P9-AGN-548

Introduction to

Second Edition

Differential Equations and Dynamical Systems

Richard E. Williamson
DARTMOUTH COLLEGE

Boston Burr Ridge, IL Dubuque, IA Madison, WI New York San Francisco St. Louis
Bangkok Bogotá Caracas Lisbon London Madrid
Mexico City Milan New Delhi Seoul Singapore Sydney Taipei Toronto

McGraw-Hill Higher Education

A Division of The ***McGraw-Hill*** *Companies*

INTRODUCTION TO DIFFERENTIAL EQUATIONS AND DYNAMICAL SYSTEMS
SECOND EDITION

Published by McGraw-Hill, an imprint of The McGraw-Hill Companies, Inc., 1221 Avenue of the Americas, New York, NY 10020.

This book is printed on acid-free paper.

1 2 3 4 5 6 7 8 9 0 VNH/VNH 0 9 8 7 6 5 4 3 2 1 0

ISBN 0-07-232573-9

Vice president and editor-in-chief: *Kevin T. Kane*
Publisher: *JP Lenney*
Senior sponsoring editor: *Maggie Rogers*
Developmental editor: *Michelle Munn*
Editorial assistant: *Allyndreth Cassidy*
Marketing manager: *Mary K. Kittell*
Project manager: *Sheila M. Frank*
Media technology lead: *Steve Metz*
Senior production supervisor: *Sandy Ludovissy*
Designer: *K. Wayne Harms*
Interior designer: *Sheilah Barrett*
Cover designer: *John Rokusek*
Senior supplement coordinator: *Audrey A. Reiter*
Compositor: *Interactive Composition Corporation*
Typeface: *10/12 Palatino*
Printer: *Von Hoffmann Press, Inc.*

The cover design is based on Figure 4.22 on page 311.

Library of Congress Cataloging-in-Publication Data

Williamson, Richard E.
Introduction to differential equations and dynamical systems / Richard E. Williamson. — 2nd ed.
p. cm.
Includes index.
ISBN 0-07-232573-9 (acid-free paper)
1. Differential equations. 2. Differentiable dynamical systems. I. Title.

QA371.W548 2001
515′.35—dc21 00–056225
CIP

www.mhhe.com

CONTENTS

†This section is not considered a prerequisite for later material and can be omitted without disrupting the flow of the material.

Chapter 3 Equations of Order Two or More 157

Chapter 4 Applications: Dynamics and Phase Space 239

Chapter 5 Introduction to Systems 317

Syllabus Suggestions

The listings by chapter and section are by no means exhaustive but display variety in emphasis to give some feeling for the book's flexibility. Sections or subsections not listed can of course be selected from the Contents at an instructor's discretion. Note in particular that either or both Chapter 8 and Chapter 9 can be covered at any point after Chapter 3.

	Basic Course	Engineering	Special Functions
Chapter		Section Number	
1: First-Order Equations	1-4	1-4	1-4
2: Applications: Dynamics & Equilibrium	1, 5	2, 3, 5	
3: Equations of Order Two or More	1-3	1-3	1-3
4: Applications: Dynamics and Phase-Space	1	1, 3	
5: Introduction to Systems	1, 2	1, 2	1, 2
6: Applications: Dynamics & Stability	1	1, 3, 4	
7: Fundamental Matrices	1		
8: Laplace Transforms	1, 2	1, 2	
9: Variable-Coefficient Methods	1, 2		1, 7
10: Partial Differential Equations	1-4	1-4	1-4
11: Sturm-Louiville Expansions			1-3

Visit www.mhhe.com/williamson for additional syllabus details including suggested homework assignments.

PREFACE

This text is written as an introduction to differential equations primarily for students in pure and applied mathematics, the physical sciences, and engineering. The main prerequisite is a solid working knowledge of calculus gained from a two- or three-semester course sequence. The approach to differential equations throughout this text is conservative in that all of the important techniques of a traditional course are included. It is flexible in that I have included extended applications in separate chapters, allowing for extensive adaptation and customizing of a course without losing the flow of the core material presented.

The primary goal of this text is to provide a variety of relevant applications of differential equations within the framework of a traditional course. This focus can be seen in both the motivational applications interwoven throughout the core material as well as in the optional extended applications found in Chapters 2, 4, 6, and 10. I have also taken great care to include application-motivated exercises within each chapter that allow a student the opportunity to further explore applied concepts throughout their homework assignments.

Another essential goal has been the development of well-rounded and extensive exercise sets. Careful care and consideration have been given to include a wide variety and range of exercises to provide flexibility in homework assignments. This adaptability can be seen through the inclusion of extensive problems that practice routine calculations necessary in the development of concepts, as well as the inclusion of more challenging and conceptual problems that help to motivate a student's continuing study.

The integration of dynamical systems coverage into an introductory course serves as a natural bridge to the use of computers in differential equations. It allows an instructor to offer a more modern course focusing on technology, yet can be omitted without any loss of continuity within the text. Content in Chapters 2, 4, 5, and 6 covers many of the ideas fundamental to the study of dynamical systems, illustrating the ideas with concrete examples. (See Suggested Syllabi within this Preface for further suggestions on a dynamical systems emphasis to differential equations.)

Topic Coverage

This text is structured to accommodate flexible course coverage. The core theory chapters (Chapters 1, 3, and 5) are organized in a fairly traditional manner. Material that is not considered a prerequisite for later topics has been indicated throughout the text with the following symbol, †. Chapters 2, 4, and 6 consist of extended applications relating to each preceding chapter, and can be chosen to suit a variety of tastes. These independent application sections can be covered or omitted without a loss of continuity within the text. A typical course would cover at most two of these sections per chapter; while some courses may cover only one. Chapter 10 is an integrated chapter containing an introduction to partial differential equations, Fourier series, and separation of variables applied to the one-dimensional heat and wave equations, and to the two-dimensional Laplace equation. Chapter 11 extends the ideas of Chapter 10 to a wider context.

The order of the chapter material follows a natural sequence for the study of differential equations. In particular, the treatment of second-order equations with nonconstant coefficients is placed closer to the chapters on partial differential equations, which is where most of the related material on special functions is really used. This arrangement also allows time for early emphasis on the study of higher-dimensional systems, along with their various applications. However, integral formulas for inverting second order constant-coefficient operators appear in Chapter 3, making it possible to deal routinely with piecewise-continuous forcing functions without resorting to Laplace transformations.

Features

Applications. This book includes a variety of applications to engineering and the physical sciences, along with sections on population dynamics. Extended applications can be found in Chapters 2, 4, 6, and also in Chapters 10 and 11. The needs and taste of the student and the instructor will determine the selection of applications used in the course. Many of these applications require a computer or calculator, usually accompanied by some paper and pencil work.

More challenging applications can be found on the text-specific website at www.mhhe.com/williamson. These applications can be used as group projects or treated as homework by assigning the accompanying exercises. A detailed applications index can also be found at www.mhhe.com/williamson.

Exercises. This text includes extensive exercise sets covering all ranges of ability. The basic material is treated repeatedly, with variations, in the exercises. The more difficult exercises have been broken down into carefully graduated steps to walk a student through the problem. Starred exercises are somewhat more demanding and usually less fundamental to an introductory course; these exercises may also be the basis for group projects.

Marginal Notes. By popular demand from my reviewers, I have included marginal notes interspersed throughout the text where appropriate. These notes include information regarding cross-referencing, prerequisite topic coverage for the extended applications, and helpful comments to the students.

Technology

The presence of computers and graphic-numeric software affect the teaching and learning of our subject in at least two related ways. One way provides immediate pleasure we can all gain from appreciating the exquisite detail in the complicated graphs, trajectories, and vector fields that are so important for our understanding of our subject. In another way, the geometric and numeric insight we can gain from computing enables us to think concretely about concepts that were considered too abstract for an introductory course until computers became widely available.

Computing Supplements. The Computing Supplements in Chapters 2, 4, and 6 are devoted to numerical methods and have exercises keyed to the applications in that specific chapter. This makes it natural, for example, to follow an extended application in Chapter 6 by the accompanying Computing Supplement section within that chapter. The formulas used in Chapters 4 and 6 are variants of those introduced in Chapter 2. I have included these sections within the appropriate applications chapter to emphasize the differences in the context of the applications, and in particular to call attention to the differences in the nature of input and output in each case. Also, the inclusion of this material within the specific chapter enables the student to see numerical methods right next to the formal algebraic techniques for each type of equation.

These Computing Supplements display some schematic descriptions of what simple computer code to implement the methods might look like. My own preference over the years has been to give students my own working programs written along these lines to let them see in detail how the computations are carried out. These programs, rewritten in Java, are accessible directly for individual use at my own website math.dartmouth.edu/~rewn/, or else through the publishers text specific website at www.mhhe.com/williamson. References are included within the text where these programs are appropriate. Computer software such as MAPLE, Mathematica, and Matlab are other possibilities, and most of the recursive numerical work can be completed using programmable hand calculators. And of course software and hand calculators can be used for symbolic work.

Technology-Recommended Exercises. Individual exercises that are appropriate for solving with the help of a computer or calculator are indicated with an icon within the text. Accompanying notebooks for Maple, Mathematica, and Matlab, keyed to each appropriate exercise, are available for downloading at www.mhhe.com/williamson.

Dynamical Systems. In addition to coverage of the traditional topics of an introductory differential equations course, this text weaves in some contemporary practice in the study of dynamical systems. Time dependent processes have been central to the application of differential equations, starting with Newton's revolutionary work on planetary motion, through Poincaré's exploitation of Gibb's phase-space ideas, and up to the current fruitful use of computer simulations in the field. This book attempts to capture the excitement of some of these ideas in Chapters 2, 4, and 6, where the study of each section is at the instructor's option. Chapter 2, Section 3 on rocket motion is an elementary introduction to the fundamental rocket equation for one space dimension. Chapter 4, Section 3 introduces the ideas of phase-space in their most elementary and accessible setting, namely the study of ordinary second order differential equations. Chapter 6, Section 4 begins with the derivation of the vector form of the two-body problem, and then shows how the method extends naturally to three or more bodies.

New to This Edition

Great thought was given to the revision of this text. Each section was analyzed in detail to ensure student accessibility to the material while providing a comprehensive and intuitive introduction to differential equations. (Also see the Note to Students from Three Students following this Preface.) In addition to improving the exposition of the text, both major and minor content revisions were made for clarity and continuity.

- **Chapter One** underwent a significant streamlining and reorganization to provide a more concise introduction to this important material. The discussion of linear equations is now introduced prior to that of exact equations. The treatment of these topics remains independent for flexibility.
- An introduction to variation of parameters is now covered in **Chapter Three.** A more comprehensive treatment of the variable coefficient theory can be found in **Chapter Nine** where the need for it is motivated by examples.
- A detailed discussion of the sliding chain problem and pendulum motion have been interwoven throughout the text in **Chapter Three** to help motivate exponential and trigonometric solutions.
- An introduction to piecewise-defined functions has been included in the Computing Supplement of **Chapter Four.**
- **Chapter Five** now contains a treatment of the eigenvalue-eigenvector method for two-dimensional systems, previously included in Chapter Seven. A new geometric introduction to generalized eigenvectors and a summary of the essential matrix algebra have also been included to clarify the concepts introduced.
- **Chapter Seven** now contains a new section on Wronskians and a new justification for the method of computing the exponential matrix.
- **Chapter Nine** now contains a new section on zeros of solutions.
- *Greater Emphasis on Visualization*—Numerous new graphs and figures have been added throughout the text to aid in visual understanding and motivation of the concepts being introduced.

Supplements

Student's Solutions Manual. This manual is available for sale to the student, and includes detailed step-by-step solutions to all odd-numbered problems throughout the text.

Instructor's Solutions Manual. This manual includes detailed solutions to all even-numbered problems throughout the text and is available to all adopters of this textbook.

Website. With the emerging importance of technology within Differential Equations instruction, a text specific website provides the opportunity to offer a technological extension of the text. At www.mhhe.com/williamson, you will find the following tools for use in your course:

Concept-driven Java programs with detailed user instructions and an introductory example to walk you through the functionality of the program. These programs are directly keyed to the text.

Computing-based Lab Projects, tied directly to each chapter's subject matter. These projects can be utilized with any computer algebra system. These projects are also available in Maple, Mathematica, and Matlab notebooks to download.

Selected numerical exercises are also available in Maple, Mathematica, and Matlab notebooks for ease in completing homework assignments. The exercises come from odd-numbered problems within the text and are indicated with a computer icon .

Further Extended Applications are also available to challenge students. These applications are extensions of the material introduced in Chapters 2, 4, and 6, and can be used as group projects or treated as homework by assigning the accompanying exercises.

For instructors using this textbook, several paths through the text are provided in our Sample Syllabi, along with accompanying homework suggestions. An Applications Index has also been provided by subject emphasis to aid in customizing the course.

Accuracy

Because of the careful checking and proofing by a number of mathematics instructors (acting independently), the author and publisher believe this book to be substantially error free. Any remaining errors or corrections will be gratefully received by the author at rewn@dartmouth.edu. Or, please contact the publisher directly via the accuracy link at www.mhhe.com/williamson.

Acknowledgments

I am grateful to my Dartmouth colleagues Jeanne Albert, Dennis Desormier, Allan Gunter, Dennis Healy, David Hemmer, John Lamperti, Tim Olson, Florin Pop, Reese Prosser, and Dana Williams for their helpful suggestions, based on their teaching experience with preliminary versions of this book. I am particularly grateful to Sara Baron, Elizabeth Hamon, and Rebecca Taxier for their careful reading of the book and for their insightful suggestions for improvement. It was always a pleasure working with them.

Special thanks go to Allan Gunter for his exhaustive attention to detail in updating the Instructor's and Student's Solutions Manuals that accompany this edition and also to Joe Good for his careful accuracy checking.

Jeanne Albert of Castleton State College, Gregory Fredericks of Lewis and Clark College, John Palmer of the University of Arizona, Craig Tracy of the University of California at Davis, and Randy Zounes of Cornell University were also very helpful. Michael Sturge of the Dartmouth Physics Department contributed several interesting exercises.

I also thank the many reviewers of the first and second editions. These reviewers have provided much helpful criticism and encouragement to me.

Seth Armstrong, Arkansas State University
Sergiu Aszicouizi, Ohio University, Athens
Leigh Atkinson, University of North Carolina, Ashville
Bruce Berndt, University of Illinois, Urbana-Champaign
Richard Black, Prairie State College
Paul Blanchard, Boston University
Carlos Borges, Naval Postgraduate School
Stewart Davidson, Georgia Southern University
John Davis, Baylor University
Paul Dawkins, Lamar University
Satyan Devadoss, Ohio State University, Columbus
John Gregory, Southern Illinois University
Victor Gummersheimer, Southeast Missouri State University
Greg Harris, Auburn University
Robert Hunt, Humboldt State University
Zdzislaw Jackiewicz, Arizona State University, Tempe
Helmut Knaust, University of Texas at El Paso
Terry Lawson, Tulane University
Anbo Le, Ohio State University, Columbus
Joyce Longman, Villanova University
Charles MacCluer, Michigan State University
George Majda, Ohio State University, Columbus
Carol Jean Martin, Dodge City Community College
Ronald Mathison, North Dakota State University, Fargo
Douglas Meade, University of South Carolina
James Meiss, University of Colorado, Boulder
Gary Meisters, University of Nebraska
Aaron Melman, University of San Francisco
Kouhestani Nader, Prairie View A&M University
Jim Peters, Weber State University
Mohammad Rammaha, University of Nebraska, Lincoln
Bill Royalty, University of Idaho
Rinaldo Schinazi, University of Colorado, Colorado Springs
Greg Suiatek, Penn State University
Louis H. Thurston, Texas A&M, Kingsville
Roberto Triggiani, University of Virginia

Larry Turyn, Wright State University

R. Lee Van de Wetering, San Diego State University

Horace Wiser, Washington State University

My association with the publisher has been a pleasure throughout the production process. Maggie Rogers, Michelle Munn, and Sheila Frank provided careful oversight for this entire project, and my thanks go as well to the copyeditor Pat Steele and the rest of the staff at McGraw-Hill for their continuing hard work on my project.

Richard E. Williamson

A Note to Students from Three Students

Everything should be made as simple as possible, but not simpler.
Albert Einstein

While our friends went off to read their history textbooks or do scientific research in a lab this past academic year, we sat down and did something fairly rare, even for students taking a math course. We really read a math book, this one on differential equations. We were reading the text to offer a student's point of view while it was being prepared for its second edition. In the process we read not only the text, but the exercises as well. We are all three freshmen at Dartmouth College, taking our first multivariable calculus course this year, none having any differential equations background.

Why were three freshmen offering their opinions on a differential equations text? Dartmouth College offers a Women in Science Program, sponsoring internships in math and science. We each chose this project from the many that were offered. Our sponsor, the author of this text, wanted to have the book edited by students so it would be clearer for students using it. Even though many math professors across the country had already read and commented on the text, we offered some new insight into what would be helpful to a student. We were the perfect subjects for this experiment since we had such limited prior knowledge of the subject.

The author gave us a little guidance as we went through the book, but mainly we were on our own, learning and understanding the material as we read. Often we found ourselves editing out excess words, substituting a word or phrase, or asking the author to add more steps in an example to clarify how to work a problem. We think our changes made the textbook more readable and student-friendly. We had the advantage of talking to the author each time we raised a question, and rest assured he clarified the text in every instance.

Though we had to balance many hours of work on the text with our other course work, we enjoyed the experience. It isn't every day that one can speak directly to the author of a book and see immediate changes in the writing. The challenge of learning about differential equations, while at the same time making the subject more accessible to other students, was fun. We wish you good fortune with your study of differential equations. Hopefully our work has made things easier and more interesting for you.

Sara Baron '03
Elizabeth Hamon '03
Rebecca Taxier '03

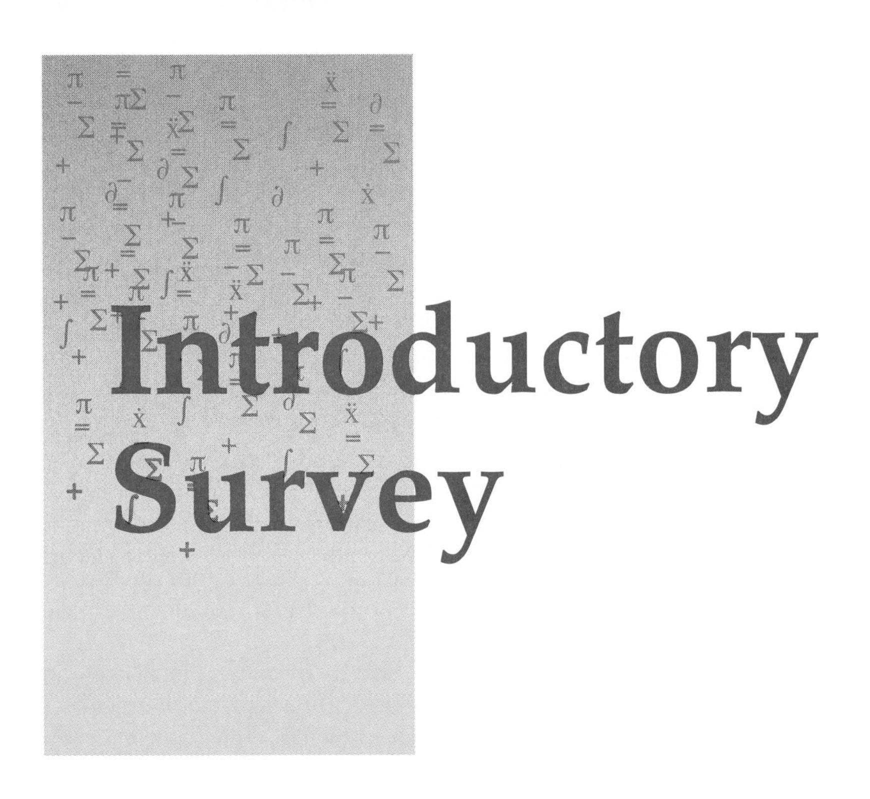

Introductory Survey

Isaac Newton's invention of what we now call differential equations has remained a cornerstone of pure and applied science, and the subject is still generating new ideas. For example, the power of numerical computation and computer graphics to clarify ideas and in particular to display chaotic behavior has contributed significantly to our understanding of many processes modeled by differential equations. Indeed, a principal aim of this book is to incorporate these methods into the study of differential equations in a useful way at an introductory level. Reading the following remarks and examples will give you some idea of how the subject is organized and what some of the applications are.

Mathematically speaking, a **dynamical system** is a description of possible evolutions over time of points, called **states,** in some space $\mathcal{S}$ called the **state space** of the system. The states in a state space might, for example, consist of number pairs that measure possible positions and velocities of some physical object, for example, a particle moving on a path in space.

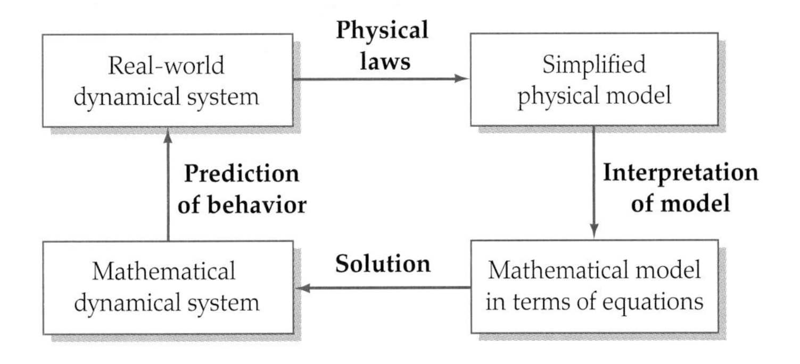

If time t is measured over an interval of the real numbers, the possible evolutions of a system will often be determined by a differential equation with t

as an independent variable. What a **differential equation** does is relate values of a function $y(t)$ to corresponding values of one or more of its derivatives. A differential equation is often at the heart of a system, so it's sometimes natural to think of this equation itself as representing the system.

Some error is usually involved in passing from one stage to another in the diagram on page 1, so a satisfactory analysis requires that a small error at one stage translates into a small error at the next stage. Such systems are essentially predictable, while a so-called "chaotic" system tends to be essentially unpredictable. Thus the principal aim in studying a chaotic system is to look for some core of overall regularity in its behavior.

It is the interplay, shown schematically here, between a differential equation and a real-world dynamical system that is the central theme of this book. In practice, specialists may find themselves focusing on this diagram at particular points. For example, a theoretical physicist might work across the top, a pure mathematician across the bottom, and an applied mathematician or scientist along the vertical directions.

The ideas sketched here also apply if time is measured in discrete jumps rather than continuously. We may then still be dealing with a dynamical system, one in which a differential equation is replaced by a discrete recurrence relation as in the Newton, Euler, and Poincaré methods of Chapters 2, 4, and 6. And equations with no designated time variable at all also play a significant role in pure and applied mathematics.

Since the specific examples in this introduction are intended to give an intuitive feeling for what the subject is like, each one is tied to a particular application. The equations are arranged according to a well-established mathematical classification, a classification that's useful because different applications often lead to the same mathematical problem. Indeed one of our aims is to show how a single mathematical idea can be used in several ways. The classification is also useful for predicting in general terms what kind of solution, if any, can be expected for a given problem. All of these introductory examples are treated at appropriate points later on. Some cases requiring approximate numerical computation using, for example, Java programs are on the companion websites to this book. We begin with an example that appears prominently in many calculus courses, and for which machine computation is unnecessary.

Java applets are available at www.mhhe.com/williamson/ or more directly at math.dartmouth.edu/~rewn/

example 1

Suppose $v = v(t)$ represents the velocity as a function of time t of an object of mass m falling in the vicinity of the earth's surface. Taking into account the pull of gravity on the object as well as the effect of air resistance will lead in some circumstances to considering a relationship of the form

$$m\frac{dv}{dt} = -kv + mg.$$

Here $a = dv/dt$ is the object's actual acceleration, g is a constant representing the acceleration due to gravity, and k is a constant that determines the force due to air resistance. If air resistance is negligible, we can assume $k = 0$, in which case we are left with just

$$\frac{dv}{dt} = g.$$

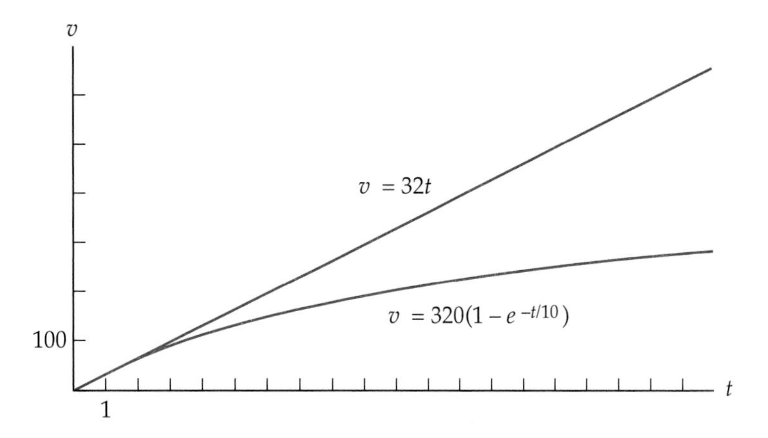

Figure IS.1 Effect of air resistance on the velocity of a falling object.

Integrating both sides with respect to t gives $v(t) = gt + v(0)$. It follows that the graph of v as a function of t, shown in Figure IS.1 for the choices $m = 1, g = 32$, and $v(0) = v_0 = 0$, is a line with slope g. This outcome predicts in particular that the velocity will continue to increase at a steady rate until the object hits the surface of the earth, which is just what the differential equation $dv/dt = g$ says; beyond that, we get a formula for velocity as a function of time. In the case where air resistance is taken into account, with resistance force proportional to velocity, we'll see in Chapter 1 that the velocity is expressible in terms of the significant parameters g, k, m, and $v(0) = v_0$ in the form

$$v(t) = \frac{mg}{k} + \left(v_0 - \frac{mg}{k}\right)e^{-kt/m}.$$

The graph of a particular one of these solutions is also shown in Figure IS.1 for the choices $m = 1, g = 32, k = 0.1$, and $v(0) = 0$. [Note that if $v_0 = mg/k$, the velocity is constant.]

example 2 A differential equation of the form

$$L\frac{d^2Q}{dt^2} + R\frac{dQ}{dt} + \frac{1}{C}Q = E(t), \quad L, R, C \text{ constant},$$

can be used to predict the time variation of electric charge $Q = Q(t)$ in a simple electric circuit when $E(t)$ represents an externally applied voltage. One of the many applications for such a circuit is to produce an enhanced response to the applied voltage in the form of a more widely oscillating charge $Q(t)$. The function $Q(t)$ is called the "response" corresponding to the "input" $E(t)$. By the choice of the constants L, R, and C a circuit can be "tuned" to produce an amplified response to the input, as shown in Figure IS.2.

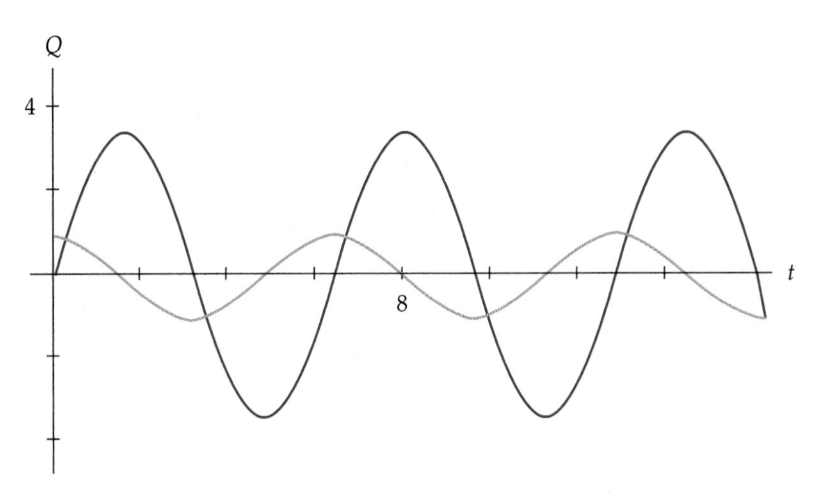

Figure IS.2 Input and amplified response from a circuit.

example 3

A horizontal beam of uniform cross section and constant density ρ per unit of length, will sag under its own weight. Let $y(x)$ represent the shape of the beam at distance x from one end. Arguments from the theory of elasticity show that under these assumptions the function $y(x)$ nearly satisfies the **Euler beam equation:**

$$EI\frac{d^4y}{dx^4} = -\rho.$$

The constant factors E and I depend respectively on the elasticity and cross-sectional configuration of the beam; their product $R = EI$ is a measure of a beam's rigidity, or resistance to bending. The shape that a beam actually takes depends on how it is supported at its two ends. Figure IS.3 shows the outline of the lower edge of a beam that is simply resting on a support at the left end and is *cantilevered* at the right end; that is, maintained more or less horizontally at that end by embedding a portion near the end in a wall. The axis scaling is chosen to give an exaggerated view of the vertical deflection. If distance were measured in feet, the maximum deflection for such a 20-foot beam might be a fraction

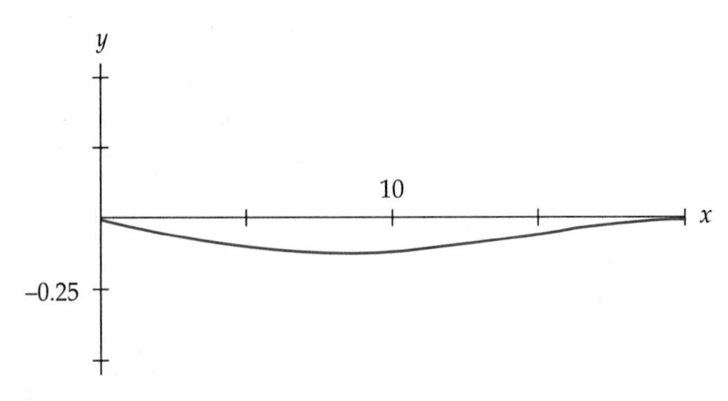

Figure IS.3 Beam profile; simply supported left end and cantilevered right end.

of an inch. Notice that the differential equation $EI\, d^4y/dx^4 = -\rho$ contains no information about what happens at the ends of the beam; this information has to be added by supplementary "boundary conditions" or "end conditions" on $y(x)$, in this case, $y(0) = 0$, $y''(0) = 0$ and $y(20) = 0$, $y'(20) = 0$. For example, $y''(0) = 0$ requires flatness at the left end, and $y'(20) = 0$ requires the beam to be horizontal at the right end. The typical solution of the differential equation contains four arbitrary constants of integration, which are then determined by the four end conditions. You can verify directly that the solution formula

$$y(x) = \frac{\rho}{EI}\left(-\frac{x^4}{24} + \frac{5x^3}{4} - \frac{500x}{3}\right)$$

satisfies the equation containing the fourth-order derivative of $y(x)$, along with the four boundary conditions. Of course, for the picture some assumption had to be made about the relation between the product EI and the density ρ per unit of length; in this example, $EI = 7000\rho$.

Linearity of a Differential Equation. The differential equations in the three examples above have an important feature in common in that they are all **linear.** Recognizing linearity when you see it is just a matter of checking that an equation has degree 1 in the unknown functions and their derivatives that occur in the equation. Another way to put the requirement is that the unknown functions and their derivatives that are in the equation should appear only in sums of multiples of each of them alone, where the multipliers depend at most on independent variables. Thus the differential equation

$$\frac{d^2y}{dx^2} + 0.3x\frac{dy}{dx} + x^2y = x^2, \text{ or } y'' + 0.3xy' + x^2y = x^2,$$

is linear since it is a polynomial of degree 1 in the symbols y, $dy/dx = y'$, and $d^2y/dx^2 = y''$. [Recall that $d^2y/dx^2 \neq (dy/dx)^2$!] But the differential equation

$$y' + yy' = x$$

is not linear since it is of second degree as a function of y and y' together. Many significant problems can be stated in terms of linear equations, and there are standard methods for providing complete solutions to lots of them, solutions that can in principle often be expressed in a single simple formula.

To say that an equation is **nonlinear** (i.e., not linear) is to say that its dependence on the functions and their derivatives that the equation determines is not of first degree. Thus $y'' + \sin y = x$ is nonlinear, since the term $\sin y$ is not a multiple of y, y', or any higher order derivative of y.

---

example 4

The differential equation

$$\frac{d^2y}{dt^2} + \frac{1}{5}\frac{dy}{dt} - y + y^3 = \frac{3}{10}\cos t$$

is nonlinear precisely because of the presence of the term y^3, which is of degree 3

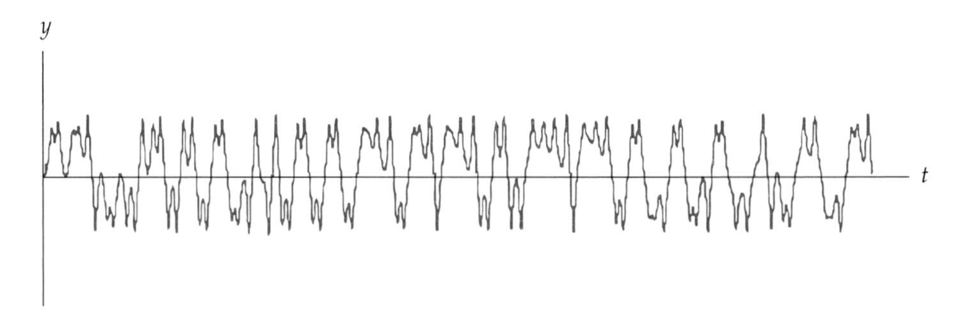

Figure IS.4 Approximate solution to the Duffing equation for $0 \leq t \leq 500$.

in y. This equation is an example of a **Duffing equation** driven by a periodic oscillation with period 2π. Figure IS.4 shows a computer simulation for $0 \leq t \leq 500$ of an evolving solution satisfying $y(0) = y'(0) = 0$. The challenge posed by this rather irregular graph is to find some regularity hidden within its apparent chaos. That issue is addressed concretely in the computing supplement Section 4D of Chapter 4.

Nonlinear equations are important because they often provide much better models for physical behavior than do the simpler linear ones. Even though solution formulas are often more difficult, or impossible, to find for nonlinear equations, the ability of computers to provide numerical or graphical solutions directly often enables us to bypass the need for a formula.

Order of a Differential Equation. The **order** of a differential equation is just the order of the highest order derivative that occurs in the equation. Thus the orders of the equations in Examples 1 through 4 are respectively 1, 2, 4, and 2. For linear equations, the number of arbitrary constants in a typical solution formula is always the same as the order of the equation. Unfortunately the corresponding situation for nonlinear equations isn't always so clear-cut, and the solutions may not be nicely indexed by a number of constants equal to the order of the equation.

In formal terms, a single ***n*th order linear differential equation** of the type we've seen so far can always be written

$$a_n(x)\frac{d^n y}{dx^n} + a_{n-1}(x)\frac{d^{n-1} y}{dx^{n-1}} + \cdots + a_1(x)\frac{dy}{dx} + a_0(x)y = f(x),$$

where the functions $a_n(x), \ldots, a_0(x)$, $f(x)$ all play a role in restricting the possible choices for solutions $y = y(x)$. We often use t, for time, as the independent variable instead of a space variable x. Depending on the application, we may use a wide variety of other letters other than y for the possible solutions.

---

example 5

The nonlinear first-order differential equation

$$\frac{dP}{dt} = kP\left(1 - \frac{P}{L}\right),$$

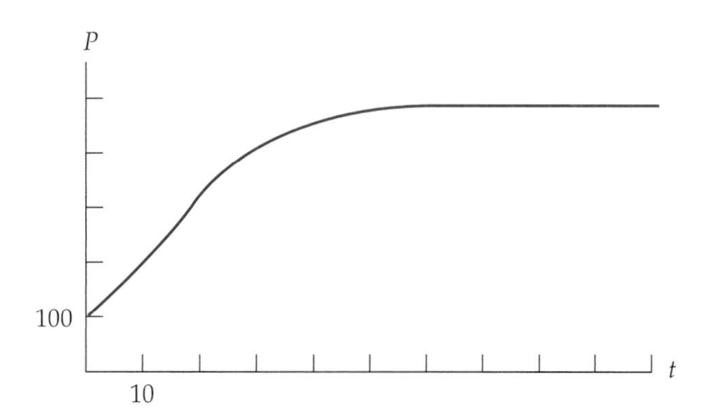

Figure IS.5 Limited population $P(t)$ with $P(0) = 100,\ L = 500$, and $k = 0.1$.

where k and L are positive constants, is an example of what is called a **logistic equation** and can be used to model the growth of certain populations that have inherent limits to their ability to increase. A typical solution has the form

$$P(t) = \frac{P_0 L}{(L - P_0)e^{-kt} + P_0},$$

where P_0 is the size of the population at time $t = 0$. The graph of this solution typically looks like the one in Figure IS.5.

example 6

The position of a pendulum oscillating back and forth in a plane can be measured by the angle $\theta = \theta(t)$ that it makes with a vertical line at time t. A discussion based on elementary physical principles shows that, if gravity is the only force acting on the pendulum, then $\theta(t)$ satisfies the second-order nonlinear differential equation

$$\frac{d^2\theta}{dt^2} = -\frac{g}{l}\sin\theta,$$

where l is the effective length of the pendulum, and g is the acceleration of gravity. Unlike all the previous examples, there are no interesting solutions to this equation expressible in terms of the elementary functions of calculus. [The exceptional solution $\theta(t) = 0$ is called a **trivial solution.**] Our approach in this book is to use numerical approximations for quantitative solutions of this equation, supplemented by general qualitative descriptions. The approximations can in principle be made to any desired degree of accuracy, and can easily be converted to graphical form. A typical graph of $\theta(t)$ is shown in Figure IS.6(b).

Higher Dimensional Equations. Each of the next two examples deals with a **system** of differential equations that imposes constraints on more than one function. A system is *linear* if each equation in it is linear, and the *order* of a system is the maximum of the orders of the equations in the system. The number of real-valued functions determined by a system is called its **dimension.**

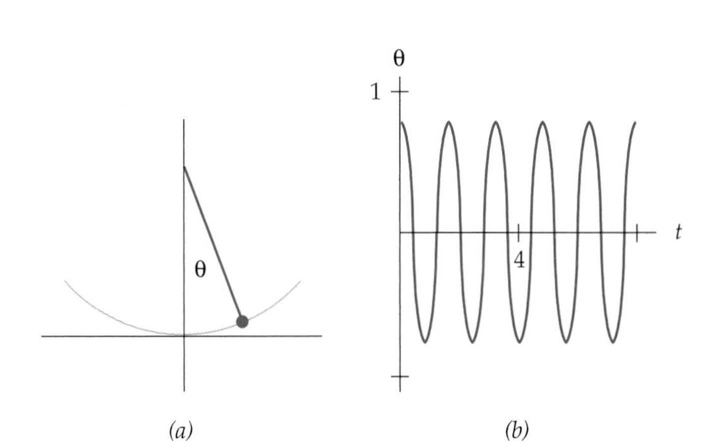

Figure IS.6 (*a*) Pendulum arc. (*b*) $\theta = \theta(t)$.

example 7

Populations of size $P_1(t)$ and $P_2(t)$ may each depend on the presence of the other for growth over time t. In the absence of the other species, each one might dwindle away according to some empirically discovered rule such as

$$\frac{dP}{dt} = -0.2P,$$

in which the negative right-hand side guarantees that $P(t)$ will decrease as time t increases. Such a decline is shown graphically in Figure IS.7(a). [The solution formula for this equation is $P(t) = P(0)e^{-0.2t}$.] On the other hand, the contribution of each population to the growth of the other can sometimes be described by a pair of linear equations, making up a first-order system of differential equations

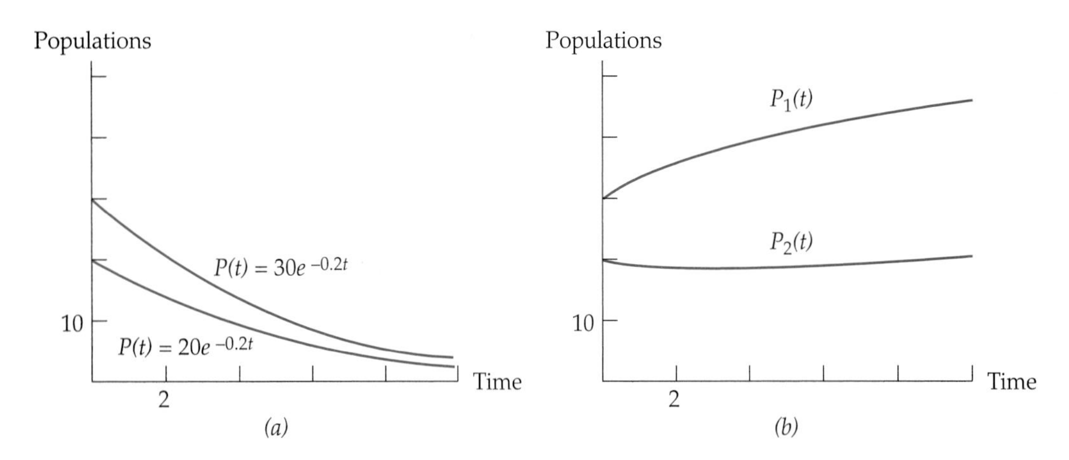

Figure IS.7 (*a*) Independent decline. (*b*) Cooperative increase.

to be satisfied by two functions $P_1(t)$ and $P_2(t)$:

$$\frac{dP_1}{dt} = -0.2P_1 + 0.5P_2, \qquad \frac{dP_2}{dt} = 0.1P_1 - 0.2P_2,$$

in which the growth rate of each population is augmented by an amount proportional to the size of the other population. The resulting behavior of the two populations over time can also be expressed in terms of exponential functions, and a typical example appears in Figure IS.7(b).

example 8

One of the earliest, and still one of the most significant, applications of differential equations is to the description of planetary motions, as determined by the fundamental ideas of Newton's mechanics. Relative to a plane rectangular coordinate system centered at the center of mass of a star, a solitary planet has position $\big(x(t),\ y(t)\big)$ at time t whose coordinates satisfy the two-dimensional, second-order system of nonlinear equations

$$\frac{d^2x}{dt^2} = -\frac{G(M+m)x}{\left(x^2+y^2\right)^{3/2}}, \qquad \frac{d^2y}{dt^2} = -\frac{G(M+m)y}{\left(x^2+y^2\right)^{3/2}}.$$

Here M is the mass of the star, m is the mass of the planet, and G is a constant associated with the inverse square law of gravitational attraction. Rather than plot solution values $x(t)$ and $y(t)$ separately as functions of t, what is often done in examples like this is to plot the **trajectory,** or **orbit,** of a solution, that is, the graph representing the path actually followed by an object. Such a path is an ellipse for a single planet, though an object that visits a star just once has a hyperbolic trajectory. Two trajectory curves are shown in Figure IS.8, with the star indicated by a dot.

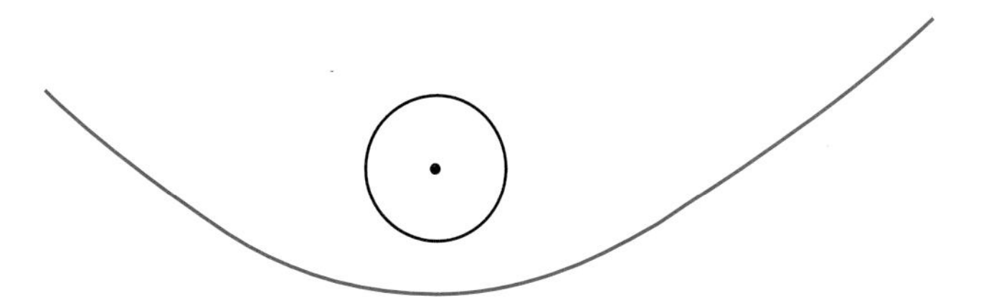

Figure IS.8 Closed planetary orbit and visitor's path.

example 9

The previous example has to do with what is called the "two-body problem," the two bodies typically being thought of as our sun and a single planet, or the earth and its moon, each pair moving in a fixed plane. The fact is that each of the three bodies just mentioned exerts some influence on the others. Taking all these effects into account in the "three-body problem" leads to a system of nine second-order nonlinear equations, since it takes three equations to determine the motion of each of the three bodies in three-dimensional space. (For the sun and nine planets, we would have 30 such equations.) In cases of any practical significance, such systems must be solved by approximate numerical methods, and it's helpful in organizing the arithmetic to write position vectors, say $\mathbf{x}_1 = (x_1, x_2, x_3)$, $\mathbf{x}_2 = (x_4, x_5, x_6)$, and $\mathbf{x}_3 = (x_7, x_8, x_9)$ for each of the three bodies. In terms of these vectors, the nine scalar equations of motion for the three body problem can be written fairly briefly as follows:

$$\frac{d^2\mathbf{x}_1}{dt^2} = \frac{Gm_2}{r_{21}^3}(\mathbf{x}_2 - \mathbf{x}_1) + \frac{Gm_3}{r_{31}^3}(\mathbf{x}_3 - \mathbf{x}_1),$$

$$\frac{d^2\mathbf{x}_2}{dt^2} = \frac{Gm_1}{r_{12}^3}(\mathbf{x}_1 - \mathbf{x}_2) + \frac{Gm_3}{r_{32}^3}(\mathbf{x}_3 - \mathbf{x}_2),$$

$$\frac{d^2\mathbf{x}_3}{dt^2} = \frac{Gm_1}{r_{13}^3}(\mathbf{x}_1 - \mathbf{x}_3) + \frac{Gm_2}{r_{23}^3}(\mathbf{x}_2 - \mathbf{x}_3),$$

where r_{jk} stands for the distance between $\mathbf{x}_j$ and $\mathbf{x}_k$, that is, $r_{jk} = |\mathbf{x}_j - \mathbf{x}_k|$. Starting with given positions and velocities for the three bodies, it is routine to predict with good accuracy where they will be, and what their velocities will be, at future times.

Partial Derivatives. So far the differential equations we've considered have had solutions that are functions of a single one-dimensional variable, for example t for time or x for distance. Differential equations of that kind are called **ordinary** differential equations because they impose conditions on only ordinary derivatives as contrasted with partial derivatives. A differential equation whose solutions are differentiated in more than one variable (e.g., time and distance, or two or three space variables, or time together with space variables) is called a **partial differential equation.**

example 10

The linear partial differential equation

$$\frac{\partial u}{\partial t} = \frac{\partial^2 u}{\partial x^2}$$

is a **heat equation** in which $u = u(x, t)$ represents the temperature at time t and position x in a one-dimensional heat-conducting medium such as a thin, heat-insulated wire. Generally speaking, partial differential equations are harder than ordinary ones to interpret directly. In this example, however, there is a direct qualitative interpretation that predicts something interesting about the solutions $u(x, t)$. Note that if $u(x, t)$ is concave up as a function of x for some value of t and some interval of x values, then $\partial^2 u(x, t)/\partial x^2 > 0$ on the interval. But the differential equation then implies that $\partial u(x, t)/\partial t > 0$, in other words, $u(x, t)$ is an increasing function of t. A careful look at the solution graph in Figure IS.9 will bear out this interpretation. Similarly, you can see that where $u(x, t)$ is concave down as a function of x, it is also decreasing as a function of t. This geometric observation is consistent with heat flowing from warmer to cooler regions.

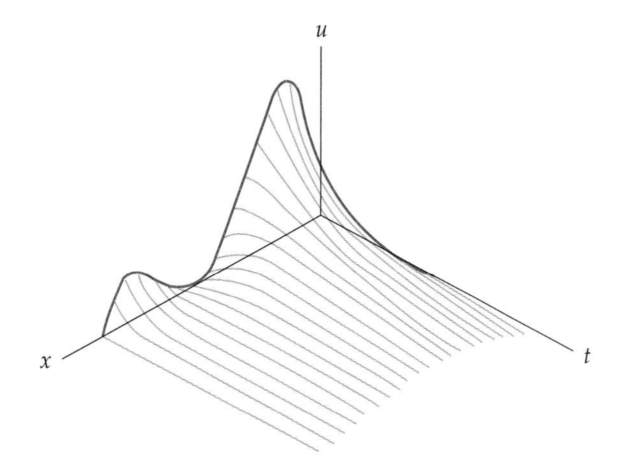

Figure IS.9 Temperature $u(x, t)$ as a function of discrete position x and continuously increasing time t.

EXERCISES

For each of the differential equations or systems of equations in Exercises 1 through 26, state (i) the order, (ii) the dimension, (iii) whether it is linear or nonlinear, and (iv) whether it is an ordinary differential equation (ODE) or a partial differential equation (PDE).

1. Equation for the height $h = h(t)$ of fluid in a vertical tank of uniform cross-sectional area A with fluid flowing from a hole of area a in the side of the tank:

$$\frac{dh}{dt} = -\frac{ka}{A}\sqrt{h}.$$

   Here k is a constant depending on the properties of the fluid.

2. Equation for the position $s = s(t)$ of a falling object at time t, measured down from initial position s_0:

$$\frac{ds}{dt} = \sqrt{v_0^2 + 2g(s - s_0)},$$

   where v_0 is the initial velocity, and g is the acceleration of gravity.

3. **Newton's law of cooling** for the surface temperature $u = u(t)$ of an object in an environment maintained at temperature u_0:

$$\frac{du}{dt} = k(u - u_0), \quad k = \text{const.}$$

4. The harmonic oscillator equation $m\,d^2y/dt^2 = -\pi r^2 \rho y$ where $y(t)$ gives the vertical motion of a cylindrical buoy of mass m and radius r floating in water of density ρ.

5. **Van der Pol equation** $d^2y/dt^2 - \epsilon(1 - y^2)\,dy/dt + \delta y = 0$ describing the behavior of periodic electrical discharges. [Note that setting $\epsilon = 0$ and $\delta = \pi r^2 \rho m$ yields the previous equation.]

6. Damped harmonic oscillator equation with externally applied periodic force:

$$\frac{d^2y}{dt^2} + k\frac{dx}{dt} + hy = \cos t.$$

7. **Catenary equation** for the shape of a chain suspended between two supports: $y'' = \sqrt{1 + (y')^2}$.

8. Equation for the velocity $v(t)$ of an object falling near the surface of the earth in a medium that offers high frictional resistance at high velocities: $dv/dt = -v - \frac{1}{2}v^2 + g$.

9. Damped oscillation of a pendulum: $d^2\theta/dt^2 = -0.2d\theta/dt - 5\sin\theta$, amplitude unrestricted.

10. Small, damped oscillation of a pendulum: $d^2\theta/dt^2 = -0.2\,d\theta/dt - 5\theta$. Here the term containing $\sin\theta$ in the preceding differential equation has been modified by the approximation $\sin\theta \approx \theta$, which may be good enough for some practical purposes if θ is small enough.

11. Relativistic differential equation for the speed v of a rocket in linear motion; v is regarded as a function of its decreasing mass m as fuel is burned:

$$m\frac{dv}{dm} + v_e\left(1 - \frac{v^2}{c^2}\right) = 0.$$

Here the constant v_e is the velocity of the expelled exhaust measured relative to the rocket, and c is the speed of light.

12. Setting $c = \infty$ in the previous (relativistic) problem yields the nonrelativistic differential equation appropriate for relating speed and mass of a rocket at lower speeds:

$$\frac{dv}{dm} = -\frac{v_e}{m}.$$

13. **Binet's equation** for the reciprocal $u = 1/r$ of the distance r from planet to star in polar coordinates: $d^2u/d\theta^2 + u = G/H^2$, G and H positive constants.

14. Relativistic equation for reciprocal $u = 1/r$ of the distance r from planet to star in polar coordinates: $d^2u/d\theta^2 + u = G/H^2 + (3G/c^2)u^2$, G and H constant. The constant c is the velocity of light. [Note that setting $c = \infty$ yields the previous equation.]

15. Equation for the shape of a wave moving with uniform speed in a narrow channel: $d^3y/dx^3 = \sigma y\, dy/dx$, σ constant.

16. **Euler beam equation** for the shape of a uniform beam subject to lengthwise compression:

$$EI\frac{d^4y}{dx^4} + F\frac{d^2y}{dx^2} = 0,$$

where F is the compression force applied at one end, and both E and I are constant.

17. **Bessel equation**

$$x^2\frac{d^2y}{dx^2} + x\frac{dy}{dx} + (x^2 - p^2)y = 0.$$

18. System of equations satisfied by the displacements from equilibrium $x_1 = x_1(t)$ and $x_2 = x_2(t)$ of two equal masses m linked by springs of equal strength k to fixed supports:

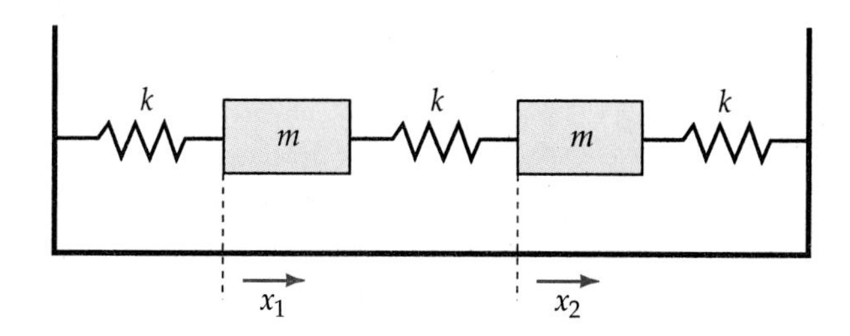

$$m\frac{d^2x_1}{dt^2} = -2kx_1 + kx_2, \qquad m\frac{d^2x_2}{dt^2} = kx_1 - 2kx_2.$$

19. **Lotka-Volterra equations** for the sizes of interacting parasite and host populations:

$$\frac{dP}{dt} = (a - bP)H, \qquad \frac{dH}{dt} = (cH - d)P, \quad a, b, c, d \text{ constant.}$$

20. **Lotka-Volterra equations** for the sizes H, K, and P of three populations interacting over time:

$$\frac{dP}{dt} = (a - bP)H, \qquad \frac{dK}{dt} = K(P + H - e), \qquad \frac{dH}{dt} = (cH - d)P.$$

Here a, b, c, d, and e are usually constant.

21. **Lorenz equations** derived by approximating partial differential equations for atmospheric flow:

$$\dot{x} = \sigma(y - x), \qquad \dot{y} = \rho x - y - xz, \qquad \dot{z} = -\beta z + xy.$$

Here σ, ρ, and β are constants. The specific choices $\sigma = 10$, $\rho = 28$, $\beta = \frac{8}{3}$ yield a system with solutions that appear to exhibit chaotic behavior.

22. One-dimensional **wave equation** $u(x, t)$:

$$\frac{\partial^2 u}{\partial t^2} = a^2\frac{\partial^2 u}{\partial x^2},$$

where a determines the constant velocity of the wave motion.

23. **Telegraph equation** for current flow in a wire:

$$A\frac{\partial^2 u}{\partial t^2} + B\frac{\partial u}{\partial t} + Cu = \frac{\partial^2 u}{\partial x^2},$$

where A, B, and C are constants. [Note that the choices $A = 1/a^2$ and $B = C = 0$ yield the previous equation.]

24. Korteweg-deVries equation

$$\frac{\partial u}{\partial t} + \frac{\partial u}{\partial x} + \beta u \frac{\partial u}{\partial x} + \frac{\partial^3 u}{\partial x^3} = 0, \quad \beta = \text{constant}$$

for a wave profile $u(x, t)$ in a narrow channel at point x and time t.

25. Vector wave equation

$$\rho(s) \frac{\partial^2 \mathbf{x}}{\partial t^2} = F \frac{\partial^2 \mathbf{x}}{\partial s^2},$$

for small vibrations of a stretched string in three dimensions, where $\mathbf{x} = \mathbf{x}(s, t)$ is the position vector of a point on the string s units from one end at time t, $\rho(s)$ is the string density as a function of distance from the same end, and F is the constant tension force on the string.

26. Black-Scholes equation

$$\frac{\partial V}{\partial t} + \frac{1}{2}\sigma^2 x^2 \frac{\partial^2 V}{\partial x^2} + rx\frac{\partial V}{\partial x} - rV = 0$$

for the value $V(x, t)$ of an option to buy or sell a stock with price x at time t. The constant r is a prevailing risk-free interest rate, for example, the rate of a government bond, and σ^2 is a measure of the volatility of the investor's return on this particular stock.

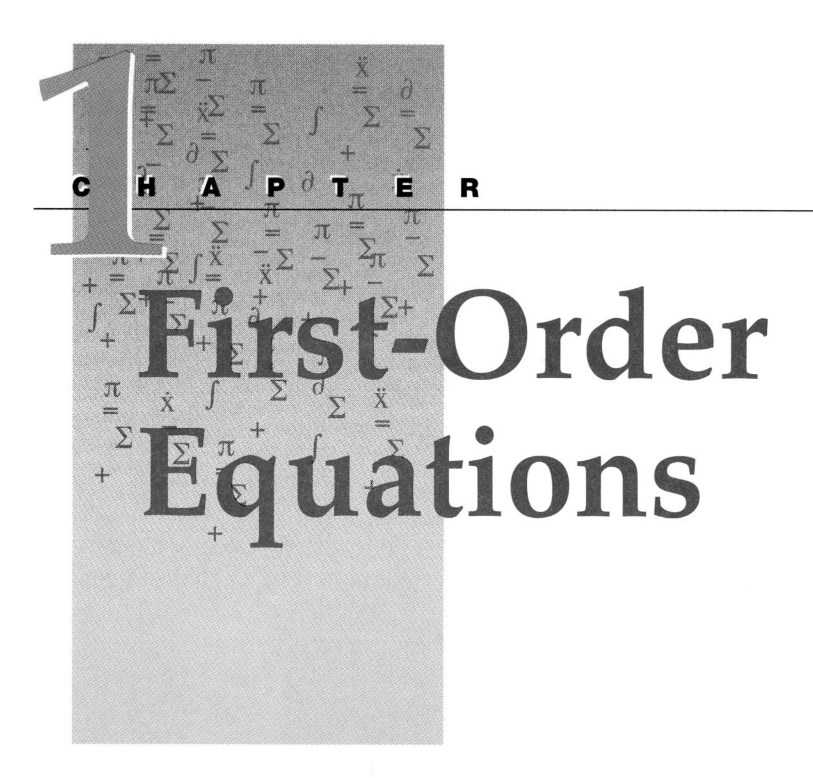

CHAPTER 1 First-Order Equations

1 SOLUTIONS AND INITIAL VALUES

It has been known for many years that in both pure and applied mathematics it is possible to establish equations relating an important function $y(x)$ to one or more of its derivatives, and then to use the equation to get information about $y(x)$. Such an equation relating a function and its derivatives is called a **differential equation.** Simple examples of differential equations are

$$y'(x) + y(x) = x \quad \text{and} \quad y'(x) = \sqrt{1 + y(x)^6}.$$

Using the alternative Leibniz notation for the derivative, and suppressing "(x)," the two equations look like

$$\frac{dy}{dx} + y = x \quad \text{and} \quad \frac{dy}{dx} = \sqrt{1 + y^6}.$$

Both of these differential equations turn out to be satisfied by infinitely many **explicitly** defined functions $y(x)$, each one called a **solution.** Since the existence of infinitely many solutions is the typical state of affairs, it will be important to try to describe the multitude of solutions in a convenient and comprehensible way. Such a description may not always be possible, but if we have some formula that contains as special cases all particular solutions $y(x)$, then it's customary to refer to that formula as the **general solution.** For example, we'll see that the first equation above has for its general solution on $-\infty < x < \infty$ the family of functions

$$y(x) = x - 1 + Ce^{-x}, \quad C \text{ constant.}$$

It turns out that by assigning the constant C an appropriate value, we can pick a particular one of these solutions whose graph passes through any given point (x_0, y_0) we choose. The second differential equation, $y' = \sqrt{1 + y^6}$, is more problematic in that a general solution can't be expressed in terms of the elementary functions of calculus or for that matter in terms of any other well-understood class of functions. The standard response to that difficulty is to study the solutions by working directly with the differential equation; in particular we'll develop numerical methods in the computing supplement to Chapter 2 for displaying the properties of particular solutions. (If we're thorough enough in our investigations, we may find that in the process we've added to our supply of well-understood functions.)

The **order** of a differential equation is naturally defined to be the same as that of the highest order derivative that appears effectively in the equation. Thus our two examples $y' + y = x$ and $y' = \sqrt{1 + y^6}$ are within the scope of the present chapter, namely, equations of order 1. Equations such as $y'' + y = x$ and $y'' = \sqrt{1 + y^6}$ are of order 2 and will be treated in detail in Chapter 3.

1A Elementary Solution Formulas

Equations are usually meant to be solved or otherwise interpreted in some way, and differential equations are no exception. A *solution* of a differential equation is a function $y = y(x)$, which, when substituted along with its derivatives into the differential equation, satisfies the equation for all x in some specified interval called the **domain** of the solution. (If we change the domain interval by increasing or decreasing it, we'll regard the resulting solution as different; such domain changes may be crucial in applications where the behavior of a solution changes decisively after some time passes.) In practice, there are sometimes elementary formulas for solutions, although by no means always. Sometimes solutions are defined in an **implicit** manner. For example, the equation $x^2 + y^2 = 1$ determines two different differentiable functions of x, namely, $y = \sqrt{1 - x^2}$ and $y = -\sqrt{1 - x^2}$, each of which turns out to satisfy the differential equation $y' = -x/y$ for $-1 < x < 1$. In the following examples, we will simply verify that certain formulas provide solutions; derivation of the solutions will be discussed later.

example 1 We'll verify that the differential equation

$$y' + y = x$$

has the solution $y(x) = x - 1$ for $-\infty < x < \infty$. We have $y'(x) = 1$, so

$$y' + y = 1 + (x - 1) = x, \quad \text{for all real } x.$$

More generally, the same differential equation has the solutions

$$y = x - 1 + ce^{-x}$$

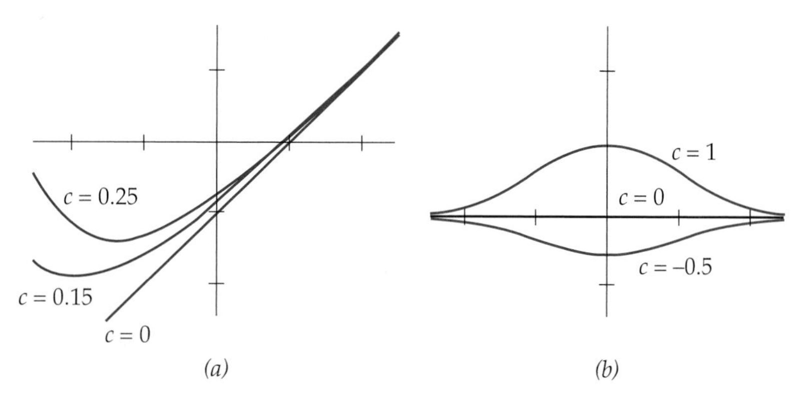

Figure 1.1 (*a*) $y = x - 1 + ce^{-x}$. (*b*) $y = ce^{-x^2/2}$.

for each choice of the constant c; just check that, for all real x, $y' = 1 - ce^{-x}$, so

$$y' + y = (1 - ce^{-x}) + (x - 1 + ce^{-x}) = x.$$

Thus there are actually infinitely many distinct solutions, and the solution $y(x) = x - 1$ can be obtained from the more general formula above by setting $c = 0$. Figure 1.1(a) shows the graphs of the solutions corresponding to $c = 0$, $c = 0.15$, and $c = 0.25$. Notice that these graphs converge together as x increases since the term ce^{-x} by which they differ tends to 0 as x tends to infinity.

example 2

The differential equation

$$y' + xy = 0$$

has the solutions $y = ce^{-x^2/2}$, one for each value of c, since we can check that

$$y' + xy = (-cxe^{-x^2/2}) + x(ce^{-x^2/2}) = 0.$$

Figure 1.1(b) shows the solution corresponding to $c = -0.5$ and $c = 1$. For $c = 0$, we get $y = 0$ identically for all x, so the graph of this solution coincides with the x axis.

example 3

Suppose $f(x)$ defines a continuous function for $-\infty < x < \infty$. The differential equation $y' = f(x)$ has as solutions

$$y = \int_0^x f(t)\,dt, \quad \text{or more generally} \quad \int_0^x f(t)\,dt + C,$$

one solution for each fixed C. In this example, the number C distinguishes the various solutions from one another by making uniform vertical shifts in the graphs. Figure 1.2(a) shows some examples for which $f(x) = 1/(1 + x^2)$.

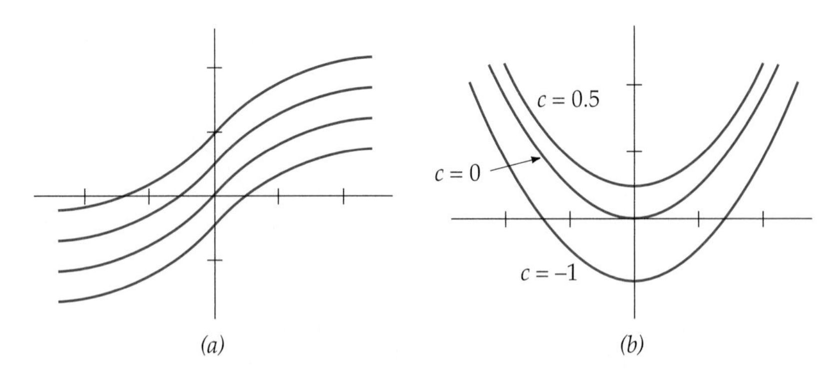

Figure 1.2 (*a*) $y = \arctan x + C$. (*b*) $y = \frac{1}{2}x^2 + C$.

Choosing $C = 0$ gives $y = \int_0^x (1 + t^2)^{-1}\,dt = \arctan x$; if $C = 1$, then $y = \int_0^x (1 + t^2)^{-1}\,dt = \arctan x + 1$.

If $f(x) = (1+x^2)^{-1}$ were replaced by some complicated function that did not have an indefinite integral in terms of an elementary formula, we could still find an approximation to the values of a solution by numerical approximation with Simpson's rule over successive intervals. (Recall that studying and computing a function such as $\arctan x$ for the first time is also a lot of work.) On the other hand, arctan is already "well known," so a solution that can be expressed in terms of arctan is often preferred over the results of an approximate computation.

1B Initial Values and Solution Families

A typical first-order differential equation has infinitely many solutions on some interval, and each of these solutions has a graph passing through infinitely many points. To distinguish one solution from another in a convenient way, we can pick a point x_0 in the common domain of several solutions. Then let $y_0 = y(x_0)$, where $y(x)$ is some particular solution, and label that solution with the pair (x_0, y_0). Geometrically, what we just did was pick a point (x_0, y_0) on the given solution curve and then use that point to specify the solution. The choice of a point (x_0, y_0) is called an **initial condition** and is often expressed in the form $y(x_0) = y_0$. General conditions under which such a condition specifies one and only one solution are taken up in Section 2C.

---

example 4 The differential equation

$$\frac{dy}{dx} = x$$

has solutions

$$y = \int_0^x t\,dt + c = \tfrac{1}{2}x^2 + c.$$

To pass through the point $(x_0, y_0) = (-1,1)$, the solution graph must satisfy

$$1 = \tfrac{1}{2}(-1)^2 + c.$$

Thus $c = \frac{1}{2}$, and the desired solution is $y = (x^2+1)/2$; this and two other solutions are shown in Figure 1.2(b) that satisfy different initial conditions. Note that in this example the graph of a solution crosses the y axis at $y = c$.

---

example 5

We saw in Example 2 that the differential equation

$$\frac{dy}{dx} = -xy$$

has solutions

$$y = Ce^{-x^2/2}.$$

For its graph to pass through the point $(x_0, y_0) = (0,2)$, a solution must satisfy

$$2 = Ce^0 = C.$$

Thus $C = 2$, and the particular solution selected by the condition $(x_0, y_0) = (0,2)$ is $y = 2e^{-x^2/2}$. As in the previous example, the value of the constant C marks the point on the y axis where the corresponding solution crosses that axis.

---

example 6

The formula $y = mx$ describes a family of lines through the origin of the xy plane, each line with a different slope m. Differentiating the equation with respect to x gives

$$\frac{dy}{dx} = m.$$

If $x \neq 0$, substituting the slope $m = y/x$ into the differential equation shows that the family of lines satisfies

$$x\frac{dy}{dx} = y.$$

In a first-order equation involving x, y, and dy/dx, a given pair of values (x, y) should ideally determine a unique value of dy/dx, which will be the slope of a solution graph passing through the point (x, y). The differential equation $x(dy/dx) = y$ in the previous example fails in that respect, because when $x = 0$ the equation specifies no particular value for dy/dx. To avoid this situation in general discussions we'll sometimes restrict attention to what are called **normal form** equations

$$y' = F(x, y),$$

in which $F(x, y)$ and hence $y'(x)$ is clearly defined on some subset of the xy plane.

If we try to single out from the solutions of a differential equation one that satisfies a given initial condition of the form $y(x_0) = y_0$, we encounter what is called an **initial-value problem.** Thus an initial-value problem for a normal form first-order differential equation $y' = F(x, y)$ on an interval $a < x < b$ makes two requirements of a solution $y = y(x)$:

(i) $y'(x) = F\left(x, y(x)\right)$ for $a < x < b$.

(ii) $y(x_0) = y_0$.

For many of the most important differential equations, the collection of solutions is large enough that a quite arbitrary initial condition can be met. In practice, this is often done by first identifying a set of solutions depending on one or more parameters, for example, a constant of integration C, and then choosing C so that the initial condition is satisfied. For a first-order equation, a set of solutions will typically depend on just one parameter, and the term **1-parameter family** is used to describe such a set. In Examples 4 and 5, the solution families $y = \frac{1}{2}x^2 + C$ and $y = Ce^{-x^2/2}$ are 1-parameter families, with parameter denoted by C. We'll see later that these two 1-parameter families describe *all* solutions of their respective differential equations. The next example is exceptional in that it has a simple 1-parameter family of solutions, all of which are physically meaningful but that still fails to describe all solutions.

example 7

Suppose an object is dropped from rest, hence with initial velocity zero, near the surface of the earth in such a way that you can ignore air resistance and slight variations in g, the acceleration of gravity. Let $s = s(t)$ be the object's vertical displacement at time t from the dropping point. From the formula $s(t) = \frac{1}{2}gt^2$, derived in calculus courses (see Exercise 47 on page 24), it follows that $s'(t) = gt$. Since $2gs = g^2t^2$, it follows that $s' = \sqrt{2gs}$. This differential equation for the dropped object is satisfied more generally by the one-parameter family

$$s(t) = \left(\sqrt{\frac{g}{2}}\,t + C\right)^2, \quad \text{with derivative} \quad s'(t) = \sqrt{2g}\left(\sqrt{\frac{g}{2}}\,t + C\right).$$

Taking the square root of the expression for $s(t)$ enables us to eliminate $\sqrt{g/2}\,t + C$ from the expression for s', giving $s' = \sqrt{2gs}$. Notice that the differential equation makes sense only when $s \geq 0$. Notice also that the differential equation itself contains significant information about the object's motion. Since s' is the object's velocity, the differential equation shows that velocity is completely determined by s, a fact not directly obtainable from the solution formula in terms of t. On the other hand, if $s(t)$ is intended to measure distance from the initial dropping point as a function of time t, the parameter C has the significance that $0 = s(0) = C^2$. Hence this initial condition requires that $C = 0$, which singles out the particular solution $s(t) = (\sqrt{g/2}\,t + 0)^2 = \frac{1}{2}gt^2$. Exercises 47 and 48 show that the first-order equation previously displayed follows from the assumption that g is the acceleration of gravity.

---

example 8

The previous example is atypical in that the initial-value problem

$$\frac{ds}{dt} = \sqrt{2gs}, \quad s(0) = 0,$$

has an additional solution, namely, $s(t) = 0$ for all t, aside from the conventional one $s(t) = \frac{1}{2}gt^2$. In cases like this we may need to resort to the physical interpretation that we place on the solution to help us decide which one to accept. Since we're talking about a falling object, the constant solution $s(t) = 0$ would seem to be ruled out, so we accept the other one. This example shows that even though a function describing a physical process can be shown logically to satisfy a differential equation, it does not follow logically that every solution of the equation gives a valid description of the physical process. The reason is that the physical assumptions used to derive the differential equation may not be satisfied by all solutions, as in the present example. The matter of nonuniqueness for the solution of an initial-value problem is taken up in detail in Section 2C.

Examples 6 and 7 show that even if we have a solution formula for a physical problem, a more complete story may still be had by concentrating directly on the associated differential equation itself. But what if we start with a family of functions; can we find a differential equation satisfied by all members of the family? In practice we can best proceed as follows:

1. Differentiate the equations that define the members of the family with respect to a chosen independent variable to produce a second 1-parameter family of equations containing derivatives of the dependent variable.
2. Eliminate the parameter using, if necessary, both families of equations to get a differential equation independent of the parameter.

---

example 9

The equations $x + cy = 0$, with parameter $c \neq 0$, describes a family of lines in the xy plane, each line with a different slope $-1/c$. Think of y as a function of x, and differentiate with respect to x to get

$$1 + c\frac{dy}{dx} = 0.$$

But $c = -x/y$ from the original equation, so the desired differential equation is

$$1 + \left(\frac{-x}{y}\right)\frac{dy}{dx} = 0 \quad \text{or} \quad x\frac{dy}{dx} = y \quad \text{or} \quad \frac{dy}{dx} = \frac{y}{x}.$$

Satisfying an initial condition $y(x_0) = y_0$ can be done by solving $x_0 + cy_0 = 0$ for c, and this works just fine unless $y_0 = 0$; in that case, the only initial condition we can satisfy is $y(0) = 0$, which is satisfied by all the lines in the family.

example 10 The 1-parameter family of curves determined by the implicit relation

$$x^2 + c^2y^2 = 1, \quad c \neq 0,$$

consists of ellipses, all passing through the four points $(x, y) = (\pm 1, 0), (0, \pm 1/c)$. If we want to regard x as independent variable, the implicitly defined functions that could serve as solutions for a differential equation for y' are

$$y(x) = (1/c)\sqrt{1 - x^2}, \quad c \neq 0.$$

However, it's easier to work with the implicit relation directly. Using implicit differentiation with respect to x, we get $2x + 2c^2yy' = 0$ or $y' = -x/(c^2y)$. From the original equation relating x and y, we find $1/c^2 = y^2/(1 - x^2)$, so substitution into $y' = -x/(c^2y)$ gives

$$y' = -(x/y)\frac{y^2}{1 - x^2} \quad \text{or} \quad y' = \frac{xy}{x^2 - 1}.$$

Note that the family of ellipses is defined only for $-1 \leq x \leq 1$ and that the vertical tangents at $(\pm 1, 0)$ correspond to points at which the right side of the differential equation is undefined. (This differential equation also has solutions valid for $|x| > 1$; methods for finding them are developed in Sections 3 and 5.)

EXERCISES

For each of the differential equations in Exercises 1 through 15, carry out the details of checking to see whether the given function is a solution on the specified interval.

1. $y' + y = 0; \ y = ce^{-x}, \ -\infty < x < \infty, \ c$ const.
2. $y' = 2y; \ y = ce^{2x}, \ -\infty < x < \infty, \ c$ const.
3. $\dfrac{dy}{dt} + 2ty = 0, \ y = be^{-t^2}, \ -\infty < t < \infty, \ b$ const.
4. $y' = y + x; \ y = ce^{-x} - x - 1, \ -\infty < x < \infty, \ c$ const.
5. $x\dfrac{du}{dx} = 1; \ u = \ln(cx), \ 0 < x, \ c$ pos. const.
6. $\dfrac{dy}{dx} = 1 + y^2; y = \tan(x - c), \ -\pi/2 + c < x < \pi/2 + c, \ c$ const.
7. $\dfrac{dz}{dx} = 2\sqrt{|z|}, \ z = x|x|, \ -\infty < x < \infty.$
8. $\dfrac{dy}{dx} = 3x^2y + 1; \ y = e^{x^3}\displaystyle\int_0^x e^{-t^3}\,dt, \ -\infty < x < \infty.$

9. $\dfrac{dy}{dx} = \sqrt{1+x^3};\ y = \int_0^x \sqrt{1+t^3}\,dt + C,\ -1 < x < \infty,\ C$ const.

10. $\dfrac{dy}{dx} = 2x\sqrt{1+x^6};\ y = \int_0^{x^2} \sqrt{1+t^3}\,dt,\ -\infty < x < \infty.$

11. $\dfrac{d^2y}{dt^2} + 4y = 0;\ y = c_1 \cos 2t + c_2 \sin 2t,\ -\infty < t < \infty,\ c_1,\ c_2$ const.

12. $\dfrac{d^2y}{dt^2} - 4y = 0;\ y = c_1 e^{2t} + c_2 e^{-2t},\ -\infty < t < \infty,\ c_1,\ c_2$ const.

13. $\dfrac{dy}{dt} = \sqrt{y};\ y = 0,\ -\infty < t < \infty.$

14. $\dfrac{dy}{dt} = \sqrt{y};\ y = \begin{cases} 0, & t < 0, \\ t^2/4, & 0 \le t, \end{cases}\ -\infty < t < \infty.$

15. $y' = (y \ln y)/x; y = e^{cx},\ x > 0,\ c = \text{const}.$

16. **(a)** The differential equation $(dx/dt)^2 + t^2 = 0$ has no real-valued solution $x = x(t)$ on a given interval $a < t < b$; explain why.

(b) The differential equation $(dx/dt)^2 + x^2 = 0$ has exactly one real-valued solution $x = x(t)$ on a given interval $a < t < b$; explain why. What is the unique solution?

For each of the one-parameter families of functions in Exercises 17 through 26, find the particular solution that satisfies the associated initial condition; then sketch the graph of that solution. Note that some solutions are implicitly defined.

17. $y = ce^{3x}; y(0) = 7.$

18. $y = ke^{-7x}; y(0) = 7.$

19. $x = \ln(ct);\ x(1) = -2.$

20. $x = c \ln t;\ x(2) = -1.$

21. $y = k \cos x; y(\pi) = 2.$

22. $y = \tan(cx);\ y(\pi) = 1.$

23. $x^2 + y^2 = c^2 > 0;\ y(1) = -1.$

24. $x^2 - c^2y^2 = 1,\ c \neq 0;\ y(2) = 1.$

25. $y^2 = cx;\ y(3) = -3.$

26. $x \ln y = k;\ y(1) = e.$

For each of the 1-parameter solution formulas in Exercises 27 through 36, find a first-order differential equation satisfied by all members of the family.

27. $y = ce^{3x}.$

28. $y = ke^{-7x}.$

29. $x = \ln(ct);\ x(1) = -2.$

30. $x = c \ln t;\ x(2) = -1.$

31. $y = k \cos x.$

32. $y = \tan(cx).$

33. $x^2 + y^2 = c^2 > 0.$

34. $x^2 - c^2y^2 = 1.$

35. $y^2 = cx.$

36. $x \ln y = k.$

It doesn't necessarily follow that because some members of a 1-parameter family satisfy a differential equation, all members do. By substitution into the differential equation in Exercises 37 through 40, find all values of the parameter r such that the corresponding function is a solution.

37. $x^2y'' + 5xy' + 4y = 0;\ y = x^r$.

38. $y'' + 5y' + 4y = 0;\ y = e^{rx}$.

39. $y' = (y \ln y)/x;\ y = re^x$.

40. $y' = 3y;\ y = re^{rx}$.

41. The definition of the exponential function is sometimes introduced along with a model for organic growth in which the size $P(t)$ of a population at time t satisfies

$$\frac{dP}{dt} = kP, \quad 0 < k = \text{const.}$$

This equation is satisfied by $P(t) = Ce^{kt}$, where C is a constant that would normally be positive given the interpretation of P as a population size.

(a) Show that $C = P(0)$.

(b) If $P(0) = 2$ and $P(2) = 6$, find k.

(c) Show that a population following this growth model, often called the **exponential growth law** increases by the factor $R > 1$ in any given time interval of length $t_R = (\ln R)/k$.

42. The **radioactive decay equation** is

$$\frac{dQ}{dt} = -kQ, \quad 0 < k = \text{const.}$$

This equation is satisfied by the quantity $Q(t)$ of a radioactive element remaining after time t, starting with some initial amount $Q(0)$. The differential equation has solutions of the form $Q(t) = Ce^{-kt}$, where C is constant.

(a) Show that $C = Q(0)$.

(b) If $Q(0) = 4$ and $Q(1) = 1$, find k.

(c) Show that a sample of radioactive element decreases by a factor r, $0 < r < 1$, in any given time interval of length $t_r = -(\ln r)/k$.

In Exercises 43 through 46 find a first-order differential equation solved by $y = y(x)$, where

43. The slope of the graph of $y(x)$ at (x, y) equals the distance from (x, y) to (0,0).

44. The tangent line to the graph of $y(x)$ at (x, y) contains the point $(x + y, x + y)$.

45. The rate of change of $y(x)$ with respect to x is proportional to x and inversely proportional to y.

46. The function $y(x)$ is constant.

47. Let g be the acceleration of gravity near the surface of the earth. ($g \approx 32.2$ feet per second per second.) By Newton's law, an object falling with negligible air resistance has acceleration

$$\frac{dv}{dt} = g,$$

where $v = v(t)$ is the velocity of the object at time t. Use integration to derive the following relations:

(a) $v(t) = gt + v_0$, where v_0 is the velocity at time 0.

(b) $s(t) = \frac{1}{2}gt^2 + v_0 t + s_0$, where $s = s(t)$ is the distance at time t of the object from the reference point $s = 0$.

48. An object dropped near the earth's surface falls a distance $s(t) = \frac{1}{2}gt^2$ in time t; specifically, we assume $s_0 = s(0) = 0$ and $v_0 = s'(0) = 0$.

(a) Show that $s = s(t)$ satisfies the first-order differential equation

$$\frac{ds}{dt} = \sqrt{2gs}.$$

(b) Show that the differential equation in part (a) has as solutions each member of the one-parameter family

$$s = \left(\sqrt{\frac{g}{2}}t + c\right)^2 = \tfrac{1}{2}gt^2 + \sqrt{2g}\,ct + c^2.$$

(c) Show that the motions described in part (b) are quite special in that they all satisfy $v(0)^2 = 2gs(0)$.

49. The general solution to the falling object problem treated in the two previous problems is

$$s = \tfrac{1}{2}gt^2 + v_0 t + s_0,$$

where v_0 is initial velocity and s_0 is initial displacement.

(a) Show that $s = s(t)$ satisfies the first-order differential equation

$$\frac{ds}{dt} = \sqrt{v_0^2 + 2g(s - s_0)}.$$

[*Hint:* Solve for t in terms of both s and ds/dt.]

(b) Show that the expression $v_0^2 + 2g(s - s_0)$ under the radical is always nonnegative, given our assumptions on s. [*Hint:* When does that expression reach its minimum as a function of t?]

50. Flow of liquid from a tank. A cylindrical tank with cross-sectional area A has an outlet hole in its side near the bottom. If $h = h(t)$ is the height of an ideal fluid above the outlet at time t, and a is the area of the outlet hole, then $V(t)$, the remaining fluid volume at time t, satisfies **Torricelli's equation**

$$\frac{dV}{dt} = -a\sqrt{2gh}.$$

An intuitive justification for the equation is to note that it depends on having the outlet velocity equal to the free-fall velocity of a drop of fluid from height h, as derived in the previous exercise; thus $-dV/dt$ equals area

a times outlet velocity $\sqrt{2gh}$. (A thoroughly scientific justification depends on principles of fluid mechanics.) Thus for an ideal fluid, the equation takes the form

$$\frac{dh}{dt} = -\frac{a}{A}\sqrt{2gh}.$$

(a) By analogy with the results of Example 7 of the text, show that the Torricelli equation has a solution of the form $h(t) = (bt + c)^2$. Then determine what the constants b and c must be.

(b) Use your answer to part (a) to find out how long it would take for the fluid height above the outlet to drop from h_0 to 0. In particular, estimate how long it would take to empty a full cylindrical tank with diameter 10 feet, height 20 feet, and circular outlet at the bottom with diameter 6 inches.

2 DIRECTION FIELDS

Just as the graph of a function can be said to be a complete geometric representation of the function, so the direction field of a first-order differential equation, described below, represents the differential equation. This fundamental idea can be used to get information about solutions without going to the trouble of finding formulas for solutions.

2A Definition and Examples

Since y' can be interpreted as a slope, a differential equation written in the **normal form**

$$y' = F(x, y)$$

assigns slope $F(x, y)$ to the point (x, y) in the xy plane. Such an assignment is called a **direction field** or **slope field.** If $y(x)$ is a solution to the differential equation, then

$$y'(x) = F\big(x, y(x)\big).$$

But since $y'(x)$ is the slope of the tangent to the graph of the solution at the point with coordinates $(x, y(x))$, the number $F\big(x, y(x)\big)$ is also equal to that slope. The slope then specifies a direction along the solution curve $y = y(x)$. From one point of view, we can think of starting with the solution $y = y(x)$ and locating a short segment of tangent line at each point of the graph of $y = y(x)$; some of these are shown in Figure 1.3(b). In Figure 1.3(a) we draw a segment of slope $F(x, y) = y$ at selected points (x, y). In this way, we get a sketch of the direction field associated with the differential equation $y' = y$. If the sketch is skillfully made, it is often possible to get a good idea of what the graphs of solutions $y = y(x)$ should look like. The motivating principle behind direction fields is that a solution graph should be tangent to the field segment located at each point that the graph passes through. Figure 1.3(b) shows such a sketch of a direction field, along with the graphs of three curves that are tangent to the segments in

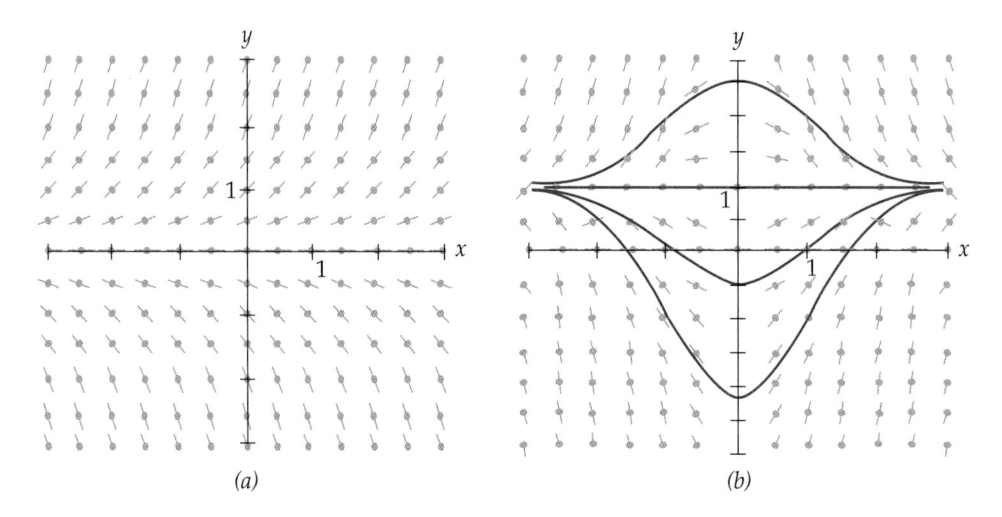

Figure 1.3 *(a)* Slope $F(x, y) = y$ at (x, y). *(b)* Slope $x - xy$ at (x, y).

the direction field. The direction field is associated with the differential equation

$$y' = x - xy.$$

Thus the function F in this case is $F(x, y) = x - xy$. In particular, $F\left(1, \frac{1}{2}\right) = \frac{1}{2}$ so that the segment through $(x, y) = \left(1, \frac{1}{2}\right)$ has slope $\frac{1}{2}$. Notice that there are apparently infinitely many possible curves that could be drawn tangent to the direction field. Of these infinitely many curves, Figure 1.3(b) shows only four; others could easily be drawn in.

***example* 1**

The differential equation

$$\frac{dy}{dx} = \frac{y^2}{x^2 + y^2}$$

defines a direction field for all (x, y) except (0,0). To sketch the field, we can first make a table of slopes. In this example, we will restrict ourselves to the first quadrant, because replacing x by $-x$ or y by $-y$ leaves the slope unchanged. We find the array of slopes shown in Table 1.1.

Looking at the table by itself doesn't convey much in the way of understanding; it's only when the information is displayed graphically that the data conveys something about the differential equation and its solutions. These points (x, y), plus many more, are plotted in Figure 1.4 along with segments having the assigned slopes. The pattern that emerges is a good guide to the general shape of the solution graphs. Although it can't be shown in the picture, it is also worth noticing what happens as $x \to \infty$ and $y \to \infty$. If $x \to \infty$ and y is held fixed, then the slopes tend to zero. If $y \to \infty$ with x held fixed, then the slopes tend to 1. Three solution curves are drawn in, showing their inclination to be consistent with the direction field. There is in fact an "elementary" solution formula for this differential equation, computable by one of the methods of Section 3C. But if what we want is simply a general picture of

Table 1.1 Direction field slopes for $dy/dx = y^2/(x^2 + y^2)$

(x, y)	$\frac{dy}{dx}$	(x, y)	$\frac{dy}{dx}$
(0,1)	1	(2,3)	$\frac{9}{13}$
(0,2)	1	(3,0)	0
(0,3)	1	(3,1)	$\frac{1}{10}$
(1,1)	$\frac{1}{2}$	(3,3)	$\frac{1}{2}$
(1,2)	$\frac{4}{5}$	(4,0)	0
(1,3)	$\frac{9}{10}$	(4,1)	$\frac{1}{17}$
(2,0)	0	(4,2)	$\frac{1}{5}$
(2,1)	$\frac{1}{5}$	(4,3)	$\frac{9}{25}$
(2,2)	$\frac{1}{2}$	(4,4)	$\frac{1}{2}$

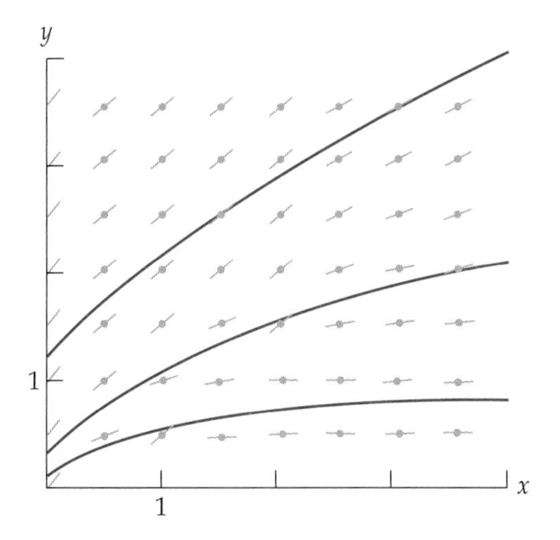

Figure 1.4 Direction field sketch for $dy/dx = y^2/(x^2 + y^2)$.

the behavior of solution graphs, we may be satisfied with a sketch of the direction field.

2B Isoclines

Sketching a direction field by hand requires a certain amount of judgment about how much information to include in the sketch. As with a caricature of a person's face, the trick is to include just enough information to enable the viewer to recognize what is being represented. Too much detail may obscure the essential features, and too little may fail to show a recognizable pattern. One way to organize a sketch is to begin by sketching some **isoclines,** that is, curves along which all slopes, or inclines, of the field are equal to each other. This organizing principle enables us to draw in parallel segments of the field along each isocline curve, making sure to include enough isoclines to ensure a good sketch of the field as a whole. Stated in terms of the function F in the differential equation $y' = F(x, y)$, an isocline is an implicitly defined level curve $F(x, y) = m$ at level m. An isocline is *not* in general the graph of a solution; hence, the isocline curves themselves should not figure prominently, if at all, in the final sketch of a direction field.

example 2

The differential equation $y' = x$ has a very simple direction field since the slopes are equal on vertical lines, that is, lines determined by equations $x = m$, where m is constant. These lines are the isoclines of the direction field, and are indicated by the vertical dotted lines in Figure 1.5(a). When $m = -1$, the slopes of the field are all -1; when $m = 0$, the slopes are all zero; when $m = 1$, the slopes are all 1; and so on. As shown in Figure 1.5(a), these slope segments are easily sketched in by hand once a single slope segment has been put in place, because the other segments on the same isocline are all parallel to the first one. Given a completed sketch of the direction field, it's an easy matter to sketch a few solution curves, which happen to be parabolas having equations of the form $y = \frac{1}{2}x^2 + C$, with C constant.

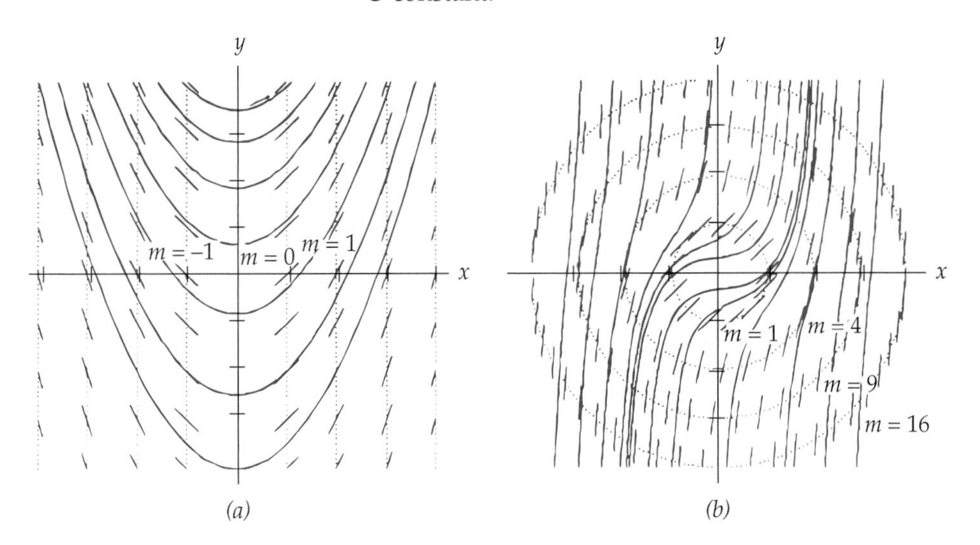

Figure 1.5 (*a*) $y' = x$. (*b*) $y' = x^2 + y^2$.

In the previous example the ultimate aim was to sketch the graphs of the solutions to $y' = x$. Since these graphs were simply parabolas, one might argue that they could have been sketched without the aid of the direction field. However, making even moderately accurate sketches of parabolas is not as easy as you might think, and the direction field is definitely helpful in doing the job by hand. In the next example, there are no available formulas for the solution graphs, so we rely more heavily on the direction field.

example 3 The differential equation $y' = x^2 + y^2$ has no elementary formula that describes its solutions. Furthermore, listing the slopes associated with a rectangular array of points requires computing many different numbers, even though noticing that the slopes are the same at (x, y) and $(-x, -y)$ is of some help. More effective is first to sketch the isocline curves $x^2 + y^2 = m$. Since these curves are circles of radius $\sqrt{m}$, it's natural to pick values such as $m = 1, 4, 9$, and 16 as in Figure 1.5(b). Then sketch in segments of slope m around the corresponding circle. Finally, sketch in the solution graphs, giving them at each point a tangent direction consistent with the nearby slope segments.

If a direction field and its isoclines are both hard to picture, there may be no alternative to the brute force method of plotting many segments. This is best done using computer graphics; here is a description of how the underlying program could be constructed. Let's assume that we want to plot equal segments of slope $F(x, y)$ centered on points (x, y) at the corners of a square grid on a rectangular region R, where s is to be the horizontal and vertical distance between plotted points. The segment from (0,0) to $\big(1, F(x, y)\big)$ has the correct slope, and we divide by its length $\sqrt{1 + F(x, y)^2}$ to get a segment of length 1 and slope $F(x, y)$ from (0,0) to

$$\left(\frac{1}{\sqrt{1 + F(x, y)^2}}, \frac{F(x, y)}{\sqrt{1 + F(x, y)^2}}\right).$$

What we do is both add to and subtract from (x, y) a small multiple of this segment to find the endpoints of our segment centered at (x, y). Taking the small multiplier to be $s/3$, the process could be formalized as follows:

```
FOR x = x1 TO x2 BY STEP s
FOR y = y1 TO y2 BY STEP s
SET d = (1 + F(x, y)^2)^(1/2)
PLOT LINE FROM (x - s/(3d), y - sF(x, y)/(3d)) TO (x + s/(3d), y + sF(x, y)/(3d))
NEXT x
NEXT y
```

Routines like this were used to plot some of the pictures in this book, and others like it have been incorporated into a number of widely available plotting routines, including the Java Applet DIRFLD at the website <http://math.dartmouth.edu/~rewn/> or through a link from <http://www.mhhe.com/williamson>. This application will plot direction fields for differential equations of the form $y' = f(x, y)$. It is necessary for the user to specify the function $f(x, y)$ along with parameters that determine the extent of the field, as well as the spacing between points where the slopes are to be evaluated.

EXERCISES

For each of the differential equations in Exercises 1 through 6, sketch the associated direction field at points (x, y) with $-2 \le x \le 2$ and $-2 \le y \le 2$. Include enough points in your sketch so that a pattern becomes apparent. Then sketch in a few solution curves.

1. $y' = x$.

2. $y' = y$.

3. $y' = -1$.

4. $y' = 0$.

5. $y' = -y$.

6. $y' = -2x$.

For each of the differential equations in Exercises 7 through 12, sketch the associated direction field at points (x, y) for which the differential equation is defined and for which $-2 \le x \le 2$ and $-2 \le y \le 2$. Include enough points in your sketch so that a pattern becomes apparent. You may want to use isoclines to help in sketching the field. Then sketch in a few solution curves.

7. $y' = 2x^2 + y^2$.

8. $y' = x^2 + y$.

9. $y' = x - y$.

10. $y' = \sqrt{x^2 + y^2}$.

11. $y' = x/(1 + y^2)$.

12. $y' = \sqrt{1 - x^2 - y^2}$.

For each of the initial-value problems in Exercises 13 through 18, sketch enough of the associated direction field to enable you to make a sketch of the solution for increasing x. Then sketch the solution.

13. $y' = y + 2x, \; y(0) = 1$.

14. $y' = (x^2 + y^2)^{-1}, \; y(1) = 0$.

15. $y' = xy, \; y(0) = 1$.

16. $y' = xy, \; y(0) = 0$.

17. $y' = \cos y, \; y(0) = 1$.

18. $y' = xy + y^2, \; y(-1) = -1$.

19. Consider the differential equation

$$y' = \begin{cases} 0, & x < 0, \\ x, & x \ge 0, \end{cases}$$

(a) Sketch the direction field for $-2 \le x \le 2$.

(b) Find a two-piece formula for the solution to the differential equation satisfying the initial condition $y(0) = 1$.

20. There is no elementary formula for solutions of the differential equation $y' = \sqrt{1 + y^4}$; use its direction field to sketch solutions passing through the points (–1,0), (0,1), and (1,1).

21. There is no elementary formula for solutions of the differential equation $y' = \sqrt{1 + y^3}$; use its direction field to sketch solutions passing through the points (–1,0), (0,1), and (1,1). What about (0,–2)?

22. There is no elementary formula for solutions of $y' = 1/\sqrt{1 + y^4}$; use its direction field to sketch solutions passing through the points (–1,0), (0,1), and (1,1).

23. The implicit formula $y + \sqrt{1+y^2} = Ce^x$, $C = \text{const.}$ determines solutions of the differential equation $y' = \sqrt{1+y^2}$, but the formula isn't very helpful for making a sketch of the solution satisfying $y(0) = 1$.

(a) Use the direction field to sketch this solution.

(b) Use implicit differentiation to verify that functions $y = y(x)$ satisfying the implicit formula given above also satisfy the differential equation $y' = \sqrt{1+y^2}$.

(c) Show that solutions of the differential equation $y' = \sqrt{1+y^2}$ also satisfy the *second-order* differential equation $y'' = y$. What can you conclude about concavity of the solution graphs of the first-order differential equation?

24. Let

$$F(y) = \begin{cases} 0, & y \le 0, \\ \sqrt{y}, & y > 0. \end{cases}$$

(a) Sketch the direction field of the differential equation $y' = F(y)$ for $-2 \le x \le 2$, $-2 \le y \le 2$.

(b) Verify that there are two distinct solutions of $y' = F(y)$ passing through each point of the form $(x_0, 0)$. [*Hint:* Consider $y = (x - x_0)^2/4$ for $x > x_0$.]

Differentiate both sides each of the differential equations in Exercises 25 through 33 with respect to x to determine those regions where their solution graphs are concave up (where $y'' > 0$) or concave down (where $y'' < 0$). Use the information about concavity as an aid in sketching some typical solution graphs.

25. $y' = y^2$.

26. $y' = \sqrt{1+y^4}$.

27. $y' = x + y$.

28. $y' = xy$.

29. $y' = \sin y$.

30. $y' = xy^2$.

31. $y' = \sin x$.

32. $y' = y/x$.

33. $y' = x/y$.

34. Show that if $F(x, y)$ has continuous partial derivatives $F_x(x, y)$ and $F_y(x, y)$ with respect to x and y in some region R of the xy plane, then solutions of the first-order equation $y' = F(x, y)$ also satisfy the second-order equation $y'' = F_x(x, y) + F_y(x, y)F(x, y)$.

35. Show that if a, b, c, and d are constants with $ad - bc \neq 0$, the differential equation

$$y' = (ax + by)/(cx + dy)$$

has lines through the origin for its isoclines. Show by example what can go wrong if $ad - bc = 0$.

36. Show that every isocline of the differential equation $y' = y/x$ is itself the graph of a solution of the differential equation.

37. Show that no isocline of the differential equation $y' = x$ is a solution graph of the differential equation.

38. Find every isocline of the differential equation $y' = x + y$ and determine which, if any, is a solution graph of the differential equation.

39. Find every isocline of the differential equation $y' = x/y$ and determine which, if any, is a solution graph of the differential equation.

Indicate by a sketch or in some other way, the regions of the xy plane where solutions of the differential equations in Exercises 40 through 45 are increasing and the regions where solutions are decreasing. Note that if $y = y(x)$ is a solution of $y' = f(x, y)$, then the graph of $y(x)$ is increasing at points (x, y) where $f(x, y) > 0$ and decreasing where $f(x, y) < 0$.

40. $y' = x - y$.

41. $y' = e^{\sin xy}$.

42. $y' = x^3 - y$.

43. $y' = x^2 - y^2$.

44. $y' = 1 - x^2 - y^2$.

45. $y' = x(x^2 - y)$.

46. Bugs in mutual pursuit Four identical bugs are on a flat table, each moving at the same constant speed. Use (x, y) coordinates on the table, and locate bugs 1 through 4 initially in respective quadrants 1 through 4, each at one of the points $(\pm 1, \pm 1)$. Bug 1 always heads directly toward bug 2, bug 2 toward bug 3, bug 3 toward bug 4, and bug 4 toward bug 1, so their paths are mutually congruent.

(a) Use the symmetry of the paths to show that at any moment the bugs are at the corners of a square, and in particular, if bug 1 is at (x, y), then bugs 2, 3, and 4 are, respectively, at $(-y, x)$, $(-x, -y)$, and $(y, -x)$.

(b) Using the notation of part (a), show that the paths of bugs 1 and 3 satisfy the differential equation

$$\frac{dy}{dx} = \frac{y - x}{x + y},$$

unless $x + y = 0$. What is the analogous differential equation for bugs 2 and 4?

(c) Sketch the direction field for the differential equation of bug 1, and use it to sketch his path starting at $(1,1)$. [*Note:* This differential equation has an elementary solution formula derivable by the methods of Section 3C.]

Use a computer graphics routine to display the direction field for each of the differential equations in Exercises 47 through 52. Then sketch in a few solution curves.

47. $y' = x^3 + y^3, \quad -2 \le x \le 2, \; -2 \le y \le 2.$

48. $y' = x^2 - y^2, \quad -2 \le x \le 2, \; -2 \le y \le 2.$

49. $y' = e^{-x^2 - y^2}, \quad -2 \le x \le 2, \; -2 \le y \le 2.$

50. $y' = \sin(x + y), \quad -4 \le x \le 4, \; -4 \le y \le 4.$

51. $y' = \ln(1 + x^4 + y^4), \quad -2 \le x \le 2,\ -2 \le y \le 2.$

52. $y' = \sqrt{1 + x^2 + y^4}, \quad -4 \le x \le 4,\ -4 \le y \le 4.$

Orthogonal Trajectories. A curve that is perpendicular to each member of a given family of curves is called an **orthogonal trajectory** to the family. If the given family of curves satisfies a first-order differential equation $y' = f(x, y)$, then an orthogonal trajectory will quite generally be the graph of a function $y(x)$ that satisfies $y' = -1/f(x, y)$; since the product of the slopes of the two fields is -1 at each point, this just says that the direction field of the orthogonal trajectories is everywhere perpendicular to that of the given family.
For each of the 1-parameter families of functions in Exercises 53 through 64, find a first-order differential equation $y' = f(x, y)$ satisfied by the members of the family. Sketch in the most convenient order (i) the direction field of the orthogonal trajectory equation $y' = -1/f(x, y)$, (ii) a few graphs of orthogonal trajectories, and (iii) some graphs of members of the given family.

53. $y = cx.$

54. $y = x + c.$

55. $xy = c.$

56. $x^2 + y^2 = c^2.$

57. $y = cx^3.$

58. $y = (x + c)^2.$

59. $y = c \sin x.$

60. $y = \cos x + c.$

61. $y = c.$

62. $y = x^2 + c.$

63. $y = c/x.$

64. $y = c/x^2.$

***65.** A point P in the xy plane starts at the origin and moves up along the positive y axis, dragging behind it (as if attached by a string of length $a > 0$) another point Q that starts at $(a, 0)$ on the positive x axis. Thus the distance between the two points remains constantly equal to a, and the tangent line to the path of the trailing point Q always passes through the leading point P on the y axis. The path $y = y(x)$ of the trailing point Q in the first quadrant is called a **tractrix.**

(a) Show that the tractrix satisfies the differential equation $y' = -\sqrt{a^2 - x^2}/x$. [*Hint:* Draw a picture.]

(b) Show that the tractrix is concave up by showing that $y'' = a^2/(x^2\sqrt{a^2 - x^2}\,) > 0$.

(c) Sketch the direction field of the differential equation for $0 < x < 1$, assuming $a = 1$.

(d) Use the results of parts (b) and (c) to sketch the tractrix that starts at $(0, 1)$.

(e) Integrating both sides of the differential equation with respect to x and assuming $y(a) = 0$ shows that a tractrix is the graph of

$$y(x) = a \ln\left(\frac{a + \sqrt{a^2 - x^2}}{x}\right) - \sqrt{a^2 - x^2}.$$

Verify this by substitution into the differential equation.

(f) Show that the length of the tractrix over $x_0 \le x \le a$ is $a \ln(a/x_0)$. [*Hint:* Use the differential equation.]

2C Continuity, Existence, and Uniqueness

Even though, for practical reasons, the sketches that we make of direction fields are incomplete, the underlying mathematical direction field is in principle a complete geometric representation of a differential equation $y' = F(x, y)$. For this reason, the underlying concept of a direction field is fundamental to understanding first-order differential equations. In relying on a sketch of a direction field for drawing solution graphs for $y' = F(x, y)$, we assume that the slope segments don't vary discontinuously from point to point. In particular, we assume that $F(x, y)$ is a continuous function of the point (x, y). (Imagine how much trouble you'd have sketching solution curves if the field of slopes varied *discontinuously*, say in some random fashion.)

It seems reasonable to expect that if the way $F(x, y)$ varies were sufficiently irregular, the corresponding differential equation might not have any solutions at all. However it can be proved (see Peano's theorem in Appendix C), that continuity alone of $F(x, y)$ in a region R is enough to guarantee the existence of at least one solution whose graph contains a given point of R. The more stringent hypotheses of Theorem 2.1 below guarantee that if $F(x, y)$ is not only continuous in a region R of the xy plane but that if also $\partial F/\partial y$ is continuous there, then $y' = F(x, y)$ has a *unique* solution graph passing through every point of R, though the domain of the solution may be smaller than the width of R in the x direction. (Put slightly differently, the assertion that there is a unique solution through a point says that two different solution graphs will never have a point in common.) Referring back to Figure 1.5 on page 29, we can conclude that the xy regions shown there are completely filled by nonintersecting solution graphs. From what we know about the parabolas $y = \frac{1}{2}x^2 + c$ this result seems fairly clear in the first picture, but is not so clear just from looking at the second one. Appendix C gives some indication of the underlying idea for a proof of existence for Theorem 2.1. For an interesting alternative approach see Chapter 2 of V. I. Arnold's *Ordinary Differential Equations*, Springer (1991). As we proceed, our solution techniques will provide existence proofs for important special cases.

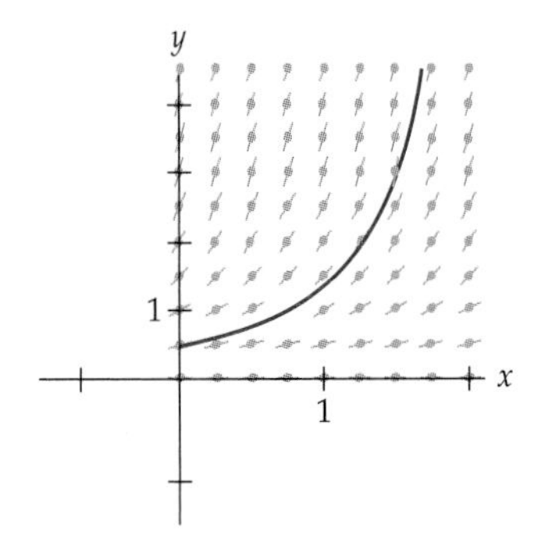

$y' = y^2$ defined for all (x, y), solution $y = 1/(2 - x)$ with $y(0) = 1/2$ defined only for $x < 2$.

2.1 EXISTENCE AND UNIQUENESS THEOREM

Assume that both $F(x, y)$ and its partial derivative $F_y(x, y)$ are continuous on an x interval I: $\alpha < x < \beta$, and for $a < y < b$. Then the initial-value problem

$y' = F(x, y)$, $y(x_0) = y_0$, with $a < y_0 < b$, has a unique solution defined on some possibly smaller subinterval J of I containing x_0. If the y interval is $-\infty < y < \infty$, and there is a constant B such that $|F_y(x, y)| \leq B$ for all x in I and for all y, then this unique solution is defined on the entire interval I.

example 4

If a boundedness condition $|F_y(x, y)| \leq B$ in Theorem 2.1 holds for all x in I and for all y, it prevents the slopes of the direction field from being so steep that the solution tends to infinity before reaching the ends of the domain interval. For example, if $F(x, y) = y$ then $F_y(x, y) = 1$ so $|F_y(x, y)| \leq B = 1$ for all real x and y. It follows from the theorem that there is a unique solution to the initial-value problem $y' = y$, $y(0) = y_0$ on the entire interval $-\infty < x < \infty$. One such solution is evidently $y = y_0e^x$, and the theorem guarantees that this is the only solution. We'll be able to prove uniqueness later in Section 4 without using the theorem.

example 5

Applying Theorem 2.1 to the differential equation $y' = 1 + y^2$, we first note that $F(x, y) = 1 + y^2$ and its partial derivative $F_y(x, y) = 2y$ are both continuous for all (x, y) with $-\infty < x < \infty$, $-\infty < y < \infty$. The conclusion of Theorem 2.1 asserts that there is a unique solution $y = y(x)$ satisfying an arbitrary initial condition $y(x_0) = y_0$ on *some* interval. Note, however, that the values of $F_y(x, y) = 2y$ are not bounded for $-\infty < y < \infty$, as the second part of Theorem 2.1 requires, so we can't conclude that the solution to an initial-value problem exists for all x in $I: -\infty < x < \infty$. Indeed, with initial condition $y(0) = 0$ you are asked to show in Exercise 23 that the unique solution is $y = \tan x$, defined and continuous only for $-\pi/2 < x < \pi/2$.

For some purposes, a better way to look at the implication of the existence and uniqueness theorem is that there is a unique solution graph through a point (x_0, y_0) and hence that no two of these graphs have a point in common. Some of the solution graphs for $y' = x^2 + y^2$ drawn in Figure 1.5(b), and while it looks as if some of them might touch each other if extended far enough up and down, Theorem 2.1 shows that they can't touch.

example 6

Some condition stronger than just continuity of $F(x, y)$ is needed to guarantee the uniqueness of solutions described in Theorem 2.1. As to how uniqueness can fail, consider the following direction field, as sketched here. Let

$$F(x, y) = \begin{cases} \sqrt{y}, & \text{if } y > 0, \\ 0, & \text{if } y \leq 0. \end{cases}$$

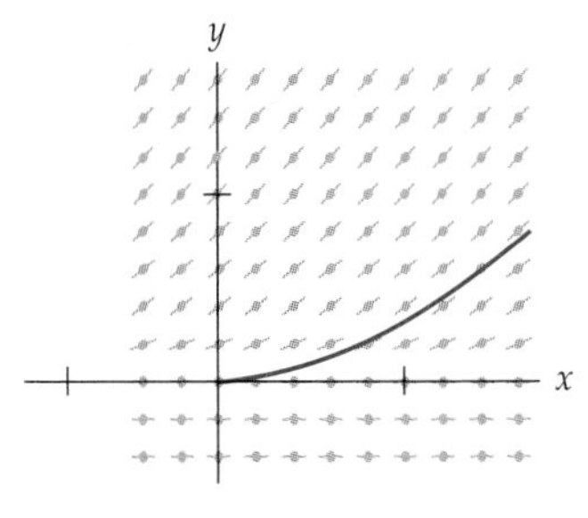

Direction field for $y' = F(x, y)$.

The function $F(x, y)$ is continuous at all points (x, y) for which $y > 0$ and also for which $y < 0$. At points $(x, 0)$ we also have continuity, because $\lim_{y \to 0} F(x, y) = 0$. However, $F_y(x, y) = \frac{1}{2}y^{-1/2}$ when $y > 0$ and fails to exist when $y = 0$. Hence Theorem 2.1 fails to apply to the equation $y' = F(x, y)$. You can verify directly that the equation has two distinct solutions satisfying the initial condition $y(0) = 0$; one of them is $y_1(x) = \frac{1}{4}x^2$ for $x \geq 0$, and the other one is $y_2(x) = 0$ for $x \geq 0$. But Exercise 22 on page 41 shows that generally speaking, failure of $F_y(x, y)$ to exist is not conclusive evidence for nonuniqueness.

example 7

Theorem 2.1 does apply to the differential equation $y' = \sqrt{y}$ of the previous example as long as we restrict consideration to points (x, y) for which $y > 0$. In that case, the partial derivative $\partial\sqrt{y}/\partial y = \frac{1}{2}y^{-1/2}$ is continuous, and we can conclude that there is a unique solution whose graph contains a specified point (x_0, y_0) with $y_0 > 0$.

example 8

If $F(x, y) = F(x)$ is independent of y and is continuous as a function of x on some interval $a < x < b$, then Theorem 2.1 applies to the differential equation $y' = F(x)$ at all points (x, y) of the infinite vertical strip containing the interval. The reason is that $\partial F(x)/\partial y$ is identically zero, and so of course is continuous. For instance, the field determined in Example 2 by the function $F(x) = x$ is of this type. This example is typical of those that are continuous but independent of y in that the existence of a unique solution through each point can be inferred without using Theorem 2.1 by using both versions of the fundamental theorem of calculus as follows. Suppose $y = y(x)$ is some solution of the initial-value problem $y' = F(x)$, $y(x_0) = y_0$. We can integrate both sides of the differential equation between x_0 and x using one version of the fundamental theorem to get

$$y(x) - y(x_0) = \int_{x_0}^{x} y'(t)\,dt = \int_{x_0}^{x} F(t)\,dt.$$

Since $y(x_0) = y_0$, it follows that $y(x)$ must be related to $F(x)$ by the equation

$$y(x) = y_0 + \int_{x_0}^{x} F(t)\,dt.$$

This establishes the uniqueness of the solution, because if $y(x)$ satisfies the differential equation, then $y(x)$ must be given by the previous formula. Regardless of where it came from, the same formula provides the existence of a solution to the initial-value problem. Just apply the other version of the fundamental theorem to differentiate both sides with respect to x to get $y'(x) = F(x)$, thus showing that the differential equation is satisfied. Then set $x = x_0$ to show that the initial condition $y(x_0) = y_0$ is also met. As a special case, the differential equation $y' = x$ of Example 2, with initial condition $y(0) = c$, yields the unique solution $y = \frac{1}{2}x^2 + c$.

EXERCISES

Each differential equation in Exercises 1 through 9 determines a direction field on a set S in the xy plane. Make a careful sketch of S in each case, and within that picture show the subset R of S where Theorem 2.1 guarantees a unique solution through each point of R.

1. $y' = xy^{4/3}$.

2. $y' = xy^{1/3}$.

3. $y' = \sqrt{xy}$.

4. $y' = \sqrt{x} + x^3y^4$.

5. $y' = y/x$.

6. $y' = x/y$.

7. $y' = \sin(x - y)$.

8. $y' = (1 + y^2)/(1 - y^2)$.

9. $y' = e^{\sin(xy)}$.

10. Does the differential equation

$$y' = \begin{cases} 0, & x < 0, \\ x, & 0 \le x, \end{cases}$$

have a unique solution with graph passing through a given point of the xy plane? Explain.

11. Define for all (x, y) the continuous function $K(x, y) = y^{2/3}$.

(a) Sketch the direction field of the differential equation $y' = K(x, y)$ on the rectangle $-2 \le x \le 2,\ -2 < y < 2$.

(b) Verify that there are *at least* three distinct solution curves through each point $(x_0, 0)$ and defined for all x. [*Hint:* Consider $y = (x - x_0)^3/27$ for $x \ge x_0$ and $y = 0$ for $x < x_0$.]

(c) Explain in detail why this example doesn't contradict Theorem 2.1.

12. Define for all (x, y) the continuous function

$$F(x, y) = \begin{cases} \sqrt{y}, & \text{if } y > 0, \\ 0, & \text{if } y \le 0. \end{cases}$$

(a) Sketch the direction field of the differential equation $y' = F(x, y)$ on the rectangle $-2 \le x \le 2,\ -2 < y < 2$.

(b) Verify that there are two distinct solution curves through each point $(x_0, 0)$ and defined for all x. [*Hint:* Consider $y = (x - x_0)^2/4$ for $x > x_0$.]

(c) Explain in detail why this example doesn't contradict Theorem 2.1.

13. A differential equation with the same essential features as the one in the previous exercise arises in mechanics.

(a) A body dropped from rest under constant gravitational acceleration g moves distance $s = \frac{1}{2}gt^2$ in time t. Show that this $s(t)$ satisfies the differential equation $ds/dt = \sqrt{2gs}$ with initial condition $s(0) = 0$.

(b) Show that the initial-value problem in part (a) has a solution other than $s = \frac{1}{2}gt^2$.

14. Define for all (x, y) the continuous function

$$G(x, y) = \begin{cases} x\sqrt{y}, & \text{if } y \geq 0, \\ 0, & \text{if } y < 0. \end{cases}$$

(a) Sketch the direction field of the differential equation $y' = G(x, y)$ on the rectangle $-2 \leq x \leq 2,\ -2 < y < 2$.

(b) Verify that there are two distinct solution curves through the point (0,0) and defined for all x. [*Hint:* Consider $y = x^4/16$ for $x > 0$.]

(c) Explain in detail why this example doesn't contradict Theorem 2.1.

15. (a) Verify that the differential equation $y' = \sqrt{1 - y^2}$ has solutions $y = \sin(x + c)$, where c is constant. Then sketch a solution graph passing through (0,0), following the convention that $\sqrt{1 - y^2} \geq 0$. Given c, for what x is $y = \sin(x + c)$ a valid solution?

(b) Sketch solution graphs for the equation in part (a) by relying entirely on a preliminary sketch of the direction field for $y' = \sqrt{1 - y^2}$. Show in particular that there is a solution graph through $(\pi/2, 1)$ given by

$$y(x) = \begin{cases} \sin x, & |x| \leq \pi/2, \\ 1, & x > \pi/2, \\ -1, & x < -\pi/2. \end{cases}$$

(c) Noting that $y(x) = 1$ is a solution with $y(\pi/2) = 1$, is there a conflict between the results in part (b) and Theorem 2.1? Explain.

16. Define for all (x, y) the continuous function

$$Q(x, y) = \begin{cases} \sqrt{1 - y^2}, & \text{if } -1 \leq y \leq 1, \\ 0, & \text{otherwise.} \end{cases}$$

(a) Sketch the direction field of the differential equation $y' = Q(x, y)$ on the rectangle $-\pi \leq x \leq \pi,\ -2 < y < 2$.

(b) Verify that there are two distinct solution curves through each point $(x_0, 1)$ and defined for all x; similarly for $(x_0, -1)$ [*Hint:* Consider $y = \sin(x + c)$ for $-\pi/2 - c \leq x \leq \pi/2 - c$; then extend the solution to have appropriate constant values.]

(c) Explain in detail why this example doesn't contradict Theorem 2.1.

17. The right-hand side of the differential equation $y' = (xy)^{3/2}$ is defined and continuous for all (x, y) in the closed quadrant $Q = \{(x, y) : x \geq 0,\ y \geq 0\}$.

(a) Determine the open set R of points (x_0, y_0) for which Theorem 2.1 guarantees the existence and uniqueness of a solution to the initial-value problem $y(x_0) = y_0$.

(b) Extend the function $F(x, y)$ to all (x, y) by

$$F(x, y) = \begin{cases} (xy)^{3/2}, & \text{for } (x, y) \text{ in } Q, \\ 0, & \text{for } (x, y) \text{ not in } Q. \end{cases}$$

Show that by applying Theorem 2.1 to the extended differential equation $y' = F(x, y)$ you can conclude that the initial-value problem $y(x_0) = y_0$ has a unique solution for all (x_0, y_0) in Q, thus avoiding the need for Q to be open for this example.

18. Suppose $y' = F(x, y)$ is a differential equation that meets the assumptions of Theorem 2.1 for all (x, y). Can it happen that the graphs of two different solutions of this differential equation have a point (x_0, y_0) in common? Explain your answer.

***19.** The region on which Theorem 2.1 guarantees a unique solution graph containing a given point need not be a rectangle. Define the continuous function $F(x, y)$ for all (x, y) by

$$F(x, y) = \begin{cases} \sqrt{1 - x^2 - y^2}, & \text{for } x^2 + y^2 < 1, \\ 0, & \text{for } x^2 + y^2 \geq 1. \end{cases}$$

(a) Show that Theorem 2.1 can be applied to the open disk $x^2 + y^2 < 1$ so as to ensure the existence of a unique solution to $y' = F(x, y)$, with graph passing through each point of the disk.

(b) Draw a conclusion analogous to that in part (a) for the open region $x^2 + y^2 > 1$.

(c) Can Theorem 2.1 be applied to $y' = F(x, y)$ on the entire xy plane? Explain.

(d) Use a sketch of the direction field of $y' = F(x, y)$ to sketch the graph of solution to the differential equation on the interval $-2 \leq x \leq 2$ and that satisfies the initial condition $y(-2) = 0$. (For solution existence see Peano's theorem, Theorem 1.3, in Appendix C.)

20. Let $a > 0$ be constant in $y' = a(1 + y^2)$.

(a) Verify that $y = \tan(ax)$ is a solution of the differential equation for $-\pi/(2a) < x < \pi/(2a)$, but for no larger interval containing $x = 0$.

(b) Make a sketch of the direction field for $y' = 1 + y^2$, and use it to account for the behavior of the solution $y = \tan x$.

21. Define $P(x)$ for all x by

$$P(x) = \begin{cases} 2x, & x \leq 0, \\ x, & x > 0. \end{cases}$$

(a) Sketch the graph of $P(x)$ for $-2 \leq x \leq 2$.

(b) Sketch the direction field of the differential equation $y' = P(x)$.

(c) Find a solution to $y' = P(x)$ satisfying $y(-1) = 0$, and sketch the graph of the solution along with your sketch of the direction field.

(d) Is the solution you found in part (c) the only one satisfying both the differential equation and the initial condition? Give reasons for your answer.

***22.** You might think that nonuniqueness for $x \geq 0$ in the initial-value problem $y' = \sqrt{y}$, $y(0) = 0$, is caused entirely by the way in which $\sqrt{y}$ fails to have a derivative at $y = 0$. The following example shows that the matter is subtler than that, and is significant for Torricelli's equation in Section 3B. Let

$$y' = \begin{cases} -\sqrt{y}, & y \geq 0, \\ 0, & y < 0, \end{cases}$$

and consider the initial-value problem with $y(0) = 0$.

(a) Show that the initial-value problem has only the trivial solution $y(x) = 0$ for $x \geq 0$. In thinking about this, it helps to sketch the direction field. Suppose there is a solution with $y(x_1) > 0$ for some $x_1 > 0$. Show that $y(x) > 0$ for $x < x_1$. Show also by Theorem 2.1 that $y(x)$ is uniquely determined for $x < x_1$ and hence that $y(0) > 0$. Use a similar argument if $y(x_1) < 0$.

(b) Show that the initial-value problem has more than one solution for $x \leq 0$.

23. Theorem 2.1 guarantees the existence of a solution only on some subinterval of the domain where the differential equation is defined, and that subinterval may be quite restricted. Verify that the initial-value problem $dy/dx = 1 + y^2$, $y(0) = 0$ has solution $y = \tan x$ defined only for $-\pi/2 < x < \pi/2$ even though the differential equation is defined everywhere. Note that the condition $|F_y(x, y)| < B$ of Theorem 2.1 fails to hold for $-\infty < y < \infty$.

24. Theorem 2.1 guarantees the existence of a solution only on some subinterval of the domain where the differential equation is defined, and that subinterval may be quite restricted.

(a) Verify that the initial-value problem $dy/dx = y^2$, $y(0) = y_0 > 0$ has solution $y = y_0/(1 - y_0 x)$ for $x < 1/y_0$, so as y_0 gets large the right endpoint of the interval on which the solution is defined tends to zero. Note that the condition $|F_y(x, y)| < B$ of Theorem 2.1 fails to hold here.

(b) Sketch the graph of the solution in part (a) for $y(0) = \frac{1}{2}$.

25. The purpose of this exercise is to show, using no prior knowledge of the exponential function, that if $y = E(x)$ solves the initial-value problem $y' = y$, $y(0) = 1$ for all real x, then $E(x)$ has the characteristic property shared by all exponentials, namely

$$E(x + z) = E(x)E(z) \quad \text{for all real } x \text{ and } z.$$

Prove this by showing that if we set $w(x) = E(x + z) - E(x)E(z)$, where z is fixed but arbitrary, then $w(x)$ solves the initial-value problem $w' = w$, $w(0) = 0$, and hence by Theorem 2.1 that $E(x + z) - E(x)E(z) = 0$ for all real x and z.

3 INTEGRATION

Broadly speaking, integration in some form or other is the basis for all solution methods for differential equations. This section is devoted to finding elementary solution formulas for some first-order differential equations that can be analyzed by fairly direct application of formal integration techniques. In practice performing an integration means one or more of the following: (i) having the integral already stored in your brain as a well-known formula; (ii) looking the integral up in a table, for example the one in Appendix D; (iii) finding the integral using a symbolic calculator; and (iv) transforming the integral so that one of the techniques of Appendix D applies.

3A $y' = f(x)$

A differential equation of the form $y' = f(x)$ can in principle be solved on any interval $a \leq x \leq b$ over which f is continuous. We can write the general solution either as

$$y(x) = \int f(x)\,dx + C \quad \text{or} \quad y(x) = \int_a^x f(u)\,du + C.$$

Differentiating either of these last two equations with respect to x shows that we've found a whole family of solutions. Indeed we've found *all* solutions, because the definite integral provides one solution, and we know that two different functions with the same derivative can differ only by an added constant of integration. Solution families of equations of this simple form always have graphs consisting of parallel curves with constant vertical distance between points on two given graphs. One way to account for this parallelism is to observe that the slope specified by the direction field of $dy/dx = f(x)$ is the same for each fixed value of x, regardless of the vertical displacement. Examples appear in Figure 1.2 on page 18.

example 1 The differential equation $y' = x$ has general solution

$$y(x) = \int x\,dx + C = \frac{1}{2}x^2 + C,$$

valid for all real x. Similarly the differential equation $y' = \sin(1 + x^2)$ has general solution that we can write as

$$y(x) = \int_0^x \sin(1 + u^2)\,du + C.$$

An important difference between this and the solution to $y' = x$ is that here we can't find an explicit elementary solution formula. Nevertheless we can get quite a bit of information about the solution by using numerical integration to produce an accurate table of values. And the derivatives of $y(x)$ are readily available, since we know $y'(x) = \sin(1 + x^2)$ implies that $y'' = 2x \cos(1 + x^2)$, and so on for higher order derivatives.

When the independent variable represents time we denote it by t rather than x, and we can interpret an equation of the form $y' = v(t)$ as a condition that the position $y(t)$ at time t of a one-dimensional motion has **velocity** $v(t)$. If $v(t)$ is differentiable then $v' = a(t)$ is called the **acceleration** of y at time t. The absolute value $|v(t)|$ is the **speed.**

---

example 2

Near the surface of the earth, the acceleration due to gravity for a falling body has a nearly constant value $g = 32.2$ feet per second per second. Thus velocity satisfies the equation $v' = g$. Integration with respect to t gives $v = gt + C_1$. The constant C_1 is determined by observing the velocity v_0 at some time t_0. If $t_0 = 0$, we find

$$v(t) = gt + v_0.$$

Since position $y(t)$ is related to velocity by $y' = v$, we can write a first order equation for $y(t)$ in the form

$$y' = gt + v_0.$$

Integration gives $y = \frac{1}{2}gt^2 + v_0 t + C_2$. If $y(0) = y_0$, as measured down from some reference point above ground level, then the constant C_2 must be y_0, so

$$y(t) = \tfrac{1}{2}gt^2 + v_0 t + y_0.$$

For example, if we were simply to drop a stone down a deep well we would have $v_0 = 0$. Since it might be natural to measure distance down from the top of the well we could set $y_0 = 0$, with the result

$$y(t) = \tfrac{1}{2}gt^2 \approx 16.1t^2 \text{ feet.}$$

If it took 4 seconds for the stone to hit the bottom of the well, we would know that it was about $16.1 \cdot 4^2 = 257.6$ feet deep.

If instead we measure up from the surface of the earth, the acceleration of gravity would act to decrease the velocity so we would replace g by $-g$ in the preceding derivation, getting $v(t) = -gt + v_0$. Thus a projectile thrown straight up from the surface of the earth ($y_0 = 0$) with velocity $v_0 = 100$ feet per second would have altitude in feet at t seconds given by

$$y(t) = -\tfrac{1}{2}gt^2 + 100t.$$

The maximum altitude y_{max} occurs when the velocity $v = y'$ is zero. Hence we find t_{max} from

$$y'(t) = -gt + 100 = 0$$

to be $t_{\max} = 100/g \approx 3.1$ seconds. Then

$$y_{\max} = -\tfrac{1}{2}g\left(\frac{100}{g}\right)^2 + 100\left(\frac{100}{g}\right) = \frac{5000}{g} \approx 155 \text{ feet}$$

at $t \approx 3.1$ seconds. Air resistance is assumed negligible here, but is taken into account in Chapter 2, Section 2.

In the previous example we solved a second-order equation, $y'' = g$ by integrating two successive first-order equations to get the solution

$$y(t) = \tfrac{1}{2}gt^2 + v_0 t + y_0;$$

this is the position at time t of an object with initial position y_0 and initial velocity v_0 that is being influenced by a constant gravitational acceleration g. The assumption at first was that distance was measured downward from some point above the attracting body, with acceleration g, say from the edge of a cliff or from the top of a deep well, and then upward from the surface of the earth, with acceleration $-g$. Whether we choose to measure distance up from below or down from above is a choice that we make depending on what seems more natural in the context of the problem. Thus the choice between using g or $-g$ for the acceleration is not required by the physical situation, but is dictated by the chosen orientation of the coordinate line used for measuring distance. Having made that choice, the decision as to whether to use g or $-g$ is dictated by whether gravity acts to increase velocity or decrease it. As shown below in Figure 1.6(a), the signed velocity $v(t)$ is always increasing, from negative values through zero to increasing positive values, so $v'(t) = g > 0$. In Figure 1.6(b), $v(t)$ always decreases, from positive values through zero to decreasing negative values, so $v'(t) = -g < 0$.

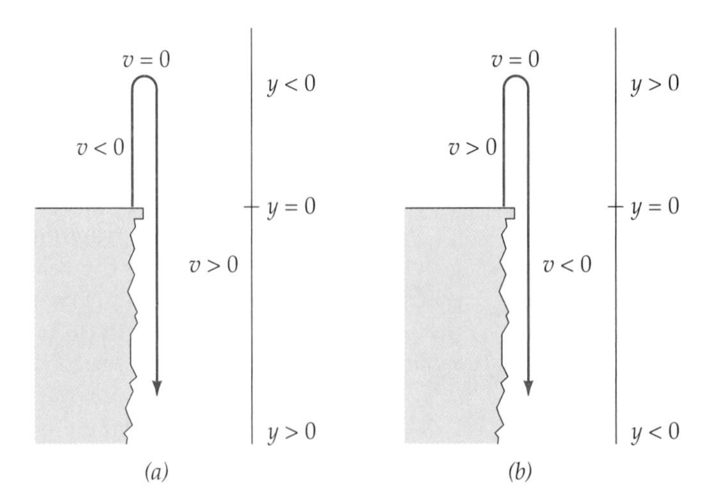

Figure 1.6 (*a*) $v' = g$. (*b*) $v' = -g$.

example **3**

A projectile is fired straight up from the surface of the earth ($y_0 = 0$) with velocity $v_0 = 1000$ feet per second. If we choose to measure distance up from the surface of the earth, then at time $t \geq 0$ we'd have $v'(t) = -g$ since gravity acts to decrease velocity. Successive integrations yield

$$\begin{aligned} v(t) = y'(t) &= -gt + v_0, \\ &\approx -32.2t + 1000 \quad \text{and} \\ y(t) &= -\tfrac{1}{2}gt^2 + v_0 t \\ &\approx -16.1t^2 + 1000t. \end{aligned}$$

Maximum height is reached when $v = 0$, at about $t_{max} \approx 1000/32.2 \approx 31$ seconds. The maximum height is about $y_{max} \approx -16.1(t_{max})^2 + 1000t_{max} \approx 15{,}528$ feet.

example **4**

A dropped object under no external influence other than constant gravity falls distance $s = \frac{1}{2}gt^2$ in time t. It's easy to show from this (see Exercise 35) that

$$\frac{ds}{dt} = \sqrt{2gs}, \qquad s(0) = 0.$$

This initial-value problem is fairly exceptional among those of applied mathematics in that it has two distinct solutions: $s = \frac{1}{2}gt^2$ and $s = 0$. Of course the differentiability assumption of the Uniqueness Theorem 2.1 of Section 2C is not satisfied by $\sqrt{2gs}$, so the theorem doesn't apply.

EXERCISES

Find the general solution of each differential equation in Exercises 1 through 10, and then find the particular solution that satisfies the associated initial condition.

1. $y' = x(1 - x)$, $y(0) = 1$.

2. $ds/dt = (t + 1)^2$, $s(1) = 2$.

3. $y' = x/(1 - x^2)$, $y(0) = 1$.

4. $du/dv = v^2 + 1$, $u(-1) = 1$.

5. $y'' = \sin x$, $y(0) = 1$, $y'(0) = 1$.

6. $y''' = 1$, $y(0) = y'(0) = y''(0) = 0$.

7. $dz/dt = te^t$, $z(0) = 1$.

8. $y' = \arctan x$, $y(0) = 0$.

9. $d^2x/dt^2 = e^t$, $x(0) = 1$, $dx/dt(0) = 0$.

10. $y'''' = x$, $y(0) = y''(0) = 0$, $y'(0) = y'''(0) = 1$.

Direction Fields. A differential equation of the special form $y' = f(x)$ has isoclines that are lines parallel to the y axis; thus to sketch the direction field you need to determine only one slope on each such line, making all slope segments centered on that line parallel to the first one. Most of the differential equations in Exercises 11 through 16 can't be solved by integration in terms of elementary functions, but you are asked only to sketch the direction field for each one and then use the field to sketch a few solution graphs.

11. $y' = x^3$.

12. $y' = \sqrt{x^2 + 1}$.

13. $y' = 1/(1 + x^4)$.

14. $y' = \sqrt[3]{1 - x^3}$.

15. $y' = e^{-x^2}$.

16. $y' = x/(1 + x^4)$.

17. A projectile is fired up from ground level with an initial velocity of 5000 feet per second. What is the maximum altitude attained, and how long does it take to get there?

18. A weight is dropped from 5000 feet above ground. How long does it take to reach the ground, and with what velocity does it hit?

19. Suppose the objects described in the previous two exercises are sent on their way at the same time and are aimed directly at each other.

(a) How long after release do they meet, and at what height above ground?

(b) What initial velocity should the projectile be given so that the two objects meet 2500 feet above ground?

20. A block of wood is kicked across a smooth ice surface with initial velocity 20 feet per second. Friction with the ice produces a constant negative acceleration a. If the block stops sliding after 4 seconds, what is the value of a? How far does the block slide?

21. A projectile is fired straight up from the surface of the earth with initial velocity v_0 feet per second and with no other force but gravity acting on it. How large must v_0 be for the projectile to achieve a height of 100 feet? [*Hint:* Solve two equations for v_0.]

22. Estimate the velocity with which an object should be thrown down from a bridge if it is to have velocity 100 feet per second when it hits the water 100 feet below.

23. An object is thrown straight up from the surface of the earth with initial velocity v_0. Show that if air resistance is neglected the maximum height is attained at $t_{\max} = v_0/g$ and is $h_{\max} = \frac{1}{2}v_0^2/g$.

24. Assuming a constant negative acceleration a for the application of an automobile's braking system; find a if the car comes to a stop within 100 feet from an initial speed of 100 feet per second.

25. A certain automobile's braking system provides an approximately constant *de*celeration of 30 feet per second per second on an asphalt pavement. Assume the brakes are applied when the car is going 60 miles per hour.

(a) About how long will it take the car to stop?

(b) Estimate the car's stopping distance.

(c) If the pavement is wet, it is observed that the car stops in 3.5 seconds from an initial speed of 60 miles per hour. Estimate the deceleration rate and stopping distance under these conditions.

26. A spherical ball of ice with uniform density ρ melts so that its mass decreases at a rate proportional to its surface area, with proportionality constant $k > 0$. Show that if the radius of the ball is initially r_0, then it will be all melted after time $\rho r_0/k$. [Recall that a ball of radius r has volume $\frac{4}{3}\pi r^3$ and surface area $(d/dr)(\frac{4}{3}\pi r^3) = 4\pi r^2$.]

27. The estimate $g \approx 32.2$ feet per second per second that is often used for the acceleration of gravity near the earth's surface is related to Newton's **inverse-square law** by the equation

$$g = \frac{GM}{R^2},$$

where M and R are respectively the mass and mean radius of the earth, and G is a universal gravitational constant. To estimate the value g_m for the corresponding acceleration near the surface of the earth's moon, we use the estimates that the moon's mass and radius respectively are about 0.012 and 0.273 times that of the earth; hence follows the estimate

$$g_m \approx g\frac{0.012}{(0.273)^2} \approx 5.18 \text{ feet/second}^2.$$

(a) Estimate the minimum initial velocity v_0 required for a jump to lift a person six inches off the earth's surface.

(b) How high a jump would the velocity found in part (a) produce on the moon?

28. A ball is dropped 100 feet above the surface of a large planet and hits the surface about 1.5 seconds later.

(a) Estimate the acceleration of gravity near the surface of the planet.

(b) Estimate the impact velocity of the ball at the surface of the planet.

29. Show that if the acceleration of gravity at the surface of a planet is g_p, then the initial vertical velocity required for a high-jumper to reach height h is $\sqrt{2hg_p}$.

30. A projectile is fired straight up from the surface of an airless planet with an initial velocity of 200 feet per second and is observed to reach a maximum height of 300 feet.

(a) Estimate the acceleration of gravity on the planet.

(b) If the acceleration of gravity on the planet were the same as that on earth, how high would the projectile have risen?

31. A stone is tossed up at the edge of a vertical cliff so that it just misses the edge on its way back down. Neglecting air resistance, show that the stone has the same speed, that is absolute value of velocity, when it passes the edge as it had from the initial toss.

32. A projectile is fired straight up with initial velocity v_0 from the surface of an airless planet where the acceleration of gravity is g_p. Show that the maximum height attained is $\frac{1}{2}v_0^2/g_p$.

33. A projectile is fired straight *down* with initial velocity v_0 from height s_0 above the surface of an airless planet where the acceleration of gravity is g_p. Show that the speed at impact is $\sqrt{v_0^2 + 2g_p s_0}$. Note that the initial conditions are different in the next problem, but that the result is the same.

34. A projectile is fired straight *up* with initial velocity v_0 from height s_0 above the surface of an airless planet where the acceleration of gravity is g_p. Show that the speed at impact is $\sqrt{v_0^2 + 2g_p s_0}$. Note that the initial conditions are different in the previous problem, but that the result is the same.

35. An object dropped near the surface of the earth under the influence of gravity alone falls distance $s = \frac{1}{2}gt^2$ in time t, where g is the acceleration of gravity.

(a) Show that $s = s(t)$ satisfies the first-order differential equation $ds/dt = \sqrt{2gs}$.

(b) Verify that the initial-value problem $ds/dt = \sqrt{2gs}$, $s(0) = 0$ has the solution $s = \frac{1}{2}gt^2$. What other solution does it have? Explain why the uniqueness statement in Theorem 2.1 of Section 2 isn't contradicted by this example.

Euler Beam Equation. Example 3 of the Introductory Survey contains a picture and some background relevant to the next six exercises in which conditions are required of solutions at two separate points called **boundary points.**

36. Suppose a uniform horizontal beam has profile shape $y = y(x)$, with x measured from one end. For a fairly rigid beam with uniform loading, $y(x)$ satisfies a fourth-order differential equation $y'''' = -P$, where $P > 0$ is a constant depending on the characteristics of the beam. If the ends of the beam are simply supported at the same level at $x = 0$ and $x = L$, then $y(0) = y(L) = 0$. The extended beam also behaves as if its profile had an inflection point at each support, so that $y''(0) = y''(L) = 0$. (Figure IS.3 in the Introductory Survey shows only the left end simply supported.)

(a) Solve the differential equation by four successive integrations and use the boundary conditions to show that the vertical deflection at point x, namely, $-y(x)$, is

$$\frac{P}{24}(x^4 - 2Lx^3 + L^3x), \quad 0 \le x \le L.$$

(b) Where does the maximum amount of vertical deflection from the level at the ends occur, and how much is the maximum deflection?

(c) It's reasonable to expect that the beam's profile will be concave up between the supports; show that this is true.

(d) Explain how the value of P can be computed for a particular beam by making appropriate measurements.

37. Suppose a uniform horizontal beam has profile shape $y = y(x)$, with x measured from one end. For a fairly rigid beam with uniform loading, $y(x)$ typically satisfies a fourth order differential equation $y'''' = -P$, where $P > 0$ is a constant depending on the characteristics of the beam. If the right end of the beam is embedded in a wall at $x = 0$ (the technical term is *cantilvered*) and simply supported at $x = L$, then $y(0) = y(L) = 0$ and $y''(0) = y'(L) = 0$. (Figure IS.3 in the Introductory Survey shows this situation.)

(a) Solve the differential equation by four successive integrations, and use the boundary conditions to show that the shape is described by the graph of

$$y(x) = -\frac{P}{48}(2x^4 - 3Lx^3 + L^3x).$$

(b) Show that the graph of $y(x)$ has an inflection on $0 < x < L$. What is the maximum vertical deflection?

(c) Make a sketch that shows the qualitative features of the graph of $y(x)$ for $0 \le x \le L$.

38. Let a horizontal beam have a profile described by the graph of $y = y(x)$. If an end of the beam are horizontally embedded at the same height in vertical walls, then $y(x)$ satisfies conditions of the form $y(0) = y(L) = 0$ and $y'(0) = y'(L) = 0$. The **Euler beam equation** for $y(x)$ has the form $y'''' = -P$, $P > 0$.

(a) Show that the solution of the above Euler equation boundary-value problem is

$$y(x) = -\frac{P}{24}(x^4 - 2Lx^3 + L^2x^2).$$

(b) Show that the beam's maximum deflection occurs at the center, and find the amount of deflection there.

(c) Find the horizontal coordinates of the inflection points of $y(x)$.

(d) Make a sketch on a scale that shows the qualitative features of the graph of $y(x)$ for $0 \le x \le L$.

*__39.__ A horizontal beam with profile graph $y = y(x)$ is embedded at both ends in such a way that that $y'(0) = y'(L) = 0$, but at different levels, so that $y(0) = 0$ and $y(L) = a \neq 0$. Here L is the distance between the vertical supports. The previous exercise treats just the case $a = 0$.

(a) Solve the Euler beam equation $y'''' = -P$ with these boundary conditions.

(b) Show that the graph of $y(x)$ has a horizontal tangent strictly between 0 and L if and only if $PL^4 > 72|a|$.

(c) Make a sketch on a scale that shows the qualitative features of the graph of $y(x)$ for the two cases $PL^4 > 72|a|$ and $PL^4 < 72|a|$.

40. If one end of a beam profile $y = y(x)$ is embedded at $x = L$ so $y(L) = y'(L) = 0$ and the other end is left hanging free at $x = 0$, the appropriate boundary conditions at the free end turn out to be $y''(0) = y'''(0) = 0$.

(a) Show that if $y(x)$ satisfies the Euler beam equation $y'''' = -P$, then the maximum deflection at the free end is $PL^4/8$.

(b) Make a sketch that shows the qualitative features of the graph of $y(x)$ for $0 \leq x \leq L$.

41. Boundary problems don't always have solutions, and there is sometimes a physical reason for this. Try to solve $y'''' = -R$, where R is a positive constant, subject to the boundary conditions $y(0) = y''(0) = 0$ simply propped up at a supported end, and $y''(L) = y'''(L) = 0$ at an unsupported end.

(a) Explain in purely mathematical terms why there is no solution.

(b) Taking into account the interpretation of the boundary conditions, explain in purely physical terms why there is no solution.

3B Separation of Variables

A differential equation that can be written in the form

$$g(y)\frac{dy}{dx} = f(x) \quad \text{or} \quad \frac{dy}{dx} = f(x)h(y)$$

can in principle always be solved by integration. Letting $y = y(x)$ be a solution of the equation, we'll look for an equation relating x and y. We try to integrate both sides of the first form of the differential equation with respect to x, getting

$$\int g(y)\frac{dy}{dx}\,dx = \int f(x)\,dx. \tag{1}$$

Now suppose we can find indefinite integrals F and G such that $F' = f$ and $G' = g$. By the chain rule,

$$\begin{aligned}\frac{dG(y)}{dx} &= G'(y)\frac{dy}{dx}\\ &= g(y)\frac{dy}{dx},\end{aligned}$$

so on the left-hand side we have

$$\begin{aligned}\int g(y)\frac{dy}{dx}\,dx &= \int \frac{d}{dx}G(y)\,dx\\ &= G(y) + C_1.\end{aligned}$$

The right-hand side of the differential equation can be written

$$\int f(x)\,dx = F(x) + C_2.$$

Hence $G(y) + C_1 = F(x) + C_2$, or

$$G(y) = F(x) + C,$$

where $C = C_2 - C_1$. In general this gives an *implicit* formula for the solution, implicit in the sense that we may have to solve the implicit formula to get one or more *explicit* solution formulas of the form $y = y(x)$. If we can solve the equation for y in terms of x, we have a formula $y = y(x, C)$ for solutions.

The process outlined is usually called **separation of variables** because it involves getting x's on one side of the equation and y's on the other. Changing variable in the integral allows us to drop the dx's on the left-hand side of equation (1). The resulting formal equation,

$$\int g(y)\,dy = \int f(x)\,dx,$$

still leads to the same solution: $G(y) = F(x) + C$, where $G' = g$ and $F' = f$. Equation (1) is sometimes written in **differential form,**

$$g(y)\,dy = f(x)\,dx,$$

which can be interpreted as either

$$g(y)\,\frac{dy}{dx} = f(x) \quad \text{or} \quad g(y) = f(x)\,\frac{dx}{dy}.$$

Unfortunately, these separated forms are not always attainable for a first-order differential equation.

example 5

In biological studies it is often observed that the size $P(t)$ of a growing bacteria population is proportional to the rate of change dP/dt. Expressing this proportionality in the form

$$\frac{dP}{dt} = kP, \quad k \text{ const.,} \tag{2}$$

gives us a first-order differential equation for P. Experience with the exponential function allows us to guess one solution. If we try

$$P(t) = Ce^{kt}, \quad C \text{ const.,}$$

we see that

$$\frac{dP}{dt}(t) = kCe^{kt} = kP(t)$$

for all real numbers t; in other words, $P = Ce^{kt}$ is a solution of the differential equation. Since we guessed this solution, the possibility remains that there are others. Under the assumption that the population size $P(t)$ is never zero, we can proceed as follows to separate variables. Divide Equation (2) by P, getting

$$\frac{1}{P}\frac{dP}{dt} = k.$$

Now integrate both sides with respect to t in the form

$$\int \frac{1}{P}\frac{dP}{dt}\,dt = \int k\,dt, \quad \text{or} \quad \int \frac{dP}{p} = \int k\,dt.$$

The integral on the left is $\ln|P| + c_1$, and the integral on the right is $kt + c_2$. Hence we can lump the constants of integration together and write

$$\ln|P| = kt + c.$$

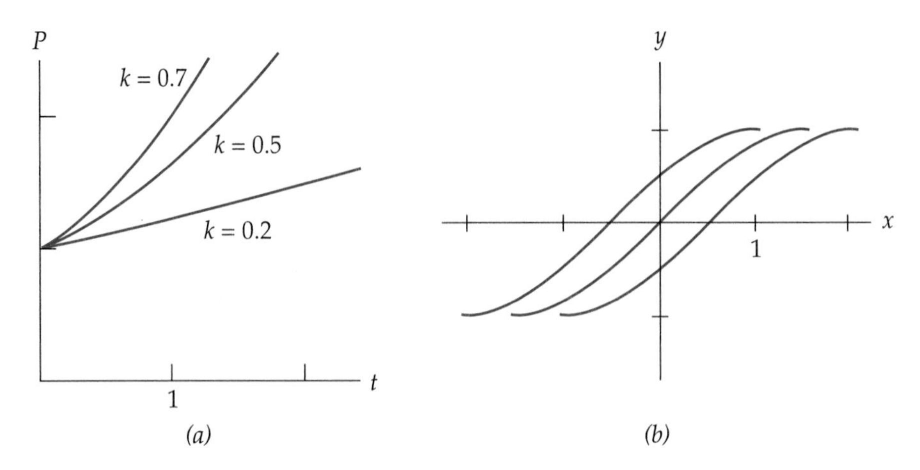

Figure 1.7 (a) $dP/dt = kP,\ P(0) = 1;\ P(t) = e^{kt}$. (b) $dy/dx = \sqrt{1 - y^2}\,;\ y = \sin(x + c)$.

Since $\exp(\ln z) = z$, we can apply the exponential function to both sides to get

$$|P(t)| = e^{kt+c}$$
$$= e^c e^{kt} = Ce^{kt},$$

where $C = e^c$ is necessarily a positive constant. But $P(t)$ is differentiable, and since it must then be continuous we must have either

$$P(t) = Ce^{kt} \quad \text{or} \quad P(t) = -Ce^{kt}.$$

Thus the constant can be either positive or negative in theory. If $P(t)$ does in fact represent a population size, then we use the first formula with $P(0) = C > 0$. Then $P(t) = P(0)e^{kt}$. Graphs of solutions satisfying $P(0) = 1$ are shown in Figure 1.7(a) for $k = 0.2$, 0.5, and 0.7. Note that this picture doesn't contradict the uniqueness result of Section 2C, because the different values of k correspond to different differential equations rather than different solutions to the same equation. If $P(0) = 0$ then $P(t) = 0$ for all t.

example 6

The differential equation $dy/dx = \sqrt{1 - y^2}$ in differential form is

$$\frac{dy}{\sqrt{1 - y^2}} = dx.$$

Integrating both sides gives $\arcsin y = x + c$. We can use any branch of the multiple-valued inverse-sine function we please, so assume we've picked the "principal branch," for which $\arcsin 0 = 0$; thus $-\pi/2 \le \arcsin y \le \pi/2$. Once a choice is made for the constant c, we then have the restriction $-\pi/2 \le x + c \le \pi/2$ on the values of x. Solving for y, we find $y = \sin(x + c)$.

Graphs of the solutions are shown in Figure 1.7(b) for $c = -1$, 0, and 1. Note that the graphs are parallel in the x direction, because the isoclines are horizontal lines. Note also that when $c = 0$, the solution graph comes to an abrupt end at $x = \pm\pi/2$. This apparent anomaly can be resolved in a geometrically natural way by extending the graph to the right with constant value 1 and to the left with value -1 (see Exercise 15).

Flow of Liquid from a Tank. Recall that $h = \frac{1}{2}gt^2$ is the distance a freely falling drop of fluid travels from rest in time t, the velocity being $v = gt$. To express velocity in terms of distance, first solve for t in terms of h to get $t = \sqrt{2h/g}$. Now multiply by g to get $gt = \sqrt{2gh}$. Since $v = gt$, the *free-fall velocity in terms of distance* is then $v = \sqrt{2gh}$. We'll use this last equation to motivate the following discussion.

Suppose a tank has an outlet hole near the bottom. Let $h = h(t)$ be the height of an ideal fluid above the outlet at time t, and let a be the area of the outlet hole. Then $V(t)$, the fluid volume remaining after time t, satisfies *Torricelli's equation*

$$\frac{dV}{dt} = -a\sqrt{2gh}.$$

The intuitive justification for the equation is that it depends on having the outlet velocity equal to the free-fall velocity $\sqrt{2gh}$ of a drop of fluid from height h; thus dV/dt should be the negative of *outlet area* a times *outlet velocity* $\sqrt{2gh}$. (A more profound justification depends on principles of fluid mechanics.) For an ordinary nonideal fluid, the equation takes the form

$$\frac{dV}{dt} = -ka\sqrt{2gh},$$

where $0 < k < 1$, the **flow constant** k, depending on properties of the fluid such as viscosity and on the shape of the hole. For a tank with cross-sectional area $A(h)$ at height h, the fluid volume V up to height h is given by an integral, and dV/dt by its derivative:

$$V(h) = \int_0^h A(y)\,dy, \quad \text{so} \quad \frac{dV}{dt} = A(h)\frac{dh}{dt}.$$

Thus Torricelli's equation appears as a differential equation to be satisfied by h:

$$A(h)\frac{dh}{dt} = -ka\sqrt{2gh}.$$

Note. Depending on what $A(h)$ is, this equation may fail to satisfy the differentiability condition in Theorem 2.1 of Section 2 that guarantees a unique solution to the initial-value problem with $h(0) = 0$. However, the condition $h(0) = 0$ means physically that we are considering draining an empty container. In addition, a look at the direction field (see Exercise 22 in Section 2C) shows that $h(t) = 0$ is the only solution for this problem when $t \geq 0$.

---

example 7

A tank in the shape of a circular cone has radius $r_0 = 5$ feet and vertical height $h_0 = 10$ feet. Hence the radius r at height h satisfies $r/h = 5/10$ or $r = h/2$ [see Figure 1.8(a)]. The cross-sectional area at height h is $A(h) = \pi r^2 = \pi h^2/4$.

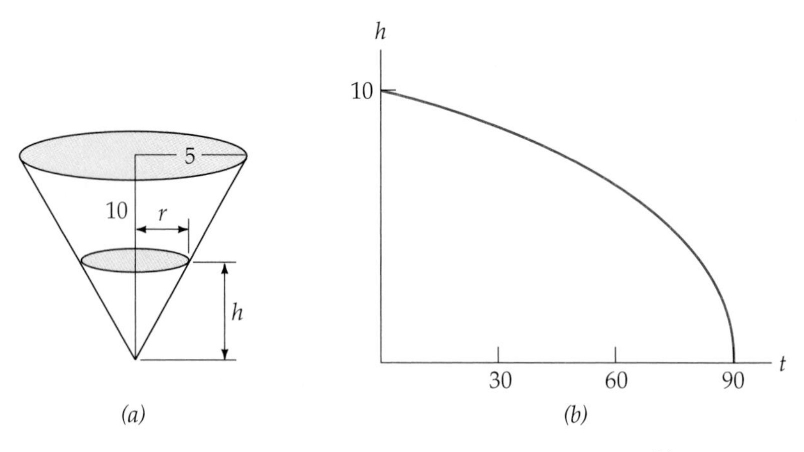

Figure 1.8 (a) $r/h = 5/10$. (b) $h(t) = \left(316.23 - 3.5t\right)^{0.4}$.

A circular outlet formed by cutting the tip of the cone off at diameter 6 inches, has area $3^2\pi$ square inches, or $a = \pi/16$ in square feet. Suppose we want to use Torricelli's equation to find out how the fluid level varies as a function of time, starting with the tank full. Torricelli says

$$\frac{\pi h^2}{4}\frac{dh}{dt} = -\frac{k\pi}{16}\sqrt{2gh} \quad \text{or} \quad h^{3/2}dh = -k\sqrt{\frac{g}{8}}\,dt.$$

Integrate the separated equation with initial condition $h(0) = 10$ to get

$$\frac{2}{5}h^{5/2} = -k\sqrt{\frac{g}{8}}\,t + \frac{2}{5}10^{5/2} \quad \text{or} \quad h(t) = \left(10^{5/2} - 5k\sqrt{\frac{g}{32}}\,t\right)^{2/5}.$$

If we were dealing with an ideal fluid we would take $k = 1$. The approximation $g \approx 32$ feet per second per second then yields $h(t) = \left(10^{5/2} - 5t\right)^{2/5}$. Under our assumptions we conclude that the tank would empty ($h = 0$) in time $t = 20\sqrt{10} \approx 63$ seconds. The empirical factor k could be determined by making an actual measurement of the time it takes to empty the tank; if this time turned out to be 90 seconds, we would solve $10^{5/2} - 5k \cdot 90 = 0$ to get $k \approx 0.7$; with this value of k, the practical numerical formula for $h(t)$ becomes $h = (316.23 - 3.5t)^{0.4}$. The graph is shown in Figure 1.8(b).

example 8

Suppose the conical tank in the previous example is modified by the addition of a valve at the bottom. With the tank full and the valve closed, someone begins to open the valve at the rate of 1/100 of a square foot per second until it reaches its fully open area of $\pi/16$ square feet. What is the height of fluid in the tank at the time the valve becomes fully open? To find out, express the area a of the outlet as a function of t. Since the rate of increase of a is 1/100 and $a(0) = 0$, we find $a(t) = t/100$; this holds until $a(t) = t/100 = \pi/16$, or until $t = 25\pi/4 \approx 19.6$ seconds. For this example, we'll assume that $g = 32$ and that the flow constant k is 1.

Torricelli's equation becomes

$$\frac{\pi h^2}{4}\frac{dh}{dt} = -\frac{8t}{100}h^{1/2} \quad \text{or} \quad h^{3/2}dh = -\frac{8t}{25\pi}\,dt.$$

Integrate the separated equation with initial condition $h(0) = 10$ to get

$$\frac{2}{5}h^{5/2} = -\frac{4t^2}{25\pi} + \frac{2}{5}10^{5/2} \quad \text{or} \quad h(t) = \left(10^{5/2} - \frac{2t^2}{5\pi}\right)^{2/5}.$$

At time $t = 25\pi/4$ seconds we find $h = (10^{5/2} - 125\pi/16)^{2/5} \approx 9.7$ feet when the valve becomes fully open. We can find subsequent behavior of $h(t)$ by an analysis like the one in the previous example.

EXERCISES

Solve each of the following equations by separating the variables. Then determine the constant of integration so that the given initial condition is satisfied and sketch the graph of that particular solution.

1. $y\,dy/dx = x^2;\, y = 1$ when $x = 2$.
2. $y' = x^2(1 + y^2);\, y = 0$ when $x = 0$.
3. $dx + e^x y\,dy = 0;\, y = 1$ when $x = 0$.
4. $s\,dt + t\,ds = 0;\, s = 1$ when $t = 1$.
5. $\cos u\,du - \sin v\,dv = 0;\, (u, v) = (\pi/2, \pi/2)$.
6. $y\sqrt{1 - x^2}\,dy = dx;\, (x, y) = (0, \pi)$.
7. Verify the indefinite integration formulas below. Also sketch the graph of the integrand and of the integral as functions of u for $u > 0$ and for $u < 0$, using the choice $\alpha = 1/2$ for part (b).

   (a) $\int (1/u)\,du = \ln|u| + C$.

   (b) $\int |u|^\alpha du = u|u|^\alpha/(\alpha + 1) + C,\ \text{if } \alpha \neq -1$.

Direction Fields. A differential equation of the special form $dy/dx = f(y)$ has isoclines that are lines parallel to the x axis; thus to sketch the direction field you need to determine only one slope on each such line, making all slope-segments centered on that line parallel to the first one. All of the following differential equations can be written in separated form, but still are not necessarily solvable by integration in terms of elementary functions. Sketch the direction field for each one and then use the field to sketch a few solution graphs.

8. $dy/dx = y^3$.
9. $dy/dx = \sqrt{y^2 + 1}$.
10. $dy/dx = 1/(1 + y^4)$.
11. $dy/dx = \sqrt[3]{1 - y^3}$.
12. $dy/dx = \sqrt{y},\ y \geq 0$.
13. $dy/dx = y/(1 + y^4)$.

14. Newton's law of cooling asserts that the rate of change of the surface temperature $u(t)$ of an object is proportional to the difference between $u(t)$ and the temperature u_0 of the surrounding medium. Thus

$$\frac{du}{dt} = n(u - u_0),$$

where the constant of proportionality n is negative if $u > u_0$, so that du/dt remains negative. Solve this equation under the assumption that u_0 is constant. Then sketch the graph of $u = u(t)$ under the assumption that $n = -0.5$, $u_0 = 10$, and $u(0)$, the initial temperature of the object, is 20.

15. In Example 6 of the text, the differential equation $dy/dx = \sqrt{1 - y^2}$ is shown to have the solution family $y = \sin(x + c)$, with the domain restrictions $-\pi/2 - c \leq x \leq \pi/2 - c$.

(a) Show that if $y = \sin(x + c)$ is extended to have value 1 for $x > \pi/2 - c$ and -1 for $x < -\pi/2 - c$, then the extended solution is valid for all real x.

(b) Explain why the extension described in part (a) is consistent with the direction field of the differential equation.

16. In Example 7 of the text we arrived at the approximate formula $h(t) = (10^{5/2} - 5t)^{2/5}$ for the height of fluid in a certain conical tank at time t. Find the corresponding volume of fluid in the tank in two ways:

(a) Use the formula $\frac{1}{3}\pi r^2 h$ for the volume of a right circular cone.

(b) Integrate Torricelli's equation for dV/dt.

17. A water tank with constant cross-sectional area A has an outlet of area a square feet at the bottom through which water is allowed to run freely, with flow constant $k = 1$. Water flows into the tank at a constant rate of q cubic feet per second. Show that if the height of the tank is less than $\frac{1}{64}q^2/a^2$, there will eventually be an overflow, and that otherwise the water level will steadily approach a fixed level. (Assume the acceleration of gravity is 32 feet per second per second.) [*Hint:* Interpret an appropriate modification of Torricelli's equation.]

18. A full water tank with constant cross-sectional area A and bottom outlet of area a empties in t_1 seconds. Assuming flow constant $k = 1$ and $g = 32$ feet per second per second, find the height of the tank.

19. Use Torricelli's equation

$$\frac{dV}{dt} = -ka\sqrt{2gh}$$

to do the following.

(a) Show that if a tank has constant cross-sectional area A, then $h(t)$ satisfies

$$\frac{dh}{dt} = -\frac{ka\sqrt{2g}}{A}h^{1/2}.$$

(b) Solve the differential equation.

(c) Use your answer to part (b) to find out how long it would take for the fluid above the outlet to drop from initial height $h = h_0$ to $h = 0$. In particular, estimate how long that would take if the tank is cylindrical with diameter 10 feet, height 20 feet, the outlet is circular with diameter 6 inches, the constant k is 1, and the tank is initially full.

(d) To estimate the value of the constant k empirically, you could measure the time it takes for the tank described in part (c) to empty and compute k from that information. Suppose the time it takes to empty the tank is 10 minutes; estimate k.

20. A tank 60 feet high and with cross-sectional area 100 square feet is to have its level lowered from full to half-full at the uniform rate of 1 foot per second. This is to be done by gradually increasing the scaled outlet area $ka(t)$ as a function of time over the required interval.

(a) Find a formula for $ka(t)$. In particular, estimate $ka(0)$ and $ka(30)$.

(b) Would it be feasible to empty the tank by continuing the method described above? Explain.

21. A full water tank h_0 feet high has constant cross-sectional area A and bottom outlet of area a. Assuming flow-constant $k = 1$ and $g = 32$ feet per second per second, show that the time it takes to empty the tank is $\frac{1}{4}A\sqrt{h_0}/a$ seconds.

22. A water tank of height h_0 feet and cross-sectional area A has two outlets of equal area a, one at the bottom and one l feet higher.

(a) Show that the time in seconds required for the tank to drain from full to level l is

$$\frac{A}{12\,al}\left[h_0^{3/2} - l^{3/2} - (h_0 - l)^{3/2}\right].$$

Assume flow constant $k = 1$ and $g = 32$ feet per second per second. [*Hint:* Find an appropriate modification of Torricelli's equation and solve it for time t.]

(b) Show that as $l \to 0$ the result of part (a) tends to one-half the result of part (a) of the previous exercise. Why does this make sense physically?

23. An upright hemispherical bowl of radius 1 foot has an outlet hole near the bottom of radius 1 inch.

(a) Express the upper surface area A of fluid in the bowl in terms of the height of the fluid.

(b) Assuming flow constant $k = \frac{1}{2}$, about how long would it take for the full bowl to empty under the influence of gravity?

24. The problem here is to design a 1-cubic-foot container, 4 feet high, with a circular cross section and an outlet at the bottom, that empties under the influence of gravity in 4 hours and in such a way that the fluid level drops at a constant rate. Assume throughout that the acceleration of gravity is 32 feet per second per second and that the outlet has flow constant $k = 1$.

(a) Use Torricelli's equation to show that the radius of the container at height h above the bottom should have the form $r = p\sqrt[4]{h}$, where p is constant.

(b) Show that the total volume requirement makes $p = \frac{1}{4}\sqrt{3/\pi}$ in part (a).

(c) Find the area of the outlet if the container is to go from full to empty in exactly 4 hours.

(d) Find the radius at the top of the container, and make a sketch of its profile as viewed from the side.

(e) Explain how the container could be made to serve as a clock that measures time throughout a 4-hour period.

25. Imagine a balloon that starts with zero radius and gets inflated at a constant rate of C cubic units per second. Express the radius of the balloon as a function of time.

26. **(a)** Velocity $v = v(t)$ of a falling object subject only to a constant negative acceleration of gravity satisfies $dv/dt = -g$. Show that under the additional assumption that $v_0 = v(0) > 0$, the velocity is zero at time $t_1 = v_0/g$.

(b) A retarding force due to air resistance can be included in the equation in part (a) to give

$$\frac{dv}{dt} = -g - kv,$$

where k is a positive constant. Show that if $v_0 = v(0) > 0$, the velocity is zero at time

$$t_2 = \frac{1}{k}\ln\left(1 + \frac{kv_0}{g}\right).$$

(c) Give an intuitive physical argument to show that the times t_1 and t_2 in the above parts (a) and (b) should satisfy $t_1 > t_2$.

(d) Give a precise mathematical argument to show that the times t_1 and t_2 in parts (a) and (b) satisfy $t_1 > t_2$. [*Hint:* Is $x > \ln(1 + x)$ for $x > 0$?]

(e) Compute directly the right-hand limit

$$\lim_{k \to 0+} \frac{1}{k}\ln\left(1 + \frac{kv_0}{g}\right).$$

Is your result reasonable on physical grounds?

27. A particular parameter in a one-parameter family of solutions can sometimes be advantageously replaced by another symbol. Here is an example.

(a) Solve $y' = -y^{3/2}$ by separating variables.

(b) Show that the equation $y' = -y^{3/2}$ has a family of solutions $y = 4/(x + c)^2$, each solution of the family being valid on intervals not containing a specified point $x = -c$. Show, however, that this family fails to contain the constant solution $y = 0$.

(c) Replace the parameter c in part (b) by $1/C$, $C \neq 0$, and multiply numerator and denominator by C^2. Show that the resulting family contains the solution missing in part (b). What else is then missing?

28. Consider the differential equation $dz/dt = -z^3$ with initial condition $z(0) = z_0$.

(a) Find a formula for the solution of the initial-value problem.

(b) Sketch the direction field of the equation.

(c) Sketch graphs for $t \geq 0$ of the solutions such that $z_0 = 2,\ 1,\ -1,$ and -2.

Differential Inequalities. Inequalities such as $y' \geq 0$ can often be used to determine qualitative properties of a function without necessarily specifying the function with as much precision as a differential equation may do. For example, $y' > 0$ on the interval $0 \leq x \leq 1$ implies only that $y(x)$ is increasing on the interval, while $y'' > 0$ implies that the graph of $y(x)$ is concave up there. Beyond that, we can sometimes integrate both sides of an inequality in such a way as to maintain inequality.

29. What conclusion can you draw about a function $y = y(x)$ under each of the following assumptions?

(a) $y'(x) \geq 1$ for $x \geq 0$, with $y(0) = 0$.

(b) $y''(x) \geq 1$ for $x \geq 0$, with $y(0) = y'(0) = 0$.

(c) $\left(e^{-x}y(x)\right)' \geq 0$ for all real x, with $y(0) = 1$.

(d) $x \leq y''(x) \leq 2x$ for all real x, with $y(0) = 0$ and $y'(0) = 1$.

(e) $y'(x) \geq y(x) > 0$ for all x, with $y(0) = 1$.

(f) $y''(x) \leq 0$ for all x with $y(0) = 0,\ y'(0) = 1$.

30. (a) Solve the differential equation $y' = \sqrt{1 - y^2}$ by separation of variables, and express y in terms of x and a constant of integration. Then sketch solution graphs passing through (0,–1), (0,0), and (0,1).

(b) Sketch the solution graphs asked for in part (a) by relying entirely on a preliminary sketch of the direction field for $y' = \sqrt{1 - y^2}$. Note that **singular solutions** $y(x) = 1$ and $y(x) = -1$ also satisfy the equation, but are not contained in the solutions of part (a).

31. (a) Solve the differential equation $y' = 1 - y^2$ by separation of variables, and express y in terms of x and a constant of integration. Then sketch solution graphs passing through (0,–2), (0,2), and (0,0).

(b) Sketch the solution graphs asked for in part (a) by relying entirely on a sketch of the direction field for $y' = 1 - y^2$.

(c) The given differential equation has the obvious singular solutions $y(x) = 1$ and $y(x) = -1$. Is either of these contained in the solution formula you derived in part (a)? Explain why, or why not.

3C Transformations †

When the variables do not separate in a first-order equation, it may still be possible to change variables in such a way that the new equation separates. After solving the new equation, then change variables back again. An appropriate substitution will be suggested by the form of the differential equation, and in that regard the choice involves some educated guessing. We'll consider two types in detail.

Linear Substitution. Suppose we have an equation that can be written in the form

$$\frac{dy}{dx} = f(ax + by + c).$$

If we let $u = ax + by + c$, then

$$\frac{du}{dx} = a + b\frac{dy}{dx}, \quad \text{so} \quad \frac{dy}{dx} = \frac{1}{b}\frac{du}{dx} - \frac{a}{b}.$$

The given equation becomes

$$\frac{1}{b}\frac{du}{dx} - \frac{a}{b} = f(u), \quad \text{or} \quad \frac{du}{dx} = bf(u) + a,$$

so integration in x can be tried; if this works let $u = ax + by + c$ in the solution.

example 9

For instance, in

$$\frac{dy}{dx} = (x + y)^2,$$

we let $u = x + y$, so $dy/dx = du/dx - 1$. Then

$$\frac{du}{dx} = u^2 + 1 \quad \text{or} \quad \frac{du}{1 + u^2} = dx.$$

Integrating, we find

$$\arctan u = x + c,$$

so

$$\arctan(x + y) = x + c, \quad \text{or} \quad y = -x + \tan(x + c).$$

Some solution curves are shown in Figure 1.9.

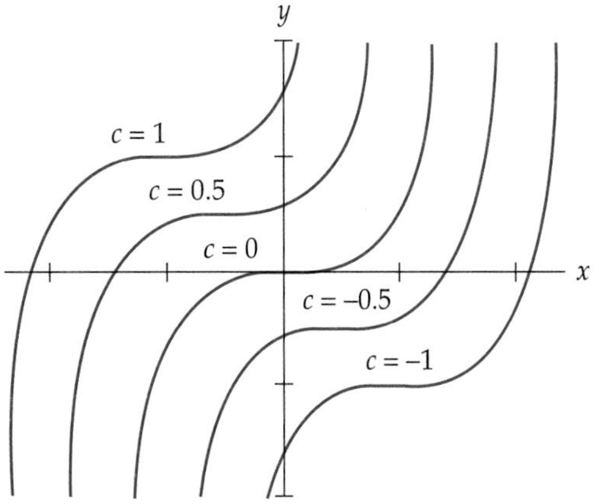

Figure 1.9 $dy/dx = (x + y)^2$;
$y = -x + \tan(x + c)$.

Quotient Substitution. Suppose we have an equation that can be written in the form

$$\frac{dy}{dx} = g\left(\frac{y}{x}\right).$$

(Equations of this type are sometimes called **homogeneous,** although a different use for that term appears in Chapter 3.) If we let $u = y/x$, then

$$\frac{du}{dx} = \frac{x(dy/dx) - y}{x^2} = \frac{1}{x}\frac{dy}{dx} - \frac{y}{x^2}.$$

We solve for dy/dx to get

$$\frac{dy}{dx} = x\frac{du}{dx} + \frac{y}{x} = x\frac{du}{dx} + u.$$

The given equation becomes

$$x\frac{du}{dx} = g(u) - u,$$

so the variables can be separated; after integration, let $u = y/x$ in the solution.

example 10 For instance, instead of

$$x^2\frac{dy}{dx} = x^2 + xy + y^2,$$

we can write

$$\frac{dy}{dx} = 1 + \frac{y}{x} + \left(\frac{y}{x}\right)^2.$$

The function $g(u)$ is evidently $g(u) = 1 + u + u^2$. Letting $u = y/x$, we transform the given equation to

$$x\frac{du}{dx} = (1 + u + u^2) - u = 1 + u^2.$$

In separated form, we find

$$\frac{du}{1 + u^2} = \frac{dx}{x}.$$

We can integrate directly to get

$$\arctan\ u = \ln|x| + c.$$

Then put $u = y/x$ back again:

$$\arctan\frac{y}{x} = \ln|x| + C.$$

To simplify the relation between y and x, we can set $c = \ln\ k, k > 0$. This gives

$$\begin{aligned}\arctan\frac{y}{x} &= \ln|x| + \ln\ k \\ &= \ln\ k|x|.\end{aligned}$$

Then $y/x = \tan(\ln\ k|x|)$, so

$$y = x\tan(\ln\ k|x|).$$

Two remarks are in order. First, it is usually impossible in integrating a separated equation to find a nice formula that expresses y in terms of x, or vice versa, as we did in the previous example; we may have to make do with an implicit formula of the form $F(x, y, C) = 0$. Second, our solution to $x^2y' = x^2 + xy + y^2$ quite naturally fails to exist when $x = 0$. Can you find other values of x, depending on k, for which $\tan(\ln k|x|)$ fails to exist?

Removing Constants from a First-Degree Quotient. A differential equation of the form

$$\frac{dy}{dx} = f\left(\frac{ax + by + m}{cx + dy + n}\right), \quad m \neq 0,$$

can't be solved using a quotient substitution $u = y/x$, because division of numerator and denominator by x leaves terms m/x and n/x above and below. However, this equation can often be transformed into the form

$$\frac{dy}{dx} = f\left(\frac{a + by/x}{c + dy/x}\right) \quad \text{if} \quad ad - bc \neq 0.$$

To remove m and n, a substitution of the form $x = z + h,\ y = w + k$, where h and k are suitably chosen constants, will often suffice. Since $dy/dx = dw/dz$, to solve the given equation we first solve the equation

$$\frac{dw}{dz} = f\left(\frac{az + bw}{cz + dw}\right),$$

and then change variables back again by $x = z + h$ and $y = w + k$. Exercise 19, part (b), examines the case $ad - bc = 0$, which turns out to be inherently simpler, because in this case a linear substitution leads directly to separation of variables.

example 11

We'll solve this differential equation, assuming $x \neq y$:

$$\frac{dy}{dx} = \frac{x + y + 1}{x - y}.$$

To use a substitution of the form $u = y/x$, we first try to find constants h and k such that letting $x = z + h$ and $y = w + k$ will make the ratio on the right homogeneous in z and w. We have

$$\frac{x + y + 1}{x - y} = \frac{z + h + w + k + 1}{z + h - w - k},$$

so we want $h + k + 1 = 0$ and $h - k = 0$. The unique solution of this pair of linear equations for h and k is $h = k = -\frac{1}{2}$. Thus we set $x = z - \frac{1}{2}$, $y = w - \frac{1}{2}$.

To solve the given differential equation, we first solve the transformed equation

$$\frac{dw}{dz} = \frac{z + w}{z - w}.$$

This can be done by rewriting it as

$$\frac{dw}{dz} = \frac{1 + w/z}{1 - w/z}$$

and substituting $w = zu$, $dw/dz = u + z(du/dz)$. The result before separating variables u and z is $u + z(du/dz) = (1 + u)/(1 - u)$. Separation and integration gives

$$\left[(1 - u)/(1 + u^2)\right] du = (1/z)\, dz, \quad \text{and} \quad 2 \arctan u - \ln(1 + u^2) = \ln(z^2) + C.$$

Then substitute back, first by $u = w/z$, then by $z = x + \frac{1}{2}$, $w = y + \frac{1}{2}$ to get

$$2 \arctan\left[(y + \tfrac{1}{2})/(x + \tfrac{1}{2})\right] = \ln\left[(x + \tfrac{1}{2})^2 + (y + \tfrac{1}{2})^2\right] + C.$$

EXERCISES

Each of the following differential equations can be written in one of two forms:

$$\frac{dy}{dx} = f(ax + by + c) \quad \text{or} \quad \frac{dy}{dx} = g\left(\frac{y}{x}\right).$$

Decide which form applies in each case and use a transformation of the type

$$u = ax + by + c \quad \text{or} \quad u = \frac{y}{x}$$

to put the equation in separated form. Then solve the equation and eliminate u from the solution.

1. $dy/dx = 2x + y + 3$.

2. $dy/dt = (t - y)^2$.

3. $x^2 dy/dx = x^2 + y^2$.

4. $t\,dx = (t + x)\,dt$.

5. $dy/dx = (y - x)/(y + x)$.

6. $dz/dv = (2z + v + 1)^2$.

7. $y' = (x - y)/(x - y + 1)$.

8. $y' = (2x + y)/(2x + y + 1)$.

None of the following equations can be written in the homogeneous form

$$\frac{dy}{dx} = g\left(\frac{y}{x}\right).$$

However, a preliminary transformation of the form $x = z + h$, $y = w + k$, where h and k are constant, gives an equation of that form in z and w. Solve the equations by first making the transformation with general h and k and then finding h and k. Then solve the equation in z, w and transform back to x, y for the final result.

9. $dy/dx = (x + y + 1)/(x + 2y + 3)$.

10. $(x + y)\,dy = (x - y + 1)\,dx$.

11. $dy/dx = (x + y + 1)/(x - y)$.

12. $dy/dx = (y - x)/(x + y + 1)$.

13. $(4x + 3y - 7)\,dx + (3x - 7y + 4)\,dy = 0$.

14. $(2x - 3y + 4)\,dx + (3x - 2y + 1)\,dy = 0$.

15. $y' = (x - y)/(x + y + 1)$.

16. $y' = (2x - y + 1)/(x - y)$.

17. **(a)** Show that for the equation $(x + 2y + 1)\,dx + (2x + 4y - 1)\,dy = 0$ it's impossible to use a transformation of the form $x = z + h$, $y = w + k$ to make the equation homogeneous.

(b) Solve the equation in part (a) by first making a substitution of the form $z = ax + by + c$.

18. Find all values of x, depending on $k > 0$, for which the solution $y = x \tan(\ln k|x|)$ to $x^2(dy/dx) = x^2 + xy + y^2$ fails to exist.

19. **(a)** Show that if $ad - bc \neq 0$, there is a unique choice for h and k in the substitution $x = z + h$, $y = w + k$ that transforms the quotient in $y' = (ax + by + m)/(cx + dy + n)$ into the homogeneous quotient $(az + bw)/(cz + dw)$.

(b) Show that if $ad - bc = 0$, then either (i) the numerator in part (a) is simply a constant multiple of the denominator or else (ii) the desired h and k fail to exist but that then letting $z = cx + dy$ permits separation of the variables y and z in the transformed differential equation.

20. Show that the differential equation

$$P(x, y)\,dx + Q(x, y)\,dy = 0$$

can be written in the form $dy/dx = g(y/x)$ if, for arbitrary t, both

$$P(tx, ty) = t^n P(x, y) \quad \text{and} \quad Q(tx, ty) = t^n Q(x, y)$$

for some number n. [*Hint:* Let $t = 1/x$.]

21. Bugs in mutual pursuit. Four identical bugs start moving at the same time on a flat table, each at the same constant speed $a > 0$. Use (x, y) coordinates on the table, and locate bugs 1 through 4 initially in respective quadrants 1 through 4, each at one of the four points $(\pm 1, \pm 1)$. Bug 1 always heads directly toward bug 2, bug 2 toward bug 3, bug 3 toward bug 4, and bug 4 toward bug 1, so their paths are mutually congruent. The next five parts ask you to examine the paths.

(a) Use the symmetry of the paths to show that at any moment the bugs are at the corners of a square, and in particular, if bug 1 is at (x, y), then 2, 3, and 4 are, respectively, at $(-y, x)$, $(-x, -y)$, and $(y, -x)$.

(b) Using the notation of part (a), show that the paths of bugs 1 and 3 satisfy the differential equation

$$\frac{dy}{dx} = \frac{y - x}{x + y},$$

unless $x + y = 0$. What is the analogous differential equation for bugs 2 and 4?

(c) Show that the coordinates of the path of bug 1 satisfy an equation of the form

$$\ln \sqrt{x^2 + y^2} + \arctan \frac{y}{x} = k.$$

(d) Show that in polar coordinates $x = r\cos\theta$, $y = r\sin\theta$, the equation can be put in the form

$$\ln r + \theta = k \quad \text{or} \quad r = Ce^{-\theta}$$

of a **logarithmic spiral.** Then sketch the first bug's path starting at (1,1).

(e) Use the formula

$$s = \int_{\theta_0}^{\theta_1} \sqrt{r^2 + \left(\frac{dr}{d\theta}\right)^2}\, d\theta$$

for arc length in polar coordinates to show that a logarithmic spiral starting at a finite point and winding in toward the origin has finite length. In particular, show that if the bugs were reduced to points they would all reach the origin from their respective starting points $(\pm 1, \pm 1)$ in time $2/a$, where a is the constant speed of each one of them.

***22.** An airplane flies horizontally with constant airspeed v miles per hour, starting at distance a east of a fixed beacon and always pointing directly at the beacon. There is a crosswind from the south with constant speed $w < v$.

(a) Locate the positive x axis along the east-west line through the beacon and with the origin at the beacon. Show that the graph $y = y(x)$ of the airplane's path satisfies

$$\frac{dy}{dx} = \frac{y}{x} - \frac{w}{v}\sqrt{1 + \left(\frac{y}{x}\right)^2}.$$

(b) Show that the airplane's path traces the graph of $y = (x/2)\left[(a/x)^{w/v} - (x/a)^{w/v}\right]$. [*Hint:* Let $y/x = u$, as in Section 3C.]

Analysis of Systems. One of the most important uses for equations of the form $dy/dx = G(x, y)/F(x, y)$, is their relation to systems of two equations such as

$$\frac{dx}{dt} = F(x, y), \qquad \frac{dy}{dt} = G(x, y).$$

The connection stems from the chain rule $dy/dt = (dy/dx)(dx/dt)$, which, slightly rewritten, leads to

$$\frac{dy}{dx} = \frac{dy/dt}{dx/dt} = \frac{G(x, y)}{F(x, y)}.$$

The single equation in x and y contains less information than the pair of equations in x, y, and t. However, the single equation is sometimes easier to solve explicitly, and its solution provides some information about the system. The next five exercises explore these ideas in some examples.

23. Show that each solution $x = x(t)$, $y = y(t)$ to the system

$$\frac{dx}{dt} = -y, \qquad \frac{dy}{dt} = x$$

traces a circle $x^2 + y^2 = r^2$ for some appropriate constant r.

24. Show that each solution $x = x(t)$, $y = y(t)$ to the system

$$\frac{dx}{dt} = y, \qquad \frac{dy}{dt} = x$$

traces a hyperbola $x^2 - y^2 = C$ for some constant C.

25. Show that each solution $x = x(t)$, $y = y(t)$ to the system

$$\frac{dx}{dt} = x, \qquad \frac{dy}{dt} = y$$

traces a path lying on a line.

26. Show that each solution $x = x(t)$, $y = y(t)$ to the system

$$\frac{dx}{dt} = x + y, \qquad \frac{dy}{dt} = x^2 - y^2$$

traces a path lying on a graph of the form $y = x - 1 + Ce^x$ for some constant C.

Generalized Solutions. A continuous function $y(x)$ that satisfies a first-order differential equation except at isolated points is sometimes called a **generalized solution.** Typically the failure of $y(x)$ to satisfy the equation at a point x_0 is due to the failure of $y'(x_0)$ to exist.

27. The direction field defined by the differential equation

$$y' = F(x, y) = \begin{cases} 0, & \text{if } x < 0, \\ 1, & \text{if } 0 \le x, \end{cases}$$

is discontinuous, though $\partial F/\partial y = 0$, so $\partial F/\partial y$ turns out to be continuous.

(a) Sketch the direction field for the equation and attempt to sketch a continuous solution graph for the differential equation on the interval $-1 \le x \le 1$ and satisfying the initial condition $y(-1) = 0$.

(b) Explain why your graph can't have a derivative everywhere on an interval containing $x = 0$.

(c) Verify that, for each constant C,

$$y(x) = \begin{cases} C, & x < 0, \\ x + C, & x \ge 0, \end{cases}$$

is (i) continuous and (ii) satisfies the differential equation for every real number $x \ne 0$, and so is a generalized solution. What choice for C satisfies $y(-1) = 0$?

28. Find a generalized solution to the equation

$$y' = \begin{cases} 1, & x < 1, \\ 2x, & x \ge 1, \end{cases}$$

that also satisfies initial condition $y(0) = 1$. Sketch the graph of your solution for $0 \le x \le 2$.

4 LINEAR EQUATIONS

The first-order differential equations we have looked at so far can nearly all be written in the form

$$\frac{dy}{dx} = F(x, y),$$

where F is some fairly simple function of x and y. If we make the requirement that F have the form $F(x, y) = -g(x)y + f(x)$, then we get what is called a **linear equation.** The term "linear" is used because the equation is of first degree in dy/dx and y. It is technically helpful to write the equation in what is called **normal form** as

4.1
$$\frac{dy}{dx} + g(x)y = f(x),$$

where *it is crucial that dy/dx have coefficient 1.* The equation is then said to be **normalized.** Linear equations are general enough that they have many interesting applications; all the same, they are special enough that we can say significant things about their solutions without specifying much about $g(x)$ and $f(x)$.

example 1

If $f(x)$ happens to be identically zero, we can find solutions $y = y(x)$ to Equation 4.1 by assuming $y \neq 0$ and rewriting the equation

$$y' + g(x)y = 0 \quad \text{or} \quad \frac{y'}{y} = -g(x).$$

Integrating both sides with respect to x, we get

$$\ln |y| = -G(x) + c,$$

where $G(x) = \int g(x)\,dx$ is an indefinite integral of $g(x)$ and c is a constant. Taking the exponential of both sides gives

$$|y| = e^c e^{-G(x)}.$$

If we remove the absolute value, we can allow for the possibility of replacing the positive constant e^c by $-e^c$, so we'll cover both cases with an arbitrary nonzero constant C:

$$y = Ce^{-G(x)} \quad \text{or} \quad y = Ce^{-\int g(x)\,dx}.$$

We check that this last form of the solution is valid by first differentiating with respect to x. We get $y' = -Cg(x)e^{\int g(x)\,dx}$, so

$$y' + g(x)y = -Cg(x)e^{\int g(x)\,dx} + Cg(x)e^{\int g(x)\,dx} = 0,$$

as required by the original differential equation.

4A Exponential Integrating Factors

The method used in Example 1 fails if the function $f(x)$ in Equation 4.1 is not zero; it also has the technical defect that it forces us to assume $y \neq 0$. [Conceivably there are solutions that take on the value zero; indeed $y(x) = 0$ is one such.] Both restrictions can be avoided at once if we use the exponential factor that occurred in the previous example. This time the product rule shows that

$$\left(e^{\int g(x)\,dx} y\right)' = e^{\int g(x)\,dx} y' + g(x)e^{\int g(x)\,dx} y = e^{\int g(x)\,dx}[y' + g(x)y].$$

This equation tells us that if we multiply the differential equation $y' + g(x)y = f(x)$ by the **exponential integrating factor**

4.2
$$M(x) = e^{\int g(x)dx},$$

the left-hand side becomes the derivative of the product $M(x)y$ and the right-hand side is $M(x)f(x)$. The next step is to integrate both sides of the resulting equation

$$(M(x)y)' = M(x)f(x).$$

Integration on the left gives $M(x)y$, so

$$M(x)y = \int M(x)f(x)\,dx + C.$$

Since $M(x)$ is an exponential, it's never zero, so we can divide by it to find all solutions $y(x)$.

example 2

To find all solutions of the linear differential equation

$$y' = xy + x,$$

we first rewrite the equation in the normal form that we used in the previous discussion:

$$y' - xy = x.$$

We then find the exponential multiplier by identifying the coefficient function $g(x) = -x$ and writing Equation 4.2 as

$$M(x) = e^{\int g(x)dx}$$
$$= e^{-\int x\,dx} = e^{-x^2/2}.$$

Multiplying the differential equation by $M(x) = e^{-x^2/2}$ gives

$$e^{-x^2/2}y' - xe^{-x^2/2}y = xe^{-x^2/2}.$$

But we know from the way $M(x)$ is chosen, or we could verify directly using the product rule, that this last equation is equivalent to

$$\frac{d}{dx}\left(e^{-x^2/2}y\right) = xe^{-x^2/2}.$$

Integrating both sides with respect to x gives

$$e^{-x^2/2}y = \int xe^{-x^2/2}\,dx + C = -e^{-x^2/2} + C.$$

Finally, multiplying by $e^{x^2/2}$ gives the family of solutions

$$y = -1 + Ce^{x^2/2}.$$

example 3

The differential equation $x^2y' + y = 1$ is not normalized, and when $x = 0$ even fails to define a value for the derivative y' at points $(0, y)$ on the y axis. Thus we can't in general expect solutions to exist when $x = 0$. To normalize the equation, divide by x^2 to get

$$y' + \frac{1}{x^2}y = \frac{1}{x^2}, \quad x \neq 0.$$

Since $g(x) = x^{-2}$ in Equation 4.2 for the exponential integrating factor we compute

$$M(x) = e^{\int x^{-2}\,dx} = e^{-x^{-1}}.$$

We multiply the normalized equation by $M(x)$ to get

$$\left(e^{-x^{-1}}y\right)' = x^{-2}e^{-x^{-1}}.$$

Integrating both sides of the equation gives

$$e^{-x^{-1}}y = \int x^{-2}e^{-x^{-1}}\,dx = e^{-x^{-1}} + C.$$

Division by $M(x)$ gives the solution formula $y = 1 + Ce^{x^{-1}}$.

Three Important Remarks about the Exponential Multiplier Method

1. The linear differential equation should be in the normal form

$$y' + g(x)y = f(x),$$

where y' has coefficient 1, before identifying the coefficient function g for the purpose of computing $M(x) = e^{\int g(x)dx}$.

2. The differential equation

$$y' + g(x)y = f(x),$$

and its multiplied forms

$$M(x)y' + M(x)g(x)y = M(x)f(x) \quad \text{or} \quad \left[M(x)y\right]' = M(x)f(x)$$

are completely equivalent to one another in the sense that any solution of one equation is also a solution of the other. The reason is that, since it's an exponential function, the multiplier $M(x)$ is never equal to zero. Hence we can multiply and divide by it as we please without altering the solutions of a given equation.

3. Integrating the previous displayed equation between x_0 and x gives

$$M(x)y(x) - M(x_0)y(x_0) = \int_{x_0}^{x} M(u)f(u)\,du.$$

Then dividing by $M(x)$ gives the unique solution to the initial-value problem with $y(x_0) = y_0$:

$$y(x) = \frac{1}{M(x)}M(x_0)y(x_0) + \frac{1}{M(x)}\int_{x_0}^{x} M(u)f(u)\,du.$$

Note that the choice of a constant of integration in the exponent of $M(x)$ doesn't spoil the uniqueness, because

$$M(x) = e^{\int g(x)\,dx + C} = e^{C}e^{\int g(x)\,dx}.$$

The result is that the nonzero constant factor e^{C} will always cancel out, since only ratios of the function M occur in the expression for $y(x)$.

EXERCISES

Find an exponential multiplier $M(x)$ for each normalized expression in Exercises 1 through 6 such that $M(x)$ times that expression can be written in the form $d[M(x)y]/dx$.

1. $y' + 2y$.

2. $dy/dx + y$.

3. $dy/dx + (2/x)y$.

4. $y' + e^x y$.

5. $y' + e^{-2x-1}y$.

6. y'.

Find the general solution of each first-order equation in Exercises 7 through 18. Then find the particular solution that satisfies the given initial condition.

7. $ds/dt + ts = t, \quad s(0) = 0$.

8. $y' = y + 1, \quad y(0) = 1$.

9. $2dy/dx = xy, \quad y(1) = 0$.

10. $t dP/dt + P = t^3, \quad P(1) = 0$.

11. $dx/dy + x = 1, \quad x(1) = 1$.

12. $dx/dt = x + t, \quad x(0) = 0$.

13. $y' + y = e^x, \quad y(0) = 0$.

14. $y' + \cos x = 0, \quad y(0) = 2$.

15. $y' + 2y = 1 + e^{-2x}, \quad y(0) = 1$.

16. $dx/dt = x - t, \quad x(1) = 1$.

17. $dx/dt = 2x - te^{-2t}, \quad x(0) = 1$.

18. $y' - y = x + 1, \quad y(0) = 1$.

Neither of the following two differential equations is linear as written, but they both become linear if the roles of dependent and independent variables are exchanged. Find solutions for each equation.

19. $\dfrac{dy}{dx} = \dfrac{1}{x + y}$.

20. $\dfrac{ds}{dt} = \dfrac{s}{t + s^2 e^{-s}}$.

21. Suppose $f(x) = 1$ for $0 \le x \le 1$ and $f(x) = 1/x$ for $1 < x$. Find the general solution of the differential equation $y' + f(x)y = x$. Try $M(x) = e^x$ if $0 \le x < 1$ and $M(x) = x$ if $1 \le x$.

22. Given $g(x)$ and $y(x)$, the exponential factor $M(x)$ is designed to make the equation $\left[M(x)y\right]' = M(x)y' + M(x)g(x)y$ valid for all differentiable $y(x)$. Expand the expression on the left and solve the resulting differential equation for $M(x)$ by separating variables to find the formula for the integrating factor.

Special Second-Order Equations. The following four differential equations of order 2 can be solved by regarding them as first-order linear equations to be solved for $z = y'$. Find the most general solution to each one by first making the appropriate substitution for y' and y'' and then integrating once to find $y(x)$.

23. $y'' + y' = 1$.

24. $y'' = y' + 1$.

25. $y'' + (1/x)y' = 1$.

26. $xy'' = 1$.

Special Third-Order Equations. The following differential equations of order 3 can be solved by regarding them as first-order linear equations to be solved for $w = y''$. Find the most general solution to each one by first making the appropriate substitution $w = y''$ and $w' = y'''$ and then integrating twice to find $y(x)$.

27. $y''' + y'' = 0$.

28. $y''' + (1/x)y'' = x, \quad x \neq 0$.

29. $2y''' + (3/x)y'' = 0, \quad x \neq 0$.

30. $y''' = 0$.

31. Show that the linear differential equation $dy/dx + g(x)y = f(x)$ with $f(x)$ and $g(x)$ continuous on an interval, satisfies the hypotheses of the general existence and uniqueness theorem of Section 2C for $dy/dx = F(x, y)$, namely that both $F(x, y)$ and its partial derivative $\partial F(x, y)/\partial y$ are continuous in a rectangle $R = \{(x, y) : a < x < b, \ c < y < d\}$.

4B Linear Models

The linear equations are rather special among the first-order equations for a number of reasons. One is that there are solution formulas for them that display the properties of the solutions in a way that makes them fairly easy to interpret. Another is that there are many processes that can be modeled very well by linear equations. In addition, what happens with a linear application can often, though not always, be used as a good indication of what will happen under somewhat more complicated assumptions. Our first illustration of these points has to do with what is called a **mixing problem:** a container holding fluid with an amount S of some substance in solution undergoes enrichment or dilution by addition and/or subtraction from S at some known rates. Thus the analysis reduces to the simple relation

$$\frac{dS}{dt} = \{\text{inflow rate for } S\} - \{\text{outflow rate for } S\}.$$

Note that, for consistency with the left-hand side, the rates on the right-hand side must be stated in terms of quantities of dissolved material, rather than volumes of solution. Figure 1.10 shows a schematic diagram with three possible

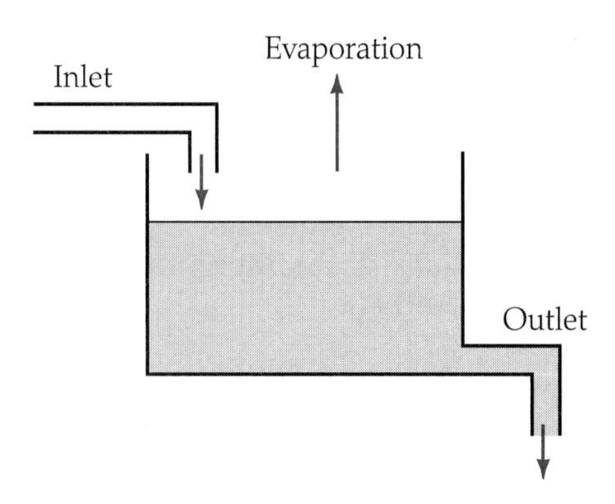

Figure 1.10 Input-Output diagram.

ways for the value of S to be influenced. To make our model a linear one, we need to assume that the solution is kept thoroughly mixed at all times, so that the concentration C of dissolved material per unit of volume V is kept homogeneous throughout the container. Under that assumption sometimes it's convenient to deal with concentrations of the substance in question. For example, in a full container of capacity V, the concentration will be $C = S/V$, though we're still free to use $S = CV$ in setting up a differential equation. It follows that if the outward flow rate of solution is r_o, then

$$\{\text{outflow rate for } S\} = r_o C = r_o \frac{S}{V}.$$

Note that the total volume V of solution in a compartment may increase or decrease depending on relative rates of inflow, outflow and evaporation. Evaporation, by changing volume V, affects the outflow rate for S.

---

example 4

Suppose that a 100-gallon vat contains 10 pounds of salt dissolved in water and that a solution of the same salt is being run into the vat at a rate of 3 gallons per minute. Assume the solution being run into the vat has a concentration in pounds per gallon that increases slowly with time according to the formula

$$C(t) = 1 - e^{-t/100}.$$

The solution is kept thoroughly mixed and the excess is drawn off, also at a rate of 3 gallons per minute. Let $S(t)$ stand for the pounds of salt in the tank at time $t \geq 0$. Since the concentration of salt in the tank running out at time t is the total amount of salt in the tank divided by the total amount of solution, we have

$$\begin{aligned}\frac{dS}{dt}(t) &= \{\text{inflow rate for } S\} - \{\text{outflow rate for } S\}\\ &= 3C(t) - 3\frac{S(t)}{100}\\ &= 3\left(1 - e^{-t/100}\right) - \frac{3}{100}S(t).\end{aligned}$$

The resulting first-order equation is linear, and in normal form it is

$$\frac{dS}{dt} + \frac{3}{100}S = 3\left(1 - e^{-t/100}\right).$$

An exponential multiplier is given by

$$M(t) = e^{\int (3/100)dt} = e^{3t/100}.$$

Multiplying the equation by M puts it in the form

$$\frac{d}{dt}\left(e^{3t/100}S\right) = 3\left(e^{3t/100} - e^{2t/100}\right).$$

Integration with respect to t gives

$$e^{3t/100}S(t) = \int 3(e^{3t/100} - e^{2t/100})\,dt + K$$
$$= 100e^{3t/100} - 150e^{2t/100} + K.$$

Then multiplication by $e^{-3t/100}$ gives

$$S(t) = 100 - 150e^{-t/100} + Ke^{-3t/100}.$$

To determine the constant K, we recall that the vat initially contains 10 pounds of salt, so $S(0) = 10$. Then setting $t = 0$ in the formula for $S(t)$ gives

$$10 = -50 + K \quad \text{or} \quad K = 60.$$

Thus the desired particular solution is

$$S(t) = 100 - 150e^{-t/100} + 60e^{-3t/100}.$$

Notice that

$$\lim_{t\to\infty} C(t) = 1,$$

so the concentration of the solution being added approaches 1 pound per gallon. From this information we could conclude on physical grounds that the total amount of salt in the 100-gallon tank should approach 100 pounds; indeed, the formula for $S(t)$ shows that

$$\lim_{t\to\infty} S(t) = 100.$$

---

example 5

Consider the following modification of the previous example: everything remains the same as before except that now the mixed solution is being drawn off at the rate of 4 gallons per minute and the salt concentration of the added solution is held constant at 1 pound per gallon. Since fresh solution is still being added at the rate of 3 gallons per minute, the tank is being emptied at 1 gallon per minute, with the result that t minutes after the process starts, the tank will contain $100 - t$ gallons of solution. In particular, after 100 minutes the tank will be empty, and the process will have effectively ended. Therefore, we'll consider the following differential equation on the time interval $0 \le t \le 100$:

$$\frac{dS}{dt}(t) = 3 - 4\frac{S(t)}{100 - t}.$$

The ratio $S(t)/(100 - t)$ is the concentration in pounds per gallon of the salt solution at time t; this varies both because $S(t)$ is changing and because the total amount of solution $(100 - t)$ is decreasing. The differential equation is solved by rewriting it in the form from which we can calculate an integrating factor:

$$\frac{dS}{dt} + \frac{4}{100 - t}S = 3, \qquad M(t) = e^{\int 4/(100-t)}dt = e^{-4\ln(100-t)} = (100 - t)^{-4}.$$

Thus we only have to integrate both sides of $d\left[(100-t)^{-4}S\right]/dt = 3(100-t)^{-4}$ to get

$$(100-t)^{-4}S = (100-t)^{-3} + K, \quad \text{or} \quad S = (100-t) + K(100-t)^4.$$

The constant K can be determined from the initial value $S(0) = 10$, as in the previous example; it now turns out that $K = -9 \cdot 10^{-7}$. From the general formula for $S(t)$ we can calculate, for example, the time at which the tank contains the maximum amount of salt, or focus on the behavior of the salt concentration $C(t) = S(t)/(100-t) = 1 + K(100-t)^3$ in pounds per gallon.

Experience tells us that hot or cold objects cool down or warm up to the temperature of their surroundings. A useful way to quantify these observations is called **Newton's law of cooling,** which asserts that the temperature $u(t)$ of an object has a rate of change proportional to the difference between $u(t)$ and the **ambient temperature** $a(t)$ of the surroundings:

$$\frac{du}{dt} = -k(u-a), \quad k = \text{const.}, \ k > 0.$$

Note that if $u > a$, the right-hand side of the equation is negative, so u will decrease, and actual cooling takes place. If $u < a$, then u will increase, and what we have is really a "law of heating." (The constant k varies with the length of the time units employed, so, for example, in changing from hours to minutes, k would be divided by 60.)

---

example 6

Suppose a beaker of water initially at 50°C. is placed in a bath of water whose temperature $a(t)$ at time t hours gradually warms from 40° to 50° for $t \geq 0$ according to $a(t) = 50 - 10e^{-t}$. With $k = \frac{1}{2}$, the temperature $u(t)$ of the beaker satisfies the equation

$$\frac{du}{dt} = -\frac{1}{2}\left[u - (50 - 10e^{-t})\right] \quad \text{or} \quad \frac{du}{dt} + \frac{1}{2}u = 25 - 5e^{-t}.$$

Multiplication by the exponential factor $M(t) = e^{t/2}$ gives

$$\frac{d}{dt}(e^{t/2}u) = 25e^{t/2} - 5e^{-t/2}.$$

Integration with respect to t gives

$$e^{t/2}u = 50e^{t/2} + 10e^{-t/2} + C \quad \text{or} \quad u = 50 + 10e^{-t} + Ce^{-t/2}.$$

At $t = 0$, we have $u(0) = 50$ so we find $C = -10$ and $u(t) = 50 + 10e^{-t} - 10e^{-t/2}$. Using this formula, you can estimate the time it would take for the beaker to cool to its minimum temperature.

EXERCISES

1. A 100-gallon tank full of copper sulfate solution contains initially 10 pounds of copper sulfate in solution. At $t = 0$, pure water is added at a rate of 2 gallons per minute, with a resulting overflow of 2 gallons per minute of copper sulfate solution. Assuming that the solution is kept thoroughly mixed at all times, find the amount of copper sulfate in the tank at time $t > 0$.

2. Ammonium nitrate solution for fertilizer enters a 100-gallon tank of initially pure water from two different sources. One source provides water containing 1 pound of ammonium nitrate per gallon at a rate of 2 gallons per minute. A second source provides 3 gallons of ammonium nitrate solution per minute at a varying concentration $C(t) = 2e^{-2t}$, measured in pounds of ammonium nitrate per gallon. Assume that the contents of the tank are kept thoroughly mixed at all times and that the solution is drawn off at a rate of 5 gallons per minute. Find the amount of ammonium nitrate in the tank at time $t > 0$.

3. The current $I(t)$ in an electric circuit satisfies the differential equation
$$L\frac{dI}{dt} + RI = E(t),$$
where L and R are positive constants representing inductance and resistance, respectively, and $E(t)$ is a variable applied voltage. Show that
$$I(t) = \frac{1}{L}e^{-Rt/L}\int_0^t E(u)e^{Ru/L}du + I(0)e^{-Rt/L}.$$

4. A pellet of mass m falling under the influence of gravity through a resisting medium has velocity $v(t)$ at time t, satisfying
$$m\frac{dv}{dt} = mg - kv.$$
Here g is the constant acceleration of gravity, and k is a positive constant that measures the resistance of the medium. Show that
$$v(t) = \left[v(0) - \frac{mg}{k}\right]e^{-kt/m} + \frac{mg}{k}.$$

5. Sketch the graph of the function $u(t)$ found at the conclusion of Example 6 in the text. What is the minimum value of $u(t)$, and when does it occur? What is $\lim_{t\to\infty} u(t)$?

6. A full 100-cubic-foot tank initially has 10 pounds of salt in solution. At a certain time additional salt solution begins to enter the tank at a rate of 1 cubic foot per hour, while thoroughly mixed solution runs out a drain at the same rate. However, the amount of salt in the added solution decreases at a constant rate from 1 pound per cubic foot initially all the way down to zero pounds per cubic foot at the end of one hour.

   (a) Find the amount of salt in the tank at a given time during the first hour. In particular, about how much salt will be in the tank at the end of one hour?

(b) Estimate the time at which the amount of salt in the tank is at a maximum during the first hour.

(c) If pure water continues to run into the tank after the first hour at the rate of 1 cubic foot per hour, how much more time will it take for the total amount of salt in the tank to reach 5 pounds?

7. Two 100-gallon mixing tanks are initially full of pure water. A solution containing one pound of calcium chloride per gallon of water pours into the first tank at the rate of one gallon per minute. Thoroughly mixed solution runs from the first tank to the second at the rate of one gallon per minute, where it too is thoroughly mixed in before draining away at 1 gallon per minute. Obviously there will always be at least as much calcium chloride in the first tank as in the second; find the maximum amount of this excess.

8. Two 100-gallon tanks X and Y are initially full of pure water. Salt solution is added to X at 1 gallon per minute (gpm) from an external source, each gallon containing 1 pound of salt. Mixed solution is pumped from X to Y at 2 gpm and runs from Y into a waste drain at 4 gpm. Let $x = x(t)$ and $y = y(t)$ be the respective amounts of salt in X and Y at time $t \geq 0$.

(a) At what times t_X and t_Y will the respective tanks first become empty?

(b) Find a pair of differential equations satisfied by $x(t)$ and $y(t)$ for $0 \leq t \leq t_X$ and $0 \leq t \leq t_Y$, respectively.

(c) Find a formula for $x(t)$ alone.

(d) Use the result of part (c) to find $y(t)$.

9. A 100-gallon tank is initially full of pure water. Salt solution is added for 10 minutes at the rate of one gallon per minute with salt content of the added solution increasing linearly over the 10 minutes from 1 pound per gallon to 2 pounds per gallon. Thoroughly mixed salt solution is drawn off at the rate of one gallon per minute. Estimate the amount of salt in the tank at the end of the 10 minutes.

10. A 100-gallon mixing vat is initially half-full of pure water. Two gallons of curing solution per minute at a concentration of 1 pound of sodium nitrite per gallon begin to flow in, while 1 gallon per minute of mixed solution flows out. Estimate the amount of sodium nitrite in the vat at the moment it begins to overflow.

11. A 100-gallon tank initially contains 50 gallons of water with a total of 10 pounds of salt dissolved in it. A drain is opened in the bottom that lets out 1 gallon of solution per minute. Simultaneously, salt solution begins to be added at 2 gallons per minute with a concentration of 2 pounds per gallon.

(a) How much salt is in the tank when it first becomes full and starts to overflow?

(b) If the process is allowed to continue with overflow at an additional outflow of 1 gallon per minute, what is the upper limit for the total amount of salt in the tank? Estimate the additional time after the start of overflow for the amount of salt in the tank to reach 175 pounds.

12. A 100-gallon mixing vat is initially full of pure water. One gallon of water begins to flow into the tank per minute, with a pound of salt in each gallon. Simultaneously 2 gallons of mixed solution per minute are pumped out. Find the maximum amount of salt in the vat, and show that the maximum occurs when the vat is half full of solution.

Evaporators

13. A 100-gallon mixing vat is initially full of pure water. One gallon per minute of salt solution with 1 pound of salt dissolved in each gallon of water flows into the tank, 1 gallon per minute of mixed solution flows out, and 1 gallon of water per minute evaporates from the vat. Estimate the amount of salt in the vat when it is half empty.

14. Suppose that, because of greater heating, the vat in the previous exercise is undergoing evaporation of pure water at a total rate of 2 gallons per minute. Estimate the amount of salt in the vat when it is half full.

15. A 100-gallon tank is initially half full of pure water. Three gallons per minute of salt solution with one pound of salt dissolved in each gallon flows into the tank, one gallon per minute of mixed solution flows out, and one gallon of water per minute evaporates from the tank. How much salt is in the tank when it begins to overflow?

Membrane Diffusion. A chemical in solution diffuses across a membrane with diffusion coefficient $k > 0$ from a compartment with known concentration $f(t)$ to a compartment with resulting concentration $u(t)$. Solve the diffusion equation

$$\frac{du}{dt} = k[f(t) - u]$$

under each of the sets of assumptions in Exercises 16 through 20. In each case, find the smallest constant B such that $u(t) \leq b$ for $t \geq 0$.

16. $u(0) = 0,\ f(t) = 10,\ k = 2.$

17. $u(0) = 0,\ f(t) = 10e^{-t},\ k = 2.$

18. $u(0) = 5,\ f(t) = 10,\ k = 2.$

19. $u(0) = 0,\ f(t) = 10(1 - e^{-t}),\ k = 2.$

20. $u(0) = 5,\ f(t) = 10,\ k(t) = (1 + t)^{-1}.$

Newton Cooling. Find the general form of a solution $u(t)$ of Newton's cooling equation $du/dt = -k(u - a)$ under each of the sets of assumptions in Exercises 21 through 25. Also, use an appropriate scale in each case to sketch the graphs of $u = u(t)$ and $u = a(t)$ relative to the same pair of axes.

21. $u(0) = 100,\ a(t) = 50,\ k = 1.$

22. $u(0) = 100,\ a(t) = 50e^{-t},\ k = 1.$

23. $u(0) = 100,\ a(t) = 50e^{-t},\ k = 2.$

24. $u(0) = 0,\ a(t) = 50 \sin t,\ k = 1.$

25. $u(0) = 100,\ a(t) = 200 + 50e^{-2t},\ k = 1.$

26. Suppose the solution $u(t) = 50 + 10e^{-t} + 40e^{-t/2}$ is derived for a cooling process from $100°$ down to $50°$. Estimate the value t for which $u(t) = 75$. [*Hint:* The relevant equation can be regarded as a quadratic equation to be solved for $e^{-t/2}$.]

27. A container of ice cream mix at 70°F is placed in a mixture of ice and brine at 30°F. Assume the validity of Newton's law described prior to Example 6 and that the ice cream has reached 40°F after 15 minutes.

- **(a)** Find an approximate value for the constant k in Newton's law.
- **(b)** When will the ice cream mix reach 35°F?

28. Suppose that an iron bar initially at 300°F is immersed in a water bath at 100°F for 30 minutes and then is transferred to another water bath at 50°F. Assume the validity of Newton's law described just before Example 6 of the text.

- **(a)** What will the temperature of the bar be after an additional 30 minutes, assuming the cooling coefficient for the iron in water is $k = 0.1$?
- **(b)** Suppose that initially the iron is cooled for 30 minutes in air at 100°F, for which the cooling coefficient is only $k = 0.07$ and is then immersed in water for 30 minutes. What will the temperature of the bar be at the end of the hour?

29. If a beaker containing s units of liquid at initial temperature u_0 has an additional unit at temperature $u_1 < u_0$ added and mixed in right away, the temperature of the mixture averages out to $(su_0 + u_1)/(s + 1)$. Assume the surrounding temperature is held at a constant $a < u_1$, and that Newton's law is valid as described before Example 6.

- **(a)** Find the resulting temperature $u(t_1)$ of the mixture if the initial s units at temperature u_0 are first allowed to cool over time period t_1, upon which a single unit at temperature $u_1 < u_0$ is mixed in.
- **(b)** Find the temperature $v(t_1)$ of the mixture if initially one unit at temperature u_1 is mixed in and then the mixture is allowed to cool over time period t_1.
- **(c)** Which of the routines described in parts (a) and (b) produces the cooler mixture? Does your intuition, informed by Newton's law, lead you to the same answer?

30. A cup of water is initially at 180°F in a 70°F room. If the water cools to 160°F in 10 minutes, estimate the constant k in Newton's law of cooling, and find the approximate temperature of the water after another 10 minutes.

Rubber Band Problems

31. A rubber band initially of length l has one end attached to a wall while the band is stretched uniformly over its entire length away from the wall by pulling the other end away from the wall at constant rate r. Note that particles of the band nearer to the fixed end move proportionately more slowly, in particular, the fixed end doesn't move at all and the other end moves at rate r.

(a) Find a differential equation satisfied by $x = x(t)$, the distance from the wall of a given particle of the band.

(b) After t time units of stretching, what is the distance $x(t)$ from the wall of a particle of rubber that is initially at distance x_0 from the wall?

32. (a) One end of a 1-foot rubber band is attached to a wall while the other end is stretched away from the wall at 1 foot per minute. Starting at the same time as the stretching, a small bug starts to crawl from the wall along the band at the constant rate of b feet per minute relative to the band. Will the bug ever reach the other end of the band, and if so how will its arrival time depend on b? Assume that the stretching is uniform throughout the length of the band.

(b) Solve the same problem under the assumptions that the band is initially l feet long, is being stretched at r feet per minute and that the bug starts out already x_0 feet from the wall.

4C From Nonlinear to Linear †

Given that there is a formula for the general solution of a first-order linear equation, and that the solutions have in general fairly simple properties, it is important to consider the question whether some nonlinear equations can for some purposes be effectively replaced by linear equations. In our initial examples, this replacement will be called **linearization**, and the resulting linear equation is called a **linearized** version of the nonlinear equation. To be reasonably representative of the nonlinear equation $y' = f(x, y)$, a linearization needs to be restricted in its application to some interval of y values containing a value y_0 of particular interest. Thus, for each x such that $f(x, y_0)$ is differentiable as a function of y, make the first-degree approximation

$$f(x, y) \approx f(x, y_0) + f_y(x, y_0)(y - y_0),$$

where f_y is the derivative of f with respect to y, holding x fixed; this approximation is geometrically equivalent to replacing the graph of f, regarded as a function of y, by its tangent line at $y = y_0$. We define the **linearization** of $y' = f(x, y)$ near y_0 to be the differential equation

4.3 $$y' = f(x, y_0) + f_y(x, y_0)(y - y_0).$$

Even if a nonlinear equation can be "solved" by an elementary formula, it may be impossible to derive a usable expression for the solution in terms of the independent variable. In that case; linearization can often provide an approximate solution formula that is accurate enough for some purposes. See, for example, Exercise 16 on Stefan's law.

example 7

In the nonlinear equation $y' = xy^2$ the function f of Equation 4.3 is $f(x, y) = xy^2$, so $f_y(x, y) = 2xy$. At $y_0 = 3$ we have $f(x, 3) = x \cdot 3^2 = 9x$ and $f_y(x, 3) = 2x \cdot 3 = 6x$. Near $y_0 = 3$ the nonlinear equation gets replaced by the linear equation $y' = 9x + 6x(y - 3)$ or $y' = 6xy - 9x$. The general solution of the linear equation is $y = \frac{3}{2} + Ce^{3x^2}$.

example 8

For the nonlinear equation $y' = f(x, y) = \tan y$, we have $f(x, 0) = 0$ and $f_y(x, 0) = \sec^2 0 = 1$. Thus the linearized Equation 4.3 at $y = 0$ becomes simply $y' = y$ with general solution $y(x) = Ce^x$. Near $y = 0$, we can impose an initial condition, say $y(0) = 0.1$, to get the particular solution $y(x) = 0.1e^x$. The given nonlinear equation can be solved by separation of variables and has the family of solutions $\ln|\sin y| = x + c$. The particular solution of this family satisfying the initial condition $y(0) = 0.1$ can be written $y(x) = \arcsin[e^x \sin(0.1)]$. This last formula is "elementary" and is after all the exact solution; on the other hand it is a bit cumbersome as compared with $y(x) = 0.1e^x$. Figure 1.11 shows that the two graphs are hard to tell apart up to $x = 1$. (The difference in values remains less than 0.003 up to that point, with a larger value for the solution of the nonlinear equation since $\tan y > y$ for $y > 0$.) By the time we get to $x = 2$ the discrepancy has increased to a little less than $\frac{1}{10}$. By comparing the differential equations you can deduce which graph has the larger values; simply note that $\tan y > y$ for $0 < y < \pi/2$.

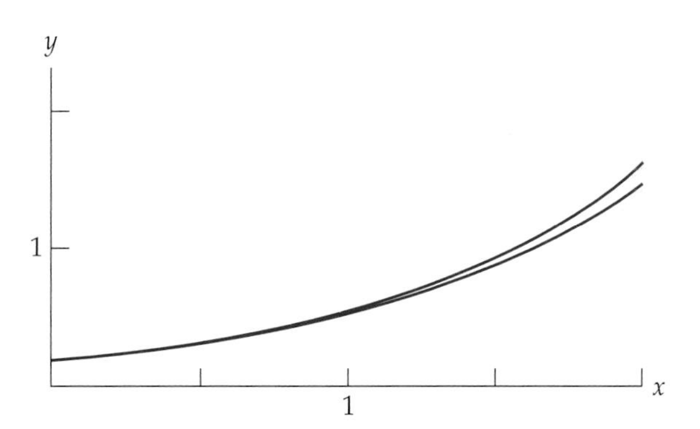

Figure 1.11 Solutions of $y' = \tan y$ and $y' = y$ with $y(0) = 0.1$.

Bernoulli Equations. The limited range of applicability of a linearized equation sometimes makes it desirable to solve the original nonlinear equation directly if at all possible. A nonlinear equation that can be transformed directly into a linear equation, thus avoiding the approximation implicit in linearization, is the **Bernoulli equation**

$$y' + g(x)y = f(x)y^n.$$

The parameter n can be any real number, but if $n = 0$ or $n = 1$ the equation is already linear so we can ignore those values in applying the following method. The trick is first to multiply the equation by y^{-n}. It then turns out that if we let $u(x) = y^{1-n}(x)$, then $(1-n)y' = y^n u'$, and the Bernoulli equation transforms to the linear equation

4.4 $$u' + (1-n)g(x)u = (1-n)f(x).$$

If you can solve this equation explicitly for $u(x)$ using the usual exponential multiplier

$$M(x) = e^{\int (1-n)g(x)\,dx},$$

all that remains to be done is set $u = y^{1-n}$ to find the solution to the Bernoulli equation.

example 9

The Bernoulli equation $y' + (1/x)y = (1/y)$ fits the general form of Equation 4.4 with exponent $n = -1$. Then the factor $1 - n$ equals 2, so let $u = y^2$. By Equation 4.4 the linear equation to be solved is

$$u' + \frac{2}{x}u = 2.$$

The appropriate exponential multiplier is $M(x) = \exp(\int (2/x)\,dx) = x^2$, and the general solution of

$$x^2u' + 2xu = 2x^2 \quad \text{is} \quad u = \frac{2}{3}x + \frac{c}{x^2}.$$

Since u and y were related by $u = y^{1-n} = y^2$ (to make the transition to a linear equation), we find $y^2 = (2/3)x + c/x^2$. The right-hand side must be positive if y is to be real-valued, so we can write the two explicit formulas $y = \pm\sqrt{(2/3)x + c/x^2}$.

EXERCISES

Find the linearization of the differential equations in Exercises 1 through 8 near the indicated point.

1. $y' = y^3 + y;\ y_0 = 0.$
2. $y' = x \arctan y;\ y_0 = 0.$
3. $y' = y^2 + 1;\ y_0 = 1.$
4. $y' = x \tan y;\ y_0 = \pi/4.$
5. $y' = -\sin y;\ y_0 = 0.$
6. $y' = -\sin y;\ y_0 = \pi/4.$
7. $y' = x^2y^2;\ y_0 = -2.$
8. $y' = x/y;\ y_0 = 1.$

9. The linearization of $y' = xy^2$ near $y = 3$ is shown in Example 1 of the text to be $y' = 6xy - 9x$.

(a) Find the solution y_1 of the nonlinear equation $y' = xy^2$ that satisfies $y_1(0) = 3$.

(b) Find the solution y_2 of the corresponding equation linearized near $y = 3$ that satisfies $y_2(0) = 3$.

(c) Sketch the graphs of the solutions y_1 and y_2 found in parts (a) and (b); how do they compare at $x = 1/2$?

10. The velocity $v = v(t)$ of a certain falling body subject to nonlinear, velocity-dependent air resistance satisfies $dv/dt = 32 - 0.01v^2$.

(a) Find the linearization of this differential equation near velocity $v_0 = 100$.

(b) Solve the linearized equation found in part (a) subject to the initial condition $v(0) = 100$.

11. The equation $m\,dv/dt = mg - kv^\alpha$, $\alpha > 0$, for a falling body with viscous damping is nonlinear for $\alpha \neq 1$, and is seldom solvable for $v = v(t)$ in elementary terms. (The cases $\alpha = 1$ and $\alpha = 2$ are exceptional.) Without solving the differential equation, show that the terminal velocity is $v_\infty = \sqrt[\alpha]{mg/k}$. [*Hint:* What happens to dv/dt as $t \to \infty$?]

12. The falling body equation $m\,dv/dt = mg - kv^2$ has solutions for which the terminal velocity is $v_\infty = \sqrt{gm/k}$.

(a) Find the linearized version of the differential equation near $v = v_\infty$.

(b) Solve the linear equation found in part (a) with initial condition $v(0) = \sqrt{gm/k}$.

13. The falling body equation $m\,dv/dt = mg - kv^\alpha$, $0 < m = \text{const.}$, has solutions for which the terminal velocity is $v_\infty = \sqrt[\alpha]{gm/k}$.

(a) Find the linearized version of the differential equation near $v = v_\infty$.

(b) Solve the linear equation found in part (a) under the assumption that $v(0) = v_\infty$.

14. (a) Find the linearization of the falling body equation $m\,dv/dt = mg - kv^2$ near $v = v_0$.

(b) Solve the linear equation found in part (a) under the assumption that $v(0) = v_0$.

15. (a) Show that the linearization of the falling body equation $m\,dv/dt = mg - kv^\alpha$ near $v = v_0$ is $m\,dv/dt = gm + k(\alpha - 1)v_0^\alpha - k\alpha v_0^{\alpha-1}v$.

(b) Show that the solution of the linear equation of part (a) under the assumption that $v(0) = v_0$ is

$$v(t) = \frac{1}{\alpha}\left(v_0 - \frac{gm}{kv_0^{\alpha-1}}\right)e^{-\alpha(kv_0^{\alpha-1}/m)t} + \frac{1}{\alpha}\left[\frac{gm}{kv_0^{\alpha-1}} + (\alpha - 1)v_0\right].$$

16. Stefan's law for the Kelvin-scale temperature $u(t)$ of a body in an environment with ambient temperature a states that $du/dt = C(a^4 - u^4)$.

(a) Show that **Newton's law of cooling,** $du/dt = k(a - u)$, can be regarded as the linearization of Stefan's law near $u = a$, with $k = 4a^3C$.

(b) Show that the solution to Newton's equation is

$$u = a + (u_0 - a)e^{-4a^3Ct},$$

where $u_0 = u(0)$.

(c) Show that the nonlinear Stefan equation has solutions that satisfy

$$\ln\left|\frac{u-a}{u+a}\right| - 2\arctan\frac{u}{a} = -4a^3Ct + c,$$

where c is an integration constant.

(d) Discuss the problem of expressing u as a function of t using the relation in part (c), and use a direction field sketch for Stefan's equation to sketch some typical solution graphs.

17. A nonlinear model for membrane transport of a chemical solution to a compartment with concentration $u = u(t)$ is $du/dt = c(k - u)^\alpha$, where c, k, and α are positive constants.

(a) Show that the linearization of the nonlinear equation is $du/dt = C(K - u)$, with $C = c\alpha(k - u_0)^{\alpha-1}$ and $K = (k - u_0 + \alpha u_0)/\alpha$, near $u = u_0$.

(b) Make a graphical comparison of the solutions $u = u(t)$ to the nonlinear equation and its linearization for the case $\alpha = 2$, $c = k = 1$ and $u_0 = 0$. In particular, describe the relative behaviors of these solutions as t gets large.

18. If $n \neq 0$ or 1, the differential equation

$$y' + g(x)y = f(x)y^n$$

is a *Bernoulli equation,* as discussed in the text. The purpose of this exercise is to go through the details of how a Bernoulli equation can be reduced to solving a linear equation.

(a) Show that if $n \neq 1$, multiplying by y^{-n} and letting $u(x) = y^{1-n}(x)$ yields the equation $u' + (1 - n)g(x)u = (1 - n)f(x)$. [*Hint:* This last equation is linear and can in principle always be solved for $u(x)$. Then replace u by y^{1-n} in the solution of the linear equation.]

(b) What problem could arise with defining u by $u = y^{1-n}$ if $y \leq 0$?

19. **(a)** Solve the Bernoulli equation $y' = y - y^4$ given $y(0) = \frac{1}{2}$. **(b)** Use the equation's direction field to sketch the solution to part (a).

20. Solve the Bernoulli equation $x^2y' - xy = y^2$ given $y(1) = 1$. Also find the solution for which $y(1) = 0$ and explain why it doesn't come out of the general method.

21. Solve the Bernoulli equation

$$\frac{dv}{dy} = \frac{1}{v} - \frac{v}{y + y_0},$$

given that y_0 is a positive constant and that $v(0) = 0$.

22. Assume $n \neq 0$ or 1 in the Bernoulli equation $y' + g(x)y = f(x)y^n$ and show that the linearization at $y = y_0$ is $y' + G(x)y = F(x)$, where $G(x) = [g(x) - ny_0^{n-1} f(x)]$ and $F(x) = (1 - n)y_0^n f(x)$. What value for y_0 must be avoided if $n = -1$?

5 EXACT EQUATIONS†

The ideas in this section provide a technique for solving some first-order differential equations. Beyond that we'll investigate the representation of solutions as level curves of a function of two variables and also see how a direction field can be modified to be the gradient field $\nabla F(x, y)$ of a function $F(x, y)$.

5A Solutions by Integration

The solutions of some differential equations are quite naturally presented in the implicit form $F(x, y) = C$ as opposed to the explicit form $y = y(x)$. For example, one particular solution of $y' = x/y$ is $y = \sqrt{1 + x^2}$, with $-1 < x < 1$, while another solution is $y = -\sqrt{1 + x^2}$, defined on the same interval. Squaring both sides in either formula gives an implicit formula for both solutions: $y^2 - x^2 = 1$, which is the equation of a hyperbola. Differentiating the more general equation $y^2 - x^2 = C$ implicitly with respect to x gives $2yy' = 2x$ or $y' = x/y$, so we see that there are many explicit solutions contained in the family of hyperbolas depending on C.

Suppose now that we start with a relation $F(x, y) = C$ that implicitly defines $y = y(x)$ as a differentiable function of x, or perhaps $x = x(y)$ as a function of y. If we differentiate $F(x, y) = C$ with respect to x using the chain rule, we get

$$F_x(x, y) + F_y(x, y)\frac{dy}{dx} = 0,$$

where the subscripts denote partial differentiation with respect to the indicated variable, while the other variable is held fixed. The resulting first-order differential equation is automatically satisfied by the function $y = y(x)$. Multiplying by dx and setting $dy = (dy/dx)\,dx$, the differential equation looks like

$$F_x(x, y)\,dx + F_y(x, y)\,dy = 0, \quad \text{or} \quad P(x, y)\,dx + Q(x, y)\,dy = 0,$$

where $P = F_x$ and $Q = F_y$. Not every equation of the second form displayed above arises as indicated from what is called the **exact differential** of a function $F(x, y)$, namely,

$$dF(x, y) = F_x(x, y)\,dx + F_y(x, y)\,dy.$$

Those differential equations that do arise that way are called **exact** differential equations. If an equation is exact, we can then try to display its solution graphs as level curves of some function $F(x, y)$ satisfying $F(x, y) = C$, with both $F_x(x, y) = P(x, y)$ and $F_y(x, y) = Q(x, y)$. Thus the problem of solving an exact differential equation is equivalent to reconstructing a function F from its gradient $\nabla F = (F_x, F_y)$. The solutions are then usually displayed in the implicit form $F(x, y) = C$.

example 1

Sometimes it's possible to make an educated guess at the exactness of a differential, and then solve the associated equation at a glance. An equation

$$g(x)\,dx + h(y)\,dy = 0,$$

with variables separated is always exact if $g(x)$ and $h(y)$ are continuous functions on respective domain intervals $a \le x \le b,\ \ c \le y \le d$. The reason is that there is always a solution of the special implicit form

$$G(x) + H(y) = C,$$

where

$$G(x) = \int g(u)\,du \quad \text{and} \quad H(y) = \int h(u)\,du.$$

The constants of integration can be combined into one, which may not be completely arbitrary. For instance, $x\,dx + y\,dy = 0$ has the implicit solution $\frac{1}{2}x^2 + \frac{1}{2}y^2 = C$, where $C > 0$. The graphs of these solutions are circles centered at the origin, and these circles are the level curves of $F(x, y) = \frac{1}{2}(x^2 + y^2)$. A circle of radius $r > 0$ corresponds to the points at level $z = r^2$ on the graph in (x, y, z) space of the equation $z = x^2 + y^2$. These are shown in Figure 1.12 below.

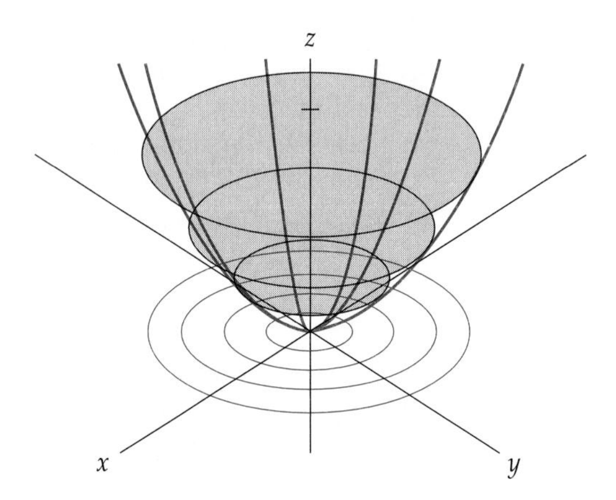

Figure 1.12

example 2

Recall that much of the technique of indefinite integration is based on our ability to recognize a function when we are given its derivative. In a slight extension of this technique, what we want to do here is recognize a function when we are given two partial derivatives. For example, since $\partial/\partial x(x \sin y) = \sin y$ and $\partial/\partial y(x \sin y) = x \cos y$, the equation

$$\sin y\,dx + x \cos y\,dy = 0$$

is easily seen to be equivalent to

$$d(x \sin y) = 0.$$

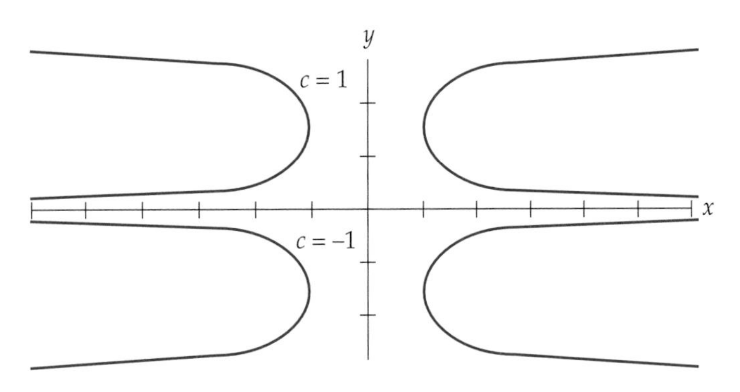

Figure 1.13 $\sin y\,dx + x\cos y\,dy = 0;\ y = \arcsin(c/x),\quad c = \pm 1.$

Thus the implicit form of the general solution to the equation

$$\sin y\,dx + x\cos y\,dy = 0$$

is $x\sin y = c$. Another representation is $x = c\csc y$. An explicit solution that we can derive by solving for y is $y = \arcsin(c/x)$. Figure 1.13 shows some graphs; the broken curves represent points on alternative branches of the arcsine function.

If you think a differential equation may be exact but you can't guess the solution, there is an integration technique for trying to find $F(x, y)$. The idea is to hypothesize the existence of F and then see what conclusions can be drawn about F. (Of course, one possible conclusion is that the equation you're looking at *isn't* exact and won't yield up *its solution* to this method.)

---

example 3

The differential equation $y' = (x + y)/(y - x)$ is equivalent, if $y \neq x$, to

$$(x + y)\,dx - (y - x)\,dy = 0 \quad \text{or} \quad (x + y)\,dx + (x - y)\,dy = 0.$$

If the left-hand side is the exact differential of some function F, that means precisely that $F_x(x, y) = x + y$ and $F_y(x, y) = x - y$. In principle, we can work with either of these last two equations; since they seem to be of roughly equal complexity, we may as well pick the first one. We integrate both sides with respect to x and find

$$\begin{aligned} F(x, y) &= \int F_x(x, y)\,dx \\ &= \int (x + y)\,dx \\ &= \tfrac{1}{2}x^2 + yx + \phi(y). \end{aligned}$$

Note the "constant" of integration $\phi(y)$; since y is held constant during the integration, we can allow for different values of the integration constant, depending on y (but *not* on x), and in fact this freedom is crucial for making the remaining steps effective. Having found that

$$F(x, y) = \tfrac{1}{2}x^2 + yx + \phi(y),$$

it now makes sense to differentiate with respect to y so that we can compare the result with our second equation $F_y = x - y$. We find

$$\begin{aligned} F_y(x, y) &= \frac{\partial}{\partial y}\left[\frac{1}{2}x^2 + yx + \phi(y)\right] \\ &= x + \phi'(y) = x - y. \end{aligned}$$

From this last equality we conclude that $\phi'(y) = -y$. [It's important that this expression for $\phi'(y)$ turns out not to depend on x; otherwise the the given equation could not have been exact.] Integration with respect to y shows that $\phi(y)$ must be of the form $-\frac{1}{2}y^2 + c$. It follows that

$$F(x, y) = \tfrac{1}{2}x^2 + yx - \tfrac{1}{2}y^2 + c,$$

and that solutions of the original differential equation must satisfy the implicit relation

$$\tfrac{1}{2}x^2 + yx - \tfrac{1}{2}y^2 + c = 0.$$

Note that this is still not quite the same as having a formula for an explicit solution; we would have to solve the quadratic equation for y, or else for x, to get that. But completing the square in the first two terms shows that the graphs of all solutions must lie on a hyperbola of the form $(x+y)^2 - 2y^2 = C$. This may be the most informative way to look at the family of solutions.

Outline of the Method. We can summarize the general form of the calculation in the previous example as follows: Since $F_x(x, y) = P(x, y)$,

$$F(x, y) = \int P(x, y)\,dx + \phi(y), \quad \text{where } \frac{\partial}{\partial y}\int P(x, y)\,dx + \phi'(y) = Q(x, y).$$

Had we started instead with $F_y(x, y) = Q(x, y)$, the calculation would look like this:

$$F(x, y) = \int Q(x, y)\,dy + \psi(x), \quad \text{where } \frac{\partial}{\partial x}\int Q(x, y)\,dy + \psi'(x) = P(x, y).$$

What we have here are methods for solving a system of the two partial differential equations $F_x = P$, $F_y = Q$ when P and Q are given. It's worth looking at these formulas to fix the ideas in mind, but as a practical matter, it's not worth memorizing them, because once you understand the method the steps are quite naturally suggested by the equations $F_x = P$ and $F_y = Q$.

5B Exactness Test

In Example 3 it was nice that in solving for $\phi'(y)$ we were able to come up with an expression independent of x. Otherwise the resulting inconsistency would have left us with $\phi(y)$ depending on x. Fortunately, there is a simple test that we can apply before going to the trouble of doing any formal integration. If the equation we're trying to solve fails the test, there's no point in continuing the solution method as outlined at the end of Section 5B. The test is easy to state and to apply.

5.1 THEOREM

Assume that $P(x, y)$ and $Q(x, y)$ have continuous first-order partial derivatives in a rectangular region R. Then

$$P(x, y)\,dx + Q(x, y)\,dy$$

is an exact differential if and only if

$$\frac{\partial Q(x, y)}{\partial x} = \frac{\partial P(x, y)}{\partial y}$$

throughout the rectangle R.

Proof. If the test fails then the differential can't be exact, because exactness means

$$P(x, y) = \frac{\partial F(x, y)}{\partial x} \quad \text{and} \quad Q(x, y) = \frac{\partial F(x, y)}{\partial y} \quad \text{for some } F(x, y).$$

The assumption that P and Q are continuously differentiable implies equality of the mixed partials F_{xy} and F_{yx}, so

$$\frac{\partial Q(x, y)}{\partial x} = \frac{\partial^2 F(x, y)}{\partial y \partial x} = \frac{\partial^2 F(x, y)}{\partial x \partial y} = \frac{\partial P(x, y)}{\partial y}.$$

This proves one of the implications of Theorem 5.1.

To verify that the exactness test $P_y(x, y) = Q_x(x, y)$ is sufficient to guarantee exactness under our hypotheses, all we have to do is walk through the routine of Section 5A in full generality. Assume the equation

$$P(x, y)\,dx + Q(x, y)\,dy = 0$$

satisfies the consistency condition

$$\frac{\partial Q(x, y)}{\partial x} = \frac{\partial P(x, y)}{\partial y}$$

in some rectangular region of the xy plane. We try to find $F(x, y)$ such that

$$F_x(x, y) = P(x, y), \quad F_y(x, y) = Q(x, y).$$

Suppose we choose to integrate the second equation with respect to y, while holding x fixed. We get

$$F(x, y) = \int Q(x, y)\,dy + \phi(x).$$

To find $\phi(x)$, differentiate this equation with respect to x and use $F_x(x, y) = P(x, y)$ to get

$$P(x, y) = \frac{\partial}{\partial x} \int Q(x, y)\,dy + \phi'(x).$$

Now $\phi'(x)$ can be expressed entirely in terms of P and Q, so we can in principle integrate again, this time with respect to x, to find a function $\phi(x)$ to put into the our equation for $F(x, y)$. It's crucial of course that our expression

$$P(x, y) - \frac{\partial}{\partial x} \int Q(x, y)\, dy$$

for $\phi'(x)$ really turns out to be independent of y. To see that it is, we'll check that its derivative with respect to y is zero. Differentiating this formula with respect to y cancels the y integration and yields

$$\frac{\partial P(x, y)}{\partial y} - \frac{\partial Q(x, y)}{\partial x}.$$

This expression is identically zero if the consistency condition $Q_x = P_y$ holds, so $\phi'(x)$ will indeed be independent of y under that assumption. If we had started with integrating $F_x = P$ instead of $F_y = Q$, everything would be essentially the same, the only alterations being the interchange of x and y and the interchange of P and Q. ■

example 4 The differential equation

$$(x - y)\, dx + (x + y)\, dy = 0$$

isn't exact because

$$\frac{\partial (x + y)}{\partial x} = 1 \neq -1 = \frac{\partial (x - y)}{\partial y}.$$

This does *not* mean that the differential equation has no solutions, but just that we can't solve it by identifying it as an exact equation. (The equation can be solved by the method of Example 10 of Section 3C.)

example 5 The differential equation

$$(e^x \cos y + 2x)\, dx - e^x \sin y\, dy = 0$$

passes the exactness test, because

$$\frac{\partial}{\partial x}(-e^x \sin y) = -e^x \sin y = \frac{\partial}{\partial y}(e^x \cos y + 2x).$$

We can expect to find a solution in the implicit form $F(x, y) = C$ by solving the equations

$$F_x(x, y) = e^x \cos y + 2x \quad \text{and} \quad F_y(x, y) = -e^x \sin y.$$

We start with whichever equation seems simpler; let's take the second equation and integrate with respect to y. We find

$$\begin{aligned} F(x, y) &= \int F_y(x, y)\, dy \\ &= \int -e^x \sin y\, dy \\ &= e^x \cos y + \phi(x). \end{aligned}$$

Here the constant of integration may turn out to depend explicitly on x. According to the equation for $F(x, y)$, we'll have

$$\begin{aligned} F_x(x, y) &= \frac{\partial}{\partial x}[e^x \cos y + \phi(x)] \\ &= e^x \cos y + \phi'(x). \end{aligned}$$

Thus $e^x \cos y + \phi'(x) = e^x \cos y + 2x$, so $\phi'(x) = 2x$. Since $\phi(x)$ must then be of the form $x^2 + c$, we find $F(x, y) = e^x \cos y + x^2 + c$. The solution can be written in implicit form as

$$e^x \cos y + x^2 = C.$$

We can then extract from this formula the explicit solution $y = \arccos [e^{-x}(C - x^2)]$.

The frequently used formulas

5.1 (**a**) $d(uv) = v\, du + u\, dv$ (**b**) $d\left(\frac{u}{v}\right) = \frac{v\, du - u\, dv}{v^2}$

can sometimes be used to make an informed guess about the solutions to some equations once they are in differential form.

---

example 6

The equation $dy/dx = -y/(x + y^2)$ can be written in differential form as

$$y\, dx + (x + y^2)\, dy = 0 \quad \text{or} \quad (y\, dx + x\, dy) + y^2\, dy = 0.$$

The first differential passes the exactness test $P_y = 1 = Q_x$, and the regrouping of terms is suggested by Equation 5.1(a). Fortunately the second term is just $d(y^3/3)$, so the entire equation can be written

$$d(xy) + d\left(\frac{y^3}{3}\right) = 0, \quad \text{with solution } xy + \frac{y^3}{3} = C.$$

---

example 7

The equation $y \cos x\, dx + \sin x\, dy = 0$ has the form $d(uv)$, where $u = \sin x$ and $v = y$. Solutions are given by $y \sin x = C$, and it's easy to check that $d(y \sin x) = y \cos x\, dx + \sin x\, dy$.

example 8

The equation $y\,dx - x\,dy = 0$ fails the exactness test since $P_y = 1 \neq -1 = Q_x$, but recalling Equation 5.1(b) suggests multiplying by y^{-2} to get

$$\frac{y\,dx - x\,dy}{y^2} = 0 \quad \text{or} \quad d\frac{x}{y} = 0.$$

Thus $x/y = c$ or $x = cy$ for the solution. The factor y^{-2} is an example of an *integrating factor* discussed previously in Section 4, and also expanded upon in Section 5C.

EXERCISES

The differential equations in Exercises 1 through 10 are either exact, separable, both, or neither. Check the consistency condition $P_y = Q_x$, and determine which of the possibilities holds. If possible, find a solution in the form $F(x, y) = C$ and find the value of C that satisfies the given initial condition.

1. $(y + 2xy^2)\,dx + (x + 2x^2y)\,dy = 0$; $y = 1$ when $x = 1$.

2. $(2y + 1)\,dy - (2x + 1)\,dx = 0$; $y = 0$ when $x = 0$.

3. $(y + e^y)\,dx + x(1 + e^y)\,dy = 0$; $y = 0$ when $x = 1$.

4. $(\cos y - y \sin x)\,dx + (\cos x - x \sin y)\,dy = 0$; $y = \pi/2$ when $x = \pi/2$.

5. $(y + 1)\,dy - xy\,dx = 0$; $y = 0$ when $x = 2$.

6. $(y - x)e^x\,dx + (1 + e^x)\,dy = 0$; $y = 1$ when $x = 1$.

7. $(y + 1/x)\,dx + (x + 1/y)\,dy = 0$; $y = 1$ when $x = 1$.

8. $4x^3y^3\,dx + 3x^4y^2\,dy = 0$; $y = 1$ when $x = 1$.

9. $(2xy + y^3)\,dx + (x^2 + 3xy^2)\,dy = 0$; $y = 1$ when $x = 1$.

10. $\ln x(y \ln y)\,dx + \left[\ln(\ln y)/x\right]dy = 0$; $y = 2$ when $x = 2$.

Determine for which values of the constant k each of the equations in Exercises 11 through 14 satisfies the consistency condition $P_y = Q_x$. Then solve the equation using those values of k.

11. $y\,dx + (kx + y^2 + 1)\,dy = 0$.

12. $\cos y\,dx + [\cos y + k(x + y)\sin y]\,dy = 0$.

13. $[(x + ky + 1)/(x + ky)]\,dx + [k/(x + ky)]\,dy = 0$.

14. $2\,dx/x + k\,dy/y = 0$.

15. Show that an equation $dy/dx = P(x)/Q(y)$ for which the variables are separable becomes exact when rewritten in separated differential form.

16. Find all functions $P(x, y)$ such that the equation $P(x, y)\,dx + (x^2y + y)\,dy = 0$ is exact.

17. Find all functions $Q(x, y)$ such that the equation $(\sin x + \cos y)\,dx + Q(x, y)\,dy = 0$ is exact.

18. **(a)** By writing both equations $P(x, y)\,dx + Q(x, y)\,dy = 0$ and $Q(x, y)\,dx - P(x, y)\,dy = 0$ in the general form y' equals some function of (x, y), show that where solution curves of the two equations intersect, they intersect perpendicularly.

(b) Show that if each of the two differential equations in part (a) is exact then two level curves $P(x, y) = C_1$ and $Q(x, y) = C_2$ intersect perpendicularly.

19. Solution curves to the equation

$$P(x, y)\,dx + Q(x, y)\,dy = \frac{2xy}{(x^2 + y^2)^2}\,dx + \left[1 + \frac{y^2 - x^2}{(x^2 + y^2)^2}\right]dy = 0$$

model the flow lines of fluid flow in the xy plane past a circular object bounded by the circle $x^2 + y^2 = 1$.

(a) Show that the differential equation is exact.

(b) Find an implicit solution formula for the equation.

(c) Show that the x axis and the circle $x^2 + y^2 = 1$ are solution curves, and sketch them along with at least two other solution curves.

(d) Use the idea in part (a) of the previous exercise to find a differential equation satisfied by curves perpendicular to the flow lines, and find equations for these perpendicular curves.

*__20.__ A basic result of multivariable calculus says that the gradient vector $\nabla F(x, y) = \big(F_x(x, y),\ F_y(x, y)\big)$ is perpendicular to the level curve of F that contains (x, y). Assume continuous second-order partials for F in what follows.

(a) Show that an implicit equation of a curve perpendicular to the level curves of F, called an *orthogonal trajectory* of the level curves, satisfies $F_y(x, y)\,dx - F_x(x, y)\,dy = 0$.

(b) Show that this last equation passes the exactness test if and only if F is a **harmonic function,** which means that $F_{xx} + F_{yy} = 0$ for all (x, y).

(c) Show that an implicit equation for a level curve of F always satisfies the exact equation $F_x(x, y)\,dx + F_y(x, y)\,dy = 0$.

*__21.__ The assumption that R is rectangular in the statement of the exactness test is used only for showing the sufficiency of the condition. It is used there just to guarantee that an integration with respect to either x or y can proceed unimpeded as we go from point to point in the region R in which the condition $Q_x = P_y$ holds.

(a) Give an example of a region R that is not rectangular, but such that the proof in the text will still work.

(b) Use Green's theorem to prove the sufficiency of the exactness test under the assumption that the region R is simply connected.

5C Integrating Factors

A first-order differential equation, whether written as $y' = f(x, y)$ or in differential form $P(x, y)\,dx + Q(x, y)\,dy = 0$, typically has associated with it a direction field, or slope field, as described in Section 2. [If $P(x, y)$ and $Q(x, y)$ are both zero at the same point (x_0, y_0), the equation doesn't say anything at (x_0, y_0), so we assume that any such points have been excluded in what follows.] The pair of functions $\big(P(x, y), Q(x, y)\big)$ is related to the slope function f by $f(x, y) = -P(x, y)/Q(x, y)$. In addition, $\big(P(x, y), Q(x, y)\big)$ generates a **vector field** that we sketch by drawing arrows from initial points (x, y) to the terminal points $\big(x + P(x, y), y + Q(x, y)\big)$, as shown in Figure 1.14(a).

Geometric Interpretation. An arrow from this vector field starting at (x, y) will have slope $Q(x, y)/P(x, y)$ and so will be perpendicular to the segment of slope $f(x, y) = -P(x, y)/Q(x, y)$ associated with the differential equation $y' = f(x, y)$ at (x, y). Thus a solution curve containing (x, y) will be perpendicular to the vector arrow starting at (x, y). However, if $P(x, y)\,dx + Q(x, y)\,dy = 0$ is an exact equation, with implicit solution formula $F(x, y) = C$, then the vector field is the **gradient field** of F, that is,

$$(P, Q) = (\partial F/\partial x, \partial F/\partial y), \quad \text{abbreviated} \quad (P, Q) = \nabla F \quad \text{or} \quad (P, Q) = \text{grad}\, F.$$

We've seen previously that exactness of a first-order equation aids in finding an implicit solution formula, with solution graphs that are level curves of some function, for example the circles in Figure 1.14(b). We now ask whether we can adjust the *lengths* of the vectors in a vector field (P, Q), possibly also reversing some directions, so that the resulting field will be the gradient of function; in other words, can we find a function $M(x, y)$ such that the vector field (MP, MQ) is a gradient field? If we can find such an M, then

$$M(x, y)P(x, y)\,dx + M(x, y)Q(x, y)\,dy = 0$$

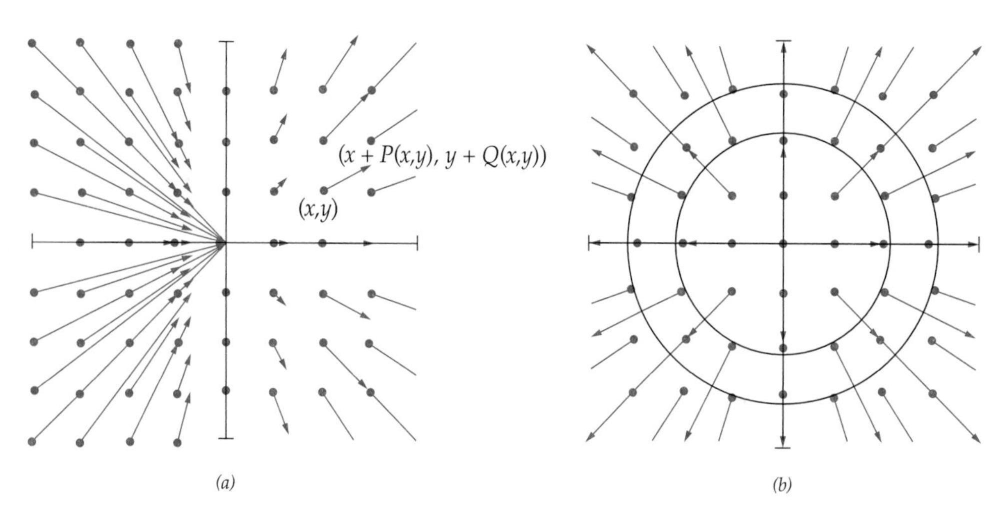

Figure 1.14 (*a*) Field not a gradient; $P(x, y) = x^2$, $Q(x, y) = xy$. (*b*) Same field modified to be a gradient; $(x^{-1})P(x, y) = x$, $(x^{-1})Q(x, y) = y$.

will be an exact equation, solvable in principle by integration, but with the same solutions as the given equation. For this reason, $M(x, y)$ is called an **integrating factor.** [Note that the related slopes in $y' = f(x, y)$ remain the same.] Though we won't give a proof, it follows from the existence and uniqueness Theorem 2.1 of Section 2C that if $P(x, y)$ and $Q(x, y)$ are continuously differentiable in a region of the xy plane, then an integrating factor $M(x, y)$ always exists in a neighborhood of a point (x_0, y_0) in the region where P and Q aren't both zero. Unfortunately it's hard to find a formula for such an $M(x, y)$ except in very special circumstances; some of these exceptions are indicated in the following examples and exercises.

example 1

The equation $x^2\,dx + xy\,dy = 0$ fails the exactness test, since letting $P(x, y) = x^2$ and $Q(x, y) = xy$ we find

$$\frac{\partial P(x, y)}{\partial y} = 0 \neq y = \frac{\partial Q(x, y)}{\partial x},$$

unless $y = 0$. Hence the pair $P(x, y)$, $Q(x, y)$ can't be a gradient. However, dividing by x, that is, multiplying by the integrating factor $M(x) = x^{-1}$ when $x \neq 0$, gives $x\,dx + y\,dy = 0$. This equation is exact, because we can let $F(x, y) = (x^2 + y^2)/2$ and note that $\partial F(x, y)/\partial x = x$ and $\partial F(x, y)/\partial y = y$. Both the inexact and the exact vector fields are shown in Figures 1.14(a) and 1.14(b), respectively. Note that the arrows at corresponding points of the two fields are parallel, but have different lengths and sometimes even point in opposite directions. Using the geometric interpretation of the differential equation, the solution graphs $F(x, y) = C$, circles in this example, will be perpendicular to vector arrows starting at points (x, y) on the graphs as in Figure 1.14(b).

example 2

The equation $y\,dx + 2x\,dy = 0$ is not exact, since $P(x, y) = y$ and $Q(x, y) = 2x$, and

$$\frac{\partial P(x, y)}{\partial y} = 1 \neq 2 = \frac{\partial Q(x, y)}{\partial x}.$$

However, the equation can be solved by separating variables: just divide by xy, so in effect $M(x, y) = (xy)^{-1}$ is an integrating factor. Now integrate $dx/x + 2dy/y = 0$ to get

$$\int \frac{dx}{x} + \int \frac{2dy}{y} = c \quad \text{or} \quad \ln|x| + \ln y^2 = c.$$

Combining logarithms we get $\ln|x|y^2 = c$ or $xy^2 = \pm e^c = C$. In this way, we found the family of implicit solutions $F(x, y) = xy^2 = C$. But now we see that this family of solutions satisfies the exact differential equation

$$F_x(x, y)\,dx + F_y(x, y)\,dy = y^2\,dx + 2xy\,dy = 0.$$

Comparison of this equation with the original differential equation enables us to make the easy guess that $N(x, y) = y$ is an even simpler integrating factor for the original equation. This observation can now be used to solve some equations of the form

$$y\,dx + \left[2x + Y(y)\right] dy = 0, \quad \text{for example,} \quad y\,dx + (2x + y^2)\,dy = 0.$$

The integrating factor $N(x, y) = y$ that we just found for the original equation now applies here to give us

$$y^2\,dx + (2xy + y^3)\,dy = 0.$$

Since $\partial y^2/\partial y = 2y = \partial(2xy + y^3)/\partial x$, we can proceed with some confidence to look for a $G(x, y)$ such that $G_x(x, y) = y^2$ and $G_y(x, y) = 2xy + y^3$. Integrating the first of these equations with respect to x gives $G(x, y) = xy^2 + g(y)$. Now differentiate this equation with respect to y and compare the result, $G_y(x, y) = 2xy + g'(y)$ with the previous expression $G_y(x, y) = 2xy + y^3$ for G_y. The conclusion is that $g'(y) = y^3$, so $g(y) = \frac{1}{4}y^4 + C$. The implicit-form solution we're looking for is then $xy^2 + \frac{1}{4}y^4 = C$.

EXERCISES

1. An integrating factor $M(x, y)$ for the equation $P(x, y)\,dx + Q(x, y)\,dy = 0$ is a function such that

$$M(x, y)P(x, y)\,dx + M(x, y)Q(x, y)\,dy = 0$$

is exact in some region.

(a) Show that a linear equation of the form $dy/dx + f(x)y = 0$, with $f(x)$ continuous on an interval $a < x < b$, can be written $f(x)y\,dx + dy = 0$.

(b) Show that multiplying by $M(x) = e^{\int f(x)\,dx}$ produces an equation that satisfies the consistency condition for exactness.

(c) Use the result of part (b) to derive the general solution to the given first-order equation:

$$y = Ce^{-\int f(x)\,dx}.$$

(d) Show that starting with the prior assumption that the solution is either always zero or else never zero, the result of part (c) can be obtained by first separating variables.

There is no universally applicable method for finding an integrating factor, though it can be proved that one always exists if the given equation satisfies certain rather mild conditions. Here are some examples for which such a factor is provided. Solve each equation in Exercises 2 through 5 using the given factor.

2. Show that $M(y) = y^{-2}$ is an integrating factor for $(y + y^2)\,dx - x\,dy = 0$.

3. Show that $M(y) = e^{-y}$ is an integrating factor for $(1 + x)e^x\,dx + (ye^y - xe^x)\,dy = 0$.

4. Show that $M(x) = 2x$ is an integrating factor for $(2x^2 + 3xy - y^2)\,dx + (x^2 - xy)\,dy = 0$.

5. Show that $M(x, y) = 1/(x^2 + y^2)$ is an integrating factor for $y\,dx - (x^2 + x + y^2)\,dy = 0$.

6. The differential equation $g(y)\,dy/dx = h(x)$ is separable type, with implicit solution $G(y) = H(x) + C$, where $G'(y) = g(y)$ and $H'(x) = h(x)$.

(a) Show that this differential equation is also exact if written in differential form.

(b) Use the method for solution of exact equations by integration to solve the equation in part (a) in terms of $G(y)$ and $H(x)$. Compare the result with the solution by the variables-separable method.

Exact Equations Using Differential Formulas. Use Equations 5.1, or simple modifications, to identify solutions to the following. In some cases you can find solutions by separating variables, though that may be more work.

7. $y\,dx + (x + y)\,dy = 0$.

8. $y\,dx - (x + y^2)\,dy = 0$.

9. $y \sin x\,dx - \cos x\,dy = 0$.

10. $y^2\,dx + 2xy\,dy = 0$.

11. $y^2\,dx - 2xy\,dy = 0$.

12. $y^2 \cos x\,dx + 2y \sin x\,dy = 0$.

13. $2xy^3\,dx + 3x^2y^2\,dy = 0$.

14. $y^2\,dx + 2(xy - y)\,dy = 0$.

Finding One-Variable Integrating Factors. It's possible to compute integrating factors for $P(x, y)\,dx + Q(x, y)\,dy = 0$ in the following two special cases.

(i) If $[P_y(x, y) - Q_x(x, y)]/Q(x, y) = R(x)$ is independent of y, then

$$M(x) = e^{\int R(x)\,dx} \text{ is an integrating factor.}$$

(ii) If $[Q_x(x, y) - P_y(x, y)]/P(x, y) = S(y)$ is independent of x, then

$$M(y) = e^{\int S(y)\,dy} \text{ is an integrating factor.}$$

Note that if the consistency condition $Q_x = P_y$ holds, then M will simply be a positive constant. Find an integrating factor for each of the following, and use it to solve the equation. Then check that your solution does satisfy the equation.

15. $y\,dx - x\,dy = 0$.

16. $(y^2 + 3x)\,dx + xy\,dy = 0$.

17. $2xy\,dx + (y^2 - x^2)\,dy = 0$.

18. $(y^2 + 4ye^x)\,dx + 2(y + e^x)\,dy = 0$.

19. $\sin y\,dx + x \cos y\,dy = 0$.

20. The one-variable integrating factors described in the preamble to the previous exercise will typically produce an exact equation.

(a) Show this for case (i) if $Q(x, y) \neq 0$.

(b) Show this for case (ii) if $P(x, y) \neq 0$.

21. Solve the nonexact equation $(x - y)\,dx + (x + y)\,dy = 0$ by using the transformation $u = y/x$ and rewriting the equation as one to be solved for $u = u(x)$. (See text Example 10 of Section 3C.)

CHAPTER REVIEW AND SUPPLEMENT

When trying to solve a first-order differential equation, it may be that an appropriate technique occurs to you right away. Otherwise, you may run mentally through a list of techniques that are familiar to you, starting with the ones that are simpler to apply and checking one at a time to see whether each one is likely to work for the problem at hand. As a rough guide to this thought process, the list below is arranged with the simpler and more commonly occurring separable and linear types at the beginning. It's important to realize that the equation you want to solve may have to be rearranged to put it in one of the forms listed below. For example, $(x + y)\,dx + (2x - 3y)\,dy = 0$ can be rewritten as $dy/dx = (-2 + 3y/x)/(1 + y/x)$, which can be solved using the quotient substitution $u = y/x$.

Variables Separable An equation may occur in the form $dy/dx = f(x)g(y)$, $dy/dx = f(x)/g(y)$, $f(x)dy/dx = g(y)$, or some variant of one of these. We try to write the equation in separated differential form

$$a(y)\,dy = b(x)\,dx, \quad \text{with solution} \quad \int a(y)\,dy = \int b(x)\,dx = C.$$

Linear Equation The normal form is $dy/dx + g(x)y = f(x)$, but it may be disguised, as, for example, in $a(x)\,dy/dx = b(x)y + c(x)$. The normal form equation can be solved by letting $M(x) = \exp[\int g(x)\,dx]$ in which case it can be solved by integrating both sides of

$$\frac{d}{dx}[M(x)y] = M(x)f(x) \quad \text{to get} \quad y = \frac{1}{M(x)}\int M(x)f(x)\,dx + \frac{C}{M(x)}.$$

Bernoulli Equation $y' + g(x)y = f(x)y^n$ is linear if n equals 0 or 1; otherwise, multiply by y^{-n} and let $u(x) = y^{1-n}(x)$. The Bernoulli equation transforms to the linear equation $u' + (1 - n)g(x)u = (1 - n)f(x)$. Replace u by y^{1-n} in the solution to the linear equation.

Quotient Substitution In an equation that can be written $dy/dx = f(y/x)$ let $u = y/x$. Then $dy/dx = u + x(du/dx)$, and the transformed equation $x(du/dx) = f(u) - u$ is separable; then let $u = y/x$ in its solution. Provided that $ad - bc \neq 0$, the equation

$$\frac{dy}{dx} = f\frac{ax + by + m}{cx + dy + n} \quad \text{becomes} \quad \frac{dw}{dz} = f\frac{a + bw/z}{c + dw/z}$$

if we choose h and k so that letting $x = z + h$ and $y = w + k$ removes the constant terms. Then let $u = w/z$ and proceed as above. Finally let $z = x - h$ and $w = y - k$.

Linear Substitution In an equation that can be written $dy/dx = f(ax + by + c)$ let $u = ax + by + c$. Then $dy/dx = (1/b)(du/dx) - (a/b)$, and the transformed equation $du/dx = bf(u) + a$ can be integrated directly; then substitute by $u = ax + by + c$ its solution.

Exact Equation An equation written in differential form $P(x, y)\,dx + Q(x, y)\,dy = 0$ is called exact if there is a function $F(x, y)$ with partial derivatives satisfying $F_x(x, y) = P(x, y)$ and $F_y(x, y) = Q(x, y)$. The equation is exact in some neighborhood of every point if and only if $P_y(x, y) = Q_x(x, y)$ there. Routines for solutions $F(x, y) = C$ to exact equations may be tried in either of two ways. (i) Integrate $F_x(x, y) = P(x, y)$ to get

$$F(x, y) = \int P(x, y)\,dx + \phi(y), \quad \text{where } \phi'(y) = Q(x, y) - \frac{\partial}{\partial y}\int P(x, y)\,dx.$$

(ii) Integrate $F_y(x, y) = Q(x, y)$ to get

$$F(x, y) = \int Q(x, y)\,dy + \psi(x), \quad \text{where } \psi'(x) = P(x, y) - \frac{\partial}{\partial x}\int Q(x, y)\,dy.$$

The equations in Exercises 1 through 16 are either linear or can be solved by separation of variables. Find all solutions that satisfy:

1. $x(dy/dx) + y - x = 0.$

2. $dy/dx = 1/[y(1-x)^2].$

3. $dx/dt = tx + e^{t^2/2}.$

4. $(1+x)y' + y = \cos x.$

5. $y^3y' = (y^4 + 1)e^x.$

6. $dy/dx = 4x^3y - y,\ y(1) = 1.$

7. $xy' + (2x-3)y = x^4.$

8. $y' = xy + y.$

9. $t(dx/dt) = -2x + t^3,\ x(2) = 1.$

10. $t(dx/dt) = 1.$

11. $dx/dt = -3x^2.$

12. $dy/dt + ty = 1.$

13. $dx/dt = (x+t)^2$. [*Hint:* Let $x + t = y$.]

14. $dy/dt = \cos^2 y.$

15. $x(dy/dx) = y - x^2e^x,\ y(1) = 1.$

16. $x(dy/dx) = y(1-x)^2.$

Each equation in Exercises 17 through 40 can be solved by at least one of the methods treated in this chapter. After considering the possibilities, pick a method for each one to find all solutions that satisfy:

17. $e^x(dx/dt) = 1 + e^x.$

18. $x\,dy + y\,dx = 0$, contains $(x, y) = (1,2)$.

19. $y\,dx - x\,dy = 0$, contains $(x, y) = (1,1)$.

20. $(t + y - 1)\,dt = dy,\ y(1) = 1.$

21. $(3zy^2 + 2z)\,dy = (3z^2 - y^3 - 2y)\,dz.$

22. $x^2y' - xy = 1/x,\ y(1) = 0.$

23. $(x^2 + y)\,dx + (x + \sin y)\,dy = 0.$

24. $(x^2 + y^2)\,dx + 2xy\,dy = 0.$

25. $(e^{-x^2} - xy)\,dx = dy.$

26. $(x + y + 1)\,dx = (x + y - 1)\,dy.$

27. $y' + xy = xy^{-3}.$

28. $y\,dx - x\,dy = y^3\,dy.$

29. $(4x + 3y + 2)\,dx + (3x + 2y + 1)\,dy = 0.$

30. $xy' + y = y^2,\ y(1) = 2.$

31. $(1 - e^{-x}\sin y)\,dx + e^{-x}\cos y\,dy = 0.$

32. $(2xy + y^2)\,dx + (x^2 + 2xy)\,dy = 0.$

33. $(x^2+y^2+1)\,dx+(x^2-2xy)\,dy=0.$ **34.** $y\,dx+(x+y)\,dy=0.$
35. $y^2e^{xy}\,dx+(1+xy)e^{xy}\,dy=0.$ **36.** $y^2\,dx+(1+2xy)\,dy=0.$
37. $y\,dx+(2x+y^{-1})\,dy=0.$ **38.** $dx+(x-t)\,dt=0.$
39. $y'+y=y^{2/3}.$ **40.** $y'=\sin(x+y),\ y(0)=0.$
41. Answer parts (a) through (d) for $dy/dx=e^{x-y}$ without first solving the equation.

(a) Find the part P of the xy plane where every solution graph is increasing while in P.
(b) Find the part Q of the xy plane where every solution graph is concave up while in Q.
(c) Is the line $y=x$ a solution graph?
(d) Is the line $y=x$ an isocline?
(e) Find an explicit solution formula for the differential equation, and try to get the information asked for in parts (a) through (d) from your formula rather than directly from the differential equation. Which approach seems simpler?

42. Consider the differential equation $dy/dx=e^{x-y}$.

(a) What conclusions can you draw from Theorem 2.1 on existence and uniqueness about solutions of this equation?
(b) Can a solution graph passing through the point $(x,y)=(0,1)$ cross the line $y=x$? Explain your reasoning.

43. Consider the family of linear equations $y'+ay=c$, with a,c constant, $a\neq 0$.

(a) Show that the isoclines of the direction field of this equation are horizontal lines, and that every such line is an isocline.
(b) Sketch the direction field associated with the differential equation $y'+2y=1$.

44. Early experiments with objects dropped from rest above the earth led to the conjecture that after an object had fallen distance s its velocity would be proportional to s. (Under the ordinary assumption that the acceleration of gravity is constant, the velocity is proportional to $\sqrt{s}$.)

(a) Is the early conjecture consistent with initial velocity zero? Explain your reasoning.
(b) Is the early conjecture consistent with positive initial velocity? How would acceleration be related to s under this assumption?

45. A 100-gallon mixing vat is initially full of pure water, whereupon 2 gallons of salt solution per minute is added, each gallon containing 1 pound of salt. Water evaporates from the tank at the rate of 1 gallon per minute, and the excess solution overflows into a drain. Find the amount of salt in the tank at time t under the given assumptions and also under the altered assumption that the tank initially contains 50 pounds of salt in solution.
46. Coffee cooling. We are presented two choices for cooling one cup of coffee over a period of 10 minutes: (i) let the coffee cool by itself for 10 minutes and then add cream at temperature c or (ii) add the same amount of cream at temperature c right away and then allow the mixture to cool for 10 minutes. Assume that mixing quantity p of liquid at temperature T_0 and quantity q at

temperature T_1 instantly results in quantity $p+q$ with average temperature $(pT_0+qT_1)/(p+q)$.

(a) Give an explanation for what your intuition tells you about whether method (i) or (ii) will result in cooler coffee at the end of 10 minutes.

(b) Show that the choice of adding the cream at the end of the waiting time will result in cooler coffee provided that the initial temperature c of the cream is strictly less than the ambient temperature a of the room.

(c) Show that the *difference* between the final temperatures attained by the two choices is independent of the initial temperature of the coffee.

47. Newton versus Stefan. Newton's law of cooling is $du/dt = n(u-a)$, where the constant n is negative. Here $u = u(t)$ is the temperature of an object in a surrounding medium that has temperature a. The purpose of this exercise is to compare Newton's law with Stefan's law, which says that $du/dt = n(u^4 - a^4)$. (Clearly both can't be precisely correct. The facts are that there is a sound empirical and theoretical basis for the validity of Stefan's law over the observable absolute temperature range and that Newton's law provides a useful linearized version of Stefan's law under certain circumstances.)

(a) Use the factorization $u^4 - a^4 = (u^3 + au^2 + a^2u + a^3)(u-a)$ to show that

$$u^4 - a^4 = 4a^3\left[\frac{(u/a)^3 + (u/a)^2 + (u/a) + 1}{4}\right](u-a).$$

(b) Show that the fractional factor in part (a) will be as close as you like to 1 if the ratio $(u-a)/a$ is small enough.

(c) Show that Newton's law in the form $du/dt = 4na^3(u-a)$ is a good approximation to Stefan's law if $u-a$ is small enough relative to a.

(d) Derive the implicit solution $\ln|(u-a)/(u+a)| - 2\arctan(u/a) = 4na^3t + c$ to Stefan's equation. Then explain why the corresponding solution $\ln|u-a| = 4na^3t + c$ to the linearized equation might be preferred for computational purposes.

48. Snowplow problem. At h_0 hours before midnight it starts snowing very heavily on bare roads, but always at a constant rate. Starting at midnight a snowplow begins clearing previously unplowed roads, its rate of progress dx/dt at a given time being always inversely proportional to the depth of snow that has fallen up to that time, hence also inversely proportional to the total time elapsed since it began snowing. The plow has cleared 2 miles of road by 2 A.M., but the snowfall is so heavy that only one additional mile is cleared by 4 A.M.

(a) Find h_0.

(b) Continuing under the same assumptions, what is the total amount cleared by 6 A.M.?

(c) What is the constant of proportionality that expresses dx/dt in inverse proportion to elapsed time?

49. The nonlinear equation $dy/dx = |y|$ is not covered by the existence and uniqueness theorem of Section 2C, because $|y|$ fails to be differentiable as a function of y at $y = 0$. Still, the equation has a unique solution graph passing through each point (x, y) of the plane: (i) Show that if a solution $y(x)$ of $y' = |y|$ is nonzero at x_0 [i.e., $y(x_0) = y_0 \neq 0$], then that solution is uniquely determined and is nonzero for all x. (ii) Explain how to conclude from (i) that a solution such that $f(x_0) = 0$ must in fact be identically zero.

50. Sometimes the area between the graph of $y = f(u)$ and the u axis over the interval from $u = a$ to $u = x$ is the same as the length of the graph over the same interval for all x. Verify that this is so if $f(u) = \cosh u = (e^u + e^{-u})/2$. Then find a differential equation satisfied by all such functions and solve the equation to find a formula for all such functions.

CHAPTER 2

Applications: Dynamics and Equilibrium

The applications discussed in Chapter 1 have been kept fairly simple to provide immediate applications of the solution techniques. This chapter contains two more detailed applications and is organized by area of application rather than by technique for solution. However, all the first-order equations needed for these applications are either linear or have variables separable. In addition, Section 3 on behavior near equilibrium complements Section 1 and 2. The examples show that interpreting a solution can be at least as complex as deriving the solution and also show that interpretation has a vital role to play in the study of dynamics.

The Computing Supplement at the end of the chapter features first-order numerical applications. You can also find an extended application to radioactive decay at http://www.mhhe.com/williamson.

1 POPULATION DYNAMICS

example 1 In Example 5 on page 51, we let $P(t)$ be the size of a population, perhaps of bacteria or elephants, at time t. We considered the consequences of the **exponential growth model**

$$\frac{dP}{dt} = rP,$$

where $r > 0$ is the **per capita growth rate.** The main conclusion is that the population will grow exponentially, more specifically that

$$P(t) = P_0 e^{rt},$$

where P_0 is the population size at time $t = 0$. Over short time intervals, these results fit certain populations, for example, bacteria, quite well. However, since r and P_0 are both positive, $P(t)$ will tend to infinity as t tends to infinity; thus without some restriction on t the model cannot be correct. For example, starting with some particular population size P_0, the model predicts that the population will double repeatedly, the first time to size $2P_0$ over time $t = D$, called the population's **doubling time.** The doubling time D is determined by $P_0 e^{rD} = 2P_0$, that is, $e^{rD} = 2$ or $D = \ln 2/r$. Since time units are the only units relevant to D, the same is true for the constant r.

---

example 2

A more realistic model than exponential growth for some populations is given by the **logistic differential equation:**

$$\frac{dP}{dt} = rP\left(1 - \frac{P}{K}\right),$$

where the constant $K > 0$ is a limiting size for the population. In this model, the per capita growth rate is $r(1 - P/K)$, and depends on population size. (At size K, the model predicts zero growth rate.) As $P(t)$ approaches K, the factor $1 - P/K$ approaches zero and so has an inhibiting effect on the growth of P.

The logistic equation can be solved by separation of variables:

$$\frac{dP}{P - P^2/K} = r\,dt, \quad P \neq 0,\ P \neq K.$$

The partial fraction identity

$$\frac{1}{P} + \frac{1/K}{1 - P/K} = \frac{1}{P - P^2/K}$$

enables us to integrate the left side, and we get

$$\ln|P| - \ln\left|1 - \frac{P}{K}\right| = rt + c, \quad \text{or} \quad \ln\left|\frac{P}{1 - P/K}\right| = rt + c.$$

Taking exponentials on both sides,

$$\frac{P}{1 - P/K} = Ce^{rt}, \qquad C = \pm e^c.$$

Solving for P gives

$$P(t) = \frac{KCe^{rt}}{K + Ce^{rt}}.$$

Setting $t = 0$, we can solve the resulting equation for C to get $C = P_0K/(K - P_0)$, where $P_0 = P(0)$. Now put this value for C in our expression for $P(t)$ and simplify to get

$$P(t) = \frac{P_0 K}{(K - P_0)e^{-rt} + P_0}.$$

The constant values $P = 0$ and $P = K$ that we excluded at the beginning of the computation are themselves constant solutions, $P(t) = 0$ and $P(t) = K$, that we can recover from the previous formula by setting $P_0 = 0$ and $P_0 = K$, respectively.

The whole point of the preceding analysis is to arrive at a general formula into which we can put known information and from which we can then derive something we didn't previously know, for example the population's doubling time from P_0, assuming $2P_0 < K$. (See Exercise 13.)

example 3

If we know r, P_0, and K in the formula for $P(t)$ in Example 2, then we can let t tend to infinity, and we see that the first term in the denominator tends to 0. Hence

$$\lim_{t \to \infty} P(t) = K.$$

This result is consistent with our original interpretation of K as a population size beyond which there is no further increase in size. The number K can be interpreted as the limiting capacity of the population's environment. We now see that, in theory the value K is never actually attained in this model unless $P_0 = K$. But if $P_0 \neq K$ the population size $P(t)$ approaches K asymptotically regardless of the size of $r > 0$. Figure 2.1(a) shows the graphs of some typical relations for $P = P(t)$.

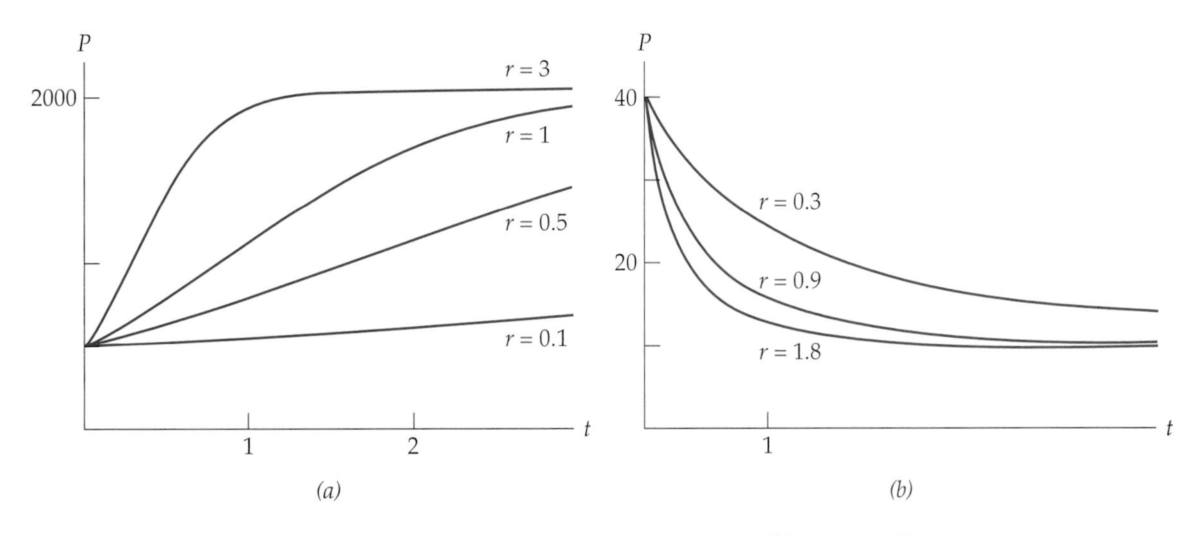

Figure 2.1 Logistic models: (a) $P_0 = 500,\ K = 2000$; (b) $P_0 = 40,\ K = 10$.

example 4

If the initial population size $P(0) = P_0$ is bigger than the constant K in the logistic model $P' = rP(1 - P/K)$, then the model predicts population decline, because the factor $(1 - P/K)$ not only starts out negative but remains so. As confirmation we note that the solution formula for $P(t)$ derived in Example 2 is valid regardless of the relationship between the constants P_0 and K. But if $P(0) = P_0 > K$, we choose to rewrite the solution formula as

$$P(t) = \frac{P_0 K}{P_0 - (P_0 - K)e^{-rt}}.$$

The denominator increases to P_0 as t tends to infinity, so the limit is still K, but the approach to the limit is now decreasing from above, as in Figure 2.1(b).

Recall that the logistic equation always has the nontrivial constant solution $P(t) = K$ and the trivial solution $P(t) = 0$. A constant solution is called an **equilibrium solution,** and while constancy is very unlikely to be maintained in practice, equilibrium solutions are still important, because they call attention to stable or unstable behavior near the equilibrium solution. These matters are discussed in detail in Section 3 of this chapter.

EXERCISES

1. How should the constants r and P_0 be chosen in the solution $P(t) = P_0e^{rt}$ to the linear growth equation $dP/dt = rP$ if $P(t)$ is to satisfy $P(10) = 100$ and $P(20) = 150$? With this choice for P_0 and r, what is $P(-10)$?

2. Show that if $P(t) = P_0e^{rt}$ for positive constants P_0 and r, then $P(t)$ always doubles in a fixed time period $D = t_2 - t_1$, where $D = \ln 2/r$. The constant D is the *doubling time* of this exponential growth law.

3. If a solution $P(t)$ for

$$\frac{dP}{dt} = rP, \quad r > 0,$$

doubles during a time period of length D, how long does $P(t)$ take to triple in size?

4. **(a)** Assume $P(t)$ differentiable for all real t and also that $P'(t) = rP(t)$ for some constant r. Show that the acceleration $P''(t)$ is $r^2P(t)$ and, more generally, that $P^{(n)} = r^nP(t)$.

(b) Show that the population acceleration under the logistic model $P' = rP(1 - P/K), r > 0$, is

$$P'' = rP\left(1 - \frac{2P}{K}\right)P' = r^2P\left(1 - \frac{P}{K}\right)\left(1 - \frac{2P}{K}\right).$$

(c) Show that the population curve for the logistic model is concave up while $P < \frac{1}{2}K$, indicating and increasing population growth rate, and with declining growth rate if $\frac{1}{2}K < P < K$.

5. If the population of the United States was 4,000,000 in 1790, 17,000,000 in 1840 and 63,000,000 in 1890, show that the size of the population from 1790 to 1890 cannot be approximated with reasonable accuracy by a formula of the form $P(t) = P_0 e^{rt}$. (Among other things, there was a lot of immigration during that period.)

6. We can regard the per capita growth rate r in the exponential growth model $dP/dt = rP$ as a *net* per capita growth rate $r = \beta - \delta$, where β is a per capita birth rate and δ is a per capita death rate. If $\delta > \beta$, then $r < 0$, and instead of considering a doubling time for a population it's appropriate to look for a **half-life,** that is, a fixed time period over which the population is cut in half. Find a formula for the half-life of a population with net growth rate $r < 0$. If the growth rate r is negative the differential equation $dP/dt = rP$ is the standard model for decay of radioactive substances.

7. We can regard the per capita growth rate $r(1 - P/K)$ in the logistic model as a *net* per capita growth rate $\beta - \delta P$, where β is a per capita birth rate, and δP is a per capita death rate that depends on population size. Derive the logistic equation using this idea and show that the constant that we interpreted as limiting capacity is $K = \beta/\delta$.

8. A natural generalization of the logistic differential equation is

$$\frac{dP}{dt} = rP\left[1 - \left(\frac{P^q}{K}\right)\right], \qquad q > 0.$$

Note: In doing this problem, the formula

$$\int \frac{dP}{P(a - P^q)} = \frac{1}{aq} \ln\left(\frac{P^q}{a - P^q}\right) + C$$

is useful; it can be derived using substitution $P^q = x$ and partial fractions.

(a) Show that the solution curves for this equation satisfy

$$\lim_{t \to \infty} P(t) = K^{1/q}.$$

(b) Find $P(t)$ if $P(0) = P_0$, noting the special case $P_0^q = K$.

9. Show that the logistic differential equation can be written in the form

$$\frac{dP}{dt} = aP - bP^2,$$

and show how the parameters a and b are related to the rate constant r and the limit population P_∞.

10. The **general logistic equation** $dP/dt = aP^\alpha(1 - P/K)^\beta$, $a > 0$, $\alpha > 0$, $\beta > 0$, has elementary solution formulas only for special values of the exponents α and β. Nevertheless, we can get qualitative information about solutions without finding formulas in terms of t.

(a) Show that solutions with initial value satisfying $0 < P(0) < K$ are strictly increasing as long as they remain less than K.

(b) Show that a solution graph has an inflection point at $P_I = [\alpha/(\alpha+\beta)]K$ if $0 < P(0) < P_I$.

11. Writing the logistic equation in the form

$$\frac{1}{P}\frac{dP}{dt} = r\left(1 - \frac{P}{K}\right)$$

shows that for this model the growth rate of the population per individual member decreases with rate $-r/K$ as P increases. For some insect populations, it turns out that the equation

$$\frac{1}{P}\frac{dP}{dt} = r\left(\frac{K-P}{K+cP}\right), \qquad c = \text{const.} > 0,$$

is more realistic.

(a) Show that if $c = 0$, we get the logistic equation.

(b) Find the general solution of this differential equation in implicit form.

(c) What form does the solution take if we determine the constant of integration by $P(0) = P_0$?

12. The **Gompertz population equation** is

$$\frac{dP}{dt} = aP(L - \ln P),$$

where a and L are positive constants. Show that solutions are

$$P(t) = Ke^{-Ce^{-at}},$$

where $K = e^L$ and $C = \ln[K/P(0)]$.

13. (a) Use the solution derived in Example 2 of the text for the logistic equation to compute the doubling time D for this model to go from $P(0) = P_0$ to $P = 2P_0$, assuming $2P_0 < K$.

(b) Show that the doubling time D found in part (a) exceeds the doubling time under the exponential growth law $P' = rP$ by the amount

$$\frac{1}{r}\ln\left(\frac{K-P_0}{K-2P_0}\right).$$

2 VELOCITY IN A RESISTING MEDIUM

One of Newton's fundamental laws of motion asserts that the sum F of all forces acting at time t on a moving body of constant mass is equal to its mass m times the acceleration of the body at that time. Since acceleration is by definition the time derivative $dv(t)/dt$ of velocity $v(t)$, the law can be written in the form $m(dv/dt) = F$. In this section, we shall consider only examples for which the motion is restricted to a single linear path with coordinate x. As a result of this simplifying assumption, force and velocity are also represented by real-valued functions. In general, F may depend not only on time t, but on position x and velocity v. However, there are several interesting applications in which F is independent of position; in that case, we can often find $v(t)$ by a single integration and then can find the position $x(t)$ by another integration.

example 1

A car propelled along a straight horizontal track experiences retarding forces due to air resistance and friction with the track. We'll assume the total retarding force is proportional to the velocity v of the car. (Thus the faster the car goes, the greater is the retarding force.) We'll also assume that the propelling force is turned off at some time that we'll call $t = 0$, leaving the car with velocity $v(0) = v_0$. It's possible to determine empirically in many applications a constant $k > 0$ such that the retarding force has the form $r(v) = -kv$, where k is a positive constant. In that case, the Newton law of motion $F = ma$ can be written as an explicit differential equation with initial condition:

$$m\frac{dv}{dt} = -kv, \quad \text{for} \quad t \geq 0 \quad \text{with} \quad v(0) = v_0 > 0.$$

This first-order linear initial-value problem can be solved by using the exponential multiplier $e^{kt/m}$ or else by separating variables. Both methods give

$$v(t) = v_0 e^{-kt/m}.$$

From the solution, we learn that the velocity drops exponentially to zero because of the retarding forces. Note also that a larger value of k (more resistance) leads to even more rapid decrease in velocity, as you can see from the position of k in the exponential. It is also interesting to note that the mass m plays a role: a larger mass corresponds to a *less rapid* decrease in velocity. (The effect of a larger mass is to increase **momentum** mv, which acts to maintain the velocity at a higher rate under the influence of a given retarding force.)

Since the car's velocity $v(t)$ at time t is the derivative of its position $x(t)$, it follows that, to within a constant of integration, position is the integral of velocity:

$$\begin{aligned} x(t) &= \int v(t)\,dt + C \\ &= \int v_0 e^{-kt/m}\,dt + C \\ &= -v_0\frac{m}{k}e^{-kt/m} + C. \end{aligned}$$

The constant of integration can be determined by observing $x(t)$ at some specific time t_1. Then

$$x(t_1) = -v_0\frac{m}{k}e^{-kt_1/m} + C.$$

Thus $C = x(t_1) + v_0 m/k e^{-kt_1/m}$, so

$$x(t) = v_0\frac{m}{k}\left(e^{-kt_1/m} - e^{-kt/m}\right) + x(t_1).$$

example 2

There is something surprising in the previous example in that the velocity never actually reaches zero, in other words the car never rolls to a stop, since the exponential is never zero. This discrepancy between normal experience and theory is due to the approximate quality of formulas for retarding forces; there is no simple formula to describe them perfectly under all circumstances; we just do the best we can in given circumstances. However, the discrepancy becomes less disturbing when we compute the distance $x = x(t)$ from the starting point, either using the formula at the end of the previous example, or else by starting over with a definite integral for the distance:

$$\begin{aligned} x(t) &= \int_0^t v(u)\,du \\ &= \int_0^t v_0 e^{-ku/m}\,du \\ &= v_0\left[-\frac{m}{k}e^{-ku/m}\right]_0^t = \frac{mv_0}{k}\left(1 - e^{-kt/m}\right). \end{aligned}$$

As t tends to infinity, the exponential term tends to zero, so the upper limit for the values of $x(t)$ is easily seen to be mv_0/k. Thus the range of motion is limited even though in theory the velocity never reaches zero.

Mass and Weight. The **mass** m of an object is a measure of the amount of matter in it, while its **weight** w is the magnitude of the gravitational force exerted on it by a standard physical object, often the earth. Thus $w = ma$, where a is the acceleration due to gravity. If weight is measured at the surface of the earth, then $w = mg$, where $g \approx 32.2$ feet/second/second, or about 9.8 meters/second/second, given that 1 meter is about 3.28 feet. (Variation in g is due to local variations in the structure of the earth.) The international standard unit of measurement for mass is the **kilogram,** and a 1-kilogram object weighs about 2.2 pounds near the surface of the earth. It is common practice to allow the term "kilogram" to denote also the *weight* of a 1-kilogram mass near the earth's surface. Thus an object of mass 1-kilogram weighs about 1 "kilogram" near the earth's surface.

example 3

A projectile of mass m fired vertically near the surface of the earth encounters not only air resistance but also the downward force mg of gravity. In terms of forces the differential equation satisfied by velocity $v(t)$ has the form

$$\begin{aligned} m\frac{dv}{dt} &= -r(v) - mg \\ &= -kv - mg, \end{aligned}$$

provided we assume air resistance proportional to the first power of velocity: $r(v) = -kv$. Again we have a first-order linear equation that can be solved by using the exponential multiplier $e^{kt/m}$ (or, since mg is constant, by separating variables):

$$\frac{d}{dt}\left(e^{kt/m}v\right) = -ge^{kt/m}.$$

The solution is

$$\begin{aligned} v(t) &= -\frac{mg}{k} + Ce^{-kt/m} \\ &= -\frac{mg}{k} + \left(\frac{mg}{k} + v_0\right)e^{-kt/m}, \end{aligned}$$

where $v_0 = v(0)$ is the initial velocity. As t gets large, the second term tends to zero, leaving the limit $v_\infty = -mg/k$. The number v_∞ is the **terminal velocity** of the projectile, a velocity never exceeded when the projectile is viewed simply as a falling body after having reached its maximum height. Figure 2.2(a) shows graphs of velocity $v(t)$ with $v_0 > 0$ (upward initial direction) and $v_0 < 0$ (downward initial direction); the graphs on the right are of the corresponding height functions, showing their tendencies toward straight lines with slopes $-mg/k$.

To find out when the projectile reaches its maximum height, we set $v(t) = 0$, and solve for time t in

$$-\frac{mg}{k} + \left(\frac{mg}{k} + v_0\right)e^{-kt/m} = 0.$$

Solving for the exponential $e^{kt/m}$ we get

$$e^{kt/m} = \left(\frac{mg}{k} + v_0\right)\left(\frac{k}{mg}\right) = 1 + \frac{kv_0}{mg}.$$

Assuming $v_0 > -mg/k$, taking logarithms gives the solution

$$t_{\max} = \frac{m}{k}\ln\left(1 + \frac{kv_0}{mg}\right).$$

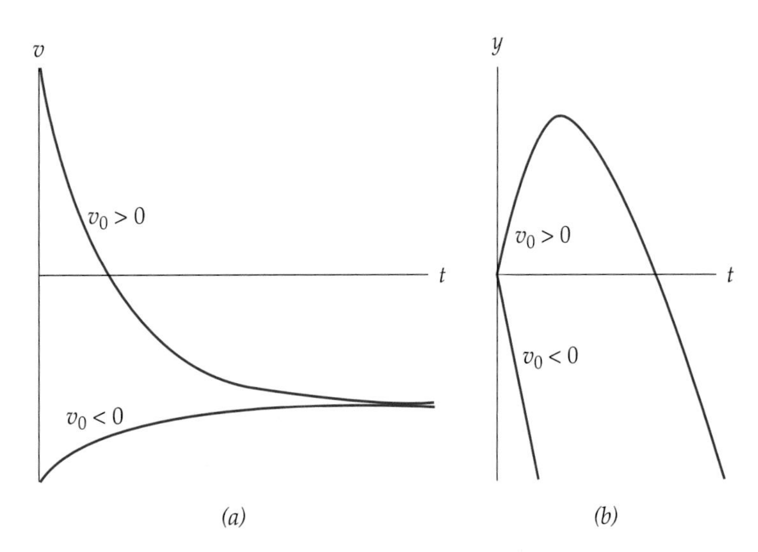

Figure 2.2 $(a)\ v = \dot{y} = -\frac{mg}{k} + \left(v_0 + \frac{mg}{k}\right)e^{-kt/m}$.
$(b)\ y = y(t)$; slope approaches $\frac{-mg}{k}$.

The height $h(t)$ attained at time t is just the integral of v from 0 to t. We leave it as an exercise to compute $h(t)$ and to show that the maximum height attained is

$$h(t_{\max}) = \frac{mv_0}{k} - \frac{m^2 g}{k^2} \ln\left(1 + \frac{kv_0}{mg}\right).$$

Typical graphs of position $y(t)$ appear in Figure 2.2(b).

---

example 4

An object of weight about $w = 2200$ pounds near the surface of the earth will have mass about $m = 2200/2.2 = 1000$ kilograms. If the velocity v of the object satisfies a differential equation of the form

$$m\frac{dv}{dt} = -kv - mg, \qquad k > 0, \qquad g = \text{acceleration of gravity},$$

then the value of the constant k will depend on the units of distance, mass, and time that we use. If we choose to deal with weight $w = mg$ instead of mass, the equation becomes

$$\frac{w}{g}\frac{dv}{dt} = -kv - \frac{w}{g}g, \quad \text{or} \quad w\frac{dv}{dt} = -kgv - wg.$$

Thus our equation has the same general form that it had when we dealt with mass instead of weight:

$$w\frac{dv}{dt} = -Kv - wg, \qquad K = kg.$$

This equation remains valid as long as w is measured in the same way at all times. It's convenient that the differential equation takes the same form whether we use mass or standardized weight, but we need to take care that the constant k or $K = kg$ is correctly matched to the units being used; since its value is usually determined empirically, the appropriate units will be built into the experiment. In particular, suppose K were to be determined in an experiment to be $K = 0.1$, using pounds w measured at the surface of the earth. If the units for distance and time were feet and seconds, then the mass m of the object would be about $2200/32.2 \approx 68.3$. The constant k would correspondingly be about $0.1/32.2 \approx 0.0031$. Switching instead to kilograms, we would have $m \approx 1000$ and $k \approx 0.0031/2.2 \approx 0.0014$.

EXERCISES

1. A 6440-pound car is guided along a track with velocity $v = v(t)$ subject to retarding force $-0.5v$ and with initial speed $v(0) = 100$ feet per second.

   (a) What is the approximate velocity at $t = 2$ seconds?

   (b) What is the upper limit of the car's progress along the track?

2. A car of mass 100 kilograms is guided along a track with velocity $v = v(t)$ subject only to a retarding force of the form $-kv$ and that $v(0) = 10$ feet per second.

(a) The car's velocity has fallen to 7 feet per second at time $t = 120$ seconds. Estimate the value of k.

(b) Assume that the car appears to have come to rest just after traveling 3600 feet. Estimate the value of k using that observation.

3. Suppose the differential equation $m(dv/dt) = -kv$ is satisfied by a moving object of mass m and that the constant k has the value 0.36 when the velocity v is measured in feet per second.

(a) What value should k have if velocity is measured in miles per second?

(b) What value should k have if velocity is measured in miles per hour?

(c) Would you get the same answers in parts (a) and (b) if the questions referred to the differential equation $m(dv/dt) = -kv^2$?

4. A body of mass m falling under the influence of gravity and with resistance force proportional to velocity v satisfies the equation

$$m\frac{dv}{dt} = gm - kv, \qquad k > 0.$$

(a) Find the most general solution of the differential equation.

(b) Express $v(t)$ in terms of g, k, m, and the initial velocity v_0.

(c) Sketch typical graphs of the solutions in part (b) for each of the three distinct cases $0 \le v_0 < mg/k$, $v_0 = mg/k$, and $mg/k < v_0$.

5. A quadratic retarding force of the form $-kv^2$ has more effect at velocities $v > 1$ than does a linear retarding force $-kv$. Suppose that $k = 0.25$ for a car of mass 100 kilograms guided on a horizontal track with only the quadratic retarding force acting and with an initial velocity of 40 feet per second.

(a) Show that $v(t) = 400/(t + 10)$.

(b) The assumption of a v^2-retarding force is usually unrealistic for low velocities; using the result of part (a), show that the car will in theory never come to rest.

6. A body of mass m having velocity $v = v(t)$ moves subject only to frictional forces F proportional to the square of the velocity: $F = -kv^2$.

(a) Find the differential equation satisfied by v.

(b) Solve the differential equation in part (a) in terms of the parameters m, k, and $v_0 = v(0)$.

(c) Find the distance covered in time t, and show that it tends to infinity as t does if $v_0 \neq 0$.

(d) Show that, for arbitrary $v_0 > 0$ and $v_1 > 0$, velocity satisfies $v(t) < v_1$, if $t > m/(kv_1)$.

7. Suppose that a 100-kilogram boat glides through the water with velocity $v = v(t)$ subject only to a retarding force of the form $-0.1v^{3/2}$.

(a) Find the velocity in terms of initial velocity $v_0 > 0$.

(b) Find the distance traveled in time t in terms of v_0.

(c) Show that the boat will theoretically never stop moving, but that there is an upper limit to its forward progress.

8. The equation $m(dv/dt) = -kv^r$ can be rewritten

$$\frac{dv}{dt} = -Kv^r.$$

(a) Assuming K and r constant, $r > 1$, find $v(t)$ in terms of K, r, and initial velocity $v(0) = v_0 > 0$.

(b) For what values of $r > 1$ is $v(t)$ bounded above?

(c) Discuss the case $0 < r < 1$ and show that the velocity falls to zero in finite time $v_0^{1-r}[K(1-r)]^{-1}$.

***9.** A baseball weighing 5 ounces is popped up vertically with an initial velocity of 161 feet per second. Assume that $g = 32.2$ feet per second per second and that the retarding force of air resistance is $-kv$, where $k = 0.5$. The units determining k are ounces and seconds. Part (c) requires you to estimate the solution of a transcendental equation.

(a) Estimate the time it takes for the ball to reach its maximum height.

(b) Estimate the maximum height of the ball.

(c) Express the height of the ball as a function of the time after maximum height, and use the result to estimate the additional time before hitting the ground and the speed of the ball when it hits the ground.

(d) Explain why the speed on returning to the ground is less than the given initial speed.

(e) Answer parts (a) through (c) under the assumption that air resistance can be neglected (i.e., $k = 0$) and explain why the differences in corresponding results are reasonable.

10. Suppose that a projectile of mass m is fired upward with initial velocity v_0 from the surface of the earth under the influence of gravitational force $-mg$ and air resistance force $-kv$, as in Example 3 of the text.

(a) Compute the height $h(t)$ attained after time t.

(b) Show that the maximum height attained is

$$h_{\max} = \frac{mv_0}{k} - \frac{m^2 g}{k^2} \ln\left(1 + \frac{kv_0}{mg}\right).$$

(c) The dependence of the maximum height attained [see part (b)] on k and m reduces to dependence on just the ratio of these two quantities. Explain how this fact can be deduced directly by looking at the differential equation satisfied by $v(t)$.

11. In Example 3 of the text, it was shown that for a projectile fired up from the surface of the earth with initial velocity v_0 and air resistance $-kv$, the time of maximum height is

$$t_{\max} = \frac{m}{k} \ln\left(1 + \frac{kv_0}{mg}\right).$$

(a) Show that $\lim_{k \to 0} t_{\max} = v_0/g$.

(b) Solve the projectile velocity problem under the assumption that $k = 0$ and compare the corresponding value for $t_{\max}$ with the result of part (a).

12. A ball of mass m is tossed up from the edge of a vertical cliff so that on the way back down it just misses the edge of the cliff. The ball is subject to acceleration of gravity g and to air resistance of magnitude $k|v|$, $k > 0$.

(a) Show that if initial velocity satisfies $|v_0| > mg/k$, then the initial speed will never again be attained.

(b) Show that if initial velocity satisfies $|v_0| < mg/k$, then the initial speed $|v_0|$ will be attained again at time

$$t_1 = \frac{m}{k} \ln\left(\frac{mg + k|v_0|}{mg - k|v_0|}\right).$$

13. An object of mass m is propelled downward with initial velocity v_0, subject to constant gravity g, and linear retarding force $-kv$.

(a) Show that if $v_0 = mg/k$, then the velocity remains constant.

(b) Show that the velocity increases if $v_0 < mg/k$ and decreases if $v_0 > mg/k$.

***14.** A projectile of mass m is fired up from the surface of the earth against the force due to gravity and against the force of air resistance in the form $-kv^2$.

(a) Show that the velocity v satisfies the equation $dv/dt = -(k/m)v^2 - g$, where $g =$ acceleration of gravity.

(b) Solve the equation in part (a), and express the time of maximum altitude in terms of initial velocity v_0.

(c) Show that the time of maximum altitude increases as v_0 is increased, but that it has upper limit $\pi\sqrt{m/(4kg)}$.

(d) Show that the maximum altitude attained is $h_{\max} = (m/2k) \ln\left[1 + (kv_0^2/mg)\right]$.

14. An object of mass m falls to earth under the force due to gravity and against an air resistance of the force of magnitude kv^2.

(a) Show that velocity v satisfies the equation $dv/dt = g - (k/m)v^2$.

(b) Rewrite the differential equation in the form

$$\frac{dv}{(gm/k) - v^2} = \frac{k}{m}\,dt,$$

and solve for v using the partial fraction identity

$$\frac{1}{a^2 - v^2} = \frac{1}{2a}\left(\frac{1}{a - v} + \frac{1}{a + v}\right),$$

to perform the integration in v.

(c) Find the terminal velocity $v_\infty = \lim_{t\to\infty} v(t)$.

3 EQUILIBRIUM SOLUTIONS

If we have a differential equation $dy/dt = f(t, y)$ and want to know what happens to the solutions $y(t)$ when t is very large, there are three methods we can use. If we can find a formula for $y(t)$ we may accept $\lim_{t\to\infty} y(t)$ for an answer. If we have no formula, we can try a numerical computation for large values t to get some idea of what happens, the drawback being that accuracy may deteriorate significantly over a long t interval. Hence this aproach is unreliable. Example 1 illustrates a third alternative that avoids finding $y(t)$ by formula or by approximation, but sometimes produces the desired information.

example 1 If an object of mass m falls under constant gravity g against resistance of the form $-kv^\alpha$, $\alpha > 0$, then the velocity $v = v(t)$ satisfies

$$m\frac{dv}{dt} = mg - kv^\alpha.$$

Depending on the value of α, solution formulas are sometimes available by separation of variables, but we can't find such a formula in terms of general $\alpha > 0$. However, if $\alpha = 1$, the differential equation is linear with exponential multiplier $M(t) = e^{kt/m}$. The general solution is

$$v = \frac{mg}{k} + Ce^{-kt/m}.$$

Note that no matter what C is in the formula, the limit of $v(t)$ as t tends to infinity is mg/k. Not only that, with $C = 0$, the constant solution $y_0(t) = mg/k$ is precisely the limit constant that we arrived at by letting t tend to infinity. However, if $\alpha = 1.1$, the limitations of integration technique won't allow us to find a formula in terms of elementary functions. Nevertheless, the equality we found between the constant solutions mg/k and limits at infinity when $\alpha = 1$ extends to the case for which α is any positive number. For low velocities such that $v < (mg/k)^{1/\alpha}$, which implies that $mg - kv^\alpha > 0$, the velocity is increasing since the differential equation then says that $dv/dt > 0$. In the opposite case, when $v > (mg/k)^{1/\alpha}$, $dv/dt < 0$ so the velocity decreases. Figure 2.3 shows the direction field for a particular choice of constants, along with a few solutions. The picture makes it plausible that the solutions actually tend to $\sqrt[\alpha]{mg/k}$; that this is true will follow from Theorem 3.1 later on. Finally, note that all constant solutions $v = v_0$ can be found without otherwise solving the differential equation. The reason is that since for a constant solution $dv_0/dt = 0$, the differential equation becomes $gm - kv_0^\alpha = 0$, so solving for v_0 gives $v_0 = \sqrt[\alpha]{mg/k}$.

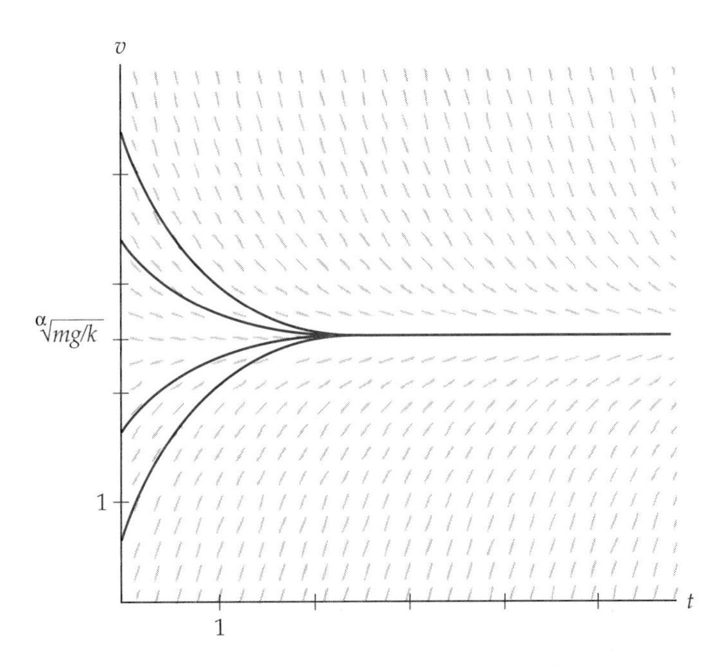

Figure 2.3 $\alpha = 1.2,\ m = 0.4,\ k = 0.1,\ g = 32.2,\ \sqrt[\alpha]{mg/k} \approx 57.3.$

The constant solution found in the previous example would model a physical process only if the retarding force has magnitude precisely kv^α and the initial velocity is precisely $v_0 = \sqrt[\alpha]{mg/k}$, an unlikely event in the real world. However, the example shows that a constant solution of a differential equation can have special significance because the graphs of other solutions may approach it. Because of the physical context in which constant solutions often appear, the term **equilibrium solution** is often used instead of "constant solution." For instance, the equilibrium solution $v = \sqrt[\alpha]{mg/k}$ in the previous example has the property that if you start out at that velocity, you maintain it. Moreover, if you start out with a nearby velocity the solution eventually gets arbitrarily close to the equilibrium value. In the present context this behavior characterizes $v = \sqrt[\alpha]{mg/k}$ as a **stable equilibrium** solution. Finding equilibrium solutions for the first-order equation $dy/dt = f(t, y)$ is in principle a simple matter. Since y is to be constant, $dy/dt = 0$, so we just have to find constant solutions y that satisfy the equation $f(t, y) = 0$ identically. Thus computationally, finding an equilibrium solution amounts to finding a single point, so the term **equilibrium point** is often used instead of equilibrium solution.

---

example 2

To find all equilibrium solutions to $dy/dt = y^2 - y$, set $dy/dt = 0$ and solve $y^2 - y = y(y-1) = 0$. The solutions are $y = 0$ and $y = 1$, so the two equilibrium solutions are $y_0(t) = 0$ and $y_1(t) = 1$. In addition, for values of y between 0 and 1, all the slopes of the direction field are negative, so solution graphs with initial values in that interval decrease, apparently toward the value 0 of the equilibrium solution $y_1(t) = 0$. See Figure 2.4(a).

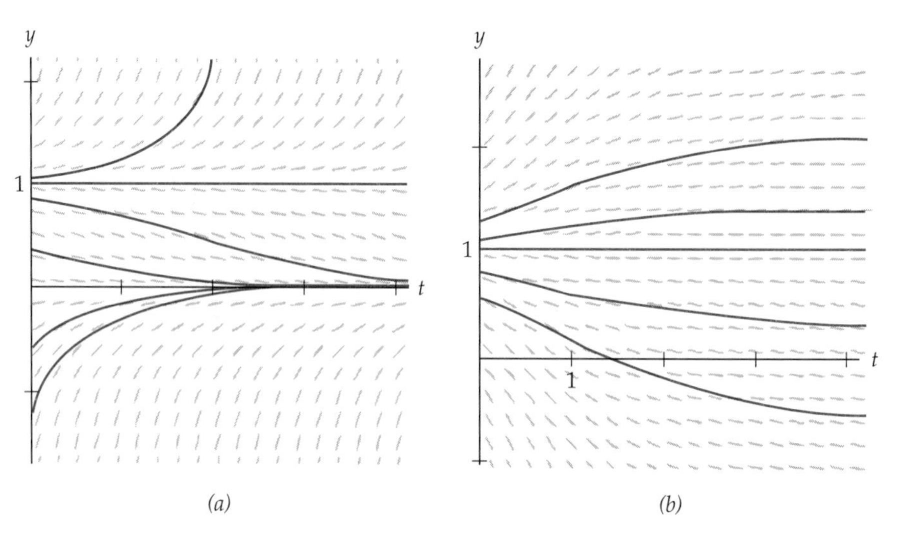

Figure 2.4 (a) $dy/dt = y^2 - y$. (b) $dy/dt = (y - 1)/(t^2 + 1)$.

example 3

The differential equation $dy/dt = (y - 1)/(t^2 + 1)$ has an equilibrium solution corresponding to each fixed value of y for which the right-hand side is identically zero, so the unique equilibrium solution is $y_0(t) = 1$. Note that the direction field has positive slopes if $y > 1$ and negative slopes if $y < 1$. Thus solutions with initial values more than 1 increase in value and solutions with initial values less than 1 decrease in value. Hence all solutions but $y_0(t) = 1$ diverge away from 1 as t increases, as in Figure 2.4(b). The solution $y_0(t) = 1$ is an example of an **unstable equilibrium,** so called because another solution that starts arbitrarily close to it can diverge away from it.

example 4

The differential equation $dy/dt = e^y$ has no equilibrium solutions at all, because the right-hand side is never zero. Hence dy/dt can't be zero, so there can't be a constant solution.

Note. If initial-value problems for a particular first-order differential equation sometimes fail to have unique solutions, the behavior of the direction field may not be an accurate indication of the relation of equilibrium solutions to the other solutions. The next example illustrates this point.

example 5

The differential equation $dy/dt = y^{2/3}$ has one equilibrium solution: $y_0(t) = 0$. Some other solutions can be found by separation of variables: $y_c(t) = (t - c)^3/27$. Still more solutions can be fitted together from pieces of these, for example

$$y = \begin{cases} (t-c)^3/27, & t < c, \\ 0, & c \le t, \end{cases} \qquad y = \begin{cases} 0, & t < c, \\ (t-c)^3/27, & c \le t. \end{cases}$$

Six of these piecewise-defined solutions are sketched in Figure 2.5 along with

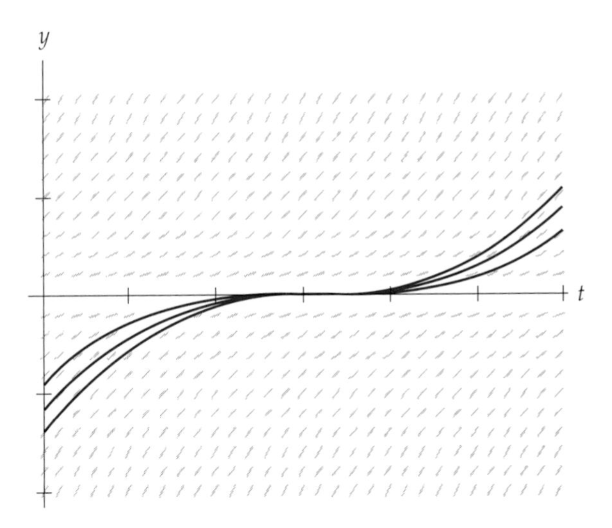

Figure 2.5 Solutions of $dy/dt = y^{2/3}$.

three conventional ones and the direction field for $0 \le t \le 6$. Note that a solution starting out negative may either cross the t axis and become positive, or else may continue with constant value 0. Similarly, a solution may start with constant 0 values and then become positive. The slopes of the direction field tend to zero as y tends to 0, so the slopes of the solution graphs flatten out as y approaches zero. However, some solutions actually cross the path of the equilibrium solution, while others become identical with it from some point on.

We've seen in the examples that a direction field can be an excellent guide for predicting stable or unstable behavior of solutions near an equilibrium solution. However, Example 5 shows that some caution is required in drawing conclusions. Theorem 3.1 provides some firm answers for equations of the special form $dy/dt = f(y)$ in Examples 1, 2, and 5 that arise often in practice. Such equations are called **autonomous** because f has no explicit dependence on the independent variable t. The following theorem enables us to draw conclusions about long-term behavior of solutions relative to equilibrium values. The result is that the direction field is a reliable guide to solution behavior except that some condition on $f(y)$ is needed to avoid the kind of behavior we saw in Example 5, where solution graphs could pass trough equilibrium values.

3.1 THEOREM

Let $y(t)$ be a solution of $dy/dt = f(y)$. Suppose that $f(y)$ has a continuous derivative in some interval containing an equilibrium point y_0 and that at some time t_0 the solution $y(t)$ enters this interval.

(i) If $f(y) > 0$ for $y(t_0) \le y < y_0$ or $f(y) < 0$ for $y_0 < y \le y(t_0)$, then $\lim_{t\to\infty} y(t) = y_0$.

(ii) If $f(y) < 0$ for $y(t_0) \le y < y_0$ or $f(y) > 0$ for $y_0 < y \le y(t_0)$, then $\lim_{t\to-\infty} y(t) = y_0$.

Remark. Geometrically the conclusion of the theorem says that if the direction field of $dy/dt = f(y)$ points a solution graph toward an equilibrium value y_0, then the the solution will in fact converge to y_0.

Proof. We'll treat only the first assumption in case (i) since the other three parts are quite similar. Because $dy/dt = f(y) > 0$ for some value t_0, $y(t)$ continues to increase for $t > t_0$ as long as $y(t) < y_0$. But $y(t) = y_0$ is impossible because the uniqueness theorem (Theorem 3.1) prevents the graphs of $y(t)$ and the constant solution y_0 from having points in common. Also, since $y(t)$ is continuous, its graph can't skip over the value y_0. Hence, $y(t)$ is strictly bounded above by y_0. The increasing function $y(t)$ thus has a finite limit $\bar{y}$ satisfying $\bar{y} \leq y_0$. To show $\bar{y} = y_0$, suppose $\bar{y} < y_0$. Since $f(y)$ is continuous,

$$\lim_{t\to\infty} y'(t) = \lim_{t\to\infty} f\big(y(t)\big) = f\big(\lim_{t\to\infty} y(t)\big) = f(\bar{y}).$$

We assumed $\bar{y} < y_0$, so $f(\bar{y}) > 0$ by (i). But if $\lim_{t\to\infty} y'(t) = f(\bar{y}) > 0$, then the integral $y(t)$ of $y'(t)$ would have to be unbounded, contradicting $y(t) < y_0$. ■

example 6

The **logistic population equation** $dy/dt = ky(c - y),\ k,\ k > 0,\ c > 0$ is a modification of the organic growth law $dy/dt = ky$. When $0 < y < c$ the factor $c - y$ has the effect of reducing the rate of increase of y as y increases toward an upper limit c. The logistic equation has equilibrium solutions $y_0(t) = 0$ and $y_c(t) = c$. Since $f(y) = ky(c - y) > 0$ in the interval $0 < y < c$, case (i) of Theorem 3.1 applies and implies that a solution $y(t)$ that starts with initial value in the interval $0 < y < c$ will tend to c as t tends to $+\infty$. For an initial value larger than c, we're in the interval $c < y < \infty$, where $f(y) = ky(c - y) < 0$. Hence the second assumption in part (i) of the theorem applies, so we still have limit c, approached from above this time. The solution $y_c(t) = c$ is an example of an **asymptotically stable equilibrium** in that every solution that has a value close enough to c has limit c as t tends to infinity.

example 7

The *logistic equation* of the previous example can be modified to include a fixed negative term $-h$ on the right-hand side: $dy/dt = ky(c - y) - h$. The constant h is usually interpreted as a plant harvest rate that might be applied to manage a population. To find equilibrium solutions we solve for y in the quadratic equation $ky(c - y) - h = 0$ or $-ky^2 + kcy - h = 0$. It turns out that there are distinct real solutions only when $h < kc^2/4$, in which case we have the two positive equilibrium solutions

$$y_m = \frac{c}{2} - \sqrt{\left(\frac{c}{2}\right)^2 - \frac{h}{k}}, \qquad y_M = \frac{c}{2} + \sqrt{\left(\frac{c}{2}\right)^2 - \frac{h}{k}}.$$

Thus $f(y)$ factors to $f(y) = -k(y - y_m)(y - y_M)$, and is positive if $y_m < y < y_M$. Part (i) of Theorem 3.1 shows that given an initial value between the two equilibrium values $y_m < y_M$, a solution will tend to the larger value y_M as t tends to infinity. Other possibilities are considered in the exercises.

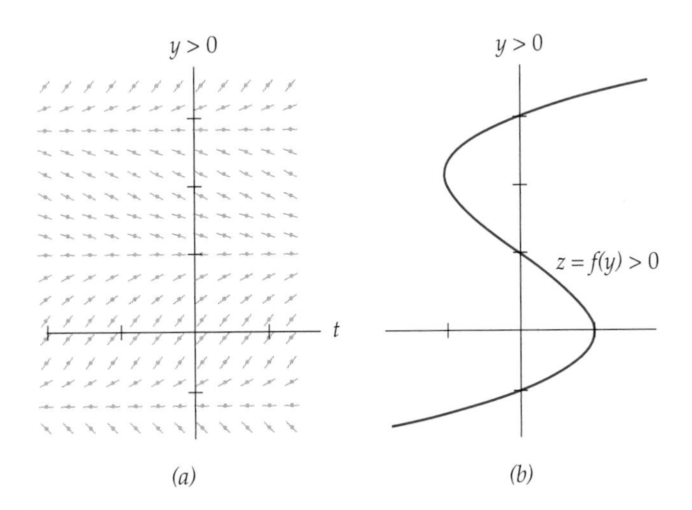

Figure 2.6 (a) Slopes $f(y)$. (b) Graph of $z = f(y)$.

Phase Curves. Here is a convenient way to display the direction field of an autonomous differential equation $y' = f(y)$ together with the graph of the function $z = f(y)$ so that you can see how the information in one relates to the other. Doing this requires us to align the positive y axes both for the direction field and for the graph of f. As shown in Figure 2.6, we have left the direction field sketch in its usual orientation in Figure 2.6(a) and oriented the graph of f in Figure 2.6(b) so that horizontal slopes of the field occur at the level at which the graph of f crosses its y axis. Similarly, the positive slopes correspond to points y where $f(y) > 0$, and negative slopes correspond to negative values of $f(y)$. The graph of $z = f(y)$ is called the **phase curve** of the differential equation $y' = f(y)$, because it is plotted in what is called **phase space,** where coordinates are position and velocity rather than position and time. Both pictures are intended to convey the same in formation, but the direction field is only a sketch, while the phase curve represents the differential equation more precisely. The vertical arrows record only the general inclination of the slopes in the field. On the other hand, solution graphs $y = y(t)$ must be plotted in the ty plane, where we sketch the direction field.

example 8

The phase curve of an autonomous first-order *linear* equation $y' = ay + b$ is a straight line, namely, the graph of $z = ay + b$. If $a \neq 0$, there is exactly one equilibrium solution, namely, $y = -b/a$.

EXERCISES

Determine which of the following differential equations has one or more equilibrium solutions, and for those that do, find all equilibrium solutions. Then determine the long-term behavior of the nonconstant solutions by applying Theorem 3.1 or by finding explicit solution formulas.

(Refer to previous page.)

1. $dy/dt = -y$.
2. $dy/dt = y^2$.
3. $dy/dt = ty$.
4. $dy/dt = \sin y$.
5. $dy/dt = 1 - y^2$.
6. $dy/dt = y - t$.
7. $du/dt = u(u - 1)$.
8. $dv/dt = 1 + 3v + v^2$.
9. $dy/dt = y(1 - e^y)$.

10. A car of mass m is propelled along a track with constant acceleration a against a frictional force of magnitude kv^3.

    (a) Find the differential equation satisfied by the velocity $v = v(t)$.

    (b) Find all equilibrium velocities.

    (c) Use Theorem 3.1 to describe the behavior of solutions to the differential equation in part (a) having positive initial velocities.

11. A body of mass m falls under the influence of constant gravitational acceleration g and with air resistance of the form $k(v^2 + 2v)$, $v \geq 0$.

    (a) Find all equilibrium velocities.

    (b) Are all your answers to part (a) physically meaningful? Explain.

12. A body of mass m falls under the influence of constant gravitational acceleration g and with air resistance of the form kv^α. If the initial velocity is equal to the equilibrium velocity, how long does it take for the body to fall distance h?

13. A body of mass m falls under the influence of constant acceleration g and with air resistance of the form $k \ln(1 + v)$. Find the equilibrium velocity.

14. A body of mass m falls under the influence of constant gravitational acceleration g and with air resistance of the form $k \tan(\alpha v)$, $\alpha > 0$.

    (a) Find the equilibrium velocity.

    (b) Show that if $v_0 = 0$, then the equilibrium velocity can be at most $\pi/(2\alpha)$, regardless of the choices for k, g, and m.

15. Define a function of velocity v by

$$F(v) = \begin{cases} v, & 0 \leq v \leq 1, \\ (v^2 + 1)/2, & 1 < v. \end{cases}$$

    Suppose an object of mass m falls under the influence of constant gravity g and retarding force $kF(v)$, $k > 0$.

    (a) Sketch the graph of $F(v)$ for $0 \leq v \leq 3$.

    (b) Find the equilibrium velocities.

16. **(a)** Newton's law of cooling is expressed by the differential equation $du/dt = -k(u - a)$, where k and a are positive constants. The equation has a single equilibrium solution; what is it?

    (b) Stefan's law is expressed by the differential equation $du/dt = -k(u^4 - a^4)$, where k and a are positive constants. The equation has a single physically realistic equilibrium solution; what is it?

17. Consider Newton's law of cooling with variable ambient temperature $a(t) = e^{-t}$: $du/dt = -k(u - e^{-t})$, $k > 0$.

(a) Show that the differential equation has no equilibrium solutions.

(b) Show that every solution of the differential equation tends to zero as t tends to infinity.

Logistic Equations

18. Consider solutions of the logistic equation $dy/dt = y(1 - y) - h$ modified by the constant harvest rate h.

(a) For what positive values of h are there two equilibrium solutions?

(b) What are the equilibrium solutions corresponding to the values of h you found in part (a)?

(c) Describe the behavior of a solution with initial value between the equilibrium values found in part (a).

(d) Describe the behavior of a solution with initial value below the lower equilibrium value found in part (a).

(e) Describe the behavior of a solution with initial value above the upper equilibrium value found in part (b).

(f) What is the special significance of the value $h = \frac{1}{4}$? In particular, would you expect this precise value to be significant in an attempt to use the differential equation to model the history of a real population?

19. Consider solutions of the logistic equation $dy/dt = ky(c - y) - h$ modified by the constant harvest rate h.

(a) For what positive values of h are there two distinct equilibrium solutions?

(b) What are the equilibrium solutions corresponding to the values of h you found in part (a)?

(c) Describe the behavior of a solution with initial value between the equilibrium values found in part (b).

(d) Describe the behavior of a solution with initial value below the lower equilibrium value found in part (b).

(e) Describe the behavior of a solution with initial value above the upper equilibrium value found in part (b).

(f) What is the special significance of the value $h = kc^2/4$? In particular, would you expect this precise value to be significant in an attempt to use the differential equation to model the history of a real population?

20. Consider solutions of the general logistic equation $dy/dt = y^\alpha(1 - y)^\beta$, α, β positive constants, $0 \le y \le 1$.

(a) What are the equilibrium solutions?

(b) Describe the behavior of a solution with initial value between the equilibrium values found in part (a).

21. Consider solutions of the general logistic equation $dy/dt = ky^{\alpha}(c - y)^{\beta} - h$, $\alpha,\ \beta > 0$, modified by the constant harvest rate h.

(a) Show that $f(y) = ky^{\alpha}(c - y)^{\beta} - h$ has a unique maximum value on the interval $0 \le y \le c$ at $y = \alpha c/(\alpha + \beta)$.

(b) For what positive values of h are there two or more distinct equilibrium solutions?

22. Example 5 preceding Theorem 3.1 shows that requiring only continuity of $f(y)$ in the statement of the theorem would be insufficient. Explain in detail how the example enables us to reach this conclusion.

23. Bifurcation. A value λ_0 for a real parameter λ in a family of differential equations is a **bifurcation point** if solutions change character in a fundamental way as λ passes through λ_0. Show that $\lambda = 0$ is a bifurcation point for $dy/dt = \lambda y - y^3$ as follows.

(a) Show that for $\lambda \le 0$ all solutions tend to the equilibrium value $y = 0$ as t tends to infinity.

(b) Show that for $\lambda > 0$, if a solution has a value $y(t_0)$ in the interval $-\sqrt{\lambda} < y < \sqrt{\lambda}$, then that solution tends away from the equilibrium value $y = 0$ as t tends to infinity.

4 VARIABLE MASS AND ROCKET PROPULSION

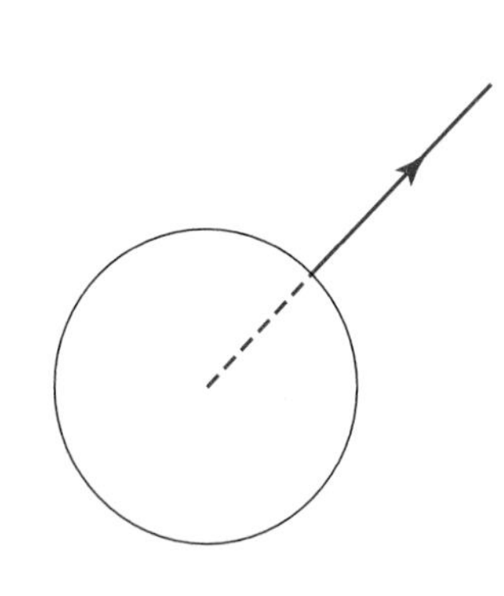

Our experience and intuition tells us that if we were to stand on a skateboard and toss a weight back over one shoulder, then the skateboard would move forward. The purpose of this section is to convert our intuition into the study of the fundamental differential equation for rocket motion. We'll consider here just the motion of a body of varying mass moving along a straight path. Such a path might lie along a line passing through the center of gravity of the earth and extending out into space. Let $F = F(t)$ be the total external force acting on the body at time t, let $m = m(t)$ be the mass of the body, and let $v = v(t)$ be the body's velocity. For the specialized case of a body of *constant* mass m, we have the familiar equation

$$F = ma, \quad \text{where } a = \frac{dv}{dt} \text{ is } \textbf{acceleration}.$$

However, in the case of an object such as a rocket, fuel is initially counted as a part of the rocket's mass and is then consumed at such a high rate that the assumption of constant mass is not applicable. Newton may not have had rockets in mind, but he formulated the relationship among F, m, and v in enough generality that it takes care of the possibility of nonconstant mass. The general **Newton second force law** is

$$F = \frac{d(mv)}{dt}.$$

Here $p = mv$ is the **momentum** of a body, defined so that the momentum of a body equals the sum of the momenta of its parts. It follows from this equation and the product rule for differentiation that if m is constant then

$$F = m\frac{dv}{dt} + v\frac{dm}{dt} = ma + 0 = ma.$$

Thus the general force law $F = d(mv)/dt$ contains the law for constant mass, $F = ma$, as a special case. In general, however, $m = m(t)$ may vary with time.

example 1

This example illustrates the principle behind rocket propulsion in simplified form. Consider a projectile of mass m_0 moving on a linear path with constant velocity $v_0 > 0$ and not being acted on by external forces. Thus the total momentum of all parts of the object remains equal to the initial constant value $m_0 v_0$. At time $t = 0$, a particle of the projectile having mass $m < m_0$ is ejected with velocity w from the rear of the projectile back along its linear path. Note that $w < 0$, because the particle is moving in the direction opposite to that of the projectile. After the ejection of mass m, the projectile has reduced its mass to $m_1 = m_0 - m$. To find the projectile's velocity v_1 after ejection, we express the constant total momentum of the system, consisting of projectile plus particle, as the sum of the momenta of the two parts: $m_0 v_0 = mw + (m_0 - m)v_1$. Solving for v_1 gives

$$v_1 = \frac{m_0 v_0 + m(-w)}{m_0 - m} = \frac{m_0}{m_0 - m} v_0 + \frac{m}{m_0 - m}(-w).$$

You can check that the projectile of reduced mass $m_0 - m$ has had its velocity increased by the difference $v_1 - v_0 = m(v_0 - w)/(m_0 - m)$. Note that increasing the mass m of the ejected particle to a higher value, still below the obvious upper limit m_0, makes for a larger increase in velocity. In a somewhat different way, increasing the speed $-w > 0$ with which the particle is ejected will also increase v_1. Continued ejection of particles will increase the projectile's velocity still more. This is the essence of how a rocket works.

To appreciate how an analysis of momentum plays a role in describing real rocket motion, we need to consider not only the velocity $v = v(t)$ of the rocket but also the velocity $w = w(t)$ of the expelled exhaust gas, which for a rocket on a straight path will have sign opposite to that of $v(t)$. In other words we need to consider the forces acting on all parts of the system consisting of both rocket and exhaust. We then delete the part associated with the exhaust to arrive at an equation for the rocket's velocity. The calculation given at the end of this section corrects the equation $F = d(mv)/dt$ that defines force by a term $-w\,dm/dt$ that accounts for the expelled exhaust gas. That calculation shows that the correct formula expressing the forces acting on the rocket alone has the smaller magnitude

$$F = \frac{d(mv)}{dt} - w\frac{dm}{dt}.$$

Note that if $v > 0$, then $w < 0$, because the exhaust will be moving in the direction opposite to that of the rocket. Also dm/dt is negative since the mass m of the rocket is decreasing, at least while the rocket is actually firing. Thus typically $w\,dm/dt > 0$. The negative correction term $-w\,dm/dt$ accounts for fuel mass ejected from the rocket.

To apply the previous displayed equation, expand the first term on the right by the product rule and recombine terms to get

$$\begin{aligned} F &= m\frac{dv}{dt} + (v - w)\frac{dm}{dt} \\ &= m\frac{dv}{dt} + v_e\frac{dm}{dt}, \end{aligned}$$

where $v_e = v - w$ is the velocity of the exhaust gas, which we note is *measured relative to the moving rocket.* Note that $v_e > 0$, since $v > 0$ and $w < 0$. In practice, v_e can often be assumed to be constant, and we'll make that assumption unless the contrary is explicitly stated. The force F also contains the sum of external forces acting on the rocket, such as the gravitational attraction of the earth or frictional air resistance. The resulting differential equation can be regarded either as a first-order equation to be solved for $m = m(t)$ or as a first-order equation to be solved for $v = v(t)$. In the latter formulation we rewrite the equation as

4.1 $$m(t)\frac{dv}{dt} = -v_e\frac{dm(t)}{dt} + F,$$

often called the **fundamental rocket equation.** Since the exhaust velocity v_e will be positive while the rocket motor is firing, and $dm/dt < 0$, the term $-v_e\,dm/dt$ will in fact be positive. In case the motor is turned off, both v_e and dm/dt will be zero and the rocket equation reduces to $m(dv/dt) = F$. The term $-v_e(dm/dt)$ in the fundamental rocket equation represents the force due to the firing of the rocket, and is called the **thrust** of the rocket motor. Note that a significant increase in velocity from the action of the motor is due not only to exhaust velocity v_e, which is often constant, but also to a rapid rate dm/dt of fuel consumption.

example 2

A rocket so remote from the effects of external forces F that they may be neglected can be assumed to satisfy an equation of the form

$$m\frac{dv}{dt} = -v_e\frac{dm}{dt},$$

or

$$\frac{dv}{dt} = -\frac{v_e}{m}\frac{dm}{dt}.$$

Assuming v_e to be constant, we can integrate both sides with respect to t to get

$$v(t) = -v_e \ln m(t) + C.$$

If we know the initial mass m_0 of rocket plus fuel, and the initial velocity v_0 of the rocket at some initial time t_0, we can determine the constant of integration

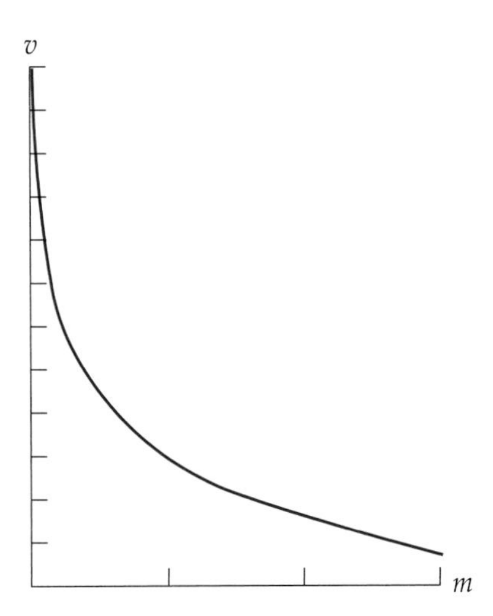

Figure 2.7 Rocket velocity as a function of mass: $v = v_e \ln(m_0/m)$.

to be

$$C = v_0 + v_e \ln m_0.$$

Hence

$$\begin{aligned} v &= v_0 + v_e \ln m_0 - v_e \ln m \\ &= v_0 + v_e \ln \left(\frac{m_0}{m} \right). \end{aligned}$$

The general form of the relation between v and m is shown in Figure 2.7. This graph shows that velocity increases quite rapidly as we decrease the mass of the rocket by burning fuel, and a more rapid rate of fuel consumption, attained, for example, by igniting more motors, will increase the velocity more rapidly. We can draw the same conclusion directly from the fundamental differential equation, which makes an assertion about the rate of change of velocity, that is the acceleration $a = dv/dt$ of the rocket. The differential equation in the absence of external forces is

$$\frac{dv}{dt} = -v_e \frac{1}{m} \frac{dm}{dt}.$$

Since m is decreasing, $dm/dt < 0$, so acceleration is always positive. The differential equation shows that acceleration $a = dv/dt$ can be made large by making m small, but doing that has the disadvantage of reducing the rocket's carrying capacity and the potential fuel supply. The alternative is to make dm/dt large in absolute value, that is, to burn the fuel very rapidly.

***example* 3**

Continuing from the previous example, we can ask for the velocity of a rocket as a function of the mass $b(t)$ of the fuel burned up to time t during the rocket's flight. Let $m(t)$ be the mass of the rocket together with its unburned fuel.

(Thus the initial mass of the rocket is $m(0) = m(t) + b(t)$.) Using the formula for $v(t)$ from the previous example, we find

$$\begin{aligned} v(t) &= v_0 + v_e \ln\left(\frac{m(t)+b(t)}{m(t)}\right) \\ &= v_0 + v_e \ln\left(1 + \frac{b(t)}{m(t)}\right). \end{aligned}$$

Thus, for given initial velocity v_0 and exhaust velocity v_e, rocket velocity at time t depends just on the ratio of burned fuel to the residual mass of the rocket, and is independent of the rate of fuel consumption, regardless of whether this rate is variable or constant.

If we use a constant rate α for fuel consumption, then the mass $b(t)$ of fuel burned up to time t is $b(t) = \alpha t$, and the total mass $m(t)$ of the rocket plus unburned fuel at time t is

$$m(t) = m_r + m_f - \alpha t, \quad 0 \le t \le \frac{m_f}{\alpha},$$

where m_f is the initial mass of fuel in the rocket, and m_r is the mass of the rocket with no fuel in it. Then the fuel is all used up when $\alpha t = m_f$, or $t = m_f/\alpha$, so $t_b = m_f/\alpha$ is called the **burnout time** for consumption rate α. Hence $m(t_b) = m_r$ and $b(t_b) = m_f$. It follows that the rocket velocity at time t_b is

$$v(t_b) = v_0 + v_e \ln\left(1 + \frac{m_f}{m_r}\right).$$

The number $v(t_b)$ is the maximum velocity attainable by firing the motor in the absence of external forces. We observed above that, even for a variable rate of fuel consumption, $v(t)$ is independent of the rate of fuel consumption, so in particular $v(t_b)$ is independent of the rate of fuel consumption.

The previous example applies only when external forces F are small enough to be ignored. The next example takes into account the variable force of gravity $F = -m(t)g$ that rocket engines have to work against in lifting off from the surface of a planet.

---

example 4

Recall that the fundamental rocket equation has the form

$$m\frac{dv}{dt} = -v_e\frac{dm}{dt} + F.$$

The only external force that we'll consider in this example is that of gravity: $F = -mg$ where g is the acceleration of gravity near the earth's surface; in particular, we're ignoring air resistance. The equation to look at is thus

$$\frac{dv}{dt} = -\frac{v_e}{m}\frac{dm}{dt} - g.$$

Integration with respect to t gives

$$v(t) = -v_e \ln m(t) - gt + c.$$

The constant c can be determined by specifying the initial velocity $v_0 = v(0)$. Suppose, for example, that the rocket is launched from the surface of the earth with initial velocity $v(0) = 0$. In that case $c = v_e \ln m(0)$, and then

$$v(t) = v_e \ln\left(\frac{m(0)}{m(t)}\right) - gt.$$

Assuming a constant rate α of fuel consumption, empty rocket mass m_r and initial fuel mass m_f, we find $m(t) = m_r + m_f - \alpha t$ whenever $0 \le t \le m_f/\alpha$. For these values of t we then have

$$v(t) = v_e \ln\left(\frac{m_r + m_f}{m_r + m_f - \alpha t}\right) - gt.$$

The parameters m_r, m_f, and v_e will typically be related so that $v(t) \ge 0$. In particular, setting $t_b = m_f/\alpha$, at which time the fuel is used up, shows after some rearrangement that the inequality $v(m_f/\alpha) \ge 0$ is equivalent to requiring that

$$v_e \ge \frac{g m_f}{\alpha \ln[1 + (m_f/m_r)]}.$$

This inequality is a requirement that the exhaust velocity v_e must meet if the rocket is to lift off the ground.

Note: For the calculations in Examples 1, 2, and 3 the mass m of the rocket plus fuel could just as well have been replaced by weight $W = mg$, where g is acceleration due to gravity. The reason is that the external force F is zero in those differential equations, and they are all of degree 1 in m and dm/dt; thus multiplying by the constant g gives us the original equation with mass m replaced by weight $W = mg$. The simple replacement of mass by weight is not always possible when $F \ne 0$, as in Example 4.

example 5

A rocket subject to frictional forces of magnitude kv and constant gravity has velocity $v(t)$ obeying

$$m\frac{dv}{dt} = -v_e\frac{dm}{dt} - kv - gm, \quad k > 0.$$

If $m = m(t)$ is known, the differential equation for v can be written in first-order linear form as

$$\frac{dv}{dt} + \frac{k}{m}v = -\frac{v_e}{m}\frac{dm}{dt} - g.$$

Now suppose there is a constant rate α of fuel consumption so that $m(t) = m_0 - \alpha t$, where m_0 is the initial mass of rocket plus fuel. The equation then becomes

$$\frac{dv}{dt} + \frac{k}{m_0 - \alpha t} v = \frac{\alpha v_e}{m_0 - \alpha t} - g.$$

The appropriate integrating factor is $(m_0 - \alpha t)^{-k/\alpha}$. With initial condition $v(0) = 0$ the solution works out to be

$$v(t) = \frac{\alpha v_e}{k} + \frac{g}{\alpha - k}(m_0 - \alpha t) - \left(\frac{\alpha v_e}{k} + \frac{g m_0}{\alpha - k}\right)\left(\frac{m_0 - \alpha t}{m_0}\right)^{k/\alpha},$$

where in practice the fuel consumption rate α is much larger than the drag constant k. Exercise 15 asks you to derive this solution, and Exercise 16 asks you to draw a conclusion from it.

Derivation of the Rocket Equation. Let $p = mv$ denote rocket momentum, exclusive of exhaust gas, at time t and let Δp, Δm, and Δv denote the changes in p, m, and v in the time interval of length Δt between t and $t + \Delta t$. With our goal to find dp/dt, we consider the momentum at time $t + \Delta t$:

$$p + \Delta p = (m + \Delta m)(v + \Delta v) + (-\Delta m)(w + \Delta w),$$

where w is the velocity of the expelled exhaust gas. The second term on the right is the change in rocket momentum due to the escape of exhaust gas. (Note that for a rocket that is burning fuel, $\Delta m < 0$, because the decrement Δm represents a decrease in mass. Also the gas velocity $w + \Delta w$ will typically be directed opposite to $v + \Delta v$, the velocity of the rocket itself. However, the assumptions are not used in the computation.) Now subtract $p = mv$ from both sides, simplify, and divide by Δt to get

$$\frac{\Delta p}{\Delta t} = m\frac{\Delta v}{\Delta t} + (v - w)\frac{\Delta m}{\Delta t} + \Delta m \frac{\Delta v}{\Delta t} - \Delta m \frac{\Delta w}{\Delta t}.$$

As Δt tends to zero, so does Δm. It follows that the last two terms tend to zero as Δt tends to zero, and we get for a limit the equation

$$\frac{dp}{dt} = m\frac{dv}{dt} + (v - w)\frac{dm}{dt}.$$

Write $v_e = (v - w)$ for the exhaust velocity relative to the rocket, and $F = dp/dt$ for the external force on the rocket. We can then write the fundamental rocket equation as

$$F = m\frac{dv}{dt} + v_e\frac{dm}{dt}.$$

All forces in this last equation are to be thought of as applied at the center of mass of the rocket. ■

It is tempting to try to derive the rocket equation just by applying the product rule to the equation $F = d(mv)/dt$:

$$F = m\frac{dv}{dt} + v\frac{dm}{dt}.$$

This equation is correct if F, m, and v are interpreted as pertaining to the entire system consisting of rocket plus fuel plus exhaust gas. The trouble is that we really want an equation in which v can be interpreted as the velocity of just the rocket together with its unburned fuel. This is precisely what the previous derivation gives us.

The derivation given above is quite valid regardless of whether m is increasing or decreasing and whether w is positive or negative. Thus the equations

$$m\frac{dv}{dt} + (v - w)\frac{dm}{dt} = F \quad \text{or} \quad \frac{d(mv)}{dt} - w\frac{dm}{dt} = F$$

can be applied when mass is not dissipating but increasing as in the final exercises for this section.

example 6

Consider a car of mass m_0 rolling with constant velocity v_0 on a straight frictionless track and with no external forces acting on the car. At time $t = 0$ the car begins to pick up sand at a constant rate r from a motionless overhead hopper. The mass of the car plus sand as a function of time is $m(t) = m_0 + rt$. The external force is $F = 0$, and the initial velocity of the increasing mass is $w = 0$, so we have

$$m\frac{dv}{dt} + (v - 0)\frac{dm}{dt} = 0,$$

or $d(mv)/dt = 0$ by the product rule for derivatives. Thus the momentum mv is constant. In particular, $m(t)v(t) = m_0 v_0$, so for $t > 0$

$$v(t) = \frac{m_0 v_0}{m_0 + rt} = \frac{v_0}{1 + rt/m_0} < v_0.$$

The velocity decreases as t increases, and indeed the acceleration is negative:

$$\frac{dv(t)}{dt} = \frac{-rm_0 v_0}{(m_0 + rt)^2} < 0.$$

EXERCISES

1. A rocket subject to a retarding force proportional to velocity has velocity $v = v(t)$ satisfying

$$m\frac{dv}{dt} = -v_e\frac{dm}{dt} - kv, \qquad k = \text{const.} > 0.$$

(a) Note that the differential equation is a first-order linear equation for v. Find the general solution under the assumption that $m(t) = m_0 - \alpha t$, where α is a constant rate of fuel consumption.

(b) Assuming initial velocity zero, show that $v(t) \le \alpha v_e/k$ for $t \ge 0$.

2. Referring to Example 1 on page 125, make the following comparisons.

(a) Show that increasing the mass of the ejected particle by a factor $\rho > 1$ causes a greater increase $v_1 - v_0$ in velocity than would increasing the relative velocity $v_0 - w$ of ejection by the same factor ρ.

(b) Show that increasing the mass of the ejected particle by a factor $\rho > 1$ causes a greater increase $v_1 - v_0$ in velocity than would multiplying the velocity w of the ejected particle by the same factor ρ.

(c) Compare the effects on $v_1 - v_0$ of increasing velocity w of ejection and increasing *relative* velocity $v_0 - w$ by the same factor $\rho > 1$.

3. (a) The discussion and graph in Example 2 on page 126 suggests that in principle a rocket might attain arbitrarily large velocity by virtue of having consumed a large enough proportion of its mass as burning fuel. What practical considerations might prevent this from happening for a rocket based on earth?

(b) Starting with initial velocity $v_0 = 0$, maintaining constant exhaust velocity v_e, and neglecting external forces, what ratio of residual rocket mass to initial mass would be required for a rocket to attain velocity c, the speed of light?

(c) The extreme upper limit for exhaust velocity from a fuel-burning rocket is about $v_e = 2.18$ miles per second. Using this value for v_e and $c = 186{,}000$ miles per second, estimate the size of the ratio found in part (b).

4. A car is propelled along a smooth track by ejecting sand from the rear of the car at a fixed rate of 20 pounds per second and with velocity relative to the car of 10 feet per second, so the car behaves somewhat like a rocket. Assume that the car together with its occupants weighs 1000 pounds and that it starts with an additional 500 pounds of sand.

(a) If the car starts from a stand-still and frictional forces are negligible, what would the car's velocity be when the sand is all used up?

(b) If the car starts from a stand-still and frictional forces retard the car by an amount $F(t) = -v(t)$, what would the car's velocity be at the instant when the sand is all used up?

(c) If the car described in part (b) continues to be subject to the same frictional forces, what would its velocity be 60 seconds after the instant at which the sand is all used up? How far would it have traveled by that time?

5. (a) Solve the fundamental rocket equation with constant exhaust velocity $v_e > 0$ and external force $F = 0$. Assume $v(0) = v_0 \geq 0$, but make no assumption about the form of $m(t)$.

(b) With the same assumptions as in part (a), assume also that there is a constant rate α of fuel consumption, so that $m(t) = m_r + m_f - \alpha t$, where m_f is the initial fuel mass and m_r the mass of the rocket itself. Show that at burnout, when $t = m_f/\alpha$, the acceleration is $\alpha v_e/m_r$.

6. To launch a rocket from a standing position on the surface of the earth, a positive initial acceleration has to be applied in order to overcome g, the acceleration of gravity. Neglecting air resistance, show that a successful launch requires $\alpha v_e > g m_0$, where α is a constant rate of fuel consumption, v_e is the exhaust velocity, and m_0 is the initial mass of rocket plus fuel. (The product αv_e is the *thrust* of the rocket motor, defined more generally in the text, and has the dimensions of force.)

7. Example 3 of the text shows that starting with velocity $v_0 = 0$ the fundamental rocket equation has solution

$$v(t) = v_e \ln\left(\frac{m_0}{m_0 - \alpha t}\right), \qquad 0 \le t \le m_f/\alpha,$$

if $m_0 = m_r + m_f$, there are no external forces, and fuel is burned at constant rate α.

(a) Assume $y(0) = 0$ and integrate velocity $v(t)$ to get distance

$$y(t) = v_e t - \frac{v_e}{\alpha}(m_0 - \alpha t)\ln\left(\frac{m_0}{m_0 - \alpha t}\right).$$

[*Hint:* $\int \ln x\, dx = x \ln x - x$.]

(b) Show that at burnout time $t_b = m_f/\alpha$, the distance covered is

$$y_b = \frac{v_e}{\alpha}\left[m_f - m_r \ln\left(1 + \frac{m_f}{m_r}\right)\right].$$

What does this equation say about the size of α if we want y_b to be large? See also the previous exercise for restrictions on the size of α.

8. (a) Solve the fundamental rocket equation for constant exhaust velocity $v_e > 0$ under the assumption that $F = -gm$, where $m(t) = m_r + m_f - at$ and $v(0) = v_0 \ge 0$.

(b) Integrate the result found in part (a) to find the height $h(t)$ of a rocket starting from the earth's surface on a line through the center of the earth, with initial velocity $v_0 = 0$.

9. A fully fueled rocket weighs 32 pounds, half of which is fuel, has exhaust velocity $v_e = 800$ feet per second. The rocket consumes all its fuel at a constant rate within 10 seconds. Suppose the rocket is launched vertically from a standing position on the surface of the earth. Take $g = 32$ feet per second per second, and neglect air resistance. Describe the entire history of the rocket from launch at $t_0 = 0$ to return to earth at t_r. In particular,

(a) Find the velocity and acceleration of the rocket for $0 \le t \le 10$.

(b) Find the height of the rocket at $t = 10$ and its position and velocity for $10 \le t \le t_r$, during which time the rocket is under the influence of gravity only. In particular, find the maximum height of the rocket above the earth and the speed of the rocket when it hits the earth.

*10. Use the data of the previous problem under the additional assumption that the rocket encounters a retarding force of air resistance proportional to its velocity, with proportionality factor $k = 0.01$.

(a) Show that if distance is measured up from the earth's surface the external force function has the form $F = -0.01v - 32m$.

(b) Find the velocity and acceleration of the rocket for $0 \le t \le 10$.

(c) Find the height of the rocket at $t = 10$ and its position and velocity for $10 \le t \le t_r$, during which time the rocket is under the influence of gravity and air resistance only. In particular find the maximum height of the rocket above the earth and the speed of the rocket when it hits the earth.

11. (a) Show that the mass $m(t)$ of a rocket plus its unburned fuel can be computed from its velocity $v(t)$ by

$$m(t) = m(0)e^{[v(0)-v(t)]/v_e}$$

under the assumption $F = 0$ for external forces.

(b) What is the corresponding result if $F = -gm(t)$, where g is a positive constant?

12. (a) Show that the fundamental rocket equation can be written

$$m\frac{dv}{dm} = -v_e$$

if there are no external forces. Solve the equation for v in terms of m and the initial mass $m_0 = m(0)$ of rocket plus fuel if $v(0) = 0$.

(b) If velocity v is significantly large relative to the speed of light c, the relativistic version of the equation in part (a) must be used:

$$m\frac{dv}{dm} = -v_e\left(1 - \frac{v^2}{c^2}\right).$$

Assume initial velocity $v(0) = 0$ and use separation of variables to solve this differential equation for v in terms of m and the initial mass $m_0 = m(0)$ of rocket plus fuel. Sketch the shape of the graph of v as a function of m on $0 \le m \le m_0$ for the special case $v_e = c/2$.
[*Note:* You can use c as the unit of measurement on the v axis and m_0 as the unit on the m axis.]

(c) Show that as c tends to infinity the solution to the problem in part (b) tends to the solution to the problem in part (a).

13. Solve the relativistic rocket equation

$$\frac{m\,dv}{dm} = -v_e\left(1 - \frac{v^2}{c^2}\right)$$

for v as a function of m with initial total mass m_0 and initial velocity v_0.

14. A multistage rocket can be thought of as a succession of single-stage rockets in which the initial velocity for each stage is the same as the final velocity for the previous stage and the sum of the total masses of the later stages are counted as part of the unexpendable rocket mass of the previous stage. A body stage is jettisoned as soon as its fuel is gone.

(a) Consider a two-stage rocket with initial velocity v_0, rocket mass M_r, fuel mass M_f, and exhaust velocity V_e for the first stage, and corresponding quantities m_r, m_f, and v_e for the second stage. Assuming no external forces, find a formula for the velocity of the second stage at burnout when all the fuel is gone. [*Hint:* Use the first part of Example 3 on page 127.]

(b) Suppose in part (a) that $M_r = m_r$, $M_f = m_f$, and $V_e = v_e$. Show that the final velocity of the second stage is greater than the final velocity attained by a single stage rocket of mass $2m_R$ and initial fuel mass $2m_f$, and compute the difference in velocities.

15. Derive the solution $v(t)$ given at the end of Example 5 of the text.

16. Compute the value of the solution $v(t)$ given at the end of Example 5 of the text at the time t_b when the fuel is used up and the remaining rocket mass is m_r. Then show that if g is large enough $v(t_b)$ will be negative.

***17.** For a rocket operating fairly near the surface of the earth, the estimate mg for the magnitude of the gravitational attraction of the earth is adequate. At greater distances Newton's **inverse square law** should be used. Let r be the earth's radius and x the distance of the rocket from the surface of the earth. Then the inverse square law gives

$$F_g = \frac{GMm}{(r+x)^2}$$

for the magnitude of the force of gravity at height x above the earth. Here M is the mass of the earth and G is the **gravitational constant,** which, of course, depends on the units of measurement being used. (Thus G can be estimated from $GMr^{-2} = g$ if we know M and r.) Using this notation, show that $x = x(t)$ satisfies this *second-order* differential equation when the rocket is above the atmosphere:

$$\frac{d^2x}{dt^2} + \frac{GM}{(r+x)^2} = -\frac{v_e}{m}\frac{dm}{dt}.$$

18. Consider a rocket with initial velocity zero and no external forces acting on it. Let m_f be the initial fuel mass and let m_r be the mass of the rocket with no fuel in it. Show that if the velocity $v(t)$ of the rocket is ever to exceed the exhaust velocity v_e it is necessary to have $m_f > (e-1)m_r \approx 1.72m_r$. [*Hint:* Integrate $dv/dt = -v_e(dm/dt)/m$.]

19. As suggested at the end of the derivation of the fundamental rocket equation in the text, it is tempting simply to apply the product rule for differentiation to the equation $d(mv)/dt = F$ to arrive at an equation for rocket motion. [This simplified derivation is incorrect because the $m(t)$ and $v(t)$ in the equation thus derived would necessarily refer to the system consisting not only of the rocket plus fuel unburned up to time t but also to the expelled exhaust.] This exercise examines the consequences of trying to predict rocket motion with no external forces by using the uncorrected equation

$$m\frac{dv}{dt} = -v\frac{dm}{dt}.$$

(a) Show that we would conclude that $v(t) = v(0)m(0)/m(t)$. What would you then conclude about motion starting with initial velocity $v(0) = 0$?

(b) Let $a_c(t)$ and $a_u(t)$ be the predicted rocket accelerations according to the corrected and uncorrected rocket equations, respectively. Show that given the same rate of fuel consumption $m'(t)$, then $a_c(t) > a_u(t)$ as long as $v_e m(t) > v(0)m(0)$.

20. Consider the rocket equation

$$m\frac{dv}{dt} = -v_e\frac{dm}{dt} + F(t),$$

with constant exhaust velocity $v_e > 0$ and external force F depending only on time.

(a) Assuming $v(t)$ known from observation and $F(t)$ identically zero, derive the formula

$$m(t) = m(0)e^{[v(0)-v(t)]/v_e}, \quad t \geq 0.$$

(b) Derive a formula analogous to the one in part (a) for the case in which $F(t)$ is continuous for $t \geq 0$.

***21.** A rocket with no forces except for a rocket motor with exhaust velocity v_e acting on it has mass m_0 when fully fueled. If fuel mass m_1 is consumed in order to accelerate the rocket from velocity v_0 to v_1, how much fuel is required to fire the motor in the opposite direction to bring the velocity down to v_2?

Motion Subject to Increasing Mass

22. A car of mass m_0 is sliding with constant velocity v_0 over a straight track with negligible friction and with no external forces acting on the car. (See Example 6 of the text.) At time $t_0 = 0$ the car begins to gain cargo having initial velocity zero at a constant rate $r > 0$.

(a) Find the distance $x(t)$ covered by the car t time units after the mass gain starts.

(b) Show that as r tends to zero the distance found in part (a) tends to the distance that would be attained if no additional mass were added after $t_0 = 0$.

23. A container of mass m_0 starts with velocity $v_0 > 0$ and moves in a straight linear path. As the container moves it absorbs mass that is streaming in the opposite direction with constant speed s.

(a) Find a differential equation relating increasing mass $m(t)$ and its velocity $v(t)$, assuming no effects on the motion other than the absorption of additional mass.

(b) Solve the equation found in part (a), and show that the velocity reaches the value zero when the mass has increased by $v_0 m_0 / s$.

(c) Show that if m is allowed to become arbitrarily large, then the velocity of the container tends to s.

24. Imagine a body of mass m_0 moving through space on a linear path with constant velocity v_0 and with no external forces acting on it. At time $t = t_0$ the body begins to absorb hitherto motionless mass, to increase its mass to $m = m(t)$.

(a) Show that after time t_0 the velocity of the body is, for some time at least, given by $v(t) = m_0 v_0 / m(t)$.

(b) Assume that $m(t) = m_0 + rt$, $r = \text{const.} > 0$, for $t \geq t_0$ and show that the body would attain velocity $v_1 < v_0$ at time $t_1 = (m_0/r)(v_0/v_1 - 1)$.

25. Imagine a body of mass m_0 moving through space on a linear path with constant velocity v_0 and with no external forces acting on it. At time $t = t_0$ the body begins to increase in mass, through uniform accretion of motionless mass, according to some formula $m = m(t)$. Note that at all times, both before and after time t_0, the velocity satisfies

$$m\frac{dv}{dt} + v\frac{dm}{dt} = 0.$$

(a) Show that after time t_0 the velocity of the body is $v(t) = m_0 v_0 / m(t)$.

(b) Find a formula for the distance traveled after time $t_0 = 0$ if the mass at time t is $m(t) = m_0 + at^2$, where a is some positive constant. Show in particular that the distance tends to a finite value s_0 as t tends to infinity, and find s_0.

26. A hailstone with zero initial velocity falling from a considerable height above the earth has an initial mass of 1 gram and is gaining mass by gathering motionless vapor according to the formula $m(t) = (1 + t/10)$, where t is measured in seconds. Assume that distance is measured down toward the earth and that $g = 32$ feet per second per second.

(a) Assuming free fall with no air resistance, find the hailstone's velocity $v(t)$ at general time $t > 0$.

(b) Assuming the air resistance term $-0.01v$, find the hailstone's velocity $v(t)$ at general time $t > 0$.

(c) Estimate the two values for $v(10)$ obtained from parts (a) and (b).

27. A hailstone falling with zero initial velocity from a considerable height above the earth has an initial mass of 1 gram and is losing mass because of evaporation according to the formula $m(t) = (1 - t/10)$, where t is measured in seconds. Assume that the vapor particles have velocity zero after they leave the hailstone. Assume also that distance is measured down toward the earth and that $g = 32$ feet per second per second.

(a) Assuming free fall with no air resistance, find the hailstone's velocity $v(t)$ at time $0 \le t < 10$.

(b) Assuming the air resistance term $-0.01v$, find the hailstone's velocity $v(t)$ at time $0 \le t < 10$.

(c) Compare the two values for $v(7)$ obtained from parts (a) and (b). Can you reconcile with reality the behavior of the two formulas for $v(t)$ as t tends to 10?

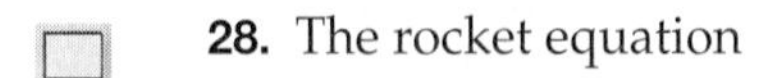

28. The rocket equation

$$m\frac{dv}{dt} = -v_e\frac{dm}{dt} + F$$

becomes a first-order equation for $v(t)$ if we specify time dependent mass $m(t) = m_0 + m_1 - at$ for $0 \le t \le m_0/a$, and $m(t) = m_1$ for $m_0/a < t$. Let the external force F be given by

$$F(t) = -gm(t) - kv^{\alpha},$$

where g is the acceleration of gravity and k and α are positive constants.

(a) Find numerical approximations to $v(t)$ in the range $0 \le t \le m_0/a$, given that $g = 32.2$, $v(0) = 0$, $m_0 = 25$, $m_1 = 75$, $a = 1$, $v_e = 3500$, and $\alpha = 1.5$. Use successively $k = 2, 1, 0.5$, and 0.1, and sketch the corresponding graphs for $v = v(t)$ using an appropriate scale.

(b) Estimate the value of k in part (a) that would produce approximately the values $v(0) = 0$, $v(5) = 17.5$, $v(10) = 40$, $v(15) = 63$, and $v(20) = 87$.

(c) Repeat part (a) for the range $0 \le t \le 50$, noting that $m(t)$ cannot be described by a single elementary function.

29. A hailstone falling with zero initial velocity from a considerable height above the earth has an initial mass of 1 gram and is gaining mass through freezing according to the formula $m(t) = (1 + t/50)^3$, where t is measured in seconds. Assume that distance is measured down toward the earth and that $g = 32$ feet per second per second.

(a) Assuming free fall with no air resistance, find a formula for the hailstone's velocity $v(t)$ for time $t > 0$.

(b) Assuming the air resistance term $-0.01v$, generate a computer plot of the hailstone's velocity $v(t)$ for $0 \le t \le 40$, and compare it with a computer plot of the graph of the solution found in part (a).

30. A hailstone falling with zero initial velocity from a considerable height above the earth has an initial mass of 1 gram and is losing mass because of melting according to the formula $m(t) = (1 - t/50)^3$, where t is measured in seconds. Assume that distance is measured down toward the earth and that $g = 32$ feet per second per second.

(a) Assuming free fall with no air resistance, find a formula for the hailstone's velocity $v(t)$ for time $t > 0$.

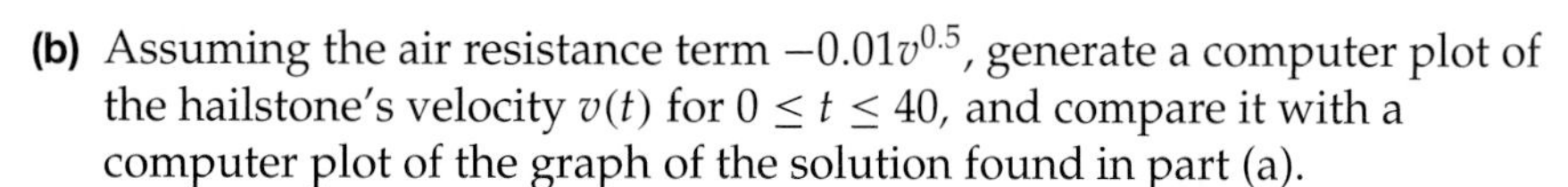

(b) Assuming the air resistance term $-0.01v^{0.5}$, generate a computer plot of the hailstone's velocity $v(t)$ for $0 \le t \le 40$, and compare it with a computer plot of the graph of the solution found in part (a).

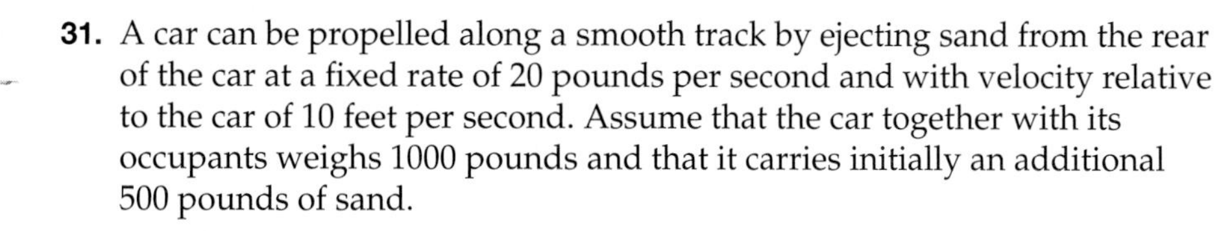

31. A car can be propelled along a smooth track by ejecting sand from the rear of the car at a fixed rate of 20 pounds per second and with velocity relative to the car of 10 feet per second. Assume that the car together with its occupants weighs 1000 pounds and that it carries initially an additional 500 pounds of sand.

(a) If the car starts with initial velocity $v(0) = 100$ feet per second and frictional forces retard the car by an amount $F(t) = -0.01v(t)^2$, write down the differential equation satisfied by $v(t)$.

(b) Find the approximate velocity of the car 25 seconds later, when the sand is all used up.

(c) Find the approximate velocity of the car 25 seconds after the instant at which the sand is all used up. Is it reasonable for this velocity to be less than 100 feet per second?

5 COMPUTING SUPPLEMENT: FIRST-ORDER EQUATIONS

In practice, a first-order differential equation

$$\frac{dy}{dx} = F(x, y) \quad \text{and solution family} \quad y = f(x, c)$$

are often expressed by formulas such as

$$y' = -2y \quad \text{and} \quad y = ce^{-2x},$$

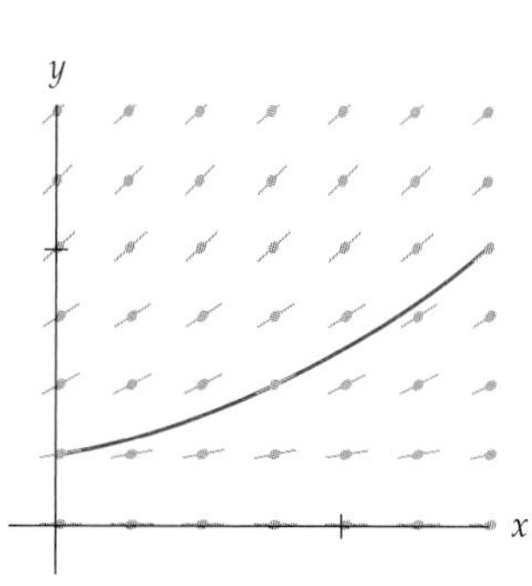

where c is a parameter that depends on an initial condition. The importance of such formulas is that they often show quite easily how the values of the solution depend not only on the independent variable x, but also on the parameter c. In our example, we know that that y tends to 0 as x tends to ∞, regardless of the value of c. However, when solution formulas are not readily available, we rely on numerical approximation. Even in examples where it's possible to find a solution formula, we may prefer to not to use it when our pupose is to generate solution graphs as we do in this section. Our methods here are strongly motivated by the idea that solution graphs take their directions from a differential equation's direction field, as in the sketch in the margin.

5A Euler Method

Suppose we are given a first-order initial-value problem that we can't solve in an elementary way, for example,

$$y' = F(x, y) = x^2 + y^2, \qquad y(-2) = -1.$$

Our problem is to find approximate values for a solution $y = y(x)$. We can make only finitely many estimates, so we restrict ourselves to some interval of values, say $x_0 \le x \le b$. We then try to approximate $y(x)$ at the equally spaced points

$$x_1 = x_0 + h, \qquad x_2 = x_0 + 2h, \ldots, x_m = x_0 + mh,$$

where m is some chosen number of subdivisions of the interval from x_0 to b, each with length

$$h = \frac{b - x_0}{m}.$$

To make our estimates, we approximate the graph of $y = y(x)$ by a sequence of straight-line segments chosen from the direction field of our differential equation

$$y' = F(x, y),$$

starting with the initial condition $y(x_0) = y_0$. The segment of the direction field through (x_0, y_0) has slope $F(x_0, y_0)$, and we extend that segment until it intersects the vertical line $x = x_1$ at a point (x_1, y_1). We then simply repeat the process with (x_1, y_1) replacing (x_0, y_0) to get (x_2, y_2), and so on to (x_m, y_m). Figure 2.8 shows such a sequence of segments along with the solution curve that they approximate. The entire sequence of segments is called an **Euler polygon.** Each segment is determined by its initial point (x_k, y_k) and the slope $F(x_k, y_k)$. The equation of the line containing the segment is thus

$$y - y_k = F(x_k, y_k)(x - x_k), \quad \text{or} \quad y = y_k + F(x_k, y_k)(x - x_k).$$

On this line, the y value corresponding to $x = x_{k+1}$ is then

$$y_{k+1} = y_k + F(x_k, y_k)(x_{k+1} - x_k),$$

or, since $h = x_{k+1} - x_k$,

$$y_{k+1} = y_k + hF(x_k, y_k).$$

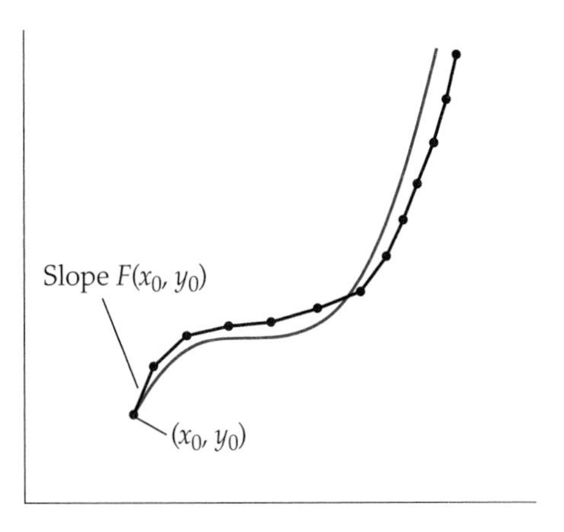

Figure 2.8 Solution and Euler polygon.

The preceding formula defines the **Euler approximation.** Starting with x_0 and y_0, we get a table of values approximating the solution $y = y(x)$.

k	x_k	$y_k = y_{k-1} + hF(x_{k-1}, y_{k-1})$
0	x_0	y_0
1	x_1	$y_1 = y_0 + hF(x_0, y_0)$
2	x_2	$y_2 = y_1 + hF(x_1, y_1)$
3	x_3	$y_3 = y_2 + hF(x_2, y_2)$
⋮	⋮	⋮

example 1

The differential equation $y' = xy$ has solutions $y = ce^{x^2/2}$ that we'll compare with the results of our numerical approximation method. We have

$$\begin{aligned} y_{k+1} &= y_k + hF(x_k, y_k) \\ &= y_k + hx_k y_k. \end{aligned}$$

With $h = 0.1$, the next table shows the comparison, with the values of the exact solution rounded off. What we get is a fair approximation for some purposes, for example, for drawing a fairly accurate graph. Notice that the accuracy deteriorates as we go down the columns, because the errors compound each other. We have started with initial values $(x_0, y_0) = (0, 1)$.

x_k	y_k	$e^{(1/2)x_k^2}$
0.0	1	1
0.1	1	1.005
0.2	1.01	1.020
0.3	1.03	1.046
0.4	1.06	1.083
0.5	1.10	1.133
0.6	1.16	1.197
0.7	1.23	1.277
0.8	1.31	1.377
0.9	1.42	1.499
1.0	1.55	1.647

A computing routine to print a more accurate pairing of x and y for x values between 0 and 1, with step size $h = 0.01$, might look like this:

```
DEFINE F(X,Y) = X * Y
SET X = 0
SET Y = 1
SET H = 0.01
DO WHILE X < 1
SET Y = Y + H * F(X, Y)
SET X = X + H
PRINT X, Y
LOOP
```

To improve accuracy, we think first of increasing m, the number of subdivisions of the interval from x_0 to b, thus making h smaller. But increasing the number of subdivisions increases the likelihood of significant round-off error accumulation in the arithmetic. Rather than recklessly increasing m, we prefer to use a simple improvement of the method. The improvement will produce a smaller error at each step without increasing the number of steps so the error won't increase at a greater rate.

5B Improved Euler Method

The improved Euler method uses a process often called **prediction-correction.** The method assigns a slope to each approximating segment that is a corrected average of the Euler slope and the next Euler slope, as it would have been predicted using the original Euler method.

We'll now use y_{k+1} to denote our improved approximate value at the $(k+1)$th step, and use p_{k+1} for the corresponding simple Euler prediction, based on a previously computed value y_k. We follow these steps to approximate the solution to $y' = F(x, y)$, $y(x_0) = y_0$:

(i) Compute the slope $F(x_k, y_k)$.
(ii) Determine a predictor estimate p_{k+1} by $p_{k+1} = y_k + hF(x_k, y_k)$.
(iii) Compute the average of the two slopes $F(x_k, y_k)$ and a predicted corrector slope $F(x_{k+1}, p_{k+1})$ and use it to determine y_{k+1} by

$$y_{k+1} = y_k + h\left[\frac{F(x_k, y_k) + F(x_{k+1}, p_{k+1})}{2}\right].$$

The formula for y_{k+1} just shown comes from writing the equation for the line through (x_k, y_k) with the average slope and then setting $x = x_{k+1}$ to get the corresponding value $y = y_{k+1}$. The general idea is illustrated in Figure 2.9. A computing routine to implement the method for the initial-value problem $y' = xy$, $y(0) = 1$, might look like this, with step size $h = 0.01$, printing values

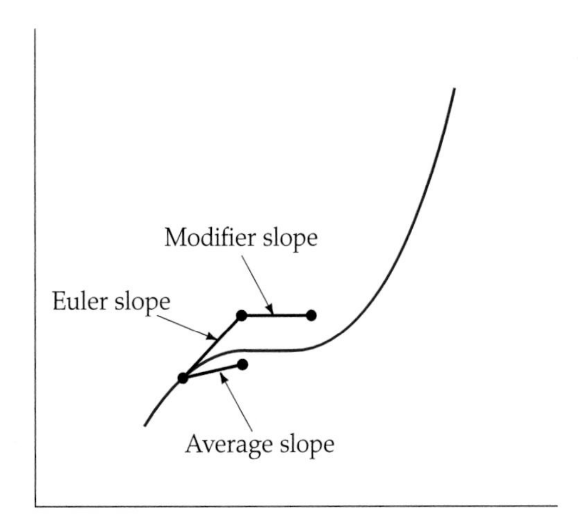

Figure 2.9 Improved Euler slope is an average.

for x between 0 and 1:

```
DEFINE F(X,Y) = X * Y
SET X = 0
SET Y = 1
SET H = 0.01
DO WHILE X < 1
SET P = Y + H * F(X,Y)
SET Y = Y + (H/2) * F(X,Y) + F(X+H, P)
SET X = X + H
PRINT X,Y
LOOP
```

At each stage the value p_k is the *prediction* and y_k is the *correction*. If $h < 0$, the approximations move from larger to smaller x values.

---

example 2

Here is the improved Euler approximation for the same problem we looked at in Example 1: $y' = xy$ with $(x_0, y_0) = (0,1)$, and step size $h = 0.1$. We have $F(x, y) = xy$, so

$$\begin{aligned} p_{k+1} &= y_k + hF(x_k, y_k) \\ &= y_k + 0.1x_k y_k \end{aligned}$$

and

$$\begin{aligned} y_{k+1} &= y_k + \frac{h}{2}[F(x_k, y_k) + F(x_{k+1}, p_{k+1})] \\ &= y_k + 0.05(x_k y_k + x_{k+1} p_{k+1}). \end{aligned}$$

x_k	p_k	y_k	$e^{(1/2)x_k^2}$
0	1	1	1
0.1	1	1.005	1.005
0.2	1.015	1.020	1.020
0.3	1.041	1.046	1.046
0.4	1.077	1.083	1.083
0.5	1.127	1.133	1.133
0.6	1.190	1.197	1.197
0.7	1.269	1.277	1.277
0.8	1.367	1.377	1.377
0.9	1.487	1.499	1.499
1.0	1.634	1.648	1.649

The values of p_k and y_k and the exact solution are rounded off to three decimal places. Notice that there is a difference between the last two columns only at $x = 1$. The arithmetic required in the preceding examples can be done quite easily with a hand calculator. However, if you want many millions of steps, a programmed digital computer is essential. It is important to realize that reducing step size h cannot indefinitely improve the accuracy of the approximations. The increased amount of arithmetic required will eventually cause enough round-off error that further reduction of h is self-defeating. It turns out that the error made at each step by using an approximate formula, called the **local formula error,** is of order h^2 for the Euler method and h^3 for the improved Euler method. Thus taking $h = 0.01$ might be expected to produce an error of order 0.0001 in one Euler step, and might be expected to produce an error of order 0.000001 in one improved Euler step. The advantage of the modification is obvious, because to get the latter accuracy from the simple Euler method would require 1000 times as many steps as the improved method over a given interval, thereby increasing by a factor of 1000 the accumulated **round-off error** inherent in the arithmetic. As a practical matter we often try a reasonable-looking step size and then accept the results if cutting the step size in half produces no significant change in the final outcome. Such checks are important, because it can happen that under adverse assumptions the accumulated error over an x-interval of length L can grow exponentially as a function of L.

EXERCISES

At the website <http://math.dartmouth.edu/~ rewn/> the Applets EULER, 1ORD, 1OR-DPLOT, and RUNKUT implement, in order, the Euler method with numeric output, the improved Euler method with numeric output, the improved Euler method with graphic output, and the fourth-order Runge-Kutta method with numeric output. (See Exercise 12.) Each of the initial-value problems 1 and 2 has a unique solution, expressible in terms of elementary functions, that satisfies the prescribed condition. First find the solution formula; then compare the values given by this formula with approximate values that you compute using the Euler method. Use at least five equally spaced steps on the given interval if you compute by hand and at least 20 steps if you use a computer.

1. $y' = y$ for $0 \le x \le 0.5$ with $y = 1$ when $x = 0$.

2. $y' = y + x$ for $0 \le x \le 0.5$ with $y = 0$ when $x = 0$.

3. and **4** For Exercises 3 and 4 substitute the improved Euler method for the Euler method in the corresponding part of Exercises 1 and 2.

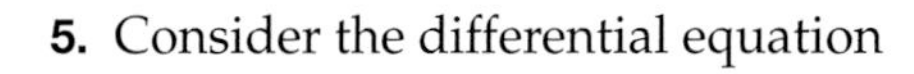

5. Consider the differential equation

$$y' = 1 + y^2 \quad \text{for} \quad -1 \le x \le 1.$$

(a) Sketch the direction field of the differential equation.

(b) Sketch a solution curve passing through $(x, y) = (0, 0)$.

(c) Apply the Euler method to approximate the solution in part (b) at points $x_k = k/5$ for $k = 1$ through $k = 5$.

(d) Use the improved Euler method to print a table of 20 values for the solution satisfying $y(0) = 0$ for $-1 \le x \le 1$. Make $h < 0$ for $x < 0$.

(e) Compare the results of (d) with the solution $y = \tan\, x$.

6. The initial-value problem

$$y' = f(x), \qquad y(x_0) = y_0$$

has the solution

$$y = y_0 + \int_{x_0}^{x} f(t)\, dt.$$

(a) Show that applying the Euler method to the differential equation is equivalent to approximating the solution integral by Riemann sums (i.e., by using the rectangle rule).

(b) Show that the improved Euler method for the differential equation is equivalent to using the trapezoid rule for approximating the solution integral.

Euler Polygon. This term is used to denote the connected sequence of line segments joining the successive points (x_k, y_k) of an Euler approximate solution of $y' = f(x, y)$. [It can be proved that for a carefully chosen sequence of step sizes tending to 0, the resulting Euler polygons will converge on some interval to the graph of a solution if $f(x, y)$ is continuous.] Plot the Euler polygon for the following initial-value problems on the indicated intervals and with the specified step size h. If possible, use a solution by formula for comparison; otherwise use a sketch of the direction field.

7. $y' = y,\ y(0) = 1, 0 \le x \le 2; h = \frac{1}{4}$.

8. $y' = x + y,\ y(0) = 0, 0 \le x \le 2; h = \frac{1}{4}$.

9. $y' = x - y,\ y(0) = 0, 0 \le x \le 4; h = \frac{1}{2}$.

10. $y' = xy,\ y(0) = 1, 0 \le x \le 1; h = \frac{1}{4}$.

11. Use computer graphics to plot a sketch of the direction field of the first-order differential equation $y' = F(x, y)$ listed below and then plot approximate solutions to initial-value problems determined by conditions of the form $y(x_0) = y_0$. Apply the routine to each of the following differential equations, plotting the direction field for points (x, y) in the rectangle $|x| \le 4$, $|y| \le 4$, and plotting the solution satisfying $y(-1) = -1$.

(a) $y' = -y + x$.
(b) $y' = xy$.
(c) $y' = \sin y$.
(d) $y' = \sin xy$.
(e) $y' = \sqrt{1 + y^2}$.
(f) $y' = \sqrt{1 + x^2}$.

12. The fourth-order **Runge-Kutta** approximations to the solution of $y' = F(x, y)$, $y(x_0) = y_0$, are computed as follows:

$$y_{k+1} = y_k + \frac{h}{6}(m_1 + 2m_2 + 2m_3 + m_4),$$

where

$$m_1 = F(x_k, y_k),$$
$$m_2 = F\left(x_k + \frac{h}{2},\ y_k + \frac{h}{2}m_1\right)$$
$$m_3 = F\left(x_k + \frac{h}{2},\ y_k + \frac{h}{2}m_2\right)$$
$$m_4 = F(x_k + h,\ y_k + hm_3).$$

(The local formula error is of order h^5.) Apply a computer algorithm to implement the routine. Then compare the results obtained from it for the problem

$$y' = xy, \qquad y(0) = 1,$$

with the results obtained from the improved Euler method, as well as with the solution formula $y = \exp(x^2/2)$. Try step sizes $h = 0.1, 0.01$, and 0.001.

13. Show that the successive application of the Taylor approximations

$$y_{n+1} \approx y_n + hy'_n + \tfrac{1}{2}h^2 y''_n \quad \text{and} \quad y'_{n+1} \approx y'_n + hy''_n$$

leads to the improved Euler formula $y_{n+1} \approx y_n + \frac{1}{2}h(y'_n + y'_{n+1})$.

14. The initial value problem

$$y' = \sqrt{y}, \qquad y(0) = 0,$$

has two distinct solutions for $x \ge 0$:

$$y_1(x) = 0 \quad \text{and} \quad y_2(x) = \frac{x^2}{4}.$$

(a) Show that direct application of the Euler method produces y_1 exactly at the points of approximation.

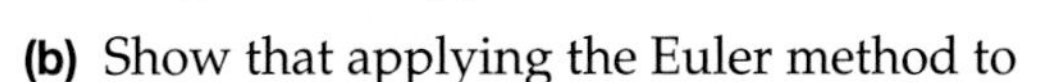

(b) Show that applying the Euler method to

$$y' = \sqrt{y}, \qquad y(x_0) = \frac{x_0^2}{4},$$

for small positive x_0 produces an approximation to y_2.

15. Consider the **generalized logistic equation**

$$\frac{dP}{dt} = kP^{\alpha}\left(1 - \frac{P^{\beta}}{m}\right).$$

(a) Let $k = 1$, $m = 5$, and $P(0) = 1$. Find numerical approximations to the solution in the range $0 \le t \le 10$ for the parameter pairs $(\alpha, \beta) = (0.5,1)$, $(0.5,2)$, $(1.5,1)$,$(1.5,2)$, $(2,2)$.

(b) Estimate a parameter pair (r, q) that yields approximately the values $P(0) = 1$, $P(2) = 2.4$, $P(4) = 2.9$.

16. A 100-gallon mixing tank is full at the time $t = 0$. A salt solution at a concentration of 1 pound per gallon is added at a decreasing rate of $e^{-0.5t}$ gallons per minute, and solution is drawn from the tank so as to maintain the 100-gallon level.

(a) Find the first-order equation satisfied by the amount $S(t)$ of salt in the tank at time t minutes.

(b) Find a numerical approximation to $S(t)$ in the range $0 \le t \le 20$, assuming that $S(0) = 0$. Sketch the graph of the solution.

17. Here is a generalized logistic equation

$$\frac{dP}{dt} = kP\left(1 - \frac{P}{L(t)}\right),$$

where $L(t)$ acts to limit population size $P(t)$. Let $k = 1$ and $P(0) = \frac{1}{2}$, and use numerical analysis to sketch the graph of $P(t)$ for $0 \le t \le 25$. In each case compare the graph of $P(t)$ with that of $L(t)$.

(a) $L(t) = 3 + \sin t$. **(b)** $L(t) = \ln(2 + t)$. **(c)** $L(t) = t + 1$.

18. A generalized logistic equation for growth of a population $P = P(t)$ is

$$\frac{dP}{dt} = KP^{\alpha}(c - P)^{\beta}, \qquad \alpha, \beta > 0.$$

With $\alpha = \beta = 2$, $K = 3$, and $c = 4$, estimate the long-term value of $P(t)$ if

(a) $P(0) = 1$.

(b) $P(0) = 5$.

(c) Explain the results of parts (a) and (b) on the basis of the behavior of the direction field of the differential equation.

19. Consider a 100-gallon tank from which mixed solution is being drawn off at the rate of 4 gallons per minute. Salt solution is being added at the rate of 3 gallons per minute and the salt concentration of the added solution is gradually increasing from time $t = 0$ according to the formula $(1 - e^{-t})$ pounds per gallon.

(a) When will the tank be effectively empty?

(b) Find a differential equation for the amount $S(t)$ of salt in the tank over the relevant time interval.

(c) Use a numerical solution method to approximate the solution satisfying the initial condition $S(0) = 10$. Then estimate the maximum total amount of salt in the tank during the process and the time at which the maximum occurs.

(d) Explain what goes wrong when you try to find an elementary solution formula for the differential equation.

20. An object weighing 1600 pounds is dropped under the influence of constant gravitational acceleration 32 feet per second per second and with air resistance force of magnitude $0.1 \ln(1 + v)$. Use a numerical method to estimate the time it takes for the velocity v to reach 500 feet per second. Recall that weight equals mass times gravitational acceleration.

21. An object weighing 640 pounds is dropped under the influence of constant gravitational acceleration 32 feet per second per second and with air resistance force of magnitude $0.1v^{1.2}$. Use a numerical method to estimate the time it takes for the velocity v to reach 500 feet per second.

5C Newton's Method

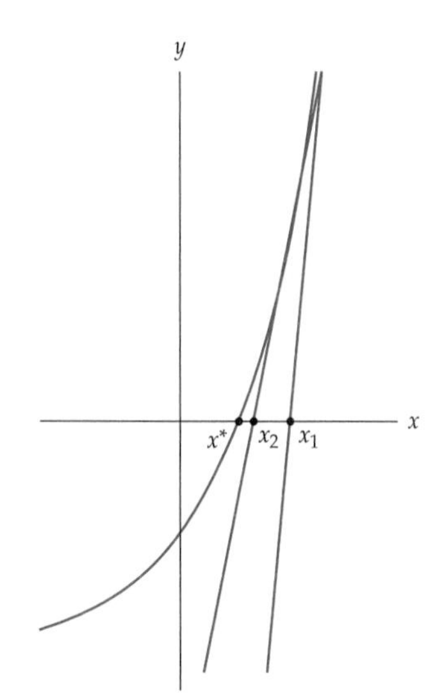

Finding approximate solutions to an equation $f(x) = 0$ is one of the most frequently occurring problems in pure and applied mathematics, and Newton's method is often the preferred approach. Geometry motivates the traditional introduction to the method, which goes as follows: To estimate a solution x^* of $f(x) = 0$, make a good initial guess x_0 and draw the tangent line to the graph of $y = f(x)$ at $(x_0, f(x_0))$. This line is the graph of

$$y = f(x_0) + f'(x_0)(x - x_0),$$

and the graph of this line crosses the x axis where $y = 0$, that is $f(x_0) + f'(x_0)(x - x_0) = 0$. Solving the last equation for $x = x_1$ gives the classical Newton formula for the next estimate

$$x_1 = x_0 - \frac{f(x_0)}{f'(x_0)}, \quad \text{and in general} \quad x_{n+1} = x_n - \frac{f(x_n)}{f'(x_n)},$$

replacing x_0 by x_1, then x_1 by x_2, and so on. The margin figure shows the graph of $y = e^x - 2$ together with the intersections of its tangent lines with the x axis at $x_1 \approx 1.27067$ and $x_2 \approx 0.831957$. Since $e^{\ln 2} = 2$, the exact solution is $x^* = \ln 2 \approx 0.693147$. The middle column of Table 1 on page 150 shows that starting with $x_0 = 2$ we get six-digit accuracy with just five steps.

Newton's method for approximating solutions to $f(x) = 0$ is a special case of the numerical solution of a differential equation as follows: To locate a point x^* such that $f(x^*) = 0$, try to find a function $x = x(t)$ such that $f\big(x(t)\big)$ converges rapidly to 0 as t tends to ∞. A way to do this is to ask that $f\big(x(t)\big) = Ce^{-t}$ for t tending to infinity. Differentiating this last equation with respect to t using the chain rule gives $f'(x)(dx/dt) = -Ce^{-t} = -f(x)$, or

5.1
$$\frac{dx}{dt} = -\frac{f(x)}{f'(x)}.$$

example 3

If $f(x) = x^2$, then $f'(x) = 2x$, and Equation 6.1 becomes

$$\frac{dx}{dt} = -\frac{x}{2}, \quad \text{with solutions} \quad x(t) = Ce^{-t/2}.$$

We started with implicit solutions to a differential equation, namely, $f(x) = Ce^{-t}$. Assuming we can't solve $f(x) = Ce^{-t}$ for x, and we want $x = x(t)$, we try applying Euler's method to Equation 5.1, with step size $h > 0$ and starting value $t_0 = 0$. The result is a formula, containing Newton's method as a special case with $h = 1$, for generating a sequence of numbers x_n tending to a solution x^* of $f(x) = 0$:

5.2
$$x_{n+1} = x_n - h\frac{f(x_n)}{f'(x_n)}.$$

The value x_n is only an approximation to a number x such that $f(x) = Ce^{-t}$ for $t_n = nh$, but since $f(x_n) \approx e^{-nh}$, and the exponential approaches zero as n tends to infinity, we can hope for rapid convergence of $f(x_n)$ to zero. Two choices remain to be made:

Larger values of h explain the chaotic behavior of iterations of the discrete logistic equation;, see Exercises 22 and 23.

1. The starting value $x(0) = x_0$, which should be somewhere near a number x^* such that $f(x^*) = 0$, a choice that determines $C = f(x_0)$, and
2. a value for h that makes nh tend to infinity fairly fast.

Customary usage has it that Newton's method uses $h = 1$, but other choices are also interesting. Making $0 < h < 1$ usually requires more steps to reach an accurate result, while making h significantly larger than 1 raises the risk of an erratic response from the method. However, we'll see later that some values of $h > 1$ can be particularly desirable.

example 4

Table 1 shows a computer printout of the values of x_n for three choices of h when $f(x) = e^x - 2$. We've used $x_0 = 2$ for the starting value each time. The exact solution to $e^x - 2 = 0$ is $\ln 2 \approx 0.693147$. Note that we arrive at this value after just five steps when $h = 1$. With $h = 2$, the numbers x_n oscillate about the true value for thousands of steps before producing even two-place accuracy.

Table 1 Newton-Euler approximations to ln 2

$h = \frac{1}{2}$	$h = 1$	$h = 2$
2	2	2
1.63534	1.27067	.541341
1.33022	.831957	.86921
1.09464	.702351	.546341
.929301	.693189	.862601
.82413	.693147	.550853
.762747	.693147	.856685
.72913	.693147	.554954
.711459	.693147	.851349
.702386	.693147	.558704
.697788	.693147	.846503
.695473	.693147	.562152
.694311	.693147	.842077
.69373	.693147	.565336
.693438	.693147	.838013
.693293	.693147	.56829
.69322	.693147	.834263
.693184	.693147	.57104
.693165	.693147	.83079
.693156	.693147	.57361
.693152	.693147	.827561

example 5

Transcendental equations such as $320t + 2200(1 - e^{-0.1t}) = 10{,}000$ come up often in applied mathematics. To apply Newton's method, let

$$f(t) = 320t + 2200(1 - e^{-0.1t}) - 10{,}000,$$

and estimate the positive values of t where $f(t) = 0$. Since $f'(t) = 320 + 220e^{-0.1t} > 0$, $f(t)$ is increasing. Also $f(0) = -10{,}000$ and $f(t) \to \infty$ as $t \to \infty$, so there must be a unique positive t^* such that $f(t^*) = 0$. The Newton iteration is

$$t_{n+1} = t_n - \frac{320t + 2200[1 - \exp(-0.1t_n)] - 10{,}000}{320 + 220\exp(-0.1t_n)}.$$

Starting with $t_0 = 1$, we arrive after six steps at the estimate $t_6 \approx 24.9426$. The Java Applet NEWTON at http://math.dartmouth.edu/~rewn/ prints the estimates, but be sure to enter the functions using the variable x.

example 6

Formal application of Newton's method without regard to the special characteristics of the function $f(x)$ can lead to anomalous results. For example, exponential functions such as $f(x) = e^{x^2/4}$ are never zero, but we can try to use Newton's method on this function anyway; this leads us to the iteration formula $x_{n+1} = x_n - 2/x_n$. Putting $x_0 = 1$ we get $x_n = (-1)^n$; this sequence oscillates between 1 and -1, and so doesn't converge to anything. We might be suspicious if we did get convergence, since there is no solution to converge to. With other starting values the behavior is no better and will usually be worse.

The exercises show other ways that Newton's method may fail to produce a sequence tending to a solution of $f(x) = 0$, though when it does what it's supposed to, it typically converges quite rapidly. Furthermore, the scope of Newton's method can be extended in a surprisingly large number of ways, for example, to finding the complex solutions of an equation, which is essentially a two-dimensional problem, and also to still higher-dimensional problems. Newton's method must be used with care, as the previous example shows, but in spite of that caveat it remains one of the most widely used recipes of contemporary numerical analysis. Finally, it has been known for many years that even when the method fails, it may do so in intriguing ways that illustrate some of the basic features of the chaotic dynamics of a **discrete dynamical system** $x_{n+1} = N(x_n)$. The "dynamics" of this system consists of the various results of starting with particular values $x = x_0$ and forming the sequences $x_0, x_1 = N(x_0), x_2 = N(x_1), x_3 = N(x_2), \ldots, x_n = N(x_{n-1}), \ldots$. An illuminating way to look at this sequence to denote the n-fold composition of N with itself by N^n; thus $N^1(x) = N(x)$, $N^2(x) = N\big(N(x)\big)$, $N^3(x) = N\Big(N\big(N(x)\big)\Big)$, and so on. In this notation, $x_n = N^n(x_0)$ when $n \geq 1$, and the particular choice for x_0 can be regarded as an initial condition.

To be specific about the relation of Newton's method to discrete dynamics, the recurrence relation $x_{n+1} = x_n - f(x_n)/f'(x_n)$ can be abbreviated $x_{n+1} = N(x_n)$, where $N(x) = x - f(x)/f'(x)$. Then the sequence of approximations to a solution of $f(x) = 0$ is $N^n(x_0)$, where x_0 is the initial estimate of the solution. Following is an example of how the method can lead to interesting chaotic dynamics.

---

example 7

The function $f(x) = x^2+1$ is never zero for real values of x, but we can still apply Newton's method to see what happens; there are, of course, purely imaginary roots at $\pm i$. In this example, the function $N(x)$ to construct is

$$\begin{aligned} N(x) &= x - \frac{f(x)}{f'(x)} \\ &= x - \frac{x^2+1}{2x} = \frac{1}{2}\left(x - \frac{1}{x}\right). \end{aligned}$$

Since $N(x)$ has only real values if x is real, we can't expect to approximate the imaginary roots by applying Newton's method with a real starting value x_0. What does happen when x_0 is real is as follows. For some starting values x_0, $N^n(x_0)$ runs repeatedly through a finite set of real values. For all other values of x_0, $N(x_0)$ wanders in an apparently aimless way through the real numbers. Some of the details of this behavior are treated in the exercises. The erratic behavior could be interpreted as a frantic and futile attempt to get at the imaginary roots while confined to the real number line. It can be shown that starting with a nonreal complex number z will take us quickly to one of the imaginary square roots $\pm i$ of -1. (What happens when you try to solve $z^3 + 1 = 0$ to approximate cube roots of -1 this way is both weird and wonderful. The mathematician

Arthur Cayley pointed this out in 1879, and it was perhaps the first observation of chaotic dynamics in the iteration of complex functions.)

Error Estimates. In numerical root-finding, the discrepancies $\epsilon_n = x_n - x^*$ between a root x^* and its successive approximations x_n are useful for describing rapidity of convergence to the root. It turns out that if $f'(x^*) \neq 0$ then a convergent Newton sequence x_n **converges quadratically** to a root of $f(x)$, which in mathematical notation says that there is a constant K such that

$$\epsilon_{n+1} \sim K\epsilon_n^2.$$

The symbol "$\sim$" is to be interpreted here to mean simply that the Taylor expansion in powers of ϵ_n of the expression on the left begins with the term on the right. If x^* is a pth-order root of $f(x)$, that is, $f(x) = (x - x^*)^p g(x)$ for some $g(x)$ not zero at x^*, we can still get quadratic convergence if we use Euler step size $h = p$. Quadratic convergence is usually desirable once we have reached a small error ϵ_n, since the next error will be to some extent determined by the square of ϵ_n, which is even smaller.

5.2 THEOREM

Suppose $f^{(p)}(x^*)$ is the first nonzero derivative of $f(x)$ at x^*, and that higher-order derivatives are continuous. If the Newton-Euler recursion

$$x_{n+1} = x_n - p\frac{f(x_n)}{f'(x_n)}$$

produces a sequence convergent to x^* for some starting value x_0, then

$$(x_{n+1} - x^*) \sim \frac{(x_n - x^*)^2}{p(p+1)} \frac{f^{(p+1)}(x^*)}{f^{(p)}(x^*)}.$$

Proof. Subtracting x^* from both sides of the Newton-Euler equation gives the error $\epsilon_{n+1} = \epsilon_n - pf(x_n)/f'(x_n)$. We now expand $f(x)$ and $f'(x)$ about x^*, in other words, in powers of $\epsilon_n = x_n - x^*$. Writing $f^{(k)}$ for $f^{(k)}(x^*)$, we get

$$f(x_n) = \frac{\epsilon_n^p f^{(p)}}{p!} + \frac{\epsilon_n^{(p+1)} f^{(p+1)}}{(p+1)!} + \frac{\epsilon_n^{(p+2)} f^{(p+2)}}{(p+2)!} + \cdots,$$

$$f'(x_n) = \frac{\epsilon_n^{(p-1)} f^{(p)}}{(p-1)!} + \frac{\epsilon_n^p f^{(p+1)}}{p!} + \frac{\epsilon_n^{(p+1)} f^{(p+2)}}{(p+1)!} + \cdots.$$

Substitute these expansions into the recursion formula and write the result as a single fraction. Because of the choice of Euler step size $h = p$, the terms of degree p in the numerator cancel out. Then cancel the common factors $(p-1)!$ and ϵ_n^{p-1} in the top and bottom. We're left with

$$\epsilon_{n+1} = \frac{\epsilon_n^2/[p(p+1)]f^{(p+1)} + \epsilon_n^3/[p(p+1)(p+2)]f^{(p+2)} + \cdots}{f^{(p)} + (1/p)\epsilon_n f^{(p+1)} + \cdots}$$

$$= \frac{\epsilon_n^2 f^{(p+1)}}{p(p+1)f^{(p)}} + \cdots.$$

Replacing ϵ_n and ϵ_{n+1} by their definitions gives the desired result. ■

The Java Applet NEWTON at website <http://math.dartmouth.edu/~rewn/> is useful for doing most of the exercises. This can also be reached through <www.mhhe.com/williamson>.

EXERCISES

Each of the equations in Exercises 1 through 6 has either a single solution or else no solution in the given interval. Find those solutions that do exist to five-place accuracy.

1. $x^3 - 24x^2 + 12x - 8 = 0, 0 \leq x < \infty.$

2. $x^3 + 24x^2 + 12x + 8 = 0, 0 \leq x < \infty.$

3. $e^{-x} = x, -\infty < x < \infty.$

4. $e^x = x, -\infty < x < \infty.$

5. $e^{x^2} = e^x, 0 < x < \infty.$

6. $x + 1 = \tan x, -\pi/2 < x < \pi/2.$

Locate all the real solutions of the equations in Exercises 7 through 12 to five-place accuracy. Use a preliminary graphical analysis to identify an appropriate starting point for applying Newton's method to each solution.

7. $x^4 - x - 1, -\infty < x < \infty.$

8. $x^3 - 4x^2 + 5x - 2, -\infty < x < \infty.$

9. $e^{-x} - \ln x, 0 < x < \infty.$

10. $\sin x - x/10, -\infty < x < \infty.$

11. $x^2 - \cos(x + 1), -\infty < x < \infty.$

12. $5 \sin 16x - \tan x, -\pi/2 < x < \pi/2.$

13. The function $f(x) = \arctan x$ has just one zero value, located at $x^* = 0$. To see the importance of having a close enough starting point x_0 in applying Newton's method, experiment with the following choices on this function.

(a) $x_0 = 1,\ h = 1.$ **(b)** $x_0 = 2,\ h = 1.$ **(c)** $x_0 = 2,\ h = \frac{1}{2}.$

14. A 10-pound weight is fired straight up from the surface of the earth with an initial velocity of 100 feet per second. There is an air-resistance force of magnitude equal to 0.1 times velocity. Use Newton's method as the last step in the process of finding out how long it takes for the weight to fall back to earth. Assume $g = 32$.

15. Since the function $f(x) = x^2 + 1$ is positive for all real x, applying Newton's method with a real number for a starting point can't find the imaginary roots at $\pm i$. Investigate what happens starting with (a) $x_0 = 1/\sqrt{3}$, (b) $x_0 = 1$, (c) $x_0 = \frac{1}{2}$.

16. Under the right circumstances choosing a step size h larger than 1 can lead to convergence.

(a) With $f(x) = x^2$, take $h = 1$ and explain what happens to Newton's method, starting at an arbitrary point x_0.

(b) With $f(x) = x^2$, take $h = 2$ and explain what happens to Newton's method, starting at an arbitrary point x_0.

(c) Let n be a positive integer. With $f(x) = x^n$, take $h = n$ and explain what happens to Newton's method, starting at an arbitrary point x_0.

17. Show that Newton's method with step size $h > 1$ replaces the tangent line approximation through $(x_n, f(x_n))$ with a line that is not as steep as the tangent. Explain the analogous conclusion for the choice $0 < h < 1$.

18. Suppose a sequence of numbers $x_1, x_2, x_3, \ldots$ is generated by successive application of the Newton formula $x_{n+1} = x_n - f(x_n)/f'(x_n)$ and that the sequence converges to a number x^*. Show that if $f'(x)$ is continuous in an open interval containing x^*, then x^* is necessarily a solution of $f(x) = 0$. [*Hint:* Let n tend to ∞ in the Newton formula.]

19. A **fixed point** of a function $F(x)$ is a domain point x^* such that $F(x^*) = x^*$.

(a) Show that if F is a real-valued function of a real variable, then F has a fixed point at just those domain points, if any, for which the graph of F crosses the line $y = x$.

(b) Assume that $f(x)$ is a differentiable real function such that $f'(x) \neq 0$, and let $N(x) = x - f(x)/f'(x)$. Show that $f(x^*) = 0$ if and only if x^* is a fixed point for $N(x)$, that is, $N(x^*) = x^*$.

(c) Assume that $f(x)$ is a differentiable real function, and let $N(x) = x - f(x)/f'(x)$. Show that if the sequence $N^n(x_0)$ converges as n tends to ∞ to a value x^*, then x^* must be a fixed point of N.

(d) Use the result of part (c) to show that if $f(x) = x^2 + 1$ and $N(x) = x - f(x)/f'(x)$, then there is no real value x_0 such that $N^n(x_0)$ can converge as $n \to \infty$.

20. **(a)** Derive the Newton iteration formula $x_{n+1} = N(x_n)$ for $f(x) = e^{x^2/4}$.

(b) Show that the formula derived in part (a) can't ever produce a convergent sequence x_n, no matter what the starting value x_0 is. [*Hint:* Suppose x_n converged to x^*, and let $n \to \infty$ in the iteration formula. Can you find x^*?]

21. Consider the iteration formula $x_{n+1} = N(x_n)$, where $N(x)$ is some continuous function of x. We can try to express the right-hand side of this formula as the result of applying Newton's method to the approximate solution of $f(x) = 0$ for some function $f(x)$. In other words, given $N(x)$, find a differentiable $f(x)$ such that

$$N(x) = x - \frac{f(x)}{f'(x)}.$$

(a) Show that $y = f(x)$ described above must satisfy the first-order linear differential equation

$$y' + \frac{y}{N(x) - x} = 0.$$

(b) Solve the differential equation in part (a) to show that if $N(x) = x^2$ then $f(x) = x/(x-1)$ is a possible choice.

(c) Repeated application of $x_{n+1} = N(x_n)$ starting with some choice for x_0 produces different results, depending on x_0. Explain these differences for $N(x) = x^2$ on the basis of the graph of the function $f(x)$ of part (b).

22. The equation $dx/dt = hx(m-x)$ is the *logistic differential equation,* and it has the well-behaved solutions described in Section 1. A popular computer recreation is to iterate the related discrete equation $x_{n+1} = hx_n(1-x_n)$ with some fixed $h \geq 1$, starting with an x_0 strictly between 0 and 1. You get a sequence of numbers

$$x_0,\ x_1 = hx_0(1-x_0),\ x_2 = hx_1(1-x_1), \ldots, x_n = hx_{n-1}(1-x_{n-1}).$$

It turns out that if $0 \leq h \leq 3$ the sequence x_n converges to $1 - 1/h$. For larger values of h the sequence exhibits *period doubling*. Thus, for h somewhat larger than 3, alternate elements of x_n converge to two limit points. For still larger choices of h, subsequences of x_n consisting of every fourth, then eighth, and in general every 2^nth value, converge to different values, with apparently *random* or *chaotic* behavior for values of h closer to $h = 4$. Here is a way to study and understand this behavior by relating it to an application of the Euler method to a family of logistic equations.

(a) Make a computer graphics simulation of the long-term behavior of $x_{n+1} = hx_n(1-x_n)$ by starting with some number x_0 between 0 and 1, for example, $x_0 = \frac{1}{2}$. Let $h = 1$, and plot x_{80} through x_{100} vertically over $h = 1$ on a horizontal h axis. Then step h ahead by successive increments of size 0.01 to get $h = 1.01,\ 1.02,\ 1.03, \ldots,$ repeating the plot of x_{80} through x_{100} until you reach $h = 3.99$. You should get a picture something like the one shown here.

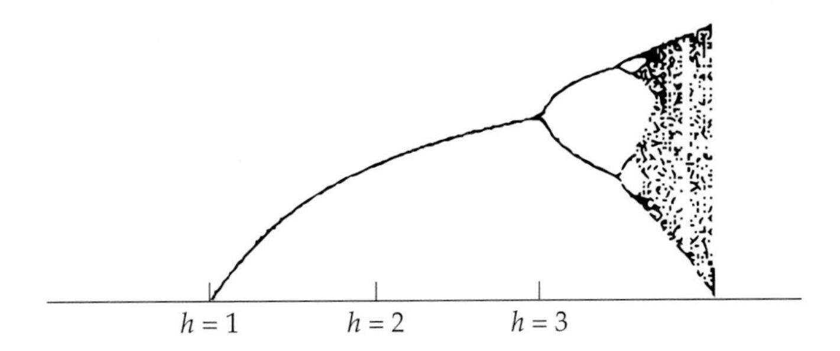

(b) Show that if the Euler method with step size h is applied to the particular case of the *logistic equation* $dx/dt = x\big((1-1/h) - x\big)$, you get the recurrence equation $x_{n+1} = hx_n(1-x_n)$ discussed above. Note that as $n \to \infty$, the sequence x_n estimates the behavior of a solution $x(t)$ to the differential equation for large values $t = nh$. (For $h > 3$ the step size is so large that the validity of the estimate is hopelessly compromised.)

(c) Each parameter value h for which branching or *period doubling* occurs in the discrete system is an example of a *bifurcation point*, that is, a point of fundamental change in system behavior. Estimate the first three such values of h from the picture, or, better still, by running a numerical simulation.

23. **(a)** Show that using Newton's method to find a root of

$$f_\lambda(x) = \left(\frac{x - (1 - 1/\lambda)}{x}\right)^{1/(\lambda-1)}$$

for $\lambda > 1$ leads to the logistic iteration formula $x_{n+1} = N(x_n) = \lambda x_n(1 - x_n)$.

(b) Explain why the iteration formula $x_{n+1} = N(x_n)$ derived in part (a) can be expected to produce a sequence converging to $1 - 1/\lambda$ for at least some values of $\lambda > 1$.

(c) In part (a) of the previous numbered exercise you are asked to demonstrate that increasing the value of λ beyond 3 leads to bifurcation of the graph and eventually chaotic behavior in the sequences generated by the relation $x_{n+1} = N(x_n)$, while using the same x_0 for each value of λ. Demonstrate that if the starting value x_0 is adjusted appropriately for each λ with root-finding in mind, the erratic behavior disappears. Thus chaotic behavior in logistic iteration can be regarded as a result of an ill-considered use of Newton's method. [*Hint:* $f_\lambda(x)$ is typically ill-defined as a real function for $x < (1 - 1/\lambda)$.]

SUPPLEMENTARY APPLICATION FOR CHAPTER 2

The website http://www.mhhe.com/williamson/ contains an application devoted to the radioactive decay equation $dy/dt = -ky$, where k is a postive constant. The underlying mathematics is the same as required for the population growth equation $dP/dt = kP$ with $k > 0$, but in practice the range of applications for the radioactive decay equation contains somewhat more complex problems such as radioactive carbon dating of archeological items.

CHAPTER 3

Equations of Order Two or More

1 INTRODUCTION

Recall that a differential equation has **order** n if n is the highest-order derivative in the equation. Thus equations of the form

$$y'' = f(x, y, y') \quad \text{and} \quad y''' = g(x, y, y', y'')$$

have orders two and three, respectively. Experience with differentiation strongly suggests that derivatives of order one or two appear most prominently in applications. The reason for this is that first and second derivatives have well-understood interpretations: slope and velocity for order one, and concavity and acceleration for order two. Put another way, a derivative of order *more* than two is typically hard to interpret directly. Interpretation of derivatives carries over to interpretation of differential equations, and second-order equations as a class turn out to be the most important ones in pure and applied mathematics, since they are in principle capable of including the four basic geometric and physical ideas mentioned above in one equation. Still, equations of order three or more arise in some applications, and understanding them is helpful for solving some second-order equations.

A large class of equations with important applications exists for which there is a generally applicable method for finding solution formulas; these are the linear constant-coefficient equations, four examples of which are

$$y'' + y = 0, \qquad y'' - 3y' + 2y = e^x, \qquad 2y'' - 3y = x, \qquad y^{(4)} - y = 1.$$

These four equations all share the defining property of what it means for a differential equation to be linear:

Definition. A differential equation is said to be **linear** if the equation has degree one in the unknown functions and their derivatives that occur in the equation.

Thus the equations $y'' + y^2 = 0$ and $y'' + yy' = 0$ are **nonlinear,** that is, not linear. We'll at first concentrate on second-order equations, partly because they are the ones that most often come up in applied mathematics, and partly because the higher-order extension doesn't involve fundamentally different ideas. Many applied examples in Section 4 of this chapter and in Chapter 4 call for an equation with independent variable t for time, for example, the **damped linear spring equation**

$$m\frac{d^2x}{dt^2} + k\frac{dx}{dt} + hx = F(t),$$

where m is the mass of an oscillating weighted spring, k is a damping coefficient, and h is a stiffness coefficient, called Hooke's constant for the spring. The function $F(t)$ represents an externally applied force and is called a **forcing function.**

Why Is Linearity Important? The practical significance of the linear form of a differential equation is as follows. We define a second-order operator L acting on twice-differentiable functions y by

$$Ly = Ay'' + By' + Cy, \qquad A \neq 0.$$

The coefficients A, B, and C may be constant or may depend on an independent variable x. It follows from the basic properties of differentiation, as well as from the definition of L, that if c_1 and c_2 are constants, and $y_1(x)$ and $y_2(x)$ are twice-differentiable, then L satisfies

1.1 $$L(c_1y_1 + c_2y_2) = c_1Ly_1 + c_2Ly_2.$$

To see this, just apply the operator L, then regroup terms by multiples of c_1 and c_2, to get

$$\begin{aligned} L(c_1y_1 + c_2y_2) &= A(c_1y_1 + c_2y_2)'' + B(c_1y_1 + c_2y_2)' + C(c_1y_1 + c_2y_2) \\ &= c_1(Ay_1'' + By_1' + Cy_1) + c_2(Ay_2'' + By_2' + Cy_2) \\ &= c_1Ly_1 + c_2Ly_2. \end{aligned}$$

Some terminology helps to fix these ideas in our minds. An operator L that satisfies Equation 1.1 is called a **linear operator,** and the sum of multiples $c_1y_1 + c_2y_2$ is called a **linear combination** of y_1 and y_2 with **coefficients** c_1 and c_2. The real importance of linearity for a differential equation lies in what it says about the general form of the solutions of a linear differential equation $Ly = 0$, namely,

1.2 SUPERPOSITION PRINCIPLE

If y_1 and y_2 are two solutions of a linear differential equation $Ly = 0$, that is, $Ly_1 = 0$ and $Ly_2 = 0$, then every linear combination $y = c_1y_1 + c_2y_2$ with arbitrarily chosen coefficients is also a solution of $Ly = 0$.

Proof. From Equation 1.1 we get $L(c_1y_1 + c_2y_2) = c_1Ly_1 + c_2Ly_2 = 0 + 0 = 0$. ■

example 1

It's easy to check that the linear equation $y'' - y = 0$ has solutions $y_1 = e^x$ and $y_2 = e^{-x}$, so the superposition principle tells us that $y = 7e^x - 2e^{-x}$ is also a specific solution. Section 1B shows that every solution of $y'' - y = 0$ is a linear combination $y = c_1 e^x + c_2 e^{-x}$.

A condition imposed on the values of a solution and its successive derivatives at a single domain point x_0 is called an **initial condition,** and the problem of finding a solution to a differential equation that also satisfies given initial conditions is called an **initial-value problem.**

---

example 2

Returning to Example 1, we can pose the initial-value problem

$$y'' - y = 0, \qquad y(0) = 5, \qquad y'(0) = 9.$$

We look for a solution among the linear combinations $y(x) = c_1 e^x + c_2 e^{-x}$ from Example 1. Because $y'(x) = c_1 e^x - c_2 e^{-x}$, the initial conditions require

$$y(0) = c_1 + c_2 = 5,$$
$$y'(0) = c_1 - c_2 = 9.$$

Adding and subtracting these two equations shows that $2c_1 = 14$ and $2c_2 = -4$. Hence $c_1 = 7$ and $c_2 = -2$, which gives us the solution $y = 7e^x - 2e^{-x}$ we noted in Example 1; Section 1B shows that this solution is uniquely determined by the initial conditions. Note, however, the definitions

$$\cosh x = \tfrac{1}{2}(e^x + e^{-x}), \qquad \sinh x = \tfrac{1}{2}(e^x - e^{-x})$$

of the **hyperbolic cosine** and the **hyperbolic sine** functions. It follows by successively adding and subtracting these two equations that

$$e^x = \cosh x + \sinh x \quad \text{and} \quad e^{-x} = \cosh x - \sinh x.$$

Hence we can write the unique solution to the initial-value problem $y'' - y = 0$, $y(0) = 5$, $y'(0) = 9$ as

$$y(x) = 7(\cosh x + \sinh x) - 2(\cosh x - \sinh x) = 5\cosh x + 9\sinh x.$$

Thus even though the solution is uniquely determined its written form is not unique.

Note that Sections 3C and 3D take care of some forcing functions not covered by Sections 3A and 3B, and can be considered optional. Both Section 3D and Chapter 8, Section 1B also cover some important discontinuous forcing functions.

The methods of the next few sections aim at giving a complete analysis of the initial-value problem for the damped linear spring equation displayed at the beginning of this introduction, as well as all constant coefficient differential equations of the same linear form. In particular, we can predict what will happen to the mechanism that obeys the damped spring equation if we know the initial position $x(t_0)$ and initial velocity $x'(t_0)$ at some time t_0. For many of the examples in this chapter, including the linear constant-coefficient ones listed in the second paragraph of this introduction, we'll be able to verify the existence and uniqueness of solutions for the appropriate initial-value problems by careful derivation of solution formulas. For other equations, such as the nonlinear equation $d^2y/dt^2 = -\sin y$, we'll refer to a general existence and uniqueness theorem in the introduction to Chapter 4. Nothing remotely like the superposition principle is valid for a nonlinear equation such as $y'' + y^2 = 0$, so the optional Section 4 shows how to find solution formulas for some second-order nonlinear equations of special types.

1A Exponential Solutions

Here is an application that we can treat thoroughly using just the ideas in Section 1.

example 3

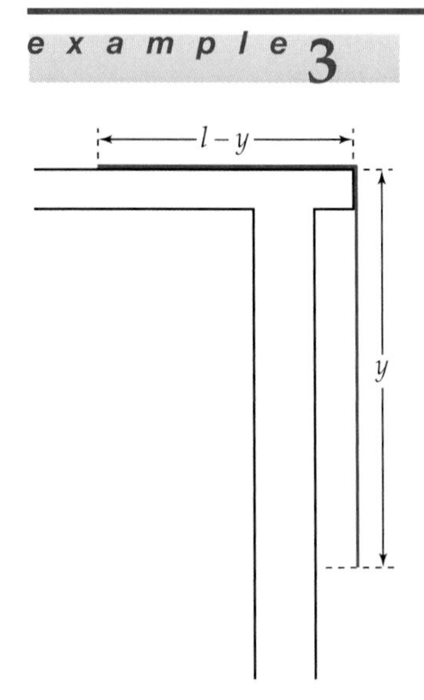

The sketch shows a uniform chain of length l and total mass m with an amount y of its length hanging over the edge of a dock. We'll assume for now that the rest of the chain, of length $l - y$, can slide over the surface of the dock with negligible retarding friction. The external force F acting to pull the chain down is the product of the varying mass of the hanging part of the chain and the acceleration of gravity g. The varying mass referred to is $(m/l)y$ since m/l is the chain's mass density per unit of its length y. Thus $F = g(m/l)y$. Note that the force acting to accelerate the chain increases as the hanging length increases.

On the other hand, from the point of view of the entire chain's motion, the definition of F is also the chain's total mass times its acceleration: $F = m\,d^2y/dt^2$. The result of equating the two expressions $F = (gm/l)y$ and $F = m\,d^2y/dt^2$ is the differential equation

$$m\frac{d^2y}{dt^2} = \frac{gm}{l}y, \quad \text{or} \quad \frac{d^2y}{dt^2} = \frac{g}{l}y.$$

Note carefully the general idea behind how we arrived at this equation, because we use the idea repeatedly: We equated the *definition* of force with a *specific* force, namely, gravitational attraction. We'll return to this equation in Section 1B as an example of the solution of an initial-value problem. For now we'll simply note how the equation relates to the equation $d^2y/dt^2 = g$, which just asserts that the the acceleration of gravity is constant. Note that while $y < l$ there is still some chain on the dock, so $(g/l)y = g(y/l) < g$. Hence the chain's acceleration is less than the acceleration of gravity g, while $y < l$. When $y = l$ the factor y/l on the left of the inequality cancels out, and the two accelerations are equal. At this point the chain becomes simply a falling body, no longer subject to the original differential equation, but with its motion governed thereafter by $d^2y/dt^2 = g$. The first exercise of several on sliding chains asks you to look at the relationship between the two differential equations.

The differential equation

$$y' - ry = 0,$$

with r constant, was treated in Sections 3B and 4 of Chapter 1. Using the exponential multiplier method of Section 5, it is easy to show that every solution has the form

$$y = ce^{rx}, \quad c \text{ constant.}$$

The reason is that multiplication of the equation by e^{-rx} gives the equivalent equation

$$e^{-rx}y' - re^{-rx}y = 0 \quad \text{or} \quad (e^{-rx}y)' = 0.$$

It follows that $e^{-rx}y = c$, so $y = ce^{rx}$. Hence it's natural to look for exponential solutions for second-order equations.

---

example 4

We look for exponential solutions to

$$y'' + 5y' + 6y = 0,$$

having the form $y = e^{rx}$, r constant. We have $y' = re^{rx}$, $y'' = r^2e^{rx}$, so to satisfy the equation we need

$$r^2e^{rx} + 5re^{rx} + 6e^{rx} = 0$$

or

$$(r^2 + 5r + 6)e^{rx} = 0.$$

The exponential factor e^{rx} is never zero, so the last equation is satisfied if and only if

$$r^2 + 5r + 6 = (r + 2)(r + 3) = 0.$$

Letting $r = -2$ and $r = -3$, we've found the respective solutions

$$y_1 = e^{-2x}, \qquad y_2 = e^{-3x}$$

for the differential equation. If we define the operator L by $Ly = y'' + 5y' + 6y$, then we've just shown that $Ly_1 = 0$ and $Ly_2 = 0$. From the Superposition Principle 1.2 it follows that the linear combination

$$y = c_1e^{-2x} + c_2e^{-3x}$$

is also a solution for arbitrary choices of the constants c_1 and c_2. We next try to choose the constants to satisfy initial conditions.

To find a solution $y(x)$ satisfying the initial conditions

$$y(0) = 1, \qquad y'(0) = 2,$$

we compute $y' = -2c_1e^{-2x} - 3c_2e^{-3x}$. Now substitute $x = 0$ in the expressions for y and y'. We find

$$y(0) = c_1 + c_2 = 1,$$
$$y'(0) = -2c_1 - 3c_2 = 2.$$

We use elimination to solve this pair of equations for c_1 and c_2. Dividing the second equation by -2 gives

$$c_1 + c_2 = 1,$$
$$c_1 + \tfrac{3}{2}c_2 = -1.$$

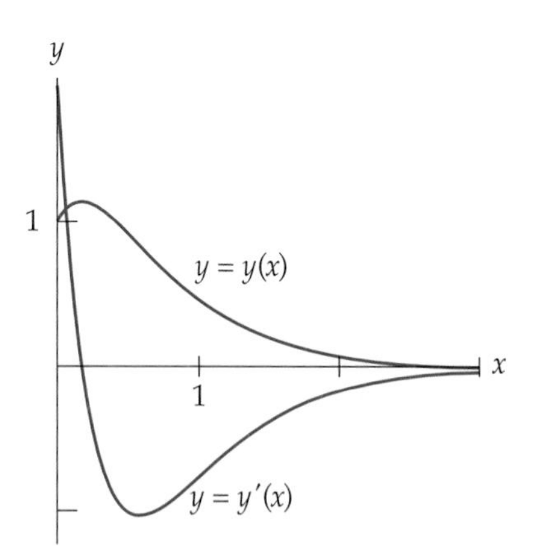

Figure 3.1 Solution $y = 5e^{-2x} - 4e^{-3x}$ and derivative $y' = -10e^{-2x} + 12e^{-3x}$.

Subtract the second equation from the first to get $-\frac{1}{2}c_2 = 2$, so $c_2 = -4$. Replacing c_2 by -4 in the first equation gives $c_1 = 5$. Hence

$$y = 5e^{-2x} - 4e^{-3x}$$

is the desired solution of the initial-value problem. The graphs of $y(x)$ and $y'(x) = -10e^{-2x} + 12e^{-3x}$ are shown in Figure 3.1. Note that the solution graph attains its maximum where the derivative's graph crosses the x axis.

example 5

The intimate relationship between a solution and its derivative will be explored more fully in Chapter 4, Section 3, in the context of *phase space*, where the basic coordinates are y and y'. For now we'll simply elaborate on the pictorial evidence in Figure 3.1, which confirms that the solution $y(x)$ found in the previous example increases when $y'(x) > 0$ and decreases when $y'(x) < 0$.

Since evidently $\lim_{x\to\infty} y(x) = 0$ the maximum value of $y(x)$ for $x \geq 0$ occurs either when $x = 0$ or when $y'(x) = 0$. The latter value of x we find by solving the equation $-10e^{-2x} + 12e^{-3x} = 0$, or $6 = 5e^x$. The solution x_0 comes from taking the natural logarithm of $e^x = \frac{6}{5}$, and is $x_0 = \ln\frac{6}{5} \approx 0.182$. Noting that $e^{-x_0} = \frac{5}{6}$, we find that

$$y(x_0) = 5\left(\tfrac{5}{6}\right)^2 - 4\left(\tfrac{5}{6}\right)^3 = \tfrac{125}{108} \approx 1.157.$$

Since $y(\ln\frac{6}{5}) > 1 = y(0)$, the maximum value of the solution for $x \geq 0$ is about 1.157.

One feature of the pair of functions $y_1(x)$, $y_2(x)$ in the previous example is important enough that it merits special attention:

Neither $y_1(x)$ nor $y_2(x)$ is a constant multiple of the other function.

Otherwise one of the two exponentials in the linear combination $c_1e^{-2x} + c_2e^{-3x}$

would be redundant, and the sum could be simplified to a single constant multiple of one function or the other. We'll restate the condition as a formal definition in a way that will generalize effectively to more than two functions when we encounter the idea later in Section 2D:

> *Two functions $y_1(x)$ and $y_2(x)$ are linearly independent on an interval I if the only constants c_1, c_2 for which the linear combination $c_1y_1(x)+c_2y_2(x)$ is identically zero on I are the obvious choices $c_1=c_2=0$.*

The two statements in italics are equivalent, so we can use whichever is more convenient, usually the former. To prove the equivalence, suppose first that $y_1 = cy_2$ on I, so $c_1y_1 - cy_2 = 0$ on I with $c_1 = 1$. Hence the functions y_1 and y_2 are not linearly independent. On the other hand, if y_1 and y_2 are not linearly independent, then $c_1y_1+c_2y_2 = 0$ on I with, for example, $c_1 \neq 0$. Hence $y_1 = -(c_2/c_1)y_2$, so y_1 is a multiple of y_2. The first of the two statements provides the simplest test for linear independence of two functions $y_1(x)$ and $y_2(x)$ on an interval I, because it just requires us to check that the ratios $y_1(x)/y_2(x)$ and $y_2(x)/y_1(x)$ are not constant.

example 6

If $r \neq s$, the nonzero functions e^{rx} and e^{sx} are linearly independent functions on every interval $a < x < b$, because if $e^{rx} = ce^{sx}$ or $e^{sx} = ce^{rx}$, then multiplying either equation by e^{-sx} implies that $e^{(r-s)x}$ is constant, which we know to be false if $r \neq s$.

To say that two functions are **linearly dependent** is to say that they are not linearly independent. Thus the functions e^x and $2e^x$ are linearly dependent because one of them is a constant multiple of the other.

example 7

The functions $\sin x$ and $\cos x$ are linearly independent functions on every interval $a < x < b$, because if $\sin x = c \cos x$ or $\cos x = c \sin x$, then the ratios $\tan x$ or $\cot x$ would have to be constant wherever they're defined on the interval, which we know to be false.

A quadratic equation $Ar^2 + Br + C = 0$ with a double root $r = a$ gives us only a single solution $y_1(x) = e^{ax}$, so it's reasonable to ask whether we can find a linearly independent solution $y_2(x)$. Theorem 1.4 in Section 1B will clear the matter up without assuming the existence of an exponential solution, but the next example shows us how to proceed given that we've made that assumption.

example 8

The equation $y'' - 2ay' + a^2y = 0$ is the generic "double-root" differential equation. When we substitute $y = e^{rx}$, we get

$$r^2e^{rx} - 2are^{rx} + a^2e^{rx} = 0, \quad \text{or} \quad (r^2 - 2ar + a^2)e^{rx} = 0.$$

Canceling the nonzero factor e^{rx} leaves us with the quadratic equation $r^2 - 2ar + a^2 = 0$, which factors to $(r - a)^2 = 0$. The resulting double root gives us just the one solution $y_1 = e^{ax}$. If there is a linearly independent solution y_2, then y_2 must be a *non*constant multiple of y_1, so $y_2(x) = u(x)e^{ax}$. [Thus if y_2 turns out to exist we know that $u(x) = y_2(x)e^{-ax}$.] Trying to find a suitable nonconstant factor $u(x)$, we substitute $y = ue^{ax}$ and the derivatives $y' = (u' + au)e^{ax}$ and $y'' = (u'' + 2au' + a^2u)e^{ax}$ into the differential equation $y'' - 2ay' + a^2y = 0$. Rearranging terms gives

$$(u'' + 2au' + a^2u - 2au' - 2a^2u + a^2u)e^{ax}, \quad \text{or} \quad u''e^{ax} = 0.$$

Now cancel e^{ax} and integrate $u'' = 0$ twice to get $u(x) = c_1x + c_2$. Hence

$$y = u(x)e^{ax} = c_1xe^{ax} + c_2e^{ax}.$$

Thus $u(x) = x$ provides a suitable nonconstant factor, and we can pick $y_2(x) = xe^{ax}$ for the other member of a pair of linearly independent solutions by setting the arbitrary constant c_2 equal to zero.

Bearing in mind that there might still be solutions not involving exponentials factors, our experience with the examples at least enables us to hope, as is proved in Theorem 1.4 of Section 1B, that to find a unique solution to an initial-value problem for

$$Ay'' + By' + Cy = 0, \qquad A \neq 0,\ A,\ B,\ C \text{ constant},$$

the thing to do is

1. Solve the associated algebraic equation $Ar^2 + Br + C = 0$, called the **characteristic equation** of the differential equation; its roots r_1 and r_2 are called the **characteristic roots** of the differential equation, and indeed these roots characterize the solutions.
2. If the roots $r = r_1$ and $r = r_2$ are distinct real numbers, then $y = c_1e^{r_1x} + c_2e^{r_2x}$ will be solutions. If r_1 is a double root, then $y = c_1e^{r_1x} + c_2xe^{r_1x}$ will be solutions.
3. We can single out specific solutions by imposing two initial conditions $y(x_0) = y_0$, $y'(x_0) = z_0$ on y to fix the values of c_1 and c_2.

---

example 9 The differential equation

$$y'' - 2y' - y = 0$$

has the characteristic equation $r^2 - 2r - 1 = 0$. The **quadratic formula** for roots of

$$Ar^2 + Br + C = 0, \quad \text{is} \quad r = \frac{-B \pm \sqrt{B^2 - 4AC}}{2A},$$

so the roots of the characteristic equation are

$$r = \frac{2 \pm \sqrt{8}}{2} = \begin{cases} 1 + \sqrt{2}, \\ 1 - \sqrt{2}. \end{cases}$$

The corresponding solutions are $y_1(x) = e^{(1+\sqrt{2})x}$ and $y_2(x) = e^{(1-\sqrt{2})x}$. A family of solutions has the form

$$y = c_1 e^{r_1 x} + c_2 e^{r_2 x}, \qquad r_1 = 1 + \sqrt{2}, \qquad r_2 = 1 - \sqrt{2}.$$

Theorem 1.4 in Section 1B will show that every solution of the differential equation is a linear combination $y(x) = c_1 e^{r_1 x} + c_2 e^{r_2 x}$, where $r_1 = 1+\sqrt{2}, r_2 = 1-\sqrt{2}$, and $c_1, \ c_2$ is a pair of arbitrarily chosen constants. Furthermore, the method of proof for Theorem 1.4 will show that c_1 and c_2 are uniquely determined by initial conditions $y(x_0) = y_0, y'(x_0) = z_0$ at a point x_0.

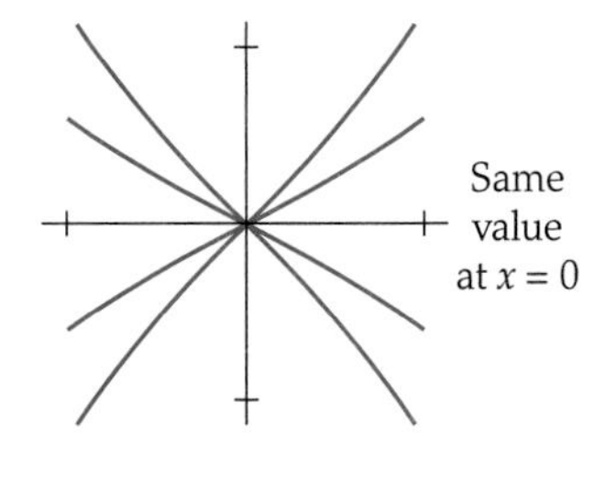

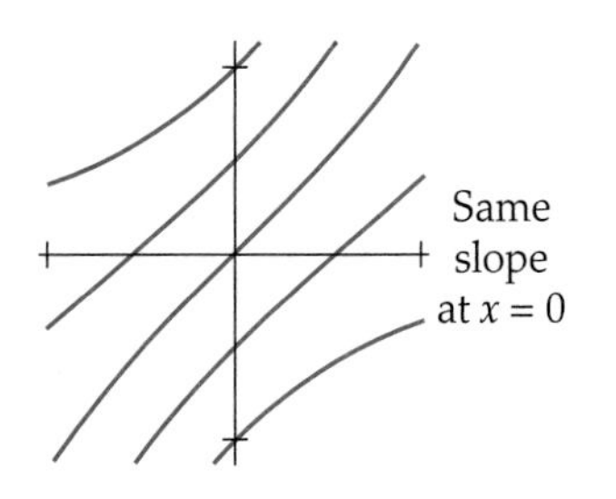

If we think of the constants c_1 and c_2 as both being allowed to vary, then the corresponding linear combinations $y(x) = c_1 y_1(x) + c_2 y_2(x)$ are an example of a **2-parameter family** with **parameters** c_1 and c_2. The initial values $y(x_0) = y_0$ and $y'(x_0) = z_0$ for a solution and its derivative at x_0 provide alternative parameters y_0 and z_0 that are often more meaningful geometrically than c_1 and c_2. If y_0 is held fixed and z_0 varies, we get solution graphs passing through the point (x_0, y_0) with different slopes. On the other hand, with z_0 fixed and y_0 varying, we get solution graphs with parallel slopes z_0 at x_0 but passing through different points when $x = x_0$.

The figures illustrate what happens if you vary just one of the two parameters x_0 and y_0. Note in particular that two uniquely determined solution graphs are not prohibited from crossing, as is the case with solutions of a first-order equation. However, Theorem 1.4 will show that tangency of solution graphs is prohibited, since the theorem asserts that a solution's value and slope of at a point together uniquely determine the solution.

1B General Solutions

If you feel that there's something unusual about multiplying differential operators, remember that in the Leibnitz notation the equation $(d/dx)(dy/dx) = d^2y/dx^2$ does exactly that.

We'll start with some basic properties of the differential operator $D = d/dx$, where D is just a simpler notation that will stand for differentiation with respect to some variable that we'll call x for now. This notation will be very useful here as well as in Sections 2 and 3, and in Chapter 5, Section 2, on systems of differential equations. We interpret $D - 2$, $D^2 - 1$, and similar expressions that we write as polynomials in D, as operations on sufficiently differentiable functions $y = y(x)$. Here are four examples:

(a)
$$(D-2)e^{2x} = De^{2x} - 2e^{2x} = 0,$$

(b)
$$\begin{aligned}(D^2+2)\sin x &= D(D\sin x) + 2\sin x \\ &= D\cos x + 2\sin x = \sin x,\end{aligned}$$

(c)
$$(D - 2)y = Dy - 2y$$
$$= y' - 2y,$$

(d)
$$(D^2 - 1)y = D^2y - y$$
$$= D(Dy) - y = y'' - y.$$

A polynomial $P(x)$ is often given meaning by thinking of letting x be a number. Analogously, a polynomial $P(D)$ is given meaning by applying it to a differentiable function.

One of the fundamental properties of polynomials $P(x)$ in x is that they can in principle be factored if their coefficients are numerical constants; the same is of course true if we replace the letter x by the letter r, the letter D, or any other letter. This fact will lead us to a complete understanding of constant-coefficient linear differential equations once we establish in the next theorem that the factorization of a polynomial operator $P(D)$ acts appropriately on suitably differentiable functions.

1.3 THEOREM

With r and s constant, when $D - r$ and $D - s$ act on twice-differentiable functions, then

$$(D - r)(D - s) = (D - s)(D - r)$$
$$= D^2 - (r + s)D + rs.$$

In particular, if r_1 and r_2 are the characteristic roots of the linear constant-coefficient differential equation $y'' + ay' + by = 0$, then

$$y'' + ay' + by = (D - r_1)(D - r_2)y$$
$$= (D - r_2)(D - r_1)y.$$

Proof. Proving that two operators are equal amounts to proving that they produce the same result when applied to a given sufficiently differentiable function $y(x)$. If r and s are constants,

$$(D - r)(D - s)y = (D - r)(y' - sy)$$
$$= D(y' - sy) - r(y' - sy)$$
$$= y'' - (r + s)y' + (rs)y$$
$$= \left[D^2 - (r + s)D + rs\right]y.$$

Interchanging r and s changes nothing in the way this last operator acts on a function, so we can conclude also that $(D - s)(D - r) = (D - r)(D - s)$, in the sense that these two operators have the same effect on twice-differentiable functions $y(x)$. ∎

We'll see that the ability to factor a constant-coefficient operator enables us to describe all solutions of the related differential equation using successive integration and also to show that an initial-value problem always has a unique solution. So far we have tacitly assumed this in our examples. In particular, we found exponential solutions by assuming there was a solution of the form e^{rx} so we could identify r and show that there was at least one solution. But it's possible that there might be solutions not of exponential form. Indeed, if the characteristic equation has a double root r, we've seen that there will be a solution

xe^{rx}. The next example is of this kind, and following it carefully will be helpful for understanding the general discussion in the proof of Theorem 1.4 on page 168.

---

example 10 We can write the differential equation

$$y'' + 2y' + y = 0 \quad \text{as} \quad (D^2 + 2D + 1)y = 0 \quad \text{or} \quad (D + 1)^2 y = 0.$$

Note that the characteristic equation is

$$r^2 + 2r + 1 = 0 \quad \text{or} \quad (r + 1)^2 = 0,$$

so there is a double root $r = -1$. By assuming an exponential solution, we saw in Section 1A that $y = e^{-x}$ solves the differential equation. Without making such an assumption, suppose that y is an arbitrary solution and let $z = (D+1)y$. Then the equation

$$(D + 1)z = (D + 1)^2 y = 0$$

gives us a first-order equation $(D + 1)z = 0$, or $z' + z = 0$, to be solved for z. Using the exponential integrating factor $M(x) = e^x$, the solutions of this first-order equation must satisfy $(e^x z)' = 0$. Hence $e^x z = c_1$, so all solutions $z(x)$ have the form

$$z = c_1 e^{-x}.$$

Since we set $z = (D + 1)y$, we now know that y must satisfy the first-order equation

$$(D + 1)y = y' + y = c_1 e^{-x}.$$

But this linear equation can be solved by multiplying by the integrating factor $M(x) = e^x$ to get $e^x y' + e^x y = c_1$, or

$$(e^x y)' = c_1.$$

Integrating both sides gives $e^x y = c_1 x + c_2$, so

$$y = c_1 x e^{-x} + c_2 e^{-x},$$

and every solution of the differential equation has the form $y = c_1 x e^{-x} + c_2 e^{-x}$.

By the **general solution** of a differential equation we mean a solution formula such that every solution is a special case of the general solution, different solutions usually depending on a choice of constants. It's worth reviewing the steps of the computation in the previous example to see that we did indeed find the general solution. The key point is that we let $y(x)$ be an arbitrary solution and allowed the differential equation to dictate the form of the solution. Using this approach we can prove the following theorem just by applying the method that we used in the previous example. In the example, we solved a second-order equation by solving succession of two first-order equations, so we refer to this widely applicable method as **order reduction**. (The numerical methods derived in the Chapter 4 computing supplement will use the same principle.) The method is the same regardless of whether the characteristic roots are equal or not, with only a small difference in detail between the two cases. Moving beyond the previous example, the proof by order reduction establishes two things in complete generality:

1. The form of a general solution's dependence on the characteristic roots and arbitrary constants
2. The unique determination of the constants in the general solution by imposing arbitrary initial conditions

1.4 THEOREM

Suppose the characteristic equation $r^2 + ar + b = 0$ of

$$y'' + ay' + by = 0$$

has real roots r_1 and r_2. There are two distinct cases.

(i) If $r_1 \neq r_2$, the general solution is $y = c_1 e^{r_1 x} + c_2 e^{r_2 x}$.
(ii) If $r_1 = r_2$, the general solution is $y = c_1 x e^{r_1 x} + c_2 e^{r_1 x}$.

The constants c_1 and c_2 are uniquely determined by prescribing arbitrary initial values $y(x_0) = y_0$ and $y'(x_0) = z_0$ of the solution and its derivative at a single point x_0.

Proof. In operator form, the differential equation is

$$(D - r_1)(D - r_2)y = 0.$$

We assume $y(x)$ is a solution and show that it has the form claimed above. Set

$$z(x) = (D - r_2)y(x) = y'(x) - r_2 y(x),$$

and substitute z for $(D - r_2)y$ in the previous displayed equation. Using the integrating factor $e^{-r_1 x}$ we solve the resulting equation $(D - r_1)z = 0$ to get $z = c_1 e^{r_1 x}$. Note that c_1 is determined by the initial conditions on y and y' since

$$c_1 = z(x_0)e^{-r_1 x_0} = [y'(x_0) - r_2 y(x_0)]e^{-r_1 x_0}.$$

Given the relation $z(x) = (D - r_2)y(x) = y'(x) - r_2 y(x)$ between y and z, the solution y must satisfy

$$(D - r_2)y = c_1 e^{r_1 x}.$$

Multiplication by $e^{-r_2 x}$ gives

$$D(e^{-r_2 x} y) = c_1 e^{(r_1 - r_2)x}.$$

Unequal Root Case. Assuming $r_1 \neq r_2$, we integrate the previous equation to get

$$e^{-r_2 x} y = \frac{c_1}{r_1 - r_2} e^{(r_1 - r_2)x} + c_2.$$

Now multiply by $e^{r_2 x}$ to get

$$y = \frac{c_1}{r_1 - r_2} e^{r_1 x} + c_2 e^{r_2 x}.$$

For neatness, we can absorb the constant $1/(r_1 - r_2)$ into c_1 to get for case (i)

$$y(x) = c_1 e^{r_1 x} + c_2 e^{r_2 x}. \qquad (*)$$

Equal Root Case. Assuming $r_1 = r_2$ in $D(e^{-r_2 x} y) = c_1 e^{(r_1 - r_2)x}$, we get $D(e^{-r_2 x} y) = c_1$. Integrating both sides gives

$$e^{-r_2 x} y = c_1 x + c_2.$$

Thus for case (ii),

$$y(x) = c_1 x e^{r_2 x} + c_2 e^{r_2 x}. \qquad (**)$$

Finally, note that once c_1 is determined from $y(x_0)$ and $y'(x_0)$ as noted above in the proof, in either case (i) or case (ii) we can determine the constant c_2 from the value $y(x_0)$ alone; just solve the appropriate equation $(*)$ or $(**)$ for c_2 with $x = x_0$. ■

Geometrically, the initial conditions described in Theorem 1.4 require the graph of a solution $y(x)$ to go through a given point (x_0, y_0) with slope $z_0 = y'(x_0)$. It's possible to extract formulas from the proof of Theorem 1.4 for determining coefficients c_1 and c_2 from the numbers x_0, y_0, and z_0, but since the formulas aren't particularly memorable it's usually just as efficient to work more directly, as in the next example.

example 11

Suppose we want the solution of $y'' - 4y = 0$ with graph passing through the point (0,1) with slope 2. In other words, we want to satisfy the initial conditions $y(0) = 1$ and $y'(0) = 2$. Since the characteristic equation of the differential equation is $r^2 - 4 = 0$ and has roots $r = \pm 2$, all solutions have the form

$$y(x) = c_1 e^{2x} + c_2 e^{-2x}.$$

The corresponding derivative formula is

$$y'(x) = 2c_1 e^{2x} - 2c_2 e^{-2x}.$$

Setting $x = 0$ in $y(x)$ and $y'(x)$, we get the two equations

$$y(0) = c_1 + c_2 = 1 \quad \text{and} \quad y'(0) = 2c_1 - 2c_2 = 2.$$

Solving for c_1 and c_2 by elimination gives the unique solution $c_1 = 1,\ c_2 = 0$, so $y(x) = e^{2x}$ is the unique solution passing through (0,1) with slope 2.

example 12

The differential equation in the initial-value problem $y'' + 2y' + y = 0$, $y(0) = 1$, $y'(0) = 0$, was solved in Example 8 with $a = -1$. The differential equation has characteristic equation $r^2 + 2r + 1 = 0$ with double characteristic root $r_1 = r_2 = -1$. Theorem 1.3 says that the general solution is $y(x) = c_1 e^{-x} + c_2 x e^{-x}$. The first of the initial conditions requires that $y(0) = c_1 = 1$. Since $y'(x) = -c_1 e^{-x} + c_2 e^{-x} - c_2 x e^{-x}$, the second initial condition requires $y'(0) = -c_1 + c_2 = 0$. Hence, $c_1 = c_2 = 1$ and the initial-value problem has the unique solution $y = e^{-x} + x e^{-x}$.

example 13

In Example 3 on page 160 we showed that if a chain of length l sliding over a dock without friction has length y hanging over the edge of the dock, then y satisfies $d^2y/dt^2 = (g/l)y$ as long as $y \le l$. Rewriting the equation as $d^2y/dt^2 - (g/l)y = 0$, we see that the characteristic equation is $r^2 - g/l = 0$ with roots $r = \pm\sqrt{g/l}$. The corresponding linear combination of linearly independent solutions is

$$y = c_1 e^{\sqrt{g/l}\,t} + c_2 e^{-\sqrt{g/l}\,t}.$$

To be specific about initial conditions, suppose $y(0) = 0$ and $y'(0) = 1$, which together mean that we're assuming the falling end of the chain starts right at the edge of the dock and is given an initial shove over the edge with velocity 1 unit per second. We use these conditions to determine c_1 and c_2 from the equations

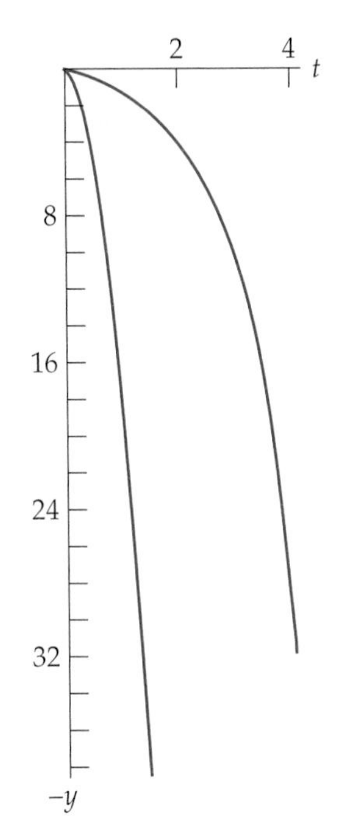

$$c_1 + c_2 = 0 \qquad \sqrt{\frac{g}{l}}\,c_1 - \sqrt{\frac{g}{l}}\,c_2 = 1.$$

From the first equation $c_2 = -c_1$, so the second equation tells us that $2\sqrt{g/l}c_1 = 1$, or $c_1 = \frac{1}{2}\sqrt{l/g}$. Hence the solution to the initial-value problem is

$$y = \frac{1}{2}\sqrt{\frac{l}{g}}\left(e^{\sqrt{g/l}\,t} - e^{-\sqrt{g/l}\,t}\right) = \sqrt{\frac{l}{g}}\,\sinh\sqrt{\frac{g}{l}}\,t,$$

where the hyperbolic sine function is defined by $\sinh x = \frac{1}{2}(e^x - e^{-x})$. Just because the chain is falling, the picture in the margin was drawn showing the graph of $-y(t)$. We assumed $g = 32$ ft./sec.^{2} and $l = 32$ ft. The graph represents the motion of the lower end of the sliding chain while $0 < y < 32$, and the more rapidly traced parabolic free-fall graph governed by $y'' = g$ is on its left.

Boundary Conditions. Suppose that in the previous examples we wanted not just to start a solution off with a given value and a given slope at the same point, but instead wanted a solution graph passing through two different points in the xy plane. Such conditions applied to a single solution at more than one point are called **boundary conditions**. The problem of finding a solution of a differential equation that also satisfies boundary conditions is called a **boundary-value problem.** Boundary-value problems are theoretically more complicated than initial-value problems, and the relevant theorems are deferred to Chapter 9, Section 3. Nevertheless many boundary-value problems are quite simple computationally, as in the next example.

example 14

Our aim in this example is to find a solution to $y'' - y = 0$ with boundary conditions $y(0) = 0$ and $y(\ln 2) = 1$. Geometrically speaking, we're looking for a solution whose graph contains the points (0,0) and (ln 2,1). We need work only with the expression

$$y(x) = c_1 e^x + c_2 e^{-x}$$

for the general solution itself, since values of $y'(x)$ aren't relevant to this problem. The resulting equations for the coefficients are $y(0) = c_1 + c_2 = 0$ and

$y(1) = e^{\ln 2}c_1 + e^{-\ln 2}c_2 = 1$. The second equation simplifies to $2c_1 + \frac{1}{2}c_2 = 1$. Though Theorem 1.4 guarantees a unique solution only for initial-value problems, the method of elimination shows that these equations do have a unique solution, namely, $c_1 = \frac{2}{3}$, $c_2 = -\frac{2}{3}$. The solution of the boundary-value problem is

$$y(x) = \tfrac{2}{3}e^{x} - \tfrac{2}{3}e^{-x} = \tfrac{4}{3}\sinh x.$$

The two previous examples suggest that the distinction between initial conditions and boundary conditions might be only a minor computational one. A more insightful way to think about the difference is as follows. An *initial-value problem* asks a solution to start at a given point with a given direction and then go where the differential equation tells it to go. Under the assumptions of Theorem 1.4, there will always be a uniquely determined graph for the solution of an initial-value problem. A *boundary-value problem* asks a solution to start at a given point and follow a solution graph to another prescribed point, but such a problem may not have a solution. ("You can't get from here to there by train.") Or there may be more than one way to go from some points to certain others by following the dictates of the differential equation. However Exercise 54 shows that second-order homogeneous equations with *real* characteristic roots form an exceptional class in that boundary-value problems specifying values at two points are always uniquely solvable for these equations.

---

example 15

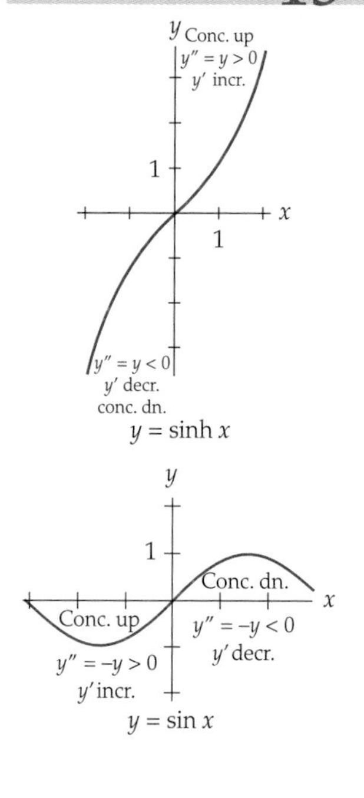

The initial-value problem $y'' = y$, $y(0) = 0$, $y'(0) = 1$, has characteristic equation $r^2 - 1 = 0$ with roots $r = \pm 1$. The unique solution to the problem is $y = \frac{1}{2}(e^x - e^{-x}) = \sinh x$, with the graph of $\sinh x$, shown in the margin. The purpose of this example is to note how the differential equation determines the shape of the graph, concave up or down depending on the sign of y. Note also that the graph of $\sinh x$ crosses the x axis just once.

On the other hand, the differential equation $y'' = -y$ has solutions that are concave down when $y > 0$ and concave up when $y < 0$. In particular, $y = \sin x$ is such a solution, since $y'' = -\sin x = -y$.

A second margin figure shows how $y'' = -y$ determines the shape of the graph of $y = \sin x$, which is profoundly different from the graph of $y = \sinh x$. In particular, the graph of $y = \sin x$ shows *oscillation* about the x axis, that is, the graph crosses the axis repeatedly. (The sine function is even periodic, with period 2π. Section 5 in Chapter 4 shows that differential equations such as $y'' = -y$ or $y'' = -y^3$ will always have periodic solutions with some positive period.) We'll see in Section 2 that $\sin x$ and $\sinh x$ are also profoundly *similar* in another way, and that the notation $\sinh x$ should not be regarded simply as shorthand for something involving exponentials.

EXERCISES

Find the characteristic equation of each of the differential equations in Exercises 1 through 8. Then solve the characteristic equation and use the roots to write a solution of the differential equation as a linear combination of exponentials with arbitrary constant coefficients.

1. $y'' + y' - 6y = 0.$

2. $2y'' - y = 0.$

3. $y'' + 2y' + y = 0.$

4. $y'' + 3y' + y = 0.$

5. $y'' - y' = 0.$

6. $y'' - 3y' - y = 0.$

7. $2y'' - 3y' + y = 0.$

8. $3y'' + 3y' = 0.$

For each differential equation in Exercises 1 through 8, find the particular solution satisfying the corresponding initial or boundary condition in Exercises 9 through 16 given below. Then sketch the graph of that particular solution.

9. $y(0) = 2, \; y'(0) = 2.$

10. $y(0) = 1, \; y'(0) = 0.$

11. $y(0) = 1, \; y'(0) = 2.$

12. $y(1) = 1, \; y'(1) = 1.$

13. $y(0) = 1, \; y(1) = 0.$

14. $y(0) = 0, \; y(1) = 0.$

15. $y(0) = 0, \; y'(0) = 0.$

16. $y(1) = 1, \; y(2) = 2.$

Sketch the graph of each function of x in Exercises 17 through 22. It's sometimes desirable to design a mechanism that produces a specified response for a solution, so first find a differential equation of the form

$$y'' + ay' + by = 0,$$

of which each function is a solution. Next write the general solution of the differential equation and check that the given function is a special case of your general solution. [*Hint:* What characteristic roots go with each solution?]

17. $xe^{-x}.$

18. $e^{x} + e^{-x}.$

19. $1 + x.$

20. $2e^{2x} - 3e^{3x}.$

21. $xe^{-x} - e^{-x}.$

22. $e^{-3x} + e^{5x}.$

With $D = d/dx$ compute the expressions in Exercises 23 through 28.

23. $(D + 1)e^{-2x}.$

24. $(D^2 + 1)e^{x}.$

25. $D^3 e^{3x}.$

26. $(D^2 + D - 1)\sin x.$

27. $(D^2 + 1)x\cos x.$

28. $D^2 e^{x^2}.$

Put each of the differential equations in Exercises 29 through 34 in operator form: $(a\,D^2 + b\,D + c)y = 0$. Then factor the operator, for example, $D^2 - 1 = (D - 1)(D + 1)$.

29. $y'' + 2y' + y = 0.$

30. $y'' - 2y = 0.$

31. $2y'' - y = 0.$

32. $y'' + 3y' = 0.$

33. $y'' = 0.$

34. $y'' - y' = 0.$

Each of the equations in Exercises 35 through 38 can be written in factored operator form:

$$(D - r_1)(D - r_2)y = f(x).$$

In each case, let $(D - r_2)y = z$ and solve $(D - r_1)z = f(x)$ for the most general possible z. Having found z, solve $(D - r_2)y = z$.

35. $D(D - 3)y = 0.$

36. $D^2 y = 1.$

37. $y'' - y = 1.$

38. $y'' + 2y' + y = x.$

39. **(a)** Find the general solution of the differential equation $y'' + 2y' + y = 0$.

(b) Find the solution singled out by the initial conditions $y(0) = 1$, $y'(0) = 2$.

(c) Describe the behavior of the solution found in part (b) as $x \to \infty$ and as $x \to -\infty$.

(d) Sketch the graph of the solution found in part (b) for $-2 \le x \le 2$, and find the coordinates of the maximum point on the graph over that interval.

40. **(a)** Find the general solution of the differential equation $y'' - 2y' + y = 0$.

(b) Find the solution singled out by the initial conditions $y(0) = 0$ and $y'(0) = 1$.

(c) Describe the behavior of the solution found in part (b) as $x \to \infty$ and as $x \to -\infty$.

(d) Sketch the graph of the solution found in part (b) for $-2 \le x \le 2$ and find the coordinates of the minimum point on the graph over that interval.

41. While we're dealing with constant-coefficient equations in this section, the coefficients in a linear differential equation needn't be constant, though factoring the operator isn't usually feasible when the coefficients depend on the independent variable. (See Chapter 9, Section 1, for an alternative.) The differential equation $y'' + (1/x)y' - (1/x^2)y = 0$, $x > 0$, can be written in operator form as $\left[D^2 + (1/x)D - 1/x^2\right]y = 0$.

(a) Show that the equation can also be written $D\big(D + 1/x\big)y = 0$.

(b) Solve the equation in part (a) by letting $z = (D + 1/x)y$ and solving a succession of first-order equations.

(c) Show that $D\big(D + 1/x\big) \neq \big(D + 1/x\big)D$.

(d) Solve $\big(D + 1/x\big)Dy = 0$.

42. Recall that the hyperbolic cosine and hyperbolic sine are defined by

$$\cosh x = \tfrac{1}{2}(e^x + e^{-x}), \qquad \sinh x = \tfrac{1}{2}(e^x - e^{-x}).$$

(a) Show that, if constants d_1 and d_2 are suitably chosen in terms of c_1 and c_2, then

$$c_1 e^{rx} + c_2 e^{-rx} = d_1 \cosh rx + d_2 \sinh rx.$$

(b) Express the general solution of

$$y'' - k^2 y = 0$$

in terms of hyperbolic functions.

43. Show that the hyperbolic cosine and hyperbolic sine functions defined by $\cosh x = \frac{1}{2}\left(e^x + e^{-x}\right)$ and $\sinh x = \frac{1}{2}\left(e^x - e^{-x}\right)$ are linearly independent on every interval $a < x < b$.

44. Show that the derivative of $\cosh x$ is $\sinh x$ and that the derivative of $\sinh x$ is $\cosh x$.

45. (a) Show that the characteristic equation of

$$Ay'' + By' + Cy = 0,$$

with A, B, and C constant, $A \neq 0$, has real roots if and only if $B^2 \geq 4AC$.

(b) Show that when $B^2 > 4AC$, the general solution of the differential equation in part (a) can be written

$$y = e^{\alpha x}(d_1 \cosh\ \beta x + d_2 \sinh\ \beta x),$$

where $\alpha = -B/2A$, $\beta = \sqrt{B^2 - 4AC}/2A$.

46. Assume $|A| < \frac{1}{4}$ in the equation $Ay'' + y' + y = 0$ and show that, as A tends to 0, and with proper choice of arbitrary constants, there are solutions of this equation tending to solutions of $y' + y = 0$ for each fixed x.

47. The differential equation $y'' - 2y' + y = 0$ has infinitely many solutions $y(x)$ with graphs passing through the point (0,1). Find the three that have slopes -1, 0, and 1 at that point and sketch their graphs.

48. Initial conditions $y(x_0) = a_0$ and $y'(x_0) = a_1$, imposed at a single point x_0, will always be satisfied by some solution of $y'' - 3y' + 2y = 0$; show that the conditions $y(\ln 2) = a_0$ and $y'(0) = a_1$, at two different points 0 and $\ln 2$, can be satisfied only if $a_0 = 2a_1$.

Sliding Chain Problems

49. Referring to text Example 3, use integration to show that for $0 < y(t) < l$,

(a) $dy(t)/dt < gt + y'(0)$. **(b)** $y(t) < \frac{1}{2}gt^2 + y'(0)t + y(0)$.

50. Here's how to get information about the sliding chain initial-value problem $y'' = (g/l)y$, $y(0) = y_0$, $y'(0) = z_0$ without solving the differential equation derived in text Example 3 for a function of t, but by instead interpreting the equation in the **state space** with variables y and y'.

(a) Multiply the equation by y' and integrate with respect to t to get $\frac{1}{2}(y')^2 = \frac{1}{2}(g/l)y^2 + C$.

(b) Use the initial conditions to determine C.

(c) Assuming either $y_0 > 0$ or $z_0 > 0$, show that when $y = l$, the speed of the falling chain is $\sqrt{gl + z_0^2 - (g/l)y_0^2}$. Note that $y_0 < l$.

51. A chain of length l with a mass m lies unattached and in a straight line on the deck of a ship in such a way that it will slide with negligible friction.

(a) Assuming the chain runs out over the side with no force acting on it other than gravity at constant acceleration g, explain in your own words why the amount y hanging over the side satisfies $d^2y/dt^2 = (g/l)y$ as long as $0 \le y \le l$. (Assume the deck is more than height l above the water.)

(b) How fast is the chain accelerating as the last link goes over the side?

(c) Find $y(t)$ if $y(0) = y_0$ and $dy/dt(0) = v_0$.

(d) Show that if the chain starts from rest with length $y_0 > 0$ hanging over the side, then the last link goes over the side at time $t_1 = \sqrt{l/g}\,\ln\left[(l + \sqrt{l^2 - y_0^2})/y_0\right]$.

52. Assume that the equation $y'' + ay' + by = 0$ has real characteristic roots. Show that a solution graph can cross the horizontal axis at most once. Consider the case of a double root separately.

53. Unstable behavior. The behavior of solutions of a differential equation can sometimes be so sensitive to changes in the initial conditions that their slightest alteration leads to drastically different behavior of the solutions; as a result, small errors in physical input data can produce very bad long-term predictions. Here is an example.

(a) Solve the differential equation $y'' - y = 0$ subject to the initial conditions $y(0) = 1,\ y'(0) = \alpha$, where α is a constant.

(b) Show that the solution $y(x)$ determined in part (a) by the choice $\alpha = -1$ satisfies $\lim_{x\to\infty} y(x) = 0$, but that any other choice for the real number α leads to $\lim_{x\to\infty} y(x) = \pm\infty$.

(c) Sketch the graphs of the solutions in part (a) for $\alpha = -1.1,\ \alpha = -1$, and $\alpha = -0.9$ in such a way as to display their differing behaviors.

54. Let $r(x)$ and $s(x)$ be differentiable functions defined on an interval $a < x < b$. Show that $\big(D - r(x)\big)\big(D - s(x)\big) = \big(D - s(x)\big)\big(D - r(x)\big)$ precisely when the functions $r(x)$ and $s(x)$ differ by a constant on the interval. [*Hint:* Show that two second-order operators are equal by showing that they give the same result when applied to an arbitrary twice differentiable function.]

55. In text Example 8 we arrived at the modified exponential solution xe^{ax} when the characteristic equation of $y'' - 2ay' + a^2y = 0$ has a as a double root. Carry out the details of the differentiation in finding a pair of independent solutions of the form $y = e^{ax}u(x)$ by substitution into the differential equation. (This technique is used in an essential way in Chapter 9 for dealing with homogeneous linear equations with nonconstant coefficients.)

56. This exercise gives a geometric clue as to why the factor x occurs in the equal-root case.

(a) Show that $(D-r)[D-(r+h)]y=0$, or $[D^2-(2r+h)D+r(r+h)]y=0$, can also be written $y''-(2r+h)y'+r(r+h)y=0$.

(b) Show that if $h\neq 0$ the general solution is $y_h=c_1e^{(r+h)x}+c_2e^{rx}$.

(c) Let $c=1/h$, $c_2=-1/h$, and show that for these choices $\lim_{h\to 0} y_h(x)=xe^{rx}$ for all x.

(d) Show that the limit in part (c) is, by definition, the derivative of the solution e^{rx} with respect to r.

***57.** Assume that the characteristic roots of the constant-coefficient differential equation $y''+ay'+by=0$ are real numbers r_1 and r_2. Show that if $x_1\neq x_2$, the *boundary-value problem* $y''+ay'+by=0,\ \ y(x_1)=y_1,\ \ y(x_2)=y_2$ always has a unique solution for given numbers y_1 and y_2, that is, show that you can always solve for the desired constants c_1 and c_2 in the general solution. Consider the cases of unequal and double characteristic roots separately.

Text Example 15 shows how a second-order equation $y''=f(y)$ can be interpreted to give information about the concavity of its solution's graphs. Without solving the equations in Exercises 58 through 61, make sketches that show the concavity of a solution satisfying the initial conditions $y(0)=0$ and $y'(0)=1$. Precise solution values aren't required.

58. $y''=y^2$.

59. $y''=y-y^2$.

60. $y''=\sin y$.

61. $y''=1-y^3$.

2 OSCILLATORY SOLUTIONS

Our prototypical application for the differential equations of Section 1 was to model the response of a chain hanging over the edge of a dock to the force of gravity. Next we'll model a different response to gravity, namely, that of a pendulum. In our example, we have a falling body whose motion is constrained in a different way, with very different results.

example 1

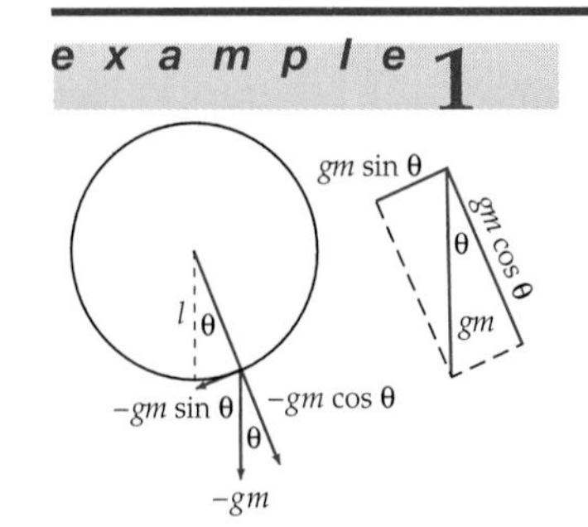

The motion of a freely swinging pendulum with no outside force but gravity acting on it is best described in terms of the angle $\theta=\theta(t)$ that the pendulum makes with a vertical line. We'll consider an **ideal pendulum** with a rod of negligible mass attaching a weight of mass m to the pivot, and with all mass concentrated at the weight's center of gravity at distance l from the pivot. The first margin figure shows the setup, where the possible positions for the center of mass describe a circle of radius l. The expected motion is an **oscillation,** which is a repeated switching back and forth from side to side of a fixed reference, in this case a vertical line.

The second figure is an enlarged analysis of the downward-directed gravitational force of magnitude mg into nonnegative components parallel and perpendicular to the pendulum rod. At any given position the gravitational force component in the direction of motion is perpendicular to the component directed along the length of the pendulum. It follows by the Pythagorean relation that the the sum of the squares of these perpendicular component magnitudes must be g^2m^2. The coordinates relative to these directions are $F = -gm \sin\theta$ and $G = -gm \cos\theta$. The force G acting along the length of the pendulum toward its end must be exactly balanced by an opposite force at the pivot, and these forces play no other role in the motion. If θ is measured in radians, distance along the circular path of motion is $y = l\theta$, so we can express the force coordinate in the direction of motion as mass times acceleration: $F = md^2(l\theta)/dt^2$. Equating our two expressions for force coordinate in the direction of motion gives the differential equation satisfied by $\theta = \theta(t)$:

$$m\frac{d^2(l\theta)}{dt^2} = -gm \sin\theta.$$

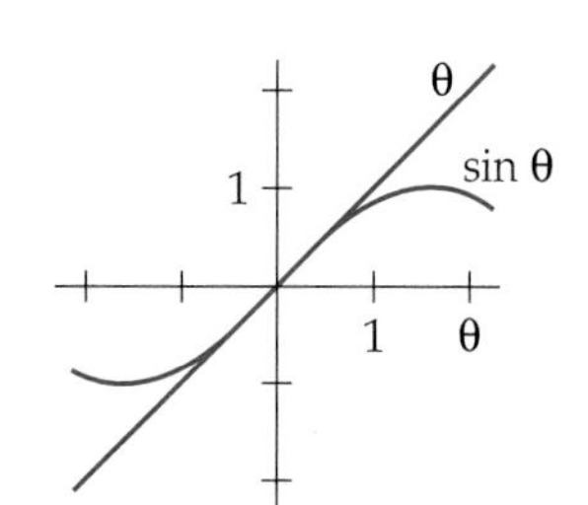

The minus sign automatically signifies that the signed velocity $d(l\theta)/dt$ is decreasing if $0 < \theta < \pi$ and increasing if $-\pi < \theta < 0$. Division by the constant lm gives the alternative form

$$\frac{d^2\theta}{dt^2} = -\frac{g}{l}\sin\theta.$$

This differential equation for $\theta(t)$ is nonlinear and has no solutions in terms of elementary functions. If θ remains small, say $|\theta| < 0.1$ radians, we may find approximate solutions that are acceptable for some purposes by using the tangent-line replacement for the graph of $\sin\theta$ near $\theta = 0$, namely, $\sin\theta \approx \theta$. This approximation, shown in the margin, leads to the linear equation

The solutions of the linearized pendulum equation are examples of harmonic oscillation, studied in detail in Section 2C. The solutions of the nonlinear pendulum equation are oscillatory also, but have their own distinct character, studied in Chapter 4, Section 3.

$$\frac{d^2\theta}{dt^2} = -\frac{g}{l}\theta.$$

It's easy to check that this differential equation has among its solutions

$$\theta = \cos\sqrt{\frac{g}{l}}\,t \quad \text{and} \quad \theta = \sin\sqrt{\frac{g}{l}}\,t.$$

See Figure IS.6 in the Introductory Survey for a sample solution graph.

Detailed study of the *nonlinear* pendulum equation appears in Section 4B of this chapter, and in Chapter 4, Section 3. More on the linearized equation is in Chapter 4, Section 2.

The process of consistently replacing nonlinear terms in an equation by linear estimates, is called **linearization** of the equation, and may or may not be acceptable in practice, depending on the demands for accuracy in the specific problem at hand. In the case of relatively small pendulum oscillations, it turns out that we can get a good idea of qualitative pendulum behavior from the linearized equation. The angle θ is related to distance y along the circular path of the moving end of the pendulum by $\theta = y/l$, so in terms of y the nonlinear and linearized pendulum equations are respectively

$$\frac{d^2y}{dt^2} = -g \sin \frac{y}{l} \quad \text{and} \quad \frac{d^2y}{dt^2} = -\frac{g}{l} y.$$

It is the linear equation on the right that we'll solve in the present section, along with others with complex characteristic roots.

2A Complex Exponentials

Leonhard Euler discovered the imaginative geometric and algebraic links that bridge the gap between the exponential and trigonometric functions, and that enable us now to unify our approach to solving the differential equations $y'' = y$ and $y'' = -y$. Euler's initial idea was this: Replace x by ix and use $i^2 = -1$ in the exponential power series

$$e^x = 1 + x + \frac{x^2}{2!} + \frac{x^3}{3!} + \frac{x^4}{4!} x^4 + \frac{x^5}{5!} \cdots \text{ to get}$$
$$e^{ix} = 1 + ix - \frac{x^2}{2!} - \frac{ix^3}{3!} x^3 + \frac{x^4}{4!} + \frac{ix^5}{5!} \cdots.$$

Appendix A reviews the geometry and arithmetic of complex numbers.

Separating this last infinite series into real and imaginary parts gives

$$e^{ix} = \left(1 - \frac{1}{2!}x^2 + \frac{1}{4!}x^4 - \cdots\right) + i\left(x - \frac{1}{3!}x^3 + \frac{1}{5!}x^5 - \cdots\right)$$
$$= \cos x + i \sin x.$$

We'll now explain the meaning and use of the resulting formula $e^{ix} = \cos x + i \sin x$.

The computations with real exponentials in Section 1 of this chapter are all valid under the more general assumption that the roots of the characteristic equation

$$r^2 + ar + b = 0 \quad \text{associated with} \quad y'' + ay' + by = 0$$

are nonreal complex numbers. To justify the computations in the new context we need a **complex exponential** function, denoted by e^{ix}, and *defined* for real numbers x by **Euler's formula:**

$$e^{ix} = \cos x + i \sin x.$$

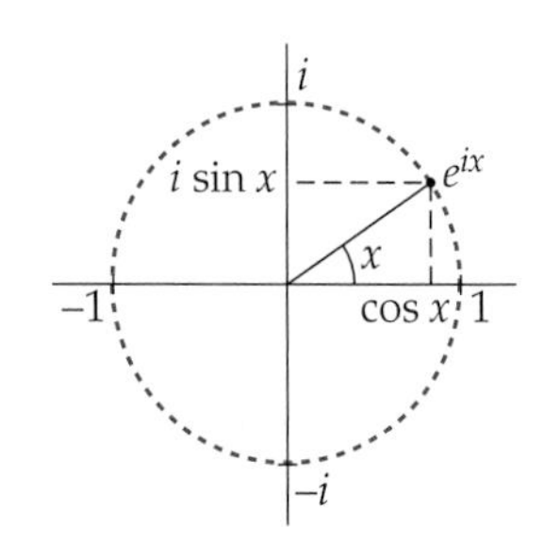

Figure 3.2 $e^{ix} = \cos x + i \sin x$.

Figure 3.2 in the margin plots the complex number e^{ix} in the complex plane by its real and imaginary parts $\cos x$ and $\sin x$. In the picture, the real number x denotes an angle measured in radians. In addition to the exponential series above, our justification for this definition of the notation e^{ix} really comes from its nice properties, properties that mimic the properties of the real exponential e^x and are established next as Equations 2.1 through 2.5.

The numbered formulas 2.1, 2.2, and 2.3, which follow, are the basic properties of e^{ix} that we use routinely. We begin with the **absolute value,** recalling that $|\alpha + i\beta| = \sqrt{\alpha^2 + \beta^2}$. Hence

2.1 $$|e^{ix}| = |\cos x + i \sin x| = \sqrt{\cos^2 x + \sin^2 x} = 1.$$

Using the addition formulas for sine and cosine shows that

$$\begin{aligned} e^{ix}e^{iz} &= (\cos x + i \sin x)(\cos z + i \sin z) \\ &= (\cos x \cos z - \sin x \sin z) + i(\cos x \sin z + \sin x \cos z) \\ &= \cos(x + z) + i \sin(x + z) = e^{i(x+z)}. \end{aligned}$$

It follows that e^{ix} has one of the characteristic properties that e^x has:

2.2 $$e^{ix}e^{iz} = e^{i(x+z)}.$$

In particular, when $z = -x$, we get $e^{ix}e^{-ix} = e^0 = 1$, so that

2.3 $$\frac{1}{e^{ix}} = e^{-ix}; \quad \text{that is,} \quad \frac{1}{\cos x + i \sin x} = \cos x - i \sin x.$$

Equations 2.2 and 2.3, aside from their usefulness, serve to justify the use of the exponential notation; the function e^{ix} behaves algebraically somewhat like the real-valued exponential e^x, for which $e^x e^z = e^{x+z}$ and $(e^x)^{-1} = e^{-x}$.

---

example 2

The function e^{ix} is in general complex valued. Since e^{ix} lies on a circle of radius 1 centered at zero in the complex plane, Figure 3.2 shows that 1 and -1 are the only real values e^{ix} assumes. In particular, if k is an integer,

$$e^{ik\pi} = \cos k\pi + i \sin k\pi = (-1)^k = \begin{cases} 1, & k \text{ even}, \\ -1, & k \text{ odd}. \end{cases}$$

The real-valued functions $\cos x$ and $\sin x$ can be represented in terms of e^{ix} as follows:

$$\begin{aligned} \frac{1}{2}(e^{ix} + e^{-ix}) &= \frac{1}{2}(\cos x + i \sin x + \cos x - i \sin x) = \cos x, \\ \frac{1}{2i}(e^{ix} - e^{-ix}) &= \frac{1}{2i}(\cos x + i \sin x - \cos x + i \sin x) = \sin x. \end{aligned}$$

It's interesting to compare these formulas with the definitions of $\sinh x$ and $\cosh x$ in Section 1.

Using complex exponentials may seem like a roundabout way to do trigonometry, but it turns out that Equation 2.2 is often a very natural and convenient substitute for the addition formulas for cosine and sine that we used to prove it. The complex exponential notation is useful not only because of the simple formulas described previously, but because it works naturally with differentiation and integration. To differentiate or integrate a complex-valued function $u(x)+iv(x)$, we have only to differentiate or integrate the real and imaginary parts. Here are the precise definitions:

$$\frac{d}{dx}[u(x)+iv(x)] = \frac{du}{dx}(x) + i\frac{dv}{dx}(x),$$

$$\int [u(x)+iv(x)]\,dx = \int u(x)\,dx + i\int v(x)\,dx.$$

Accordingly, we have

$$\begin{aligned}\frac{d}{dx}e^{ix} &= \frac{d}{dx}(\cos x + i\sin x)\\ &= -\sin x + i\cos x = i(\cos x + i\sin x) = ie^{ix}.\end{aligned}$$

The end result is

$$\frac{d}{dx}e^{ix} = ie^{ix},$$

just as if i were for a real constant. Similarly,

$$\int e^{ix}\,dx = \frac{1}{i}e^{ix} + C,$$

where C is an arbitrary real or complex constant. More generally, we define

$$e^{(\alpha+i\beta)x} = e^{\alpha x}e^{i\beta x}$$

and compute

2.4 $$\frac{d}{dx}e^{(\alpha+i\beta)x} = (\alpha+i\beta)e^{(\alpha+i\beta)x}$$

and

2.5 $$\int e^{(\alpha+i\beta)x}\,dx = \frac{1}{(\alpha+i\beta)}e^{(\alpha+i\beta)x} + C, \quad C \text{ real or complex},$$

assuming α and β are not both zero in Equation 2.5. These computations are left as exercises; they are simple but it's important to understand them by working through them yourself.

2B Complex Characteristic Roots

Given the facts about complex exponentials just described, the derivation of the general solution of

$$y'' + ay' + by = 0,$$

given in Section 1, proceeds word for word the same if the roots of the characteristic equation are complex. The only thing that remains to be done then is to interpret the solution when the coefficients a and b are real numbers. Rather than go through the identical general computation again, we'll illustrate it with an example and then state the general result.

example 3

The differential equation $y'' + y = 0$ in operator form is $(D^2 + 1)y = 0$, and in factored form this is

$$(D - i)(D + i)y = 0.$$

If $y(x)$ is some solution, let

$$(D + i)y = z.$$

We then need to solve

$$(D - i)z = 0.$$

As in the real case treated in Section 1, we multiply by an exponential factor designed to make the left side the derivative of a product. The correct factor is

$$e^{\int -i\,dx} = e^{-ix},$$

and we observe that by the product rule $e^{-ix}(D - i)z = D(e^{-ix}z)$. To solve $D(e^{-ix}z) = 0$, all we have to do is integrate with respect to x and then multiply by e^{ix}. Since $e^{ix}e^{-ix} = 1$, we get

$$e^{-ix}z = c_1, \quad \text{so} \quad z = c_1 e^{ix}.$$

Substitution of this result into the equation for y gives

$$(D + i)y = c_1 e^{ix}.$$

This time the multiplier is e^{ix}, and we find

$$D(e^{ix}y) = c_1 e^{2ix}.$$

Since the integral of e^{2ix} is $(1/2i)e^{2ix}$, after integration we have

$$e^{ix}y = \frac{1}{2i}c_1 e^{2ix} + c_2.$$

Now multiply by e^{-ix} to get

$$y = c_1 e^{ix} + c_2 e^{-ix},$$

where we have absorbed the factor $1/(2i)$ into the constant c_1. The roots of the characteristic equation $r^2 + 1 = 0$ are $\pm i$, so the same rule for writing down the solution works here as it does in the case of real roots.

It remains only to extract the real-valued solutions from this general complex-valued solution. This is done simply by using the definition of $e^{\pm ix}$ in terms of $\cos x$ and $\sin x$:

$$\begin{aligned}
y &= c_1 e^{ix} + c_2 e^{-ix} \\
&= c_1(\cos x + i \sin x) + c_2(\cos x - i \sin x) \\
&= (c_1 + c_2)\cos x + i(c_1 - c_2)\sin x \\
&= d_1 \cos x + d_2 \sin x.
\end{aligned}$$

To simplify the formula we have set $d_1 = c_1 + c_2$ and $d_2 = i(c_1 - c_2)$. The notational change doesn't involve a loss of generality in the constants, because we can always get the original c's back in terms of the d's. Indeed it's easy to check that

$$c_1 = \tfrac{1}{2}(d_1 - id_2) \quad \text{and} \quad c_2 = \tfrac{1}{2}(d_1 + id_2).$$

Note that if the d's are to be real numbers, which is the important case in most applications, then the c's will in general be complex.

example 4 Continuing from the previous example, we find a formula for the solution satisfying initial conditions $y(0) = 1$ and $y'(0) = 2$. We compute

$$y(x) = d_1 \cos x + d_2 \sin x, \quad \text{so} \quad y(0) = d_1 = 1,$$
$$y'(x) = -d_1 \sin x + d_2 \cos x, \quad \text{so} \quad y'(0) = d_2 = 2.$$

Thus the required solution is

$$y = \cos x + 2 \sin x.$$

Whenever the equation $y'' + ay' + by = 0$ has coefficients that are real numbers, the characteristic roots will be complex if $a^2 - 4b < 0$. This follows from the quadratic formula

$$r = \frac{-a \pm \sqrt{a^2 - 4b}}{2}, \quad \text{so} \quad r = -\frac{a}{2} \pm \frac{i}{2}\sqrt{4b - a^2} \quad \text{when} \quad a^2 - 4b < 0.$$

Complex numbers of the form $\alpha + i\beta$, $\alpha - i\beta$, β, β real, $\beta \neq 0$, are said to be **conjugate** to each other, and if characteristic roots of $y'' + ay' + by = 0$ are complex conjugate, then we can write solutions in the form

$$y = c_1 e^{(\alpha + i\beta)x} + c_2 e^{(\alpha - i\beta)x}, \qquad \alpha, \beta \text{ real}.$$

We can always rewrite this formula as follows:

$$\begin{aligned} y &= e^{\alpha x}(c_1 e^{i\beta x} + c_2 e^{-i\beta x}) \\ &= e^{\alpha x}[c_1(\cos \beta x + i \sin \beta x) + c_2(\cos \beta x - i \sin \beta x)] \\ &= e^{\alpha x}[(c_1 + c_2) \cos \beta x + i(c_1 - c_2) \sin \beta x] \\ &= d_1 e^{\alpha x} \cos\ \beta x + d_2 e^{\alpha x} \sin \beta x. \end{aligned}$$

Recall that a function $f(x)$ has period $p \neq 0$ if $f(x + p) = f(x)$ for all x. Section 2C reviews periodicity and discusses related matters.

The special importance of this last form of the solution is that the basic solutions $e^{\alpha x} \cos\ \beta x$ and $e^{\alpha x} \sin \beta x$ are products of the exponentials and periodic functions of period $p = 2\pi/\beta$.

The proof of Theorem 2.6 is formally the same as that of Theorem 1.3 in the previous section, so we collect a dividend from our development of the complex exponential in that we don't have to repeat the main part of the proof. The only difference here is that we can interpret the solutions as being complex-valued, though of course real-valued solutions are included as special cases as we just saw.

2.6 THEOREM

The differential equation

$$y'' + ay' + by = 0, \qquad a, b \text{ constant},$$

has for its general solution a linear combination

$$y = c_1 e^{r_1 x} + c_2 e^{r_2 x}, \qquad r_1 \neq r_2,$$

$$y = c_1 x e^{r_1 x} + c_2 e^{r_1 x}, \qquad r_1 = r_2,$$

where r_1 and r_2 are the roots of $r^2 + ar + b = 0$. Assuming a and b are real with $a^2 - 4b < 0$, we can write $r_1 = \alpha + i\beta$ and $r_2 = \alpha - i\beta$, with $\beta \neq 0$. The general solution can then be written

$$y = c_1 e^{\alpha x} \cos \beta x + c_2 e^{\alpha x} \sin \beta x.$$

Initial conditions $y(x_0) = y_0$ and $y'(x_0) = z_0$ can always be satisfied by a unique choice of c_1 and c_2.

Classification of Solutions. The solutions of the differential equation

$$Ay'' + By' + Cy = 0, \quad A \neq 0,$$

can be classified according to the sign and relative size of the constants A, B, and C. The characteristic equation is

$$Ar^2 + Br + C = 0$$

with roots

$$r_1, r_2 = \frac{-B \pm \sqrt{B^2 - 4AC}}{2A}.$$

1. If $B^2 - 4AC > 0$, the roots are real and unequal, so the solutions are

$$y = c_1 e^{r_1 x} + c_2 e^{r_2 x}.$$

2. If $B^2 - 4AC = 0$, the roots are equal and real, so solutions are all of the form

$$y = c_1 x e^{r_1 x} + c_2 e^{r_1 x}.$$

3. If $B^2 - 4AC < 0$, the solutions have the form

$$y = c_1 e^{\alpha x} \cos \beta x + c_2 e^{\alpha x} \sin \beta x$$

where $\alpha = -B/2A$ and $\beta = \sqrt{4AC - B^2}/2A$.

The case $B^2 - 4AC = 0$ is critical in the sense that it represents the division between the real exponential solutions covered in Section 1 ($B^2 - 4AC > 0$) and solutions involving sine and cosine ($B^2 - 4AC < 0$). Sections 2, 3, and 4 of Chapter 4 develop the geometric and physical significance of this classification in detail.

example 5

The differential equation $y'' + 2y' + 5y = 0$ has characteristic equation $r^2 + 2r + 5 = 0$ with roots $r_1 = -1 + 2i, r_2 = -1 - 2i$. The real forms of the general solution and its derivative are

$$\begin{aligned} y(x) &= c_1 e^{-x} \cos 2x + c_2 e^{-x} \sin 2x, \\ y'(x) &= c_1(-e^{-x} \cos 2x - 2e^{-x} \sin 2x) + c_2(-e^{-x} \sin 2x + 2e^{-x} \cos 2x) \\ &= (-c_1 + 2c_2)e^{-x} \cos 2x + (-2c_1 - c_2)e^{-x} \sin 2x. \end{aligned}$$

Initial conditions $y(0) = 1$ and $y'(0) = -1$ require that $y(0) = c_1 = 1$ and $y'(0) = -c_1 + 2c_2 = -1$ with unique solution $c_1 = 1$ and $c_2 = 0$. Hence the unique solution satisfying the initial conditions is $y = e^{-x} \cos 2x$. The graph is shown in Figure 3.3 and it's helpful to think about the action of the two factors e^{-x} and $\cos 2x$ on each other. The cosine factor by itself lies between -1 and 1, and oscillates with period π, since $\cos 2(x + \pi) = \cos 2x$ for all x. The decreasing exponential factor squashes the cosine factor between the two exponential graphs of $y = \pm e^{-x}$ that lie above and below it.

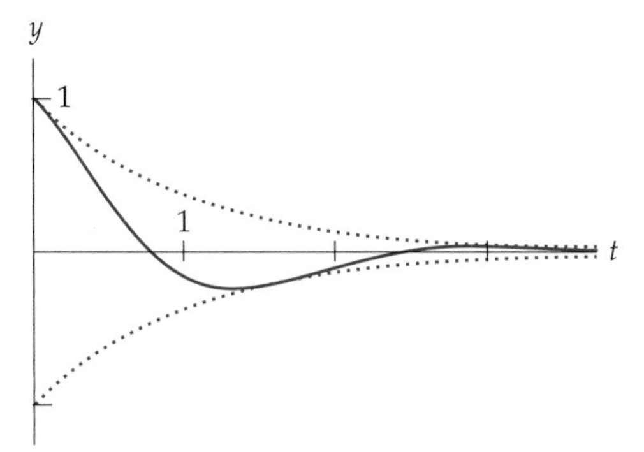

Figure 3.3 $y = e^{-x} \cos 2x$.

example 6

The linearized pendulum equation $d^2\theta/dt^2 + (g/l)\theta = 0$ has characteristic equation $r^2 + g/l = 0$ with roots $r = \pm i\sqrt{g/l}$. The solutions are linear combinations of the basic oscillations $\cos \sqrt{g/l}\, t$ and $\sin \sqrt{g/l}\, t$:

$$\theta(t) = c_1 \cos \sqrt{\frac{g}{l}}\, t + c_2 \sin \sqrt{\frac{g}{l}}\, t.$$

If the pendulum's motion is started by holding it at angle θ_0, $0 < \theta_0 < \pi$, and then just letting it go with initial velocity zero, the initial conditions take the form $\theta(0) = \theta_0$, $d\theta/dt(0) = 0$. Since

$$\frac{d\theta(t)}{dt} = -c_1\sqrt{\frac{g}{l}} \sin \sqrt{\frac{g}{l}}\, t + c_2\sqrt{\frac{g}{l}} \cos \sqrt{\frac{g}{l}}\, t,$$

applying these conditions to our solution $\theta(t)$ gives

$$c_1 = \theta_0 \quad \text{and} \quad c_2\sqrt{\frac{g}{l}} = 0.$$

Hence $\theta(t) = \theta_0 \cos \sqrt{g/l}\, t$, while $(d\theta/dt)(t) = -\theta_0\left(\sqrt{g/l}\right) \sin \sqrt{g/l}\, t$. These two periodic formulas describe the future states in terms of the corresponding

Examples 5 and 6 on pages 226 and 227 discuss a correction factor used for larger values of θ_0 in the linearized pendulum equation. The effect of friction on the linearized equation is taken up in Chapter 4, Section 1.

initial states $\theta(0) = \theta_0$, $(d\theta/dt)(0) = 0$, and they enable us to answer many questions. For example, the time t_1 it takes for the pendulum's initial fall to the vertical position $\theta(t_1) = 0$ satisfies $\theta_0 \cos \sqrt{g/l}\, t_1 = 0$, so $\sqrt{g/l}\, t_1 = \pi/2$. Thus $t_1 = (\pi/2)\sqrt{l/g}$. Note that this formula predicts that t_1 doesn't depend on the initial angle θ_0, but a moment's thought tells us this is unlikely to be correct: the formula asserts that the time it takes for the pendulum to fall from a high initial position is the same as from a lower one. Recall, however, that we've been working with the linearized pendulum equation that's reasonably accurate only for small values of θ_0.

EXERCISES

Show that each of the complex numbers 1 through 4 has absolute value 1. Then find a real number x such that the complex number can be written in the form $\cos x + i \sin x = e^{ix}$; for example

$$\frac{\sqrt{3}+i}{2} = \cos\frac{\pi}{6} + i\sin\frac{\pi}{6} = e^{i\pi/6}.$$

1. i.

2. $(1+i)/\sqrt{2}$.

3. $(1-i)/\sqrt{2}$.

4. $(\sqrt{3}-i)/2$.

Given a pair d_1, d_2 of real numbers, find complex numbers c_1, c_2 such that

$$c_1 e^{(\alpha+i\beta)x} + c_2 e^{(\alpha-i\beta)x} = e^{\alpha x}(d_1 \cos\beta x + d_2 \sin\beta x).$$

5. $d_1 = 1,\ d_2 = 0$.

6. $d_1 = 4,\ d_2 = -2$.

7. $d_1 = 0,\ d_2 = \pi$.

8. $d_1 = 1,\ d_2 = 1$.

9. Recall that a function is periodic, with period p, if $f(x+p) = f(x)$ for all x in the domain of f,

(a) Show that e^{ix} has period 2π.

(b) Show that $e^{i\beta x}$ is periodic for β real, and find the smallest positive period in terms of $\beta \neq 0$.

10. Show that $e^{-i\beta x} = \cos\beta x - i\sin\beta x$. What properties of cos and sin are used here?

11. **(a)** Verify Equation 2.4.

(b) Verify Equation 2.5.

Solve each of the differential equations in Exercises 12 through 15 by factoring the differential operator associated with it and then successively solving a pair of first-order linear equations.

12. $y'' + y = 1$.

13. $y'' + 2y' + 2y = 0$.

14. $y'' + 2y = 0$.

15. $y'' + y' = x$.

16. Find all real or complex solutions of $y'' + iy' = 0$.

Find the roots of the characteristic equation of each of the differential equations in Exercises 17 through 24. Then write the general solution of the differential equation, replacing complex exponentials by $e^{\alpha x}\cos\beta x$ and $e^{\alpha x}\sin\beta x$ where it's appropriate.

17. $y'' + 2y = 0$.

18. $2y'' + 3y' = 0$.

19. $y'' - 2y' + 2y = 0$.

20. $y'' - y' + y = 0$.

21. $2y'' + y' - y = 0$.

22. $y'' + y' = 2y$.

23. $2y'' + y' + y = 0$.

24. $3y'' - y' + y = 0$.

For each of the differential equations in Exercises 17 through 24, find the solution satisfying the corresponding initial-value or boundary-value conditions of Exercises 25 through 32; do this by solving a pair of linear equations for the unknown constants.

25. $y(0) = 0,\ y'(0) = 1$.

26. $y(0) = 1,\ y'(0) = 0$.

27. $y(\pi) = 0,\ y'(\pi) = 0$.

28. $y(0) = 2,\ y'(0) = -1$.

29. $y(0) = 0,\ y'(0) = 2$.

30. $y(0) = 0,\ y(1) = 1$.

31. $y(0) = 0,\ y'(0) = 0$.

32. $y(0) = 0,\ y(1) = 0$.

33. The differential equation $y'' + y = 0$ may fail to have a unique solution if we require the solution to satisfy certain critically chosen boundary conditions. [Exercise 57 in Section 1 shows that if the constant-coefficient equation $y'' + ay' + by = 0$ has real characteristic roots, then the associated boundary-value problem $y(x_1) = y_1,\ y(x_2) = y_2$ always has a unique solution.]

(a) Show that the boundary-value problem $y'' + y = 0,\ y(0) = 1,\ y(\pi) = 1$ has no solution. [*Hint:* You know all solutions of $y'' + y = 0$.]

(b) Show that the boundary-value problem $y'' + y = 0, y(0) = 1,\ y(2\pi) = 1$ has infinitely many solutions.

34. (a) Show that $(c_1\cos\beta x + c_2\sin\beta x)$ can be written in the form $A\cos(\beta x - \phi)$, where $A = \sqrt{c_1^2 + c_2^2}$ if ϕ satisfies $\cos\phi = c_1/A$ and $\sin\ \phi = c_2/A$.

(b) The result of part (a) is useful because it shows that

$$c_1\cos\beta x + c_2\sin\beta x$$

has a graph that is the same as that of $\cos\beta x$ shifted by a **phase angle** ϕ and multiplied by an **amplitude** A. Sketch the graph of

$$\cos 2x + \sqrt{3}\sin 2x$$

by first finding ϕ and A.

35. (a) Show that $y(x) = c_1\cos\beta x + c_2\sin\beta x$ can be written in the form $A\sin(\beta x + \theta)$ where $A = \sqrt{c_1^2 + c_2^2}$ is the amplitude of $y(x)$ if θ satisfies $\sin\theta = c_1/A$ and $\cos\theta = c_2/A$. (This is an alternative to the more usual transformation discussed in the previous exercise and in Section 2D that uses cosine instead of sine as the basic reference function.)

(b) The result of part (a) says that

$$c_1 \cos \beta x + c_2 \sin \beta x = A \sin(\beta x + \theta)$$

for appropriately chosen real numbers A and θ. Use this result to sketch the graph of $y(x) = \cos 2x + \sqrt{3} \sin 2x$, by first finding A and θ.

(c) Show that

$$A \sin(\beta x + \theta) = A \cos(\beta x - \phi),$$

where $\phi = \pi/2 - \theta$. The number ϕ (or sometimes $-\phi$) is called a **phase shift,** and A is the amplitude of the trigonometric function $A \sin(\beta x + \theta)$.

36. For us, Euler's formula is really a definition of the notation e^{ix}, so we're not in a position to prove it. However, the power series computation we used to motivate the definition is valid if you take the trouble to clarify the meaning of convergence for complex series. Break the full infinite series

$$\sum_{k=1}^{\infty} \frac{(ix)^k}{k!}$$

into real and imaginary parts, and identify these parts with the corresponding series expansions for $\cos x$ and $\sin x$.

Differentiation rules for complex-valued functions $f(x) = u(x) + iv(x)$ and $g(x) = s(x) + it(x)$ obey the same basic rules relative to differentiation that real-valued functions do. Use the corresponding relations for real-valued functions to show that the following formulas hold on an interval $a < x < b$ on which both f and g are differentiable complex-valued functions.

37. $(f + g)' = f' + g'$.

38. $(cf)' = cf'$, c constant.

39. $(fg)' = fg' + f'g$.

40. $(f/g)' = (f'g - fg')/g^2$, $g \neq 0$.

41. The nonlinear pendulum equation $d^2\theta/dt^2 = -(g/l) \sin \theta$ has the constant solution $\theta = \pi$.

(a) Explain the physical significance of this solution, in particular the difficulty of exhibiting it in practice.

(b) What other constant solution does this equation have, and what is their physical meaning?

42. **(a)** Solve the linear initial-value problem $d^2\theta/dt^2 = -(g/l)\theta$, $\theta(0) = \theta_0$, $d\theta/dt(0) = \psi_0$.

(b) Text Example 6 shows that a solution of the linear equation will give a poor approximation to a solution of the nonlinear equation if θ_0 isn't close to zero. What can you say about ψ_0 in this regard?

43. **(a)** Show that if $y = y(x)$ is a solution to the constant-coefficient equation $y'' + ay' + by = 0$ and c is a constant, then the function y_c defined by

$y_c(x) = y(x + c)$ is also a solution. [*Hint:* It's not necessary to know the form of $y(x)$ in terms of elementary functions.]

(b) Generalize the result of part (a) to a solution $y(x)$ of the nth order constant-coefficient equation

$$y^{(n)} + a_{n-1}y^{(n-1)} + \cdots + a_1 y' + a_0 y = 0.$$

(c) Generalize the result of part (a) to a solution $y(x), a < x < b$, of a second-order equation, linear or nonlinear, of the form $y'' = F(y, y')$. [*Hint:* $y_c(x)$ will in general be defined on an interval different from $a < x < b$.]

44. Let $y(x)$ be a solution of a second-order equation $y'' + ay' + by = 0$. Could constants a and b have been chosen so that $y(x) = x \cos x$? What about $\sin x + \cos 2x$? Justify your answers.

Find second-order differential equations $y'' + ay' + by = 0$ that have Exercises 45 through 52 as solutions. [*Hint:* What are the characteristic roots associated with each solution?]

45. $\sin 2x$.

46. $\cos 2x$.

47. $e^x \cos 2x$.

48. $e^{-x} \cos(x/3)$.

49. $\sin(x/2)$.

50. xe^{2x}.

51. $\sin 3x - \cos 3x$.

52. $x - 7$.

53. For fixed α and fixed $\beta \neq 0$, show directly that $e^{\alpha x} \cos \beta x$ and $e^{\alpha x} \sin \beta x$ are linearly independent. What if $\beta = 0$?

Bifurcation Points. If a family of differential equations depends on a real parameter a and the qualitative behavior of the solutions to these equations is different for when a passes from one side of a value a_0 to the other then a_0 is called a **bifurcation point** or "splitting point" for the family. The following exercise is about a bifurcation point.

54. The family of differential equations $y'' + ay' + y = 0$ has $a_0 = 0$ as a bifurcation point for the parameter a. Show this as follows.

(a) Show that if $a < 0$, then there is always a solution $y_a(x)$ that assumes arbitrarily large values for large values of x.

(b) Show if $a > 0$, then every solution tends to zero as x tends to infinity.

2C Sinusoidal Oscillations †

The linearized pendulum equation $y'' + (g/l)^2 y = 0$ has the family of solutions

$$y(x) = c_1 \cos \beta x + c_2 \sin \beta x,$$

where $\beta = \sqrt{g/l}$. Such families occur often in the description of *oscillation*, a term we defined in the introduction to this section to mean the repeated switching back and forth from one side to another of some reference curve, in this example the x axis. The oscillation displayed above is called **harmonic** because of the central role it plays in the analysis of musical tones. An equation $y'' + \beta^2 y = 0$ is called a **harmonic oscillator** equation if $\beta \neq 0$. A function $f(x)$ is called **periodic** with period $2p$, if there is a number $p \neq 0$ such that $f(x + 2p) = f(x)$ for all x. Thus the number $2p = 2\pi/\beta$ is a **period** of the oscillation $y(x)$, since for its two terms we have

$$\cos \beta\left(x + \frac{2\pi}{\beta}\right) = \cos(\beta x + 2\pi) = \cos \beta x,$$

$$\sin \beta\left(x + \frac{2\pi}{\beta}\right) = \sin(\beta x + 2\pi) = \sin \beta x, \quad \text{for all } x.$$

The **frequency** or number of oscillations per x unit of a periodic function $f(x)$ is just the reciprocal of the period. For our harmonic oscillation $y(x)$ the frequency will be $\nu = \beta/(2\pi)$. (The number β itself is called the **circular frequency** of the oscillation and is equal to the number of oscillations per 2π units.)

example 7

The harmonic oscillation

$$y(x) = \cos 2x - 3 \sin 2x$$

has period π and frequency $1/\pi$, since both $\cos 2x$ and $\sin 2x$ have period π. In addition, $y(x)$ has other periods $n\pi$ for integer n, but the period π, the smallest positive period, is singled out as the fundamental period. The function

$$f(x) = \cos 2x - 3 \sin 4x$$

is periodic with period π, but is not a harmonic oscillation, because the fundamental periods of the two terms are unequal.

Since the parameters c_1 and c_2 in our general harmonic oscillation

$$y(x) = c_1 \cos \beta x + c_2 \sin \beta x$$

often don't have clear meanings, it's sometimes preferable to write $y(x)$ in terms of other parameters that do have useful interpretations. We choose an angle ϕ, shown in the figure, called a *phase angle*, such that

$$\cos \phi = \frac{c_1}{\sqrt{c_1^2 + c_2^2}}, \qquad \sin \phi = \frac{c_2}{\sqrt{c_1^2 + c_2^2}}.$$

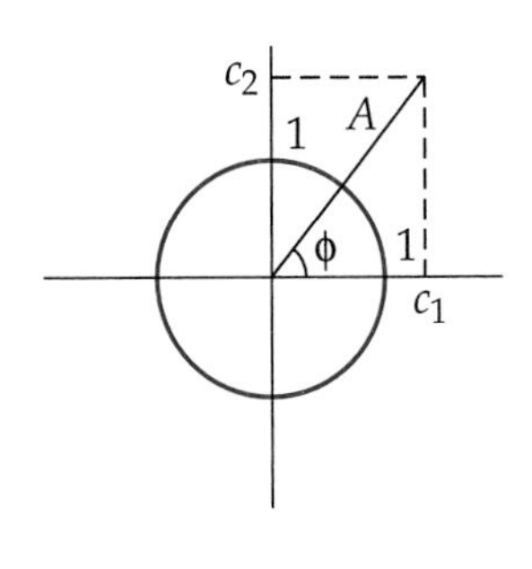

Such angles ϕ always exist. The reason is that the sum of the squares of the expressions for $\cos \phi$ and $\sin \phi$ is 1 so these two numbers represent the coordinates of a point that determines an angle ϕ with the positive horizontal axis, as in the margin. Now set $A = \sqrt{c_1^2 + c_2^2}$, so that $c_1 = A \cos \phi$ and $c_2 = A \sin \phi$. Using an addition formula for the cosine we can then write

$$y(x) = c_1 \cos \beta x + c_2 \sin \beta x = A(\cos \phi \cos \beta \mathbf{x} + \sin \phi \sin \beta x)$$
$$= A \cos(\beta x - \phi) = A \cos \beta(x - \phi/\beta).$$

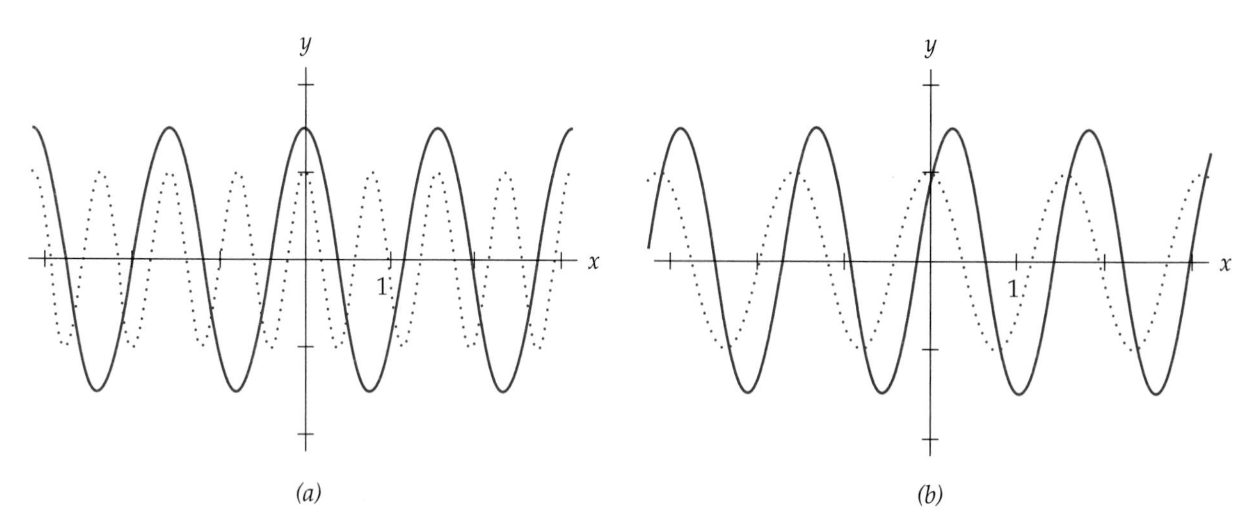

Figure 3.4 (*a*) $y = \cos 8x$; $y = \frac{3}{2}\cos 4x$. (*b*) $y = \cos 4x$; $y = \frac{3}{2}\cos(4x - \pi/3)$.

Thus our original 2-parameter family has been rewritten in terms of two new parameters A and ϕ. Furthermore, the graph of $y(x)$ is really just a horizontal shift of the simple oscillation $A\cos\beta x$. Note that if $\phi/\beta > 0$, the graph of $A\cos\beta x$ gets shifted to the right by ϕ/β. The number A is the **amplitude** of the oscillation about the value $y=0$, and the actual width of the oscillation from side to side is consequently $2A$. Figure 3.4(a) shows two graphs of oscillations with different periods and different amplitudes. In Figure 3.4(b) $\cos 4x$ is shifted to the right by the phase angle $\pi/3$ and is also amplified by the factor $\frac{3}{2}$. The graph of $y = \frac{3}{2}\cos(4x - \pi/3)$ is shifted to the right as compared with $y = \frac{3}{2}\cos 4x$, because after the shift, x has to increase to get to the same point on the graph.

As for finding the phase angle ϕ, it satisfies $\tan\phi = c_2/c_1$, so in general $\phi = \arctan(c_2/c_1)$. Note that if c_1 and c_2 are both positive then the point $(\cos\phi, \sin\phi)$ lies in the first quadrant, in which case we can pick a unique ϕ so that $0 < \phi < \pi/2$. Specifically, we can determine ϕ by $\phi = \arctan(c_2/c_1)$, where the principal branch of the arctangent function is assumed. In other cases, it's better to check directly that ϕ is in the correct quadrant; for example, if c_1 and c_2 are both negative then we can choose to have $\pi < \phi < 3\pi/2$. To summarize, $y(x) = c_1\cos\beta x + c_2\sin\beta x = A\cos(\beta x - \phi)$, where

$$\textbf{2.7} \qquad A = \sqrt{c_1^2 + c_2^2}, \quad \phi = \arctan\frac{c_2}{c_1}, \quad \text{and} \quad c_1 = A\cos\phi, \quad c_2 = A\sin\phi.$$

To determine ϕ when $c_1 = 0$, we set $\phi = \pi/2$ if $c_2 > 0$ and $\phi = -\pi/2$ if $c_2 < 0$.

example 8

The trigonometric function $y(x) = 3\cos 4x + 3\sin 4x$ has period $2\pi/4 = \pi/2$ and hence a frequency of $2/\pi \approx 0.64$ complete oscillations per x unit. The amplitude is $A = \sqrt{3^2 + 3^2} = 3\sqrt{2}$. The phase angle is $\phi = \arctan(3/3) = \pi/4$. Hence the given function can be written as

$$y(x) = A\cos(\beta x - \phi) = 3\sqrt{2}\cos\left(\frac{4x - \pi}{4}\right).$$

Note that the phase angle is incorporated into $\cos\beta x$, and if $\phi > 0$ the angle indicates that the graph of $y(x)$ is a shift to the right of the graph of $\cos\beta x$ by the amount ϕ/β. Because the cosine function has period 2π, the phase angle ϕ can always be altered by adding an integer multiple of 2π.

example 9

To analyze $\cos 4x - \sqrt{3}\sin 4x$, note that the phase angle is $\phi = \arctan(-\sqrt{3}) = -\pi/3$ and that the amplitude is $A = \sqrt{1^2 + (-\sqrt{3})^2} = 2$. Then express the oscillation in the form $A\cos(\beta x - \phi) = 2\cos(4x + \pi/3)$. Thus the given combination of sine and cosine has the same graph as $2\cos 4(x + \pi/12)$, which itself is a shift to the *left* of the graph of $2\cos 4x$ by $\pi/12$.

Two cosine oscillations $A\cos(\beta x - \phi)$ and $B\cos(\gamma x - \psi)$, having a common period $2p$ are said to be **in phase** if $\phi - \psi$ is an integer multiple of 2π; otherwise they are shifted **out of phase** relative to each other.

example 10

Figure 3.4(b) shows the graphs of $\cos 4x$ and $\frac{3}{2}\cos(4x - \pi/3)$. The oscillations have the same period, $\pi/2$, and are out of phase with each other, the wider oscillation being shifted to the right by less than the period $\pi/2$.

The functions that arise as solutions of second-order constant-coefficient equations $y'' + ay' + by = 0$ are purely harmonic oscillations

$$y(x) = c_1\cos\beta x + c_2\sin\beta x = A\cos(\beta x - \phi),$$

only in the special case $a = 0,\ b = \beta^2 > 0$ that corresponds to the case of purely imaginary conjugate roots for the characteristic equation. In all other cases, the solutions don't oscillate at all, or else oscillate but without a constant amplitude. The latter case, nonperiodic oscillation, occurs when $a \neq 0$ and $a^2 - 4b < 0$, which corresponds to complex conjugate roots with nonzero real part: $\alpha \pm i\beta$. The solutions then look like

$$\begin{aligned} y(x) &= e^{\alpha x}(c_1\cos\beta x + c_2\sin\beta x) \\ &= Ae^{\alpha x}\cos(\beta x - \phi). \end{aligned}$$

Thus the contrast between harmonic and nonharmonic oscillation of these solutions boils down to the behavior of the exponential factor $e^{\alpha x}$. Because $-1 \leq \cos(\beta x - \phi) \leq 1$, the graphs of these oscillations always lie between the graphs of $-Ae^{\alpha x}$ and $Ae^{\alpha x}$.

example 11 Here are two examples of nonharmonic oscillation:

$$y_1(x) = e^{-2x/5}\cos 6x, \qquad y_2(x) = \tfrac{3}{2}e^{x/5}\cos\left(5x + \tfrac{2}{5}\right).$$

Note that the harmonic factors in each example are periodic with respective periods $\pi/3$ and $2\pi/5$, but that the function as a whole is not periodic. Since $\cos x$ takes the value zero whenever $6x = \left(k + \frac{1}{2}\right)\pi$, k an integer, it follows that $y_1(x)$ is zero when $x = \left(k/6 + \frac{1}{12}\right)\pi$. Similarly $y_2 = 0$ when $x = \left(k/5 + \frac{1}{10}\right)\pi - \frac{2}{25}$. However, the precise locations of the successive maxima and minima depend on the exponential factor and the shift. See Figure 3.5 and Exercise 32.

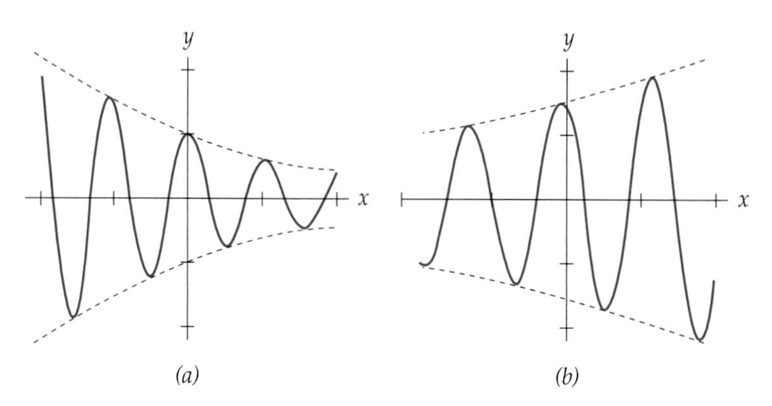

Figure 3.5 (*a*) $y_1(x) = e^{-2x/5}\cos 6x$. (*b*) $y_2(x) = \left(\frac{3}{2}\right)e^{x/5}\cos\left(5x + \frac{2}{5}\right)$.

The function $y_1(x)$ is a solution of a homogeneous equation with characteristic roots $-\frac{2}{5} \pm 6i$, while $y_2(x)$ stems from characteristic roots $\frac{1}{5} \pm 5i$.

EXERCISES

Determine (i) the smallest positive period $2p$, (ii) the frequency ν, and (iii) the amplitude A of each of the periodic functions in Exercises 1 through 6.

1. $\sin 3x$.

2. $\sin 3x + 2\cos 3x$.

3. $\sin 3x - 2\cos 3x$.

4. $\sin\sqrt{2}x$.

5. $\sin 3(x+1) + \cos 3(x+1)$.

6. $\sin 4x - 4\cos 4x$.

Sketch the graphs of the oscillatory functions in Exercises 7 through 14. Also state whether it is a solution of a constant coefficient differential equation $y'' + ay' + by = 0$ and, if so, what the equation is.

7. $y(x) = 2\cos(x - \pi/2)$.

8. $y(x) = 3\cos(x + \pi/2) + 1$.

9. $y(x) = \cos(2x - \pi/3)$.

10. $y(x) = 3\cos(x/2 + \pi/4) - 1$.

11. $y(x) = e^{-x}\cos(x - \pi)$.

12. $y(x) = -e^{x}\cos(x + \pi/3)$.

13. $y(x) = 2e^{-x/2}\cos(x - \pi/3) + 2$.

14. $y(x) = 3\sin x + \cos x$.

Phase Angles. Each of the trigonometric functions in Exercises 15 through 22 can be written in the form $A\cos(\beta x - \phi)$. In each case, find the amplitude A and a phase angle ϕ. Then sketch the graph.

15. $y(x) = \cos x + \sqrt{3}\sin x$.

16. $y(x) = \sqrt{2}(\cos x + \sin x)$.

17. $y(x) = 3\sin 2x$.

18. $y(x) = 2\sqrt{2}(\cos 5x - \sin 5x)$.

19. $y(x) = -\cos 4x$.

20. $y(x) = 2\sqrt{3}\cos x + 2\sin x$.

21. $y(x) = -3\cos 2x + 3\sqrt{3}\sin 2x$.

22. $y(x) = \sin x$.

Initial Values. For each of the functions in Exercises 23 through 30 find the initial-value problem that the function satisfies; this means find (i) a constant-coefficient differential equation $y'' + ay' = by = 0$ and (ii) $y(0)$ and $y'(0)$.

23. $y(x) = 3\cos(2x - \pi/2)$.

24. $y(x) = -\cos(3x + \pi/2)$.

25. $y(x) = 4\cos(x + \pi/3)$.

26. $y(x) = \cos(\sqrt{2}x + \pi/4)$.

27. $y(x) = e^{-x}\cos(2x + \pi)$.

28. $y(x) = -e^{-2x}\cos(3x + \pi/3)$.

29. $y(x) = e^{x}\cos(x - \pi/2)$.

30. $y(x) = e^{2x}\cos(2x - \pi/4)$.

31. Show that the function $y(x) = e^{\alpha x}\cos\beta x$ is never periodic unless $\alpha = 0$.

32. Show that the function $y(x) = e^{\alpha x}\cos\beta x$ has its maxima and minima at points with x coordinate $(1/\beta)\arctan(\alpha/\beta) + k\pi/\beta$ for integer k.

***33.** Let $0 < \beta < \gamma$ and define $f(x) = A\sin\beta x + B\sin\gamma x$, with A and B nonzero constants.

(a) Show that if γ/β is a rational number, then $f(x)$ has period $2n\pi/\beta$ for some integer n.

(b) Show that if $f(x)$ is periodic with period $\alpha > 0$, then γ/β must be a rational number. [*Hint:* Show that $f(\alpha) = 0$ and $f''(\alpha) = 0$. Then conclude that $\sin\beta\alpha = 0$ and $\sin\gamma\alpha = 0$.]

***34.** Let $f(x)$ be differentiable for all real x. Show that if f has a sequence of periods α_n tending to zero, then f must have derivative zero everywhere and so must be constant. Conclude that if f is nonconstant and periodic then it has a least positive period.

***35.** Let $f(x)$ have least positive period α. Show that if f also has period $\beta > \alpha$, then β is an integer multiple of α. [*Hint:* If not then $n\alpha < \beta < (n+1)\alpha$ for some integer n. Show $(n+1)\alpha - \beta < \alpha$ is also a period.]

***36.** Assume that the differential equation $y'' + ay' + by = 0$ has conjugate complex roots $r_1 = \alpha + i\beta$ and $r_2 = \alpha - i\beta$.

(a) Show that if $x_1 < x_2$, the boundary-value problem

$$y'' + ay' + by = 0, \quad y(x_1) = y_1, \qquad y(x_2) = y_2$$

always has a unique solution for given y_1 and y_2, provided that $\beta(x_2 - x_1)$ is not an integer multiple of π. [*Hint:* Use explicit solution formulas, and show that you can solve for the desired constants c_1 and c_2 in the general solution.]

(b) Show that the boundary-value problem $y'' + y = 0,\ \ y(0) = y(\pi) = 1$ has no solution.

(c) Show that the boundary-value problem $y'' + y = 0,\ \ y(0) = y(2\pi) = 1$ has infinitely many solutions.

37. The differential equation $y'' + \beta^2 y = 0$ has solutions $A\cos(\beta x - \phi),\ \beta > 0$. Initial conditions $y(0) = y_0 \neq 0$ and $y'(0) = z_0$ determine a unique solution as follows.

(a) Solve the equations $A\cos\phi = y_0,\ \ \beta A\sin\phi = z_0$ to get $A = \sqrt{y_0^2 + z_0^2/\beta^2}$ and $\phi = \arctan[z_0/(\beta y_0)]$.

(b) Show that if $y_0 = 0$, the formulas in part (a) can reasonably be interpreted to mean $A = |z_0|/\beta$ and $\phi = \pi/2$ when $z_0 > 0$, $\phi = -\pi/2$ when $z_0 < 0$.

2D Order More Than Two

The general nth-order constant-coefficient homogeneous linear equation is $L(y) = 0$, where the differential operator L looks like

$$L = D^n + a_{n-1}D^{n-1} + \cdots + a_1 D + a_0.$$

In applications, the coefficients a_k are almost always real, though that's not essential to the theory. Understanding the solutions to the nth-order equation is significant for understanding some special types of second-order nonhomogeneous equations $y'' + ay' + by = f(x)$, treated in Section 3. Mainly second-order equations arise directly in applications, though there are exceptions of order 3 and 4, because it's the first two derivatives of a function that have well-understood interpretations.

example 12

If $L = D^3 + 2D^2 + 1$, the associated homogeneous differential equation is $L(y) = y''' + 2y'' + y = 0$, whose solutions model the buckling of a column under compressive force. (See Exercises 46 and 47.) If $L = D^4 + 3D^2$, the differential equation is $L(y) = y^{(4)} + 3y'' = 0$, sometimes written $y^{(4)} + 3y'' = 0$.

The order-reduction method used in the second-order case, with resulting solutions explained in the statement of Theorem 2.6, applies successively to produce the complete solution to the nth-order equation, provided that we can find the roots of the corresponding characteristic equation

$$r^n + a_{n-1}r^{n-1} + \cdots + a_1 r + a_0 = 0.$$

In complete generality, finding characteristic roots can be a big technical problem whose practical solution requires numerical approximation techniques such as Newton's method, treated in Chapter 2, Section 4C. Nevertheless, we can deal with a number of important examples fairly simply.

The significance of the next theorem is that it guarantees not only the existence but the general form of all solutions of the appropriate initial-value problem for the nth-order constant-coefficient equation $L(y) = 0$. Each of these solutions can be expressed uniquely as a sum of constant multiples of n basic functions of the form $x^l e^{\alpha x} \cos \beta x$, or $x^l e^{\alpha x} \sin \beta x$. As a consequence of this unique representation, we show in Theorem 2.9 that these n functions form, in a precise technical sense explained below, a basis for the set of all solutions to $L(y) = 0$.

2.8 THEOREM

The nth-order constant-coefficient equation $L(y) = 0$ has for its general solution a linear combination

$$c_1 y_1(x) + \cdots + c_n y_n(x)$$

of solutions $y_k(x)$. The constants c_k can be prescribed arbitrarily; in particular, they are uniquely prescribed by n initial conditions

$$y(x_0) = z_0, \qquad y'(x_0) = z_1, \ldots, y^{n-1}(x_0) = z_{n-1}.$$

If the roots of the characteristic equation $r^n + a_{n-1}r^{n-1} + \cdots + a_1 r + a_0 = 0$ are $r_1, \ldots, r_n$, and m is the multiplicity of the root r_k, then some m of the functions $y_k(x)$ in the solution have the form $x^l e^{r_k x}$, $l = 0, 1, \ldots, m-1$. If roots $\alpha + i\beta$ and $\alpha - i\beta$ occur in complex conjugate pairs, then the corresponding pair of exponential solutions can be replaced by

$$x^l e^{\alpha x} \cos \beta x, \qquad x^l e^{\alpha x} \sin \beta x.$$

The order-reduction method argument used in the proof of Theorem 2.8 involves no additional ideas beyond those used in proving Theorems 1.2 and 2.6, so we omit the proof in favor of some examples. The constants of integration in the solution formulas in Theorem 2.8 appear so often that we extend our application of the term linear combination to cover expressions of the form

$$c_1 y_1 + c_2 y_2 + \cdots + c_n y_n.$$

As before, the numbers c_k are regarded as parameters, and the linear combination displayed above is an example of an n-**parameter family** of functions.

example 13

The differential equation $y^{(4)} - y = 0$ has characteristic equation $r^4 - 1 = 0$. To find the roots, we note that since $r^4 = 1$, either $r^2 = 1$ or else $r^2 = -1$. Hence the roots are $r_1 = 1$, $r_2 = -1$, $r_3 = i$, and $r_4 = -i$. The first two roots provide the solutions e^x and e^{-x}. The second pair provides e^{ix} and e^{-ix} or the equivalent real-valued pair $\cos x$ and $\sin x$. The complete solution is then

$$y(x) = c_1 e^x + c_2 e^{-x} + c_3 \cos x + c_4 \sin x.$$

Initial conditions $y(0) = 0$, $y'(0) = 1$, $y''(0) = 2$, and $y'''(0) = 1$ impose conditions on the constants c_k. For example,

$$y'(x) = c_1 e^x - c_2 e^{-x} - c_3 \sin x + c_4 \cos x,$$

so $y'(0) = c_1 - c_2 + c_4 = 1$. The complete set of four conditions reduces to

$$c_1 + c_2 + c_3 = 0,$$
$$c_1 - c_2 + c_4 = 1,$$
$$c_1 + c_2 - c_3 = 2,$$
$$c_1 - c_2 - c_4 = 1.$$

Straightforward elimination starts with adding the first and second, then the third and fourth, to show that $c_1 = 1,\ c_2 = 0,\ c_3 = -1$, and $c_4 = 0$. Thus the particular solution that satisfies the initial conditions is $y(x) = e^x - \cos x$.

example 14

The third-order differential equation $(D-1)^3 y = 0$ is $y''' - 3y'' + 3y' - y = 0$ when written without operator notation and has the single characteristic root $r = 1$ with multiplicity 3. The general solution is

$$y = c_1 e^x + c_2 x e^x + c_3 x^2 e^x.$$

If we notice that initial conditions $y(0) = y'(0) = 0$ and $y''(0) = 1$ are satisfied by $y(x) = \frac{1}{2}x^2 e^x$, then the uniqueness of the solution to this initial-value problem guarantees that we have the correct solution without doing any more work.

example 15

The equation $y^{(4)} + \lambda y'' = 0$ is satisfied under certain conditions by a function $y(x)$ that describes the lateral deflection, measured x units from one end, of a uniform column under a vertical compressive force. (The constant $\lambda = P/\rho$ depends on the structure of the column and on the vertical load P applied to it.) The characteristic equation is $r^4 + \lambda r^2 = 0$, or $r^2(r^2 + \lambda) = 0$. With $\lambda > 0$, the roots are $r_1 = r_2 = 0$ and $r_3 = \sqrt{\lambda}\, i$, $r_4 = -\sqrt{\lambda}\, i$. So the general solution is

$$y(x) = c_1 + c_2 x + c_3 \cos \sqrt{\lambda}\, x + c_4 \sin \sqrt{\lambda}\, x.$$

Initial conditions at a single point x_0 are physically uninteresting in this problem. What is usually done is to impose boundary conditions on $y(x)$ and $y'(x)$, or else on $y(x)$ and $y''(x)$ at points corresponding to the two ends of the column, say at $x = 0$ and $x = L$. The existence of a unique solution depends critically on the value of λ. These matters are taken up in the following exercises and also following Chapter 9, Section 3.

Finding the roots of an nth-degree characteristic equation is in general a difficult problem. The analogues of the quadratic formula for solutions of quadratic equations exist for equations of degree 3 and 4, but are somewhat complicated. The following examples illustrate some fairly simple special cases.

example 16

The cubic equation $r^3 + 2r^2 + 5r = 0$ can be written $r(r^2 + 2r + 5) = 0$. Apart from the obvious root $r_1 = 0$, the quadratic equation $r^2 + 2r + 5 = 0$ contributes roots

$r_2 = -1 + 2i$ and $r_3 = -1 - 2i$. The corresponding solutions of the differential equation

$$y''' + 2y'' + 5y' = 0$$

are

$$y = c_1 + c_2 e^{-x} \cos 2x + c_3 e^{-x} \sin 2x.$$

---

example 17

The fourth-degree, or quartic, equation $r^4 - 13r^2 + 36 = 0$ is a quadratic equation in r^2, with solutions $r^2 = (13 \pm 5)/2$. Since $r^2 = 4$ or $r^2 = 9$, the four distinct roots are $r = \pm 2, \ \pm 3$. Hence the solutions to

$$y^{(4)} - 13y'' + 36y = 0$$

consist of all members of the 4-parameter family

$$y = c_1 e^{2x} + c_2 e^{-2x} + c_3 e^{3x} + c_4 e^{-3x}.$$

It will be helpful in Section 3 to apply the idea behind Theorem 2.8 in reverse order, starting with solutions and arriving at a differential equation that has those solutions.

---

example 18

To find a constant coefficient equation $L(y) = 0$ of least possible order having e^x, e^{2x}, and e^{-2x} for solutions, we write the linear operator equation

$$(D - 1)(D - 2)(D + 2)y = 0.$$

Clearly e^{-2x} is a solution, because $(D + 2)e^{-2x} = 0$. Since the operator factors can be written in any order, and $(D - 2)e^{2x} = 0$ and $(D - 1)e^x = 0$, the three functions e^x, e^{2x}, and e^{-2x} are solutions of the differential equation

$$(D - 1)(D - 2)(D + 2)y = 0 \quad \text{or} \quad y''' - y'' - 4y' + 4y = 0.$$

---

example 19

Let us find a constant coefficient equation $L(y) = 0$ of least possible order having $\cos x$ and $\sin 2x$ for solutions. These solutions would have arisen from characteristic roots $\pm i$ and $\pm 2i$, respectively. We write the linear operator equation

$$(D^2 + 1)(D^2 + 4)y = 0.$$

Clearly $\sin 2x$ is a solution, because $(D^2 + 4) \sin 2x = 0$. Since the operator factors can be written in any order, and $(D^2 + 1) \cos x = 0$, both $\sin 2x$ and $\cos x$ are solutions of the differential equation. Linear combinations $y = c_1 \cos x + c_2 \sin x + c_3 \cos 2x + c_4 \sin 4x$ constitute the general solution.

In Section 1 we defined linear independence for just two functions.

Independence of Basic Solutions. The unique determination of the constants in Theorem 2.8 has as a consequence the linear independence of the associated solutions. The definition of linear independence extended to cover n functions is this: $y_1(x), y_2(x), \ldots, y_n(x)$ *are* **linearly independent** *on an interval if the identity*

$$c_1y_1(x) + c_2y_2(x) + \cdots + c_ny_n(x) = 0$$

holds on the interval only with the obvious choice $c_1 = c_2 = \cdots = c_n = 0$ *for the constants.* You are asked in Exercise 49 to show that a set of two or more functions is linearly independent if and only if none of them is expressible as a linear combination of the others using constant coefficients. If one function of a set is expressible as a linear combination of the others, the set is said to be linearly dependent. Thus independence ensures that there is no redundancy in the expression of a given function as a linear combination of functions from an independent subset. (In the terminology of linear algebra, an independent subset is called a **basis** for the set of linear combinations of its elements.) That every solution of $L(y) = 0$ is a linear combination of the solutions $x^le^{\alpha x}\cos\beta x,\ x^le^{\alpha x}\sin\beta x$ listed at the end of Theorem 2.8 is the main conclusion of the theorem; that these solutions are indeed linearly independent is an additional consequence of the theorem, as follows.

2.9 THEOREM

The solutions $x^le^{r_kx}$ in Theorem 2.8, with real forms $x^le^{\alpha x}\cos\beta x$, $x^le^{\alpha x}\sin\beta x$, are linearly independent; hence these solutions form a basis for the set of all solutions of $L(y) = 0$.

Proof. Suppose a linear combination $y(x)$ of these n solutions is identically zero:

$$c_1y_1(x) + \cdots + c_ny_n(x) = 0.$$

To prove independence we need to show that all $c_k = 0$. The linear combination is a solution of a linear homogeneous differential equation $L(y) = 0$, and since this solution is identically zero, it obviously satisfies initial conditions $y(x_0) = y'(x_0) = \cdots = y^{(n-1)}(x_0) = 0$ at a given point x_0. Since choosing all the $c_k = 0$ suffices to produce this solution, and since Theorem 2.8 guarantees uniqueness of the numbers c_k, all c_k must be zero. ■

EXERCISES

Find the general solution to each of the differential equations in Exercises 1 through 10.

1. $y''' - y = 0.$

2. $y^{(4)} - 2y'' + y = 0.$

3. $y''' - 2y' = 0.$

4. $y^{(4)} - y'' = 0.$

5. $y''' - 8y = 0.$

6. $y^{(4)} - 4y'' + 4y = 0.$

7. $(D^2 - 4)(D^2 - 1)y = 0.$

8. $D(D - 1)^3y = 0.$

9. $y''' - 2y'' = 0.$

10. $y^{(4)} - y = 0.$

Find constant-coefficient linear differential equations with real coefficients and of the smallest possible order that have the functions in Exercises 11 through 20 as solutions. [*Hint:* Find the characteristic roots associated with each equation.]

11. $y(x) = \cos 4x.$

12. $y(x) = x \cos 4x.$

13. $y(x) = x^2 \cos 4x.$

14. $y(x) = x \sin 4x.$

15. $y(x) = xe^x \sin x.$

16. $y(x) = x^3 \cos 4x.$

17. $y(x) = x^5.$

18. $y(x) = \cos 4x + \sin 3x.$

19. $y(x) = e^{-x} \cos x.$

20. $y(x) = x \cos x + \cos 2x.$

Listed in Exercises 21 through 30 are families of solutions to some linear constant-coefficient differential equations. In each case, find a differential equation of least possible order satisfied by the family.

21. $c_1 + c_2 x + c_3 e^x.$

22. $c_1 \cos x + c_2 \sin x + c_3 + c_4 x.$

23. $c_1 \cos x + c_2 \sin x + c_3 e^x.$

24. $c_1 \cos 2x + c_2 \sin 2x + c_3 e^{-x}.$

25. $c_1 + c_2 x + c_3 x^2.$

26. $c_1 \cos 3x + c_2 \sin 3x + c_3.$

27. $c_1 e^x + c_2 e^{-x} + c_3 \cos x + c_4 \sin x.$

28. $c_1 e^x \cos 2x + c_2 e^x \sin 2x + c_3.$

29. $c_1 \cos x + c_2 \sin x.$

30. $c_1 \cos x + c_2 \sin x + c_3 \cos 3x + c_4 \sin 3x.$

Each of the solution families listed in Exercises 31 through 38 can satisfy the corresponding set of conditions listed below with appropriate values for the c_ks. Find the c_ks for Exercises 31 through 38.

31. $y(0) = 1,\ y'(0) = 2,\ y''(0) = 1.$

32. $y(0) = y'(0) = 2,\ y''(0) = y'''(0) = -1$

33. $y(0) = 2,\ y'(0) = y''(0) = 0.$

34. $y(0) = 2,\ y'(0) = y''(0) = -3.$

35. $y(0) = y'(0) = 1,\ y''(0) = 2.$

36. $y(0) = 1,\ y'(0) = 3,\ y''(0) = -9.$

37. $y(0) = y'(0) = 1,\ y''(0) = -1,\ y'''(0) = 3.$

38. $y(0) = 2,\ y'(0) = 3,\ y''(0) = 1.$

Find the general solution to each equation in Exercises 39 through 44. [*Hint:* In 42, $r^4 + 1 = 0$, so $r^2 = \pm i = \pm e^{i\pi/2}$. Then $r = \pm e^{i\pi/4}$ or $r = \pm e^{3i\pi/4}$.]

39. $y^{(6)} - y^{(4)} = 0.$

40. $y^{(4)} - 2y''' = 0.$

41. $y^{(5)} - y' = 0.$

42. $y^{(4)} + y = 0.$

43. $(D^2 - 4)(D^2 - 16)y = 0.$

44. $D^2(D - 1)^3 y = 0.$

45. Explain why $y(x) = \cos x + \sin 2x$ can't be the solution to a constant coefficient equation of the form $y'' + ay' + by = 0$ with a and b real. Find a real equation of order greater than 2 that $y(x)$ does satisfy.

46. The general solution $y(x) = c_1 + c_2x + c_3 \cos \sqrt{\lambda}\, x + c_4 \sin \sqrt{\lambda}\, x$ to $y^{(4)} + \lambda y'' = 0$ is derived in Example 15 of the text; use it to do the following.

(a) If $\lambda = 4\pi^2$, find the infinitely many solutions that satisfy the boundary conditions $y(0) = y(1) = 0,\ y'(0) = y'(1) = 0$.

(b) Under the assumption that $y(x)$ represents the horizontal deflection of a column under a vertical compressing force, we can interpret the boundary conditions in part (a) to mean that the ends of the column are rigidly embedded in floor and ceiling. Sketch some typical solutions, assuming small displacements.

(c) Show that if $\lambda = \pi^2$, then the only solution satisfying the boundary conditions in part (a) is the identically zero solution.

47. The general solution $y(x) = c_1 + c_2x + c_3 \cos \sqrt{\lambda}\, x + c_4 \sin \sqrt{\lambda}\, x$ to $y^{(4)} + \lambda y'' = 0$ is derived in Example 15 of the text; use it to do the following.

(a) If $\lambda = \pi^2$, find the infinitely many solutions that satisfy the boundary conditions $y(0) = y(1) = 0,\ y''(0) = y''(1) = 0$.

(b) Under the assumption that $y(x)$ represents the horizontal deflection of a column under a vertical compressing force, we can interpret the boundary conditions in part (a) to mean that the ends of the column are hinged at both floor and ceiling. Sketch some typical solutions, assuming small displacements.

(c) Show that if $\lambda = \pi^2/4$, then the only solution satisfying the boundary conditions in part (a) is the identically zero solution.

48. A differential equation for the lateral displacement $y = y(x)$ at distance x from one end of a uniform rotating shaft is

$$y^{(4)} - \lambda y = 0,$$

where the constant $\lambda > 0$ is proportional to the speed of rotation.

(a) Show that the most general solution to the differential equation can be written

$$y = c_1 \cosh \sqrt[4]{\lambda}\, x + c_2 \sinh \sqrt[4]{\lambda}\, x + c_3 \cos \sqrt[4]{\lambda}\, x + c_4 \sin \sqrt[4]{\lambda}\, x,$$

where $\cosh u = (e^u + e^{-u})/2$ and $\sinh u = (e^u - e^{-u})/2$.

(b) Let $\lambda = n^4\pi^4$, where n is a positive integer, and find solutions of the differential equation subject to boundary conditions $y(0) = y''(0) = 0$, $y(1) = y''(1) = 0$.

(c) Sketch the graphs of the solutions found in part (b) for $n = 1,\ 2,\ 3$.

(d) Show that if λ is not of the form prescribed in part (b) then the only solution to the problem posed there is the identically zero solution.

49. Show that a set of two or more functions is linearly independent if and only if no one of them is expressible as a linear combination of the others.

50. **(a)** Show without using explicit solution formulas that if $y = y(x)$ is a solution to $y''' + ay'' + by' + cy = 0$, where a, b, and c are constant, then for any fixed constant c the function $y = y(x + c)$ is also a solution of the same equation. This property of the entire set of solutions is called **translation invariance.**

(b) Is the set of solutions of an nth-order constant-coefficient linear equation $L(y) = 0$ translation invariant? Explain.

(c) What can you say about translation-invariance of the set of solutions of equations such as $y^{(n)} = f(y, y', \ldots, y^{(n-1)})$ if f is not itself explicitly dependent on x?

(d) Explain why translation invariance of the set of solutions fails to hold for equations with nonconstant coefficients, even, for example, for a first-order equation such as $y' + g(x)y = 0$. [*Hint:* Consider the direction field of a first-order linear equation.]

51. A uniform horizontal beam sags by an amount $y = y(x)$ at a distance x measured from one end. For a fairly rigid beam with uniform loading, $y(x)$ typically satisfies a fourth-order differential equation $y^{(4)} = R$, where R is a constant depending on the load being carried and on the characteristics of the beam itself. If the ends of the beam are supported at $x = 0$ and at $x = L$, then $y(0) = y(L) = 0$. The extended beam also behaves as if its profile had an inflection point at each support, so that $y''(0) = y''(L) = 0$.

(a) Use the multiple characteristic root of the associated homogeneous equation to find the general solution of the homogeneous equation.

(b) Show that the vertical deflection at x is $\frac{1}{24}R(x^4 - 2Lx^3 + L^3x)$, $0 \leq x \leq L$.

3 NONHOMOGENEOUS LINEAR EQUATIONS

3A General Solution

We began Sections 1 and 2 with the respective linear differential equations $y'' - (g/l)y = 0$ and $y'' + (g/l)y = 0$, then proceeded to generalize to $y'' + ay' + by = 0$. Here we generalize again, to allow for supplementary terms $f(x)$ or forces $F(t)$ applied in either of the forms

$$(D^2 + a\,D + b)y = f(x), \quad \text{or} \quad \frac{d^2y}{dt^2} + a\frac{dy}{dt} + by = F(t).$$

Recall from Section 1 that whether the coefficients a, b, and c depend on the independent variable or not, the real significance of linearity for a differential operator $L = D^2 + a\,D + b$, even if a and b depend on the independent variable, is that,

$$L(c_1y_1 + c_2y_2) = c_1Ly_1 + c_2Ly_2,$$

for every pair of constants c_1, c_2 and every pair of twice-differentiable functions y_1, y_2. As we saw in Section 1, verification of the linearity of this second-order operator L is a direct consequence of the linearity of the differential operator $D = d/dx$ and of the arithmetical operations involved. The linearity of L has

The use of the term *homogeneous* here is standard, but it unfortunately conflicts with another standard use in Chapter 1, Section 3C. The contexts differ by enough that misunderstanding is unlikely.

already enabled us to state, under the *superposition principle*, that $c_1y_1 + c_2y_2$ is a solution of a linear **homogeneous equation** $Ly = 0$, which for our immediate application is a constant-coefficient equation.

$$y'' + ay' + by = 0,$$

whenever y_1 and y_2 are solutions of the same equation. The linearity of L will play a part in describing the general solution of the **nonhomogeneous equation** $Ly = f$, or

$$y'' + ay' + by = f(x),$$

where $f(x)$ is defined on some interval. We'll see that the solution has two parts, only one of which depends on $f(x)$. The function f is often called a *forcing function* because of its interpretation in mechanical and electrical problems.

3.1 THEOREM

Let y_p be some particular solution of the nonhomogeneous equation

$$Ly = f,$$

where L is a linear operator. Then the general solution of $Ly = f$ is a sum

$$y = y_h + y_p,$$

where y_h is the general solution of the associated homogeneous equation $Ly = 0$.

Proof. We assume that $Ly_p = f$ and that y is some solution of $Ly = f$. Then

$$L(y - y_p) = Ly - Ly_p = f - f = 0.$$

Thus $y - y_p = y_h$ for some solution y_h of $Ly = 0$. It follows that $y = y_h + y_p$. Since y stood for an arbitrary solution, all solutions have the form $y_p + y_h$. ■

The only linear operators we've discussed are differential operators, but linear integral operators appear in Section 3C, and also in Chapter 8. Theorem 3.1 applies to these operators as well. For now we'll use Theorem 3.1, along with Theorem 2.6, to deduce the following.

3.2 COROLLARY

The general solution of

$$y'' + ay' + by = f(x) \quad \text{has the form} \quad y = c_1y_1(x) + c_2y_2(x) + y_p(x),$$

where y_p is some particular solution, and y_1 and y_2 are linearly independent solutions of the associated homogeneous equation $y'' + ay' + by = 0$.

example 1

The equation

$$y'' - y = x$$

has $y'' - y = 0$ for its associated homogeneous equation, with general solution

$$y_h = c_1e^x + c_2e^{-x}.$$

We can verify directly that $y_p = -x$ is a solution of the nonhomogeneous equation. (We'll see shortly how to discover such a solution by integration.) It follows from Theorem 3.1 that every solution has the form

$$y = y_h + y_p = c_1 e^x + c_2 e^{-x} - x.$$

To find the particular solution satisfying initial conditions

$$y(x_0) = y_0, \qquad y'(x_0) = z_0,$$

just compute $y'(x) = -1 + c_1 e^x - c_2 e^{-x}$ and solve the two equations

$$c_1 e^{x_0} + c_2 e^{-x_0} - x_0 = y_0$$
$$c_1 e^{x_0} - c_2 e^{-x_0} - 1 = z_0$$

for the correct constants c_1 and c_2. For example, if $x_0 = y_0 = 0$ and $z_0 = 1$, we solve $c_1 + c_2 = 0$ and $c_1 - c_2 = 2$ to get $c_1 = 1,\ c_2 = -1$, so $y = -x + e^x - e^{-x}$ is the solution of the initial-value problem with initial conditions $y(0) = 0$ and $y'(0) = 1$.

3B Undetermined-Coefficient Method

Understanding this traditional method for solving many nonhomogeneous equations requires some careful thought. The alternatives in Sections 3C and 3D are somewhat more mechanical in practice.

The previous example has the advantage of simplicity, but it is unsatisfying because it used a solution $y_p(x) = -x$ that was just an educated guess, or as we said, was found "by inspection." The present subsection puts the inspection method on a more formal and routine footing for a special class of equations. This routine is the most efficient way to solve many second-order constant-coefficient equations $Ly = f(x)$, but fails to apply to some functions $f(x)$, notably discontinuous ones, to which the methods of the optional Sections 3C and 3D apply.

The undetermined coefficient method consists of finding trial linear combinations containing enough terms to contain a solution $y_p(x)$ to $Ly = f(x)$. The constant coefficients in the linear combination are then determined by substitution into the differential equation. We'll see as we go along that the method works when the forcing function $f(x)$ is itself a solution of a homogeneous constant-coefficient differential equation $Ny = 0$. As we know, these solutions are also linear combinations, with constant coefficients, of functions of either of the two basic types

3.3 $\quad x^n e^{\alpha x} \cos \beta x \quad \text{or} \quad x^n e^{\alpha x} \sin \beta x, \qquad n \geq 0$ an integer.

In dealing with linear differential operators such as $L = D^2 + aD + b$, a differential equation

$$Ly = a_1 f_1(x) + \cdots + a_m f_m(x), \qquad a_k \text{ constant},$$

can be solved by first finding a solution $y_k(x)$ to each of the m related equations

$$Ly = f_1(x), \ldots, Ly = f_m(x).$$

The linearity of L then implies that the function

$$y(x) = a_1 y_1(x) + \cdots + a_m y_m(x)$$

is a solution to the given equation. This consequence of linearity is an extension to nonhomogeneous equations of the *superposition principle* discussed in Section 1 for homogeneous equations.

example 2

To solve $y'' - y = 4e^{2x} + 5e^{-3x}$, we can use linearity to split the problem into the solution of the two equations

$$y'' - y = e^{2x} \quad \text{and} \quad y'' - y = e^{-3x}.$$

The associated homogeneous equation $y'' - y = 0$ is of course the same for both equations, with general solution $y_h(x) = c_1e^x + c_2e^{-x}$ in common. All that remains to be done is to find single solutions $y_1(x)$ and $y_2(x)$ to each of the two simpler equations and then write the general solution in the form $y = y_h + 4y_1 + 5y_2$.

For the first equation, $y'' - y = e^{2x}$, since the second derivative of e^{2x} is $4e^{2x}$, a reasonable guess is that there might be a solution $y_1(x) = Ae^{2x}$, for some constant A. Indeed, we compute $y_1' = 2Ae^{2x}$, $y_1'' = 4Ae^{2x}$, and substitute:

$$\begin{aligned} y_1'' - y_1 &= 4Ae^{2x} - Ae^{2x} \\ &= 3Ae^{2x} = e^{2x}. \end{aligned}$$

To satisfy the differential equation, we need $3A = 1$, so $A = \frac{1}{3}$ gives us $y_1 = \frac{1}{3}e^{2x}$. To solve $y'' - y = e^{-3x}$, a trial solution of a similar general form, $y_2 = Ae^{-3x}$, leads to $y_2' = -3Ae^{-3x}$, $y_2'' = 9Ae^{-3x}$ and thus to $y_2'' - y_2 = 8Ae^{-3x} = e^{-3x}$. Hence $A = \frac{1}{8}$, and $y_2 = \frac{1}{8}e^{-3x}$. Adding $4y_1 + 5y_2$ together with y_h gives the general solution

$$y = c_1e^x + c_2e^{-x} + \tfrac{4}{3}e^{2x} + \tfrac{5}{8}e^{-3x}.$$

Why Did Our Guesses Work? Our guesses were certainly adequate for the form of the two particular solutions that we needed in the previous example. Now let's look at a procedure that will lead to correct guesses automatically and will also work as well in more complicated examples. The general setup was

$$Ly = (D - r_1)(D - r_2)y = e^{ax},$$

with $r_1 = 1$, $r_2 = -1$, and either $a = 2$ or $a = -3$. Recall that we already know how to solve constant-coefficient *homogeneous* equations if we know the roots of the characteristic equation. Consider what happens if we apply the operator $(D - a)$ to both sides of the previous equation:

$$(D - a)(D - r_1)(D - r_2)y = (D - a)e^{ax} = ae^{ax} - ae^{ax} = 0.$$

The operator $N = (D - a)$ is called an **annihilator** of e^{ax}, since $Ne^{ax} = 0$. We now have two relevant facts at our disposal:

(i) All solutions of the above third-order homogeneous equation have the form

$$y = c_1e^{r_1x} + c_2e^{r_2x} + c_3e^{ax},$$

provided that a is different from the two unequal numbers r_1 and r_2. If it happens that $a = r_2$, then the third-order equation has the general solution

$$y = c_1 e^{r_1 x} + c_2 e^{r_2 x} + c_3 x e^{ax}.$$

(ii) Because $(D - a)e^{ax} = 0$, a solution of $(D - r_1)(D - r_2)y = e^{ax}$ also solves

$$(D - a)(D - r_1)(D - r_2)y = 0.$$

Since the functions in the 2-parameter family $y_h = c_1 e^{r_1 x} + c_2 e^{r_2 x}$ are solutions of the associated *homogeneous* linear second-order equation, that is, $Ly_h = 0$, they can't by themselves provide a solution of the *nonhomogeneous* equation. Hence a solution to the nonhomogeneous equation must be among the functions $c_3 e^{ax}$, or else $c_3 x e^{ax}$ if $a = r_2$. These ideas suggest the following.

3.4 METHOD

Steps for finding the terms in a linear combination for a trial solution y_p to $Ly = f(x)$, where $f(x)$ is a basic function of the form $x^n e^{\alpha x} \cos \beta x$ or $x^n e^{\alpha x} \sin \beta x$:

0. As a preliminary step, solve the associated homogeneous equation $Ly = 0$ to get $y_h(x)$.

1. Include in the trial linear combination y_p, the **derivative set** of $f(x)$, consisting of the linearly independent sets of functions of which $f(x)$ and its successive derivatives are linear combinations. For example, the derivative set of $f(x) = 2x + 3x \sin x$ consists of two parts, $\{x, 1\}$ and $\{x \sin x, x \cos x, \sin x, \cos x\}$.

2. If a term from step 1 satisfies the homogeneous equation $Ly = 0$, multiply that term and all terms in its derivative set by the lowest power x^k such that the resulting terms are not homogeneous solutions. The exponent k needs to be at most the order of the original operator L, so for a second-degree equation if we have to multiply at all it will be by x or x^2.

3. Form a linear combination, with undetermined constant coefficients, of the terms from step 2, and find the the coefficients by substitution into $Ly = f(x)$.

Step 1 includes the derivative set of $f(x)$, because if N annihilates $f(x)$ we need to include all solutions of $Ny = 0$. For example, if $f(x) = x \sin x$, then $N = (D^2 + 1)^2$, and the derivative set of $x \sin x$ is $\{x \sin x, x \cos x, \sin x, \cos x\}$, which is the independent solution set of $Ny = 0$.

example 3

Here are examples of functions $f(x)$ in $Ly = f(x)$ and trial solutions $y_p(x)$ formed from the derivative set of $f(x)$, assuming that no term in $y_p(x)$ satisfies $Ly = 0$. Note that N annihilates both $f(x)$ and the derivative terms in $y_p(x)$.

$f(x) = x^2$	$N = D^3$	$y_p(x) = Ax^2 + Bx + C$
$f(x) = e^{rx}$	$N = D - r$	$y_p(x) = Ae^{rx}$
$f(x) = xe^{rx}$	$N = (D - r)^2$	$y_p(x) = (Ax + B)e^{rx}$
$f(x) = x^2 e^{rx}$	$N = (D - r)^3$	$y_p(x) = (Ax^2 + Bx + C)e^{rx}$
$f(x) = \cos \beta x$	$N = D^2 + \beta^2$	$y_p(x) = A \cos \beta x + B \sin \beta x$
$f(x) = x \sin \beta x$	$N = (D^2 + \beta^2)^2$	$y_p(x) = (Ax + B) \cos \beta x + (Cx + D) \sin \beta x$

example 4

Finding an annihilator N for $f(x) = e^{\alpha x} \cos \beta x$ or $f(x) = e^{\alpha x} \sin \beta x$ is a little more complicated than for the functions listed in the previous example. The characteristic roots of N are $\alpha \pm i\beta$ so

$$N = [D - (\alpha + i\beta)][D - (\alpha - i\beta)] = D^2 - 2\alpha D + (\alpha^2 + \beta^2).$$

If $\alpha \pm i\beta$ aren't roots of L, the associated trial solutions for $Ly = f(x)$ are just

$$y_p = Ae^{\alpha x} \cos \beta x + Be^{\alpha x} \sin \beta x.$$

If L also has characteristic roots $\alpha \pm i\beta$, then in step 2 we use

$$y_p = Axe^{\alpha x} \cos \beta x + Bxe^{\alpha x} \sin \beta x.$$

example 5

To solve $y'' + 4y = 3\cos 2x$, note that the equation has the associated homogeneous solution $y_h = c_1 \cos 2x + c_2 \sin 2x$. The derivative set of $\cos 2x$ is $\{\cos 2x, \sin 2x\}$, but the choice $y_p = A\cos 2x + B\sin 2x$ for $y_p(x)$ is inadequate, because substitution into the left-hand side produces zero, not a multiple of $\cos 2x$. (This is why we need step 0, to make us aware of the solutions of the homogeneous equation before finding a trial solution.) For this example, step 2 in Method 3.4 tells us to multiply by x, so we use instead

$$y_p(x) = Ax\cos 2x + Bx\sin 2x.$$

Successive differentiation shows that

$$y_p'(x) = A\cos 2x + B\sin 2x + 2Bx\cos 2x - 2Ax\sin 2x,$$
$$y_p''(x) = -4A\sin 2x + 4B\cos 2x - 4Ax\cos 2x - 4Bx\sin 2x.$$

We find $y_p'' + 4y_p = -4A\sin 2x + 4B\cos 2x$. Since we want $y_p'' + 4y_p = 3\cos 2x$, equating coefficients in the two formulas on the right gives $A = 0$ and $B = \frac{3}{4}$. The general solution is $y = y_h + y_p = c_1 \cos 2x + c_2 \sin 2x + \frac{3}{4}x\sin 2x$.

example 6

The equation $y'' + 2y' + y = 3e^{-x} + 2x$ has $y_h(x) = c_1 e^{-x} + c_2 xe^{-x}$ for its homogeneous solution. Solving first the equation $y'' + 2y' + y = 2x$, we note the derivative set of x is $\{x, 1\}$ and that neither of these is a homogeneous solution. We set $y_1(x) = Ax + B$ and find $y_1'(x) = A$, $y_1''(x) = 0$. Substitution into the equation with $2x$ on the right gives

$$0 + 2A + Ax + B = 2x.$$

Equating coefficients of x gives $A = 2$. The constant terms satisfy $2A + B = 0$, so $B = -2A = -4$. Thus $y_1(x) = 2x - 4$.

To solve $y'' + 2y' + y = 3e^{-x}$, we note that the derivative set of e^{-x} is just $\{e^{-x}\}$. But e^{-x} and xe^{-x} are homogeneous solutions, so step 2 with $k = 2$ suggests the trial solution

$$y_2(x) = Ax^2 e^{-x}.$$

Differentiating twice with the product rule, we find

$$y_2'(x) = A(2x - x^2)e^{-x} \quad \text{and} \quad y_2''(x) = A(2 - 4x + x^2)e^{-x}.$$

Substitution into the equation with $3e^{-x}$ on the right gives

$$A(2 - 4x + x^2)e^{-x} + 2A(2x - x^2)e^{-x} + Ax^2e^{-x} = 3e^{-x}.$$

The terms with x and x^2 as factors all cancel out and we are left with $2A = 3$. Thus $A = \frac{3}{2}$, so $y_2(x) = \frac{3}{2}x^2e^{-x}$. The general solution of the original equation is then

$$\begin{aligned} y &= y_1 + y_2 + y_h \\ &= 2x - 4 + \tfrac{3}{2}x^2e^{-x} + c_1e^{-x} + c_2xe^{-x}. \end{aligned}$$

We could have found the complete trial y_p by using a single trial solution

$$y_p = Ax + B + Cx^2e^{-x},$$

which would probably be more efficient. But splitting the problem into two parts does have the effect of reducing the number of undetermined coefficients you need to work with at one time.

Summary. Because it involves several steps, Method 3.4 sometimes looks a bit complicated in practice, so it's important to keep in mind the main idea behind it. Understanding the method depends on the observation that we made earlier in special cases: If $y(x)$ is a solution of

$$Ly = f(x),$$

where $f(x)$ is itself a solution of a homogeneous equation $Ny = 0$, then

$$N(Ly) = N(f(x)) = 0.$$

Thus $y(x)$ must be among the solutions of the homogeneous equation

$$N(Ly) = 0.$$

If you've found an annihilator N of $f(x)$, the steps in solving $Ly = f(x)$ are as follows:

0. Solve $Ly = 0$ to get the general homogeneous solution $y_h(x)$.
1. Find the general solution $y_N(x)$ of $NLy = 0$.
2. Substitute $y(x) = y_N(x) - y_h(x)$ into $Ly = f(x)$ to determine the remaining coefficients for $y_p(x)$. The general solution to $Ly = f(x)$ is then $y_h(x) + y_p(x)$.

example 7

The differential equation $y''' + y' = x - x^2$ has characteristic equation $r^3 + r = 0$ with 0 and $\pm i$ for roots. Hence the homogeneous solution is

$$y_h = c_1 \cos x + c_2 \sin x + c_3.$$

An annihilator for $x - x^2$ is $N = D^3$, so all solutions of the given equation are among the solutions of

$$D^3(D^3 + D)y = D^4(D^2 + 1)y = 0.$$

These solutions are

$$y = c_1x^3 + c_2x^2 + c_3x + c_4 + c_5 \cos x + c_6 \sin x.$$

The last three terms are homogeneous solutions, so we try $y_p = Ax^3 + Bx^2 + Cx$. At this point we note that the derivative sets of x and x^2 are $\{x, 1\}$ and $\{x^2, x, 1\}$. Since 1 is a homogeneous solution of the given equation, we multiply the elements of the derivative sets by x and keep only the larger set $\{x^3, x^2, x\}$. Thus using Method 3.4 we arrive at the same trial $y_p = Ax^3 + Bx^2 + Cx$. In any case, we compute

$$y_p' = 3Ax^2 + 2Bx + C \quad \text{and} \quad y_p''' = 6A.$$

Hence we want

$$y_p''' + y_p' = 6A + C + 2Bx + 3Ax^2 = x - x^2.$$

Equating coefficients of like powers of x on both sides gives $C + 6A = 0, 2B = 1$, and $3A = -1$, so $A = -\frac{1}{3}$, $B = \frac{1}{2}$, and $C = 2$. Hence the general solution is

$$y = c_4 + c_5 \cos x + c_6 \sin x + 2x + \tfrac{1}{2}x^2 - \tfrac{1}{3}x^3.$$

EXERCISES

Show that each of the following operators L in 1 through 4 is linear by verifying that

$$L(c_1y_1 + c_2y_2) = c_1Ly_1 + c_2Ly_2$$

for arbitrary constants c_1, c_2 and sufficiently differentiable functions y_1, y_2.

1. $Ly = D^2y.$

2. $Ly = D^2y + Dy.$

3. $Ly = 2Dy + xy.$

4. $Ly = 2D^2y + xDy.$

Which of the operators L in Exercises 5 through 8 is linear?

5. $Ly = xDy.$

6. $Ly = xD^2y + xy^2.$

7. $Ly = x^2D^2y + y^2.$

8. $Ly = y^2.$

Each of the linear differential equations in Exercises 9 through 12 has the given function y_p as a solution. Verify this and find the most general solution.

9. $y'' + y = x;\ y_p = x.$

10. $y'' - 2y' + y = 1;\ y_p = 1.$

11. $2y'' - y = e^x;\ y_p = e^x.$

12. $xy' + y = 1;\ y_p = 1.$

Find homogeneous differential equations $Ny = 0$ of least possible order for which the functions in Exercises 13 through 20 are solutions. In other words, find an annihilator N of least possible order for these functions.

13. $e^x + 2e^{2x}$.

14. $e^x \cos x - e^x \sin x$.

15. $x + 1$.

16. $xe^x - 2e^x$.

17. $x \sin 3x$.

18. $x^2 \cos 4x$.

19. $xe^x \sin x$.

20. $x^3 e^{-x} \cos 2x$.

Without computing the constants, find the appropriate form for a trial solution y_p for each of the equations in Exercises 21 through 26.

21. $y'' - y = \cos x$.

22. $y'' + y = \cos x$.

23. $y'' - y = e^x$.

24. $y'' - y = xe^x$.

25. $y'' - 2y' + y = xe^x$.

26. $y'' = x^5$.

For each of the equations in Exercises 27 through 40:

(a) Find the form of a trial particular solution, leaving the coefficients undetermined.

(b) Find the general solution of the equation.

(c) Find the particular solution satisfying the initial conditions $y(0) = 0$, $y'(0) = 1$.

(d) Sketch the graph of the solution found in part (c).

27. $y'' - y = e^{2x}$.

28. $y'' - y = 3e^x$.

29. $y'' + 2y' + y = e^x$.

30. $y'' - 3y = x$.

31. $y'' - y = e^x + x$.

32. $y'' - 2y = \cos 2x$.

33. $y'' + y = \cos x$.

34. $y'' = \cos x + \sin x$.

35. $y'' + y = x \cos x$.

36. $y'' - y = xe^x$.

37. $y'' + 4y' + 4y = 3x$.

38. $y'' - y' - 12y = 2e^{4x}$.

39. $y'' + 2y' + 2y = e^x$.

40. $y'' - y' = x$.

The associated homogeneous equations in Exercises 41 through 50 are the same as in Exercises 1 through 10 of Section 2D. Find the general solution to each of the nonhomogeneous equations in Exercises 41 through 50.

41. $y''' - y = e^x$.

42. $y^{(4)} - 2y'' + y = 5$.

43. $y''' - 2y' = 4$.

44. $y^{(4)} - y'' = x$.

45. $y''' - 8y = e^{2x}$.

46. $y^{(4)} - 4y'' + 4y = \sinh 2x$.

47. $(D^2 - 4)(D^2 - 1)y = \cosh x$.

48. $D(D - 1)^3 y = x - x^2$.

49. $y''' - 2y'' = x^3$.

50. $y^{(4)} - y = \sin x$.

***51.** Let $f(x)$ be a basic homogeneous solution 3.3: $x^n e^{\alpha x} \cos x$ or $x^n e^{\alpha x} \sin x$. Let N be the annihilator of lowest order for $f(x)$. Verify that the derivative set of $f(x)$ is the same as the standard set of basic solutions of $Ny = 0$.

3C Variation of Parameters †

This topic is treated in the context of variable-coefficient equations in Chapter 9, Section 2.

We can always find a pair of linearly independent solutions y_1 and y_2 of the associated homogeneous equation of the constant-coefficient equation $y'' + ay' + by = f(x)$. Reduced to simplest terms these pairs of solutions take one of the two forms

$$y_1(x) = e^{r_1 x}, \quad y_2(x) = e^{r_2 x}, \; r_1 \neq r_2, \quad \text{or} \quad y_1(x) = e^{r_1 x}, \quad y_2(x) = xe^{r_1 x}.$$

If in the former case $r_1 = \alpha + i\beta$, $r_2 = \alpha - i\beta$, then it's sometimes convenient to write $y_1(x) = e^{\alpha x}\cos\beta x$, $y_2(x) = e^{\alpha x}\sin\beta x$. We'll now show that we can in principle always find functions $u_1(x)$ and $u_2(x)$ such that

$$y(x) = u_1(x)y_1(x) + u_2(x)y_2(x)$$

is a particular solution of the nonhomogeneous equation. If u_1 and u_2 were constant parameters, $y(x)$ would be a solution of the associated homogeneous equation, but the additional freedom we gain by allowing u_1 and u_2 to be functions of x enables us to satisfy the nonhomogeneous equation. It's this additional freedom that motivates using the term *variation of parameters.*

The steps leading to the form of the solution described above involve some algebra that we'll carry out once and for all. Then we perform two integrations that are specific to each special problem. To standardize the solution formula, we work with the **normalized equation** in which the coefficient of y'' is 1:

$$y'' + ay' + by = f(x).$$

If y_1, y_2 are independent homogeneous solutions, we form the linear combination

$$y(x) = u_1(x)y_1(x) + u_2(x)y_2(x),$$

where our primary condition on $u_1(x)$ and $u_2(x)$ will be that $y(x)$ should be a solution of the nonhomogeneous equation. We also include a secondary condition that simplifies the computation of $u_1(x)$ and $u_2(x)$.

3.5 THEOREM
Let $y_1(x)$ and $y_2(x)$ be linearly independent solutions of $y'' + ay' + by = 0$. If the functions $u_1(x)$ and $u_2(x)$ are chosen so that their derivatives satisfy

$$\begin{aligned} y_1(x)u_1'(x) + y_2(x)u_2'(x) &= 0, \\ y_1'(x)u_1'(x) + y_2'(x)u_2'(x) &= f(x), \end{aligned}$$

then $y_p(x) = u_1(x)y_1(x) + u_2(x)y_2(x)$ will be a solution of the nonhomogeneous equation $y'' + ay' + by = f(x)$.

Proof. Preparing to force $y_p(x)$ to satisfy the differential equation $y_p(x)$ into the differential equation, we compute the derivative

$$y_p' = y_1'u_1 + y_1u_1' + y_2'u_2 + y_2u_2'.$$

Since we have two functions to determine, we can impose a second condition on u_1 and u_2, namely, that

$$y_1u_1' + y_2u_2' = 0,$$

so that y'_p simplifies to $y'_p = y'_1 u_1 + y'_2 u_2$. Now compute y''_p from this simplified expression for y'_p. We find

$$y''_p = y''_1 u_1 + y''_2 u_2 + y'_1 u'_1 + y'_2 u'_2.$$

Substitution into $y'' + ay' + by = f$ and regrouping terms gives

$$(y''_1 + ay'_1 + by_1)u_1 + (y''_2 + ay'_2 + by_2)u_2 + y'_1 u'_1 + y'_2 u'_2 = f.$$

Since y_1 and y_2 satisfy the associated homogeneous equation, the first two terms equal zero. Thus $y'_1 u'_1 + y'_2 u'_2 = f$ for our second requirement. ■

---

example 8

The equation $y'' - y = e^x$ has homogeneous solutions $y_1(x) = e^x$ and $y_2(x) = e^{-x}$. Since $f(x) = e^x$, $y'_1(x) = e^x$, and $y'_2(x) = -e^{-x}$, the two conditions of Theorem 3.5 translate into

$$e^x u'_1 + e^{-x} u'_2 = 0,$$
$$e^x u'_1 - e^{-x} u'_2 = e^x.$$

Adding these two equations gives $2e^x u'_1 = e^x$, so $u'_1 = \frac{1}{2}$. We integrate and take $u_1 = \frac{1}{2}x$. Subtracting the second equation from the first gives $2e^{-x} u'_2 = -e^x$, so $u'_2 = -\frac{1}{2}e^{2x}$. We integrate and take $u_2 = -\frac{1}{4}e^{2x}$. According to Theorem 3.5, we'll get a solution of the nonhomogeneous differential equation by writing

$$\begin{aligned} y_p(x) &= u_1(x)y_1(x) + u_2(x)y_2(x) \\ &= \left(\tfrac{1}{2}x\right)(e^x) + \left(-\tfrac{1}{4}e^{2x}\right)(e^{-x}) = \tfrac{1}{2}xe^x - \tfrac{1}{4}e^x. \end{aligned}$$

We recognize the term $-\frac{1}{4}e^x$ as a part of the general homogeneous solution, so we write the general solution of the nonhomogeneous equation as

$$y = \tfrac{1}{2}xe^x + c_1 e^x + c_2 e^{-x}.$$

The previous example was chosen for simplicity. We could have worked through the example by just starting with the observation that a trial particular solution should have the form $y = Axe^x$ and then finding the constant A by substitution into the differential equation. Before illustrating the variation of parameters method with a more compelling example we'll outline in general terms the steps that follow the application of Theorem 3.5. The equations

3.6

$$y_1 u'_1 + y_2 u'_2 = 0,$$
$$y'_1 u'_1 + y'_2 u'_2 = f,$$

are the result of applying the theorem. We can solve this system of equations for u'_1 and u'_2 by elimination or by Cramer's rule (Appendix B, Theorem 2.9). Either way we find

$$u_1'(x) = \frac{\begin{vmatrix} 0 & y_2 \\ f & y_2' \end{vmatrix}}{\begin{vmatrix} y_1 & y_2 \\ y_1' & y_2' \end{vmatrix}} = \frac{-y_2(x)f(x)}{y_1(x)y_2'(x) - y_2(x)y_1'(x)},$$

$$u_2'(x) = \frac{\begin{vmatrix} y_1 & 0 \\ y_1' & f \end{vmatrix}}{\begin{vmatrix} y_1 & y_2 \\ y_1' & y_2' \end{vmatrix}} = \frac{y_1(x)f(x)}{y_1(x)y_2'(x) - y_2(x)y_1'(x)}.$$

The expression in the denominators is the same in both formulas and we can write it as the two-by-two determinant

$$w(x) = \begin{vmatrix} y_1(x) & y_2(x) \\ y_1'(x) & y_2'(x) \end{vmatrix} = y_1(x)y_2'(x) - y_2(x)y_1'(x),$$

called the **Wronskian determinant** of y_1, y_2. Since we're dealing with second-order constant-coefficient equations, we can check directly that $w(x)$ is never zero:

(i) If $y_1 = e^{r_1 x}$, $y_2 = e^{r_2 x}$ if $r_1 \neq r_2$, then $w(x) = (r_2 - r_1)e^{(r_1+r_2)x}$.
(ii) If $y_1 = e^{r_1 x}$, $y_2 = xe^{r_1 x}$, then $w(x) = e^{2r_1 x}$.

If the solutions y_1 and y_2 were not independent, one of them would be a constant multiple of the other, so $w(x)$ would be zero for all x.

To complete the solution, integrate the formulas for u_1', u_2' to find u_1 and u_2, and then combine with y_1, y_2 to get a particular solution

3.7
$$\begin{aligned} y_p(x) &= y_1(x)u_1(x) + y_2(x)u_2(x) \\ &= y_1(x)\int \frac{-y_2(x)f(x)}{w(x)}\,dx + y_2(x)\int \frac{y_1(x)f(x)}{w(x)}\,dx. \end{aligned}$$

Because Equations 3.6 are easier to remember than Equation 3.7, people often prefer to start with them in each problem and carry out the rest of the computation to arrive at Equation 3.7. Because of the way $f(x)$ enters Equations 3.6 and 3.7, it's essential that the differential equation be in normalized form if Equations 3.6 and 3.7 are to be valid.

example 9

The equation $y'' - 2y' + y = e^x/x$ can't be solved by the undetermined coefficient method, because $f(x) = e^x/x$ is not itself a solution of constant-coefficient homogeneous differential equation. The characteristic equation $r^2 - 2r + 1 = 0$ has $r = 1$ for a double root, so we can take $y_1(x) = e^x$ and $y_2(x) = xe^x$ for a convenient pair of independent solutions of the associated homogeneous equation. We have

$$w(x) = \begin{vmatrix} e^x & xe^x \\ e^x & e^x + xe^x \end{vmatrix} = e^{2x} + xe^{2x} - xe^{2x} = e^{2x}.$$

Hence Equation 3.7 becomes

$$\begin{aligned}
y_p(x) &= y_1(x)\int \frac{-y_2(x)f(x)}{w(x)}\,dx + y_2(x)\int \frac{y_1(x)f(x)}{w(x)}\,dx \\
&= e^x \int \frac{-xe^x(e^x/x)}{e^{2x}}\,dx + xe^x\int \frac{e^x(e^x/x)}{e^{2x}}\,dx. \\
&= -xe^x + xe^x \ln|x|.
\end{aligned}$$

The first term in our expression for $y_p(x)$ satisfies the associated homogeneous equation, so we can absorb that term into the general homogeneous solution. The general solution of the nonhomogeneous equation is then

$$y = c_1e^x + c_2xe^x + xe^x\ln|x|,\ x \neq 0.$$

EXERCISES

Use variation of parameters to find a particular solution $y_p(x)$ for each of the following constant coefficient equations.

1. $y'' - y = e^x$.

2. $y'' + 2y + y = e^x$.

3. $4y'' + y = 1$.

4. $y'' + 6y' + 9y = e^{5x}$.

5. $y'' - 2y' + y = e^x/x^2$.

6. $y'' - 2y' + y = e^x/x^3$.

7. $y'' + y = \sec x$.

8. $y'' + y = \csc^2 x$.

9. $y'' + y = \tan x$.

10. $y'' + 3y' + 2y = (1 + e^x)^{-1}$.

11. Find a particular solution $y'' + 4y = \sin^2 x$ in two ways.

(a) Use variation of parameters.

(b) Use the identity $\sin^2 x = \frac{1}{2}(1 - \cos 2x)$ and the method of undetermined coefficients.

12. Verify that the assertions in the text that

(a) if $y_1(x) = e^{r_1x}$ and $y_2(x) = e^{r_2x}$ with $r_1 \neq r_2$, then $w(x) = (r_2 - r_1)e^{(r_1+r_2)x}$.

(b) if $y_1(x) = e^{r_1x}$ and $y_2(x) = xe^{r_1x}$ with $r_1 = r_2$, then $w(x) = e^{2r_1x}$.

13. Assume $y_1(x)$ and $y_2(x)$ are differentiable on some interval and that $y_2(x) = cy_1(x)$ for some constant c. Show that

$$w(x) = \begin{vmatrix} y_1(x) & y_2(x) \\ y_1'(x) & y_2'(x) \end{vmatrix} = 0$$

for all x in the interval.

3D Integral Transform Formulas †

The proof of the following Theorem 3.8 just amounts to solving $(D - r_1)(D - r_2)y = f(x)$ by successive integrations as in Section 1, and then finding a formula for the solution in terms of r_1, r_2 and f. We'll defer the proof in favor of

Because they require only integration, the integration methods are often preferred to finding a correct trial solution and computing undetermined coefficients. Variation of parameters treated in Section 3C, and in Chapter 9, Section 2, for equations with nonconstant coefficients, applies in principle when the forcing function $f(x)$ is of a more general type than the ones to which the undetermined coefficient method applies. The method of this section has the additional advantages that it applies to discontinuous forcing functions, for example functions representing on–off inputs.

an interpretation and some examples. *Note that the differential equation must first be normalized so that the coefficient of y'' equals 1.*

3.8 THEOREM

Assume $f(x)$ is continuous on an interval containing x_0, and let $y_p(x)$ be the solution of the normalized constant coefficient equation

$$y'' + ay' + by = f(x)$$

satisfying $y_p(x_0) = y_p'(x_0) = 0$. If r_1, r_2 are the roots of the characteristic equation $r^2 + ar + b = 0$, then

(i) if r_1, r_2 are real and unequal,

$$y_p(x) = \int_{x_0}^{x} \frac{e^{r_1(x-t)} - e^{r_2(x-t)}}{r_1 - r_2} f(t)\, dt,$$

(ii) if $r_1 = r_2$,

$$y_p(x) = \int_{x_0}^{x} (x-t)e^{r_1(x-t)} f(t)\, dt,$$

(iii) if $r_1, r_2 = \alpha \pm i\beta$, $\beta \neq 0$,

$$y_p(x) = \int_{x_0}^{x} \frac{1}{\beta} e^{\alpha(x-t)} \sin \beta(x-t) f(t)\, dt.$$

In each case (i), (ii), and (iii) we use an integral operator L^{-1} defined by

$$L^{-1} f(x) = \int_{x_0}^{x} g(x-t) f(t)\, dt,$$

where the variable x is held fixed during the integration. The operator L^{-1} acts as an **inverse integral operator** to the differential operator $D^2 + aD + b$ with initial conditions $y(x_0) = y'(x_0) = 0$. The function g has one of the three forms

$$\text{(i)} \quad g(u) = \frac{1}{r_1 - r_2}(e^{r_1 u} - e^{r_2 u}), \qquad r_1 \neq r_2,$$

$$\text{(ii)} \quad g(u) = ue^{r_1 u}, \qquad r_1 = r_2,$$

$$\text{(iii)} \quad g(u) = \frac{1}{\beta} e^{\alpha u} \sin \beta u, \qquad r_1 = \alpha + i\beta,\ r_2 = \alpha - i\beta,\ \beta \neq 0.$$

An integral of the form $L^{-1} f(x)$ displayed above is an example of an **integral transform** of a type called a **convolution integral** that converts the function $f(x)$ into $y_p(x)$. Since the variable x that occurs in the upper limit and in the difference $x - t$ must be treated as a constant with respect to integration in the other variable, some care is necessary in doing the integration. The examples show, however, that if you want to you can separate the x variable from the integration process. For example, the addition formula for the sine function enables us to write

$$\int \frac{1}{\beta} e^{\alpha(x-t)} \sin \beta(x-t) f(t)\, dt = \frac{1}{\beta} e^{\alpha x} \sin \beta x \int e^{-\alpha t} \cos(\beta t)\, f(t)\, dt$$
$$- \frac{1}{\beta} e^{\alpha x} \cos \beta x \int e^{-\alpha t} \sin(\beta t)\, f(t)\, dt.$$

example 10

Here is an equation in which the right-hand side is itself a solution of the associated homogeneous equation:

$$y'' + 2y' + y = e^{-x}.$$

The homogeneous solution is $y_h = c_1 xe^{-x} + c_2 e^{-x}$, corresponding to a double root $r = -1$ of the equation $r^2 + 2r + 1 = 0$. The function $g(x-t) = (x-t)e^{-(x-t)}$. Since the differential equation is normalized, we can compute the solution satisfying $y(0) = y'(0) = 0$ from

$$y_p(x) = \int_0^x (x-t)e^{-(x-t)}e^{-t}\,dt.$$

Note that integration is with respect to t, so x is temporarily fixed. We find

$$\begin{aligned} y_p(x) &= \int_0^x xe^{-(x-t)}e^{-t}\,dt - \int_0^x te^{-(x-t)}e^{-t}\,dt \\ &= xe^{-x}\int_0^x dt - e^{-x}\int_0^x t\,dt \\ &= xe^{-x}[t]_0^x - e^{-x}\left[\tfrac{1}{2}t^2\right]_0^x \\ &= xe^{-x}(x-0) - e^{-x}\left(\tfrac{1}{2}x^2 - 0\right) \\ &= x^2e^{-x} - \tfrac{1}{2}x^2e^{-x} = \tfrac{1}{2}x^2e^{-x}. \end{aligned}$$

This is the solution $y_p(x)$ with $y_p(0) = y_p'(0) = 0$. By Corollary 3.2 the general solution is

$$y(x) = y_h(x) + y_p(x) = c_1 xe^{-x} + c_2 e^{-x} + \tfrac{1}{2}x^2e^{-x}.$$

To satisfy initial conditions of the form $y(0) = y_0$, $y'(0) = z_0$ we use the flexibility inherent in the general homogeneous part of the solution $y_h(x)$. Since the particular solution $y_p(x)$ given by the integral transform formula satisfies $y_p(0) = 0$, $y_p'(0) = 0$, we just need to determine values for c_1 and c_2 in $y_h(x)$ so that the independent solutions $y_1(x) = xe^{-x}$ and $y_2(x) = e^{-x}$ satisfy

$$c_1 y_1(0) + c_2 y_2(0) = y_0 \quad \text{and} \quad c_1 y_1'(0) + c_2 y_2'(0) = z_0.$$

For this example, we have $y_1(0) = 0$, $y_2(0) = 1$, $y_1'(0) = 1$, and $y_2'(0) = -1$. Hence we solve

$$0c_1 + c_2 = y_0 \quad \text{and} \quad c_1 - c_2 = z_0$$

to get $c_2 = y_0$ and $c_1 = y_0 + z_0$. Thus our solution is $y(x) = x^2e^{-x} + (y_0 + z_0)xe^{-x} + y_0 e^{-x}$.

example 11

Here is an example similar to the previous one, but involving trigonometric functions:

$$y'' + 9y = \sin 3x.$$

The characteristic roots of the associated homogeneous equation are $r_1 = 3i$, $r_2 = -3i$. The function $g(x-t)$ is then $g(x-t) = \frac{1}{3}\sin 3(x-t)$, so the integrand in the integral transform formula is then $\frac{1}{3}\sin 3(x-t)\sin 3t$. The integration with respect to t follows from using Formula 40 of the table of integrals in Appendix D, or more directly using the identity that proves Formula 40: $\sin A \sin B = \frac{1}{2}\cos(A - B) - \frac{1}{2}\cos(A + B)$. Thus the particular solution satisfying $y_p(0) = y_p'(0) = 0$ is

$$\begin{aligned} y_p &= \tfrac{1}{3}\int_0^x \sin 3(x-t)\sin 3t\,dt \\ &= \tfrac{1}{6}\int_0^x \big[\cos(3x-6t) - \cos 3x\big]\,dt \\ &= \tfrac{1}{6}\big[-\tfrac{1}{6}\sin(3x-6t) - t\cos 3x\big]_0^x \\ &= \tfrac{1}{18}\sin 3x - \tfrac{1}{6}x\cos 3x. \end{aligned}$$

The next example can't be carried through by undetermined coefficients except by breaking the problem into two separate parts and getting new initial conditions for the second part from the terminal values from the previous part. The key to avoiding this nuisance is to use the integral transform by breaking the integration interval at points a where the form of the integrand changes. Then treat the integration limits this way:

$$\int_{x_0}^x g(x-t)f(t)\,dt = \int_{x_0}^a g(x-t)f(t)\,dt + \int_a^x g(x-t)f(t)\,dt.$$

Care has to be taken to distinguish the values of the upper limit x that lie in the different intervals.

---

example 12 The equation $y'' - y = f(x)$ has the associated homogeneous solution $y_h = c_1e^x + c_2e^{-x}$, so

$$g(x-t) = \tfrac{1}{2}\big[e^{(x-t)} - e^{-(x-t)}\big].$$

Suppose $f(x)$ is given by

$$f(x) = \begin{cases} 0, & x < 1, \\ \frac{1}{2}, & 1 \le x. \end{cases}$$

The solution satisfying $y(0) = y'(0) = 0$ is

$$y(x) = \tfrac{1}{2}\int_0^x \big[e^{(x-t)} - e^{-(x-t)}\big]f(t)\,dt.$$

Since $f(t) = 0$ for $t < 1$, we clearly have $y(x) = 0$ for $x < 1$. When $x \ge 1$, we find

$$\begin{aligned} y(x) &= \tfrac{1}{2}\int_1^x \big[e^{(x-t)} - e^{-(x-t)}\big]\big[\tfrac{1}{2}\big]\,dt \\ &= \tfrac{1}{4}e^x\int_1^x e^{-t}\,dt - \tfrac{1}{4}e^{-x}\int_1^x e^t\,dt \\ &= \tfrac{1}{4}e^x\big[-e^{-t}\big]_1^x - \tfrac{1}{4}e^{-x}\big[e^t\big]_1^x \\ &= -\tfrac{1}{4} + \tfrac{1}{4}e^{(x-1)} - \tfrac{1}{4} + \tfrac{1}{4}e^{-(x-1)} = \tfrac{1}{2}\cosh(x-1) - \tfrac{1}{2}. \end{aligned}$$

Just as $f(x)$ is defined by different formulas on different intervals, so is the solution $y(x)$:

$$y(x) = \begin{cases} 0, & x < 1, \\ \frac{1}{2}\cosh(x-1) - \frac{1}{2}, & 1 \le x, \end{cases}$$

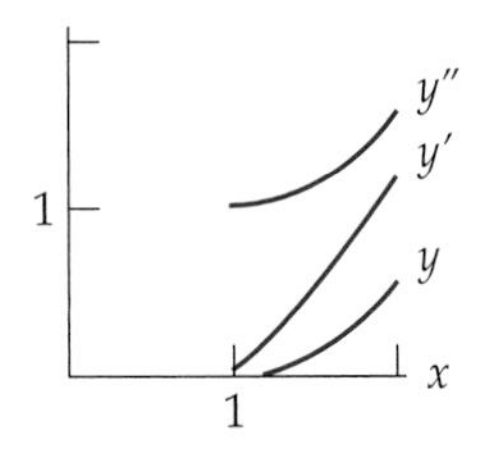

Since the forcing function $f(x)$ is discontinuous in the previous example, we expect $y''(x)$ to be discontinuous at $x = 1$. The functions $y(x)$, $y'(x)$, and $y''(x)$, with graphs as shown, are identically zero for $x \le 0$. The two pieces of $y(x)$ fit together at $x = 1$ in such a way that $y(x)$ has one continuous derivative, though the graph of $y'(x)$ has a sharp corner at $x = 1$. Examples like this prompt the broadening of the idea of "solution" to "generalized solution," discussed in the exercises.

Similar problems are often treated using Laplace transforms, as in Chapter 8, Section 1B. If what you really want is only a solution graph or a table of values rather than a formula, then the numerical method of Chapter 4, Section 5B, may be preferable.

DEFERRED PROOF OF THEOREM 3.8

We factor the operator L defined by $Ly = y'' + ay' + by$ and write

$$(D - r_1)(D - r_2)y = f(x), \quad (D - r_2)y = z.$$

Then solve $(D - r_1)z = f(x)$ by first multiplying by $e^{-r_1 x}$. The result is

$$e^{-r_1 x} z = \int e^{-r_1 x} f(x)\, dx + c.$$

We choose the constant c so that $z(x_0) = 0$. This can be done by using a definite integral, and we get

$$z(x) = e^{r_1 x} \int_{x_0}^{x} e^{-r_1 s} f(s)\, ds.$$

Now solve $(D - r_2)y = z$ by the same method:

$$\begin{aligned} y(x) &= e^{r_2 x} \int_{x_0}^{x} e^{-r_2 t} z(t)\, dt \\ &= e^{r_2 x} \int_{x_0}^{x} e^{(r_1 - r_2)t} \left[\int_{x_0}^{t} e^{-r_1 s} f(s)\, ds \right] dt. \end{aligned}$$

Here we have replaced x by t in the integral for z. Setting $x = x_0$ shows that $y(x_0) = 0$ and $z(x_0) = 0$. Since $z(x_0) = y'(x_0) - r_2 y(x_0)$, it follows that $y'(x_0) = 0$ also.

Now integrate by parts in t using $u = \int_{x_0}^{t} e^{-r_1 s} f(s)\,ds$, $dv = e^{(r_1-r_2)t}\,dt$. Assuming $r_1 \neq r_2$, we get

(i)
$$y(x) = \frac{1}{r_1 - r_2} e^{r_1 x} \int_{x_0}^{x} e^{-r_1 s} f(s)\,ds - \frac{1}{r_1 - r_2} e^{r_2 x} \int_{x_0}^{x} e^{-r_2 t} f(t)\,dt$$
$$= \frac{1}{r_1 - r_2} \int_{x_0}^{x} \left[e^{r_1(x-t)} - e^{r_2(x-t)}\right] f(t)\,dt.$$

A similar integration by parts when $r_1 - r_2 = 0$ (see Exercise 29) gives

(ii)
$$y(x) = \int_{x_0}^{x} (x-t) e^{r_1(x-t)} f(t)\,dt.$$

If $r_1 = \alpha + i\beta$ and $r_2 = \alpha - i\beta$, then $r_1 - r_2 = 2i\beta$, and

$$\frac{1}{r_1 - r_2}\left[e^{r_1(x-t)} - e^{r_2(x-t)}\right] = \frac{1}{\beta} e^{\alpha(x-t)} \sin \beta(x-t).$$

Substituting this into (i) gives part (iii) of Theorem 3.8.

EXERCISES

For each of the differential equations in Exercises 1 through 8 find the function $g(x-t)$ for an initial-value problem $y(x_0) = y'(x_0) = 0$. Then write the integral transform formula for the solution.

1. $y'' - 4y = f(x)$.

2. $y'' + 9y = f(x)$.

3. $y'' - 5y' + 6y = f(x)$.

4. $y'' + 2y' + 3y = f(x)$.

5. $y'' - 2y' + y = f(x)$.

6. $y'' - 4y' + 13y = f(x)$.

7. $2y'' - 3y' + 2y = f(x)$.

8. $-y'' + 4y' - 29y = f(x)$.

For each differential equation in Exercises 9 through 16, first find the general solution of the associated homogeneous equation; then find the associated function $g(x-t)$ and use it to find the particular solution satisfying the given initial condition of the special form $y(x_0) = y'(x_0) = 0$.

9. $y'' - y = x;\ x_0 = 0$.

10. $y'' + y = x;\ x_0 = 0$.

11. $y'' + 2y' + y = e^{2x};\ x_0 = 0$.

12. $y'' + 2y' + y = 0;\ x_0 = 1$.

13. $y'' - 4y = e^{x};\ x_0 = -1$.

14. $y'' - 3y' + 2y = e^{-x};\ x_0 = 2$.

15. $y'' + 9y = \sin x;\ x_0 = 0$.

16. $y'' + 3y' + 2y = e^{-x};\ x_0 = 0$.

In Exercises 17 through 22, find the particular solution to $y'' - 4y = f(x)$ satisfying $y(0) = y'(0) = 0$ with $f(x)$ as given.

17. $f(x) = \begin{cases} 1, & 0 \le x, \\ 0, & x < 0. \end{cases}$

18. $f(x) = \begin{cases} 0, & 0 \le x, \\ 1, & x < 0. \end{cases}$

19. $f(x) = \begin{cases} 1, & 0 \le x, \\ -1, & x < 0. \end{cases}$

20. $f(x) = \begin{cases} x, & 0 \le x, \\ 0, & x < 0. \end{cases}$

21. $f(x) = \begin{cases} x - 1, & 1 \le x, \\ 0, & x < 1. \end{cases}$

22. $f(x) = x^2 e^{2x}$.

23. In $d^2x/dt^2 + a(dx/dt) + bx = f(t)$, the letter t usually stands for time.

(a) Show that the integral transform solution can be written
$x_p(t) = \int_{t_0}^{t} g(t-u) f(u)\, du.$

(b) Use part (a) to solve $d^2x/dt^2 + x = \sin t$, if $x(0) = 0$ and $dx/dt(0) = 1$.

24. Consider $d^2x/du^2 + a(dx/du) + bx = f(u)$.

(a) Find suitable variables for the general integral transform formula for a solution $x(u)$ satisfying $x(0) = dx/du(0) = 0$.

(b) Use the result of part (a) to solve $d^2x/du^2 - x = e^u$, if $x(0) = 0$ and $dx/du(0) = 1$.

25. Since the forcing function of the equation $y'' + 3y' + 2y = (1 + e^x)^{-1}$ is not itself a solution of a homogeneous constant-coefficient linear equation, the undetermined-coefficient method won't work. Find the particular solution satisfying $y(0) = y'(0) = 0$. $\left[\textit{Hint: } e^{2t}/(1+e^t) = e^t(1 - 1/(1+e^t)).\right]$

26. Since the right-hand side of the equation $y'' - 2y' + y = e^x/x$ is not itself a solution of a homogeneous constant-coefficient linear equation, the undetermined-coefficient method won't work. Note also that the equation has no solution on an interval containing $x = 0$, since the right-hand side blows up there.

(a) Find the particular solution satisfying $y(1) = y'(1) = 0$ for $0 < x$.

(b) Find the particular solution satisfying $y(-1) = y'(-1) = 0$ for $x < 0$.

27. Find the general solution of each of the following differential equations by writing the equation in the form

$$(D - r_1)(D - r_2)y = f(x),$$

letting $z = (D - r_2)y$ and then solving in succession the two first-order equations $(D - r_1)z = f(x)$ and $(D - r_2)y = z$.

(a) $y'' - y' - 2y = e^x$.

(b) $y'' - 4y = e^{3x}$.

28. **(a)** Show that if $r_1 = \alpha + i\beta$, $r_2 = \alpha - i\beta$, then

$$\frac{1}{r_1 - r_2}\left(e^{r_1 u} - e^{r_2 u}\right) = \frac{1}{\beta} e^{\alpha u} \sin \beta u.$$

(b) Show that

$$\lim_{\beta \to 0} \frac{1}{\beta} e^{\alpha u} \sin \beta u = u e^{\alpha u}.$$

(c) Let r_1 and r_2 be real-valued. Show that

$$\lim_{r_1 \to r_2} \frac{1}{r_1 - r_2}(e^{r_1 u} - e^{r_2 u}) = u e^{r_2 u}.$$

The limit relations in parts (b) and (c) make the form of the function $g(x-t)$ plausible for the case of equal roots, though they don't qualify immediately as proofs of the solution formulas.

29. To complete the derivation of the function $g(x-t)$ for the case of equal roots, use integration by parts to show that

$$e^{rx} \int_{x_0}^{x} \left[\int_{x_0}^{t} e^{-rs} f(s)\, ds \right] dt = \int_{x_0}^{x} (x-t) e^{r(x-t)} f(t)\, dt.$$

[*Hint:* Let $u = \int_{x_0}^{t} e^{-rs} f(s)\, ds$, $dv = dt$.]

***30.** There are integral transform formulas for equations of order higher than 2. Here it is for normalized third-order constant-coefficient equations $Ly = f(x)$ with initial conditions $y(x_0) = y'(x_0) = y''(x_0) = 0$ and distinct characteristic roots r_1, r_2, r_3:

$$y_p(x) = \int_{x_0}^{x} \left[\frac{e^{r_1(x-t)}}{(r_1-r_2)(r_1-r_3)} + \frac{e^{r_2(x-t)}}{(r_2-r_1)(r_2-r_3)} + \frac{e^{r_3(x-t)}}{(r_3-r_1)(r_3-r_2)} \right] f(t)\, dt.$$

The derivation follows the same lines as that given in the text for second-order equations.

(a) Express the third order formula in terms of r_1, α, and β in case r_1 is real, $r_2 = \alpha + i\beta$, and $r_3 = \alpha - i\beta$, $\beta > 0$.

(b) What do you think the analogous formula is in case $r_1 \neq r_2 = r_3$?

Generalized Solutions. A differentiable function that has one continuous derivative and fails to have a second derivative at some points of definition, but which otherwise satisfies a second-order equation is an example of a **generalized solution.** (This term should not be confused with the term *general solution* that we sometimes use for an entire family of solutions.)

31. Verify that the solution obtained in Example 12 of the text has a continuous first derivative at all points, but that its second derivative fails to exist at $x = 1$.

32. Consider the differential equation $y'' = \begin{cases} -1, & x < 0, \\ 1, & 0 \le x, \end{cases}$ for all real x. Verify that $y = \begin{cases} -x^2/2, & x < 0, \\ x^2/2, & 0 \le x. \end{cases}$ is a generalized solution: (i) $y(x)$ satisfies the differential equation everywhere but at $x = 0$; (ii) $y'(x)$ is continuous for all x; (iii) $y''(0)$ fails to exist.

33. The purpose of this exercise is to show that if $f(x)$ is continuous on some interval containing a, then the integral transform formula $y(x) = \int_a^x g(x-t) f(t)\, dt$ represents a function with two continuous derivatives on the interval.

(a) Consider first the case $g(x-t) = (e^{r_1(x-t)} - e^{r_2(x-t)})/(r_1 - r_2)$, breaking up the integral formula for $y(x)$ so that the variable x doesn't occur under an integral sign.

(b) Treat as in part (a) the case $g(x-t) = (x-t)e^{r(x-t)}$.

34. What is the form of the integral transform operator for the second-order initial-value problem $y'' = f(x)$, $y(0) = y'(0) = 0$?

35. What is the form of the integral solution operator for the general first-order initial-value problem $y' - ry = f(x)$, $y(0) = 0$, where r is constant?

36. Show that the inverse integral operator L^{-1} defined in connection with Theorem 3.8 is linear as defined by Equation 1.1 in Section 1.

37. The function $g(x-t)$ associated with each linear differential operator L is easy to remember because $g(x)$ is a solution of $Ly = 0$. Show that the function $g(x - t_0)$ associated with a differential operator $L = D^2 + aD + b$, and with t_0 fixed, satisfies the homogeneous differential equation $Ly = 0$ along with the initial conditions $y(t_0) = 0$, $y'(t_0) = 1$. Do this for each of the three cases separately: (a) $r_1,\ r_2$ both real, $r_1 \neq r_2$; (b) $r_1 = r_2$; and (c) $r_1 = \alpha + i\beta$, $r_2 = \alpha - i\beta$ complex conjugate.

Interpretation. Theorem 3.8 then says that the solution $y_p(x)$ of the nonhomogeneous equation is an integrated combination of homogeneous solutions shifted by t over the interval between x_0 and x. Such an integral can be regarded as a kind of generalized linear combination of homogeneous solutions.

4 NONLINEAR EQUATIONS†

The general second-order differential equation of the form $d^2y/dt^2 = f(t, y, dy/dt)$ has as special cases many second-order equations other than the constant-coefficient linear ones discussed earlier in the chapter. In particular, the equation may be nonlinear, or it may be linear but with nonconstant coefficients. Even though the computer-driven numerical methods treated in the next chapter are quite adequate for some purposes, there are often good reasons for pursuing general solution formulas. One is that the dependence of a solution on parameters is clarified when it is displayed in a formula. Another is that solution formulas often enable us to settle questions about the behavior of solutions as the independent variable tends to infinity.

We use t for the independent variable in this section because it will represent time in most of the examples. Thus $y = y(t)$ will denote a solution, and Newton's convenient overdot notation $\dot{y}(t)$ and $\ddot{y}(t)$ will denote its first and second time derivatives. Apart from numerical approximations, there aren't any universally applicable methods for solving $\ddot{y} = f(t, y, \dot{y})$ explicitly. Section 4A uses order reduction, already used in Sections 1 and 2 for constant-coefficient linear equations, under the assumption that the differential equation is not explicitly dependent of the position variable y. In Section 4B the equation is assumed to be *autonomous*, which means that the equation is not explicitly dependent on the variable t, for example, the pendulum equation $\ddot{y} = -y$. In these two subsections, the standard integral tables can sometimes be used to good advantage.

4A Independence of Position: $\ddot{y} = f(t,\dot{y})$

The yz space is an example of a "state space" in which present states determine future states if $f(t, y)$ and $f_y(t, y)$ are continuous; for historical reasons the term "phase space" also occurs, for example, in Chapter 4, Section 3.

Let's assume we have an equation that doesn't involve the unknown function y directly, but contains only t, $\dot{y}$, and $\ddot{y}$ explicitly. The trick is again to use order reduction and introduce the new function $z = z(t)$ by setting $\dot{y} = z$. Then $\ddot{y} = \dot{z}$, and we replace $\ddot{y}$ and $\dot{y}$ in the original equation to get the pair of equations

$$\dot{y} = z,$$
$$\dot{z} = f(t, z).$$

This is a first-order system of equations, but of a very special kind. Most significantly, the second equation $\dot{z} = f(t, z)$ doesn't contain y. So we try to solve this equation for z by treating it as a linear equation if it is one, or by any other method, for example, separation of variables, that seems appropriate. Once $z(t)$ is known, then we can find y by integrating both sides of $\dot{y} = z(t)$. Note that $\dot{y}$ and $\ddot{y}$ are unaltered by adding a constant to y. Since $\ddot{y} = f(t, \dot{y})$ is independent of y, adding a constant to a solution always gives another solution, and solution formulas in the following examples will consequently show an arbitrary additive constant. Geometrically this means that shifting a solution graph up or down along the y axis always produces another solution graph.

example 1

In the linear equation

$$t\ddot{y} + \dot{y} = 0$$

we set $\dot{y} = z$, so $\ddot{y} = \dot{z}$ to get

$$t\dot{z} + z = 0.$$

We can then treat this as a standard first-order linear equation. Or we may recognize directly from the product rule that

$$t\dot{z} + z = \frac{d}{dt}tz,$$

so integration gives

$$tz = c_1.$$

Then using our substitution $\dot{y} = z$ in the other direction gives

$$t\dot{y} = c_1 \quad \text{or} \quad \dot{y} = c_1 t^{-1}.$$

Another integration gives the solution

$$y = c_1 \ln|t| + c_2, \quad t \neq 0.$$

example 2

Consider the nonlinear equation

$$\ddot{y} - 2t(\dot{y})^2 = 0.$$

Since y is absent, we set $\dot{y} = z$ to get $\ddot{y} = \dot{z}$ and $\dot{z} = 2tz^2$. The variables separate giving

$$z^{-2}\dot{z} = 2t.$$

One integration gives

$$-z^{-1} = t^2 + c_1 \quad \text{or} \quad z = \frac{-1}{t^2 + c_1}.$$

Since $\dot{y} = z$, we integrate again to get

$$y = -\int \frac{dt}{t^2 + c_1} + c_2.$$

Consulting a table of indefinite integrals shows that there are three separate cases. Graphs of each type appear in Figure 3.6, where we have assumed $c_2 = 0$.

$$c_1 > 0 : y = \frac{-1}{\sqrt{c_1}} \arctan\left(\frac{t}{\sqrt{c_1}}\right) + c_2,$$

$$c_1 = 0 : y = \frac{1}{t} + c_2,$$

$$c_1 < 0 : y = \frac{1}{2\sqrt{|c_1|}} \ln\left|\frac{\sqrt{|c_1|} + t}{\sqrt{|c_1|} - t}\right| + c_2.$$

If $y(0) = 0$, the formula with $c_1 > 0$ provides solutions that pass through points in the second and fourth quadrants, and the one with $c_1 < 0$ provides the rest. The appearance of the first of these two formulas could be improved by renaming the positive constant $\sqrt{c_1}$ simply c_1, and the other one by renaming the positive constant $\sqrt{|c_1|}$ simply $-c_1$. However, that would complicate the simple relationship between c_1 and the initial values for $\dot{y}$. For example, as things stand, $\dot{y}(0) = -1/c_1$ as long as $c_1 \neq 0$.

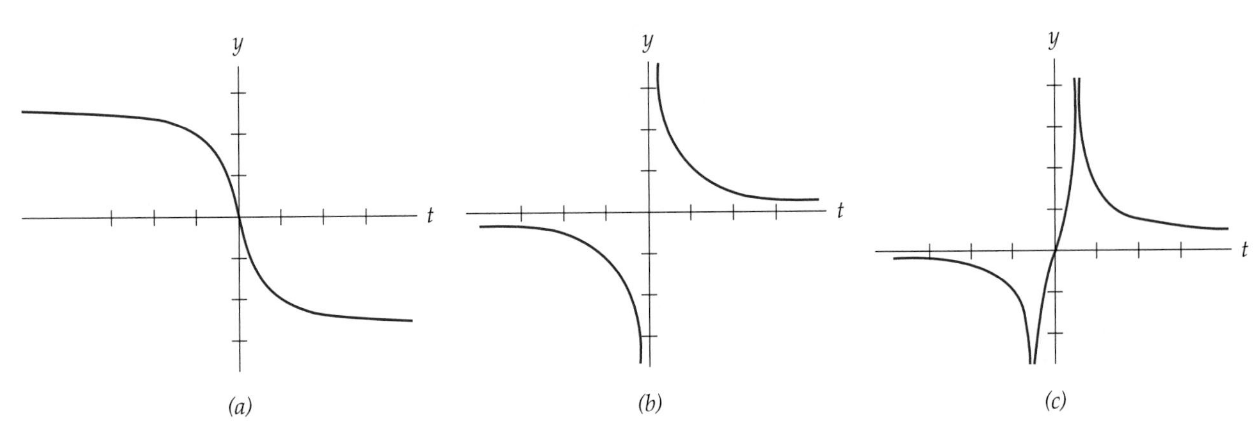

Figure 3.6 (a) $c_1 > 0$. (b) $c_1 = 0$. (c) $c_1 < 0$.

4B Independence of Time: $\ddot{y} = f(y, \dot{y})$

An equation in which the independent variable is not explicitly present is called **autonomous.** A solution to such an equation is still a function of the independent variable t, and is defined on one or more intervals. Because the equation itself is independent of t, shifting a solution graph right or left along the t axis will always produce another solution graph. The idea here is again to think of the "velocity" $\dot{y}$ as a quantity of primary interest and to give it a name, say $\dot{y} = z$, or $\dot{y} = v$ if $\dot{y}$ really can be interpreted as a velocity. Then $\ddot{y} = \dot{z}$, so the equation $\ddot{y} = f(y, \dot{y})$ can be written $\dot{z} = f(y, z)$. While this is a first-order equation, it is awkward, because it contains not only y and z but also a derivative with respect to a third variable t. The way around this is to use the chain rule to eliminate explicit reference to t:

$$\begin{aligned} \dot{z} &= \frac{dz}{dt} \\ &= \frac{dy}{dt}\frac{dz}{dy} = z\frac{dz}{dy}. \end{aligned}$$

Replacing $\dot{z}$ by $z(dz/dy)$ enables us to eliminate explicit use of t in the velocity equation $\dot{z} = f(y, z)$:

$$z\frac{dz}{dy} = f(y, z).$$

If we can solve this equation with an implicit relation $G(y, z, C) = 0$, then we can use the definition of z as $z = dy/dt$, and try to solve the first-order equation $G(y, \dot{y}, C) = 0$. The equation $G(y, \dot{y}, C) = 0$ is called a **first integral** of $\ddot{y} = f(y, \dot{y})$, because in principle it takes just one more integration to solve the original equation $\ddot{y} = f(y, \dot{y})$. Thus if a first integral can be rewritten explicitly as $dy/dt = h(y, C)$, then the variables separate easily, and we are in fact faced with a single integration problem:

$$\int \frac{dy}{h(y, C)} = \int dt.$$

Actually carrying out the integration still leaves us only with t as a function of y, and we then have the problem of inverting that relation to find $y = y(t)$.

Summary. To solve $\ddot{y} = f(y, \dot{y})$:

(i) Set $dy/dt = z$, $d^2y/dt^2 = z(dz/dy)$, and solve the first-order equation $z(dz/dy) = f(y, z)$ to get a relation $G(y, z, C) = 0$ between y and z.
(ii) Set $z = dy/dt$ and solve the first-order equation $G(y, \dot{y}, C) = 0$.

---

example 3

To solve $d^2y/dt^2 = y(dy/dt)$ with initial conditions $y(0) = 0$, $\dot{y}(0) = \frac{1}{2}$, let $dy/dt = z$ and $d^2y/dt^2 = z(dz/dy)$ to get

$$z\frac{dz}{dy} = yz.$$

Note that $z(0) = \dot{y}(0) = \frac{1}{2} \neq 0$, so we can assume $z(t) \neq 0$ for t near 0. We divide by z and are left with $dz/dy = y$. Integrating both sides with respect to y gives the first integral $z = \frac{1}{2}y^2 + C_1$. The initial conditions show that $z = \frac{1}{2}$ and $y = 0$ at $t = 0$, so $\frac{1}{2} = 0 + C_1$ and $C_1 = \frac{1}{2}$. Thus $z = dy/dt = \frac{1}{2}(y^2 + 1)$, so the additional integration is

$$\int \frac{dy}{y^2 + 1} = \int \frac{1}{2}\,dt.$$

This second integration gives $\arctan y = \frac{1}{2}t + C_2$. Since $y = 0$ when $t = 0$, we find that $C_2 = 0$, so

$$y = \tan\left(\frac{t}{2}\right)$$

is our solution to the initial-value problem.

example 4

If instead of the initial values for $\ddot{y} = y\dot{y}$ given in the previous example, we had $y(0) = 0$, and $\dot{y}(0) = -\frac{1}{2}$, the value for the first integration constant would have been $C_1 = -\frac{1}{2}$. In that case, we would have had a different integration problem:

$$\int \frac{dy}{y^2 - 1} = \int \frac{1}{2}\,dt, \quad \text{with result} \quad \frac{1}{2}\ln\left|\frac{y-1}{y+1}\right| = \frac{1}{2}t + C_2.$$

Solving for y gives

$$y = \frac{1 + Ke^t}{1 - Ke^t}, \quad \text{where} \quad K = \pm e^{2C_2} \neq 0.$$

To satisfy the remaining initial condition, $y(0) = 0$, we make $K = -1$, in which case $y = (1 - e^t)(1 + e^t)^{-1}$. Figure 3.7 shows not only the two solutions we've just computed, but also some others with $y(0) = 0$ and the slope $\dot{y}(0)$ at the origin

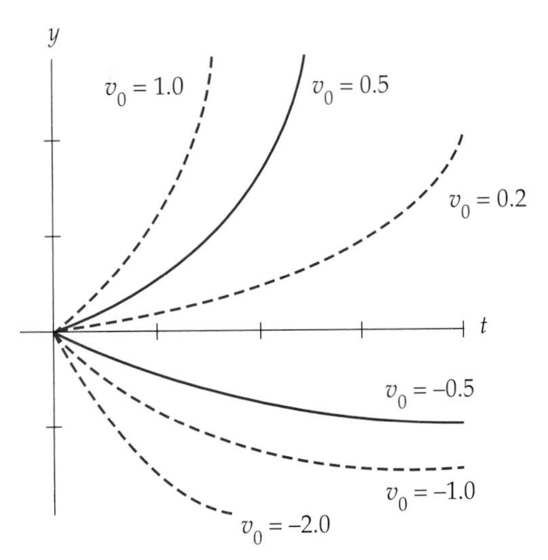

Figure 3.7 Solutions of $\ddot{y} = y\dot{y},\ y(0) = 0,\ \dot{y}(0) = v_0$.

taking on various values. Note that if $dy/dt(0) = 0$ we get the constant solution $y(t) = 0$; this one could be regarded as having come from dropping the condition $z \neq 0$ that we made near the beginning of Example 3.

One thing that sets the family of solutions to the previous nonlinear example apart from those of linear constant-coefficient equations $\ddot{y} = -k\dot{y} - hy$ is that simply by changing the initial value $\dot{y}(0)$ we can pass from a solution that tends to infinity in finite time to one that tends to a limit as t tends to infinity. No such behavior occurs with $\ddot{y} = -k\dot{y} - hy$.

---

example 5

The undamped pendulum equation, derived in Section 2 of this chapter, is $\ddot{y} = -(g/l)\sin y$, where y is the angular displacement of the pendulum from the vertical position. We write the equation using $z = \dot{y}$ as $z(dz/dy) = (g/l)\sin y$. A first integral is $\frac{1}{2}z^2 = -(g/l)\cos y + C$ or $\frac{1}{2}\dot{y}^2 = (g/l)\cos y + C$. The constant C could be determined by a variety of conditions, one of which is that the velocity $\dot{y}$ is zero when y reaches its maximum value $y_{\max} = \eta > 0$. Thus $0 = (g/l)\cos\eta + C$, so $C = -(g/l)\cos\eta$ and $\frac{1}{2}\dot{y}^2 = (g/l)\big(\cos y - \cos\eta\big)$. We are faced with the problem of integrating the separated equation

$$\int \frac{dy}{\sqrt{\cos y - \cos\eta}} = \pm\sqrt{\frac{2g}{l}}\int dt.$$

The integral on the left cannot be evaluated in terms of elementary functions but is an *elliptic integral* that can be rewritten in a standard form with tabulated values. Note that we haven't actually solved the equation in the sense of having found functions $y(t)$ that satisfy it. However, we can use the elliptic integral formula to determine the time required for a complete cycle from position η to $-\eta$ and back to η again, that is, the time period T of oscillation. Because of symmetry in the motion of y, we see that

$$T = 4\int_0^{T/4} dt = 4\sqrt{\frac{l}{g}}\int_0^{\eta} \frac{dy}{\sqrt{2(\cos y - \cos\eta)}}, \quad 0 < \eta < \pi.$$

At this point it's instructive to compare the above formula for T with the linear equation $\ddot{y} = -(g/l)y$ derived by linearizing $\ddot{y} = -(g/l)\sin y$ near $y = 0$. The complete solution to the linearized version of the problem posed above is $y = \eta\cos(\sqrt{g/l}\,t)$. Under that assumption, the period is $T = 2\pi\sqrt{l/g}$. Two things stand out right away when we compare the two expressions for T: (i) The dependence of T on length l and gravity g is proportional to $\sqrt{l/g}$ in both versions and (ii) the linearized version predicts a period that is independent of the maximum amplitude η, while the nonlinear version predicts a period that depends on η. The nonlinear version is the correct one, and the nature of T's dependence on η is investigated in the next example.

example 6

The period of an undamped pendulum with angular amplitude η was determined in the previous example to be $T = 4\sqrt{l/g}\, I(\eta)$, where

$$I(\eta) = \int_0^{\eta} \frac{dy}{\sqrt{2(\cos y - \cos \eta)}}.$$

Numerical evaluation of this integral for η in the interval $0 \leq \eta < \pi$ enables us to draw the graph of $I(\eta)$ shown in Figure 3.8. It turns out that $I(0+) = \pi/2$ and that $I(\pi-) = +\infty$. The picture shows that for small values of the maximum amplitude η the number $I(\eta)$ is only slightly larger than $\pi/2$; this makes the period only slightly larger than $T = 4\sqrt{l/g}\,(\pi/2) = 2\pi\sqrt{l/g}$, which is the period predicted by the solution $y = \eta\cos(\sqrt{g/l}\,t)$ of the linearized equation $\ddot{y} = -(g/l)\sin y$. But with $\eta = 1$ radian, about 57°, the linear estimate $\pi/2$ is about 6% too small when compared with $I(1)$. The deterioration of the linearized estimate's accuracy is obviously much more severe thereafter, becoming unbounded as η approaches π.

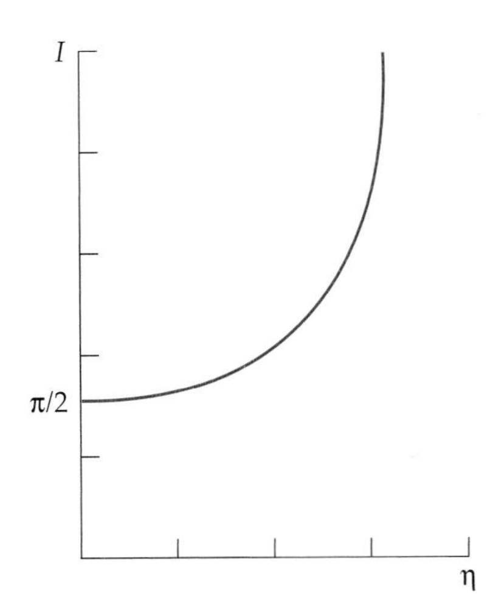

Figure 3.8 Graph of elliptic integral $I(\eta)$ for $0 < \eta < \pi$.

example 7

Finding approximate values of $I(\eta)$, introduced in the previous example, by using Simpson's rule is awkward because the integrand is unbounded as y approaches η. This difficulty can be avoided by changing the variable of integration in such a way as to write $I(\eta)$ in terms of the **complete elliptic integral**

$$K(k) = \int_0^{\pi/2} \frac{d\phi}{\sqrt{1 - k^2 \sin^2 \phi}} \qquad k^2 < 1.$$

It's left as an exercise to show that a change of variable gives

$$I(\eta) = \int_0^{\eta} \frac{dy}{\sqrt{2(\cos y - \cos \eta)}} = K(k), \quad \text{where} \quad k = \sin(\eta/2).$$

Note that the interval of integration $0 \leq y \leq \eta$ changes to the fixed interval $0 \leq \phi \leq \pi/2$. Tables of this integral are usually printed for values of a parameter α between $0°$ and $90°$ such that $k = \sin \alpha°$.

For a more detailed, but still brief, treatment of elliptic integrals see M. L. Boas, *Mathematical Methods in the Physical Sciences,* 2nd ed., Wiley, 1983.

EXERCISES

Find solutions for the equations in Exercises 1 through 6 by the method of Section 4A.

1. $t\ddot{y} - \dot{y} = 0.$ **2.** $t^2\ddot{y} + \dot{y}^2 = 0.$ **3.** $\ddot{y} + \dot{y}^2 = 0.$

4. $t\ddot{y} + \dot{y} = 0.$ **5.** $t\ddot{y} + \dot{y} = t^3.$ **6.** $\ddot{y} + \cos t = 0.$

Find solutions for the equations in Exercises 7 through 12 by the method of Section 4B.

7. $y\ddot{y} - \dot{y}^2 = 0.$ **8.** $y^2\ddot{y} + \dot{y}^3 = 0.$ **9.** $\ddot{y} = \dot{y}^3.$

10. $\ddot{y} + \dot{y}^2 = 0.$ **11.** $\ddot{y} + \dot{y}^2 = 1.$ **12.** $\ddot{y} = y.$

Find solutions for the equations in Exercises 13 through 16 that satisfy the given conditions.

13. $\ddot{y} - 2t = 0;\ y(0) = 0,\ \dot{y}(0) = 1.$

14. $t\ddot{y} + 2\dot{y} = 0;\ y(1) = 1,\ \dot{y}(1) = 2.$

15. $\ddot{y} = y\dot{y};\ y(0) = 0,\ \dot{y}(0) = 1.$

16. $\ddot{y} = 2(y^3 - y);\ y(0) = 0,\ \dot{y}(0) = 1.$

17. Show that the differential equation in Exercise 2 has more than one solution satisfying $y(0) = 0$ and $\dot{y}(0) = 0$.

18. Show that the differential equation in Exercise 5 has precisely one solution satisfying $y(0) = 0$ and $\dot{y}(0) = 0$.

19. The **catenary equation** $d^2y/dx^2 = k\sqrt{1 + (dy/dx)^2}$, $0 < k =$ const., is satisfied by the shape $y(x)$ of a chain suspended between two supports. The constant k depends on the location of the fixed ends and the length of the chain.

(a) Solve the equation with the initial conditions $y(0) = 1/k,\ y'(0) = 0$, and sketch the graph of this solution.

(b) Use the differential equation directly to show that the curvature $\kappa = y''/[1 + (y')^2]^{3/2}$ of all solutions is at most k.

(c) Show that the catenary equation can also be written $d^2y/dx^2 = k(ds/dx)$, where s is arc-length measured along the chain, and hence that if the catenary equation satisfies the boundary conditions $y(a) = \alpha,\ y(b) = \beta,\ a < b$, then $ks = y'(b) - y'(a)$, where s is the length of the chain. [*Hint:* Don't solve the differential equation.]

20. The formula

$$y = \frac{1}{2\sqrt{|c_1|}} \ln \left| \frac{\sqrt{|c_1|} + t}{\sqrt{|c_1|} - t} \right|$$

appears in Example 2 of the text as a solution formula for solutions to the nonlinear differential equation $\ddot{y} = 2t(\dot{y})^2$. This equation is valid for solutions whose graphs contain points in the first and third quadrants of the ty plane. Show that for $|t| < \sqrt{|c_1|}$ the formula can be expressed by the **inverse hyperbolic tangent** function as

$$y = \frac{1}{\sqrt{|c_1|}} \tanh^{-1}\left(\frac{t}{\sqrt{|c_1|}}\right),$$

and that for $|t| > \sqrt{|c_1|}$ it can be expressed by the **inverse hyperbolic cotangent** function as

$$y = \frac{1}{\sqrt{|c_1|}} \coth^{-1}\left(\frac{t}{\sqrt{|c_1|}}\right).$$

Here $\tanh x = (e^x - e^{-x})/(e^x + e^{-x})$ and $\coth x = 1/\tanh x$. The point is to express their inverse functions in terms of the natural logarithm function.

21. This exercise shows that applying the method of Section 4B to an equation can produce an extraneous "solution" that fails to satisfy the equation.

(a) Show that the solution $y = y(t)$ of $\ddot{y} = -y,\ y(0) = 1,\ \dot{y}(0) = 0$ also satisfies an equation of the form $(\dot{y})^2 = -y^2 + C$, and that with $C = 1$ this last equation has the constant solution $y(t) = 1$.

(b) Verify that $y(t) = 1$ fails to satisfy $\ddot{y} = -y$ but that the first-order and second-order equations in part (a) have in common the solution $y(t) = \cos t$ satisfying the conditions $y(0) = 1$, $\dot{y}(0) = 0$.

22. Integration of equations such as $\ddot{y} = f(y)$ can be done as follows.

(a) To integrate $\ddot{y} = f(y)$, multiply both sides by $\dot{y}$ to get $\dot{y}\ddot{y} = f(y)\dot{y}$. Show that if there is a function $F(y)$ such that $F'(y) = f(y)$, then $\frac{1}{2}\dot{y}^2 = F(y) + C_1$.

(b) Apply the technique of part (a) to show that the solution of the initial-value problem $\ddot{y} = -y$, $y(0) = 1$, $\dot{y}(0) = 0$ is $y = \sin(t + \pi/2) = \cos t$.

(c) Apply the technique of part (a) to show that a solution $y = y(t)$ of the undamped pendulum equation $\ddot{y} = -\sin y$ satisfies

$$\dot{y}^2 = 2\cos y + C,$$

where C is constant.

(d) Find the constant C in part (c) if $\ddot{y} = -\sin y$ is subject to initial conditions $y(0) = y_0$ and $\dot{y}(0) = z_0$.

(e) Letting $z = \dot{y}$, use the results of parts (c) and (d) to sketch the curves $z^2 = 2\cos y + C$ in the yz plane, for $C = -1,\ C = 0,\ C = 1$, and $C = 2$. What is the physical significance of the value $C = -2$? What can you say about the case $C = -3$?

***23.** **(a)** Find all solutions of the differential equation $\ddot{y} = 2y\dot{y}$. (This problem is of a different type from the one in Example 2 of the text. Nevertheless, the indefinite integrals needed for that example can be used here also.)

(b) Sketch the graph of a typical solution of each kind from part (a).

24. The equation $\ddot{y} = -1/y^2$ is a normalized version of the inverse-square law of gravitational attraction $\ddot{y} = -GM/y^2$ for the distance $y(t)$ from an attracting body of mass M.

(a) Find the linearized equation near the value $y_0 = 1$, and find the solution of the linearized equation subject to the initial conditions $y(0) = 1,\ \dot{y}(0) = 0$.

(b) Show that under the same initial conditions as in part (a), the solution $y(t)$ to the nonlinear equation satisfies $\dot{y} = -\sqrt{(2-2y)/y}$.

(c) Show that $y(t)$ satisfies

$$\arccos\sqrt{y} + \sqrt{y(1-y)} = \sqrt{2}t.$$

Estimate the value of t corresponding to $y = 0$.

(d) Sketch the graph of the equation in part (c) to show the relation between y and t.

25. Follow the steps of parts (b) and (c) of the previous exercise for the more general equation $\ddot{y} = -GM/y^2$, with initial conditions $y(0) = y_0,\ \dot{y}(0) = 0$, to show that

$$y_0 \arccos\sqrt{y/y_0} + \sqrt{y(y_0 - y)} = \sqrt{2GM/y_0}\, t.$$

Use this equation to show that the time it takes to fall from y_0 to 0 is $\pi y_0^{3/2}/(2\sqrt{2GM})$.

26. A spherical raindrop falls through the air under the influence of the earth's gravitational acceleration g and subject to negligible air resistance. The raindrop increases in mass m through the accumulation of additional moisture in such a way that its rate of change with respect to distance y fallen is proportional to the square of its radius ($dm/dy = kr^2$). Assume that the raindrop starts with radius zero and velocity zero at $t = 0$.

(a) Show that radius r is proportional to distance y.

(b) Combining Newton's second law (Chapter 2, Section 4.) with the constant gravitational acceleration g gives $d/dt(mv) = gm$. Show that this equation is equivalent to

$$v\frac{d}{dy}(y^3 v) = gy^3.$$

(c) Solve the differential equation in part (b) with initial conditions $y = v = 0$. [*Hint:* Multiply both sides by y^3.]

(d) Find how far the raindrop falls in time t and show that its acceleration is a constant $g/7$.

27. The second order linear equation

$$y'' + b(x)y' + c(x)y = 0$$

with coefficients $b(x)$ and $c(x)$ that are not necessarily constant can sometimes be solved by making the substitution $y' = zy$.

(a) Show that this substitution leads to the associated **Riccati equation**

$$\frac{dz}{dx} + z^2 + b(x)z + c(x) = 0.$$

(b) Show that if $z = z(x)$ is a solution of the Riccati equation, then a solution of the related second order linear equation is given by

$$y(x) = Ce^{Z(x)},$$

where $Z(x)$ is some indefinite integral of $z(x)$.

28. Show that if y is a solution of the linear equation

$$a(x)y'' + [a(x)b(x) - a'(x)]y' + a^2(x)c(x)y = 0,$$

then $z = y'/(a(x)y)$ is a solution of the general **Riccati equation**

$$\frac{dz}{dx} + a(x)z^2 + b(x)z + c(x) = 0.$$

29. **(a)** Use $\cos y = 1 - 2\sin^2(y/2)$ to show that the pendulum period factor $I(\eta)$ is

$$I(\eta) = \int_0^\eta \frac{dy}{\sqrt{2(\cos y - \cos\eta)}} = \int_0^\eta \frac{\frac{1}{2}dy}{\sqrt{\sin^2(\eta/2) - \sin^2(y/2)}}.$$

(b) Use part (a) and the change-of-variable relation $\sin(y/2) = k\sin\phi$ with $k = \sin(\eta/2)$ to show that

$$I(\eta) = \int_0^{\pi/2} \frac{d\phi}{\sqrt{1 - k^2\sin^2\phi}}, \qquad 0 < k^2 < 1.$$

30. The complete elliptic integral is

$$K(k) = \int_0^{\pi/2} \frac{d\phi}{\sqrt{1 - k^2\sin^2\phi}}.$$

Show that (a) $K(0) = \pi/2$ and (b) $K(k)$ is strictly increasing for $0 \le k < 1$.

31. Make a computer plot of the results of a Simpson's rule computation of the elliptic integral $K(k)$, as defined in the text, for $0 \le k \le 0.9$. Use at least 90 equally spaced values of k and at least 20 Simpson-rule intervals in the interval $[0, \pi/2]$. Note that $K(0) = \pi/2$.

***32.** The complete elliptic integral

$$K(k) = \int_0^{\pi/2} \frac{d\phi}{\sqrt{1 - k^2\sin^2\phi}}, \qquad 0 \le k < 1,$$

tends to $+\infty$ as k tends to 1 from the left.

(a) Show that

$$K(k) \ge \frac{1}{\sqrt{2}} \int_0^{\pi/2} \frac{d\phi}{\sqrt{1 - k \sin \phi}}.$$

(b) Make the substitution $x = \sin \phi$ in the integral in part (a) to conclude that

$$K(k) \ge \frac{1}{2} \int_0^1 \frac{dx}{\sqrt{1-x}\sqrt{1-kx}} \ge \frac{1}{2} \int_0^1 \frac{dx}{1-kx}.$$

(c) Evaluate the last integral in part (b), and show that it tends to infinity as $k \to 1$ from the left.

33. The elliptic integral $K(k)$ defined in the text is more specifically called the **complete elliptic integral of the first kind.** The **complete elliptic integral of the second kind** is defined by

$$E(k) = \int_0^{\pi/2} \sqrt{1 - k^2 \sin^2 \phi}\, d\phi, \qquad 0 \le k \le 1.$$

The term "elliptic integral" comes from the application of the integral of the second kind to computing the arc length of an ellipse.

(a) Show that the length l of the ellipse parametrized by $x = a \cos t,\ y = b \sin t$, with $0 \le t \le 2\pi$, can be expressed by

$$l = \int_0^{2\pi} \sqrt{a^2 \sin^2 t + b^2 \cos^2 t}\, dt.$$

(b) Assume $a \le b$ in part (a), and show that $l = 4bE(k)$, where $k = \sqrt{1 - (a^2/b^2)}$.

34. Make a computer plot of the results of a Simpson's rule computation of the elliptic integral $E(k)$, as defined in the previous exercise, for $0 \le k \le 1$. Use at least 10 equally spaced values of k and at least 10 Simpson-rule intervals in the interval $[0, \pi/2]$. Note that $E(0) = \pi/2$ and $E(1) = 1$.

CHAPTER REVIEW AND SUPPLEMENT

Constant-Coefficient Equations. The main business of this chapter is learning to solve second-order initial-value problems of the form $Ay'' + By'' + Cy = f(x)$, $y(x_0) = y_0$, and $y'(x_0) = z_0$, where A, B, and C are real constants, and $f(x)$ is a continuous function. In broad outline, we usually solve this problem in three steps:

1. Solve the *associated homogeneous equation* $Ay'' + By' + Cy = 0$ by finding the *characteristic roots* of the quadratic equation $Ar^2 + Br + C = 0$ and using the roots r_1, r_2 to write the *general homogeneous solution* in the form $y_h(x) = c_1 y_1(x) + c_2 y_2(x)$ as

$$y_h(x) = c_1 e^{r_1 x} + c_2 e^{r_2 x}, \quad \text{if } r_1 \ne r_2 \quad \text{or} \quad y_h(x) = c_1 e^{r_1 x} + c_2 x e^{r_1 x}, \quad \text{if } r_1 = r_2.$$

2. The general solution to the nonhomogeneous equation, where $f(x)$ is not identically zero, is the sum $y = y_h + y_p$, where y_p is an arbitrary solution of the nonhomogeneous equation. If $f(x)$ is itself a solution of a homogeneous constant-coefficient equation $Ny = 0$, then a suitable y_p can be found by undetermined coefficient trial substitution. (See Section 3B for two different desriptions of the process.) Regardless of the form of $f(x)$ we can try integration to find a $y_p(x)$ by either of two options:

(i) The variation of parameters formula

$$y_p(x) = y_1(x)\int \frac{-y_2(x)f(x)}{w(x)}\,dx + y_2(x)\int \frac{y_1(x)f(x)}{w(x)}\,dx,$$

where $w(x) = y_1(x)y_2'(x) - y_2(x)y_1'(x)$.

(ii) Compute an integral $\int_{x_0}^{x} g(x-t)f(t)\,dt$, where $g(u)$ depends on the characteristic roots:

$$g(u) = \frac{e^{r_1 u} - e^{r_2 u}}{r_1 - r_2}, \qquad r_1 \neq r_2; \qquad g(u) = e^{r_1 u}, \qquad r_1 = r_2;$$

$$g(u) = \frac{1}{\beta}e^{\alpha u}\sin \beta u, \qquad r_1, r_2 = \alpha \pm i\beta.$$

3. An important point about the basic homogeneous solutions $y_1(x)$, $y_2(x)$ of step 1 is that they are *linearly independent*, meaning that neither one is a constant multiple of the other. [If, on the other hand, $y_1(x)$ and $y_2(x)$ were *linearly dependent* so that, for example $y_1(x) = cy_2(x)$ for some constant c, then the general homogeneous solution $c_1y_1(x) + c_2y_2(x)$ would reduce to a constant multiple of a single function, making it impossible to satisfy arbitrary initial conditions of the form $y(x_0) = y_0$, $y'(x_0) = z_0$.] As it is, we can always satisfy such conditions using the linearly independent homogeneous solutions by solving for c_1 and c_2 in the equations

$$c_1y_1(x_0) + c_2y_2(x_0) = y_0 - y_p(x_0),$$
$$c_1y_1'(x_0) + c_2y_2'(x_0) = z_0 - y_p'(x_0).$$

Find all solutions that satisfy:

1. $y'' + 2y' + y = e^{-x} + 3e^{x}$.

2. $y'' + y = x\sin x$.

3. $y'' - y = \sin x$.

4. $y'' - y' - y = 1,\ y(0) = y'(0) = 1$.

5. $y'' + 2y' + 3y = 1$.

6. $(D-1)^2y = x^3 - x$.

7. $y'' + 9y = \sin 3x,\ y(0) = 1,\ y'(0) = 0$.

8. $y''' = x$.

9. $(D^2 + 4)y = \cos 3x$ (using Section 3C).

10. $y'''' = 81y$.

11. $y'' + y = 0,\ y(0) = -1,\ y(\pi) = 1$.

12. $y'' + y = 0,\ y(0) = 0,\ y(\pi/2) = 2$.

Find the general form for a trial solution y_p for each of the following. (For example, for $y'' - y = e^x$, choose $y_p = Axe^x$.) You need not determine any coefficient values.

13. $y'' - 4y = xe^{2x} + e^{2x}$.

14. $y'' + y = x^2 \cos x$.

15. $y'' - 5y' + 6y = xe^{2x} + e^{3x}$.

16. $y'' + 4y = x^2 \cos 2x - 2\sin 2x$.

17. $y'' - 4y = e^{2x} + 5\cos x$.

18. $y'' + y = 3x\sin(x - 3)$.

19. $y'' - y' = x^2 + 2e^x$.

20. $y''' - y = e^{x/2}\sin\sqrt{3}x$.

21. $y''' = 1 + x + x^3$.

22. $y'' + y = x^{99}\cos x$.

The simplest real forms for basic solution pairs to the constant-coefficient equation $y'' + ay' + by = 0$ are $\{e^{r_1 x}, e^{r_2 x}\}$, $\{e^{r_1 x}, xe^{r_1 x}\}$, and $\{e^{\alpha x}\cos\beta x, e^{\alpha x}\sin\beta x\}$.

23. Make a corresponding complete list of triples for the equation $y''' + ay'' + by' + cy = 0$.

24. Make a corresponding complete list of quadruples for $y'''' + ay''' + by'' + cy' + dy = 0$.

25. Derive from scratch the fundamental sinusoidal solutions to the harmonic oscillator problem $\ddot{y} + y = 0,\ y(0) = 0,\ \dot{y}(0) = 1$. Our earlier derivations were made using complex exponentials:

(a) Multiply the equation by $\dot{y}$. Integrate with respect to t to get $\frac{1}{2}\dot{y}^2 + \frac{1}{2}y^2 = C$.

(b) Find C, solve for $\dot{y}$, and solve the resulting first order equation to get $y(t) = \sin(t + c)$.

26. Suppose that $y_1(x)$ and $y_2(x)$ are real-valued functions defined for all real x and you know that $y_1(x)$ is not a constant multiple of $y_2(x)$. Are the two functions necessarily linearly independent? Explain your answer, using an example if necessary.

27. Let the functions $y_1,\ y_2$ be defined for all real x by $y_1(x) = e^{rx}$ and $y_2(x) = e^{sx}$, where r and s are unequal complex numbers. Show that y_1 and y_2 are linearly independent even if complex constants $c_1,\ c_2$ are allowed in $c_1 y_1(x) + c_2 y_2(x)$.

28. For what values of the constant b do the nonidentically zero solutions of $y'' + y' + by = 0$ oscillate as functions of x?

29. The current $I(t)$ flowing through a certain electric circuit at time t satisfies

$$I'' + RI' + I = \sin t,$$

where $R > 0$ is a constant resistance. The equation has a solution of the form $I(t) = A\sin(t - \alpha)$ for certain constants $A > 0$ and α. Find A and α.

30. Show that the functions e^x and e^{-x}, are linearly independent on an interval $a < x < b$ by showing directly that neither function can be a constant multiple of the other on such an interval.

31. (a) Show that the functions e^x and e^{-x}, are linearly independent on an interval $a < x < b$ by showing directly that the equation

$$c_1 e^x + c_2 e^{-x} = 0$$

can't be satisfied for all real x unless the constants c_1 and c_2 are both zero.

(b) Show that the equation $2e^x - 3e^{-x} = 0$ is satisfied for exactly one real x and that $2e^x + 3e^{-x} = 0$ is satisfied for no real x.

(c) For what *complex* values of x is $2e^x + 3e^{-x} = 0$ satisfied?

32. Suppose that

$$\frac{d^2y}{dt^2} = -y.$$

(a) Find the general solution $y(t)$ of this equation.

(b) Let

$$z(t) = \frac{dy}{dt}(t).$$

Show that the parametrized curve $\big(y(t), z(t)\big)$ traces either a clockwise circular path or else reduces to a single point.

33. Theorem 2.8 implies that there is a one-to-one correspondence between the set of all n-tuples of initial values $(z_0, z_1, \ldots, z_{n-1})$ and and all solutions to the nth-order homogeneous equation $L(y) = 0$. Explain how this conclusion follows.

34. Theorem 2.8 implies that there is a one-to-one correspondence between the set of all n-tuples of initial values $(z_0, z_1, \ldots, z_{n-1})$ and all n-tuples $(c_1, c_2, \ldots, c_n)$ of coefficients of linear combination. Explain how this implication follows from the theorem.

In Exercises 35 to 38, assume only that L is a linear operator such that $L(y_1) = w_1$, $L(y_2) = w_2$, and $L(y_3) = w_3$. Using just this information, find linear combinations z of y_1, y_2, and y_3 so that

35. $L(z) = 2w_1 - 3w_2$.

36. $L(z) = w_1 + 2w_2 - 4w_3$.

37. $L(z) = L(2z) + w_1 + w_2$.

38. $L(z) = 0$.

39. **(a)** Show that the family of differential equations $y'' - (2r + h)y' + r(r + h)y = 0$, depending on the parameter h, is the same as $(D - r)[D - (r + h)]y = 0$.

(b) Show that the equations of part (a) have solutions $y = c_1 e^{(r+h)x} + c_2 e^{rx}$ if $h \neq 0$.

(c) Let $c_1 = 1/h$, $c_2 = -1/h$, and show that for each fixed x, the resulting solution $y_h(x)$ tends as $h \to 0$ to

$$y = \frac{d}{dr}e^{rx} = xe^{rx}.$$

40. What is the form of the integral formula, as described in Section 3D, for the initial-value problem $y'' = f(x)$, $y(0) = y'(0) = 0$?

41. Show that if $y(x)$ is a solution of $y'' + ay + b = 0$, a, b constant, then $y(x) = e^{-ax/2}v(x)$, where $v(x)$ is a solution of $v'' + (b - \frac{1}{4}a^2)v = 0$.

42. A ball of mass m, and hence weight $w = mg$, is dropped from a point 200 feet above the surface of the earth so that its initial velocity is 0. Assume that the drag force is insignificant.

(a) When does the ball hit the ground?

(b) What is the impact velocity at ground level?

(c) From what height should the ball be dropped to get an impact velocity of 100 feet per second?

(d) Is the mass of the ball relevant to any of the preceding questions?

43. Suppose in Exercise 42 that the weight of the ball is 32.2 pounds, so we can take $m = 1$, and that the drag force $-kv$ is significant, with drag constant $k = 0.1$.

(a) Show that the ball hits the ground after time t_1, where t_1 is the positive solution of the equation $t/10 + e^{-t/10} = 171/161$, with $g = 32.2$.

(b) Show that t_1 in part (a) is about 3.74 seconds.

(c) Compare the result obtained in part (b) with the answer obtained by using the free-fall formula $y(t) = gt^2/2$.

(d) Find the terminal velocity of the ball.

(e) About how long does it take for the ball to reach velocity 75 feet per second?

44. Suppose a falling body is subject to a drag constant k and has mass m.

(a) If the initial velocity is 0, show that the distance covered in time t is

$$y(t) = \frac{mg}{k}t - \frac{m^2 g}{k^2}(1 - e^{-kt/m}).$$

(b) Show that the formula for $y(t)$ in part (a) satisfies $y(t) \leq \frac{1}{2}gt^2$ for $t \geq 0$. [*Hint:* $\ddot{y} = g - (k/m)\dot{y} \leq g$. Now integrate from 0 to t.]

(c) Find the analogue of the formula given in part (a) for the case of initial velocity $\dot{y}(0) = v_0$.

45. (a) For a falling body of mass m subject to linear drag with constant k, show that initial velocity v_0, leads to velocity at time t given by

$$\dot{y}(t) = \frac{mg}{k} + \left(v_0 - \frac{mg}{k}\right)e^{-kt/m}.$$

(b) Find the limit as k tends to 0 of the formula for $\dot{y}(t)$ in part (a). Does this agree with the free-fall formula $v_0 + gt$?

46. The drag constant k can be estimated by using an observed value of the terminal velocity $v_\infty = mg/k$.

(a) Find k if a body of weight $w = 100$ pounds achieves terminal velocity $v_\infty = 180$ feet per second.

(b) How far must the body in part (a) fall to obtain the velocity of 150 feet per second?

47. Determine the amplitude A of the displacement $y(t)$ in terms of $\theta(0)$ and $\dot{\theta}(0)$ of an undamped pendulum satisfying the linear equation $\ddot{\theta} + (g/l)\theta = 0$. What is its period?

48. How long should an undamped pendulum be if its displacement angle satisfies $\ddot{\theta} + (g/l)\theta = 0$ and it is to have a period of 2 seconds?

49. **(a)** Given the relation $y(t) = l\theta(t)$, express the equation $\ddot{\theta} + (k/m)\dot{\theta} + (g/l)\theta = 0$ in terms of y, k, l, and m.

(b) What condition must k satisfy for the pendulum described by the differential equation in part (a) to oscillate?

(c) Show that oscillations described by the differential equation in part (a) undergo an increase in the time between successive vertical positions if l increases slightly. What is the effect of an increase in m?

50. Consider the equation

$$\ddot{\theta} + \frac{k}{m}\dot{\theta} + \frac{g}{l}\theta = 0,$$

where m, k, g, and l are positive constants.

(a) Show that if the roots of the characteristic equation are real then these roots are both negative, and that if they are complex then they both have negative real parts.

(b) What can you conclude about the oscillation of a pendulum from part (a)?

(c) If $k^2/m^2 \geq 4(g/l)$, explain why $\theta(t)$ is nonoscillatory.

51. Show that the general solution of

$$\ddot{\theta} + \frac{k}{m}\dot{\theta} + \frac{g}{l}\theta = 0,$$

in the case $k^2/m^2 < 4(g/l)$, has the alternative form

$$\theta(t) = Ae^{-kt/2m}\cos\left(\sqrt{\frac{B}{l}}t + \alpha\right),$$

where $B = g - lk^2/(4m^2)$. [*Hint:* Use the addition formula for the cosine.]

52. Solve the equation

$$\ddot{y} + \dot{y} + y = a_0 \cos wt, \qquad a_0 \neq 0,\ \omega \neq 0,\ \sqrt{2}/3,$$

for the oscillation of a pendulum with externally applied force $a_0 \cos wt$. First find a particular solution $y_p(t)$ by the method of undetermined coefficients using the trial solution $y_p(t) = A\cos wt + B\sin wt$.

53. Let $y = y(t)$ be a periodic solution of

$$y'' = -\frac{g}{l}\sin y,$$

and let y_0 be the maximum value of $y(t)$, assuming $0 < y_0 < \pi/2$.

(a) Show that if $y(t)$ satisfies

$$y' = \sqrt{\frac{2g}{l}(\cos y - \cos y_0)},$$

then $y(t)$ satisfies the second-order pendulum equation.

(b) Show that the time T required for $y(t)$ to go from $-y_0$ to y_0 is

$$T = \sqrt{\frac{2l}{g}} \int_0^{y_0} \frac{dy}{\sqrt{\cos y - \cos y_0}}.$$

Thus the period of the solution is $p = 2T$.

54. A cylindrical object floating in liquid with its central axis vertical satisfies a differential equation for its vertical displacement $y(t)$ from equilibrium:

$$\ddot{y} + \frac{k}{m}\dot{y} + A\frac{w}{m}y = 0,$$

where m is the mass of the object, A is its circular cross-sectional area, w is the density of the liquid, and k is a positive friction constant.

(a) Show that the differential equation has oscillatory solutions if and only if $k^2 < 4Awm$.

(b) Discuss the effect on $y(t)$ of independently increasing m and w. Assume $k^2 < 4Awm$. Note that as m increases, frequency alters critically at

$$k^2 = 2Awm.$$

55. A 2-foot undamped pendulum is released from angle $\theta = 0.1$ radian (about 5.7°) from the vertical.

(a) Express θ as a function of time t after release.

(b) About how long does it take for the pendulum to reach the vertical position $\theta = 0$ for the first time?

(c) Find the angular speed $|\dot{\theta}|$ when $\theta = 0$.

56. A 2-foot undamped pendulum has maximum linear velocity 0.5 feet per second at its tip. What is the maximum angular displacement?

57. A pendulum 6 inches long is pushed down from an angular displacement of $\frac{1}{10}$ radian with angular speed $\frac{1}{2}$ radian per second. Find the period, the maximum amplitude, and the maximum angular velocity.

58. Leave everything in the previous problem the same except that the pendulum is initially pushed up instead of down. Do any of the answers change? Explain why.

4 CHAPTER

Applications: Dynamics and Phase Space

This chapter is about **dynamics,** meaning change over time t of a quantity $y(t)$ determined by its relation to dy/dt and d^2y/dt^2. Fundamental to much of our discussion is the simplest form of **Newton's second law of motion,** according to which force F is equal to constant mass m times acceleration a. If y denotes displacement, then it follows from $a = d^2y/dt^2$ that

$$F = ma = m\frac{d^2y}{dt^2}.$$

It's often convenient to follow Newton's convention of writing time derivatives using overdots. Thus $\dot{y} = dy/dt$, and $\ddot{y} = d^2y/dt^2$. Thus many of the differential equations we'll be looking at take the form

$$m\ddot{y} = f(t, y, \dot{y}).$$

The overdot notation is standard in the field of dynamical systems, and is convenient because it enables us to write $\dot{y}^2$ instead of notations such as $(dy/dt)^2$, or $(y')^2$, or y'^2.

In this chapter, the mathematical *state space* or *phase space,* as it is called in certain classical settings, will be 2-dimensional, consisting of coordinate pairs of the form $(y, \dot{y})$. Section 3 focuses directly on this state space by reducing second-order equations to a standard first-order form. The emphasis on states of a system is fundamental to our study whenever we can be assured that *future states of a system governed by a second-order equation will be determined by both position y and velocity $\dot{y}$ at some fixed time t_0.* Up to this point, we've usually been able to guarantee determination of future states by integration, or else successive integration using order reduction. The nonlinear pendulum equation is an exception, and for that equation and many others we'll sometimes need to refer to the following theorem that provides conditions under which this unique determination is guaranteed.

The existence and uniqueness theorem is most conveniently viewed as a special case of Theorem 1.1 in Chapter 5, page 342.

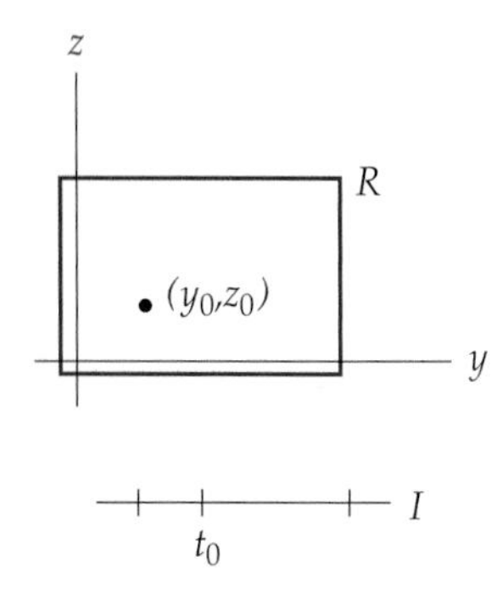

1.1 Existence and Uniqueness Theorem
Assume the function of three variables $f(t, y, z)$ and its two first-order partial derivatives $f_y(t, y, z)$ and $f_z(t, y, z)$ are continuous for t in an interval I and for all (y, z) in an open rectangle R containing (y_0, z_0). Then the initial-value problem

$$\ddot{y} = f(t, y, \dot{y}), \qquad y(t_0) = y_0, \qquad \dot{y}(t_0) = z_0$$

has a unique solution on some subinterval J of I containing t_0. If also there is constant B such that $|f_y(t, y, z)| \leq B$ and $|f_z(t, y, z)| \leq B$ for all x in I and all (y, z) in $\mathcal{R}^2$, then the unique solution is defined on the entire interval I.

Supplemental project material on gravitational forces, pendulum motion, and energy is available at the website <http://math.dartmouth.edu/~rewn/>.

1 MECHANICAL OSCILLATORS

A second-order, constant-coefficient linear differential equation

1.2
$$m\frac{d^2x}{dt^2} + k\frac{dx}{dt} + hx = F(t)$$

has solutions that can be neatly classified into distinct types, depending on the relations between the constants $m > 0, k \geq 0, h > 0$, and the external forcing function $F(t)$. The most important class consists of the *oscillations*, namely, those solutions whose values persist in switching back and forth from one side to the other of some fixed reference. The external forcing function $F(t)$ can be regarded as an "input" to the underlying mechanism modeled by the left-hand side of Equation 1.2. The solution $y(t)$ is the resulting "output."

Harmonic Oscillation. This mode of oscillation is called "harmonic" because of its association with musical tones, as described in Chapter 10. Here we assume that $k = 0$ and $F(t) = 0$ in Equation 1.2. The assumption $k = 0$ represents an ideal frictionless process that is hard to realize in practice. Nevertheless, harmonic oscillation has fundamental importance as the limiting case as k tends to zero. Our assumptions $F(t) = 0$ and $k = 0$ enables us to conclude that every nontrivial solution is **periodic** with a **period** $2p$ such that $x(t+2p) = x(t)$ for all t. The reason is that if $F(t) = 0$ and $k = 0$, Equation 1.2 becomes

$$\frac{d^2x}{dt^2} + \frac{h}{m}x = 0, \qquad \frac{h}{m} > 0,$$

and all solutions consist of the periodic trigonometric functions

$$\begin{aligned} x(t) &= c_1\cos\left(\sqrt{h/m}\,t\right) + c_2\sin\left(\sqrt{h/m}\,t\right), \\ &= c_1\cos\omega t + c_2\sin\omega t, \quad \text{where } \omega = \sqrt{h/m}. \end{aligned}$$

Expressions of this last form occur often in physical settings, and a process that can be represented in this way is called a **harmonic oscillator.** The analysis described in Chapter 3, Section 2C, is helpful for interpreting oscillations, where the arbitrary constants c_1 and c_2 in the general solution formulas are replaced by constants that have some direct physical significance. We'll review the essentials briefly in terms of time t. We choose an angle ϕ such that

$$\cos\phi = \frac{c_1}{\sqrt{c_1^2 + c_2^2}}, \qquad \sin\phi = \frac{c_2}{\sqrt{c_1^2 + c_2^2}}.$$

Then set $A = \sqrt{c_1^2 + c_2^2}$, so that $c_1 = A\cos\phi$ and $c_2 = A\sin\phi$. Using the addition formula for the cosine of a difference we can then write

$$\begin{aligned} x(t) &= c_1 \cos\left(\sqrt{\frac{h}{m}}\,t\right) + c_2 \sin\left(\sqrt{\frac{h}{m}}\,t\right), \\ &= A(\cos\phi\cos\omega t + \sin\phi\sin\omega t) \\ &= A\cos(\omega t - \phi). \end{aligned}$$

Since the cosine factor is between -1 and 1, the number $A > 0$ is the maximum deviation of the oscillation from the value $x = 0$, and A is the *amplitude* of the oscillation; the total width of the oscillation is then $2A$. In the expression $\omega t - \phi$, the number ϕ is the **phase angle** of the oscillation. If $\phi > 0$, it indicates that the graph of $x(t)$ is a shift to the right of the graph of $\cos\omega t$ by the amount ϕ/ω. Using periodicity, we can always assume $0 \le \phi < 2\pi/\omega$.

We see that the solutions of $\ddot{x} + (h/m)x = 0$ exhibit the *circular frequency* $\omega = \sqrt{h/m}$ that depends only on h and m. On the other hand, the amplitude A and phase angle ϕ depend on the constants c_1 and c_2, which in turn depend on initial conditions as well as on h and m. It's impossible to find a mechanical oscillator that's not subject to a significant amount of frictional force and that will keep on oscillating with no external driving force, but we'll see in Chapter 10 that harmonic oscillators arise in the solution of partial differential equations. While the harmonic oscillator model represents an ideal system in the ordinary physical world, the model comes close to mimicking a massive marine navigation buoy that has lots of inertial energy and is floating in calm water, as in the next example.

example 1

If an object such as a marine navigation buoy floats in water, it has a tendency to oscillate vertically with a steady rhythm. To understand how this happens we use the **Archimedes principle,** which states that an object immersed partly or entirely in a fluid is buoyed up by a force equal to the weight of the displaced fluid. (The reason is that the buoyant force of the water acts on the floating object just as it would have acted to buoy up the displaced fluid; legend has it that Archimedes shouted "Eureka" when this clever explanation occurred to him.) In particular, an object of mass m floating at equilibrium will be buoyed up by a force equal to its own weight $w = mg$ as it is pulled down by gravity. Thus, with no motion at equilibrium, the magnitude of the buoyant force equals the weight of the floating object, which in this case is also the weight of displaced fluid. If the object is depressed below its equilibrium level, the additional buoyant force will equal the weight of the additional displaced fluid. On the other hand, in a position elevated above equilibrium the force of gravity acting down exceeds the buoyant force by the excess of the object's weight over the weight of the displaced fluid.

We'll discuss a simple case in which the surface of the water remains flat and a buoy floats as a vertical cylinder. Three positions of the same buoy are shown

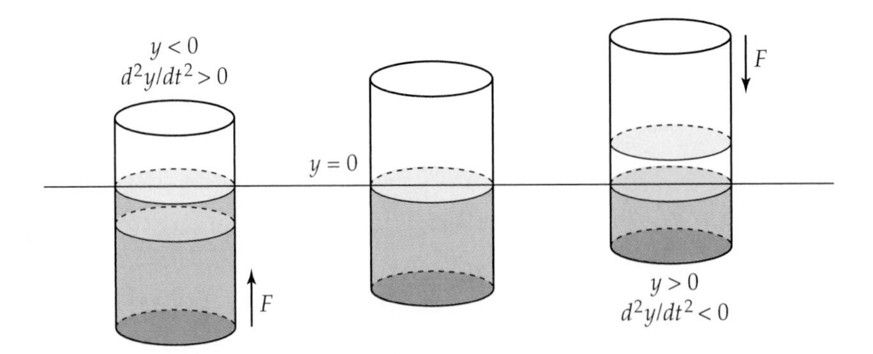

Figure 4.1 Floating buoy.

in Figure 4.1. Choose a vertical coordinate line with origin $y = 0$ at the water's surface and with positive y values directed up. In a depressed position ($y < 0$) of the buoy's equilibrium waterline, the additional displaced water has weight $-\rho Sy$, where ρ is the weight density of water and S is the cross-sectional area of the cylindrical buoy. The resulting upward directed force F can be equated to mass times acceleration:

$$m\frac{d^2y}{dt^2} = -\rho Sy.$$

When $y > 0$, the term on the right is negative, consistent with the expected excess downward force of gravity, so the differential equation is valid in both cases. The solutions of this equation all have the form

$$\begin{aligned} y(t) &= c_1 \cos\sqrt{\frac{\rho S}{m}}\,t + c_2 \sin\sqrt{\frac{\rho S}{m}}\,t \\ &= A\cos\left(\sqrt{\frac{\rho S}{m}}\,t - \phi\right). \end{aligned}$$

The amplitude $A = \sqrt{c_1^2 + c_2^2}$ of the oscillations is completely determined by initial conditions. Since $\sqrt{\rho S/m}\,[t + 2\pi\sqrt{m/(\rho S)}] = \sqrt{\rho S/m}\,t + 2\pi$, the period of oscillation will be $T = 2\pi\sqrt{m/(\rho S)}$ time units. Rapid oscillation goes with a small period, which is what we get with a fluid of high density or a buoy of large cross section S. On the other hand, a relatively massive buoy will oscillate fairly slowly.

Note that since the buoy's mass m and the mass density ρ of water appear on both sides of the differential equation, we can just as well use units of weight, as measured in air, for both of these. Since water weighs about 62.5 pounds per cubic foot, a 2000-pound buoy with a 2-foot radius will have a natural oscillation period of about $2\pi\sqrt{2000/(62.5 \cdot 4\pi)} \approx 10$ seconds.

Damped Oscillation. A typical mechanism that can be analyzed using a constant-coefficient equation appears in cross section in Figure 4.2. The working parts consist of a piston that oscillates in a cylinder that generates frictional forces, and a spring that extends and compresses.

A spring exerts a force with magnitude that is a multiple of its extension or compression from its equilibrium position, which we'll denote by 0 on a fixed

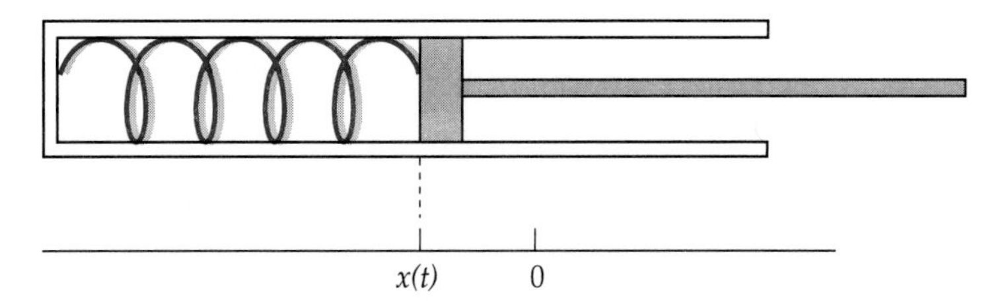

Figure 4.2 Spring-loaded piston.

scale. Thus, if x is the amount of displacement from 0, then the force F_s exerted by the spring is representable, according to *Hooke's law,* with Hooke's stiffness constant h, by

$$F_s = -hx, \qquad h > 0.$$

For an ideal spring, it is assumed that Hooke's coefficient h is constant. The minus sign in front of the factor h comes about because a positive extension of the spring, $x > 0$, induces a restoring force that *decreases* the velocity dx/dt, making $d^2x/dt^2 < 0$; likewise compression, $x < 0$, induces an increase in velocity, making $d^2x/dt^2 > 0$. The schematic diagrams in Figure 4.3 track a spring through two stages each of extension and compression. Note carefully the sign changes in x, dx/dt, and d^2x/dt^2 from one stage to the next.

The retarding frictional force F_f in the mechanism may be due to the viscosity

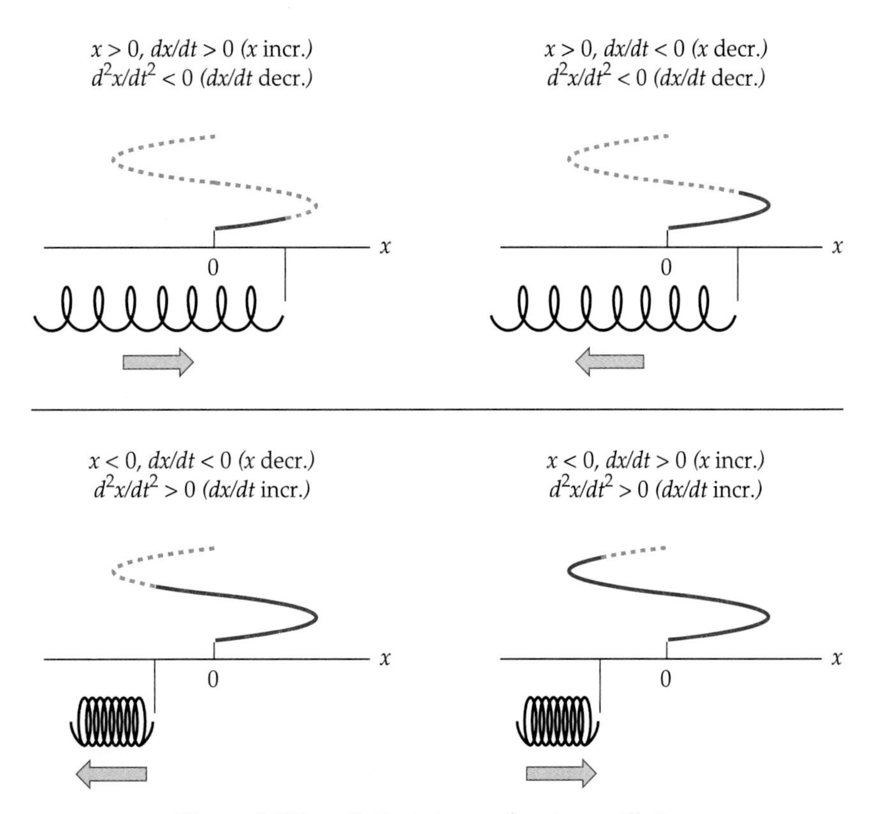

Figure 4.3 Four distinct stages of spring oscillation.

of some surrounding fluid or to friction in mechanical parts of the mechanism; we assume that the sum of these forces is proportional to the velocity:

$$F_f = -k\frac{dx}{dt}, \qquad k > 0.$$

A time-dependent external force $F_e = F(t)$ may act independently of F_s and F_f, so the total force F_T acting parallel to the measuring scale is

$$F_T = F_s + F_f + F_e = -hx - k\frac{dx}{dt} + F(t).$$

On the other hand, general physical principles assert that this force must also be equal to the mass m of the moving parts times the acceleration d^2x/dt^2, so that $F_T = md^2x/dt^2$. Thus

$$m\frac{d^2x}{dt^2} = -k\frac{dx}{dt} - hx + F(t), \quad \text{or} \quad m\frac{d^2x}{dt^2} + k\frac{dx}{dt} + hx = F(t).$$

The external forcing function $F(t)$ can be thought of as an "input" to the mechanism, and a solution $x(t)$ can be thought of as a "response" that the mechanism makes to the input. For our calculations, the mechanism is replaced by the differential equation.

The assumption that there is a constant $h > 0$ such that hx describes the force required to extend or compress a spring by an amount x is called **Hooke's law.** If Hooke's law holds for a spring, then an extension x requires force hx, and an extension $x + \Delta x$ requires force $h(x + \Delta x)$. It follows that the incremental force required to extend from position x to $x + \Delta x$ is $h\Delta x$. An additional consequence is that, if Hooke's law applies, the coordinate x can be chosen to have its zero value located at an arbitrary position of the extension or compression. Of course, there are always practical limits to the applicability of Hooke's law; in extreme cases, overextension could even break a spring and overcompression could buckle it. Thus it's customary to make various restrictive assumptions about the nonnegative constants m, k, and h, and also about the externally applied force $F(t)$.

The piston shown in Figure 4.2 exerts a damping force that can be varied by changing the viscosity of the medium in which the piston moves. If we continue to assume that $F(t) = 0$ in Equation 1.2, then we have to deal with the differential equation

$$\frac{d^2x}{dt^2} + \frac{k}{m}\frac{dx}{dt} + \frac{h}{m}x = 0.$$

The characteristic equation is

$$r^2 + \frac{k}{m}r + \frac{h}{m} = 0,$$

which has roots

$$r_1 = \frac{1}{2m}\left(-k + \sqrt{k^2 - 4mh}\right), \qquad r_2 = \frac{1}{2m}\left(-k - \sqrt{k^2 - 4mh}\right).$$

We can distinguish three distinct cases depending on the discriminant $k^2 - 4mh$.

Overdamping, $k^2 - 4mh > 0$. Because we have assumed that k and h are both positive, this case occurs when $k > 2\sqrt{mh}$. Physically, this inequality means that the friction constant k exceeds the constant $\sqrt{mh}$, depending on the spring stiffness h and the mass m, by a factor more than 2. The effect of the assumption $k > 2\sqrt{mh}$ is to make the roots of the characteristic equation satisfy $r_2 < r_1 < 0$. As a result, the general solution is

$$x(t) = c_1 e^{r_1 t} + c_2 e^{r_2 t},$$

where the exponentials decrease as t increases. It follows that in the case of overdamping, the solution really fails to oscillate at all in the sense described at the beginning of the section.

example 2

In the previous discussion we'll assume $m = 2, h = 2$, and $k = 5$, so that $k > 2\sqrt{mh}$. The characteristic equation is $2r^2 + 5r + 2 = 0$, with roots $r_1 = -\frac{1}{2}, r_2 = -2$, so that the displacement from equilibrium at time t is

$$x(t) = c_1 e^{-t/2} + c_2 e^{-2t}.$$

A typical graph is shown in Figure 4.4(a). The maximum displacement occurs at just one point, after which the displacement tends steadily to 0 as t increases.

Underdamping, $k^2 - 4mh < 0$. This case occurs when $k < 2\sqrt{mh}$, so that, relative to $\sqrt{mh}$, the friction constant k is small. A typical solution is illustrated in Figure 4.4(b). The characteristic roots are now complex conjugates of one another:

$$r_1 = \frac{1}{2m}(-k + i\sqrt{4mh - k^2}), \qquad r_2 = \frac{1}{2m}(-k - i\sqrt{4mh - k^2}).$$

The general form of the displacement function is then

$$\begin{aligned} x(t) &= e^{-kt/2m}\left(c_1 \cos\frac{\sqrt{4mh - k^2}}{2m}t + c_2 \sin\frac{\sqrt{4mh - k^2}}{2m}t\right) \\ &= Ae^{-kt/2m}\cos\left(\frac{\sqrt{4mh - k^2}}{2m}t - \phi\right). \end{aligned}$$

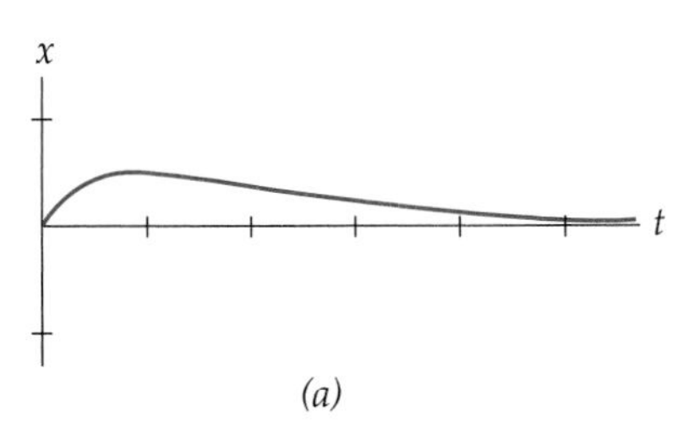

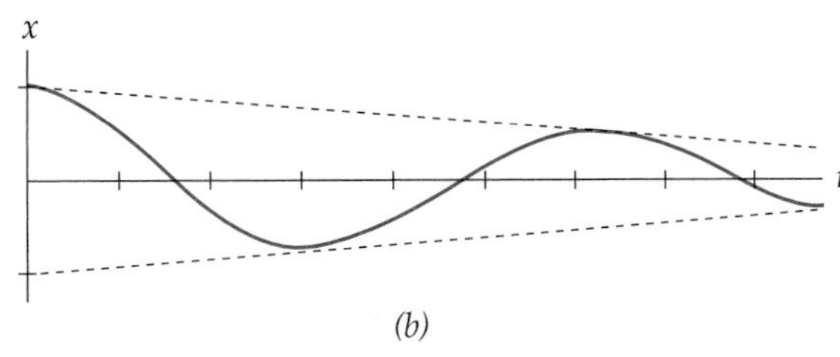

Figure 4.4 (a) $x = e^{-t/2} - e^{-2t}$. (b) $x = e^{-t/10}\cos t$.

example 3

If we take $h = 2$, $m = 1$, and $k = 2$, then $k < 2\sqrt{mh}$, and the displacement at time t is

$$x(t) = Ae^{-t}\cos(t - \phi).$$

Figure 4.4(b) shows the graph of such a function with $A = 1$ and $\phi = 0$. You can check by substitution that this choice for the constants A and ϕ gives a solution satisfying the initial conditions

$$x(0) = 1, \qquad \frac{dx}{dt}(0) = -1.$$

Critical Damping, $k = 2\sqrt{mh}$. This case lies between overdamping and underdamping, and it is unstable in the sense that an arbitrarily small change in one of the parameters k, m, or h will disturb the equality $k = 2\sqrt{mh}$ and produce one of the other two cases. Numerically, the case of critical damping is distinguished by the equality of the characteristic roots: $r_1 = r_2 = -k/2m$. It follows that the displacement function is given by

$$x(t) = c_1te^{-kt/2m} + c_2e^{-kt/2m}.$$

example 4

If we take $m = h = 1$ and $k = 2$, then

$$x(t) = (c_1t + c_2)e^{-t}.$$

If $x(0) = x_0$ and $dx/dt(0) = v_0$, then

$$c_1 = (v_0 + x_0), \qquad c_2 = x_0,$$

so that

$$x(t) = [(v_0 + x_0)t + x_0]e^{-t}.$$

Figure 4.5 shows four possibilities, depending on the size of the initial velocity v_0.

A critically damped displacement is like an overdamped one in that there is no oscillation from one side to the other of the equilibrium position. But with fixed mass m and spring constant h, the critical viscosity value $k = 2\sqrt{mh}$ generally produces a relatively efficient return toward equilibrium: the exponential decay rate for critical damping and underdamping is $\frac{1}{2}k/m$, while for overdamping it is $\frac{1}{2}(k + \sqrt{k^2 - 4mh})/m$. In physical terms, a higher viscosity than the critical value $k = 2\sqrt{mh}$ produces a more sluggish return from a nonequilibrium value, and lower viscosity allows oscillation.

Forced Oscillation. In the specific instances considered so far, the differential equation

$$m\frac{d^2x}{dt^2} + k\frac{dx}{dt} + hx = F(t)$$

has been subject to initial conditions $x(0) = x_0$, $dx/dt(0) = v_0$, but the external forcing function $F(t)$ has been assumed to be identically zero. The resulting free oscillation is described by a solution of a homogeneous differential

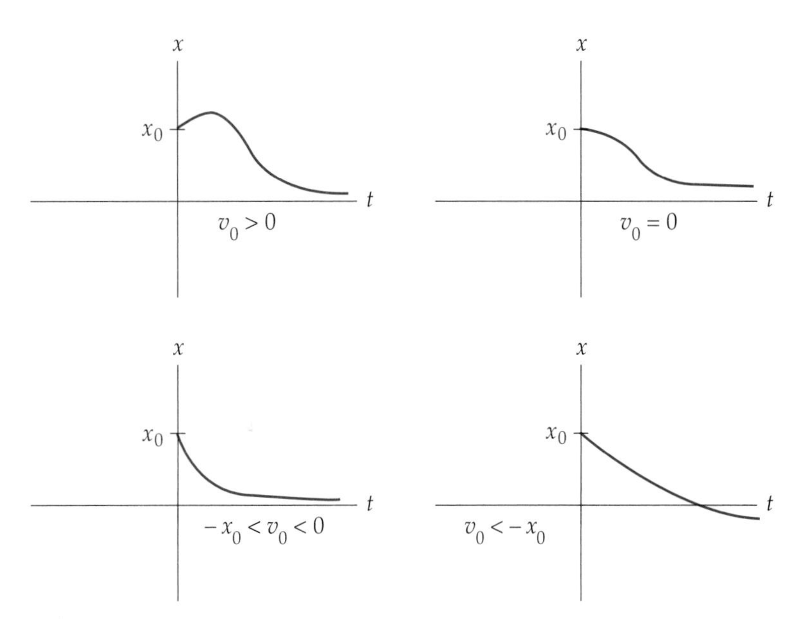

Figure 4.5 Critically damped states.

equation. (Note that "free" does not mean "undamped.") When $F(t)$ is not identically zero, we speak of **forced oscillation.** From a purely mathematical point of view, there is no reason why the function $F(t)$ on the right-hand side of the preceding differential equation cannot be chosen to be, say, an arbitrary continuous function for $t \geq 0$. However, a forcing function $F(t)$ that assumes large values could easily drive the oscillations quite outside the range in which we can maintain the original assumptions used to derive the differential equation. (For example, stretching a spring too far might change its characteristics to the point of destroying its elasticity altogether.) For this reason, the functions $F(t)$ chosen in the examples and exercises have a fairly restricted range of values. In every example, of course, we can use the decomposition of a solution $x(t)$ into homogeneous and particular parts,

$$x(t) = x_h(t) + x_p(t).$$

The homogeneous solution $x_h(t)$ was already discussed earlier in this section for various choices of m, k, and h in the homogeneous differential equation. What remains to be done now is to discuss the effect of adding a solution $x_p(t)$ of the nonhomogeneous equation.

example 5 As we know from Chapter 3, Section 2, the sinusoidally forced harmonic oscillator

$$\frac{d^2x}{dt^2} + \beta^2 x = A\cos(\omega t - \phi), \quad \omega \neq \beta,$$

has solutions

$$x_h(t) + x_p(t) = B\cos(\beta t - \psi) + C\cos(\omega t - \eta),$$

which are sums of periodic terms with different periods $2\pi/\beta$ and $2\pi/\omega$. In the exceptional case $\omega = \beta$, we've seen in Chapter 3, Section 3B, that these two terms coalesce to produce a solution in which the second term in the general form of the solution is amplified by the increasing factor t:

$$x(t) = B\cos(\beta t - \psi) + Ct\cos(\beta t - \psi).$$

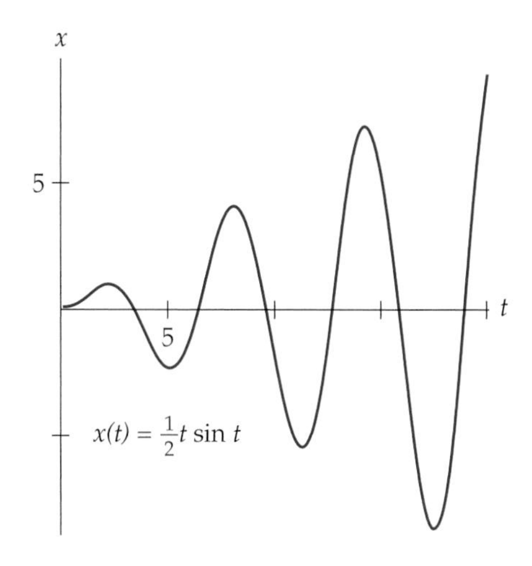

We can look at an amplified part of the solution as a prototype for the phenomenon of *resonance*, which we'll consider in a more realistic context in Examples 6 and 7. For now we'll just illustrate with the initial-value problem

$$\frac{d^2x}{dt^2} + x = \cos t, \qquad x(0) = 0, \qquad \frac{dx}{dt}(0) = 0.$$

We find the solution by the methods of Chapter 3, Section 3B or Section 3C, to be

$$x(t) = \tfrac{1}{2}t\sin t.$$

Qualitatively this unbounded solution suggests that if the forcing oscillation has a frequency very close to the frequency of the unforced oscillator, then the result may be a response of large amplitude. In practice, it's impossible to tune two frequencies to be exactly equal, so we'll examine this suggestion more realistically in the following examples. You are asked in the exercises to investigate the responses of different oscillators $d^2x/dt^2 + \beta^2 x$ to the fixed input $F(t) = \cos t$.

If $k > 0$, the analysis given in the earlier examples shows that every homogeneous solution tends to zero as fast as an exponential of the form $e^{-kt/2m}$. Thus, for values of t that make $\frac{1}{2}kt/m$ moderately large, the addition of the homogeneous solution has a negligible effect. Such an effect is called **transient,** and the particular solution is therefore called the **steady-state** solution, because that's what persists after the transient part has become negligible.

---

example 6

If $F(t) = a_0 \cos \omega t$, then the differential equation

$$m\frac{d^2x}{dt^2} + k\frac{dx}{dt} + hx = a_0 \cos \omega t, \qquad k > 0,$$

has a particular solution of the form

$$x_p(t) = A \cos \omega t + B \sin \omega t,$$

which can be found by the method of undetermined coefficients. Substitution of x_p into the equation yields

$$(h - \omega^2 m)A + \omega k B = a_0,$$
$$\omega k A - (h - \omega^2 m)B = 0.$$

Solving this pair of equations for A and B gives

$$A = \frac{(h - \omega^2 m)a_0}{(h - \omega^2 m)^2 + \omega^2 k^2}, \qquad B = \frac{\omega k a_0}{(h - \omega^2 m)^2 + \omega^2 k^2},$$

so the particular solution we get is

$$x_p(t) = \frac{a_0}{(h - \omega^2 m)^2 + \omega^2 k^2}[(h - \omega^2 m) \cos \omega t + \omega k \sin \omega t]$$
$$= \frac{a_0}{\sqrt{(h - \omega^2 m)^2 + \omega^2 k^2}} \cos(\omega t - \phi),$$

where $\phi = \arctan[\omega k/(h - \omega^2 m)]$. What we have found is just one of many solutions, each satisfying a different set of initial conditions. However, our particular solution is the steady-state solution. Notice that $x_p(t)$ is very much like the external force $F(t) = a_0 \cos \omega t$. In fact, the two amplitudes differ only by the factor $\rho(\omega) = [(h - \omega^2 m)^2 + k^2\omega^2]^{-1/2}$.

There are two problems that often occur in the design of a damped oscillator. One problem is making the amplitude of the steady-state response $x_p(t)$, called the **response amplitude,** large relative to the amplitude of the external forcing amplitude a_0. For example, in the design of a sensitive seismograph the mechanism needs to be able to "amplify" a small vibration, that is, to increase its amplitude to the point where it can be detected mechanically or electrically. The other problem is that we may want to design a detector that can receive an input $F(t)$ of large amplitude and then reduce or "attenuate" the amplitude of the output $x_p(t)$ to manageable size while maintaining the other characteristics of $F(t)$. In either case, a key formula is the factor $\rho(\omega)$ derived in the previous example. To amplify we make ρ large, and to attenuate we make ρ small.

Note that the frequency ω of the steady-state response $x_p(t)$ is the same as the frequency of the external forcing function $a_0 \cos \omega t$, so that in the long run, all frequencies are to a large extent transmitted by the mechanism. The phenomenon called **resonance** is a relationship between a mechanism and an external oscillatory force $F(t)$ such that the steady-state response $x_p(t)$ of the mechanism oscillates with a significantly large amplitude. The test for what constitutes "significantly large" depends on the physical characteristics of the mechanism.

example 7 The expression

$$\rho(\omega) = \frac{1}{\sqrt{(h - \omega^2 m)^2 + k^2\omega^2}}$$

in the previous example represents the ratio of the amplitudes of $x_p(t)$ and $F(t)$. It's reasonable to say that resonance occurs for the mechanism if the ratio $\rho(\omega)$ is larger than 1, in which case we have **amplification,** that is, an increase in amplitude. More particularly, if we choose $\omega = \sqrt{h/m}$, then the first term inside the square root becomes zero, and we are left with

$$\rho(\omega) = \frac{1}{\omega k} = \frac{1}{k}\sqrt{\frac{m}{h}}.$$

If $\rho(\omega) > 1$, that is, if $\sqrt{m} > k\sqrt{h}$, we would say that ω is a *resonance frequency* for the mechanism.

Some people prefer to use the term "resonance" only in cases for which the response amplitude is maximized under various conditions, but the usual usage is our less restrictive one in which the output amplitude is greater than the input amplitude. Nevertheless, the problem of maximizing response amplitude by maximizing the factor $\rho(\omega)$ is a significant one and is taken up in Exercises 46 and 47 on pages 257 and 258. Figure 4.6 shows the graphs of the amplification factor $\rho(\omega) = 1/\sqrt{(h - \omega^2 m)^2 + k^2\omega^2}$ with the choice $h = m = 1$ for four different values of k.

Finally, if $\omega = \sqrt{h/m}$ and $\rho < 1$, we get reduced response amplitude as compared with input amplitude, called **attenuation,** rather than amplification.

Beat Frequencies. The term **beat** refers to a phenomenon in which a relatively high frequency oscillation has an amplitude that varies in such a way that the successive local maximum and minimum values of the oscillation partially delineate another oscillation of lower frequency. Figure 4.7 shows the graph of such an oscillation, on which we have superimposed smooth dotted curves that outline the paths of the extreme values. If the underlying "carrier" oscillation represents a sonic vibration, these outlining or "enveloping" oscillations can sometimes be heard as a distinct lower frequency sound, in which case a listener is said to be "hearing a beat." In electrical circuits, there is an analogue to beats that often goes by the name "amplitude modulation."

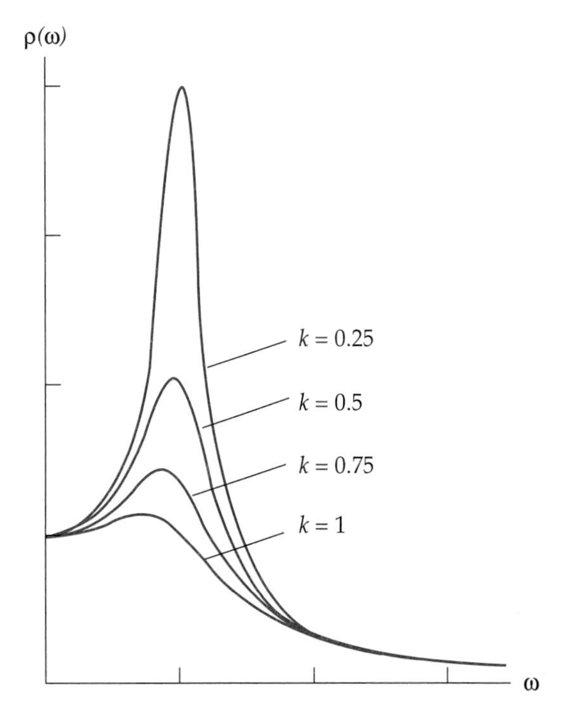

Figure 4.6 $\rho(\omega) = 1/\sqrt{(1-\omega^2)^2 + k^2\omega^2}$.

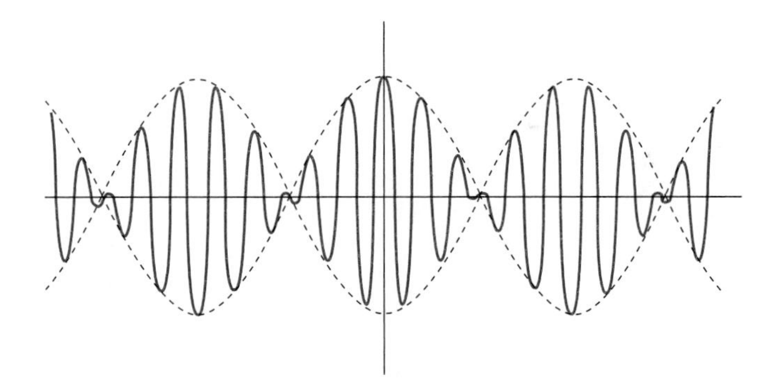

Figure 4.7 Graph of $\frac{1}{2}\cos 22t + \frac{1}{2}\cos 18t = \cos 20t \cos 2t$.
Carrier frequency $10/\pi$; beat frequency $1/\pi$.

Beats can be generated by simultaneously producing two harmonic oscillations with relatively high but distinct frequencies. For example, the identity

$$\sin \omega_0 t - \sin \omega t = 2 \sin \tfrac{1}{2}(\omega_0 - \omega)t \, \cos \tfrac{1}{2}(\omega_0 + \omega)t$$

exhibits the difference of two sine oscillations, with circular frequencies ω_0 and ω, as a single cosine oscillation with large circular frequency $\frac{1}{2}(\omega_0 + \omega)$, and modified by an amplitude $A(t) = 2 \sin \frac{1}{2}(\omega_0 - \omega)t$ that varies with time. (A suggestion for deriving this and similar identities is given in Exercise 53.) The frequency of the beat oscillation is just $\frac{1}{2}(\omega_0 - \omega)$. Thus a low frequency (i.e., a low note in the case of an audible sound) will be produced by making $\omega_0 - \omega$ small. It could happen that the high-frequency oscillation, with frequency $\frac{1}{2}(\omega_0 + \omega)$, produces a sound outside anyone's hearing range, but that the beat tone is audible. The next example describes an idealized mechanism for producing beats.

example 8

Consider an undamped oscillator driven by an external force $F(t)$:

$$m\frac{d^2x}{dt^2} + hx = F(t).$$

The solutions of the associated homogeneous equation all have the form

$$\begin{aligned} x_h &= c_1 \cos\sqrt{\frac{h}{m}}\,t + c_2 \sin\sqrt{\frac{h}{m}}\,t \\ &= c_1 \cos\omega_0 t + c_2 \sin\omega_0 t, \quad \text{where } \omega_0 = \sqrt{\frac{h}{m}}. \end{aligned}$$

We'll assume $\omega \neq \omega_0$ and let $F(t) = a_0 \cos\omega t$. You can check by substitution that with $h = m\omega_0^2$, a particular solution of the differential equation is

$$x_p = \frac{a_0}{m\left(\omega_0^2 - \omega^2\right)} \cos\omega t.$$

It follows that the general solution can be written in the form $x = x_h + x_p$, or

$$x(t) = c_1 \cos\omega_0 t + c_2 \sin\omega_0 t + \frac{a_0}{m\left(\omega_0^2 - \omega^2\right)} \cos\omega t.$$

We choose initial conditions $x(0) = 0$ and $\dot{x}(0) = 0$ and check that these conditions require

$$c_1 = \frac{-a_0}{m\left(\omega_0^2 - \omega^2\right)}, \quad \text{and} \quad c_2 = 0.$$

The unique solution to the initial-value problem is then

$$x(t) = \frac{a_0}{m\left(\omega^2 - \omega_0^2\right)}(\cos\omega_0 t - \cos\omega t).$$

This formula expresses the response to the driving force as the difference of two periodic functions with the same amplitude but different circular frequencies ω_0 and ω. The appropriate trigonometric identity for understanding this oscillation is treated in Exercise 49:

$$\cos\omega_0 t - \cos\omega t = 2\sin\tfrac{1}{2}(\omega - \omega_0)t\ \sin\tfrac{1}{2}(\omega + \omega_0)t;$$

using it we find

$$x(t) = \frac{2a_0}{m\left(\omega^2 - \omega_0^2\right)} \sin\frac{1}{2}(\omega - \omega_0)t\ \sin\frac{1}{2}(\omega + \omega_0)t.$$

If ω is close to ω_0, we get a low-frequency "beat tone" of the form

$$b(t) = \frac{2a_0}{m(\omega^2 - \omega_0^2)} \sin \frac{1}{2}(\omega - \omega_0)t,$$

which modifies or "modulates" the higher frequency oscillation $\sin \frac{1}{2}(\omega_0 + \omega)t$.

EXERCISES

Each of the differential equations in Exercises 1 through 6 represents a free oscillation. Classify them according to type: harmonic, overdamped, underdamped, or critically damped. [*Note:* $\dot{x} = dx/dt$, $\ddot{x} = d^2x/dt^2$.]

1. $d^2x/dt^2 + 2dx/dt + x = 0.$

2. $d^2x/dt^2 + 2dx/dt + 2x = 0.$

3. $\ddot{x} + 9x = 0.$

4. $\ddot{x} + 3\dot{x} + x = 0.$

5. $\ddot{x} + a\dot{x} + a^2x = 0, a > 0.$

6. $d^2x/dt^2 + \frac{1}{4}dx/dt + \frac{1}{8}x = 0.$

7.–12. Find the general solution for each differential equation in Exercises 1–6.

For each of the general solutions in Exercises 13 through 17 of some second-order constant-coefficient equation, find the choice of the arbitrary constants that satisfies the corresponding initial conditions. Sketch the solution that you find. Also find a differential equation that the solution satisfies.

13. $x(t) = c_1 \cos 2t + c_2 \sin 2t$; $x(0) = 0,\ dx/dt(0) = 1.$

14. $x(t) = A\cos(3t - \phi)$; $x(0) = 1,\ dx/dt(0) = -1.$

15. $x(t) = c_1 te^{-2t} + c_2 e^{-2t}$; $x(0) = 0,\ dx/dt(0) = -1.$

16. $x(t) = Ae^{-t}\cos(2t - \phi)$; $x(0) = 2,\ dx/dt(0) = 2.$

17. $x(t) = c_1 e^{-2t} + c_2 e^{-4t}$; $x(0) = -1,\ dx/dt(0) = -1.$

Find the steady-state solution to the differential equations in Exercises 18 and 19. Also estimate the earliest time beyond which the transient solution remains less than 0.01, assuming initial conditions $x(0) = 1,\ \dot{x}(0) = 0.$

18. $d^2x/dt^2 + 2dx/dt + 2x = 2\cos 3t$. [*Hint:* Show $|x(t)| \le \sqrt{2}e^{-t}$.]

19. $d^2x/dt^2 + 3dx/dt + 2x = \cos t.$

20. There is a well-known analogy between the behavior of a damped mass-spring system and of an LRC electrical circuit. Here L is the inductance (analogue of mass) of a coil, R is the resistance (analogue of friction constant) in the circuit, and C is the capacitance (analogue of reciprocal of spring stiffness) or ability of a capacitor to store a charge. The differential equation satisfied by the charge $Q(t)$ on the capacitor at time t is

$$L\frac{d^2Q}{dt^2} + R\frac{dQ}{dt} + \frac{1}{C}Q = E(t),$$

where $E(t)$ is the voltage impressed on the circuit from an external source.

(See Figure 4.8(c) on page 262.) The charge $Q(t)$ is related to the current flow $I(t)$ by $I = dQ/dt$.

(a) Derive the relations that must hold between R, L, and C so that the response of $Q(t)$ should be respectively underdamped, critically damped, and overdamped.

(b) Show that if $C = \infty$ (capacitor is absent) the equation for $I(t)$ is $L\,dI/dt + RI = E(t)$.

(c) Solve the equation in part (b) when $E(t) = E \sin \omega t$ and $I(0) = 0$, and show that, if t is large enough, the current response differs negligibly from

$$\frac{E}{Z}\sin(\omega t - \theta),$$

where $Z = \sqrt{R^2 + \omega^2 L^2}$ and $\cos\theta = R/Z$. The function $Z(\omega)$ is called the *impedance* of the circuit in response to the sinusoidal input of frequency ω.

(d) Show that, if $0 < C < \infty$, a long-term current response to input voltage $E(t) = E \sin \omega t$ is $E(t - \alpha)/Z(\omega)$, where $Z(\omega) = \sqrt{R^2 + [\omega L - 1/(\omega C)]^2}$.

21. The range of validity of Hooke's law for a given spring, and the corresponding value of a Hooke constant h, can be determined as follows. Hang the spring with known weights $W_j = m_j g$, $j = 1, \ldots, n$, attached to the free end. If the additional weight is always proportional to the additional extension, then Hooke's law is valid for this range of extensions, and h is the constant of proportionality. A similar procedure applies to compression.

(a) Assume distance units are in feet. A spring with a 5-pound weight appended has length 6 inches, but with an 8-pound weight appended has length 1 foot. If the spring satisfies Hooke's law with constant h, find h.

(b) What if distance is measured in meters and force in kilograms in part (a)? [*Note*: There are about 3.28 feet in a meter and 2.2 pounds in a kilogram.]

(c) Suppose we know Hooke's constant to be $h = 120$ for a certain spring. We observe that between hanging a 20-pound weight from it and then a larger weight we get an additional extension of 6 inches. How big is the larger weight?

(d) A spring is compressed to length 20 centimeters by a force of 5 kilograms, to length 10 centimeters by a force of 6 kilograms, and to 5 centimeters by a force of 7 kilograms. Discuss the possible validity of Hooke's law given this information.

Forced Harmonic Oscillator

22. Make computer plots of the solutions to the initial-value problem

$$\frac{d^2x}{dt^2} + hx = \cos t, \qquad x(0) = \frac{dx}{dt}(0) = 0$$

over the interval $0 \leq t \leq 36$ and for the values $h = 0.25, 0.5, 1.25$, and 1.5. What do the results suggest about these values of h as compared with $h = 1$?

23. Prove that $t \sin t$ assumes arbitrarily large positive and negative values as t tends to infinity.

24. Make computer plots of the solutions to the initial-value problem

$$\frac{d^2x}{dt^2} + hx = \cos t, \qquad x(0) = \frac{dx}{dt}(0) = 0$$

over the interval $0 \leq t \leq 36$ and for the values $h = 0.8, 1$, and 1.2. What do the results suggest about values $h \neq 1$?

25. Use the results of text Example 5 to prove that for each solution $x(t)$ of $d^2x/dt^2 + hx = \cos t$ there is a constant K such that $|x(t)| \leq K$ for all t.

26. The equation $\ddot{x} + hx = \cos \omega t$ has solutions $x_p(t) = (h - \omega^2)^{-1} \cos \omega t$ if $\omega^2 \neq h$. [This follows from Example 6 if we extend it to the case $k = 0$.]

(a) If $h = 2$, what is a value of ω that will produce a response of amplitude 4 in x_p?

(b) If $\omega = 2$, what is a value of h that will produce a response of amplitude 5 in x_p?

(c) If $h = 2$, what is the unique positive value of ω for which solutions to $\ddot{x} + hx = \cos \omega t$ become unbounded as $t \to \infty$?

(d) If $\omega = 10$, what is the range of h values that will produce response amplitudes for x_p above 4?

Answer questions 27 through 30 about solutions $x(t)$ to the damped and unforced equation $m\ddot{x} + k\dot{x} + hx = 0$, where m, k, and h are positive constants.

27. If $m = 2$, how should h and k be related so that the nontrivial solutions will be oscillatory?

28. If the friction and Hooke constants satisfy $h = k = 1$, how should the mass m be chosen so that all nontrivial solutions will oscillate?

29. If $m = h = 1$, how should k be chosen so that $x(t)$ will have a factor with circular frequency $\frac{1}{2}$?

30. If $m = h = 1$ how should k be chosen so that $x(t)$ is oscillatory?

Answer questions 31 through 34 about solutions $x(t)$ of the damped and forced equation $m\ddot{x} + k\dot{x} + hx = \cos \omega t$, where $h > 0$ and $k > 0$.

31. If $m = k = h = 1$, how should ω be chosen so that the amplitude of the steady-state solution will be 1?

32. If $k = \omega = 1$, what relation must hold between m and h so that the steady-state amplitude will be 1?

33. If $m = h = \omega = 1$, how should k be chosen so that the amplitude of the steady-state solution will be 2?

34. What polynomial relation must hold among m, h, k, and ω if the frequency of the transient solution is to be the same as the steady-state frequency? As a special case, show that if the homogeneous solutions have circular frequency ω, and also $h = m\omega^2$, then $k = 0$, so there is no damping and no nontrivial transient solution.

Find the amplitude A, the frequency $\omega/2\pi$ and a phase-angle ϕ for the periodic functions in Exercises 35 through 37.

35. $2\cos t + 3\sin t$. **36.** $-2\cos 2t + 3\sin 2t$. **37.** $\sin \pi t + 2\cos \pi t$.

By how much are each of the pairs of oscillations in Exercises 38 through 40 shifted apart? You can decide this by expressing each pair in the general form $A\cos(\omega t - \phi)$, $B\cos(\omega t - \psi)$.

38. $(\sqrt{3}/2)\cos t + (1/2)\sin t,\ \cos t$.

39. $(1/2)\cos t + (\sqrt{3}/2)\sin t,\ \cos t$.

40. $(1/\sqrt{2})\cos t + (1/\sqrt{2})\sin t,\ \sin t$.

41. A weight of mass $m = 1$ is attached by horizontal springs with Hooke constants h_1 and h_2 to two fixed vertical supports. The weight oscillates along a horizontal line with negligible friction. By analyzing the force due to each spring, show that the displacement $x = x(t)$ of the weight from equilibrium satisfies $\ddot{x} = -(h_1 + h_2)x$.

42. A weight of mass $m = 1$ is attached by springs with Hooke constants h_1 and h_2 to two fixed vertical supports. The weight oscillates along a horizontal line with negligible friction. Let the respective unstressed lengths of the two springs be l_1 and l_2, and let b denote the distance between the supports.

(a) By analyzing the force due to each spring, show that the displacement $x = x(t)$ of the weight from the support attached to the first spring satisfies

$$\ddot{x} = -(h_1 + h_2)x + h_1 l_1 + h_2(b - l_2).$$

(b) Use the result of part (a) to show that the constant equilibrium solution for x is

$$x_e = [h_1 l_1 + h_2(b - l_2)]/(h_1 + h_2).$$

(c) Show that the constant value x_e of part (b) is the value of the solution to the differential equation that satisfies the initial conditions $x(0) = x_e,\ \dot{x}(0) = 0$.

***43.** A clock pendulum is often suspended from its fixed support by a short segment of flat spring. Let h be the Hooke constant applicable to the distance of deflection of such a pendulum along the path of motion of its center of gravity.

(a) Show that the circular frequency of oscillation is approximately $\sqrt{(h/m) + (g/l)}$, where m and l are the mass and length of the pendulum.

(b) Let the motion of such a pendulum be subject to linear damping with damping constant $k > 0$. Show that if oscillation is to take place, it would be desirable to have $4[hm + (g/l)m^2] > k^2$.

44. The recoil mechanism of an artillery piece is designed containing a linearly damped spring mechanism. The spring stiffness h and the damping factor k should be chosen so that after firing the gun barrel will tend to its original position before firing without additional oscillation. We'll assume a given initial velocity V_0 and mass m for the gun barrel during recoil, and also a fixed maximum recoil distance E, always attained by the barrel.

(a) How should h and k be chosen so that under these conditions the gun barrel undergoes critical damping after firing?

(b) Show that the differential equation for displacement $x(t)$ can be written so as to display dependence only on the parameters V_0 and E.

***45.** During construction of a suspension bridge, two towers have been erected, and a 10-ton weight is suspended between the towers by a cable anchored to both towers. Because of the elastic properties of the cable and the towers, it takes a $\frac{1}{2}$-ton force to move the weight sideways by 0.1 feet. An earthquake moves the base of each tower sideways with identical displacements of the form $0.25 \cos 6t$ feet in t seconds. Assume a linear model for the lateral force on the weight and that damping is negligible.

(a) Find Hooke's constant h and the natural unforced frequency of oscillation for the weight.

(b) Find the amplitude of the quake motion as a function of t, and compare it with the amplitude of the forced oscillation of the weight.

46. The differential equation

$$m\ddot{x} + k\dot{x} + hx = a_0 \sin \omega t$$

determines the displacement $x(t)$ of a damped spring with external forcing $F(t) = a_0 \sin \omega t$ as a function of time t. A frequency ω that maximizes the amplitude of $x(t)$ is called a *maximum resonance frequency*. This exercise asks you to investigate the maximum resonance frequency for fixed ω, and under various assumptions about the mechanism.

(a) Consider the mechanically ideal case where $k = 0$. Show that choosing ω to equal the natural circular frequency $\omega_0 = \sqrt{h/m}$ produces a response $x(t)$ that contains the factor t and hence has deviations from the equilibrium position $x = 0$ that become arbitrarily large as t increases. (Thus there is no theoretical maximum resonance frequency in this case, though in practice the maximum response will be limited by the structural capacity of the mechanism to accommodate wide deviations from equilibrium.)

(b) Assume that k is a fixed positive number. Show that the steady-state displacement $x_p(t)$ has maximum amplitude when h and m are chosen so that $\sqrt{h/m} = \omega$ and that the maximum amplitude is $a_0(\omega k)^{-1}$. (Making such choices for h and m can be thought of as **tuning** the mechanism for maximum response.)

(c) Continuing with the ideas of part (b), show that a small response amplitude to a given forcing frequency ω can be achieved by making $|h - \omega^2 m|$ large.

47. In the previous exercise, we assumed the input frequency to be fixed and considered the effect of varying h and m in the differential equation. Suppose now that h, m, and k are fixed positive numbers and that we want to choose ω so as to maximize the amplitude of the response $x_p(t)$.

(a) Show that the amplitude ratio

$$\rho(\omega) = [(h - \omega^2 m)^2 + k^2\omega^2]^{-1/2}$$

of $x_p(t)$ over $F(t)$ is maximized when the function $F(\omega) = (h - \omega^2 m)^2 + k^2\omega^2$ is minimized.

(b) Show that if $k^2 \leq 2hm$, then $F'(\omega) = 0$ when $\omega^2 = (2hm - k^2)/(2m^2)$. Conclude from this that in this case the maximum response amplitude

$$\frac{2m|a_0|}{k\sqrt{4hm - k^2}} \quad \text{occurs for } \omega = \omega_0\sqrt{1 - \frac{k^2}{2hm}},$$

where $\omega_0 = \sqrt{h/m}$ is the natural circular frequency of the undamped $(k = 0)$ mechanism.

(c) Use the result of part (b) to show that maximum response amplitude is greater than input amplitude (i.e., there is amplification) if $2m > k\sqrt{4mh - k^2}$. [Here we assume that $4mh - k^2 > 0$, which is equivalent to assuming that the transient response $x_h(t)$ is oscillatory. Why are these two assumptions equivalent?]

(d) Show that if $k^2 \geq 2hm$, then $F'(\omega) = 0$ only when $\omega = 0$ and that $F(\omega)$ is strictly increasing for $\omega \geq 0$. Hence conclude that in this case maximum response is $|a_0|/h$, and occurs only for the constant forcing function $F(t) = 0$.

48. The purpose of this exercise is to show that if m, k, and h are positive constants, then each particular solution of

$$m\ddot{x} + k\dot{x} + hx = b_0 \sin \omega t$$

remains bounded as $t \to \infty$.

(a) Show that the solutions of the associated homogeneous equation all tend to zero as $t \to \infty$. (These are the transient solutions.)

(b) Show that

$$x_p(t) = \frac{b_0}{k^2\omega^2 + (h - m\omega^2)^2}[-k\omega \cos \omega t + (h - m\omega^2) \sin \omega t]$$

is a particular solution and that it satisfies $|x_p(t)| \leq 1/\sqrt{k^2\omega^2 + (h - m\omega^2)^2}$. [*Hint:* See text Example 6.]

(c) Use the inequality $|x_p + x_h| \leq |x_p| + |x_h|$ to conclude from the results of (a) and (b) that all solutions are bounded.

49. The purpose of this exercise is to observe the effect on the individual solutions of the initial-value problem

$$\ddot{x} + hx = \sin \omega t, \qquad x(0) = \dot{x}(0) = 0$$

of letting the parameter ω approach the positive constant $\sqrt{h}$. (The differential equation represents a highly idealized situation from a physical point of view, because there is no damping term.)

(a) Show that the unique solution to the initial-value problem with $\omega \neq \sqrt{h}$ is

$$x(t) = \frac{1}{h - \omega^2}\left(\sin \omega t - \frac{\omega}{\sqrt{h}} \sin \sqrt{h}t\right),$$

and that the solution satisfies

$$|x(t)| \leq \frac{1 + \omega/\sqrt{h}}{|h - \omega^2|} \quad \text{for all values of } t.$$

(b) Show that as ω approaches $\sqrt{h}$, the solution-values found in part (a) approach

$$\frac{1}{2h}(-\sqrt{h}t \cos \sqrt{h}t + \sin \sqrt{h}t).$$

Show also that in contrast to the inequality in part (a), this function oscillates with arbitrarily large amplitude as t tends to infinity.

(c) Find an initial-value problem that has the function obtained in part (b) as a solution. [*Hint:* What happens to the original differential equation as $\omega \to \sqrt{h}$?]

50. Suppose that an undamped, but forced, oscillator has the form

$$\ddot{x} + 2x = \sum_{k=0}^{n} a_k \cos kt.$$

(a) Use the linearity of the differential equation to show that it has the particular solution

$$x_p(t) = \sum_{k=0}^{n} \frac{a_k}{2 - k^2} \cos kt.$$

(The trigonometric sum on the right in the differential equation is an example of a *Fourier series*, discussed in general in Chapter 10. An extension of such a sum to an infinite series can be used to represent a very general class of functions.)

(b) How does the solution in part (a) change if the left side of the differential equation is replaced by $\ddot{x} + 4x$ and the right side remains the same?

Floating Buoys

51. A 64-pound cylindrical buoy with a 2-foot radius floats in water with its axis vertical. Assume the buoy is depressed $\frac{1}{2}$ foot from equilibrium and then released.

(a) Find the differential equation for the displacement from equilibrium during undamped motion.

(b) Find the amplitude and period of the motion.

52. A cylindrical buoy with a 1-foot radius floats in water with its axis vertical. Assume the buoy is depressed from equilibrium, released, and allowed to oscillate undamped. If the period of oscillation is 4 seconds, how much does the buoy weigh?

53. (a) Prove the identity $\cos\alpha - \cos\beta = -2\sin\frac{1}{2}(\alpha+\beta)\sin\frac{1}{2}(\alpha-\beta)$, used in the analysis of beats. [*Hint:* Apply addition formula for cosine to $\cos\big((\alpha+\beta)/2 + (\alpha-\beta)/2\big) = \cos\alpha$ and to $\cos\big((\alpha+\beta)/2 - (\alpha-\beta)/2\big) = \cos\beta$.]

(b) Prove the identity $\sin\alpha - \sin\beta = 2\sin\frac{1}{2}(\alpha-\beta)\cos\frac{1}{2}(\alpha+\beta)$, used in the analysis of beats. [*Hint:* Apply addition formula for cosine to $\cos\big((\alpha+\beta)/2 + (\alpha-\beta)/2\big) = \cos\alpha$ and to $\cos\big((\alpha+\beta)/2 - (\alpha-\beta)/2\big) = \cos\beta$.]

54. Consider the differential equation

$$\ddot{x} + 25x = 16\cos 3t.$$

(a) Show that the equation has general solution

$$x(t) = c_1\cos 5t + c_2\sin 5t + \cos 3t.$$

(b) Show that the particular solution satisfying $x(0) = 0,\ \dot{x}(0) = 0$ is $x_p(t) = \cos 3t - \cos 5t$.

(c) Show that $\cos 3t - \cos 5t = 2\sin 4t\sin t$. [*Hint:* Use an identity from the previous problem.]

(d) Use the result of part (c) to sketch the graph of the particular solution found in part (a) for $0 \le t \le 2\pi$.

55. (a) The phase difference between $\cos(\omega t - \alpha)$ and $\cos(\omega t - \beta)$ is $\alpha - \beta$. What is the *time* shift required to put the two oscillations in phase?

(b) What is the phase difference between $\cos(\omega t - \alpha)$ and $\sin(\omega t - \beta)$? [*Hint:* Express the second one in terms of cosine.]

56. Let $F(t) = \sin\alpha t + \sin\beta t$, where α and β are positive numbers.

(a) Show that if $\beta = r\alpha$ for some rational number r then $F(t)$ is periodic for some period $p > 0$, that is, $F(t+p) = F(t)$ for all t. Show also that p can be expressed as (possibly different) integer multiples of both π/α and π/β.

***(b).** Prove that if an $F(t)$ of the form given above is periodic with period $p > 0$ then $\beta = r\alpha$ for some rational number r. Thus, for example, $\sin t + \sin\sqrt{2}t$ can't be periodic. [*Hint:* Check that $F(p) = 0$ and $F''(p) = 0$. Then conclude that $(\alpha^2 - \beta^2)\sin\alpha p = (\alpha^2 - \beta^2)\sin\beta p = 0$, so that either $\alpha = \pm\beta$ or else αp and βp are integer multiples of π.]

57. Suppose a physical process is accurately modeled by a differential equation of the form

$$m\frac{d^2x}{dt^2} + k\frac{dx}{dt} + hx = 0,$$

with m, k, and h positive constants. It may be possible by observation to draw conclusions about the parameters in the underlying process. Given each of the following sets of information about the constants and a solution, find the implications for the other constants.

(a) $m = 1, \ x(t) = e^{-3t}\cos 6t$.

(b) $h = 1, \ x(t) = e^{-t}\sin 5t$.

(c) $k = 1, \ x(t) = e^{-t/2}\cos(t/2)$.

(d) $k = 3, \ h = 2, \ x(t) = e^{-4t}\sin 4t$.

58. Suppose we want to construct a damped harmonic oscillator with Hooke constant $h = 2$ and damping constant $k = 3$. What is the lower limit m_0 for the mass m such that oscillatory solutions are possible? Does oscillation occur for $m = m_0$?

59. A weight of mass m is attached to an elastic cord of length l that obeys Hooke's law with constant h. The other end of the cord is attached to the top of a tower from which the weight is then dropped. Assume negligible air resistance. (In practice this is sometimes done with a long *bungee cord*.)

(a) What is the velocity of the weight when the cord first begins to take the strain of the fall?

(b) Show that from the time the cord begins to stretch until the weight bounces back up again the cord's extension beyond its relaxed state is given by

$$y(t) = \frac{gm}{h}\left(1 - \cos\sqrt{\frac{h}{m}}t\right) + \sqrt{\frac{2glm}{h}}\sin\sqrt{\frac{h}{m}}t.$$

(c) Show that the maximum extension of the cord from its relaxed state is

$$y_{\max} = \frac{gm + \sqrt{g^2m^2 + 2glmh}}{h}.$$

[*Hint:* Express the result of part (b) in terms of phase and amplitude.]

(d) Show that the maximum stretching force experienced is $\sqrt{g^2m^2 + 2glmh}$.

(e) For some types of elastic cord material, Hooke's constant h is inversely proportional to the length l of the cord. Show that, for cords made with a given type of such material, the maximum extension resulting from the experiment described above is proportional to l. Show also that the maximum stretching force experienced with cords of different unstressed lengths l is independent of l. The latter fact has made it possible for people to make a stunt out of the experiment by jumping off a tower while tied to a very long cord, but with limited physical stress to the jumper.

2 ELECTRIC CIRCUITS

The study of electric circuits, and of the more general networks taken up in the supplement to Chapter 6, has to do with time-dependent flow of **current** $I(t)$ in conductors. Electric circuits discussed in this section can all be viewed as consisting of a single wire loop containing a (possibly time dependent) source of voltage, or electric force, with **voltage** $E(t)$, a resistor with **resistance** R, a capacitor capable of storing electric charge $Q(t)$, and having **capacitance** C, and an inductor coil with **inductance** L that induces additional voltage in response to a decreasing current as well as inducing a decrease in voltage in response to an increasing current. (Thus an inductor has the effect of smoothing out the current and making it more nearly constant.) Direction of current flow will be described relative to a fixed loop direction; in our single-loop diagrams, such as the ones in Figure 4.8, this direction has been consistently chosen to be clockwise. With this convention, current flowing in the clockwise direction will be designated **positive**, and the reverse flow will be **negative.** A circuit containing resistance, inductance and capacitance, so that L, R, and C are all positive, is sometimes called an ***LRC* circuit;** we'll name the relevant units for these quantities later when we introduce numerical values for them.

The relation between circuit elements that enables us to predict the values $I(t)$ of current is **Kirchhoff's loop law:** *The sum of voltage differences across the elements of a closed loop is equal to the voltage source in the loop.*

The **voltage differences** referred to in the loop law are due to the presence of resistance, inductance, or capacitance in the loop. The differences are related to current I in the following ways:

Voltage difference $E_R = IR$ is due to resistance R.

Voltage difference $E_L = L\,dI/dt$ is due to inductance L.

Voltage difference $E_C = Q/C$ is due to capacitance C, where Q is charge on the capacitor, and charge is related to current by

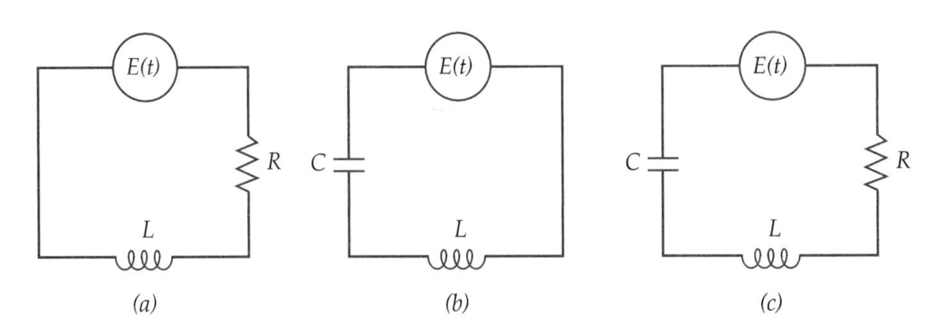

Figure 4.8 (a) LR circuit. (b) LC circuit. (c) LRC circuit.

2.1

$$I = \frac{dQ}{dt}.$$

Remarks

1. Equation 2.1 expresses the definition of *current* $I(t)$ in a loop as the rate of change of positive charge $Q(t)$ through a cross section of wire in the loop. (According to the standard convention, electrons, which have negative charge, move opposite to current.)
2. The constants L, R, and C are never negative; making one of them zero is equivalent to asserting that the corresponding type of circuit element is not included in the loop.
3. Note that for a steady, that is, constant, current, we have $dI/dt = 0$ so that there is no voltage difference across an inductor in that case. For this reason, it makes sense to include an inductor only in circuits in which the current varies over time, the particular case of most interest being a circuit driven by an alternating voltage source, for example, $E(t) = E_0 \sin \omega t$.
4. Many people denote current by i rather than I, and then write $\sqrt{-1} = j$ instead of $\sqrt{-1} = i$. We've kept the convention $\sqrt{-1} = i$, and so use I for current.

example 1

The loop in Figure 4.8(a) contains, in addition to the voltage source, only resistance and inductance, so the loop law dictates that

$$L\frac{dI}{dt} + RI = E.$$

This first-order equation has an elementary solution if L, R, and E are constant:

$$I = \frac{E}{R} + c_1 e^{-Rt/L}.$$

The constant c_1 can be determined by an initial condition of the form $I(0) = \text{const}$. Given that R and L are positive constants, the solution formula predicts that the current will approach the value E/R exponentially as t tends to infinity. Note that if $L = 0$ the differential equation reduces to a purely algebraic relation from which we deduce that $I = E/R$ for all values of t. This last equation is the statement of **Ohm's law.**

example 2

Computing $I(t)$ was particularly simple in the previous example because we assumed that E was constant. However, if we have variable voltage input, say from a microphone, of the form $E(t) = E_0 \cos \omega t$, then we resort to finding an exponential multiplier to solve the resulting first-order linear equation in the standard form

$$\dot{I} + \frac{R}{L}I = \frac{E_0}{L}\cos \omega t.$$

We find $M(t) = e^{Rt/L}$, with the result that

$$\frac{d}{dt}\left(e^{Rt/L} I\right) = \frac{E_0}{L} e^{Rt/L} \cos \omega t.$$

The integral of the function of t on the right is computed using two integrations by parts and is available from tables or symbolic computing programs:

$$e^{Rt/L} I = \frac{(E_0/L)e^{Rt/L}}{R^2/L^2 + \omega^2}\left(\frac{R}{L} \cos \omega t + \omega \sin \omega t\right) + c_1.$$

Hence,

$$I(t) = \frac{E_0}{R^2 + L^2\omega^2}(R \cos \omega t + L\omega \sin \omega t) + c_1 e^{-Rt/L}.$$

What we learn from this formula is that, just as with a constant input voltage, there is a **transient** term, that is, one that dies off to zero as t tends to infinity. However, the other term is periodic with the same period ω as the input voltage. If we write the formula for current in terms of a phase angle α, we get

$$I(t) = \frac{E_0}{\sqrt{R^2 + L^2\omega^2}} \cos(\omega t - \alpha) + c_1 e^{-Rt/L}, \quad \text{where } \alpha = \arctan\left(\frac{L\omega}{R}\right).$$

From this formula we learn how the phase angle α depends on the design of the circuit: to increase α, increase L or reduce R. Similar interpretations are taken up in the exercises.

---

example 3

The loop in Figure 4.8(b) contains all the basic circuit elements except resistance. In classical circuits such a state of affairs is impossible; there is always some resistance. However, we can think of such a circuit as being "superconducting" in the sense that the resistance is negligible for our purposes. Furthermore, the simple model we get helps to elucidate the more general one discussed in the following example.

The loop law asserts that

$$L\dot{I} + \frac{1}{C}Q = E(t).$$

In order to solve this equation for I, or for Q, we need some other relation between those two quantities. That relation is provided by Equation 2.1, $I = \dot{Q}$, referred to above, and we can use it to eliminate I from the loop-law equation, getting

$$L\frac{d^2Q}{dt^2} + \frac{Q}{C} = E.$$

We assume L and C are positive constants, so the characteristic equation $Lr^2 + C^{-1} = 0$ has roots $r_1, r_2 = \pm i/\sqrt{LC}$. The resulting solutions of the homogeneous equation all have the form

$$Q_h = c_1 \cos \frac{t}{\sqrt{LC}} + c_2 \sin \frac{t}{\sqrt{LC}}$$
$$= c_1 \cos \omega_0 t + c_2 \sin \omega_0 t.$$

In the previous line and hereafter, we use the abbreviation $\omega_0 = 1/\sqrt{LC}$, because this **natural circular frequency** ω_0 appears in a number of fundamental formulas. The voltage source often supplies an alternating current that can be characterized by voltage $E(t) = E_0 \cos \omega t$. If ω happened to equal the natural circular frequency ω_0, we would expect the nonhomogeneous equation to have unbounded oscillatory solutions containing the factor t. Since this phenomenon can't be realized for very long without exceeding the capacity of a real physical mechanism, we'll assume $\omega \neq \omega_0$; in that case you can check that a particular solution is

$$Q_p = \frac{E_0}{L(\omega_0^2 - \omega^2)} \cos \omega t.$$

It follows that the general solution can be written in the form $Q = Q_h + Q_p$, or

$$Q(t) = c_1 \cos \omega_0 t + c_2 \sin \omega_0 t + \frac{E_0}{L(\omega_0^2 - \omega^2)} \cos \omega t.$$

Initial conditions for $Q(t)$ that are of particular interest are $Q(0) = 0$ (this says that there is no initial charge on the capacitor) and $\dot{Q}(0) = 0$ [since $I(0) = \dot{Q}(0) = 0$, this says that the initial current is zero]. You can check that these initial conditions require

$$c_1 = \frac{-E_0}{L(\omega_0^2 - \omega^2)}, \qquad c_2 = 0.$$

The unique solution is then

$$Q(t) = \frac{E_0}{L(\omega_0^2 - \omega^2)} (\cos \omega t - \cos \omega_0 t).$$

The previous formula describes the response to the input voltage as the difference of two periodic functions with the same amplitude but different frequencies ω_0 and ω. The current, which is the time derivative of the charge, is then

$$I(t) = \frac{E_0}{L(\omega_0^2 - \omega^2)} (\omega_0 \sin \omega_0 t - \omega \sin \omega t).$$

An analysis of this last formula in Exercise 3 leads to a description of the phenomenon of "amplitude modulation."

The loop in Figure 4.8(c) contains all the basic circuit elements. According to the loop law, the sum of the three voltage difference E_R, E_L, and E_C equals the voltage source $E(t)$, so we have

2.2
$$L\frac{dI}{dt} + RI + \frac{Q}{C} = E(t).$$

As in the previous example, we eliminate I in favor of Q using Equation 2.1: $I = dQ/dt$. We get

2.3
$$L\frac{d^2Q}{dt^2} + R\frac{dQ}{dt} + \frac{Q}{C} = E(t).$$

If we then manage to find $Q(t)$, we can later find $I(t)$ by using Equation 2.1 again; just differentiate $Q(t)$ to find $I(t)$. But if what we really want is just $I(t)$, and we know that L, R, and C are constant, we can differentiate both sides of Equation 2.2 and then use Equation 2.1 to get a differential equation satisfied by $I(t)$:

2.4
$$L\frac{d^2I}{dt^2} + R\frac{dI}{dt} + \frac{1}{C}I = \dot{E}(t).$$

This last equation has the disadvantage that satisfying initial conditions requires not only a knowledge of $I(0)$, which can be measured, but also of $\dot{I}(0)$, which is hard to determine.

---

example 4

When L, R, and C are constant in Equation 2.4, the roots of the characteristic equation

$$Lr^2 + Rr + \frac{1}{C} = 0$$

are important in determining the precise behavior of the circuit under an applied voltage $E(t)$. If $E(t)$ is also constant, then $\dot{E}(t) = 0$, and Equation 2.4 is homogeneous. The roots of the characteristic equation are

$$r_1, r_2 = -\frac{1}{2}\frac{R}{L} \pm \frac{1}{2}\sqrt{\left(\frac{R}{L}\right)^2 - \frac{4}{LC}}.$$

The corresponding solutions will be oscillatory whenever the discriminant $(R/L)^2 - 4/(LC)$ is negative. This is the most interesting case for applications, and we'll assume it holds for the purposes of this example. We assume also that the constants R, L, and C are all positive. Thus the homogeneous equation, with $\dot{E}(t) = 0$, has solutions of the form

$$I_h = e^{-Rt/2L}(c_1 \cos\omega_0 t + c_2 \sin\omega_0 t), \quad \text{where } \omega_0 = \frac{1}{2}\sqrt{\frac{4}{LC} - \left(\frac{R}{L}\right)^2}.$$

These solutions are transient because the exponential factor makes them tend rapidly to zero as time goes on. It may be that they die out so rapidly that we can safely ignore them in computing the response of the circuit to an input voltage. On the other hand, it may be the initial response of the circuit that we are particularly interested in. And, of course, the solutions of the homogeneous equation have a role to play in determining the effect of initial conditions on the general solution.

It is important to remember that the solution $I_h(t)$ we have just been discussing represents current as a function of time. If we want then to find the charge $Q(t)$ on the capacitor, we solve the first-order equation $\dot{Q} = I(t)$. To do this explicitly we, of course, need an initial value $Q(0)$, which may be difficult to measure in practice.

example 5

If the transient solutions found in the previous example are considered negligible, we may want to make specific calculations only for a solution of the related nonhomogeneous equation. Let's assume an alternating voltage source $E(t) = E_0 \sin \omega t$, where $E_0 > 0$. Hence $\dot{E}(t) = \omega E_0 \cos \omega t$, and to find $I(t)$ we solve

$$L\ddot{I} + R\dot{I} + \frac{1}{C}I = \omega E_0 \cos \omega t.$$

We use a trial solution of the form $I_p = A \cos \omega t + B \sin \omega t$. (There is no chance that this could turn out to satisfy the homogeneous equation, because all homogeneous solutions contain the exponential factor $e^{-Rt/2L}$ found in the previous example.) Next we compute $\dot{I}$ and $\ddot{I}$, substitute into the differential equation and equate coefficients of $\cos \omega t$ and $\sin \omega t$. The result is a pair of equations to be solved for A and B:

$$\begin{aligned}(C^{-1} - \omega^2 L)A + R\omega B &= \omega E_0\\ -R\omega A + (C^{-1} - \omega^2 L)B &= 0.\end{aligned}$$

Solving for A and B, we find

$$I_p(t) = \frac{(C^{-1} - \omega^2 L)\omega E_0}{(C^{-1} - \omega^2 L)^2 + R^2\omega^2} \cos \omega t + \frac{R\omega^2 E_0}{(C^{-1} - \omega^2 L)^2 + R^2\omega^2} \sin \omega t.$$

The two terms combine into one cosine term (see Chapter 3, Section 2C) to introduce a phase-angle α, so

$$I_p(t) = \frac{E_0}{\sqrt{[L\omega - (C\omega)^{-1}]^2 + R^2}} \cos(\omega t - \alpha),$$

where $\alpha = \arctan[R\omega/(C^{-1} - L\omega^2)]$, $0 \leq \alpha \leq \pi$. This solution is called the *steady-state* solution, because it differs from the general solution only by the transient solution, which itself tends exponentially to zero as t tends to infinity.

What we can immediately conclude from all this is that if the input voltage is represented by a pure sine of frequency ω and amplitude E_0, then the current also tends to vary with frequency ω. The factor

$$Z(\omega) = \sqrt{[L\omega - (C\omega)^{-1}]^2 + R^2}$$

in the denominator of the expression for I_p is called the **impedance** of the circuit; it is a function not only of the parameters R, L, and C but also of the input frequency ω. Note that a large impedance produces an attenuated output amplitude. On the other hand, minimizing $Z(\omega)$ produces a maximum value for the output amplitude. The value ω_0 for which this maximum is attained is called the **maximum resonance frequency** of the circuit.

The discussion in the previous example is reminiscent of the analysis of vibrations of a spring mechanism in Section 1. The analogy is in fact complete, and is helpful in thinking about circuits if you have some intuition about what happens in a spring mechanism. Table 1 shows the correspondences in parallel columns.

Table 1

Spring $m\ddot{x} + k\dot{x} + hx = F(t)$		**Circuit** $L\ddot{Q} + R\dot{Q} + \frac{1}{C}Q = E(t)$	
Position	x	Charge	Q
Velocity	$\dot{x}$	Current	$\dot{Q} = I$
Mass	m	Inductance	L
Damping	k	Resistance	R
Elasticity	$1/h$	Capacitance	C
Applied force	$F(t)$	Applied voltage	$E(t)$

Notice particularly that the relationship between stiffness h and capacitance C is reciprocal. A very strong spring has a large h and hence small elasticity, which in the correspondence table is analogous to small capacitance. Put the other way, large capacitance is analogous to a very elastic spring, that is, one with a small h. You may find it illuminating to review the preceding circuit examples with the mechanical analogy in mind.

Before proceeding with numerical values for L, R, and C, standard units of measurement should be agreed on for the quantities we have to deal with. Voltage is in **volts,** current is in **amperes,** and charge is in **coulombs.** Because of Equation 2.1 relating current and charge, there is a relation between amperes and coulombs; it turns out that 1 ampere equals 1 coulomb per second. Resistance is measured in **ohms,** inductance in **henrys,** and capacitance in **farads.** Just as with a mechanical system, it is the balance between the sizes of these quantities that determines the inherent qualities of an electric circuit. In practice, capacitance C is usually rather small, often on the order of 10^{-6} farads. The resistance R is usually measured in tens, hundreds, or thousands of ohms. If inductance L is in the range 0.1 to 1.0 henrys, the natural circular frequency $\omega_0 = 1/\sqrt{LC}$ will be on the order of 1000 to 3000 cycles per second.

---

example 6 Analog computers can be designed to mimic the operation of mechanical systems by using the analogy described above. The parameters L, R, and C are analogues to mass m, damping k, and elasticity $1/h$. Suppose we want to model the displacement $x(t)$ of a spring mechanism with a mass of 1000 pounds, damping factor $k = 24$, and stiffness $h = 200$. If there is an externally applied force $F(t) = 220 \cos 95t$, with t measured in seconds, an appropriate LRC circuit model might have inductance 1000 henrys, resistance 24 ohms, and capacitance $1/200 = 0.005$ farads. The external alternating current source would be characterized by voltage $E(t) = 220 \cos 95t$, having maximum voltage 220 volts, period $2\pi/95$, and hence frequency $95/2\pi \approx 15.1$ cycles per second. Remember that for any LRC circuit, it is the charge $Q(t)$ on the capacitor that models the physical displacement. To determine a particular solution satisfying given initial conditions on $x(0)$ and $\dot{x}(0)$, just start the circuit with $Q(0) = x(0)$ and $I(0) = \dot{x}(0)$.

Alternatively, we could scale the constants down by the factor 100, since this would just amount to dividing the associated differential equation by 100. The resulting circuit would have $L = 10$ henrys, $R = 0.24$ ohms, and $C = 0.00005$ farads (or 50 microfarads) and maximum input voltage 2.2 volts, still with frequency ≈ 15.1 cycles per second. This low-voltage circuit has the advantage that it is simpler to operate.

example 7

Let's turn the previous example around. Suppose we observe an electrical response x from some system that looks like $x(t) = 2.2e^{-31t} \sin(3t - 2)$. Can we design a circuit that gives precisely this response to some voltage input and initial conditions? In other words, can we find a differential equation having $x(t)$ as a solution? The answer is yes, of course, since a solution of that form is associated with a homogeneous equation with characteristic roots $\alpha \pm i\beta = -31 \pm 3i$. The associated differential operator is

$$(D + 31 - 3i)(D + 31 + 3i) = D^2 + 62D + 970.$$

Hence the choice $L = 1$, $R = 62$, and $C = 1/970 \approx 0.001$ will suffice for the circuit elements. The circuit should be started out with $Q(0) = x(0) = 0$ and $I(0) = \dot{x}(0) \approx 59.27$ amperes in order to produce something like the given oscillations. The voltage across the resistor required to produce that much initial current is $62 \cdot 59.27 \approx 3675$ volts.

example 8

Figure 4.9(a) is a diagram of an RC circuit. This, of course, is really just an LRC circuit with negligible inductance present. (The mechanical analog would be a massless spring with damping, something that would be impossible using a real spring.) The differential equation for the charge Q on the capacitor is a special case, with $L = 0$, of the one in Equation 2.3, and we'll assume $E(t) = E_0 \sin \omega t$ there. Thus we want to solve

$$R\dot{Q} + \frac{1}{C}Q = E_0 \sin \omega t.$$

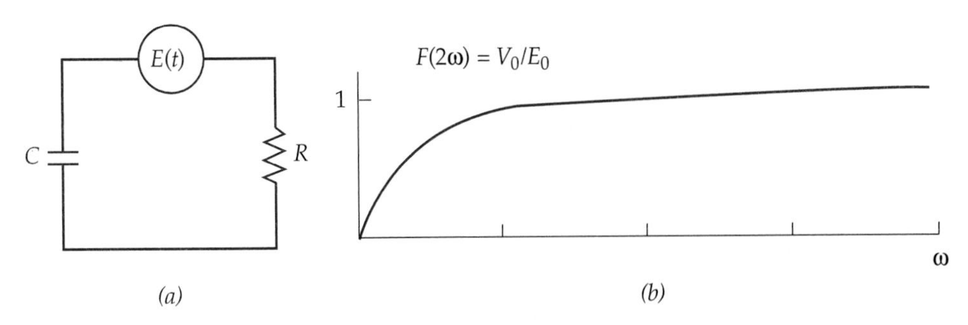

Figure 4.9 (*a*) RC circuit. (*b*) $V_0/E_0 = RC\omega/\sqrt{1+R^2C^2\omega^2}$.

We can solve this equation from scratch, or rely on the result of Example 5 as follows. We introduce the defining relation $I = \dot{Q}$ into Example 5 and also set $L = 0$. Integrating the steady-state solution $I_p(t) = \dot{Q}(t)$ of Example 5 with respect to t yields

$$Q_p(t) = \frac{E_0}{\sqrt{C^{-2} + R^2\omega^2}} \sin(\omega t - \alpha) = \frac{CE_0}{\sqrt{1 + R^2C^2\omega^2}} \sin(\omega t - \alpha),$$

where $\alpha = \arctan RC\omega$. Transient solutions have the form $Q_h(t) = Ke^{-t/RC}$.

It's interesting to compare the amplitude E_0 of the input voltage with the amplitude of the output voltage $V(t)$ as measured across the resistance R in the circuit. According to Ohm's law, $V(t) = RI(t) = R\dot{Q}(t)$. Computing $\dot{Q}$ from the above solution formula, we find

$$V(t) = \frac{RC\omega E_0}{\sqrt{1 + R^2C^2\omega^2}} \cos(\omega t - \alpha).$$

Thus the voltage amplitude as a function of ω is $V_0 = RC\omega E_0/\sqrt{1 + R^2C^2\omega}$, and the ratio of output amplitude to input amplitude is

$$\frac{V_0}{E_0} = \frac{RC\omega}{\sqrt{1 + R^2C^2\omega^2}}.$$

What we see here is that $V_0/E_0 = F(RC\omega)$, where $F(x) = x/\sqrt{1+x^2}$. The function F is increasing as x increases, it tends to 1 as x tends to infinity, and $F(0) = 0$. With R and C fixed, the same can be said about $F(RC\omega)$ as a function of ω. Figure 4.9(b) shows the graph of V_0/E_0 as a function of ω for values of R and C such that $RC = 2$.

When the amplitude ratio is near 1, in other words, when ω is large, the output amplitude is almost as large as the input amplitude. On the other hand, small values of ω produce a response of small amplitude as compared with the amplitude of the input voltage. For this reason, an RC circuit can be used as a "high-pass filter" that tends to "filter out" low-frequency input and to "passthrough" high-frequency input.

EXERCISES

1. Consider a circuit with inductance $L = 1/5$ henrys, resistance $R = 2$ ohms, and capacitance $C = 1/1285$ farads. See Example 5 of the text.

(a) Find the general formula for the capacitor charge $Q(t)$ if $E(t) = 0$.

(b) Use the result of part (a) to find $Q(t)$ if $Q(0) = 1$ and $I(0) = 0$.

(c) Find $Q(t)$ under the initial conditions of part (b) if $E(t) = e^{-t}$.

(d) Find $Q(t)$ under the initial conditions of part (b) if $E(t) = \sin 20t$.

2. Referring to Equation 2.4 of the text, show that if the voltage source is constant and $CR^2 > 4L$, then all solutions $I(t)$ are transient and tend to zero exponentially without oscillation as t tends to infinity.

3. In Example 3 of the text we derived the formula

$$Q(t) = \frac{E_0}{L\left(\omega_0^2 - \omega^2\right)}(\cos \omega t - \cos \omega_0 t).$$

for the response of a superconducting LC circuit to input voltage $E(t) = E_0 \cos \omega t$ in which $\omega_0 = 1/\sqrt{LC}$ is the natural circular frequency of the circuit.

(a) Show that

$$\cos \omega t - \cos \omega_0 t = 2 \sin \tfrac{1}{2}(\omega_0 - \omega)t \, \sin \tfrac{1}{2}(\omega_0 + \omega)t.$$

[*Hint:* Let $\alpha = \frac{1}{2}(\omega - \omega_0)t$ and $\beta = \frac{1}{2}(\omega + \omega_0)t$, and use $\cos(\alpha \pm \beta) = \cos \alpha \cos \beta \mp \sin \alpha \sin \beta$.]

(b) Assume that ω_0 is fairly large and that ω is close to ω_0 with $\omega \neq \omega_0$. Show that $Q(t)$ can then be interpreted as a low frequency oscillation [with frequency $(\omega_0 - \omega)/(4\pi)$] and rapidly varying amplitude [varying with frequency $(\omega_0 + \omega)/(4\pi)$]. (In practice the low frequency oscillation transmits information and the high frequency factor is thought of as a **carrier wave** subject to **amplitude modulation** by the low-frequency oscillation; see Figure 4.8.)

4. (a) Work out the details for finding the solution

$$Q(t) = \frac{E_0}{L\left(\omega_0^2 - \omega^2\right)}(\cos \omega t - \cos \omega_0 t)$$

(given in Example 3 of the text) to the problem $L\ddot{Q} + Q/C = E_0 \cos \omega t$, with $Q(0) = \dot{Q}(0) = 0$. (The natural frequency ω_0 is equal to $1/\sqrt{LC}$.)

(b) Show that for fixed t, as ω approaches ω_0, the solution of part (a) approaches

$$Q_0(t) = \frac{E_0}{2L\omega_0} t \sin \omega_0 t.$$

5. In Example 8 of the text we got the steady-state solution of the RC circuit equation simply by specializing from the solution of the general

LRC equation. If we want to find the general solution for the current in an RC circuit, we need to proceed differently.

(a) Show that in an RC circuit the current satisfies

$$R\dot{I} + \frac{1}{C}I = \dot{E}(t).$$

(b) Find the general solution of the differential equation in part (a) for the case $E(t) = E_0 \cos \omega t$. (You already know the steady-state solution from Example 8; can you use that to save some work?)

(c) Use the result of part (b) to find $I(t)$ for general R and C and $E(t) = E_0 \cos \omega t$ if $I(0) = 0$.

6. In Example 8 of the text the phase angle is $\alpha = \arctan RC\omega$. Show that as ω tends to infinity the oscillations of the input and steady-state output voltages tend to be more nearly in phase. Show on the other hand that as ω tends to zero the corresponding oscillations become more nearly out of phase by $\pi/2$. [*Hint:* Output voltage is R times the time derivative of capacitor charge Q.]

7. Show that the resonant frequency that maximizes the amplitude of the output response of an LRC circuit to a sinusoidal input $E_0 \sin \omega t$ is $\omega_M = 1/\sqrt{LC}$ and that the resulting impedance is $Z(\omega_M) = R$. See Example 5 of the text.

8. Show that if the voltage source in a circuit loop is turned off at some time t_0, then the current $I(t)$ and charge $Q(t)$ will both satisfy the same second-order homogeneous differential equation from that time on.

9. An LRC circuit loop with no voltage source ($E(t) = 0$) is closed at time $t = 0$ when the charge on the capacitor is $Q_0 > 0$]. Show that if $CR^2 = L$, then $Q(t) = Q_0[1 + Rt/(2L)]e^{-Rt/(2L)}$ and that $Q(t)$ eventually tends decreasingly to zero as t tends to infinity.

10. A circuit contains elements with parameters $L = 1$, $R = 1000$, and $C = 4 \cdot 10^{-6}$. At $t = 0$ voltage $E(t) = 110 \sin 50\pi t$ is applied. Find $I(t)$ assuming $Q(0) = I(0) = 0$.

11. A circuit containing elements with parameters $L = 1$, $R = 1000$, and $C = 6.25 \cdot 10^{-6}$ has constant voltage $E = 24$ applied at $t = 0$.

(a) Find the steady-state charge Q_∞ on the capacitor.

(b) At what time does the charge on the capacitor reach half of its steady-state value Q_∞?

(c) What is the maximum value $I_{\max}$ attained by the current?

12. A circuit in which $Q(0) = I(0) = 0$ has elements with parameters $L = 0.02$, $R = 250$, and $C = 2 \cdot 10^{-6}$. At time $t = 0$ constant voltage $E = 28$ is applied. How long does it take for $Q(t)$ to reach 10% of its steady-state value?

13. A circuit in which $Q(0) = 1$ and $I(0) = 0$ has elements with parameters $L = 0.05$, $R = 1$, and $C = 1/505$. Find the charge Q on the capacitor and the current I when $t = 0.01$.

14. Find the steady-state current in a circuit with $L = 10$, $R = 20$, $C = 100$, and $E(t) = 5\cos t$.

15. Let $I(t)$ be the current in an LRC circuit in which the charge on the capacitor is $Q(t)$ and the voltage is $E(t)$. Show that $\dot{I}(0) = \left[E(0) - RI(0) - Q(0)/C\right]/L$. Thus the observable quantities $E(0)$, $I(0)$, and $Q(0)$ determine $\dot{I}(0)$.

3 PHASE SPACE AND EQUILIBRIUM

Our geometric point of view toward second-order differential equations and their solutions undergoes a fundamental shift in this section. This shift requires us to maintain our current thinking about differential equations and then blend in a new point of view. Section 3A introduces a geometric setting that originated in applied mathematics and was exploited by Henri Poincaré with profound effect on the theory and application of differential equations. In Section 3B we identify a simple property that ensures the existence of periodic solutions to a wide variety of second-order differential equations, including the equation $\ddot{y} = -y^3$ and the nonlinear pendulum equation $\ddot{y} = -\sin y$.

3A Phase Space

The differential equation for the displacement $y = y(t)$ of a linearly damped spring-loaded piston (see Section 1) is

$$m\ddot{y} + k\dot{y} + hy = 0.$$

Previously, our pictorial view of a solution $y = y(t)$ of such an equation came simply from looking at the graph of $y(t)$ in the ty plane. What we'll do here is consider simultaneously with $y(t)$ the new dependent variable $z(t) = \dot{y}(t)$, and plot the curve in the yz plane represented parametrically by $(y,z) = \big(y(t), z(t)\big)$. The yz plane in this context is called the **phase space** or **state space** of the differential equation, and an image curve represented by $(y,z) = \big(y(t), z(t)\big)$ is called a **phase curve** or **phase path.** Note that the time variable t is not explicitly present in the plot of a phase curve, so a phase plot does lack some information about solutions. But phase plots do two important things for us:

1. A phase plot enables us to focus attention on the relation between the two quantities that determine future states of an ongoing process, namely position y and velocity $z = \dot{y}$. The Existence and Uniqueness Theorem 1.1 in the introduction to this chapter gives standard conditions under which the determination of future states is ensured and under which distinct phase curves can't touch each other.
2. A phase plot often enables us to make a graphical study of long-term behavior of solutions without having to plot points for extremely large values of t extended along a time axis. For example, we'll be interested in studying solutions as their values approach, for large t, the value of a constant solution $y(t) = y_0$ having velocity $\dot{y}(t)$ identically zero. A constant solution is also called an **equilibrium solution.**

We could, in principle, study phase curves of nonlinear second-order equations of the form $\ddot{y} = f(t, y, \dot{y})$. However, phase space analysis yields the most useful results for the **autonomous** equation $\ddot{y} = f(y, \dot{y})$, from which the time variable t is explicitly absent. (See Exercise 34 to understand why.) The first example is both simple and familiar.

example 1

The undamped linear oscillator equation $\ddot{y} + \omega^2 y = 0$, $\omega \neq 0$, was introduced in Chapter 3, Section 2. There we found that the solutions $y(t) = c_1 \cos \omega t + c_2 \sin \omega t$ could all be written in the form $y(t) = A\cos(\omega t - \phi)$, and from this it follows that the derivative $\dot{y}(t) = z(t)$ is $z(t) = -A\omega \sin(\omega t - \phi)$. The parametric representation

$$\big(y(t), z(t)\big) = \big(A\cos(\omega t - \phi),\ -A\omega \sin(\omega t - \phi)\big)$$

describes ellipses with semiaxes A and $|\omega|A$ centered at the origin of the yz plane. To see this, eliminate t between the equations for y and z to get

$$y^2 + \frac{z^2}{\omega^2} = A^2.$$

Phase curves for three different values of ω are shown in Figure 4.10(a). Note that the velocity z attains both its maximum and its minimum when the position y is zero. Similarly, the position y is at its maximum and minimum when the velocity z is zero. Furthermore, it's clear from this finite picture that the motion is periodic, not just in the sense that the position y repeatedly traces over the same interval from $-A$ to A, but also that a given position y always goes with the same velocity z. Note, however, that some information about solutions in

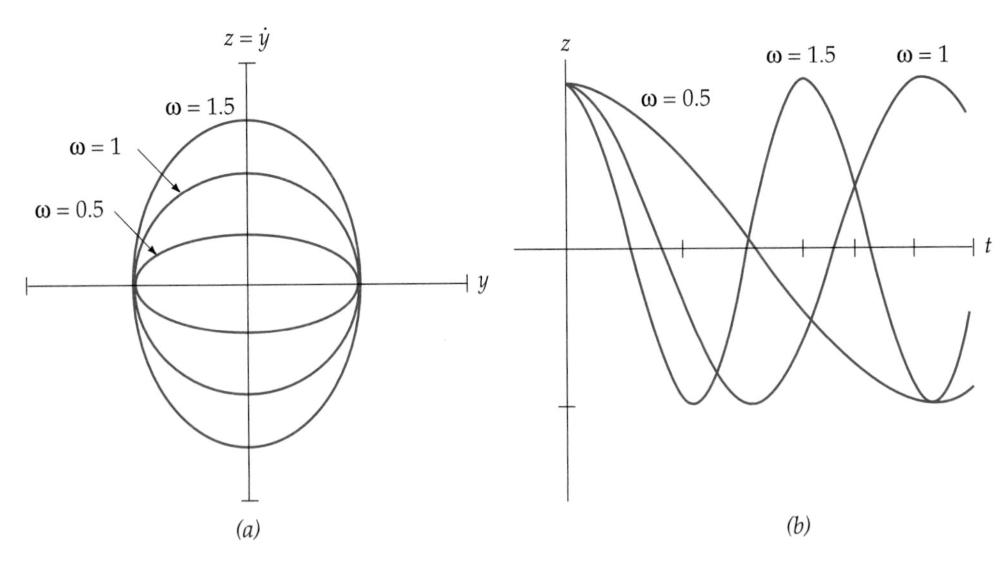

Figure 4.10 $\ddot{y} + \omega^2 y = 0,\ y(0) = 1,\ \dot{y}(0) = 0;\ \omega = 0.5,\ 1,\ 1.5.$
(*a*) Curves $(y, z) = (y(t), \dot{y}(t))$ for $\omega = 0.5, 1, 1.5$. (*b*) Graphs $y = y(t)$ for $\omega = 0.5, 1, 1.5$.

terms of t is lost by going to this picture. Specifically, Figure 4.10(a) contains little information about how y and $z = \dot{y}$ are each related to specific times t, the notable exception being that as t increases the ellipses are traced *clockwise* regardless of whether $\omega > 0$ or $\omega < 0$. For example, at a point in the first and fourth quadrants, y must be increasing, because $z = \dot{y}$ is positive there. Similarly, y decreases in the second and third quadrants, because $z = \dot{y}$ is negative there. Figure 4.10(b) shows the corresponding graphs of some solutions; note that a steeper graph goes with increased velocity and increased vertical elongation of a phase curve.

The phase curves of an autonomous second order equation $\ddot{y} = f(y, \dot{y})$ are determined by an *order-reduction* technique to obtain the first-order equation

3.1
$$z\frac{dz}{dy} = f(y, z).$$

To see this, use the chain rule, and the relations $\dot{y} = z$, $\dot{z} = f(y, z)$, to write

$$\frac{dz}{dy}\frac{dy}{dt} = \frac{dz}{dt}, \quad \text{or} \quad z\frac{dz}{dy} = f(y, z).$$

---

example 2

The oscillator equation $\ddot{y} + \omega^2 y = 0$ leads, as described above, to the first-order equation

$$z\frac{dz}{dy} + \omega^2 y = 0, \quad \text{or in separated form,} \quad z\,dz = -\omega^2 y\,dy.$$

The solutions are found by integration to be

$$\tfrac{1}{2}z^2 = -\tfrac{1}{2}\omega^2 y^2 + C, \quad \text{or} \quad z^2 + \omega^2 y^2 = 2C.$$

These are just the equations of the ellipses found in the previous example by starting with explicit solutions in terms of t. These elliptical phase curves characterize harmonic oscillation.

Finding the constant, or equilibrium, solutions of a second-order equation $\ddot{y} = f(y, \dot{y})$ is straightforward. A constant solution y must satisfy the two equations $\dot{y} = 0$ and $\ddot{y} = 0$. In terms of $f(y, z)$ with phase variables y and z, these two requirements say that $z = 0$ and $f(y, z) = 0$. Thus we have the criterion: An equilibrium solution y_0 of $\ddot{y} = f(y, \dot{y})$ is represented in yz phase space as an **equilibrium point** $(y_0, 0)$. This point is on the y axis with y_0 determined by the equation $f(y, 0) = 0$.

example 3

To locate in phase space the equilibrium solutions of the equation $\ddot{y} + \omega^2 y = 0$ in the previous example, we can proceed directly from the relations

$$\dot{y} = z \quad \text{and} \quad \dot{z} = -\omega^2 y.$$

If y is to be constant, both expressions must be zero. The only solution is $(y, z) = (0, 0)$. We could also have drawn this conclusion from the explicit solution formula obtained in Example 1.

example 4

As discussed in Chapter 3, Section 2, the damped linear oscillator equation $\ddot{y} + k\dot{y} + y = 0$, $0 < k < 2$, has solutions of the form

$$y(t) = e^{-kt/2}(c_1 \cos \omega t + c_2 \sin \omega t), \qquad \omega = \sqrt{1 - (k/2)^2}\,.$$

These solutions can be rewritten as $y(t) = Ae^{-kt/2} \cos(\omega t - \phi)$. Hence, we can compute the derivative $z(t) = \dot{y}(t)$ to be

$$z(t) = Ae^{-kt/2}\left[-\left(\frac{k}{2}\right)\cos(\omega t - \phi) - \omega \sin(\omega t - \phi)\right].$$

This we can rewrite using an addition formula for the cosine as $z(t) = Be^{-kt/2} \cos(\omega t - \psi)$. Since $y(t)$ and $z(t)$ have a common exponential factor, the phase curves may be represented parametrically by

$$\big(y(t), z(t)\big) = e^{-kt/2}\big(A \cos(\omega t - \phi), -B \sin(\omega t - \psi)\big).$$

If k were zero, the exponential factor would be absent, and we would typically get elliptic phase curves. But since the exponential tends to zero as t tends to infinity, each curve spirals in toward the origin as t increases as is shown in Figure 4.11(a).

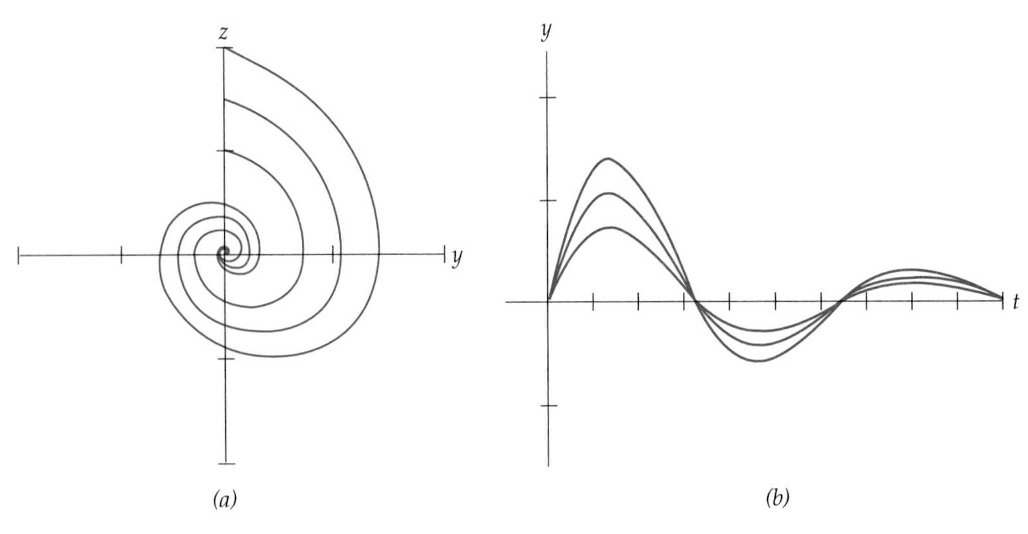

Figure 4.11 $\ddot{y} = -0.5\dot{y} - y$, $y(0) = 0$, $\dot{y}(0) = 1,\ 1.5,\ 2$. (*a*) Phase portrait. (*b*) Graphs of $y = y(t)$.

Getting used to switching back and forth between ty space and yz space requires some careful thought at first; studying the correspondence between the curves in parts (a) and (b) of Figure 4.11 will help and is well worth the effort.

Typical phase curves are shown in Figure 4.11(a) for $y(0) = 0$ and for various values of $z(0) = \dot{y}(0)$. Graphs of corresponding solutions are shown in Figure 4.11(b). This collection of pictures is not strictly analogous to the ones in Figure 4.10(a). In Figure 10 the phase curves and graphs illustrate responses to changes in the differential equation itself, through changes in the parameter ω, rather than to changes in initial conditions as in Figure 4.11. The result of plotting several phase curves, clearly spaced, of a *single* differential equation is called a **phase portrait** for the differential equation. Note that Figure 4.10(a) can't be a phase portrait for a single differential equation satisfying the conditions of the Existence and Uniqueness Theorem 1.1, because the distinct curves touch each other at $(y,z) = (0,\pm 1)$. Figures 4.11(a) and (b) compare phase plots with solution graphs for $y = y(t)$ for just a single differential equation.

example 5

The equilibrium solutions of the equation $\ddot{y} = -k\dot{y} - y$ in the previous example are the solutions in the yz plane of the pair of equations

$$z = 0 \quad \text{and} \quad -kz - y = 0.$$

The only solution is therefore $(y, z) = (0, 0)$.

The next example is about a nonlinear equation and is particularly important because it illustrates how a phase space analysis can easily provide information about solutions even when elementary solution formulas aren't available.

example 6

Assuming there are no retarding frictional forces, we saw in Chapter 3, Section 2, that a pendulum of length l has angular displacement $y(t)$ from the vertical at time t that satisfies

$$\ddot{y} = -\frac{g}{l}\sin y, \quad g = \text{gravitational acceleration.}$$

We can't find an elementary formula for $y(t)$, so we proceed as we did in Example 2 to find an equation satisfied by y and $z = \dot{y}$. We replace $\ddot{y}$ by $z(dz/dy)$ as in Equation 3.1 to get

$$z\frac{dz}{dy} = -\frac{g}{l}\sin y.$$

Integrating both sides with respect to y gives

$$\frac{1}{2}z^2 = \frac{g}{l}\cos y + C \quad \text{or} \quad z = \pm\sqrt{\frac{2g}{l}\cos y + 2C}\,.$$

In Figure 4.12(a), where $l = g$, the values of $2C$ for which $|2C| < 2$ correspond to disjoint loops that represent ordinary swings of a pendulum, somewhat like

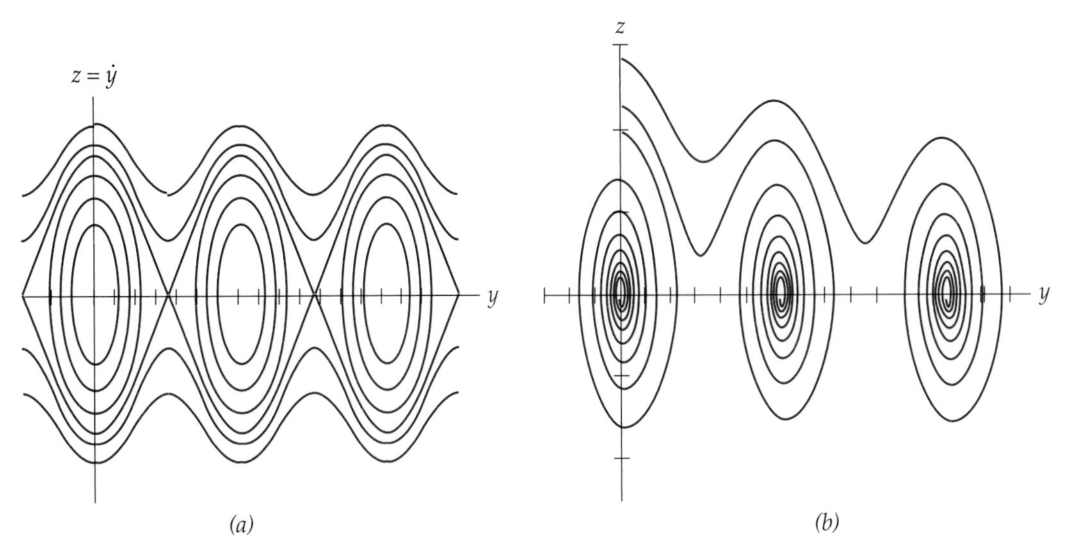

Figure 4.12 Phase portraits. (a) Undamped pendulum. $\ddot{y} = -\sin y,\ z^2 = 2\cos y + 2C$; $C = -1/2,\ 0,\ 1/2,\ 1,\ 6/5,\ 2$. (b) Damped pendulum. $\ddot{y} = -0.1\dot{y} - \sin y,\ y(0) = 0;\ \dot{y}(0) = 2,\ 2.3,\ 2.8$.

the oscillations shown in Figure 4.10(b). If $2C > 2$, the velocity is never zero, so the pendulum goes "over the top" repeatedly. Hence there are two distinct classes of phase path that are periodic as functions of y, one above and one below the y axis in Figure 4.12(a). The upper paths are linked to increasing y and clockwise rotation of the pendulum, the lower ones to counterclockwise rotation.

We can sketch the graphs of the phase paths for various values of the integration constant C using the observation that the curves are tangent to the direction field of $dz/dy = \dot{z}/\dot{y} = -\sin y/z$. (See Exercises 35 through 43 on page 283.) Another alternative is to use computer graphics based on numerical solution techniques such as those described in Section 4. (The Java applet 2ORDPLOT at <http://math.dartmouth.edu/~rewn/> will do this.)

Finally, the equilibrium solutions are just the points on the y axis satisfying $-\sin y = 0$, in other words the points $(y, z) = (k\pi, 0)$, where k is an integer. These points lie at the centers of the closed loops in Figure 4.12(a).

Approaching alternate equilibrium points we get phase curves that *appear* in Figure 4.12(a) to cross each other on the x axis. A curve of this kind is called a **separatrix** because it separates phase curves representing two distinct behaviors: periodic, called **oscillation,** or sometimes **libration,** where the pendulum swings back and forth, and nonperiodic, called **rotation,** where the pendulum swings over the top. In fact, these separating curves themselves don't actually cross on the y axis, but they do have the odd integer multiples $y = (2k + 1)\pi$ as limits on the y axis. Each separatrix approaches an equilibrium point without reaching it, and the equilibrium point itself represents a precarious balancing of a pendulum, quite distinct from the separatrix. Each equilibrium point is itself a kind of degenerate phase curve.

The transition from "back and forth" pendulum swings to the "over the top" behavior described in the previous example can sometimes be achieved by making very small changes in initial conditions. For example, a pendulum may swing so close to the top that a minute boost in velocity will send it over the top. The corresponding change in phase curve is from a closed loop to a wavelike curve as you can see from Figure 4.12(a). This behavior is an example of "instability," some aspects of which are taken up in detail in the next section and in Chapter 6, Section 8.

example 7

The nonlinear pendulum equation with linear damping is

$$\ddot{y} = -k\dot{y} - \frac{g}{l}\sin y.$$

We have no way to find formulas for either the solutions or the phase curves of this equation. In both instances, we need the numerical methods described in Section 4 to get phase portraits. Figure 4.12(b) shows a sample. This picture, particularly the curve starting at $(y, z) = (0, 2)$, should be compared with the one in Figure 4.12(a). Note that the damping factor k for Figure 4.11(a) is five times as large as it is for Figure 4.12(b), so the curves in the latter picture spiral in less quickly toward their equilibrium position.

example 8

One form of the **Van der Pol equation** is

$$\ddot{y} = -k(y^2 - 1)\dot{y} - y, \qquad k > 0,$$

and it can be regarded as a modification of the harmonic oscillator equation $\ddot{y} = -y$ by a term $-k(y^2 - 1)\dot{y}$ that has a damping effect when $|y| > 1$ and an amplifying effect when $|y| < 1$. Realized in an electric circuit, this term has the effect of driving an output signal toward periodic oscillation, an oscillation that will be displayed as a closed loop in a phase portrait, because the entire state $(y, \dot{y})$ repeats over the time period of the oscillation. A larger value of k will force a more rapid transition toward periodicity, as you can see from Figure 4.13.

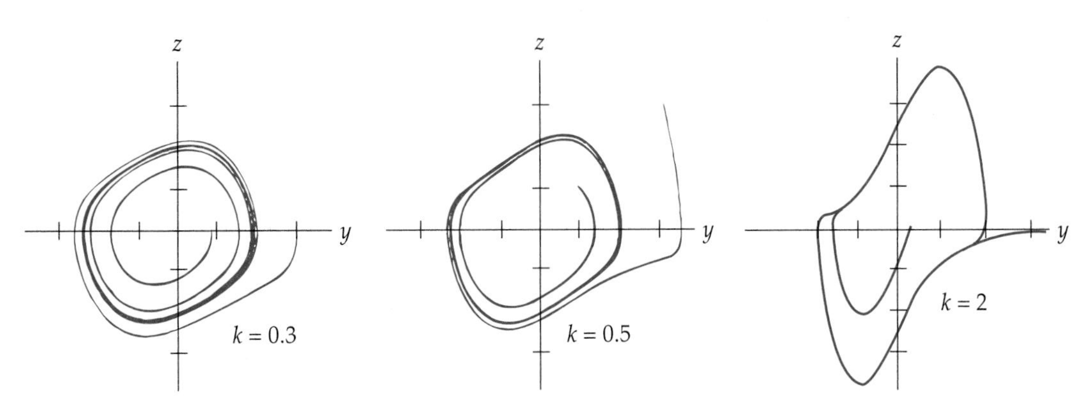

Figure 4.13 Phase curves of the Van der Pol equation $\ddot{y} = -k(y^2 - 1)\dot{y} - y,\ k > 0$.

These pictures were made using a numerical method described in Section 4. Each picture contains just three phase curves, one that spirals out toward a closed loop from the inside, and one that spirals in toward the same loop. The closed loops toward which the phase curves tend are called **limit cycles** and are themselves phase curves of solutions. A limit cycle trajectory is an example of an **equilibrium set** to be contrasted with a single equilibrium point; the latter is necessarily static in character while a limit cycle is a dynamically active trajectory.

EXERCISES

Phase Space versus Space Time. Make a sketch in the (y, z) phase space of the solution to each of the following initial-value problems. [You can do this either directly using expressions for $y(t)$ and $z(t)$ or by eliminating t to get a single equation in y and z.] Also make a separate sketch relative to (t, y) axes of the solution graph itself.

1. $\ddot{y} + y = 1,\ y(0) = 2,\ \dot{y}(0) = 0.$

2. $\ddot{y} - y = 0,\ y(0) = 1,\ \dot{y}(0) = 0.$

3. $\ddot{y} = 1,\ y(0) = \dot{y}(0) = 0.$

4. $\ddot{y} + y = 0,\ y(0) = 2,\ \dot{y}(0) = -1.$

5. $\ddot{y} + \dot{y} = 0,\ y(0) = 2,\ \dot{y}(0) = 0.$

6. $\ddot{y} - \dot{y} = 0,\ y(0) = 1,\ \dot{y}(0) = 0.$

Phase Portraits. Sketch at least three different phase curves for each differential equation below by first finding a family of equations relating y and $z = \dot{y}$ to a constant depending on initial conditions. You can find a suitable equation by eliminating t from the general formulas for $y(t)$ and $z(t) = \dot{y}(t)$ and, if necessary, rewriting the equation so that it displays only one arbitrary constant.

7. $\ddot{y} + y = 1.$

8. $\ddot{y} - y = 0.$

9. $\ddot{y} = 1.$

10. $\ddot{y} + y = 0.$

11. $\ddot{y} + \dot{y} = 0.$

12. $\ddot{y} - \dot{y} = 0.$

13. What are the phase curves of the equations $\ddot{y} = \dot{y}$ and $\ddot{y} = -\dot{y}$?

14. Some differential equations have constant solutions, which are called **equilibrium solutions,** and some don't. Find all equilibrium solutions of the differential equations in Exercises 1 through 6.

15. The differential equation $\ddot{y} + \dot{y} - y + y^3 = 0$ has three distinct constant solutions. What are they?

***16.** The differential equation $\ddot{y} = -y - k\,\mathrm{sgn}(\dot{y})$ depends on the discontinuous real function sgn defined by

$$\mathrm{sgn}(z) = \begin{cases} 1, & z > 0, \\ 0, & z = 0, \\ -1, & z < 0. \end{cases}$$

The differential equation is a harmonic oscillator equation modified by a friction term depending only on the direction of motion and not on the magnitude of its velocity. This assumption about friction is sometimes appropriate for motion at low speed with high resistance, and is called **Coulomb friction.**

(a) Show that the phase paths are patched together from two families of semicircles, one family centered at $y = -k$ for $\dot{y} \geq 0$ and the other family centered at $y = k$ for $\dot{y} < 0$.

(b) Let $k = 1$ and plot the phase curves starting at (i) $(y, \dot{y}) = (-6,0)$, (ii) $(y, \dot{y}) = (-4,0)$, and (iii) $(y, \dot{y}) = (-2,0)$.

(c) Show by using properties of the phase plane that the motion stops completely in finite time regardless of the initial conditions.

Behavior Near Equilibrium. A first-order autonomous equation $\dot{y} = f(y)$ has in its phase portrait just a single curve, one that is particularly easy to identify: It is the graph in the yz plane of $z = f(y)$. The equilibrium points are the points where this graph crosses the horizontal axis. Sketch the phase curve of each of the following equations, and mark the intervals that have an equilibrium point at one or both ends. Then for an arbitrary initial point y_0 in each such interval, decide how the corresponding solution will behave relative to each equilibrium endpoint; will it move toward such a point or away from it?

17. $\dot{y} = -y + 1$.

18. $\dot{y} = -y + y^3$.

19. $\dot{y} = y^2$.

20. $\dot{y} = y - y^3$.

21. $\dot{y} = 1 - y^2$.

22. $\dot{y} = \sin y$.

23. (a) Show that the general solution $y(t) = Ae^{-kt/2}\cos(\omega t - \alpha)$, $A > 0$, of a damped linear oscillator equation has derivative $\dot{y}(t)$ that can be written in the form $z(t) = -Be^{-kt/2}\sin(\omega t - \beta)$, $B > 0$. In particular, find the constant B in terms of the constants A, k, and ω.

***(b).** Show that $\big(y(t), z(t)\big) = \big(A\cos(\omega t - a), -B\sin(\omega t - \beta)\big)$, A, B nonzero constants, is a parametric representation of an ellipse unless $\alpha - \beta$ is an odd multiple of $\pi/2$, in which case it degenerates to a line segment.

24. Show as in text Example 1 that each phase curve in the yz plane of a solution of $\ddot{y} = \omega^2 y$ satisfies an equation $\omega^2 y^2 - z^2 = C$ for some appropriate constant C. Sketch these curves for $\omega = \frac{1}{2}$, 1 and $\frac{3}{2}$ if the curves are to pass through $(y,z) = (1,0)$. This should produce a picture analogous to Figure 4.10(a), but containing hyperbolas instead of ellipses.

25. The phase curves $z^2 = 2(g/l)\cos y + 2C$ with $C = g/l$ (see Example 6 of the text) relate to two distinct types of pendulum behavior.

(a) A single point of the form $(y,z) = (k\pi,0)$, k an integer, is called an *equilibrium point*, and corresponds to a constant solution of $\ddot{y} = -(g/l)\sin y$. What physical states do these solutions describe? In particular, what is the physical distinction between odd and even values of k?

(b) If $z^2 = 2(g/l)(\cos y + 1)$ and $z > 0$ the pendulum is in a very special kind of motion. Describe that motion in words, and do the same for the case $z < 0$.

*26. The pendulum equation $\ddot{y} = -\sin y$ has been integrated once in Example 6 of the text to produce the relation $\frac{1}{2}\dot{y}^2 = \cos y + C$ between y and $\dot{y}$.

(a) Requiring $y(0) = 0$, $\dot{y}(0) = 2$, show that the specific relation satisfied under those initial conditions is $\dot{y}^2 = 2(1 + \cos y)$.

(b) The equation found in part (a) has a graph in phase space called a *separatrix* because it separates the back-and-forth phase curves from the over-the-top rotational ones. Taking square-roots, separating variables and using $\cos(y/2) = \pm\sqrt{(1 + \cos y)/2}$, show that a separatrix satisfies a first-order equation $\sec(y/2)\, dy = \pm 2\, dt$.

(c) Integrate the differential equation in part (b) to show that a separatrix for the equation $\ddot{y} = -\sin y$ satisfies $1 + \sin(y/2) = e^{\pm t}\cos(y/2)$. (There is no such elementary formula relating time and position for the rotating and oscillating solutions.)

27. By studying Figure 4.12(b) carefully you can sketch the phase curves for the damped pendulum equation $\ddot{y} = -0.1\dot{y} - \sin y$ passing through the distinct points

$$(y, z) = (0, -2),\ (0, -2.3),\ (0, -2.8).$$

Make your three sketches relative to a single set of y, z axes, and explain in words how the corresponding pendulum motions differ from the ones represented in Figure 4.12(a). [*Hint:* You'll need to use more of the negative y axis than Figure 4.12(b) does.]

*28. A differential equation of the form $\ddot{y} = \phi(y)$ expresses the acceleration of a point on a line as a function of position y on the line.

(a) Show that the equation is equivalent to the pair of equations $\dot{y} = z$, $\dot{z} = \phi(y)$.

(b) Show that the phase curves of $\ddot{y} = \phi(y)$ satisfy $dz/dy = \phi(y)/z$ for $z \neq 0$.

(c) Show that the differential equation of part (b) has solutions $\frac{1}{2}z^2 = \Phi(y) + \text{const.}$, where $\Phi'(y) = \phi(y)$.

(d) Assume $\phi(y) = -y^{-2}$ and sketch the phase curve through $(y, z) = (2, -1)$.

(e) Find the solution of $\ddot{y} = -y^{-2}$ under the assumption that $(y, z) = (2, -1)$ when $t = 0$. Sketch the graph of this solution for $0 \le t \le \frac{4}{3}$.

29. (a) Find all equilibrium points for the damped pendulum equation $\ddot{y} = -k\dot{y} - (g/l)\sin y$.

(b) Based on what you know about a physical pendulum, separate the equilibrium points found in part (a) into two classes: (i) those for which a solution starting near the equilibrium point tends toward it and (ii) those for which a solution starting near the equilibrium point tends away from it.

30. The equation $\ddot{y} + k(y^2 + \dot{y}^2 - 1)\dot{y} + y = 0$, $k > 0$, is a modification of the Van der Pol equation $\ddot{y} + k(y^2 - 1)\dot{y} + y = 0$. The modified equation exhibits damping when $y^2 + \dot{y}^2 > 1$ and amplification when $y^2 + \dot{y}^2 < 1$.

(a) Show that the circle of radius one centered at the origin is a phase curve for the equation.

(b) Find every solution to this modified equation that has a phase plot that lies on the circle of part (a).

(c) What are the equilibrium solutions of the modified equation?

31. In all our examples, phase curves of $\ddot{y} = f(y, \dot{y})$ that wind repeatedly around an equilibrium point, for example the origin in the yz plane, wind in the *clockwise* direction. Explain why this winding is always clockwise. [*Hint:* Show first that $z = \dot{y}$ must change sign as the path winds around the point.]

32. The reason for restricting phase space analysis for second-order equations to the autonomous case $\ddot{y} = f(y, \dot{y})$ is that for a nonautonomous equation $\ddot{y} = f(t, y, \dot{y})$ there are often many phase curves passing through a given point of the yz plane, with the result that a phase portrait becomes too cluttered to be useful.

(a) Show that the equation $\ddot{y} = 6t$ has phase curves parametrized by equations of the form $y(t) = t^3 + c_1 t + c_2$, $z(t) = 3t^2 + c_1$, where c_1 and c_2 are arbitrary constants.

(b) Show that distinct curves of the form found in part (a) satisfy the initial conditions $y(t_0) = 0$, $z(t_0) = 0$ for arbitrary choices of t_0.

(c) Sketch three different phase curves near $(y, z) = (0, 0)$ and passing through that point.

Plotting Phase Curves Using Direction Fields and Isoclines. Letting $\dot{y} = z$, we've seen that phase curves of the autonomous second-order equation $\ddot{y} = f(y, \dot{y})$ satisfy the equation $dz/dy = f(y, z)/z$. These curves can sometimes be sketched using the direction field of the first-order equation. (See Chapter 1, Section 2B, where it is explained how to do this using isoclines.) Use this method, with or without isoclines, to sketch phase portraits for the following equations.

33. $\ddot{y} = -y$.

34. $\ddot{y} = y + 1$.

35. $\ddot{y} = y^2$.

36. $\ddot{y} = y$.

37. $\ddot{y} = -y^2$.

38. $\ddot{y} = y + \dot{y}$.

39. $\ddot{y} = -\sin y$.

40. $\ddot{y} = \dot{y}$.

41. $\ddot{y} = -|y|y$.

[*Hint:* Show for 41 that phase curves for $\ddot{y} = -|y|y$ are symmetric in the z axis.]

42. The inverse-square law equation $\ddot{y} = -y^{-2}$, $y > 0$, has paths in phase space that are symmetric about the y axis and become unbounded as y tends to zero.

(a) Sketch the ones passing through the points $(y,z) = (1,1),\ (1,2),\ (2,2)$. Also indicate the directions that $(y(t), \dot{y}(t))$ traces these curves as t tends to $+\infty$.

(b) Show that a phase curve of $\ddot{y} = -y^{-2}$ crosses the y axis if and only if it contains a point (y_0, z_0) for which $y_0 z_0^2 < 2$.

43. In Example 6 of the text it is shown that a solution $y = y(t)$ of the undamped pendulum equation $\ddot{y} = -(g/l)\sin y$ satisfies $\frac{1}{2}\dot{y}^2 = (g/l)\cos y + C$, $C = \text{const.}$

(a) If $y(0) = y_0$ and $\dot{y}(0) = z_0$, find the value of C for the resulting solution.

(b) Given that the initial angle is $y(0) = y_0$, find out how large the initial speed $|z_0|$ must be so that the pendulum goes over the top. [*Hint:* You must have $\dot{y}^2(t) > 0$ for all t.]

44. Calvin has asked Hobbes for a push on a swing with ropes 9 feet long. If Hobbes pushes the swing when it's stationary at its lowest position, what angular velocity does he need to give it to make it go over the top. Neglect friction.

45. A certain second-order differential equation has, among others, the phase curves shown below. Answer the following questions about the specified positions $y(t)$ and their corresponding velocities $\dot{y}(t)$. The axis ticks are spaced 1.5 units apart and are indicated at $y = \dot{y} = 3$.

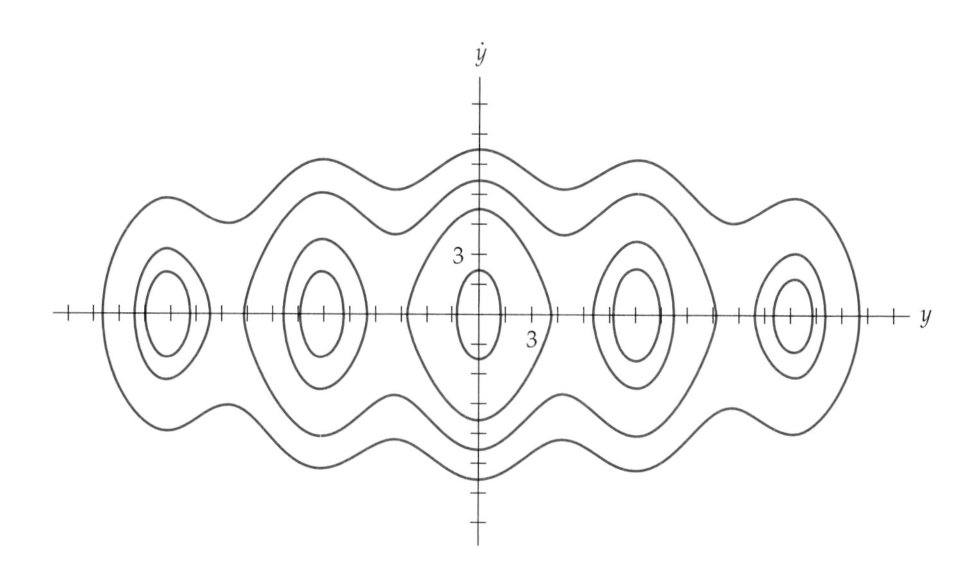

(a) For a solution for which the velocity is 4.5 when the position is 21, estimate the velocity when the position is 10.

(b) Does the solution referred to in part (a) appear to be periodic?

(c) For a solution for which the position is 7 when the velocity is 2.5, estimate the maximum and minimum velocities.

(d) Estimate the maximum and minimum positions attained by the solution referred to in part (c).

(e) Would you expect a solution satisfying $y(0) = \dot{y}(0) = 1$ to be periodic?

(f) The differential equation has constant solutions; try to identify the ones that appear in the phase portrait.

(g) If a phase curve were parametrized by $\big(y(t), \dot{y}(t)\big)$, would it be traced clockwise or counterclockwise?

46. A certain second-order differential equation has, among others, the phase curves shown below. Answer the following questions about the specified positions $y(t)$ and their corresponding velocities $\dot{y}(t)$.

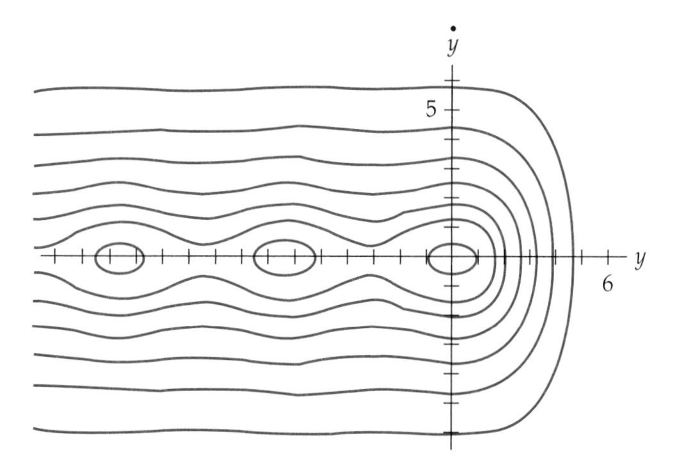

(a) For a solution for which the velocity is 3.4 when the position is 0, estimate the velocity when the position is 3.

(b) Does the solution referred to in part (a) appear to be periodic?

(c) For a solution for which the position is 1 when the velocity is 1, estimate the maximum and minimum velocities.

(d) Estimate the maximum position attained by the solution referred to in part (c); what can you say about the minimum position?

(e) Would you expect a solution satisfying $y(0) = \dot{y}(0) = -4$ to be periodic?

(f) The differential equation has constant solutions; try to identify the ones that appear in the phase portrait.

(g) Try to guess what the differential equation is.

3B Periodic Solutions†

Periodic phenomena are important because they exhibit a kind of dynamic equilibrium, repeating themselves over and over. Note that we can study a periodic phenomenon by confining our attention to a domain interval with length equal to the length of the period, for example, an interval of length 2π for the sine function. That the pendulum equation $\ddot{y} = -\sin y$ has periodic solutions is not obvious and, as we'll see, depends in no way on the periodicity of the sine function. (Indeed the nonlinear pendulum equation has lots of nonperiodic solutions of the rotational or "over the top" variety, and the harmonic oscillator equation $\ddot{y} = -y$ has all its nontrivial solutions periodic with period 2π even though $-y$ is not periodic.) To guarantee the existence of at least *some* periodic solutions to $\ddot{y} = -\sin y$, or more generally to $\ddot{y} = -f(y)$, requires that the function $f(y)$ should be positive in some interval $0 < y < a$ and negative in some interval $-b < y < 0$. As a special case, this will happen if $f(y)$ is positive for positive y and if also $f(y)$ is an *odd* function, meaning $f(-y) = -f(y)$, as in the first two parts of Figure 4.14. Thus, using phase space ideas, the second-order equations $\ddot{y} = -y^3$, $\ddot{y} = -\sin^5 y$, and $\ddot{y} = -ye^y$ will all be seen to have at least some periodic solutions, just as does the equation $\ddot{y} = -y$.

To understand why the claims in the paragraph above are valid, we'll turn our attention from the second-order equation $\ddot{y} + f(y) = 0$ to an equivalent first-order system in yz phase space that we obtain by an order-reduction method described also in Chapter 3; Section 4B.

$$\frac{dy}{dt} = z, \qquad \frac{dz}{dt} = -f(y).$$

We can use the chain rule relation $(dz/dt)/(dy/dt) = dz/dy$ to draw some conclusions about periodicity from the related first-order equation that we get by dividing dz/dt by $dy/dt = z$:

$$\frac{dz}{dy} = \frac{-f(y)}{z}.$$

Assuming $f(y)$ is continuous will allow us to make a geometric argument showing that the path of a solution $y = y(t)$, when looked at in the yz phase space with $\big(y(t), z(t)\big) = \big(y(t), \dot{y}(t)\big)$, traces a closed curve. Thus we identify certain paths that loop back to an initial point (y_0, z_0) and start over again with the same initial values on the same path. Since the function $f(y)$ in $\ddot{y} + f(y) = 0$ is not explicitly dependent on t, the solution will be periodic, and the period is just the time it takes to complete one loop of the closed curve. We can summarize with the following qualitative statement: *periodic solutions correspond to closed loops in phase space.*

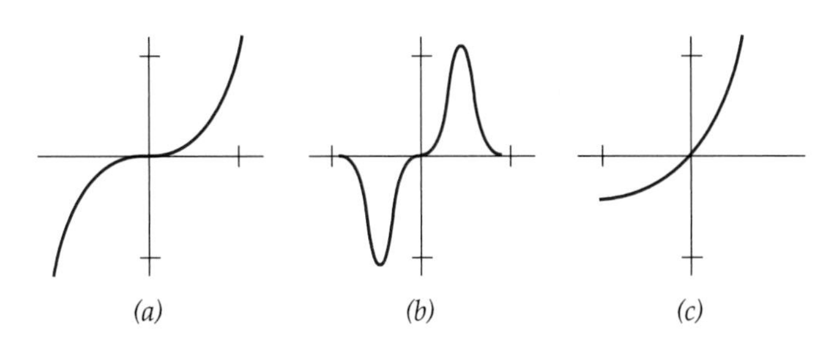

Figure 4.14 (*a*) $f(y) = y^3$ (odd). (*b*) $f(y) = \sin^5 y$ (odd). (*c*) $f(y) = ye^y$ (not odd).

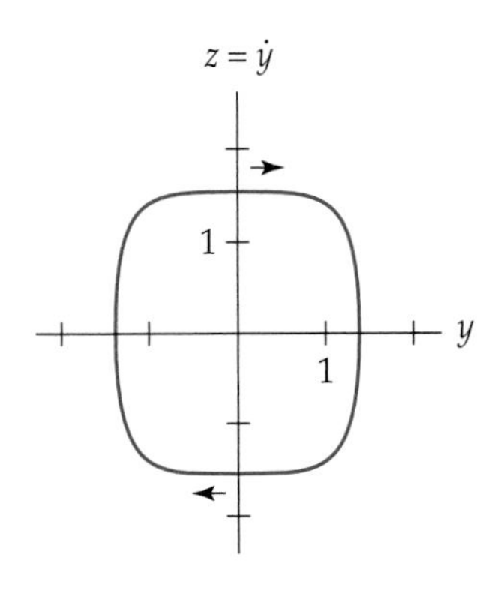

3.2 PERIODICITY THEOREM

Suppose that $f(y)$ is continuous and is positive for $0 < y \leq a$ and negative for $-a \leq y < 0$. Let $y = y(t)$ be the solution to a second-order initial-value problem $\ddot{y} + f(y) = 0$ with $y(0) = y_0$, $\dot{y}(0) = z_0$. For initial points (y_0, z_0) sufficiently close to (0,0) in the yz phase space, the phase paths are nonintersecting closed loops traced clockwise around (0,0); hence the corresponding solutions $y(t)$ are periodic functions of t. We defer the proof to the end of the section in favor of some examples.

example 9

Let y stand for the extension ($y > 0$) or compression ($y < 0$) from the unstressed position of a spring for which the linear relation $F = -hy$ of Hooke's law is replaced by $F = -\gamma y^3$, $\gamma > 0$. In the absence of retarding forces, the differential equation of motion is $\ddot{y} = -\gamma y^3$. The motion in phase space is governed by the pair of equations

$$\dot{y} = z, \qquad \dot{z} = -\gamma y^3.$$

Since $dz/dy = \dot{z}/\dot{y}$, the phase curves satisfy

$$\frac{dz}{dy} = \frac{-\gamma y^3}{z},$$

with solutions $\frac{1}{2}z^2 + (\gamma/4)y^4 = C$, C = const. In this example, all the phase paths are bounded loops, corresponding to periodic oscillations of the spring. Figure 4.15 shows phase plots and corresponding solution graphs for the choice $\gamma = 1$. The graphs of solutions of $dz/dy = -y^3/z$ shown in Figure 4.15(a) were made using the numerical methods described in Chapter 2. The pictures of solutions $y = y(t)$ were made using the numerical methods of Section 4 of

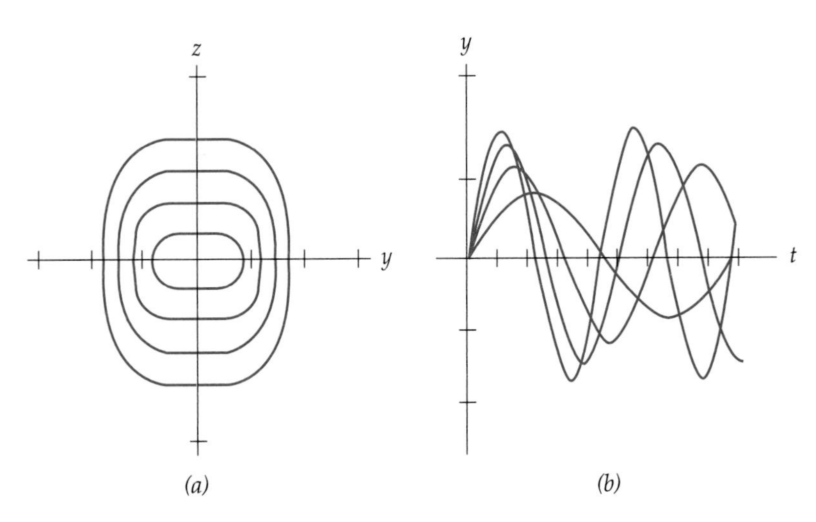

Figure 4.15 (a) Phase portrait. $dz/dy = -y^3/z$.
(b) Graphs of $y = y(t)$. $\ddot{y} + y^3 = 0$.

this chapter. The successive initial values (y_0, z_0) correspond to larger and larger phase curves along with steeper and steeper initial slopes to the solution graphs. All of the graphs in Figure 4.15(b) are periodic, but the period of oscillation varies, with larger period corresponding to smaller initial velocity; these contrast with the solutions of the linear harmonic oscillator equation $\ddot{y} = -\gamma y$ for which all solutions have $\sqrt{\gamma}$ for a common period.

Computing the period of a periodic solution of the type guaranteed by the periodicity theorem is most often done using the numerical techniques of the next section. A few examples where hand computation suffices are described in the exercises and in the following example.

example 10

The harmonic oscillator equation $\ddot{y} = -\alpha^2 y$ has nothing but periodic solutions. In particular, the initial conditions $y(0) = 0$ and $\dot{y}(0) = z_0$ specify the unique solution $y(t) = (z_0/\alpha)\sin \alpha t$ with period $2\pi/\alpha$. A related example is a differential equation with a two-piece right side:

$$\ddot{y} = \begin{cases} -\alpha^2 y, & \text{if } y \geq 0, \\ -\beta^2 y, & \text{if } y < 0. \end{cases}$$

We can think of this equation as representing a spring that behaves linearly under both extension and compression, but with different stiffness for each of the two modes. It is left as an exercise to show that one complete oscillation of the solution satisfying initial conditions $y(0) = 0,\ \dot{y}(0) = z_0$ is

$$y(t) = \begin{cases} \dfrac{z_0}{\alpha}\sin \alpha t, & 0 \leq t \leq \dfrac{\pi}{\alpha}, \\ -\dfrac{z_0}{\beta}\sin \beta(t - \pi/\alpha), & \dfrac{\pi}{\alpha} \leq t \leq \dfrac{\pi}{\alpha} + \dfrac{\pi}{\beta}. \end{cases}$$

Thus the period of the oscillation is $\pi/\alpha + \pi/\beta$. The shapes of the phase curves and solution graph are taken up in Exercises 6 and 7.

Proof of the Periodicity Theorem 3.2. The equation $\ddot{y} + f(y) = 0$ is equivalent to the system $\dot{y} = z,\ \dot{z} = -f(y)$. The phase curves satisfy the separated first-order equation $z\,dz + f(y)\,dy = 0$. When this is integrated we get

$$\tfrac{1}{2}z^2 + F(y) = C, \quad \text{where } F(y) = \int_0^y f(u)\,du.$$

Note that $F(y)$ is strictly increasing on the interval $0 \leq y \leq a$, because the integrand $f(u)$ is positive there. Thus each solution path in phase space must lie on the graph of an equation $z^2/2 + F(y) = C$, with distinct non-intersecting paths corresponding to different values of the constant C. Pick a curve through $(y, z) = (0, z_0)$ with $z_0 > 0$, so that $\frac{1}{2}z_0^2 < F(a)$. Then $C = \frac{1}{2}z_0^2$. Then as y increases toward a, $F(y)$ increases to $\frac{1}{2}z_0^2$, and z decreases to 0. The reason y is strictly increasing along this part of the path is that $dy/dt = z > 0$. Note that the path has a vertical tangent where it crosses the positive y-axis, because $dy/dt = 0$ and

$dz/dt < 0$ there; similarly, the path has a horizontal tangent where it crosses the positive z-axis, because $dz/dt = -f(0) = 0$ and $dy/dt > 0$ there.

The equation $\ddot{y} = -f(y)$ is also equivalent to the "reversed" system $\dot{y} = -z$, $\dot{z} = f(y)$, in which the phase paths are traced in the opposite direction. If we now allow y to *decrease* from $y = 0$, we get a piece of a phase path of the reversed system for which z decreases as y decreases. The reason is that $f(y)$ is negative for $-a < y < 0$, so its integral,

$$F(y) = \int_0^y f(u)\,du = \int_y^0 -f(u)\,du,$$

increases as y decreases. By choosing z_0, perhaps smaller than before, so that $\frac{1}{2}z_0^2 < F(-a)$, we guarantee that z decreases to 0 as y decreases. (If we needed a smaller z_0 for the left side, we agree to use that one on the right also.) So far, we have a phase path that lies above the y-axis and joins a point $(y,z) = (-B,0)$ to a point $(y,z) = (A,0)$. The lower half of the path, starting with $y = A$, $z = 0$, traces the graph of $\frac{1}{2}z^2 + F(y) = \frac{1}{2}z_0^2$ right to left, since here $\dot{y} = z < 0$, and so completes a closed loop. ■

EXERCISES

1. The differential equation $\ddot{y} = -y + y^2$ has constant equilibrium solutions at two distinct points. What are they? Does the equation have periodic solutions?

2. The closed loop shown in the margin next to Theorem 3.2 is a phase curve of $\ddot{y} = -y^5$ that contains the point $(y,z) = (0,\frac{4}{3})$. Find the implicit equation of the form $G(y,z) = C$ for this curve.

3. **(a)** The Periodicity Theorem 3.2 guarantees the existence of periodic solutions to the pendulum initial-value problem $\ddot{y} = -(g/l)\sin y$, $y(0) = y_0$, $\dot{y}(0) = z_0$ if (y_0, z_0) is near phase space points of the form $(2k\pi, 0)$ but not necessarily if $2k\pi$ is replaced by an odd multiple of π. Explain this difference on the basis of the statement of the periodicity theorem.

(b) Explain the difference described in part (a) on physical grounds.

4. The **Morse model** for the displacement y from equilibrium of the distance between the two atoms of a diatomic molecule is $\ddot{y} = K(e^{-2ay} - e^{-ay})$, where K and a are positive constants. Show that the equation has periodic solutions.

5. A nonconstant periodic solution $y(t)$, along with its derivative $\dot{y}(t)$, to the nonautonomous (i.e., time dependent) second-order equation $\ddot{y} = g(t, y, \dot{y})$ always generates a closed loop in $(y, \dot{y})$ phase space.

(a) Verify the assertion above.

(b) Verify that the equation $\ddot{y} - \dot{y} + e^{2t}y = 0$ has solutions of the form $y(t) = c_1\cos(e^t) + c_2\sin(e^t)$. Show that these solutions are generally not periodic functions of t, and that they do not trace closed loops in $(y, \dot{y})$ phase space.

(c) Assume that the autonomous differential equation $\ddot{y} = g(y, \dot{y})$ has a unique solution satisfying each initial condition of the form $y(t_0) = y_0$, $\dot{y}(t_0) = z_0$. Show that every solution that traces a closed loop in $(y, \dot{y})$ phase space is periodic, with period τ equal to the time it takes to traverse the loop once.

6. The differential equation in Example 10 of the text is $\ddot{y} + f(y) = 0$, where

$$f(y) = \begin{cases} \alpha^2 y, & \text{if } 0 \le y, \\ \beta^2 y, & \text{if } y < 0. \end{cases}$$

Show that a phase curve passing through $(y, z) = (0, z_0)$, with $z_0 > 0$, consists of a closed loop composed of right and left halves of two ellipses, one passing through $(y, z) = (z_0/\alpha, 0)$ and the other passing through $(y, z) = (-z_0/\beta, 0)$. This can be done by considering the first-order equation $dz/dy = -f(y)/z$ in two separate pieces and then coordinating the results.

7. Verify directly that the solution given in the second part of Example 10 of the text is in fact a solution to the oscillator equation with two-piece right-hand side and that the solution can be extended periodically. Note that this is not simply a matter of plugging some formulas into the differential equation; it's also necessary to check that the slopes of the solution graphs match up properly at $t = 0$, π/α, and $\pi/\alpha + \pi/\beta$.

8. If the function $f(y)$ in Theorem 3.2 is an odd function, that is $f(-y) = -f(y)$, then the phase paths of the equation $\ddot{y} = -f(y)$ will be not only symmetric about the horizontal y axis but also about the z axis. Explain why this is so.

9. The differential equation $\ddot{y} = -y + \delta y^3$ is a harmonic oscillator equation if $\delta = 0$, and is said to model a **hard spring** if $\delta < 0$ and a **soft spring** if $\delta > 0$.

(a) Show that in all cases the equation has periodic solutions with initial conditions near enough to $(y_0, z_0) = (0, 0)$. Do you expect to find periodic solutions near the other two equilibria when $\delta > 0$? Explain.

(b) Explain why the terms "hard" and "soft" are appropriate.

10. Show that the **Duffing oscillator** $\ddot{y} = -y^3 + y$ has constant equilibrium solutions $y(t) = 1$ and $y(t) = -1$ and that a solution with initial values $y(0) = y_0$, $\dot{y}(0) = z_0$ close enough to $(1, 0)$ or $(-1, 0)$ is periodic, with phase curve looping around $(1, 0)$ or $(-1, 0)$.

4 COMPUTING SUPPLEMENT: SECOND-ORDER EQUATIONS

Second-order differential equations are either linear, of the form

$$\frac{d^2y}{dt^2} + a(t)\frac{dy}{dt} + b(t)y = f(t),$$

or nonlinear, for example, the damped pendulum equation

$$\frac{d^2y}{dt^2} + k\frac{dy}{dt} + \frac{g}{l}\sin y = 0.$$

Equations of both types can be written in the very general form

$$\frac{d^2y}{dt^2} = F\left(t, y, \frac{dy}{dt}\right).$$

It is this form that we will treat here along with **initial conditions** of the form

$$y(t_0) = y_0, \qquad \frac{dy}{dt}(t_0) = z_0,$$

where t_0, y_0, and z_0 are given constants. The Existence and Uniqueness Theorem 1.1 described in the introduction to this chapter guarantees a unique solution to this problem if $F(t, y, z)$ and its partial derivatives with respect to y and z are continuous on some interval containing t_0 and in some rectangle containing the point (y_0, z_0). It is important to realize that the only type of problem for which we so far have a universally effective method of actually displaying such solutions is the linear equation with $a(t)$ and $b(t)$ both constant. Even then, if $f(t)$ is not a function that can be integrated formally, we may have trouble finding a formula for a particular solution. What we may then settle for is a numerical approximation y_k to the value $y(t_k)$ of the true solution at a discrete set of points t_k. Such approximations were treated in the Chapter 2 Computing Supplement for first-order equations, and the methods used here for second-order equations are simple modifications of the first-order methods.

If the purpose in solving a differential equation is just to obtain numerical values or a graph for some particular solution, then a purely numerical approach may be more efficient than first finding a solution formula and then finding the desired graphic or numeric results from the formula. On the other hand, if what you want is to display the nature of a solution's dependence on certain parameters in the differential equation, or on initial conditions, then solution by formula is preferable if at all possible. Beyond that, detailed properties of a solution, such as whether it is periodic or only approximately so, can be hard to get from a numerical approximation and easy to get from a formula such as $y(t) = 2\cos 3t$.

4A Euler Methods

Numerical methods for first-order equations can be motivated by using the interpretation of the first derivative as a slope. Rather than trying to make something of the interpretation of the second derivative in a second-order equation, what we will do is find a pair of first-order equations equivalent to a given second-order equation and then apply first-order methods to the simultaneous solution of the pair of equations. The principle is easiest to understand in the general case

$$y'' = F(t, y, y'), \qquad y(t_0) = y_0, \qquad y'(t_0) = z_0.$$

There are many ways to find an equivalent pair of first-order equations, but the most natural is usually to introduce the first derivative y' as a new unknown function z. We write $z = y'$, $z' = y''$, so we can replace $y'' = F(t, y, y')$ by $z' = F(t, y, z)$. The pair to be solved numerically is then

$$y' = z, \qquad y(t_0) = y_0,$$
$$z' = F(t, y, z), \qquad z(t_0) = z_0.$$

Since $y'(t_0) = z(t_0) = z_0$, there is an initial condition that goes naturally with each equation. To find an approximate solution, we can do what we would do with a single first-order equation, except that at each step we find new approximate values for both unknown functions y and z, and then use these values to compute new approximations in the next step. The iterative formulas are as follows for the simple **Euler method,** with step size h:

$$y_{k+1} = y_k + hz_k,$$
$$z_{k+1} = z_k + hF(t_k, y_k, z_k),$$

where $t_k = t_0 + kh$. The starting values y_0 and z_0 come from the initial conditions.

---

example 1

The initial-value problem

$$\ddot{y} = -\sin y, \qquad y(0) = \dot{y}(0) = 1, \quad \text{for } 0 \le t < 20$$

describes the motion of a pendulum with fairly large amplitude, so large that the linear approximation $\ddot{y} = -y$ would be inadequate. We go ahead to solve the system

$$\dot{y} = z, \qquad y(0) = 1,$$
$$\dot{z} = -\sin y, \qquad z(0) = 1.$$

The essence of the computation can be expressed as follows.

```
DEFINE F(T,Y,Z) = -SIN(Y)
SET T=0
SET Y=1
SET Z=1
SET H=0.01
DO WHILE T<20
    SET S=Y
    SET Y=Y+H*Z
    SET Z=Z+H*F(T,S,Z)
    SET T=T+H
    PRINT T,Y
LOOP
```

Note the command SET S=Y, saving the current value of y for use two lines later; without this precaution, the advanced value $y + hz$ would be used, which is not correct. The printout results in 2000 values of t from 0.01 to 20 by steps of size 0.01 along with the corresponding y values. We could also print the z

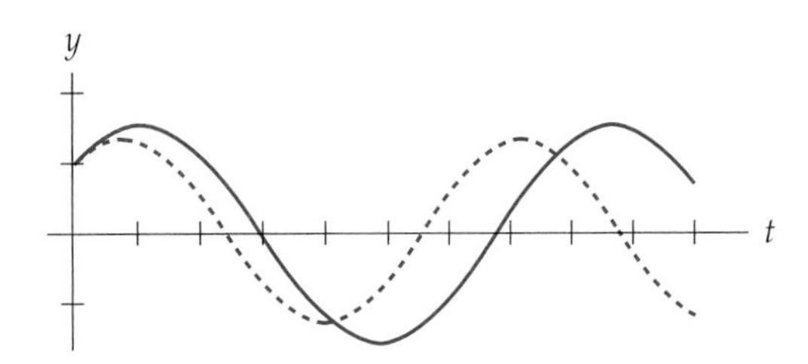

Figure 4.16 Solutions $y(t)$ to $\ddot{y} = -\sin y$ (—) and $\ddot{y} = -y$ $(\cdots)$ with $y(0) = y'(0) = 1$.

values, which are approximations to the values of the derivative $\dot{y}$. This is useful in making a phase-space plot of the solution, plotting approximations to the points $\big(y(t), \dot{y}(t)\big)$. In the formal routine listed above, the line PRINT T,Y would be replaced by something like PLOT Y,Z for a phase plot.

Rather than displaying a table with 2000 entries, Figure 4.16 shows as an unbroken curve the graph of y for the initial conditions $y(0) = y'(0) = 1$. The replacement of $\sin y$ by y in the differential equation is inappropriate in this instance, because the values of y that occur are too large to make the approximation a good one. The linearized initial-value problem $\ddot{y} = -y$, $y(0) = y'(0) = 1$, has solution $\cos t + \sin t$, whose graph is shown in Figure 4.16 as a dotted curve. It's clear that solution to the linearized problem differs drastically from the shape and period shown by the solution to the true nonlinear pendulum problem.

The **improved Euler method** results from applying the single-variable version of the method to each the equations $y' = z$ and $z' = F(t, y, z)$, as we did for the Euler method:

$$\begin{aligned}
p_{k+1} &= y_k + hz_k, \\
q_{k+1} &= z_k + hF(t_k, y_k, z_k), \\
y_{k+1} &= y_k + \frac{h}{2}(z_k + q_{k+1}), \\
z_{k+1} &= z_k + \frac{h}{2}[F(t_k, y_k, z_k) + F(t_{k+1}, p_{k+1}, q_{k+1})].
\end{aligned}$$

Here p_k and q_k provide the simple Euler estimates that are then used to compute the final estimates for y_k and $z_k = \dot{y}_k$; $t_k = t_0 + kh$ as before. (As with the simple Euler method, the value y_k has to be kept for use in computing z_{k+1} and cannot be replaced by y_{k+1} without significant error.)

The advantage of the modification is that the error in the final estimates is substantially reduced without adding much complexity to the computation on a digital computer.

example 2

The initial-value problem

$$y'' + y = 0, \qquad y(0) = 0, \qquad y'(0) = 1$$

has solution $y = \sin x$. Table 2 presents values that compare the Euler method, the improved Euler method, and the correct value rounded to six decimal places.

Table 2

x	Euler y	Improved Euler y	$y \approx \sin x$
0.1	0.099838	0.099834	0.099834
0.2	0.198689	0.198669	0.198669
0.3	0.295564	0.295520	0.295520
0.4	0.389496	0.389418	0.389418
0.5	0.479545	0.479426	0.479426
0.6	0.564812	0.564643	0.564642
0.7	0.644443	0.644218	0.644218
0.8	0.717643	0.717356	0.717356
0.9	0.783679	0.783327	0.783327
1.0	0.841892	0.841471	0.841471
1.1	0.891697	0.891208	0.891207
1.2	0.932598	0.932039	0.932039
1.3	0.964185	0.963558	0.963558
1.4	0.986140	0.985450	0.985450
1.5	0.998243	0.997495	0.997495
1.6	1.000370	0.999574	0.999574
1.7	0.992508	0.991665	0.991665
1.8	0.974725	0.973848	0.973848
1.9	0.947200	0.946300	0.946300
2.0	0.910297	0.909297	0.909297
2.1	0.864117	0.863209	0.863209
2.2	0.809387	0.808496	0.808496
2.3	0.746564	0.745705	0.745705
2.4	0.676275	0.675463	0.675463
2.5	0.599221	0.598472	0.598472
2.6	0.516173	0.515501	0.515501
2.7	0.427958	0.427379	0.427379
2.8	0.335458	0.334988	0.334988
2.9	0.239597	0.239249	0.239249
3.0	0.141333	0.141119	0.141120

The value of h used is 0.001, but only every hundredth value is given. There are only three discrepancies between the final digits in the last two columns.

An algorithm to produce the first, third, and fourth columns in the previous table might look like this. (The routine produces only every hundredth row of the computed values.)

```
DEFINE F(T,Y,Z)= -Y

SET T=0

SET Y=0

SET Z=1

SET H=.001

FOR J=1 TO 30
```

```
FOR K=I TO 100
    SET P=Y+H*Z
    SET Q=Z+H*F(T,Y,Z)
    SET S=Y
    SET Y=Y+.5*H*(Z+Q)
    SET Z=Z+.5*H*(F(T,S,Z)+F(T+H,P,Q))
    SET T=T+H
    NEXT K
    PRINT T, Y, SIN(T)
NEXT J
```

Recall that in this algorithm we are dealing with a pair of equations of the form

$$\dot{y} = z \qquad \dot{z} = F(t, y, z).$$

The Applets 2ORD and 2ORDPLOT at website <http://math.dartmouth.edu/~rewn/> use this routine. Accuracy in the Euler methods can be improved, within reasonable limits, by decreasing the step size h. This requires more steps to reach a given value of t and may produce more approximate solution values than is convenient. Thus we'd print results only after m steps of calculation. For example, $h = 0.001$ and $m = 10$ would produce approximate values with argument differences of 0.01. The Applets referred to above allow for this feature.

4B Stiffness

We remarked in Computing Supplement of Chapter 2 that the local formula errors due to using the Euler and improved Euler methods are respectively of order h and h^2. In addition to the formula error and the round-off error that we have to deal with, for some second-order equations, an additional difficulty occurs due to what is called the **stiffness** of a differential equation. The problem is that in a two-parameter family of solutions, a small error at some point may lead to approximation of a solution that is very different from the one we started on. Here is a simple example.

example 3

The differential equation $y'' - 100y = 0$ has the general solution

$$y(t) = c_1 e^{10t} + c_2 e^{-10t}.$$

The initial conditions $y(0) = 1$ and $y'(0) = -10$ single out the solution $y_1(t) = e^{-10t}$. But the "nearby" initial conditions $y(0) = 1$, $y'(0) = -10 + 2\epsilon$, for small $\epsilon > 0$, single out the solution

$$y_2(t) = 0.1\epsilon e^{10t} + (1 - 0.1\epsilon)e^{-10t}.$$

The behavior of these two solutions is very different for only moderately large values of t. In particular, $y_1(t)$ tends to zero as t tends to infinity, but $y_2(t)$ gets large very soon unless the "error" ϵ is required to be small to an impractical degree. Figure 4.17 shows comparative behaviors, with $\epsilon = 0.00001$. Solutions

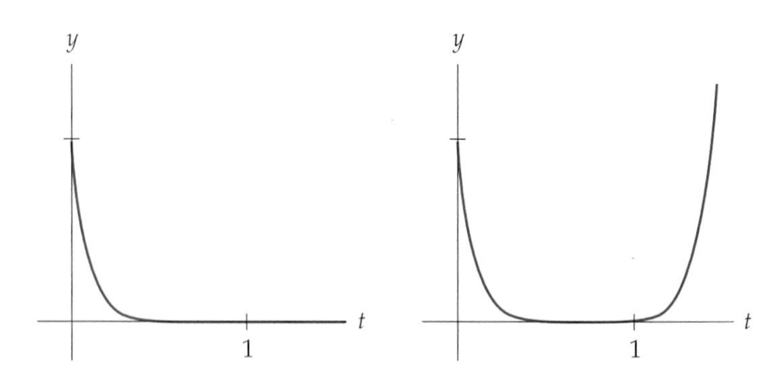

Figure 4.17 Initially close solutions that diverge later.

of nonlinear equations may exhibit the same kind of behavior, but decreasing the step size is inherently nonproductive for getting accuracy in stiff problems; one way to deal with the problem is suggested in the exercises.

EXERCISES

The Applets 2ORD, 2ORDPLOT, and PHASEPLOT at the website <http://math.dartmouth.edu/~rewn> that can be used to do the graphic and numeric exercises in this section. They can also be reached via <http://www.mhhe.com/williamson>. The function H(t) is available for use in these Applets and is used in Section 4C.

1. Consider the **Airy equation** $\ddot{y} + ty = 0$ for $0 \leq t \leq 12$.

  (a) Estimate the location of t values corresponding to the zero values of the solution satisfying $y(0) = 0$ and $\dot{y}(0) = 2$. What changes if you replace the condition $\dot{y}(0) = 0$ by $\dot{y}(0) = a$ for various choices of a? [*Hint:* Look for successive approximations to y with opposite sign.]

  (b) What can you say about the questions in part (a) on an interval $n \leq t \leq 0$?

  (c) Estimate the location of the maxima and minima of the solution satisfying $y(0) = 0$ and $\dot{y}(0) = 2$. [*Hint:* Look for successive approximations to $z = \dot{y}$ with opposite sign.]

2. The **Bessel equation** of order zero (see Chapter 9, Section 7B) is

$$x^2y'' + xy' + x^2y = 0.$$

For $1 \leq x \leq 40$, estimate the location of the zero values of a solution satisfying $y(1) = 1$, $y'(1) = 0$.

3. Perform a numerical comparison of the solution of $\ddot{y} = -\sin y$, $y(0) = 0$, $\dot{y}(0) = 1$, with the solution of $\ddot{y} = -y$ using the same initial conditions. In particular, estimate the discrepancies between the location of successive zero values for the two solutions, one of which is just $y = \sin t$.

4. The nonlinear equation

$$\frac{d^2y}{dx^2} + k\frac{dy}{dx} + hy^2 = 0, \qquad h, k, \text{ constant,}$$

has solutions defined near $x = 0$. Compare the behavior of numerical solutions of the initial value problem $y(0) = 0$, $y'(0) = 1$ with the corresponding behavior when the nonlinear term hy^2 is replaced by a linear term hy. To do this you should investigate the result of choosing several different values of $k > 0$ and $h > 0$.

5. The linear equation

$$\frac{d^2y}{dx^2} + a(x)\frac{dy}{dx} + b(x)y = 0,$$

with continuous coefficients $a(x)$ and $b(x)$ occurs often with rather special choices for the coefficient functions. (Chapters 9 and 11 contain several examples.) For other choices of the coefficients, numerical methods are used. Study the behavior of numerical solutions of the initial value problem $y(0) = 0$, $y'(0) = 1$ as follows.

(a) Let $a(x) = \sin x$ and $b(x) = \cos x$ for $0 \le x \le 2\pi$.

(b) Let $a(x) = e^{-x/2}$ and $b(x) = e^{-x/3}$ for $0 \le x \le 1$.

6. Nonuniqueness. The initial-value problem $\ddot{y} = 3\sqrt{y}$, $y(0) = 0$, $\dot{y}(0) = 0$ has the identically zero solution.

(a) Verify that $y(t) = \frac{1}{16}t^4$ is also a solution.

(b) Discuss the application of Euler methods to the problem.

Stiff Equations

7. The initial-value problem

$$y'' = 25y, \qquad y(0) = 1, \qquad y'(0) = -5$$

has solution $y(t) = e^{-5t}$. Try to approximate this solution using the improved Euler method with step size $h = 0.001$ over the interval $0 \le t \le 10$. Interpret your results.

Note. The next two exercises deal with order reductions of somewhat different forms from those that occur in the previous exercises. Ready-made computer routines may have to be altered accordingly.

8. The second-order initial-value problem in the previous exercise can be converted to a pair of first-order equations by means of substitutions other than $y' = z$.

(a) Let $y' = -5y + u$ [i.e., $(D + 5)y = u$]. Then $u' = 5u$ [i.e., $(D - 5)u = 0$] gives $y'' = 25y$. Show that the initial conditions $y(0) = 1$ and $y'(0) = -5$ are converted to $y(0) = 1$ and $u(0) = 0$. Then approximate y and u by the improved Euler method over the interval $0 \leq t \leq 10$ and compare the results with those of Exercise 7.

(b) Let $y' = wy$, which is called a **Riccati substitution,** and show that $y'' = 25y$ becomes $w' = 25 - w^2$, provided that $y \neq 0$. Then show that $y(0) = 1$, $y'(0) = -5$ is equivalent to $y(0) = 1$, $w(0) = -5$, and approximate the solution to the system

$$y' = wy, \qquad y(0) = 1,$$
$$w' = 25 - w^2, \qquad w(0) = -5.$$

9. Show that the *Riccati substitution* $y' = wy$ reduces the general linear equation

$$\frac{d^2y}{dt^2} + a(t)\frac{dy}{dt} + b(t)y = 0$$

to the system of equations

$$y' = wy,$$
$$w' = -a(t)w - b(t) - w^2$$

provided that $y \neq 0$. Show also that the initial conditions $y(t_0) = y_0$, $y'(t_0) = z_0$ get converted by the same substitution into $y(t_0) = y_0$, $w'(t_0) = z_0/y_0$.

Falling Objects

10. Suppose that the displacement $y(t)$ of a falling object is subject to a nonlinear friction force

$$m\ddot{y} = -k\dot{y}^\alpha + mg.$$

(a) Find numerical approximations to $y(t)$ in the range $0 \leq t \leq 20$, with $g = 32.2$, $y(0) = 0$, $\dot{y}(0) = 0$, $m = 1$, and $\alpha = 1.5$. Use the values $k = 0$, 0.1, 0.5, and 1, and sketch the graphs of $y = y(t)$ using an appropriate scale.

(b) Estimate the value of k that, along with $\dot{y}(0) = 0$ and the other parameter values in part (a), produces approximately the values $y(0) = 0$, $y(5) = 66$, $y(10) = 137$, and $y(15) = 208$.

11. A typical nonlinear model for an object of mass 1 dropped from rest with frictional drag is $\ddot{y} = -k\dot{y}^\alpha + g$, $y(0) = \dot{y}(0) = 0$; the model is linear if $\alpha = 1$. In numerical work assume $g = 32$ feet per second squared.

(a) If the terminal velocity for the linear model is 36 feet per second, what is k?

(b) Estimate the time it takes for the linear model to reach velocity 35.99 feet per second. Use the value of k found in (a).

(c) If the terminal velocity for the nonlinear model with $\alpha = 1.1$ is 36 feet per second, estimate k. This requires trial and error.

(d) Estimate the time it takes for the nonlinear model of part (c) to reach velocity 35.99 feet per second.

12. **Bead on a wire.** Exercise 24 in the chapter supplement on energy and stability considers a bead sliding without friction under constant vertical gravity along a wire bent into the shape of the twice differentiable graph $y = f(x)$; the result is that x obeys $\ddot{x} = -[g + f''(x)\dot{x}^2]f'(x)/[1 + f'(x)^2]$. With $f(x) = -x^3 + 4x^2 - 3x$, $g = 32$, and $x(0) = 0$, estimate how large $\dot{x}(0) > 0$ should be for the bead to overcome periodic oscillation and go over the hump in the wire.

Pendulum

13. **(a)** Use the Euler method for

$$y'' = F(t, y, y'), \qquad y(t_0) = y_0, \qquad y'(t_0) = z_0,$$

and apply it to the pendulum equation with $F(t, y, y') = -16 \sin y$, $y(0) = 0$, $y'(0) = 0.5$.

(b) Do part (a) using the improved Euler method.

14. For small oscillations of y, the approximation $\sin y \approx y$ is fairly good, leading to the replacement of the pendulum equation $y'' + (g/l) \sin y = 0$ by the linearized equation $y'' + (g/l)y = 0$, with solutions of the form

$$y_a(t) = c_1 \cos \sqrt{\frac{g}{l}}\, t + c_2 \sin \sqrt{\frac{g}{l}}\, t.$$

Assuming $g/l = 16$, compare $y_a(t)$ with the improved Euler approximation to the solution of the nonlinear equation under initial conditions.

(a) $y(0) = 0, \ y'(0) = 0.1.$

(b) $y(0) = 0, \ y'(0) = 4.$

15. Consider the damped pendulum equation $\ddot{\theta} = -(g/l) \sin \theta - (k/m)\dot{\theta}$, with $g = 32.2, l = 20, k = 0.03$, and $m = 5$. For the solution with $\theta(0) = 0$ and $\dot{\theta}(0) = 0.2$.

(a) Estimate the maximum angles θ for $0 \le t \le 15$.

(b) Estimate the successive times between occurrences of the value $\theta = 0$ for $0 \le t \le 15$.

(c) Repeat part (b), but with initial conditions $\theta(0) = 0$ and $\dot{\theta}(0) = 2.0$.

16. Consider the following modification of the pendulum equation $ml\ddot{\theta} = -gm \sin \theta$, written here in terms of forces. If the pendulum pivot is moved vertically from its usual fixed position at level 0, so that at time t it is at $f(t)$, with $f(0) = 0$, the additional vertical force component is $m\ddot{f}(t)$. Thus the vertical force due to gravity alone is replaced by $[-gm + m\ddot{f}(t)] \sin \theta$. It follows that the equation for displacement angle $\theta = \theta(t)$ becomes

$$ml\ddot{\theta} = -gm \sin \theta + m\ddot{f}(t) \sin \theta \quad \text{or} \quad \ddot{\theta} = \frac{1}{l}[-g + \ddot{f}(t)] \sin \theta.$$

(a) Show that if $f(t) = at$, with a constant, then there is no change in acceleration as compared with the fixed-pivot case and hence that the equation for the displacement angle θ remains the same: $\ddot{\theta} = -(g/l)\sin\theta$.

(b) Show that in the case of a general twice-differentiable $f(t)$, the position of the pendulum weight at time t is $\mathbf{x}(t) = \big(l\sin\theta(t),\ f(t) - l\cos\theta\big)$, where θ satisfies either of the differential equations displayed above.

(c) Let $g = 32$, $l = 5$, $m = 1$, and $f(t) = 2\sin 4t$. Plot $\big(t, \theta(t)\big)$ for t values 0.01 apart between 0 and 100, assuming $\theta(0) = 0$ and $\dot{\theta}(0) = 0.01,\ 0.001,\ 0.0001$. Note the long-term deviation in behavior as compared with the identically zero solution corresponding to $\dot{\theta}(0) = 0$.

(d) Plot the path of the pendulum weight under the assumptions in part (c).

Oscillators and Phase Space

17. An unforced oscillator displacement $x = x(t)$ satisfies $m\ddot{x} + k\dot{x} + hx = 0$, where k and h may depend on time t.

(a) Suppose that $k(t) = 0.2(1 - e^{-0.1t})$, $h = 5$, and $m = 1$ and that $x(0) = 0$ and $\dot{x}(0) = 5$. Compute a numerical approximation to $x(t)$ on the range of $0 \le t \le 20$. Then sketch the graph of $x = x(t)$.

(b) Do part (a) using instead $k = 0$ and $h(t) = 5(1 - e^{-0.2t})$.

18. The **spherical pendulum** equation is discussed in the exercises for the chapter supplement on energy and stability.

$$\ddot{y} = -\frac{g}{l}\sin y + M^2\frac{\cos y}{\sin^3 y},$$

where M is a constant velocity of rotation of the pendulum about a vertical line through the pendulum's pivot.

(a) Rewrite the equation as a first-order system suitable for numerical solution and phase plots.

(b) Assume $l = g$ and compare solution graphs in the ty plane for the two cases $M^2 = 0.5$ and $M^2 = 0$. Discuss the effect of rotation on the pendulum's motion.

(c) Assume $l = g$ and make a phase portrait for the case $M^2 = 2$.

(d) Assume $l = g$ and make a phase portrait for the case $M^2 = \frac{9}{8}$, showing clearly the behavior near the stable equilibrium points at $y = \pm\pi/3$.

19. Make phase-plots of the **soft spring oscillator equation** $\ddot{y} = -\gamma y^3 + \delta y$ under each of the following assumptions.

(a) $\gamma = \delta = 1$. **(b)** $\gamma = 1,\ \delta = 2$. **(c)** $\gamma = 2,\ \delta = 1$.

20. Make solution graphs and $(y, \dot{y})$ phase plots for the periodically driven **hard spring oscillator equation** $\ddot{y} = -y - y^3 + k\dot{y} + \frac{3}{10}\cos t$ and use the results to detect long-term approach to periodic behavior for some $k < 0$.

21. Time-dependent linear spring mechanism. The equation $\ddot{y} + k(t)\dot{y} + h(t)y = f(t)$ represents an oscillator externally forced by $f(t)$, damped by factor $k(t)$ and with stiffness $h(t)$. Use initial conditions $y(0) = 0$ and $\dot{y}(0) = 1$ to plot solutions with the following choices for $k(t)$ and $h(t)$.

(a) $k(t) = 0,\ h(t) = e^{t/10}$.

(b) $k(t) = 0,\ h(t) = e^{-t}$.

(c) $k(t) = \frac{1}{10},\ h(t) = e^{t/10}$.

(d) $k(t) = (1 - e^{-t}),\ h(t) = 1$.

(e) $k(t) = 1,\ h(t) = 1/(1 + t^2)$.

(f) $k(t) = 0,\ h(t) = t/(1 + t^2)$.

22. Plot closed periodic phase paths for the *Morse model* of displacement y from equilibrium of the distance between the two atoms of a diatomic molecule:

$$\ddot{y} = K(e^{-2ay} - e^{-ay}), \qquad K = a = 1.$$

23. Consider the nonlinear oscillator equation $m\ddot{x} + k\dot{x}|\dot{x}|^{\beta} + hx = 0$, $0 \le \beta = \text{const}$. Let $m = 1$, $k = 0.2$, and $h = 5$ and suppose that $x(0) = 0$ and $\dot{x}(0) = 5$. Compute numerical approximations to $x(t)$ on the range $0 \le t \le 20$ for $\beta = 0, 0.5, 1$. Sketch the resulting graphs using computer graphics.

24. Chaos. The nonlinear *Duffing oscillator* considered in the previous section generalizes to the periodically driven, damped initial-value problem

$$\ddot{y} + k\dot{y} - y + y^3 = A\cos\omega t, \qquad y(0) = \tfrac{1}{2}, \qquad \dot{y}(0) = 0.$$

(a) Make a computer plot of the solution $y(t)$ for $0 \le t \le 500$ for the parameter choices $k = 0.2$, $A = 0.3$, $\omega = 1$, and initial conditions $y(0) = \dot{y}(0) = 0$. Compare your result graphically with Figure IS.4 in the Introductory Survey at the beginning of the book. You will probably find that while your picture has the same general sort of irregularity as the one in the book, it will not agree in detail beyond some t value. This behavior of the damped, periodically driven Duffing equation is often described as "chaotic," which in practical terms means unpredictable. In particular, the specific output that you get will depend significantly not only on the parameter and initial values, but on the choice of numerical method and step size, and even on the internal arithmetic of the machine used to generate the output. For this reason, it seems impossible to describe accurately the global shape of the output from the damped, periodically driven Duffing oscillator. To find some order in the chaos we can resort to the method of time sections described in Section 4D.

(b) Change just the damping constant in part (a) to $k = 0$, and make both a plot of the solution. Comment on the qualitative changes that you see as compared with the output in part (a).

(c) Make several phase plane plots, starting at different points in the $y\dot{y}$ plane, of solutions of the Duffing equation. Use the parameter values $k = 0.2$, $A = 0.3$, and $\omega = 1$.

(d) Experiment with part (c) by trying your own choices for the three parameter values.

4C Functions Defined Piecewise

Step functions appear in Example 12 of Chapter 3, Section 3D, and in the exercises for that section. Some of the examples discussed there and in this section can also be treated using Laplace transforms, as shown in Chapter 8, Section 2B, but the methods used there apply only to constant-coefficient linear differential equations. To get graphic and numeric output resulting from discontinuous inputs for nonlinear equations or linear equations with variable coefficients you will probably need to use the ideas in the present section.

A **step function** is a function whose domain is a union of intervals on each of which the function's value is constant. Figure 4.18 shows the graphs of three step functions. Step functions are often used to model on-or-off phenomena that exhibit abrupt changes, and they are also useful in a technical way for numerical work, in particular representing functions that are patched together from pieces of elementary functions. In the step function graphs shown in Figure 4.18 the presence of a function value on one side of a jump is indicated by a dot, even though in all of our applications the precise values right at each jump don't matter from a basic mathematical point of view. Nevertheless, since here we're planning to use step functions in computer work, we have to be aware that computer programs need us to be precise about such matters, so to help make this precision routine we'll often follow a convention that makes a consistent choice, namely, to have step functions agree with their right-hand limits at jumps; such functions are called **right continuous** at the jumps. The three functions with graphs shown in Figure 4.18 are right continuous everywhere. What's more important is to make sure that a function is assigned one, and only one, value at each point where it needs to be defined. Indeed the integral formulas of Chapter 3, Section 3D, show that the precise value of a forcing function at a jump is irrelevant for the values of a solution, because the integral is unaffected by a change of value at a single point.

example 4

For the graphs in Figures 4.18(a), 4.18(b), and 4.18(c), we have

$$f(t) = \begin{cases} 1, & -1 \le t < 1, \\ -1, & 1 \le t < 2, \\ 1, & 2 \le t < 3, \end{cases} \qquad g(t) = \begin{cases} -1, & 0 \le t < 1, \\ \frac{3}{2}, & 1 \le t < 2, \\ \frac{1}{2}, & 2 \le t < 4, \end{cases}$$

$$r(t) = \begin{cases} 0, & t < 0, \\ 1, & 0 \le t < 1, \\ 0, & 1 \le t. \end{cases}$$

The simplest of step functions is the **Heaviside function**

$$\mathrm{H}(t) = \begin{cases} 0, & t < 0, \\ 1, & 0 \le t, \end{cases}$$

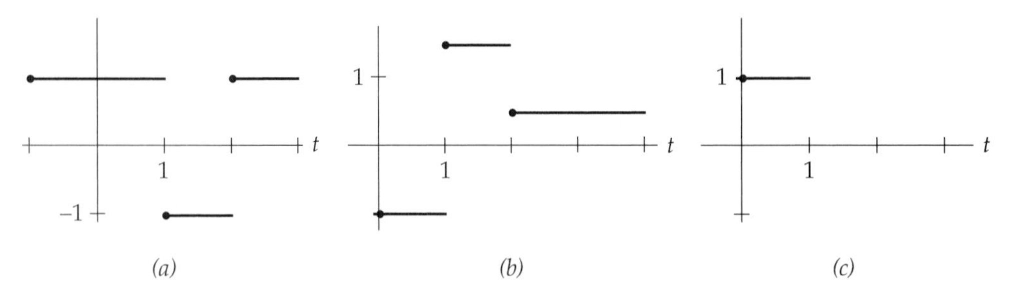

Figure 4.18 (a) $y = f(t)$. (b) $y = g(t)$. (c) $y = r(t) = \mathrm{H}(t) - \mathrm{H}(t-1)$.

and the related functions

$$H(t-a) = \begin{cases} 0, & t < a, \\ 1, & a \le t, \end{cases}$$

that shift the Heaviside function's jump from $t = 0$ to $t = a$. From the functions $H(t-a)$ we can construct a formula for an arbitrary right-continuous step function. For example, the function $r(t)$ in Figure 4.18(c) is just

$$H(t-a) - H(t-b) = \begin{cases} 0, & t < a, \\ 1, & a \le t < b, \\ 0, & b \le t, \end{cases}$$

with $a = 0$ and $b = 1$. The function $H(t-a) - H(t-b)$ is sometimes called the **indicator function** of the interval $a \le t < b$, taking value 1 on that interval and 0 elsewhere.

example 5

It's easy to check that the function $y(t) = (t^2/2)H(t)$ has a continuous first derivative $\dot{y}(t) = (t)H(t)$ and that $\ddot{y}(t) = H(t)$ if $t \neq 0$; this last equation holds everywhere but at $t = 0$ since $\ddot{y}(0)$ fails to exist. Thus $y(t)$ is a generalized solution to the initial-value problem $\ddot{y} = H(t)$, $y(0) = \dot{y}(0) = 0$, in the sense discussed in Chapter 3, Section 3D. Generalized solutions are the typical output from second-order problems with forcing functions that have jump discontinuities, the underlying reason being that a function can't have a derivative with a jump discontinuity. The reason for discussing this point in some detail here is that a numerical solution with graphic output from a discontinuous forcing function typically produces such a smooth-looking graph that we might be inclined to think that there was nothing mathematically unusual about the underlying solution. Still, generalized solutions provide reasonable results in applications, in particular to electric circuit problems where a strictly on–off model is fairly appropriate for current.

example 6

Suppose we want a precise input formula for the function S defined by

$$S(t) = \begin{cases} -1, & t < -\pi/2, \\ \sin t, & -\pi/2 \le t < \pi/2, \\ 1, & \pi/2 \le t. \end{cases}$$

You can check that for $t < -\pi/2$ we can write $S(t)$ as $H(t+\pi/2) - 1$, that between $-\pi/2$ and $\pi/2$ we can write $\sin t\big(H(t+\pi/2) - H(t-\pi/2)\big)$, and that for $\pi/2 \le t$ we can write $H(t-\pi/2)$. Notice that each of these three formulas has the property that it has value zero wherever the other two are not zero. Hence, we can just add them up to get

$$S(t) = \big[H(t+\pi/2) - 1\big] + \big[\sin t\big(H(t+\pi/2) - H(t-\pi/2)\big)\big] + \big[H(t-\pi/2)\big],$$

with graph sketched in Figure 4.19(a).

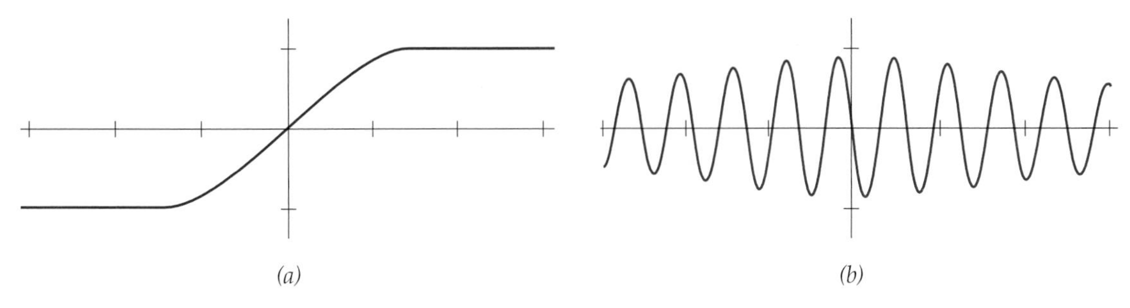

(a) (b)

Figure 4.19 (*a*) $y = S(t)$. (*b*) $\ddot{y} = -100y - 0.4S(t)\dot{y}$, $y(-3) = -\frac{1}{2}$, and $\dot{y}(-3) = \frac{1}{2}$.

The formula for $S(t)$ in terms of $\mathrm{H}(t)$ in the previous example looks a little hard to interpret as compared to the bracketed cases we used to define it at first, but the formula will be easily read by a computer program. (If you're writing the program yourself you can make a definition by cases similar to the one we gave first.) The next example shows the use of $S(t)$ as a coefficient in a linear differential equation.

example 7

The initial-value problem $\ddot{y} + 0.4S(t)\dot{y} + 100y = 0$, $y(-3) = -\frac{1}{2}$, $\dot{y}(-3) = \frac{1}{2}$, represents the action of a forced oscillator that is subject to velocity-dependent damping when $t > 0$ and amplification when $t < 0$. When $t > \pi/2$ or $t < -\pi/2$, $S(t)\dot{y} = \pm 0.4\dot{y}$, so we're dealing there with constant-coefficient equations that are easy to solve in terms of elementary functions. On the interval $-\pi/2 < t < \pi/2$, the term $0.4S(t)\dot{y} = 0.4(\sin t)\dot{y}$ provides a smooth transition from one mode to the other, but for $-\pi/2 < t < \pi/2$ the equation is not solvable in terms of elementary functions.

Even if we did have a formula for the solution in the middle interval, we'd be faced with the nuisance of adjusting constants to piece together smoothly a solution from separate solutions on the two intervals to the right of $-\pi/2$. Putting the differential equation in the form for numerical solution, we let $\dot{y} = z$, $\dot{z} = -4y - S(t)z$, with $y(-3) = -\frac{1}{2}$, $z(-3) = \frac{1}{2}$. The solution graph plotted using the modified Euler method with step size $h = 0.001$ is shown in Figure 4.19(b).

Routine for Piecewise Definitions. Example 5 suggests the following way to represent functions defined in pieces.

1. For $f(t)$ on a finite interval $a \le t < b$, write $f(t)\big(\mathrm{H}(t-a) - \mathrm{H}(t-b)\big)$, for $g(t)$ on $b \le t < c$, write $g(t)\big(\mathrm{H}(t-b) - \mathrm{H}(t-c)\big)$, and so on.
2. For $f(t)$ on an infinite interval $-\infty < t < b$, write $f(t)\big(1 - \mathrm{H}(t-b)\big)$, and for $g(t)$ on an infinite interval $c \le t < \infty$, write $g(t)\mathrm{H}(t-c)$.
3. Having taken care that none of the intervals overlap, just add up the separate pieces. Depending on the specific functions f, g, and so on, some simplification may suggest itself.

example 8

Following the routine just outlined, we note that the functions $F(t)$ and $G(t)$, defined below, are based on the same union of four intervals:

$$F(t) = \begin{cases} 2, & t < -1, \\ -1, & -1 \le t < 0, \\ 2, & 0 \le t < 1, \\ -2, & 1 \le t, \end{cases} \qquad G(t) = \begin{cases} 1, & t < -1, \\ -t, & t \le t < 0, \\ t, & 0 \le t < 1, \\ 1, & 1 \le t. \end{cases}$$

They can be written

$$\begin{aligned} F(t) &= 2[1 - \mathrm{H}(t+1)] + (-1)[\mathrm{H}(t+1) - \mathrm{H}(t)] \\ &\quad + 2[\mathrm{H}(t) - \mathrm{H}(t-1)] + (-2)\mathrm{H}(t-1) \\ &= 2 - 3\mathrm{H}(t+1) + 3\mathrm{H}(t) - 4\mathrm{H}(t-1), \end{aligned}$$

$$\begin{aligned} G(t) &= (1)[1 - \mathrm{H}(t+1)] + (-t)[\mathrm{H}(t+1) - \mathrm{H}(t)] \\ &\quad + t[\mathrm{H}(t) - \mathrm{H}(t-1)] + (1)\mathrm{H}(t-1) \\ &= 1 - (t+1)\mathrm{H}(t+1) + (2t)\mathrm{H}(t) - (t-1)\mathrm{H}(t-1). \end{aligned}$$

The routine often helps to keep your thinking orderly, but can be bypassed to arrive directly at the simplified form after a little careful thought; just take account of the change that takes place when you pass from one interval to the next as you work from left to right.

example 9

The **square-wave function** $\mathrm{sqw}(t)$ whose graph is shown in Figure 4.20(a) for $0 \le t < 8$, is periodic with period 2 in the limited sense that $\mathrm{sqw}(t+2) = \mathrm{sqw}(t)$ if $t \ge 0$. For most applications it's enough to represent the function just for $t \ge 0$, and that's what we'll do here. To synthesize the function using $\mathrm{H}(t)$ it's more direct to deviate from the general method outlined above. We start with $\mathrm{H}(t) = 1$ for $t \ge 0$. Then subtract $2\mathrm{H}(t-1)$, which gives us 1 for $0 \le t < 1$ and -1 for $t \ge 1$. Next add $2\mathrm{H}(t-2)$, which takes us back up to 1 for $t \ge 2$. Continue in this way, adding terms $(-1)^k\mathrm{H}(t-k)$. Since we're restricting to $t \ge 0$, we replace the

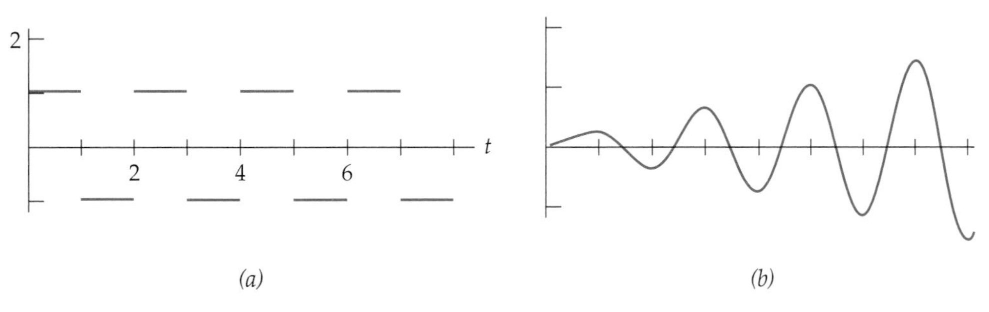

Figure 4.20 (*a*) Square wave $\mathrm{sqw}(t)$. (*b*) $\ddot{y} + \pi^2 y = \mathrm{sqw}(t)$, $y(0) = \dot{y}(0) = 0$.

first term H(t) by 1 and get the infinite sum

$$\mathrm{sqw}(t) = 1 + 2\sum_{k=1}^{\infty}(-1)^k \mathrm{H}(t-k), \qquad t \geq 0.$$

Since $\mathrm{H}(t-k) = 0$ whenever $k > t$, all but finitely many terms of the series are zero, so the infinite series converges for each fixed t. Exercise 19 shows that if $[t]$ denotes the greatest integer less than or equal to t, then the equation $\mathrm{sqw}(t) = (-1)^{[t]}$ extends sqw(t) to have period 2 for $-\infty < t < \infty$. This is useful for numerical work, because the greatest integer function is often a built-in part of software packages, for example our Java Applets will recognize it as int(t). If you want to stretch or compress the graph to have period $2p$ instead, just use $\mathrm{sqw}(t/p)$.

example 10

The linear differential equation $\ddot{y} + \beta^2 y = \mathrm{sqw}(t)$, where sqw($t$) is the square wave function illustrated in Figure 4.20(a), has exact solutions in terms of trigonometric functions and H(t). (See Exercise 22.) The simplest initial-value problem requires $y(0) = \dot{y}(0) = 0$. Before implementing the modified Euler method, we make a decision about what domain of t values we want to work over, and then include enough terms in our expression for sqw(t) in the definition of $\ddot{y} = -\beta^2 y + \mathrm{sqw}(t)$ to cover that domain. A computer plot of the solution for $0 \leq t < 8$ when $\beta = \pi$ is shown in Figure 4.20(b).

example 11

This example concerns the design of a pendulum clock, which in essence can be regarded as a slightly damped pendulum with an additional mechanism called an **escapement** that imparts with each swing to one side or the other a brief acceleration designed to overcome the damping effect. The basic externally driven pendulum equation we consider is

$$\ddot{y} = -\frac{g}{l}\sin y - \frac{k}{m}\dot{y} + \frac{1}{lm}F(y, \dot{y}),$$

where $y(t)$ is angular displacement from the vertical, g is the acceleration of gravity, l is the pendulum length, m is the pendulum mass, and $k > 0$ is a damping constant. When F is identically zero the equation is solved via linear approximation in Section 2 of this chapter and is also discussed in greater generality in Example 5 on page 216.

The typical escapement imparts its brief impulse near the vertical position, where $y(t)$ is close to zero. We start by modeling the force of the impulse with $\alpha\big(\mathrm{H}(y+p) - \mathrm{H}(y-p)\big)$, where $\alpha > 0$ is the absolute magnitude of the applied force. This term applies force α over the interval $-p \leq y < p$, of angular measure $2p$, where p is a small positive constant. However, to enhance the existing motion when $y(t)$ is decreasing rather than increasing, we still need to multiply this by -1 just when $\dot{y}(t) < 0$. This change we obtain with the factor $2\mathrm{H}(\dot{y}) - 1$, which is 1 when $\dot{y} > 0$ and -1 when $\dot{y} < 0$. Thus we choose

$$F(y, \dot{y}) = \alpha\big(\mathrm{H}(y+p) - \mathrm{H}(y-p)\big)\big[2\mathrm{H}(\dot{y}) - 1\big].$$

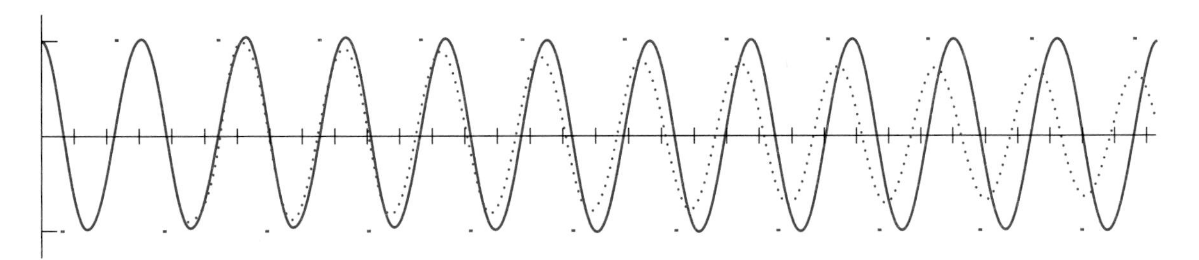

Figure 4.21 Damped pendulum (dotted) and forced pendulum with impulse intervals.

For the purposes of this model, an escapement is completely determined by the two positive numbers p and α, and these values can be built into a mechanical device.

What we have in the escapement-driven pendulum is an example of a rather crude but effective **feedback control** mechanism, crude because the monitoring of behavior that determines the application of the feedback force takes place only over short, widely spaced intervals of the pendulum's motion. Figure 4.21 shows plots relative to the same axes of $y(t)$, both for the damped pendulum without feedback, and for the one with feedback using $p = 0.08$ and $\alpha = 0.5$; parameters common to both plots are $g = 32.2$, $l = 7$, $m = 4$, and $k = 0.1$. The time intervals where feedback occurs are indicated by horizontal segments of length $2p = 0.16$ with vertical displacements alternately -1 and 1. The vertical scale is enlarged to display the difference between the two curves more clearly. You are asked in Exercise 23 to show, using numerical approximation, that once the escapement parameters p and α are adjusted to overcome damping and maintain a certain amplitude A of motion, the escapement mechanism will drive the pendulum toward amplitude $A = 1$ starting with a wide variety of initial conditions. For a free pendulum there is a precise relationship between A and the time period $2T$ for a full oscillation. (See Exercise 7 in Section 2, page 272 of this chapter.) However, the combined effect of damping and escapement impulse is complex enough that final tuning to a particular period is usually done by adjusting the length l.

EXERCISES

Note: The functions $\mathrm{H}(t)$ and $\mathrm{int}(t)$ are accepted as input to the Java Applets for graphic and numeric computation.

In Exercises 1 through 8 sketch the graph of the given function.

1. $\mathrm{H}(t-2) - \mathrm{H}(t-4)$.

2. $\mathrm{H}(t-2) + \mathrm{H}(t-4)$.

3. $2\mathrm{H}(t-2) - 3\mathrm{H}(t-4)$.

4. $-\mathrm{H}(t-2) + 2\mathrm{H}(t-4)$.

5. $(t-2)\mathrm{H}(t-2) - (t-4)\mathrm{H}(t-4)$.

6. $(t-1)\mathrm{H}(t-2) - 2\mathrm{H}(t-3)$.

7. $\sin \pi t\big(\mathrm{H}(t) - \mathrm{H}(t-1)\big) + \mathrm{H}(t-1)$.

8. $t\mathrm{H}(t) + (t-2)\mathrm{H}(t-1) - (t-1)\mathrm{H}(t-2)$.

9. Represent the function $f(t)$ in Example 4 of the text in terms of $\mathrm{H}(t)$, and simplify the result.

10. Represent the function $g(t)$ in Example 4 of the text in terms of $\mathrm{H}(t)$, and simplify the result.

11. Sketch the graph of the function $F(t)$ in Example 8 of the text and make a computer plot of the solution to $\ddot{y} + \sin y = F(t)$, $y(-2) = \dot{y}(-2) = 0$.

12. Sketch the graph of the function $G(t)$ in Example 8 of the text and make a computer plot of the solution to $\ddot{y} + y = F(t)$, $y(-2) = \dot{y}(-2) = 0$.

13. **(a)** Sketch the graph of the continuous function

$$c(t) = \begin{cases} \cos t, & t \leq \pi, \\ -1, & \pi < t, \end{cases}$$

then represent $c(t)$ using $\mathrm{H}(t)$.

(b) Make a computer plot of the solution to the initial-value problem $\ddot{y} + \pi^2 y = c(t)$, $y(0) = \dot{y}(0) = 0$.

14. **(a)** Sketch the graph of the continuous function

$$R(t) = \begin{cases} 0, & t < 0, \\ t, & 0 \leq t < 1, \\ 1, & 1 \leq t, \end{cases}$$

then represent $R(t)$ using $\mathrm{H}(t)$.

(b) Make a computer plot of the solution to the initial-value problem $\ddot{y} + ty = R(t)$, $y(-1) = \dot{y}(-1) = 0$.

15. **(a)** Sketch the graph of the discontinuous function

$$d(t) = \begin{cases} 0, & t < 0, \\ 1 - t, & 0 \leq t < 1, \\ 1, & 1 \leq t, \end{cases}$$

then represent $d(t)$ using $\mathrm{H}(t)$.

(b) Make a computer plot of the solution to the initial-value problem $\ddot{y} + t^2 y = d(t)$, $y(0) = \dot{y}(0) = 0$.

16. **(a)** Sketch the graph of the discontinuous function

$$D(t) = \begin{cases} -1, & t < \pi, \\ \cos t, & \pi \leq t < 2\pi, \\ 0, & 2\pi \leq t, \end{cases}$$

then represent $D(t)$ using $\mathrm{H}(t)$.

(b) Make a computer plot of the solution to the initial-value problem $\ddot{y} + e^{-t} y = D(t)$, $y(0) = \dot{y}(0) = 0$.

17. Find a formula in terms of the function $\mathrm{H}(t)$ that extends the formula derived in Example 9 of the text so as to extend the square wave to be left continuous for $t < 0$. [*Hint:* Try using $\mathrm{H}(-t)$.]

18. Show that the function $S(t)$ defined in Example 6 has a continuous derivative $S'(t)$ for all t, but that $S''(t)$ fails to exist at $t = \pm\pi/2$.

19. Consider the equation $\text{sqw}(t) = (-1)^{[t]}$, where $[t]$ stands for the greatest integer less than or equal to. (For example, $[\frac{1}{2}] = 0$ and $[-\frac{1}{2}] = -1$.)

(a) Show that $\text{sqw}(t) = (-1)^{[t]}$ with the function $\text{sqw}(t)$ as defined in the text for $t \geq 0$.

(b) Show that the equation in part (a) extends $\text{sqw}(t)$ to $-\infty < t < 0$ so that the extended function is right continuous for all t and is periodic of period 2. Show also that $\text{sqw}(t/p)$, $p > 0$, has period $2p$.

20. (a) Make a computer plot of the solution of

$$\ddot{y} + \pi^2 y = \text{sqw}(t), \qquad y(0) = \dot{y}(0) = 0$$

for $0 \leq t \leq 10$, where $\text{sqw}(t)$ is the square wave function illustrated in Figure 4.20(a) and described in Example 9 of the text.

(b) Repeat part (a) with 1 replacing π^2 in the differential equation.

21. (a) Make a computer plot of the solution of

$$\ddot{y} + \tfrac{1}{10}\dot{y} + \pi^2 y = \text{sqw}(t), \qquad y(0) = \dot{y}(0) = 0$$

for $0 \leq t \leq 10$, where $\text{sqw}(t)$ is the square wave function illustrated in Figure 4.20(a) and described in Example 9 of the text.

(b) Repeat part (a) with 1 replacing π^2 in the differential equation.

***22.** Consider the linear initial-value problem

$$\ddot{y} + \beta^2 y = \text{sqw}(t), \qquad y(0) = \dot{y}(0) = 0,$$

where $\text{sqw}(t)$ is the square wave function illustrated in Figure 4.20(a).

(a) Show that the problem has solution

$$y(t) = \frac{1}{\beta^2}(1 - \cos\beta t) + \frac{2}{\beta^2}\sum_{k=1}^{\infty}(-1)^k H(t-k)\big[1 - \cos\beta(t-k)\big].$$

[*Hint:* Using the expression derived for $\text{sqw}(t)$ in Example 10 of the text, solve the problem for a generic term depending on k by using an integral formula. Then use the linearity of the differential equation.]

(b) Show that when $\beta = \pi$ the solution can be written

$$y(t) = \frac{1}{\pi^2}(1 - \cos\pi t) + \frac{2}{\pi^2}\left[\sum_{j=1}^{\infty} H(t-2j)\right](1 - \cos\pi t)$$

$$- \frac{2}{\pi^2}\left[\sum_{j=0}^{\infty} H(t-2j-1)\right](1 + \cos\pi t).$$

(c) Show how to conclude from part (b) that when $\beta = \pi$, the successive local maxima and minima of $y(t)$ occur alternately at odd and even integer t values respectively and tend alternately to $+\infty$ and $-\infty$ as t tends to ∞.

(d) Show that the solution $y(t) = \big(1/(2\pi^2)\big) \sin \pi t - \big(1/(2\pi)\big)t \cos \pi t$ to the related problem $\ddot{y} + \pi^2 y = \sin \pi t,\ y(0) = \dot{y}(0) = 0$, also has the properties described in part (c).

23. The initial-value problem $\ddot{y} = -(g/l) \sin y,\ y(0) = y_0,\ \dot{y}(0) = z_0$, produces a natural oscillation amplitude A in $y(t)$ for each choice of the four parameters g, l, y_0, z_0. An escapement impulse, detailed in text Example 11 can be regarded as an attempt to restore this natural amplitude when A is degraded by adding the damping term $-k\dot{y}$ to the right-hand side of the differential equation. For instance, the choices $g = 32.2, l = 7, y_0 = 1$, and $z_0 = 0$ in Example 11 produce a free oscillation with $A = 1$. When the damping term $k\dot{y} = 0.1\dot{y}$ is included, the escapement choices $p = 0.08$ and $\alpha = 0.5$ are an attempt to repeatedly nudge the amplitude back to $A = 1$.

(a) Using step size $h = 0.001$ for numerical approximation to the solution of the initial-value problem $\ddot{y} = -(32.2/7) \sin y,\ y(0) = 1,\ \dot{y}(0) = 0$, demonstrate experimentally that the natural amplitude of free oscillation is $A = 1$.

(b) Estimate the natural amplitude A when the initial conditions in part (a) are changed to $y(0) = 0$ and $\dot{y}(0) = 1$.

(c) Using the data in Example 11 of the text except for the initial conditions, experiment with reasonable initial conditions to show that the clock model tends toward compliance with the amplitude $A = 1$.

24. (a) The escapement $F(y, \dot{y})$ described in Example 11 of the text delivers its impulse of size α to the pendulum uniformly over the interval $-p \le y < p$. Design an escapement $G(y, \dot{y})$ that delivers its impulse over the interval $-p \le y < 0$ when y is increasing and over the interval $0 \le y < p$ when y is decreasing.

(b) By conducting numerical experiments find parameters α and p for $G(y, \dot{y})$ that will drive a damped pendulum toward amplitude $A = 1$, assuming $g = 32.2, l = 5, k = 0.1$, and $m = 3$.

4D Poincaré Sections

Some solution graphs and phase plots for nonautonomous equations $\ddot{y} = f(t, y, \dot{y})$ are so irregular that they display very little apparent order. This irregularity is sometimes just a symptom of an underlying inability of numerical approximations to predict the actual time evolution of a solution that is theoretically completely determined. (See Exercise 12.) To uncover a more stable structure within a continuous-time system it may be helpful to sample a numerically plotted phase curve at equally spaced discrete times, with fixed time-difference τ. Thus we look as at a discrete-time "subsystem." A plot of the resulting sample is called a **Poincaré time section** and consists of a sequence $\mathbf{x}_n = \big(y(n\tau), z(n\tau)\big)$ of points in phase space. Thus we have an example of a **discrete dynamical system,** that is, a system in which the system is observed at

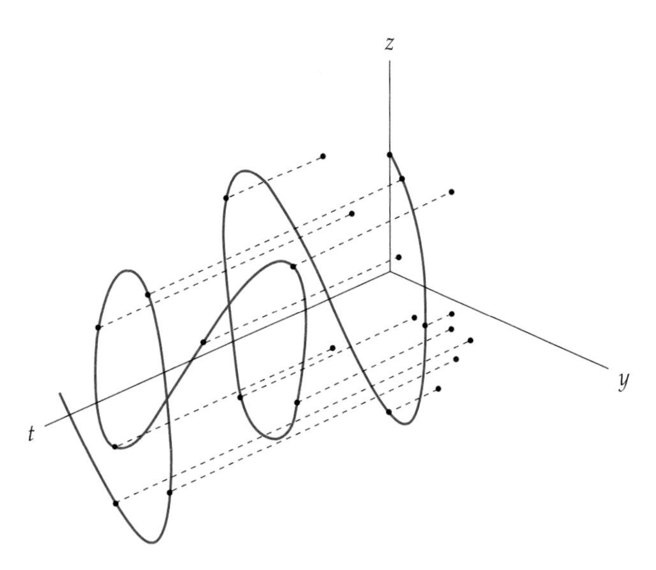

Figure 4.22 Projecting equally time-spaced points onto the yz plane.

discrete times that are τ units apart. The term **stroboscopic section,** named after a visual sampling procedure for physical processes, is often used instead of time section. Interesting results come mainly from periodically driven systems of the form $\ddot{y} = f(y, \dot{y}) + A\cos\omega t$.

We'll typically be monitoring a time variable t that increases by a fixed amount h at each step in a numerical approximation. We want to check to see just which times make t close to an integer multiple of the sample step τ; to do this we can check to see when $t - m\tau < h$, where m is the largest integer with $m\tau < t$, and h is the step size of the numerical method in use. [This criterion may be written in software routines as $\text{mod}(t, \tau) < h$, where $\text{mod}(t, \tau)$ is the "remainder" on division of t by τ, i.e., $t = m\tau + \text{mod}(t, \tau)$.] We plot the corresponding phase point $\mathbf{x} = (y,z)$ when the criterion is met. Figure 4.22 shows a few points on a solution graph (*not* on a trajectory) chosen to be equally spaced in time and are then projected parallel to the time axis onto the yz phase space. Here the differential equation isn't specified, but projecting 3200 times produced Figure 23(a) for the equation specified there. As with any numerical approximation, it should be understood that the results here are not completely accurate. In this procedure, an additional inaccuracy occurs because we don't sample precisely at integer multiples of τ; if you try that, given the inherent numerical inaccuracies, you're likely get no picture at all. The Java Applet TIMESECT at website <http://math.dartmouth.edu/~rewn> is designed to produce such pictures.

A second-order differential equation $\ddot{y} = f(t, y, \dot{y})$ is **time-periodic** with period $\tau > 0$, if $f(t + \tau, x, y) = f(t, x, y)$ for all (t, x, y). If $\ddot{y} = f(t, y, \dot{y})$ is time-periodic then its period τ is a likely choice for the fixed time difference τ for a Poincaré time section. In the next example $\tau = 2\pi$.

***example* 12**

The periodically driven Duffing oscillator $\ddot{y} + k\dot{y} - y + y^3 = A\cos\omega t$ is treated in the previous set of exercises. See also Figure IS.4 on page 6 in the Introductory Survey for a solution graph. (Theorem 8.9 of Chapter 6, Section 8B, explains that

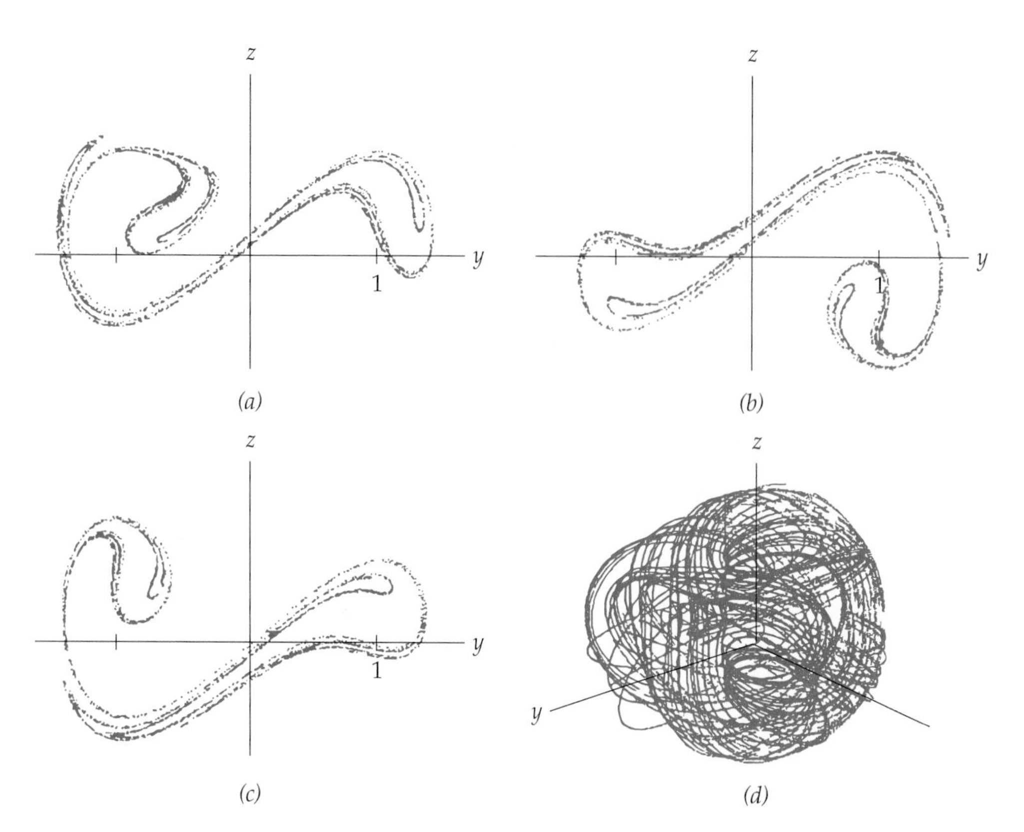

Figure 4.23 $\dot{y} = z,\ \dot{z} = y - y^3 - \frac{1}{5}z + \frac{3}{10}\cos t,\ y(0) = z(0) = 0$. (*a*) Sampling at $t = 2n\pi$. (*b*) Sampling at $t = 2n\pi + 2$. (*c*) Sampling at $t = 2n\pi + 5.4$. (*d*) Cylindrical image, $0 \le t \le 500$.

the long-term limit behavior for the unforced equation, with $A = 0$, is confined to a few fairly simple types.) In contrast, if $A \neq 0$ the problem of analyzing solution behavior gains an added dimension of complexity that we can study using time sections. We choose $k = \frac{1}{5}$, $A = \frac{3}{10}$, $\omega = 1$, and the initial values $y(0) = \dot{y}(0) = 0$, and then convert to a first-order system by setting $\dot{y} = z$. Figure 4.23(a) shows a resulting phase-space sample of 3200 points starting at $t = 0$ and using sample step $\tau = 2\pi$ because the forcing function $\cos t$ has period 2π. Parts (b) and (c) show the qualitatively similar pictures obtained by advancing the sampling times by 2 and 5.4, respectively. These samplings convey a sense of structure that is hard to quantify but nevertheless is quite striking. The exercises contain suggestions for supplementing these pictures to show more structure. As the pictures evolve point by point you can see succeeding points hop around the general pattern, seemingly in a highly "random" manner.

Cylindrical Representation. Here is a way to think about these time sections that is in some ways more satisfying than simply regarding them as projections parallel to a t axis. We interpret time t as a cylindrical coordinate angle, measuring this angle about the vertical z axis. We then plot $\big(y(t)\cos t,\ y(t)\sin t,\ z(t)\big)$ in three-dimensional space for an extended range of t values. In this way we transfer an extended image of a solution in rectangular (t, y, z) space to an image in a more confined region. Figure 4.23(d) is a perspective drawing of just such a transferred image, which we'll call a **cylindrical representation** of the solution. The preceding three pictures are cross sections of this 3-dimensional image taken in planes containing the vertical coordinate axis and set at angles of 0, 2, and 5.4 radians to the left hand horizontal axis in part (d). Extending the 3-dimensional trace in Figure 4.23(d) significantly beyond $t = 500$ results, at this scale, in a blob that conveys very little information. In contrast, the time-section pictures span the much larger time interval $0 \leq t \leq (3200)(2\pi) \approx 20,106$. Taken altogether, the four pictures convey some feeling for what the solution is like. The analytical details are described in the exercises under cylindrical representation, and the Java Applet Cylinder Rep at website <http://math.dartmouth.edu/~rewn/> or via <http://www.mhhe.com/williamson> will do the graphics.

Caution. Unless you have an exceptionally fast computer, you may need some patience in plotting a satisfactory time section. The reason is that typically an enormous amount of computed information is discarded between recorded samples. In this regard, note that if you're computing a periodic, or very nearly periodic, solution with least positive period τ, and you're recording only a phase space time section with sample step τ, you can expect to get just repetition of a single dot in your picture.

EXERCISES

The Java Applets TIMESECT and CYLINDER REP referred to in the text can be used to do some of these exercises.

1. Suppose you're making a phase plot of a solution having least positive period τ, but are recording only a time section with sample step τ/n, where n is a fixed positive integer. How many dots would you expect your time-section to display?

2. The unforced Duffing oscillator $\ddot{y} = y - y^3$ appears in the exercises for Section 3, where you're asked to show that solutions starting close enough to phase-space equilibrium are periodic.

- **(a)** Plot some typical graphs of the periodic solutions referred to above, and also make separate plots of the corresponding phase curves.
- **(b)** Plot some higher-energy solutions and corresponding phase-plots that loop around both equilibrium points.

3. Time sections of the solution to $\ddot{y} = y - y^3 - 0.2\dot{y} + 0.3\cos t$, $y(0) = \dot{y}(0) = 0$, are considered in Example 12 of the text. The sampling is done at times $2n\pi,\ 2n\pi + 2,\ 2n\pi + 5$ for nonnegative integer n. Make additional time sections at (a) $2n\pi + 1$, (b) $2n\pi + 3$, and (c) $2n\pi + 4$. Explain the relation of each picture you make to Figure 4.23(d). Note that sampling times $m\tau + \alpha$ for integer m can be estimated by selecting times for which $\text{mod}(t - \alpha,\ \tau) < h$. For example, plot a point only when $\text{mod}(t - 4,\ 2\pi) < h$ in part (c).

4. Changing the forcing term in the initial-value problem in the previous exercise to $0.3\cos(t/2)$ results in a differential equation with time period 4π instead of 2π. Investigate graphically what this change does to the solution. In particular, what characteristics does a time section have after the change?

***5.** The time sequences used in making Figures 4.23(b) and 4.23(c) differ by 3.4. Note that the resulting pictures are nearly symmetric to each other about the origin in the yz plane, where $z = \dot{y}$. Since the differential equation is time-periodic with period 2π, one might guess that with a sampling difference of exactly π, the corresponding pictures will be precisely symmetric to each other. This could turn out to be true for example if for all positive integers n a sampled solution $y(t)$ satisfied

$$\big(y(2n\pi), z(2n\pi)\big) = \big(-y((2n+1)\pi)), -z\big((2n+1)\pi)\big).$$

(Numerical experiment shows the guess to be false.) Given initial conditions $y(0) = z(0) = 0$, show that our guess is strictly true if and only if the constants k and A in $\ddot{y} = y - y^3 - k\dot{y} + A\cos t$ are chosen so that also $y(\pi) = z(\pi) = 0$.

6. **(a)** Show that if $y - y^3$ is replaced by $-y$ in the periodically driven Duffing equation with $k > 0$ and $A > 0$ to get $\ddot{y} = -y - k\dot{y} + A\cos t$, then all solutions of the resulting linear equation are oscillatory and tend to the periodic function $y_p(t) = (A/k)\sin t$ as t tends to infinity.

(b) Explain qualitatively what time-sections for this equation can be expected to look like.

7. Figure IS.4 in the Introductory Survey at the beginning of the book shows a numerically simulated time evolution for the initial-value problem

$$\frac{d^2y}{dt^2} + \frac{1}{5}\frac{dy}{dt} - y + y^3 = \frac{3}{10}\cos t, \;\; y(0) = \dot{y}(0) = 0.$$

Such simulations are fairly satisfying in a qualitative way, but tend to be quantitatively imprecise, because in the long run, the sensitivity of the differential equation to small changes in input data magnifies the inevitable error that occurs at each step. Plot solutions $y(t)$ to this differential equation for $0 \le t \le 500$, replacing the initial values $\big(y(0),\ \dot{y}(0)\big) = (0, 0)$ successively by

(a) $(0,\ 0.01),\ (0,\ 0.001),\ (0,\ 0.0001)$, and so on.

(b) $(0.01,\ 0),\ (0.001,\ 0),\ (0.0001,\ 0)$, and so on.

Record in general terms whatever long-term discrepancies you see among these pictures.

8. Make a time-section portrait for the Duffing initial-value problem

$$\frac{d^2y}{dt^2} + \frac{1}{5}\frac{dy}{dt} - y + y^3 = \frac{3}{10}\cos t, \;\; y(0) = -1, \;\; \dot{y}(0) = 1,$$

using time difference $\tau = 2\pi \approx 6.283185307$ and starting at $t = 0$. Compare the result with Figure 4.23(a), noting the lack of sensitivity to even a large change in initial conditions. This kind of experimental evidence of "stability" is one reason for the interest in time sections.

9. The differential equation $\ddot{y} = -y^3 - \frac{1}{20}\dot{y} + 7\cos t$ can be thought of as a periodically driven, slightly damped spring that is very soft for small displacements and very hard for larger ones. Use a step size $h \leq 0.01$. Assume below that $y(0) = z(0) = 0$.

(a) Let $\dot{y} = z$ to get a first-order system.

(b) Plot a time section at times $t \approx 2n\pi$, for example, $\text{mod}(t, 2\pi) < h$.

(c) Plot a time section at times $t \approx 2n\pi + 1$, for example, $\text{mod}(t - 1, 2\pi) < h$.

(d) Plot a time-section at times $t \approx 2n\pi + 4$, for example, $\text{mod}(t - 4, 2\pi) < h$.

10. The differential equation $\ddot{y} = -y^3 - \frac{1}{5}\dot{y} + 7\cos t$ is a more strongly damped version of the one in the previous exercise. Assume $y(0) = z(0) = 0$.

(a) Plot a time section at times $t = 2n\pi$, and compare the result with the corresponding picture from the previous exercise.

(b) Is the comparison in part (a) consistent with the increased damping factor? Explain.

11. In contrast to the pictures in Figure 4.23, time sections for linear second-order equation tend to be more regulated.

(a) Show that the time-section plot with sample step $\tau = 2\pi$ for the linear initial-value problem

$$\ddot{y} + \omega^2 y = (\omega^2 - 1)\sin t, \qquad y(0) = 0, \qquad \dot{y}(0) = \omega$$

lies on the phase-space ellipse $\omega^2 y^2 + (z - 1)^2 = (\omega - 1)^2$ if $\omega \neq 0$.

(b) Plot this ellipse for $\omega = \frac{1}{3}$ and for $\omega = \frac{4}{3}$. On each ellipse plot the corresponding time-sections with $\tau = 2\pi$: $\big(y(2\pi n),\ z(2\pi n)\big)$.

(c) Plot the time section for $\omega = \sqrt{2}/2$. (The next exercise is particularly relevant to this part.)

***12.** Since digital computing is carried out entirely with rational numbers, the distinction between rational and irrational numbers becomes blurred in numerical work. The reason is that an irrational number can only be represented by a rational approximation with as large a denominator as the computer can accommodate. It follows from Exercise 33 in Chapter 3, Section 2C, that oscillatory solutions such as $y_\omega(t) = \sin \omega t + \sin t$ are periodic if and only if ω is rational, in which case, if $\omega = p/q$, then $2\pi q$ is a period.

(a) Show that as an irrational ω is replaced by increasingly more accurate rational approximations $\omega = r$, the least positive period of the resulting function $y_r(t)$ tends to infinity.

(b) Show that a time section with sample step $\tau = 2\pi$ for y_ω consists of only finitely many points if ω is rational, and, in theory, infinitely many different points if ω is irrational.

13. When choosing a sample step τ for sampling in the making a time section it's important to take account of the natural time-periodicity in $\ddot{y} = f(t, y, \dot{y})$. For example, in the forcing term $A \cos \omega t$ in the Duffing equation of text Example 12 the natural period is $2\pi/\omega$. Thus with $\omega = 1$ the natural period is 2π so we used sample step $\tau = 2\pi$ in making Figures 4.23(a), 4.23(b), and 4.23(c).

(a) Make a time-section plot with the specific choices $k = \frac{1}{5}$, $A = 0.3$, and $\omega = 1$ in the Duffing equation of Example 12 and using sample step $\tau = \pi$ starting at $t = 0$. The pattern of your plot should appear to contain the one in Figure 4.23(a) as a subset; explain why.

(b) Make a time-section plot with the specific choices $k = \frac{1}{5}$, $A = 0.3$, and $\omega = 1$ in the Duffing equation of Example 12 and using sample step $\tau = 4$ starting at $t = 0$. Note that the resulting pattern is relatively featureless as compared with Figure 4.23(a), where the choice of τ is keyed to the period of the system. (Note that we have perversely assumed in our choice $\tau = 4$ that we're dealing with a period 4 system.)

Cylindrical Representation of Solution Trajectories. The picture in Figure 4.23(d) was made by plotting $\big(y(t)\cos t,\, y(t)\sin t,\, z(t)\big)$ in three-dimensional perspective for $0 \leq t \leq 500$, where $\big(y(t), z(t)\big)$ is a phase-plane point at time t. More generally, we can do a perspective plot of

$$\big(y(t)\cos(2\pi t/\tau),\ y(t)\sin(2\pi t/\tau),\ z(t)\big),$$

where $y(t)$ is a solution of an equation $\ddot{y} = f(t, y, \dot{y})$, $z(t) = \dot{y}(t)$, and τ is the constant time between samples. As explained in the text, the Poincaré time sections are just the intersections of the cylindrical representation with planes containing its vertical z axis. (Note that we synchronize the cylindrical plotting with the sampling interval to maintain this intersection property.) The phase plots are sampled at times $n\tau + \sigma$, n a positive integer, having time τ between samples and perhaps a constant shift σ.

14. Reproduce Figure 4.23(d) using the above prescription with $\tau = 2\pi$.

15. Make a cylindrical representation of the solution of the initial-value problem $\ddot{y} = -y^3 - \frac{1}{20}\dot{y} + 7\cos t$, $y(0) = \dot{y}(0) = 0$.

16. Make a cylindrical representation of the solution of the initial-value problem $\ddot{y} = -y^3 - \frac{1}{5}\dot{y} + 7\cos t$, $y(0) = \dot{y}(0) = 0$. What change do you notice in comparison to the result of the previous exercise?

SUPPLEMENTARY APPLICATIONS FOR CHAPTER 4

The website http://www.mhhe.com/williamson/ contains two supplementary applications: Gravitational Forces, and Energy and Stability. The first of these explores in the context of second-order equations some of the issues raised earlier using first-order equations. The second explores the connection between kinetic and potential energy on the one hand, and stable and unstable behavior on the other.

CHAPTER 5

Introduction to Systems

The earlier chapters were restricted to single differential equations for which the solution to an initial-value problem was a single real-valued function of a real variable. But most real-world problems involve interactions, for example among forces, between moving parts of a mechanism, among chemical solutions flowing between compartments, among populations of living beings. Suppose now each of the subparts of a physical system can be quantified by one or more real-valued time-dependent variable. If we establish equations among the variables and their derivatives, then we have a *system* of differential equations. Section 1 contains simple examples of such systems along with some useful notation, geometric background, and simple applications. Section 2 describes solution techniques for some special systems. The more complex examples are gathered together in Chapter 6, but Examples 7, 8, and 9 in the Introductory Survey give some idea of what these examples are about. Solution techniques using matrices are taken up in Section 3 and also in Chapter 7.

To deal with mutual interactions quantitatively we use systems of two or more real-valued differential equations, each involving one or more of the relevant variables. As a conceptual aid we may prefer to recast such systems as a single vector differential equation whose solutions are vector-valued functions of a real variable. This point of view is also helpful in writing computer code, but the main reason for making the step to vector systems is that many phenomena in geometry and applied mathematics are most naturally presented in vector form. When writing vector equations we'll at first follow the convention of writing rows of coordinates for vectors and abbreviating them with boldface letters when the dimension is clearly specified. Thus $\mathbf{x} = (x, y)$ means the same thing as the frequently used notation $x\mathbf{i} + y\mathbf{j}$, and $\mathbf{x} = (x, y, z)$ means the same thing as $x\mathbf{i} + y\mathbf{j} + z\mathbf{k}$.

Most of the systems we'll study will be written using time t as independent variable, so it will be convenient to use Newton's overdot notation for derivatives with respect to time:

$$\dot{x} = \frac{dx}{dt}, \qquad \dot{y} = \frac{dx}{dt}, \qquad \ddot{x} = \frac{d^2x}{dt^2}, \qquad \ddot{y} = \frac{d^2y}{dt^2}.$$

1 VECTOR EQUATIONS

1A Geometric Setting

Recall that a solution of a single first-order differential equation of the form

$$\dot{x} = F(t, x)$$

is a real-valued function $x = x(t)$ defined on some interval $a < t < b$ and such that

$$\dot{x}(t) = F\big(t, x(t)\big), \qquad a < t < b.$$

For example, $\dot{x} = x + t$ has general solution $x = Ce^t - t - 1$, defined for all real t. A pair of differential equations

$$\dot{x} = F_1(t, x, y),$$
$$\dot{y} = F_2(t, x, y),$$

for unknown functions $x(t)$ and $y(t)$ is called a **system of dimension 2,** and a solution will have the form

$$x = x(t),$$
$$y = y(t), \qquad a < t < b.$$

As t increases from a to b, the point $\big(x(t), y(t)\big)$ in the xy plane will trace out some path, perhaps like the one in the figure. Such a path, together with its direction of traversal, is called a **trajectory** of the system. Planetary orbits and particle paths in fluid flows have historically been prime examples of solution

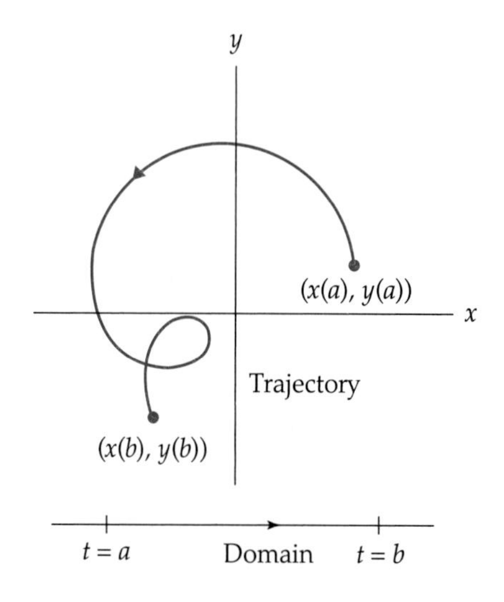

trajectories, so the terms **orbit** and **flow line** are often used instead of trajectory. The idea of solution trajectory for a first-order system $\dot{x} = F(t,x)$ of dimension 1 makes sense, but the trajectories are all confined to the x axis so it's hard to interpret the picture of a trajectory; this is why we always put the independent domain variable t in the picture and draw graphs of solutions for this equation rather than trajectories.

If the coordinates in a problem are chosen so that each equation in a system contains only one unknown function and its derivatives, then the system is called **uncoupled.** In an uncoupled system, the equations operate independently so we can try solving these equations separately. Section 3 of this chapter explains how in a given system we can sometimes change coordinates so as to find an uncoupled system that is **equivalent** in that it has the same solutions in different notation. In the next example, we are presented with an uncoupled system, and each coordinate equation of the system is easy to solve.

example 1

The uncoupled system

$$\dot{x} = -x, \qquad \dot{y} = -2y$$

is particularly simple; each unknown function occurs in just one equation. Thus we can solve each equation by itself to get the general solution

$$x = c_1 e^{-t}, \qquad y = c_2 e^{-2t}.$$

At $t = 0$, we have $x(0) = c_1$ and $y(0) = c_2$, so the initial conditions $x(0) = 3$ and $y(0) = 1$ require $c_1 = 3$ and $c_2 = 1$. Hence

$$x = 3e^{-t}, \qquad y = e^{-2t}.$$

Squaring the first equation and comparing the result with the second equation shows that the trajectory satisfies the equation of a parabola, $y = e^{-2t} = (x/3)^2$. Since the exponential functions are both positive, the trajectory obeys the restrictions $x > 0$ and $y > 0$. Figure 5.1(a) shows the *trajectory* for $0 \le t < \infty$, where the arrow point shows the direction of traversal for increasing t. Figure 5.1(b) shows the *graph* of the solution in txy space, along with its projections into the three-coordinate planes. These projections into the tx and ty planes are the graphs of

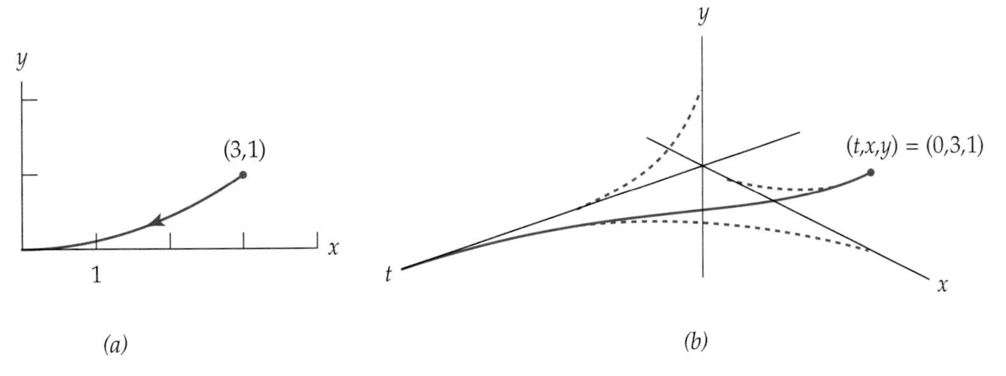

Figure 5.1 (*a*) Directed trajectory. (*b*) Solution graph and projections.

the coordinate functions $x(t)$ and $y(t)$ of the solution. In the xy plane we get the graph of the trajectory.

---

example 2

To solve the coupled system

$$\dot{x} = -y, \qquad \dot{y} = x,$$

note that since $\dot{x} = -y$, then $\ddot{x} = -\dot{y} = -x$, so $\ddot{x} + x = 0$. The solutions of this equation for x are linear combinations of $\cos t$ and $\sin t$. Since $y = -\dot{x}$, we have

$$x = c_1 \cos t + c_2 \sin t, \qquad y = c_1 \sin t - c_2 \cos t.$$

Initial conditions $x(0) = 1$ and $y(0) = 0$ require $c_1 = 1$ and $c_2 = 0$. Hence the initial-value problem has solution

$$x = \cos t, \qquad y = \sin t.$$

Figure 5.2(a) shows the circular trajectory directed by increasing t. The circle is traced infinitely often for $0 \leq t < \infty$. The graph of the solution is a helix, shown in Figure 5.2(b) just for $0 \leq t \leq 4\pi$, along with its projections into the three-coordinate planes. These projections into the tx and ty planes are the graphs of $\cos t$ and $\sin t$. In the xy plane, we get the solution trajectory, traced repeatedly as t varies. Note that a smaller scale is used on the t axis.

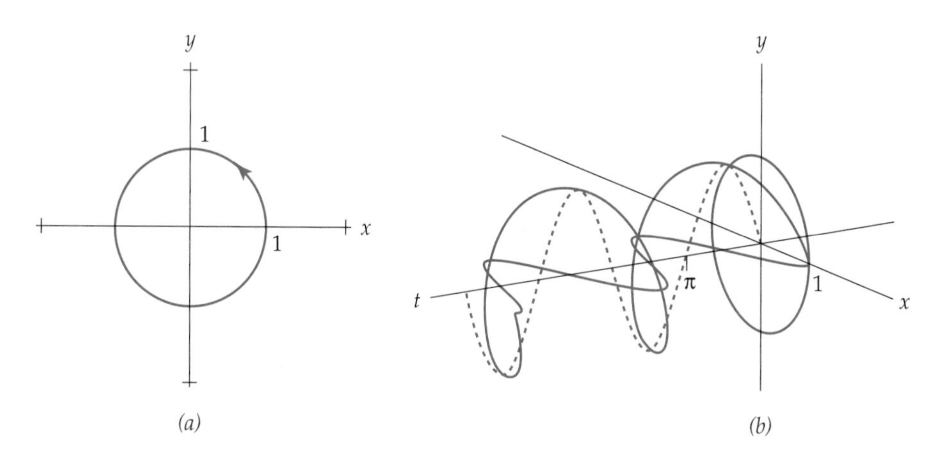

Figure 5.2 (*a*) Directed trajectory. (*b*) Solution graph and projections.

It's important to understand that the solution trajectories in Figures 5.1(a) and 5.2(a) by themselves fail to give a complete geometric description of the solution; the correspondence between t values and points on the trajectory path is not made explicit by the trajectory alone. Having enhanced our picture with a t axis perpendicular to the x and y axes, we got pictures that showed the correspondence between t and $(x(t), y(t))$ for the intervals in question. We often want only the trajectory; for example, the trajectory may represent the path followed by some physical object, whereas the solution graph may have no direct physical significance. However, the concept of a *vector field* introduced later in Section 1B enables us to recover very important information directly in the trajectory plane, namely the speed at which the trajectories are traversed.

Here is a simple application in which the system concept is essential for quantifying the problem.

example 3

Administering Medication

This model serves in many contexts, but we'll make it specific here. A drug taken orally passes through two primary stages, first the digestive system and then the blood circulatory system. Our basic assumption is that the natural rate of change of drug concentration out across the boundary of each stage is proportional to the concentration present in that stage. Letting $x = x(t)$ stand for the drug concentration in the digestive system and $y = y(t)$ the concentration in the circulatory system, the pair of differential equations

$$\frac{dx}{dt} = -ax + f(t), \qquad x(0) = 0,$$

$$\frac{dy}{dt} = cax - by, \qquad y(0) = 0,$$

is a model for determining a drug's presence in the two primary systems at time t, starting from an initial state in which no drug is present. The function $f(t)$ represents the rate at which the drug concentration is increased in the digestive system by external dosage, a rate that might be continuous or discontinuous. The rate constant a determines the rate $r_1 = ax$ at which the drug concentration is lowered in the digestive system and the factor c, $0 < c < 1$ represents its dilution as it enters the blood stream. The rate constant b serves similarly for the rate $r_2 = by$ at which the concentration is lowered in the circulatory system as the drug passes on to do its work elsewhere in the body. The coupling in this system is only partial, since the first equation can in principle be solved alone as a first-order linear equation. The resulting solution $x(t)$ is then plugged into the second equation, which can be similarly solved. You are asked to do this in Exercise 45 and, with a variation, in Exercise 46. Just the same, writing the equations as a system is important for displaying the nature of the total process.

We often use vector notation to describe a differential equation when we want to emphasize geometric ideas.

example 4

We can write the system of differential equations in Example 1 using vector notation by letting

$$\mathbf{x} = (x, y), \qquad \frac{d\mathbf{x}}{dt} = \dot{\mathbf{x}} = (\dot{x}, \dot{y}), \quad \text{and} \quad \mathbf{F}(x, y) = (-x, -2y).$$

Then equating corresponding coordinates gives

$$(\dot{x}, \dot{y}) = (-x, -2y) \quad \text{or} \quad \frac{d\mathbf{x}}{dt} = \mathbf{F}(\mathbf{x}) \quad \text{or} \quad \dot{\mathbf{x}} = \mathbf{F}(\mathbf{x}).$$

example 5

For a time-dependent example, we can write the nonautonomous system

$$\dot{x} = x + y + t, \qquad \dot{y} = x - y - t,$$

as $\dot{\mathbf{x}} = \mathbf{F}(t,\mathbf{x})$, where $\mathbf{x} = (x, y)$ and $\mathbf{F}(t,\mathbf{x}) = (x + y + t, x - y - t)$.

While the systems encountered in applications are often low dimensional, perhaps two or three, systems of very high dimension do occur, so it's important to use notation that accommodates arbitrary dimension. For example, the motion of three bodies subject only to their own mutual gravitational attraction are described by the solution of an 18-dimensional first-order system, discussed in Chapter 6, Section 4. (For our own solar system we need dimension at least 60.) When we speak of a **first-order initial-value problem of dimension n in normal form,** we mean a first-order initial-value problem of the form

$$\begin{aligned}
\dot{x}_1 &= F_1(t, x_1, x_2, \ldots, x_n), \qquad & x_1(t_0) &= a_1,\\
\dot{x}_2 &= F_2(t, x_1, x_2, \ldots, x_n), \qquad & x_2(t_0) &= a_2,\\
&\vdots\\
\dot{x}_n &= F_n(t, x_1, x_2, \ldots, x_n), \qquad & x_n(t_0) &= a_n.
\end{aligned}$$

Normal form is important for two reasons. One is that under the conditions described in Theorem 1.1 of Section 1E on page 342, a normal-form initial-value problem will always have a solution. The other related reason is that numerical solution of systems is most conveniently carried out with a normal-form system, as we'll see in Chapter 6, Section 5.

Using the vector notations

$$\mathbf{x} = (x_1, x_2, \ldots, x_n), \qquad \dot{\mathbf{x}} = (\dot{x}_1, \dot{x}_2, \ldots, \dot{x}_n) \qquad \mathbf{a} = (a_1, \ldots, a_n),$$

we can write the problem more concisely for some purposes as

$$\dot{\mathbf{x}} = \mathbf{F}(t,\mathbf{x}), \qquad \mathbf{x}(t_0) = \mathbf{a} \quad \text{where} \quad \mathbf{F}(t,\mathbf{x}) = \big(F_1(t,\mathbf{x}), \ldots, F_n(t,\mathbf{x})\big).$$

A solution is then a vector-valued function $\mathbf{x} = \mathbf{x}(t)$ defined on an interval $a < t < b$. Verifying that $\mathbf{x}(t)$ really is a solution amounts to checking that

$$\dot{\mathbf{x}} = \mathbf{F}(t, \mathbf{x}(t)), \quad \text{for } a < t < b.$$

The concept of phase space appears also in Chapter 4, Section 3. Aside from the concept itself there is no effective overlap between that section and this chapter. The reason is that in Chapter 4 the mathematical technique focuses directly on the simplest approach to the equations $y'' = f(y, y')$ and their reduction to the first-order equations $z(dy/dz) = f(y, z)$. After widening our scope in this chapter, the required techniques will involve matrix algebra and analysis of vector fields.

We can display trajectory curves for n-dimensional systems only for $n = 2$ and $n = 3$, but the basic idea is still valid in higher dimensions. The space where trajectories of a *first-order* system are traced is the **state space** or **phase space** of the associated system. Under the assumptions of Theorem 1.1 in Section 1E, specifying a state will determine a unique trajectory starting at that state that typically trace a differentiable path as the independent variable increases. Figures 5.1(a) and 5.2(a) depict trajectories in two-dimensional space. A trajectory in 3-dimensional space appears in Figure 5.3(a). Thus one advantage of the vector interpretation of a system is that a solution $\mathbf{x}(t)$ can be interpreted as a position, at time t, of a point in a space of some dimension, and we can even draw a picture of this space in the case of dimension 2 or 3.

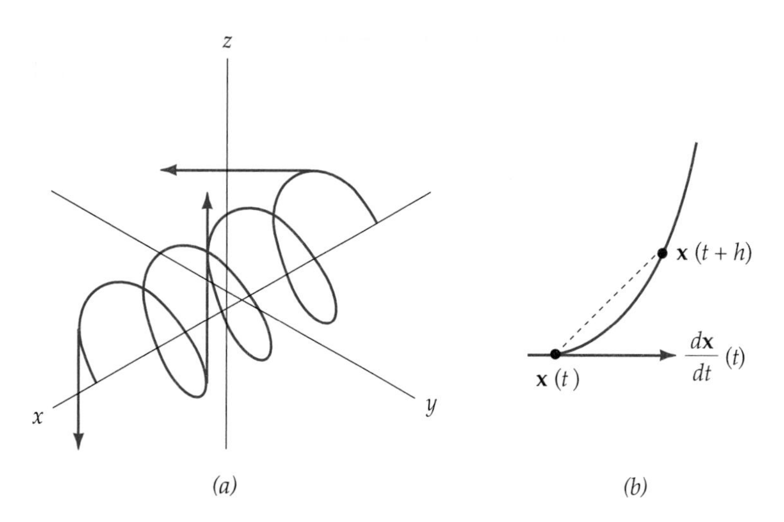

Figure 5.3 (a) Trajectory with tangent vectors.
(b) Limiting position of tangent.

A second advantage of the vector interpretation is that the vector derivative $\dot{\mathbf{x}}$ has a geometric meaning that the coordinate derivatives $\dot{x}_k$ do not have when considered separately: $\dot{\mathbf{x}}$ can be interpreted as a tangent vector to the trajectory $\mathbf{x} = \mathbf{x}(t)$. Furthermore, if t is interpreted as a time variable, then $\dot{\mathbf{x}}(t)$ is a **velocity vector** in the sense that its length is the **speed** $v(t)$ of traversal of the trajectory, and the direction of the tangent is the direction of motion along the trajectory at time t. The following discussion shows how the formal definitions of tangent and velocity are motivated by the intuitive ideas behind them.

Figure 5.3(b) shows the points $\mathbf{x}(t)$ and $\mathbf{x}(t+h)$. The parallelogram law for addition of vectors shows that, if $\mathbf{z}$ is the vector arrow joining $\mathbf{x}(t)$ to $\mathbf{x}(t+h)$, then $\mathbf{z} + \mathbf{x}(t) = \mathbf{x}(t+h)$. It follows that $\mathbf{z} = \mathbf{x}(t+h) - \mathbf{x}(t)$. Now multiply this vector by the number $1/h$ to get a parallel vector that approaches the tangent direction to the trajectory at $\mathbf{x}(t)$ as h tends to 0 from the right. The result is easier to read if we write the entries of $\mathbf{x}(t)$ in columns:

$$\lim_{h\to 0} \frac{\mathbf{x}(t+h) - \mathbf{x}(t)}{h} = \begin{pmatrix} \lim_{h\to 0} \dfrac{x_1(t+h) - x_1(t)}{h} \\ \vdots \\ \lim_{h\to 0} \dfrac{x_n(t+h) - x_n(t)}{h} \end{pmatrix} = \begin{pmatrix} \dfrac{dx_1(t)}{dt} \\ \vdots \\ \dfrac{dx_n(t)}{dt} \end{pmatrix} = \dot{\mathbf{x}}(t).$$

We define the **tangent vector** or *velocity vector* to the trajectory at $\mathbf{x}(t)$ to be $\dot{\mathbf{x}}(t)$. Geometrically, $\dot{\mathbf{x}}$ appears as an arrow with its initial point at $\mathbf{x}(t)$, tangent to the trajectory and pointing in the direction of increasing t along the trajectory. The length of the vector $\dot{\mathbf{x}}(t)$ is the speed $v(t)$ of $\mathbf{x}(t)$ at time t,

$$|\dot{\mathbf{x}}| = \sqrt{\dot{x}_1^2 + \cdots + \dot{x}_n^2}.$$

This is just the length of the velocity vector, where the **length** of a vector $\mathbf{y} = (y_1, y_2, \ldots, y_n)$ is defined as usual by $|\mathbf{y}| = \sqrt{y_1^2 + \cdots + y_n^2}$. The second time derivative $\ddot{\mathbf{x}}(t)$ is the **acceleration vector** of the trajectory, and along with $\dot{\mathbf{x}}(t)$ it plays a fundamental role in classical mechanics. The reason is that the **force vector** $\mathbf{F}(t)$ that we think of as altering the motion of a body of constant mass m along its path is determined by $\mathbf{F}(t) = m\ddot{\mathbf{x}}(t)$.

example 6

Suppose a projectile is fired over level ground at an angle of elevation θ, where $0 < \theta < \pi/2$ and with initial speed v_0. If we neglect air resistance, there is no horizontal acceleration or deceleration, so the x coordinate of the projectile's position $\mathbf{x} = (x, y)$ satisfies $\ddot{x} = 0$. In the vertical direction, there is just the downward acceleration $-g$ of gravity, so $\ddot{y} = -g$. We integrate the equations $\ddot{x} = 0$ and $\ddot{y} = -g$ twice with respect to t, getting first

$$\begin{aligned}\dot{x} &= c_1, \\ \dot{y} &= -gt + c_3,\end{aligned} \quad \text{then} \quad \begin{aligned}x &= c_1 t + c_2, \\ y &= -\tfrac{1}{2}gt^2 + c_3 t + c_4.\end{aligned}$$

If we choose coordinates so that the projectile starts at $t = 0$ with position $x_0 = 0$ and $y_0 = 0$, then the right-hand equations at $t = 0$ imply that $c_2 = c_4 = 0$. Ordinarily we give the other two initial conditions in terms of $\dot{x}(0) = z_0$ and $\dot{y}(0) = w_0$, but for projectile problems it's customary to specify instead the parameters over which we have direct control; these are the speed v_0 of the projectile as it leaves the gun, called the **muzzle velocity,** and the **elevation angle** θ, which is the angle the gun barrel makes with the horizontal line pointing from the gun to the target. Since $v_0 = \sqrt{z_0^2 + w_0^2}$, the elementary definitions of cosine and sine show that $\cos\theta = z_0/v_0$ and $\sin\theta = w_0/v_0$. Hence $z_0 = v_0\cos\theta$ and $w_0 = v_0 \sin\theta$. Thus the left-hand equations at $t = 0$ require $c_1 = v_0\cos\theta$ and $c_3 = v_0\sin\theta$. Altogether we have

$$\begin{aligned}\dot{x} &= v_0\cos\theta, \\ \dot{y} &= -gt + v_0\sin\theta,\end{aligned} \quad \text{and} \quad \begin{aligned}x &= (v_0\cos\theta)t, \\ y &= -\tfrac{1}{2}gt^2 + (v_0\sin\theta)t.\end{aligned}$$

Eliminating t between the equations for x and y shows that the trajectory lies on a downward pointing parabola. The maximum point on the trajectory occurs when $\dot{y} = 0$, that is when $t = v_0 \sin\theta/g$. Figure 5.4 shows two trajectories, the initially steeper one parabolic, with endpoints 75 miles apart. (See Example 7 and Exercise 37 for the other trajectory.)

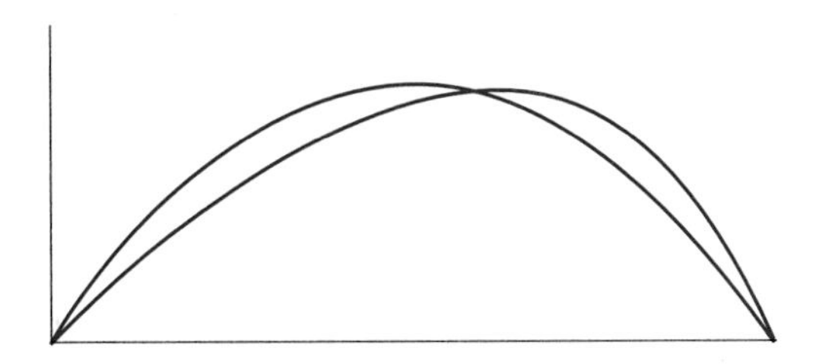

Figure 5.4 (a) $\theta = 54.4^\circ$, $v_0 = 3659$ feet per second. (b) $k/m = 0.007$, $\theta = 43.4^\circ$, $v_0 = 5250$ feet per second.

example 7

We can write system of equations in the previous example in vector form if we denote the gravitational acceleration vector $(0, g)$ by $\mathbf{g}$. With $\mathbf{x} = (x, y)$, we get simply $\ddot{\mathbf{x}} = -\mathbf{g}$. If a retarding force of air resistance is proportional to velocity is included, we have to write the equation in terms of force rather than acceleration, and so take the projectile mass into account. Thus

$$m\ddot{\mathbf{x}} = -k\dot{\mathbf{x}} - m\mathbf{g} \quad \text{or} \quad \ddot{\mathbf{x}} = -\frac{k}{m}\dot{\mathbf{x}} - \mathbf{g}, \qquad k > 0.$$

In terms of rectangular xy coordinates the vector equation becomes the pair of equations

$$\ddot{x} = -\frac{k}{m}\dot{x}, \qquad \ddot{y} = -\frac{k}{m}\dot{y} - g.$$

An asymmetrical trajectory with $k/m = 0.007$ appears with the parabolic trajectory in Figure 5.4. You are asked in Exercise 34 to solve the system using the data in the part (b) caption of Figure 5.4.

In the general first-order system

$$\frac{d\mathbf{x}}{dt} = \mathbf{F}(t, \mathbf{x}),$$

the vector-valued function **F** is understood to depend explicitly on the independent variable t, so different tangent vectors may be assigned to a trajectory if it happens to pass through the same point **x** at two different times, as in the figure on page 318. If $\mathbf{F}(t, \mathbf{x})$ is independent of t, the system is called **autonomous** and can be written

$$\frac{d\mathbf{x}}{dt} = \mathbf{F}(\mathbf{x}).$$

Examples 1, 2, 4, and 6 above are autonomous systems, whereas Examples 3 and 5 are **nonautonomous** and time-dependent.

1B Vector Fields

For an autonomous system $\dot{\mathbf{x}} = \mathbf{F}(\mathbf{x})$, the vector $\mathbf{F}(\mathbf{x})$, necessarily equals the tangent vector $\dot{\mathbf{x}}$ to a trajectory at **x**. For an autonomous system, this tangent vector is always the same regardless of what time it is when the trajectory passes through **x**. Such an assignment of vectors $\mathbf{F}(\mathbf{x})$ to points **x** is called a **vector field.** Figure 5.5(a) shows a sketch, using vector arrows, of a vector field derived from the next example. To make this sketch, a vector $\mathbf{F}(\mathbf{x})$, rather than starting at the origin, is translated parallel to itself so that its tail is located at the point of tangency **x**. In other words, *the arrow is drawn from the point* **x** *to the point* $\mathbf{x} + \mathbf{F}(\mathbf{x})$.

A sketch of a vector field $\mathbf{F}(\mathbf{x})$ can't include all the arrows in the field. Similarly, we can't draw all the solution trajectories of the associated first-order system $\dot{\mathbf{x}} = \mathbf{F}(\mathbf{x})$. But we can draw representative collection of trajectories of a first-order system, called a **phase portrait.** Figure 5.5(b) shows a phase portrait for the system in Example 8. Unfortunately phase portraits are often comprehensible only for 2-dimensional systems; Figure 5.10 in Section 1E shows an attempt at a 3-dimensional example.

Chapter 4, Section 3, discusses 2-dimensional phase portraits in the context of single second-order equations.

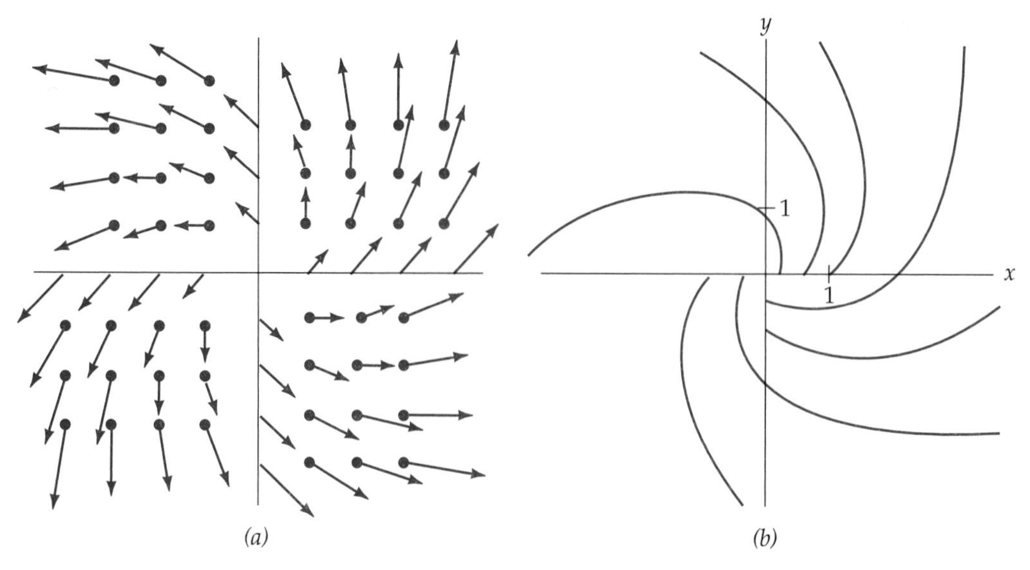

Figure 5.5 (*a*) Vector field. (*b*) Phase portrait.

example 8

We'll see in Section 2 how to find a general solution formula for the coupled pair of differential equations

$$\dot{x} = x - y, \qquad \dot{y} = x + y.$$

For now we'll look at a geometric interpretation of this system of equations and its solutions. Figure 5.5(a) shows a vector field sketch as an array of tangent vectors $\mathbf{x} = (x - y, x + y)$ at $\mathbf{x} = (x, y)$ to solution trajectories of the system. You can imagine superimposing Figure 5.5(a) on the picture of solution trajectories in the phase portrait in Figure 5.5(b). If the axes are aligned, the arrows will be tangent to the trajectories. The arrows in the vector field indicate the direction of traversal of the trajectories, and longer arrows farther from the origin show that the speed of traversal of a trajectory (see Example 9.) increases as the path moves away from the origin; this information you can't get just from looking at the trajectories in the phase portrait.

example 9

The speed of traversal at a point (x, y) on a trajectory of the system $\dot{x} = x - y$, $\dot{y} = x + y$ in the previous example is the length of the velocity vector $(\dot{x}, \dot{y})$, and so is

$$\sqrt{\dot{x}^2 + \dot{y}^2} = \sqrt{(x - y)^2 + (x + y)^2} = \sqrt{2}\sqrt{x^2 + y^2}.$$

example 10

The autonomous system

$$\frac{dx}{dt} = -y, \qquad \frac{dy}{dt} = x,$$

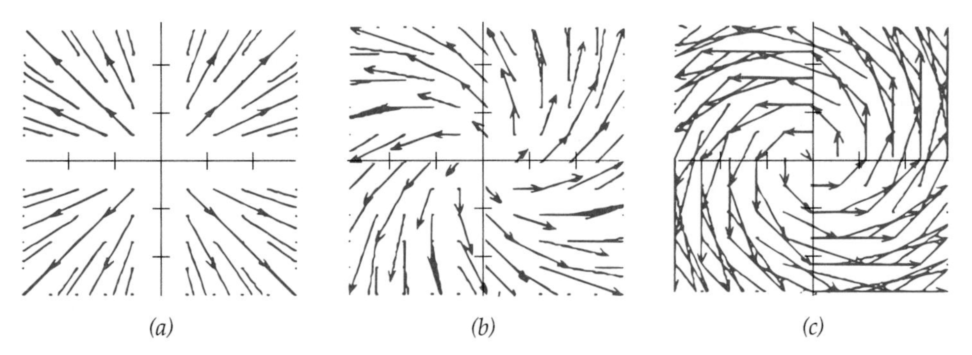

Figure 5.6 (*a*) $\mathbf{F}(0,x,y) = (x,y)$. (*b*) $\mathbf{F}(\frac{1}{2},x,y) = \left(\frac{1}{2}(x-y), \frac{1}{2}(x+y)\right)$. (*c*) $\mathbf{F}(1,x,y) = (-y,x)$.

can be written $d\mathbf{x}/dt = \mathbf{F}(\mathbf{x})$, where $\mathbf{x} = (x,y)$ and $\mathbf{F}(\mathbf{x}) = (-y,x)$. The geometric interpretation of the system that the tangent vector to a solution trajectory through the point (x,y) has the direction of the vector $(-y,x)$. The tangent then has slope $-x/y$, while the line joining the origin to the point (x,y) has slope y/x. It follows that these two directions are perpendicular. Figure 5.6(c) shows a sketch of the vector field. The picture suggests that the trajectories might be spirals aimed at the origin, but they're actually circles centered at the origin. However, we can verify that the initial-value problem with $x(0) = u$, $y(0) = v$ is satisfied by the family of solutions

$$x(t) = u\cos t - v\sin t, \qquad y(t) = u\sin t + v\cos t,$$

and that $x(t)^2 + y(t)^2$ equals the constant $u^2 + v^2$. Thus we get circular trajectories. Section 2 contains an elementary routine for deriving such solution formulas.

For the general time-dependent system $d\mathbf{x}/dt = \mathbf{F}(t,\mathbf{x})$, the function $\mathbf{F}(t,\mathbf{x})$ specifies a tangent vector to a trajectory through $\mathbf{x}$ that may be different for different t. The stationary vector field of the kind pictured in Figure 5.5(a) is no longer appropriate, but it can be replaced by a sequence of "snapshots" taken at different times. Each snapshot will be a sketch of a single vector field, but will show changes in the arrows as time t varies. By making the changes quickly and in small steps, you can produce the illusion of a continuous evolution of one vector field into another.

---

example **11** The system

$$\frac{dx}{dt} = (1-t)x - ty, \qquad \frac{dy}{dt} = tx + (1-t)y$$

is determined by the time-dependent vector field

$$\mathbf{F}(t,x,y) = ((1-t)x - ty, tx + (1-t)y).$$

Figure 5.6 sketches of the vector field for three values of t.

Comparison Vector Field and Direction Field. A vector field plays a somewhat different role for a first-order system than a direction field, or slope field, does for a scalar equation $\dot{x} = F(t,x)$. In a *direction field,* defined in Chapter 1, Section 2, the axis of the domain variable t is in the picture, the line segments all have the same arbitrary but convenient length, and the segments are tangent to the *graph* of a solution. In a direction field the speed is indicated by the inclination of the segments relative to the horizontal axis, and the direction is the same as the positive direction along that axis. (This description of direction field holds also in higher dimensions, where direction fields have considerable theoretical importance.) In a *vector field,* defined in this section, the domain axis is *not* in the picture, while each arrow is tangent to a *trajectory* (i.e., an image curve) of a solution, and speed and direction are indicated by the length and direction of the arrows.

1C Order Reduction and Normal Form

It's often a useful trade-off to replace a differential equation of order higher than one by a first-order system at the expense of increasing the number of unknown functions. This is done in a useful way, for example, in Chapter 3, Section 2, and at several places in the applications of Chapter 6.

example 12

In the general second-order equation

$$\frac{d^2y}{dt^2} = f\left(t, y, \frac{dy}{dt}\right),$$

we can let $dy/dt = z$. Then $d^2y/dt^2 = dz/dt$, so the two equations

$$\frac{dy}{dt} = z, \qquad \frac{dz}{dt} = f(t,y,z),$$

form a system equivalent to the original second-order equation. This is essentially the technique we used to establish the general solution formulas for second-order constant coefficient equations, and for extending the Euler numerical methods from first-order to second-order equations.

Notice that a solution of the system, being a pair of functions, $x(t)$, $y(t)$ produces not only a solution $y(t)$ of the given equation but also its derivative $dy/dt = z(t)$.

example 13

Here are some second-order equations and their companion systems.

$$(a) \quad \ddot{y} + y = 0; \quad \begin{cases} \dot{y} = z, \\ \dot{z} = -y; \end{cases}$$

$$(b) \quad \ddot{y} = -\sin y; \quad \begin{cases} \dot{y} = z, \\ \dot{z} = -\sin y; \end{cases}$$

$$(c) \quad \ddot{y} + t\dot{y} + t^2y = 0; \quad \begin{cases} \dot{y} = z, \\ \dot{z} = -tz - t^2y. \end{cases}$$

Initial conditions for the second-order equations have the form $y(t_0) = y_0$, $dy/dt(t_0) = z_0$. Since $dy/dt = z$, these initial conditions translate directly into the conditions $y(t_0) = y_0$, $z(t_0) = z_0$ for the corresponding systems.

While the general form of the systems we came up with in the previous example may require a little more writing than the second-order equations they came from, it often turns out to be conceptually and computationally simpler to deal with the first-order vector derivative $\dot{\mathbf{x}} = (\dot{y}, \dot{z})$ in the system rather than with the pair $(\dot{y}, \ddot{y})$ containing derivatives of different orders. A similar remark applies to the four-dimensional vector derivative $\dot{\mathbf{x}} = (\dot{x}, \dot{y}, \dot{u}, \dot{v})$ introduced in the next example. The basic space in which the position trajectories lie is called the **configuration space** to distinguish it from the higher-dimensional space that incorporates derivative coordinates, which we have called the *state space* or *phase space.* Figures 4.12 and 4.13 of Chapter 4 (pages 278 and 279) show phase portraits of first-order systems specifically derived from second-order equations. It's important to understand that it is the *state* of a system at a given time t that is used to determine future states through the action of a system.

---

example 14 Second-order systems of the form

$$\ddot{x} = f(t, x, y, \dot{x}, \dot{y})$$
$$\ddot{y} = g(t, x, y, \dot{x}, \dot{y})$$

occur often in applications, for example, in Chapter 6 in the study of vibrating mechanisms and in the study of planetary motion. Dealing instead with an equivalent first-order system made it quite natural to extend the numerical methods of Chapter 2, Section 4, to second-order systems, and often provides some additional geometric insight into the problem under study. The idea in this case is to introduce two new dependent variables $u = \dot{x}$ and $v = \dot{y}$. It follows then that we have $\ddot{x} = d\dot{x}/dt = \dot{u}$ and $\ddot{y} = d\dot{y}/dt = \dot{v}$. The configuration space consists of points (x, y) and the state space of points (x, y, u, v). The original system can then be rewritten as the system of four first-order equations

$$\dot{x} = u,$$
$$\dot{y} = v,$$
$$\dot{u} = f(t, x, y, u, v),$$
$$\dot{v} = g(t, x, y, u, v).$$

Initial conditions for the second-order system specify positions and velocities for x and y, and might look, for example, something like $x(0) = x_0$, $\dot{x}(0) = u_0$, $y(0) = y_0$, $\dot{y}(0) = v_0$. In terms of $x, y, u,$ and v these conditions convert into $x(0) = x_0$, $u(0) = u_0$, $y(0) = y_0$, and $v(0) = v_0$.

The best-known system of the type described above is not explicitly dependent on the velocities $\dot{x}$ and $\dot{y}$; it is the system of equations, derived in Chapter 6, Section 4, that governs the planar motion of a single planet of mass m_1 relative to a fixed star of mass m_2:

$$\ddot{x} = \frac{-G(m_1 + m_2)x}{(x^2 + y^2)^{3/2}}, \qquad x(0) = x_0, \qquad \dot{x}(0) = u_0,$$

$$\ddot{y} = \frac{-G(m_1 + m_2)y}{(x^2 + y^2)^{3/2}}, \qquad y(0) = y_0, \qquad \dot{y}(0) = u_0.$$

The number G is the universal gravitational constant, which depends on the units of measurement being used. Knowing the initial conditions means knowing the positions (x_0, y_0) and velocities (u_0, v_0) of the planet at some time we call $t = 0$. Simple solution formulas are impossible to find in general, though it's known that all trajectories are elliptical in shape. In particular, you can check that there are solution trajectories of radius a centered at the center of mass of the star. Indeed these solutions have the form $x = a \cos \omega t$, $y = a \sin \omega t$, and it's left as an exercise to find the correct uniform angular speed w for each radius a that makes these formulas satisfy the differential equations.

For doing approximate computations as well as for geometric interpretation, it is important to be able to reduce a system to the standardized *first-order normal form*

$$\frac{d\mathbf{x}}{dt} = \mathbf{F}(t, \mathbf{x})$$

containing only first-order derivatives isolated on the left, even if the original system contains higher order derivatives.

example 15 The first two systems are in normal form, but the third is not, for two reasons, one for each equation in the system.

$$\begin{cases} \dfrac{dx}{dt} = y + t, \\ \dfrac{dy}{dt} = x + t, \end{cases} \qquad \begin{cases} \dfrac{dx}{dt} = xy, \\ \dfrac{dy}{dt} = x^2 + t, \end{cases} \qquad \begin{cases} \dfrac{dx}{dt} = \dfrac{dy}{dt} + y, \\ 2\dfrac{dy}{dt} = x - t. \end{cases}$$

example 16 To convert the second-order system

$$\frac{d^2x}{dt^2} + 2\frac{dy}{dt} = t,$$

$$\frac{d^2y}{dt^2} - \frac{dx}{dt} + y = 0,$$

to first-order normal form, let $dx/dt = z$ and $dy/dt = w$. Then $d^2x/dt^2 = dz/dt$

and $d^2y/dt^2 = dw/dt$. Substitution gives us altogether four equations

$$\frac{dx}{dt} = z, \qquad \frac{dy}{dt} = w,$$
$$\frac{dz}{dt} = -2w + t, \qquad \frac{dw}{dt} = z - y.$$

The reduction to normal form in this example depended first on our ability to solve for the second derivatives in terms of x and y and their first derivatives.

example 17 An example of a first-order system *not* in normal form is

$$\dot{x} + \dot{y} = 2x + 4y,$$
$$2\dot{x} + 3\dot{y} = 2x + 6y.$$

This system can be put in normal form by applying simple elimination to the derivative terms. Multiply the first equation by 2 and subtract from the second to get $\dot{y} = -2x - 2y$. Now subtract this equation from the first one to get $\dot{x} = 4x + 6y$. The result is a system in first-order normal form:

$$\dot{x} = 4x + 6y,$$
$$\dot{y} = -2x - 2y.$$

If the coefficients 2 and 3 in the second of the original equations were both replaced by 1, we wouldn't have been able to solve for $\dot{x}$ and $\dot{y}$, so the reduction to normal form would have been impossible.

EXERCISES

The uncoupled systems in Exercises 1 through 4 can be solved by treating each equation separately. Find the general solution, and then find the particular solution that satisfies the given initial conditions.

1. $dx/dt = x + 1,\ x(0) = 1,$
$dy/dt = y,\ y(0) = 2.$

2. $dx/dt = t,\ x(1) = 0,$
$dy/dt = y,\ y(1) = 0.$

3. $dx/dt = x,\ x(0) = 0,$
$dy/dt = \frac{1}{2}y,\ y(0) = 1,$
$dz/dt = \frac{1}{3}z,\ z(0) = -1.$

4. $dx/dt = x + t,\ x(0) = 0,$
$dy/dt = y - t,\ y(0) = 0,$
$dz/dt = z,\ z(0) = 1.$

5.–8. For each of the systems in Exercises 1 through 4, there is a vector-valued function $\mathbf{F}(t, \mathbf{x})$, with $\mathbf{x} = (x, y)$ or (x, y, z) such that the system can be written in the form $d\mathbf{x}/dt = \mathbf{F}(t, \mathbf{x})$.

(i) Find $\mathbf{F}$ in each case.

(ii) Find the speed of a trajectory through $\mathbf{x}$ at time t.

9. The trajectory in the figure on page 318 belongs to the system

$$\frac{dx}{dt} = -\beta y - (1 - \beta)\sin t, \qquad \frac{dy}{dt} = \beta x + (1 - \beta)\cos t,$$

for the parameter choice $\beta = 0.35$. Verify that the system has solutions of the form $x = \cos t + c \cos \beta t,\ y = \sin t + c \sin \beta t,\ c = \text{const.}$

Sketch the vector fields in Exercises 10 through 13, associated with the systems of Exercise 1, by drawing a few arrows for $\mathbf{F}(\mathbf{x})$ or $\mathbf{F}(t,\mathbf{x})$ with their tails at selected points $\mathbf{x}$ of the form (x,y) or (x,y,z). In Exercises 11 and 13 make separate sketches for $t = -1,\ t = 0$, and $t = 1$.

10. $\mathbf{F}(x,y) = (x+1,\ y)$.

11. $\mathbf{F}(t,x,y) = (t,y)$.

12. $\mathbf{F}(x,y,z) = (x,\ \frac{1}{2}y,\ \frac{1}{3}z)$.

13. $\mathbf{F}(t,x,y,z) = (x+t,\ y-t,\ z)$.

14. Sketch the vector field $\mathbf{F}(x,y) = (-y,\ x)$. Then sketch the trajectory curve tangent to arrows in the field sketch, starting at $(x,y) = (1,0)$.

***15.** Show that the system $\dot{x} = -ty,\ \dot{y} = tx$ has circular solution trajectories of radius $r > 0$, traced with increasing speed rt as time increases. [*Hint:* Show that $x\dot{x} + y\dot{y} = 0$.]

For Exercises 16 and 17, note that computer software for plotting 2-dimensional vector fields is widely available, for example, the Java Applet VECFIELD at <http://math.dartmouth.edu/~rewn/>, which also plots trajectories. Scaling the lengths of the arrows up or down by a constant factor often makes a better picture; normalization to uniform length is another possibility, though with loss of indications of relative speed at different points.

16. Sketch the vector field $\mathbf{F}(x,y) = (y-x,\ y+x)$.

17. Sketch the vector field $\mathbf{F}(x,y) = (xy,\ x+y)$.

Let $\dot{y} = x$ in Exercises 18 and 19. (a) Express the second-order differential equation as a first-order system of dimension 2, (b) find the corresponding initial conditions for $x(0)$ and $y(0)$, and (c) solve either the second-order equation or the system, whichever seems simpler.

18. $\ddot{y} + \dot{y} + y = 0,\ y(0) = 1,\ \dot{y}(0) = 1$.

19. $\ddot{y} + t\dot{y} = t,\ y(0) = 0,\ \dot{y}(0) = 1$.

Find a first-order system equivalent to each of Exercises 20 through 23.

20. $d^2x/dt^2 + (dx/dt)^2 + x^2 = e^t$. [*Hint:* Let $dx/dt = y$.]

21. $d^2x/dt^2 = x\,dx/dt$.

22. $d^3x/dt^3 = (d^2x/dt^2)^2 - x\,dx/dt - t$. [*Hint:* Let $dx/dt = y,\ dy/dt = z$.]

23. $d^3x/dt^3 = 12x\,dx/dt$.

Reduce each system in Exercises 24 through 27 to normal form, with each first derivative by itself on the left side.

24. $dx/dt + dy/dt = t,$
$dx/dt - dy/dt = y.$

25. $dx/dt + dy/dt = y,$
$dx/dt + 2dy/dt = x.$

26. $2dx/dt + dy/dt + x + 5y = t,$
$dx/dt + dy/dt + 2x + 2y = 0.$

27. $dx/dt - dy/dt = e^{-t},$
$dx/dt + dy/dt = e^t.$

28. Consider the 2-dimensional coupled system $\dot{x} = x + y$, $\dot{y} = 4x + y$.

(a) Use $x = z + w$, $y = 2z - 2w$, to change coordinates from (x, y) to (z, w) and show that this change results in the uncoupled system $\dot{z} = 3z$, $\dot{w} = -w$.

(b) Solve the uncoupled system in part (a) for z and w, then change back to xy coordinates to solve for x and y. Then verify by substitution that your solution for x and y satisfies the given system.

Finding Trajectories from a Single Equation. Suppose the 2-dimensional system $dx/dt = F(t, x, y)$, $dy/dt = G(t, x, y)$ is such that $G(t, x, y)/F(t, x, y) = R(x, y)$ is independent of t. This would occur in particular if neither F nor G depended explicitly on t. Since the chain rule allows us to write

$$\frac{dy}{dx} = \frac{dy/dt}{dx/dt}$$

under fairly general conditions, we can sometimes conclude that there are trajectory curves of the system satisfying the differential equation

$$\frac{dy}{dx} = R(x, y).$$

If this equation can be solved, we have a way to plot trajectories without finding solutions $x(t)$ and $y(t)$. For example, $dx/dt = ty$, $dy/dt = -tx$ leads us to consider

$$\frac{dy}{dx} = -\frac{x}{y},$$

which has solutions $x^2 + y^2 = c$, representing circular trajectories. Using this method, sketch some trajectories for the systems in Exercises 29 through 32. Looking at the vector field will tell you how a trajectory is traced and let you find constant solutions, for which $\dot{x} = \dot{y} = 0$.

29. $dx/dt = x - y$, $dy/dt = x^2 - y^2$.

30. $dx/dt = e^{2y}$, $dy/dt = e^{x+y}$.

31. $dx/dt = e^t y$, $dy/dt = e^t x$.

32. $dx/dt = xy + y^2$, $dy/dt = x + y$.

Projectile Trajectories

33. The 1-dimensional equation $\ddot{y} = -g$ is used to determine the motion of an object moving perpendicularly to a large attracting body and subject to no other forces. In Example 6 of the text, the range of motion is extended to a vertical plane with horizontal coordinate x. If there are no forces acting horizontally, the single equation for y is replaced by the 2-dimensional uncoupled system

$$\ddot{x} = 0, \qquad \ddot{y} = -g.$$

(a) Solve the 2-dimensional system, subject to the four initial conditions

$$x(0) = 0, \qquad y(0) = 0, \qquad \dot{x}(0) = z_0 > 0, \qquad \dot{y}(0) = w_0 > 0.$$

(b) Show that the trajectory of the solution found in part (a) follows a parabolic path.

(c) Show that the maximum height is attained when the horizontal displacement is $z_0 w_0/g$ and that the maximum height is $\frac{1}{2}w_0^2/g$.

(d) Show that the horizontal distance traversed before returning to height $y(0) = 0$ is $2z_0 w_0/g$. Show also that for a given initial speed $v_0 = \sqrt{z_0^2 + w_0^2}$, this horizontal distance is maximized by having $z_0 = w_0$.

34. A **projectile fired against air resistance** proportional to velocity satisfies the uncoupled system

$$\ddot{x} = -\frac{k}{m}\dot{x},$$
$$\ddot{y} = -\frac{k}{m}\dot{y} - g, \qquad k > 0.$$

(a) Solve the 2-dimensional system, subject to the four initial conditions

$$x(0) = 0, \qquad y(0) = 0, \qquad \dot{x}(0) = z_0 > 0, \qquad \dot{y}(0) = w_0 > 0.$$

(b) Show that the trajectory of the solution found in part (a) rises to a unique maximum at time $t_{\max} = (m/k)\ln(1 + kw_0/mg)$.

(c) Show that the position of maximum height has coordinates

$$x_{\max} = \frac{mz_0 w_0}{mg + kw_0}, \qquad y_{\max} = \frac{mw_0}{k} - \frac{m^2 g}{k^2}\ln\left(1 + \frac{kw_0}{mg}\right).$$

(d) Show that as k tends to zero the maximum height tends to $\frac{1}{2}w_0^2/g$.

35. Suppose you want to kick a ball over an h-foot-high vertical fence a feet away from you in such a way that the ball just barely clears the fence and lands b feet away from it on the other side. Assuming no air resistance, what should the angle of elevation θ and initial speed v_0 of the kick be?

36. Find the general solution of the uncoupled system $\ddot{x} + (k/m)\dot{x} = 0$, $\ddot{y} + (k/m)\dot{y} = -g$ of Example 6 of the text. Then use the data from the part (b) caption of Figure 5.4 to find a particular solution.

***37. Big Bertha.** During World War I Paris was bombarded by a gun located 75 miles away using shells that arrived in Paris 186 seconds after firing. Air resistance was negligible during a substantial part of the 25-mile-high trajectory. Using the gun's maximum powder charge, a shell could attain a muzzle velocity as high as 5500 feet per second. Since the shells had to be relatively small to attain the necessarily high muzzle velocity, they didn't cause an enormous amount of physical damage, but the Germans hoped that the considerable psychological damage would make the war-weary French capitulate.

(a) Neglecting air resistance altogether, and using $g = 32$ feet per second squared, formulate a pair of second-order differential equations for the coordinates x and y of a shell's trajectory.

(b) Use part (a) to estimate the gun's elevation angle θ, the muzzle velocity v_0, and the maximum trajectory height.

(c) Assume a retarding force of magnitude equal to a function $a(y) > 0$ of altitude y above ground times projectile speed $v = \sqrt{\dot{x}^2 + \dot{y}^2}$.

Formulate a pair of differential equations satisfied by $x(t)$ and $y(t)$, denoting the mass of a shell by m.

(d) Solve the equations in part (c) assuming that $a = a(y)$ is constant, muzzle velocity is $v_0 > 0$, and elevation angle θ satisfies $0 < \theta < \pi/2$.

38. A target is suspended over level ground at height h and is released to fall with acceleration $-g$. At the same time as the release, a gun aimed directly at the suspended target is fired from a point at ground level a feet from a point directly below the suspended target. Assuming negligible air resistance, show that the bullet will hit the target if and only if the initial speed v_0 of the bullet satisfies $2v_0^2 h > g(a^2 + h^2)$.

39. Here is an outline of a derivation of the **pendulum equation** $\ddot{\theta} = -(g/l)\sin\theta$ using a system of differential equations.

(a) Show that if x and y are rectangular coordinates and θ is the angle formed by the vector (x, y), measured counterclockwise from the downward vertical direction, then $x = l\sin\theta$ and $y = -l\cos\theta$. [*Hint:* These equations are a slight modification of the usual polar coordinate relations.]

(b) Show that $\ddot{x} = -l\sin\theta\dot{\theta}^2 + l\cos\theta\ddot{\theta}$ and $\ddot{y} = l\cos\theta\dot{\theta}^2 + l\sin\theta\ddot{\theta}$.

(c) Use the representation $\ddot{x} = 0$, $\ddot{y} = -g$ for the coordinates of the acceleration of gravity, together with the result of part (b), to derive the pendulum equation. (This derivation safely ignores the lengthwise, or radial, force on the pendulum, since that force is always perpendicular to the path of motion.) [*Hint:* Eliminate terms containing $\dot{\theta}^2$.]

***40.** **Pumping on a swing.** A playground swing with someone "pumping" on it can be looked at as a pendulum of varying length, because pumping up and down has the effect of alternately raising and lowering the center of mass, thus changing the effective length of the pendulum. Here is an outline of the derivation of the differential equation satisfied by $\theta = \theta(t)$, the angle the swing makes with the vertical direction, where $l = l(t)$ is the effective length of the swing at time t. The Java applet SWING at website <http://math.dartmouth.edu/~rewn/> plots trajectories of a pumped swing using the differential equation for θ derived in part (b).

(a) Let $x = l\sin\theta$ and $y = -l\cos\theta$ be the rectangular coordinates of the center of mass, with pivot at $x = 0$ and $y = 0$. Use the chain rule to derive the two equations

$$\ddot{x} = l(\cos\theta)\ddot{\theta} - l(\sin\theta)\dot{\theta}^2 + 2\dot{\theta}\dot{l}\cos\theta + \ddot{l}\sin\theta,$$

$$\ddot{y} = l(\sin\theta)\ddot{\theta} + l(\cos\theta)\dot{\theta}^2 + 2\dot{\theta}\dot{l}\sin\theta - \ddot{l}\cos\theta.$$

(b) By eliminating $\ddot{l}$ and then $\ddot{\theta}$ between the two relations $\ddot{x} = 0$ and $\ddot{y} = -g$, derive the equation

$$\ddot{\theta} = -\frac{g}{l}\sin\theta - \frac{2\dot{\theta}\dot{l}}{l},$$

which governs the angular motion of the swing.

(c) The effective length $l(t)$ can be controlled through the swinger's application of radial force to alternately shift the center of mass of this variable "pendulum." The length $l(t)$, controlled in this way, then determines a second-order equation for $\theta = \theta(t)$; find this nonlinear differential equation if $l(t) = 10 + \sin t$.

*41. A **spherical pendulum** is one that pivots freely about a fixed point in 3-dimensional space. (Note that the end of the pendulum moves on a sphere centered at the pivot.) Let l be the effective length of the pendulum, let θ be the angle the pendulum makes with the downward vertical direction, and let ϕ be the angle a vertical plane containing the pendulum makes with a positive horizontal axis through the pivot. The purpose of this exercise is to show that the angles θ and ϕ satisfy the following system of differential equations:

$$\ddot{\theta} = -\frac{g}{l}\sin\theta + \dot{\phi}^2 \sin\theta\cos\theta,$$

$$\ddot{\phi} = -2\dot{\theta}\dot{\phi}\cot\theta, \qquad \theta \neq k\pi,\ k \text{ integer}.$$

The restriction $\theta \neq k\pi$ has a physical significance: if the pendulum is ever in a vertical position ($\theta = k\pi$), then the motion is confined to a vertical plane and is governed by the single equation to which the system reduces if ϕ is constant, namely $\ddot{\theta} = -(g/l)\sin\theta$. The Java applet SPHEREPEND at website <http://math.dartmouth.edu/~rewn/> plots trajectories of a spherical pendulum bob, based on the equation derived here.

(a) Show that if x, y, and z are rectangular coordinates of the center of mass of the pendulum, then

$$x = l\sin\theta\cos\phi,$$

$$y = l\sin\theta\sin\phi,$$

$$z = -l\cos\theta.$$

(b) The gravitational acceleration vector has coordinates $\ddot{x} = 0$, $\ddot{y} = 0$, and $\ddot{z} = -g$. Show that

$$\sin\phi\ddot{x} - \cos\phi\ddot{y} = -l\ddot{\phi}\sin\theta - 2l\dot{\theta}\dot{\phi}\cos\theta = 0,$$

and use this to derive the second equation of the spherical pendulum system.

(c) Show that these two equations are valid:

$$\cos\phi\ddot{x} + \sin\phi\ddot{y} = l\ddot{\theta}\cos\theta - l\dot{\theta}^2\sin\theta - l\dot{\phi}^2\sin\theta = 0,$$

$$\ddot{z} = l\ddot{\theta}\sin\theta + l\dot{\theta}^2\cos\theta = -g.$$

Then derive the first equation of the spherical pendulum system from these two equations.

42. **Heat exchange.** The temperatures $u(t) \geq v(t)$ of two bodies in thermal contact with each other may be governed for the warmer body by Newton's law of cooling and for the cooler body by the analogous heating law:

$$\frac{du}{dt} = -p(u - v), \qquad \frac{dv}{dt} = q(u - v),$$

where p and q are positive constants and the equations are subject to initial conditions $u(0) = u_0,\ v(0) = v_0$. [Note that p may be different from q if the two bodies have different capacities to absorb heat.] This is a coupled system, but because of its simple form it can easily be solved as follows.

(a) Show that $q\,du/dt + p\,dv/dt = 0$. Then integrate with respect to t to show that $qu(t) + pv(t) = c_0$, where $c_0 = qu_0 + pv_0$.

(b) Use the relation between u and v derived in part (a) together with the given system to derive a single differential equation satisfied by $u(t)$. Then solve this equation using the initial conditions.

(c) Find a formula for $v(t)$.

(d) Find out how long it takes for the initial temperature difference between the two bodies to be cut in half.

43. **Bugs in mutual pursuit.** Four identical bugs are on a flat table, each moving at the same constant speed v. Use xy coordinates on the table, and locate bugs 1 through 4 initially in respective quadrants 1 through 4, each at one of the points $(\pm 1, \pm 1)$. Bug 1 always heads directly toward bug 2, bug 2 toward bug 3, bug 3 toward bug 4, and bug 4 toward bug 1, so their paths are mutually congruent.

(a) Use the symmetry of the paths to show that at any moment the bugs are at the corners of a square, and in particular, if bug 1 is at (x, y), then bugs 2, 3, and 4 are respectively at $(-y, x)$, $(-x, -y)$ and $(y, -x)$.

(b) Show that for bug 1 $\dot{y}/\dot{x} = (y - x)/(x + y)$ and that $\dot{x}^2 + \dot{y}^2 = v^2$.

(c) Use part (b) to show that the path $(x, y) = \big(x(t), y(t)\big)$ followed by bug 1 satisfies the system

$$\frac{dx}{dt} = \frac{-v}{\sqrt{2}}\frac{x+y}{\sqrt{x^2+y^2}}, \qquad \frac{dy}{dt} = \frac{v}{\sqrt{2}}\frac{x-y}{\sqrt{x^2+y^2}}.$$

Note. Solution formulas for bug 1, with $0 \le t \le 2/v$, are

$$x(t) = \sqrt{2}\left(1 - \frac{vt}{2}\right)\cos\left(\ln\left(1 - \frac{vt}{2}\right) + \frac{\pi}{4}\right),$$

$$y(t) = \sqrt{2}\left(1 - \frac{vt}{2}\right)\sin\left[\ln\left(1 - \frac{vt}{2}\right) + \frac{\pi}{4}\right].$$

Another description of the trajectories is obtained in Exercise 12 of Chapter 1, Section 3C.

44. An airplane flies horizontally with constant airspeed v, pointing always at a fixed nondirectional radio beacon at position $\mathbf{x} = 0$. Wind with constant velocity vector $\mathbf{w}$ is blowing horizontally.

(a) Show that if $\mathbf{x}(t)$ is the airplane's position at time t, then $\dot{\mathbf{x}} = -(v/|\mathbf{x}|)\mathbf{x} + \mathbf{w}$.

(b) If $\mathbf{x} = (x, y)$ and $\mathbf{w} = (w_1, w_2)$ show that the system is written

$$\dot{x} = \frac{-vx}{\sqrt{x^2+y^2}} + w_1, \qquad \dot{y} = \frac{-vy}{\sqrt{x^2+y^2}} + w_2.$$

(c) See Exercise 7(b) in Chapter 6, Section 5, for numerical solutions.

*45. **Administering a drug.** With reference to Example 3 of Section 1A:

(a) Solve the system

$$\frac{dx}{dt} = -ax + f(t), \qquad x(0) = 0,$$
$$\frac{dy}{dt} = ax - by, \qquad y(0) = 0,$$

under the assumption that the rate constants satisfy $b > a > 0$ and that $f(t) = 1 - \cos \gamma t$, γ a positive constant.

(b) For $0 \le t \le 15$, sketch the graphs of $f(t)$, $x(t)$, and $y(t)$, assuming $a = 1$, $b = 2$, and $\gamma = \frac{1}{2}$.

*46. A version of the drug delivery equations in text Example 3 is

$$\frac{dx}{dt} = -cx + f(t), \qquad x(0) = 0,$$
$$\frac{dy}{dt} = ax - by, \qquad y(0) = 0,$$

under the assumption that the rate constants satisfy $c > b > a > 0$ and that $f(t) = 1 - \cos \gamma t$, γ a positive constant. In this model some portion of the drug is eliminated from the digestive tract without passing into the circulatory system.

(a) Solve the system under the assumption that $f(t) = 1 - \cos \gamma t$, γ a positive constant.

(b) For $0 \le t \le 15$, sketch the graphs of $f(t)$, $x(t)$, and $y(t)$, assuming $a = 1$, $b = 2$, $c = 3$, and $\gamma = \frac{1}{2}$.

47. **Planetary orbits**

(a) Find the relation between a and ω that enables $x = a \cos \omega t$, $y = a \sin \omega t$ to satisfy the orbit equations in Example 14 of the text.

(b) Find, in terms of a and $G(m_1 + m_2)$, four initial conditions on $x(t)$ and $y(t)$ at $t = 0$ that determine the circular orbit of radius a.

1D Equilibrium Solutions

The stability of equilibrium for a given differential equation may be very vulnerable to slight perturbations of its associated initial conditions. See Chapter 6, Section 6, and Chapter 7, Section 4.

There are special solutions of autonomous systems that seem rather unimportant at first sight but that turn out to be quite significant in practice; these are the constant solutions of the autonomous vector equation

$$\frac{d\mathbf{x}}{dt} = \mathbf{F}(\mathbf{x}).$$

A solution $\mathbf{x}(t) = \mathbf{x}_0$ that is constant for all t is called an **equilibrium solution** because it is invariant for all time.

The trajectory of a constant equilibrium solution is always the single point in state space representing its constant value; once we have identified that point, the trajectory has in principle been completely determined. It is the behavior of other trajectories near the equilibrium point that then becomes the main focus of attention. The direct approach to finding the equilibrium solutions of the system $d\mathbf{x}/dt = \mathbf{F}(\mathbf{x})$ is to note that a constant solution $\mathbf{x}(t) = \mathbf{x}_0$ necessarily satisfies $d\mathbf{x}/dt = 0$ for all t in the domain of $\mathbf{x}(t)$. It follows that the equilibrium solutions are just the constant solutions $\mathbf{x} = \mathbf{x}_0$ of the equation $\mathbf{F}(\mathbf{x}) = 0$. Note that this last equation is not a differential equation; in practice, it may be a purely algebraic equation, although there is no guarantee that it will be easy to solve.

---

example **18**

If the system

$$\begin{aligned}\dot{x} &= 2x + y,\\ \dot{y} &= x + y + 1\end{aligned}$$

is to have constant solutions $x(t)$ and $y(t)$, then for such solutions we must have zero derivatives $\dot{x} = \dot{y} = 0$ identically in t. It follows that an equilibrium solution must satisfy the purely algebraic system

$$\begin{aligned}2x + y &= 0,\\ x + y + 1 &= 0.\end{aligned}$$

Subtract the first equation from the second to get $x = 1$, from which it follows that $y = -2$. Thus there is just one equilibrium solution: $(x, y) = (1, -2)$.

---

example **19**

The second-order equation $\ddot{x} = -(g/l)\sin x$ governs the motion of an undamped pendulum of length l subject to gravitational acceleration of magnitude g. In this equation, $x = x(t)$ is the angle the pendulum makes with a vertical line directed down from the pendulum's fixed pivot point. Letting $y = \dot{x}$, so that $\dot{y} = \ddot{x}$, we arrive at the equivalent first-order system

$$\begin{aligned}\dot{x} &= y,\\ \dot{y} &= -\frac{g}{l}\sin x.\end{aligned}$$

This reduction to a first-order system is necessary, at least in principle, so that the role of $\dot{x}$ will be clearly displayed. [Note that $\dot{x}$ doesn't occur explicitly in $\ddot{x} = -(g/l)\sin x$.] Setting $\dot{x} = \dot{y} = 0$, we conclude that the equilibrium solutions in the state space must satisfy $\sin x = 0$ and $y = 0$. Thus the infinitely many equilibrium points are located at $(x, y) = (k\pi, 0)$, where k is an integer, positive, negative, or zero. Even values of k correspond to motionless downward vertical positions of the pendulum. Odd values of k correspond to upward vertical positions that are very precariously balanced.

---

example **20**

The system

$$\begin{aligned}\dot{x} &= xy,\\ \dot{y} &= x - y - 1,\end{aligned}$$

has equilibrium solutions that must satisfy,

$$xy = 0,$$
$$x - y - 1 = 0.$$

The first equation is satisfied by taking $x = 0$, so $y = -1$ follows from the second equation. Taking $y = 0$ in the first equation makes $x = 1$ in the second. Hence the equilibrium solutions are $(x, y) = (0, -1)$ and $(x, y) = (1, 0)$.

EXERCISES

Find all equilibrium solutions of the following systems.

1. $dx/dt = 2x - 3y + 1$,
   $dy/dt = x + y$.

2. $dx/dt = x(y - 1)$,
   $dy/dt = (x - 1)y$.

3. $dx/dt = x - 2xy$,
   $dy/dt = tx$.

4. $dx/dt = x \sin y$,
   $dy/dt = y \sin x$.

5. $dx/dt = x - 2xy$,
   $dy/dt = x - z$,
   $dz/dt = x + y$.

6. $dx/dt = 1 - e^{(x-y)}$,
   $dy/dt = \sin(x - y)$,
   $dz/dt = xyz$.

7. The system

$$\dot{x} = 1 - x^2 - y^2,$$
$$\dot{y} = x + cy,$$

   has two different equilibrium solutions for each value of the constant c; what are they, and on what curve in state space do they all lie?

8. Show that the general system of linear constant-coefficient equations

$$\dot{x} = ax + by + h, \qquad \dot{y} = cx + dy + k,$$

   has either a unique equilibrium solution, or an entire line full of them, or else none at all.

9. Find the equilibrium solutions of the **Lotka-Volterra system**

$$\frac{dH}{dt} = (a - bP)H, \qquad \frac{dP}{dt} = (cH - d)P, \qquad a, b, c, d \text{ positive constants.}$$

   This system models the behavior of interacting host H and parasite P population sizes.

10. The Lotka-Volterra equations can be refined to take account of fixed limits $L > H(t)$ and $M > P(t)$ to the growth of the host and parasite populations as follows:

$$\frac{dH}{dt} = (a - bP)H(L - H), \qquad \frac{dP}{dt} = (cH - d)P(M - P),$$

   where a, b, c, and d are positive constants, with $L > d/c > 1$ and $M > a/b > 1$. Find all equilibrium solutions. The Lotka-Volterra system is discussed in detail in Chapter 6, Section 2.

11. Find all equilibrium solutions of the system of nonlinear equations

$$\dot{x} = x(3 - y), \qquad \dot{y} = y(x + z - 3), \quad \dot{z} = z(2 - y).$$

12. In the **Lorenz system** $\dot{x} = \sigma(y - x)$, $\dot{y} = -xz + \rho x - y$, $\dot{z} = xy - \beta z$, the constants β, ρ, and σ are assumed to be positive. Show that the only equilibrium solutions are $(0,0,0)$ and, if $\rho > 1$, $(\pm\sqrt{\beta(\rho - 1)}, \pm\sqrt{\beta(\rho - 1)}, \rho - 1)$.

13. The position $\mathbf{x} = (x, y)$ and velocity $\mathbf{v} = (z, w)$ vectors of a single planet orbiting a fixed sun are shown in Chapter 6, Section 4, to obey the four-dimensional system

$$\dot{x} = z, \qquad \dot{y} = w, \qquad \dot{z} = \frac{-kx}{(x^2 + y^2)^{3/2}}, \qquad \dot{w} = \frac{-ky}{(x^2 + y^2)^{3/2}},$$

where k is a positive constant. Does the system have equilibrium solutions (x_0, y_0, z_0, w_0)?

The systems in Exercises 14 and 15 have curves or surfaces of which every point is an equilibrium solution. Identify all such solutions and sketch the relevant curves or surfaces.

14. $\dot{x} = (1 - x^2 - y^2)(x + y)$, $\dot{y} = (1 - x^2 - y^2)(x^2 - y^2)$.

15. $\dot{x} = y(z - x - y)$, $\dot{y} = y(x - y)$, $\dot{z} = x^2y - y^3$.

16. The 1-dimensional linear equation $\dot{x} = tx$ is **equivalent** to the 2-dimensional system of nonlinear autonomous equations $\dot{x} = xy$, $\dot{y} = 1$ in the sense that knowing all solutions of one system allows us to find all solutions of the other.

(a) Find all solutions of $\dot{x} = tx$. What are the constant solutions, if any?

(b) Find all solutions of the nonlinear system $\dot{x} = xy$, $\dot{y} = 1$. Show that this system has no constant solutions. Thus the single equation has an equilibrium solution but the associated autonomous system does not. Note that the given equation is linear but the system isn't.

In Exercises 17 to 19, letting $x_{n+1} = t$ and incorporating the single real equation $\dot{x}_{n+1} = 1$ into an n-dimensional nonautonomous vector system $\dot{\mathbf{x}} = \mathbf{F}(\mathbf{x}, t)$ produces an $n + 1$-dimensional autonomous system $\dot{\mathbf{y}} = \mathbf{F}(\mathbf{y})$, where $\mathbf{y} = (\mathbf{x}, x_{n+1}) = (x_1, x_2, \ldots, x_n, x_{n+1})$. Find such an autonomous system for

17. $\dot{y} = \sin(ty)$.

18. $\dot{x} = x + y + t$, $\dot{y} = x - y + t^2$.

19. $\dot{x} = ty$, $\dot{y} = tz$, $\dot{z} = tx$.

1E Existence, Uniqueness, and Flows†

This section gives a precise description of what a dynamical system is in the mathematical sense. We begin by noting that reduction of a system to first-order normal form isn't always possible, but when it is, and when the vector field is continuously differentiable, we can apply the following theorem that establishes not only existence and uniqueness of solutions, but also continuous differentiability of the flow lines. For an illuminating discussion of the theorem see Chapter 2 of V. I. Arnold's *Ordinary Differential Equations*, Springer (1991).

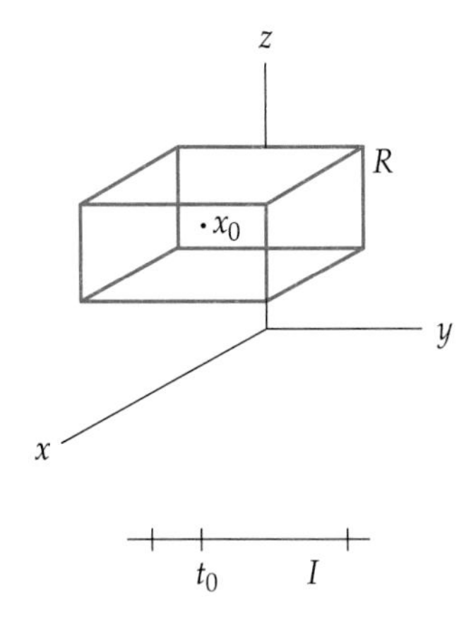

1.1 EXISTENCE AND UNIQUENESS THEOREM

Assume $\mathbf{F}(t, \mathbf{x})$ and its first-order vector partial derivatives with respect to the coordinate variables x_k of $\mathbf{x}$ are continuous for all t in an interval I containing t_0 and for all $\mathbf{x}$ in an open rectangle R in $\mathcal{R}^n$ containing $\mathbf{x}_0$. Then the initial-value problem $\dot{\mathbf{x}} = \mathbf{F}(t, \mathbf{x})$, $\mathbf{x}(t_0) = \mathbf{x}_0$ has a unique solution on some subinterval J of I containing t_0. Furthermore, the value $\mathbf{x}(t, \mathbf{x}_0)$ of this solution is a continuously differentiable function not only of t but also the initial point $\mathbf{x}_0$. If in addition the values of the first-order partial derivatives $\mathbf{F}_{x_k}(t,\mathbf{x})$ are confined to a finite interval for all $\mathbf{x}$ in $\mathcal{R}^n$, then the above conclusions are valid for t in the entire interval I.

The case $n = 3$ is illustrated in the figure, with $\mathbf{x} = (x, y, z)$ and assuming $\mathbf{x}_0 = (a_0, b_0, c_0)$. The partial derivatives of $\mathbf{F} = (F_1, F_2, F_3)$ are then calculated with respect to x, y, and z.

For an autonomous system $\dot{\mathbf{x}} = \mathbf{F}(\mathbf{x})$ satisfying the hypotheses of the theorem, the uniqueness part of the theorem implies that two distinct solution trajectories can't have an isolated point $\mathbf{x}_0$ in common, as stated in Corollary 1.2 below; in particular, trajectories can't cross or be tangent at an isolated point $\mathbf{x}_0$.

1.2 COROLLARY

If an autonomous system $\dot{\mathbf{x}} = \mathbf{F}(\mathbf{x})$ satisfies the conditions of Theorem 1.1, and two solution trajectories of the system have a point $\mathbf{x}_0$ in common, then the two trajectories coincide on either side of $\mathbf{x}_0$.

Proof. If two trajectories agree at $\mathbf{x}_0$, this common value taken as initial value for each one dictates that the trajectories are the same over some time interval J. ■

example 21 You can verify by substitution that the system $\dot{x} = -y$, $\dot{y} = x$ has circular trajectories of radius $\sqrt{u^2 + v^2}$ traced by $x(t) = u\cos t - v\sin t$, $y(t) = u\sin t + v\cos t$ and satisfying $x(0) = u$, $y(0) = v$. These are shown in Figure 5.7(a), and the picture shows that distinct trajectories fail to intersect, but are traced repeatedly if t is not restricted.

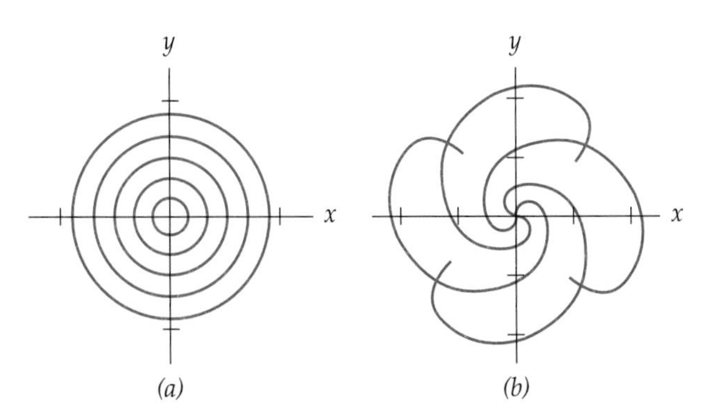

Figure 5.7 (*a*) Autonomous system trajectories.
(*b*) Nonautonomous system trajectories.

example 22

Figure 5.7(b) shows some computer plots of trajectories for the *nonautonomous* system $\dot{x} = (1-t)x - ty$, $\dot{y} = tx + (1-t)y$. Each one of the four trajectories is shown crossing one of the others.

Remark. Comparison with the statement of the Existence and Uniqueness Theorem 2.1 of Chapter 1, Section 2C, raises the question of why we restrict ourselves in Corollary 1.2 to autonomous equations. The reason is that in Chapter 1 we were talking about the *graph* of a solution, while here we are looking at the *trajectory* of a solution, which is a customary object of attention for the solution of a system. If Corollary 1.2 were stated in terms of solution graphs, we could admit nonautonomous equations in the statement. One trajectory of a nonautonomous system may very well cross another, or even intersect itself at a nonzero angle, because in arriving at the same point in state space at different times, the direction of the vector field may turn out to be different. Snapshots of the vector field of the system in Example 22 are shown in Figure 5.6. Note, however, that graphs of different solutions in the txy space will have no points in common, because the first coordinate t will be different at each distinct point of the graph. This idea can be used to show how Corollary 1.2, which applies directly to autonomous systems, can be applied to time-dependent systems. See Exercise 12 for the details.

Flows. A trajectory of an autonomous system $\dot{\mathbf{x}} = \mathbf{F}(\mathbf{x})$ we earlier called a **flow line** of the vector field $\mathbf{F}$, and we can think of their paths, as shown for example in Figure 5.7(a), as the possible paths followed by fluid particles in a steady fluid flow with velocity vector $\mathbf{F}(\mathbf{x})$ at $\mathbf{x}$. In what follows, we'll assume that the autonomous vector field $\mathbf{F}$ satisfies the conditions of Theorem 1.1 in some region B in $\mathcal{R}^n$, thus guaranteeing (i) that there is a unique flow line through each $\mathbf{x}$ in B and (ii) that distinct flow lines have no points in common. We associate with each such vector field $\mathbf{F}$ a family of **flow transformations** T_t from B to B defined by

1.3 $T_t(\mathbf{x}) = \mathbf{y}(t)$, where $\mathbf{y}(t)$ solves $\dot{\mathbf{x}} = \mathbf{F}(\mathbf{x})$ with initial value $\mathbf{y}(0) = \mathbf{x}$.

In words, $T_t(\mathbf{x})$ is the point on the flow line of $\mathbf{F}$ starting at $\mathbf{x}$ that is reached after time t. Thus a family $\{T_t\}$ is a formal representation of a continuous-time **dynamical system** described less formally at the beginning of the Introductory Survey of the entire book as the determination of possible time-evolutions of states of a system. The families $\{T_t\}$ are important in applications, because they, rather than systems of equations, are ultimately what we use to model real-world phenomena.

example 23

The system $\dot{x} = -y$, $\dot{y} = x$ has circular trajectories, as in Figure 5.7(a). Thus a flow line of radius A for the vector field $\mathbf{F}(x,y) = (-y,x)$ can be parametrized by $x(t) = A\cos(t+\alpha)$, $y(t) = A\sin(t+\alpha)$. To start one of these flow lines at a fixed point (u,v) when $t=0$, we note that $A = \sqrt{u^2+v^2}$ and write

$$T_t(u,v) = \left(\sqrt{u^2+v^2}\cos(t+\alpha),\ \sqrt{u^2+v^2}\sin(t+\alpha)\right),$$

where α is an angle that the radius from the origin to (u,v) makes with the positive x axis. If (u,v) is in the first quadrant, then $\alpha(u,v) = \arctan(v/u)$ if

$u \neq 0$, and $\alpha(0,v)$ is $\pi/2$ if $v \neq 0$; $\alpha(0,0)$ is not defined. Figure 5.8 illustrates these ideas.

The function $\phi: \mathcal{R}^3 \to \mathcal{R}^2$ defined by $\phi(t,u,v) = T_t(u,v)$ in the previous example is obviously, for each fixed (u,v), not one-to-one as a function of t unless t is somehow restricted. However, for fixed $t = t_0$, T_{t_0} turns out not only to be one-to-one but to have a nice inverse, namely T_{-t_0}. Indeed, it's a straightforward exercise to show that T_{t_0} is just a rotation about the origin through angle t_0. Hence the inverse of T_{t_0} is T_{-t_0}, which is a rotation through angle $-t_0$. This simple relationship between T_t and its inverse holds very generally, as described below.

The transformations T_t introduced above have the **composition property:**

1.4 $\qquad T_t T_s = T_{t+s}$, in other words, $T_t\big(T_s(\mathbf{x})\big) = T_{t+s}(\mathbf{x})$,

whenever all three transformations are defined. Equation 1.4 holds because the system

$$\frac{d\mathbf{x}}{dt} = \mathbf{F}(\mathbf{x}),$$

of which $\mathbf{y}(t) = T_t(\mathbf{x})$ is a solution, has a unique solution starting at $T_t(\mathbf{x}_0)$ whose value at s time units later must coincide with the unique solution value achieved by starting at $\mathbf{x}_0$ and running for time $t + s$. Note that t and s can be negative in Equation 1.4, since they can decrease from zero. Another interpretation for negative time-values is based on the *reversed system,*

$$\frac{d\mathbf{x}}{dt} = -\mathbf{F}(\mathbf{x}),$$

which has solution trajectories traced in the direction $-\mathbf{F}(\mathbf{x})$ exactly opposite to that of the solutions of $d\mathbf{x}/dt = \mathbf{F}(\mathbf{x})$. We can use solutions of the reversed system to define T_t for $t < 0$ by $T_t(\mathbf{x}_0) = \mathbf{z}(t)$, where $\mathbf{z}(t)$ satisfies

$$\frac{d\mathbf{z}}{dt} = -\mathbf{F}(\mathbf{z}), \qquad \mathbf{z}(0) = \mathbf{x}_0.$$

It follows that each of T_{-t} and T_t is an **inverse operator** to the other, that is

1.5 $$T_{-t}T_t = T_tT_{-t} = I,$$

where I is an identity operator that leaves points fixed.

---

example 24 The circular flow defined by

$$T_t(u,v) = \left(\sqrt{u^2+v^2}\cos(t+\alpha),\ \sqrt{u^2+v^2}\sin(t+\alpha)\right), \qquad \alpha = \arctan(v/u),$$

is discussed in Example 23. While the parameter t is customarily interpreted as the time it takes for point $\mathbf{x}$ on a flow line to move to $T_t(\mathbf{x})$, in this special example t is also the angle subtended at the origin from $\mathbf{x}$ to $T_t(\mathbf{x})$. Two such

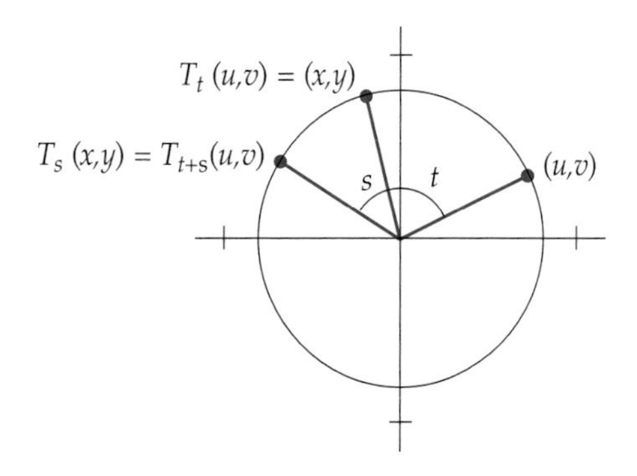

Figure 5.8 $T_{t+s}(u,v) = T_s\big(T_t(u,v)\big)$.

angles are shown in Figure 5.8, t from (u,v) to $T_t(u,v)$, and s from (x,y) to $T_s(x,y)$. If $(x,y) = T_t(u,v)$, as shown in the figure, then Equation 1.4 will hold, since both times and angles are to be added.

example 25 You can check that the uncoupled system $\dot{x} = x$, $\dot{y} = 2y$ has the solution $x(t) = ue^t$, $y = ve^{2t}$ with initial values $x(0) = u$, $y(0) = v$. The flow generated by the vector field $\mathbf{F}(x,y) = (x,2y)$ is therefore

$$T_t(u,v) = (ue^t, ve^{2t}).$$

The (x,y) coordinates of the flow line of the flow starting at (u,v) satisfy $x = ue^t$, $y = ve^{2t}$, so $u^2y = vx^2$. Thus as $t > 0$ increases the flow lines trace parabolas heading away from the origin unless u or v is zero, in which case we get segments of the axes. See Figure 5.9(a), where the initial points (u,v) shown all lie on a circle of radius $\frac{1}{2}$ centered at the origin. With decreasing negative t we get the dotted parts of the parabolas heading toward the origin but never reaching it. The origin is an equilibrium solution, since $T_t(0,0) = (0,0)$ for all t, so the origin is itself a flow line, though a trivial one.

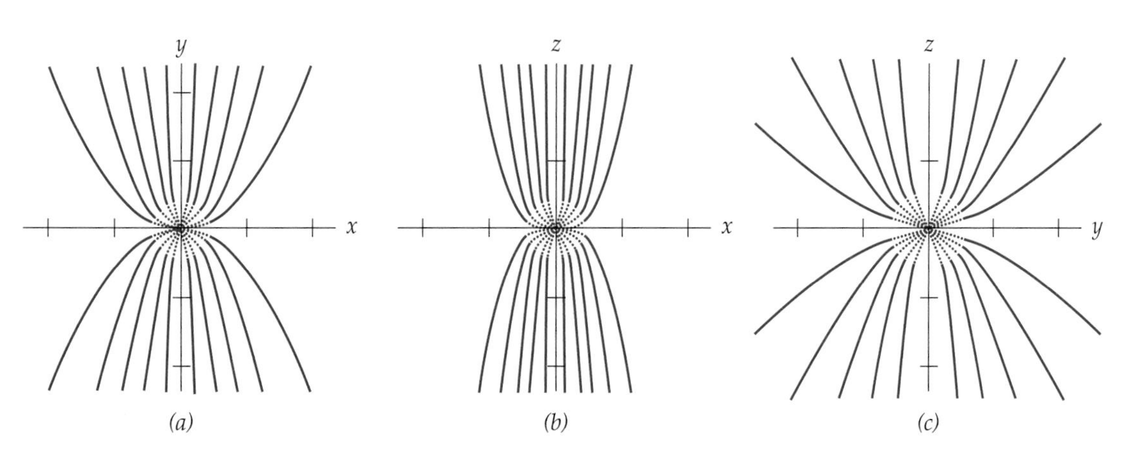

Figure 5.9 (a) $u^2y = vx^2$. (b) $u^3z = wx^3$. (c) $v^3z^2 = w^2y^3$.

example 26

The 3-dimensional system $\dot{x} = x$, $\dot{y} = 2y$, $\dot{z} = 3z$ has the solution $x(t) = ue^t$, $y = ve^{2t}$, $z(t) = we^{3t}$ with initial values $x(0) = u$, $y(0) = v$, $z(0) = w$. The flow generated by the vector field $\mathbf{G}(x,y,z) = (x,2y,3z)$ has a family $\{T_t\}$ of flow transformations such that

$$T_t(u,v,w) = (ue^t, ve^{2t}, we^{3t}).$$

An individual flow line of the flow is either one of the positive or one of the negative axes or else has the shape of a curve called a "twisted cubic." Plotting any one of these curves is not difficult, but a representative sample of them in 3-dimensional perspective looks confusing, so as an alternative Figure 5.9 shows projections of such a sample into each of the three coordinate planes of $\mathcal{R}^3$. These projections are traced away from the origin starting at distance $\frac{1}{2}$ from there. In particular Figure 5.9(a) is the same picture used to illustrate the flow lines of the field $\mathbf{F}(x,y) = (x,2y)$ in the previous example.

example 27

The 3-dimensional initial-value problem

$$\dot{x} = -z, \qquad \dot{y} = \tfrac{1}{2}, \qquad \dot{z} = x, \qquad x(0) = u, \qquad y(0) = v, \qquad z(0) = w$$

has the solution $x(t) = u\cos t - w\sin t$, $y = \frac{1}{2}t + v$, $z(t) = w\cos t + u\sin t$. The flow ψ generated by the vector field $\mathbf{G}(x,y,z) = (-z,\frac{1}{2},x)$ has a family $\{T_t\}$ of flow transformations such that

$$T_t(u,v,w) = (u\cos t - w\sin t,\ \tfrac{1}{2}t + v,\ w\cos t + u\sin t).$$

The second coordinate, $y = \frac{1}{2}t + v$, makes the flow move in the general direction of the positive y axis. The x and z coordinates satisfy

$$x^2 + z^2 = (u\cos t - w\sin t)^2 + (w\cos t + u\sin t)^2 = u^2 + w^2.$$

Hence the projection on the xz plane of a flow line through (u,v,w) is a circle of radius $\sqrt{u^2 + v^2}$ centered at the origin. It follows that each flow line is a helix, a type of spiral curve, traced as t increases in the direction of the positive y axis. See Figure 5.10.

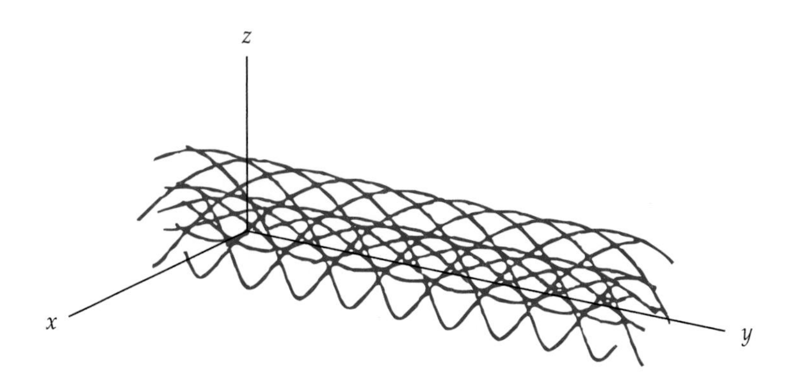

Figure 5.10 Helical flow lines.

EXERCISES

1. The domain of t values for which the solution of even an autonomous system is valid may be quite restricted. Illustrate this point by deriving the explicit solution to the 1-dimensional initial-value problem $\dot{x} = ax^2$, $x(0) = 1$, where a is a positive constant.

2. Theorem 1.1 on existence and uniqueness of solutions fails to apply everywhere to the 1-dimensional equation defined by $\dot{x} = \sqrt{x}$, $x \geq 0$ and $\dot{x} = 0$, $x < 0$.

   (a) Explain why the theorem doesn't apply if the initial condition is $x(0) = 0$.

   (b) Find two distinct solutions to the equation, both satisfying $x(0) = 0$.

3. The flow transformation of the vector field $\mathbf{F}(x, y) = (x, 2y)$ in Example 25 of the text is $T_t(u, v) = (ue^t, ve^{2t})$. Verify directly for this example that $T_{t+s}(u, v) = T_s\big(T_t(u, v)\big)$ for all t and all (u, v).

4. Consider the system $\dot{x} = y$, $\dot{y} = x$.

   (a) Verify that the system's flow is

$$T_t(u, v) = \left(\tfrac{1}{2}(u + v)e^t + \tfrac{1}{2}(u - v)e^{-t},\ \tfrac{1}{2}(u + v)e^t - \tfrac{1}{2}(u - v)e^{-t}\right).$$

   (b) Show that the flow line through (u, v) lies on a straight line if $|u| = |v|$ and otherwise traces one branch of a hyperbola as t varies.

   (c) Verify directly that $T_{t+s}(u, v) = T_s\big(T_t(u, v)\big)$ for all t and (u, v).

5. The flow transformation of the vector field $\mathbf{F}(x, y) = (-y, x)$ in Example 23 of the text is $T_t(u, v) = \big(u \cos t - v \sin t,\ u \sin t + v \cos t\big)$. Verify by direct substitution for this example that $T_{t+s}(u, v) = T_s\big(T_t(u, v)\big)$ for all t and (u, v).

6. What are the flow "lines" of the identically zero vector field on $\mathcal{R}^2$?

Hamiltonian Systems

Hamiltonian systems are discussed in more detail in Chapter 6, Section 6C, on stability.

7. A 2-dimensional **Hamiltonian system** has the form $\dot{x} = \partial H/\partial y,\ \dot{y} = -\partial H/\partial x$, where the real-valued **Hamiltonian function** $H(x, y)$ is assumed to be twice continuously differentiable.

   (a) Show that the system $\dot{x} = -y,\ \dot{y} = x$ is Hamiltonian, and find the Hamiltonian function $H(x, y)$ for this system.

   (b) Show that the flow lines of a 2-dimensional Hamiltonian system follow level curves of the associated Hamiltonian function.

8. Show that the second-order equation $\ddot{x} = -f(x)$ can be looked at as a first-order system by setting $y = \dot{x}$. Then show that the first-order system can be looked at as a Hamiltonian system, as defined in the previous exercise, with Hamiltonian

$$H(x, y) = \tfrac{1}{2}y^2 + U(x), \quad \text{where} \quad U'(x) = f(x).$$

The function $U(x)$ is the **potential energy** of the system, and $H(x,y)$ is the **total energy.**

*9. Consider the flow transformation $T_t(u,v) = \big(x(t),\, y(t)\big)$ generated by solutions of a system $\dot{x} = F(x,y)$, $\dot{y} = G(x,y)$ with initial conditions $x(0) = u$, $y(0) = v$. Assume that F and G have continuous first order partials.

(a) For fixed t let $J_t(u,v)$ be the Jacobian determinant of $T_t(u,v)$ with respect to u and v. Show that

$$\frac{d}{dt} J_t(u,v) = F_u y_v - F_v y_u + G_v x_u - G_u x_v.$$

(b) Apply the chain rule, for example $F_u = F_x x_u + F_y y_u$, to the partials of F and G in part (a) to show that $(d/dt)J_t = (F_x + G_y)J_t$, where F_x and G_y are evaluated at $\big(x(t),\, y(t)\big)$.

(c) Noting that T_0 is an identity transformation, so that $J_0 = 1$, solve the differential equation in part (b) to show that $J_t = e^{t(F_x+G_y)}$. The real-valued function $\operatorname{div}(F,G) = F_x + G_y$ is called the **divergence** of the vector field with coordinate functions F and G.

(d) Use the result of part (c) together with the change-of-variable theorem for double integrals to show that a flow transformation acting in a region R sends a set of positive area into a set of smaller area, the same area, or larger area depending on whether $\operatorname{div}(F,G)$ is negative, zero, or positive throughout R.

10. A 2-dimensional **gradient system** has the form $\dot{x} = \partial U/\partial x$, $\dot{y} = \partial U/\partial y$, where the real-valued **potential function** $U(x,y)$ is assumed to be twice continuously differentiable. Show that the flow lines of a gradient system are perpendicular to the level curves of its potential function.

*11. Consider the 2-dimensional uncoupled system $\dot{x} = x^3, \;\; \dot{y} = y^3$.

(a) Sketch the vector field of the system near the origin.

(b) For each fixed t, compute the Jacobian determinant J_t of the flow transformation T_t of the system.

(c) Let B be a region of positive area in $\mathcal{R}^2$. Use the result of part (b) to show that the area of the image of B under T_t is bigger than $A(B)$, the area of B, if $t > 0$ and less than $A(B)$ if $t < 0$.

(d) Can you draw the same conclusion as in part (c) if the original system is replaced by $\dot{x} = x^2, \;\; \dot{y} = y^2$? Explain your reasoning.

12. Corollary 1.2 is relevant to nonautonomous systems because an n-dimensional nonautonomous system $\dot{\mathbf{x}} = \mathbf{F}(\mathbf{x},t)$ can be regarded as an $n+1$-dimensional autonomous system by setting $t = x_{n+1}$ in $\mathbf{F}(\mathbf{x},t)$ and adding the equation $\dot{x}_{n+1} = 1$ to the system.

(a) Show how the system $\dot{x}_1 = tx_2, \;\; \dot{x}_2 = tx_1$ would look after this conversion.

(b) Show however that while the system given in part (a) has $(x_1, x_2) = (0,0)$ for an equilibrium solution, the augmented 3-dimensional system has no equilibrium solutions.

13. The purpose of this exercise is to show that if two twice-differentiable functions $S(t)$ and $C(t)$ solve the initial-value problem

$$S' = C, \qquad S(0) = 0,$$
$$C' = -S, \qquad C(0) = 1,$$

then S and C together satisfy the addition formulas for sine and cosine:

$$S(t+s) = S(t)C(s) + C(t)S(s),$$
$$C(t+s) = C(t)C(s) - S(t)S(s).$$

Thus the derivative formulas for sine and cosine, together with their values at zero, imply the addition formulas. Show this by setting

$$x(t) = S(t+s) - S(t)C(s) - C(t)S(s),$$
$$y(t) = C(t+s) - C(t)C(s) + S(t)S(s),$$

for fixed but arbitrary s. Check that $x(t)$ and $y(t)$ satisfy the given system with $S = x, C = y$, and initial conditions $x(0) = y(0) = 0$. Then apply the uniqueness part of Theorem 1.1 to prove the addition formulas for S and C, which must then be sine and cosine since they too solve the initial-value problem.

2 LINEAR SYSTEMS

2A Definitions

Example 1 below shows four linear systems with functions $x(t)$, $y(t)$, and $z(t)$ to be determined. To establish general properties and techniques for such systems we need some definitions. Recall that a single differential equation for a single real or complex valued function $x = x(t)$ is called a **linear differential equation** if it is a polynomial of degree 1 in x and one or more of its derivatives, with coefficients that may be functions of an independent variable t. Thus $\ddot{x} + t^2\dot{x} = \sin t$ and $e^t\ddot{x} - tx = t$ are linear, while $\ddot{x} + x^2 = 0$ and $x\ddot{x} + tx = 0$ are nonlinear. If we define a linear differential operator $P(D)$ by $P(D) = a_n D^n + a_{n-1}D^{n-1} + \cdots + a_1 D + a_0$, then a linear differential equation of order n can be written $P(D)x = f(t)$. A coefficient a_k may be constant or may be a function of the independent variable t alone.

A **linear system** that determines solutions $\mathbf{x} = (x(t), y(t), \ldots)$ is a set of equations in which each scalar-valued equation can be written in the form

$$P_1(D)x + P_2(D)y + \cdots = f(t),$$

where each polynomial operator $P_k(D)$ is a linear differential operator. If $f = 0$ in each equation of such a system, the system is called **homogeneous,** otherwise the system is **nonhomogeneous.** In applying these definitions, the written order of the terms can always be changed. (For example the equation $dx/dt = ty$

is equivalent to $dx/dt - ty = 0$.) Thus a homogeneous linear system is one for which both sides of each equation consists entirely of a sum of multiples of the unknown functions or their derivatives. The system becomes nonhomogeneous if we add a nonzero term $f(t)$ to at least one of the equations; replacement of all such terms by zero then produces the **associated homogeneous system.**

example 1

Here are some linear systems.

$$\text{(a)}\quad \frac{dx}{dt} + \frac{dy}{dt} = e^t, \qquad \frac{dx}{dt} - \frac{dy}{dt} = e^{-t}.$$

$$\text{(b)}\quad \frac{dx}{dt} - ty = \sin t, \qquad \frac{d^2y}{dt^2} + t^2 x = 0.$$

$$\text{(c)}\quad \frac{d^2x}{dt^2} + \frac{dx}{dt} + \frac{dy}{dt} = 0, \qquad x + \frac{dy}{dt} + y = 0.$$

$$\text{(d)}\quad \frac{dx}{dt} = x + y + z, \qquad \frac{dy}{dt} = x + y - z, \qquad \frac{dz}{dt} = x - y + z.$$

Only (c) and (d) are homogeneous systems. The associated homogeneous systems in (a) and (b) are obtained by replacing the exponential and sine terms by zero.

The following systems are nonlinear. The property of homogeneity isn't really significant for nonlinear systems, because the decomposition of solutions into homogeneous and particular parts isn't valid for nonlinear systems.

$$\text{(e)}\quad \frac{dx}{dt} + \left(\frac{dy}{dt}\right)^2 = t, \qquad \frac{dx}{dt} - \frac{dy}{dt} = 0.$$

$$\text{(f)}\quad \frac{dx}{dt} - ty = 0, \qquad \frac{d^2y}{dt^2} + \sin x = 0.$$

2B Elimination Method

Matrix methods for linear systems are taken up in Section 3 and also in Chapter 7.

This method is often the most direct if a system isn't too complicated. By adding nonzero multiples of one equation to another, we can sometimes rewrite a system so that it becomes easier to interpret and solve.

example 2

The sum and difference of the two equations in Example 1(a) are, respectively,

$$2\frac{dx}{dt} = e^t + e^{-t}, \qquad 2\frac{dy}{dt} = e^t - e^{-t}.$$

Integrating both equations with respect to t and dividing by 2 gives

$$x = \tfrac{1}{2}(e^t - e^{-t}) + c_1, \qquad y = \tfrac{1}{2}(e^t + e^{-t}) + c_2.$$

We can also do a part of the computation using the hyperbolic cosine and sine functions $\cosh t = (e^t + e^{-t})/2$, $\sinh t = (e^t - e^{-t})/2$. We then write the uncoupled system as

$$\frac{dx}{dt} = \cosh t, \qquad \frac{dy}{dt} = \sinh t.$$

The solutions can also be written $x(t) = \sinh t + c_1$, $y(t) = \cosh t + c_2$; this follows, for example, from using two differentiation formulas for hyperbolic functions: derivative of $\sinh t$ is $\cosh t$ and vice versa.

Differential operators are helpful if the coefficients in a linear system are constant, as the following example shows.

---

example 3

The vector equation

$$\frac{d}{dt}(x, y) = (2x + 4y + 2,\ x - y + 4)$$

represents the system

$$\frac{dx}{dt} = 2x + 4y + 2,$$
$$\frac{dy}{dt} = x - y + 4.$$

Using $D = d/dt$, we can write the system as

$$(D - 2)x - 4y = 2,$$
$$-x + (D + 1)y = 4.$$

Since $x(t)$ and $y(t)$ must be at least once differentiable, the first equation in the form $\dot{x} = 2x + 4y + 2$ shows that $x(t)$ must be twice differentiable, since $\ddot{x} = 2\dot{x} + 4\dot{y}$. A similar remark applies to $y(t)$. If D were a number, we could add multiples of one equation to another so as to eliminate either x or y and then substitute back to get the other variable. The algebra of differential operators allows us to do something similar here. Operate on the second equation with $(D - 2)$:

$$(D - 2)x - 4y = 2,$$
$$-(D - 2)x + (D - 2)(D + 1)y = (D - 2)4.$$

Noting that $(D - 2)4 = -8$, we add the first equation to the second to get

$$(D - 2)(D + 1)y - 4y = -6.$$

But this equation is the same as

$$(D^2 - D - 6)y = -6.$$

The characteristic equation is $r^2 - r - 6 = (r+2)(r-3) = 0$. The roots are $r_1 = -2$, $r_2 = 3$; so the solution of the homogeneous equation for y is

$$y_h = c_1 e^{-2t} + c_2 e^{3t}.$$

By inspection we find a particular solution $y_p = 1$, so $y(t)$ must be of the form

$$y = y_h + y_p = c_1 e^{-2t} + c_2 e^{3t} + 1.$$

Now solve the second of the original two equations for x:

$$x = (D+1)y - 4.$$

Substitution of the formula for $y(t)$ gives

$$x(t) = -c_1 e^{-2t} + 4c_2 e^{3t} - 3.$$

The vector solution is then

$$(x, y) = (-c_1 e^{-2t} + 4c_2 e^{3t} - 3,\ c_1 e^{-2t} + c_2 e^{3t} + 1).$$

Initial conditions, for example $\big(x(0), y(0)\big) = (2,1)$, require that

$$(-c_1 + 4c_2 - 3,\ c_1 + c_2 + 1) = (2,1),$$

or $-c_1 + 4c_2 = 5$, $c_1 + c_2 = 0$. Solving these equations gives $c_1 = -1$, $c_2 = 1$ for the particular solution

$$(x_p, y_p) = (e^{-2t} + 4e^{3t} - 3,\ -e^{-2t} + e^{3t} + 1).$$

2C General Form of Solutions

A step-by-step analysis of the computation in the previous example shows that we have indeed found the most general solution and that it contains exactly the right number of arbitrary constants. Extraneous constants may turn up in a solution formula when applying an extra operator increases the order of the system, and so increases the number of constants. To find relations among the extra constants just substitute the solution back into the given system. And to determine in advance what the correct number of arbitrary constants is, we appeal to the theory of Chapter 7, Sections 1A and 1D; Theorem 1.2 and Theorem 1.11 there show for constant-coefficient systems, and more generally for variable-coefficient systems, that *if the system is in first-order normal form,* then its solution contains a number of arbitrary constants equal to the number of equations in the system. Thus we have the following:

Rule of Thumb. To determine the correct number of arbitrary constants for the solution of a linear system, attempt to put the system in first-order normal form. If your attempt succeeds, the correct number of arbitrary constants will equal the number of equations in your normal-form system.

It's important to know whether a system of purely algebraic linear equations has solutions, and how many, or else is inconsistent with no solutions. Exercises 28 through 31 for this section show analogously that not every linear system is equivalent to one in normal form. In particular, a system that is not in normal form may be inconsistent, having no solutions, or may have a nonstandard number of arbitrary constants in its general solution. In the latter case, we still have recourse to substitution back into the system for finding relations among constants.

---

example 4 The second-order system

$$\frac{d^2x}{dt^2} = x + 2y + t,$$
$$\frac{d^2y}{dt^2} = 3x + 2y$$

can be written in operator form as

$$(D^2 - 1)x - 2y = t,$$
$$-3x + (D^2 - 2)y = 0.$$

If we multiply the second equation by 2 and operate on the first with $(D^2 - 2)$, then adding the resulting equations eliminates y:

$$(D^2 - 2)(D^2 - 1)x - 6x = -2t \quad \text{or} \quad (D^4 - 3D^2 - 4)x = -2t.$$

The characteristic equation is

$$r^4 - 3r^2 - 4 = (r^2 + 1)(r^2 - 4) = 0,$$

and has roots $r_1 = i, r_2 = -i, r_3 = 2$, and $r_4 = -2$. The homogeneous solution for x is then

$$x_h(t) = c_1 \cos t + c_2 \sin t + c_3 e^{2t} + c_4 e^{-2t}.$$

By inspection, we find a particular solution $x_p = \frac{1}{2}t$. Thus

$$x = x_h + x_p = c_1 \cos t + c_2 \sin t + c_3 e^{2t} + c_4 e^{-2t} + \tfrac{1}{2}t.$$

To find y we could start over again, eliminating x between the two equations and finding an expression for $y_h + y_p$ containing four additional constants c_5, c_6, c_7, and c_8. The original system written in first-order normal form contains just four equations, so there will be some relations among the constants that enable us to reduce to just four that can have arbitrarily assigned values. This could be done by substitution into the original system. However, it's easier in this example not to start the elimination process over and solve a differential equation for y alone, but to proceed as follows. Just solve the first of the given differential equations for y in terms of x, and then substitute the formula we just found for $x(t)$. A straightforward calculation gives

$$y = \tfrac{1}{2}(D^2 - 1)x - \tfrac{1}{2}t,$$

and y turns out to be

$$y = -c_1 \cos t - c_2 \sin t + \tfrac{3}{2}c_3 e^{2t} + \tfrac{3}{2}c_4 e^{-2t} - \tfrac{3}{4}t.$$

Putting this together with the formula we found above for x gives the complete solution. The four constants can be determined by imposing four initial conditions such as $x(0) = x_0,\ \dot{x}(0) = x_1,\ y(0) = y_0$, and $\dot{y}(0) = y_1$.

For both technical and conceptual reasons it's sometimes helpful to break a complicated problem into pieces, solve each piece separately, and then combine the results into a complete solution. As an example, recall the single linear equation $\ddot{y}+4y=1$. This solution procedure can be separated into (i) finding the complete solution of the homogeneous equation $\ddot{y}+4y=0$, for which $y_h = c_1 \cos 2t + c_2 \sin 2t$, and (ii) finding a single particular solution of the nonhomogeneous equation, for which we can in this instance simply guess the constant solution $y_p = \frac{1}{4}$. Because the equation is linear, the complete solution for y is then $y = y_h + y_p$. The decomposition of solutions $x(t) = x_h(t) + x_p(t)$ that holds for a 1-dimensional linear equation is valid in the vector form $\mathbf{x}(t) = \mathbf{x}_h(t) + \mathbf{x}_p(t)$ for linear systems also. The essence of the proof is simply a vector reinterpretation of the proof of Theorem 3.1 in Chapter 3, Section 3A, where we proved it for dimension 1. We'll give another proof below using coordinates.

2.1 THEOREM

If $\mathbf{x}(t) = (x(t), y(t), \dots)$ is a given solution of a nonhomogeneous linear system and $\mathbf{x}_p(t) = \big(x_p(t), y_p(t), \dots\big)$ is an arbitrary particular solution of the same system, then

$$\mathbf{x}(t) = \mathbf{x}_h(t) + \mathbf{x}_p(t) = (x_h(t) + x_p(t), y_h(t) + y_p(t), \dots)$$

for some appropriately chosen solution $\mathbf{x}_h(t) = \big(x_h(t), y_h(t), \dots\big)$ of the associated homogeneous system. Thus every solution of a linear system can be expressed as the sum of a fixed particular solution and a solution of the associated homogeneous system, so the number of arbitrary constants in the general solution of the system is the same as the number of arbitrary constants in the solution of the associated homogeneous system.

Proof. Suppose a nonhomogeneous system consists of m equations

$$\sum_{k=1}^{n} P_{jk}(D)x_k = f_j(t), \quad j = 1, \dots, m,$$

with solutions $\mathbf{x}(t) = (x_1(t), x_2(t), \dots)$ and $\mathbf{x}_p(t) = (p_1(t), p_2(t), \dots)$. Since each operator $P_{jk}(D)$ acts linearly on a real or complex function, $P_{jk}(D)(x_k - p_k) = P_{jk}(D)x_k - P_{jk}(D)p_k$. Hence

$$\begin{aligned}\sum_{k=1}^{n} P_{jk}(D)(x_k - p_k) &= \sum_{k=1}^{n} P_{jk}(D)x_k - \sum_{k=1}^{n} P_{jk}(D)p_k \\ &= f_j(t) - f_j(t) = 0, \qquad j = 1, \dots, m.\end{aligned}$$

Thus $\mathbf{x}(t) - \mathbf{x}_p(t) = \mathbf{x}_h(t)$ for some solution $\mathbf{x}_h(t)$ of the associated homogeneous system. So $\mathbf{x}(t) = \mathbf{x}_h(t) + \mathbf{x}_p(t)$. ■

The previous theorem enables us to break the solution of a linear system into homogeneous and particular parts that can be computed by first finding the homogeneous part and then a particular solution. For now, finding the particular part will involve the informed guessing of Chapter 3, although this feature can be avoided by using an integral formula described in Chapter 7, Section 3.

example 5

Here is a nonhomogeneous system and the associated homogeneous system:

$$\begin{cases} \dot{x} = -y + \cos t \\ \dot{y} = x + \sin t, \end{cases} \qquad \begin{cases} \dot{x} = -y \\ \dot{y} = x. \end{cases}$$

The homogeneous system is most easily solved by noting that $\ddot{x} = -\dot{y} = -x$, or $\ddot{x} + x = 0$; this has solutions

$$x_h = c_1 \cos t + c_2 \sin t.$$

The equation $\dot{x} = -y$ says $y = -\dot{x}$, from which we get

$$y_h = c_1 \sin t - c_2 \cos t.$$

These last two displayed equations give us the complete solution to the homogeneous system. To find a particular solution to the system, we use our experience with a single constant-coefficient equation as a guide. Multiples of $\cos t$ and $\sin t$ won't suffice, because they are already identified as solutions of the homogeneous system. So we try

$$x_p = At \cos t + Bt \sin t \quad \text{and} \quad y_p = Ct \cos t + Dt \sin t.$$

Substitute these expressions for x_p and y_p into the nonhomogeneous system to get

$$A \cos t - At \sin t + B \sin t + Bt \cos t = -Ct \cos t - Dt \sin t + \cos t,$$

$$C \cos t - Ct \sin t + D \sin t + Dt \cos t = At \cos t + Bt \sin t + \sin t.$$

Now equate coefficients of like terms in the first equation to get $A = 1$ and $B = 0$, so $x_p = t \cos t$. Similarly, from the second equation we get $C = 0$ and $D = 1$, so $x_p = t \sin t$. The complete solution is

$$x = x_h + x_p = c_1 \cos t + c_2 \sin t + t \cos t,$$

$$y = y_h + y_p = c_1 \sin t - c_2 \cos t + t \sin t.$$

Systems of the type considered in the next example occur fairly often, so it's worth noting that there is a simple formula for computing the characteristic roots.

example 6

First-order constant-coefficient homogeneous systems of the form

$$\begin{matrix} \dot{x} = ax + by \\ \dot{y} = cx + dy \end{matrix} \quad \text{can be written} \quad \begin{matrix} (D - a)x - by = 0 \\ -cx + (D - d)y = 0. \end{matrix}$$

To solve for x, operate on the first equation by $(D - d)$ and multiply the second equation by b. Adding the results gives

$$(D - d)(D - a)x - bcx = 0, \quad \text{or} \quad \big(D^2 - (a + d) + (ad - bc)\big)x = 0.$$

The characteristic roots r_1 and r_2 thus satisfy $r^2 - (a + d)r + (ad - bc) = 0$, so x has the form $x = c_1 e^{r_1 t} + c_2 e^{r_2 t}$, or $x = c_1 e^{r_1 t} + c_2 t e^{r_1 t}$ if $r_1 = r_2$. Multiplying the first equation by c and operating on the second by $(D - a)$, then adding, turns out to produce the same characteristic equation. Hence y has the same general form as x, with constants c_3 and c_4. We determine dependence of c_3 and c_4 on $c_1\, c_2$ by substitution into the given system. The common characteristic equation for x and y is easy to remember if you note that we can write it using a 2-by-2 determinant:

$$\begin{vmatrix} a - r & b \\ c & d - r \end{vmatrix} = 0.$$

The meaning of this determinant is clarified in the geometric discussion in the next section.

EXERCISES

Classify each of the systems in Exercises 1 through 6 as linear or nonlinear.

1. $dx/dt = t + x^2 + y,$
   $dy/dt = t^2 + x + y.$
2. $dy/dt = t^2 + z,$
   $dz/dt = t^3 + y.$
3. $dx/dt = t^2x - y + e^t,$
   $dy/dt = 1.$
4. $d^2x/dt^2 = tx + y.$
   $dy/dt = x + ty.$
5. $dx/dt + dz/dt = 1,$
   $dx/dt - t(dz/dt) = x.$
6. $d^2x/dt^2 + dy/dt = 0,$
   $y + t(d^2y/dt^2) = t.$

Use elimination by operator multiplication to get rid of one of the dependent variables in the systems in Exercises 7 through 10. Solve the resulting equation for the remaining variable, and then determine the general solution $(x(t), y(t))$.

7. $dx/dt = 6x + 8y,$
   $dy/dt = -4x - 6y.$
8. $dx/dt = x + 2y,$
   $dy/dt = -2x + y.$
9. $dx/dt = x + 2y,$
   $dy/dt = x + y + t.$
10. $dx/dt = -y - t,$
    $dy/dt = x + t.$
11. It is in general not possible to find closed-form solutions for linear systems with nonconstant-coefficient functions. Here is one that can nevertheless be solved fairly easily by solving a second-order equation for y. Find the general solution.

$$\dot{x} = (t^{-1} - t)x - t^2 y, \qquad t > 0,$$

$$\dot{y} = x + ty.$$

Reduce each system in Exercises 12 through 15 to normal form with a single first derivative having coefficient one on the left side.

12. $dx/dt + dy/dt = t,$
    $dx/dt - dy/dt = x.$
13. $dx/dt + dy/dt = y,$
    $dx/dt + 2dy/dt = x.$
14. $2dx/dt + dy/dt + x + 5y = t,$
    $dx/dt + dy/dt + 2x + 2y = 0.$
15. $dx/dt + dy/dt = \sin t,$
    $dx/dt - dy/dt = \cos t.$

Having put each system in Exercises 12 through 15 in normal form, solve each system, and then determine the constants in your solution so as to satisfy the corresponding initial conditions in Exercises 16 through 19.

16. $x(0) = 1$, $y(0) = -1$.

17. $x(0) = 0$, $y(0) = 5$.

18. $x(0) = -1$, $y(0) = 0$.

19. $x(\pi) = 1$, $y(\pi) = 2$.

Use elimination by operator multiplication to get rid of one of the dependent variables in the systems in Exercises 20 through 23. Solve the resulting equation for the remaining variable and then determine the general solution of the system. Substitution may be necessary to find relations among constants. Then determine the constants so that the initial conditions are satisfied.

20. $d^2x/dt^2 - x + dy/dt + y = 0$,
$dx/dt - x + d^2y/dt^2 + y = 0$, $x(0) = y(0) = 0$, $\dot{x}(0) = 0$, $\dot{y}(0) = 1$.

21. $d^2x/dt^2 - dy/dt = 0$,
$dx/dt + d^2y/dt^2 = 0$, $x(0) = 1$, $y(0) = 0$, $\dot{x}(0) = \dot{y}(0) = 0$.

22. $d^2x/dt^2 - dy/dt = t$, $dx/dt + dy/dt = x + y$, $x(0) = y(0) = 0$, $\dot{x}(0) = 1$.

23. $d^2x/dt^2 - y = e^t$,
$d^2y/dt^2 + x = 0$, $x(0) = y(0) = \dot{x}(0) = \dot{y}(0) = 0$.

24–27. By introducing new dependent variables, $u = \dot{x}$ and $v = \dot{y}$, attempt to reduce each system in Exercises 20 through 23 to first-order normal form. If the reduction fails, explain why.

None of the linear systems in Exercises 28 through 31 is equivalent to a first-order system in normal form. Discuss their solutions, or lack thereof.

28. $dx/dt + dy/dt = 0$,
$dx/dt + dy/dt = 1$.

29. $dx/dt + dy/dt = 0$,
$dx/dt + dy/dt = x$.

30. $dx/dt + dy/dt = t$,
$dx/dt + dy/dt = x$.

31. $dx/dt + dy/dt = y$,
$dx/dt + dy/dt = x$.

32. Suppose that you have used elimination on the system

$$\dot{x} + x - y = 0, \qquad x + \dot{y} + 3y = e^t$$

to find a differential equation for y alone and that you have solved that equation for $y(t)$.

(a) Explain which equation it is then better to substitute $y(t)$ back into to finish solving the system.

(b) Explain why you generate an extraneous constant of integration if you make the less good choice for your back substitution, and explain what you should do to eliminate the extra constant.

33. (a) Find a first-order system of dimension 4 equivalent to the second-order system

$$\ddot{y} - 3x - 2y = 0, \qquad \ddot{x} - y + 2x = 0.$$

(b) Write the system found in part (a) in normal form.

(c) Solve the system in part (a).

34. (a) Find a two-dimensional system of order 2 satisfied by the x and y coordinates of the solutions to $\dot{x} = z + w$, $\dot{y} = z - w$, $\dot{z} = x - y$, and $\dot{w} = x + y$. Then solve the second-order system and use its solution to solve the given system.

(b) Find a two-dimensional system of order 2 satisfied by the z and w coordinates of the solutions to the system in part (a).

35. Consider the harmonic oscillator equation $\ddot{x} + \omega^2 x = 0$, $\omega > 0$.

(a) Show that a phase portrait of the associated first-order normal form system consists of ellipses.

(b) Show that the solution graphs of the associated first-order normal form system are elliptical helices.

***36.** Consider the motion of a particle located at point (x, y) on a rotating turntable with its own (x, y) coordinates fixed to the turntable and having constant angular velocity ω about $(x, y) = (0,0)$. The particle's motion can be regarded as governed by a **centrifugal acceleration** $\omega^2(x, y)$ directed away from the center of rotation, and a **Coriolis acceleration** perpendicular to the instantaneous direction of the particle's motion and given by $2\omega(\dot{y}, -\dot{x})$. Under these assumptions, show the following, which justifies introducing the two accelerations defined above.

(a) Show that x and y satisfy the system $\ddot{x} = \omega^2 x + 2\omega\dot{y}$, $\ddot{y} = \omega^2 y - 2\omega\dot{x}$. Operators and elimination can be used to solve such systems, but the method suggested below is neater for this special system.

(b) Show that if we let z denote the complex number $x + iy$, the system in part (a) can be rewritten as the single complex equation $\ddot{z} + 2i\omega\dot{z} - \omega^2 z = 0$.

(c) Show that the equation in part (b) has complex solutions $z(t) = (c_1 + c_2 t)e^{-i\omega t}$ and hence that the initial value problem $x(0) = R$, $y(0) = 0$, $\dot{x}(0) = u$, $\dot{y}(0) = v$ for the system in part (a) has solutions in real form given by

$$x(t) = (R + ut)\cos\omega t + (v + \omega R)t\sin\omega t,$$

$$y(t) = -(R + ut)\sin\omega t + (v + \omega R)t\cos\omega t.$$

(d) Use either the complex or else the real form of the solution found in part (c) to show that the motion with respect to the turntable can be regarded as a linear motion having constant velocity vector **v**, but with a correction for the rotation of the turntable. What is the constant velocity vector **v** in terms of ω, R, u, and v?

3 MATRIX OPERATORS

The elimination method for systems described in Section 2 is a natural outgrowth of the methods of Chapter 3. However, the elimination method can be fairly cumbersome if the dimension of an equivalent normal-form system is greater than two. In addition to providing computational efficiency, using the matrix methods of this section will give us some geometric insight that enables us to see that many examples of linear systems are really equivalent to uncoupled systems. Carrying matrix methods a step further, we'll see in Chapter 7 that we can always solve the constant-coefficient normal-form systems in vector form by a method analogous to the exponential multiplier method of Chapter 1, Section 4, that works for a single 1-dimensional equation $\dot{x} = x + f(t)$.

All the facts about matrix algebra we'll use in this chapter are explained in detail in Appendix B, but we'll provide here a review of the operations that are essential for this section. Throughout the chapter we'll need matrix operations only for single-column matrices, which we interpret as vectors, and for square matrices, which we'll describe in context now. Our focus will be on homogeneous linear systems with constant coefficients in normal form, such as

$$\begin{aligned}\dot{x} &= 2x + 3y\\ \dot{y} &= 4x - y,\end{aligned} \quad \text{and} \quad \begin{aligned}\dot{x} &= -x + 2y\\ \dot{y} &= 4x - 4y - 3z\\ \dot{z} &= 2x - 2y + 3z.\end{aligned}$$

Because we'll assume the systems are in normal form and the variables are in alphabetical order, writing the coefficients on the right in matrix form determines the two systems completely:

$$\begin{pmatrix}\dot{x}\\ \dot{y}\end{pmatrix} = \begin{pmatrix}2 & 3\\ 4 & -1\end{pmatrix}\begin{pmatrix}x\\ y\end{pmatrix} \quad \text{and} \quad \begin{pmatrix}\dot{x}\\ \dot{y}\\ \dot{z}\end{pmatrix} = \begin{pmatrix}-1 & 2 & 0\\ 4 & -4 & -3\\ 2 & -2 & 3\end{pmatrix}\begin{pmatrix}x\\ y\\ z\end{pmatrix}.$$

In abbreviated form the two systems look like $\dot{\mathbf{x}} = A\mathbf{x}$ and $\dot{\mathbf{x}} = B\mathbf{x}$, where we understand that in the abbreviated systems,

$$\dot{\mathbf{x}} = \begin{pmatrix}\dot{x}\\ \dot{y}\end{pmatrix}, \quad A = \begin{pmatrix}2 & 3\\ 4 & 1\end{pmatrix}, \quad \text{and} \quad \mathbf{x} = \begin{pmatrix}x\\ y\end{pmatrix},$$

$$\dot{\mathbf{x}} = \begin{pmatrix}\dot{x}\\ \dot{y}\\ \dot{z}\end{pmatrix}, \quad B = \begin{pmatrix}-1 & 2 & 0\\ 4 & -4 & -3\\ 2 & -2 & 3\end{pmatrix}, \quad \text{and} \quad \mathbf{x} = \begin{pmatrix}x\\ y\\ z\end{pmatrix}.$$

We've used columns to write the vectors that we wrote $\mathbf{x} = (x, y)$ or $\mathbf{x} = (x, y, z)$ in Section 1 because of the way matrix products are usually defined using **row-by-column multiplication.** By this definition of matrix multiplication, the product AB of a matrix $A = (a_{ij})$ and a matrix $B = (b_{ij})$ has for its entry in the ith row and jth column the dot product of the ith row A with the jth column of B. For example, the products displayed above are exactly what are required by the systems in their original forms, since

$$\begin{pmatrix}2 & 3\\ 4 & -1\end{pmatrix}\begin{pmatrix}x\\ y\end{pmatrix} = \begin{pmatrix}2x + 3y\\ 4x - y\end{pmatrix}$$

$$\text{and} \quad \begin{pmatrix}-1 & 2 & 0\\ 4 & -4 & -3\\ 2 & -2 & 3\end{pmatrix}\begin{pmatrix}x\\ y\\ z\end{pmatrix} = \begin{pmatrix}-x + 2y\\ 4x - 4y - 3z\\ 2x - 2y + 3z\end{pmatrix}.$$

We'll also need scalar multiples and sums of matrices. If c is a scalar and A is a matrix then cA is the matrix A modified by multiplying each entry by c. Similarly the sum of matrices of the same dimensions is the matrix formed by adding corresponding entries. Thus

$$c\begin{pmatrix} x \\ y \end{pmatrix} = \begin{pmatrix} cx \\ cy \end{pmatrix} \quad \text{and} \quad \begin{pmatrix} 2 & 3 \\ 4 & 5 \end{pmatrix} + \begin{pmatrix} 2 & -3 \\ 0 & 1 \end{pmatrix} = \begin{pmatrix} 4 & 0 \\ 4 & 6 \end{pmatrix}.$$

Linear combinations of column vectors occur often, and by the definitions just given we have for example

$$c_1\begin{pmatrix} 1 \\ 2 \end{pmatrix} + c_2\begin{pmatrix} 1 \\ 3 \end{pmatrix} = \begin{pmatrix} c_1 + c_2 \\ 2c_1 + 3c_2 \end{pmatrix}.$$

A useful way to regard the product of a matrix and a column vector is to write it as linear combinations of the columns of the matrix with the vector entries as coefficients. In our first two examples, these would be respectively

$$\begin{pmatrix} 2x + 3y \\ 4x - y \end{pmatrix} = x\begin{pmatrix} 2 \\ 4 \end{pmatrix} + y\begin{pmatrix} 3 \\ -1 \end{pmatrix},$$

$$\begin{pmatrix} -x + 2y \\ 4x - 4y - 3z \\ 2x - 2y + 3z \end{pmatrix} = x\begin{pmatrix} -1 \\ 4 \\ 2 \end{pmatrix} + y\begin{pmatrix} 2 \\ -4 \\ -2 \end{pmatrix} + z\begin{pmatrix} 0 \\ -3 \\ 3 \end{pmatrix}.$$

Furthermore, matrices act linearly on linear combinations of column vectors, for example

$$A\left(c_1\begin{pmatrix} 1 \\ 2 \end{pmatrix} + c_2\begin{pmatrix} 1 \\ 3 \end{pmatrix}\right) = c_1A\begin{pmatrix} 1 \\ 2 \end{pmatrix} + c_2A\begin{pmatrix} 1 \\ 3 \end{pmatrix}.$$

A related operational rule that we use is $(A + B)\mathbf{x} = A\mathbf{x} + B\mathbf{x}$.

For some square matrices A there is a unique square matrix A^{-1} such that $AA^{-1} = A^{-1}A = I$, where I is the **identity matrix**, of the same size as A, having ones along the main diagonal from upper left to lower right, and zeros elsewhere. The matrix A^{-1} is called the **inverse** of A, and A is said to be **invertible** if it has an inverse. If A is invertible then the matrix equation $A\mathbf{x} = \mathbf{b}$ has a unique solution; just multiply by A^{-1} on the left to get $A^{-1}A\mathbf{x} = I\mathbf{x} = A^{-1}\mathbf{b}$, so $\mathbf{x} = A^{-1}\mathbf{b}$. For an invertible 2 by 2 matrix A the inverse is computed as follows, as discussed in Appendix B:

$$\text{If} \quad A = \begin{pmatrix} a & b \\ c & d \end{pmatrix} \quad \text{then} \quad A^{-1} = \frac{1}{ad - bc}\begin{pmatrix} d & -b \\ -c & a \end{pmatrix}, \quad ad - bc \neq 0.$$

For example, if

$$A = \begin{pmatrix} 1 & 1 \\ 1 & 3 \end{pmatrix}, \quad \text{then} \quad A^{-1} = \tfrac{1}{2}\begin{pmatrix} 3 & -1 \\ -1 & 1 \end{pmatrix},$$

$$\text{so} \quad \begin{pmatrix} 1 & 1 \\ 1 & 3 \end{pmatrix}\begin{pmatrix} \frac{3}{2} & -\frac{1}{2} \\ -\frac{1}{2} & \frac{1}{2} \end{pmatrix} = \begin{pmatrix} 1 & 0 \\ 0 & 1 \end{pmatrix}.$$

The expression $\det A = ad - bc$ is the **determinant** of the 2 by 2 matrix A displayed above. For square matrices larger than 2 by 2 there are computational rules and determinant formulas for inverses described in the last part of Appendix B. In this connection we'll note here the important fact that *a square* *matrix A has an inverse A^{-1} if and only if its determinant* det A *is not zero.* A closely related fact is that, viewed as vectors, *the columns of a square matrix A are linearly independent if and only if* det A *is not zero,* where the definition of linear independence for column vectors is formally the same as for scalar-valued functions. In particular, two vectors **u** and **v** are linearly independent if and only if neither one is a scalar multiple of the other.

3A Eigenvalues and Eigenvectors

Many of the normal form systems in Sections 1 and 2 of this chapter are examples of the vector differential equation $\dot{\mathbf{x}} = A\mathbf{x}$. These examples show that if a constant-coefficient system has dimension 2 or 3, then exponentials arise naturally in the solutions. In Chapter 1, for the case of dimension 1, we saw that $\dot{x} = ax$ has solutions $x(t) = ce^{at}$. Consequently, we may hope to find solutions

3.1 $$\mathbf{x}(t) = e^{\lambda t}\mathbf{u},$$

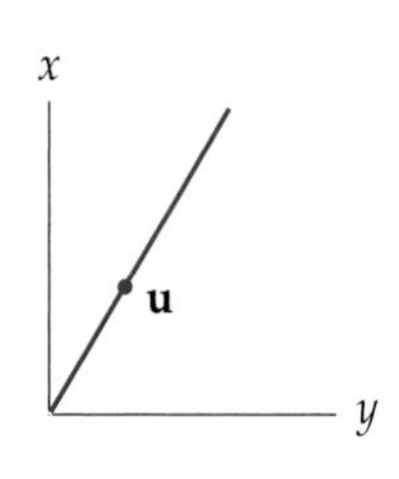

where, as is customary, we now put the scalar factor $e^{\lambda t}$ in front of the constant vector **u**. When $n = 2$, and u, v and λ are nonzero real numbers, the trajectory of such a solution $\begin{pmatrix} x \\ y \end{pmatrix} = e^{\lambda t}\begin{pmatrix} u \\ v \end{pmatrix}$ would just be a line emanating from the origin with slope v/u. By forming linear combinations of such solutions, and allowing complex values for the parameters u, v, and λ, we'll be able to produce a wide variety of solutions and trajectories.

To determine values for u,v, and λ we substitute from Equation 3.1 into the differential equation $\dot{\mathbf{x}} = A\mathbf{x}$. Differentiation of $\mathbf{x}(t)$ in Equation 3.1 gives

$$\frac{d\mathbf{x}}{dt} = \lambda e^{\lambda t}\mathbf{u}.$$

As we saw in Section 1, the justification for this differentiation is that derivatives of vector-valued functions are defined so that the operation can be carried out one entry at a time. For example, if

$$\mathbf{x}(t) = e^{\lambda t}\begin{pmatrix} u \\ v \end{pmatrix} = \begin{pmatrix} e^{\lambda t}u \\ e^{\lambda t}v \end{pmatrix},$$

then

$$\frac{d\mathbf{x}(t)}{dt} = \begin{pmatrix} \lambda e^{\lambda t}u \\ \lambda e^{\lambda t}v \end{pmatrix} = \lambda e^{\lambda t}\begin{pmatrix} u \\ v \end{pmatrix} = \lambda e^{\lambda t}\mathbf{u}.$$

Since $e^{\lambda t}$ is a scalar, we can factor it past the matrix A, so $A\mathbf{x} = A(e^{\lambda t}\mathbf{u}) = e^{\lambda t}A\mathbf{u}$. To solve the differential equation $d\mathbf{x}/dt = A\mathbf{x}$, we equate our expressions for $d\mathbf{x}/dt = \lambda e^{\lambda t}\mathbf{u}$ and $A\mathbf{x} = e^{\lambda t}\mathbf{u}$, with the result that

$$e^{\lambda t}A\mathbf{u} = \lambda e^{\lambda t}\mathbf{u}.$$

Dividing by the nonzero factor $e^{\lambda t}$ reduces the problem of finding an exponential solution to solving by elimination a linear algebraic system involving λ and **u**, namely,

3.2 $$A\mathbf{u} = \lambda\mathbf{u}.$$

The choice $\mathbf{u} = 0$ doesn't provide useful solutions to the original system, so we set that possibility aside and make the following definition.

Definition. A real or complex number λ is an **eigenvalue** of the square matrix A with corresponding **eigenvector** $\mathbf{u} \neq 0$ if $\mathbf{u}$ and λ satisfy Equation 3.2, namely, $A\mathbf{u} = \lambda\mathbf{u}$. Thus Equation 3.1 provides a nontrivial solution to the vector differential equation whenever $\mathbf{u}$ is an eigenvector of A with eigenvalue λ. The problem of solving the vector differential equation has been reduced to an algebraic problem to which we can apply an idea from matrix algebra. This is done by rewriting Equation 3.2 as follows. We note that if I denotes the identity matrix of the same dimensions as A, then

$$A\mathbf{u} - \lambda\mathbf{u} = A\mathbf{u} - \lambda I\mathbf{u} = (A - \lambda I)\mathbf{u}.$$

We now rewrite Equation 3.2 using $A\mathbf{u} - \lambda\mathbf{u} = (A - \lambda I)\mathbf{u}$ to get

3.3 $$(A - \lambda I)\mathbf{u} = 0.$$

Equation 3.3 poses the **eigenvalue-eigenvector problem** of finding all real or complex numbers λ and corresponding nonzero vectors $\mathbf{u}$ in a slightly different form from the one stated in the definition. While Equation 3.2 is equivalent to Equation 3.3, the latter equation has the technical advantage that it displays the vector equation in a more convenient form for solution. If the square matrix $A - \lambda I$ should happen to have an inverse, we could apply $(A - \lambda I)^{-1}$ to both sides of Equation 3.3 and conclude that $\mathbf{u}$ must be 0. But $\mathbf{u} = 0$ is not useful in the context of solving differential equations, so we have to choose λ so that $(A - \lambda I)^{-1}$ fails to exist. Theorem 2.8 of Appendix B tells us that the matrix $(A - \lambda I)$ fails to have an inverse precisely when the matrix has determinant zero, that is, when λ satisfies the **characteristic equation**

3.4 $$\det(A - \lambda I) = 0.$$

The determinant $P(\lambda) = \det(A - \lambda I)$ is a polynomial, called the **characteristic polynomial** of A, and the roots of $P(\lambda)$ will be the eigenvalues of A. For example, if $A = \begin{pmatrix} 2 & 1 \\ 3 & 2 \end{pmatrix}$, then Equation 3.4 says

$$\det\left(\begin{pmatrix} 2 & 1 \\ 3 & 2 \end{pmatrix} - \lambda\begin{pmatrix} 1 & 0 \\ 0 & 1 \end{pmatrix}\right) = \det\begin{pmatrix} 2-\lambda & 1 \\ 3 & 2-\lambda \end{pmatrix} = 0.$$

The 2 by 2 determinant equation works out to be

$$(2 - \lambda)^2 - 3 = \lambda^2 - 4\lambda + 1 = 0.$$

This characteristic equation is a condition on λ alone; having found all values of λ that satisfy it, we can then put each root in Equation 3.3 to find a corresponding eigenvector $\mathbf{u}$. Note that if $\mathbf{u}$ is a nonzero solution of Equation 3.2, or the equivalent Equation 3.3 then so is $c\mathbf{u}$ for an arbitrary nonzero scalar c, because c can be factored out and canceled. As we already know from Section 2, if the homogeneous linear differential equation $\dot{\mathbf{x}} = A\mathbf{x}$ has solutions of the form $e^{\lambda_k t}\mathbf{u}$ and $e^{\lambda_k t}\mathbf{v}$, then solutions are also given by writing a **linear combination** of the vector functions $e^{\lambda_1 t}\mathbf{u}$ and $e^{\lambda_2 t}\mathbf{v}$:

$$\mathbf{x}(t) = c_1 e^{\lambda_1 t}\mathbf{u} + c_2 e^{\lambda_2 t}\mathbf{v}.$$

example 1

To solve the system

$$\begin{pmatrix} \dfrac{dx}{dt} \\ \dfrac{dy}{dt} \end{pmatrix} = \begin{pmatrix} x + y \\ 4x + y \end{pmatrix} = \begin{pmatrix} 1 & 1 \\ 4 & 1 \end{pmatrix}\begin{pmatrix} x \\ y \end{pmatrix},$$

try to find nonzero vectors $\mathbf{u} = (u, v)$ that satisfy the eigenvector equation

$$\begin{pmatrix} 1 & 1 \\ 4 & 1 \end{pmatrix}\begin{pmatrix} u \\ v \end{pmatrix} = \lambda\begin{pmatrix} u \\ v \end{pmatrix}$$

for some number λ. In other words, we find numbers λ such that the equation

$$\begin{pmatrix} 1-\lambda & 1 \\ 4 & 1-\lambda \end{pmatrix}\begin{pmatrix} u \\ v \end{pmatrix} = \begin{pmatrix} 0 \\ 0 \end{pmatrix}$$

has nonzero vector solutions $\mathbf{u}$. If the 2-by-2 matrix above is invertible, then only the solution $(u, v) = (0, 0)$ exists. If the matrix $A - \lambda I$ fails to have an inverse we must have

$$\det\begin{pmatrix} 1-\lambda & 1 \\ 4 & 1-\lambda \end{pmatrix} = 0.$$

For our specific example this means

$$\begin{aligned}(1-\lambda)^2 - 4 &= \lambda^2 - 2\lambda - 3 \\ &= (\lambda - 3)(\lambda + 1) = 0.\end{aligned}$$

The only solutions for the eigenvalues are $\lambda = 3$ and $\lambda = -1$.

Case (1): $\lambda = 3$. We want nonzero vectors $\mathbf{u} = (u, v)$ such that

$$\begin{pmatrix} -2 & 1 \\ 4 & -2 \end{pmatrix}\begin{pmatrix} u \\ v \end{pmatrix} = \begin{pmatrix} 0 \\ 0 \end{pmatrix} \quad \text{or} \quad \begin{aligned} -2u + v &= 0, \\ 4u - 2v &= 0. \end{aligned}$$

The two numerical equations reduce to just one, $v = 2u$, so from the infinitely many possible solutions we can choose for simplicity $u = 1$, $v = 2$. Thus, since $\lambda = 3$,

$$\mathbf{x}_1(t) = e^{3t}\begin{pmatrix} 1 \\ 2 \end{pmatrix}$$

is a solution, along with any numerical multiple of it.

Case (2): $\lambda = -1$. We want nonzero vectors $\mathbf{v} = (u,v)$ such that

$$\begin{pmatrix} 2 & 1 \\ 4 & 2 \end{pmatrix}\begin{pmatrix} u \\ v \end{pmatrix} = \begin{pmatrix} 0 \\ 0 \end{pmatrix}, \quad \text{or} \quad \begin{aligned} 2u + v &= 0, \\ 4u = 2v &= 0. \end{aligned}$$

The two equations reduce to $v = -2u$, satisfied for example by $u = 1, v = -2$, so

$$\mathbf{x}_2(t) = e^{-t}\begin{pmatrix} 1 \\ -2 \end{pmatrix}$$

is a solution, as well as any numerical multiple. Forming a linear combination of our two solutions we get a more general solution of the vector differential equation:

$$\begin{aligned} \mathbf{x}(t) &= c_1 e^{3t}\begin{pmatrix} 1 \\ 2 \end{pmatrix} + c_2 e^{-t}\begin{pmatrix} 1 \\ -2 \end{pmatrix} \\ &= \begin{pmatrix} c_1 e^{3t} + c_2 e^{-t} \\ 2c_1 e^{3t} - 2c_2 e^{-t} \end{pmatrix}. \end{aligned}$$

Geometric Interpretation. The purely algebraic computations in the previous example are effective for finding solution formulas, but by themselves they're a bit abstract. Understanding the geometric significance of the result in the example is helpful here, so we'll review the results with reference to Figure 5.11. Note first that the two eigenvectors

$$\mathbf{u} = \begin{pmatrix} 1 \\ 2 \end{pmatrix}, \qquad \mathbf{v} = \begin{pmatrix} 1 \\ -2 \end{pmatrix}$$

are not multiples of each other, so any point in the plane can be written as a sum of multiples of $\mathbf{u}$ and $\mathbf{v}$, in other words any vector $\mathbf{x}$ can be written using zw coordinates as

$$\mathbf{x} = z\mathbf{u} + w\mathbf{v}.$$

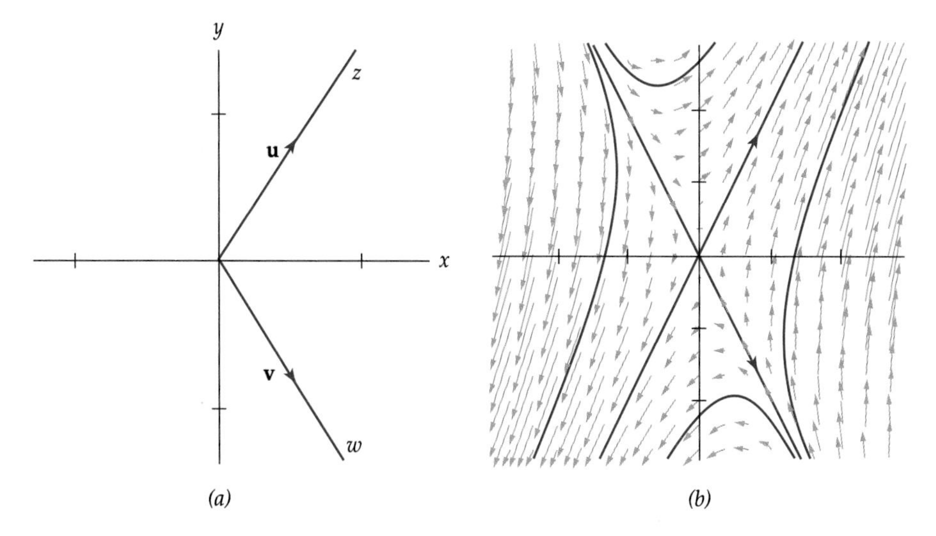

Figure 5.11 (a) Eigenvector coordinates. (b) Vector field and trajectories.

By the rules for multiplication by a matrix A,

$$\begin{aligned} A\mathbf{x} &= Az\begin{pmatrix}1\\2\end{pmatrix} + Aw\begin{pmatrix}1\\-2\end{pmatrix} \\ &= zA\begin{pmatrix}1\\2\end{pmatrix} + wA\begin{pmatrix}1\\-2\end{pmatrix}. \end{aligned}$$

Figure 5.11(a) shows the eigenvectors, and Figure 5.11(b) shows how naturally these vectors relate geometrically to the vector field and some trajectories of the system. In particular, the axes determined by $\mathbf{u}$ and $\mathbf{v}$ partition the plane into four regions in which the field and its trajectories show quite distinct behaviors.

Since $\mathbf{u}$ and $\mathbf{v}$ are eigenvectors, with eigenvalues 3 and -1, respectively, we have $A\mathbf{u} = 3\mathbf{u}$ and $A\mathbf{v} = -\mathbf{v}$. In other words, A has the effect of multiplying the first vector $\mathbf{u}$ by 3 and the second vector $\mathbf{v}$ by -1. Hence,

$$\begin{aligned} A\mathbf{x} &= zA\mathbf{u} + wA\mathbf{v} \\ &= 3z\,\mathbf{u} - w\,\mathbf{v}. \end{aligned}$$

Since we've also seen that $\mathbf{x} = z\mathbf{u} + w\mathbf{v}$, we can differentiate this equation to get

$$\frac{d\mathbf{x}}{dt} = \frac{dz}{dt}\mathbf{u} + \frac{dw}{dt}\mathbf{v}.$$

To make $d\mathbf{x}/dt = A\mathbf{x}$ we equate the coefficients of $\mathbf{u}$ and $\mathbf{v}$ on the right sides of the last two displayed equations to get

$$\begin{aligned} \frac{dz}{dt} &= 3z, \\ \frac{dw}{dt} &= -w, \end{aligned} \qquad \text{or in matrix form,} \qquad \begin{pmatrix}\dfrac{dz}{dt}\\ \dfrac{dw}{dt}\end{pmatrix} = \begin{pmatrix}3 & 0\\ 0 & -1\end{pmatrix}\begin{pmatrix}z\\ w\end{pmatrix}.$$

The underlying reason for introducing coordinates relative to the eigenvectors is that the new system with variable z and w is particularly simple to solve, because each involves only one unknown function. In other words, the new system is *uncoupled*, and we see immediately that

$$z(t) = c_1e^{3t}, \qquad w(t) = c_2e^{-t}.$$

Since $\mathbf{x} = z\mathbf{u} + w\mathbf{v}$, we can express our solution in terms of the original xy coordinates:

$$\mathbf{x}(t) = c_1e^{3t}\mathbf{u} + c_2e^{-t}\mathbf{v} = c_1e^{3t}\begin{pmatrix}1\\2\end{pmatrix} + c_2e^{-t}\begin{pmatrix}1\\-2\end{pmatrix}.$$

The procedure in the previous example can be tried in any number of dimensions, but we'll restrict ourselves primarily to dimension $n = 2$ in this section, since higher dimensional problems are generally handled more completely and efficiently using the methods of Chapter 7. If A has complex eigenvalues as in the next example, the method outlined above is the same, with complex exponentials in the solutions, or else their trigonometric equivalent.

example 2 The system

$$\frac{dx}{dt} = x - y, \quad \frac{dy}{dt} = x + y \qquad \text{has matrix} \quad A = \begin{pmatrix} 1 & -1 \\ 1 & 1 \end{pmatrix}.$$

The eigenvalues are solutions of

$$\det \begin{pmatrix} 1-\lambda & -1 \\ 1 & 1-\lambda \end{pmatrix} = 0,$$

which works out to be

$$(1-\lambda)^2 + 1 = \lambda^2 - 2\lambda + 2 = 0.$$

Using the quadratic formula we find solutions $\lambda_1 = 1 + i$ and $\lambda_2 = 1 - i$.

Case (1): $\lambda = 1 + i$. The eigenvectors are the nonzero solutions of

$$\begin{pmatrix} -i & -1 \\ 1 & -i \end{pmatrix} \begin{pmatrix} u \\ v \end{pmatrix} = \begin{pmatrix} 0 \\ 0 \end{pmatrix}, \quad \text{or} \quad \begin{aligned} -iu - v &= 0, \\ u - iv &= 0. \end{aligned}$$

Multiplying the first equation by i shows that the two equations together are equivalent to $u = iv$, so we can choose $\begin{pmatrix} u \\ v \end{pmatrix} = \begin{pmatrix} 1 \\ -i \end{pmatrix}$.

Case (2): $\lambda = 1 - i$. The eigenvectors are the nonzero solutions of

$$\begin{pmatrix} i & -1 \\ 1 & i \end{pmatrix} \begin{pmatrix} u \\ v \end{pmatrix} = \begin{pmatrix} 0 \\ 0 \end{pmatrix}, \quad \text{or} \quad \begin{aligned} iu - v &= 0, \\ u + iv &= 0. \end{aligned}$$

This pair of equations is equivalent to $v = iu$ so we can choose $\begin{pmatrix} u \\ v \end{pmatrix} = \begin{pmatrix} 1 \\ i \end{pmatrix}$. The general solution of the differential equation can be written using $e^{(1\pm i)t} = e^t(\cos t \pm i \sin t)$:

$$\begin{aligned} \mathbf{x}(t) &= c_1 e^{(1+i)t} \begin{pmatrix} 1 \\ -i \end{pmatrix} + c_2 e^{(1-i)t} \begin{pmatrix} 1 \\ i \end{pmatrix} \\ &= c_1 e^t \begin{pmatrix} \cos t + i \sin t \\ -i \cos t + \sin t \end{pmatrix} + c_2 e^t \begin{pmatrix} \cos t - i \sin t \\ i \cos t + \sin t \end{pmatrix} \\ &= \begin{pmatrix} (c_1 + c_2)e^t \cos t + i(c_1 - c_2)e^t \sin t \\ -i(c_1 - c_2)e^t \cos t + (c_1 + c_2)e^t \sin t \end{pmatrix}. \end{aligned}$$

If we rename the constants so that $c_1 + c_2 = d_1$ and $i(c_1 - c_2) = d_2$, then

$$\begin{aligned} \mathbf{x}(t) &= \begin{pmatrix} d_1 e^t \cos t + d_2 e^t \sin t \\ -d_2 e^t \cos t + d_1 e^t \sin t \end{pmatrix} \\ &= d_1 \begin{pmatrix} e^t \cos t \\ e^t \sin t \end{pmatrix} + d_2 \begin{pmatrix} e^t \sin t \\ -e^t \cos t \end{pmatrix}. \end{aligned}$$

Because $c_1 = (d_1 - id_2)/2$ and $c_2 = (d_1 + id_2)/2$, either pair of constants, $c_1,\ c_2$ or $d_1,\ d_2$ can be chosen arbitrarily, in which case the other pair is determined by these equations. In particular, d_1 and d_2 can be chosen to be real numbers.

Summary of the Eigenvector Method. To solve the constant-coefficient equation

$$\frac{d\mathbf{x}}{dt} = A\mathbf{x},$$

where A is an n by n matrix, proceed as follows:

1. Find eigenvalues of A by solving the polynomial equation $\det(A - \lambda I) = 0$.
2. For each eigenvalue λ_k, find an eigenvector $\mathbf{u}_k$ by solving

$$(A - \lambda_k I)\mathbf{u} = 0.$$

3. Form linear combinations of solutions $\mathbf{x}_k(t) = e^{\lambda_k t}\mathbf{u}_k$. If there are n *linearly independent* eigenvectors (i.e., none of them is replaceable by a linear combination of the others), then step 3 produces the most general solution.

The exponential matrix method of Chapter 7 offers the double advantage that it dispenses with the unnecessary computing of eigenvectors altogether and applies with only minor computational change if the characteristic equation has repeated roots.

If an eigenvector $\mathbf{u}$ is *linearly dependent* on others (i.e., $\mathbf{u}$ is replaceable by a linear combination of other vectors), the procedure outlined above produces solutions but not the most general one. This deficiency of solutions can occur only if the characteristic equation has repeated roots, and may not occur even then. In case there aren't enough independent eigenvectors, we can use the elimination method explained in Section 2 or, for low-dimensional sytems, the generalized eigenvectors of part B of this section.

3B Generalized Eigenvectors†

If the characteristic equation of a matrix A has a repeated root, there may or may not be sufficiently many eigenvectors to represent all possible solutions.

example 3

The system

$$\begin{aligned}\dot{x} &= x + y \\ \dot{y} &= \quad\;\; y\end{aligned} \quad \text{or} \quad \begin{pmatrix}\dot{x}\\ \dot{y}\end{pmatrix} = \begin{pmatrix}1 & 1\\ 0 & 1\end{pmatrix}\begin{pmatrix}x\\ y\end{pmatrix},$$

is only partially coupled, so we can solve the second equation for $y(t) = c_1 e^t$ and put the result into the first equation. Applying the exponential multiplier method of Chapter 1, Section 4, the resulting first-order equation $\dot{x} - x = c_1 e^t$ has solutions $x(t) = c_1 t e^t + c_2 e^t$. In vector form the solutions look like

$$\begin{pmatrix}x\\ y\end{pmatrix} = c_1\begin{pmatrix}te^t\\ e^t\end{pmatrix} + c_2\begin{pmatrix}e^t\\ 0\end{pmatrix}.$$

This is the easiest way to solve this system, but trying to follow the eigenvector method will be instructive at this point so we'll do it. The characteristic equation is $(1 - \lambda)^2 = 0$ with $\lambda = 1$ as a double root. The equation to solve for an eigenvector is

$$\begin{pmatrix}1-\lambda & 1\\ 0 & 1-\lambda\end{pmatrix}\begin{pmatrix}u\\ v\end{pmatrix} = \begin{pmatrix}0 & 1\\ 0 & 0\end{pmatrix}\begin{pmatrix}u\\ v\end{pmatrix} = \begin{pmatrix}0\\ 0\end{pmatrix}.$$

This vector equation requires only that $v = 0$, so all the eigenvectors have the form $\mathbf{u} = \begin{pmatrix} u \\ 0 \end{pmatrix}$, where $u \neq 0$. The only solutions we get for the original system are consequently $\begin{pmatrix} x \\ y \end{pmatrix} = e^t \begin{pmatrix} u \\ 0 \end{pmatrix} u \begin{pmatrix} e^t \\ 0 \end{pmatrix}$, so we have to proceed differently if we want to find the remaining solutions using matrix methods.

We saw at the beginning of the previous example that to express the general solution we need to include terms not only involving e^t but also te^t. Suppose we want to solve a system $\dot{\mathbf{x}} = A\mathbf{x}$ that has a repeated eigenvalue λ and associated eigenvector $\mathbf{u}$, but that fails to have an associated eigenvector $\mathbf{v}$ independent of $\mathbf{u}$. Guided by the form of the solution we found, we look for additional independent solutions of the form

3.5
$$\mathbf{x}(t) = te^{\lambda t}\mathbf{u} + e^{\lambda t}\mathbf{v},$$

where $\mathbf{u}$ is an eigenvector with eigenvalue λ. We'll see that by choosing the auxiliary vector $\mathbf{v}$ correctly the given system will become, not uncoupled, but only *partially* coupled, like the system in the previous example. To achieve this partial coupling all we'll have to do is express the system using coordinates z and w with respect to the vectors $\mathbf{u}$ and $\mathbf{v}$, so that $\mathbf{x} = z\mathbf{u} + w\mathbf{v}$. To do this effectively, $\mathbf{u}$ and $\mathbf{v}$ will have to be linearly independent.

To find a suitable vector $\mathbf{v}$ we assume the existence of a solution to $\dot{\mathbf{x}} = A\mathbf{x}$ of the form $\mathbf{x} = te^{\lambda t}\mathbf{u} + e^{\lambda t}\mathbf{v}$, where $\mathbf{u}$ is an eigenvector with repeated eigenvalue λ. We multiply this linear combination by A, pulling the scalars out to get

$$\begin{aligned} A\mathbf{x} &= te^{\lambda t}A\mathbf{u} + e^{\lambda t}A\mathbf{v} \\ &= \lambda te^{\lambda t}\mathbf{u} + e^{\lambda t}A\mathbf{v}, \end{aligned}$$

and using $A\mathbf{u} = \lambda\mathbf{u}$ in the second step. Using the product rule for derivatives and noting that $\mathbf{u}$ and $\mathbf{v}$ are constant we find from Equation 3.5 that

$$\dot{\mathbf{x}} = \lambda te^{\lambda t}\mathbf{u} + e^{\lambda t}\mathbf{u} + \lambda e^{\lambda t}\mathbf{v}.$$

When we equate $A\mathbf{x}$ and $\dot{\mathbf{x}}$, the common term $\lambda te^{\lambda t}\mathbf{u}$ cancels to give

$$e^{\lambda t}A\mathbf{v} = e^{\lambda t}\mathbf{u} + \lambda e^{\lambda t}\mathbf{v} \quad \text{or} \quad A\mathbf{v} = \mathbf{u} + \lambda\mathbf{v},$$

since the factor $e^{\lambda t}$ is never zero. This last equation is like Equation 3.3 for finding eigenvectors in that it is purely algebraic. We rewrite the equation as we did Equation 3.3 for convenience in solving for the eigenvector $\mathbf{u}$. A **generalized eigenvector v** associated with eigenvalue λ is a vector $\mathbf{v}$ satisfying

3.6
$$(A - \lambda I)\mathbf{v} = \mathbf{u},$$

where $\mathbf{u}$ is an eigenvector with eigenvalue λ. Since $\mathbf{u} \neq 0$, a solution $\mathbf{v}$ can't be zero and can't be an eigenvector with eigenvalue λ, because the left side of Equation 3.6 would then be zero. Hence $\mathbf{v}$ can't be a scalar multiple of $\mathbf{u}$, so $\mathbf{u}$ and $\mathbf{v}$ will be independent vectors.

Geometric Interpretation. Suppose $\mathbf{u}$ is an eigenvector with eigenvalue λ for a square matrix A and $\mathbf{v}$ is a generalized eigenvector satisfying Equation 3.6. We can represent vectors $\mathbf{x}$ in terms of the independent vectors $\mathbf{u}$ and $\mathbf{v}$ by using coordinates z and w. Thus for an arbitrary vector $\mathbf{x}$ we have $\mathbf{x} = z\mathbf{u} + w\mathbf{v}$. Then since $A\mathbf{v} = \lambda\mathbf{v} + \mathbf{u}$ by Equation 3.6, and $A\mathbf{u} = \lambda\mathbf{u}$, we have

$$\begin{aligned} A\mathbf{x} &= zA\mathbf{u} + wA\mathbf{v} = \lambda z\,\mathbf{u} + w(\lambda\mathbf{v} + \mathbf{u}) \\ &= (\lambda z + w)\mathbf{u} + \lambda w\mathbf{v}. \end{aligned}$$

But since $\dot{\mathbf{x}} = \dot{z}\mathbf{u} + \dot{w}\mathbf{v}$, equating coefficients of $\mathbf{u}$ and $\mathbf{v}$ in $\dot{\mathbf{x}}$ and $A\mathbf{x}$ gives us, not an uncoupled system, but a *partially* coupled system that is easy to solve by successive integration:

$$\begin{aligned} \dot{z} &= \lambda z + w, \\ \dot{w} &= \lambda w, \end{aligned} \quad \text{or} \quad \begin{pmatrix} \dot{z} \\ \dot{w} \end{pmatrix} = \begin{pmatrix} \lambda & 1 \\ 0 & \lambda \end{pmatrix} \begin{pmatrix} z \\ w \end{pmatrix}.$$

example 4

We saw in Example 3 that the system

$$\begin{pmatrix} \dot{x} \\ \dot{y} \end{pmatrix} = \begin{pmatrix} 1 & 1 \\ 0 & 1 \end{pmatrix} \begin{pmatrix} x \\ y \end{pmatrix}$$

had only the eigenvectors $\mathbf{u} = \begin{pmatrix} u \\ 0 \end{pmatrix}$, $u \neq 0$, all of which lie along the x axis, and which generated the solutions $\mathbf{x}_1(t) = u\begin{pmatrix} e^t \\ 0 \end{pmatrix}$ to the system. To find independent solutions, we solve Equation 3.6 for a generalized eigenvector $\mathbf{v}$ with $\lambda = 1$. For $\mathbf{u}$ we'll choose the eigenvector $\mathbf{u} = \begin{pmatrix} 1 \\ 0 \end{pmatrix}$. Thus Equation 3.6 becomes

$$\left(\begin{pmatrix} 1 & 1 \\ 0 & 1 \end{pmatrix} - 1\begin{pmatrix} 1 & 0 \\ 0 & 1 \end{pmatrix}\right)\mathbf{v} = \begin{pmatrix} 1 \\ 0 \end{pmatrix}, \quad \text{or} \quad \begin{pmatrix} 0 & 1 \\ 0 & 0 \end{pmatrix}\begin{pmatrix} u \\ v \end{pmatrix} = \begin{pmatrix} 1 \\ 0 \end{pmatrix}.$$

This matrix equation is equivalent to $v = 1$ and $0 = 0$, so u can be chosen arbitrarily. For simplicity we'll let $u = 0$, so $\mathbf{v} = \begin{pmatrix} 0 \\ 1 \end{pmatrix}$. Our independent solution determined by Equation 3.5 is then

$$\mathbf{x}_2(t) = te^t\begin{pmatrix} 1 \\ 0 \end{pmatrix} + e^t\begin{pmatrix} 0 \\ 1 \end{pmatrix} = \begin{pmatrix} te^t \\ e^t \end{pmatrix}.$$

Linear combinations of our basic solutions are

$$\mathbf{x}(t) = c_1\begin{pmatrix} e^t \\ 0 \end{pmatrix} + c_2\begin{pmatrix} te^t \\ e^t \end{pmatrix},$$

in accord with the solution we found by successive integration at the beginning of Example 3. Note that when $c_2 = 0$, we get just one family of solution trajectories that lie on a line, namely, the horizontal axis. For all other choices of c_2 the trajectories are curved, as shown in Figure 5.12(b). Compare this picture with Figure 5.11(b), where there are two solution families with straight-line trajectories.

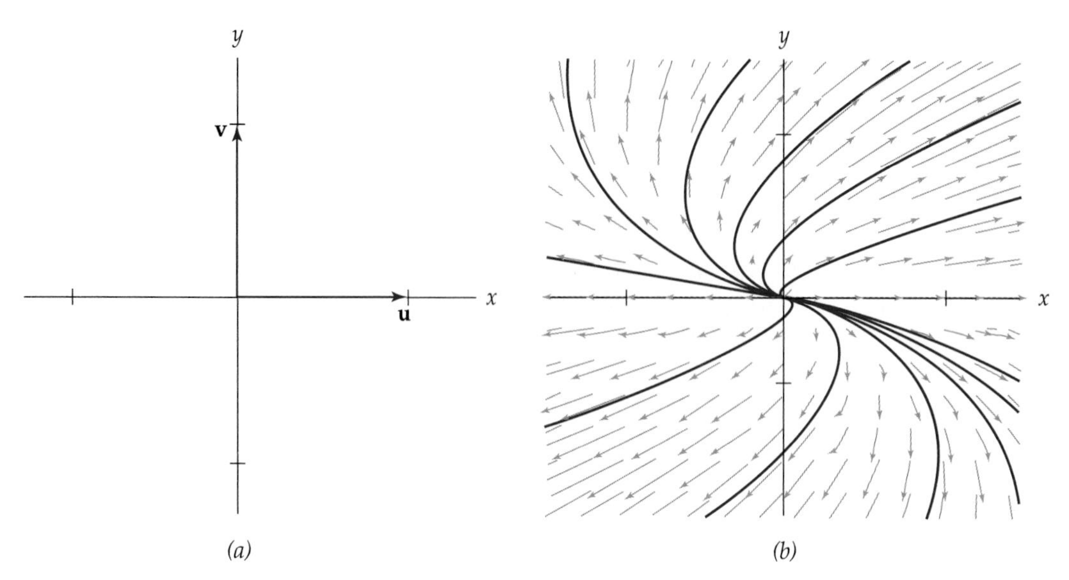

Figure 5.12 (a) Eigenvector: **u**; generalized: **v**. (b) Vector field and trajectories.

EXERCISES

The matrix $A = \begin{pmatrix} 1 & 12 \\ 3 & 1 \end{pmatrix}$ has eigenvalues 7 and −5. Which of the vectors in Exercises 1 through 5 is an eigenvector of A? For those that are eigenvectors, what is the corresponding eigenvalue?

1. $\begin{pmatrix} 2 \\ 1 \end{pmatrix}$.

2. $\begin{pmatrix} -2 \\ 1 \end{pmatrix}$.

3. $\begin{pmatrix} -4 \\ -2 \end{pmatrix}$.

4. $\begin{pmatrix} -2 \\ 2 \end{pmatrix}$.

5. $\begin{pmatrix} 1 \\ 1 \end{pmatrix}$.

Find all the eigenvalues of each of the matrices in Exercises 6 through 9.

6. $\begin{pmatrix} 1 & 4 \\ 1 & 1 \end{pmatrix}$.

7. $\begin{pmatrix} 0 & 4 \\ 1 & 0 \end{pmatrix}$.

8. $\begin{pmatrix} 2 & 4 \\ 1 & 2 \end{pmatrix}$.

9. $\begin{pmatrix} 1 & 0 & 0 \\ 2 & 1 & 0 \\ 0 & 1 & 2 \end{pmatrix}$.

10.–13. In Exercises 10 through 13, for each matrix in Exercises 6 through 9, find an eigenvector corresponding to each eigenvalue.

Each of the two-dimensional systems in Exercises 14 through 17 has the form

$$\frac{d\mathbf{x}}{dt} = A\mathbf{x}$$

in which the matrix A has constant entries. In each example, find the eigenvalues of A, and for each eigenvalue find a corresponding eigenvector.

14. $\begin{pmatrix} \dot{x} \\ \dot{y} \end{pmatrix} = \begin{pmatrix} -3 & 2 \\ -4 & 3 \end{pmatrix} \begin{pmatrix} x \\ y \end{pmatrix}$.

15. $\dot{x} = 3x,$
$\dot{y} = 2y.$

16. $\dot{x} = x + 4y,$
$\dot{y} = 5y.$

17. $\begin{pmatrix} \dot{x} \\ \dot{y} \end{pmatrix} = \begin{pmatrix} 2 & -1 \\ 1 & 2 \end{pmatrix} \begin{pmatrix} x \\ y \end{pmatrix}.$

18.–21. Use the eigenvalues and eigenvectors of each of the systems in Exercises 14 through 17 to write the general solution of the system in the form

$$\mathbf{x}(t) = c_1 e^{\lambda_1 t}\mathbf{u}_1 + c_2 e^{\lambda_2 t}\mathbf{u}_2.$$

In the case of complex eigenvalues, convert the solution to real form.

In Exercises 22 through 25, for the systems of equations in Exercises 14 through 17, solve for the arbitrary constants to find the particular solution that satisfies the corresponding condition.

22. $(x(0), y(0)) = (1,3).$

23. $(x(0), y(0)) = (-1,0).$

24. $(x(0), y(0)) = (1,1).$

25. $(x(0), y(0)) = (0,0).$

26. **(a)** Find the general solution of the system $\dot{\mathbf{x}} = A\mathbf{x}$, where

$$A = \begin{pmatrix} 1 & -1 & 4 \\ 3 & 2 & -1 \\ 2 & 1 & -1 \end{pmatrix}.$$

(b) Find a constant vector $\mathbf{x}_p$ that is a particular solution of the system $\dot{\mathbf{x}} = A\mathbf{x} + \mathbf{c}$, where $\mathbf{c}$ is the column vector with entries 1, 0, and 2, top to bottom.

(c) Find, by any method, the particular solution $\mathbf{x}$ of the system in part (b) that satisfies the condition $\mathbf{x}(0) = (1,1,2)$.

27. Show that, if A is n-by-n with invertible eigenvector matrix U, and the eigenvalues of A have negative real parts, then all solutions of $d\mathbf{x}/dt = A\mathbf{x}$ tend to zero as t tends to $+\infty$.

***28.** The second-order, constant-coefficient, differential equation

$$\frac{d^2y}{dt^2} + a\frac{dy}{dt} + by = 0$$

is equivalent to the first-order system

$$\begin{pmatrix} \dot{x} \\ \dot{y} \end{pmatrix} = \begin{pmatrix} -a & -b \\ 1 & 0 \end{pmatrix} \begin{pmatrix} x \\ y \end{pmatrix}.$$

Show that the eigenvalues of the matrix are the same as the characteristic roots of the second-order differential equation.

29. Let U be an invertible n-by-n matrix with columns $\mathbf{u}_1, \ldots, \mathbf{u}_n$. Let D be the diagonal matrix with diagonal entries $\lambda_1, \ldots, \lambda_n$, and define the n-by-n matrix A by $A = UDU^{-1}$.

(a) Show that $A\mathbf{u}_k = \lambda_k \mathbf{u}_k$, for $k = 1, \ldots, n$.

(b) Find the 2-by-2 matrix that has eigenvectors $\mathbf{u}_1 = (2,3)$, $\mathbf{u}_2 = (1,1)$, and corresponding eigenvalues $\lambda_1 = 3$ and $\lambda_2 = 1$.

(c) Show that the system $d\mathbf{x}/dt = A\mathbf{x}$ has solutions $e^{\lambda_k t}\mathbf{u}_k$, for $k = 1, \ldots, n$.

(d) Find a two-dimensional system having

$$\mathbf{x}(t) = c_1 e^{3t}\begin{pmatrix} 2 \\ 3 \end{pmatrix} + c_2 e^{t}\begin{pmatrix} 1 \\ 1 \end{pmatrix}$$

as its general solution.

In applications of linear systems, we're sometimes concerned mainly with the general behavior of solutions. By finding the eigenvalues $\lambda = a + ib$ of each matrix A for the systems in Exercises 30 through 33, find out whether the solutions of $d\mathbf{x}/dt = A\mathbf{x}$ oscillate ($b \neq 0$), tend to 0 as t tends to $+\infty$ ($a < 0$), or become unbounded ($a > 0$). Of course, a single equation may have solutions exhibiting more than one kind of behavior.

30. $\dfrac{d}{dt}\begin{pmatrix} x \\ y \end{pmatrix} = \begin{pmatrix} -1 & -2 \\ 1 & 0 \end{pmatrix}\begin{pmatrix} x \\ y \end{pmatrix}.$

31. $\dfrac{d}{dt}\begin{pmatrix} x \\ y \end{pmatrix} = \begin{pmatrix} 0 & -1 \\ 1 & 0 \end{pmatrix}\begin{pmatrix} x \\ y \end{pmatrix}.$

32. $\dfrac{d}{dt}\begin{pmatrix} x \\ y \end{pmatrix} = \begin{pmatrix} 2x - y \\ 3x - y \end{pmatrix}.$

33. $dx/dt = 2x + y, dy/dt = 3x + y.$

***34.** The second-order system $\ddot{x} = ax + by$, $\ddot{y} = cx + dy$, is equivalent to the first-order system

$$\begin{pmatrix} \dot{x} \\ \dot{y} \\ \dot{z} \\ \dot{w} \end{pmatrix} = \begin{pmatrix} 0 & 0 & 1 & 0 \\ 0 & 0 & 0 & 1 \\ a & b & 0 & 0 \\ c & d & 0 & 0 \end{pmatrix}\begin{pmatrix} x \\ y \\ z \\ w \end{pmatrix}.$$

(a) Verify this statement and show by calculating the determinant of $A - \lambda I$ that the characteristic equation of the first-order system is $\lambda^4 - (a + d)\lambda^2 + (ad - bc) = 0$.

(b) Assume that a, b, c, and d are real numbers. Show that all eigenvalues of the system are purely imaginary if and only if both $a + d < 0$ and $ad - bc > 0$, with $(a + d)^2 > 4(ad - bc)$. [*Hint:* Find out when $\lambda^2 < 0$.]

(c) Show under the conditions of part (b) that all nontrivial solutions of the system are sums of periodic oscillations.

35. Verify that if $\mathbf{x}_1(t)$ and $\mathbf{x}_2(t)$ are solutions of $\dot{\mathbf{x}} = A\mathbf{x}$, then so is $c_1\mathbf{x}_1(t) + c_2\mathbf{x}_2(t)$ for arbitrary constants c_1 and c_2.

36. Let A be an n-by-n matrix that has 0 for an eigenvalue. Explain why A can't have an inverse.

37. Finding a matrix with given eigenvalues and eigenvectors. Suppose we want an n-by-n matrix A that has $\lambda_1, \ldots, \lambda_n$ and $\mathbf{u}_1, \ldots, \mathbf{u}_n$ as eigenvalues and corresponding eigenvectors: $A\mathbf{u}_k = \lambda_k\mathbf{u}_k$. Let Λ be the square matrix with the λs on the main diagonal, and let U be the square matrix with the vectors $\mathbf{u}_k$ as columns, in the same order as the λs. Show that if U^{-1} exists, then $A = U\Lambda U^{-1}$ is the desired matrix.

38. Use the method described in the previous exercise to find a matrix A with the following eigenvalues and corresponding eigenvectors.

(a) $\lambda_1 = 2, \lambda_2 = 3, \mathbf{u}_1 = \begin{pmatrix} 1 \\ 2 \end{pmatrix}, \mathbf{u}_2 = \begin{pmatrix} 2 \\ 5 \end{pmatrix}$.

(b) $\lambda_1 = 2, \lambda_2 = 3, \ \lambda_3 = 3, \mathbf{u}_1 = \begin{pmatrix} 1 \\ 2 \\ 3 \end{pmatrix}, \mathbf{u}_2 = \begin{pmatrix} 2 \\ 5 \\ 0 \end{pmatrix}, \mathbf{u}_3 = \begin{pmatrix} -1 \\ 0 \\ -14 \end{pmatrix}$.

39. The matrix

$$A = \begin{pmatrix} a & 0 \\ 0 & b \end{pmatrix} \begin{pmatrix} \cos\theta & -\sin\theta \\ \sin\theta & \cos\theta \end{pmatrix} = \begin{pmatrix} a\cos\theta & -a\sin\theta \\ b\sin\theta & b\cos\theta \end{pmatrix}, \quad a > 0, b > 0,$$

represents a counterclockwise rotation about the origin in $\mathcal{R}^2$ through angle θ, followed by dilation or contraction in two perpendicular directions.

(a) Assume $a = b = 1$. Show then that A has a real eigenvalue if and only if $\theta = k\pi$ for some integer k. What are the possible real or complex eigenvalues and corresponding eigenvectors?

(b) Show that A has distinct real eigenvalues if and only if $(a + b)^2 \cos^2\theta > 4ab$.

40. Let $A = \begin{pmatrix} a & b \\ c & d \end{pmatrix}$, where the entries are real numbers and $bc \geq 0$.

(a) Show that all eigenvalues of A are real.

(b) Show that both eigenvalues of A are positive when $a + d > 0$ and $ad - bc > 0$, both are negative when $a + d < 0$ and $ad - bc > 0$, and have opposite signs when $ad - bc < 0$.

41. Show that the independent solutions obtained at the end of Example 2 of the text can be obtained just from Case (1) by taking real and imaginary parts of $e^{(1+i)t}\begin{pmatrix} 1 \\ -i \end{pmatrix}$.

42. Show that if the square matrix A has eigenvalue $\lambda = 0$ with corresponding eigenvector $\mathbf{u} \neq 0$, the the system $\dot{\mathbf{x}} = A\mathbf{x}$ has constant solutions $\mathbf{x} = c\mathbf{u}$.

Matrices with Repeated Eigenvalues. If an n-by-n matrix A has an eigenvalue of multiplicity more than one there may or may not be n independent eigenvectors. When $n \geq 3$ the case of repeated eigenvalues is efficiently handled using the exponential matrix of Chapter 7. The following exercises illustrate the possible variety of solutions when $n = 2$.

43. Consider the system $\begin{pmatrix} \dot{x} \\ \dot{y} \end{pmatrix} = \begin{pmatrix} a & b \\ 0 & a \end{pmatrix} \begin{pmatrix} x \\ y \end{pmatrix}, b \neq 0$.

(a) Show that $\lambda = a$ is a double root of the characteristic equation and that the only eigenvectors are $\begin{pmatrix} u \\ 0 \end{pmatrix}, u \neq 0$. [Since every eigenvector is a nonzero multiple of (1,0), we can't find two independent eigenvectors.]

(b) Solve the partially coupled system by first solving for $y(t)$, then using the result to solve for $x(t)$.

(c) Solve the system by first finding a generalized eigenvector.

44. Consider the system $\begin{pmatrix} \dot{x} \\ \dot{y} \end{pmatrix} = \begin{pmatrix} a & 0 \\ 0 & a \end{pmatrix} \begin{pmatrix} x \\ y \end{pmatrix}$.

(a) Show that the system is uncoupled and that initial-value problem with $x(0) = x_0$, $y(0) = y_0$ has solutions $\begin{pmatrix} x \\ y \end{pmatrix} = e^{at} \begin{pmatrix} x_0 \\ y_0 \end{pmatrix}$.

(b) Show that $\lambda = a$ is a double root of the characteristic equation and that every nonzero vector (u, v) is an eigenvector.

(c) Pick two independent eigenvectors and use them to arrive at the solutions in part (a).

45. For the system $\begin{pmatrix} \dot{x} \\ \dot{y} \end{pmatrix} = \begin{pmatrix} -1 & 4 \\ -1 & 3 \end{pmatrix} \begin{pmatrix} x \\ y \end{pmatrix}$,

(a) Find an eigenvector $\mathbf{u}$, and show that every other eigenvector is a scalar multiple of $\mathbf{u}$.

(b) Find a generalized eigenvector $\mathbf{v}$.

(c) Express the solutions $\mathbf{x}(t)$ of the system in terms of $\mathbf{u}$ and $\mathbf{v}$.

(d) Use computer graphics to sketch the vector field of the system along with some typical solution trajectories.

46. Solve the equations

$$\dot{y} = -\tfrac{1}{25}y + \tfrac{1}{25}z, \qquad \dot{z} = +\tfrac{1}{100}y - \tfrac{1}{25}z$$

using the methods of this section. Then find the solution such that $y(0) = 10$ and $z(0) = 20$.

47. Show that if a 2-dimensional system $\dot{\mathbf{x}} = A\mathbf{x}$ has $\lambda = 0$ as a repeated eigenvalue then every solution trajectory lies on a straight line.

48. (a) Find by successive integration the general solution of the system

$$\frac{dx}{dt} = -x + y, \qquad \frac{dy}{dt} = -y.$$

(b) Show that the eigenvectors of the related matrix $\begin{pmatrix} -1 & 1 \\ 0 & -1 \end{pmatrix}$ are insufficient for the application of the method of Section 1A.

(c) Solve the system using the optional method of Section 3B.

49. (a) Find by successive integration the general solution of the system

$$\frac{dx}{dt} = x, \qquad \frac{dy}{dt} = y + z, \qquad \frac{dz}{dt} = z.$$

(b) Show that the eigenvectors of the related matrix $\begin{pmatrix} 1 & 0 & 0 \\ 0 & 1 & 1 \\ 0 & 0 & 1 \end{pmatrix}$ are insufficient for the application of the method of Section 1A.

(c) Solve the system using the optional method of Section 3B for the eigenvalue $\lambda = 1$.

50. If the characteristic equation of A has a triple root λ_1, it may be necessary to find, in addition to an eigenvalue $\mathbf{u}$ satisfying $(A - \lambda_1 I)\mathbf{u} = 0$ and an independent generalized eigenvalue $\mathbf{v}$ such that $(A - \lambda_1 I)\mathbf{v} = \mathbf{u}$, an additional generalized eigenvector $\mathbf{w}$ such that $(A - \lambda_1 I)\mathbf{w} = \mathbf{v}$.

(a) Show that such a vector $\mathbf{w}$ is not a linear combination of $\mathbf{u}$ and $\mathbf{v}$.

(b) Show that in addition to $\mathbf{x}_1 = e^{\lambda t}\mathbf{u}$ and $\mathbf{x}_2 = te^{\lambda t}\mathbf{u} + e^{\lambda t}\mathbf{v}$, as shown in the text, the differential equation $\mathbf{x} = A\mathbf{x}$ has solution $\mathbf{x}_3 = \frac{1}{2}t^2 e^{\lambda t}\mathbf{u} + te^{\lambda t}\mathbf{v} + e^{\lambda t}\mathbf{w}$.

(c) Find $\mathbf{u}$, $\mathbf{v}$, and $\mathbf{w}$ if $A = \begin{pmatrix} 1 & 1 & 0 \\ 0 & 1 & 1 \\ 0 & 0 & 1 \end{pmatrix}$. Then form a linear combination of the solutions of part (b).

(d) Solve $\dot{\mathbf{x}} = A\mathbf{x}$ by successive integration, and compare the result with part (c).

4 STABILITY FOR AUTONOMOUS SYSTEMS†

When an explicit solution formula $\mathbf{x}(t)$ is available for a vector differential equation

$$\frac{d\mathbf{x}}{dt} = \mathbf{F}(t,\mathbf{x}), \qquad t_0 \leq t < \infty,$$

it may be possible to determine whether $\lim_{t\to\infty} \mathbf{x}(t)$ exists or whether $\mathbf{x}(t)$ exhibits some other kind of behavior. Even when solution formulas are not available, we can sometimes still get information about the long-term behavior of solutions. We'll make the simplifying assumption that $\mathbf{F}(t,\mathbf{x}) = \mathbf{F}(\mathbf{x})$ does not depend explicitly on t; such a system is called **autonomous.** We'll deal at first only with the case in which $\mathbf{x} = (x, y)$ is two-dimensional and the system is linear. For nonlinear and 2-dimensional systems see Section 4B. Chapter 7, Section 4, treats higher-dimensional systems.

4A Linear Systems

Throughout this subsection we'll be dealing with normal-form autonomous systems

4.1

$$\begin{aligned} \frac{dx}{dt} &= ax + by, \\ \frac{dy}{dt} &= cx + dy, \qquad a, b, c, d \text{ real numbers.} \end{aligned}$$

In practice, the system might have arisen initially from an order reduction of a second-order equation for x to a first-order system by letting $\dot{x} = y$. As we did in Chapter 4, Section 3, we call the xy plane the **phase plane** of the second-order equation, and the same term has come to be used for the xy plane of Equations 4.1 also.

In Section 2 we reduced the solution of such systems to solving single second-order equations which we could then treat by the characteristic root method of Chapter 3. In Section 3 we introduced eigenvectors and eigenvalues λ_1 and λ_2 to solve the same system. Whichever approach we use, the basic solutions to Equations 6.1 are of course the same, namely,

4.2
$$x(t) = c_1 e^{\lambda_1 t} + c_2 e^{\lambda_2 t},$$
$$y(t) = c_3 e^{\lambda_1 t} + c_4 e^{\lambda_2 t},$$

where only two of the four constants can be chosen arbitrarily. If $\lambda_1 = \lambda_2$, then $e^{\lambda_2 t}$ will be replaced by $te^{\lambda_1 t}$. Using either approach the numbers λ_1 and λ_2 are the **roots** of the **characteristic equation,** in this case a quadratic equation, that's expressible as a 2-by-2 determinant, namely,

$$\begin{vmatrix} a-\lambda & b \\ c & d-\lambda \end{vmatrix} = 0, \quad \text{or} \quad \lambda^2 - (a+d)\lambda + (ad - bc) = 0.$$

There will in general be some relation between the constants c_k, determined by substitution into the system.

The long-term behavior of a constant solution $x(t) = x_0$, $y(t) = y_0$ as t tends to infinity is very simple. The solution simply remains constant, and such a solution is called an **equilibrium solution.** Viewed as a single point, (x_0, y_0) is called an **equilibrium point.** Substitution shows that Equations 4.1 always have $x(t) = 0$, $y(t) = 0$ for an equilibrium solution. The following theorem describes conditions under which all solution trajectories that start close enough to (0,0) will remain close to the equilibrium point as $t \to \infty$, in which case (0,0) is called a **stable equilibrium.** If instead there are trajectories starting arbitrarily close to (0,0) that are infinitely often a fixed distance away from (0,0), then (0,0) is called an **unstable equilibrium.**

4.3 THEOREM

Let λ_1 and λ_2 be the characteristic roots that occur in Solution 4.2 to System 4.1. We have the following possibilities:

(i) If the real parts of both λ_1 and λ_2 are negative, then $\lim_{t\to\infty}(x(t), y(t)) = (0,0)$, for all choices of the constants c_k, and (0,0) is called an **asymptotically stable equilibrium.**

(ii) If λ_1 and λ_2 are conjugate and purely imaginary $(\lambda_1 = iq, \lambda_2 = -iq, q \neq 0)$, then all solutions 4.2 are periodic of period $2\pi/q$, and (0,0) is a **stable** equilibrium.

(iii) If either λ_1 or λ_2 has positive real part, then there are choices of the constants c_k for which $x(t)$ and $y(t)$ are unbounded, and (0,0) is an **unstable** equilibrium.

(iv) If either or both of λ_1 and λ_2 is zero, all solution trajectories are confined to straight lines and may even reduce to single points.

Proof. The proof consists of examining the form that the solutions take in the four cases:

(i) With $\lambda_1 = -p + iq$, $\lambda_2 = -p - iq$, $p > 0$, the solutions are linear combinations of $e^{-pt}\cos qt$ *and* $e^{-pt}\sin qt$, or else of e^{-pt} and te^{-pt}. Since these basic functions tend to zero as t tends to ∞, all solutions tend to zero also.

(ii) With $\lambda_1 = iq, \lambda_2 = -iq, q \neq 0$, the solutions are linear combinations of $\cos qt$ and $\sin qt$ and so are either periodic or zero.

(iii) With $\lambda_1 = p + iq, \lambda_2 = p - iq, p > 0$, the solutions are linear combinations of $e^{pt} \cos qt$ and $e^{pt} \sin qt$, or else of e^{pt} and te^{pt}. Thus nonzero linear combinations can assume arbitrarily large values as $t \to \infty$; if the coefficients are zero, the solutions are identically zero.

(iv) With $\lambda_1 = 0$ and $\lambda_2 = r \neq 0$, the solutions are of the form $x = c_1 + c_2 e^{rt}$, $y = c_3 + c_4 e^{rt}$. If $r = 0$, we get instead $x = c_1 + c_2 t$, $y = c_3 + c_4 t$. If both c_2 and c_4 are zero, the trajectory is a single point. Otherwise, if for instance $c_4 \neq 0$, we can eliminate e^{rt} or t and find straight-line trajectories. ■

The description that Theorem 4.3 gives for the behavior of solutions of System 4.1 fails to capture the interesting variety of geometric behaviors that solutions can exhibit; the best way to understand this is to look at some pictures of what can happen. Since everything depends on the roots λ_1 and λ_2 of the characteristic equation, it helps to derive the characteristic equation once and for all. Writing the system in operator form, we have

$$\begin{aligned} (D - a)x - by &= 0, \\ -cx + (D - d)y &= 0. \end{aligned}$$

Eliminating y gives

$$(D - d)(D - a)x - bcx = 0,$$

or

$$\left[D^2 - (a + d)D + (ad - bc)\right]x = 0.$$

Thus the **characteristic roots** λ_1 and λ_2 are the roots of the **characteristic equation**

$$\lambda^2 - (a + d)\lambda + (ad - bc) = 0;$$

this equation is easy to recall by writing it using a 2-by-2 determinant:

$$\begin{aligned} \begin{vmatrix} a - \lambda & b \\ c & d - \lambda \end{vmatrix} &= (a - \lambda)(d - \lambda) - bc \\ &= \lambda^2 - (a + d)\lambda + (ad - bc) = 0. \end{aligned}$$

It's from the determinant form of the equation that the roots arise as eigenvalues of the system.

---

example 1 In the system

$$\begin{aligned} \frac{dx}{dt} &= -3x + 2y, \\ \frac{dy}{dt} &= -4x + 3y, \end{aligned}$$

$a + d = 0$ and $ad - bc = -1$, so the characteristic equation is

$$\lambda^2 - 1 = 0.$$

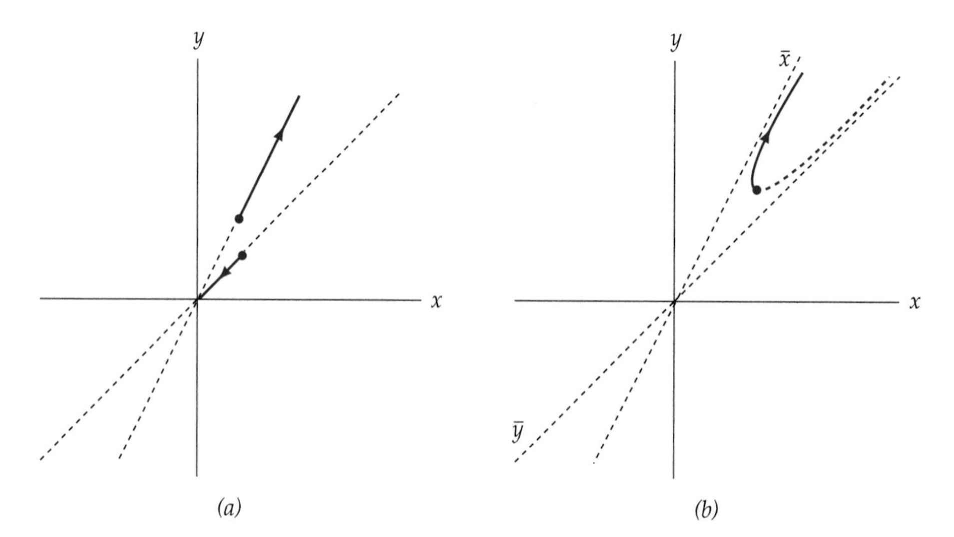

Figure 5.13 Directed trajectories.

The roots are $\lambda_1 = -1$, $\lambda_2 = 1$, and substitution of Equations 4.2 into either one of the differential equations shows that $c_3 = c_1$ and $c_4 = 2c_2$. Hence the general solution is

$$x(t) = c_1e^{-t} + c_2e^t, \qquad y(t) = c_1e^{-t} + 2c_2e^t.$$

The choice $c_1 = 0$, $c_2 = 1$ gives us

$$x(t) = e^t, \qquad y(t) = 2e^t.$$

The trajectory satisfies $2x = y$, which is the equation of a line. Since $x(t) \to \infty$ and $y(t) \to \infty$ as $t \to \infty$, the positive functions in $\big(x(t), y(t)\big) = (e^t, e^{2t})$ trace the upper trajectory in the first quadrant, shown in Figure 5.13(a). On the other hand, choosing $c_1 = c_2 = 1$ gives us

$$x(t) = e^{-t} + e^t, \qquad y(t) = e^{-t} + 2e^t.$$

Since $y(t) - 2x(t) = -e^{-t}$ and $y(t) - x(t) = e^t$ the trajectory satisfies $(y - x)(y - 2x) = -1$, which is a hyperbola. Again $x(t) \to \infty$ and $y(t) \to \infty$ as $t \to \infty$. Introducing new coordinates $(\bar{x}, \bar{y})$ by $\bar{x} = y - x$, $\bar{y} = y - 2x$, the hyperbola becomes $\bar{x}\bar{y} = -1$, shown in Figure 5.13(b). The directed trajectory is shown for $t > 0$. Finally, If we choose $c_1 = 1$ and $c_2 = 0$, the trajectory satisfies $x = y$ and is traced toward the origin by $\big(x(t), y(t)\big) = (e^{-t}, e^{-t})$ on the lower trajectory in Figure 5.13(a).

A **phase portrait** for a system consists of directed sketches of enough trajectories to convey a good idea of the variety of possible solution behaviors. Typical portraits organized by type of characteristic root are on pages 379 and 380. The first sketch in each case shows the simplest possible behavior relative to x and y axes. A more typical portrait is also shown for each type along with the conventional term for the displayed type. The entire portrait display is intended as a reference guide, and you should convince yourself, one type at a time, that each picture looks reasonable relative to the given characteristic values.

I. Unstable Node: $0 < \lambda_1 < \lambda_2$

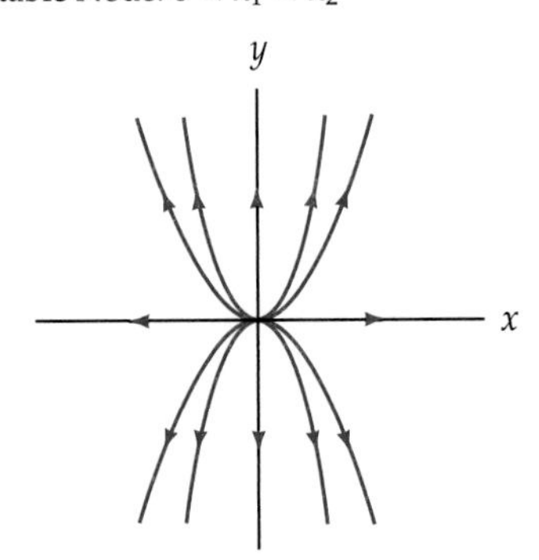

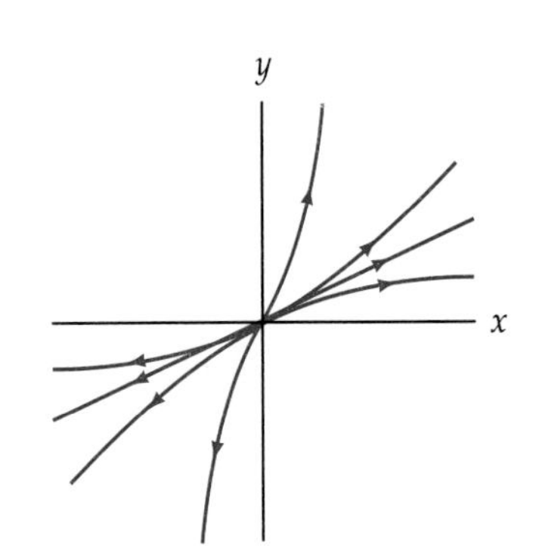

Example: $x = 2e^t$, $y = e^{3t}$; $8y = x^3$.

II. Saddle (Unstable): $\lambda_1 < 0 < \lambda_2$

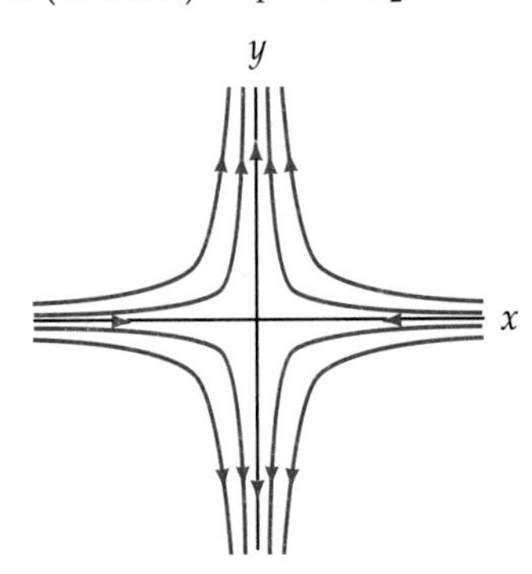

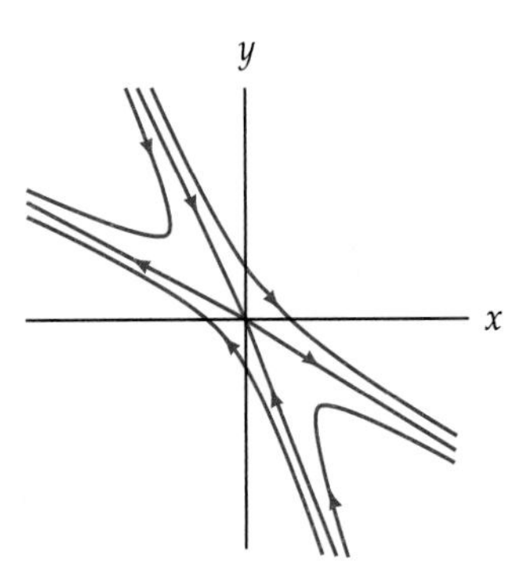

Example: $x = e^{-t}$, $y = 3e^t$; $xy = 3$.

III. Asymptotically Stable Node: $\lambda_1 < \lambda_2 < 0$

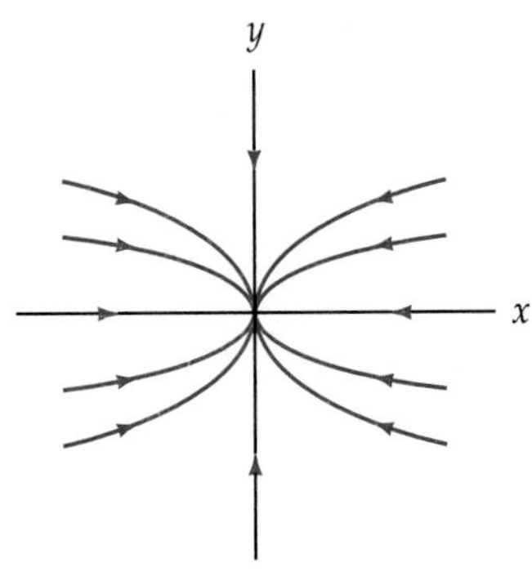

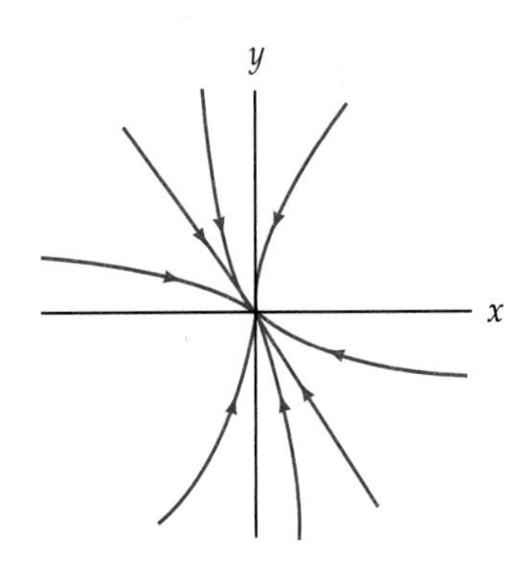

Example: $x = 2e^{-3t}$, $y = e^{-t}$; $x = 2y^3$.

IV. Unstable Spiral: $\lambda_1 = p + iq$, $\lambda_2 = p - iq$, $p > 0, q \neq 0$

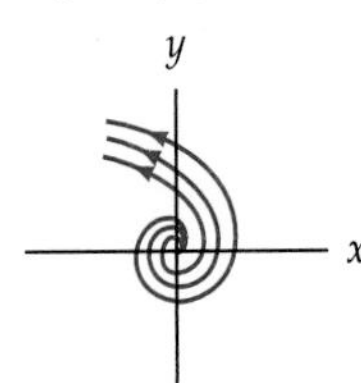

Example: $x = e^t \cos 2t$, $y = e^t \sin 2t$; $x^2 + y^2 = e^{\arctan y/x}$.

V. Stable Center: $\lambda_1 = iq, \lambda_2 = -iq, q \neq 0.$

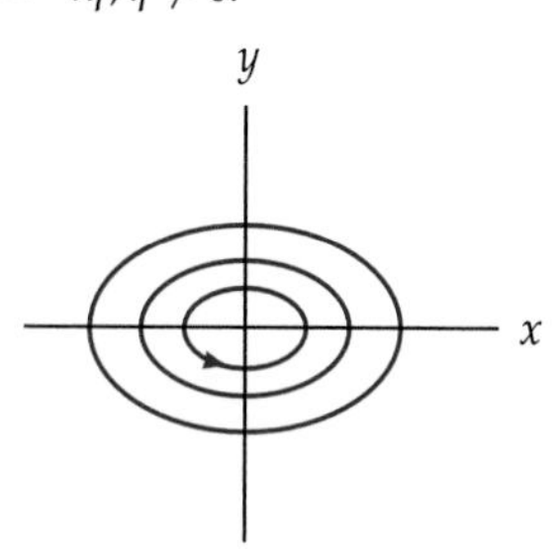

Example: $x = 2\cos t,\ y = \sin t;\ x^2 + 4y^2 = 4.$

VI. Asymptotically Stable Spiral: $\lambda_1 = -p + iq, \lambda_2 = -p - iq, p > 0, q \neq 0$

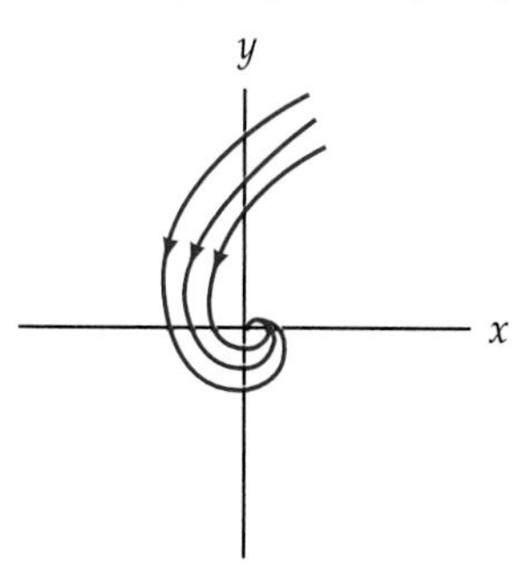

Example: $x = e^{-t}\cos t,\ y = e^{-t}\sin t;\ x^2 + y^2 = e^{-2\arctan y/x}.$

VII. Unstable Star: $\lambda_1 = \lambda_2 > 0$

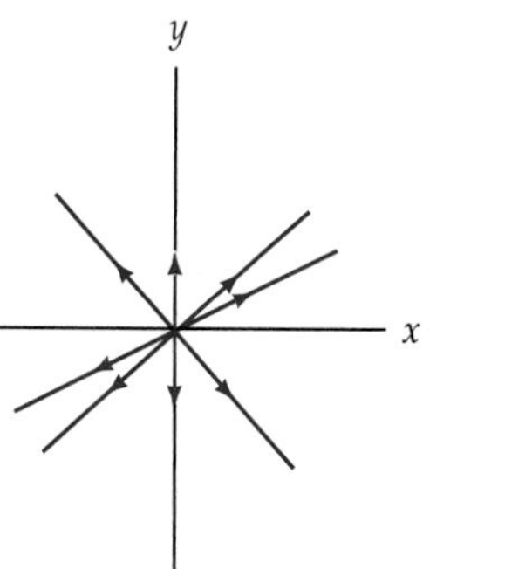

Example: $x = 2e^t,\ y = 3e^t,\ 3x = 2y.$

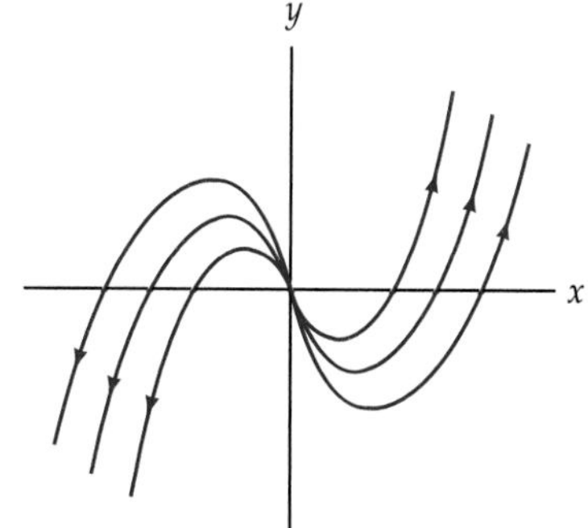

Example: $x = e^t,\ y = te^t,\ y = x\ln x.$

VIII. Asymptotically Stable Star: $\lambda_1 = \lambda_2 < 0$

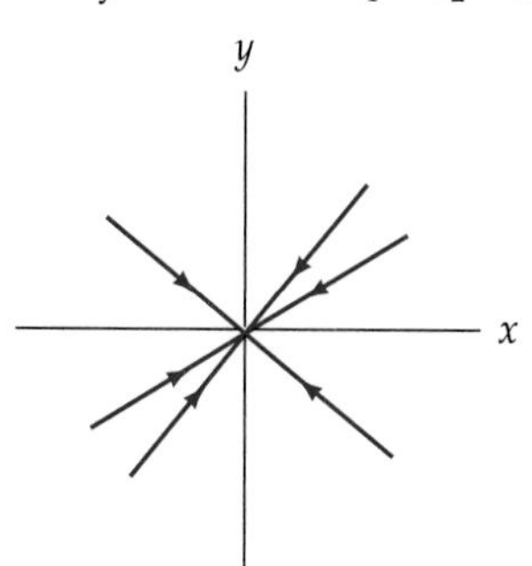

Example: $x = e^{-t},\ y = 3e^{-t},\ 3x = y.$

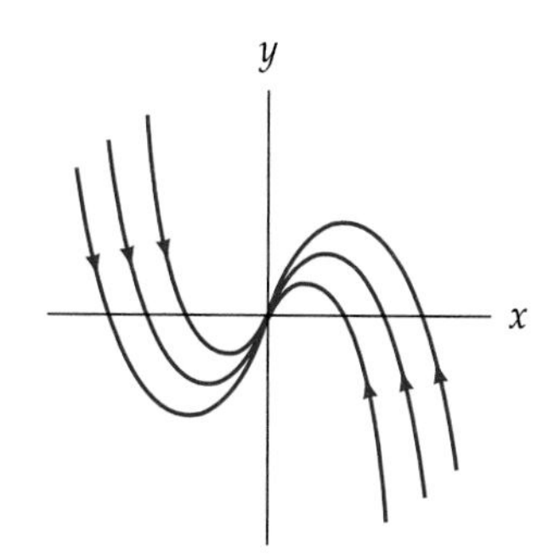

Example: $x = e^{-t},\ y = te^{-t},\ y = -x\ln x.$

example 2

The trajectories in the phase portrait in Figure 5.14. All appear to be spiraling toward the origin, so we might conclude tentatively that we're looking at an asymptotically stable spiral. If the underlying system were linear, the characteristic roots would have the form $\lambda_1 = -p + iq$, $\lambda_2 = -p - iq$, $p > 0$, $q \neq 0$. Indeed, the portrait belongs to the system $\dot{x} = -x + y$, $\dot{y} = -x - y$, which has roots $-1 \pm i$.

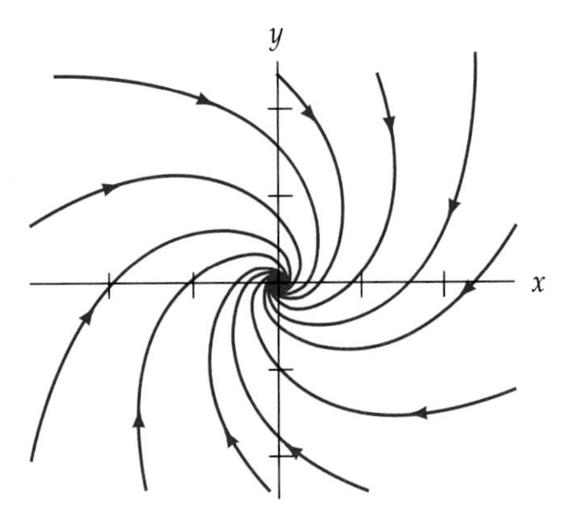

Figure 5.14 Asymptotically stable spiral.

example 3

The damped oscillator equation $\ddot{x} + k\dot{x} + hx = 0$ is equivalent to the system $\dot{x} = y$, $\dot{y} = -hx - ky$. We find characteristic roots $\lambda_{1,2} = -(k/2) \pm \sqrt{(k/2)^2 - h}$. Assuming as usual that $k > 0$, we see that if $h < (k/2)^2$, then both roots are negative, so the origin is an asymptotically stable node in the xy phase plane. If $h > (k/2)^2$ the roots are complex, but still with negative real parts, so we get an asymptotically stable spiral. If $h = (k/2)^2$ the roots are negative and equal, so we get an asymptotically stable star.

The characteristic equation of Equations 4.1 is

$$\lambda^2 - (a + d)\lambda + (ad - bc) = 0$$

and has $\lambda_1 = 0$ for a root precisely when $ad - bc = 0$. This is case (iv) in Theorem 4.3 and is not covered by the preceding stability classification; this case is called **degenerate** for two reasons. One is that the solutions $c_1 + c_2 e^{\lambda_2 t}$, or $c_1 + c_2 t$ if $\lambda_2 = 0$, contain at most one nonconstant function. The other reason is that $ad - bc = 0$ precisely when there are infinitely many equilibrium points for the system. The equilibrium points (x, y) occur when

$$ax + by = 0$$
$$cx + dy = 0$$

both hold in Equations 4.1. Exercise 26(b), asks you to show that the equilibrium points cover at least an entire line when $ad - bc = 0$. If a, b, c, and d are all zero, every point is an equilibrium point.

example 4 The system

$$\frac{dx}{dt} = x + y,$$
$$\frac{dy}{dt} = 2x + 2y$$

has equilibrium points (x, y) satisfying

$$x + y = 0,$$
$$2x + 2y = 0.$$

The solutions (x, y) lie on the line $x + y = 0$. The characteristic roots satisfy $\lambda^2 - 3\lambda = 0$, and are $\lambda_1 = 0$, $\lambda_2 = 3$. The solutions of the system are

$$x(t) = c_1 + c_2 e^{3t},$$
$$y(t) = -c_1 + 2c_2 e^{3t}.$$

Elimination of e^{3t} between these two equations shows that the trajectories of the nonequilibrium solutions are parallel lines with slope 2.

EXERCISES

Find the characteristic roots λ_1 and λ_2 associated with each system in Exercises 1 through 8. Then classify each system by name, according to the types categorized in I through VIII of the text.

1. $dx/dt = -3x + 2y,$
$dy/dt = -4x + 3y.$

2. $dx/dt = x + 4y,$
$dy/dt = 5y.$

3. $dx/dt = 2x - y,$
$dy/dt = x + 2y.$

4. $dx/dt = x,$
$dy/dt = 3x + y.$

5. $dx/dt = 2x - 3y,$
$dy/dt = 2x - 2y.$

6. $dx/dt = -x + y,$
$dy/dt = -4x - y.$

7. $dx/dt = x,$
$dy/dt = 2x.$

8. $dx/dt = x + y,$
$dy/dt = x + y.$

Each of the second-order equations in Exercises 9 through 12 is equivalent to a first-order system obtained by letting $dx/dt = y$. Classify by name, according to the listing I–VIII, each equation (or equivalently each system) by finding the characteristic roots of the given second-order equation directly.

9. $d^2x/dt^2 - x = 0.$

10. $d^2x/dt^2 + dx/dt - x = 0.$

11. $d^2x/dt^2 + dx/dt + x = 0.$

12. $d^2x/dt^2 + x = 0.$

The second-order equations in Exercises 13 through 16 can be thought of as determining the development $x(t)$ of a spring system subject to a frictional force $k\,dx/dt$, where $k \geq 0$ is constant. Find conditions on k under which the equation belongs to any or all of the classes in the standard I–VIII.

13. $d^2x/dt^2 + k\,dx/dt + x = 0.$

14. $d^2x/dt^2 + k\,dx/dt + 3x = 0.$

15. $2d^2x/dt^2 + k\,dx/dt + x = 0.$

16. $m\,d^2x/dt^2 + k\,dx/dt + 3x = 0,\ m > 0.$

17. Sketch typical phase plots in the xy plane for each of the systems in Exercises 1 through 8.

18. Sketch typical phase plots in the xy plane, where $y = dx/dt$, for each of the second-order equations in Exercises 9 through 12.

19. Show that the system

$$\frac{dx}{dt} = ax + by, \qquad \frac{dy}{dt} = cx + dy$$

has infinitely many constant equilibrium solutions if and only if $ad - bc = 0$. [*Hint:* Try to solve $ax + by = cx + dy = 0$.]

20. Consider the general solution (Equations 4.2 of the text) of a linear system in which $\lambda_1 < \lambda_2 < 0$, and for which (0,0) is an asymptotically stable node.

(a) Show that, if c_2 and c_4 are not both zero in Equations 4.2, then as $t \to \infty$ the slope $dy/dx = (dy/dt)/(dx/dt)$ of the associated trajectory approaches c_4/c_2, interpreted as a vertical slope if $c_2 = 0$.

(b) Show that if $c_2 = c_4 = 0$, but c_1 and c_3 are not both zero, then all trajectories are straight lines with slope c_3/c_1, interpreted as vertical slope if $c_1 = 0$.

21. (a) Show that the nonlinear system

$$\frac{dx}{dt} = A - Bx - x + x^2y, \qquad A \neq 0,$$
$$\frac{dy}{dt} = Bx - x^2y$$

has a single constant solution $x(t) = A$, $y(t) = B/A$. Recall that the one-point trajectory of a constant solution is called an *equilibrium point.*

(b) Assume that $A > 0$ and $B > A^2 + 1$. Trajectories starting sufficiently near, but not at, the equilibrium point exhibit *limit cycle* behavior in that they approach a closed trajectory. Investigate this claim by using a graphical-numerical method.

22. What difference, if any, is there between the phase portraits of the system $\dot{x} = f(x,y)$, $\dot{y} = g(x,y)$ and the system $\dot{x} = -f(x,y)$, $\dot{y} = -g(x,y)$?

23. Match the portraits (a), (b), (c), and (d) with the systems

(a) $\dot{x} = x,\ \dot{y} = x - y.$

(b) $\dot{x} = x,\ \dot{y} = x + y.$

(c) $\dot{x} = x - y,\ \dot{y} = x + y.$

(d) $\dot{x} = 2y,\ \dot{y} = -x.$

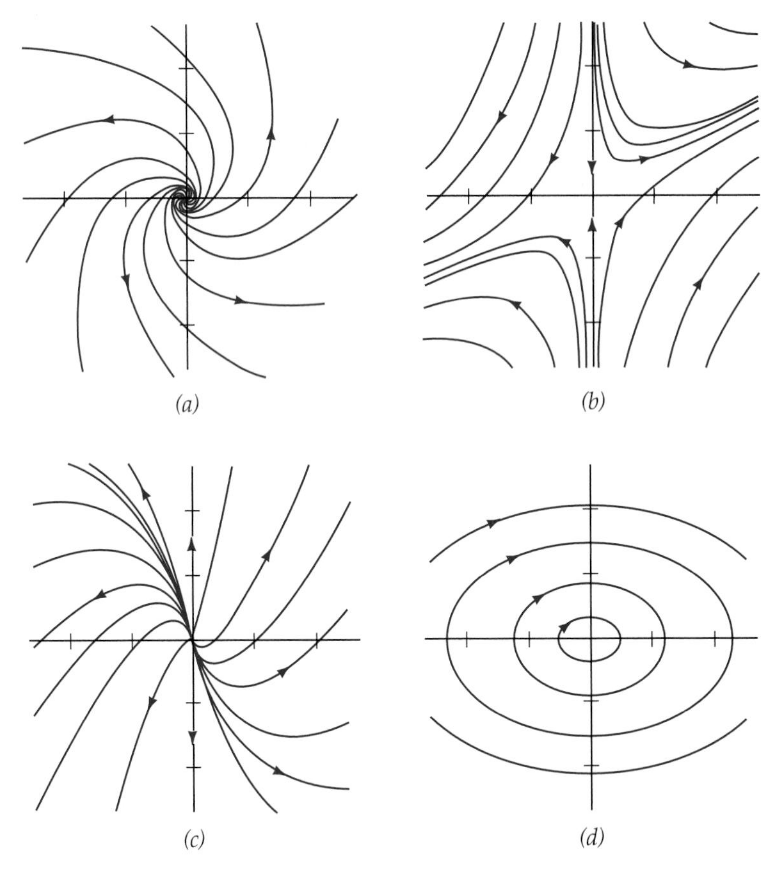

24. Show that if the system $\dot{x} = ax + by$, $\dot{y} = cx + dy$ has real characteristic roots then it has an asymptotically stable node at the origin if and only if $a + d < 0$ and $ad - bc > 0$.

25. **(a)** Solve the system $\dot{x} = y$, $\dot{y} = -x - y$, $\dot{z} = x - z$.

(b) Show that every solution tends to the equilibrium point (0,0,0), which is thus asymptotically stable.

(c) What changes if the last equation is replaced by $\dot{z} = x + z$?

26. **(a)** Show that Equations 4.1 of the text have (0,0) as their only equilibrium point precisely when $ad - bc \neq 0$.

(b) Show that if $ad - bc = 0$ in Equations 4.1, then there are infinitely many equilibrium points, covering an entire line or even the entire plane.

4B Nonlinear Systems and Linearization

Methods for determining the long term ($t \to \infty$) stability or instability of solutions of the 2-dimensional nonlinear system

4.4
$$\frac{dx}{dt} = f(x,y), \qquad \frac{dy}{dt} = g(x,y)$$

are useful because explicit solution formulas are often impossible to find, and approximate numerical solutions are necessarily always limited to some finite time interval $t_0 \le t \le t_1$. The 2-dimensional case is singled out first for detailed attention because of its relevance to the system $\dot{x} = y$, $\dot{y} = f(t,x,y)$ associated with the widely applicable second-order equation $\ddot{x} = f(x,\dot{x})$. The most extreme instability occurs for systems of dimension more than two, and systems of higher dimension appear in the following Section 4C and also in Chapter 7, Section 4. First we'll consider the behavior of solution trajectories near an *equilibrium point* of the system 4.4, that is, a point (x_0, y_0) having the property that $f(x_0, y_0) = g(x_0, y_0) = 0$. Associated with each such point is a constant *equilibrium solution* $x(t) = x_0$, $y(t) = y_0$. We will be concerned with finding out whether or not a solution that passes near such a point tends asymptotically to the point or exhibits some other kind of behavior. Given the nature of the general criteria we'll develop, we'll mainly restrict ourselves to considering an **isolated equilibrium point,** that is, an equilibrium point that does not have other equilibrium points arbitrarily close to it. Thus the kind of solution behavior described as "degenerate" in Section 4A will not be included.

Let (x_0, y_0) be an isolated equilibrium point of System 4.4. We are interested in solution behavior near (x_0, y_0), but we change variable, replacing x by $x + x_0$ and y by $y + y_0$, to reduce to the case $(x_0, y_0) = (0,0)$ of Section 4A. The system then looks like

4.5
$$\frac{dx}{dt} = f(x + x_0, y + y_0), \qquad \frac{dy}{dt} = g(x + x_0, y + y_0),$$

and the equilibrium point of interest is now at $x = 0$, $y = 0$. Assume f and g are continuously differentiable, and use a first-degree **Taylor expansion:** thus we replace System 4.5 by the **linearized system** in terms of partial derivatives f_x, f_y, g_x, and g_y of f and g, namely,

4.6
$$\frac{dx}{dt} = f_x(x_0, y_0)x + f_y(x_0, y_0)y,$$
$$\frac{dy}{dt} = g_x(x_0, y_0)x + g_y(x_0, y_0)y.$$

It can be shown that System 4.6 is in a precise sense the best of all linear systems to examine near (x_0, y_0) instead of the nonlinear System 4.5. [If 4.5 were itself a homogeneous linear system, then 4.6 would be exactly the same system as 4.5. For instance, if $f(x,y) = ax + by$, then $f_x(x_0, y_0) = a$ and $f_y(x_0, y_0) = b$, for all x_0, y_0.]

example 5

The equilibrium points of the nonlinear system

$$\frac{dx}{dt} = 2x - y^2,$$
$$\frac{dy}{dt} = x + y$$

are found by solving the algebraic system

$$2x - y^2 = 0,$$
$$x + y = 0.$$

The only solutions are $(x_0, y_0) = (0,0)$ and $(x_1, y_1) = (2,-2)$, so we have two isolated equilibrium points. Near (x_0, y_0) we consider the linearized system

$$\frac{dx}{dt} = 2x - (2y_0)y,$$
$$\frac{dy}{dt} = x + y.$$

When $y_0 = 0$, the system has solutions

$$x(t) = c_1 e^{2t},$$
$$y(t) = c_1 e^{2t} + c_2 e^t.$$

The two positive characteristic roots show that the linear system has an unstable node at (0,0). Near $(x_1, y_1) = (2,-2)$ the linearization has the distinct characteristic roots $\lambda_1 = (3 + \sqrt{17})/2$, $\lambda_2 = (3 - \sqrt{17})/2$. Thus the solutions exhibit saddle-type behavior as $t \to \infty$, since $\lambda_1 > 0$ and $\lambda_2 < 0$.

The next important step in Example 5 would be to draw some conclusion about the behavior of solutions of the given nonlinear system from what we know about the solutions of the linearized system at each equilibrium point. To state the connection clearly and succinctly, we will use some standard terminology. Let $x_0(t) = x_0$, $y_0(t) = y_0$ be an equilibrium solution of an autonomous system of differential equations, linear or nonlinear. Then (x_0, y_0) is **asymptotically stable** if every solution $(x(t), y(t))$ of the system that has a trajectory point close enough to (x_0, y_0) satisfies

$$\lim_{t\to\infty} \big(x(t), y(t)\big) = (x_0, y_0).$$

For the linear systems in Section 4A, the examples numbered III, VI, and VIII show asymptotically stable behavior at (0,0). Special types of asymptotic stability are the **spiral,** in which the trajectory winds around the equilibrium point as it approaches it, and the **node,** in which the tangent to each trajectory approaches a limiting position. A less strict condition on the system at (x_0, y_0) is that (x_0, y_0) be simply **stable,** which means that every trajectory that has a point close enough to (x_0, y_0) will remain within a preassigned distance of (x_0, y_0). According to the definitions, if (x_0, y_0) is asymptotically stable for a system, then it is also stable. Number V in Section 4A is stable without being asymptotically stable. All other equilibrium points are called **unstable;** these include types I, II, IV, and VII of Section 4A. In particular, type II, the **saddle point,** can be characterized for the nonlinear system as having two lines through it, one of which is tangent to a trajectory approaching the point and the other of which is tangent to a trajectory that moves away from the point.

4.7 THEOREM

Let $f(x,y)$ and $g(x,y)$ be continuously differentiable in a region R of the xy plane, and suppose the point (x_0, y_0) in R is an isolated equilibrium point of the system

$$\frac{dx}{dt} = f(x,y), \qquad \frac{dy}{dt} = g(x,y).$$

Suppose that the associated linearized system at (x_0, y_0),

$$\frac{dx}{dt} = f_x(x_0, y_0)x + f_y(x_0, y_0)y, \qquad \frac{dy}{dt} = g_x(x_0, y_0)x + g_y(x_0, y_0)y,$$

is **nondegenerate:**

$$f_x(x_0, y_0)g_y(x_0, y_0) - f_y(x_0, y_0)g_x(x_0, y_0) \neq 0.$$

Then, at (x_0, y_0),

(i) Asymptotic stability for the linearized system at (0,0) implies asymptotic stability for the given system at (x_0, y_0), with nodal and spiral behavior corresponding similarly.

(ii) Instability for the linearized system at (0,0) implies instability for the given system at (x_0, y_0). In particular, if (0,0) is a saddle point for the linearized system, it is a saddle point for the given system, while nodal and spiral behavior correspond similarly.

(iii) If (0,0) is a stable center for the linearized system, then (x_0, y_0) may be an unstable spiral, an asymptotically stable spiral, or perhaps merely stable for the nonlinear system.

The proof of parts (i) and (ii) involves some technical analysis that we won't undertake. Proving part (iii) just amounts to looking at some appropriate examples, and this is asked for in Exercise 12 at the end of the section.

example 6

Continuing with Example 5, we conclude from Theorem 4.7, part (ii), above, that the nonlinear system has an unstable equilibrium at $(x_0, y_0) = (0,0)$, since (0,0) is an unstable node for the linearized system. The same conclusion holds for the point $(2,-2)$, at which the linearized system has a saddle point. More specifically, $(2,-2)$ is a saddle point for the nonlinear system since $\lambda_1 = (3 - \sqrt{17})/2 < 0 < \lambda_2 = (3 + \sqrt{17})/2$.

example 7

The undamped nonlinear pendulum equation $\ddot{\theta} = -(g/l)\sin\theta$ is derived in Section 2 of Chapter 3. Here $\theta = \theta(t)$ is the displacement angle from the downward-oriented vertical. Letting $x = \theta$ and $y = \dot{\theta}$, the equivalent system is

$$\frac{dx}{dt} = y, \qquad \frac{dy}{dt} = -\frac{g}{l}\sin x.$$

Mechanical intuition suggests that the equilibrium solutions are those for which $y = \dot{\theta} = 0$ and $x = \theta = n\pi$, where n is an integer. When n is even the pendulum hangs straight down, and when n is odd, it is precisely balanced in a straight-up position, the latter obviously being mechanically unstable. Solving the equations $y = 0, -(g/l)\sin x = 0$ for $(x, y) = (0, n\pi)$ confirms some of our intuition. For example, consider the linearized system at $(x_n, y_n) = (0, n\pi)$. We find

$$\frac{dx}{dt} = y,$$
$$\frac{dy}{dt} = -\left(\frac{g}{l}\right)(\cos n\pi)x.$$

The characteristic roots satisfy

$$\lambda^2 + (-1)^n \frac{g}{l} = 0,$$

since $\cos n\pi = (-1)^n$. If n is even (pendulum down), the roots are $\pm i\sqrt{g/l}$. If n is odd, the roots are $\pm\sqrt{g/l}$. In the first case, we have a center for the linear system, so Theorem 4.7 fails to confirm or deny our intuition. It takes a more detailed analysis (see Exercise 11) to show that the expected stability does occur near $(x_0, y_0) = (2m\pi, 0)$. When n is odd, the linear system has a saddle point at $((2m+1)\pi, 0)$. This tells us that, while the vertical positions with zero velocity are unstable, there is in principle a path in the (x, y) phase space tending toward, but not reaching, a straight-up balanced position.

Figure 4.14 of Chapter 4, Section 3, shows the result of some numerical computations of phase-space trajectories. The effect of frictional damping on the nature of equilibrium points is taken up in Exercise 13 below and in Section 4 of Chapter 4.

There may be an equilibrium point of an autonomous system that is isolated from any other specific trajectory of the system, though we've seen in the previous section that this situation is far from typical. (Indeed the stable center numbered V in the display is the only one in the entire list of 2-dimensional linear types that has that property.) The following theorem gives conditions that imply the complementary case, in which an equilibrium point $\mathbf{x}_0$ has at least one other trajectory with points arbitrarily close to $\mathbf{x}_0$.

4.8 THEOREM

Let $F(\mathbf{x})$ be a continuous vector field and let $\mathbf{x}(t)$ be a solution of the autonomous system $\dot{\mathbf{x}} = F(\mathbf{x})$. If either of the limits

$$\lim_{t\to\infty} \mathbf{x}(t) \quad \text{or} \quad \lim_{t\to-\infty} \mathbf{x}(t),$$

exists then the limit vector is an equilibrium point for the system.

Proof. Denoting the limit of $\mathbf{x}(t)$ as $t \to \infty$ by $\mathbf{x}_0$, we just need to show that $F(\mathbf{x}_0) = 0$ to guarantee that $\mathbf{x}_0$ is an equilibrium point. Since $F(\mathbf{x})$ is continuous at $\mathbf{x}_0$,

$$\lim_{t\to\infty} F(\mathbf{x}(t)) = F\left(\lim_{t\to\infty} \mathbf{x}(t)\right) = F(\mathbf{x}_0).$$

If $F(\mathbf{x}_0) \neq 0$, it must have some coordinate $a_k \neq 0$. Then for all large enough t,

$$|F_k(\mathbf{x}(t)) - a_k| < \tfrac{1}{2}|a_k|, \quad \text{that is} \quad -\tfrac{1}{2}|a_k| + a_k < F_k(\mathbf{x}(t)) < \tfrac{1}{2}|a_k| + a_k.$$

If $a_k > 0$ then $\frac{1}{2}a_k < F_k(\mathbf{x}(t)) = \dot{x}_k(t)$ for all large enough t, so $x_k(t)$ is unbounded above. If $a_k < 0$ then $\frac{1}{2}a_k > F_k(\mathbf{x}(t)) = \dot{x}_k(t)$ for all large enough t, so $x_k(t)$ is unbounded below. In either case $\mathbf{x}(t)$ can't have a limit, so $F(\mathbf{x}_0) = 0$ is the only possibility. If $t \to -\infty$ the argument is essentially the same. ■

example 8

Figure 5.15(a) is a scaled-down sketch of the vector field of the system $\dot{x} = -x(x^2 + y^2 - 1)$, $\dot{y} = -y(x^2 + y^2 + 1)$. When $y \neq 0$ the second coordinate of the field can't be zero, so all equilibrium points must be on the x axis, where $y = 0$. In that case we just have to find solutions of $-x(x^2 - 1) = 0$; these are $x = -1,\ 0,\ 1$, so the equilibrium points are at $(-1,0)$, $(0,0)$, and $(1,0)$. The sketch suggests that the points $(\pm 1,0)$ are asymptotically stable equilibrium points, because the nearby arrows point at them. Similarly, $(0,0)$ appears to be a saddle point. Establishing the truth of these statements using Theorem 4.7 is left as an exercise. Our main interest here is in illustrating the conclusion of Theorem 4.8. It appears from the sketch of the vector field that there is a trajectory lying on the y-axis that approaches $(0,0)$ from above and also one from below. In addition there appears to be a trajectory on the x axis extending from $(0,0)$ to $(1,0)$ and another from $(0,0)$ to $(-1,0)$. The precise meaning of this last statement is that a solution assuming a value $\mathbf{x}_0$ on the line segment between $(-1,0)$ and $(0,0)$ tends along the segment to $(-1,0)$ as $t \to \infty$ and to $(0,0)$ as $t \to -\infty$. This, as well as our other assertions about trajectories can be verified by direct computation or by appealing to Theorem 3.1 of Chapter 2.

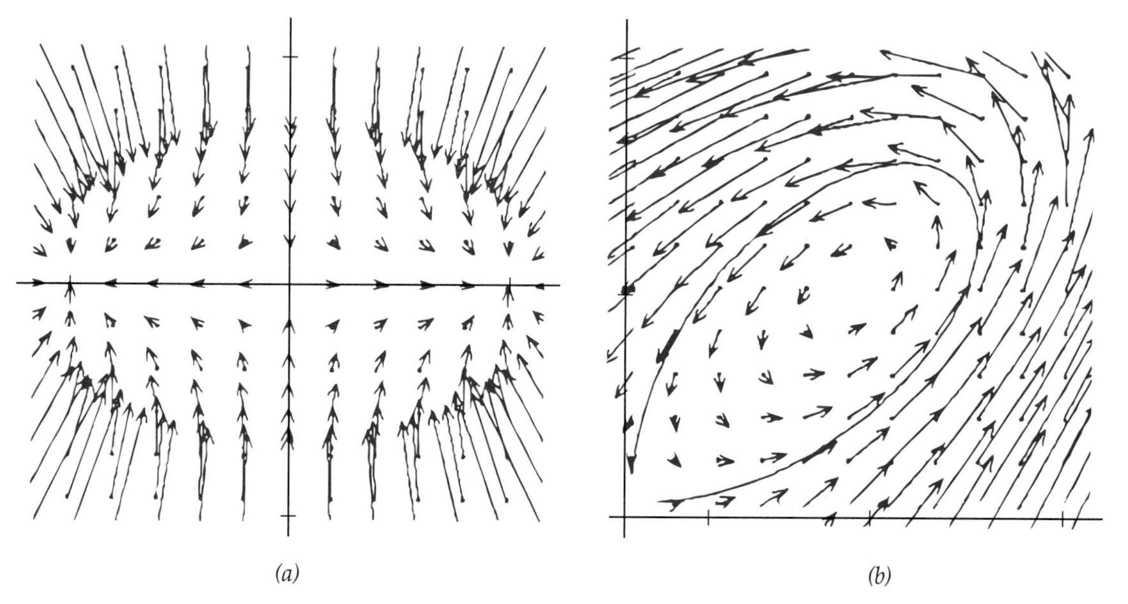

Figure 5.15 (a) $\dot{x} = -x(x^2 + y^2 - 1)$, $\dot{y} = -y(x^2 + y^2 + 1)$. (b) $\dot{x} = x - y^2$, $\dot{y} = -y + x^2$.

example 9

The system $\dot{x} = x - y^2$, $\dot{y} = -y + x^2$ has a scaled vector field shown for the first quadrant in Figure 5.15(b), along with a trajectory extending counterclockwise away from the origin and back again. The equilibrium point at the origin is evidently of saddle type, since the linearization there is $\dot{x} = x$, $\dot{y} = -y$. The trajectories of the nonlinear system satisfy

$$\frac{dy}{dx} = \frac{\dot{y}}{\dot{x}} = \frac{-y + x^2}{x - y^2}.$$

In differential form this equation is $(y - x^2)dx + (x - y^2)dy = 0$ and is exact with solution $3xy - x^3 - y^3 = C$. The choice $C = 0$ is an algebraic curve called the **folium of Descartes,** which in addition to the equilibrium point (0,0) and the loop in the first quadrant contains unbounded trajectories in the second and fourth quadrants. The equilibrium at (1,1) looks like a stable center, and it is one.

If a solution trajectory fails to converge to an equilibrium point as described in Theorem 4.8, it remains for us to describe other possibilities for relatively stable limiting behavior as t tends to $+\infty$ or to $-\infty$. Let $\mathbf{x} = \mathbf{x}(t)$ be a solution of an autonomous system $\dot{\mathbf{x}} = F(\mathbf{x})$. A point $\mathbf{z}$ is said to be an **ω-limit point** of the solution trajectory if there is a sequence $\{t_k\}$ of times tending to $+\infty$ such that $\lim_{k\to\infty} \mathbf{x}(t_k) = \mathbf{z}$. The set of all ω-limit points of $\mathbf{x}$ is called the **ω-limit set** of $\mathbf{x}$ and is denoted by $\omega(\mathbf{x})$. The **α-limit set** of $\mathbf{x}$ is defined analogously with $\{t_k\}$ tending to $-\infty$. Thus if a solution $\mathbf{x} = \mathbf{x}(t)$ has a unique limit $\mathbf{z}_0$ as $t \to \infty$, then $\omega(\mathbf{x})$ consists of the single point $\mathbf{z}_0$. The next two examples are typical of the other possible limit sets for 2-dimensional systems.

---

example 10

The system $\dot{x} = -y$, $\dot{y} = x$ has counterclockwise circular trajectories except for a single equilibrium point at the origin. The system

$$\dot{x} = x(1 - x^2 - y^2) - y, \qquad \dot{y} = x + y(1 - x^2 - y^2)$$

is a modification that preserves the single equilibrium point and just one circular trajectory, the one with radius 1. All other trajectories either spiral in to the circle from outside it or else spiral out to the circle from inside it, winding around infinitely many times while approaching arbitrarily close to the circle but never

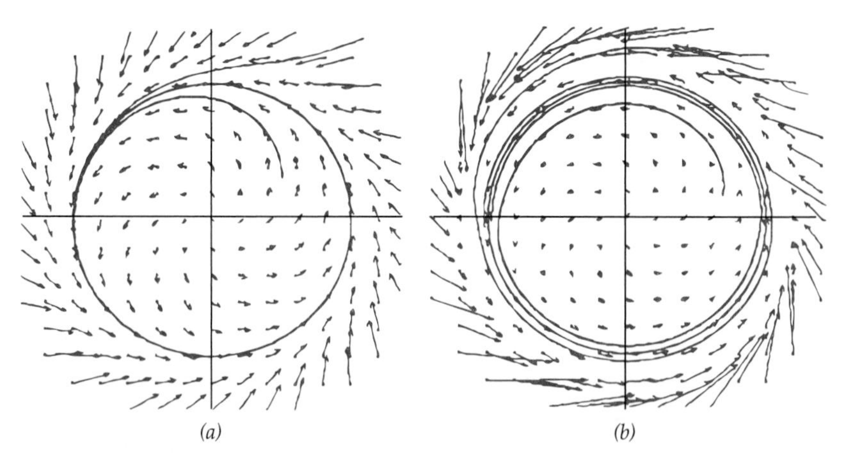

Figure 5.16 (a) $\dot{x} = x(1 - x^2 - y^2) - y$, $\dot{y} = x + y(1 - x^2 - y^2)$.
(b) $\dot{x} = x(1 - x^2 - y^2)^3 - y(1 - x^2 - y^2)^2 - y^3$,
$\dot{y} = x(1 - x^2 - y^2)^2 + y(1 - x^2 - y^2)^3 + xy^2$.

touching it. Thus the circular trajectory is the ω-limit set of every trajectory except the equilibrium point. The α-limit set of every trajectory strictly inside the circle is the equilibrium point (0,0). Figure 5.16(a) shows a sketch of the vector field and some trajectories that strongly suggests that these statement are valid. The analytical details are left as an exercise.

The circular trajectory in the previous example is an instance of a *limit cycle,* which is by definition a closed, nonconstant trajectory that is also a limit set of some other trajectory. The van der Pol equation as treated in Example 8 of Chapter 4, Section 3, provides another example of limit-cycle behavior. The next example is similar to the previous one, because it exhibits the circle $x^2 + y^2 = 1$ as a limit set. While this time the limit set is still a circle, it is not a single cyclic trajectory but decomposes into two semicircular trajectories and two equilibrium points.

---

example 11 The vector field of the system

$$\dot{x} = x(1 - x^2 - y^2)^3 - y(1 - x^2 - y^2)^2 - y^3,$$
$$\dot{y} = x(1 - x^2 - y^2)^2 + y(1 - x^2 - y^2)^3 + xy^2$$

is obtained by multiplying the one in the previous example by $(1 - x^2 - y^2)^2$ and then adding the vector $y^2(-y,x)$. Comparing Figures 5.16(a) and 5.16(b) shows some geometric similarities. in particular, the modified system turns out to have the circle $x^2 + y^2 = 1$ as an ω-limit set for all but a few of its trajectories. One striking difference is that, in addition to a common equilibrium point at (0,0), the modified system has equilibrium points at (−1,0) and (1,0) that aren't shared by the simpler system. Since these additional equilibrium points lie on the unit circle, the circle is not a limit cycle as in the previous example. Indeed, this limit set consists of two semicircular trajectories separated by equilibrium points. The semicircle $x^2 + y^2 = 1$, $y > 0$, is a trajectory with α-limit point (1,0) and ω-limit point (−1,0). The semicircle $x^2 + y^2 = 1$, $y < 0$, is a trajectory with α-limit point (−1,0) and ω-limit point (1,0). Verifying some of the details is left for the exercises.

The next theorem gives conditions under which an ω-limit set of a 2-dimensional system must be "tame," restricted to being either an equilibrium point, a limit cycle, or else the type considered in the previous example. The core of the theorem is called the Poincaré-Bendixon Theorem. For a proof see S. Wiggins, *Applied Nonlinear Dynamical Systems and Chaos*, Springer (1990). Higher dimensional systems satisfying the same conditions may have a much wider range of behavior, in particular "chaotic" behavior. See Chapter 4, Section 4D, where the third dimension is time t.

4.9 POINCARÉ-BENDIXON THEOREM
Assume a 2-dimensional autonomous system has a continuously differentiable vector field, and let R be a region with at most finitely many equilibrium points of the system such that if $\mathbf{x}_0$ is in R, then the trajectory $\mathbf{x}(t)$ with $\mathbf{x}(0) = \mathbf{x}_0$ remains in R for $t \geq 0$. Then the only possible ω-limit set for a trajectory starting in R is either (i) an equilibrium point, (ii) a limit cycle, or (iii) a connected set consisting of equilibrium points joined pairwise by trajectories.

A region R such that trajectories starting there remain there for $t \geq 0$ is said to be **positively invariant.** Verifying positive invariance may be difficult to do precisely in general, but for 2-dimensional systems a reasonably accurate sketch of the relevant vector field may be convincing enough. For instance, in the vector fields shown in Figure 5.16, it's clear that the annular region $\frac{1}{2} < x^2 + y^2 < 2$ around the circle $x^2 + y^2 = 1$ is positively invariant for both systems. In Example 10 this region contains no equilibrium points at all, so the only possible ω-limit set is a limit cycle, of type (ii) in Theorem 4.9. In Example 11 the unit circle is the prime candidate to be a limit set. Since it contains equilibrium points, the circle can't be a limit cycle, so it must be of type (iii).

EXERCISES

Find all the equilibrium points (x_0, y_0) for each of the systems in Exercises 1 through 8. Then find the associated linearized system for each such point. Use Theorem 4.7 to draw what conclusions you can about the behavior of solution trajectories near the equilibrium point. Warning: If the linearized system is degenerate $(ad = bc)$, Theorem 4.7 does not apply.

1. $\dot{x} = y,$
$\dot{y} = x + y^2.$

2. $\dot{x} = y,$
$\dot{y} = y + x^2.$

3. $\dot{x} = -y,$
$\dot{y} = x + x^2.$

4. $\dot{x} = x + y^2,$
$\dot{y} = x^2 - y.$

5. $\dot{x} = x - xy^2,$
$\dot{y} = x - y^3.$

6. $\dot{x} = e^x - 1,$
$\dot{y} = x + y.$

7. $\dot{x} = x(1 - x^2 - y^2) - y,\ \dot{y} = x + y(1 - x^2 - y^2).$

8. $\dot{x} = x(1 - x^2 - y^2)^3 - y(1 - x^2 - y^2)^2 - y^3,$
$\dot{y} = x(1 - x^2 - y^2)^2 + y(1 - x^2 - y^2)^3 + xy^2.$

Each of the systems in Exercises 9 and 10 has infinitely many equilibrium points. In each case, single out all isolated equilibrium points for which the linearized system is nondegenerate, and draw what conclusions you can about solution trajectories near these equilibrium points.

9. $\dot{x} = 1 - y,$
$\dot{y} = \sin x.$

10. $\dot{x} = x(1 - x^2 - y^2),$
$\dot{y} = y(1 - x^2 - y^2).$

11. In Example 7 of the text we looked at the system

$$dx/dt = y, \qquad dy/dt = -(g/l)\sin x,$$

derived from the nonlinear pendulum equation. Since $dy/dx = (dy/dt)/(dx/dt)$, the solutions are related by

$$\frac{dy}{dx} = \frac{-g \sin x}{ly}.$$

(a) Use separation of variables to show that solution trajectories satisfy

$$y^2 = \frac{2g}{l} \cos x + c, \quad \text{where } c \text{ is constant.}$$

(b) Show that, if the constant c in part (a) satisfies $|c| < 2g/l$, then the trajectories are closed circuits about the equilibrium points $(x, y) = (2m\pi, 0)$, corresponding to stable oscillations.

(c) Show that if $c > 2g/l$, the trajectories are periodic graphs, unbounded in the x direction. These trajectories correspond to a perpetual winding motion of the pendulum.

(d) Show that if $c = 2g/l$, the trajectories represent critical paths of motion that extend from one unstable equilibrium point to another. [Hence one such trajectory fits the limit behavior described in Theorem 4.8 for two distinct equilibrium points.]

12. Consider the family of nonlinear systems

$$\frac{dx}{dt} = y + \alpha x(x^2 + y^2),$$
$$\frac{dy}{dt} = -x + \alpha y(x^2 + y^2),$$

where α is constant. Linearized analysis is inadequate for this system.

(a) Show that the only equilibrium point is $(x_0, y_0) = (0,0)$, regardless of the value of α.

(b) Show that the linearized system associated with (0,0) is $\dot{x} = y,\ \dot{y} = -x$, and that the origin is a stable center as $t \to \infty$. Note that (0,0) is also a stable center for the given system when $\alpha = 0$.

(c) Show that in polar coordinates the given system takes the form $\dot{r} = \alpha r^3, \dot{\theta} = -1$.
[*Hint:* Apply d/dt to the equations $x = r \cos\theta$ and $y = r \sin\theta$.]

(d) Solve the polar-form system in part (c), and show that if $\alpha > 0$ the equilibrium point is unstable, and that if $\alpha < 0$ it is stable. Thus the parameter value $\alpha = 0$ is called a **bifurcation point** for the system, because the stability of the system at the equilibrium point changes in a fundamental way as α passes through zero.

13. A nonlinear pendulum with frictional damping has a displacement angle $\theta = \theta(t)$ that satisfies $\ddot{\theta} + [k/(lm)]\dot{\theta} + (g/l)\sin\theta = 0$, where $k > 0$ is constant.

(a) Show that the equation for θ is equivalent, with $x = \theta$, $y = \dot{\theta}$, to the first-order system

$$\dot{x} = y, \qquad \dot{y} = -\frac{g}{l}\sin x - \frac{k}{lm}y.$$

(b) Show that the equilibrium points of the system in part (a) are independent of k.

(c) Show that the unstable equilibrium points are all saddles.

(d) Show that the stable equilibrium points are nodes if $k^2 > 4glm^2$ and asymptotic spirals if $k^2 < 4glm^2$. Why does this last distinction make sense physically?

14. The *Lotka-Volterra equations* are discussed in Chapter 6, Section 2.

$$\frac{dH}{dt} = (a - bP)H, \qquad \frac{dP}{dt} = (cH - d)P, \qquad a, b, c, d > 0.$$

(a) Show that for $H > 0$ and $P > 0$, the only equilibrium point is $(H_0, P_0) = (d/c, a/b)$, and find the associated linearized system.

(b) Show that the equilibrium solution of the linearized system is a stable center.

(c) Explain why the result of part (b) is consistent with the results of text Example 5 of Section 2 of Chapter 6.

(d) Discuss the equilibrium solution of the nonlinear system at $(H,P) = (0,0)$. Do your conclusions make sense, given the interpretation of P and H as sizes of parasite and host populations respectively?

15. Discuss the nature of the equilibrium point of the system in Exercise 21 of Section 4A, under the assumptions $A > 0$, $B > A^2 + 1$.

16. The nonlinear system $\dot{x} = 2ye^{x^2+y^2}$, $\dot{y} = -2xe^{x^2+y^2}$ is shown in Example 15 of the next section to have a stable equilibrium at (0,0). Show that an analysis of the linearized system at (0,0) fails to lead to this conclusion.

17. Find all equilibrium points of $\dot{x} = -y(1 - x^2 - y^2)$, $\dot{y} = x(1 - x^2 - y^2)$. Then show that all other trajectories are circles and that no trajectory converges to an equilibrium point. [*Hint:* $x\,dx + y\,dy = 0$.]

18. Find all equilibrium points of $\dot{x} = x(2 - x - y)$, $\dot{y} = y(x - 1)$ and discuss their stability. Then, by hand or using computer graphics, make a sketch of some typical trajectories near the equilibrium points.

19. Compute the equilibrium solutions of $\dot{x} = -x(x^2 + y^2 - 1)$, $\dot{y} = -y(x^2 + y^2 + 1)$ and discuss their stability.

20. The system $\dot{x} = -x(x^2 + y^2 - 1)$, $\dot{y} = -y(x^2 + y^2 + 1)$ discussed in Example 8 of the text reduces to $\dot{x} = -x(x^2 - 1)$ on the x axis and to $\dot{y} = -y(y^2 + 1)$ on the y axis.

(a) Solve the y equation explicitly to show that a trajectory starting at y_0 on the y axis converges to the origin as $t \to \infty$.

(b) Solve the x equation explicitly to show that if $|x_0| < 1$ a trajectory starting at x_0 on the x axis converges to either (1,0) or (−1,0) as $t \to \infty$ and converges to the origin as $t \to -\infty$.

21. The system $\dot{x} = -x(x^2 + y^2 - 1)$, $\dot{y} = -y(x^2 + y^2 + 1)$ discussed in Example 8 of the text reduces to $\dot{x} = -x(x^2 - 1)$ on the x-axis and to $\dot{y} = -y(y^2 + 1)$ on the y-axis. Use Theorem 3.1 of Chapter 2, Section 3, to establish the following.

(a) Show that a trajectory starting at y_0 on the y axis converges to the origin as $t \to \infty$.

(b) Show that if $|x_0| < 1$, a trajectory starting at x_0 on the x axis converges to either (1,0) or (−1,0) as $t \to \infty$ and converges to the origin as $t \to -\infty$.

22. In Example 10 of the text we considered the system

$$\dot{x} = x(1 - x^2 - y^2) - y, \qquad \dot{y} = x + y(1 - x^2 - y^2).$$

(a) Show the system has a circular orbit traced by the solution $x = \cos t$, $y = \sin t$.

(b) Show that all solutions satisfy $(d/dt)(y/x) = 1 + (y/x)^2$.

(c) Use part (b) to show the polar angle θ of a point on a nonconstant trajectory satisfies $\theta = \arctan(y/x) = t + c$ and hence that all such trajectories wind counterclockwise infinitely often around the origin.

(d) Show that all solutions satisfy $x\dot{x} + y\dot{y} = (x^2 + y^2)(1 - x^2 - y^2)$.

(e) Use part (d) to show that the polar radius r of a point on a trajectory satisfies $dr/dt = r(1 - r^2)$, with solutions $r = ke^t/\sqrt{k^2e^{2t} \pm 1}$, the sign depending on whether $0 < r < 1$ or $r > 1$. Show that these solutions have as ω-limit set the circular trajectory $x^2 + y^2 = 1$.

23. In Example 10 of the text we considered the system

$$\dot{x} = x(1 - x^2 - y^2) - y, \qquad \dot{y} = x + y(1 - x^2 - y^2).$$

(a) Show that all solutions satisfy $x\dot{x} + y\dot{y} = (x^2 + y^2)(1 - x^2 - y^2)$ and hence that the polar radius r of a point on a trajectory satisfies $dr/dt = r(1 - r^2)$.

(b) Show that the xy plane with (0,0) deleted is positively invariant for the system by integrating the result of part (a) from 0 to t.

(c) Explain how to conclude from Theorem 4.9 that the circle $x^2 + y^2 = 1$ is a limit cycle.

24. In Example 11 of the text we considered the system

$$\dot{x} = x(1 - x^2 - y^2)^3 - y(1 - x^2 - y^2)^2 - y^3,$$
$$\dot{y} = x(1 - x^2 - y^2)^2 + y(1 - x^2 - y^2)^3 + xy^2.$$

(a) Show that all solutions satisfy $x\dot{x} + y\dot{y} = (x^2 + y^2)(1 - x^2 - y^2)^3$ and hence that the polar radius r of a point on a trajectory satisfies $dr/dt = r(1 - r^2)^3$.

(b) Show that the xy plane with (0,0) deleted is positively invariant for the system by integrating the result of part (a) from 0 to t.

(c) Use Theorem 4.9 to show that the circle $x^2 + y^2 = 1$ is an ω-limit set.

(d) Show that the system has trajectories on the unit circle satisfying $\dot{x} = -y^3$, $\dot{y} = xy^2$. Explain why the unit semicircle with $y > 0$ is a trajectory with $(1,0)$ as α-limit point and $(-1,0)$ as ω-limit point. What can you say about the semicircle with $y < 0$?

25. The van der Pol equation $\ddot{x} + \alpha(x^2 - 1)\dot{x} + x = 0$ is equivalent to the system $\dot{x} = y$, $\dot{y} = -x - \alpha(x^2 - 1)y$.

(a) Find the linearization of the system near $(x_0, y_0) = (0,0)$.

(b) Discuss the behavior of solutions near $(x_0, y_0) = (0,0)$ and their dependence on the constant $\alpha > 0$. What happens if $\alpha = 0$ or if $\alpha < 0$?

(c) Make a computer-aided sketch of the vector field of the system for $\alpha = 0.5$, use it to identify a positively invariant region R of the xy plane that excludes $(0,0)$, and conclude that this van der Pol oscillator has a limit cycle. For a sketch of some trajectories see Figure 4.14 in Chapter 4, Section 3.

26. It's true, though we won't prove it, that if a 2-dimensional system has a closed (i.e., periodic) solution trajectory and the system's vector field is also defined and continuous everywhere inside this trajectory, then the system has an equilibrium point inside the trajectory curve.

(a) Illustrate this principle with an example.

(b) Find an example of a 2-dimensional system with an equilibrium point, but which has no closed trajectories.

27. (a) Show that the van der Pol equation $\ddot{x} + \alpha(x^2 - 1)\dot{x} + x = 0$ is equivalent to the system $\dot{x} = -\alpha(x^3/3 - x) + y$, $\dot{y} = -x$.

(b) Find the linearization at the equilibrium point $(x_0, y_0) = (0,0)$ of the system in part (a), and compare it with the linearization found in Exercise 25.

28. (a) The system

$$\dot{x} = \tfrac{1}{2}y\left[\tfrac{39}{16} - \tfrac{3}{4}(x^2 + y^2)\right],$$
$$\dot{y} = -\tfrac{1}{2}x\left[\tfrac{39}{16} - \tfrac{3}{4}(x^2 + y^2)\right] + \tfrac{9}{16}$$

has three equilibrium solutions. One is $(-2,0)$; where are the others?

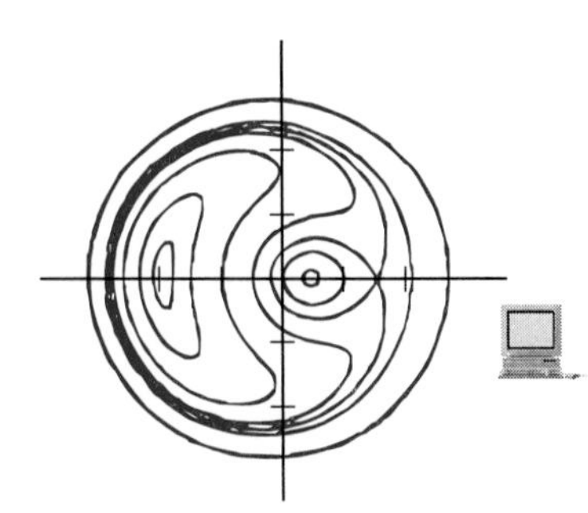

(b) Use a numerical system plotter to plot typical solution curves for the system in part (a), paying particular attention to direction of traversal and behavior near equilibrium points. Before adding direction arrows, your picture should look something like the one in the margin. At one point in the picture a trajectory appears to cross itself. Does this really happen? Explain what does happen.

29. Use **Green's theorem** in the form

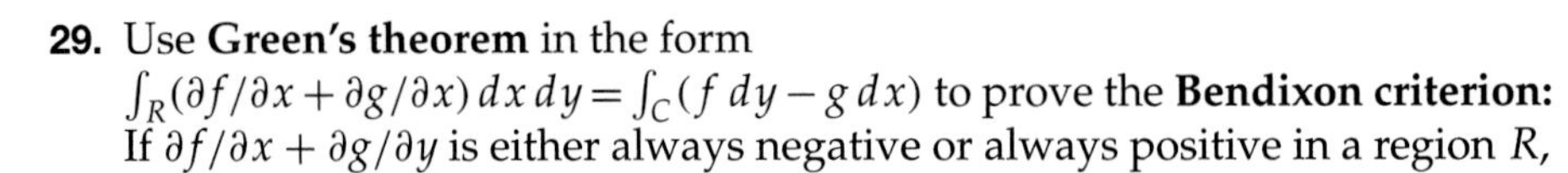

$\int_R(\partial f/\partial x + \partial g/\partial x)\,dx\,dy = \int_C(f\,dy - g\,dx)$ to prove the **Bendixon criterion:** If $\partial f/\partial x + \partial g/\partial y$ is either always negative or always positive in a region R, then R can't contain a closed trajectory C of the system $\dot{x} = f(x, y)$, $\dot{y} = g(x, y)$. Assume the partial derivatives are continuous on R. [*Hint:* Show the line integral is zero.]

4C Liapunov's Method†

The method of Section 4B for determining trajectory behavior near an equilibrium point $\mathbf{x}_0$ depends on comparing the behavior of a nonlinear system to that of a related linear system. The Liapunov method described below is sometimes called "direct," because it avoids reference to an intervening auxiliary system. Furthermore, the method enables us to determine a specific region of initial points such that solutions starting there tend to a given equilibrium point $\mathbf{x}_0$; the set of *all* such points is called the **basin of attraction** of $\mathbf{x}_0$. The geometric idea behind the Liapunov method is quite simple. Given a first-order system $\dot{\mathbf{x}} = \mathbf{F}(x)$ with an equilibrium point $\mathbf{x}_0$, suppose you can find a real-valued function $V(\mathbf{x})$ on $\mathcal{R}^n$ that has a strict local minimum at $\mathbf{x} = \mathbf{x}_0$ and such that $-\nabla V(\mathbf{x}) \cdot \mathbf{F}(\mathbf{x}) > 0$ in some neighborhood of $\mathbf{x}_0$. Recall that the *negative* of the gradient vector $\nabla V(\mathbf{x})$ points in the direction of maximum *decrease* of V at $\mathbf{x}$. On a solution trajectory $\mathbf{x} = \mathbf{x}(t)$, we have $\dot{\mathbf{x}} = \mathbf{F}(\mathbf{x})$, so the vector $\mathbf{F}(\mathbf{x})$ points in the direction of the trajectory. But the relations

$$\begin{aligned} -\nabla V(\mathbf{x}) \cdot \mathbf{F}(\mathbf{x}) &= -\nabla V(\mathbf{x}) \cdot \dot{\mathbf{x}} \\ &= |\nabla V(\mathbf{x})||\dot{\mathbf{x}}| \cos\theta > 0 \end{aligned}$$

show that the cosine of the angle θ between the trajectory and the direction of maximum decrease satisfies $\cos\theta > 0$. In other words, we must have $-\pi/2 < \theta < \pi/2$, so the trajectory points in a "downhill" direction relative to the graph of V. It is reasonable to expect that maintaining this general direction near a strict minimum $V(\mathbf{x}_0)$ for V will keep the trajectory from wandering away from $\mathbf{x}_0$; this is stable behavior, which may even result in asymptotically stable behavior with $\mathbf{x}(t)$ tending to $\mathbf{x}_0$. Figure 5.17 shows level curves of $V(\mathbf{x})$ for an example in two dimensions, with $-\nabla V(\mathbf{x})$ perpendicular to a level curve and $\mathbf{F}(\mathbf{x})$ pointing

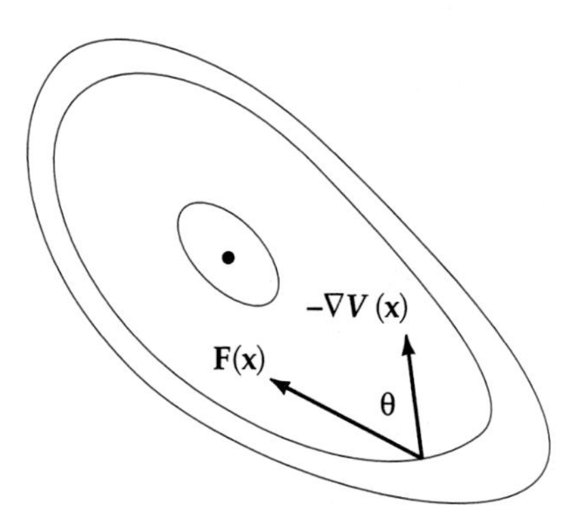

Figure 5.17 Acute angle between $F(\mathbf{x})$ and $-\nabla V(\mathbf{x})$.

in the general direction of the minimizing point for $V(\mathbf{x})$. The expression $\nabla V \cdot \mathbf{F}$ occurs often enough that the abbreviation $\dot{V}$ is often used for it:

$$\dot{V}(\mathbf{x}) = \nabla V(\mathbf{x}) \cdot \mathbf{F}(\mathbf{x}).$$

The notation is fairly natural; if $\mathbf{x}(t)$ is a solution of $\dot{\mathbf{x}} = \mathbf{F}(\mathbf{x})$, then the chain-rule then gives us

$$\frac{d}{dt}V(\mathbf{x}) = \nabla V(\mathbf{x}) \cdot \dot{\mathbf{x}} = \nabla V(\mathbf{x}) \cdot \mathbf{F}(\mathbf{x}) = \dot{V}(\mathbf{x}).$$

Note. The important thing to observe about $\dot{V}(\mathbf{x})$ is that if $\dot{V}(\mathbf{x}) < 0$ for $\mathbf{x}$ in some region R, then $V(\mathbf{x})$ decreases along solution trajectories of $\dot{\mathbf{x}} = \mathbf{F}(\mathbf{x})$ that lie in R; similarly $V(\mathbf{x})$ increases if $\dot{V}(\mathbf{x}) > 0$ in R.

4.10 LIAPUNOV STABILITY THEOREM

Let $\dot{\mathbf{x}} = \mathbf{F}(\mathbf{x})$ have an equilibrium solution $\mathbf{x}_0$ in a region R in which $\mathbf{F}(\mathbf{x})$ is continuous. Let $V(\mathbf{x})$ be real-valued and continuously differentiable on a neighborhood $B_r \subset R$ centered at $\mathbf{x}_0$ such that $V(\mathbf{x})$ has a strict minimum at $\mathbf{x} = x_0$. Let m be the minimum of $V(\mathbf{x})$ on the boundary of B_r, and let S be the subset of B_r for which $V(\mathbf{x}) < m$.

(i) If $\dot{V}(\mathbf{x}) \leq 0$ on B_r then $\mathbf{x}_0$ is a stable equilibrium point, and a trajectory that enters S stays in S.
(ii) If $\dot{V}(\mathbf{x}) < 0$ on B_r except at $\mathbf{x}_0$, then $\mathbf{x}_0$ is an asymptotically stable equilibrium point, and S is contained in the basin of attraction of $\mathbf{x}_0$.

Definition. A function V with an isolated minimum at $\mathbf{x}_0$ is called a **Liapunov function** for $\dot{x} = \mathbf{F}(\mathbf{x})$ if $\dot{V}(\mathbf{x}) = \nabla V(\mathbf{x}) \cdot F(\mathbf{x})$ satisfies the assumption in (i) and is called a **strict Liapunov function** if $\dot{V}(\mathbf{x})$ satisfies the assumption in (ii).

Proof. By a shift of coordinates we can assume that $\mathbf{x}_0 = 0$, and that $V(0) = 0$. Suppose r is chosen so that $V(\mathbf{x})$ is continuous and positive for $0 < |\mathbf{x}| \leq r$. Let m be the minimum of $V(\mathbf{x})$ on the sphere $|\mathbf{x}| = r$ in $\mathcal{R}^n$. Now suppose a solution $\mathbf{x}(t)$ satisfies $0 < V(\mathbf{x}) < m$ for some value $t = t_0$. In order for the trajectory to leave the region S, the value of $V(\mathbf{x})$ would have to become at least as large as m. But this is impossible. The reason is that $V(\mathbf{x}) < m$ at one point of the trajectory, and since

$$\begin{aligned}\frac{d}{dt}V(\mathbf{x}(t)) &= \nabla V(\mathbf{x}(t)) \cdot \dot{\mathbf{x}}(t) \\ &= V(\mathbf{x}(t)) \cdot \mathbf{F}(\mathbf{x}) = \dot{V}(\mathbf{x}(t)) \leq 0,\end{aligned}$$

$V(\mathbf{x})$ is nonincreasing thereafter along the trajectory. Hence $\mathbf{x}_0$ is stable, and the trajectory stays inside S. This proves conclusion (i).

To prove (ii), observe that since now $\dot{V}(\mathbf{x}) < 0$, we must have $V(\mathbf{x})$ strictly decreasing along a trajectory unless the trajectory reaches $\mathbf{x}_0 = 0$. We'll first prove that $\dot{V}(\mathbf{x})$ tends to zero on a trajectory starting in S. Suppose not, and that there is a solution $\mathbf{x}(t)$ and a number b such that $0 < b < V(\mathbf{x}(t))$ for $t \geq t_0$. Choose a positive $s < r$ such that $V(\mathbf{x}) < b$ for $|\mathbf{x}| < s$. It follows that $|\mathbf{x}(t)| > s$ for $t \geq t_0$. Since $\dot{V}(\mathbf{x})$ has an isolated zero value at 0, there is a $\mu > 0$ such that $\dot{V}(\mathbf{x}) \leq -\mu < 0$ when $s \leq |\mathbf{x}| \leq r$. Applying the fundamental theorem of calculus to the previous displayed equation then gives

$$V(\mathbf{x}(t)) - V(\mathbf{x}(t_0)) = \int_{t_0}^{t} \dot{V}(\mathbf{x}(u))\,du \leq -\mu(t - t_0).$$

The left side of the inequality is bounded below by $b - V(\mathbf{x}(t_0))$ as t increases, but the right side is not. This contradiction negates our assumption, so $V(\mathbf{x})$ must tend to zero along the trajectory. To show that $\mathbf{x}(t)$ tends to $\mathbf{x}_0$, let ρ be an arbitrarily small positive number. We'll show that the trajectory stays in the neighborhood B_ρ of 0. Let m_ρ be the minimum of $V(\mathbf{x})$ on the boundary of B_ρ. Since $V(\mathbf{x}(t))$ tends to 0, we can find a t_ρ such that $V(\mathbf{x}(t_\rho)) < m_\rho$. It follows from part (i) that the trajectory stays in B_ρ for $t > t_\rho$. ■

example 12 The van der Pol equation in electric circuit theory,

$$\ddot{x} + \alpha(1 - x^2)\dot{x} + x = 0,$$

is equivalent to the system

$$\dot{x} = y,$$
$$\dot{y} = -x - \alpha y(1 - x^2).$$

Thus we're dealing with $\dot{\mathbf{x}} = \mathbf{F}(\mathbf{x})$, where $\mathbf{F}(x,y) = (y, -x - \alpha y(1 - x^2))$. [The equilibrium point at $(x_0, y_0) = (0,0)$ can be characterized for $\alpha \neq 0$ using the method of Section 4B, but using the Liapunov method will give us some additional information about the region of stability.] A Liapunov function should have a strict minimum at $(x,y) = (0,0)$. A choice of the form $V(x,y) = ax^2 + by^2$, with positive a and b has the required minimum and enables us some latitude to choose a and b so that $\dot{V}(x,y) = \nabla V(x,y) \cdot \mathbf{F}(x,y) < 0$. We find that

$$\begin{aligned}\dot{V}(x,y) &= \nabla V(x,y) \cdot \mathbf{F}(x,y) \\ &= (2ax, 2by) \cdot (y, -x - \alpha(1 - x^2)y) \\ &= 2axy - 2bxy - 2b\alpha y^2(1 - x^2).\end{aligned}$$

We choose $a = b$ to make the first two terms cancel, since those terms could contribute positive values. Assume $\alpha > 0$. Then $\dot{V}(x,y) = -2b\alpha y^2(1-x^2) \leq 0$ for $(x,y) \neq (0,0)$ and for any $b > 0$, whenever $|x| \leq 1$. Let $b = 1$. Evidently $\dot{V}(x,y) \leq 0$ when $x^2 + y^2 \leq 1$, so a solution trajectory starting in or on a circle of radius 1 about the origin must stay in that disk. Also $\dot{V}(x,y) < 0$ when $x^2 + y^2 < 1$. Thus a solution trajectory starting in the interior of the disk must tend to (0,0). What the Liapunov theorem guarantees is that every trajectory starting inside the circle not only stays there but tends to (0,0). The actual region of asymptotic stability is somewhat larger, as the computer plot in Figure 5.18 shows. In the case $\alpha = 0$, $\dot{V}(x,y) = 0 \leq 0$ for all (x,y), so it follows that (0,0) is stable, though this follows independently from the elementary solution of the system.

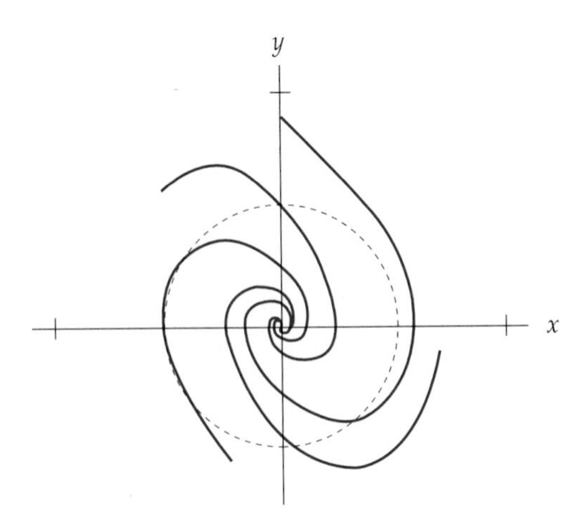

Figure 5.18 Asymptotically stable van der Pol equilibrium.

example 13 The system

$$\dot{x} = y(z+1), \qquad \dot{y} = -x(z+1), \qquad \dot{z} = -z^3$$

has a single equilibrium point $(x_0, y_0, z_0) = (0,0,0)$. We try to find a Liapunov function of the form $V(x,y,z) = ax^2 + by^2 + cz^2$ with a, b, c positive. Then

$$\begin{aligned}\dot{V}(x,y,z) &= \nabla V(x,y,z) \cdot \mathbf{F}(x,y,z) \\ &= (2ax, 2by, 2cz) \cdot \big(y(z+1), -x(z+1), -z^3\big) \\ &= 2axy(z+1) - 2bxy(z+1) - 2cz^4.\end{aligned}$$

Choosing $a = b = c = 1$, we have a Liapunov function with $\dot{V}(x,y,z) = -2z^4 \le 0$; this function is not strict because $\dot{V}(x,y,0) = 0$ not only at $(0,0,0)$ but also at every point of the xy plane. However, we can conclude that the origin is a stable equilibrium, and that every solution trajectory that enters the neighborhood $x^2 + y^2 + z^2 \le r^2$ will stay there from that time on.

The previous examples show that finding an appropriate Liapunov function is not necessarily an automatic routine. However, there are two important classes of system for which there are sometimes"ready-made" Liapunov functions. One of these classes consists of systems of the form $\dot{\mathbf{x}} = -\nabla V(\mathbf{x})$ where $V(\mathbf{x})$ is a continuously differentiable function; such a system is called a **gradient system.** The solution trajectories of the system are perpendicular to the level sets of V and point in the direction of maximum decrease of V at each point $\mathbf{x}$. An equilibrium point of a gradient system is a critical point $\mathbf{x}_0$ of V, and if $V(\mathbf{x}_0)$ is an isolated local minimum value for V, then V itself will be a Liapunov function for the system at $\mathbf{x}_0$. The reason is that

$$\begin{aligned}\dot{V}(\mathbf{x}) &= \nabla V(\mathbf{x}) \cdot \big(- \nabla V(\mathbf{x})\big) \\ &= -|\nabla V(\mathbf{x})|^2 \le 0.\end{aligned}$$

To say that $\mathbf{x}_0$ is an *isolated* minimum point for V, means that there is a neighborhood B of $\mathbf{x}_0$ such that $\nabla V(\mathbf{x}) \neq 0$ throughout B except at $\mathbf{x}_0$ itself. Thus $\dot{V}(\mathbf{x}) < 0$ except at $\mathbf{x} = \mathbf{x}_0$, so $\mathbf{x}_0$ will be an asymptotically stable equilibrium point.

example 14

Suppose $V(x, y) = e^{x^2+y^2}$. The associated gradient system is

$$\dot{x} = -2xe^{x^2+y^2}, \qquad \dot{y} = -2ye^{x^2+y^2}.$$

Taking $V(x, y)$ itself for a Liapunov function at the unique critical point (0,0) we find

$$\dot{V}(\mathbf{x}) = -|\nabla V(\mathbf{x})|^2 = -4(x^2 + y^2)e^{2(x^2+y^2)}.$$

Thus $\dot{V}(x, y) < 0$ except at the critical equilibrium point. It follows from part (ii) of the Liapunov theorem that (0,0) is asymptotically stable, but the theorem says even more. The minimum of $V(x, y) = e^{x^2+y^2}$ on the circle $x^2 + y^2 = r^2$ centered at (0,0) is $m = e^{r^2}$. But $V(x, y) < m$ on the disk S inside the circle. Thus every such disk is in the basin of attraction of (0,0), so the basin consists of all points (x, y) in $\mathcal{R}^2$.

Apart from gradient systems, there is another kind of system, called a **Hamiltonian system,** that sometimes has a ready-made Liapunov function. These systems necessarily have even dimension $2n$. We start with a continuously differentiable function $H(x, y) = H(x_1, \ldots, x_n, y_1, \ldots, y_n)$ of $2n$ real variables, called the **Hamiltonian** of the system and form the system of $2n$ first-order equations

$$\dot{x}_k = \frac{\partial H}{\partial y_k}, \qquad k = 1, \ldots, n$$

$$\dot{y}_k = -\frac{\partial H}{\partial x_k}, \qquad k = 1, \ldots, n.$$

[Hamiltonian systems first became important in the study of classical mechanics in which H represents the total energy of a mechanical system.] What is important for us here is this: $\nabla H(\mathbf{x}, \mathbf{y})$ of H is perpendicular to the $2n$-dimensional vector field of the Hamiltonian system. Hence the solution trajectories of the system lie on level sets of H. To see this, take as a Liapunov function at an equilibrium point the Hamiltonian function, as we do in the next theorem.

4.11 THEOREM

If $H(\mathbf{x}, \mathbf{y})$ is continuously differentiable, then the dot product of ∇H and the vector field of the associated Hamiltonian system is zero, that is, $\dot{H}(\mathbf{x}, \mathbf{y}) = 0$. As a result the solution trajectories of a Hamiltonian system lie on level sets of H.

Proof.

$$\begin{aligned}\dot{H}(\mathbf{x}, \mathbf{y}) &= \left(\frac{\partial H}{\partial x_1}, \ldots, \frac{\partial H}{\partial x_n}, \frac{\partial H}{\partial y_1}, \ldots, \frac{\partial H}{\partial y_n}\right) \cdot \left(\frac{\partial H}{\partial y_1}, \ldots, \frac{\partial H}{\partial y_n}, -\frac{\partial H}{\partial x_1}, \ldots, -\frac{\partial H}{\partial x_n}\right) \\ &= \frac{\partial H}{\partial x_1}\frac{\partial H}{\partial y_1} + \cdots + \frac{\partial H}{\partial x_n}\frac{\partial H}{\partial y_n} - \frac{\partial H}{\partial y_1}\frac{\partial H}{\partial x_1} - \cdots - \frac{\partial H}{\partial y_n}\frac{\partial H}{\partial x_n} = 0.\end{aligned}$$

The second statement follows from the previous equation, since then $(d/dt)H(\mathbf{x}(t), \mathbf{y}(t)) = \dot{H}(\mathbf{x}(t), \mathbf{y}(t)) = 0$. Hence $H(x, y)$ is constant on the path. ■

4.12 COROLLARY

If a critical point of $H(\mathbf{x}, \mathbf{y})$ gives an isolated minimum $H(\mathbf{x}_0, \mathbf{y}_0)$, then part (i) of the Liapunov theorem applies, and we conclude that $(\mathbf{x}_0, \mathbf{y}_0)$ is a stable equilibrium for the associated Hamiltonian system.

example 15

Let $H(x, y) = e^{x^2+y^2}$. The associated Hamiltonian system is

$$\dot{x} = 2ye^{x^2+y^2}, \qquad \dot{y} = -2xe^{x^2+y^2}.$$

Try $V(x, y) = H(x, y)$ for a Liapunov function. Since $\dot{V}(x, y) = 0$ for all (x, y), we just have to note that V has an isolated minimum at (0,0) to conclude from part (i) of the Liapunov theorem that (0,0) is a stable equilibrium. Every disk centered at (0,0) is a stability region. The reason is that the minimum of $V(x, y) = e^{x^2+y^2}$ on the circle $x^2 + y^2 = r^2$ centered at (0,0) is $m = e^{r^2}$. But $V(x, y) < m$ on the disk S inside the circle. Thus by the Liapunov theorem every trajectory that enters such a disk stays there from that time on.

Every second-order system of dimension n is equivalent to a $2n$-dimensional first-order system and so meets the even-dimensionality requirement for Hamiltonian systems. Given the usual method for reducing the order of $\ddot{\mathbf{x}} = \mathbf{F}(\mathbf{x})$ by setting $\dot{\mathbf{x}} = \mathbf{y}$, the only additional requirement is that $\mathbf{F}(\mathbf{x}) = -\nabla U(\mathbf{x})$ for some continuously differentiable real-valued function $U(\mathbf{x})$. [In other words, the original second-order system should be $\ddot{\mathbf{x}} = -\nabla U(\mathbf{x})$, a **conservative system.**] It follows that the equivalent first-order system $\dot{\mathbf{x}} = \mathbf{y}$, $\dot{\mathbf{y}} = -\nabla U(\mathbf{x})$ is Hamiltonian with Hamiltonian function

$$\begin{aligned} H(\mathbf{x}, \mathbf{y}) &= \tfrac{1}{2}|\mathbf{y}|^2 + U(\mathbf{x}) \\ &= \tfrac{1}{2}(y_1^2 + \cdots + y_n^2) + U(x_1, \ldots, x_n). \end{aligned}$$

Thus $H = T + U$ can be regarded as the total energy, kinetic plus potential, of the system, and by Theorem 4.11, trajectories lie on level sets $H = C$ called **energy shells.** The Hamiltonian function has a strict minimum at $(0, \mathbf{x}_0)$ if and only if U has a strict minimum at $\mathbf{x}_0$. This observation together with the Liapunov theorem proves a stability test to the effect that $\ddot{\mathbf{x}} = -\nabla U(\mathbf{x})$ *has a stable equilibrium at a strict minimum of* $U(\mathbf{x})$.

example 16

Let $U(x, y) = x^2y^2$. The second-order conservative system $\ddot{x} = -2xy^2$, $\ddot{y} = -2x^2y$ is equivalent to the first-order system

$$\begin{aligned} \dot{x} &= z, \\ \dot{y} &= w, \\ \dot{z} &= -2xy^2, \\ \dot{w} &= -2x^2y. \end{aligned}$$

This system is Hamiltonian with Hamiltonian function $H(x,y,z,w) = x^2y^2 + \frac{1}{2}z^2 + \frac{1}{2}w^2$. The function H has an isolated minimum at the equilibrium point $(x,y,z,w) = (0,0,0,0)$. By the previous remark $\dot{H}$ is identically zero, so H is a Liapunov function for the system at the equilibrium, and we can conclude that the equilibrium is stable.

If an attempt to show that a function $V(\mathbf{x})$ is a Liapunov function fails because $\dot{V}(\mathbf{x}) > 0$ near an equilibrium point $\mathbf{x}_0$, it may be possible to use V to show that $\mathbf{x}_0$ is unstable by using the following test.

4.13 INSTABILITY TEST

Let $\dot{\mathbf{x}} = \mathbf{F}(\mathbf{x})$ have an equilibrium solution $\mathbf{x}_0$ in a region R in which $\mathbf{F}(\mathbf{x})$ is continuous. Let $V(\mathbf{x})$ be real-valued and continuously differentiable on a neighborhood $B_r \subset R$ centered at $\mathbf{x}_0$ such that $V(\mathbf{x})$ has a strict minimum at $\mathbf{x} = \mathbf{x}_0$ and such that $\dot{V}(\mathbf{x}) > 0$ on B_r except at $\mathbf{x}_0$. Then $\mathbf{x}_0$ is an unstable equilibrium.

Proof. Let B_a, with $0 < a < r$, be an arbitrary neighborhood of radius a centered at $\mathbf{x}_0$. We'll show that if $\mathbf{x}_1$ is in B_a, $\mathbf{x}_1 \neq \mathbf{x}_0$, then the system trajectory with $\mathbf{x}(0) = \mathbf{x}_1$ must leave B_a. Let $0 < \epsilon < |\mathbf{x}_1|$, and let $\mu > 0$ be the minimum of $\dot{V}$ for $\epsilon \leq |\mathbf{x}| \leq a$. Then

$$V(\mathbf{x}(t)) - V(\mathbf{x}(0)) = \int_0^t \dot{V}(\mathbf{x}(u))\,du \geq \mu t.$$

It follows that $V(\mathbf{x}(t))$ is unbounded on this trajectory if it remains in B_a, in particular would become larger than the maximum of $V(\mathbf{x})$ for $|\mathbf{x}| \leq a$. The only way to avoid the contradiction is to have the trajectory leave B_a. ■

---

example 17

The system

$$\dot{x} = 2xe^{x^2+y^2}, \qquad \dot{y} = 2ye^{x^2+y^2}$$

is the gradient system associated with the function $-e^{x^2+y^2}$. However, the latter function has its maximum rather than its minimum at the unique equilibrium point (0,0). If we try letting $V(x,y) = e^{x^2+y^2}$, we find

$$\begin{aligned}\dot{V}(x,y) &= \left(2xe^{x^2+y^2}, 2ye^{x^2+y^2}\right) \cdot \left(2xe^{x^2+y^2}, 2ye^{x^2+y^2}\right)\\ &= 4(x^2+y^2)e^{x^2+y^2} > 0,\end{aligned}$$

except at $(x,y) = (0,0)$. By the instability test, the equilibrium is unstable.

EXERCISES

For each of the systems in Exercises 1 through 6, determine whether the constants a and b can be chosen so that $V(x,y) = ax^2 + by^2$ is (i) a Liapunov function for the system at (0,0), or (ii) a strict Liapunov function.

1. $\dot{x} = -x^3,\ \dot{y} = -y^3.$

2. $\dot{x} = -x^2,\ \dot{y} = -y^2.$

3. $\dot{x} = -\sin x,\ \dot{y} = -\sin y.$

4. $\dot{x} = -\sin y,\ \dot{y} = -\sin x.$

5. $\dot{x} = -xy^2,\ \dot{y} = -x^2 y.$

6. $\dot{x} = -x^2 y,\ \dot{y} = -xy^2.$

For each of the systems in Exercises 7 through 12, determine whether the constants a, b, and c can be chosen so that $V(x,y,z) = ax^2 + by^2 + cz^2$ is (i) a Liapunov function for the system at (0,0,0) (ii) a strict Liapunov function.

7. $\dot{x} = -x^3,\ \dot{y} = -y^3,\ \dot{z} = -z^3.$

8. $\dot{x} = -x,\ \dot{y} = -y,\ \dot{z} = -z.$

9. $\dot{x} = -yz,\ \dot{y} = -xz,\ \dot{z} = -xy$

10. $\dot{x} = -xe^y,\ \dot{y} = -ye^z,\ \dot{z} = -ze^x.$

11. $\dot{x} = -x,\ \dot{y} = -y,\ \dot{z} = 0.$

12. $\dot{x} = -x,\ \dot{y} = -y^3,\ \dot{z} = x^2 z.$

13. **(a)** Show that the system $\dot{x} = -x - 3y^2,\ \dot{y} = xy - y^3$ has a unique equilibrium point at $(x,y) = (0,0)$.

(b) Choose a and b so that $V(x,y) = ax^2 + by^2$ is a strict Liapunov function at (0,0). Then deduce that (0,0) is asymptotically stable for the system, with the entire xy plane as basin of attraction.

14. The system $\dot{x} = y(z+1),\ \dot{y} = -x(z+1),\ \dot{z} = -z^3$ is shown in Example 13 of the text to have a stable equilibrium at (0,0,0) using the Liapunov method. The question is left open there as to whether the equilibrium is in fact asymptotically stable.

(a) Solve the third equation of the system for $z = z(t)$.

(b) Show that the trace of a nonequilibrium solution $\big(x(t), y(t), z(t)\big)$ in the xy plane satisfies $x^2 + y^2 = x_0^2 + y_0^2$, and so is either a circle or point. [*Hint:* $dy/dx = -x/y$.]

(c) Explain why the equilibrium at (0,0,0) is not asymptotically stable.

15. The system $\dot{x} = yz,\ \dot{y} = -xz,\ \dot{z} = -z$ is similar to the one treated in Example 13 of the text using the Liapunov method.

(a) Find all equilibrium points of the system.

(b) Solve the third equation of the system for $z = z(t)$.

(c) Show that the trace of a nonequilibrium solution $\big(x(t), y(t), z(t)\big)$ in the xy plane satisfies $x^2 + y^2 = x_0^2 + y_0^2$, and so is either a circle or point. [*Hint:* $dy/dx = -x/y$.]

(d) Show that (0,0,0) is a stable equilibrium and that all others are unstable.

16. The system $\dot{x} = yz,\ \dot{y} = -xz,\ \dot{z} = -z^3$ is similar to the one treated in Example 13 of the text using the Liapunov method.

(a) Find all equilibrium points of the system.

(b) Solve the third equation of the system for $z = z(t)$.

(c) Show that the trace of a nonequilibrium solution $\big(x(t), y(t), z(t)\big)$ in the xy plane satisfies $x^2 + y^2 = x_0^2 + y_0^2$, and so is either a circle or point. [*Hint:* $dy/dx = -x/y$.]

(d) Show that (0,0,0) is a stable equilibrium, but is not asymptotically stable, and that all other equilibrium points are unstable.

Hamiltonian Systems

17. The Hamiltonian system

$$\dot{x} = 2ye^{x^2+y^2},\ \dot{y} = -2xe^{x^2+y^2}$$

is shown in Example 15 of the text to have a stable equilibrium at (0,0). Show that an analysis of the linearized system at (0,0) fails to lead to this conclusion.

18. Show that the solution trajectories of a two-dimensional Hamiltonian system $\dot{x} = \partial H/\partial y$, $\dot{y} = -\partial H/\partial x$ are tangent to level sets $H(\mathbf{x}, \mathbf{y}) = C$ at nonequilibrium points.

19. Consider the **Lotka-Volterra system** $\dot{x} = x(a - by)$, $\dot{y} = y(cx - d)$, where a, b, c, and d are positive constants, and let $H_0(x, y) = cx - d \ln x + by - a \ln y$.

(a) Show that each solution trajectory satisfies an implicit relation $H_0(x, y) = C$. [*Hint:* Compute dy/dx and observe that the variables separate.]

(b) Show that $H_0(x, y)$ has a strict minimum at the equilibrium $(d/c, a/b)$.

(c) Compute $\dot{H}_0(x, y)$ to show that the equilibrium is stable.

(d) Show that if the vector field of the Lotka-Volterra system is multiplied by $\rho(x, y) = -(xy)^{-1}$, the result is the Hamiltonian system with Hamiltonian $H_0(x, y)$.

20. Let $H(x, y) = x^d y^a e^{-cx-by}$, a, b, c, d positive, and consider the Hamiltonian system $\dot{x} = H_y(x, y)$, $\dot{y} = -H_x(x, y)$.

(a) Show that the Hamiltonian system has the same solution trajectories as the Lotka-Volterra system of the previous exercise, but that solutions with the same initial points are not the same as functions of t.

(b) Show that $H(x, y)$ has a strict maximum at the equilibrium $(d/c, a/b)$, making it unsuitable as a Liapunov function for the system.

21. Let $H(x, y)$ be continuously differentiable. Suppose that for each solution trajectory of a given two-dimensional system $\dot{x} = f(x, y)$, $\dot{y} = g(x, y)$ there is a constant C such that $H(x, y) = C$ on the trajectory. Show that the Hamiltonian system

$$\dot{x} = H_y(x, y), \qquad \dot{y} = -H_x(x, y)$$

has the same trajectories as the given system.

22. Let $U(x)$ be a continuously differentiable function of one variable, and consider the differential equation $\ddot{x} = -U'(x)$.

(a) Show that the second-order equation is equivalent to the first-order system

$$\dot{x} = y, \qquad \dot{y} = -U'(x).$$

(b) Show that the system in part (a) is Hamiltonian with Hamiltonian function $H(x, y) = U(x) + \frac{1}{2}y^2$, with the conclusion that if $U(x_0)$ is an isolated minimum for $U(x)$, then $(x_0, 0)$ is a stable equilibrium point for the system and hence for $\ddot{x} = -U'(x)$.

(c) Noting that the system in part (a) is equivalent to the 1-dimensional conservative system $\ddot{x} = -U'(x)$, show that the Hamiltonian $H(x, y)$ is constant on trajectories $\big(x(t), y(t)\big)$ of the given system.

23. Consider the second-order linear system

$$\ddot{x} = -ax - by, \qquad \ddot{y} = -bx - cy,$$

where a, b, c, and d are constants.

(a) Show that the system is equivalent to a first-order Hamiltonian system, with Hamiltonian function $H(x, y, z, w) = \frac{1}{2}(ax^2 + 2bxy + cy^2 + z^2 + w^2)$.

(b) Show that the system has a stable equilibrium at the origin if $a > 0$ and $ac > b^2$.

24. (a) If the vector field of a two-dimensional Hamiltonian system $\dot{x} = f(x, y)$, $\dot{y} = g(x, y)$ is continuously differentiable, show that the **divergence** $\text{div}(f, g) = \partial f/\partial x + \partial g/\partial y$ is identically zero.

(b) Verify that the vector field of the system $\dot{x} = x^2 y$, $\dot{y} = -xy^2$ has divergence identically zero, and find a Hamiltonian function for the system. [*Hint:* Solve $H_y(x, y) = x^2 y$.]

Nonlinear Oscillators

25. Consider the nonlinear oscillator equation $\ddot{x} + k\dot{x} + hx + \alpha x^2 + \beta x^3 = 0$ with linear damping and h, α, and β positive. Let $V(x, y) = ax^2 + bx^3 + cx^4 + dy^2$.

(a) By letting $\dot{x} = y$, find an equivalent first-order system.

(b) Show that if $k \geq 0$, the constants a, b, c, and d can be chosen so that V is a Liapunov function for the equilibrium solution $x = 0$, $\dot{x} = 0$, and hence that this equilibrium is stable.

26. Consider the nonlinear systems $\dot{x} = y + \alpha x(x^2 + y^2)$, $\dot{y} = -x + \alpha y(x^2 + y^2)$, where α is constant. Linearized analysis is inadequate for this system. Let $V(x, y) = x^2 + y^2$.

(a) Show that the only equilibrium point is $(x_0, y_0) = (0,0)$, regardless of the value of α.

(b) Show that V is a Liapunov function if $\alpha < 0$, and that the origin is asymptotically stable.

(c) Use V to show that the origin is unstable if $\alpha > 0$.

(d) What can you say about stability if $\alpha = 0$?

27. Suppose $\mathbf{x} = \mathbf{x}(t)$ satisfies the gradient system $\dot{\mathbf{x}} = -\nabla V(\mathbf{x})$. Use the chain rule to show that the tangent vectors $\dot{\mathbf{x}}(t)$ (i) are perpendicular to the level sets of V, defined implicitly by equations $V(\mathbf{x}) = c$, and

(ii) point in the direction of maximum decrease of V at $\mathbf{x}(t)$.
[*Hint:* $\nabla V(\mathbf{x}) \cdot \dot{\mathbf{x}} = |\nabla V(\mathbf{x})||\dot{\mathbf{x}}| \cos \theta$.]

28. Suppose $\mathbf{x} = \mathbf{x}(t)$ traces the solution trajectory γ of a system $\dot{\mathbf{x}} = \mathbf{F}(\mathbf{x})$ and that $V(\mathbf{x})$ is a real-valued, continuously differentiable function defined on γ.

(a) Show that if $\nabla V(\mathbf{x}) \cdot \mathbf{F}(\mathbf{x}) = 0$ for all $\mathbf{x}$ on γ, then γ lies on a level set for V, that is, there is a constant C such that $V(\mathbf{x}) = C$ for all $\mathbf{x}$ on γ.

(b) Apply the result of part (a) to the system $\dot{x} = -y$, $\dot{y} = x$ with $V(x, y) = x^2 + y^2$ to show that all nonequilibrium trajectories are circular.

(c) Apply the result of part (a) to the system $\dot{x} = yz$, $\dot{y} = -2xz$, $\dot{z} = xy$ with $V(x, y, z) = x^2 + y^2 + z^2$ to show that each nonequilibrium trajectory lies on a sphere.

(d) What conclusions can you draw from part (a) about trajectories of Hamiltonian systems?

29. Suppose a 2-dimensional vector field is continuously differentiable on an open set and that the field is both Hamiltonian with Hamiltonian $H(x, y)$ and also a gradient field with potential $G(x, y)$.

(a) Show that the pair $G(x, y)$, $H(x, y)$ satisfies the **Cauchy-Riemann partial differential equations** $\partial G/\partial x = \partial H/\partial y$ and $\partial G/\partial y = -\partial H/\partial x$.

(b) Show that if the vector field is twice continuously differentiable then the functions $G(x, y)$ and $H(x, y)$ are **harmonic,** that is, that they both satisfy $\partial^2 u/\partial x^2 + \partial^2 u/\partial y^2 = 0$.

CHAPTER REVIEW AND SUPPLEMENT

Basic Definitions A *first-order normal form vector system,* may be *autonomous* $[\dot{\mathbf{x}} = \mathbf{F}(\mathbf{x})]$ or *nonautonomous* $[\dot{\mathbf{x}} = \mathbf{F}(t, \mathbf{x})]$. Each *solution* $\mathbf{x} = \mathbf{x}(t)$ has a *trajectory,* which is the curve traced by the values of $\mathbf{x}(t)$. Attached to each point $\mathbf{x}(t)$ of a trajectory is the *velocity,* or *tangent,* vector $\dot{\mathbf{x}}(t)$, and if the solution is twice differentiable, the *acceleration* vector $\ddot{\mathbf{x}}(t)$. The tangent vectors $\dot{\mathbf{x}}(t)$ of a solution $\mathbf{x}(t)$ coincide with vectors $\mathbf{F}(\mathbf{x}(t))$, or $\mathbf{F}(t, \mathbf{x}(t))$ of the *vector field* of the system. Constant solutions are called *equilibrium solutions.*

Reduction to First-Order Normal Form This form is important for two reasons. One reason is that it brings the geometry of the vector field to the fore. Another reason is that it is the form to which the existence and uniqueness theorem applies most directly. Higher order systems are always reducible to first-order systems by introducing new variable names for the lower order derivatives in the system, thus increasing the dimension of the system. For example $\mathbf{F}(\mathbf{x}, \dot{\mathbf{x}}, \ddot{\mathbf{x}}) = 0$ could be replaced by the system $\dot{\mathbf{x}} = \mathbf{y}$, $\mathbf{F}(\mathbf{x}, \mathbf{y}, \dot{\mathbf{y}}) = 0$, of higher dimension. Additional reduction to normal form is a matter of algebraic manipulation in practice, and may not always be possible.

Existence and Uniqueness Theorem (1.1 in Section 1E.) Assume $\mathbf{F}(t, \mathbf{x})$ and its first-order vector partial derivatives with respect to the coordinate variables x_k of $\mathbf{x}$ are continuous for all t in an interval I containing t_0 and for all $\mathbf{x}$ in an open rectangle R in $\mathcal{R}^n$ containing $\mathbf{x}_0$. Then the initial-value problem $\dot{\mathbf{x}} = \mathbf{F}(t,\mathbf{x})$, $\mathbf{x}(t_0) = \mathbf{x}_0$, has a unique solution on some subinterval J of I containing t_0. Furthermore, the value $\mathbf{x}(t,\mathbf{x}_0)$ of this solution is a continuously differentiable function not only of t but also the initial point $\mathbf{x}_0$. If also the first-order partial derivatives $\mathbf{F}_{x_k}(t,\mathbf{x})$ are bounded for all $\mathbf{x}$ in $\mathcal{R}^n$ then the above conclusions are valid for t in the entire interval I.

Solution Techniques Section 2: Elimination method. Section 3: λ is an *eigenvalue* with corresponding *eigenvector* $\mathbf{u} \neq 0$ for the square matrix A if $A\mathbf{u} = \lambda\mathbf{u}$. The eigenvalues of A satisfy the polynomial equation $\det(A - \lambda I) = 0$, and $\mathbf{x}(t) = e^{\lambda t}\mathbf{u}$ will be a solution of the system $\dot{\mathbf{x}} = A\mathbf{x}$.

EXERCISES

Solve the initial-value problems in Exercises 1 through 12.

1. $dx/dt = x^2 + 1$, $x(0) = 1$,
   $dy/dt = y$, $y(0) = 2$.
2. $dx/dt = t$, $x(1) = 0$,
   $dy/dt = x + y$, $y(1) = 0$.
3. $dx/dt = y + 1$, $x(0) = 1$,
   $dy/dt = x$, $y(0) = 2$.
4. $dx/dt = -y$, $x(1) = 0$,
   $dy/dt = -x$, $y(1) = 0$.
5. $\dot{x} = 3x - 4y$, $x(0) = 1$,
   $\dot{y} = 4x - 7y$, $y(0) = 2$.
6. $\dot{x} = 3x - 5y$, $x(0) = 0$,
   $\dot{y} = x - y$, $y(0) = 1$.
7. $\dot{x} = y + t$, $x(0) = 1$,
   $\dot{y} = 4x - 1$, $y(0) = 2$.
8. $\dot{x} = 2$, $x(0) = 0$,
   $\dot{y} = x + y$, $y(0) = 1$.
9. $\ddot{x} = -3x + y$, $x(0) = 3$, $\dot{x}(0) = 0$,
   $\ddot{y} = 2x - 2y$, $y(0) = 3$, $\dot{y}(0) = 0$.
10. $\ddot{x} = y$, $x(0) = 0$, $\dot{x}(0) = 1$,
    $\ddot{y} = x$, $y(0) = 1$, $\dot{y}(0) = 0$.
11. $dx/dt = -y$, $x(0) = 0$,
    $dy/dt = x$, $y(0) = 1$,
    $dz/dt = z$, $z(0) = -1$.
12. $dx/dt = t$, $x(0) = 0$,
    $dy/dt = y$, $y(0) = 0$,
    $dz/dt = y + z$, $z(0) = 1$.

13. Suppose that the autonomous differential equation $\dot{\mathbf{x}} = F(\mathbf{x})$ has solution $\mathbf{x} = \mathbf{x}(t)$ satisfying $\mathbf{x}(0) = \mathbf{x}_0$ and $\mathbf{x}(1) = \mathbf{x}_1$. What can you say about a solution to $\dot{\mathbf{x}} = -F(\mathbf{x})$?

14. Consider the system $(\dot{x},\dot{y},\dot{z}) = (-\omega y,\omega x,\sigma)$, where ω and σ are nonzero constants.

    (a) Without solving the system, show that the acceleration vector of a nonzero solution trajectory (i) is always perpendicular to its velocity vector and (ii) is parallel to the xy plane, with length equal to ω^2 times the length of the corresponding position vector (x,y,z).

    (b) Solve the system. Then sketch the trajectory passing through (1,0,0), if $\omega = \sigma = 1$.

15. Let $f(\mathbf{x}) = f(x,y)$ be a continuously differentiable function of two real variables. The **gradient vector field** $\nabla f(x,y) = \left(f_x(x,y), f_y(x,y)\right)$ generates

the autonomous vector differential equation $\dot{\mathbf{x}} = \nabla f(\mathbf{x})$, called a *gradient system*. (Subscripts denote partial derivatives.)

(a) Show that the solution trajectories of a gradient system are perpendicular to the level curves $f(x,y) = c$ of f. [*Hint:* Show $(d/dt)f\big(x(t), y(t)\big) = 0$.]

(b) Illustrate part (a) using the example $f(x,y) = x^2 + y^2$.

(c) Suppose that $\mathbf{x}(t)$ is a solution of a gradient system satisfying $\mathbf{x}(t_0) = \mathbf{x}_0$ and $\mathbf{x}(t_1) = \mathbf{x}_1$, where $\mathbf{x}_0$ and $\mathbf{x}_1$ lie on the same level curve of f, that is, $f(\mathbf{x}_0) = f(\mathbf{x}_1)$. Show that $\int_{t_0}^{t_1} |\dot{\mathbf{x}}(t)|^2\,dt = 0$ and hence that the solution $\mathbf{x}(t)$ must reduce to a constant equilibrium solution.

16. Let $H(\mathbf{x}) = H(x,y)$ be a continuously differentiable function of two real variables. The vector field $\mathbf{H}(x,y) = \big(H_y(x,y), -H_x(x,y)\big)$ is called a **Hamiltonian field,** and the autonomous vector differential equation $\dot{\mathbf{x}} = \mathbf{H}(\mathbf{x})$, called a *Hamiltonian system.*

(a) Show that the solution trajectories of a Hamiltonian system are level curves of $H(x,y)$. [*Hint:* Use part (a) of the previous exercise.]

(b) Illustrate part (a) using the example $H(x,y) = x^2 - y^2$.

17. Using only geometric terms, describe in general (a) the trajectory and (b) the graph of an equilibrium solution for a 2-dimensional system.

CHAPTER 6

Applications: Linear and Nonlinear

The examples in this chapter illustrate ideas and techniques that appear often in applied mathematics, all involving mutual interactions. Step one is finding a system of differential equations that the interacting quantities satisfy under clearly defined assumptions. We'll be able to give complete solutions for some systems in terms of elementary functions; for others, we'll be able to give qualitative descriptions of the solutions, then use the numerical methods described in Section 5 for more accurate quantitative descriptions. The dimension of the state space will vary depending on the details of a particular application.

1 MULTICOMPARTMENT MIXING

Chapter 1, Section 4B, has a discussion of analogous problems for a single container.

When fluid flows back and forth among several tanks or compartments of a physical system, the amounts $x(t)$, $y(t)$, ... of a particular dissolved substance in two or more tanks X, Y, ... will typically vary as a function of time. Given enough information about the processes involved, we may be able to find a system of differential equations that has the functions $x(t)$, $y(t)$, ... as the coordinates of its solution, and the logic behind the derivations is analogous to what we used in Chapter 1, Section 4B. Setting up the system will usually require for each compartment that we express the rate of change of a typical amount x as the difference between rate of inflow and rate of outflow of the dissolved substance $x(t)$:

$$\frac{dx}{dt} = \{\text{inflow rate to tank X}\} - \{\text{outflow rate from tank X}\}.$$

There will be one such equation for each compartment. Note that the rates on the right-hand side must also be stated in the same terms as the rate on the left, namely, quantities of dissolved material rather than volumes of solution. An underlying assumption is that the fluid in each compartment is kept thoroughly mixed all the time. Because ordinary salt is so familiar, we may think of that dissolved in water, though other soluble materials will do as well.

example 1

Figure 6.1 shows two 50-gallon tanks connected by flow pipes and with inlets and outlets all having the rates of flow as marked in gallons per minute (g/m). The flow rates are arranged so that each tank is maintained at its capacity at all times. We suppose that each tank initially contains salt solution at a concentration in pounds per gallon that we leave unspecified for the moment, that the left-hand tank is receiving salt solution at a concentration of 1 pound per gallon, and that the right-hand tank is receiving pure water. The problem is to find out what happens to the amount of salt over time in each tank. We assume that tanks are kept thoroughly mixed at all times so that the salt concentration is uniform throughout each tank at any given time. In the left-hand tank, with salt content $x(t)$, the rate of change of the amount of salt is, by definition, dx/dt. On the other hand, because of the various flow rates, we can break this rate of change into three parts

$$\frac{dx}{dt} = -4\frac{x}{50} + 3\frac{y}{50} + 1,$$

where $x/50$ is the concentration of salt in the left tank and $y/50$ the concentration in the right tank, both in pounds per gallon. The term $-4(x/50)$ is the rate of outflow of salt, and the remaining terms represent the rate of inflow. Similarly,

$$\frac{dy}{dt} = 2\frac{x}{50} - 3\frac{y}{50}.$$

Thus we have a system of differential equations that we can write as

$$\begin{aligned} \frac{dx}{dt} &= -\frac{4}{50}x + \frac{3}{50}y + 1, \\ \frac{dy}{dt} &= \frac{2}{50}x - \frac{3}{50}y. \end{aligned}$$

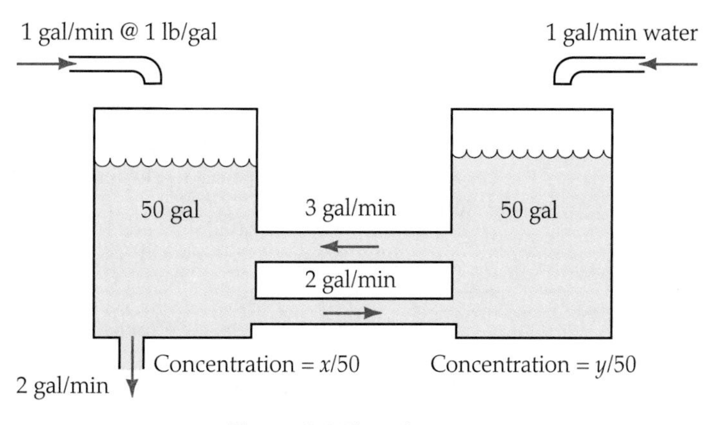

Figure 6.1 Flow diagram.

To solve it we use the elimination method, first writing the system in the form

$$\left(D+\frac{4}{50}\right)x-\frac{3}{50}y=1,$$
$$-\frac{2}{50}x+\left(D+\frac{3}{50}\right)y=0.$$

Multiply the first equation by 2/50 and operate on the second by $(D+\frac{4}{50})$. Adding the two equations eliminates x and gives

$$\left(D+\frac{4}{50}\right)\left(D+\frac{3}{50}\right)y-\frac{6}{(50)^2}y=\frac{2}{50}, \quad \text{or} \quad \left(D^2+\frac{7}{50}D+\frac{6}{(50)^2}\right)y=\frac{2}{50}.$$

We find the roots of the characteristic equation in this case by the factorization

$$r^2+\frac{7}{50}r+\frac{6}{(50)^2}=\left(r+\frac{1}{50}\right)\left(r+\frac{6}{50}\right).$$

Thus $r_1=-\frac{1}{50}$ and $r_2=-\frac{6}{50}$. A constant particular solution is $y_p(t)=\frac{50}{3}$. Thus, in general,

$$y(t)=c_1e^{-(1/50)t}+c_2e^{-(6/50)t}+\tfrac{50}{3}.$$

Using the second equation of the system to write $x(t)$ in terms of $y(t)$, we find

$$\begin{aligned} x(t) &= \tfrac{50}{2}\left(D+\tfrac{3}{50}\right)y(t) \\ &= c_1e^{-(1/50)t}-\tfrac{3}{2}c_2e^{-(6/50)t}+\tfrac{50}{2}. \end{aligned}$$

Thus the general solution is

$$\begin{aligned} x(t) &= c_1e^{-(1/50)t}-\tfrac{3}{2}c_2e^{-(6/50)t}+\tfrac{50}{2}, \\ y(t) &= c_1e^{-(1/50)t}+c_2e^{-(6/50)t}+\tfrac{50}{3}. \end{aligned}$$

From these equations, we see immediately that

$$\begin{aligned} \lim_{t\to\infty} x(t) &= \tfrac{50}{2}, \\ \lim_{t\to\infty} y(t) &= \tfrac{50}{3}. \end{aligned}$$

In other words, the concentration, in pounds per gallon, in the left tank approaches $\frac{1}{2}$ and in the right tank approaches $\frac{1}{3}$.

The constants c_1 and c_2 depend on the initial values $x(0)$ and $y(0)$. Thus the equations

$$\begin{aligned} x(0) &= c_1-\tfrac{3}{2}c_2+\tfrac{50}{2}, \\ y(0) &= c_1+c_2+\tfrac{50}{3} \end{aligned}$$

determine c_1 and c_2 when $x(0)$ and $y(0)$ are known. In particular, if $x(0)=y(0)=0$, subtracting these equations shows that $c_2=\frac{10}{3}$, from which it follows by substitution in either equation that $c_1=-20$. The graphs of x and y are

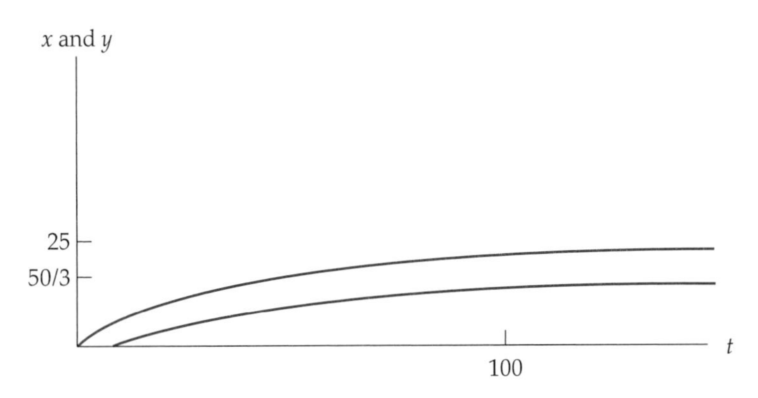

Figure 6.2 Amounts $x(t)$ and $y(t)$ of salt.

plotted in Figure 6.2 using the same vertical axis for both x and y. The constants c_1 and c_2 are determined if we know $x(t_1)$ and $y(t_1)$ at an arbitrary time $t_1 > 0$. See Exercise 2, part (c).

In the previous example, the constant equilibrium solution $x = \frac{50}{2}$, $y = \frac{50}{3}$ could have been found without solving the entire system; just set $dx/dt = dy/dt = 0$ and solve the resulting equations

$$-\tfrac{4}{50}x + \tfrac{3}{50}y = -1$$
$$\tfrac{2}{50}x - \tfrac{3}{50}y = 0.$$

Beyond that, all we need to show that all solutions tend to equilibrium is to note that both characteristic roots are negative, leading to exponential solutions tending to zero.

---

example 2

Two tanks, one of capacity 100 gallons, the other of capacity 200 gallons are initially half-full of pure water. Mixed fluid flows through a pipe from the 100-gallon tank to the other tank at 1 gallon per minute. Mixed fluid flows in the opposite direction through another pipe at 2 gallons per minute. In addition, salt solution at a concentration of 1 pound per gallon runs from an external source into the 100-gallon tank at the rate of 1 gallon per minute. The entire process is stopped as soon as either tank starts to overflow or becomes empty. To analyze what happens it's very helpful to draw and relabel your own version of Figure 6.1, and you should do this now for practice before reading the rest of this example.

We first determine the stopping time. The 100-gallon tank gains 3 gallons per minute and loses 1 gallon per minute for a total gain of 2 gallons per minute; this tank will start to overflow after it has gained 50 gallons, or $50/2 = 25$ minutes. The 200-gallon tank loses 2 gallons per minute and gains 1 gallon per minute for a total loss of 1 gallon per minute; thus this tank would lose its initial 100 gallons after 100 minutes. It follows that we need consider the process only over the shorter time interval $0 \le t \le 25$ minutes. Let $x(t)$ and $y(t)$ be the respective amounts of salt in the smaller and larger tanks, respectively. The amount of solution in the small tank at time t is $50 + 2t$ gallons and in the large tank is $100 - t$ gallons. Then

$$\frac{dx}{dt} = -\frac{x}{50+2t} + 2\frac{y}{100-t} + 1$$
$$\frac{dy}{dt} = \frac{x}{50+2t} - 2\frac{y}{100-t}.$$

Since there is initially no salt in either tank, the corresponding initial conditions are $x(0) = y(0) = 0$. We see that there are no equilibrium solutions by setting $dx/dt = 0$ and $dy/dt = 0$ and trying to solve the resulting equations for x and y. Solutions to the initial-value problem do exist according to Theorem 1.1 in Chapter 5, Section 1E, but not in terms of elementary functions, so the solution depends in practice on the numerical methods of the type discussed in the Computing Supplement for this chapter. Since salt is being added to the system at 1 pound per minute, after 25 minutes the total amount of salt in the system will be 25 pounds. Numerical calculation will show that when the process stops $x(25) \approx 22.2$, $y(25) \approx 2.8$.

EXERCISES

1. Suppose that two initially full 100-gallon tanks of salt solution contain amounts of salt $y(t)$ and $z(t)$ at time t. Suppose that the solution in the y tank is flowing to the z tank at a rate of 1 gallon per minute, and that the solution in the z tank is flowing to the y tank at the rate of 4 gallons per minute. Suppose also that the overflow from the y tank goes down the drain, whereas the z tank is kept full by the addition of fresh water. Assume that each tank is kept thoroughly mixed at all times.

(a) Find a linear system satisfied by $y(t)$ and $z(t)$.

(b) Find the general solution of the system in part (a) and then determine the constants in it so that the initial values will be $y(0) = 10$ and $z(0) = 20$.

(c) Draw the graphs of the particular solutions found in part (b) and interpret the results.

2. In Example 1 of the text, the general solution to the system of differential equations is found to be

$$x(t) = c_1 e^{-(1/50)t} - \tfrac{3}{2} c_2 e^{-(6/50)t} + \tfrac{50}{2},$$
$$y(t) = c_1 e^{-(1/50)t} + c_2 e^{-(6/50)t} + \tfrac{50}{3}.$$

(a) Find values for the constants c_1 and c_2 so that the initial conditions $x(0) = 25$ and $y(0) = \frac{2}{3}$ are satisfied.

(b) Show that it is possible to choose c_1 and c_2 so that an arbitrary initial condition $(x(0), y(0)) = (x_0, y_0)$ is satisfied. Is such a condition always meaningful for the application we have in mind here?

(c) Show that it is possible to choose c_1 and c_2 so that an arbitrary condition $(x(t_1), y(t_1)) = (x_1, y_1)$ is satisfied. Can you conclude from this result that it is possible to find the initial condition $(x(0), y(0)) = (x_0, y_0)$ that would lead to these values at $t = t_1$?

3. Two mixing vats of the same 1-gallon size containing a dissolved chemical are maintained at full capacity by the following process. Solution is pumped from vat X to vat Y at rate $a > 0$ and also in the opposite direction at rate a.

(a) Find the system that the respective amounts $x(t)$, $y(t)$ of chemical in solution in the two vats satisfy.

(b) Show that every solution of the system in part (a) is either a constant equilibrium solution or else tends to one exponentially as t tends to infinity, without exhibiting any oscillatory behavior. [*Hint:* Show that the relevant characteristic roots are nonpositive.]

4. Three mixing vats of the same 1-gallon size containing salt solution stay at full capacity because there is no gain or loss in any of the vats. Solution is pumped cyclically from vat X to vat Y at rate a, from Y to Z at rate a, and from Z to X at rate a.

(a) Find the system that the respective amounts $x(t)$, $y(t)$, $z(t)$ of chemical in solution in the vats satisfy.

(b) Show that $\dot{x} + \dot{y} + \dot{z} = 0$, and interpret this equation.

(c) Show that $x(t) = y(t) = z(t) =$ const. is always a solution of the system.

(d) By examining the relevant characteristic roots, show that the nonconstant solutions exhibit strongly damped oscillatory behavior.

5. Two tanks, one of capacity 100 gallons and the other of capacity 200 gallons, are initially full of liquid. The 100-gallon tank starts with nothing but pure water, but the other tank starts out with 10 pounds of salt dissolved in the water. Solution flows through a pipe from the 100-gallon tank to the other tank at 1 gallon per minute. Solution flows in the opposite direction through another pipe at 2 gallons per minute. Both tanks are allowed to overflow their thoroughly mixed solutions into a drain if necessary, and the entire process is stopped if either tank becomes empty.

(a) Which tank empties first, and at what time?

(b) Write down a system of differential equations and initial conditions whose solution describes this process as a function of time.

(c) Show that the rate of decrease of the total amount of salt in the two tanks is equal to the concentration of salt in the 100-gallon tank.

6. Two tanks, one of capacity 100 gallons and the other of capacity 200 gallons, are each initially half-full of liquid. The 100-gallon tank starts with nothing but pure water, but the other tank starts out with 10 pounds of calcium chloride dissolved in the water. Solution flows through a pipe from the 100-gallon tank to the other tank at 1 gallon per minute. Solution flows in the opposite direction through another pipe at 2 gallons per minute. The entire process is stopped if either tank becomes empty or either tank overflows.

(a) How long does it take for the process to stop?

(b) Write down a system of differential equations and initial conditions whose solution describes the process as a function of time.

(c) During the process what is the total change in calcium chloride content of the two tanks together?

7. Two tanks, one of capacity 100 gallons and the other of capacity 200 gallons, are each initially half-full of liquid. The 100-gallon tank starts with nothing but pure water, but the other tank starts out with 10 pounds of salt dissolved in the water. Solution flows through a pipe from the 100-gallon tank to the other tank at 2 gallons per minute. Solution flows in the opposite direction through another pipe at 1 gallon per minute. The entire process is stopped if either tank becomes empty or either tank overflows.

(a) How long does it take for the process to stop?

(b) Write down a system of differential equations and initial conditions whose solution describes the process as a function of time.

(c) How much liquid is in each tank when the process stops?

8. Three tanks, all of capacity 100 gallons, are initially full of liquid. There is an equal exchange of solution between Z and Y at the rate of 2 gallons per minute each way. There is also an equal exchange between X and Y at the rate 2 gallons per minute, but no *direct* exchange between X and Z.

(a) Find a system of differential equations whose solution describes this process as a function of time.

(b) Show that the system has equilibrium solutions with all three salt concentrations equal.

***(c)** Solve the system by elimination, assuming tanks X and Y start out with nothing but pure water while tank Z starts out with 1 pound of salt dissolved in water.

9. Two 100-gallon tanks X and Y are initially full of pure water. Copper sulfate solution is added to X from an external source at 1 gallon per minute (gpm), each gallon containing 1 pound of copper sulfate. Mixed solution is pumped from X to Y at 2 gpm and drains from Y at 3 gpm. Let $x = x(t)$ and $y = y(t)$ be the respective amounts of copper sulfate in X and Y at time $t \geq 0$.

(a) At what time t_1 will X or Y first be empty?

(b) Find a system of differential equations satisfied by $x(t)$ and $y(t)$ for $0 \leq t \leq t_1$.

(c) Solve the system found in part (b).

10. Two 100-gallon tanks X and Y contain initially 50 and 100 gallons, respectively, of pure water. From an external source, salt solution is added to Y at 1 gallon per minute (gpm), each gallon containing 1 pound of salt. Mixed solution flows from Y to X at 2 gpm and from X to Y at 1 gpm. Let $x = x(t)$ and $y = y(t)$ be the respective amounts of salt in X and Y at time $t \geq 0$. [*Note:* You are not asked to solve any differential equations for this question.]

(a) At what time t_1 will X begin to overflow? Express the total amount of salt in the two tanks as a function of t while $0 \leq t \leq t_1$.

(b) Find a system of differential equations satisfied by $x(t)$ and $y(t)$ for $0 \leq t \leq t_1$.

(c) Find a system of differential equations satisfied by $x(t)$ and $y(t)$ for $t_1 \leq t$, while X is overflowing.

11. Two 100-gallon tanks X and Y are initially full of salt solution, with x_0 pounds of salt in X and y_0 pounds of salt in Y. Mixed solution is pumped from X to Y at 2 gallons per minute and from Y to X at 3 gallons per minute. Pure water evaporates from X at 2 gallons per minute. Let $x = x(t)$ and $y = y(t)$ be the respective amounts of salt in X and Y at time $t \geq 0$.

(a) At what time t_1 will one of the tanks first overflow or become empty?

(b) Find, but don't solve, a system of differential equations satisfied by $x(t)$ and $y(t)$ for $0 \leq t \leq t_1$.

(c) Show that $x(t) + y(t)$ remains constant and that the amount of salt in each tank is in equilibrium whenever $x_0 = \frac{3}{2} y_0$.

(d) Assume that $x(0) = 10$ and $y(0) = 20$. Use the equation $x(t) + y(t) = 30$ to solve the system you found in part (b).

12. Two tanks, one of capacity 100 gallons, the other of capacity 200 gallons are each initially half-full of liquid. The 100-gallon tank starts with nothing but pure water, but the other tank starts out with 10 pounds of salt dissolved in the water. Solution flows from the 100-gallon tank to the other tank at 2 gallons per minute. Solution flows in the opposite direction at 1 gallon per minute. Pure water is added to the 100-gallon tank at 1 gallon per minute. The entire process is stopped if either tank becomes empty or either tank overflows.

(a) How long does it take for the process to stop?

(b) Write down the system of differential equations and initial conditions whose solution describes the process as a function of time. As a check, notice that the total amount of salt present in the system remains unchanged.

(c) Use the check in part (b) to find a first-order linear initial-value problem for $x(t)$ alone.

(d) Solve the initial-value problem in part (c) for $x(t)$. Then find $y(t)$, and estimate the amount of salt in each tank when the process stops.

2 INTERACTING POPULATIONS

The simple population growth equation $dP/dt = kP$, $k > 0$, was modified in Chapter 2, Section 1, to take account of the inhibiting effect of larger population size on the growth rate. One such modification resulted in the nonlinear equation $dP/dt = kP(1 - P/K)$. The factor $(1 - P/K)$ approaches zero as P approaches K and tends to reduce the growth rate as $P(t)$ approaches K from below. [If $P(t) > K$, the growth rate is negative, and the population declines.] When populations of varying size interact mutually, the possibilities for expressing the interactions are more interesting. The general qualitative behavior of these models is much more significant than specific numeric or graphic representations, though the latter are very useful for illustrative purposes.

---

example 1

Linear Competition

The system

$$\frac{dP}{dt} = kP - mQ, \qquad k > 0,\ m > 0,$$
$$\frac{dQ}{dt} = -nP + lQ, \qquad l > 0,\ n > 0,$$

includes growth-inhibition of both populations P and Q by simply subtracting from the growth rate of each population a term proportional to the size of the other population. Thus increasing the growth factor Q inhibits the growth of P more and vice versa. The size of the coefficients m and n might reflect how much the associated populations eat, or how much living space they occupy.

In operator form we write the above system as

$$(D - k)P + mQ = 0$$
$$nP + (D - l)Q = 0.$$

To eliminate Q, operate on the first equation by $(D - l)$, multiply the second by m, and subtract to get

$$(D - l)(D - k)P - mnP = 0 \quad \text{or} \quad \left[D^2 - (k + l)D + (kl - mn)\right] P = 0.$$

The characteristic equation $r^2 - (k + l)r + (kl - mn) = 0$ has roots

$$r_1, r_2 = \tfrac{1}{2}\left[(k + l) \pm \sqrt{(k + l)^2 - 4kl + 4mn}\,\right] = \tfrac{1}{2}\left[(k + l) \pm \sqrt{(k - l)^2 + 4mn}\,\right].$$

Since $(k - l)^2 + 4mn > 0$, the roots are unequal real numbers, so there is no oscillatory behavior, and P has the form $P(t) = c_1e^{r_1t} + c_2e^{r_2t}$. Solving the first equation in the original system for $Q = (kP - \dot{P})/m$ shows that $Q(t) = ac_1e^{r_1t} + bc_2e^{r_2t}$, where a and b are constants depending on k, l, m and n, where $a = (k - r_1)/m$, $b = (k - r_2)/m$. Figure 6.3 shows typical behaviors of $P(t)$ and $Q(t)$ plotted using the same vertical axis. Depending on the four parameters, either graph could be P or Q.

The graphs in Figure 6.3 suggest qualitatively that, depending on the parameters of a computation, both populations may prosper, or that one or the other may die out regardless of which one starts out ahead. (See also Figure IS.7 in the

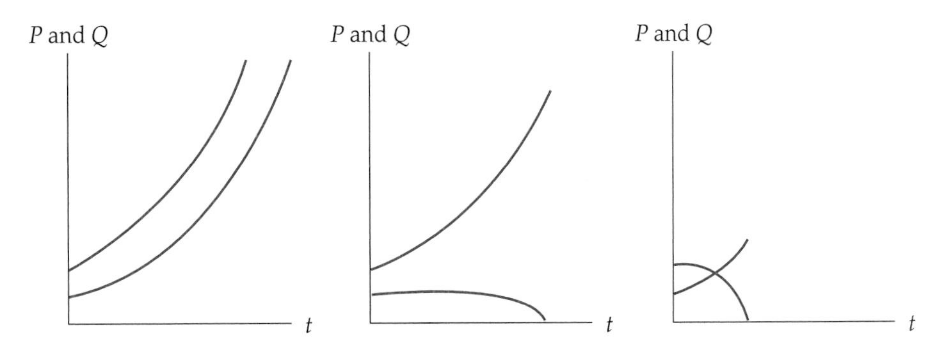

Figure 6.3 Possible population interactions.

Introductory Survey.) The next example presents a nonlinear refinement based on the logistic model of Chapter 2, Section 1.

example 2

Nonlinear Competition

Suppose each of two population sizes $P(t)$ and $Q(t)$ over time obey a logistic law in the absence of the other population; this would mean that P and Q satisfy the uncoupled version of the system

$$\frac{dP}{dt} = r_1 P\left(\frac{K_1 - a_2 Q - P}{K_1}\right) \quad \text{and} \quad \frac{dQ}{dt} = r_2 Q\left(\frac{K_2 - a_1 P - Q}{K_2}\right),$$

where $a_1 = a_2 = 0$. The positive constants K_1 and K_2 are the respective **carrying capacities** of the environment for populations P and Q. The constant $a_2 \geq 0$ represents the per capita aggressiveness with which Q competes for the part of P's environment that Q also consumes, while P's analogous aggressiveness factor is $a_1 \geq 0$. Put another way, increasing a_2 degrades the effective carrying capacity of P's environment, while increasing a_1 degrades Q's environment.

There are only three equilibrium solutions that exist for all admissible choices of the constants K_1, K_2, a_1, and a_2, namely $(P, Q) = (0,0), (K_1,0), (0,K_2)$. The only other significant equilibrium solution is the first-quadrant solution, if there is one, of the linear algebraic system we get by setting only the second factor on the right in each differential equation equal to zero:

$$P + a_2 Q = K_1, \qquad a_1 P + Q = K_2,$$

with solution

$$(P_E, Q_E) = \left(\frac{K_1 - a_2 K_2}{1 - a_1 a_2}, \frac{K_2 - a_1 K_1}{1 - a_1 a_2}\right), \quad \text{if} \quad a_1 a_2 \neq 1.$$

The algebraic signs of the top and bottom in the above ratios imply relations among the constants that have intuitively appealing interpretations that are borne out by closer examination. For example, if $a_1 K_1 > K_2$ and $a_1 a_2 < 1$, then $Q_E < 0$, so the only equilibrium points are the three on the axes. If in addition $P_E > 0$, so $K_1 > a_2 K_2$, then population P with carrying capacity K_1 has a clear competitive advantage over Q: not only does P's consumption $a_1 K_1$ at carrying capacity exceed Q's carrying capacity K_2, but P's carrying capacity exceeds Q's consumption at its carrying capacity. Graphic analysis of the system's vector field (see Exercise 10) shows that under these conditions P will inevitably drive Q toward extinction. The next two examples discuss the two distinct configurations that arise when (P_E, Q_E) represents an equilibrium in the positive quadrant. [In the event that $a_1a_2 = 1$ either there are no solutions (P_E, Q_E), or, if also $K_1 = a_2K_2$, then all points on the line $P + a_2Q = K_1$ are equilibrium solutions, a situation of no practical interest.]

example 3

Coexistence

Here we assume $K_1 > a_2K_2$, $K_2 > a_1K_1$, and $a_1a_2 < 1$ in the system of Example 2. Thus neither population's resource consumption at carrying capacity exceeds the carrying capacity of the other population. The third inequality suggests that even if one population is an aggressive competitor, the other will be proportionately less so. Figure 6.4(a) shows a regime in which the positive equilibrium is $P_E \approx 3.2$, $Q_E \approx 2.1$. Every trajectory that starts in the positive quadrant tends toward this equilibrium as t tends to infinity. Exercise 12 asks you to try other choices of constants K_j and a_j satisfying the three inequalities above. You can also vary the rate constants $r_1 = 0.7$ and $r_2 = 0.6$ that were used for the figures. For clarity the arrows in Figure 6.4 were all scaled to have the same length.

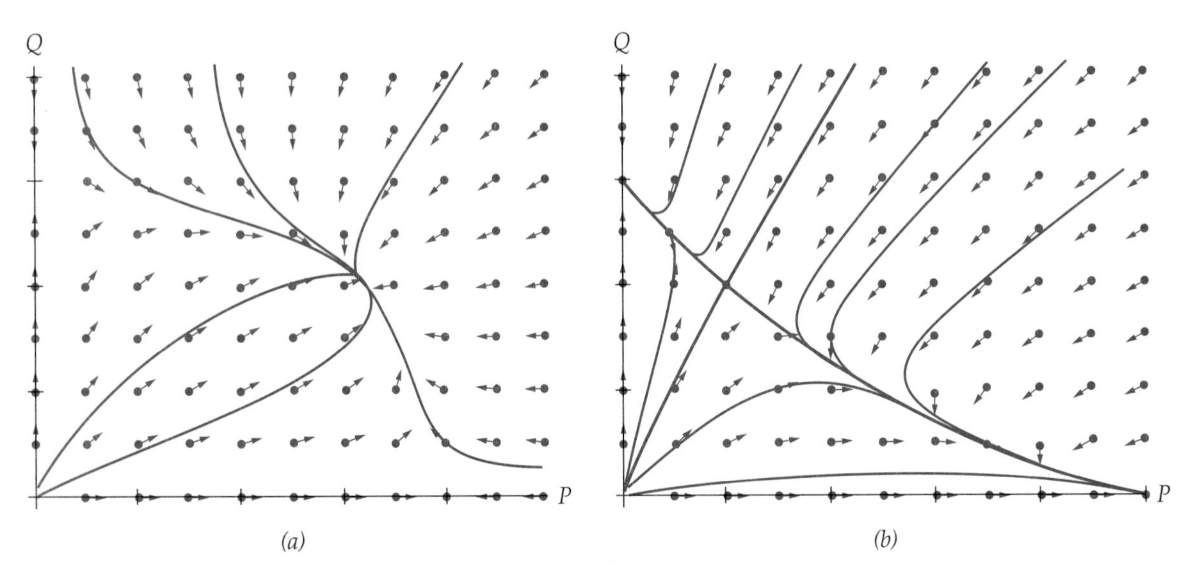

Figure 6.4 *(a)* Coexistence. $(K_1, K_2, a_1, a_2) = (4,3,0.3,0.4)$. *(b)* Selective extinction. $(K_1, K_2, a_1, a_2) = (5,3,1,2)$. The vector fields have been normalized so all arrows are equally long.

example 4

Selective Extinction

Here we assume $K_1 < a_2K_2$, $K_2 < a_1K_1$, and $a_1a_2 > 1$ in the system of Example 2. The resource consumption of each population at carrying capacity exceeds the carrying capacity of the other population, and aggressive consumption by one population is not necessarily balanced by corresponding restraint on the part of the other. Two trajectories, each one called a **separatrix,** together with the equilibrium point, cut the positive PQ quadrant into four regions. Depending on the initial population sizes, one of two things will happen with overwhelming likelihood. Points in the right-hand region in Figure 6.4(b) initiate trajectories that lead to extinction for Q, while trajectories that start in the left-hand region lead to extinction for P. (Passing from one region to the other is impossible. Why?) Exercise 13 asks you to try other choices of constants K_j and a_j satisfying the three inequalities at the beginning of the example. There is a vanishingly small likelihood that the initial point lies on a separatrix, in which case the populations would tend to P_E and Q_E along the separatrix.

The most widely known nonlinear population model involves parasite-host, or alternatively predator-prey, populations in an asymmetric relationship described in detail in the next example.

---

example 5

Lotka-Volterra Model

Let $P = P(t)$ denote the size at time t of a parasite population that preys on a host population of size $H = H(t)$. If the parasites are absent, the host population may grow for some time according to $\dot{H} = aH,\ a > 0$. However, in general, the growth of H will be inhibited by the increasingly destructive effect of a larger parasite population P, since the death rate per host will be larger. We assume these negative effects to take the form $-bPH,\ b > 0$, when added to the right-hand side of the expression for $\dot{H}$. This leads to the first displayed equation below. If the host population is absent, the parasites can be expected to die out rapidly according to $\dot{P} = -dP,\ d > 0$, with additional growth provided only by $H > 0$ in the form of the term $cHP,\ c > 0$. This leads to the second equation below.

The equations

$$\dot{H} = (a - bP)H, \qquad \dot{P} = (cH - d)P$$

are nonlinear unless $b = c = 0$. We assume here that $a, b, c,$ and d are positive constants, though they could in principle depend on t. The factor $a - bP$ acts to increase the size of H if $P(t)$ is below the critical level a/b, and to decrease H in the opposite case. Similarly, the factor $cH - d$ acts to increase the size of P if $H(t)$ is above the critical level d/c, and to decrease P in the opposite case.

The equilibrium solutions satisfy $(a - bP)H = 0$ and $(cH - d)P = 0$, and the solutions are $(H,P) = (0,0)$ and $(H,P) = (d/c, a/b)$. The $(0,0)$ solution is not interesting from the point of view of population dynamics, since it represents a total absence of both species and it turns out not even to be a limit of solutions with both populations present. To understand the significance of the equilibrium at $H = d/c,\ P = a/b$, we examine the behavior near that point. To that end, let $H = H_0 + d/c$ and $P = P_0 + a/b$, where H_0 and P_0 are assumed to be small. Substitution into the system for H and P gives

$$\dot{H}_0 = (a - bP_0 - a)\left(H_0 + \frac{d}{c}\right) = -\left(\frac{bd}{c}\right)P_0 - bP_0H_0,$$

$$\dot{P}_0 = (cH_0 + d - d)\left(P_0 + \frac{a}{b}\right) = \frac{ac}{b}H_0 + cH_0P_0.$$

This system is no simpler than the original one, but since H_0 and P_0 are assumed small, say less than 1 on some scale, the product H_0P_0 will be even smaller.

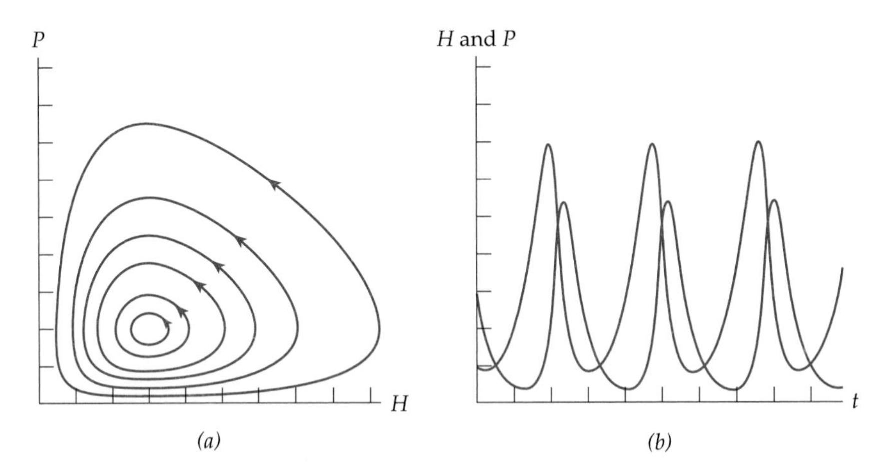

Figure 6.5 Nonlinear Lotka-Volterra solutions. (a) HP trajectories. (b) Graphs: $H(0) = 1,\ P(0) = 3$.

Removal of the last two terms in the modified version of the system results in the linear system

$$\dot{H}_0 = -\left(\frac{bd}{c}\right) P_0, \qquad \dot{P}_0 = \frac{ac}{b} H_0.$$

The replacement of the original nonlinear system by an approximate system of linear form is called **linearization.** The idea is taken up in more generality in Section 4B of this chapter.

The solution of the linearized system by elimination is

$$H_0(t) = b\sqrt{d}\, A\cos(\sqrt{ad}\, t + \phi), \qquad P_0(t) = c\sqrt{a}\, A\sin(\sqrt{ad}\, t + \phi),$$

where the constants A and ϕ are to be determined by initial conditions. The point $(H_0(t), P_0(t))$ traces a family of ellipses centered at (0,0) in the (H_0, P_0) plane. Shifting back to the (H,P) plane, we get ellipses centered at the equilibrium point $(d/c, a/b)$.

It is a remarkable fact that the closed, periodic orbits exhibited by the linearized system have as their counterparts closed orbits for the original nonlinear system, which implies that $H(t)$ and $P(t)$ are actually periodic. You are asked to show in Exercise 16 that this is true for every trajectory of the nonlinear system in the positive (H,P) quadrant. The sketches in Figure 6.5(a) are scaled-down versions associated with the choice $a = 2$, $d = 1$, and $b = c = 1$, and they show that, with the axes labeled as they are, the trajectories circulate counterclockwise about the equilibrium point $(H,P) = (3,2)$. Figure 6.5(b) shows parametrizing graphs for the trajectory through $(H,P) = (1,3)$.

The Lotka-Volterra model serves generally as a predator-prey model, with P the predator and H the prey. The closed orbits are much too neat to be numerically accurate predictors, though they do suggest a roughly synchronized qualitative behavior of two population sizes. In real life, there will always be perturbations of population models, perhaps large enough to send P to extinction with $P(t_1) = 0$ at some time t_1.

EXERCISES

The Java applet VECFIELD at the website <http://math.dartmouth.edu/~rewn/>, or via <http:/www.mhhe.com/williamson>, is adequate for doing the computer exercises below, but many other software packages are also available.

1. In text Example 1, let $k = l = 1$, $m = 2$, and $n = 3$.

(a) Find explicit solution formulas for $P(t)$ and $Q(t)$ in terms of initial parameters $P_0 = P(0) > 0$ and $Q_0 = Q(0) > 0$.

(b) Investigate to what extent, if any, the relative sizes of P_0 and Q_0 affect the general nature of the solutions.

2. Assume $k = n$ and $l = m$ in text Example 1.

(a) Find explicit solution formulas for $P(t)$ and $Q(t)$ in terms of initial parameters $P_0 = P(0) > 0$ and $Q_0 = Q(0) > 0$.

(b) Investigate to what extent, if any, the relative sizes of P_0 and Q_0 affect the general nature of the solutions.

3. If in text Example 1 the growth and inhibition constants are essentially the same, we may assume $k = l$ and $m = n$. Thus the linear system is

$$\frac{dP}{dt} = kP - mQ,$$

$$\frac{dQ}{dt} = -mP + kQ, \qquad k > 0,\ m > 0.$$

(a) Show that the characteristic roots are $r_1 = k + m$ and $r_2 = k - m$.

(b) Conclude from part (a) that

$$P(t) = c_1 e^{(k+m)t} + c_2 e^{(k-m)t}, \qquad Q(t) = -c_1 e^{(k+m)t} + c_2 e^{(k-m)t}.$$

(c) Show that with initial conditions $P(0) = P_0$ and $Q(0) = Q_0$, the solutions become

$$P(t) = \tfrac{1}{2}(P_0 - Q_0)e^{(k+m)t} + \tfrac{1}{2}(P_0 + Q_0)e^{(k-m)t},$$

$$Q(t) = \tfrac{1}{2}(Q_0 - P_0)e^{(k+m)t} + \tfrac{1}{2}(P_0 + Q_0)e^{(k-m)t}.$$

(d) Show that a gradual exponential decline in either population is impossible in this model except in the critical circumstance that $P_0 = Q_0$, and that in that unlikely event, the relation $k < m$ between the growth and inhibition rates implies exponential decline in both P and Q.

4. Assume in text Example 1 that $k = n$ and $l = m$. How are $P(t)$ and $Q(t)$ related in general?

5. Verify the relations $a = (k - r_1)/m$ and $b = (k - r_2)/m$ that are asserted near the end of text Example 1.

6. In part (c) of Exercise 3 the equations

$$P(t) = \tfrac{1}{2}(P_0 - Q_0)e^{(k+m)t} + \tfrac{1}{2}(P_0 + Q_0)e^{(k-m)t},$$
$$Q(t) = \tfrac{1}{2}(Q_0 - P_0)e^{(k+m)t} + \tfrac{1}{2}(P_0 + Q_0)e^{(k-m)t},$$

describe the history of two competing populations that have common growth and inhibition rates k and m.

(a) Show that if $Q_0 > P_0$, then P will reach extinction, that is, $P(t_e) = 0$, at time

$$t_e = \frac{1}{2m} \ln\left(\frac{Q_0 + P_0}{Q_0 - P_0}\right).$$

(b) Show that if $Q_0 > P_0$ and $k > m$, then $Q(t)$ increases steadily while $P(t)$ is heading for extinction as shown in part (a).

(c) Show that if $Q_0 > P_0$ and $k < m$, then $Q(t)$ decreases steadily up to time

$$t_1 = \frac{1}{2m} \ln\left[\frac{(m-k)(Q_0 + P_0)}{(m+k)(Q_0 - P_0)}\right]$$

and increases thereafter.

7. The general solution of the linear model in Example 1 of the text is shown to be

$$P(t) = c_1 e^{r_1 t} + c_2 e^{r_2 t},$$
$$Q(t) = c_1 \frac{(k - r_1)}{m} e^{r_1 t} + c_2 \frac{(k - r_2)}{m} e^{r_2 t},$$

where $r_1 = \frac{1}{2}\left[(k+l) + \sqrt{(k-l)^2 + 4mn}\,\right] > r_2 = \frac{1}{2}\left[(k+l) - \sqrt{(k-l)^2 + 4mn}\,\right]$.

(a) Show that $mn = (r_1 - k)(r_1 - l) = (r_2 - k)(r_2 - l)$.

(b) Use the result of part (a) to show that the solutions can be written in the more symmetrical form

$$P(t) = -c_1 \frac{(k - l + \sqrt{d})}{2n} e^{r_1 t} + c_2 e^{r_2 t},$$

$$Q(t) = c_1 e^{r_1 t} + c_2 \frac{(k - l + \sqrt{d})}{2m} e^{r_2 t},$$

where the quadratic discriminant $d = (k-l)^2 + 4mn$. [Remember that c_1 and c_2 are *arbitrary* constants.]

(c) Use the result of part (b) to show that if $k = l$, then the initial-value problem $P(0) = P_0,\ Q(0) = Q_0$ has solution

$$P(t) = \frac{1}{2}\left(P_0 - \sqrt{\frac{m}{n}}\, Q_0\right) e^{(k+\sqrt{mn})t} + \frac{1}{2}\left(\sqrt{\frac{m}{n}}\, Q_0 + P_0\right) e^{(k-\sqrt{mn})t},$$

$$Q(t) = \frac{1}{2}\left(Q_0 - \sqrt{\frac{n}{m}}\, P_0\right) e^{(k+\sqrt{mn})t} + \frac{1}{2}\left(Q_0 + \sqrt{\frac{n}{m}}\, P_0\right) e^{(k-\sqrt{mn})t}.$$

(d) Use the result of part (c) to show that the two populations decline exponentially to zero if and only if both $\sqrt{n}\,P_0 = \sqrt{m}\,Q_0$ and $k^2 < mn$ are true.

8. Derive the solutions for P_E and Q_E given in text Example 2.

9. Verify the parenthetical statement at the end of text Example 2.

10. Use computer graphics to draw a first-quadrant phase portrait for the system of text Example 2 under some choice of constants satisfying $a_1K_1 > K_2$, $K_1 > a_2K_2$, and $a_1a_2 < 1$. Your portrait should indicate extinction for Q from a variety of initial points in the positive quadrant.

11. Use computer graphics to draw a first-quadrant phase portrait for the system of text Example 2 under some choice of constants satisfying $a_2K_2 > K_1$, $K_2 > a_1K_1$, and $a_1a_2 < 1$. Your portrait should indicate extinction for P from a variety of initial points in the positive quadrant.

12. Use computer graphics to illustrate the coexistence regime of text Example 3 with a phase portrait.

13. Use computer graphics to illustrate the selective extinction regime of text Example 4 with a phase portrait.

14. Mutual benefit. Bees and flowering plants are famous for benefitting each other.

(a) Explain how the equations

$$\frac{dP}{dt} = r_1P\left(\frac{K_1 + b_2Q - P}{K_1}\right) \quad \text{and} \quad \frac{dQ}{dt} = r_2Q\left(\frac{K_2 + b_1P - Q}{K_2}\right),$$

model the interaction of two such species assuming the parameters K_j and b_j are positive.

(b) Find all equilibrium solutions of the system in part (a) if $b_1b_2 \neq 1$.

(c) Discuss the possible equilibrium solutions if $b_1b_2 = 1$.

15. Use computer graphics to illustrate the system in the previous exercise with phase portraits under the following assumptions: (a) $b_1b_2 < 1$ and (b) $b_1b_2 > 1$.

***16.** The Lotka-Volterra equations

$$\frac{dH}{dt} = (a - bP)H, \qquad \frac{dP}{dt} = (cH - d)P,$$

where a, b, c, and d are positive constants, are used to describe the size relationship of parasite $P(t)$ and host $H(t)$ populations at time t.

(a) Show that if $P(t) > a/b$, then $H(t)$ decreases, and that if $H(t) < d/c$, then $P(t)$ decreases. Describe the implications of these results for the host and parasite populations.

(b) Show that the parameterized solution curves $(H,P) = (H(t), P(t))$ satisfy

$$\frac{dH}{dP} = \frac{(a - bP)H}{(cH - d)P}.$$

Then solve this equation by separation of variables to get

$$\left(H^d e^{-cH}\right)\left(P^a e^{-bP}\right) = k, \qquad k \text{ constant.}$$

(c) Show that $f(H) = H^d e^{-cH}$ has the unique local maximum value $f(d/c) = (d/ec)^d$ and tends to zero as H tends to infinity. Similarly, show that $g(P) = P^a e^{-bP}$ has unique local maximum $g(P) = (a/eb)^a$ and tends to zero as P tends to infinity.

(d) Show graphically that for $k > 0$ the curves found in part (c) are closed circuits in the first quadrant of the P, H plane. [*Hint:* For each value of $H \neq d/c$ there are two corresponding values of P. An analogous statement holds with P and H exchanged.]

***17.** Show that the closed circuits identified as orbits of the Lotka-Volterra equations are concave right in the H direction if $H < d/c$ and concave left if $H > d/c$ by showing that

$$\frac{d^2H}{dP^2} = \frac{1}{d - cH}\left[\frac{aH}{P^2} + \frac{d}{H}\left(\frac{dH}{dP}\right)^2\right].$$

Thus, the sign of the second derivative is determined by the first factor: negative when $H > d/c$, positive when $H < d/c$. (The dH over dP denotes a first derivative.)

18. The Lotka-Volterra equations can be refined to take account of fixed resource or living-space limitations, $L > H(t)$ and $M > P(t)$, to the growth of the host and parasite populations as follows:

$$\frac{dH}{dt} = (a - bP)H(L - H),$$
$$\frac{dP}{dt} = (cH - d)P(M - P),$$

where $a, b, c,$ and d are positive constants, with $L > d/c > 1$ and $M > a/b > 1$. The equations make no sense if the upper limits on H and P are not respected.

(a) Show that if $P(t) > a/b$, then $H(t)$ decreases, and that if $H(t) < d/c$, then $P(t)$ decreases. Describe the implications of these results for the host and parasite populations.

(b) Show that the parameterized solution curves $(H,P) = (H(t), P(t))$ satisfy

$$\frac{dH}{dP} = \frac{(a - bP)H(L - H)}{(cH - d)P(M - P)}.$$

Then solve this equation by separation of variables to get

$$\left[H^{d/L}(L-H)^{c-d/L}\right]\left[P^{a/M}(M-P)^{b-a/M}\right] = f(H)g(P) = k, \quad k \text{ constant.}$$

(c) Show that $f(H)$ in part (b) has its unique local maximum value at $H = d/c$ and is equal to zero at $H = 0$ and $H = L$. Similarly, show that $g(P)$ has its unique local maximum at $P = a/b$ and is zero at $P = 0$ and $P = M$.

***(d)** Show that for $k > 0$, the curves found in part (b) are closed circuits in the first quadrant of the P, H plane. [*Hint:* For each value of $H \neq c/d$ there are two corresponding values of P. An analogous statement holds with P and H exchanged.]

19. If $H(t)$ and $K(t)$ are the sizes of two populations that are preyed on by a common parasite population of size $P(t)$, an analogue of the Lotka-Volterra system extended to cover the three species is

$$\frac{dH}{dt} = (a_1 - b_1 P)H,$$
$$\frac{dP}{dt} = (c_1 H + c_2 K - d)P,$$
$$\frac{dK}{dt} = (a_2 - b_2 P)K,$$

where the a_k, b_k, c_k, and d are positive.

(a) Verify that if either the H population or the K population is absent, the system reduces to the form of the two-species Lotka-Volterra system. Then find the corresponding constant equilibrium solutions.

(b) The system for three species given above has equilibrium points with strictly positive coordinates only under certain conditions. Find these conditions, and find all the equilibrium points with positive coordinates. There may, for some choices of a_k and b_k, be infinitely many equilibrium points.

20. The three-species system described in the previous exercise can be further modified to take into account the direct effect of the growth or decline of $K(t)$ on $H(t)$ and vice versa:

$$\frac{dH}{dt} = (a_1 - b_1 P - b_3 K)H,$$
$$\frac{dP}{dt} = (c_1 H + c_2 K - d)P,$$
$$\frac{dK}{dt} = (a_2 - b_2 P - b_4 H)K.$$

(a) Show that if the constants a_k, b_k, c_k, and d are positive then the system has finitely many equilibrium solutions.

(b) Find conditions on the constants a_k, b_k, c_k, and d under which all equilibrium solutions have nonnegative coordinates.

21. Disease epidemic. Let $H(t)$ be the size of a population of healthy but susceptible animals and $P(t)$ the size of a mingled population of infected animals that can spread the disease by contact. The system

$$\frac{dH}{dt} = -bHP, \qquad \frac{dP}{dt} = bHP - cP,$$

where b and c are positive constants, is a special case of the Lotka-Volterra system that models the development of an epidemic. The term bHP increases as the likelihood of contact between healthy and infected animals increases. The term $-cP$ is the rate of removal of animals from the infective population because of death or survival with immunity. Note that every point $(H,0)$ represents an equilibrium point.

(a) Show that $\dot{H} + \dot{P} = (c/b)\dot{H}/H$.

(b) Integrate the equation in part (a) with respect to t to get a one-parameter family of equations satisfied by the system trajectories.

(c) Show that the graph of $P = P(H)$ is concave down in the first quadrant with maximum value when $H = c/b$. The sketch the graphs of some typical trajectories using computer plotting.

22. The average value of a function $P(t)$ time interval $t_0 \le t \le t_1$ is

$$\bar{P} = \frac{1}{t_1 - t_0} \int_{t_0}^{t_1} P(t)\,dt.$$

For a coordinate of a periodic trajectory such as a Lotka-Volterra orbit with time period $T > 0$ we can just as well integrate from 0 to T.

(a) Solve the Lotka-Volterra equation $\dot{H} = (a - bP)H$, and show that the average of $P(t)$ over a period interval is equal to the equilibrium coordinate $\bar{P} = a/b$.

(b) Solve the Lotka-Volterra equation $\dot{P} = (cH - d)P$, and show that the average of $H(t)$ over a period interval is equal to the equilibrium coordinate $\bar{H} = d/c$.

(c) Show that if the Lotka-Volterra equations are modified to

$$\dot{H} = (a - bP)H - hH, \qquad \dot{P} = (cH - d)P - kP, \qquad 0 < h < a, 0 < k,$$

where h and k are per capita harvesting rates, then $\bar{H} = (d + k)/c > d/c$, and $\bar{P} = (a - h)/b < a/b$, so H fares better and P fares worse, whatever the harvest rates.

3 MECHANICAL OSCILLATIONS

3A Masses in Linear Motion

Consider weights of mass m_1 and m_2 tied by springs to each other and also to fixed walls. Suppose the springs have stiffness constants h_1, h_2, and h_3, as shown in Figure 6.6, where the two fixed spring ends are b units apart. By Hooke's law,

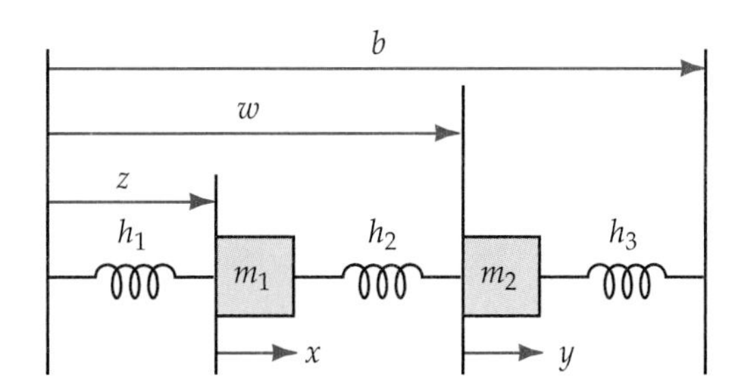

Figure 6.6 Spring-linked masses.

the magnitude of the restoring force for the jth spring is proportional to the absolute value of its displacement from rest, with proportionality constant h_j. Let x and y be displacements from equilibrium of the first and second bodies, so that equilibrium occurs when $x = y = 0$. (The labels z and w in Figure 6.6 are there for comparison in Example 3.) The force acting on the first body is by definition equal to $m_1(d^2x/dt^2)$. In calculating the force induced by the springs on m_1 or m_2 at any instant, we have to take account of the relative positions of both bodies. Note that the force exerted by h_3 on m_1 and h_1 on m_3 is transmitted through the middle spring. Thus

$$m_1\frac{d^2x}{dt^2} = -h_1x + h_2(y - x).$$

The choice of sign in each term is dictated by whether a displacement x or $y - x$ causes an increase or decrease in the velocity dx/dt. For example, if $y - x > 0$, the middle spring is extended and so acts to increase dx/dt, making $d^2x/dt^2 > 0$. Similarly,

$$m_2\frac{d^2y}{dt^2} = -h_2(y - x) - h_3y.$$

In deriving both equations, we have neglected frictional forces. We can rewrite the system in the form

$$\frac{d^2x}{dt^2} = -\frac{h_1 + h_2}{m_1}x + \frac{h_2}{m_1}y,$$
3.1
$$\frac{d^2y}{dt^2} = \frac{h_2}{m_2}x - \frac{h_2 + h_3}{m_2}y.$$

---

example 1

If the weights are equal, say $m_1 = m_2 = 1$, and also $h_1 = h_2 = h_3 = 1$, then the system 3.1 becomes

$$\frac{d^2x}{dt^2} = -2x + y,$$

$$\frac{d^2y}{dt^2} = x - 2y.$$

In operator form this is

$$(D^2 + 2)x - y = 0$$
$$-x + (D^2 + 2)y = 0.$$

Operating on the second equation with $(D^2 + 2)$ and adding gives

$$(D^4 + 4D^2 + 3)y = (D^2 + 1)(D^2 + 3)y = 0.$$

We can read off the characteristic roots at a glance, $(\pm i, \ \pm i\sqrt{3})$, so

$$y(t) = c_1 \cos t + c_2 \sin t + c_3 \cos \sqrt{3}t + c_4 \sin \sqrt{3}t.$$

Using the second of the pair of equations to find x gives

$$\begin{aligned} x(t) &= (D^2 + 2)y(t) \\ &= c_1 \cos t + c_2 \sin t - c_3 \cos \sqrt{3}t - c_4 \sin \sqrt{3}t. \end{aligned}$$

The constants $c_1, c_2, c_3,$ and c_4 are determined by initial displacements and velocities. The special choice $x(0) = -1$, $y(0) = 1$, $\dot{x}(0) = -\sqrt{3}$, and $\dot{y}(0) = \sqrt{3}$ gives

$$\begin{aligned} x(t) &= -\cos \sqrt{3}t - \sin \sqrt{3}t & \qquad y(t) &= \cos \sqrt{3}t + \sin \sqrt{3}t \\ &= -\sqrt{2} \cos\left(\sqrt{3}t - \frac{\pi}{4}\right), & &= \sqrt{2} \cos\left(\sqrt{3}t - \frac{\pi}{4}\right) \end{aligned}$$

If a homogeneous system has oscillatory solutions expressible as a sum of terms with distinct frequencies, then a term with one of these frequencies is called a **normal mode** of oscillation for the system. Thus the oscillations $\cos \sqrt{3}\,t$ and $\sin \sqrt{3}\,t$ with circular frequency $\sqrt{3}$ in the previous example express the same normal mode of the system. Other normal modes in the same example, for instance, $\cos t$ or $2 \sin t$, have circular frequency 1. The typical normal mode of a constant-coefficient system contains a factor $\cos \mu t$ or $\sin \mu t$ with circular frequency μ. The normal modes are clearly important characteristics of an oscillatory system, and an efficient way to find them using different methods appears in Chapter 7. In the special case $m\ddot{x} + hx = 0$ of dimension 1 that we treated in detail in Chapter 4, the normal modes all have circular frequency $\omega_0 = \sqrt{h/m}$.

---

example 2

Suppose the third spring is removed altogether from our previous example, so that $h_3 = 0$. We're left with

$$\ddot{x} = -2x + y,$$
$$\ddot{y} = x - y,$$

or, in operator form,

$$(D^2 + 2)x - y = 0$$
$$-x + (D^2 + 1)y = 0.$$

Elimination of x proceeds as in the previous example, but this time the differential equation for y is $(D^4 + 3D^2 + 1)y = 0$, with characteristic equation $r^4 + 3r^2 + 1 = 0$. Regarding the left as quadratic as a function of r^2, we find $r^2 = (-3 \pm \sqrt{5})/2$. Both values are negative, so the characteristic roots are $\pm i\sqrt{(3+\sqrt{5})/2} \approx \pm 1.62i$ and $\pm i\sqrt{(3-\sqrt{5})/2} \approx \pm 0.62i$. The normal modes from which solutions are constructed have circular frequencies

$$\mu_1 = \sqrt{(3+\sqrt{5})/2} \quad \text{and} \quad \mu_2 = \sqrt{(3-\sqrt{5})/2}.$$

Note that for Hooke's law to remain valid in practice, the condition $x(t) < y(t)$ has to be maintained with some leeway. Nevertheless, it's quite possible to have this condition violated under certain purely mathematical initial conditions.

In the previous example, it was essential to know in advance the equilibrium positions of the two masses to establish the precise location of each mass relative to the other and to the spring supports. Thus our additional problem is to find an **equilibrium solution** to the appropriate equations of motion, that is, a constant solution, for which all time derivatives are zero. For the equations derived in the previous example, it's easy to check that the unique equilibrium solution is $x(t) = 0$, $y(t) = 0$. Indeed we chose our coordinates so that these would be the equilibrium solutions. The next example shows how to avoid making any assumptions about the coordinates.

Instead of measuring the locations of the two masses shown in Figure 6.6 from their equilibrium positions we can measure both displacements from the same point. If we know the unstressed (i.e., relaxed) lengths l_1, l_2, and l_3 of the three springs, and the distance b between the supports, this approach enables us to determine the precise location of the equilibrium positions. (This information had to be assumed in our earlier analysis.) Let z and w be the respective distances of masses m_1 and m_2 from the left end, as shown in Figure 6.6. It's now the (positive) extension or (negative) compression of the springs beyond their unstressed lengths that is of critical interest; for the first spring, this is $z - l_1$; for the second, it is $(w - z) - l_2$; for the third, it is $(b - w) - l_3$. Taking into account the force direction of each spring action on the adjacent mass, we find

$$m_1 \frac{d^2 z}{dt^2} = -h_1(z - l_1) + h_2\big[(w - z) - l_2\big],$$

$$m_2 \frac{d^2 w}{dt^2} = -h_2\big[(w - z) - l_2\big] + h_3\big[(b - w) - l_3\big].$$

These equations simplify to the system

3.2

$$m_1 \frac{d^2 z}{dt^2} = -(h_1 + h_2)z + h_2 w + h_1 l_1 - h_2 l_2,$$

$$m_2 \frac{d^2 w}{dt^2} = h_2 z - (h_2 + h_3)w + h_2 l_2 - h_3 l_3 + h_3 b.$$

These equations are almost the same as the ones derived in the previous example except for the presence of additional constant terms on the right side. Thus they constitute a nonhomogeneous system rather than a homogeneous one. Since equilibrium solutions are constant, the second derivatives $\ddot{z}$ and $\ddot{w}$ are identically zero. Consequently, to find the equilibrium positions, all we have to do is set the right-hand sides of the differential equations equal to zero and solve for z and w.

example 3

If springs governed by Equations 3.2 have equal stiffness $h_1 = h_2 = h_3$, and equal relaxed lengths $l_1 = l_2 = l_3$, then the h_k and l_k all cancel from the purely algebraic system obtained by setting $\ddot{z} = \ddot{w} = 0$. The system to be solved is then

$$-2z + w = 0,$$
$$z - 2w = -b.$$

The unique solution is the particular constant solution $z_e = b/3$, $w_e = 2b/3$. [A moment's reflection shows that this result is what you would expect on physical grounds: If all three springs have the same properties, then the equilibrium points will be equally-spaced between the fixed supports.] Moreover, the general solution to the system can be derived from the solution to the homogeneous system by adding a particular solution. Since the homogeneous system is the same as the system in the previous example, we have $z(t) = x(t) + z_e$, $w(t) = y(t) + w_e$.

Taking linear damping into account is a simple matter in principle. For example, the two-body system we started with becomes

$$m_1 \frac{d^2x}{dt^2} = -(h_1 + h_2)x + h_2 y - k_1 \dot{x},$$
$$m_2 \frac{d^2y}{dt^2} = h_2 x - (h_2 + h_3)y - k_2 \dot{y}.$$

The major technical difference is that the introduction of first-order terms means that the equation for the characteristic roots r can no longer be regarded as a quadratic equation in r^2.

3B Nonlinear Motion in Space

Analysis of linear motion of physical bodies linked by springs satisfying Hooke's law leads to a system of second-order linear differential equations with dimension equal to the number n of bodies under consideration. If the motion takes place in a plane, or else in 3-dimensional space, it takes respectively two or three coordinates to specify the position of each body. Hence the relevant system of differential equations will then involve $2n$, or else $3n$, coordinates if we're keeping track of n bodies. At first sight, it might seem that just setting up the system of differential equations is a formidable problem.

We shall describe a simple way to proceed using vector ideas and notation. The Figure 6.7 shows two objects with masses m_1 and m_2 linked by three springs to each other, and to two points **a** and **b**. Depending on how the motion starts, it might in fact be confined to the line joining **a** and **b**, or to some plane containing

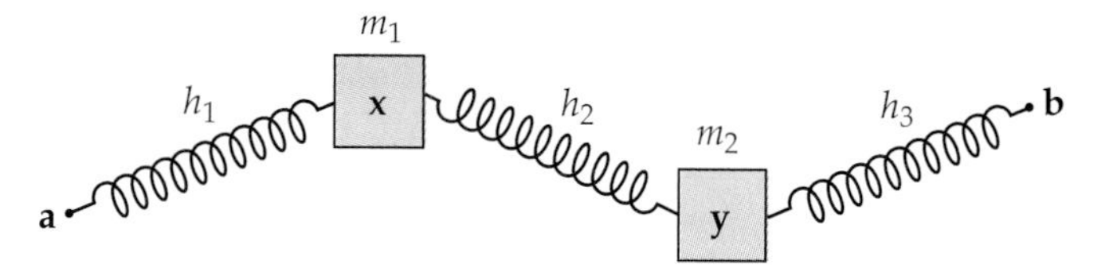

Figure 6.7 Two masses in space.

those two points, or in the most general circumstances the motion of each body might be three-dimensional in an essential way. Our derivation of the differential equations will include all these cases at once. Along with the stiffness constants h_1, h_2, and h_3 we need to take into account the relaxed (i.e., unstressed) lengths of the springs, which we'll denote by l_1, l_2, and l_3, respectively. Denoting the position at time t of the body of mass m_1 by $\mathbf{x} = \mathbf{x}(t)$, the total force acting on the body can be expressed as $m_1\ddot{\mathbf{x}}$, that is, mass times the acceleration vector. On the other hand, this same total force can be expressed as the sum of the two forces acting on the body via the two springs attached to it. The first spring will have extended, or maybe compressed, length $|\mathbf{a} - \mathbf{x}|$. The difference $|\mathbf{a} - \mathbf{x}| - l_1$, if it is positive, measures the extension of the spring beyond its relaxed length, and, if this difference is negative, it measures the amount of compression of the spring below its relaxed length. The numerical value of the force due to the first spring is then $h_1(|\mathbf{a} - \mathbf{x}| - l_1)$. Since an extended spring pulls the body toward $\mathbf{a}$, we can express the force vector by its numerical value times the vector of length one pointing from the body toward the point $\mathbf{a}$. This unit vector can be written as the vector $\mathbf{a} - \mathbf{x}$ divided by its length: $(\mathbf{a} - \mathbf{x})/|\mathbf{a} - \mathbf{x}|$. The resulting force vector is then the first of the two vectors

$$h_1(|\mathbf{a} - \mathbf{x}| - l_1)\frac{(\mathbf{a} - \mathbf{x})}{|\mathbf{a} - \mathbf{x}|}, \qquad h_2(|\mathbf{y} - \mathbf{x}| - l_2)\frac{(\mathbf{y} - \mathbf{x})}{|\mathbf{y} - \mathbf{x}|}.$$

Denoting the position of the second body by $\mathbf{y} = \mathbf{y}(t)$, we similarly get the second of these two vectors for the other spring force acting on the first body. The differential equation expressing $m_1\ddot{\mathbf{x}}$ as the sum of these forces is then

$$m_1\ddot{\mathbf{x}} = h_1(|\mathbf{a} - \mathbf{x}| - l_1)\frac{(\mathbf{a} - \mathbf{x})}{|\mathbf{a} - \mathbf{x}|} + h_2(|\mathbf{y} - \mathbf{x}| - l_2)\frac{(\mathbf{y} - \mathbf{x})}{|\mathbf{y} - \mathbf{x}|}.$$

A parallel derivation yields the equation for the force on the second body:

$$m_2\ddot{\mathbf{y}} = h_2(|\mathbf{x} - \mathbf{y}| - l_2)\frac{(\mathbf{x} - \mathbf{y})}{|\mathbf{x} - \mathbf{y}|} + h_3(|\mathbf{b} - \mathbf{y}| - l_3)\frac{(\mathbf{b} - \mathbf{y})}{|\mathbf{b} - \mathbf{y}|}.$$

For motion in three dimensions, each of these two vector equations is equivalent to three real coordinate equations, making up a system of six equations governing the motion of the entire mass-spring system. One striking feature of this system is that it is in general nonlinear. The nonlinearity is due to the presence of the expressions $|\mathbf{b} - \mathbf{y}|$, $|\mathbf{y} - \mathbf{x}|$, and $|\mathbf{a} - \mathbf{x}|$. For example, letting $\mathbf{y} = (y_1, y_2, y_3)$ and $\mathbf{b} = (b_1, b_2, b_3)$, we note that

$$|\mathbf{b} - \mathbf{y}| = \sqrt{(b_1 - y_1)^2 + (b_2 - y_2)^2 + (b_3 - y_3)^2},$$

an obviously nonlinear function of the coordinates y_1, y_2, and y_3.

Though the two-body mass-spring system is in general nonlinear, the next example shows that the linear motion of Section 3A is a special case of the nonlinear one.

---

example 4

For two bodies moving on a line, we may as well take the line of motion to be a coordinate axis. We place the fixed points $\mathbf{a} = a$ and $\mathbf{b} = b$ and the variable points $\mathbf{x} = x$ and $\mathbf{y} = y$ that designate the locations of the two masses in order on the line: $a < x < y < b$. Then the vector lengths can all be interpreted as absolute values, and we can use the order relations among the points to compute their values. For example, $|a - x| = x - a$ since $a < x$. In particular, the unit vectors that we use to determine direction of force all reduce to either 1 or -1. For example, $(a - x)/|a - x| = -1$, since $a < x$ implies $a - x < 0$. Using such relations, the two-mass vector system becomes

$$m_1\ddot{x} = -h_1(x - a - l_1) + h_2(y - x - l_2),$$

$$m_2\ddot{y} = -h_2(y - x - l_2) + h_3(b - y - l_3).$$

Rearranging terms yields a pair with the nonhomogeneous terms at the end:

$$m_1\ddot{x} = -(h_1 + h_2)x + h_2 y + h_1(a + l_1) - h_2 l_2,$$

$$m_2\ddot{y} = h_2 x - (h_2 + h_3)y + h_2 l_2 + h_3(b - l_3).$$

The system is clearly linear. The reason these equations don't look just like the ones we derived earlier for the special case of motion restricted to a line is that then we used two different locations on the line as the respective zero-values for x and y. However, the new system contains more information than the one we found earlier; now we can find the equilibrium values for x and y from the new equations by setting their right-hand sides equal to zero and solving for x and y. The analogue for nonlinear systems requires numerical analysis techniques not covered in this book.

---

example 5

Equilibrium solutions of the above pair of second-order equations for two masses are those for which there is no motion, that is, $x(t)$ and $y(t)$ are both constant. It follows that $m_1\ddot{x} = m_1\ddot{y} = 0$, so the pair of differential equations reduces to the pair of purely algebraic equations

$$-(h_1 + h_2)x + h_2 y + h_1(a + l_1) - h_2 l_2 = 0,$$

$$h_2 x - (h_2 + h_3)y + h_2 l_2 + h_3(b - l_3) = 0.$$

For example, with $h_1 = h_2 = 1$, $h_3 = 2$, $l_1 = 2$, $l_2 = 3$, $l_3 = 1$, and $a = 0$, and $b = 7$ we get

$$2x - y = -1,$$

$$-x + 3y = 15.$$

The unique solution is $x = \frac{12}{5} = 2.4$, $y = \frac{29}{5} = 5.8$. Consequently, starting the system with initial conditions $x(0) = \frac{12}{5}$, $\dot{x}(0) = 0$, $y(0) = \frac{29}{5}$, and $\dot{y}(0) = 0$

yields the constant solution we just computed. In physical terms, the forces are precisely balanced at these positions, so there is no motion.

example 6

The linear case treated in the previous example is the only one that has solutions that can be expressed in terms of elementary functions. If we want to resort to numerical approximations to the true solutions, it's sometimes desirable to display the differential equations in coordinate form rather than vector form. Written out, the six real coordinate equations for a two-body mass-spring system might use the following coordinates: $\mathbf{x} = (x_1, x_2, x_3)$, $\mathbf{y} = (y_1, y_2, y_3)$, $\mathbf{a} = (a_1, a_2, a_3)$, and $\mathbf{b} = (b_1, b_2, b_3)$. Each of the two vector equations becomes three coordinate equations, making six in all. To avoid repeating expressions for vector length such as $|\mathbf{x} - \mathbf{y}| = \sqrt{(x_1 - y_1)^2 + (x_2 - y_2)^2 + (x_3 - y_3)^2}$ more often than necessary, we rearrange the equations slightly as follows:

$$m_1\ddot{x}_j = h_1\big(1 - l_1/|\mathbf{a} - \mathbf{x}|\big)(a_j - x_j) + h_2\big(1 - l_2/|\mathbf{y} - \mathbf{x}|\big)(y_j - x_j), \qquad j = 1, 2, 3,$$

$$m_2\ddot{y}_j = h_2\big(1 - l_2/|\mathbf{x} - \mathbf{y}|\big)(x_j - y_j) + h_3\big(1 - l_3/|\mathbf{b} - \mathbf{y}|\big)(b_j - y_j), \qquad j = 1, 2, 3.$$

Linear damping would simply add terms of the form $-k_1\dot{x}_j$ and $-k_2\dot{y}_j$, respectively, to the two right-hand sides of our equations for the undamped motion, with positive damping constants k_1 and k_2.

example 7

There is nothing in the derivation of the vector equations of motion that requires **a** and **b** to be fixed in space, as we tacitly regarded them at the time. For example, we could hold one of them fixed at $\mathbf{a}(t) = (0,0,0)$ and specify a periodic motion for the other one using three coordinates, for example by $\mathbf{b}(t) = (2\sin t, \cos t, \sin 2t)$. Note that the **configuration space** or **position space** for a two-body system is 6-dimensional, with three coordinates for each body. The *state space*, or *phase space*, including six additional velocity coordinates, is 12-dimensional.

At website <http://math.dartmouth.edu/~rewn/> or via <http://www.mhhe.com/williamson> the Java Applets SPRINGS2-1, for two springs attached to one mass, SPRINGS3-1, for three springs attached to one mass, and SPRINGS3-2, for two masses linked in series by three springs, enable you to choose masses m, spring constants h, and spring lengths l for perspective graphic simulations of nonlinear mass-spring systems in 3-dimensional space.

EXERCISES

1. In Example 1 of the text, the system

$$(D^2 + 2)x - y = 0,$$
$$-x + (D^2 + 2)y = 0$$

is shown to have the general solution

$$x(t) = c_1 \cos t + c_2 \sin t - c_3 \cos \sqrt{3}t - c_4 \sin \sqrt{3}t,$$
$$y(t) = c_1 \cos t + c_2 \sin t + c_3 \cos \sqrt{3}t + c_4 \sin \sqrt{3}t,$$

where $x(t)$ and $y(t)$ are interpreted as the displacements at time t of two masses in a mass-spring physical system.

(a) Show that the initial conditions $x(0) = 0$, $\dot{x}(0) = 1$ and $y(0) = 1$, $\dot{y}(0) = 0$ can be satisfied by choosing the constants properly in the general solution.

(b) Show that initial conditions of the form $x(0) = x_0$, $y(0) = y_0$, $\dot{x}(0) = u_0$, $\dot{y}(0) = v_0$ can always be satisfied.

Normal Modes. Calculate the circular frequencies of the various constituent oscillations associated with the system 3.1 of the text under the following assumptions.

2. $m_1 = m_2 = 1,\ h_1 = h_2 = h_3 = 1.$

3. $m_1 = m_2 = 1,\ h_1 = h_2 = 1,\ h_3 = 2.$

4. $m_1 = 1,\ m_2 = 2,\ h_1 = 1,\ h_2 = h_3 = 4.$

5. How do the normal modes of the system 3.1 of the text compare with those of the system 3.2?

6. Suppose that the middle spring is removed from the system governed by Equations 3.2.

(a) Show that the system becomes uncoupled.

(b) What are the normal modes?

7. Double pendulum. An ideal pendulum of mass m_1 and length l_1 with another one of mass m_2 and length l_2 attached to its free end moves in a single plane with respective displacement angles θ_1 and θ_2 shown in the Figure. The linearized system for small oscillations is

$$\ddot{\theta}_1 = -(\rho + 1)\frac{g}{l_1}\theta_1 + \rho\frac{g}{l_1}\theta_2, \qquad \ddot{\theta}_2 = (\rho + 1)\frac{g}{l_2}\theta_1 - (\rho + 1)\frac{g}{l_2}\theta_2,$$

where $\rho = (m_2/m_1)$. Show that if $m_1 = m_2$ and $l_1 = l_2$, then the normal modes have frequencies $\left[(2 - \sqrt{2})g/l_1\right]^{1/2} \approx 0.765\sqrt{g/l_1}$ and $\left[(2 + \sqrt{2})g/l_1\right]^{1/2} \approx 1.848\sqrt{g/l_1}$.

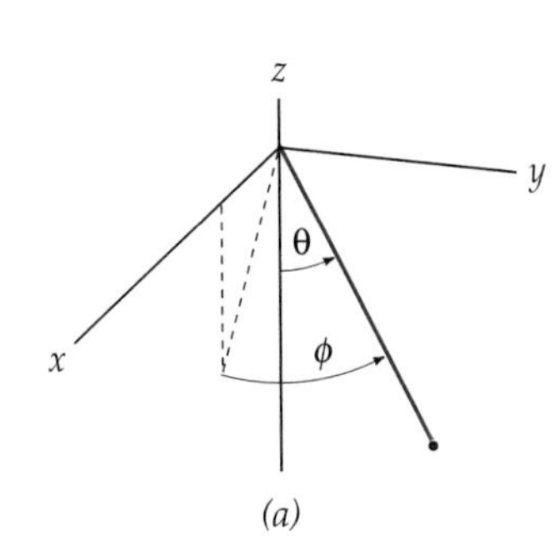

Spherical pendulum.

8. A 2-dimensional mechanical system $m_1\ddot{x} = f(x, y)$, $m_2\ddot{y} = g(x, y)$ is called **conservative** if there is a **potential function** $U(x, y)$ such that

$$\frac{\partial U(x, y)}{\partial x} = -f(x, y) \quad \text{and} \quad \frac{\partial U(x, y)}{\partial y} = -g(x, y).$$

(a) Show that the general two-body system

$$m_1\ddot{x} = -(h_1 + h_2)x + h_2 y,$$
$$m_2\ddot{y} = h_2 x - (h_2 + h_3)y,$$

is conservative by computing a potential.

(b) The **kinetic energy** of the system is $T = \frac{1}{2}(m_1\dot{x}^2 + m_2\dot{y}^2)$. Show that for a general conservative system of the type considered here, the **total energy** $T + U$ is constant. [*Hint:* Multiply the $\ddot{x}$ equation by $\dot{x}$, the $\ddot{y}$ equation by $\dot{y}$, add the two equations and integrate.]

9. Suppose two bodies at $x(t) < y(t)$ on a line are joined to each other and to two fixed points l units apart by springs of equal stiffness, so $h_1 = h_2 = h_3$. Find the equilibrium points for the system in terms of the unstressed lengths l_1, l_2, and l_3 of the springs and the distance l. Assume no forces are acting except that of the springs. [*Hint:* The answer is intuitively obvious in case $l_1 = l_2 = l_3$, and your general answer should be consistent with this special case.]

10. Suppose two bodies of mass 1 at $0 < x(t) < y(t) < l$ on a line are joined to each other and to two fixed points l units apart by springs of equal stiffness h and unstressed lengths $l_1 = l_2 = l_3$. Here x and y are both measured from the same origin, whereas in the first text example each was measured from a different point. Assume no forces are acting except that of the springs.

(a) Show that x and y satisfy

$$\ddot{x} = -2hx + hy,$$
$$\ddot{y} = hx - 2hy + hl.$$

(b) Find the constant equilibrium solutions for the system in terms of length l. These solutions remain at the fixed points from which the mass displacements are measured in the text example.

11. Suppose two bodies of mass m_1 and m_2 at $0 < x(t) < y(t) < l$ on a line are joined to each other and to two fixed points l units apart by springs of respective stiffnesses h_1, h_2, and h_3 and unstressed lengths l_1, l_2, and l_3. Here x and y are both measured from the same origin, as in Equations 3.2, rather than from different points as in Equations 3.1. Assume no forces are acting except that of the springs.

(a) Assume the middle spring is both twice as strong and half as long as the other two. Show that x and y satisfy a system of the form

$$m_1\ddot{x} = -3hx + 2hy,$$
$$m_2\ddot{y} = 2hx - 3hy + hl,$$

where h is some positive constant.

(b) Find the constant equilibrium solutions for the system in terms of l.

(c) Find the natural circular frequencies of the system, assuming $m_1 = m_2 = 1$.

12. Derive the vector equation of motion for two bodies of mass m_1 and m_2, respectively, if they are joined sequentially by springs to two fixed points and to each other, and assuming a constant gravitational acceleration of magnitude g in the direction of a fixed unit vector $\mathbf{u}$.

13. Find a first order system in standard form that is equivalent to the second order system for two bodies and three springs given in Equations 3.2.

14. (a) Derive the vector equation of motion for a single body of mass m at $\mathbf{x}(t)$ attached by springs with lengths l_1 and l_2 and respective Hooke constants h_1 and h_2 to two fixed points **a**, and **b**.

(b) Express the equation of part (a) in (x,y,z) coordinates assuming $\mathbf{a} = (0,0,0)$ and $\mathbf{b} = (b,0,0)$.

(c) Show that if the motion remains on the segment joining $(0,0,0)$ and $(b,0,0)$, the system is linear.

15. Derive the vector equations of motion for a single body of mass m at $\mathbf{x}(t)$ attached by springs with lengths l_1, l_2, and l_3 and respective Hooke constants h_1, h_2, and h_3 to three fixed points **a**, **b**, and **c**.

***16.** On physical grounds the linear system in Equations 3.1 should in general have all its solutions representable as linear combinations of undamped sinusoidal oscillations; equivalently, all characteristic roots arising in their solution should be purely imaginary. Prove this as follows.

(a) Note that Equations 3.1 have the form

$$\ddot{x} = -px + qy,$$
$$\ddot{y} = rx - sy, \qquad p, q, r, s \text{ positive constants.}$$

Use the elimination method to show that the squares of the characteristic roots of this more general system are always real.

(b) Show that assuming $p = (h_1 + h_2)/m_1$, $q = h_2/m_1$, $r = h_2/m_2$, $s = (h_2 + h_3)/m_2$ makes the squares of the characteristic roots negative, so the roots themselves are imaginary.

17. Example 1 of the text discusses a system with normal modes of vibration having respective circular frequencies $\omega_1 = 1$ and $\omega_2 = \sqrt{3}$. Show that for oscillation strictly in the first of these two modes, movement of the masses is always in the same direction and that in the second mode movement is always in opposite directions. [*Hint:* Compare $\dot{x}$ and $\dot{y}$ for each of the two modes.]

18. Replacing parallel springs by a single spring of equivalent strength. Consider two parallel springs with separate equilibrium lengths l_1 and l_2 and respective Hooke constants h_1 and h_2. Show that they can be replaced by a single spring with Hooke constant $h = h_1 + h_2$ and equilibrium length $(h_1 l_1 + h_2 l_2)/(h_1 + h_2)$. [Note that the Hooke constants are additive; in this respect parallel springs behave like electric resistances in series.]

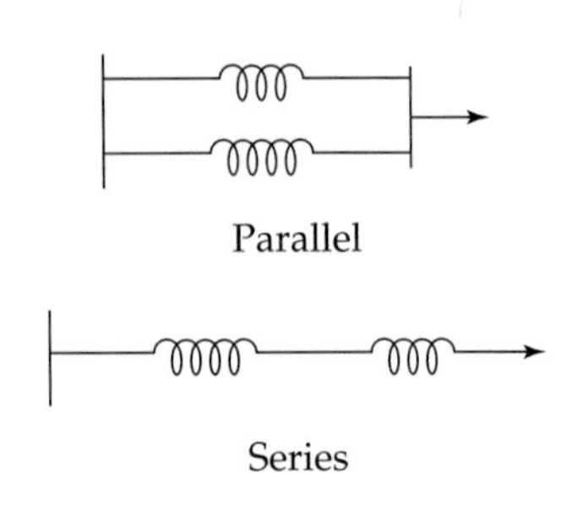

19. Replacing springs in series by a single spring of equivalent strength. Consider two springs in series with respective equilibrium lengths l_1 and l_2 and respective Hooke constants h_1 and h_2. Show that they can be replaced by a single spring with Hooke constant $h = h_1 h_2/(h_1 + h_2)$ and equilibrium length $l_1 + l_2$. [Note that the reciprocals of the Hooke constants are additive; in this respect springs in series behave like parallel electric resistances.]

4 INVERSE-SQUARE LAW

Let $\mathbf{x}_1 = \mathbf{x}_1(t)$ and $\mathbf{x}_2 = \mathbf{x}_2(t)$ represent the positions at time t of two bodies in space such that each acts on the other by the inverse square law of gravitational attraction, with no other forces considered. If m_1 and m_2 are the respective masses of the two bodies, the magnitude of the mutually attractive force is then

$$F = \frac{Gm_1m_2}{r^2},$$

where $r = |\mathbf{x}_1 - \mathbf{x}_2|$ is the distance between $\mathbf{x}_1$ and $\mathbf{x}_2$ (i.e., the length of the vector between them). The **gravitational constant** G is about $6.673 \cdot 10^{-11}$ if the relevant units are meters, kilograms, and seconds. The normalized vectors

$$\mathbf{u}_2 = \frac{\mathbf{x}_1 - \mathbf{x}_2}{|\mathbf{x}_1 - \mathbf{x}_2|}, \qquad \mathbf{u}_1 = -\frac{\mathbf{x}_1 - \mathbf{x}_2}{|\mathbf{x}_1 - \mathbf{x}_2|}$$

have length 1 and point respectively from the second body to the first, and vice versa. Thus the vectors that describe the force acting on each body are the product of magnitude F and a normalized direction unit vector $\mathbf{u}$; the vector $F\mathbf{u}_1$ acts on the first body and $F\mathbf{u}_2$ acts on the second body. Since these forces can also be described by Newton's second law as mass times acceleration, we have

$$m_1\ddot{\mathbf{x}}_1 = F\mathbf{u}_1, \qquad m_2\ddot{\mathbf{x}}_2 = F\mathbf{u}_2.$$

Figure 6.8 shows the positions and force vectors. The acceleration vectors, which actually govern the motion, are depicted as if m_1 is much larger than m_2. Written out in more detail these **Newton equations** are

4.1 $$\ddot{\mathbf{x}}_1 = -\frac{Gm_2}{|\mathbf{x}_1 - \mathbf{x}_2|^3}(\mathbf{x}_1 - \mathbf{x}_2), \qquad \ddot{\mathbf{x}}_2 = -\frac{Gm_1}{|\mathbf{x}_1 - \mathbf{x}_2|^3}(\mathbf{x}_2 - \mathbf{x}_1),$$

where m_1 has been canceled from the first equation and m_2 from the second. Subtracting the second equation from the first gives

$$\ddot{\mathbf{x}}_1 - \ddot{\mathbf{x}}_2 = -\frac{G(m_1 + m_2)(\mathbf{x}_1 - \mathbf{x}_2)}{|\mathbf{x}_1 - \mathbf{x}_2|^3}.$$

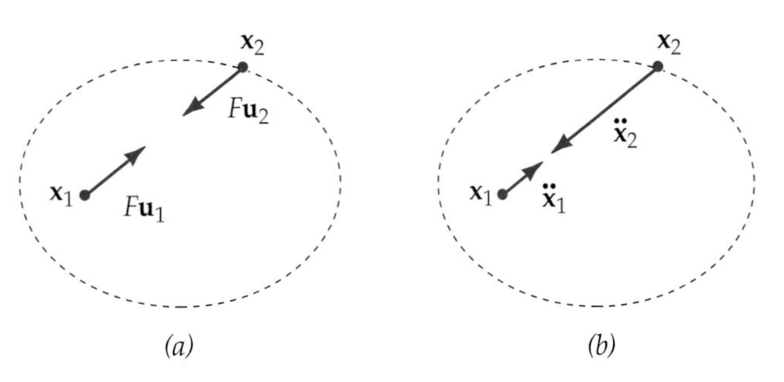

Figure 6.8 (*a*) Equal force magnitudes. (*b*) Unequal accelerations: $|\ddot{\mathbf{x}}_1| = F/m_1 < |\ddot{\mathbf{x}}_2| = F/m_2$.

Equations 4.1 form a system of vector equations for the motions of the two bodies relative to some coordinate system. If a moving coordinate system has its origin maintained at the center of mass of one of the bodies, say, the second, we can let $\mathbf{x} = \mathbf{x}_1 - \mathbf{x}_2$ and consider only the equation of relative motion for the first body:

4.2
$$\ddot{\mathbf{x}} = -\frac{G(m_1 + m_2)}{|\mathbf{x}|^3}\mathbf{x}.$$

Writing $\mathbf{x} = (x, y, z)$ and $G(m_1 + m_2) = k$, we get three scalar-valued equations:

4.3
$$\ddot{x} = \frac{-kx}{(x^2 + y^2 + z^2)^{3/2}}, \quad \ddot{y} = \frac{-ky}{(x^2 + y^2 + z^2)^{3/2}}, \quad \ddot{z} = \frac{-kz}{(x^2 + y^2 + z^2)^{3/2}}.$$

A solution of this nonlinear system will describe a trajectory of the first body relative to the second, or vice versa. We can eliminate the third equation from consideration by choosing (x, y, z) coordinates so that initial conditions on z are $z(0) = \dot{z}(0) = 0$. By Theorem 1.1 of Chapter 5, Section 1E, if $\mathbf{x}(0) \neq 0$ the system then has a unique solution with third coordinate $z = z(t)$ identically zero. Thus the system can be viewed as 2-dimensional (see also Exercise 18, part a):

$$\ddot{x} = \frac{-kx}{(x^2 + y^2)^{3/2}}, \qquad \ddot{y} = \frac{-ky}{(x^2 + y^2)^{3/2}}.$$

There are no simple formulas for the solution of the remaining two equations of 4.3. The classical approach to the problem is to derive characteristic properties of the solutions. These properties are usually stated as **Kepler's laws of planetary motion,** laws that were discovered empirically for closed orbits before Newton's work.

1. The path described by a solution $(x(t), y(t))$ is an ellipse with the sun (fixed body) at one focus.
2. The radius from the sun to the planet sweeps out equal areas in equal time periods.
3. If P is the time required to complete one orbit and a is the mean distance from the planet to the sun, then $P^2 = 4\pi^2 a^3/[G(m_1 + m_2)]$.

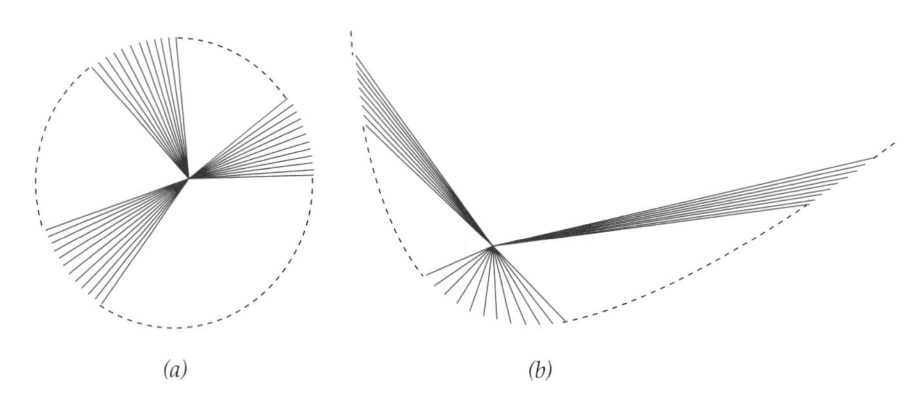

Figure 6.9 Equal areas in equal times. (a) Elliptic orbit. (b) Hyperbolic orbit.

These beautiful laws are derived in many calculus and physics texts; one approach is outlined in the exercises at the end of this section. In such derivations, it is often assumed that one of the bodies has negligible mass relative to the other, so that only the larger mass appears in the equation. For example, our sun is about 333,434 times as massive as the earth, so for some purposes the mass of the earth can be neglected.

example 1

Escape Speed

The closed elliptic orbits predicted by the first Kepler law occur only if the speed of separation of the two bodies is not too large. Nonreturning comets follow unbounded hyperbolic orbits for which the second Kepler law holds. To find the relative speed of separation required for a hyperbolic trajectory, suppose the relative motion of the two bodies is restricted to a fixed line, which we take to be an x axis. Thus y and z are always zero, so that all that survives of Equations 4.3 is

$$\ddot{x} = -\frac{k}{x^2}, \qquad k = G(m_1 + m_2).$$

To solve this equation we can multiply both sides by $\dot{x}$ to get

$$\dot{x}\ddot{x} = -kx^{-2}\dot{x}.$$

Integrating the left side with respect to t gives

$$\int \dot{x}\ddot{x}\,dt = \int \frac{d}{dt}\left(\tfrac{1}{2}\dot{x}^2\right) dt = \tfrac{1}{2}\dot{x}^2.$$

Integrating the right-hand side we get

$$-k\int x^{-2}\frac{dx}{dt}\,dt = -k\int x^{-2}\,dx = \frac{k}{x} + C.$$

Equating the results of the two integrations gives

$$\frac{1}{2}\dot{x}^2 = \frac{k}{x} + C.$$

Suppose now that at some definite time t_0 the values for x and $\dot{x} = v$ are x_0 and v_0, respectively. Then

$$C = \frac{v_0^2}{2} - \frac{k}{x_0}.$$

Thus the equation relating v and x can be written

$$\frac{v^2}{2} - \frac{v_0^2}{2} = \frac{k}{x} - \frac{k}{x_0}.$$

This relation between speed and distance shows that speed decreases as the separation distance x increases; it also enables us to find the **escape speed** of two bodies relative to each other; that is, the speed v_0 that must be attained at distance x_0 so that the speed v will always remain strictly positive thereafter.

To achieve this speed we must have

$$0 < v^2 = v_0^2 + \frac{2k}{x} - \frac{2k}{x_0}.$$

Since $k/x \to 0$ as $x \to \infty$, the only way the inequality can hold forever is to have

$$v_0^2 - \frac{2k}{x_0} > 0.$$

Thus the critical escape speed that must be exceeded at distance x_0 must satisfy $v_0^2 = 2k/x_0$. Since $k = G(m_1 + m_2)$, we find that the escape speed is

$$v_0 = \sqrt{\frac{2G(m_1 + m_2)}{x_0}}.$$

The conclusions we have reached are valid even if the relative motion is not restricted to a linear path.

It is useful to know what the constant G is in terms of various units of measurement.

Metric Units. Using meters, kilograms, and seconds, it has been calculated from careful observation that $G \approx 6.67 \cdot 10^{-11}$.

Planetary Units. Using the radius of the earth (about 6378 kilometers or 3963 miles) as the distance unit, the mass of the earth (about $5.98 \cdot 10^{24}$ kilograms) as the unit of mass and one hour as the time unit, it turns out that $G \approx 19.91$.

example 2

Using the results of the previous example we can compute escape speed relative to the earth. We'll use planetary units and assume the mass of the other body is negligible compared with the earth's mass of $5.98 \cdot 10^{24}$ kilograms. Thus we assume $m_1 + m_2 \approx 1$ in planetary units. With $G = 19.91$, we find $\sqrt{2G(m_1 + m_2)} = \sqrt{2 \cdot 19.91} \approx 6.31$. Escape speed is then about $v_0 = 6.31 x_0^{-1/2}$.

It is correct to measure the distance x_0 between the centers of mass of the two spherically homogeneous bodies, and for an object starting at the surface of the nearly spherical earth it is approximately correct to take x_0 to be 1 earth radius. In this case, $v_0 \approx 6.31$ earth radii per second, or about 25,000 miles per hour. Starting at distance $x_0 = 2$ earth radii, which would be about 4000 miles above the earth's surface, we could get away with $v_0 = 6.31/\sqrt{2} \approx 4.46$ earth radii per hour, or about 17,700 miles per hour.

example 3

A circular orbit of one body about another is a theoretical possibility, but in practice is a very unlikely state of affairs. Nevertheless, some orbits are nearly circular, so understanding circular ones is helpful for getting a qualitative feeling for what the nearby ones are like. You are asked to show in Exercise 14 that the orbital speed v_1 of one body in a circular orbit of radius x_0 about another is

$$v_1 = \sqrt{\frac{G(m_1 + m_2)}{x_0}}.$$

For example, since the moon's mass is about 1.2% of the earth's, for the moon to have a circular orbit of radius x_0 the orbital speed would have to be about $\sqrt{19.91 \cdot 1.012/x_0} \approx 4.49/\sqrt{x_0}$, where the orbital radius x_0 is measured in earth radius units (er). The mean distance of the moon from the earth in its actual elliptical orbit is about 384,404 kilometers, or $384{,}404/6378 \approx 60.27$ er. Since the moon's orbit is not very eccentric, its mean orbital velocity would be about $4.49/\sqrt{60.27} \approx 0.5783$ earth radii per hour, or about 3688 kilometers per hour.

The same sort of argument that leads to Equations 4.3 for the motion of two bodies can be applied to the case of three or more bodies with masses m_k and centers of mass located at $\mathbf{x}_k(t)$, $k = 1, 2, 3, \ldots, n$. Let $r_{ij} = r_{ji} = |\mathbf{x}_i - \mathbf{x}_j|$. Acceleration vectors are added, just as forces are, so the second-order vector equations for three bodies turn out to be

4.4
$$\begin{aligned}
\ddot{\mathbf{x}}_1 &= \frac{Gm_2}{r_{21}^3}(\mathbf{x}_2 - \mathbf{x}_1) + \frac{Gm_3}{r_{31}^3}(\mathbf{x}_3 - \mathbf{x}_1),\\
\ddot{\mathbf{x}}_2 &= \frac{Gm_1}{r_{12}^3}(\mathbf{x}_1 - \mathbf{x}_2) + \frac{Gm_3}{r_{32}^3}(\mathbf{x}_3 - \mathbf{x}_2),\\
\ddot{\mathbf{x}}_3 &= \frac{Gm_1}{r_{13}^3}(\mathbf{x}_1 - \mathbf{x}_3) + \frac{Gm_2}{r_{23}^3}(\mathbf{x}_2 - \mathbf{x}_3).
\end{aligned}$$

To convert these into equations among forces rather than accelerations we would multiply the kth equation by the mass m_k.

Equations 4.4 have their analogues for n bodies, with $6n$-dimensional *state space* vectors $(\mathbf{x}_1, \ldots, \mathbf{x}_n, \dot{\mathbf{x}}_1, \ldots, \dot{\mathbf{x}}_n)$. The more general system consists of the n equations

4.5
$$\ddot{\mathbf{x}}_k = \sum_{j=1,\ j\neq k}^{n} \frac{Gm_j}{|\mathbf{x}_j - \mathbf{x}_k|^3}(\mathbf{x}_j - \mathbf{x}_k), \qquad k = 1, 2, \ldots, n.$$

To solve such systems we need to know $2n$ initial vectors $\mathbf{x}_k(0)$ and $\dot{\mathbf{x}}_k(0)$. Since each of these vectors is 3-dimensional, the *state space* $\mathcal{S}$ has dimension $6n$; in particular, the three body problem governed by Equations 4.4 is 18-dimensional.

Notice that there is a term excluded from the kth sum in Equations 4.4, namely $Gm_k(\mathbf{x}_k - \mathbf{x}_k)/|\mathbf{x}_k - \mathbf{x}_k|^3$. This omitted term would not be well defined because the denominator is zero. As an alternative we could have a formal agreement that the ill-defined term should stand for zero, the value that makes the equation correct. (This is a useful definition to make when writing the equations for insertion into a numerical solution program.)

Except for rather special cases, there is no known set of characteristic properties such as the Kepler laws for describing the solutions of Equation 4.5, even for the three-body case governed by Equations 4.4. The numerical methods described in Section 5 can often be used to provide adequate graphical and numerical descriptions of solutions to the initial-value problems. However, an arbitrary choice of velocity coordinates can lead to nonzero velocity for the center of mass of the entire system, causing it eventually to go off to infinity. To avoid this, we can determine the velocity vector of the most massive body from the other $n-1$ by using the *conservation of momentum equation* (see Exercise 6)

4.6
$$\sum_{k=1}^{n} m_k \ddot{\mathbf{x}}_k(t) = 0.$$

EXERCISES

1. If position as a function of time is given by the vector $\mathbf{x}(t) = (x(t), y(t), z(t))$, then the magnitude of the acceleration vector is $a = \sqrt{\ddot{x}^2 + \ddot{y}^2 + \ddot{z}^2}$. Use Equations 4.3 to show that $a = k/r^2$, where $r = \sqrt{x^2 + y^2 + z^2}$, for a single planet orbiting a star.

2. The radius of the earth's atmospheric shell is about 6500×10^6 meters, and the earth's mass is about $5976 \cdot 10^{21}$ kilograms. With $G = 6.673 \times 10^{-11}$, estimate the escape speed required near the surface of the shell for a projectile of mass 100 kilograms. How is your answer affected if the projectile mass becomes 1000 kilograms? How about 10^{22} kilograms?

3. Letting $z = 0$ in Equations 4.3, we get a system of two second-order equations for x and y. Find the equivalent first-order system obtained by letting $\dot{x} = z$, $\dot{y} = w$.

4. Equations 4.4 for three bodies are equivalent to nine second-order equations for the coordinates (x_1, y_1, z_1), (x_2, y_2, z_2), and (x_3, y_3, z_3) of the three bodies. Write out these equations in terms of x_k, y_k, z_k, and $r_{ij} = \sqrt{(x_i - x_j)^2 + (y_i - y_j)^2 + (z_i - z_j)^2}$.

5. Write the four second-order vector equations for four bodies of mass m_k at $\mathbf{x}_k(t)$, $k = 1, 2, 3, 4$.

6. The **conservation of momentum law** states that, taking into account all forces acting on a physical system of n masses, the sum

$$\sum_{k=1}^{n} m_k \dot{\mathbf{x}}_k(t)$$

of the **momentum vectors** $m_k \dot{\mathbf{x}}_k(t)$ of the system is constant. Verify that conservation of momentum holds for a system governed by

(a) Equations 4.1. [*Hint:* Show that the time derivative of the sum is zero as in Equation 4.6.]

(b) More generally, for Equations 4.4.

(c) Most generally, for Equations 4.5.

Remark. Conservation of momentum may seem to fail in Equation 4.2, since $\ddot{\mathbf{x}} \neq 0$, but here $\mathbf{x}$ describes only the relative positions of two bodies, not their actual motion.

7. Derive the analogue of Equation 4.2 assuming an inverse pth power gravitation law.

8. Let g be the acceleration of gravity at the surface of a homogeneous spherical body of mass M and radius R. Use the inverse-square law to show that $g = GM/R^2$, where G is the gravitational constant in appropriate units of measurement.

9. (a) Use the equation established in the previous exercise to estimate the gravitational constant using a measured value of 9.8 meters per second per second for the acceleration of gravity near the surface of the earth. (Use the values for the mass and radius of the earth given in the text.)

(b) Estimate the acceleration of gravity near the surface of the earth using the value $6.67 \cdot 10^{-11}$ for the gravitational constant G.

10. Falling body problems in the absence of air resistance are often modeled by the equation $\ddot{x} = -g$, with $v_0 = 0$ and the initial height of the body x_0.

(a) Show that $\dot{x} = \sqrt{2g(x_0 - x)}$ under these assumptions, with distance measure down from the initial position x_0.

(b) Assuming instead acceleration $\ddot{x} = -k/x^2$, show that $\dot{x} = \sqrt{2k(x_0 - x)/(x_0 x)}$.

11. Suppose at some time that two bodies subject only to their mutual gravitational attraction are at distance r_0 apart and are receding from each other along a fixed line at a certain fraction q of escape velocity, where $0 < q < 1$. Show that their separation velocity reaches zero, and the bodies start to "fall" toward each other, when their distance apart becomes $r_0/(1 - q^2)$.

12. This exercise is a reminder that there would be no such thing as escape velocity for a body of constant mass if the acceleration of gravity were really constant. For linear motion away from the attracting body, we would have $\ddot{x} = -g$, for some positive constant g. Show that no matter how large $x_0 = x(0) > 0$ and $v_0 = \dot{x}(0) > 0$ are, $x(t)$ has a finite maximum.

13. (a) Use Kepler's second law, equal areas swept out in equal times, to show that a circular orbit must have constant speed.

(b) Use Kepler's third law, $P^2 = 4\pi^2 a^3/[G(m_1 + m_2)]$, together with the result of part (a), to show that a circular orbit of radius a has constant orbital speed $v = \sqrt{G(m_1 + m_2)/a}$. [*Hint:* Express v in terms of the period P.]

14. The **uniform orbital speed** of a satellite of mass m_1 at distance x_0 from an attracting body of mass m_2 is the speed v_1 that the satellite must attain to keep it in a uniform circular orbit.

(a) Show that the orbit $\mathbf{x} = x_0\big(\cos(v/x_0)t,\ \sin(v/x_0)t\big)$ represents circular motion of radius x_0 with uniform speed v and acceleration $\ddot{\mathbf{x}}$ toward the origin of magnitude v^2/x_0. This acceleration vector is called **centripetal acceleration.** [*Hint:* Compute $|\mathbf{x}|$, $|\dot{\mathbf{x}}|$, and $|\ddot{\mathbf{x}}|$.]

(b) Show that if gravitational acceleration $G(m_1 + m_2)/x_0^2$ is to provide precisely the centripetal acceleration of the circular orbit found in part (a), then the uniform orbital speed will be $v_1 = \sqrt{G(m_1 + m_2)/x_0}$.

(c) How is uniform orbital speed related to escape speed?

(d) How many days would there be in a month if the earth's moon had a uniform circular orbit of radius equal to 384,404 kilometers, the mean distance of the moon from the earth? (The actual number is about 27.32 days.)

***15.** The **synchronous orbit** of a body of mass m about a uniformly rotating body of mass $M > m$ is the one that maintains the orbiting body directly over one point on the rotating one. Assume the mass of each body to be concentrated at its center. (See Exercise 33.)

(a) Show that a synchronous orbit must lie in the plane of the rotating body's equator. Then use the first two Kepler laws and the polar area formula $\int \frac{1}{2}r^2\,d\theta$ to show that a synchronous orbit must be circular.

(b) Use the third Kepler law to show that if P is the period of rotation of the larger body, then the radius of the synchronous orbit is $R = K\,P^{2/3}$, where $K = \sqrt[3]{G(M+m)/4\pi^2}$.

(c) Show that the synchronous orbit about the earth for a small satellite has radius approximately 6.622846 times the radius of the earth, or 26,360 miles. (Continuing orbital correction of communication satellites is required because of uneven mass concentrations on earth and the influence of other bodies such as the sun and the moon.)

16. The Newton equations for orbits of a single planet of mass m_2 relative to a fixed sun of mass m_1 have the form

$$\ddot{x} = \frac{-kx}{(x^2 + y^2)^{3/2}}, \qquad \ddot{y} = \frac{-ky}{(x^2 + y^2)^{3/2}},$$

where $k = G(m_1 + m_2)$.

(a) Find the relationship that must hold between the positive constants a and ω so that these differential equations will have solutions with circular orbits described by $x(t) = a\cos\omega t$, $y(t) = a\sin\omega t$.

(b) Show that the relationship described in part (a) expresses the third Kepler law.

(c) Show that the orbit

$$\mathbf{x} = (a\cos\omega t,\ a\sin\omega t), \quad \omega = \text{const.} > 0$$

obeys the second Kepler law.

17. A vector system $\ddot{\mathbf{x}} = -\mathbf{F}(\mathbf{x})$ is called *conservative* if there is a real-valued *potential energy function* $U(\mathbf{x})$ such that $\mathbf{F}(\mathbf{x}) = \nabla U(\mathbf{x})$. [For a 1-dimensional vector field the relation is just $F(x) = U'(x)$; note that a potential function is determined only up to an additive constant.]

(a) Verify that the Newtonian vector field

$$-\mathbf{F}(x, y) = \left(\frac{-kx}{(x^2 + y^2)^{3/2}}, \frac{-ky}{(x^2 + y^2)^{3/2}} \right)$$

has $U(x, y) = -k(x^2 + y^2)^{-1/2}$ as potential; that is, $\nabla U = \mathbf{F}$.

(b) The *kinetic energy* of a body of mass 1 following a path $(x, y) = \big(x(t), y(t)\big)$ is $T = \frac{1}{2}(\dot{x}^2 + \dot{y}^2)$, and the *total energy* of motion in the Newtonian field is

$$E = T + U = \frac{1}{2}(\dot{x}^2 + \dot{y}^2) - \frac{k}{\sqrt{x^2 + y^2}}.$$

Verify that the total energy E is constant for the motion in the vector field of part (a). [*Hint:* Show that $dE/dt = 0$, and use the Newton equations of motion.]

(c) Verify that for motion governed by the equation $\ddot{\mathbf{x}} = -\mathbf{F}(\mathbf{x})$ the total energy E is constant if the vector field is conservative: $F(\mathbf{x}) = \nabla U(\mathbf{x})$.

The next five exercises establish the validity of Kepler's laws.

18. We've seen that the orbit of one body relative to a second always lies in a fixed plane containing both bodies. This is often shown as follows.

(a) Show that if a body of mass m has a path of motion that obeys the inverse-square law $m\ddot{\mathbf{x}} = -\big(k/|\mathbf{x}|^3\big)\mathbf{x}$, then the motion is confined to a plane through the center of attraction determined by the initial position and the velocity vectors. [*Hint:* Establish the relation $d/dt(\mathbf{x} \times \dot{\mathbf{x}}) = \mathbf{x} \times \ddot{\mathbf{x}}$ to show that the plane containing $\mathbf{x}$ and $\dot{\mathbf{x}}$ is perpendicular to a fixed vector.]

(b) A **central force law** is one such that motion is governed by an equation of the form $\ddot{\mathbf{x}} = G\mathbf{x}$, where G is a real-valued function of some unspecified variables, for example $G = G(\mathbf{x})$. Show that motion subject to a central force law is confined to a plane.

19. The **angular momentum** of a planet at position $\mathbf{x}$ in its plane orbit about the sun is the vector $\mathbf{l} = \mathbf{x} \times m\dot{\mathbf{x}}$, that is, $\mathbf{l}$ is the cross-product of the position vector $\mathbf{x}$ with the linear momentum vector $m\dot{\mathbf{x}}$.

(a) Introduce rectangular coordinates x, y in the plane of motion so that $\mathbf{x} = (x, y, 0)$ to show that the length $l = |\mathbf{l}|$ of angular momentum can be expressed as $l = m|x\dot{y} - y\dot{x}|$.

(b) Show that in terms of polar coordinates $x = r\cos\theta$, $y = r\sin\theta$, the angular momentum is $mr^2\dot{\theta}$, if $\dot{\theta} > 0$.

(c) Kepler's second law of planetary motion decrees that the radius joining a planet to the sun sweeps out equal areas in equal times. Use the Kepler law together with the formula

$$A = \frac{1}{2}\int_{\theta_1}^{\theta_2} r^2\, d\theta$$

for area in polar coordinates to show that the angular momentum $mr^2\dot{\theta}$ is constant on an orbit. [*Hint:* Express area swept out along an orbit as an integral with respect to time t between t and $t+\tau$.]

20. Kepler's second law (radius vector from sun to planet sweeps out equal areas in equal times) holds for any *central force law,* that is, a force law expressible in the form $\ddot{\mathbf{x}} = G\mathbf{x}$, where G is some real-valued function. This includes as special cases the inverse-square law of attraction, where $G(\mathbf{x}) = -k|\mathbf{x}|^{-2}$, $k > 0$, and the **Coulomb repulsion law,** where $G(\mathbf{x}) = k|\mathbf{x}|^{-2}$, $k > 0$; the latter governs interaction of particles bearing electric charges of the same sign.

(a) Assuming planar motion and using rectangular coordinates (x, y) for $\mathbf{x}$, show that a central force law can be written

$$\ddot{x} = Gx, \qquad \ddot{y} = Gy,$$

and conclude that $x\ddot{y} - y\ddot{x} = 0$ for a motion governed by a central force law.

(b) Use integration by parts with respect to t in the conclusion of part (a) to show that $x\dot{y} - y\dot{x} = h$ for some constant h.

(c) Change to polar coordinates by $x = r\cos\theta$ and $y = r\sin\theta$ to show that $x\dot{y} - y\dot{x} = r^2\dot{\theta}$, and hence, using part (b), show that $r^2\dot{\theta} = h$ for some constant h. [This result can be interpreted as saying that for motion in a central force field, the angular velocity $\dot{\theta}$ is inversely proportional to the square of the distance from the center of the field.]

(d) Use the result of part (c) and a computation of area in polar coordinates to prove Kepler's second law for a central force field by showing that, as a function of time t, area swept out has the form $A = \frac{1}{2}ht + c$. Explain why this proves Kepler's second law under the given assumptions.

***(e)** Apply Green's theorem to the equation $x\dot{y} - y\dot{x} = h$ derived in part (b) to show directly, without using polar coordinates, that Kepler's second law holds.

21. If $a \geq b$, an ellipse $x^2/a^2 + y^2/b^2 = 1$ has its **focus points** at $(\pm c, 0)$, where $c^2 = a^2 - b^2$. Parametrize the ellipse by $(x, y) = (a\cos u, b\sin u)$, $0 \leq u \leq 2\pi$ to show that the semimajor axis a is also the average, or mean, distance from points on the ellipse to a focus.

***22.** A single planet with position $\mathbf{x} = \mathbf{x}(t)$ obeying $\ddot{\mathbf{x}} = -(k/|\mathbf{x}|^3)\mathbf{x}$ follows an elliptic, parabolic, or hyperbolic path. Here is an outline of a way to show this by deriving a linear differential equation from the vector equation.

(a) Use $\mathbf{x} = (r\cos\theta, r\sin\theta)$ to express the vector equation of motion in the two polar coordinate equations $\ddot{r} - r\dot{\theta}^2 = -k/r^2, r\ddot{\theta} + 2\dot{r}\dot{\theta} = 0$.

(b) Show that the second equation derived in part (a) implies $r^2\dot{\theta} = h$ for some constant h, and use this to write the other equation in the form $\ddot{r} = h^2r^{-3} - kr^{-2}$. In particular, show that if $h = 0$ the motion is confined to a line and results either in collision or escape.

(c) Use the results of part (b) to show that if $h \neq 0$,

$$\frac{1}{r^2}\frac{d^2r}{d\theta^2} - 2\frac{1}{r^3}\left(\frac{dr}{d\theta}\right)^2 = \frac{1}{r} - \frac{k}{h^2}.$$

[*Hint:* Use the chain rule to express $\dot{r}$ and $\ddot{r}$ in terms of derivatives with respect to θ.]

(d) Let $r = 1/u$, so that $dr/d\theta = -u^{-2}du/d\theta$ and $d^2r/d\theta^2 = -u^{-2}d^2u/d\theta^2 + 2u^{-3}(du/d\theta)^2$, to show that the equation in part (c) can be written as the second-order linear equation $d^2u/d\theta^2 + u = k/h^2$, sometimes called **Binet's equation.**

(e) Show that the solution $u = 1/r = A\cos(\theta + \alpha) + k/h^2$ to the previous equation represents an ellipse, parabola, or hyperbola in polar coordinates according as $|A| < k/h^2$, $|A| = k/h^2$, or $|A| > k/h^2$. [*Hint:* Change to x, y coordinates.]

(f) Each **focus** of an ellipse lies on the major axis at distance c from the center where $c^2 = a^2 - b^2$ and a and b are the semiaxes. The **eccentricity** is $e = c/a$. Show that for an elliptic orbit, the center of attraction is at one focus and the eccentricity is $|A|h^2/k$. Then show that the polar equation for an orbit can be written

$$kr(1 + e\cos\theta) = h^2.$$

[*Hint:* For the first part, convert the polar equation, with $\alpha = 0$, to rectangular coordinates.]

(g) Assume that the orbit in part (f) is elliptic, with $0 \leq e < 1$. Show that the time for one complete revolution is $P = 2\pi ab/h$. Then show that $h^2/k = b^2/a$ to derive the third Kepler law $P^2 = 4\pi^2a^3/k$. [*Hint:* The sum of the maximum and minimum values for r is equal to $2a$.]

23. Suppose a projectile is fired directly away from and at distance x_0 from the center of mass of a planet with initial speed z_0. If z_0 is less than the escape speed, show that the maximum additional distance attained from the center of mass of the planet is

$$\frac{x_0^2z_0^2}{2GM - x_0z_0^2},$$

where M is the sum of the masses of the two bodies.

24. A simplified model of a **black hole** is a body such that the escape speed from its surface exceeds c, the speed of light.

(a) How small should the radius r of a solid homogeneous ball of mass m be if it is to become a black hole?

(b) Estimate the radius r of part (a) for the earth and for the sun. Assume that the earth and sun are homogeneous solid balls, that the earth has mass $6 \cdot 10^{24}$ kilograms and that the sun is 334,000 times as massive as the earth; in units of meters, kilograms, and seconds, $G \approx 6.67 \cdot 10^{-11}$. Also $c \approx 300{,}000$ kilometers per second.

25. Equations 4.1 with $\mathbf{x}_1 = (x_1, y_1, z_1)$ and $\mathbf{x}_2 = (x_2, y_2, z_2)$ are equivalent to six second-order scalar equations. Write out these equations in terms of coordinates x_k, y_k, and z_k and $r = \sqrt{(x_1 - x_2)^2 + (y_1 - y_2)^2 + (z_1 - z_2)^2}$.

26. The distance $x(t)$ between two bodies moving on a fixed line obeys a repelling Coulomb law $\ddot{x} = k/x^2$, $k > 0$. Show that $\dot{x}(t)$ always increases if $\dot{x}(0) > 0$.

27. Suppose two stars of identical mass m at $\mathbf{x}_1$ and $\mathbf{x}_2$ form a binary system with a fixed center of mass at $\frac{1}{2}(\mathbf{x}_1 + \mathbf{x}_2)$. Show that they can move in the same circular orbit of radius r and that their common orbital speed will then be $v = \frac{1}{2}\sqrt{Gm/r}$. [*Hint:* Use Equations 4.1.]

28. Explain the derivation of Equations 4.4 for three bodies. Be sure to include an explanation of why the differences of vectors $\mathbf{x}_i$ and $\mathbf{x}_j$ appear in the order they do.

29. Explain the derivation of Equations 4.5 for n bodies. Be sure to include an explanation of why the differences of vectors $\mathbf{x}_i$ and $\mathbf{x}_j$ appear in the order they do.

***30.** The principal result in Example 1 of the text was established under the assumption that the relative distance separating two bodies subject only to the forces of mutual gravitational attraction could always be measured along the same fixed line. The purpose of this problem is to show that this last assumption isn't needed, and that in fact the relative speed v and distance r are always related by

$$\frac{v^2}{2} - \frac{v_0^2}{2} = \frac{k}{r} - \frac{k}{r_0}.$$

Here the constant is $k = Gm$, where m is the sum of the two masses, G is the gravitational constant, and v_0 and r_0 are initial speed and distance.

Starting with the Newton vector equation $\ddot{\mathbf{x}} = -k\mathbf{x}/|\mathbf{x}|^3$, we form the dot product of both sides by $\dot{\mathbf{x}}$ to get the scalar equation

$$\dot{\mathbf{x}} \cdot \ddot{\mathbf{x}} = -k\frac{\mathbf{x} \cdot \dot{\mathbf{x}}}{|\mathbf{x}|^3}.$$

(a) Show that the left side of the previous equation is equal to

$$\frac{d}{dt}\frac{v^2}{2},$$

where $v = \sqrt{\dot{\mathbf{x}} \cdot \dot{\mathbf{x}}} = |\dot{\mathbf{x}}|$.

(b) Show that the right-hand side of that same equation is equal to

$$k\left(\nabla \frac{1}{|\mathbf{x}|}\right) \cdot \dot{\mathbf{x}},$$

where ∇ is the gradient operator: $\nabla f = (\partial f/\partial x, \partial f/\partial y, \partial f/\partial z)$.

(c) Show that the result of part (b) can be written $k(d/dt)|\mathbf{x}|^{-1}$, and conclude that

$$\frac{d}{dt}\frac{v^2}{2} = k\frac{d}{dt}\frac{1}{r}.$$

(d) Integrate the previous equation between 0 and an arbitrary positive time t to get the equation relating v, v_0, r, and r_0.

31. Write out the notational details of the derivation of the vector rocket equation, following the scalar equation derivation given at the end of Section 4 of Chapter 2. This is mainly a matter of replacing scalar notation by vector notation, but you should verify that these replacements are meaningful and correct.

32. [The results of this exercise are used in numerical solution of the planetary motion equations.] The **center of mass c** of a system of n bodies with masses m_k concentrated respectively at $\mathbf{x}_k$ is

$$\mathbf{c} = \frac{1}{m}\sum_{k=1}^{n} m_k \mathbf{x}_k, \quad \text{where} \quad m = \sum_{k=1}^{n} m_k.$$

(a) Show that if the *conservation of momentum* holds, that is $\dot{\mathbf{c}}$ is constant, then the center of mass stays on a line. Show also that by subtracting $\mathbf{c}$ from all vectors $\mathbf{x}_k$, the center of mass can be placed at the origin.

(b) Show that if the conservation of momentum Equation 4.6 holds, then $\dot{\mathbf{c}}$ is constant. Show that if also the center of mass is to stay at the origin, then the initial velocity vectors $\dot{\mathbf{x}}_k(0)$ can't be chosen arbitrarily but must satisfy

$$\sum_{k=1}^{n} m_k \dot{\mathbf{x}}_k(0) = 0.$$

***33. When can you assume that the mass of a ball is concentrated at its center?** A solid ball B_R of radius R is **spherically homogeneous** if its density is constant on every spherical shell with center at the center of B_R. It's important for the application of Newton's inverse-square law to planetary motions that each of the most significant celestial bodies is fairly near to being a spherically homogeneous ball. (Among bodies that diverge widely from this norm, erratic motion has been observed that is quite uncharacteristic of the orbital motion of a typical planet.) The purpose of this exercise is to establish Newton's result that the gravitational attraction of a spherically homogeneous ball acting at a point $\mathbf{a}$ is the same as it would be if all the mass of the ball were concentrated at its center, unless $\mathbf{a}$ is inside the ball, in which case the part of the ball at distance from the center greater than $|\mathbf{a}|$ can be neglected. More specifically, if B_R is centered at the origin and has density $\rho(|\mathbf{x}|)$ at $\mathbf{x}$, **Newton's formula** for the attracting force vector acting on a particle of mass 1 at $\mathbf{a}$ is given by G times the 3-dimensional vector integral

$$\int_{B_R} \rho(|\mathbf{x}|)\frac{\mathbf{x}-\mathbf{a}}{|\mathbf{x}-\mathbf{a}|^3}\,dV_{\mathbf{x}} = -\frac{M_{\mathbf{a}}}{|\mathbf{a}|^3}\mathbf{a},$$

where $M_{\mathbf{a}}$ is the mass of the part of B_R that lies within distance $|\mathbf{a}|$ of its center.

(a) Choose perpendicular x, y, z axes with origin at the center of B_R and positive z axis passing through $\mathbf{a} = (0,0,a)$. Show, without computing any indefinite integrals, that the x and y coordinates of the vector integral are zero and that the z coordinate is given in spherical coordinates by

$$2\pi \int_0^R r^2\rho(r)\left[\int_0^{\pi} \frac{r\cos\phi - a}{(r^2 - 2ar\cos\phi + a^2)^{3/2}}\sin\phi\,d\phi\right]dr.$$

(b) Let $\cos\phi = u$ and integrate by parts to show that the inner integral in part (a) is

$$\int_{-1}^{1} (ru - a)(r^2 + a^2 - 2aru)^{-3/2}\,du = \begin{cases} -2/a^2, & a > r, \\ 0, & a < r. \end{cases}$$

(c) Show that the mass of B_S is

$$4\pi \int_0^S r^2\rho(r)\,dr,$$

and then use the previous results to prove Newton's formula for the attracting force.

(d) Use the results of parts (a) and (b) to show that matter distributed in a spherically homogeneous way between two concentric spheres exerts no gravitational attraction at points inside the inner sphere.

(e) Specialize the results of parts (a) and (b) to the case of a homogeneous ball with *constant* density ρ to show that the gravitational attraction of the ball on a unit point-mass inside the ball and a units from the center has magnitude proportional to a with constant of proportionality $\frac{4}{3}\pi\rho G$.

5 COMPUTING SUPPLEMENT: SYSTEMS OF EQUATIONS

The numerical methods illustrated here apply to a first-order vector equation $\dot{\mathbf{x}} = \mathbf{F}(t, \mathbf{x})$ with initial condition $\mathbf{x}(t_0) = \mathbf{x}_0$, and include a wide variety of systems, both linear and nonlinear. As in Chapter 2, Section 4, and Chapter 4, Section 4, the methods approximate continuous-time systems by discrete-time systems.

5A Euler's Method

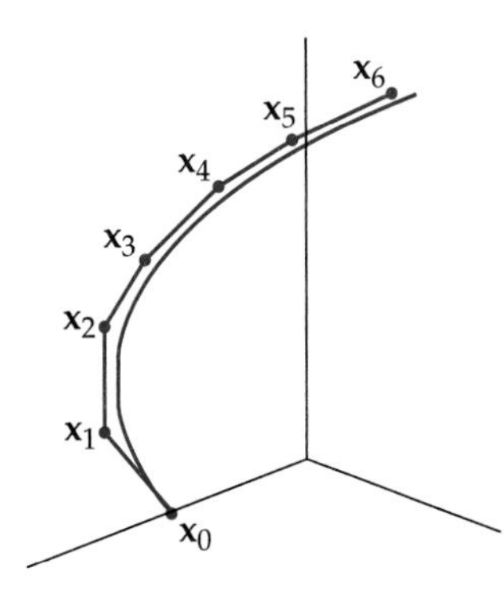

We choose a step of size h and find successive approximations $\mathbf{x}_k$ to the true values $\mathbf{x}(t_0 + kh)$ of the solution $\mathbf{x}(t)$. The idea is to multiply the difference-quotient approximation to the vector derivative by h so that

$$\frac{\mathbf{x}(t+h) - \mathbf{x}(t)}{h} \approx \mathbf{F}(t,\mathbf{x}) \quad \text{becomes} \quad \mathbf{x}(t+h) \approx \mathbf{x}(t) + h\mathbf{F}(t,\mathbf{x}).$$

Start with $\mathbf{x}(t_0) = \mathbf{x}_0$, and having found $\mathbf{x}_k$ corresponding to $t_k = t_0 + kh$, we define the approximation $\mathbf{x}_{k+1}$ at $t_{k+1} = t_0 + (k+1)h$ by

$$\mathbf{x}_{k+1} = \mathbf{x}_k + h\mathbf{F}(t_k, \mathbf{x}_k).$$

The sketch shows a 3-dimensional example.

example 1

For a two-dimensional system with $\mathbf{x} = (x, y)$,

$$\dot{x} = F(t,x,y), \qquad x(t_0) = x_0,$$
$$\dot{y} = G(t,x,y), \qquad y(t_0) = y_0.$$

The initial vector is $\mathbf{x}_0 = (x_0, y_0)$. Then

$$x_1 = x_0 + hF(t_0, x_0, y_0),$$
$$y_1 = y_0 + hG(t_0, x_0, y_0).$$

Next, with $t_1 = t_0 + h$,

$$x_2 = x_1 + hF(t_1, x_1, y_1),$$
$$y_2 = y_1 + hG(t_1, x_1, y_1).$$

In general, with $t_k = t_{k-1} + h = t_0 + kh$, we get

$$x_{k+1} = x_k + hF(t_k, x_k, y_k),$$
$$y_{k+1} = y_k + hG(t_k, x_k, y_k).$$

The basic loop for sequential software would save the initial x with SET S = X. Then

```
SET X = X + H * F(T, S, Y)
SET Y = Y + H * G(T, S, Y)
SET T = T + H
PRINT T, X, Y
```

We really need the saved value S only in the second step to avoid using the newly computed X from the first step, but we really need S in the second step to ensure that F and G are evaluated at the same points.

example 2 The system

$$\dot{x} = y, \qquad x(0) = 0, \qquad \dot{y} = -tx, \qquad y(0) = 1$$

is equivalent to the second-order equation $\ddot{x} = tx$. Together with step size $h = 0.01$, initial values $x(0) = 0$, $y(0) = 1$, and definitions

```
DEFINE F(T,X,Y) = Y
DEFINE G(T,X,Y) = -T * X
```

the loop above will produce a table of approximate values of t, $x(t)$, and the derivative $y(t) = \dot{x}(t)$ at t intervals of 0.01. Replacing the print command by PLOT T,X produces a graph of $x(t)$ shown in Figure 6.10(a). Similarly, PLOT T,Y produces a graph of $y(t) = \dot{x}(t)$ shown in Figure 6.10(b).

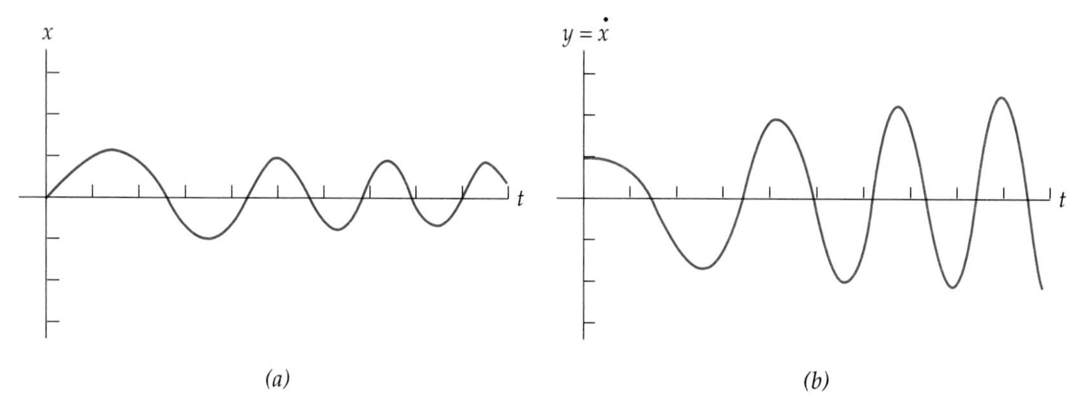

Figure 6.10 $\dot{x} = y$, $\dot{y} = -tx$, $x(0) = 0$, $y(0) = 1$.

5B Improved Euler Method

The next simplest numerical method for a vector equation is an improvement of the Euler method in which, instead of using the tangent vector $\mathbf{F}(t_k, \mathbf{x}_k)$ to find the next value, we use the vector average

$$\tfrac{1}{2}[\mathbf{F}(t_k, \mathbf{x}_k) + \mathbf{F}(t_k + h, \mathbf{p}_{k+1})],$$

where $\mathbf{p}_{k+1}$ is the value that the Euler method would have predicted, that is,

$$\mathbf{p}_{k+1} = \mathbf{x}_k + h\mathbf{F}(t_k, \mathbf{x}_k).$$

Using the predicted value $\mathbf{p}_{k+1}$, the **improved Euler approximation** to $\mathbf{x}(t_k, x_k)$ is given by

$$\mathbf{x}_{k+1} = \mathbf{x}_k + \frac{h}{2}[\mathbf{F}(t_k, \mathbf{x}_k) + \mathbf{F}(t_k + h, \mathbf{p}_{k+1})].$$

example 3

For a two-dimensional system,

$$\dot{x} = F(t,x,y), \qquad x(t_0) = x_0,$$
$$\dot{y} = G(t,x,y), \qquad y(t_0) = y_0,$$

we start with (x_0, y_0). Then letting $\mathbf{p} = (p,q)$, we compute the Euler approximation

$$p_1 = x_0 + hF(t_0, x_0, y_0),$$
$$q_1 = y_0 + hG(t_0, x_0, y_0),$$

followed by the corrected approximation

$$x_1 = x_0 + \frac{h}{2}[F(t_0, x_0, y_0) + F(t_0 + h, p_1, q_1)],$$
$$y_1 = y_0 + \frac{h}{2}[G(t_0, x_0, y_0) + G(t_0 + h, p_1, q_1)].$$

At the $(k+1)$th step, we compute $t_k = t_{k-1} + h = t_0 + kh$, and

$$p_{k+1} = x_k + hF(t_k, x_k, y_k),$$
$$q_{k+1} = y_k + hG(t_k, x_k, y_k),$$
$$x_{k+1} = x_k + \frac{h}{2}[F(t_k, x_k, y_k) + F(t_{k+1}, p_{k+1}, q_{k+1})],$$
$$y_{k+1} = y_k + \frac{h}{2}[G(t_k, x_k, y_k) + G(t_{k+1}, p_{k+1}, q_{k+1})].$$

One should be careful to avoid using the value x_{k+1} in the fourth equation instead of x_k; in sequential implementation x_k should be stored for use later.

An algorithmic loop might look like this:

```
SET P = X+H * F(T,X,Y)
SET Q = Y+H * G(T,X,Y)
SET S = X
SET X = X+(H/2) * (F(T,X,Y)+F(T+H,P,Q))
SET Y = Y+(H/2) * (G(T,S,Y)+G(T+H,P,Q))
```

For three equations $\dot{x} = F_1(t,x,y,z)$, $\dot{y} = F_2(t,x,y,z)$, $\dot{z} = F_3(t,x,y,z)$ recursive formulas for an algorithm can be written in similar style, but for these, and still higher dimensional systems it's more efficient to introduce vector arrays in which a vector variable X is assigned an appropriate dimension n along with coordinates X(1), . . . , X(n). Coordinate functions F_k can be similarly assigned to a vector function **F**, with the result that the entire computational loop may be given the form SET **X** = **X** + H $*$ **F**(T,**X**)

example 4

An example of equations of Lotka-Volterra type for three species is

$$\dot{x} = x(3 - y), \qquad \dot{y} = y(x + z - 3), \qquad \dot{z} = z(2 - y).$$

One option is to plot the relationship between two species at a time, winding up with three related pictures. Alternatively, we can plot the trajectory of a solution relative to three-dimensional axes, as shown in Figure 6.11. If either $x = 0$ or $z = 0$, this system reduces to dimension two, for example $\dot{x} = x(3 - y)$, $\dot{y} = y(x-3)$, if $z = 0$. The associated trajectories are then the closed loops as discussed in Section 2.

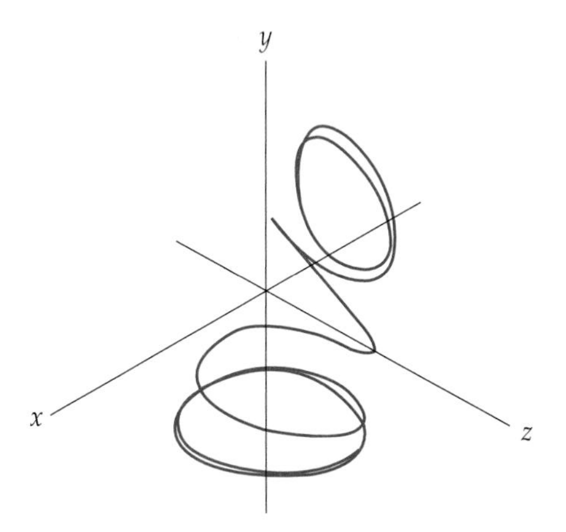

Figure 6.11 Three-species competition.

example 5

Recall that Newton's equations of planetary motion for a star with one planet, with the sum of their masses equal to M, are

$$\ddot{x} = -\frac{GMx}{(x^2 + y^2)^{3/2}}, \qquad x(0) = x_0, \qquad \dot{x}(0) = z_0,$$
$$\ddot{y} = -\frac{GMy}{(x^2 + y^2)^{3/2}}, \qquad y(0) = y_0, \qquad \dot{y}(0) = w_0.$$

These equations are derived in Section 4, where we remarked that the second-order system is equivalent to the first-order system

$$\dot{x} = z, \qquad x(0) = x_0,$$
$$\dot{y} = w, \qquad y(0) = y_0,$$
$$\dot{z} = -\frac{GMx}{(x^2 + y^2)^{3/2}}, \qquad z(0) = z_0,$$
$$\dot{w} = -\frac{GMy}{(x^2 + y^2)^{3/2}}, \qquad w(0) = w_0.$$

Because of the way $4\pi^2$ enters Kepler's third law, we sometimes choose units of measurement so that $GM = 4\pi^2$, although for some purposes we could just as well choose units so that GM has some other value. In choosing initial values for position and velocity, recall that we get an elliptic orbit only if the orbital

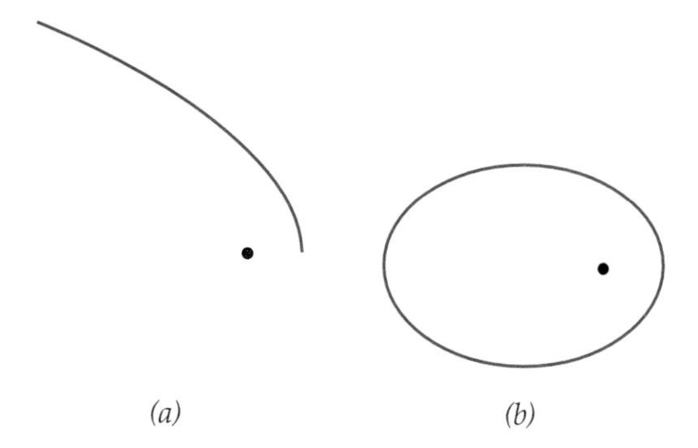

Figure 6.12 (a) Hyperbolic trajectory.
(b) Elliptical trajectory.

speed is less than the escape speed, $V_e = \sqrt{2GM/r}$. In other words, we want

$$z_0^2 + w_0^2 < \frac{2GM}{\sqrt{x_0^2 + y_0^2}}.$$

Thus, if $GM=2$ and $(x_0, y_0) = (1,0)$, we should choose (z_0, w_0) so that $z_0^2 + w_0^2 < 4$ to get a closed trajectory. The orbit in Figure 6.12(a) was made using the improved Euler method, having chosen $GM = 2$, initial position $(x_0, y_0) = (1,0)$, and initial velocity $(z_0, w_0) = (0, 2.01)$. The step size was $h = 0.01$, but the result is plotted only for every 10 steps. The closed orbit in Figure 6.12(b) is the result of changing w_0 to 1.8.

If we were to choose constants $M = 1/G$, $x_0 = 1$, $y_0 = 0$, $z_0 = 0$, and $w_0 = 1$ in this example, then the equations would be satisfied by the solution $(x(t), y(t)) = (\cos t, \sin t)$, which has a circular orbit for its trajectory. We could then use this elementary solution as a check on the accuracy of our method of numerical approximation.

The remarks about error accumulation at the end of Chapter 2, Section 4B, apply to systems also. However, one outstanding source of error in any numerical method for the approximate solution of differential equations is the evaluation of functions. If, for a single first-order equation $y' = F(x, y)$, the Euler method requires a single evaluation of $F(x, y)$ at each step, the improved Euler method applied to the same equation requires two evaluations, $F(x_k, y_k)$ and $F(x_{k+1}, p_{k+1})$, at the kth step. For an n-dimensional system, the improved Euler method typically requires $2n$ evaluations at the kth step:

$$F_j(t_k, \mathbf{x}_k),\ F_j(t_{k+1}, \mathbf{p}_{k+1}), \qquad j = 1, 2, \ldots, n,$$

where $F_j(t, \mathbf{x})$ is the jth coordinate function of the system. Since errors made at each step are then fed back into the functions F_j at the next step, it is clear that the possibility for error accumulation will be significantly enhanced for a large system.

The distinction between error at a single step and accumulated error is such that the two can be treated separately. If $\mathbf{F}(t, \mathbf{x})$ is continuously differentiable in $\mathbf{x}$, then the one-step error using the Euler method has order h^2, where h is the step size. Using the improved Euler method, the one-step error has order h^3 provided $\mathbf{F}(t,\mathbf{x})$ is twice continuously differentiable in $\mathbf{x}$. The accumulated errors for the Euler and improved Euler methods have order h and h^2, respectively; it is only proportionality factors that may increase as the dimension increases. These ideas are introduced here only as a rough guide; deeper understanding requires a thorough study of numerical analysis.

The matter of existence and uniqueness of solutions is particularly pertinent to finding approximate solutions of systems for which analytic solution formulas don't exist. For example, a numerical method may produce numbers that purport to approximate a nonexistent solution, or there may exist multiple solutions for which a numerical method attempts approximate just one. Theorem 1.1 of Chapter 5, Section 1E, shows that $\dot{\mathbf{x}} = \mathbf{F}(t,\mathbf{x})$, $\mathbf{x}(t_0) = \mathbf{x}_0$ has a unique solution if $\mathbf{F}$ and its partial derivatives with respect to the $\mathbf{x}$ variables x_k are continuous for t an interval containing t_0 and for $\mathbf{x}$ in an open set in $\mathcal{R}^n$ containing $\mathbf{x}_0$.

EXERCISES

1. The first-order autonomous system $\dot{x} = y$, $\dot{y} = x$ is equivalent to the single equation $\ddot{y} = y$, via the relation $\dot{y} = x$.

(a) Show that the system has solutions $x(t) = c_1e^t + c_2e^{-t}$, $y(t) = c_1e^t - c_2e^{-t}$.

(b) Find the particular solution satisfying $x(0) = 1$ and $y(0) = 2$.

(c) Compute a table of approximations to the particular solution found in part (b). Do this computation on the interval $0 \le t \le \frac{1}{2}$ in steps of size $h = 0.1$, both by computing values from the explicit exponential solution formula and by a direct numerical solution of the system using either the Euler method or its corrected modification.

2. Find a table of approximations to the solution $x(t)$, $y(t)$ of the system

$$\dot{x} = x + y^2, \qquad \dot{y} = x^2 + y + t,$$

with initial condition $x(0) = 1$, $y(0) = 2$. Use a step of size $h = 0.1$ on the interval $0 \le t \le \frac{1}{2}$ and make the approximation with (a) Euler method, and (b) improved Euler.

3. (a) Show that the second-order equation $\ddot{y} + 2\dot{y} + y = t$, with initial conditions $y(0) = 1$, $\dot{y}(0) = 2$, is equivalent to the first-order system

$$\dot{x} = -2x - y + t, \qquad x(0) = 2,$$

$$\dot{y} = x, \qquad y(0) = 1.$$

(b) Find a numerical approximation to the solution of the system in part (a) for the interval $0 \le t \le 1$.

(c) Solve the given second-order equation by using its characteristic equation, and compare the solution with the numerical results of part (b).

4. (a) Find a first-order system equivalent to $y'' - ty' + y = t$, $y(0) = 1$, and $y'(0) = 2$.

(b) Find a numerical approximation to the solution of the system in part (a) for $0 \leq t \leq 1$.

5. Apply the improved Euler method to the equation $dy/dt = t^2 + y^2$, $y(0) = 1$.

***6.** A basic result of multivariable calculus says that the gradient vector $\nabla f(x,y) = \big(f_x(x,y),\, f_y(x,y)\big)$ is perpendicular to the level curve of f that contains (x,y); consequently, $\big(-f_y(x,y),\, f_x(x,y)\big)$ is tangent to a level set, which is then a trajectory of the system $\dot{x} = -f_y(x,y)$, $\dot{y} = f_x(x,y)$. Suppose below that $f(x,y) = x^2 + \frac{1}{2}y^2$.

(a) Use these ideas to make a computer-graphics plot of the elliptic level curves of f.

(b) Make a computer-graphics plot of some **orthogonal trajectories,** that is, curves perpendicular to the level sets of the function f.

(c) Identify the well-known family of orthogonal trajectory curves by solving the relevant uncoupled system analytically.

7. Make computer plots of the solution to these problems in Chapter 5, Section 1C.

(a) The *bug-pursuit* problem of Exercise 43 with $v = 1$.

(b) The *airplane* problem of Exercise 44 with $v = 3$ and both $\mathbf{w} = (1,2)$ and $\mathbf{w} = (-1,2)$.

8. The fourth-order **Runge-Kutta approximations** to the solution of $\dot{\mathbf{x}} = \mathbf{F}(t,\mathbf{x})$, $\mathbf{x}(t_0) = \mathbf{x}_0$, are computed as follows:

$$\mathbf{x}_{k+1} = \mathbf{x}_k + \frac{h}{6}(\mathbf{m}_1 + 2\mathbf{m}_2 + 2\mathbf{m}_3 + \mathbf{m}_4),$$

$$\text{where} \quad \mathbf{m}_1 = \mathbf{F}(t_k, \mathbf{x}_k), \qquad \mathbf{m}_2 = \mathbf{F}(t_k + \tfrac{1}{2}h,\, \mathbf{x}_k + \tfrac{1}{2}h\mathbf{m}_1)$$

$$\mathbf{m}_3 = \mathbf{F}(t_k + \tfrac{1}{2}h,\, \mathbf{x}_k + \tfrac{1}{2}h\mathbf{m}_2), \qquad \mathbf{m}_4 = \mathbf{F}(t_k + h,\, \mathbf{x}_k + h\mathbf{m}_3).$$

[The one-step formula error is of order h^5.] Apply a computer algorithm to implement the routine. Then compare the results obtained from it for the problem

$$\dot{x} = -y, \qquad x(0) = 1, \qquad \dot{y} = x, \qquad y(0) = 0$$

with the results obtained from the improved Euler method, as well as with the solution formula $x = \cos t$, $y = \sin t$. Try step sizes $h = 0.1, 0.01$, and 0.001.

9. Apply the Runge-Kutta method of the previous exercise to a 2-dimensional system equivalent to $\ddot{x} = -tx$, $x(0) = 0$, $\dot{x}(0) = 1$, using $h = .01$. Compare with improved Euler.

Mixing

10. Tank 1 at capacity 100 gallons and tank 2 at capacity 200 gallons are initially full of salt solution. Tank 1 has 5 gallons per minute of salt solution at 1 pound per gallon running in while mixed solution is drawn off, also at 5 gallons per minute, with an additional 3 gallons per minute flowing out to tank 2. Tank 2 has 2 gallons per minute of pure water running in and 3 gallons per minute being drawn off, while 2 gallons per minute flow out to tank 1.

(a) Find a system of differential equations satisfied by the salt contents of the tanks up to the time when one is empty.

(b) Make a computer plot that compares the graphs of the components of the solution to part (a), assuming tank 1 has initially 10 pounds of salt and tank 2 has 20 pounds. Estimate the maximum amount of salt in each tank and the times when these are attained.

11. Tank 1 at capacity 100 gallons and tank 2 at capacity 200 gallons are initially half-full of salt solution. Tank 1 has 5 gallons per minute of pure water running in while mixed solution is drawn off at 4 gallons per minute, with an additional 3 gallons per minute pumped to tank 2. Tank 2 has 2 gallons per minute of salt solution at 1 pound per gallon running in and 1 gallon per minute being drained off, while 3 gallons per minute are pumped to tank 1.

(a) Find a system of differential equations satisfied by the salt contents of the tanks up to the time when one is full.

(b) Make a computer plot that compares the graphs of the components of the solution to part (a), assuming tank 1 has initially 10 pounds of salt and tank 2 has 20 pounds. Estimate the minimum amount of salt in tank 1 and the time when this is attained.

Gravitational Attraction

12. The system of Newton equations $\ddot{x} = -x(x^2 + y^2)^{-3/2}$, $\ddot{y} = -y(x^2 + y^2)^{-3/2}$ with initial conditions $x(0) = 1$, $y(0) = 0$, $\dot{x}(0) = 0$, $\dot{y}(0) = v_0$, has a solution with a closed trajectory if $v_0 < \sqrt{2}$. Use the improved Euler method to make an approximate computation of the trajectory of a single orbit if

(a) $v_0 = 0.35$. **(b)** $v_0 = 0.7$. **(c)** $v_0 = 1.4$.

13. To implement a routine that applies the improved Euler method to the three body problem, it's helpful to start with the center of mass of the physical system at the origin:

$$m_1\mathbf{x}_1 + m_2\mathbf{x}_2 + m_3\mathbf{x}_3 = 0.$$

Show that this equation is valid for all time since *conservation of momentum* holds for the Newton equations for three bodies, that is,

$$m_1\dot{\mathbf{x}}_1 + m_2\dot{\mathbf{x}}_2 + m_3\dot{\mathbf{x}}_3 = 0.$$

Thus if m_1 is the largest mass, it is appropriate to let the initial values for $\mathbf{x}_1(0)$ and $\dot{\mathbf{x}}_1(0)$ be determined by the corresponding initial values for the other two bodies. Because of the law of conservation of momentum, the previous two equations will be maintained forever.

14. When the inverse-square law $F = m_1 m_2 r^{-2}$ is replaced by $F = m_1 m_2 r^{-p}$, where $0 < p$, the Newton equations for the orbit of a single planet about a fixed sun take the form

$$\ddot{x} = -kx(x^2 + y^2)^{-(p+1)/2}, \qquad \ddot{y} = -ky(x^2 + y^2)^{-(p+1)/2}.$$

Assume $k = 1$ and make a pictorial comparison of the orbits with initial conditions $x(0) = 1$, $y(0) = 0$, $\dot{x}(0) = 0$, and $\dot{y}(0) = 0.5$ for the choices $p = 1.9$, $p = 2$, and $p = 2.1$. Discuss the differences among the three cases.

Oscillatory Systems

15. The equations $\ddot{\theta} = -(g/l)\sin\theta + \dot{\phi}^2 \sin\theta\cos\theta$, $\ddot{\phi} = -2\dot{\theta}\dot{\phi}\cot\theta$, $\theta \neq k\pi$, k integer, govern the **spherical pendulum** described in Exercise 41 of Chapter 5, Section 1C. Let $g/l = 1$.

(a) Plot the trajectory in θ, ϕ space for a solution if $\theta(0) = \dot{\theta}(0) = 1$, $\phi(0) = 0$, and $\dot{\phi}(0) = 1$.

(b) Make a 3-dimensional perspective plot of the x, y, z path of the bob if the initial conditions are as in part (a). Use the coordinate relations given in Exercise 41 referred to above.

16. The **van der Pol equation** is $\ddot{x} - \alpha(1 - x^2)\dot{x} + x = 0$, where α is a positive constant.

(a) Let $y = \dot{x}$ and write down the first-order system in x and y that is equivalent to the van der Pol equation.

(b) Use the improved Euler method to plot numerical solutions to the system found in part (a), using initial values $x(0) = 2$, $y(0) = 0$, while successively letting $\alpha = 0.1, 1.0$, and 2.0. Plot for $0 < t < t_1$, where t_1 is in each plot large enough that the trajectory of $(x(t), y(t))$ appears to be a closed loop.

(c) Repeat the three experiments in part (b) with initial values $x(0) = 1$, $y(0) = 0$, and with $x(0) = 3$, $y(0) = 0$. The closed loops being approximated in part (b) are each an example of a **limit cycle,** described in Chapter 4, Section 3, Example 8.

In Exercises 17 through 19, use a numerical system solver to plot graphical solutions of the nonlinear equations of motion for mass-spring systems in dimension 3. For two masses and three springs the vector equations were derived in Section 4:

$$m_1\ddot{\mathbf{x}} = h_1(|\mathbf{a} - \mathbf{x}| - l_1)\frac{(\mathbf{a} - \mathbf{x})}{|\mathbf{a} - \mathbf{x}|} + h_2(|\mathbf{y} - \mathbf{x}| - l_2)\frac{(\mathbf{y} - \mathbf{x})}{|\mathbf{y} - \mathbf{x}|},$$

$$m_2\ddot{\mathbf{y}} = h_2(|\mathbf{x} - \mathbf{y}| - l_2)\frac{(\mathbf{x} - \mathbf{y})}{|\mathbf{x} - \mathbf{y}|} + h_3(|\mathbf{b} - \mathbf{y}| - l_3)\frac{(\mathbf{b} - \mathbf{y})}{|\mathbf{b} - \mathbf{y}|}.$$

Run your solver with graphic output and with parameter and initial values chosen to show typical nonlinear, and if possible linear behavior.

17. **(a)** Plot typical solutions for masses m_1 and m_2 joined to each other and to two fixed points **a** and **b** by springs with unstressed lengths l_1, l_2, and l_3 and Hooke constants h_1, h_2, and h_3.

(b) What is the effect of making $h_3 = 0$ in part (a)? Run your system solver with that assumption included.

(c) Assume initial conditions that will restrict the motion in part (a) to a line and run your system solver under that assumption.

18. Derive the equations for a single mass m joined to two fixed points **a** and **b** by springs with unstressed lengths l_1 and l_2 and Hooke constants h_1 and h_2. (See Exercise 14(a) in Section 3B.)

19. **(a)** Derive the equations for a single mass m joined to three fixed points **a**, **b**, and **c** by springs with unstressed lengths l_1, l_2, and l_3 and Hooke constants h_1, h_2, and h_3.

(b) Then plot typical graphical solutions for some specific constants and initial values.

20. The general **Lorenz system** is $\dot{x} = \sigma(y - x)$, $\dot{y} = \rho x - y - xz$, $\dot{z} = -\beta z + xy$, where β, ρ, and σ are positive constants. For certain values of the parameters, in particular $\beta = \frac{8}{3}$, $\rho = 28$, $\sigma = 10$, solution trajectories exhibit an often-studied type of "chaotic" oscillation. Plot the orbits with these parameter choices and initial value $(x,y,z) = (2,2,21)$. A particularly good view is obtained by projecting on the plane through the origin perpendicular to the vector $(-2,3,1)$. Observe the effect of small changes in the initial vector on the successive numbers of circuits in each spiral configuration. This is another manifestation of "chaotic" behavior.

Interacting Populations

21. The special Lotka-Volterra system $\dot{H} = (3 - 2P)H$, $\dot{P} = (\frac{1}{2}H - 1)P$ is discussed in general terms in Section 2.

(a) Using the initial conditions $H(0) = 3$, $P(0) = \frac{3}{2}$, compute sufficiently many approximate values of the solutions $(H(t), P(t))$ so the values nearly return to the initial values.

(b) Using the approximate data obtained in part (a), sketch the graphs of $H(t)$ and $P(t)$ using the same vertical axis and the same horizontal t axis.

(c) Sketch an approximate trajectory in the HP plane for the solution found in part (a).

22. The Lotka-Volterra system $\dot{H} = (3 - 2P)H$, $\dot{P} = (\frac{1}{2}H - 1)P$ has solutions $\big(H(t), P(t)\big)$ that are periodic, because a given orbit always returns to its initial point in some finite time t_1; thus $H(t + t_1) = H(t)$ and $P(t + t_1) = P(t)$ for all t. The time period t_1 depends on the particular orbit, however. Make numerical estimates of t_1 for the system on orbits with (a) $H(0) = 3$, $P(0) = \frac{1}{2}$; (b) $H(0) = 3$, $P(0) = \frac{3}{2}$; and (c) $H(0) = 3$, $P(0) = 2$.

23. A modification of the Lotka-Volterra equations is $\dot{H} = (a - bP)H(L - H)$, $\dot{P} = (cH - d)P(M - P)$, with $a, b, c,$ and d positive constants, and it's designed to accommodate fixed limits $L > H(t)$ and $M > P(t)$ to the growth of the host and parasite populations. Assume $L > d/c > 1$ and $M > a/b > 1$. The restrictions L and M are typically imposed by lack of sufficient habitat or food supply. Use $a = 3$, $b = c = 2$, and $d = 1$ and $L = 4$ and $M = 3$ to plot the trajectories for (a) $H(0) = 3$, $P(0) = \frac{1}{2}$; (b) $H(0) = 3$, $P(0) = \frac{3}{2}$; and (c) $H(0) = 3$, $P(0) = 2$.

24. Estimate the time it takes for an orbit to close under each of the three sets of initial conditions proposed in the previous exercise.

25. The three-species system $\dot{x} = x(3 - y)$, $\dot{y} = y(x + z - 3)$, $\dot{z} = z(2 - y)$ discussed in Example 4 of the text has equilibrium solutions at (3,3,0) and (0,2,3), as well as the trivial one at (0,0,0). Use initial conditions $(x, y, z) = (0.001, 3, 2)$ to plot the following. (a) The orbit of $\big(x(t), y(t)\big)$. (b) The orbit of $\big(y(t), z(t)\big)$. (c) The orbit of $\big(x(t), z(t)\big)$.

Drug Delivery

26. Referring to text Example 3 in Chapter 5, Section 1, consider the system

$$\frac{dx}{dt} = -ax + f(t), \qquad x(0) = 0,$$

$$\frac{dy}{dt} = ax - by, \qquad y(0) = 0,$$

under the assumption that the rate constants satisfy $b > a > 0$ and that

$$f(t) = \begin{cases} 2, & \text{if } 1 \le t < 2 \text{ or } 4 \le t < 5, \\ 0, & \text{otherwise.} \end{cases}$$

Plot the solution for $0 \le t \le 15$ under the assumption that $a = 1$ and $b = 2$. Note that the function $f(t)$ can be represented using the Heaviside function $\mathrm{H}(t) = \begin{cases} 1, & t \ge 0, \\ 0, & t < 0, \end{cases}$ described in Chapter 4, Section 4C.

27. Modify the previous exercise so that the drug input function $f(t)$ remains the unchanged for $0 \le t < 15$ and is extended periodically thereafter.

SUPPLEMENTARY APPLICATIONS FOR CHAPTER 6

The website http://www.mhhe.com/williamson/ contains additional applications: Electric Networks, Energy, and Calculus of Variations. The first of these extends the analysis of simple loop circuits taken up in Chapter 2 to combinations of such loops. The second describes the Lagrange method for deriving differential equations and systems of differential equations using the concept of energy.

CHAPTER 7

Fundamental Matrices

1 MATRIX EXPONENTIALS

In this chapter we'll discuss an efficient method using matrix exponentials for solving normal-form linear systems $\dot{\mathbf{x}} = A\mathbf{x} + \mathbf{b}(t)$. The relative efficiency of this method compared to the alternative elimination method or the eigenvector method increases rapidly as the dimension of the system increases. In particular, the exponential matrix method enables us to avoid computing eigenvectors altogether. Of course, for some special systems it's possible to write down complete solutions at a glance, which should then be the preferred method. The key step for constant-coefficient systems is to define the exponential e^A of a square matrix A. The ordinary numerical exponential e^x can be defined in several ways, and the simplest one on which to model our matrix definition is the infinite sum

$$e^x = 1 + \frac{x}{1!} + \frac{x^2}{2!} + \frac{x^3}{3!} + \cdots + \frac{x^k}{k!} + \cdots.$$

This formula is valid for all real or complex values of x, and while it is not usual to do so, all the properties of e^x can be derived directly from it. For example, notice that the formal derivative of the infinite sum is equal to the sum itself, which suggests that e^x equals its derivative, as we know. Also, letting $x = 0$ shows that $e^0 = 1$. From these two properties alone it follows that e^x has the familiar properties such as $e^{x+z} = e^x e^z$. (See Exercise 25, Chapter 1, Section 2C.) Using matrix arguments, we'll be able to establish analogous properties for e^{tA}, where A is a square matrix and t is a real variable. The exponential matrices e^{tA} provide the foundation for efficient solution of constant coefficient nonhomogeneous systems $\dot{\mathbf{x}} = A\mathbf{x} + \mathbf{b}(t)$ of arbitrary dimension, and are the prime examples of what we'll define in Section 1D to be *fundamental matrices* for these systems.

The initial approach using an infinite series of matrices enables us to see most directly the general properties of the solutions we are looking for. For the purpose of computing explicit formulas, the exponential matrix is readily computable directly from the definition only in special cases, so we'll introduce another method for computing e^{tA} in Section 2. The payoff will be that we'll have an effective way to compute solutions for normal-form constant-coefficient systems, $\dot{\mathbf{x}} = A\mathbf{x}+\mathbf{b}(t)$ expressed in a notation that provides an instant reminder of significant properties of the solutions.

1A Definition and Properties

Matrix multiplication is associative, so we can raise a square matrix to an integer power without regard to the order of the multiplications, for example,

$$\begin{pmatrix} 2 & 1 \\ 0 & 3 \end{pmatrix}^2 = \begin{pmatrix} 2 & 1 \\ 0 & 3 \end{pmatrix}\begin{pmatrix} 2 & 1 \\ 0 & 3 \end{pmatrix} = \begin{pmatrix} 4 & 5 \\ 0 & 9 \end{pmatrix},$$

$$\begin{pmatrix} 2 & 1 \\ 0 & 3 \end{pmatrix}^3 = \begin{pmatrix} 2 & 1 \\ 0 & 3 \end{pmatrix}\begin{pmatrix} 4 & 5 \\ 0 & 9 \end{pmatrix} = \begin{pmatrix} 8 & 19 \\ 0 & 27 \end{pmatrix}.$$

Thus if A has dimensions n-by-n and $A^0 = I$ is the n-by-n identity matrix, we can write the sum

$$I + A + \frac{A^2}{2!} + \cdots + \frac{A^k}{k!} = \sum_{j=0}^{k} \frac{A^j}{j!}.$$

This linear combination of n-by-n matrices is also an n-by-n matrix. We define the **exponential** of A by

$$e^A = \lim_{k\to\infty} \sum_{j=0}^{k} \frac{A^j}{j!} = \sum_{j=0}^{\infty} \frac{A^j}{j!},$$

where the existence of the matrix limit is understood to mean that the limit exists in each of the n^2 entries in the matrix. It is sometimes convenient to use the notation $\exp A$ for e^A. For example, if

$$A = \begin{pmatrix} 2 & 0 \\ 0 & 3 \end{pmatrix},\ A^2 = \begin{pmatrix} 2^2 & 0 \\ 0 & 3^2 \end{pmatrix}, \ldots, A^j = \begin{pmatrix} 2^j & 0 \\ 0 & 3^j \end{pmatrix},$$

then

$$\exp\begin{pmatrix} 2 & 0 \\ 0 & 3 \end{pmatrix} = \lim_{k\to\infty} \sum_{j=0}^{k} \frac{1}{j!}\begin{pmatrix} 2^j & 0 \\ 0 & 3^j \end{pmatrix}$$

$$= \begin{pmatrix} \sum_{j=0}^{\infty} \frac{2^j}{j!} & 0 \\ 0 & \sum_{j=0}^{\infty} \frac{3^j}{j!} \end{pmatrix} = \begin{pmatrix} e^2 & 0 \\ 0 & e^3 \end{pmatrix}.$$

It is a remarkable and useful fact that the exponential of a square matrix always exists and has many of the properties of the ordinary real or complex exponential function. The justification for this statement is in the following fundamental theorem, in which the derivative of a matrix is defined to be simply the matrix obtained by differentiating the individual entries, for example,

$$\frac{d}{dt}\begin{pmatrix} 1 & t \\ 0 & e^t \end{pmatrix} = \begin{pmatrix} 0 & 1 \\ 0 & e^t \end{pmatrix}.$$

1.1 THEOREM

If A is an n-by-n real or complex matrix, the matrix series

$$\sum_{j=0}^{\infty} \frac{t^j A^j}{j!}$$

converges to an n-by-n matrix e^{tA} having the properties

(a) $\frac{d}{dt}e^{tA} = Ae^{tA} = e^{tA}A$.
(b) $e^{(t+s)A} = e^{tA}e^{sA} = e^{sA}e^{tA}$, for all real or complex numbers t and s.
(c) $e^{0A} = I$, and e^{tA} is invertible, with $e^{-tA}e^{tA} = e^{tA}e^{-tA} = I$.

Proof. If $A = (a_{ij})$, choose a positive number b such that $|a_{ij}| \le b$ for $i, j = 1, \dots, n$. Then the absolute values of the entries in tA are at most $b|t|$. Since the entries in A^2 are of the form

$$a_{i1}a_{1j} + \cdots + a_{in}a_{nj},$$

it follows that the entries in t^2A^2 are at most $n\,b^2|t|^2$ in absolute value. Proceeding inductively, the entries in t^kA^k are at most $n^{k-1}b^k|t|^k$ in absolute value. It follows that each entry in e^{tA} is defined by an absolutely convergent power series dominated by the convergent series

$$1 + b|t| + \frac{nb^2|t|^2}{2!} + \cdots + \frac{1}{j!}n^{j-1}b^j|t|^j + \cdots = \frac{1}{n}e^{nb|t|}.$$

Hence all entries exist and e^{tA} is well-defined by the series definition.

To prove property (a), we note that we can try to compute the derivative of e^{tA} from the infinite series by differentiating term by term, as follows:

$$\frac{d}{dt}e^{tA} = \frac{d}{dt}\sum_{j=0}^{\infty}\frac{t^jA^j}{j!}$$
$$= \sum_{j=1}^{\infty}\frac{jt^{j-1}A^j}{j!} = A\sum_{j=1}^{\infty}\frac{t^{j-1}A^{j-1}}{(j-1)!} = Ae^{tA}.$$

Note that the factor A could be taken out on the right just as well as on the left. The term-by-term differentiation of series is justified because, as we saw above, the series in each matrix entry is a power series that converges for all t.

We could prove property (b) similarly, but instead we'll rely on the Existence and Uniqueness Theorem 1.1 in Chapter 5, Section 1E. By that theorem the initial-value problem $\dot{\mathbf{x}} = A\mathbf{x}, \mathbf{x}(0) = \mathbf{c}$ has a unique solution for every constant n-dimensional vector $\mathbf{c}$. Let $\mathbf{x}(t) = e^{(t+s)A}\mathbf{c} - e^{tA}e^{sA}\mathbf{c}$. With the number s and vector **c** both fixed but arbitrary, we use part (a) and the chain rule to get

$$\frac{d}{dt}[e^{(t+s)A}\mathbf{c} - e^{tA}e^{sA}\mathbf{c}] = Ae^{(t+s)A}\mathbf{c} - Ae^{tA}e^{sA}\mathbf{c}$$
$$= A(e^{(t+s)A} - e^{tA}e^{sA})\mathbf{c}.$$

We have $e^{0A} = I$ by the definition of e^{tA}, so our choice for $\mathbf{x}(t)$ solves the initial-value problem $\dot{\mathbf{x}} = A\mathbf{x}$, $\mathbf{x}(0) = 0$. By the uniqueness theorem, $\mathbf{x}(t)$ must equal the obvious identically zero solution. Since **c** is arbitrary, it follows that $e^{(t+s)A} - e^{tA}e^{sA}$ is identically zero, so $e^{(t+s)A} = e^{tA}e^{sA}$. Interchanging t and s doesn't change the entry values in the matrix on the left, so we have also proved the second equality in property (b).

Property (c) follows from (b) on taking $s = -t$. Since $e^{0A} = I$, we have $I = e^{tA}e^{-tA} = e^{-tA}e^{tA}$, so e^{tA}. Hence e^{tA} and e^{-tA} are inverses to each other. ■

1B Solving Systems

The best way to see the importance of e^{tA} is to show that

$$\mathbf{x}(t) = e^{tA}\mathbf{x}_0$$

defines the solution of

$$\frac{d\mathbf{x}}{dt} = A\mathbf{x}, \qquad \mathbf{x}(0) = \mathbf{x}_0.$$

First, to differentiate the vector $e^{tA}\mathbf{x}_0$, we need only differentiate each entry in the matrix e^{tA}, because $\mathbf{x}_0$ has constant entries. Hence

$$\frac{d}{dt}e^{tA}\mathbf{x}_0 = Ae^{tA}\mathbf{x}_0$$

by part (a) of the previous theorem. Thus the differential equation is satisfied. Second, to show that the initial condition is satisfied, note that

$$\mathbf{x}(0) = I\mathbf{x}_0 = \mathbf{x}_0.$$

More generally,

$$\mathbf{x}(t) = e^{(t-t_0)A}\mathbf{x}_0$$

satisfies the initial condition $\mathbf{x}(t_0) = \mathbf{x}_0$.

1.2 THEOREM

If A is a constant n-by-n matrix, the general solution of $d\mathbf{x}/dt = A\mathbf{x}$ is

$$\mathbf{x}(t) = e^{tA}\mathbf{c} = c_1\mathbf{x}_1(t) + \cdots + c_n\mathbf{x}_n(t),$$

where $\mathbf{x}_k(t)$ is the kth column of e^{tA} and **c** is a constant column vector with kth entry c_k. The unique solution of the initial-value problem with $\mathbf{x}(t_0) = \mathbf{x}_0$ is $\mathbf{x}(t) = e^{tA}\mathbf{x}_0$.

Proof. We could apply the uniqueness theorem here, but we can prove the theorem directly by using the results of Theorem 1.1. Multiply the differential equation by e^{-tA} to get

$$e^{-tA}\frac{d\mathbf{x}}{dt} - e^{-tA}A\mathbf{x} = 0.$$

The product rule for derivatives (Theorem 1.1, Formula 5, in Appendix B) shows that this is just $d/dt(e^{-tA}\mathbf{x}) = 0$, so $e^{-tA}\mathbf{x} = \mathbf{c}$, for some constant vector $\mathbf{c}$. Multiplying by e^{tA} gives $\mathbf{x}(t) = e^{tA}\mathbf{c}$, which by row-by-column multiplication is a linear combination of the columns of the exponential matrix. Finally $\mathbf{c} = e^{0A}\mathbf{c} = \mathbf{x}(0) = \mathbf{x}_0$. Finally, replace the constant vector $\mathbf{x}_0$ by $e^{-t_0A}\mathbf{x}_0$. We get the more general solution

$$\mathbf{x}(t) = e^{tA}e^{-t_0A}\mathbf{x}_0 = e^{(t-t_0)A}\mathbf{x}_0,$$

which satisfies $\mathbf{x}(t_0) = \mathbf{x}_0$. ■

---

example 1

The system

$$\dot{x} = x,$$
$$\dot{y} = -y$$

can be written in matrix form as $d\mathbf{x}/dt = A\mathbf{x}$, where

$$A = \begin{pmatrix} 1 & 0 \\ 0 & -1 \end{pmatrix}.$$

Hence the system has general solution

$$\mathbf{x}(t) = e^{tA}\mathbf{c},$$

where $\mathbf{c}$ is the 2-dimensional column vector with constant entries c_1 and c_2. To compute the exponential matrix, write

$$\begin{aligned} e^{t\begin{pmatrix} 1 & 0 \\ 0 & -1 \end{pmatrix}} &= e^{\begin{pmatrix} t & 0 \\ 0 & -t \end{pmatrix}} \\ &= \sum_{j=0}^{\infty}\frac{1}{j!}\begin{pmatrix} t & 0 \\ 0 & -t \end{pmatrix}^j = \sum_{j=0}^{\infty}\frac{1}{j!}\begin{pmatrix} t^j & 0 \\ 0 & (-t)^j \end{pmatrix} \\ &= \begin{pmatrix} \sum_{j=0}^{\infty}\frac{1}{j!}t^j & 0 \\ 0 & \sum_{j=0}^{\infty}\frac{1}{j!}(-t)^j \end{pmatrix} = \begin{pmatrix} e^t & 0 \\ 0 & e^{-t} \end{pmatrix}. \end{aligned}$$

Hence the general solution is

$$\begin{aligned} \mathbf{x}(t) &= \begin{pmatrix} e^t & 0 \\ 0 & e^{-t} \end{pmatrix}\begin{pmatrix} c_1 \\ c_2 \end{pmatrix} \\ &= \begin{pmatrix} c_1e^t \\ c_2e^{-t} \end{pmatrix}. \end{aligned}$$

Of course this solution can easily be found by a glance at the original system; we derive it using the exponential matrix because it provides a very simple example of the definition and use of the exponential.

example 2 The system

$$\begin{aligned} \dot{x} &= x + y, \\ \dot{y} &= \quad\; y \end{aligned}$$

can be written in matrix form $\dot{\mathbf{x}} = A\mathbf{x}$ as

$$\begin{pmatrix} \dot{x} \\ \dot{y} \end{pmatrix} = \begin{pmatrix} 1 & 1 \\ 0 & 1 \end{pmatrix} \begin{pmatrix} x \\ y \end{pmatrix}.$$

We compute

$$A = \begin{pmatrix} 1 & 1 \\ 0 & 1 \end{pmatrix}, A^2 = \begin{pmatrix} 1 & 2 \\ 0 & 1 \end{pmatrix}, A^3 = \begin{pmatrix} 1 & 3 \\ 0 & 1 \end{pmatrix}, \ldots, A^k = \begin{pmatrix} 1 & k \\ 0 & 1 \end{pmatrix}, \ldots.$$

Then

$$\begin{aligned} e^{tA} &= \sum_{k=0}^{\infty} \frac{t^k}{k!} \begin{pmatrix} 1 & k \\ 0 & 1 \end{pmatrix} = \begin{pmatrix} \sum_{k=0}^{\infty} \frac{t^k}{k!} & \sum_{k=1}^{\infty} \frac{t^k}{(k-1)!} \\ 0 & \sum_{k=0}^{\infty} \frac{t^k}{k!} \end{pmatrix} \\ &= \begin{pmatrix} e^t & te^t \\ 0 & e^t \end{pmatrix}. \end{aligned}$$

Hence the solution with initial conditions $x(0) = 2$, $y(0) = -3$ is

$$\begin{aligned} \begin{pmatrix} x(t) \\ y(t) \end{pmatrix} &= \begin{pmatrix} e^t & te^t \\ 0 & e^t \end{pmatrix} \begin{pmatrix} 2 \\ -3 \end{pmatrix} \\ &= \begin{pmatrix} 2e^t - 3te^t \\ -3e^t \end{pmatrix}. \end{aligned}$$

Note that instead of computing eigenvectors we computed powers of A; we'll see in the next section that we really only need to do this for finitely many exponents. Since this system is partially uncoupled, it's most easily solved by integration.

1C Relationship of e^{tA} to Eigenvectors

The connection between matrix exponential solutions and the eigenvectors of a square matrix A is as follows. Form the diagonal matrix

$$\Lambda_t = \begin{pmatrix} e^{\lambda_1 t} & 0 & \ldots & 0 \\ 0 & e^{\lambda_2 t} & \ldots & 0 \\ \vdots & \vdots & \ddots & \vdots \\ 0 & 0 & \ldots & e^{\lambda_n t} \end{pmatrix},$$

where the numbers λ_k are the eigenvalues of A in an arbitrary but henceforth fixed order. Let $\mathbf{u}_k$ be an eigenvector corresponding to the eigenvalue λ_k. Form a matrix U with the column vector $\mathbf{u}_k$ as the kth column of U. For example, if $\mathbf{u}_1 = \begin{pmatrix} 1 \\ 2 \end{pmatrix}$ and $\mathbf{u}_2 = \begin{pmatrix} 3 \\ 4 \end{pmatrix}$, then $U = \begin{pmatrix} 1 & 3 \\ 2 & 4 \end{pmatrix}$. If $\mathbf{c}$ is a column vector with entries

$c_1, c_2, \ldots, c_n$, then $\Lambda_t\mathbf{c}$ is the column vector with entries $c_1e^{\lambda_1 t}, c_2e^{\lambda_2 t}, \ldots, c_ne^{\lambda_n t}$, for example, $\begin{pmatrix} e^{\lambda_1 t} & 0 \\ 0 & e^{\lambda_2 t} \end{pmatrix}\begin{pmatrix} c_1 \\ c_2 \end{pmatrix} = \begin{pmatrix} c_1e^{\lambda_1 t} \\ c_2e^{\lambda_2 t} \end{pmatrix}$. It follows from the definition of row-by-column multiplication that if we form the product $U\Lambda_t\mathbf{c}$, then

$$\mathbf{x}(t) = U\Lambda_t\mathbf{c} = c_1e^{\lambda_1 t}\mathbf{u}_1 + c_2e^{\lambda_2 t}\mathbf{u}_2 + \cdots + c_ne^{\lambda_n t}\mathbf{u}_n,$$

where the column vector $\mathbf{u}_k$ is the kth column of U. Differentiation and multiplication by A both act linearly on $\mathbf{x}(t)$, so since each term of $\mathbf{x}(t)$ is a solution the sum is also a solution. However, if the n eigenvectors $\mathbf{u}_k$ are linearly independent then the matrix U is invertible with inverse U^{-1}, so we can write the vector $\mathbf{c}$ as $U^{-1}\mathbf{x}_0$ for an arbitrarily specified vector $\mathbf{x}_0$. Then $\mathbf{x}(t)$ becomes

$$\mathbf{x}(t) = U\Lambda_tU^{-1}\mathbf{x}_0.$$

Note that $\Lambda_0 = I$, so $\mathbf{x}(0) = U\Lambda_0U^{-1}\mathbf{x}_0 = UU^{-1}\mathbf{x}_0 = \mathbf{x}_0$. Hence $\mathbf{x}(t) = U\Lambda_tU^{-1}\mathbf{x}_0$ solves the initial-value problem $\mathbf{x}(0) = \mathbf{x}_0$ for the equation $\dot{\mathbf{x}} = A\mathbf{x}$. Since

$$\mathbf{x}(t) = e^{tA}\mathbf{x}_0$$

solves the same problem, we are faced with the question of whether the two solutions are the same. But by Theorem 1.2 there is only one solution satisfying $\mathbf{x}(0) = \mathbf{x}_0$. Hence $e^{tA}\mathbf{x}_0 = U\Lambda_tU^{-1}\mathbf{x}_0$ for all t. It follows, since $\mathbf{x}_0$ is arbitrary, that when U is invertible

1.3 $$e^{tA} = U\Lambda_tU^{-1}.$$

Thus if the eigenvector matrix U is invertible, or equivalently when the columns of U are linearly independent, we can use U to compute the exponential matrix, as in the next example.

example 3

In Example 1 of Chapter 5, Section 3, we solved the system

$$\begin{pmatrix} \dot{x} \\ \dot{y} \end{pmatrix} = \begin{pmatrix} 1 & 1 \\ 4 & 1 \end{pmatrix}\begin{pmatrix} x \\ y \end{pmatrix}$$

by finding the eigenvalues $\lambda_1 = 3$, $\lambda_2 = -1$ and corresponding eigenvectors (1,2), (1,–2) of the 2-by-2 matrix A of the system. Thus

$$U = \begin{pmatrix} 1 & 1 \\ 2 & -2 \end{pmatrix}, \qquad \Lambda_t = \begin{pmatrix} e^{3t} & 0 \\ 0 & e^{-t} \end{pmatrix}, \qquad U^{-1} = \begin{pmatrix} \frac{1}{2} & \frac{1}{4} \\ \frac{1}{2} & -\frac{1}{4} \end{pmatrix}$$

and

$$e^{tA} = U\Lambda_tU^{-1} = \begin{pmatrix} \frac{1}{2}e^{3t} + \frac{1}{2}e^{-t} & \frac{1}{4}e^{3t} - \frac{1}{4}e^{-t} \\ e^{3t} - e^{-t} & \frac{1}{2}e^{3t} + \frac{1}{2}e^{-t} \end{pmatrix}.$$

As a partial check on the computation, notice that $e^{0A} = I$. This example shows that, if we know the eigenvectors of A and an eigenvector matrix U of A is invertible, it may be easier to use U and Λ to compute e^{tA} than to use the matrix power series definition. However, the method for finding e^{tA} explained in Section 2 doesn't require us to find eigenvectors and works even if the eigenvectors aren't linearly independent, in which case U^{-1} fails to exist.

EXERCISES

Find the exponential e^{tA} of each of the matrices in Exercises 1 through 4 by first computing the successive terms $I, tA, t^2A^2/2!, \ldots$ that appear in the series definition.

1. $A = \begin{pmatrix} -1 & 0 \\ 0 & 1 \end{pmatrix}$.

2. $A = \begin{pmatrix} i & 0 \\ 0 & -i \end{pmatrix}$.

3. $A = \begin{pmatrix} 1 & 0 \\ 1 & 0 \end{pmatrix}$.

4. $A = \begin{pmatrix} 1 & 0 & 0 \\ 0 & 2 & 0 \\ 0 & 0 & 4 \end{pmatrix}$.

5. Text Example 2 shows that

$$\exp\left(t\begin{pmatrix} 1 & 1 \\ 0 & 1 \end{pmatrix}\right) = \begin{pmatrix} e^t & te^t \\ 0 & e^t \end{pmatrix}.$$

Verify directly, by using this equation, that

(a) $\exp\left(t\begin{pmatrix} 1 & 1 \\ 0 & 1 \end{pmatrix}\right)$ and $\exp\left(-t\begin{pmatrix} 1 & 1 \\ 0 & 1 \end{pmatrix}\right)$ are inverse matrices to each other.

(b) $\exp\left(t\begin{pmatrix} 1 & 1 \\ 0 & 1 \end{pmatrix}\right)\exp\left(s\begin{pmatrix} 1 & 1 \\ 0 & 1 \end{pmatrix}\right) = \exp\left((t+s)\begin{pmatrix} 1 & 1 \\ 0 & 1 \end{pmatrix}\right)$.

(c) $\dfrac{d}{dt}\exp\left(t\begin{pmatrix} 1 & 1 \\ 0 & 1 \end{pmatrix}\right) = \begin{pmatrix} 1 & 1 \\ 0 & 1 \end{pmatrix}\exp\left(t\begin{pmatrix} 1 & 1 \\ 0 & 1 \end{pmatrix}\right)$.

(d) What is the solution of the system

$$\begin{pmatrix} \dot{x} \\ \dot{y} \end{pmatrix} = \begin{pmatrix} 1 & 1 \\ 0 & 1 \end{pmatrix}\begin{pmatrix} x \\ y \end{pmatrix} \quad \text{that satisfies} \quad \begin{pmatrix} x(0) \\ y(0) \end{pmatrix} = \begin{pmatrix} -1 \\ 2 \end{pmatrix}?$$

6. If A is a 1-by-1 matrix with entry a, what is e^{tA}?

7. Let

$$A = \begin{pmatrix} -1 & -2 & -2 \\ 0 & 1 & -1 \\ 0 & 0 & 2 \end{pmatrix}.$$

Using methods developed in Section 3, we will be able to show quite easily that

$$e^{tA} = \begin{pmatrix} e^{-t} & e^{-t} - e^t & e^{-t} - e^t \\ 0 & e^t & e^t - e^{2t} \\ 0 & 0 & e^{2t} \end{pmatrix}.$$

(a) Replace t by $-t$ in e^{tA} and verify directly that $e^{-tA}e^{tA} = I$ and that $e^{0A} = I$.

(b) If $B(t) = (d/dt)e^{tA}$, verify that $B(0) = A$.

8. (a) Find the general solution of the system

$$\begin{pmatrix} \dot{x} \\ \dot{y} \end{pmatrix} = \begin{pmatrix} 9 & -4 \\ 4 & 1 \end{pmatrix} \begin{pmatrix} x \\ y \end{pmatrix}$$

by the method of elimination.

(b) Use the result of part (a) to compute the matrix e^{tA}, where

$$A = \begin{pmatrix} 9 & -4 \\ 4 & 1 \end{pmatrix}.$$

[*Hint:* Find solutions such that $\mathbf{x}_1(0) = \begin{pmatrix} 1 \\ 0 \end{pmatrix}$ and $\mathbf{x}_2(0) = \begin{pmatrix} 0 \\ 1 \end{pmatrix}$.]

9. Let A be an n-by-n matrix with real entries. The matrix e^{itA} is defined by

$$e^{itA} = \sum_{k=0}^{\infty} \frac{1}{k!}(i)^k t^k A^k.$$

Define $\cos tA$ to be the real part of the series and $\sin tA$ to be the imaginary part, so that

$$e^{itA} = \cos tA + i \sin tA.$$

Show that the matrices $\cos tA$ and $\sin tA$ satisfy

(a) $\cos(-t)A = \cos tA$, $\sin(-t)A = -\sin tA$.

(b) $(d/dt)\cos tA = -A \sin tA$, $(d/dt)\sin tA = A \cos tA$. [*Hint:* Express $\cos tA$ and $\sin tA$ in terms of e^{itA}.]

(c) $(\cos tA)^2 + (\sin tA)^2 = I$, where I is the n-by-n identity matrix.

10. Let A be the 2-by-2 matrix

$$\begin{pmatrix} 1 & 1 \\ 0 & 1 \end{pmatrix}.$$

Define $\cos tA$ and $\sin tA$ as in Exercise 9, and verify the formulas given in (a), (b), and (c).

11. Show that if A is an n-by-n matrix then a system of the form

$$\frac{d^2\mathbf{x}}{dt^2} + A^2\mathbf{x} = 0$$

has solutions of the form

$$\mathbf{x}(t) = (\cos tA)\mathbf{c}_1 + (\sin tA)\mathbf{c}_2,$$

where $\mathbf{c}_1$ and $\mathbf{c}_2$ are constant n-dimensional vectors, whereas $\cos tA$ and $\sin tA$ are the n-by-n matrices defined in Exercise 9. Is this always the most general solution?

12. For each of the following matrices A, compute e^{tA} by using Equation 1.3. Then find the inverse matrix e^{-tA}, and check your original computation by showing that the derivative of e^{tA} at $t = 0$ is equal to A.

(a) $\begin{pmatrix} 0 & 1 \\ -6 & 5 \end{pmatrix}$. **(b)** $\begin{pmatrix} -4 & 4 \\ -6 & 6 \end{pmatrix}$. **(c)** $\begin{pmatrix} -1 & 0 & 0 \\ 0 & \frac{3}{2} & -\frac{1}{2} \\ 0 & -\frac{1}{2} & \frac{3}{2} \end{pmatrix}$.

13. Use eigenvector matrices U to find e^{tA} for each of the following matrices A.

(a) $A = \begin{pmatrix} -3 & 2 \\ -4 & 3 \end{pmatrix}$. **(b)** $A = \begin{pmatrix} 3 & 0 \\ 0 & 2 \end{pmatrix}$.

(c) $A = \begin{pmatrix} 1 & 4 \\ 0 & 5 \end{pmatrix}$. **(d)** $A = \begin{pmatrix} 2 & -1 \\ 1 & 2 \end{pmatrix}$.

14. Let A be the n-by-n matrix with all entries equal to 1.

(a) Show that $A^2 = nA$, and more generally that $A^k = n^{k-1}A$ for integer $k \geq 1$.

(b) Show that $e^{tA} = I + (1/n)(e^{nt} - 1)A$.

(c) Find the four entries in e^{tA} when $n = 2$.

15. It can be shown that if A and B are n-by-n matrices then $e^A e^B = e^{A+B}$ if and only if $AB = BA$. Find 2-by-2 matrices A and B for which the exponential equation fails to hold.

16. It can be shown that $e^{tA}e^{sB} = e^{sB}e^{tA}$ if A and B commute, that is, if $AB = BA$. Find 2-by-2 matrices A and B such that $e^{tA}e^{sB} \neq e^{sB}e^{tA}$.

17. Prove that if $B(t)$ is a 2-by-2 matrix and $\mathbf{x}(t)$ is an n-dimensional column vector, both with differentiable entries, then $(d/dt)B(t)\mathbf{x}(t) = \dot{B}(t)\mathbf{x}(t) + B(t)\dot{\mathbf{x}}(t)$.

1D Independent Solutions†

Theorem 1.2 of Section 1A shows that, for every equation $d\mathbf{x}/dt = A\mathbf{x}$ with constant square matrix A, there is an exponential solution formula $\mathbf{x}(t) = e^{tA}\mathbf{c}$. In Example 2 of Section 1B we had

$$e^{tA}\mathbf{c} = \begin{pmatrix} e^t & te^t \\ 0 & e^t \end{pmatrix} \begin{pmatrix} c_1 \\ c_2 \end{pmatrix} = c_1 \begin{pmatrix} e^t \\ 0 \end{pmatrix} + c_2 \begin{pmatrix} te^t \\ e^t \end{pmatrix}.$$

Since c_1 and c_2 are arbitrary constants, it's important to avoid the redundancy that would occur in the formula in case one of the two columns in the matrix is a constant multiple of the other. We will see that this cannot happen in general, but to state the general result we need to look more closely at what is meant by linear independence of vector functions. Let $\mathbf{x}_1(t), \mathbf{x}_2(t), \ldots, \mathbf{x}_m(t)$ be n-dimensional column vectors whose entries are functions on some common interval $a < t < b$. (It is not ruled out of course that some or all of the entries may happen to be constant.) Vector functions $\mathbf{x}_k(t)$, $k = 1, \ldots, m$ defined on a t interval are said to be **linearly independent** if whenever

$$c_1\mathbf{x}_1(t) + c_2\mathbf{x}_2(t) + \cdots + c_m\mathbf{x}_m(t) = 0$$

Formally these definitions look the same as the ones we made in Chapter 3, Sections 1 and 2D, for scalar-valued functions, but the definitions need to be understood now in the wider context of vector-valued functions.

for all t, then the constant coefficients c_k are all zero. When we have only two functions ($m = 2$), asserting linear independence is the same as saying that neither function is a constant multiple of the other. The reason is that if either c_1 or c_2 is not zero we could divide by it and express one vector as a multiple of the other. Similarly if $m > 2$, the linear independence of m vector functions means that none of them can be replaced by a sum of scalar multiples, or **linear combination,** of the others. The negation of linear independence of a set of vector functions is called **linear dependence,** and it means simply that at least one of the vector functions is a linear combination of the others.

example 4

The vector functions

$$\mathbf{x}_1(t) = \begin{pmatrix} e^t \\ 0 \end{pmatrix}, \qquad \mathbf{x}_2 = \begin{pmatrix} te^t \\ e^t \end{pmatrix}$$

in the columns of the exponential matrix in Example 1 are linearly independent. Note that

$$c_1 \begin{pmatrix} e^t \\ 0 \end{pmatrix} + c_2 \begin{pmatrix} te^t \\ e^t \end{pmatrix} = \begin{pmatrix} 0 \\ 0 \end{pmatrix}, \qquad -\infty < t < \infty$$

is the same as

$$\begin{aligned} c_1 e^t + c_2 te^t &= 0, \\ c_2 e^t &= 0. \end{aligned}$$

It follows that $c_2 = 0$. Hence $c_1 = 0$. This conclusion holds for any fixed value of t, so in particular the constant vectors $\begin{pmatrix} 1 \\ 0 \end{pmatrix}, \begin{pmatrix} 0 \\ 1 \end{pmatrix}$ are linearly independent. Just set $t = 0$. Theorem 1.4 will show that the columns of an exponential matrix e^{tA} are always independent vector functions and are also independent for each t. (But see Exercise 8.)

example 5

Consider the vector functions

$$\mathbf{x}_1(t) = \begin{pmatrix} e^t \\ 0 \\ 0 \end{pmatrix}, \qquad \mathbf{x}_2(t) = \begin{pmatrix} 0 \\ e^t \\ te^t \end{pmatrix}, \qquad \mathbf{x}_3(t) = \begin{pmatrix} 0 \\ e^t \\ t^2e^t \end{pmatrix}.$$

The check for independence for $-\infty < t < \infty$ is to solve

$$c_1 \begin{pmatrix} e^t \\ 0 \\ 0 \end{pmatrix} + c_2 \begin{pmatrix} 0 \\ e^t \\ te^t \end{pmatrix} + c_3 \begin{pmatrix} 0 \\ e^t \\ t^2e^t \end{pmatrix} = \begin{pmatrix} 0 \\ 0 \\ 0 \end{pmatrix}$$

for c_1, c_2, and c_3. This vector equation is equivalent to the three equations

$$c_1 e^t = 0, \qquad c_2 e^t + c_3 e^t = 0, \qquad c_2 te^t + c_3 t^2 e^t = 0.$$

The first equation shows that $c_1 = 0$. The middle equation implies $c_2 = -c_3$, so the last equation says $c_2 t - c_2 t^2 = 0$ for all t. Thus $c_2 = c_3 = 0$, so the vector functions are independent as defined on the interval $-\infty < t < \infty$. Note, however, that when $t = 0$ we get

$$\mathbf{x}_1(0) = \begin{pmatrix} 1 \\ 0 \\ 0 \end{pmatrix}, \qquad \mathbf{x}_2(0) = \begin{pmatrix} 0 \\ 1 \\ 0 \end{pmatrix}, \qquad \mathbf{x}_3(0) = \begin{pmatrix} 0 \\ 1 \\ 0 \end{pmatrix},$$

and these constant vectors are linearly dependent. Similarly, we have dependence when $t = 1$. This shows that vector functions may be linearly independent while their restrictions to some smaller domain, in this example one or two points, may be linearly dependent.

Here is the theorem that guarantees independence of the columns of an exponential matrix for any and all values of t. Thus a matrix whose columns are the vector functions in the previous example can't be an exponential matrix.

1.4 THEOREM

Let A be an n-by-n matrix with constant entries and let $\mathbf{x}_k(t)$ be the kth column of the exponential matrix e^{tA}. Then the vector functions $\mathbf{x}_1(t), \ldots, \mathbf{x}_n(t)$ are linearly independent over an arbitrary set S of t values.

Proof. We want to show that if all t values in some set S the vector equation

$$c_1\mathbf{x}_1(t) + \cdots + c_n\mathbf{x}_n(t) = 0$$

holds, then all the c_k must be zero. By the definition of the vectors $\mathbf{x}_k(t)$ as columns of e^{tA}, the displayed equation is equivalent to the vector equation $e^{tA}\mathbf{c} = 0$, where $\mathbf{c}$ is the column vector with the c_k as entries. Now multiplying this last equation by the inverse matrix e^{-tA} we get $e^{-tA}e^{tA}\mathbf{c} = \mathbf{c} = 0$. Hence all the c_k are zero. ■

A square matrix whose columns form a linearly independent set of solutions of a homogeneous system $\dot{\mathbf{x}} = A(t)\mathbf{x}$ is called a **fundamental matrix** for the system, because, as we'll see in Section 3, a fundamental matrix for $\dot{\mathbf{x}} = A\mathbf{x}$ enables us to generate all solutions to the nonhomogeneous system $\dot{\mathbf{x}} = A\mathbf{x} + \mathbf{b}(t)$. Theorem 1.4 shows that if A is a square matrix with constant entries then e^{tA} is a fundamental matrix for the system $\dot{\mathbf{x}} = A\mathbf{x}$.

EXERCISES

1. Each of the given matrices is the exponential matrix of some constant matrix A. In each case, express the vector function $\mathbf{x}(t) = e^{tA}\mathbf{c}$ as a linear combination of the columns of e^{tA}.

(a) $e^{tA} = \begin{pmatrix} 1 & 0 \\ 0 & e^t \end{pmatrix}$.

(b) $e^{tA} = \begin{pmatrix} e^t & 0 \\ 0 & e^{2t} \end{pmatrix}$.

(c) $e^{tA} = \begin{pmatrix} e^t & 0 & 0 \\ 0 & e^{2t} & 0 \\ 0 & 0 & e^{3t} \end{pmatrix}$.

(d) $e^{tA} = \begin{pmatrix} e^t & te^t & 0 \\ 0 & e^t & 0 \\ 0 & 0 & e^{2t} \end{pmatrix}$.

2. For each of the exponential matrices in Exercise 1 find A by computing the derivative of e^{tA} at $t = 0$. Then verify that each column of e^{tA} is a solution of the equation $\dot{\mathbf{x}} = A\mathbf{x}$.

3. Not every square matrix with linearly independent columns is an exponential matrix. For example, an exponential matrix e^{tA} must equal I when $t = 0$. Show that each of the following matrices has linearly independent columns, but is not an exponential matrix.

(a) $\begin{pmatrix} e^t & 1 \\ 0 & e^t \end{pmatrix}$. **(b)** $\begin{pmatrix} e^{2t} & t^2e^t \\ 0 & e^t \end{pmatrix}$. **(c)** $\begin{pmatrix} e^t & te^{2t} \\ 0 & e^t \end{pmatrix}$.

4. Theorem 1.4 is a simple consequence of the following more general theorem: If $A(t)$ is an invertible square matrix for some t in an interval $a < t < b$, then the columns of $A(t)$ are linearly independent vector functions on this interval. Show how to prove this theorem using the ideas in the proof of Theorem 1.4.

5. Let D be the n-by-n diagonal matrix with entries $d_1, \ldots, d_n$ on the main diagonal and zeros elsewhere. Show that e^{tD} is the diagonal matrix with entries $e^{d_1t}, \ldots, e^{d_nt}$.

6. Prove that a set $\{\mathbf{x}_1(t), \mathbf{x}_2(t), \ldots, \mathbf{x}_m(t)\}$ of vector-valued functions of the same dimension is linearly independent on a t interval if and only if no one of them is a linear combination of the others.

7. A system $\dot{\mathbf{x}} = A\mathbf{x}$ has solution $\mathbf{x} = (c_1e^{2t} + c_2e^{3t},\ c_1e^{2t} - c_2e^{3t})$. Find the matrix A by first finding e^{tA}.

8. Theorem 1.4 shows that if $M(t)$ is an exponential matrix, then not only are its columns independent vector functions of t, but in addition the columns are independent vectors for each *fixed* t.

(a) Show that the matrix $\begin{pmatrix} e^t & te^t \\ te^t & e^t \end{pmatrix}$ has independent vector functions for its columns.

(b) Show that the matrix in part (a) can't be an exponential matrix, because there is a value $t = t_0$ for which the columns are *not* independent vectors.

9. The purpose of this exercise is to show independently of the exponential matrix e^{tA} that square matrices $U\Lambda_tU^{-1}$ behave in some ways like the exponential function e^t. Let U be an invertible n-by-n matrix and let $\lambda_1, \ldots, \lambda_n$ be n arbitrarily chosen complex numbers. Let Λ_t be the n-by-n matrix with entries $e^{\lambda_1t}, \ldots, e^{\lambda_nt}$ on the main diagonal and zeros elsewhere. Define $E(t) = U\Lambda_tU^{-1}$. Prove the following properties of $E(t)$.

(a) $E(0) = I$.

(b) $E(s + t) = E(s)E(t) = E(t)E(s)$.

(c) $E(t)E(-t) = I$.

1E Wronskians†

This section extends the preceding discussion of independence for solutions from the constant-coefficient equation $\dot{\mathbf{x}} = A\mathbf{x}$ to the variable-coefficient equation $\dot{\mathbf{x}} = A(t)\mathbf{x}$. The material is relatively theoretical in that when it comes to concrete examples there are no widely applicable techniques for finding solution formulas for variable-coefficient systems. The following existence and uniqueness theorem, which follows from Theorem 1.1 in Chapter 5, Section 1E, applies whenever $A(t)$ has entries that are continuous functions of t, and does guarantee that the ensuing discussion applies in theory to a very large class of equations; this class includes of course the case of constant A, which for all practical purposes is completely taken care of by the exponential matrix e^{tA}.

1.5 EXISTENCE AND UNIQUENESS THEOREM

Suppose the square matrix $A(t)$ has continuous entries $a_{jk}(t)$ on a t interval J. Then for every t_0 in J there is a unique solution on J to the initial-value problem $\dot{\mathbf{x}} = A(t)\mathbf{x}$, $\mathbf{x}(t_0) = \mathbf{x}_0$.

We begin with a determinant criterion for the linear independence of n constant vectors $\mathbf{x}_1, \mathbf{x}_2, \ldots, \mathbf{x}_n$, each of dimension n. Write the vectors $\mathbf{x}_k$ as columns, and form a

$$\mathbf{x}_1 = \begin{pmatrix} x_{11} \\ x_{21} \\ \vdots \\ x_{n1} \end{pmatrix}, \mathbf{x}_2 = \begin{pmatrix} x_{12} \\ x_{22} \\ \vdots \\ x_{n2} \end{pmatrix}, \ldots, \mathbf{x}_n = \begin{pmatrix} x_{1n} \\ x_{2n} \\ \vdots \\ x_{nn} \end{pmatrix},$$

square matrix X from the columns. Now let $\mathbf{c}$ be an arbitrary column vector with entries $c_1, c_2, \ldots, c_n$. Then

$$X\mathbf{c} = \begin{pmatrix} x_{11} & x_{12} & \cdots & x_{1n} \\ x_{21} & x_{22} & \cdots & x_{2n} \\ \vdots & \vdots & & \vdots \\ x_{n1} & x_{n2} & \cdots & x_{nn} \end{pmatrix} \begin{pmatrix} c_1 \\ c_2 \\ \vdots \\ c_n \end{pmatrix} = c_1 \begin{pmatrix} x_{11} \\ x_{21} \\ \vdots \\ x_{n1} \end{pmatrix} + c_2 \begin{pmatrix} x_{12} \\ x_{22} \\ \vdots \\ x_{n2} \end{pmatrix} + \cdots + c_n \begin{pmatrix} x_{1n} \\ x_{2n} \\ \vdots \\ x_{nn} \end{pmatrix}.$$

Thus the vector equation to check for independence of the $\mathbf{x}_k$ is related to X as follows:

$$c_1\mathbf{x}_1 + c_2\mathbf{x}_2 + \cdots + c_n\mathbf{x}_n = X\mathbf{c} = 0.$$

1.6 THEOREM

The constant columns of a square matrix X are independent if and only if det $X \neq 0$, in which case X^{-1} exists; the corresponding statement holds for the rows of X also.

Proof. There are two possibilities for det X; if det $X \neq 0$, then X is invertible and the only possibility for $\mathbf{c}$ is $\mathbf{c} = 0$, in which case the vectors $\mathbf{x}_k$ are independent. Otherwise, if det $X = 0$, X is not invertible, $X\mathbf{c} = 0$ has a solution $\mathbf{c} \neq 0$, and the n vectors are linearly dependent. Transposing a square matrix across its main diagonal interchanges rows and columns, and since the determinant of a transposed matrix equals the determinant of the matrix, $\det X$ serves as a criterion for independence of rows as well. ■

example 6

Of the two matrices below, the columns in (i) are independent (the same for the rows), and the columns in (ii) are dependent (the same for the rows).

$$\text{(i)}\quad \det\begin{pmatrix}1 & 2 & 3\\ 2 & 3 & 2\\ 3 & 1 & 1\end{pmatrix} = -12, \qquad \text{(ii)}\quad \det\begin{pmatrix}1 & 2 & 3\\ 4 & 5 & 6\\ 7 & 8 & 9\end{pmatrix} = 0.$$

For 2-by-2 matrices you can often tell at a glance whether you have row or column dependence or not, but for these two 3-by-3 examples that's not so easy. Note that for (ii) the determinant criterion gives no information about the specific dependency relations; we'd have to solve the related homogeneous set of linear equations to find such relations.

Our main interest here is in independence of column vectors $\mathbf{x}_k(t)$ that are solutions of first-order systems of linear homogeneous differential equations of the form $\dot{\mathbf{x}} = A(t)\mathbf{x}$, where $A(t)$ is an n-by-n matrix with entries that are continuous functions of t. An n-by-n matrix $X(t)$ whose columns are solutions of the system is called a **solution matrix** of the system. Thus a determinant $\det X(t)$ formed from these columns will in general also be a function of t. If we allow the entries in the matrix to be arbitrary differentiable functions of t on some interval $a < t < b$, then $\det X(t)$ is called a **Wronskian determinant,** or a **Wronskian** for short. There are now three possibilities: (i) $\det X(t) = 0$ for all t in the interval, (ii) $\det X(t) \neq 0$ for all t in the interval, and (iii) $\det X(t) = 0$ for some t in the interval while $\det X(t) \neq 0$ for the other t values in the interval. We'll see below that when the columns of $X(t)$ are solution vectors of a system $\dot{\mathbf{x}} = A(t)\mathbf{x}$, case (iii) can't occur. To fix the ideas in mind we'll illustrate all three cases.

example 7

The three columns in each of the matrices below are possible candidates for being sets of independent solutions to a 3-dimensional, first-order, linear homogeneous system of differential equations.

$$\text{(i)}\quad \det\begin{pmatrix}1 & 1 & t\\ 1 & t & t^2\\ 1 & t^2 & t^3\end{pmatrix} = 0, \qquad \text{(ii)}\quad \det\begin{pmatrix}e^t & te^t & 0\\ 0 & e^t & 0\\ 0 & 0 & e^{3t}\end{pmatrix} = e^{5t},$$

$$\text{(iii)}\quad \det\begin{pmatrix}e^t & 0 & 0\\ 0 & e^t & e^t\\ 0 & te^t & t^2e^t\end{pmatrix} = t\,(t-1)e^{3t}.$$

By Theorem 1.6 the columns in matrix (i) are dependent for each t, with coefficients of combination that may possibly depend on t. However, Theorem 1.6 doesn't tell us whether or not the columns are independent as functions of t

or not. Although the matrix is invertible for no t values at all, the columns are independent as functions on an arbitrary t interval, as you can see by showing that an identically zero linear combination must have all coefficients c_1, c_2, and c_3 equal to zero. (See Exercise 1.)

The columns in the matrix $X(t)$ in (ii) are independent for each t. They are also independent as functions over an arbitrary t interval, because $\det X(t) \neq 0$ enables us to multiply the equation $X(t)\mathbf{c} = 0$ on the left by $X^{-1}(t)$ to show that $\mathbf{c} = 0$.

For $t = 0$ and $t = 1$ the columns in (iii) are dependent; indeed the second and third columns are identical for each of those two values. Direct examination (see Exercise 3) shows that as functions of t over an arbitrary interval the columns are independent.

The previous example shows that the determinant criterion is not by itself completely effective in determining independence of vector functions. However, in the situation we're mainly concerned with, the vector functions in question are solutions of an equation $\dot{\mathbf{x}} = A(t)\mathbf{x}$ with continuous entries in $A(t)$ on an interval $a < t < b$. In this case, the following theorem is an interesting step toward a decisive criterion.

1.7 ABEL'S THEOREM

Let $A(t)$ be an n-by-n matrix with entries $a_{jk}(t)$ that are continuous functions on a common interval $a < t < b$. Form an n-by-n matrix $X(t)$ using as columns known solutions $\mathbf{x}_k(t)$, $k = 1, \ldots, n$ to the differential equation $\dot{\mathbf{x}} = A(t)\mathbf{x}$. Then for some constant C,

$$\det X(t) = Ce^{\int (a_{11}(t)+\cdots+a_{nn}(t))\,dt}.$$

It follows that for $a < t < b$, $\det X(t)$ is either never zero, if $C \neq 0$, or else identically zero, if $C = 0$.

Proof. We'll show that $y(t) = \det X(t)$ satisfies the differential equation $\dot{y} = \operatorname{tr} A(t) y$, where $\operatorname{tr} A(t) = a_{11}(t) + \cdots + a_{nn}(t)$ denotes the sum of the main diagonal elements of $A(t)$ and is called the **trace** of the matrix. Denoting the jth column entry in $\mathbf{x}_k$ by x_{jk}, we can write the statement that $\mathbf{x}_k$ satisfies the vector differential equation as

$$\dot{x}_{jk} = \sum_{l=0}^{n} a_{jl} x_{lk} \qquad j = 1, \ldots, n.$$

It is a simple consequence of the chain rule for functions of several variables (see Exercise 9), that the derivative of a determinant is the sum of the determinants obtained by differentiating one row at a time. Thus the derivative of $\det X$ with respect to t becomes

$$\frac{d}{dt}\det X = \begin{vmatrix} \dot{x}_{11} & \dot{x}_{12} & \cdots & \dot{x}_{1n} \\ x_{21} & x_{22} & \cdots & x_{2n} \\ \vdots & \vdots & & \vdots \\ x_{n1} & x_{n2} & \cdots & x_{nn} \end{vmatrix} + \begin{vmatrix} x_{11} & x_{12} & \cdots & x_{1n} \\ \dot{x}_{21} & \dot{x}_{22} & \cdots & \dot{x}_{2n} \\ \vdots & \vdots & & \vdots \\ x_{n1} & x_{n2} & \cdots & x_{nn} \end{vmatrix} + \cdots + \begin{vmatrix} x_{11} & x_{12} & \cdots & x_{1n} \\ x_{21} & x_{22} & \cdots & x_{2n} \\ \vdots & \vdots & & \vdots \\ \dot{x}_{n1} & \dot{x}_{n2} & \cdots & \dot{x}_{nn} \end{vmatrix}.$$

Now replace the entries $\dot{x}_{jk}$ by the previously displayed sum to get

$$\begin{vmatrix} \sum_{l=1}^{n} a_{1l}x_{l1} & \sum_{l=1}^{n} a_{1l}x_{l2} & \cdots & \sum_{l=1}^{n} a_{1l}x_{ln} \\ x_{21} & x_{22} & \cdots & x_{2n} \\ \vdots & \vdots & & \vdots \\ x_{n1} & x_{n2} & \cdots & x_{nn} \end{vmatrix} + \begin{vmatrix} x_{11} & x_{12} & \cdots & x_{1n} \\ \sum_{l=1}^{n} a_{2l}x_{l1} & \sum_{l=1}^{n} a_{2l}x_{l2} & \cdots & \sum_{l=1}^{n} a_{2l}x_{ln} \\ \vdots & \vdots & & \vdots \\ x_{n1} & x_{n2} & \cdots & x_{nn} \end{vmatrix} + \cdots$$

$$+ \begin{vmatrix} x_{11} & x_{12} & \cdots & x_{1n} \\ x_{21} & x_{22} & \cdots & x_{2n} \\ \vdots & \vdots & & \vdots \\ \sum_{l=1}^{n} a_{nl}x_{l1} & \sum_{l=1}^{n} a_{nl}x_{l2} & \cdots & \sum_{l=1}^{n} a_{nl}x_{ln} \end{vmatrix}.$$

In the first of these determinants subtract a_{12} times the second row from the first, leaving the value of the determinant unchanged, then a_{13} times the third row from the first, and finally, a_{1n} times the last row from the first. This leaves the first row with entries $a_{11}x_{11}, a_{11}x_{12}, \ldots, a_{11}x_{1n}$. Factoring a_{11} from the first row then leaves $a_{11} \det X$. Now treat each of the other $n-1$ determinants analogously, leaving successively $a_{22} \det X, \ldots, a_{nn} \det X$. Adding all n of these gives $(d/dt) \det X = (a_{11} + \cdots + a_{nn}) \det X$. Thus $\det X$ solves the first-order linear differential equation we claimed it would, and solving using an exponential multiplier for an integrating factor gives the formula for $\det X(t)$ as stated. ■

example 8

If the matrix A in $\dot{\mathbf{x}} = A\mathbf{x}$ has constant entries, or even just diagonal entries that are constant, then Abel's formula becomes $\det X(t) = Ce^{(a_{11}+\cdots+a_{nn})t}$.

example 9

Part (ii) of Example 7 features an $X(t)$ with column solutions in the exponential matrix of the matrix

$$A = \begin{pmatrix} 1 & 1 & 0 \\ 0 & 1 & 0 \\ 0 & 0 & 3 \end{pmatrix}, \quad \text{where} \quad e^{tA} = \begin{pmatrix} e^t & te^t & 0 \\ 0 & e^t & 0 \\ 0 & 0 & e^{3t} \end{pmatrix}.$$

Abel's Theorem tells us that $\det e^{tA} = Ce^{(1+1+3)t} = Ce^{5t}$. Since $e^{0A} = I$ and $\det I = 1$, C must equal 1. Interchanging two columns in the exponential solution matrix gives us another matrix $X(t)$ with $\det X(t) = -e^{tA}$, so $C = -1$ for this example.

Recall that a square matrix whose columns are linearly independent vector solutions of a vector equation $\dot{\mathbf{x}} = A(t)\mathbf{x}$ is called a *fundamental matrix* for the differential equation. If the matrix A has only constant entries then we saw in Section 1D that the matrix e^{tA} is a fundamental matrix for the equation. Interchanging two columns in e^{tA} would give us another fundamental matrix $X(t)$

for the same system, but this one would lack the nice algebraic properties of the exponential matrix such as $e^{tA}e^{sA} = e^{(t+s)A}$. At any rate we've seen that the general solution to $\dot{\mathbf{x}} = A\mathbf{x}$ when A has constant entries has the form $\mathbf{x}(t) = e^{tA}\mathbf{c}$. The next theorem shows that a similar result holds for fundamental matrices of an equation $\dot{\mathbf{x}} = A(t)\mathbf{x}$.

1.8 THEOREM

Let $X(t)$ be a fundamental matrix for an equation $\dot{\mathbf{x}} = A(t)\mathbf{x}$ with continuous coefficients on an interval $a < t < b$. Then (a) $X^{-1}(t)$ exists for each t in the interval and (b) every solution of the equation has the form $\mathbf{x}(t) = X(t)\mathbf{c}$, where $\mathbf{c}$ is a uniquely determined constant vector.

Proof. The matrix $X(t)$ has columns that are independent as functions of t, but we need to know that they are independent for each fixed t_0. Suppose not, and that for some t_0 there is a vector $\mathbf{c} \neq 0$ such that $X(t_0)\mathbf{c} = 0$. Then $\mathbf{x}(t) = X(t)\mathbf{c}$ solves the initial-value problem $\dot{\mathbf{x}} = A(t)\mathbf{x}$, $\mathbf{x}(t_0) = 0$; but the identically zero solution solves it too, so by the Uniqueness Theorem 1.1 of Chapter 5, Section 1E, $X(t)\mathbf{c}$ must be identically zero, contradicting the independence of the columns of $X(t)$ as functions of t. Hence $X(t)$ is invertible for each t, with inverse denoted by $X^{-1}(t)$.

To prove that given a solution $\mathbf{x}(t)$ there is a unique $\mathbf{c}$ such that $\mathbf{x}(t) = X(t)\mathbf{c}$ on the interval $a < t < b$, let t_0 be a point in the interval and define $\mathbf{c} = X^{-1}(t_0)\mathbf{x}(t_0)$. Then $X(t_0)\mathbf{c} = \mathbf{x}(t_0)$, and $X(t)\mathbf{c}$ is a solution that agrees with the given solution $\mathbf{x}(t)$ at t_0. By the uniqueness theorem again, $\mathbf{x}(t) = X(t)\mathbf{c}$ for all t in the interval. ■

example 10

The columns of the matrix (i) in Example 7 can't form a fundamental matrix for an equation $\dot{\mathbf{x}} = A(t)\mathbf{x}$. Even though the columns are independent as functions of t, they are not independent for each fixed t. Thus $X(t)$ is not invertible, which contradicts Theorem 1.8 for fundamental matrices.

We're now in a position to understand the full significance of the Wronskian of a square matrix whose column vectors are solutions of an equation $\dot{\mathbf{x}} = A(t)\mathbf{x}$.

1.9 THEOREM

If $X(t)$ is an arbitrary solution matrix for an equation $\dot{\mathbf{x}} = A(t)\mathbf{x}$ with continuous coefficients on an interval $a < t < b$, then $X(t)$ is a fundamental matrix for the equation if and only if the Wronskian satisfies $\det X(t_0) \neq 0$ for some t_0, and hence by Abel's theorem for all t in the interval.

Proof. If $\det X(t_0) \neq 0$ then Abel's theorem implies that $\det X(t) \neq 0$ for all t in the interval. Since $X(t)$ is then invertible, the equation $X(t)\mathbf{c} = 0$ has only the solution $\mathbf{c} = 0$, and the columns of $X(t)$ are independent as functions of t on the interval. Conversely, if $X(t)$ is a fundamental matrix, then by Theorem 1.8, every solution $\mathbf{x}(t)$ satisfies $\mathbf{x}(t) = X(t)\mathbf{c}$ for a unique constant vector $\mathbf{c}$. Hence $X(t)$ is invertible, and $\det X(t) \neq 0$. ■

example 11 We return to the three matrices in Example 7. There is no interval on which the matrix $X(t)$ in Example 7 (i) can be a fundamental matrix of a system $\dot{\mathbf{x}} = A(t)\mathbf{x}$, because det $X(t)$ is identically zero for all real t, thus contradicting Theorem 1.9, and, more immediately, contradicting Abel's Theorem 1.7.

On the other hand, Theorem 1.9 guarantees that the matrix (ii) in Example 7 is the fundamental matrix of a first-order linear system; indeed this matrix is an exponential matrix. (See Exercise 2.)

By Theorem 1.8, the matrix (iii) in Example 7 would be the fundamental matrix of a system $\dot{\mathbf{x}} = A(t)\mathbf{x}$ on each interval not containing either $t = 0$ or $t = 1$ if we could establish that $X(t)$ is actually a solution matrix. The next theorem settles this point affirmatively.

1.10 THEOREM

Assume $X(t)$ is a square matrix with continuously differentiable entries on an interval $a < t < b$ such that det $X(t) \neq 0$ on the interval. Then $X(t)$ is a fundamental matrix for the vector differential equation $\dot{\mathbf{x}} = A(t)\mathbf{x}$, where $A(t) = \dot{X}(t)X^{-1}(t)$.

Proof. Let $\mathbf{x}_k$ be the kth column of X. Then $X^{-1}\mathbf{x}_k = \mathbf{e}_k$, the column vector with 1 for kth entry and 0 elsewhere. Hence $A\mathbf{x}_k = \dot{X}X^{-1}\mathbf{x}_k = \dot{X}\mathbf{e}_k = \dot{\mathbf{x}}_k$, showing that $\mathbf{x}_k$ satisfies the differential equation $\dot{\mathbf{x}} = A\mathbf{x}$ with $A = \dot{X}X^{-1}$. ■

Our final theorem proves the existence of a fundamental matrix $X(t)$ of independent solutions for a system $\dot{\mathbf{x}} = A(t)\mathbf{x}$. Exercise 12 shows that every other fundamental matrix for the system has the form $X(t)C$ for some invertible constant matrix C.

1.11 THEOREM

If the n-by-n matrix $A(t)$ has continuous entries on an interval $a < t < b$, then the equation $\dot{\mathbf{x}} = A(t)\mathbf{x}$ has an n-by-n fundamental matrix $X(t)$. Every solution $\mathbf{x}(t)$ of $\dot{\mathbf{x}} = A(t)$ has the form $X(t)\mathbf{c} = c_1\mathbf{x}_1(t) + c_2\mathbf{x}_2(t) + \cdots + c_n\mathbf{x}_n(t)$, where $\mathbf{c}$ is the column vector with constant entries $c_1, c_2, \ldots, c_n$.

Proof. Let t be a point in the interval, and for $k = 1, \ldots, n$, let $\mathbf{e}_k$ be the n-dimensional column vector with 1 for the kth entry and 0 elsewhere. Theorem 1.5 on page 477 guarantees the existence of a solution $\mathbf{x}_k(t)$ to each initial-value problem $\dot{\mathbf{x}} = A(t)\mathbf{x}$, $\mathbf{x}(t_0) = \mathbf{e}_k$. Let $X(t)$ be the matrix with $\mathbf{x}_k(t)$ as kth column. Then det $X(t_0) = 1$, so Theorem 1.10 implies that $X(t)$ is a fundamental matrix for the system. ■

EXERCISES

1. For the matrix $X(t)$ in part (i) of Example 7:

   (a) Show that $X(t)$ has determinant identically equal to zero, so there is no t value for which $X^{-1}(t)$ exists.

   (b) Show that when $t = 1$, the columns are linearly dependent.

   (c) Show that the columns are linearly independent as vector-valued functions of t over an arbitrary interval $a < t < b$; that is, show that if a linear combination of the columns with constant coefficients is identically zero, then all the coefficients are zero.

2. Verify that the matrix $X(t)$ in part (ii) of Example 7 has the form e^{tA} for some constant matrix A by first identifying A as $\dot{X}(0)$ and then computing e^{tA}.

3. Consider the matrix $X(t)$ in part (iii) of Example 7.

(a) Show by using $X^{-1}(t)$ that the columns $X(t)$ are linearly independent as vector-valued functions of t over each of the three intervals $-\infty < t < 0, 0 < t < 1$, and $1 < t < \infty$, and that $X(t)$ is the fundamental matrix of some linear system over each interval.

(b) Show further that the columns of $X(t)$ are linearly independent vector functions over $-\infty < t < \infty$. Show, however, that $X(t)$ is not a fundamental matrix over $-\infty < t < \infty$.

4. (a) Prove that if a square matrix $X(t)$ has columns that are linearly independent for each fixed t on an interval, then the columns are linearly independent as functions of t on the interval.

(b) Show by example that the converse statement is false in general.

(c) Other than assuming the matrix is constant, what additional hypothesis about $X(t)$ makes the converse statement true?

5. Consider the system

$$\begin{pmatrix} \dot{x} \\ \dot{y} \end{pmatrix} = \begin{pmatrix} 0 & 0 \\ -(1-t)^{-1} & (1-t)^{-1} \end{pmatrix} \begin{pmatrix} x \\ y \end{pmatrix}.$$

(a) By solving the system, find a fundamental matrix $X(t)$ for the system on every t interval that doesn't contain $t = 1$.

(b) Compute the Wronskian of the result of part (a) by Abel's formula, and also directly from the solutions.

6. Consider the system

$$\begin{pmatrix} \dot{x} \\ \dot{y} \end{pmatrix} = \begin{pmatrix} 1 & 0 \\ t & 1 \end{pmatrix} \begin{pmatrix} x \\ y \end{pmatrix}.$$

(a) By solving the system, find a fundamental matrix $X(t)$ for the system for $-\infty < t < \infty$. Your choice for $X(t)$ may look vaguely like an exponential matrix e^{tA} for some constant matrix A, so show that it is not one.

(b) Compute the Wronskian of the result of part (a) by Abel's formula, and also directly from the solutions.

7. Show that if $X(t)$ is an n-by-n matrix with continuously differentiable entries that is also invertible for each t in an interval $a < t < b$, then $X(t)$ is a fundamental matrix for the system $\dot{\mathbf{x}} = A(t)\mathbf{x}$, where $A(t) = \dot{X}(t)X^{-1}(t)$. (This technique shows that it's relatively easy to make up systems with a given set of independent solutions, though the equations may not be so easy to solve a priori.) [*Hint:* Observe that if $\mathbf{x}_k(t)$ is the kth column of $X(t)$, then $X^{-1}(t)\mathbf{x}_k(t)$ is the vector with 1 in the kth entry and 0 elsewhere.]

8. **(a)** Show that if $X(t)$ is an arbitrary fundamental matrix for the system $\dot{\mathbf{x}} = A(t)\mathbf{x}$, then $A(t) = \dot{X}(t)X^{-1}(t)$. [*Hint:* Show that if $\mathbf{x}_k$ is the kth column of X, then $\dot{\mathbf{x}}_k = \dot{X}X^{-1}\mathbf{x}_k$ and hence that $\dot{X} = AX$.]

(b) Illustrate part (a) twice, using the fundamental matrices $\begin{pmatrix} e^t & te^t \\ 0 & e^t \end{pmatrix}$ and $\begin{pmatrix} te^t & e^t \\ e^t & 0 \end{pmatrix}$.

9. The derivative of a determinant $\det X(t)$ with differentiable entries is the sum of the determinants obtained by differentiating the successive rows of $\det X(t)$ one at a time. Prove this by replacing t by $u_j(t)$ in the jth row of $X(t) = \left(a_{jk}(t)\right)$, applying the chain rule for a function of several variables, then replacing $u_j(t)$ by t after carrying out each partial differentiation in

$$\frac{d}{dt}\det\left(a_{jk}(u_j(t))\right) = \sum_j \frac{\partial}{\partial u_j}\det\left(a_{jk}(u_j(t))\right)\frac{du_j(t)}{dt}.$$

10. Write out in detail the proof of Abel's theorem 1.7 for the special case of a 2 by 2 matrix $X(t)$.

11. Show that Abel's formula can be written

$$\det X(t) = \det X(t_0)e^{\int_{t_0}^t \operatorname{tr}A(u)\,du},$$

where $\operatorname{tr}A(t)$, called the *trace* of $A(t)$, is the sum of the main diagonal elements of $A(t)$.

12. **(a)** Let $X(t)$ and $Y(t)$ be two fundamental matrices for a system $\dot{\mathbf{x}} = A(t)\mathbf{x}$, where $A(t)$ has entries continuous on an interval $a < t < b$. Show that there is a constant invertible matrix C such that $Y(t) = X(t)C$. [*Hint:* If $\mathbf{y}_k(t)$ is the kth column of $Y(t)$, then $\mathbf{y}_k(t) = X(t)\mathbf{c}_k$ for some constant vector $\mathbf{c}_k$.]

(b) Illustrate part (a) by finding a constant matrix C that relates the two matrices in Exercise 8, part (b).

13. Show how Theorem 1.1 on existence and uniqueness of solutions in Chapter 5, Section 1E, implies that the initial-value problem $\dot{\mathbf{x}} = A(t)\mathbf{x}$, $\mathbf{x}(t_0) = \mathbf{x}_0$ has a unique solution on an interval $a < t < b$ containing t_0 whenever the entries in the matrix $A(t)$ are continuous functions on the interval.

14. Wronskians sometimes appear in the study of linear differential equations of the form $\ddot{y} + a(t)\dot{y} + b(t)y = 0$. (See Chapter 9, Section 2.) We first rewrite the equation as a system, letting $\dot{y} = z$:

$$\begin{aligned} \dot{y} &= z \\ \dot{z} &= -b(t)y - a(t)z, \end{aligned} \qquad \text{or} \qquad \begin{pmatrix} \dot{y} \\ \dot{z} \end{pmatrix} = \begin{pmatrix} 0 & z \\ -b(t)y & -a(t)z \end{pmatrix}.$$

(a) Show that the Wronskian of two solutions of the system is $w(t) = \det\begin{pmatrix} y_1(t) & y_2(t) \\ \dot{y}_1(t) & \dot{y}_2(t) \end{pmatrix}$, called the **Wronskian** of two scalar solutions $y_1(t)$ and $y_2(t)$ of the second-order differential equation.

(b) Show that in this context Abel's formula becomes $w(t) = ce^{-\int a(t)\,dt}$ for c constant.

(c) Use part (b) to show that $w(t) = w(t_0)e^{-\int_{t_0}^{t} a(u)\,du}$ for fixed t_0 in the domain of $a(t)$ and $b(t)$.

2 COMPUTING e^{tA} IN GENERAL

The method described below is usually simpler than appealing directly to the definition of e^{tA}. Also, the method works with little change if the matrix A has multiple eigenvalues λ_k satisfying $\det(A - \lambda I) = 0$, and in particular doesn't require us to compute eigenvectors as we did in Chapter 5, Section 3. Our goal is finding numerical coefficient functions $b_j(t)$, depending on the eigenvalues $\lambda_1, \ldots, \lambda_n$ of an n-by-n matrix A, and such that

2.1
$$e^{tA} = b_0(t)I + b_1(t)A + b_2(t)A^2 + \cdots + b_{n-1}(t)A^{n-1}.$$

For example, if A is a 3-by-3 matrix, we'll be able to express the exponential matrix in the form $e^{tA} = b_0(t)I + b_1(t)A + b_2(t)A^2$, where $A^0 = I$ is the 3-by-3 identity matrix. An intuitive justification for Equation 2.1 follows from the following remarkable theorem, proved in Appendix B, Section 2, as Theorem 2.10. This theorem allows us to express nth and higher powers of A in terms of lower powers.

2.2 CAYLEY-HAMILTON THEOREM

If A is an n-by-n matrix with nth degree characteristic polynomial

$$P(\lambda) = \det(A - \lambda I),$$

then the matrix polynomial obtained by substituting A^k for λ^k in $P(\lambda)$ satisfies $P(A) = 0$, with the understanding that $A^0 = I$ replaces $\lambda^0 = 1$ in the substitution.

example 1

The Cayley-Hamilton theorem shows that a 2-by-2 matrix A satisfies a quadratic equation. Suppose $A = \begin{pmatrix} 1 & 1 \\ 0 & 1 \end{pmatrix}$, with characteristic polynomial

$$\det(A - \lambda I) = (1 - \lambda)^2 = \lambda^2 - 2\lambda + 1.$$

The Cayley-Hamilton theorem asserts that $A^2 - 2A + I$ is the zero matrix, or equivalently that $A^2 = 2A - I$. Multiplying the last equation by A and using the expression for A^2 gives

$$A^3 = 2A^2 - A = 2(2A - I) - A = 3A - 2I.$$

Continuing in this way, all higher powers of A are also expressible as first-degree polynomials in A. We'll see in Theorem 2.3 that for every n-by-n matrix A the power series for e^{tA} can also be reduced to an $(n-1)$-degree polynomial in A as in Equation 2.1.

The most important point about Equation 2.1 is that the infinite sum that defines e^{tA} reduces to a finite sum. Using the Cayley-Hamilton theorem the successive additions to coefficients of terms involving the lower powers $I, A, \dots, A^{n-1}$ accumulate to finite sums in the form of Equation 2.1. Rather than pursuing this approach directly, the next theorem shows how to compute the coefficients $b_j(t)$ in Equation 2.1 by solving a single system of n nonhomogeneous linear equations.

2.3 THEOREM

The coefficient functions $b_j(t)$ in the representation

$$e^{tA} = b_0(t)I + b_1(t)A + b_2(t)A^2 + \cdots + b_{n-1}(t)A^{n-1}$$

satisfy the equations

$$e^{t\lambda_k} = b_0(t) + b_1(t)\lambda_k + b_2(t)\lambda_k^2 + \cdots + b_{n-1}(t)\lambda_k^{n-1}, \qquad k = 1, 2, \dots, n, \tag{1}$$

where the numbers λ_k are the eigenvalues of the matrix A. If some m of the eigenvalues λ_k are equal then for each such multiple eigenvalue, say $\lambda_1 = \cdots = \lambda_m$, we include in the system to be solved $m-1$ additional equations evaluated with λ equal to the repeated eigenvalue after each differentiation:

$$\frac{d^\mu}{d\lambda^\mu} e^{t\lambda} = \frac{d^\mu}{d\lambda^\mu}\left[b_0(t) + b_1(t)\lambda + b_2(t)\lambda^2 + \cdots + b_{n-1}(t)\lambda^{n-1}\right],$$
$$\text{for } \mu = 1, \dots, m-1. \tag{2}$$

REMARK 1

The representation of e^{tA} in Equation 2.1 is

$$e^{tA} = b_0(t)I + b_1(t)A + b_2(t)A^2 + \cdots + b_{n-1}(t)A^{n-1},$$

so we obtain the basic Equations (1) in the theorem by deleting the identity matrix I and successively replacing A by its eigenvalues λ_k.

REMARK 2

The coefficients $b_j(t)$ are evidently determined just by a knowledge of the eigenvalues of A and their multiplicities, so these $b_j(t)$ serve equally well for representing e^{tB} if B has the same eigenvalues and corresponding multiplicities that A has.

Proof. We're looking for conditions on the coefficients $b_j(t)$ of a polynomial in λ that we'll write as $E(t,\lambda) = \sum_{j=0}^{n-1} b_j(t)\lambda^j$ so as to make $e^{tA} - E(t,A) = 0$ for all t. Note that whatever the $b_j(t)$ are, the difference $h(t,\lambda) = e^{t\lambda} - E(t,\lambda)$ is a power series in λ that converges for all λ, with convergence as rapid as that of the exponential power series; the reason is that subtracting the polynomial $E(t,\lambda)$ from $e^{t\lambda}$ affects only finitely many terms in the infinite series. We'd like to factor the characteristic polynomial $P(\lambda)$ of A out of $h(t,\lambda)$ to get

$$h(t,\lambda) = e^{t\lambda} - E(t,\lambda) = H(t,\lambda)P(\lambda).$$

Having done this, we would then replace λ by A throughout to get

$$e^{tA} - E(t,A) = H(t,A)P(A) = 0,$$

since $P(A)$ is the zero matrix by the Cayley-Hamilton theorem. This is our desired result, but the factorization is valid if and only if $H(t,\lambda)$ is finite at all roots $\lambda = \lambda_k$ of $P(\lambda_k)$. Thus we require that the quotient $\left[e^{t\lambda} - E(t,\lambda)\right]/P(\lambda)$ remain finite at the roots λ_k of $P(\lambda)$. For a roots of multiplicity 1, our requirement is just Equations (1), which say that the numerator is zero when λ equals a λ_k. In the case of a multiple root λ_1 of multiplicity m, we need to cancel the multiple root in the denominator with a multiple zero value in the numerator. This is achieved by requiring the numerator's derivatives of order 1 through $m-1$ to be zero at λ_1. Thus the Taylor expansion of the numerator $e^{t\lambda} - E(t,\lambda)$ about λ_1 will start with the power $(\lambda - \lambda_1)^m$. The result is that this power will cancel the same factor in the denominator. But the condition that these derivatives of $e^{t\lambda} - E(t,\lambda)$ be zero at $\lambda = \lambda_1$ is just a rewriting of Equations (2). ■

example 2

The matrix $A = \begin{pmatrix} 1 & 1 \\ 4 & 1 \end{pmatrix}$ has characteristic equation $\lambda^2 - 2\lambda - 3 = 0$, as shown in Example 1 of Chapter 5, Section 3. The eigenvalues are the roots $\lambda_1 = -1$, $\lambda_2 = 3$. Equations (1) of Theorem 2.3 are then

$$\begin{aligned} e^{-t} &= b_0(t) - b_1(t), \\ e^{3t} &= b_0(t) + 3b_1(t). \end{aligned}$$

Subtract the first equation from the second to get $b_1(t) = -\frac{1}{4}e^{-t} + \frac{1}{4}e^{3t}$. Then $b_0(t) = b_1(t) + e^{-t} = \frac{3}{4}e^{-t} + \frac{1}{4}e^{3t}$. Plugging these coefficient functions into Equation 2.1 along with A gives

$$\begin{aligned} e^{tA} &= \left(\tfrac{3}{4}e^{-t} + \tfrac{1}{4}e^{3t}\right)\begin{pmatrix} 1 & 0 \\ 0 & 1 \end{pmatrix} + \left(-\tfrac{1}{4}e^{-t} + \tfrac{1}{4}e^{3t}\right)\begin{pmatrix} 1 & 1 \\ 4 & 1 \end{pmatrix} \\ &= \begin{pmatrix} \frac{1}{2}e^{-t} + \frac{1}{2}e^{3t} & -\frac{1}{4}e^{-t} + \frac{1}{4}e^{3t} \\ -e^{-t} + e^{3t} & \frac{1}{2}e^{-t} + \frac{1}{2}e^{3t} \end{pmatrix}. \end{aligned}$$

As a partial check on the accuracy of our computation we can verify that our expression for e^{tA} equals the identity matrix I when $t = 0$. Note that we didn't need to have computed eigenvectors as we did when we computed e^{tA} in Example 3 of Section 1C.

example 3

In Example 2 of Chapter 5, Section 3A we saw that the matrix $\begin{pmatrix} 1 & -1 \\ 1 & 1 \end{pmatrix}$ has eigenvalues $\lambda_1 = 1 + i$ and $\lambda_2 = 1 - i$. Equations (1) of Theorem 2.3 are then

$$\begin{aligned} e^{(1+i)t} &= b_0(t) + b_1(t)(1+i), \\ e^{(1-i)t} &= b_0(t) + b_1(t)(1-i). \end{aligned}$$

Subtracting the second equation from the first gives

$$2ib_1(t) = e^t(e^{it} - e^{-it}), \quad \text{so} \quad b_1(t) = \frac{1}{2i}e^t(e^{it} - e^{-it}) = e^t \sin t.$$

Substituting for $b_1(t)$ in the first equation gives

$$b_0(t) = -(1+i)e^t \sin t + e^t(\cos t + i \sin t) = e^t(\cos t - \sin t).$$

Substituting A for λ gives

$$\exp\left(t\begin{pmatrix} 1 & -1 \\ 1 & 1 \end{pmatrix}\right) = e^t(\cos t - \sin t)\begin{pmatrix} 1 & 0 \\ 0 & 1 \end{pmatrix} + e^t \sin t \begin{pmatrix} 1 & -1 \\ 1 & 1 \end{pmatrix}$$
$$= \begin{pmatrix} e^t \cos t & -e^t \sin t \\ e^t \sin t & e^t \cos t \end{pmatrix}.$$

example 4

The matrix $A = \begin{pmatrix} 0 & 1 \\ -4 & -4 \end{pmatrix}$ has characteristic equation $\lambda^2 + 4\lambda + 4 = 0$ with eigenvalues $\lambda_1 = \lambda_2 = -2$. Equations (1) of Theorem 2.3 are identical:

$$e^{t\lambda} = b_0(t) + b_1(t)\lambda, \qquad \text{at } \lambda = -2, \quad \text{or} \quad e^{-2t} = b_0(t) - 2b_1(t).$$

Hence we need another equation satisfied by the coefficients. To find Equation (2) of Theorem 2.3, we differentiate with respect to λ in the first equation above and then set $\lambda = -2$ to get

$$te^{t\lambda} = b_1(t), \quad \text{at } \lambda = -2, \quad \text{or} \quad te^{-2t} = b_1(t).$$

We see from the last two displayed equations that $b_0(t) = e^{-2t} + 2te^{-2t}$. Equation 2.1 then becomes

$$e^{tA} = b_0(t)\begin{pmatrix} 1 & 0 \\ 0 & 1 \end{pmatrix} + b_1(t)\begin{pmatrix} 0 & 1 \\ -4 & -4 \end{pmatrix}$$
$$= (1+2t)e^{-2t}\begin{pmatrix} 1 & 0 \\ 0 & 1 \end{pmatrix} + te^{-2t}\begin{pmatrix} 0 & 1 \\ -4 & -4 \end{pmatrix}$$
$$= e^{-2t}\begin{pmatrix} 1+2t & t \\ -4t & 1-2t \end{pmatrix} = \begin{pmatrix} (1+2t)e^{-2t} & te^{-2t} \\ -4te^{-2t} & (1-2t)e^{-2t} \end{pmatrix}.$$

example 5

Let $A = \begin{pmatrix} 2 & 1 & 0 \\ 0 & 2 & 1 \\ 0 & 0 & 2 \end{pmatrix}$. The characteristic equation is $(2-\lambda)^3 = 0$, so there is a triple root $\lambda_1 = 2$. Equations (1) of Theorem 2.3 reduce to

$$e^{t\lambda} = b_0(t) + b_1(t)\lambda + b_2(t)\lambda^2, \quad \text{at } \lambda = 2, \quad \text{or} \quad e^{2t} = b_0(t) + 2b_1(t) + 4b_2(t).$$

We get two additional relations among the $b_k(t)$ by differentiating the first equation above twice with respect to λ and then setting $\lambda = 2$ after differentiation. Differentiating twice gives first $te^{t\lambda} = b_1(t) + 2b_2(t)\lambda$ and then $t^2e^{t\lambda} = 2b_2(t)$. Then setting $\lambda = 2$ gives

$$te^{2t} = b_1(t) + 4b_2(t), \qquad t^2e^{2t} = 2b_2(t).$$

Solving the last three displayed equations in reverse order for the $b_k(t)$ gives

$$b_2(t) = \tfrac{1}{2}t^2e^{2t}, \qquad b_1(t) = (t - 2t^2)e^{2t}, \qquad b_0(t) = (1 - 2t + 2t^2)e^{2t}.$$

Equation 2.1 is then $e^{tA} = b_0(t)I + b_1(t)A + b_2(t)A^2$, or

$$e^{tA} = (1 - 2t + 2t^2)e^{2t}\begin{pmatrix} 1 & 0 & 0 \\ 0 & 1 & 0 \\ 0 & 0 & 1 \end{pmatrix} + (t - 2t^2)e^{2t}\begin{pmatrix} 2 & 1 & 0 \\ 0 & 2 & 1 \\ 0 & 0 & 2 \end{pmatrix} + \frac{1}{2}t^2e^{2t}\begin{pmatrix} 4 & 4 & 1 \\ 0 & 4 & 4 \\ 0 & 0 & 4 \end{pmatrix}$$

$$= e^{2t}\begin{pmatrix} 1 & t & \frac{1}{2}t^2 \\ 0 & 1 & t \\ 0 & 0 & 1 \end{pmatrix}.$$

EXERCISES

Solve $\dot{\mathbf{x}} = A\mathbf{x}$ for the matrices in Exercises 1 through 6 by finding e^{tA}; the general solution is then a linear combination of the columns of e^{tA}.

1. $A = \begin{pmatrix} 8 & -3 \\ 10 & -3 \end{pmatrix}$.

2. $A = \begin{pmatrix} 8 & 9 \\ -4 & -4 \end{pmatrix}$.

3. $A = \begin{pmatrix} 1 & 1 & 2 \\ 0 & 1 & -1 \\ 0 & 0 & 2 \end{pmatrix}$.

4. $A = \begin{pmatrix} 1 & 1 & 1 \\ 0 & 1 & 0 \\ 0 & 0 & 2 \end{pmatrix}$.

5. $A = \begin{pmatrix} 1 & 1 & 1 & 0 \\ 0 & 0 & -1 & 0 \\ 1 & 1 & 2 & 0 \\ 1 & 0 & -1 & 1 \end{pmatrix}$.

6. $A = \begin{pmatrix} 1 & 0 & 0.5 & -0.5 \\ 1 & 0 & -1 & 0 \\ 0 & 2 & 2.5 & 0.5 \\ -1 & 1 & 0.5 & 1.5 \end{pmatrix}$.

Find the solutions to the equations $\dot{\mathbf{x}} = A\mathbf{x}$ determined by the matrices in Exercises 1 through 6 that satisfy the corresponding initial conditions in Exercises 7 through 12.

7. $\mathbf{x}(0) = \begin{pmatrix} 2 \\ -3 \end{pmatrix}$.

8. $\mathbf{x}(1) = \begin{pmatrix} 1 \\ 1 \end{pmatrix}$.

9. $\mathbf{x}(0) = \begin{pmatrix} 0 \\ 1 \\ 0 \end{pmatrix}$.

10. $\mathbf{x}(0) = \begin{pmatrix} 2 \\ 1 \\ 1 \end{pmatrix}$.

11. $\mathbf{x}(0) = \begin{pmatrix} 0 \\ 1 \\ 0 \\ 1 \end{pmatrix}$.

12. $\mathbf{x}(0) = \begin{pmatrix} 2 \\ 0 \\ 1 \\ 0 \end{pmatrix}$.

13. Find the matrix of the following normal-form system, then solve the system.

$$\frac{dx}{dt} = 2x \qquad + z,$$
$$\frac{dy}{dt} = -x + 3y + z,$$
$$\frac{dz}{dt} = -x \qquad + 4z.$$

[*Hint:* $\lambda = 3$ is a triple eigenvalue.]

14. Find the matrix of the following normal-form system, then solve the system

$$\begin{aligned}\frac{dx}{dt} &= 2x + z,\\ \frac{dy}{dt} &= y + w,\\ \frac{dz}{dt} &= 2z + w,\\ \frac{dw}{dt} &= -y + w.\end{aligned}$$

15. Let $A = \begin{pmatrix} \alpha & 1 \\ 0 & \beta \end{pmatrix}$, where α and β are real numbers.

(a) Show that if $\alpha \neq \beta$ then A has two linearly independent eigenvectors.

(b) Show that if $\alpha = \beta$ then the only eigenvectors of A are of the form $\mathbf{u} = \begin{pmatrix} c \\ 0 \end{pmatrix}$, $c \neq 0$.

(c) Compute e^{tA} in each of the two cases.

16. Show by using the definition of e^{tA} as a matrix power series that $e^{tI} = e^t I$.

17. Theorem 2.3 shows that the coefficients $b_k(t)$ in an expansion

$$e^{tA} = \sum_{k=0}^{n-1} b_k(t) A^k$$

are completely determined by the eigenvalues of the n-by-n matrix A. For example, the matrices $\begin{pmatrix} 1 & \beta \\ 0 & 2 \end{pmatrix}$ and $\begin{pmatrix} 2 & \beta \\ 0 & 1 \end{pmatrix}$ both have characteristic polynomial with 1 and 2 as roots. Hence the $b_k(t)$ are the same for all these matrices regardless of the value of the matrix entry β.

(a) Compute $b_0(t)$ and $b_1(t)$ for the two matrices above, and use them to find the corresponding exponential matrices, each depending on the parameter β.

(b) Compute the exponential matrix for $\begin{pmatrix} \alpha & \beta \\ 0 & \alpha \end{pmatrix}$.

18. Use Theorem 2.2 to compute A^2 and A^3 if $A = \begin{pmatrix} 1 & 2 \\ 3 & 4 \end{pmatrix}$.

19. Here is an alternative way to derive Equations (1) of Theorem 2.3. Let $\mathbf{v}_k$ be an eigenvector of A corresponding to eigenvalue λ_k: $A\mathbf{v}_k = \lambda_k \mathbf{v}_k$, $\mathbf{v}_k \neq 0$.

(a) Apply the matrix sum on the right side of Equation 2.1 to $\mathbf{v}_k$ to get

$$e^{tA}\mathbf{v}_k = \sum_{j=0}^{n-1} b_j(t) A^j \mathbf{v}_k = \left(\sum_{j=0}^{n-1} b_j(t) \lambda_k^j \right) \mathbf{v}_k.$$

(b) Apply the matrix power series for e^{tA} to $\mathbf{v}_k$ to get another expression for $e^{tA}\mathbf{v}_k$:

$$e^{tA}\mathbf{v}_k = \lim_{N\to\infty} \sum_{j=0}^{N} \frac{t^j}{j!}(A^j\mathbf{v}_k) = e^{t\lambda_k}\mathbf{v}_k.$$

(c) Conclude from (a) and (b) that Equations (1) hold.

20. Show that Equations (2) of Theorem 2.3 can be written more directly for computing, though not so easy to remember, as

$$t^{\mu}e^{t\lambda} = \sum_{j=\mu}^{n-1} \frac{j!}{(j-\mu)!} b_j(t)\lambda^{j-\mu}, \qquad \text{at } \lambda = \lambda_1 \quad \text{for} \quad \mu = 1, \ldots, m-1.$$

21. Equations 3.1 of Chapter 6, Section 3A, govern the linear motions of two spring-linked masses. The system specializes to $\ddot{x} = -2x + y$, $\ddot{y} = x - 2y$ in text Example 1 of that section.

(a) Let $\dot{x} = z$ and $\dot{y} = w$, and write the system in first-order normal form.

(b) Identify the matrix A of the system in part (a).

(c) Find the eigenvalues of the matrix A, for example using expansion of $\det(A - \lambda I)$ by the first row.

(d) Use Theorem 2.3 of the present section to find the general solution of the 4-dimensional system for x, y, z, and w. Then compare the results with Example 1 referred to above.

3 NONHOMOGENEOUS SYSTEMS

3A Solution Formula

To develop efficient methods for solving nonhomogeneous systems, we need a formula for the derivative of a matrix product, or, more particularly, the product of a matrix and a vector. The rule is similar to the usual product rule for real derivatives and is easily proved using that special case. The result, listed as Formula 5 in Theorem 1.1 of Appendix B, is as follows.

3.1
$$\frac{d}{dt}[A(t)B(t)] = \left[\frac{d}{dt}A(t)\right]B(t) + A(t)\left[\frac{d}{dt}B(t)\right].$$

Recall that the derivative of a matrix is just the matrix obtained by differentiating the entries. However, the order of the factors on the right is important because it involves matrix multiplication, which is not in general commutative.

To solve the nonhomogeneous vector equation

$$\frac{d\mathbf{x}}{dt} = A\mathbf{x} + \mathbf{b}(t),$$

where A is a constant matrix, we mimic what we did in the case where $\mathbf{x}$, A, and $\mathbf{b}$ were scalar-valued, with $\mathbf{b}(t)$ continuous. We first write

$$\frac{d\mathbf{x}}{dt} - A\mathbf{x} = \mathbf{b}(t).$$

Then multiply by the exponential matrix integrating factor e^{-tA} to get

$$e^{-tA}\frac{d\mathbf{x}}{dt} - e^{-tA}A\mathbf{x} = e^{-tA}\mathbf{b}(t).$$

But by the product rule for differentiation of matrices, this is the same as

$$\frac{d}{dt}(e^{-tA}\mathbf{x}) = e^{-tA}\mathbf{b}(t).$$

Integration of both sides gives, with constant vector $\mathbf{c}$,

$$e^{-tA}\mathbf{x}(t) = \int e^{-tA}\mathbf{b}(t)\,dt + \mathbf{c}.$$

In specific examples the integral of the vector $e^{-tA}\mathbf{b}(t)$ is performed one entry at a time. Recalling that e^{tA} is the inverse of e^{-tA}, we multiply by it to get the **general solution**

3.2 $$\mathbf{x}(t) = e^{tA}\int e^{-tA}\mathbf{b}(t)\,dt + e^{tA}\mathbf{c}.$$

A review of the steps used in deriving Equation 3.2 shows that the term "general solution" is justified: if $\mathbf{x}(t)$ is some solution of the given equation, then it must have the form of Equation 3.2 for some choice of the indefinite vector integral and n-dimensional constant vector $\mathbf{c}$.

---

example **1**

In Example 2 of Section 1, we saw that the homogeneous system

$$\begin{pmatrix} dx/dt \\ dy/dt \end{pmatrix} = \begin{pmatrix} 1 & 1 \\ 0 & 1 \end{pmatrix}\begin{pmatrix} x \\ y \end{pmatrix}$$

had associated with it the fundamental matrix exponential

$$e^{tA} = \begin{pmatrix} e^t & te^t \\ 0 & e^t \end{pmatrix}.$$

As an application Equation 3.2, we'll use this exponential matrix to find a particular solution of the nonhomogeneous system

$$\begin{pmatrix} dx/dt \\ dy/dt \end{pmatrix} = \begin{pmatrix} 1 & 1 \\ 0 & 1 \end{pmatrix}\begin{pmatrix} x \\ y \end{pmatrix} + \begin{pmatrix} e^t \\ e^{-t} \end{pmatrix}.$$

This system arises from adding a term to the right side of the homogeneous system. Using Equation 3.2, we compute the particular solution

$$\begin{aligned}
e^{tA}\int e^{-tA}\begin{pmatrix} e^t \\ e^{-t} \end{pmatrix}dt &= \begin{pmatrix} e^t & te^t \\ 0 & e^t \end{pmatrix}\int \begin{pmatrix} e^{-t} & -te^{-t} \\ 0 & e^{-t} \end{pmatrix}\begin{pmatrix} e^t \\ e^{-t} \end{pmatrix}dt \\
&= \begin{pmatrix} e^t & te^t \\ 0 & e^t \end{pmatrix}\int \begin{pmatrix} 1 - te^{-2t} \\ e^{-2t} \end{pmatrix}dt \\
&= \begin{pmatrix} e^t & te^t \\ 0 & e^t \end{pmatrix}\begin{pmatrix} t + \frac{1}{2}te^{-2t} + \frac{1}{4}e^{-2t} \\ -\frac{1}{2}e^{-2t} \end{pmatrix} \\
&= \begin{pmatrix} te^t + \frac{1}{4}e^{-t} \\ -\frac{1}{2}e^{-t} \end{pmatrix}.
\end{aligned}$$

Adding the particular solution found to the general homogeneous solution we found in Example 2 of Section 1 gives for the general solution

$$\mathbf{x}(t) = \begin{pmatrix} e^t & te^t \\ 0 & e^t \end{pmatrix} \begin{pmatrix} c_1 \\ c_2 \end{pmatrix} + \begin{pmatrix} te^t + \frac{1}{4}e^{-t} \\ -\frac{1}{2}e^{-t} \end{pmatrix}$$
$$= \begin{pmatrix} c_1e^t + c_2te^t + te^t + \frac{1}{4}e^{-t} \\ c_2e^t - \frac{1}{2}e^{-t} \end{pmatrix}.$$

Assuming $\mathbf{b}(t)$ is continuous on an interval $\alpha \leq t \leq \beta$, we can solve the initial-value problem with $\mathbf{x}(t_0) = \mathbf{x}_0$, where t_0 is in the interval. We make a specific choice of the integral to get

3.3
$$\mathbf{x}(t) = \int_{t_0}^{t} e^{(t-u)A}\mathbf{b}(u)\,du + e^{(t-t_0)}\mathbf{x}_0.$$

3B Nonconstant Coefficients†

Theorem 1.1 in Appendix B shows that matrix multiplication acts linearly on n-dimensional columns $\mathbf{x}$ in the sense that

$$A(c_1\mathbf{x}_1 + c_2\mathbf{x}_2) = c_1A\mathbf{x}_1 + c_2A\mathbf{x}_2.$$

This purely algebraic relation holds even if A, $\mathbf{x}_1$, and $\mathbf{x}_2$ are functions of t. In addition, d/dt acts linearly:

$$\frac{d}{dt}(c_1\mathbf{x}_1 + c_2\mathbf{x}_2) = c_1\frac{d}{dt}\mathbf{x}_1 + c_2\frac{d}{dt}\mathbf{x}_2, \quad \text{if } c_1, c_2 \text{ are constant.}$$

Allowing for the possibility that $A = A(t)$ has functions of t for entries, we define the operator $D - A(t)$ by

$$\big(D - A(t)\big)\mathbf{x} = \frac{d\mathbf{x}}{dt} - A(t)\mathbf{x}.$$

It follows that the difference $D - A(t)$ of these two operators acts linearly on vector functions $\mathbf{x}(t)$. This is helpful to know here because of what Theorem 2.1 of Chapter 5 says about solutions $\mathbf{x}(t)$ of the linear system

$$\frac{d\mathbf{x}}{dt} - A(t)\mathbf{x} = \mathbf{b}(t).$$

The theorem says in this instance that to find $\mathbf{x} = \mathbf{x}(t)$ we find a particular solution $\mathbf{x}_p(t)$ and add it to the general solution $\mathbf{x}_h(t)$ of the *homogeneous* equation

$$\frac{d\mathbf{x}}{dt} - A(t)\mathbf{x} = 0.$$

Thus $\mathbf{x}(t) = \mathbf{x}_p(t) + \mathbf{x}_h(t)$. Equation 3.2 of Section 3A is an instance of this general form for solutions when A has constant entries. Equation 3.4 on page 495 shows how to find $\mathbf{x}_p(t)$ if we already have $\mathbf{x}_h(t)$. General methods for finding $\mathbf{x}_h(t)$ when $A = A(t)$ has nonconstant entries are not available. We can in that case resort to the numerical methods of Chapter 6, Section 5.

It's interesting that even in the case of a nonconstant, continuous n-by-n matrix $A(t)$ there is a formula for a particular solution of

$$\frac{d\mathbf{x}}{dt} = A(t)\mathbf{x} + \mathbf{b}(t)$$

in terms of solutions of the associated homogeneous equation. Suppose $\mathbf{x}_1(t), \ldots, \mathbf{x}_n(t)$ is a set of n linearly independent solutions of

$$\frac{d\mathbf{x}}{dt} = A(t)\mathbf{x}.$$

Solutions would be independent, for example, if their values for some $t = t_0$ were independent vectors. We now form the n-by-n *fundamental matrix*

$$X(t) = (\mathbf{x}_1(t), \ldots, \mathbf{x}_n(t)),$$

whose columns are these independent vector solutions of the homogeneous equation $\dot{\mathbf{x}} = A\mathbf{x}$. If A has constant entries, the columns of e^{tA} will suffice; just set $\mathbf{x}_k(t) = e^{tA}\mathbf{e}_k$.

---

example 2

If A is a constant matrix, then the matrix $X(t) = e^{tA}$ is an example of a fundamental matrix, because its columns are linearly independent solutions of $d\mathbf{x}/dt = A\mathbf{x}$. In Example 1, a fundamental matrix is

$$X_1(t) = e^{tA} = \begin{pmatrix} e^t & te^t \\ 0 & e^t \end{pmatrix}.$$

Another fundamental matrix for the same system is

$$X_2(t) = \begin{pmatrix} te^t & e^t \\ e^t & 0 \end{pmatrix},$$

although $X_2(t)$ can't be an exponential matrix, because $X_2(0) \neq I$.

To find a particular solution to the nonhomogeneous equation we try to find a vector-valued function $\mathbf{v}(t)$ such that

$$\mathbf{x}_p(t) = X(t)\mathbf{v}(t)$$

is a solution of the nonhomogeneous equation. It turns out that this can be done as follows. Using the product rule for differentiation, we substitute $X(t)\mathbf{v}(t)$ into the nonhomogeneous equation to get

$$\frac{dX(t)}{dt}\mathbf{v}(t) + X(t)\frac{d\mathbf{v}(t)}{dt} = A(t)X(t)\mathbf{v}(t) + \mathbf{b}(t).$$

Since each column of $X(t)$ is a solution of the homogeneous equation, we have

$$\frac{dX(t)}{dt} = A(t)X(t).$$

Therefore, the first term cancels on each side, leaving

$$X(t)\frac{d\mathbf{v}(t)}{dt} = \mathbf{b}(t).$$

If $X(t)$ is invertible, we can multiply both sides of this last equation by $X^{-1}(t)$ to get

$$\frac{d\mathbf{v}(t)}{dt} = X^{-1}(t)\mathbf{b}(t).$$

Integration gives a formula for $\mathbf{v}(t)$ (we need only one solution $\mathbf{x}_p$ in $\mathbf{x}_h + \mathbf{x}_p$, so we omit the constant of integration):

$$\mathbf{v}(t) = \int X^{-1}(t)\mathbf{b}(t)\,dt.$$

Finally,

3.4

$$\begin{aligned}\mathbf{x}_p(t) &= X(t)\mathbf{v}(t)\\ &= X(t)\int X^{-1}(t)\mathbf{b}(t)\,dt.\end{aligned}$$

Notice that this formula is the same as the one previously derived in the constant-coefficient case, with e^{tA} now replaced by the more general $X(t)$. This process for finding $\mathbf{x}_p$ is a crucial step in extending Equation 3.2 to the case of a matrix $A(t)$ with nonconstant entries, but the real difficulty lies in finding the columns of $X(t)$.

In specific examples we can always verify that a fundamental matrix is invertible simply by finding the inverse matrix. We know that in the special case of an exponential matrix e^{tA} that the inverse is e^{-tA}. Theorem 1.8 in the optional Section 1E shows that all fundamental matrices are invertible.

example 3

The homogeneous system associated with

$$\frac{d\mathbf{x}}{dt} = \begin{pmatrix} 1 & e^t \\ 0 & 1 \end{pmatrix}\mathbf{x} + \begin{pmatrix} e^t \\ e^{2t} \end{pmatrix}$$

is uncoupled, so successive integration shows that it has independent solutions

$$\mathbf{x}_1(t) = \begin{pmatrix} e^t \\ 0 \end{pmatrix}, \qquad \mathbf{x}_2(t) = \begin{pmatrix} e^{2t} \\ e^t \end{pmatrix}.$$

We form a fundamental matrix $X(t)$ and its inverse:

$$X(t) = \begin{pmatrix} e^t & e^{2t} \\ 0 & e^t \end{pmatrix}, \qquad X^{-1}(t) = \begin{pmatrix} e^{-t} & -1 \\ 0 & e^{-t} \end{pmatrix}.$$

Formula 3.3 gives the particular solution

$$\begin{aligned}\mathbf{x}_p(t) &= \begin{pmatrix} e^t & e^{2t} \\ 0 & e^t \end{pmatrix}\int \begin{pmatrix} e^{-t} & -1 \\ 0 & e^{-t} \end{pmatrix}\begin{pmatrix} e^t \\ e^{2t} \end{pmatrix} dt\\ &= \begin{pmatrix} e^t & e^{2t} \\ 0 & e^t \end{pmatrix}\int \begin{pmatrix} 1 - e^{2t} \\ e^t \end{pmatrix} dt\\ &= \begin{pmatrix} e^t & e^{2t} \\ 0 & e^t \end{pmatrix}\begin{pmatrix} t - \frac{1}{2}e^{2t} \\ e^t \end{pmatrix} = \begin{pmatrix} te^t + \frac{1}{2}e^{3t} \\ e^{2t} \end{pmatrix}.\end{aligned}$$

The general solution is then $\mathbf{x}(t) = c_1\mathbf{x}_1(t) + c_2\mathbf{x}_2(t) + \mathbf{x}_p(t)$, or

$$\mathbf{x}(t) = c_1\begin{pmatrix} e^t \\ 0 \end{pmatrix} + c_2\begin{pmatrix} e^{2t} \\ e^t \end{pmatrix} + \begin{pmatrix} te^t + \frac{1}{2}e^{3t} \\ e^{2t} \end{pmatrix}$$

$$= \begin{pmatrix} c_1e^t + c_2e^{2t} + te^t + \frac{1}{2}e^{3t} \\ c_2e^t + e^{2t} \end{pmatrix}.$$

In this example we could have found the same solution by applying successive integration directly to the uncoupled nonhomogeneous system.

If the matrix $A(t)$ in $\dot{\mathbf{x}} = A(t)\mathbf{x} + \mathbf{b}$ is not constant, the practical significance of the fundamental matrix $X(t)$ as compared with the special case of an exponential matrix e^{tA} is that the inverse exponential matrix is obtainable simply by replacing t by $-t$ to get e^{-tA}. Since the columns of a fundamental matrix $X(t)$ will be independent for each t, the matrix $X^{-1}(t)$ will exist for each t, but its entries will typically have to be computed by matrix inversion.

3C Summary of Methods

For linear systems in the standard form $d\mathbf{x}/dt = A\mathbf{x} + \mathbf{b}$, and hence for the systems and equations reducible to this form, we usually proceed as follows:

1. Find the general solution of the homogeneous equation $d\mathbf{x}/dt = A\mathbf{x}$, either by elimination, by the eigenvector method in Chapter 5, Section 3, or by finding e^{tA} directly. In the constant-coefficient case the homogeneous solution is always of the form $\mathbf{x}_h(t) = e^{tA}\mathbf{c}$, where $\mathbf{c}$ is a constant vector. If A is not constant, there is no general method for finding $\mathbf{x}_h(t)$, and we will very likely have to use numerical methods.
2. Find a particular solution to the nonhomogeneous equation. We can do this by the elimination method, by undetermined coefficients if applicable, or by Formula 3.2 or 3.3.
3. Write the general solution as $\mathbf{x}(t) = \mathbf{x}_h(t) + \mathbf{x}_p(t)$.

EXERCISES

1. Use Formula 3.2 to solve the following equations of the form $d\mathbf{x}/dt = A\mathbf{x} + \mathbf{b}(t)$. The associated homogeneous equations $d\mathbf{x}/dt = A\mathbf{x}$ have exponential matrices e^{tA} as shown.

(a) $$\begin{pmatrix} \dot{x} \\ \dot{y} \end{pmatrix} = \begin{pmatrix} -3 & 2 \\ -4 & 3 \end{pmatrix}\begin{pmatrix} x \\ y \end{pmatrix} + \begin{pmatrix} e^{2t} \\ 1 \end{pmatrix}; e^{tA} = \begin{pmatrix} -e^t + 2e^{-t} & e^t - e^{-t} \\ -2e^t + 2e^{-t} & 2e^t - e^{-t} \end{pmatrix}.$$

(b) $$\begin{pmatrix} \dot{x} \\ \dot{y} \end{pmatrix} = \begin{pmatrix} 3 & 0 \\ 0 & 2 \end{pmatrix}\begin{pmatrix} x \\ y \end{pmatrix} + \begin{pmatrix} e^t + 1 \\ e^{-t} \end{pmatrix}; e^{tA} = \begin{pmatrix} e^{3t} & 0 \\ 0 & e^{2t} \end{pmatrix}.$$

(c) $$\begin{pmatrix} \dot{x} \\ \dot{y} \end{pmatrix} = \begin{pmatrix} 1 & 4 \\ 0 & 5 \end{pmatrix}\begin{pmatrix} x \\ y \end{pmatrix} + \begin{pmatrix} 1 \\ e^t \end{pmatrix}; e^{tA} = \begin{pmatrix} e^t & e^{5t} - e^t \\ 0 & e^{5t} \end{pmatrix}.$$

(d) $\begin{pmatrix} \dot{x} \\ \dot{y} \end{pmatrix} = \begin{pmatrix} 2 & -1 \\ 1 & 2 \end{pmatrix} \begin{pmatrix} x \\ y \end{pmatrix} + \begin{pmatrix} e^{2t} \\ 2e^{2t} \end{pmatrix}; e^{tA} = \begin{pmatrix} e^{2t}\cos t & -e^{2t}\sin t \\ e^{2t}\sin t & e^{2t}\cos t \end{pmatrix}.$

2. Find a particular solution of each of the differential equations in Exercise 1 that satisfies the following corresponding initial condition:

(a) $\mathbf{x}(0) = (-1, -1)$.

(b) $\mathbf{x}(0) = (0, 0)$.

(c) $\mathbf{x}(0) = (0, 1)$.

(d) $\mathbf{x}(0) = (-1, -1)$.

3. (a) Show that for a solution of the form

$$\mathbf{x}(t) = e^{tA}\mathbf{c}, \quad \mathbf{c} \text{ constant},$$

to satisfy the condition $\mathbf{x}(t_0) = \mathbf{x}_0$, we must have $\mathbf{c} = e^{-t_0 A}\mathbf{x}_0$.

(b) Show that if $X(t)$ is an n-by-n matrix with linearly independent columns, in particular if $X(t)$ is a fundamental matrix, then for

$$\mathbf{x}(t) = X(t)\mathbf{c}, \quad \mathbf{c} \text{ constant}$$

to satisfy $\mathbf{x}(t_0) = \mathbf{x}_0$ we must have $\mathbf{c} = X^{-1}(t_0)\mathbf{x}_0$.

4. Let $X(t)$ be a fundamental matrix such that $X(t)\mathbf{c}$ is the general solution to the homogeneous equation

$$\dot{\mathbf{x}} = A(t)x.$$

(a) Show that if $\mathbf{x}_p(t)$ is a particular solution of the nonhomogeneous system

$$\dot{\mathbf{x}} = A(t)\mathbf{x} + \mathbf{b}(t),$$

then the general solution of the nonhomogeneous system can be written $\mathbf{x}(t) = \mathbf{x}_p(t) + X(t)\mathbf{c}$, where $\mathbf{c}$ is a constant.

(b) Show that, if the general solution in part (a) is to satisfy an initial condition $\mathbf{x}(t_0) = \mathbf{x}_0$, then $\mathbf{c}$ should be chosen so that $\mathbf{c} = X^{-1}(t_0)(\mathbf{x}_0 - \mathbf{x}_p(t_0))$.

5. The systems in Exercise 1 can be solved by the method of **undetermined coefficients** as follows: Form linear combinations of the terms, and their derivatives, that occur in each entry of the nonhomogeneous part of the differential equation, taking care to include appropriate multiples by powers of t for terms that are also homogeneous solutions. Then substitute into the equation to determine the coefficients of combination. Use this method on each of the parts of Exercise 1.

6. Each of the following systems has the *homogeneous* solutions shown. Verify that these are linearly independent solutions. Find a particular solution of the nonhomogeneous equation, using Equation 3.3.

(a) $\begin{pmatrix} \dot{x} \\ \dot{y} \end{pmatrix} = \begin{pmatrix} 3/(2t) & -1/2 \\ -1/(2t^2) & 1/(2t) \end{pmatrix} \begin{pmatrix} x \\ y \end{pmatrix} + \begin{pmatrix} t^3 \\ 3t^2 \end{pmatrix}; \begin{pmatrix} t \\ 1 \end{pmatrix}, \begin{pmatrix} -t^2 \\ t \end{pmatrix}.$

(b) $\begin{pmatrix} \dot{x} \\ \dot{y} \end{pmatrix} = \begin{pmatrix} t/(t-1) & -1/(t-1) \\ 1 & 0 \end{pmatrix} \begin{pmatrix} x \\ y \end{pmatrix} + \begin{pmatrix} 1-t \\ 1-t^2 \end{pmatrix}; \begin{pmatrix} 1 \\ t \end{pmatrix}, \begin{pmatrix} e^t \\ e^t \end{pmatrix}.$

7. Find the general solution of the system of text Example 1 in Chapter 6, Section 1, using the methods of Sections 2 and 3 of this chapter.

8. Find the general solution of the system of text Example 1 in Chapter 6, Section 1, using the methods of Sections 1 and 3 of this chapter.

9. **(a)** Let $\mathbf{x}_1(t), \ldots, \mathbf{x}_n(t)$ be continuously differentiable functions, linearly independent for each t, and taking values in $\mathcal{R}^n$. Let $X(t)$ be the n-by-n matrix with columns $\mathbf{x}_1(t), \ldots, \mathbf{x}_n(t)$. Show that if we define

$$A(t) = X'(t)X^{-1}(t),$$

then the system $d\mathbf{x}/dt = A(t)\mathbf{x}$ has $\mathbf{x}_1(t), \ldots, \mathbf{x}_n(t)$ as solutions and thus has $X(t)$ as a fundamental matrix.

(b) Find a first-order homogeneous linear system of the form $d\mathbf{x}/dt = A(t)\mathbf{x}$ having

$$\mathbf{x}_1(t) = \begin{pmatrix} e^t \\ 2e^{2t} \end{pmatrix}, \qquad \mathbf{x}_2(t) = \begin{pmatrix} 1 \\ e^t \end{pmatrix}$$

as solutions. Are these two solutions linearly independent?

10. Let A be an n-by-n invertible matrix of constants, and let $\mathbf{b}$ be a fixed vector in $\mathcal{R}^n$. Show that the equation

$$\frac{d\mathbf{x}}{dt} = A\mathbf{x} + \mathbf{b}$$

always has $\mathbf{x}_p = -A^{-1}\mathbf{b}$ for a particular solution.

11. Show that if A is a constant n-by-n matrix then the equation

$$\frac{d\mathbf{x}}{dt} = A\mathbf{x}$$

has $X(t) = e^{tA}$ for its fundamental matrix of independent column solutions with $X(0) = I$.

12. Let $A(t)$ be a square matrix with entries that are differentiable on some interval.

(a) Show that

$$\frac{dA^2}{dt} = 2A\frac{dA}{dt}$$

when A and dA/dt **commute,** that is, when $(dA/dt)A = A(dA/dt)$.

(b) Show that

$$\frac{de^A}{dt} = e^A\frac{dA}{dt}$$

when A and dA/dt commute.

13. **(a)** Write out in complete detail the derivation of an analogue to Equation 3.2 to show that, for A constant and $\mathbf{b}(t)$ continuous, the initial-value problem

$$\frac{d\mathbf{x}}{dt} = A\mathbf{x} + \mathbf{b}(t), \qquad \mathbf{x}(t_0) = \mathbf{x}_0$$

has a unique solution of the form

$$\mathbf{x}(t) = e^{tA} \int_{t_0}^{t} e^{-uA}\mathbf{b}(u)\, du + e^{(t-t_0)A}\mathbf{x}_0.$$

[*Hint:* Integrate from t_0 to t instead of using an indefinite integral.]

(b) Use the result of part (a) to show that the solution to the initial-value problem with the special choice $\mathbf{x}(t_0) = 0$ is

$$\mathbf{x}(t) = \int_{t_0}^{t} e^{(t-u)A}\mathbf{b}(u)\, du.$$

The matrix factor $e^{(t-u)A}$ is a **Green's function** for the initial value problem, and plays a role analogous to that of the integral formula introduced in Chapter 3, Section 3C.

***14.** It's not true that every square matrix $M(t)$ with entries that are differentiable functions of t arises as an e^{tA}. For example, it must be the case that (i) $M(0) = I$, that (ii) $M(t+s) = M(t)M(s)$, and that (iii) $M'(t) = AM(t)$. The purpose of this exercise is to characterize those differentiable matrices $M(t)$ that are exponential matrices by showing that if (i) and (ii) hold for all real t and s, then there is a constant matrix A such that $M(t) = e^{tA}$.

(a) Show that if (ii) holds and $M(t_0)$ is invertible for some t_0, then (i) holds also.

(b) Show that if (i) and (ii) both hold and $M(t)$ has differentiable entries, then (iii) holds for some constant matrix A. [*Hint:* Apply (i) and (ii) to $(1/h)\big(M(t+h) - M(t)\big)$, and let $h \to 0$.]

(c) Show that if (i) and (iii) hold for some constant matrix A, then $M(t) = e^{tA}$. [*Hint:* Multiply both sides of (iii) by e^{-tA}, and show that $d/dt\big(e^{tA}M(t)\big)$ is the zero matrix.]

15. (a) Prove that if $F(t)$ is a differentiable real-valued function and A is a constant matrix, then

$$\frac{d}{dt}e^{F(t)A} = F'(t)Ae^{F(t)A}.$$

(b) Prove that the initial-value problem $\dot{\mathbf{x}} = f(t)A\mathbf{x}$, $x(t_0) = \mathbf{x}_0$, where A is a constant square matrix, has solution $\mathbf{x}(t) = e^{F(t)A}\mathbf{x}_0$, where $F(t) = \int_{t_0}^{t} f(u)\, du$.

4 STABILITY FOR AUTONOMOUS SYSTEMS

In Section 6 of Chapter 6 there is a detailed examination of possible behaviors for 2-dimensional autonomous systems depending on the nature of the roots of the associated characteristic polynomials. A similar analysis of higher dimensional systems quickly leads to an overwhelmingly large number of possibilities. What we'll do here instead is just establish conditions that distinguish among various

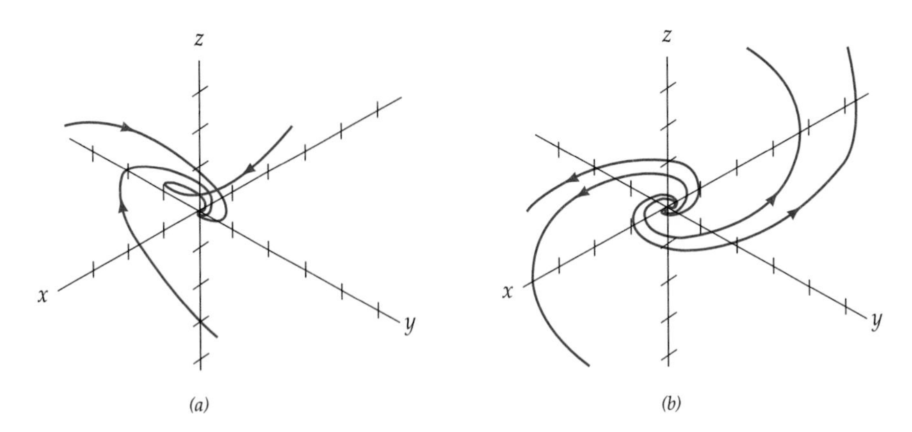

Figure 7.1 (*a*) Asymptotically stable at $\mathbf{x}_0 = 0$. (*b*) Unstable at $\mathbf{x}_0 = 0$.

kinds of stable and unstable behavior (Figure 7.1) of solutions $\mathbf{x}(t)$ to $\dot{\mathbf{x}} = \mathbf{F}(\mathbf{x})$ near an equilibrium point $\mathbf{x}_0$:

(i) $\mathbf{x}_0$ is **asymptotically stable** if there is a number $d_0 > 0$ such that every solution starting within distance d_0 of $\mathbf{x}_0$ tends to $\mathbf{x}_0$ as t tends to infinity.

(ii) $\mathbf{x}_0$ is **stable** if there is a $d_0 > 0$ such that all solutions starting within some distance $d_1 < d_0$ from $\mathbf{x}_0$ remain within distance d_0 of $\mathbf{x}_0$. Clearly asymptotic stability near an equilibrium point $\mathbf{x}_0$ implies stability near $\mathbf{x}_0$.

(iii) An equilibrium point $\mathbf{x}_0$ that is not stable is called **unstable.**

4A Linear Systems

We begin by relating stability for the solutions of an autonomous vector system $\dot{\mathbf{x}} = A\mathbf{x} + \mathbf{b}$, directly to the eigenvalues of the square matrix A. Note that the constant $\mathbf{x} = 0$ is always an equilibrium solution for the homogeneous system $\dot{\mathbf{x}} = A\mathbf{x}$, since both sides of the equation are zero if $\mathbf{x}(t) = 0$.

4.1 THEOREM

Let A be an n-by-n constant matrix. The equilibrium solution $\mathbf{x}_0 = 0$ for the homogeneous system $\dot{\mathbf{x}} = A\mathbf{x}$ is asymptotically stable if every eigenvalue of A has negative real part and is unstable if A has at least one eigenvalue with positive real part. The same conclusions hold for stability of an equilibrium solution for $\dot{\mathbf{x}} = A\mathbf{x} + \mathbf{b}$, where $\mathbf{b}$ is an n-dimensional constant vector. If all eigenvalues of A have real part zero, we can draw no conclusion for then an equilibrium solution $\mathbf{x}_0$ may be stable or may be unstable.

Proof. The method for computing the exponential matrix described in Section 2 shows that every entry in the fundamental solution matrix e^{tA} has the form $e^{\lambda t}Q(t)$, where $Q(t)$ is a polynomial and λ is an eigenvalue of A. If every such λ has negative real part, then the entries, and hence all solutions, tend to zero as t tends to $+\infty$. On the other hand, if the eigenvalue λ, with eigenvector $\mathbf{v}$, has positive real part, then the solution $\mathbf{x}(t) = \delta e^{\lambda t}\mathbf{v}$ is unbounded in every nonzero coordinate as t tends to $+\infty$, regardless of how small the positive number δ is chosen. However, if $\lambda < 0$, then $\delta e^{\lambda t}\mathbf{v} \to 0$.

If the nonhomogeneous equation has $\mathbf{x}_0$ for an equilibrium solution the homogeneous plus particular form $\mathbf{x}(t) = e^{tA}\mathbf{c} + \mathbf{x}_0$ of the general solution shows that the same conclusions hold for the nonhomogeneous equation. The last statement of the theorem is settled by checking out the two examples in Exercise 15. ■

If all eigenvalues of the matrix A have negativereal parts, then all solutions tend exponentially to 0, and the origin is accordingly called a **sink** for the system $\dot{\mathbf{x}} = A\mathbf{x}$. On the other hand, if all eigenvalues have positive real parts, the origin is called a **source,** with all nonzero trajectories traced away from the origin. If some eigenvalues are positive and all others are negative, the equilibrium at $\mathbf{x}_0 = 0$ is unstable and is called a **saddle point.**

example 1

The 3-dimensional system $\dot{x} = y,\ \dot{y} = -x - y,\ \dot{z} = x - z$ has a unique equilibrium point at the origin (0,0,0). In matrix form the system is

$$\dot{\mathbf{x}} = \begin{pmatrix} 0 & 1 & 0 \\ -1 & -1 & 0 \\ 1 & 0 & -1 \end{pmatrix} \mathbf{x}, \quad \text{with characteristic equation}$$

$$\det \begin{pmatrix} -\lambda & 1 & 0 \\ -1 & -1-\lambda & 0 \\ 1 & 0 & -1-\lambda \end{pmatrix} = 0.$$

This last equation is just $\lambda^3 + 2\lambda^2 + 2\lambda + 1 = 0$. By inspection we see that $\lambda_1 = -1$ is a root. Factoring out $\lambda + 1$ leaves the quadratic $\lambda^2 + \lambda + 1$ with roots $\lambda_2 = (-1 + \sqrt{3}i)/2$, $\lambda_3 = (-1 - \sqrt{3}i)/2$. Since all roots have negative real part, either -1 or $-\frac{1}{2}$, the equilibrium is an asymptotically stable sink by Theorem 4.1. The general solution can be computed without much trouble, but it isn't needed if we're only checking equilibrium stability.

Showing the existence of one positive eigenvalue is enough to guarantee instability.

example 2

If we replace the third equation in the previous example by $\dot{z} = x + z$, the only equilibrium point is still (0,0,0). The characteristic equation becomes

$$P(\lambda) = \det \begin{pmatrix} -\lambda & 1 & 0 \\ -1 & -1-\lambda & 0 \\ 1 & 0 & 1-\lambda \end{pmatrix} = 0 \quad \text{or} \quad P(\lambda) = -\lambda^3 + 1 = 0.$$

Clearly $\lambda_1 = 1$ is a root. Thus there is a positive eigenvalue, and by Theorem 4.1 the equilibrium at the origin is unstable for this system. Dividing $\lambda^3 - 1$ by $\lambda - 1$ leaves the quadratic $\lambda^2 + \lambda + 1$, which has roots $-\frac{1}{2}(1 \pm \sqrt{3}i)$. Since all eigenvalues have either positive or negative real parts, the origin is an unstable equilibrium point.

example 3

Setting the right sides equal to zero in the system

$$\dot{x} = x + y + 1,$$
$$\dot{y} = x - y - 1$$

yields the unique equilibrium solution $(x_0, y_0) = (0, -1)$. The matrix A of the system is $\begin{pmatrix} 1 & 1 \\ 1 & -1 \end{pmatrix}$ with characteristic equation $(1-\lambda)(-1-\lambda) - 1 = \lambda^2 - 2 = 0$. The roots are real eigenvalues $\pm\sqrt{2}$ of opposite sign, so according to Theorem 4.1 the equilibrium at $(0, -1)$ unstable and more particularly is of saddle type.

4B Nonlinear Systems

Building on the eigenvalue analysis for linear systems, we consider equilibrium points $\mathbf{x}_0 = (a_1, \ldots, a_n)$ of a nonlinear system $\dot{\mathbf{x}} = \mathbf{F}(\mathbf{x})$, that is, points such that $\mathbf{F}(\mathbf{x}_0) = 0$. The first step is to *linearize* the system at each equilibrium point $\mathbf{x}_0$, replacing each real-valued equation $\dot{x}_k = F_k(x_1, \ldots, x_n)$ in the system by its own linearization

$$\dot{x}_k = \sum_{j=1}^{n} \frac{\partial F_k}{\partial x_j}(\mathbf{x}_0)(x_j - a_j), \qquad k = 1, \ldots, x_n.$$

The resulting linear autonomous system is called the **linearization** of $\dot{\mathbf{x}} = \mathbf{F}(\mathbf{x})$ at $\mathbf{x}_0$, and we can write it as $\dot{\mathbf{x}} = F'(\mathbf{x}_0)(\mathbf{x} - \mathbf{x}_0)$ using the **derivative matrix** or **Jacobian matrix**

$$F'(\mathbf{x}_0) = \begin{pmatrix} \dfrac{\partial F_1}{\partial x_1}(\mathbf{x}_0) & \ldots & \dfrac{\partial F_1}{\partial x_n}(\mathbf{x}_0) \\ \vdots & & \vdots \\ \dfrac{\partial F_n}{\partial x_1}(\mathbf{x}_0) & \ldots & \dfrac{\partial F_n}{\partial x_n}(\mathbf{x}_0) \end{pmatrix}.$$

We'll see that the eigenvalues of the matrix $\mathbf{F}'(\mathbf{x}_0)$ are the key to our criteria, and for this reason we can work with the simpler homogeneous equation $\dot{\mathbf{x}} = \mathbf{F}'(\mathbf{x}_0)\mathbf{x}$ at $\mathbf{x}_0$, as we did in Section 3B and in Chapter 6, Section 4B.

example 4

The system

$$\dot{x} = -\sigma x + \sigma y,$$
$$\dot{y} = \rho x - y - xz,$$
$$\dot{z} = xy - \beta z$$

is a family of systems (see also Example 6) called the **general Lorenz system.** The equilibrium solutions are just the solutions to the system obtained by setting the right-hand sides equal to zero. Looking at the special case $\beta = \sigma = 1$, $\rho = 2$, we solve

$$-x + y = 0,$$
$$2x - y - xz = 0,$$
$$xy - z = 0.$$

Noting that $x = y$, it's easy to see that there are just three solutions (1,1,1), (0,0,0) and $(-1,-1,1)$. The derivative matrix $\mathbf{F}'(x,y,z)$ for the linearization at (x,y,z) is

$$\begin{pmatrix} \frac{\partial(-x+y)}{\partial x} & \frac{\partial(-x+y)}{\partial y} & \frac{\partial(-x+y)}{\partial z} \\ \frac{\partial(2x-y-xz)}{\partial x} & \frac{\partial(2x-y-xz)}{\partial y} & \frac{\partial(2x-y-xz)}{\partial z} \\ \frac{\partial(xy-z)}{\partial x} & \frac{\partial(xy-z)}{\partial y} & \frac{\partial(xy-z)}{\partial z} \end{pmatrix} = \begin{pmatrix} -1 & 1 & 0 \\ 2-z & -1 & -x \\ y & x & -1 \end{pmatrix}.$$

Evaluating this matrix at $(x,y,z) = (1,1,1)$ gives

$$\mathbf{F}'(1,1,1) = \begin{pmatrix} -1 & 1 & 0 \\ 2-z & -1 & -x \\ y & x & -1 \end{pmatrix}_{(1,1,1)} = \begin{pmatrix} -1 & 1 & 0 \\ 1 & -1 & -1 \\ 1 & 1 & -1 \end{pmatrix}.$$

Similarly, the linearization matrices at (0,0,0) and $(-1,-1,1)$ are obtained by evaluating the same derivative matrix at these two additional points:

$$\mathbf{F}'(0,0,0) = \begin{pmatrix} -1 & 1 & 0 \\ 2 & -1 & 0 \\ 0 & 0 & -1 \end{pmatrix}, \quad \mathbf{F}'(-1,-1,1) = \begin{pmatrix} -1 & 1 & 0 \\ 1 & -1 & 1 \\ -1 & -1 & -1 \end{pmatrix}.$$

The next theorem enables us to draw conclusions about equilibrium stability from the eigenvalues of the matrices in the previous example. We omit the proof.

4.2 THEOREM

Assume that the real-valued coordinate functions F_k of $\mathbf{F}$ are continuously differentiable and that the system $\dot{\mathbf{x}} = \mathbf{F}(\mathbf{x})$ has an equilibrium point at $\mathbf{x}_0$. The equilibrium solution $\mathbf{x}_0$ for the system is asymptotically stable if every eigenvalue of the derivative matrix $\mathbf{F}'(\mathbf{x}_0)$ has negative real part, and is unstable if $\mathbf{F}'(\mathbf{x}_0)$ has at least one eigenvalue with positive real part. If all eigenvalues of $\mathbf{F}'(\mathbf{x}_0)$ have real part zero, we can draw no definite conclusion, and the equilibrium $\mathbf{x}_0$ may be stable or may be unstable.

---

example 5

The special Lorenz system treated in the previous example has an equilibrium at (1,1,1) with characteristic polynomial $P(\lambda) = \det(\mathbf{F}'(1,1,1) - \lambda I)$. From Example 3 we see that

$$P(\lambda) = \det \begin{pmatrix} -1-\lambda & 1 & 0 \\ 1 & -1-\lambda & -1 \\ 1 & 1 & -1-\lambda \end{pmatrix}.$$

Computing the determinant, we get $P(\lambda) = -[(\lambda+1)^3+1]$, and we see by inspection that $\lambda_1 = -2$ is a root. Division by $\lambda+2$ gives $P(\lambda) = -(\lambda+2)(\lambda^2+\lambda+1)$. The roots of the quadratic factor are $\lambda_2 = (-1+\sqrt{3}i)/2$ and $\lambda_3 = (-1-\sqrt{3}i)/2$. Thus the real parts of all three eigenvalues are negative, so we conclude from Theorem 4.2 that (1,1,1) is an asymptotically stable equilibrium solution.

example 6

Continuing with the special Lorenz system, we examine the equilibrium at (0,0,0). The relevant characteristic polynomial is $P(\lambda) = \det(\mathbf{F}'(0,0,0) - \lambda I)$, or

$$\det\begin{pmatrix} -1-\lambda & 1 & 0 \\ 2 & -1-\lambda & 0 \\ 0 & 0 & -1-\lambda \end{pmatrix}.$$

The determinant is computed to be $-(\lambda+1)^3+2\lambda+2 = -(\lambda+1)(\lambda^2+2\lambda-1)$. The roots are $\lambda_1 = -1$, $\lambda_2 = -1-\sqrt{2}$, and $\lambda_3 = -1+\sqrt{2}$. Since $\lambda_3 > 0$ we conclude from Theorem 4.2 that (0,0,0) is an unstable equilibrium. This point is in fact a saddle point, since there are two negative eigenvalues that contribute to making the other basic solutions tend to zero. Checking out the equilibrium at $(-1,-1,1)$ is left as an exercise.

example 7

The *general Lorenz system* is

$$\dot{x} = \sigma(y-x), \qquad \dot{y} = \rho x - y - xz, \qquad \dot{z} = -\beta z + xy,$$

$$\beta, \rho, \sigma \text{ positive constants.}$$

This system has been studied extensively with the aim of understanding trajectories such as the one shown in Figure 7.2. With the choice of parameters shown there, the equilibrium points, aside from the obvious one at the origin, are at $(\pm 6\sqrt{2}, \pm 6\sqrt{2}, 27)$. The trajectory shown in the figure has initial point (2,2,21). It winds around in the area of one equilibrium an apparently random number of times, then switching to the other equilibrium with similar behavior. The eigenvalues of the linearizations are the same at the two equilibrium points; they are approximately as follows: $\lambda_1 \approx -13.85$, $\lambda_2, \lambda_3 \approx 0.09 \pm 10.19i$. Thus these two points are saddle points, and it can be shown that each one has a surface containing it on which all trajectories gradually spiral away from the point, as

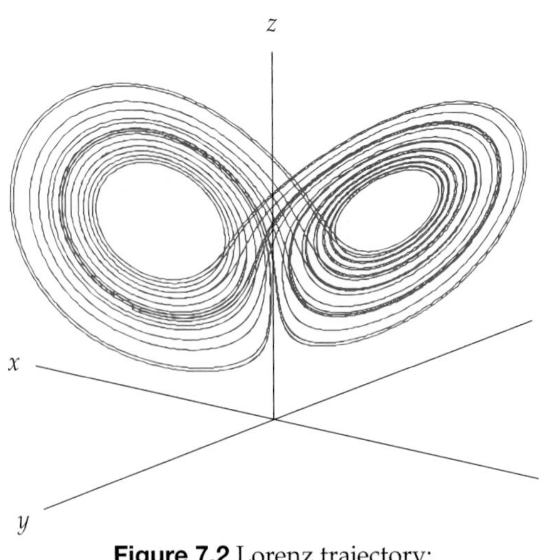

Figure 7.2 Lorenz trajectory: $\beta = \frac{8}{3}$, $\rho = 28$, $\sigma = 10$.

well as a trajectory transverse to the surface that converges to the point. The typical trajectory behavior lies somewhere between these extremes, winding away from one equilibrium until it is attracted by the other, then reversing. The number of circuits about each point, and the path taken, is very sensitive to minute changes in the initial conditions. This issue is taken up numerically in Exercise 18 of Chapter 6, Section 5.

Edward Lorenz began the study of this system by using it in the approximation of more complicated differential equations in the study of weather patterns; hence the interest in the system's sensitivity to initial conditions, sometimes called the butterfly effect. For more details about the Lorenz system see Colin Sparrow, *The Lorenz Equations: Bifurcations, Chaos, and Strange Attractors*, Springer (1982).

EXERCISES

Find the eigenvalues of the square matrices A in Exercises 1 through 4. Then use that information to decide about the stability or instability of the zero-solution to the associated linear system $\dot{\mathbf{x}} = A\mathbf{x}$.

1. $\begin{pmatrix} 2 & 1 \\ 1 & -3 \end{pmatrix}$.

2. $\begin{pmatrix} 4 & -3 \\ 3 & -4 \end{pmatrix}$.

3. $\begin{pmatrix} 1 & 1 & -1 \\ 0 & 1 & 3 \\ 0 & 0 & -1 \end{pmatrix}$.

4. $\begin{pmatrix} -6 & -1 & -2 \\ -2 & -4 & -2 \\ 2 & -1 & -2 \end{pmatrix}$.

Find all equilibrium points of the systems $\dot{\mathbf{x}} = \mathbf{F}(\mathbf{x})$ in Exercises 5 through 8. For each equilibrium $\mathbf{x}_0$, compute the derivative matrix $\mathbf{F}'(\mathbf{x}_0)$. Using the eigenvalues of $\mathbf{F}'(\mathbf{x}_0)$ try to determine the stability or instability of each equilibrium.

5. $\dot{x} = x - y^2 + 1,$
$\dot{y} = x^2 - 1.$

6. $\dot{x} = x - y,$
$\dot{y} = x^2 + y^2 - 1.$

7. $\dot{x} = y,$
$\dot{y} = e^x - 1.$

8. $\dot{x} + \dot{y} = y,$
$\dot{x} - \dot{y} = x.$

Find all equilibrium points of the systems $\dot{\mathbf{x}} = \mathbf{F}(\mathbf{x})$ in Exercises 9 and 10. For each equilibrium $\mathbf{x}_0$, compute the derivative matrix $\mathbf{F}'(\mathbf{x}_0)$. Using the eigenvalues of $\mathbf{F}'(\mathbf{x}_0)$ try to determine the stability or instability of each equilibrium.

9. $\dot{x} = -x - y^2,$
$\dot{y} = -y,$
$\dot{z} = -z - x^2.$

10. $\dot{x} = x - y,$
$\dot{y} = x^2 + y^2 - 1,$
$\dot{z} = z.$

11. Find the characteristic polynomial of the derivative matrix $\mathbf{F}'(-1,-1,1)$ associated with the equilibrium $(-1,-1,1)$ of the special Lorenz system in Example 3 of the text. Then find the eigenvalues of $\mathbf{F}'(-1,-1,1)$ and use the information to decide about the stability of $(-1,-1,1)$.

12. The general Lorenz system is

$$\dot{x} = \sigma(y - x), \qquad \dot{y} = \rho x - y - xz, \qquad \dot{z} = -\beta z + xy,$$
$$\beta, \rho, \sigma \text{ positive constants.}$$

(a) Show that (i) if $0 < \rho \leq 1$, the only equilibrium point is (0,0,0), and (ii) if $\rho > 1$, there are three equilibrium points: (0,0,0) and $\left(\pm\sqrt{\beta(\rho - 1)}, \pm\sqrt{\beta(\rho - 1)}, \rho - 1\right)$.

(b) Show that at the equilibrium points $\left(\pm\sqrt{\beta(\rho - 1)}, \pm\sqrt{\beta(\rho - 1)}, \rho - 1\right)$, $\rho > 1$, the characteristic polynomial of the linearized system is

$$P(\lambda) = \lambda^3 + (\beta + \sigma + 1)\lambda^2 + \beta(\rho + \sigma)\lambda + \beta(\rho - 1).$$

Explain why there must be at least one negative eigenvalue.

(c) Use Newton's method or a careful graphical analysis to show that with the choices $\beta = \frac{8}{3}$, $\rho = 28$, $\sigma = 10$, $P(\lambda)$ has a negative root λ_1 at about -13.85.

(d) Use the result of part (c) to show that $P(\lambda)$ has complex conjugate roots with positive real part about 0.09. Conclude that these two nonzero equilibrium points are saddle points. [*Hint:* Divide $P(\lambda)$ by $\lambda - \lambda_1$.]

13. Consider the second-order system $\ddot{x} = -4x + y^2$, $\ddot{y} = -y + 2xy$.

(a) Show that the assumption $y(t) = 0$ for all t enables us to infer the existence of periodic solutions for some initial conditions.

(b) Let $\dot{x} = z$, and $\dot{y} = w$ to get an equivalent first-order system. Find all equilibrium solutions $\big(x(t), y(t)\big)$ of the second-order system.

(c) Use linearization to show instability of the nonzero equilibrium point.

(d) Show that the origin is stable for the first-order system. Do this by showing that the system is Hamiltonian, with Hamiltonian function

$$H(x, y, z, w) = \tfrac{1}{2}(4x^2 + y^2 + z^2 + w^2) - xy^2,$$

having the origin as an isolated minimum. (See Chapter 5, Section 4C on Liapunov's method for this part.)

14. The second-order **Hénon-Heilas system** is

$$\ddot{x} = -x - 2xy, \qquad \ddot{y} = -y - x^2 + y^2.$$

(a) Show that the assumption $x(t) = 0$ for all t enables us to infer the existence of periodic solutions for some initial conditions.

(b) Let $\dot{x} = z$ and $\dot{y} = w$ to get an equivalent first-order system. Find all equilibrium solutions $\big(x(t), y(t)\big)$ of the second-order system.

(c) Use linearization to show instability of the nonzero equilibrium points.

(d) Show that the origin is stable for the first-order system. Do this by showing that the system is Hamiltonian, with Hamiltonian function

$$H(x, y, z, w) = \tfrac{1}{2}(x^2 + y^2 + z^2 + w^2) + x^2 y - \tfrac{1}{3}y^3,$$

having the origin as an isolated minimum. (See Chapter 5, Section 4C on Liapunov's method.)

15. Use the systems $\dot{\mathbf{x}} = A\mathbf{x}$ with the respective matrices

$$\begin{pmatrix} 0 & 1 & 0 & 0 \\ 0 & 0 & 1 & 0 \\ 0 & 0 & 0 & 1 \\ -1 & 0 & -2 & 0 \end{pmatrix} \quad \text{and} \quad \begin{pmatrix} 0 & 1 & 0 & 0 \\ 0 & 0 & 1 & 0 \\ 0 & 0 & 0 & 1 \\ -4 & 0 & -5 & 0 \end{pmatrix}$$

to confirm the last statement in Theorem 4.1.

16. Do Exercise 12 in Chapter 5, Section 4B, to confirm the last statement in Theorem 4.2.

17. An equilibrium point $\mathbf{x}_0$ of a system $\dot{\mathbf{x}} = \mathbf{F}(\mathbf{x})$ is called **hyperbolic** if the eigenvalues of the linearization at $\mathbf{x}_0$ all have nonzero real parts; if in addition eigenvalues of both signs occur and all are either positive or neagative, then $\mathbf{x}_0$ is called a **saddle point.** Of the types I through VIII in Chapter 5, Section 4, (a) which are of hyperbolic type and which are also of saddle type? (b) Which are sources and which are sinks?

CHAPTER REVIEW AND SUPPLEMENT

This chapter concerns solutions of linear systems of the form $\dot{\mathbf{x}} = A(t)\mathbf{x} + \mathbf{b}(t)$, where $A(t)$ is a square matrix and $\mathbf{b}(t)$ is a column vector, both of which have entries that are continuous functions of t on a fixed interval $\alpha \le t \le \beta$. A *fundamental matrix* for the system is by definition a square matrix $X(t)$ of the the same size as $A(t)$ such that the columns of $X(t)$ are linearly independent vector solutions of the associated *homogeneous system* $\dot{\mathbf{x}} = A(t)\mathbf{x}$.

Fundamental matrices are most widely applicable when the matrix A of the system has all its entries constant. In that case the exponential matrix defined by

$$e^{tA} = I + \frac{1}{1!}tA + \frac{1}{2!}t^2A^2 + \frac{1}{3!} + \cdots + \frac{1}{k!}t^kA^k + \cdots$$

is a fundamental matrix for $\dot{\mathbf{x}} = A\mathbf{x}$. This infinite series of matrices converges in each entry for all real and complex values of t, with the following consequences:

(a) $\dfrac{d}{dt}e^{tA} = Ae^{tA} = e^{tA}A.$

(b) $e^{(t+s)A} = e^{tA}e^{sA} = e^{sA}e^{tA}$, for all real or complex numbers t and s.

(c) $e^{0A} = I$, and e^{tA} is invertible, with $e^{-tA}e^{tA} = e^{tA}e^{-tA} = I$.

(d) The initial-value problem $\dot{\mathbf{x}} = A\mathbf{x}$, $\mathbf{x}(t_0) = \mathbf{x}_0$ has the unique solution $\mathbf{x}(t) = e^{(t-t_0)A}\mathbf{x}_0$.

(e) If $\mathbf{b}(t)$ is continuous the initial-value problem $\dot{\mathbf{x}} = A\mathbf{x} + \mathbf{b}(t)$, $\mathbf{x}(t_0) = \mathbf{x}_0$ has the unique solution $\mathbf{x}(t) = e^{(t-t_0)A}\mathbf{x}_0 + \int_{t_0}^{t} e^{(t-u)A}\mathbf{b}(u)\,du$.

Computing e^{tA}. This computation is important enough that over a dozen distinct ways of doing it are known; we discuss only three, of which the first two have rather limited applicability. The third method is quite general and is useful both for hand computation and for computer algebra implementation.

1. If the powers of A exhibit some fairly regular pattern it may be possible to compute e^{tA} directly from the infinite series definition, as in Example 1 of Section 1B.
2. If you can compute not only the eigenvalues $\lambda_1, \lambda_2, \dots, \lambda_n$ of A but can also find corresponding linearly independent eigenvectors $\mathbf{u}_1, \mathbf{u}_2, \dots, \mathbf{u}_n$, then the matrix U with the eigenvectors as columns is invertible. If Λ_t is the diagonal matrix with diagonal entries $e^{\lambda_1 t}, e^{\lambda_2 t}, \dots, e^{\lambda_n t}$, as described in Section 1C, then $e^{tA} = U\Lambda_t U^{-1}$.
3. If you can compute the eigenvalues $\lambda_1, \lambda_2, \dots, \lambda_n$ of A, then the exponential matrix can always be represented, as described in Section 2, as a sum of n terms

$$e^{tA} = b_0(t)I + b_1(t)A + b_2(t)A^2 + \cdots + b_{n-1}(t)A^{n-1},$$

where the coefficients $b_j(t)$ must satisfy the closely related equations

$$e^{t\lambda_k} = b_0(t) + b_1(t)\lambda_k + b_2(t)\lambda_k^2 + \cdots + b_{n-1}(t)\lambda_k^{n-1}, \quad k = 1, 2, \dots, n.$$

If some m of the eigenvalues are equal, say $\lambda_1 = \lambda_2 = \cdots = \lambda_m$, then $m - 1$ additional equations satisfied by the coefficients $b_j(t)$ replace the redundant equations. We obtain these additional $m - 1$ equations by differentiating the equation

$$e^{t\lambda} = b_0(t) + b_1(t)\lambda + b_2(t)\lambda^2 + \cdots + b_{n-1}(t)\lambda^{n-1}$$

successively $m - 1$ times with respect to λ and then replacing λ by λ_1 after each differentiation. Now solve the resulting system of n nonhomogeneous linear equations for the coefficients $b_j(t)$ in the finite sum expression for e^{tA}.

EXERCISES

For each of the following matrices A: (a) Find the exponential matrix e^{tA} and (b) characterize the equilibrium solution $\mathbf{x}_0 = 0$ of $\dot{\mathbf{x}} = A\mathbf{x}$.

1. $\begin{pmatrix} 1 & -1 \\ 1 & 1 \end{pmatrix}$ **2.** $\begin{pmatrix} 2 & 1 \\ 1 & 2 \end{pmatrix}$ **3.** $\begin{pmatrix} 2 & -1 \\ 3 & -2 \end{pmatrix}$

4. $\begin{pmatrix} 4 & -3 \\ 8 & -6 \end{pmatrix}$ **5.** $\begin{pmatrix} 0 & 1 \\ -3 & 4 \end{pmatrix}$ **6.** $\begin{pmatrix} 1 & 0 & 0 \\ 0 & 2 & 1 \\ 0 & 0 & 1 \end{pmatrix}$

7. $\begin{pmatrix} 1 & 0 & 0 \\ 0 & 1 & 1 \\ 0 & 0 & 1 \end{pmatrix}$ **8.** $\begin{pmatrix} 1 & 0 & 0 \\ 0 & 2 & 0 \\ 0 & 0 & 3 \end{pmatrix}$ **9.** $\begin{pmatrix} 0 & 0 & 0 \\ 1 & 0 & 1 \\ 1 & 0 & 0 \end{pmatrix}$

10. Explain why $\begin{pmatrix} 1 & te^t \\ te^{-t} & 1 \end{pmatrix}$ can't be an exponential matrix e^{tA}.

11. Solve the system $\dot{\mathbf{x}} = A\mathbf{x} + \mathbf{b}(t)$, where A is the matrix in Exercise 1 and $\mathbf{b}(t) = \begin{pmatrix} 0 \\ t \end{pmatrix}$.

12. Solve the system $\dot{\mathbf{x}} = A\mathbf{x} + \mathbf{b}$, where A is the matrix in Exercise 6 and $\mathbf{b}$ is a constant vector with each entry equal to 1.

13. Let $A = \begin{pmatrix} a & 1 \\ 0 & a \end{pmatrix}$ where a is an arbitrary real number.

(a) Compute A^2 and show inductively that $A^n = \begin{pmatrix} a^n & na^{n-1} \\ 0 & a^n \end{pmatrix}$.

(b) Use part (a) and the series definition of e^{tA} to show that

$$e^{tA} = \begin{pmatrix} e^{at} & te^{at} \\ 0 & e^{at} \end{pmatrix}.$$

CHAPTER 8 Laplace Transforms

1 BASIC PROPERTIES

We used the techniques described in Chapter 3 to find formulas for the solutions of initial-value problems such as

$$\ddot{y} + a\dot{y} + by = f(t), \qquad y(0) = y_0, \qquad \dot{y}(0) = z_0,$$

where the coefficients a and b are constant. We use the overdot notation for time derivatives, because the independent variable t stands for time in the most frequent areas of application for the method described in this chapter. The operator $D^2 + aD + b$ associated with the left side of the differential equation doesn't itself depend on the time variable t, so the operator is called **time-invariant.** It is these time-invariant operators that the Laplace transform method was designed to deal with, and for a reason that we'll illustrate in Example 4 the method has become particularly popular in electrical and mechanical engineering. Recall from Chapter 3 that solving the initial-value problem was broken into three steps:

(i) By assuming there is an exponential solution e^{rt} to the associated homogeneous equation, we transformed the problem of solving the homogeneous equation into the algebraic problem of finding the roots of the differential operator's *characteristic polynomial*

$$P(r) = r^2 + ar + b.$$

(ii) We found a particular solution of the nonhomogeneous equation by judiciously guessing its form and computing coefficients, or optionally by computing an integral transform

$$y_p(t) = \int_0^t g(t-u)f(u)\,du.$$

(iii) Finally we added the results of (i) and (ii), and adjusted the coefficients in the homogeneous solution to satisfy the initial conditions.

If the forcing function $f(t)$ is integrable on every finite interval $0 \le t \le T$, and if $f(t)$ and the desired solution $y(t)$ don't grow too rapidly as t tends to ∞, we can use an integral transform, the *Laplace transform*, that replaces a function $f(t)$ by a function $F(s)$ of a real or complex variable s. Furthermore, the Laplace transform method will convert the entire initial-value problem at $t = 0$ into an equation involving functions of s, an equation that involves both the characteristic polynomial $P(s)$ and the Laplace transform $Y(s)$ of the desired solution $y(t)$. It will always be easy to solve the resulting equation for $Y(s)$, so the final step will then be a matter of using a table of transforms to identify the solution $y(t)$ having transform $Y(s)$. We'll see that counterparts of steps (i), (ii), and (iii) from Chapter 3 are all recognizable in the Laplace transform method.

Definition. If for some real or complex numbers s, the integral

$$\mathcal{L}[f](s) = \int_0^\infty e^{-st} f(t)\,dt = \lim_{T\to\infty} \int_0^T e^{-st} f(t)\,dt$$

converges to a function $\mathcal{L}[f](s)$ depending on s, then $F(s) = \mathcal{L}[f](s)$ is called the **Laplace transform** of the function f.

If a is constant, the function $f(t) = e^{at}$ has a special role to play, as it does in the method of Chapter 3 outlined above, so we'll begin by calculating its Laplace transform:

1.1
$$\mathcal{L}[e^{at}](s) = \int_0^\infty e^{-st} e^{at}\,dt = \frac{1}{s-a}, \quad \text{if } s > a.$$

Note that both a and s are held fixed during the integration with respect to t. To calculate the integral over the infinite interval, we calculate the partial integral

$$\begin{aligned}\int_0^T e^{-(s-a)t}\,dt &= \left[-\frac{1}{s-a} e^{-(s-a)t}\right]_0^T \\ &= -\frac{1}{s-a} e^{-(s-a)T} + \frac{1}{s-a}.\end{aligned}$$

We assumed $s - a > 0$, so the first term tends to zero as $T \to \infty$, leaving $1/(s-a)$.

Transforming Differential Equations. The key to using the Laplace transform to solve differential equations is the following formula:

1.2
$$\int_0^\infty e^{-st} \dot{y}(t)\,dt = s \int_0^\infty e^{-st} y(t)\,dt - y(0), \quad \text{or}$$
$$\mathcal{L}[\dot{y}](s) = s\mathcal{L}[y](s) - y(0).$$

We prove this under two assumptions: (a) the improper integral on the right is convergent and (b) $\lim_{t\to\infty} e^{-st} y(t) = 0$. Indeed, integration by parts of the partial integral from 0 to T gives

$$\int_0^T e^{-st}\dot{y}(t)\,dt = [e^{-st}y(t)]_0^T + s\int_0^T e^{-st}y(t)\,dt$$
$$= e^{-sT}y(T) - y(0) + s\int_0^T e^{-st}y(t)\,dt.$$

By our assumptions, the terms on the right tend to $-y(0) + s\mathcal{L}[y](s)$ as $T \to \infty$. Hence $\mathcal{L}[\dot{y}](s)$ exists for those values of s for which $\mathcal{L}[y](s)$ exists, and Equation 1.2 is valid for the same values of s. Assumptions (a) and (b) raise a significant question, because we may not know whether a solution $y(t)$ satisfies these conditions or not until we've found the solution. A standard way to deal with the question is to assume that the forcing function $f(t)$ of a constant-coefficient equation $L(y) = f(t)$ has **exponential order** as t tends to infinity, which means that there are constants γ and C such that $|f(t)| \leq Ce^{\gamma t}$ as t tends to $+\infty$. The assumption that $f(t)$ has exponential order guarantees that the solution $y(t)$ has exponential order and hence that both $\mathcal{L}[f](s)$ and $\mathcal{L}[y](s)$ are finite for large enough s. (See Exercises 38 and 39.) The alternative approach that people often use to deal with the question is to ignore it temporarily, find a solution and observe that it's correct. Such checks should be routine anyway.

example 1

The example

$$\dot{y} + 2y = 0, \qquad y(0) = 3,$$

is too simple to show the real advantages of using Laplace transforms, but it does illustrate the general principles involved. Step one is to express the differential equation in terms of Laplace transforms by multiplying both sides by e^{-st} and then integrating from 0 to ∞ with respect to t. We can write the result as

$$\int_0^\infty e^{-st}\dot{y}\,dt + 2\int_0^\infty e^{-st}y\,dt = 0.$$

Equation 1.2 enables us to rewrite the first integral, obtaining

$$-3 + s\int_0^\infty e^{-st}y\,dt + 2\int_0^\infty e^{-st}y\,dt = 0.$$

Here we have used the assumption that $y(0) = 3$. We also rely on our knowledge of the exponential nature of the solution to justify assumptions (a) and (b) needed for the application of Equation 1.2. We solve the previous equation for the Laplace transform of the solution $y(t)$ to get

$$\int_0^\infty e^{-st}y(t)\,dt = \frac{3}{s+2}. \tag{1}$$

Thus we have found not the solution $y(t)$, but its Laplace transform. However, if we set $a = -2$ in Equation 1.1 and then multiply by 3, we get

$$\int_0^\infty e^{-st}3e^{-2t}\,dt = \frac{3}{s+2}. \tag{2}$$

Since our solution $y(t)$ and the functions $3e^{-2t}$ have the same Laplace transform, it's natural to want to conclude that $y(t) = 3e^{-2t}$. We'll see from Theorem 1.3 below that this conclusion is justified.

When we use formulas such as Equation 1.1 to identify solutions from their transforms, it will be important for us to know in general that if a solution $y(t)$ has Laplace transform $Y(s)$, then $y(t)$ is the only function with transform $Y(s)$. Thus $y(t)$ will necessarily be the desired solution. This conclusion follows from the following theorem, along with the observation that solutions of differential equations are differentiable and therefore continuous.

1.3 UNIQUENESS THEOREM
If a continuous function $y(t)$ has Laplace transform $Y(s)$, then $y(t)$ is the only continuous function that has Laplace transform $Y(s)$. (Exercise 40 describes just one of several ways to prove this theorem.)

1A Solving Equations

The essence of the Laplace transform method is to use the transform to reduce the solution of an initial-value problem to some routine algebraic manipulations, as we did in Chapter 3. Example 1 already showed that a distinctive feature of the method is that it enables us to include initial values in the calculations from the very start.

To apply the Laplace transform to differential equations with order higher than one, we need a simple extension of Equation 1.2. Applying Equation 1.2 to computing $\mathcal{L}[\ddot{y}](s)$, the Laplace transform of $\ddot{y}$, gives

$$\mathcal{L}[\ddot{y}](s) = s\mathcal{L}[\dot{y}] - \dot{y}(0).$$

Now apply Equation 1.2 again, to the first term, to get

$$\begin{aligned}\mathcal{L}[\ddot{y}](s) &= s\{s\mathcal{L}[y](s) - y(0)\} - \dot{y}(0)\\ &= s^2\mathcal{L}[y](s) - \dot{y}(0) - sy(0).\end{aligned}$$

The same formal routine leads after n steps to the formula

1.4 $\mathcal{L}[y^{(n)}](s) = s^n\mathcal{L}[y](s) - y^{(n-1)}(0) - sy^{(n-2)}(0) - \cdots - s^{n-1}y(0).$

Of course, assumptions (a) and (b) used for Equation 1.2 have to be increased. We assume

(A) The integrals $\int_0^\infty e^{-st}y^{(k)}(t)\,dt$ are convergent for $k = 1, 2, \ldots, n$.
(B) $\lim_{t\to\infty} e^{-st}y^{(k)}(t) = 0$ for $k = 0, 1, \ldots, n-1$.

For simplicity in solving equations, we'll continue to denote the Laplace transform of $y(t)$ by $Y(s)$, of $f(t)$ by $F(s)$, and so on. Thus $Y(s) = \mathcal{L}[y](s)$ and $F(s) = \mathcal{L}[f](s)$. Since the most frequently occurring differential equations encountered in applying Laplace transforms are of order 2, we display the relevant versions of Equation 1.4 as they are commonly used:

1.4′ $\mathcal{L}[\dot{y}](s) = sY(s) - y(0)$, and $\mathcal{L}[\ddot{y}](s) = s^2Y(s) - \dot{y}(0) - sy(0)$.

Integration from 0 to ∞ and multiplication by e^{-st} both act linearly on functions $y(t)$, so from Equations 1.4′ it follows that the initial-value problem $\ddot{y}+a\dot{y}+by = f(t)$, $y(0) = y_0$, $\dot{y}(0) = z_0$ transforms into

$$P(s)Y(s) = B(s) + F(s),$$

where $P(s)=s^2+as+b$ is the characteristic polynomial of the operator D^2+aD+b, $F(s)$ is the transform of the forcing function $f(t)$, and $B(s)$ is a first-degree polynomial in s depending on the initial conditions. We'll carry this computation out step by step in the next example. (See also Exercise 34.)

---

example 2

We can solve the initial-value problem

$$\ddot{y} - \dot{y} - 2y = 3e^t, \qquad y(0) = 1, \qquad \dot{y}(0) = 0,$$

by applying the Laplace transform to both sides. Using Equations 1.4′ for $Y(s) = \mathcal{L}[y](s)$ and $\mathcal{L}[\dot{y}](s)$, together with the initial conditions, we get

$$\mathcal{L}[\dot{y}] = sY(s) - y(0) = sY(s) - 1,$$
$$\mathcal{L}[\ddot{y}] = s^2Y(s) - \dot{y}(0) - sy(0) = s^2Y(s) - s.$$

We can write the equation

$$\mathcal{L}[\ddot{y} - \dot{y} - 2y] = \mathcal{L}[3e^t] \quad \text{as} \quad \mathcal{L}[\ddot{y}] - \mathcal{L}[\dot{y}] - 2\mathcal{L}[y] = 3\mathcal{L}[e^t].$$

The expressions for $\mathcal{L}[y]$, $\mathcal{L}[\dot{y}]$, and $\mathcal{L}[\ddot{y}]$ in terms of Y displayed above, together with Equation 1.1, enable us to write this last equation as

$$[s^2Y(s) - s] - [sY(s) - 1] - 2[Y(s)] = 3\frac{1}{s-1}.$$

Rearrangement separates out $Y(s)$ times the characteristic polynomial $P(s)$ of the differential operator $D^2 - D + 2$ in the original differential equation, namely

$$(s^2 - s - 2)Y(s) = \frac{3}{s-1} + s - 1 = \frac{s^2 - 2s + 4}{s-1}.$$

Hence

$$Y(s) = \frac{s^2 - 2s + 4}{(s-1)(s^2 - s - 2)}.$$

Having found an expression for $Y(s)$, our problem is now to identify precisely the solution $y(t)$ that satisfies $\mathcal{L}[y](s) = Y(s)$. Because $Y(s)$ is a rational function, we can theoretically always break it down according to the partial fraction decomposition associated with the computation of indefinite integrals. Thus the decomposition is possible because the denominator of $Y(s)$ factors. We need to determine the coefficients A, B, and C in

$$\frac{s^2 - 2s + 4}{(s-1)(s+1)(s-2)} = \frac{A}{s-1} + \frac{B}{s+1} + \frac{C}{s-2}.$$

Multiplying through by $(s-1)$, and then setting $s=1$, gives $A=-\frac{3}{2}$. Similarly, we multiply by $(s+1)$ and then set $s=-1$ to get $B=\frac{7}{6}$. Finally, we multiply by $(s-2)$ and then set $s=2$ to get $C=\frac{4}{3}$. As a result, we have

$$Y(s) = -\frac{\frac{3}{2}}{s-1} + \frac{\frac{7}{6}}{s+1} + \frac{\frac{4}{3}}{s-2}.$$

Equation 1.1 now allows us to identify $y(t)$ as

$$y(t) = -\tfrac{3}{2}e^{t} + \tfrac{7}{6}e^{-t} + \tfrac{4}{3}e^{2t}.$$

For example, the first term comes from using Equation 1.1 with $a = 1$. As a check on the computation, you can verify that $y(0)=1$ and $\dot{y}(0)=0$.

In the examples given previously, we used the linearity of the Laplace transform operator $\mathcal{L}$. This property follows from the linearity of the operation of integration, and is formally expressed by the equation

1.5 $$\mathcal{L}[c_1y_1 + c_2y_2] = c_1\mathcal{L}[y_1] + c_2\mathcal{L}[y_2], \qquad c_1, c_2 \text{ constant.}$$

Furthermore, since the transform $\mathcal{L}$ acts linearly, so does the **inverse transform,** denoted $\mathcal{L}^{-1}$, that takes us back from $Y(s)$ to $f(t)$. Thus, for example, if c_1 and c_2 are constants we have

1.6 $$\mathcal{L}^{-1}[c_1Y_1(s) + c_2Y_2(s)] = c_1y_1(t) + c_2y_2(t).$$

Equations 1.5 and 1.6, together with Equations 1.4′ for the Laplace transforms of $\dot{y}$ and $\ddot{y}$, need to be supplemented in practice by the calculation of Laplace transforms of specific functions, as is done in Equation 1.1, which asserts that $\mathcal{L}[e^{at}](s) = 1/(s-a)$. Table 1 is a specialized integral table containing enough entries to do the exercises.

It's important to note that the tables of Laplace transforms intended for solving differential equations are meant to be used in both directions, from function $f(t)$ to transform $\mathcal{L}[f](s) = F(s)$ and also back the other way. Some entries that may appear superfluous when reading the table left to right for finding a transform turn out to be quite useful when reading right to left for finding an inverse transform. More elaborate tables may contain several hundred entries. Since tables are used in both directions, the entry

$$\mathcal{L}[t^n] = \frac{n!}{s^{n+1}}, \qquad n = 0, 1, 2, \ldots$$

not only provides the transform of $f(t) = t^n$, it also provides, after division by $n!$, the inverse transform, $\mathcal{L}^{-1}[s^{-(n+1)}] = t^n/n!$, $n = 0, 1, 2\ldots.$ All entries in Table 1 are computable using elementary integration techniques, sometimes using the **partial fraction decomposition** described in general below. Computer algebra applications such as *Maple* follow the rules automatically.

In finding inverse transforms, it is sometimes essential to decompose a rational function $P(s)/Q(s)$, with the degree of $P(s)$ less than the degree of $Q(s)$,

Table 1 Laplace Transforms

$f(t)$	$F(s) = \int_0^\infty e^{-st} f(t)\,dt$
1. 1	$\dfrac{1}{s}$
2. t	$\dfrac{1}{s^2}$
3. t^n	$\dfrac{n!}{s^{n+1}}, \quad n = 0, 1, 2, \ldots$
4. e^{at}	$\dfrac{1}{s-a}$
5. te^{at}	$\dfrac{1}{(s-a)^2}$
6. $t^n e^{at}$	$\dfrac{n!}{(s-a)^{n+1}}, \quad n = 0, 1, 2, \ldots$
7. $\sin bt$	$\dfrac{b}{s^2+b^2}$
8. $\cos bt$	$\dfrac{s}{s^2+b^2}$
9. $t \sin bt$	$\dfrac{2bs}{(s^2+b^2)^2}$
10. $t \cos bt$	$\dfrac{s^2-b^2}{(s^2+b^2)^2}$
11. $e^{at} \sin bt$	$\dfrac{b}{(s-a)^2+b^2}$
12. $e^{at} \cos bt$	$\dfrac{s-a}{(s-a)^2+b^2}$
13. $e^{at} - e^{bt}$	$\dfrac{a-b}{(s-a)(s-b)}$
14. $ae^{at} - be^{bt}$	$\dfrac{(a-b)s}{(s-a)(s-b)}$
15. $(b-c)e^{at} + (c-a)e^{bt} + (a-b)e^{ct}$	$\dfrac{(a-b)(b-c)(a-c)}{(s-a)(s-b)(s-c)}$
16. $\sin bt - bt \cos bt$	$\dfrac{2b^3}{(s^2+b^2)^2}$

into a sum of terms, according to the following rules:

1. If the denominator $Q(s)$ has the factor $(s-a)^m$ as the highest power of $s-a$ that divides $Q(s)$, then include in the decomposition of $P(s)/Q(s)$ the fractions of the form

$$\frac{A_j}{(s-a)^j}, \quad j = 1, 2, \ldots, m.$$

2. If the denominator $Q(s)$ has the factor $(s^2 + ps + q)^n$ as the highest power of $s^2 + ps + q$, irreducible over the reals, that divides $Q(s)$, then include in the decomposition of $P(s)/Q(s)$ all fractions of the form

$$\frac{B_k s + C_k}{(s^2 + ps + q)^k}, \qquad k = 1, 2, \dots, n.$$

example 3

To find the function $f(t)$ having Laplace transform

$$F(s) = \frac{s+1}{(s-1)^2(s^2+1)},$$

we decompose the function into a sum of fractions as follows:

$$\frac{s+1}{(s-1)^2(s^2+1)} = \frac{A}{s-1} + \frac{B}{(s-1)^2} + \frac{Cs+D}{s^2+1}.$$

We clear the equation of fractions by multiplying by the left-hand side's denominator:

$$s + 1 = A(s-1)(s^2+1) + B(s^2+1) + Cs(s-1)^2 + D(s-1)^2.$$

Writing the right-hand side in powers of s gives

$$s + 1 = (A + C)s^3 + (-A + B - 2C + D)s^2 + (A + C - 2D)s + (-A + B + D).$$

Equating coefficients on both sides gives the four equations

$$A + C = 0, \qquad -A + B - 2C + D = 0, \qquad A + C - 2D = 1, \qquad -A + B + D = 1.$$

Since $A + C = 0$, the third equation yields $D = -\frac{1}{2}$, and the two remaining equations become $A + B = \frac{1}{2}, -A + B = \frac{3}{2}$. Then $B = 1$ and $A = -C = -\frac{1}{2}$.

Of course our solution of this 4-dimensional system was a bit ad hoc. It could have been done by methodical row operations if the system had been more complicated. In outline, the method is very straightforward: (i) clear out the fractions, (ii) equate coefficients of like powers of s, and (iii) solve the resulting linear equations for the desired coefficients. This method works for an arbitrary rational function of s, and it is the method employed in computer algebra programs with the same goal.

Using ad hoc ideas we can often simplify the computation, or at least make it less tedious and error-prone. We'll illustrate using the same example:

$$\frac{s+1}{(s-1)^2(s^2+1)} = \frac{A}{s-1} + \frac{B}{(s-1)^2} + \frac{Cs+D}{s^2+1}.$$

To compute B, we can multiply through by $(s-1)^2$ and then set $s = 1$. We get $B = 1$. The same kind of trick does not apply directly to the other coefficients, but if we subtract $1/(s-1)^2$ from both sides, we find we can cancel $(s-1)$ on the left to get

$$\frac{-s^2 + s}{(s-1)^2(s^2+1)} = \frac{-s}{(s-1)(s^2+1)} = \frac{A}{s-1} + \frac{Cs+D}{s^2+1}.$$

Now multiply by $(s-1)$ and then set $s=1$ to get $A=-\frac{1}{2}$. As a result,

$$\frac{-s}{(s-1)(s^2+1)}=\frac{-\frac{1}{2}}{s-1}+\frac{Cs+D}{s^2+1}.$$

To find C and D, we can multiply through by $(s-1)(s^2+1)$ to get

$$-s=-\tfrac{1}{2}(s^2+1)+(s-1)(Cs+D).$$

Rearranging the powers of s gives

$$\tfrac{1}{2}s^2-s+\tfrac{1}{2}=Cs^2+(D-C)s-D.$$

We equate coefficients of like powers on both sides and find that $C=\frac{1}{2}$ and $D=-\frac{1}{2}$. The result is that

$$\frac{s+1}{(s-1)^2(s^2+1)}=\frac{-\frac{1}{2}}{s-1}+\frac{1}{(s-1)^2}+\frac{\frac{1}{2}s}{s^2+1}-\frac{\frac{1}{2}}{s^2+1}.$$

From the table of transforms, we conclude that

$$\mathcal{L}^{-1}[F(s)]=-\tfrac{1}{2}e^t+te^t+\tfrac{1}{2}\cos t-\tfrac{1}{2}\sin t.$$

The examples show that the most arduous part of solving a differential equation using Laplace transforms is the passage from the transform $Y(s)$ of the solution to the solution $y(t)$: $\mathcal{L}^{-1}[Y(s)](t)=y(t)$. However, it often happens in applications to mechanical and electrical systems, as, for example, in Sections 1 and 2 of Chapter 4, and Section 3A of Chapter 6, that all we need is some qualitative information that we can get about a solution by focusing attention directly on $Y(s)$. The convenience of the type of analysis shown in the next example of what make Laplace transforms so popular in engineering applications, where we're often interested only in long-term behavior of solutions.

example 4 The equation $\ddot{x}+5\dot{x}+6x=\sin\omega t$ represents a damped harmonic oscillator driven externally by a sinusoidal oscillation. We expect this equation to have oscillatory solutions because $\sin\omega t$ is oscillatory. Suppose we're just looking for long-term behavior of a variety of solutions, in which case we don't expect initial conditions to be decisive, so we assume the simplifying initial conditions $x(0)=\dot{x}(0)=0$. The Laplace transform of this particular initial-value problem is $(s^2+5s+6)Y(s)=\omega/(s^2+\omega^2)$. The characteristic polynomial $P(s)=s^2+5s+6$ factors into $(s+2)(s+3)$, so

$$Y(s)=\frac{\omega}{(s^2+\omega^2)(s^2+5s+6)}=\frac{\omega}{(s+i\omega)(s-i\omega)(s+2)(s+3)}.$$

A partial fraction decomposition of $Y(s)$ would consist of a linear combination of terms $1/(s-a)$, where $a=i\omega$, $-i\omega$, 2, or 3. Entry 4 in Table 1 shows that the corresponding solutions will be linear combinations of $e^{i\omega t}$, $e^{-i\omega t}$, e^{-2t}, and e^{-3t}. The first two terms contribute sinusoidal oscillations. The last two terms are *transient*, decaying to zero as time increases. After sufficient time the transient terms become negligible, so for practical purposes we're left with only the oscillations, called *steady-state* terms. This much information may be enough for us, allowing us to skip the formal calculation of the complete solution.

Solving Linear Systems. Laplace transforms are applicable to linear systems with constant coefficients, as in the next example.

example 5

We'll solve the initial-value problem

$$\ddot{x} + \dot{y} = 4x, \qquad x(0) = 0, \qquad \dot{x}(0) = 1,$$
$$4\dot{x} - \ddot{y} = 9y, \qquad y(0) = -1, \qquad \dot{y}(0) = 2.$$

Denote $\mathcal{L}[x]$ and $\mathcal{L}[y]$ by X and Y, respectively. Then

$$\left[s^2 X - sx(0) - \dot{x}(0)\right] + \left[sY - y(0)\right] = 4X,$$
$$\left[4sX - x(0)\right] - \left[s^2 Y - sy(0) - \dot{y}(0)\right] = 9Y.$$

Inserting the specific initial values gives

$$\left(s^2 X - 1\right) + \left(sY + 1\right) = 4X,$$
$$4sX - \left(s^2 Y + s - 2\right) = 9Y.$$

Now rearrange into a form suitable for solving for X and Y:

$$\left(s^2 - 4\right)X + sY = 0,$$
$$4sX - \left(s^2 + 9\right)Y = s - 2.$$

Suppose we choose to eliminate Y. Multiply the first equation by $(s^2 + 9)$, the second by s, and add to get

$$(s^4 + 9s^2 - 36)X = s^2 - 2s, \quad \text{or} \quad X(s) = \frac{s^2 - 2s}{s^4 + 9s^2 - 36}.$$

The denominator in the expression for X is quadratic in s^2 with roots $s^2 = 3, -12$ so we can factor and find the partial fraction decomposition

$$X(s) = \frac{s^2 - 2s}{(s^2 - 3)(s^2 + 12)}$$
$$= \frac{2}{15}\frac{s}{s^2 + 12} + \frac{4}{5}\frac{1}{s^2 + 12} - \frac{2 + \sqrt{3}}{30}\frac{1}{s + \sqrt{3}} - \frac{2 - \sqrt{3}}{30}\frac{1}{s - \sqrt{3}}.$$

Reading from right to left in Table 1, we find

$$x(t) = \frac{2}{15}\cos\sqrt{12}\,t + \frac{4}{5\sqrt{12}}\sin\sqrt{12}\,t - \frac{2 + \sqrt{3}}{30}e^{-\sqrt{3}t} - \frac{2 - \sqrt{3}}{30}e^{\sqrt{3}t}.$$

To find $Y(s)$, and then $y(t)$ we can return to the pair of linear equations for X and Y. Substituting the factored expression above for $X(s)$ in the first of those two equations gives

$$(s^2 - 4)\frac{s^2 - 2s}{(s^2 - 3)(s^2 + 12)} + sY(s) = 0 \quad \text{or} \quad Y(s) = -\frac{(s - 2)(s^2 - 4)}{(s^2 - 3)(s^2 + 12)}.$$

Partial fraction decomposition yields

$$Y(s) = -\frac{16}{15}\frac{s}{s^2+12} + \frac{32}{15}\frac{1}{s^2+12} + \frac{2+\sqrt{3}}{30\sqrt{3}}\frac{1}{s+\sqrt{3}} - \frac{2-\sqrt{3}}{30\sqrt{3}}\frac{1}{s-\sqrt{3}}.$$

Using Table 1 gives

$$y(t) = -\frac{16}{15}\cos\sqrt{12}\,t + \frac{32}{15\sqrt{12}}\sin\sqrt{12}\,t + \frac{2+\sqrt{3}}{30\sqrt{3}}e^{-\sqrt{3}t} - \frac{2-\sqrt{3}}{30\sqrt{3}}e^{\sqrt{3}t}.$$

Transforms of Periodic Functions. Table 1 lists the transforms of two periodic functions, $\mathcal{L}[\sin bt](s) = b/(s^2+b^2)$ and $\mathcal{L}[\cos bt](s) = s/(s^2+b^2)$, and these are routinely computed using integration by parts. For many functions of period $T > 0$, it's more efficient to reduce the computation to an integral over the interval $0 \le t \le T$. In the context of the Laplace transform, functions $f(t)$ need to be defined only for $t > 0$, to say that $f(t)$ has **period** T will sometimes mean in this chapter only that $f(t+T) = f(t)$ for $t > 0$. For all such functions having a Laplace transform, we have the following computational aid.

1.7 THEOREM

If $f(t)$ has a Laplace transform and has period T, then

$$\mathcal{L}[f](s) = \frac{1}{1-e^{-Ts}}\int_0^T e^{-st} f(t)\,dt.$$

Proof.

$$\begin{aligned}
\mathcal{L}[f](s) &= \int_0^T e^{-st} f(t)\,dt + \int_T^\infty e^{-st} f(t)\,dt && \text{[Break up integral.]}\\
&= \int_0^T e^{-st} f(t)\,dt + \int_0^\infty e^{-s(u+T)} f(u+T)\,du && \text{[Let } t = u+T.\text{]}\\
&= \int_0^T e^{-st} f(t)\,dt + \int_0^\infty e^{-s(u+T)} f(u)\,du && \text{[Use periodicity of } f.\text{]}\\
&= \int_0^T e^{-st} f(t)\,dt + e^{-sT}\mathcal{L}[f](s). && \text{[Pull out factor } e^{-Ts}.\text{]}
\end{aligned}$$

Now solve for $\mathcal{L}[f](s)$. ■

The forcing function $f(t)$ may be a **step function,** namely a function whose domain is a union of intervals on each of which the function's value is constant. Section 1B treats step functions in more detail, but we can treat the next example directly using Theorem 1.7.

example 6

The **square-wave step function** is

$$\text{sqw}(t) = \begin{cases} 1, & 2k \le t < 2k+1, \\ -1, & 2k+1 \le t < 2k+2;\ k = 0, 1, 2, \ldots. \end{cases}$$

This function has period 2, and its graph is sketched in Figure 8.1. Setting $T = 2$ in the right side of the equation in Theorem 1.7, we note that the integral

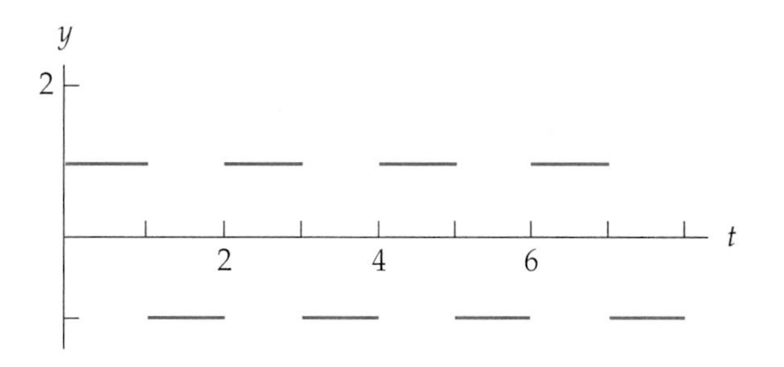

Figure 8.1 Square wave.

decomposes into a sum of two pieces:

$$\int_0^1 1 \cdot e^{-st}\,dt + \int_1^2 (-1) \cdot e^{-st}\,dt = \left[\frac{e^{-st}}{-s}\right]_0^1 - \left[\frac{e^{-st}}{-s}\right]_1^2$$

$$= \frac{-1}{s}(e^{-s} - 1) - \frac{-1}{s}(e^{-2s} - e^{-s})$$

$$= \frac{1}{s}(1 - 2e^{-s} + e^{-2s}).$$

According to Equation 1.7, the transform of $f(t)$ equals this last expression divided by $(1 - e^{-2s})$:

$$\mathcal{L}[\text{sqw(t)}](s) = \frac{1 - 2e^{-s} + e^{-2s}}{s(1 - e^{-2s})}.$$

---

example 7 Suppose we want the Laplace transform of the function

$$g(t) = \begin{cases} \sin bt, & 0 \le t \le 2\pi/b,\ b > 0, \\ 0, & 2\pi/b < t. \end{cases}$$

The graph of this function consists of one complete cycle of $\sin bt$ on $0 \le t \le 2\pi/b$ and coincides with the t axis elsewhere. We can compute the transform using integration by parts, but it's simpler first to note that

$$\mathcal{L}[g](s) = \int_0^\infty e^{-st} g(t)\,dt = \int_0^{2\pi/b} e^{-st} \sin bt\,dt.$$

Apply Theorem 1.7 and Entry 7 in Table 1 to the periodic function $f(t) = \sin bt$ to get

$$\mathcal{L}[f](s) = \frac{b}{s^2 + b^2} = \frac{1}{1 - e^{-2\pi s/b}}\mathcal{L}[g](s).$$

Solving for $\mathcal{L}[g]$ gives us

$$\mathcal{L}[g](s) = \frac{b\left(1 - e^{-2\pi s/b}\right)}{s^2 + b^2}.$$

EXERCISES

1. (a) Verify that

$$\mathcal{L}[e^{at}](s) = \int_0^\infty e^{-st}e^{at}\,dt = \frac{1}{s-a}, \quad \text{for } s > a.$$

(b) Use integration by parts to verify that

$$\mathcal{L}[t](s) = \int_0^\infty te^{-st}\,dt = \frac{1}{s^2}, \quad \text{for } s > 0.$$

2. Use Equations 1.4′ to show that, if $y(0) = 2$ and $\dot{y}(0) = 3$, then

$$\int_0^\infty e^{-st}\ddot{y}(t)\,dt = s^2Y(s) - 2s - 3,$$

where $Y(s)$ is the Laplace transform of $y(t)$.

By computing the appropriate integral, or by using the table of Laplace transforms, compute $\mathcal{L}[f](s)$, for $f(t)$ in Exercises 3 through 8.

3. $t \sin 2t$.

4. $\cos t + 2\sin t$.

5. $t^2 + 2t - 1$.

6. $\cos(t + a)$.

7. $(2t + 1)e^{3t}$.

8. $e^t + e^{-t}$.

Use Table 1 to find the inverse Laplace transforms of functions $F(s)$ in Exercises 9 through 14; that is, find $y(t)$ such that $\mathcal{L}[y](s) = F(s)$.

9. $1/(s^2 - 1)$.

10. $s/(s^2 - 4)$.

11. $2/(s^2 + 4)$.

12. $4s/(s^2 + 4)^2$.

13. $1/[(s - 2)^2 + 9]$.

14. $1/s^2 - 1/(s - 1)^2$.

Use the Laplace transform to solve initial-value problems in Exercises 15 through 26. Check by substitution.

15. $\dot{y} - y = t,\ y(0) = 2$.

16. $\dot{y} + 2y = 1,\ y(0) = 1$.

17. $\dot{y} + 3y = \cos 2t,\ y(0) = 0$.

18. $\ddot{y} + y = e^{-t} + 1,\ y(0) = -1,\ \dot{y}(0) = 1$.

19. $2\ddot{y} - \dot{y} = 2\cos 3t,\ y(0) = 0,\ \dot{y}(0) = 2$.

20. $y''' = t,\ y(0) = \dot{y}(0) = 0,\ \ddot{y}(0) = 1$.

21. $\dot{x} = 6x + 8y,\ x(0) = 0,$
$\dot{y} = -4x - 6y,\ y(0) = 1.$

22. $\dot{x} = x + 2y,\ x(0) = 0,$
$\dot{y} = -2x + y,\ y(0) = 1.$

23. $dx/dt = x + 2y,\ x(0) = 1,$
$dy/dt = x + y + t,\ y(0) = -1.$

24. $dx/dt = -y - t,\ x(0) = 2,$
$dy/dt = x + t,\ y(0) = 0.$

25. $\ddot{x} - \dot{y} = -t + 1,\ x(0) = 0,\ \dot{x}(0) = 1,$
$\dot{x} - x + 2\dot{y} = 4e^t,\ y(0) = 0.$

26. $\ddot{x} = y,\ x(0) = \dot{x}(0) = 0,$
$\ddot{y} = x,\ y(0) = 0,\ \dot{y}(0) = 1.$

27. Define $\mathrm{H}(t) = \begin{cases} 0, & \text{if } t < 0 \\ 1, & \text{if } 0 \le t. \end{cases}$

(a) Show that $\mathcal{L}[\mathrm{H}(t-a)](s) = (1/s)e^{-as}$.

(b) Show that if $g(t) = \mathrm{H}(t-a)f(t)$ for $0 < t$, and $a \geq 0$, then $\mathcal{L}[g](s) = e^{-as}\mathcal{L}[f(t+a)](s)$.

(c) Sketch the graph of $\mathrm{H}(t-a)$ for $a = 1$.

(d) Solve the differential equation $\ddot{y} = \mathrm{H}(t-a)$, $0 < a$, with conditions $y(0) = 1$, $\dot{y}(0) = 0$.

28. **(a)** Show that if $f(t)$ is defined and differentiable only for $t > 0$ (instead of $t \geq 0$), then

$$\mathcal{L}[f'](s) = -f(0+) + s\mathcal{L}[f](s), \quad \text{where} \quad f(0+) = \lim_{t\to 0+} f(t).$$

(b) Show that, if the limits $f^k(0+)$ all exist, then Equation 1.4 generalizes to

$$\mathcal{L}[f^{(n)}](s) = -f^{(n-1)}(0+) - \cdots - s^{(n-1)}f(0+) + s^n\mathcal{L}[f](s).$$

29. There are many functions $f(t)$ for which $\mathcal{L}[f(t)](s)$ is infinite for every value of s; show this for $f(t) = e^{t^2}$. [*Hint:* $e^{(t-s/2)^2} \geq 1$ for all t and s.]

Transforms of Periodic Functions. Sketch the graph of each of the functions in the next four exercises; then use Theorem 1.7 to verify the correctness of the given transform.

30. Graph the function of period 1 defined for $0 \leq t < 1$ by $f(t) = t$; verify

$$\mathcal{L}[f](s) = \frac{1 - se^{-s} - e^{-s}}{s^2(1-e^{-s})}.$$

31. Graph the function $f_2(t)$ that agrees with $f(t)$ in the previous exercise for $0 \leq t \leq 2$ and is equal to zero for $2 \leq t$; verify

$$\mathcal{L}[f_2](s) = \frac{(1 - se^{-s} - e^{-s})(1-e^{-2s})}{s^2(1-e^{-s})} = \frac{(1 - e^{-s} - se^{-s})(1+e^{-s})}{s^2}.$$

32. Graph the function of period 2 defined for $0 \leq t < 2$ by $g(t) = \begin{cases} 1, & 0 \leq t < 1, \\ 0, & 1 \leq t < 2. \end{cases}$ Show that $\mathcal{L}[g](s) = (1-e^{-s})/\left[s(1-e^{-2s})\right]$.

33. Graph the function $g_4(t)$ that agrees with $g(t)$ in the previous exercise for $0 \leq t \leq 4$ and is equal to zero for $4 \leq t$; verify

$$\mathcal{L}[g_4](s) = \frac{(1-e^{-s})(1-e^{-4s})}{s(1-e^{-2s})} = \frac{1 - e^{-s} + e^{-2s} - e^{-3s}}{s}.$$

34. Verify that if $y(0) = y_0$ and $\dot{y}(0) = z_0$, then $A\ddot{y} + B\dot{y} + Cy = f(t)$, transforms to $P(s)Y(s) = B(s) + F(s)$, where $P(s)$ is the characteristic polynomial of the operator $AD^2 + BD + C$ and $B(s)$ is a first-degree polynomial.

35. Let $P(D)$ be an nth-order, linear, constant-coefficient differential operator. Show that

$$\mathcal{L}[P(D)y](s) = P(s)\mathcal{L}[y](s) + Q(s),$$

where $Q(s)$ is some polynomial of degree $n-1$. Use induction.

36. Make an analysis of long-term behavior of solutions analogous to the one in Example 4 of the text but for the equation $\ddot{x} + 2\dot{x} + x = \sin t$.

37. Make an analysis of long-term behavior of solutions analogous to the one in Example 4 of the text but for the equation $\ddot{x} + x = e^{-t} \sin t$.

38. Assume that the forcing function $f(t)$ of a first-order constant-coefficient equation $L(y) = f(t)$ is of **exponential order** γ, which means that there are constants γ and C such that $|f(t)| \le Ce^{\gamma t}$ for all large enough t.

(a) Show that the initial value problem $\dot{y} - ay = f(t)$, $y(0) = y_0$ has solution

$$y(t) = \int_0^t e^{a(t-u)} f(u)\, du + y_0 e^{at}.$$

(b) Use the result of part (a) along with the general integral inequality $|\int_0^t F(u)\, du| \le \int_0^t |F(u)|\, du$ to prove that if $f(t)$ has order γ then $y(t)$ has order δ for some $\delta \ge \gamma$.

(c) Show that both $\mathcal{L}[f(t)](s)$ and $\mathcal{L}[y(t)](s)$ are finite for large enough s.

*__39.__ Assume that the forcing function $f(t)$ of a second-order constant-coefficient equation $L(y) = f(t)$ is of *exponential order* γ, which means that there are constants γ and C such that $|f(t)| \le Ce^{\gamma t}$ for all large enough t.

(a) Show that the Chapter 3, Section 3D, function $g(t - u)$ associated with a second-order constant-coefficient differential operator L satisfies $|g(t - u)| \le e^{\alpha(t-u)}$ for some constant α and for $0 \le u \le t$.

(b) Use the result of part (a) along with the general integral inequality $|\int_0^t F(u)\, du| \le \int_0^t |F(u)|\, du$ to prove that if $f(t)$ has order γ then $y(t)$ has order δ for some $\delta \ge \gamma$.

(c) Show that both $\mathcal{L}[f(t)](s)$ and $\mathcal{L}[y(t)](s)$ are finite for large enough s.

40. Theorem 1.3 implies in particular that a continuous function is completely determined by its Laplace transform. One way to prove this is to derive an **inversion formula** that takes us from $F(s)$ back to $f(t)$. An elementary inversion formula involving only real variables is

$$\lim_{k \to \infty} \frac{(-1)^k}{k!} \left(\frac{k}{t}\right)^{k+1} F^{(k)}\left(\frac{k}{t}\right) = f(t), \quad \text{wherever } f(t) \text{ is continuous,}$$

and $F^{(k)}(s)$ is the kth derivative of $F(s)$. (These derivatives always exist for large enough s.) Thus there can't be two essentially different piecewise continuous functions with the same transform, because both would be equal at points of continuity. The proof is fairly sophisticated; see page 288 of D. V. Widder's, *The Laplace Transform*, Princeton (1946).

(a) Apply this inversion formula to recover $f(t)$ if $F(s) = (s - a)^{-1}$.

(b) The inversion formula above works for a concrete example only if you know $F^{(k)}(s)$ for all large positive integers k. Carry out the inversion for entry 6 in Table 1, and explain what difficulty you have with later entries in the table. A more widely applicable inversion formula involves line integrals of complex-valued functions.

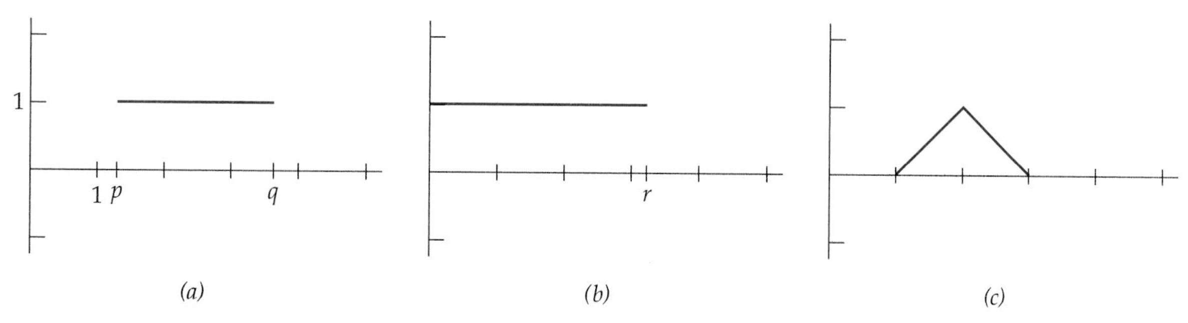

Figure 8.2 (*a*) $H(t-p) - H(t-q)$. (*b*) $1 - H(t-r)$. (*c*) $(t-1)\left[H(t-1) - H(t-2)\right] + (3-t)\left[H(t-2) - H(t-3)\right]$.

1B Piecewise-Defined Functions

In Chapter 4, Section 4C, a *step function* is defined to be a function whose domain is a union of intervals on each of which the function's value is constant. It would be worthwhile for you to take a look at the first few examples displayed there, where they are used to model intermittent processes that exhibit abrupt changes in behavior such as the turning on or off of an electric current. Since our intent here is to apply Laplace transforms it's natural to pay primary attention to functions whose domain is the entire interval $0 \le t < \infty$. Our basic device depends on the **Heaviside function** $H(t) = \begin{cases} 0, & t < 0, \\ 1, & 0 \le t, \end{cases}$ and its shifts by a positive number p: $H(t-p) = \begin{cases} 0, & t < p, \\ 1, & p \le t. \end{cases}$ From two of these shifts we can construct the basic **indicator function** of an interval $p \le t < q$, which is defined to be the function that takes the value 1 on the interval and 0 elsewhere. Examining the cases $t < p$, $p \le t < q$, and $q \le t$ shows that

$$H(t-p) - H(t-q) = \begin{cases} 1, & p \le t < q, \\ 0, & \text{elsewhere.} \end{cases}$$

The indicator function for the infinite interval $q \le t < \infty$ is just $H(t-q)$. A typical indicator function, also called a **rectangular impulse function,** appears in Figure 8.2(a).

example 8 The formula for the roof function illustrated in Figure 8.2(c) is fairly easy to interpret geometrically, because the factors $(t-1)$ and $(3-t)$ multiply indicator functions of disjoint intervals. But for algebraic treatment it's simpler to collect terms that multiply distinct shifts $H(t-p)$; doing so for this example gets us

$$(t-1)H(t-1) - 2(t-2)H(t-2) + (t-3)H(t-3).$$

To interpret this three-term formula we have to add terms as t increases past $t = 2$ and $t = 3$, but we'll see that the formula cooperates well with the transform operator $\mathcal{L}$.

Method for Piecewise Definitions. Using indicator functions we can construct essentially arbitrary step functions as well as other functions defined piecewise on nonoverlapping intervals, as in the previous and following examples. The method is this: (i) multiply the indicator function of an interval by a function, possibly constant, with the desired values on the interval; (ii) add up the results for all the intervals at hand; and (iii) then simplify the final result down to a linear combination of functions $H(t-p)$ with function coefficients.

A fairly typical initial-value problem with a piecewise-defined forcing function has the form

$$\ddot{y} + a\dot{y} + by = H(t-p)f(t) + H(t-q)g(t), \qquad y(0) = y_0, \qquad \dot{y}(0) = z_0,$$

so it's helpful to have the next theorem as an aid in finding the Laplace transform of each of the two terms on the right.

1.8 SHIFT THEOREM

If $F(s) = \mathcal{L}[f(t)](s)$ and $p > 0$, then

(a) $\mathcal{L}[H(t-p)f(t)](s) = e^{-ps}\mathcal{L}[f(t+p)](s)$.
(b) $\mathcal{L}[H(t-p)f(t-p)](s) = e^{-ps}F(s)$.
(c) $\mathcal{L}^{-1}[e^{-ps}F(s)](t) = H(t-p)f(t-p)$.

Proof. For part (a), note that the effect of $H(t-p)$ in the integrand is to make the lower limit effectively p. Then change variable, replacing t by $t+p$, to get

$$\mathcal{L}[f(t)H(t-p)](s) = \int_p^\infty e^{-st} f(t)\,dt = e^{-ps}\int_0^\infty e^{-st} f(t+p)\,dt.$$

For part (b), apply part (a) to $f(t-p)$ in place of $f(t)$. Part (c) is the result of applying $\mathcal{L}^{-1}$ to both sides of part (b). ■

example 9

Consider the simple initial-value problem $\ddot{y} = H(t-1)$, with $y(0) = \dot{y}(0) = 0$. The Laplace transform of $\ddot{y}$ is $s^2Y(s)$, and the Laplace transform of $H(t-1)$ is, by direct computation

$$\int_0^\infty e^{-st}H(t-1)\,dt = \int_1^\infty e^{-st}\,dt = \frac{e^{-s}}{s}.$$

Alternatively we can use part (a) of Theorem 1.8 with $f(t) = 1$ to find this transform. In any case, the initial-value problem transforms to $s^2Y(s) = e^{-s}/s$, or $Y(s) = e^{-s}/s^3$. By part (c) of Theorem 1.8, using $p = 1$ and $\mathcal{L}[\frac{1}{2}t^2](s) = s^{-3}$ from Table 1, we get

$$y(t) = \tfrac{1}{2}(t-1)^2H(t-1) = \begin{cases} 0, & 0 \le t < 1, \\ \tfrac{1}{2}(t-1)^2, & 1 \le t. \end{cases}$$

This problem is simple enough that we could have solved it easily by solving it first on the interval from 0 to 1, then solving the problem $\ddot{y} = 1$, $y(1) = \dot{y}(1) = 0$

and fitting the two pieces together. However, using Laplace transforms here does display quite simply the main technical features of applying them to similar, though more complicated, problems.

example 10

An initial-value problem $\ddot{y} + 4y = r(t)$, $y(0) = \dot{y}(0) = 0$, is determined by the continuous ramp function

$$r(t) = \begin{cases} 0, & 0 \le t < 1, \\ 3(t-1), & 1 \le t < 2, \\ 3, & 2 \le t. \end{cases}$$

In terms of $\mathrm{H}(t)$ the function $r(t)$ can first be written, and then simplified, as

$$\begin{aligned} r(t) &= 3(t-1)\big[\mathrm{H}(t-1) - \mathrm{H}(t-2)\big] + 3\mathrm{H}(t-2) \\ &= 3(t-1)\mathrm{H}(t-1) + (-3t+3+3)\mathrm{H}(t-2) \\ &= 3(t-1)\mathrm{H}(t-1) - 3(t-2)\mathrm{H}(t-2). \end{aligned}$$

Next we apply the Laplace transform, using Theorem 1.8, part (b), to

$$\ddot{y} + 4y = 3(t-1)\mathrm{H}(t-1) - 3(t-2)\mathrm{H}(t-2).$$

The result is

$$s^2Y(s) + 4Y(s) = 3\frac{e^{-s}}{s^2} - 3\frac{e^{-2s}}{s^2}, \quad \text{so} \quad Y(s) = 3\frac{e^{-s}}{s^2(s^2+4)} - 3\frac{e^{-2s}}{s^2(s^2+4)}.$$

A partial fraction expansion of the rational part of these expressions yields

$$\frac{3}{s^2(s^2+4)} = \frac{\frac{3}{4}}{s^2} - \frac{\frac{3}{4}}{s^2+4}.$$

Using this expansion in each term of $Y(s)$ gives

$$Y(s) = \frac{3}{4}\left(\frac{e^{-s}}{s^2} - \frac{e^{-s}}{s^2+4} - \frac{e^{-2s}}{s^2} + \frac{e^{-2s}}{s^2+4}\right).$$

Now use Theorem 1.8, part (c), and Table 1 to find the inverse transform of each term in the expression for $Y(s)$. We find four terms in $y(t)$, and then simplify to get

$$\begin{aligned} y(t) &= \tfrac{3}{4}\big[(t-1) - \tfrac{1}{2}\sin 2(t-1)\big]\mathrm{H}(t-1) \\ &\quad - \tfrac{3}{4}\big[(t-2)\mathrm{H}(t-2) - \tfrac{1}{2}\sin 2(t-2)\big]\mathrm{H}(t-2) \\ &= \big[\tfrac{3}{4}(t-1) - \tfrac{3}{8}\sin 2(t-1)\big]\mathrm{H}(t-1) - \big[\tfrac{3}{4}(t-2) - \tfrac{3}{8}\sin 2(t-2)\big]\mathrm{H}(t-2). \end{aligned}$$

The solution displayed by cases is

$$y(t) = \begin{cases} 0, & 0 \le t < 1, \\ \frac{3}{4}(t-1) - \frac{3}{8}\sin 2(t-1), & 1 \le t < 2, \\ \frac{3}{4} - \frac{3}{8}\sin 2(t-1) + \frac{3}{8}\sin 2(t-2), & 2 \le t. \end{cases}$$

The response to the increasing part of $r(t)$ is confined to $0 \le t < 1$, after which the solution becomes periodic. Note that all multiples of the two Heaviside shifts

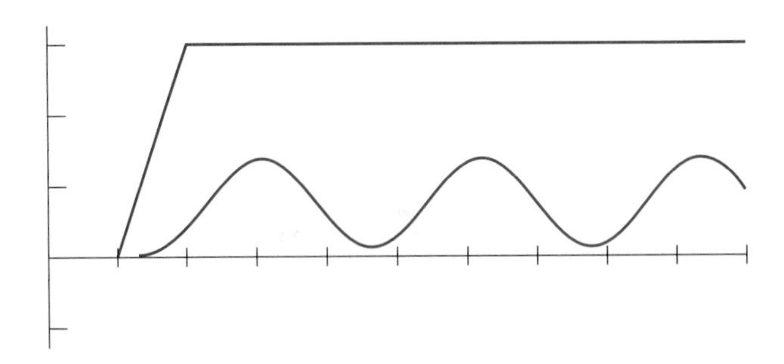

Figure 8.3 Input ramp $r(t)$ and response $y(t)$ to $\ddot{y} + 4y = r(t)$, $y(0) = \dot{y}(0) = 0$.

get added together when $t \geq 2$, and that none are included when $t < 1$. The graphs of the input $r(t)$ and the response $y(t)$ are plotted relative to the same axes in Figure 8.3; the drawing for $y(t)$ was done using the numerical method of Chapter 4, Section 4C. As a check on the correctness of our computation, note that $y(t)$ satisfies the differential equation on each of the three intervals, and that $y(t)$ is continuous at $t = 1$ and $t = 2$. It's also easy to verify directly that $y(t)$ has two continuous derivatives for $t > 0$, which it should have as a consequence of the continuity of $r(t)$. (See Exercise 23.)

1C Summary of Properties

Table 2 lists the most frequently used general properties of the Laplace transform. The entries that have not already been discussed can easily be proved using elementary calculus techniques, under the assumption that the indicated transform values are finite.

Table 2 Laplace Transform Properties

1. $\mathcal{L}[af + bg] = a\mathcal{L}[f] + b\mathcal{L}[g]$, a, b constant
2. $\mathcal{L}[f'](s) = s\mathcal{L}[f](s) - f(0)$
3. $\mathcal{L}[f^{(n)}](s) = s^n\mathcal{L}[f](s) - s^{n-1}f(0) - s^{n-2}f'(0) - \cdots - f^{(n-1)}(0)$
4. $\mathcal{L}\left[\int_0^t f(u)\,du\right](s) = \frac{1}{s}\mathcal{L}[f](s)$
5. $\mathcal{L}[e^{at}f(t)](s) = \mathcal{L}[f](s-a)$
6. $\mathcal{L}[f(t)\mathrm{H}(t-p)](s) = e^{-ps}\mathcal{L}[f(t+p)](s)$, where $\mathrm{H}(t) = \begin{cases} 0, & t < 0, \\ 1, & 1 \leq t \end{cases}$
7. $\mathcal{L}[f(t-p)\mathrm{H}(t-p)](s) = e^{-ps}F(s)$
8. $\mathcal{L}^{-1}[e^{-ps}F(s)](t) = f(t-p)\mathrm{H}(t-p)$
9. If $F(s) = \mathcal{L}[f(t)](s)$, then $F^{(n)}(s) = (-1)^n\mathcal{L}[t^n f(t)](s)$

example 11

Formula 4 in Table 2 is proved by appealing to Formula 2, already proved in Section 1B. To do this define $h(t) = \int_0^t f(u)\,du$ and note that $h(0) = 0$. Now apply Formula 2 to $h(t)$. To apply Formula 4 to computing an inverse transform, we start with

$$\mathcal{L}[\sin t](s) = \frac{1}{s^2+1}.$$

Taking $f(t) = \sin t$ in Formula 4 gives

$$\begin{aligned}\frac{1}{s(s^2+1)} &= \frac{1}{s}\mathcal{L}[\sin t](s)\\ &= \mathcal{L}\left[\int_0^t \sin u\,du\right](s) = \mathcal{L}[-\cos t + 1].\end{aligned}$$

Hence

$$\mathcal{L}^{-1}\left[\frac{1}{s(s^2+1)}\right] = -\cos t + 1.$$

Repeating the application of Formula 4 gives

$$\begin{aligned}\frac{1}{s^2(s^2+1)} &= \mathcal{L}\left[\int_0^t (-\cos u + 1)\,du\right]\\ &= \mathcal{L}[-\sin t + t].\end{aligned}$$

example 12

Starting with the formula

$$\mathcal{L}[\cos t](s) = \frac{s}{s^2+1},$$

we can apply Formula 9 in Table 2 to get

$$\begin{aligned}\mathcal{L}[t\cos t](s) &= -\frac{d}{ds}\frac{s}{s^2+1}\\ &= \frac{s^2-1}{(s^2+1)^2}.\end{aligned}$$

Another application of the same formula gives

$$\mathcal{L}[t^2\cos t](s) = -\frac{d}{ds}\frac{s^2-1}{(s^2+1)^2} = \frac{2s^2-6s}{(s^2+1)^3}.$$

example 13

To apply Formula 6 of Table 2 to the function $f(t) = t$, we define $f(t) = 0$ for $t < 0$. The graphs of $f(t)$ and $f(t-p)\mathrm{H}(t-p)$ are indicated in Figure 8.4 for $p = 1$, $p = 2$, and $p = 3$. Each function is zero where it is not positive, which is the point of including the factor $\mathrm{H}(t-a)$ in Table 2, Formula 6. From Formula 6, we find

$$\begin{aligned}\mathcal{L}[f(t-1)\mathrm{H}(t-1)](s) &= e^{-s}\mathcal{L}[f(t)](s)\\ &= e^{-s}\mathcal{L}[t](s) = e^{-s}\frac{1}{s^2}.\end{aligned}$$

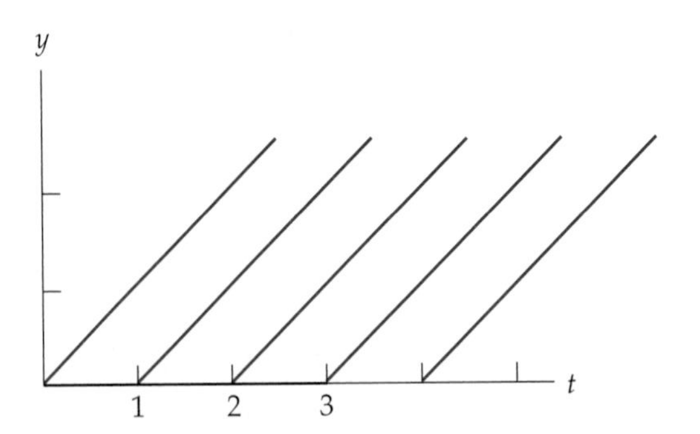

Figure 8.4 Shifted graph of $t\,\mathrm{H}(t)$.

Similarly,

$$\begin{aligned}\mathcal{L}[f(t-2)\mathrm{H}(t-2)](s) &= e^{-2s}\mathcal{L}[f(t)](s)\\ &= e^{-2s}\mathcal{L}[t](s) = e^{-2s}\frac{1}{s^2}.\end{aligned}$$

EXERCISES

In Exercises 1 through 6 sketch graphs of functions $f(t)$ defined using the Heaviside function $\mathrm{H}(t)$. Then use Theorem 1.8 to compute the Laplace transform of $f(t)$.

1. $f(t) = \mathrm{H}(t-1)) - \mathrm{H}(t-2)$.

2. $f(t) = \mathrm{H}(t) + 2\mathrm{H}(t-1)$.

3. $f(t) = \mathrm{H}(t) + \mathrm{H}(t-1) - 2\mathrm{H}(t-2)$.

4. $f(t) = t\,[\mathrm{H}(t) - \mathrm{H}(t-1)] + \mathrm{H}(t-1)$.

5. $f(t) = t\,[\mathrm{H}(t) - \mathrm{H}(t-1)] - t\,[\mathrm{H}(t-1) - \mathrm{H}(t-2)]$.

6. $f(t) = \cos(t)[\mathrm{H}(t) - \mathrm{H}(t-\pi/2)] + \sin(t)[\mathrm{H}(t-\pi) - \mathrm{H}(t-2\pi)]$.

In Exercises 7 through 12 sketch the graphs of the functions $g(t)$ for $t \geq 0$ and write the function as a sum of multiples of shifted Heaviside functions $\mathrm{H}(t-p)$, using coefficient multipliers that may depend on t. Then find the Laplace transform of $g(t)$.

7. $g(t) = \begin{cases} 1, & 1 \leq t < 2, \\ 0, & \text{elsewhere.} \end{cases}$

8. $g(t) = \begin{cases} 1, & 1 \leq t < 2, \\ -2, & 2 \leq t < 3, \\ 0, & \text{elsewhere.} \end{cases}$

9. $g(t) = \begin{cases} \sin t, & 0 \leq t < \pi, \\ 1 + \cos t, & \pi \leq t < 2\pi, \\ 0, & \text{elsewhere.} \end{cases}$

10. $g(t) = \begin{cases} t, & 0 \leq t < 2, \\ t-2, & 2 \leq t < 4, \\ 0, & \text{elsewhere.} \end{cases}$

11. $g(t) = \begin{cases} 0, & 0 \leq t < 1, \\ (t-1)^2, & 1 \leq t < 2, \\ e^{t-1}, & \text{elsewhere.} \end{cases}$

12. $g(t) = \begin{cases} 0, & 0 \le t < 1, \\ 1 - t, & 1 \le t < 2, \\ -1, & \text{elsewhere.} \end{cases}$

13. Solve the initial-value problem $\ddot{y} + \dot{y} = \text{H}(t-1)$, $y(0) = 0$, $\dot{y}(0) = 1$, expressing $y(t)$ by cases depending on t.

14. Consider the initial-value problem $\dot{y} = r(t)$, $y(0) = 0$, where $r(t)$ is the roof function of Figure 8.2(c).

(a) Note that $y(t)$ is the integral of $r(t)$ from $t = 0$. On geometric grounds, what value does $y(t)$ have for $t \ge 3$?

(b) Solve the initial-value problem for $0 \le t < 3$.

(c) Sketch the graphs of the input $r(t)$ and its integral $y(t)$ relative to the same set of axes.

15. (a) Solve the initial-value problem $\ddot{y} = r(t)$, $y(0) = \dot{y}(0) = 0$, where $r(t)$ is the roof function of Figure 8.2(c).

(b) Sketch the graphs of the input $r(t)$ and the second integral $y(t)$ relative to the same set of axes.

16. Solve the initial-value problem $\ddot{y} + \pi^2 y = g(t)$, $y(0) = \dot{y}(0) = 0$, where $g(t) = \begin{cases} 1, & 1 \le t < 2, \\ 0, & \text{elsewhere.} \end{cases}$ Then sketch the graphs of $g(t)$ and $y(t)$ relative to the same axes.

17. Solve the initial-value problem $\ddot{y} + \pi^2 y = g(t)$, $y(0) = \dot{y}(0) = 0$, where $g(t) = \begin{cases} t, & 0 \le t < 1, \\ 1, & 1 \le t. \end{cases}$ Then sketch the graphs of $g(t)$ and $y(t)$ relative to the same axes.

18. Solve the initial-value problem $\ddot{y} + y = t + (2 - t)\text{H}(t-1)$, $y(0) = \dot{y}(0) = 0$. Then sketch the graphs of the forcing function and the solution relative to the same axes.

19. Prove directly, not using Theorem 1.8, that $\mathcal{L}[\text{H}(t-p)](s) = s^{-1}e^{-ps}$, $p \ge 0$.

20. Prove directly, not using Theorem 1.8, $\mathcal{L}[t\,\text{H}(t-p)](s) = (1 + ps)s^{-2}e^{-ps}$, $p \ge 0$.

21. The gamma function, denoted by Γ, can be defined by

$$\Gamma(\nu) = \int_0^\infty t^{\nu-1}e^{-t}\,dt, \qquad \nu > 0.$$

(a) Use integration by parts to show that

$$\Gamma(\nu + 1) = \nu\Gamma(\nu).$$

(b) Deduce from part (a) that $\Gamma(n + 1) = n!$, for $n = 0, 1, 2, \ldots$.

(c) Show that if $\nu > -1$ then

$$\mathcal{L}[t^\nu](s) = \frac{\Gamma(\nu + 1)}{s^{\nu+1}}.$$

22. Derive Formula 9 in Table 2 by repeated differentiation with respect to s.

23. (a) Verify that the solution $y(t)$ found in Example 9 of the text has only one continuous derivative.

***(b)** Do Exercise 33 in Chapter 3, Section 3D, where you are asked to prove that if the forcing function is continuous then the solution of a second-order constant-coefficient equation has two continuous derivatives.

2 CONVOLUTION†

Let us review the solution of the second-order differential equation

$$\ddot{y} + a\dot{y} + by = f(t), \qquad y(0) = 0, \qquad \dot{y}(0) = 0.$$

Taking the Laplace transform of both sides gives

$$(s^2 + as + b)Y(s) = F(s) + \dot{y}(0) + (s + a)y(0).$$

The polynomial factor $P(s) = s^2 + as + b$ on the left is the **characteristic polynomial** of the operator $D^2 + aD + b$. Assuming as we did that the initial values at $t = 0$ are both zero, the reciprocal $G(s) = 1/P(s)$ is called the **transfer function** of the operator. We multiply by $G(s)$, or equivalently divide by $P(s)$, to get the formula

$$Y(s) = \frac{F(s)}{P(s)} = G(s)F(s)$$

for the Laplace transform $Y(s) = \mathcal{L}[y](s)$. Thus the "transfer" that is effected in multiplying by $G(s)$ takes us from the transform $F(s)$ of the input function $f(t)$ to the transform $Y(s)$ of the output $y(t)$: $Y(s) = G(s)F(s)$. The remaining step is then to find the inverse transform $y(t) = \mathcal{L}^{-1}[Y(s)](t)$, where $y(0) = \dot{y}(0) = 0$.

It's significant that, except for the letter used for the variable, the polynomial $P(s)$ is the same as the polynomial $P(r) = r^2 + ar + b$ whose roots we used in Chapter 3 to find characteristic roots for writing homogeneous solutions y_h to the same differential equation. Note that broadly speaking what we do with either the transform method or the characteristic root method is replace a calculus computation by an algebra computation. It might seem at first sight that when it comes to finding particular solutions the two methods differ considerably, but there is an underlying connection as we'll see now.

In Chapter 3, Section 3D, we introduced the concept of an integral transform for solving the nonhomogeneous initial-value problem for the particular solution with both initial values assumed to be zero. At that point we were using different variable names, but using our present notation we wrote the solution formula

$$y_p(t) = \int_0^t g(t-u)f(u)\,du, \quad \text{where} \quad y(0) = \dot{y}(0) = 0,$$

and called the integral a **convolution integral.** It turns out that convolution integrals have a signicant connection to the use of transforms. The most important of these connections answers the following question: If $F(s)$ and $G(s)$ are the Laplace transforms of $f(t)$ and $g(t)$, respectively, what function has Laplace transform equal to the product $F(s)G(s)$? It turns out that under rather general hypotheses in Theorem 2.1 below the answer is given by the convolution integral

$$f * g(t) = \int_0^t f(u)g(t-u)\,du.$$

Regardless of any relation to a differential equation, given two functions $f(u)$ and $g(u)$ integrable for $0 \le u \le t$, the function denoted by $f * g(t)$ is called the **convolution** of f and g. The notation suggests that $f * g$ may be regarded as a kind of product of f and g. Two basic facts about convolutions are as follows.

2.1 CONVOLUTION THEOREM

Let $f(t)$ and $g(t)$ be integrable on $0 \le t \le T$ for every positive T; then $f * g$ and $g * f$ both exist and are equal; thus convolution is commutative:

$$f * g = g * f.$$

If in addition $\mathcal{L}[|f|](s)$ and $\mathcal{L}[|g|](s)$ are both finite then $\mathcal{L}[f * g](s)$ is finite, and

$$\mathcal{L}[f * g](s) = (\mathcal{L}[f](s))(\mathcal{L}[g](s)) = F(s)G(s).$$

Note: The extra conditions on the transforms of $|f|$ and $|g|$ hold for large enough s if f and g are of exponential order.

Proof. We prove the first statement by changing the variable in the definition of $f * g$. Holding t fixed and replacing u by $t - v$ in the integral, we get

$$\begin{aligned} f * g(t) &= \int_0^t f(u)g(t-u)\,du = -\int_t^0 f(t-v)g(v)\,dv \\ &= \int_0^t g(v)f(t-v)\,dv = g * f(t). \end{aligned}$$

For the second statement, since $\mathcal{L}[|f|](s)$ and $\mathcal{L}[|g|](s)$ are finite, so are $F(s)$ and $G(s)$. (See Exercise 25.) To simplify writing limits of integration, we can extend both $f(t)$ and $g(t)$ to have the value 0 for $t < 0$, so we can write

$$\begin{aligned} F(s)G(s) &= \int_{-\infty}^{\infty} e^{-su} f(u)\,du \int_{-\infty}^{\infty} e^{-sv} g(v)\,dv \\ &= \int_{-\infty}^{\infty} f(u) \left[\int_{-\infty}^{\infty} e^{-s(u+v)} g(v)\,dv\right] du. \end{aligned}$$

We make the change of variable $v = t - u$ in the inner integral to get

$$F(s)G(s) = \int_{-\infty}^{\infty} f(u) \left[\int_{-\infty}^{\infty} e^{-st} g(t-u)\,dt\right] du.$$

Under the hypotheses of the theorem, we can interchange the order of integration. (We will assume the required theorem, called Fubini's theorem.) Then we have

$$F(s)G(s) = \int_{-\infty}^{\infty} e^{-st} \left[\int_{-\infty}^{\infty} f(u)g(t-u)\,du\right] dt.$$

Because we've defined $f(t)$ and $g(t)$ to be zero for $t < 0$, the inner integrand is zero for $t < 0$. It follows that the t integration is needed only for $0 \le t \le \infty$. Similarly, the u integration is needed only for $0 \le u \le t$. Thus we get our conclusion

$$F(s)G(s) = \int_0^\infty e^{-st}\left[\int_0^t f(u)g(t-u)\,du\right] dt$$
$$= \mathcal{L}[f * g](s).$$

■

example 1

In this example we return to the formula that motivated our introduction of the convolution, namely, the integral transform that we now denote by $y(t) = g * f(t)$:

$$y(t) = g * f(t) = \int_0^t g(t-u)f(u)\,du.$$

We're now in a position to show, using Laplace transforms, that with the proper choice of the function $g(t)$ this integral solves the time-invariant initial-value problem

$$\ddot{y} + a\dot{y} + by = f(t), \qquad y(0) = 0, \qquad \dot{y}(0) = 0.$$

Note that the left-hand side is normalized so that $\ddot{y}$ has coefficient 1. At the beginning of the section we saw that via the Laplace transform the problem converts to

$$Y(s) = G(s)F(s) \quad \text{where} \quad G(s) = \frac{1}{s^2 + as + b},$$

and $s^2 + as + b$ is the characteristic polynomial of the operator $D^2 + aD + b$. We now recognize that $G(s)$ is the Laplace transform of a function that depends on the coefficients a and b and hence, in factored form, on the roots r_1 and r_2:

$$G(s) = \begin{cases} \dfrac{1}{(s-r_1)(s-r_2)}, & r_1 \neq r_2, \\ \dfrac{1}{(s-r_1)^2}, & r_1 = r_2, \\ \dfrac{1}{(s-\alpha)^2 + \beta^2}, & r_1 = \alpha + i\beta, r_2 = \alpha - i\beta. \end{cases}$$

The third of these forms follows from multiplying out $[s - (\alpha + i\beta)][s - (\alpha - i\beta)]$. Entries 13, 5, and 11 in transform Table 1, show that

$$\mathcal{L}^{-1}[G(s)](t) = g(t) = \begin{cases} \dfrac{1}{r_1 - r_2}(e^{r_1 t} - e^{r_2 t}), & r_1 \neq r_2, \\ te^{r_1 t}, & r_1 = r_2, \\ \dfrac{1}{\beta}e^{\alpha t}\sin \beta t, & r_1 = \alpha + i\beta, r_2 = \alpha - i\beta. \end{cases}$$

You can check (Exercise 21) that for each fixed number u the function $g(t-u)$ is a solution of the homogeneous differential equation. It follows from the convolution formula that the solution $y(t)$ of the nonhomogeneous equation is a weighted average of homogeneous solutions, with weight density $f(u)$ for u between 0 and t. Such a weighted average can be regarded as a kind of generalized linear combination of homogeneous solutions. Note that this interpretation is not at all apparent from the final form of the solution $y(t) = \frac{1}{2}e^{3t} - e^{2t} + \frac{1}{2}e^{t}$ of the initial-value problem $\ddot{y} - 3\dot{y} + 2y = e^{3t}$, $y(0) = \dot{y}(0) = 0$.

example 2

From Table 1 in Section 1, we see that $\mathcal{L}[t](s) = 1/s^2$ and $\mathcal{L}[\sin t](s) = 1/(s^2+1)$. It follows from Theorem 2.1 that

$$\frac{1}{s^2} \cdot \frac{1}{s^2+1} = \mathcal{L}\left[\int_0^t (t-u)\sin u\, du\right](s).$$

But holding t fixed, we can use integration by parts to show that

$$\int_0^t (t-u)\sin u\, du = [-(t-u)\cos u]_0^t - \int_0^t \cos u\, du$$
$$= t - \sin t.$$

The alternative method involves computing a partial fraction decomposition of the form

$$\frac{1}{s^2} \cdot \frac{1}{s^2+1} = \frac{A}{s} + \frac{B}{s^2} + \frac{Cs+d}{s^2+1},$$

and then finding the inverse transform of each term.

EXERCISES

Find the convolution $f * g$ of the following pairs of functions in Exercises 1 through 4:

1. $f(t) = t, g(t) = e^{-t}, t \geq 0.$

2. $f(t) = t^2, g(t) = (t^2+1), t \geq 0.$

3. $f(t) = 1, g(t) = t, t \geq 0.$

4. $f(t) = t^2, g(t) = \sin t.$

Use the convolution of two functions to find the inverse Laplace transform of each of the products in Exercises 5 through 8 of Laplace transforms.

5. $1/\left[s^2(s+1)\right].$

6. $1/\left[(s-1)(s-2)\right].$

7. $s^{-3}e^{-2s}.$

8. $1/(s^2+1)^2.$

By using the formulas in Tables 1 and 2, find the inverse Laplace transforms of Exercises 9 through 14.

9. $1/\left[s(s+3)^2\right].$

10. $e^{-2s}/\left[s(s^2+4)\right].$

11. $1/(s^2+2s+2).$

12. $1/(s^2+1).$

13. $(e^{-s}+1)/s.$

14. $s/(s^2+10).$

Solve the initial-value problems in Exercises 15 through 18. Check by substitution.

15. $\ddot{y} - y = \sin 2t + 1$, $y(0) = 1$, $\dot{y}(0) = 1$.

16. $\ddot{y} + 2y = t$, $y(0) = 0$, $\dot{y}(0) = 1$.

17. $\ddot{y} + \dot{y} = t + e^{-t}$, $y(0) = 2$, $\dot{y}(0) = 1$.

18. $\ddot{y} + 2\dot{y} + y = t$, $y(0) = \dot{y}(0) = 0$.

19. Solve the equation

$$\dot{y} + y = \int_0^t y(u)\,du + t,$$

given that $y(0) = 1$.

20. (a) Use Formula 4 in Table 2 of Section 1C to show that

$$\mathcal{L}\left[\int_0^t \left(\int_0^{t_1} \left(\cdots \int_0^{t_n} f(t_{n+1})\,dt_{n+1}\right)\cdots dt_2\right)dt_1\right](s) = \frac{1}{s^{n+1}} F(s).$$

(b) Use Convolution Theorem 1.2 to show that the n-fold iterated integral is equal to

$$\frac{1}{n!}\int_0^t (t-u)^n f(u)\,du.$$

21. Show that the function $g(t - u_0)$ associated with a differential operator $L = D^2 + aD + b$, and with u_0 fixed, satisfies the homogeneous differential equation $L(y) = 0$ along with the initial conditions $y(u_0) = 0$, $\dot{y}(u_0) = 1$. Do this for each of the three cases separately: (a) r_1, r_2 both real, $r_1 \neq r_2$; (b) $r_1 = r_2$; and (c) r_1, r_2 complex conjugate.

22. Prove that the convolution product is **associative:** $f * (g * h) = (f * g) * h$.

23. Prove that the convolution product is **distributive** over addition: $f * (g + h) = f * g + f * h$.

24. Show that the condition in the Convolution Theorem 2.1 that $\mathcal{L}[f](s)$ be finite will be satisfied if $s > \alpha$ and $f(t)$ is of exponential order α, that is $|f(t)| \leq Ce^{\alpha t}$ for some constant C and all sufficiently large t.

25. Prove that if $\int_0^\infty |h(t)|\,dt$ is finite then $\int_0^\infty h(t)\,dt$ exists and is finite, in three steps.

(a) Show that

$$\int_0^N (h(t) + |h(t)|)\,dt \leq 2\int_0^\infty |h(t)|\,dt.$$

(b) Use the inequality $0 \leq h(t) + |h(t)|$ to prove that $\int_0^\infty (h(t) + |h(t)|)\,dt$ exists and is finite.

(c) Prove the existence and finiteness of

$$\int_0^\infty (h(t) + |h(t)|)\,dt - \int_0^\infty |h(t)|\,dt = \int_0^\infty h(t)\,dt.$$

(d) Let $h(t) = e^{-st} f(t)$ to prove that if $\mathcal{L}[|f|](s)$ is finite, then $\mathcal{L}[f](s)$ exists and is finite.

*26. The conditions on the absolute values of f and g in Theorem 2.1 that $\mathcal{L}[|f|](s)$ and $\mathcal{L}[|g|](s)$ both be finite is needed to allow interchange of integration order in the second part of the proof. These hypotheses are usually easy to verify in ordinary practice, but to see that it really is a restriction we use a highly oscillatory example.

(a) Let

$$f(t) = e^{(t+e^t)} \sin e^{(e^t)}.$$

Show that

$$\int_0^\infty e^{-st} f(t)\,dt = \int_{e^e}^\infty \frac{\sin u}{(\ln u)^s}\,du,$$

and that this is finite for all $s > 0$. [*Hint:* Express the second integral as an alternating infinite series.]

(b) Show that $\int_0^\infty e^{-st}|f(t)|\,dt = +\infty$ for all s. [*Hint:* Compare the analogue of the second integral above with a smaller, but divergent, infinite series.]

3 DIRAC DELTA FUNCTION

3A Definition of δ(t)

The solution of the initial-value problem

$$\ddot{y} + a\dot{y} + by = f(t), \qquad y(0) = y_0, \qquad \dot{y}(0) = z_0,$$

is often wanted when $f(t)$ is an **impulse function,** that is, a function with values representing force, that is identically zero outside some small time interval. A quick hammer blow behaves like this. Figure 8.5(b) shows the graphs of rectangular impulse functions $(1/\tau)\Delta_\tau(t - t_0)$ assuming the constant value $1/\tau$ on

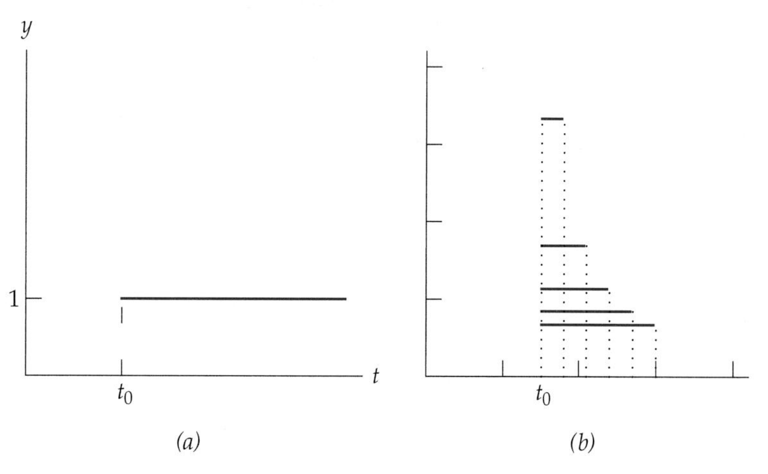

Figure 8.5 (a) $y = \mathrm{H}(t - t_0)$.
(b) $y = (1/\tau)\Delta_\tau(t - t_0)$ for incrementally varying τ.

intervals $t_0 \le t < t_0+\tau$ as τ decreases. The effect of decreasing τ is to concentrate the impulse closer to the fixed point t_0, simultaneously increasing the size of the force. The function $\Delta_\tau(t-t_0)$ is the *indicator function* of the interval $t_0 \le t < t_0+\tau$:

$$\Delta_\tau(t-t_0) = \mathrm{H}(t-t_0) - \mathrm{H}(t-\tau-t_0) = \begin{cases} 1, & t_0 \le t < t_0+\tau, \\ 0, & \text{elsewhere.} \end{cases}$$

Here $\mathrm{H}(t) = \begin{cases} 1, & t \ge 0, \\ 0, & t < 0, \end{cases}$ sketched in Figure 8.5(a), is the Heaviside function discussed in Section 1B of this chapter and in Chapter 4, Section 4C. The integrals of the related impulse functions are

$$\int_{-\infty}^{\infty} \frac{1}{\tau}\Delta_\tau(t-t_0)\,dt = \int_{t_0}^{t_0+\tau} \frac{1}{\tau}\,dt = 1.$$

Physical Interpretation. If $F(t) = m(dv/dt)$ represents a force exerted over the time interval $t_0 \le t < t_0+\tau$, the integral of $F(t)$ over the interval is called the **total impulse** of the force over the interval. Thus in general the total impulse is

$$\int_{t_0}^{t_0+\tau} F(t)\,dt = \int_{t_0}^{t_0+\tau} m\frac{dv}{dt}\,dt = mv(t_0+\tau) - mv(t_0);$$

this is just the change in momentum $mv(t)$ due to the impulse function. Interpreted this way, we see that the forces $(1/\tau)\Delta_\tau(t-t_0)$ are designed so that they all cause the same total impulse, or change in momentum. As τ tends to zero, the force is more concentrated on the decreasing time intervals of length τ and tends to infinity while the total impulse stays constantly equal to 1.

Mathematical Interpretation. The impulse function $(1/\tau)\Delta_\tau(t-t_0)$ that we discussed above is also the difference quotient that we would consider if we wanted to differentiate the shifted Heaviside function $\mathrm{H}(t-t_0) = \begin{cases} 1, & t \ge t_0, \\ 0, & t < t_0, \end{cases}$ with respect to t. We find

$$\frac{d}{dt}\mathrm{H}(t-t_0) = \lim_{\tau\to 0}\frac{\mathrm{H}(t-t_0) - \mathrm{H}(t-\tau-t_0)}{\tau} = \begin{cases} +\infty, & t = t_0;\ \tau \to 0,\ \tau > 0, \\ 0, & \text{otherwise.} \end{cases}$$

To see this, note that if $t = t_0$ and $\tau > 0$, then $\Delta_\tau(t-t_0) = \mathrm{H}(0) - \mathrm{H}(-\tau) = 1$. If $t_0 \ne 0$, then $\Delta_\tau(t-t_0) = 0$, at least for small enough τ. Thus we have computed $(d/dt)\mathrm{H}(t-t_0)$. It is often said that functions with infinite values are not functions in the ordinary sense, which is true enough if by ordinary function we mean a function with values expressible using only real or complex numbers. And indeed $\mathrm{H}(t-t_0)$ is not differentiable at $t = t_0$. However, there are many situations in which it turns out to be useful to admit functions with values $\pm\infty$, particularly if they are subject to integration. Integration sometimes has the effect of taming such functions, as in the case of improper integrals over a finite interval, and is one of the reasons the δ function, defined on page 539, is often introduced in the context of Laplace transforms. The physical interpretation in which we focused on the total impulse integral rather than impulse function itself as $\tau \to 0$ is an example of this use of integration.

With this in mind, we'll compute the limit as $\tau \to 0$ of the total impulse integral of a force $(1/\tau)\Delta_\tau(t - t_0)$ modified by a continuous factor $g(t)$; as we'll see, this turns out to be

$$\lim_{\tau \to 0} \int_{-\infty}^{\infty} \frac{1}{\tau} \Delta_\tau(t - t_0) g(t)\, dt = g(t_0).$$

To prove this we define $G(t)$ to be an integral of $g(t)$, so that $G'(t) = g(t)$. Then

$$\begin{aligned} \int_{-\infty}^{\infty} \frac{1}{\tau} \Delta_\tau(t - t_0) g(t)\, dt &= \int_{t_0}^{t_0+\tau} \frac{1}{\tau} g(t)\, dt \\ &= \frac{1}{\tau}\left[\int_0^{t_0+\tau} g(t)\, dt - \int_0^{t_0} g(t)\, dt\right] \\ &= \frac{1}{\tau}\left[G(t_0 + \tau) - G(t_0)\right]. \end{aligned}$$

As $\tau \to 0$, this last expression tends to $G'(t_0) = g(t_0)$, which is what we wanted to show. If we were to take the limit under the integral sign, we would get the product of $g(t_0)$ and what is historically called the **Dirac δ function,**

$$\delta(t - t_0) = \lim_{\tau \to 0} \frac{1}{\tau} \Delta_\tau(t - t_0) = \begin{cases} +\infty, & t = t_0 \\ 0, & t \neq t_0 \end{cases}.$$

The way we arrived at these results strongly suggests that we should extend the definition of integral to the product $\delta(t - t_0)g(t)$ and define the **δ function integral operator** by

$$\int_{-\infty}^{\infty} \delta(t - t_0) g(t)\, dt = g(t_0)$$

whenever $g(t)$ is continuous at $t = t_0$. If we take $g(t) = 1$, we see that $\delta(t - t_0)$ acts like a force that generates total impulse 1, but with all the force concentrated at the instant $t = t_0$.

example 1

Using the definition of integral involving the δ-function stated above, we have $\int_0^\infty \delta(t-2)e^{-t}\, dt = e^{-2}$. More generally, we can compute the Laplace transform of $\delta(t - t_0)$ by

$$\mathcal{L}[\delta(t - t_0)](s) = \int_0^\infty \delta(t - t_0) e^{-st}\, dt = e^{-t_0 s}.$$

In particular, with $t_0 = 0$, we get $\mathcal{L}[\delta(t)](s) = 1$.

example 2

The *convolution* of δ with f is

$$f * \delta(t) = \int_0^t f(t - u)\delta(u)\, du = f(t), \qquad t > 0.$$

Example 1 shows that $\mathcal{L}[\delta](s) = 1$. Hence the rule $\mathcal{L}[f * \delta](s) = \mathcal{L}[f](s)\mathcal{L}[\delta](s)$, shown already when δ is replaced by an ordinary function, still holds for convolution of δ with a *continuous* function.

---

example 3

To give additional meaning to the δ function, consider the following. We have elsewhere defined the *Heaviside function* $H(t)$ by

$$H(t) = \begin{cases} 0, & t < 0, \\ 1, & t \geq 0 \end{cases}.$$

Its graph for a generic value of t_0 is shown in Figure 8.5(a). Note that $H(t - t_0)$ has the property that its derivative $(d/dt)H(t)$ is zero everywhere but at $t = t_0$, but fails to exist at that one point. Given the abruptness of the jump in $H(t - t_0)$ at t_0, it may not be very surprising that there is a sense in which $(d/dt)H(t - t_0) = \delta(t - t_0)$. To justify this equation, we need to show that it makes sense in the context in which $\delta(t - t_0)$ makes sense. Thus let $g(t)$ be a continuously differentiable function tending rapidly enough to zero as $t \to \infty$ that the improper integrals

$$\int_0^\infty g(t)\,dt \quad \text{and} \quad \int_0^\infty g'(t)\,dt$$

exist. For example, the functions $g(t) = e^{-pt}$ satisfy these conditions if $p > 0$. We now do the following formal integration by parts, assuming $t_0 > 0$:

$$\begin{aligned}\int_0^\infty H'(t - t_0)g(t)\,dt &= H(t - t_0)g(t)\Big|_0^\infty - \int_0^\infty H(t - t_0)g'(t)\,dt \\ &= H(\infty)g(\infty) - H(-t_0)g(0) - \int_{t_0}^\infty g'(t)\,dt.\end{aligned}$$

But $H(\infty) = \lim_{t\to\infty} H(t) = 1$ by definition, and $g(\infty) = \lim_{t\to\infty} g(t) = 0$ by assumption. Also $H(-t_0) = 0$ since $-t_0 < 0$. Hence

$$\int_0^\infty H'(t - t_0)g(t)\,dt = -\int_{t_0}^\infty g'(t)\,dt = -g(\infty) + g(t_0) = g(t_0).$$

Comparing this result with our previous definition of the way $\delta(t)$ operates under an integral sign, namely,

$$\int_0^\infty \delta(t - t_0)g(t)\,dt = g(t_0),$$

we see that $(d/dt)H(t - t_0)$ behaves just like $\delta(t - t_0)$ in the same context. So we define an extension of the concept derivative for the Heaviside function by

$$\int_0^\infty \frac{d}{dt}H(t - t_0)g(t)\,dt = \int_0^\infty \delta(t - t_0)g(t)\,dt.$$

Important Note. The equation $(d/dt)\mathrm{H}(t-t_0) = \delta(t-t_0)$ has meaning only in the context of the previous equation, where it is quite legitimate.

example **4**

We saw in Example 1 that

$$\mathcal{L}[\delta(t-t_0)] = \int_0^{\infty} \delta(t-t_0)e^{-st}\,dt = e^{-t_0 s}.$$

Thus $\mathcal{L}[\delta(t-t_0)](s) = e^{-t_0 s}$. On the other hand,

$$\begin{aligned}\mathcal{L}[\mathrm{H}(t-t_0)](s) &= \int_0^{\infty} \mathrm{H}(t-t_0)e^{-st}\,dt = \int_{t_0}^{\infty} e^{-st}\,dt = -\frac{1}{s}e^{-st}\Big|_{t_0}^{\infty} \\ &= \frac{1}{s}e^{-t_0 s}.\end{aligned}$$

It follows that if $t_0 > 0$,

$$\mathcal{L}[\delta(t-t_0)](s) = s\mathcal{L}[\mathrm{H}(t-t_0)](s) - \mathrm{H}(-t_0) = s\mathcal{L}[\mathrm{H}(t-t_0)](s).$$

The rule $\mathcal{L}[f'(t)](s) = s\mathcal{L}[f(t)](s)$ has already been shown to hold with $\mathrm{H}(t-t_0)$ replaced by an ordinary differentiable function $f(t)$ for which $f(0) = 0$. Thus identification of $(d/dt)\mathrm{H}(t-t_0)$ and $\delta(t-t_0)$ cooperates with a basic rule of transform calculus.

3B Solving Equations

example **5**

Here is a direct solution of $\ddot{y} = \delta(t-t_0)$. We recall that $(d/dt)\mathrm{H}(t-t_0) = \delta(t-t_0)$, so

$$\begin{aligned}\dot{y} &= \mathrm{H}(t-t_0) + c_1, \\ y &= \int_0^t \mathrm{H}(u-t_0)\,du + c_1 t + c_2, \\ &= \begin{cases} 0 + c_1 t + c_2, & t < t_0 \\ (t-t_0) + c_1 t + c_2, & t \ge t_0. \end{cases}\end{aligned}$$

Since $\mathrm{H}(t-t_0) = 0$ for $t < t_0$, $y(t)$ can be written

$$y = (t-t_0)\mathrm{H}(t-t_0) + c_1 t + c_2.$$

If we also require $y(0) = \dot{y}(0) = 0$, we get the solution

$$y = (t-t_0)\mathrm{H}(t-t_0).$$

As a check, we calculate, using the product rule,

$$\begin{aligned}\dot{y} &= \mathrm{H}(t-t_0) + (t-t_0)(d/dt)\mathrm{H}(t-t_0) \\ &= \mathrm{H}(t-t_0) + (t-t_0)\delta(t-t_0).\end{aligned}$$

But the second term simply acts like 0, because

$$\int_0^\infty (t - t_0)\delta(t - t_0)g(t)\,dt = (t_0 - t_0)g(t_0) = 0,$$

for every g. It follows that $\ddot{y} = (d/dt)\mathrm{H}(t - t_0) = \delta(t - t_0)$, as we wanted to show.

example 6

We can use the Laplace transform to solve the problem $\ddot{y} = \delta(t - t_0)$, $y(0) = 0$, $\dot{y}(0) = 0$, more quickly. We find

$$\mathcal{L}[\ddot{y}](s) = s^2 Y(s) - sy(0) - \dot{y}(0) = s^2 Y(s).$$

Since $\mathcal{L}[\delta(t - t_0)](s) = e^{-t_0 s}$, we have $Y(s) = e^{-t_0 s}/s^2$. Since $\mathcal{L}[t](s) = 1/s^2$, reference to entry 6 in Table 2 shows that

$$y(t) = (t - t_0)\mathrm{H}(t - t_0).$$

The effect that $\delta(t - t_0)$ has in the equation $\ddot{y} = \delta(t - t_0)$ follows from the formula for $\dot{y}$ we derived previously:

$$\dot{y} = \mathrm{H}(t - t_0) + c_1.$$

Thus the presence of $\delta(t - t_0)$ produces a jump of height 1 in $\dot{y}$ at the point t_0.

example 7

To solve the initial-value problem

$$\ddot{y} + 2\dot{y} + y = 2\delta(t - 1), \qquad y(0) = 1, \qquad \dot{y}(0) = 1,$$

apply $\mathcal{L}$ to both sides:

$$s^2 Y(s) + 2sY + Y - 3 - s = 2e^{-s}.$$

Then

$$\begin{aligned} Y(s) &= \frac{3}{(s+1)^2} + \frac{s}{(s+1)^2} + 2\frac{e^{-s}}{(s+1)^2} \\ &= \frac{2}{(s+1)^2} + \frac{1}{s+1} + 2\frac{e^{-s}}{(s+1)^2}. \end{aligned}$$

From Tables 1 and 2, we find

$$y(t) = 2te^{-t} + e^{-t} + 2(t - 1)e^{-(t-1)}\mathrm{H}(t - 1).$$

Notice that $y(t)$ is continuous at $t = 1$. However, its derivative $\dot{y}(t)$ jumps up by 2 at $t = 1$.

The advantage of using the Laplace transform in Example 6 lies partly in the natural way in which it handles the δ function. Additional advantages are that the method incorporates initial conditions at $t = 0$ in a routine way and also allows for the easy handling of discontinuous forcing functions $f(t)$. In practice, these advantages are sometimes overshadowed by the inability of the Laplace transform method to deal with the nonlinear problems that are so important in contemporary applied mathematices. The following example illustrates this point.

example 8

In Example 8 of the optional Section 4C of Chapter 4, we discuss an escapement mechanism for a pendulum clock. The escapement is the mechanism that regulates the transfer of energy from the clockworks to the pendulum. The nonlinear model we use there imparts an impulse to the pendulum's motion when it is near the vertical position, that is, near where the displacement angle $y(t)$ is zero. If we're to analyze the mechanism using Laplace transforms we can consider only an idealization of that mechanism in which the undamped nonlinear pendulum equation is linearized by replacing $\sin y$ by y. We'll assume a periodic escapement force of size a applied by instantaneous impulses. A very much simplified version of the initial-value problem is thus

$$\ddot{y} + y = a \sum_{j=1}^{\infty} (-1)^j \delta(t - j\pi), \qquad y(0) = \dot{y}(0) = 0.$$

The timing of the impulses at times $\tau_j = j\pi$ is dictated by the desire to have them occur when the unforced pendulum has angular displacement zero. The alternating signs in the impulse function are designed to produce a decreasing $y(t)$ at $t = \pi$, then enhance an increasing $y(t)$ as the pendulum swings back through $y = 0$, and so on. (For all practical purposes, we could stop the forcing after finitely many terms, but there is no technical advantage to doing so.) Applying $\mathcal{L}$ to both sides gives

$$s^2 Y(s) + Y(s) = a \sum_{j=1}^{\infty} (-1)^j e^{-j\pi s}, \quad \text{or} \quad Y(s) = a \sum_{j=1}^{\infty} (-1)^j \frac{e^{-j\pi s}}{s^2 + 1}.$$

Taking inverse transforms, we get

$$y(t) = a \sum_{j=1}^{\infty} (-1)^j \sin(t - j\pi) \mathrm{H}(t - j\pi).$$

The series converges for each fixed t, because $\mathrm{H}(t - j\pi) = 0$ as soon as $j\pi > t$. The graph of the first few terms, with $a = 1$, appears in Figure 8.6; note its

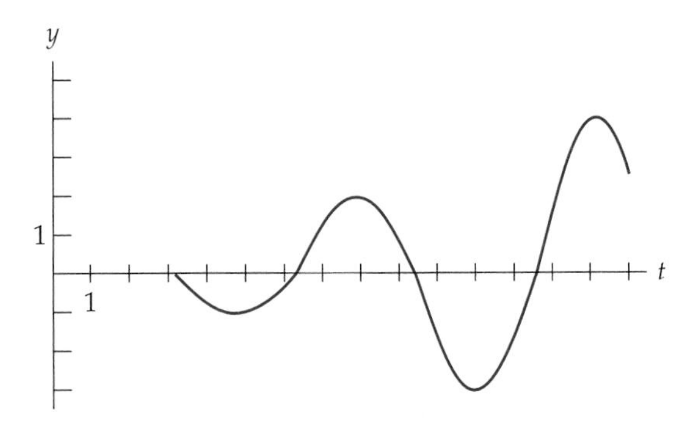

Figure 8.6 Linear undamped pendulum response to impulses $a = \pm 1$ at $t = j\pi$.

slightly disjointed appearance at the axis crossings, indicating jerky motion due to the abrupt application of impulses at the crossing instants.

example 9

In Chapter 3, Section 3D, we introduced an integral formula for the solution of the second-order initial-value problem

$$\ddot{y} + a\dot{y} + by = f(t), \qquad y(0) = \dot{y}(0) = 0, \quad \text{namely} \quad y(t) = \int_0^t g(t-u) f(u)\,dt.$$

By letting $f(t) = \delta(t - t_0)$ for fixed t_0 we find that for $t > t_0$ we get $y(t) = g(t - t_0)$. This gives us an interpretation for the function $g(t - u)$. For a fixed value u, this function of t is the response by the system governed by the operator $D^2 + aD + b$ to an instantaneous impulse of size 1 at the single point u. Indeed, this property of the function is sometimes taken as the definition of $g(t - u)$, but then of course you still have the problem of calculating a formula for it. For example, if the operator is $D^2 + 2D + 1$, the Green's function is $g(t - u) = (t - u)e^{-(t-u)}$, so the response to a unit impulse at the instant $t = 2$ is $y(t) = (t - 2)e^{-(t-2)}\mathrm{H}(t - 2)$.

EXERCISES

1. Sketch the graphs of $\Delta_1(t)$, $\Delta_1(t - 2)$, and $\mathrm{H}(t - 4)$.

2. Verify that if $\Delta_\tau(t) = 1$ for $0 \le t < \tau$ and $\Delta_\tau(t) = 0$ otherwise, then

$$\Delta_\tau(t) = [\mathrm{H}(t) - \mathrm{H}(t - \tau)].$$

Conclude then that

$$\Delta_\tau(t - t_0) = [\mathrm{H}(t - t_0) + \mathrm{H}(t - \tau - t_0)].$$

Evaluate the integrals in Exercises 3 through 8.

3. $\int_{-\infty}^{\infty} \delta(t - \pi) \cos t\, dt.$

4. $\int_{-1}^{1} \delta(t) \cos t\, dt.$

5. $\int_2^3 \delta(t - \pi)(t - 1)\, dt.$

6. $\int_{-2}^{2} \delta(t)(t^2 - 1)\, dt.$

7. $\int_2^4 \delta(t - \pi)(t^2 + 1)\, dt.$

8. $\int_{-2}^{2} \delta(t) e^{(t^2-1)}\, dt.$

Compute the Laplace transforms in Exercises 9 through 14.

9. $\int_0^\infty e^{-st}\delta(t - \pi)\, dt.$

10. $\int_0^\infty e^{-st}\delta(t - \pi) \cos t\, dt.$

11. $\int_0^\infty e^{-st}\mathrm{H}(t - \pi)\, dt.$

12. $\int_0^\infty e^{-st}\mathrm{H}(t - 1) \sin 2t\, dt.$

13. $\int_0^\infty e^{-st}\delta(t - \pi)\, dt.$

14. $\int_0^\infty e^{-st}\delta(t - 1) \sin 2t\, dt.$

15. In discussing the physical interpretation of the integrals $\int_{t_0}^{t_0+\tau} (1/\tau)\Delta_\tau(t - t_0)\, dt$ we saw that total impulse equals change in momentum under the assumption of constant mass in $F(t) = m(dv/dt)$.

Show that the relation between impulse and momentum still holds if mass is allowed to vary and the more general definition of force holds: $F(t) = (d/dt)(mv)$.

16. Solutions to second-order equations forced by δ functions are notably nonsmooth. The following example is typical: $\ddot{y} = \delta(t-1)$, $y(0) = \dot{y}(0) = 0$. Solve this initial-value problem, sketch the graph of the solution, and show that $\dot{y}(1)$ fails to exist.

17. Consider the convolution integral with $g(u) = 0$ if $u < 0$:

$$\Delta_\tau * g(t) = \int_0^t \Delta_\tau(u) g(t-u)\, du.$$

(a) By making simple changes of variable in the integral, show that this convolution can be written

$$\Delta_\tau * g(t) = \int_{t-\tau}^t g(u)\, du, \quad \text{if } t \geq \tau \geq 0.$$

(b) Show that the convolution can also be written

$$\Delta_\tau * g(t) = \int_0^t g(u)\, du, \quad \text{if } \tau \geq t \geq 0.$$

18. (a) Solve $\dot{y} = \Delta_\tau(t - t_0)$ under the initial condition $y(0) = 0$.

(b) Sketch the graph of the solution $y_\tau(t)$ found in part (a) when $t_0 = 1$.

(c) Find $\lim_{\tau \to 0+} y_p(t)$, the limit of $y_\tau(t)$ as t approaches zero from the right, and compare the result with $H(t - t_0)$.

19. Express the solution of the initial-value problem

$$\ddot{y} = \delta(t - t_0), \qquad y(0) = \dot{y}(0) = 0,$$

in terms of $H(t - t_0)$ if $t_0 > 0$.

20. (a) Solve the initial-value problems

$$\ddot{y} + y = \delta(t - \pi), \qquad y(0) = \dot{y}(0) = 0,$$

and

$$\ddot{y} + y = \delta(t - 2\pi), \qquad y(0) = \dot{y}(0) = 0.$$

(b) Use the results of part (a) to solve

$$\ddot{y} + y = c\delta(t - \pi) + d\delta(t - 2\pi), \quad y(0) = \dot{y}(0) = 0,$$

where c and d are constants.

(c) Show that if $c = d$ in part (b) then the solution is not only 0 for $0 < t < \pi$, but also for $2\pi < t$.

21. Verify that $\mathcal{L}[\delta(t - t_0) f(t)] = e^{-t_0 s} f(t_0)$. Then use this result to solve

$$\ddot{y} + 2\dot{y} + y = \delta(t - 1)t, \qquad y(0) = \dot{y}(0) = 0.$$

22. Verify that $\mathcal{L}[\mathrm{H}(t-t_0)f(t)] = e^{-t_0 s}\mathcal{L}[f(t+t_0)](s)$. Then use this result to solve

$$\ddot{y} + 2\dot{y} + y = \mathrm{H}(t-1)t, \qquad y(0) = \dot{y}(0) = 0.$$

Solve the initial-value problems in Exercises 23 though 28 using transforms or Green's functions, whichever seems more convenient.

23. $\ddot{y} + \dot{y} = \delta(t-1)$, $y(0) = 0$, $\dot{y}(0) = 0$.

24. $\ddot{y} + \dot{y} = \delta(t-1)$, $y(0) = 1$, $\dot{y}(0) = 0$.

25. $\ddot{y} - y = 2\delta(t-2)$, $y(0) = 0$, $\dot{y}(0) = 0$.

26. $\ddot{y} + 4y = \delta(t-\pi)$, $y(0) = 0$, $\dot{y}(0) = 0$.

27. $2\ddot{y} + y + 2y = \delta(t-1) - \delta(t-2)$, $y(0) = 0$, $\dot{y}(0) = 1$.

28. $\ddot{y} - 4y = \delta(t-3)$, $y(0) = 1$, $\dot{y}(0) = -1$.

29. In Example 3 of the text we identified $(d/dt)\mathrm{H}(t-t_0)$ with $\delta(t-t_0)$ by using formal integration by parts. Using a similar argument, show that if $t_0 > 0$ then

$$\int_0^\infty \delta'(t-t_0)g(t)\,dt = -g'(t_0).$$

30. Verify the statement about $\dot{y}(t)$ at the end of Example 7. The differentiation must be done by the cases $t < 1$ and $1 \le t$; it is not legitimate in this context to differentiate $\mathrm{H}(t-1)$ to get $\delta(t-1)$,

31. Referring to Example 8 of the text, make a direct determination of the constants c_1 and c_2 in

$$y(t) = c_1 t e^{-t} + c_2 e^{-t} + 2(t-1)e^{-(t-1)}\mathrm{H}(t-1)$$

so as to satisfy $y(0) = \dot{y}(0) = 1$.

32. **(a)** Referring to Example 8 of the text, show that if the factors $(-1)^j$ are removed from the forcing function then the solution $y(t)$ to the otherwise unaltered initial-value problem comes to a dead stop at $t = 2\pi$, and then repeats its behavior over $0 \le t < 2\pi$ periodically thereafter.

(b) Sketch the resulting graph for part (a), and give a physical explanation for its behavior.

33. Example 8 of the text describes a very much simplified pendulum model with periodic instantaneous forcing based on minimal feedback. Here we make the initial-value problem slightly more elaborate by introducing linear damping with damping constant $k > 0$:

$$\ddot{y} + k\dot{y} + y = a\sum_{j=1}^{\infty}(-1)^j\delta(t-\tau_j), \qquad y(0) = \dot{y}(0) = 0.$$

The time constants $\tau_j = \tau_j(k)$ depend on k and are determined to be those times at which the unforced pendulum would have zero displacement if it were started with initial conditions $y(0) = 0$, $\dot{y}(0) = z_0 \neq 0$. (Chapter 3, Section 2, shows that the τ_j are integer multiples of a fixed number, even though the damped oscillation isn't periodic overall.) The point of the impulses is to overcome the damping effect.

(a) Solve the associated homogeneous equation to determine the values of the times τ_j, $j = 1, 2, 3, \ldots$.

(b) Solve the given initial-value problem in terms of shifted Heaviside functions, using the impulse times τ_j found in part (a).

(c) Make a sketch of the solution $y(t)$ assuming $a = 1$ and $k = 0.75$; this is best done using a computer.

9 CHAPTER

Variable-Coefficient Methods

The tie that binds this chapter together is the development of techniques that go beyond those for the second-order constant-coefficient linear equations that have predominated in our study so far. Here we deal with linear equations with continuously variable coefficients. In addition, the solution of boundary-value problems in Section 3 makes it appropriate to take up in Section 4 the shooting method for numerical solution of such problems, both linear and nonlinear.

The discussion in Chapter 3 shows that if $f(x)$ is a continuous function and a and b are constants, the linear differential equation

$$y'' + ay' + by = f(x)$$

is such that all its solutions can be obtained by specializing the constants c_1 and c_2 in a formula that's valid on the domain interval of $f(x)$, namely,

$$y = c_1 y_1(x) + c_2 y_2(x) + y_p(x).$$

Much of this chapter is devoted to seeing in detail that a similar solution formula is valid even if $a = a(x)$ and $b = b(x)$ are allowed to be continuous functions of x defined on an interval I. The most obvious difference will be that we can no longer expect $y_1(x)$ and $y_2(x)$ to have the simple forms $x^l e^{\alpha x} \cos \beta x$ and $x^l e^{\alpha x} \sin \beta x$ that they have when a and b are constant. We've seen by direct computation that for the constant-coefficient case an initial-value problem for the homogeneous second-order equation has a unique solution defined on I. If the coefficients are continuous functions the same conclusion holds, and that case is covered by Theorem 1.1 in the introduction to Chapter 4. We restate the theorem here, specialized to linear equations.

1.1 EXISTENCE AND UNIQUENESS THEOREM
Let $a(x)$ and $b(x)$ be continuous on an interval I. Then for every x_0 in I the initial-value problem

$$y'' + a(x)y' + b(x)y = 0, \qquad y(x_0) = y_0, \qquad y'(x_0) = z_0$$

has a unique solution defined on I.

Our method for constructing solution formulas to the general linear equation with continuous coefficients will be reduced to showing first that finding just one solution to the corresponding homogeneous equation is sufficient to generate all solutions to the homogeneous equation. We'll then take up nonhomogeneous equations in Section 2.

1 INDEPENDENT SOLUTIONS AND THEIR ZEROS

1A Independent Solutions

Recall from Chapter 3, Section 1, that two functions y_1 and y_2 defined on the same interval are **linearly independent** if whenever a linear combination $c_1y_1 + c_2y_2$ is identically zero the constant coefficients c_1 and c_2 must both be zero. It is a routine exercise to show that y_1 and y_2 are linearly independent if and only if neither one is a constant multiple of the other; correspondingly, y_1 and y_2 are **linearly dependent** if one of them *is* a constant multiple of the other, say $y_2 = ky_1$. When y_1 and y_2 are linearly dependent, the general linear combination $c_1y_1+c_2y_2$ then reduces to a multiple of a single function. For example, if $y_2 = ky_1$, then $c_1y_1 + c_2y_2 = (c_1 + c_2k)y_1$.

example 1 A glance at the graphs of $y_1(x) = e^x$ and $y_2(x) = e^{-x}$ in Figure 9.1(a) shows that there is no interval on which one of these functions is a constant times the other, because one of them increases, the other decreases, and both are positive. Hence

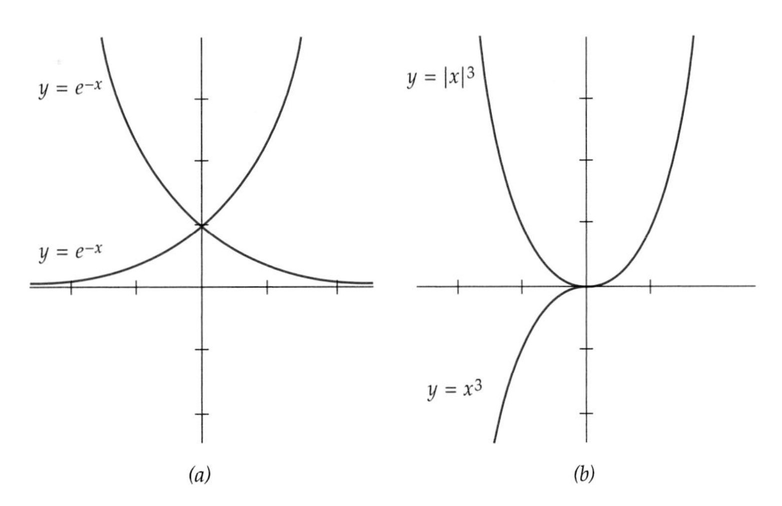

Figure 9.1 (*a*) Independent on every interval.
(*b*) Independent on some intervals.

the two functions are linearly independent. The same is true of any two exponentials $e^{\alpha x}$ and $e^{\beta x}$ where $\alpha \neq \beta$. For if $e^{\alpha x} = ce^{\beta x}$, then $e^{(\alpha-\beta)x} = c$. But $e^{(\alpha-\beta)x}$ is a constant only if $\alpha = \beta$. Hence $e^{\alpha x}$ and $e^{\beta x}$ are independent unless they are equal.

example 2

The formulas $y_1(x) = x^3$ and $y_2(x) = |x|^3$ define independent functions on some intervals and dependent functions on other intervals. On any interval consisting of only positive numbers, the two functions are equal and so are dependent. On a negative interval, $y_2(x) = -y_1(x)$, as shown in Figure 9.1(b). This is again a dependence relation. However, over an interval $a < x < b$ with $a < 0 < b$, neither function is a unique constant multiple of the other, so the functions thus defined are linearly independent.

The routine we used to solve constant-coefficient linear equations in Chapter 3 doesn't apply in general if an operator's coefficients aren't constant. (If we could factor variable-coefficient operators, we could proceed routinely as we did in Chapter 3 to perform successive integrations to find independent solutions, but we have no general method for factoring such linear operators.) However, it sometimes happens that we have a method for finding one **nontrivial solution** y_1, that is, one that's not identically zero for all values of x in the interval I, where the homogeneous equation

$$y'' + a(x)y' + b(x)y = 0$$

is defined. For example, we may be able to make a shrewd guess at a solution formula. A more likely possibility is that Theorem 1.1 guarantees the existence of a solution that we can study using the numerical methods of Chapter 4, Section 4. Power series representations discussed in Sections 6 and 7 of this chapter are also an option if the coefficients $a(x)$ and $b(x)$ are also representable by power series. In any case, to be able to write solutions as linear combinations

$$y = c_1 y_1 + c_2 y_2,$$

we will need to find a linearly independent solution y_2. For y_2 to be independent of y_1, there must be a *nonconstant* factor $u(x)$ such that

$$y_2(x) = y_1(x)u(x).$$

It's not at all obvious that we should be able to find such a factor $u(x)$ if we don't already know $y_2(x)$, but Theorem 1.2 provides a formula for $u(x)$ in terms of $y_1(x)$ and the coefficient $a(x)$. Example 8 in Chapter 3, Section 1A, illustrates the method, but here we'll be dealing with an equation with nonconstant coefficients.

example 3

It's easy to check that the linear differential equation

$$y'' - \frac{2}{x}y' + \frac{2}{x^2}y = 0, \qquad x > 0,$$

has $y_1(x) = x$ for a solution. To find an independent solution, try

$$y(x) = y_1(x)u(x) = xu(x).$$

We have

$$y = xu, \qquad y' = xu' + u, \qquad y'' = xu'' + 2u'.$$

When we substitute these expressions into the differential equation, we get

$$xu'' + 2u' - \frac{2}{x}(xu' + u) + \frac{2}{x^2}xu = 0.$$

This simplifies to $xu'' = 0$. Since we're assuming $x > 0$ we can divide x out, leaving $u'' = 0$. The solution is

$$u(x) = c_1x + c_2.$$

So, since $y = uy_1$, and $y_1 = x$, we get

$$y(x) = y_1(x)(c_1x + c_2) = x(c_1x + c_2).$$

We'll see that this is the general solution, but we can pick out a solution independent of $y_1(x) = x$ by setting $c_1 = 1$ and $c_2 = 0$, giving $y_2(x) = x^2$. The result of applying the method for finding a second independent solution differs from the explicit proof in Theorem 1.4 of Chapter 3, Section 1B, that showed we had found the general solution. But here we haven't yet proved that just because we have two independent solutions, that we've found the most general solution. For that we'll need the help of Theorem 1.3 proved later in the section.

The routine we applied in Example 3 works quite generally, and rather than go through it repeatedly we can condense the general result into a single theorem. Because the technique that leads to the next theorem involves solving only a first-order equation, it is sometimes described as an **order reduction method.**

1.2 THEOREM

Assume $y'' + a(x)y' + b(x)y = 0$ has continuous coefficient functions $a(x)$ and $b(x)$ on an interval I. If $y_1(x)$ is a solution of the equation and is nonzero on a subinterval J of I, then

$$y_2(x) = y_1(x)\int \frac{\exp\left[-\int a(x)\,dx\right]}{y_1^2(x)}\,dx$$

defines a solution on J that is independent of $y_1(x)$. The domain of $y_2(x)$ extends to the entire interval I.

Proof. We look for a nonconstant factor $u(x)$ such that $y_2(x) = y_1(x)u(x)$. Define $y = y_1u$ on J. Then

$$y' = y_1'u + y_1u', \quad \text{and} \quad y'' = y_1''u + 2y_1'u' + y_1u''.$$

Substitution of these expressions for y, y', and y'' into the differential equation enables us to collect the terms containing u on the right:

$$y_1u'' + \left[2y_1' + a(x)y_1\right]u' + \left[y_1'' + a(x)y_1' + b(x)y_1\right]u = 0.$$

The factor $\left(y_1'' + a(x)y_1' + b(x)y_1\right)$ is zero because y_1 is a solution of the original equation. The remaining terms form a differential equation of first order for $u'(x)$, namely

$$(u')' + \left(\frac{2y_1'}{y_1} + a(x)\right)u' = 0,$$

valid on every interval where $y_1(x) \neq 0$. An exponential integrating factor for this first-order linear equation is the exponential of

$$\int \left(\frac{2y_1'(x)}{y_1(x)} + a(x)\right) dx = 2\log|y_1(x)| + \int a(x)\,dx.$$

We find

$$u'(x) = c_1 \exp\left(-2\log|y_1(x)| - \int a(x)\,dx\right) = c_1 \frac{\exp(-\int a(x)\,dx)}{y_1^2(x)}.$$

Integration gives

$$u(x) = c_1 \int \frac{\exp(-\int a(x)\,dx)}{y_1^2(x)}\,dx + c_2.$$

We take $c_1 = 1$ and $c_2 = 0$ to get the solution $y_2 = y_1 u$ as stated in the theorem. This formula defines a solution independent of y_1 on J, because for the factor $u(x)$ to be constant, the integrand $y_1^{-2}\exp(-\int a\,dx)$ would have to be identically zero on I; that's impossible, because the integrand is the reciprocal of the continuous nonzero function $y_1^2 \exp(\int a\,dx)$. Finally, y_2 extends uniquely to all of I by the Existence and Uniqueness Theorem 1.1. ■

Remark. Because of the specific role played by $a(x)$, the formula in Theorem 1.2 applies only when the differential equation is normalized to have leading coefficient 1.

---

example 4 This example illustrates how the apparent problem with zero-values of y_1 in the denominator resolves itself in practice. The **Bessel equation** of index $\frac{1}{2}$ can be written

$$x^2y'' + xy' + (x^2 - \tfrac{1}{4})y = 0.$$

Suitably normalized, the equation is

$$y'' + \frac{1}{x}y' + \left(1 - \frac{1}{4x^2}\right)y = 0.$$

Direct substitution shows that the formula

$$y_1(x) = \frac{\sin x}{\sqrt{x}}, \qquad x > 0,$$

provides a solution. Since $a(x) = 1/x$, Theorem 1.2 gives

$$y_2(x) = \frac{\sin x}{\sqrt{x}} \int \frac{x}{\sin^2 x} e^{-\ln x}\, dx = \frac{\sin x}{\sqrt{x}} \int \frac{dx}{\sin^2 x}$$
$$= -\frac{\sin x}{\sqrt{x}} \cot x = -\frac{\cos x}{\sqrt{x}}.$$

We see directly that the solutions y_1 and y_2 are independent, because neither one is a constant multiple of the other, just as is the case with $\cos x$ and $\sin x$. (Alternatively, note that y_1 is bounded near zero, while y_2 is unbounded.) By Theorem 1.3 on the following page the general solution for $x > 0$ will then be $y(x) = c_1 x^{-1/2} \sin x + c_2 x^{-1/2} \cos x$.

The order reduction method produces a formula for a second, independent solution of a second-order linear differential equation if we already have one solution. Finding that first solution may be fairly easy, involving only a little intelligent guessing. At the other extreme, we may begin with no particularly handy formula for a solution and have to rely on an existence theorem to tell us that there is a solution. From there we may proceed to study some particular solution, using a combination of numerical methods, together with the form of the differential equation, to deduce properties of the solution. Once we are familiar enough with this first solution to give it a suitable name, we can then proceed to derive an independent solution by the order reduction method or some other way. Solutions arrived at this way constitute the great majority of the so-called "special functions," also called "higher transcendental functions," as opposed to the "elementary" solutions of Example 4. Prime examples of higher transcendental functions are the index ν Bessel functions of Section 7B.

We've seen by direct computation that for the constant-coefficient case $y'' + ay' + by = 0$, two independent solutions y_1 and y_2 to a homogeneous second-order equation suffice to represent all solutions as linear combinations $c_1 y_1(x) + c_2 y_2(x)$. If the coefficients $a(x)$ and $b(x)$ are continuous functions the same result still holds. Consequently, satisfying arbitrary initial conditions $y(x_0) = y_0$ and $y'(x_0) = z_0$ amounts to being able to determine c_1 and c_2 uniquely from the equations

$$c_1 y_1(x_0) + c_2 y_2(x_0) = y_0,$$
$$c_1 y_1'(x_0) + c_2 y_2'(x_0) = z_0.$$

An attempt to solve these equations for c_1 and c_2 by elimination (see Exercise 30) shows that there is a unique solution if and only if the **Wronskian determinant** $w(x) = w[y_1, y_2](x)$ is not zero at x_0, where $w(x)$ is defined by a 2-by-2 determinant:

$$w[y_1, y_2](x) = \begin{vmatrix} y_1(x) & y_2(x) \\ y_1'(x) & y_2'(x) \end{vmatrix} = y_1(x) y_2'(x) - y_2(x) y_1'(x).$$

We need this observation along with Theorem 1.1 to prove the next theorem. The theorem follows also from Theorem 7.11 in the optional Section 1E of Chapter 7.

1.3 THEOREM

Let $a(x)$ and $b(x)$ be continuous on an interval I. If y_1 and y_2 are linearly independent solutions of the homogeneous equation $y'' + a(x)y' + b(x)y = 0$ on I, then (i) the general solution is given by the 2-parameter family $y = c_1 y_1(x) + c_2 y_2(x)$; (ii) the Wronskian satisfies $w(x) \neq 0$ for all x in I; and in addition, (iii) arbitrary initial conditions $y(x_0) = y_0$, $y'(x_0) = z_0$ at a point x_0 in I will be satisfied by a unique choice of c_1 and c_2.

Proof. Let $y(x)$ be some solution defined on I, and define $Y(x)$ by

$$Y(x) = c_1 y_1(x) + c_2 y_2(x) - y(x).$$

Let x_0 be an arbitrary point in I. By the remark preceding the theorem, the equations

$$Y(x_0) = c_1 y_1(x_0) + c_2 y_2(x_0) - y(x_0) = 0,$$
$$Y'(x_0) = c_1 y_1'(x_0) + c_2 y_2'(x_0) - y'(x_0) = 0$$

enable us to solve uniquely for c_1 and c_2 if and only if $w(x_0) \neq 0$ is true. Assuming for the moment that the condition $w(x_0) \neq 0$ is satisfied, the uniqueness part of Theorem 1.1 implies that $Y(x)$ must equal the obvious identically zero solution on I, since the zero solution and its derivative are also zero at x_0. This would prove the first assertion of the theorem. But if $w(x_0) = 0$, the equations

$$c_1 y_1(x_0) + c_2 y_2(x_0) = 0,$$
$$c_1 y_1'(x_0) + c_2 y_2'(x_0) = 0$$

would have solutions with c_1 and c_2 not both zero. It follows, again from the Uniqueness Theorem, that with these values for c_1 and c_2 the solution $c_1 y_1(x) + c_2 y_2(x)$ is identically zero. This contradicts linear independence of the solutions $y_1(x)$ and $y_2(x)$, so $w(x_0) \neq 0$. Since x_0 is an arbitrary point in I, (ii) is true. Finally, since $w(x_0) \neq 0$ the equations for initial conditions

$$c_1 y_1(x_0) + c_2 y_2(x_0) = y_0,$$
$$c_1 y_1'(x_0) + c_2 y_2'(x_0) = z_0$$

have a unique solution for c_1 and c_2. ■

EXERCISES

Sketch the relevant graphs, and explain why, or why not, the pairs of functions in Exercises 1 through 6 are linearly independent on their indicated intervals.

1. $\cos x, \sin x; -\infty < x < \infty$.

2. $x, x^2; -\infty < x < \infty$.

3. $\tan x, \sin x; -\pi/2 < x < \pi/2$.

4. $x^5, |x|^5; -\infty < x < \infty$.

5. $x^5, |x|^5; -\infty < x < 0$.

6. $e^x, xe^{2x}; -\infty < x < \infty$.

7. As defined in Chapter 3, two functions $y_1(x)$ and $y_2(x)$ are *linearly independent* on an x interval if the identity $c_1 y_1(x) + c_2 y_2(x) = 0$ holds only for coefficients $c_1 = c_2 = 0$. Show that y_1 and y_2 are linearly independent on an interval if and only if neither one is a constant multiple of the others.

Each of the equations in Exercises 8 through 12 has the given function as a solution. Verify this, and find a linearly independent solution using Theorem 1.2.

8. $y'' - 4y' + 4y = 0$; $y_1(x) = e^{2x}$.

9. $xy'' - (2x+1)y' + (x+1)y = 0$; $y_1(x) = e^x$.

10. $(x-1)y'' - xy' + y = 0$; $y_1(x) = e^x$.

11. $x^2y'' - 2xy' + 2y = 0$; $y_1(x) = x$, $x > 0$.

12. $(x^2-1)y'' - 2xy' + 2y = 0$; $y_1(x) = x$.

Each of the equations in Exercises 13 and 14 is of the kind that can be treated using the method of this section, but because the term in y is already absent, it's more straightforward to solve each equation directly by treating it as a first-order equation in the unknown function y', and then find y by integration. For each equation, find the general solution in the form $c_1y_1 + c_2y_2$.

13. $y'' + xy' = 0$.

14. $y'' + (1/x)y' = 0$.

Use Theorem 1.2 to derive a solution independent of $y_1(x)$ for each equation in Exercises 15 and 16, assuming $\alpha \neq 0$ is constant.

15. $y'' - 2\alpha y' + \alpha^2 y = 0$; $y_1(x) = e^{\alpha x}$.

16. $y'' + \alpha^2 y = 0$; $y_1(x) = \cos \alpha x$.

17. The *Legendre equation* of index 1 (see Example 5 in Section 6) is

$$(1-x^2)y'' - 2xy' + 2y = 0$$

and has a solution $y_1(x) = x$.

(a) Use Theorem 1.2 to find an independent solution in terms of elementary functions.

(b) Find the solution $y(x)$ to the Legendre equation such that $y(0) = 1$ and $y'(0) = 2$.

18. The **Hermite equation** of index 1 is

$$y'' - 2xy' + 2y = 0$$

and has a solution $y_1(x) = x$. Find an independent solution in power series form by using Theorem 1.2 and the expansion $e^{x^2} = \sum_{k=0}^{\infty} x^{2k}/k!$, and then integrating term by term.

19. Let $L(y) = x^2y'' - 2xy' + 2y$. Verify that $y_1(x) = x$ is a solution of $L(y) = 0$. Then find a solution of the nonhomogeneous equation $L(y) = x^2$ in the form $y(x) = y_1(x)u(x) = xu(x)$. [*Hint:* Substitute $y = xu(x)$ into the equation to be solved and then solve for u.]

20. Theorem 1.2 contains the formula $e^{-\int a(x)\,dx}$ that arose from an integrating factor used to solve a first-order linear equations. The same formula arises in part (b) of this exercise. Suppose y_1 and y_2 are solutions of the same homogeneous equation:

$$y_1'' + a(x)y_1' + b(x)y_1 = 0,$$
$$y_2'' + a(x)y_2' + b(x)y_2 = 0.$$

(a) Multiply the first equation by y_2 and the second by y_1, and subtract to show that

$$w = y_1 y_2' - y_2 y_1' \quad \text{satisfies} \quad w' + a(x)w = 0.$$

(b) Solve the first-order equation in part (a) to get

$$y_1 y_2' - y_2 y_1' = ce^{-\int a(x)\,dx}.$$

This result is called *Abel's formula.*

(c) Show that Abel's formula can be written using determinants as

$$\begin{vmatrix} y_1(x) & y_2(x) \\ y_1'(x) & y_2'(x) \end{vmatrix} = ce^{-\int a(x)\,dx}.$$

The determinant is called the *Wronskian determinant* of y_1 and y_2. (Wronskians are discussed more generally in Chapter 7, Section 1E.)

(d) Show that the Wronskian of solutions y_1 and y_2 of $y'' + b(x)y = 0$ is constant.

(e) Show that the Wronskian of the two solutions y_1 and y_2 is either identically zero or else never zero.

21. Verify that, if $L = D^2 + a(x)D + b(x)$ on some interval $x_1 < x < x_2$, then L acts linearly on twice-differentiable functions y_1 and y_2 on that interval. That is, show that

$$L(c_1 y_1 + c_2 y_2) = c_1 L(y_1) + c_2 L(y_2), \qquad c_1, c_2 \text{ constant.}$$

22. (a) Show that the **Euler differential equation** (see also Section 7A)

$$x^2 y'' + axy' + by = 0, \qquad a, b \text{ real constants,}$$

can be solved as follows, assuming $x > 0$. Let $y = x^\mu$, so that $y' = \mu x^{\mu-1}$, and $y'' = \mu(\mu - 1)x^{\mu-2}$. Show that for y to solve the differential equation, μ must satisfy the *indicial equation* $\mu^2 + (a - 1)\mu + b = 0$.

(b) Show that if the indicial equation has real roots $\mu_1 \neq \mu_2$ then, $y_1 = x^{\mu_1}$ and $y_2 = x^{\mu_2}$ are solutions.

(c) Show that if the indicial equation has complex conjugate roots $\mu_1 = \alpha + i\beta$ and $\mu_2 = \alpha - i\beta$, then $y_1 = x^\alpha \cos(\beta \ln x)$ and $y_2 = x^\alpha \sin(\beta \ln x)$ are solutions. Note that, by definition, $x^{\alpha+i\beta} = x^\alpha e^{i\beta \ln x}$ for $x > 0$.

(d) Show that if μ_1 is a double root of the indicial equation then $y_1 = x^{\mu_1}$ and $y_2 = x^{\mu_1} \ln x$ are solutions. Use Theorem 1.2 for this.

Use the results of the previous exercise to solve the Euler equations in Exercises 23 through 26.

23. $x^2 y'' + xy' - y = 0.$

24. $x^2 y'' + 4xy' + y = 0.$

25. $x^2 y'' + 3xy' + y = 0.$

26. $x^2 y'' + xy' + y = 0.$

27. Find independent solutions of the general Euler equation of Exercise 22 for $x > 0$ as follows.

(a) Let $t = \ln x$ and define $u(t) = y(e^t)$. Use the chain rule to show that $dy/dx = x^{-1}\dot{u}$ and $d^2y/dx^2 = x^{-2}(\ddot{u} - \dot{u})$.

(b) Show that $x^2 y'' + axy' + by = \ddot{u} + (a-1)\dot{u} + bu$ and hence that the Euler equation has solution $e^{\mu_0 t}x^{\mu_0}$, where μ_0 is a root of $\mu^2 + (a-1)\mu + b = 0$.

(c) Use Theorem 1.2 to find a solution independent of x^{μ_0}. Note particularly the case $\mu_0 = \frac{1}{2}(1-a)$.

28. The following systems have solutions that can be found by first eliminating one of the unknown functions of t. Find the general solution.

(a) $\dot{x} = x - (t-1)y,\ \dot{y} = 0.$

(b) $\dot{x} = -2t^{-2}y,\ t > 0,\ \dot{y} = x$. [*Hint:* Use Exercise 22.]

29. Let $f(x)$ and $g(x)$ be twice differentiable functions on an interval $a < x < b$ on which $f(x)g'(x) - f'(x)g(x) \neq 0$.

(a) Show that the 3-by-3 determinant equation

$$\begin{vmatrix} y & f & g \\ y' & f' & g' \\ y'' & f'' & g'' \end{vmatrix} = 0$$

is a second-order homogeneous linear equation on the interval $a < x < b$ having $f(x)$ and $g(x)$ as solutions.

(b) Find a second-order homogeneous linear equation having $f(x) = \sin x$ and $g(x) = x \sin x$ as solutions for $0 < x < \pi$.

(c) Find a second-order homogeneous linear equation having $f(x) = x$ and $g(x) = e^x$ as solutions for all $x \neq 1$.

30. Use the elimination method to show that the equations

$$Ac_1 + Bc_2 = y_0,$$

$$Cc_1 + Dc_2 = z_0,$$

have a unique solution for c_1 and c_2 if and only if $AD - BC \neq 0$.

31. It is sometimes erroneously inferred from experience with solutions of differential equations that the *Wronskian determinant*

$$w[f,g](x) = \begin{vmatrix} f(x) & g(x) \\ f'(x) & g'(x) \end{vmatrix}$$

of two linearly independent differentiable functions f and g can't equal 0. Put this idea to rest by computing $w[f,g](x)$ for all real x when $f(x) = x^2$, $g(x) = x|x|$. Note that you need to use the definition of derivative as a limit of a difference quotient to compute $g'(0)$.

32. The previous exercise may seem to contradict the assertion in Theorem 1.3 that $w[y_1, y_2](x) \neq 0$. Why doesn't it?

33. Solving the general second-order equation $y'' + a(x)y' + b(x)y = 0$ can be reduced to the problem of solving $u'' + q(x)u = 0$, with

$$q(x) = -\tfrac{1}{2}a'(x) - \tfrac{1}{4}a^2(x) + b(x) \quad \text{and} \quad y(x) = u(x)e^{-1/2\int a(x)\,dx}.$$

(a) Verify this assertion by substituting the expression for $y(x)$ into the more general linear equation. [*Hint:* Denote the exponential factor by $v(x)$ and note that $v' = -\frac{1}{2}av$ and $v'' = -\frac{1}{2}a'v + \frac{1}{4}a^2v$.]

(b) The *Bessel equation* of index $\frac{1}{2}$ can be written for $x > 0$ as

$$y'' + \frac{1}{x}y' + \left(1 - \frac{1}{4x^2}\right)y = 0.$$

Show that for this equation $q(x)$ is constant. Then use the approach outlined in part (a) to find two independent solutions.

34. Assume that $y_1(x)$ and $y_2(x)$ are linearly independent solutions of the homogeneous equation $y'' + a(x)y' + b(x)y = 0$, with coefficients continuous on an interval I. The purpose of this exercise is to show that $y_1(x)$ and $y_2(x)$ can't both be zero at a point x_0 in I. (See Theorem 1.4 in Section 1B for an alternative proof.)

(a) Suppose $y_1(x_0) = y_2(x_0) = 0$ for some x_0 in I. Use Theorem 1.1 to show that $y_1'(x_0) \neq 0$ and $y_2'(x_0) \neq 0$.

(b) Conclude from part (a) that there is a nonzero constant c such that $y_1'(x_0) = cy_2'(x_0)$ and hence from Theorem 1.1 that $y_1(x) = cy_2(x)$ for all x in I, thus contradicting independence of $y_1(x)$ and $y_2(x)$.

(c) Use solutions of the equation $y'' + y = 0$ to show that independent solutions can have a common value other than zero at a single point x_0.

1B Zeros of Solutions†

A point $x = x_0$ where a nontrivial solution $y(x)$ of $y'' + a(x)y' + b(x)y = 0$ assumes the value zero is called a **zero** of the solution. In geometric terms a zero is a point where the graph of a solution $y(x)$ crosses the x axis. It might seem that, along with crossing, tangency of the graph to the axis might be possible. However, tangency is impossible under the assumptions Theorem 1.1, because if x_0 is a point of tangency then we have not only $y(x_0) = 0$ but also $y'(x_0) = 0$. Thus by Theorem 1.1, $y(x)$ would have to be the trivial identically zero solution. Thus tangency to the x axis is ruled out. The same argument shows that solution graphs for a second-order equation can cross, but cannot be tangent to each other. (The analogous conclusion for first-order equations is that solution graphs of equations satisfying the assumptions of a first-order uniqueness theorem can have no points in common at all.) A related result is that a solution of a homogeneous second-order equation that has infinitely many zeros in a closed, bounded interval must be identically zero; to see why look at Exercise 11.

example 5

It's clear where the zeros of the two solutions $\sin x/\sqrt{x}$ and $\cos x/\sqrt{x}$ in Example 4 in Section 1A are located, because we know that $\sin x$ and $\cos x$, which are solutions of $y'' + y = 0$, have their zeros respectively at $n\pi$ and $(n + \frac{1}{2})\pi$ for integer n. More generally, $y'' + \beta^2 y = 0$ has solutions $\sin \beta x$ and $\cos \beta x$ with zeros at $n\pi/\beta$ and $(n + \frac{1}{2})\pi/\beta$.

For most equations there are no simple formulas for locating the zeros of a solution, so the following theorem is helpful for organizing an orderly numerical search, for example, using Newton's method. Briefly, the theorem says two things about solutions of homogeneous linear equations: (i) *two independent solutions can't have a common zero* and (ii) *the zeros of two independent solutions occur alternately.* The second statement has real content only if at least one solution has at least two zeros.

1.4 SEPARATION THEOREM

Assume $y'' + a(x)y' + b(x)y = 0$ has continuous coefficients $a(x)$ and $b(x)$ on an interval I and has independent solutions $y_1(x)$ and $y_2(x)$ on I. Then (i) the two solutions can't have a common zero and (ii) between each two consecutive zeros of one solution there is exactly one zero of the other solution.

Remark. Since independent solutions, for example, $\sin x$ and $\cos x$, are allowed to have values other than zero at the same point, part (i) of the conclusion shows that zeros play a very special role for solutions of homogeneous linear equations.

Proof. If we first single out $y_1(x)$ for attention, then Theorems 1.2 and 1.3 together show that all solutions independent of $y_1(x)$ have the form

$$y_2(x) = c_1 y_1(x) \int \frac{\exp(-\int a(x)\,dx)}{y_1^2(x)}\,dx + c_2 y_1(x),$$

where $c_1 \neq 0$. Suppose that $y_1(x) \neq 0$ on an interval $\alpha < x < \beta$, so on that interval we can write

$$\frac{y_2(x)}{y_1(x)} = c_1 \int \frac{\exp(-\int a(x)\,dx)}{y_1^2(x)}\,dx + c_2.$$

Using the quotient rule we differentiate this equation, then multiply by $y_1^2(x)$, to get

$$y_1(x)y_2'(x) - y_2(x)y_1'(x) = c_1 e^{-\int a(x)\,dx}.$$

This is called **Abel's formula,** and since all the functions in it are continuous on I the formula is valid on all of I, even at points where $y_1(x) = 0$. (For alternative proofs see Exercise 20 of Section 1A, and in a wider context, Exercise 14 in Chapter 7, Section 1E.) Since $c_1 \neq 0$, Abel's formula shows that y_1 and y_2 can't both be zero at $x = \alpha$, at $x = \beta$, or at any other point in I, thus proving (i).

To prove (ii) assume α and β are consecutive zeros of $y_2(x)$. Again from Abel's formula, $y_1(\alpha) \neq 0$, $y_2'(\alpha) \neq 0$, and $y_1(\beta) \neq 0$, $y_2'(\beta) \neq 0$. Since α and β are consecutive zeros, $y_2(x)$ must be increasing at one of the two points and decreasing at the other, so $y_2'(\alpha)$ and $y_2'(\beta)$ have opposite signs. Since $y_2(\alpha) = y_2(\beta) = 0$, $y_1(\alpha)y_2'(\alpha)$, and $y_1(\beta)y_2'(\beta)$ have the same sign by Abel's formula, so $y_1(\alpha)$ and $y_1(\beta)$ must have opposite signs. Thus the continuous graph of $y_1(x)$ crosses the

x axis for some x_0 with $\alpha < x_0 < \beta$. Finally, reversing the roles of y_1 and y_2, we conclude that $y_2(x)$ is zero between consecutive zeros of $y_1(x)$. If $y_1(x)$ had more than one zero between α and β, then α and β could not be consecutive zeros of $y_2(x)$, and vice versa. Hence zeros of the two solutions alternate. ■

example 6

We've already remarked in Example 5 that the zeros alternate for the solutions $y_1(x) = \cos\beta x$ and $y_2(x) = \sin\beta x$ of $y'' + \beta^2 y = 0$. Of course there are other pairs of independent solutions for this equation. Exercise 1 asks you to verify that all independent pairs have the form

$$y_3(x) = a\cos\beta x + b\sin\beta x \quad \text{and} \quad y_4(x) = c\cos\beta x + d\sin\beta x,$$

where $ad - bc \neq 0$, so under that condition the zeros of y_3 and y_4 alternate. Since these linear combinations both have the form $A\sin\beta(x - x_0)$, it follows that for every x_0 there are constants a and b such that $y_3(x_0 + n\pi/\beta) = 0$ for all integers n. This observation is used to implement Theorem 1.5 below, and it will be important that, while β^2 fixes the space between zeros at π/β, there will be a solution of $y'' + \beta^2 y = 0$ with one of its zeros located at an arbitrary point x_0.

The next theorem enables us to draw conclusions about the zeros of solutions of one equation if we know enough about the zeros of solutions of a related equation, both equations being of the special form $y'' + q(x)y = 0$.

1.5 COMPARISON THEOREM

Let $y_1(x)$ and $y_2(x)$ be respective solutions of the two equations

$$y'' + q_1(x)y = 0 \quad \text{and} \quad y'' + q_2(x)y = 0,$$

where $q_1(x)$ and $q_2(x)$ are continuous on an interval I. Suppose that $q_1(x) \geq q_2(x)$ everywhere on I with strict inequality at least on some subinterval. Then $y_1(x)$ has at least one zero strictly between consecutive zeros of $y_2(x)$.

Remark. An aid in remembering which way the implication goes is to recall that solutions of the harmonic oscillator equation $y'' + \beta^2 y = 0$ have more zeros as β^2 increases.

Proof. Let α and β be consecutive zeros of $y_2(x)$, and suppose in contradiction to the statement of the theorem that $y_1(x) \neq 0$ for $\alpha < x < \beta$. Replacing y_1 or y_2 by its negative does nothing to alter the zeros, so we'll assume that both solutions are nonnegative for $\alpha < x < \beta$, noting that then y_2 is increasing at α and decreasing at β. Let

$$w(x) = y_1(x)y_2'(x) - y_2(x)y_1'(x).$$

Since $y_2(\alpha) = y_2(\beta) = 0$, and $y_2'(\alpha) > 0$ while $y_2'(\beta) < 0$, we find that

$$w(\alpha) = y_1(\alpha)y_2'(\alpha) \geq 0 \quad \text{while} \quad w(\beta) = y_1(\beta)y_2'(\beta) \leq 0.$$

Thus under our assumptions $w(x)$ can't increase overall as x goes from α to β. However, using the differential equations we see that

$$\begin{aligned} w'(x) &= y_1(x)y_2''(x) - y_2(x)y_1''(x) \\ &= -y_1(x)q_2(x)y_2(x) + y_2(x)q_1(x)y_1(x) \\ &= y_1(x)y_2(x)\left[q_1(x) - q_2(x)\right] \geq 0. \end{aligned}$$

It follows that $w(x)$ is not only nondecreasing between a and β but is strictly increasing on whatever subinterval $q_1(x) > q_2(x)$. The contradiction forces us to give up our temporary assumption that $y_1(x) \neq 0$ on $\alpha < x < \beta$. ■

example 7

That a nontrivial solution $y_1(x)$ of $y'' + x^2 y = 0$ has infinitely many zeros for both $x < 0$ and $x > 0$ follows from application of the Comparison Theorem 1.5 using the harmonic oscillator equation $y'' + y = 0$ and its solution $\sin x$. Indeed, using the notation of Theorem 1.5, on the intervals $x < 1$ and $1 < x$, we have $q_1(x) = x^2 > 1 = q_2(x)$. Theorem 1.5 also implies that the space between consecutive zeros of $y_1(x)$ tends to zero as $|x|$ increases. To see this note that if $|x| > \beta > 0$ then $x^2 > \beta^2$, so between each successive pair $\{n\pi/\beta, (n+1)\pi/\beta\}$ of zeros of $\sin \beta x$ there is a zero of $y_1(x)$. As β tends to infinity the length π/β of the interval between zeros of $\sin \beta x$ shrinks to zero.

It seems at first sight that the differential equations in Theorem 1.5 are too special for the theorem to be widely applicable. However Exercise 8 shows that solutions $y(x)$ and $u(x)$ of the two equations

1.6 $$y'' + a(x)y' + b(x)y = 0 \quad \text{and} \quad u'' + q(x)y = 0$$

are related by

1.7 $$y(x) = u(x)e^{1/2\int a(x)\,dx} \quad \text{if} \quad q(x) = b(x) - \tfrac{1}{4}a^2(x) - \tfrac{1}{2}a'(x).$$

Equation 1.7 shows that on an interval where $a(x)$ is continuous, $u(x)$ and $y(x)$ have the same zeros, so the equation for $u(x)$ is adequate for locating the zeros of $y(x)$.

example 8

The normalized Bessel equation of order ν is

$$y'' + \frac{1}{x}y' + \left(1 - \frac{\nu^2}{x^2}\right)y = 0.$$

Equations 1.6 and 1.7 show that for $x > 0$ solutions $y(x)$ satisfy

$$y(x) = \frac{u(x)}{\sqrt{x}}, \quad \text{where } u(x) \text{ satisfies } u'' + \left(1 + \frac{1 - 4\nu^2}{4x^2}\right)u = 0.$$

When $\nu = \frac{1}{2}$ the basic solutions are the elementary functions shown in Example 4 of Section 1A. (See Exercise 5.) We'll deal in this example with the case $0 \le \nu < \frac{1}{2}$, showing that a nontrivial solution of the Bessel equation has an infinite sequence of zeros with consecutive differences tending to π from below. The approach will be by comparison of the differential equation for $u(x)$ with harmonic oscillator equations $u'' + \beta^2 u = 0$ having solution zeros separated by π/β. We'll compare solutions using Theorem 1.5 in each of the two possible directions.

1. *Bessel solutions have infinitely many consecutive zeros spaced less than π apart.* With $0 \le \nu < \frac{1}{2}$ we have $q_1(x)$ tending decreasingly to 1 as x gets bigger, so

$$q_1(x) = 1 + (1 - 4\nu^2)/(4x^2) > 1 = q_2(x) \quad \text{for} \quad 0 < x.$$

Zeros of solutions of the harmonic oscillator equation $y'' + y = 0$ occur in consecutive pairs $\{x_0, x_0 + \pi\}$. Let x_0 be a zero of a Bessel solution. By Theorem 1.5 a Bessel solution must have a zero x_1 strictly between every pair $\{x_0, x_0 + \pi\}$. Let x_1 be the first Bessel zero larger than x_0 so $x_0 < x_1 < x_0 + \pi$. Hence $0 < x_1 - x_0 < \pi$, in other words Bessel zeros are spaced less than π apart.

2. *Spacing of consecutive Bessel zeros tends to π.* Suppose

$$q_1(x) = \beta^2 > 1 + (1 - 4\nu^2)/(4x^2) = q_2(x) \quad \text{for} \quad x > x_1.$$

We can achieve the inequality with $\beta > 1$, and β arbitrarily close to 1 if x_1 is large enough. To see that spaces between successive Bessel zeros tend to π as x gets large, note that there are harmonic oscillator solutions with no zero in arbitrarily located intervals of length π/β. Theorem 1.5 implies that consecutive Bessel zeros larger than x_1 can't be closer than $\pi/\beta < \pi$. Now let β tend to 1 as x_1 tends to infinity, so π/β tends to π from below. Hence the spacing of Bessel zeros tends to π along with π/β as β tends to π from above.

example 9

When $\nu > \frac{1}{2}$, a nontrivial solution of the Bessel equation has infinitely many zeros separated by more than π, and this spacing tends to π as the zeros tend to infinity. The proof uses the same ideas used in the previous example and is outlined in Exercise 10.

EXERCISES

1. Prove that all pairs of independent solutions of $y'' + \beta^2 y = 0$ have the form $y(x) = a\cos\beta x + b\sin\beta x$, $z(x) = c\cos\beta x + d\sin\beta x$, where $ad - bc \ne 0$.

2. **(a)** Given a real number x_0 prove that the constants in solutions $y(x) = a\cos\beta x + b\sin\beta x$ of $y'' + \beta^2 y = 0$ can be chosen so $y(x) = 0$ precisely when $x = x_0 + n\pi/\beta$ for all integers n.

   (b) Find all zeros of the solutions $\cos x - \sin x$ and $\cos x + \sin x$ of $y'' + y = 0$, and show that the zeros of the two solutions alternate.

   (c) Find all zeros of the solutions $2\cos x - \sin x$ and $3\cos x + \sin x$ of $y'' + y = 0$, and show that the zeros of the two solutions alternate.

3. Use the Separation Theorem 1.4 to prove that if $y'' + a(x)y' + b(x)y = 0$ has continuous coefficients for $x > 0$ and has a nontrivial solution with infinitely many positive zeros then every solution has infinitely many positive zeros.

4. Use the Separation Theorem 1.4 to prove that if $y'' + a(x)y' + b(x)y = 0$ has continuous coefficients for $x > 0$ and has a nontrivial solution with n zeros on an interval I then every nontrivial solution has at most $n + 1$ zeros on I.

5. Prove that all zeros of a nontrivial solution of the Bessel equation of order $\nu = \frac{1}{2}$ have the form $x_0 + n\pi$ for a fixed number x_0 depending on the solution and for all integers n.

6. Show that a nontrivial solution of Airy's equation $y'' + xy = 0$ has infinitely many zeros for $x > 0$ and at most one for $x < 0$. [*Hint:* Compare with a harmonic oscillator and with $y'' = 0$.]

7. Prove that if $\beta > \gamma > 0$ then $\sin \beta x$ has at least one zero between every consecutive pair of zeros of $\sin \gamma x$.

8. In $y'' + a(x)y' + b(x)y = 0$ let $a(x)$ be continuously differentiable and $b(x)$ continuous.

(a) Show that letting $y = uv$ in the differential equation we get an equation of the form $u'' + q(x)u = 0$ to be satisfied by $u(x)$ if we choose

$$v(x) = ce^{-1/2 \int a(x)\, dx} \quad \text{and} \quad q(x) = \frac{v''(x) + a(x)v'(x) + b(x)v(x)}{v(x)}.$$

(b) Show that $q(x) = b(x) - \frac{1}{4}a^2(x) - \frac{1}{2}a'(x)$ and hence that if $u(x)$ is a nontrivial solution of $u'' + q(x)u = 0$ then $y(x) = u(x)v(x)$ is a nontrivial solution of the given equation.

9. Assume that $q(x)$ is continuous for $x > x_0$ in the differential equation $y'' + q(x)y = 0$.

(a) Show that if there is a constant $\delta > 0$ such that $q(x) \geq \delta$ for $x > 0$ then every nontrivial solution of the equation has infinitely many zeros for $x > 0$.

(b) Use the equation $y'' + \frac{1}{4}x^{-2}y = 0$ to show that the conclusion in part (a) is false if the condition on $q(x)$ is replaced by the weaker condition $q(x) > 0$. [*Hint:* Try $y = x^{\mu}$.]

***10.** Use the Comparison Theorem 1.5 to prove the following facts about zeros of nontrivial solutions of the Bessel equation of order $\nu > \frac{1}{2}$. The ideas are modifications of the ones used in Example 8 of the text.

(a) Show that a solution has at most one zero for $0 < x < \frac{1}{2}\sqrt{4\nu^2 - 1}$.

(b) For $0 < x$, let $q_1(x) = 1 > 1 + (1 - 4\nu^2)/(4x^2) = q_2(x)$. Show that the zeros of a nontrivial solution of the Bessel equation are spaced apart by more than π.

(c) For $x > x_1 > 0$, let $q_1(x) = 1 + (1 - 4\nu^2)/(4x^2) > \beta^2 = q_2(x)$. Show that a solution has an infinite sequence of zeros with an upper bound for their spacing tending to π from above as x_1 tends to infinity. Conclude that the zero spacing of the Bessel solutions tends to π from above.

11. Assume $y'' + a(x)y' + b(x)y = 0$ has continuous coefficients on an interval I and has a solution $y(x)$ with distinct zeros z_n such that $\lim_{n\to\infty} z_n = x_0$, where x_0 is a finite point in I.

(a) Use the continuity of $y(x)$ to show that $y(x_0) = 0$.

(b) Use the definition of $y'(x_0)$ as a limit to show also that $y'(x_0) = 0$, and hence by Theorem 1.1 that $y(x)$ must be identically zero on I.

(c) Assume the basic property of the real numbers that if an infinite set S of numbers is contained in a closed, bounded interval $\alpha \le x \le \beta$ then there is a sequence with distinct values z_n in S that converges to some point z_0 in the interval. Using this property, show that a solution of the homogeneous differential equation that has infinitely many zeros in a closed, bounded interval must be the zero solution.

2 NONHOMOGENEOUS EQUATIONS

2A Variation of Parameters

The technique used in Section 1 to find an independent solution of a homogeneous differential equation can be applied just as well to solving a nonhomogeneous equation. We let

$$y(x) = y_1(x)u(x) \quad \text{in} \quad y'' + a(x)y' + b(x)y = f(x),$$

where $y_1(x)$ is a single nontrivial solution of the associated homogeneous equation, for which $f(x) = 0$. Because the multiplicative parameter u in the trial solution is allowed to be a function of x, the procedure is sometimes called *variation of parameters*, though that term is usually reserved for a procedure that starts with two independent solutions. This substitution works for the same reason it worked for the homogeneous equation in Section 1A: it leaves us with a first-order linear equation that we can solve for u'.

example 1

It's easy to check that the equation

$$x^2y'' - 2xy' + 2y = x^4$$

has $y_1(x) = x$ for a *homogeneous* solution. Letting $y = y_1(x)u(x) = xu(x)$, we compute

$$y = xu, \qquad y' = xu' + u, \qquad y'' = xu'' + 2u'.$$

Substitution followed by simplification of the differential equation yields

$$x^2(xu'' + 2u') - 2x(xu' + u) + 2xu = x^3u'' = x^4.$$

The last part of this equation reduces to $u'' = x$. Integrating this equation twice gives $u = \frac{1}{6}x^3 + c_1x + c_2$, so

$$y = y_1(x)u(x) = xu(x) \quad \text{or} \quad y = \tfrac{1}{6}x^4 + c_1x^2 + c_2x.$$

Finding the homogeneous solution $y_1(x) = x$ was done by guessing, but lacking a correct guess we could have tried to find a solution in power series form using methods to be described in Sections 6 and 7. (See also Exercise 11 of Section 1A.)

The number of integrations required above can be reduced from two to one if we already know two independent solutions y_1 and y_2 of the associated homogeneous equation. The routine is complicated enough that rather than repeat it for every application we'll standardize it, along with the final result, as follows. Thus suppose $a(x)$ and $b(x)$ are continuous on some interval I and that $f(x)$ is integrable. We work with the **normalized equation** in which the coefficient of y'' is divided out throughout the equation:

$$y'' + a(x)y' + b(x)y = f(x).$$

If y_1 and y_2 are independent homogeneous solutions, we form the linear combination

$$y(x) = y_1(x)u_1(x) + y_2(x)u_2(x),$$

where $u_1(x)$ and $u_2(x)$ are to be determined so that $y(x)$ is a solution of the nonhomogeneous equation. (Note that if u_1 and u_2 were constant, $y(x)$ would be a solution of the associated homogeneous equation.) What we do now is much like what we did in the previous example: compute the derivatives y' and y'' and substitute into the nonhomogeneous equation. Then rearrange the terms as follows:

$$(y_1'' + ay_1' + by_1)u_1 + (y_2'' + ay_2' + by_2)u_2 + (y_1u_1' + y_2u_2')' \\ + a(y_1u_1' + y_2u_2') + (y_1'u_1' + y_2'u_2') = f.$$

The first two collections of terms are zero because y_1 and y_2 are homogeneous solutions. What remains of the equation will be satisfied if we can choose u_1 and u_2 so that the two equations

2.1
$$\begin{aligned} y_1u_1' + y_2u_2' &= 0, \\ y_1'u_1' + y_2'u_2' &= f \end{aligned}$$

hold identically for all x on the interval in question. We can solve this system of equations for u_1' and u_2' by elimination or by Cramer's rule. (Appendix B, Theorem 2.9.) Either way we find

$$u_1'(x) = \frac{-y_2(x)f(x)}{y_1(x)y_2'(x) - y_2(x)y_1'(x)},$$

$$u_2'(x) = \frac{y_1(x)f(x)}{y_1(x)y_2'(x) - y_2(x)y_1'(x)}.$$

The expression in the denominators is the same in both formulas and we can write it as the 2-by-2 determinant

$$w(x) = \begin{vmatrix} y_1(x) & y_2(x) \\ y_1'(x) & y_2'(x) \end{vmatrix} = y_1(x)y_2'(x) - y_2(x)y_1'(x),$$

called the *Wronskian determinant* of y_1 and y_2. Theorem 2.6 in Section 2C shows that $w(x)$ is never zero if y_1 and y_2 are linearly independent solutions of the homogeneous equation, so the examples we consider here will satisfy $w(x) \neq 0$. To complete the solution, integrate the formulas for u_1' and u_2' to find u_1 and u_2, and then combine with y_1 and y_2 to get a particular solution

$$\begin{aligned} y_p(x) &= y_1(x)u_1(x) + y_2(x)u_2(x) \\ &= y_1(x)\int \frac{-y_2(x)f(x)}{w(x)}\,dx + y_2(x)\int \frac{y_1(x)f(x)}{w(x)}\,dx. \end{aligned} \tag{2.2}$$

Because Equations 2.1 are easier to remember than Equation 2.2, people often prefer to start with them in each problem and carry out the rest of the computation to arrive at Equation 2.2. The next example will be done that way. Because of the way $f(x)$ enters Equations 2.1 and 2.2, it's essential that differential equation be in normalized form if Equation 2.2 is to be valid.

example 2

The equation $x^2y'' - 2xy' + 2y = x^3$ is normalized, for $x \neq 0$, to be

$$y'' - \frac{2}{x}y' + \frac{2}{x^2}y = x, \quad \text{for} \quad x > 0, \quad \text{or} \quad x < 0,$$

and has homogeneous solutions $y_1(x) = x$ and $y_2(x) = x^2$ in either case. We guess $y_1(x)$, and the second solution follows from the first by the order-reduction method of Section 1A. For this example, Equations 2.1 become

$$\begin{aligned} xu_1' + x^2u_2' &= 0, \\ u_1' + 2xu_2' &= x. \end{aligned}$$

Multiplying the second equation by x and then subtracting the first equation from it gives $x^2u_2' = x^2$, or $u_2' = 1$. It then follows from the first equation that $u_1' = -x$. Integrating to find u_1, u_2 gives

$$u_1(x) = -\tfrac{1}{2}x^2, \qquad u_2(x) = x.$$

A particular solution is

$$\begin{aligned} y_p &= y_1u_1 + y_2u_2 \\ &= x\left(-\tfrac{1}{2}x^2\right) + x^2(x) = \tfrac{1}{2}x^3. \end{aligned}$$

Adding constants of integration to u_1 and u_2 would only add a linear combination of homogeneous solutions to y_p. Theorem 2.6 in Section 2C will show that the general solution is

$$y = c_1x + c_2x^2 + \tfrac{1}{2}x^3.$$

We'll now solve a generalization of the equation in Example 2, but using Equation 2.2 instead of Equation 2.1.

example 3 In normalized form, we consider

$$y'' - \frac{2}{x}y' + \frac{2}{x^2}y = f(x), \qquad x \neq 0.$$

We find homogeneous solution $y_1(x) = x$ by guessing, and $y_2(x) = x^2$ by guessing or the method of Section 1. The Wronskian determinant of y_1 and y_2 is

$$w(x) = \begin{vmatrix} x & x^2 \\ 1 & 2x \end{vmatrix} = 2x^2 - x^2 = x^2.$$

Equation 2.2 reduces to

$$\begin{aligned} y_p(x) &= x\int \frac{-x^2 f(x)}{x^2}\,dx + x^2\int \frac{xf(x)}{x^2}\,dx \\ &= -x\int f(x)\,dx + x^2\int \frac{f(x)}{x}\,dx. \end{aligned}$$

To make the integration fairly easy, we can use the example $f(x) = x\cos x$. For this choice, we use integration by parts to get

$$\begin{aligned} y_p(x) &= -x\int x\cos x\,dx + x^2\int \cos x\,dx \\ &= -x\left(x\sin x - \int \sin x\,dx\right) + x^2\int \cos x\,dx \\ &= -x(x\sin x + \cos x) + x^2\sin x \\ &= -x\cos x. \end{aligned}$$

Theorem 2.6 will show that the general solution is $y = c_1x + c_2x^2 - x\cos x$.

2B Green's Functions

A slight modification of Equation 2.2 gives us the generalization to the case of variable coefficients of the integral formula that we found in Chapter 3, Section 3D, for the initial-value problem $y(x_0) = y'(x_0) = 0$ associated with a constant-coefficient operator. All we need to do to get a formula is to convert the integrals in Equation 2.2 to definite integrals over the interval from x_0 to x:

$$y_p(x) = y_1(x)\int_{x_0}^{x} \frac{-y_2(t)f(t)}{w(t)}\,dt + y_2(x)\int_{x_0}^{x} \frac{y_1(t)f(t)}{w(t)}\,dt,$$

2.3

$$y_p(x) = \int_{x_0}^{x} \frac{y_1(t)y_2(x) - y_2(t)y_1(x)}{w(t)} f(t)\,dt,$$

where $w(t) = y_1(t)y_2'(t) - y_2(t)y_1'(t)$ is the Wronskian determinant of $y_1(t), y_2(t)$.

Setting $x = x_0$ in Equation 2.3 gives $y_p(x_0) = 0$. It is left as Exercise 24 to show that $y_p'(x_0) = 0$ also. Thus the *Green's function*

$$G(x,t) = \frac{y_1(t)y_2(x) - y_2(t)y_1(x)}{w(t)}$$

generates the solution

$$y_p(x) = \int_{x_0}^{x} G(x,t)\, f(t)\, dt$$

to the initial-value problem

$$y'' + a(x)y' + b(x)y = f(x), \qquad y(x_0) = y'(x_0) = 0.$$

Note that the numerator of $G(x,t)$ equals the Wronskian with $y_1'(t)$ and $y_2'(t)$ replaced by $y_1(x)$ and $y_2(x)$, respectively.

---

example 4

The equation $xy'' - (x+1)y' + y = x^2e^x$ has homogeneous solutions $y_1(x) = e^x$ and $y_2(x) = x + 1$. (Both solutions are easy to guess; if you guess only y_1, then y_2 follows from Theorem 1.2 of Section 1.) The Wronskian is $w[y_1, y_2](x) = -xe^x$, so the Green's function for the normalized equation $y'' - (1+1/x)y' + (1/x)y = xe^x$ is

$$G(x,t) = \frac{e^t(x+1) - (t+1)e^x}{-te^t} = \frac{-(x+1)}{t} + \left(1 + \frac{1}{t}\right)e^{(x-t)}.$$

Then Equation 2.3 for the normalized equation gives us $y_p(x)$ with $y_p(x_0) = y_p'(x_0) = 0$:

$$\begin{aligned}
y(x) &= \int_{x_0}^{x} \left[-\frac{x+1}{t} + \left(1 + \frac{1}{t}\right)e^{(x-t)}\right] te^t\, dt \\
&= \int_{x_0}^{x} \left[-(x+1)e^t + (t+1)e^x\right] dt \\
&= \left[-(x+1)e^t + \left(\frac{1}{2}t^2 + t\right)e^x\right]_{x_0}^{x} \\
&= \frac{1}{2}x^2e^x + e^{x_0}(x+1) - \left(\frac{1}{2}x_0^2 + x_0 + 1\right)e^x,
\end{aligned}$$

the second and third terms being homogeneous solutions.

Aside from the neatness with which Equation 2.3 displays a solution, it is convenient for calculating solutions when the forcing function $f(x)$ is defined piecewise.

---

example 5

The equation

$$y'' - \frac{2x}{x^2 - 1}y' + \frac{2}{x^2 - 1}y = f(x)$$

has homogeneous solutions $y_1(x) = x$ and $y_2(x) = x^2 + 1$. Their Wronskian is

$$w(x) = \begin{vmatrix} x & x^2 + 1 \\ 1 & 2x \end{vmatrix} = x^2 - 1.$$

The Green's function for an initial-value problem is then

$$G(x,t) = \frac{t(x^2+1) - (t^2+1)x}{t^2-1}.$$

Suppose we want the solution with forcing function

$$f(x) = \begin{cases} 0, & -1 < x < 0, \\ 1, & 0 \le x \le \frac{1}{2}, \\ 0, & \frac{1}{2} < x < 1, \end{cases}$$

and satisfying $y(0) = y'(0) = 0$. Note that if $x > \frac{1}{2}$ then $\int_0^x G(x,t) f(t)\, dt = \int_0^{1/2} G(x,t)\, dt$, so

$$y_p(x) = \int_0^x G(x,t) f(t)\, dt = \begin{cases} 0, & -1 < x < 0, \\ \int_0^x G(x,t)\, dt, & 0 \le x \le \frac{1}{2}, \\ \int_0^{1/2} G(x,t)\, dt, & \frac{1}{2} < x < 1. \end{cases}$$

For $0 \le x < \frac{1}{2}$ we compute

$$\begin{aligned} \int_0^x G(x,t)\, dt &= (x^2+1) \int_0^x \frac{t}{t^2-1}\, dt - x \int_0^x \frac{t^2+1}{t^2-1}\, dt \\ &= \frac{1}{2}(x^2+1)\ln(1-x^2) - x\left[x + \ln\left(\frac{1-x}{1+x}\right)\right]. \end{aligned}$$

The complete solution is then

$$y_p(x) = \begin{cases} 0, & -1 < x < 0, \\ \frac{1}{2}(x^2+1)\ln(1-x^2) - x\ln\big[(1-x)/(1+x)\big] - x^2, & 0 \le x \le \frac{1}{2}, \\ \frac{1}{2}(\ln\frac{3}{4})(x^2+1) - \left(\frac{1}{2} - \ln 3\right)x, & \frac{1}{2} < x < 1. \end{cases}$$

The first and third lines are of course solutions of the homogeneous equation on their respective intervals, because $f(x) = 0$ there. Finding the solution that satisfies more general initial conditions, $y(0) = y_0$, $y'(0) = z_0$, is just a matter of solving for c_1 and c_2 using

$$\begin{aligned} y(x) &= c_1 x + c_2(x^2+1) + y_p(x), \\ y'(x) &= c_1 + 2c_2 x + y_p'(x). \end{aligned}$$

But $y_p(0) = y_p'(0) = 0$, so we find c_1 and c_2 from equations $c_2 = y_0$ and $c_1 = z_0$.

2C Summary

This section and the previous one show how, given one solution to the homogeneous equation $y'' + a(x)y' + b(x)y = 0$, we can find solution formulas

2.4 $$y(x) = c_1 y_1(x) + c_2 y_2(x) + y_p(x)$$

for differential equations of the form

2.5 $$y'' + a(x)y' + b(x)y = f(x).$$

We already proved in Chapter 3 that if $a(x)$ and $b(x)$ are constant then Equation 2.5 has its most general solution of the form 2.4. Furthermore, we saw there that by choosing c_1 and c_2 properly we could satisfy arbitrary initial conditions of the form $y(x_0) = y_0$, $y'(x_0) = z_0$. This last possibility followed simply from the meaning of c_1 and c_2 as arbitrary constants of integration. If $a(x)$ and $b(x)$ are continuous functions, the analogous theorem still holds. The general results can be summarized in an extension of Theorem 1.1 in Section 1:

2.6 THEOREM

Let $a(x)$ and $b(x)$ be continuous and $f(x)$ integrable on an interval $\alpha < x < \beta$. Then the general solution of $y'' + a(x)y' + b(x)y = f(x)$ can be written in the form $y = c_1y_1 + c_2y_2 + y_p$, where y_1 and y_2 are independent solutions of the associated homogeneous equation and y_p is some particular solution of the nonhomogeneous equation. The Wronskian

$$w(x) = y_1(x)y_2'(x) - y_2(x)y_1'(x) \neq 0 \quad \text{is nonzero for} \quad \alpha < x < \beta,$$

and every pair of initial conditions $y(x_0) = y_0$, $y'(x_0) = z_0$ on the interval can be satisfied by a unique choice of c_1 and c_2.

Proof. We need to appeal to Theorem 1.1 of Section 1 for the existence and form of the general solution to the homogeneous equation. Then Theorem 3.1 of Chapter 3, Section 3A, proved there for an arbitrary linear equation, guarantees the form of the solution to the nonhomogeneous equation, with y_p given by Equation 2.3 or by variation of parameters. To prove that the Wronskian $w(x)$ appearing in Equation 2.3 is never zero, suppose on the contrary that for some x_0 in the interval we have $w(x_0) = 0$. But $w(x_0)$ is the determinant of the system

$$c_1y_1(x_0) + c_2y_2(x_0) = 0,$$
$$c_1y_1'(x_0) + c_2y_2'(x_0) = 0.$$

We know from Theorem 1.1 that the system has a unique solution for c_1 and c_2. Hence the determinant $w(x_0)$ can't be zero. Initial conditions for the nonhomogeneous equation can be written

$$c_1y_1(x_0) + c_2y_2(x_0) = y_0 - y_p(x_0),$$
$$c_1y_1'(x_0) + c_2y_2'(x_0) = z_0 - y_p'(x_0).$$

Since the determinant $w(x_0) \neq 0$ there is a unique solution for c_1 and c_2 here also. ■

example 6

It's easy to verify that $x^2y'' - 2xy' + 2y = 0$ has independent solutions $y_1(x) = x$, $y_2(x) = x^2$. The Wronskian is $w(x) = x^2 \neq 0$ except at $x = 0$. The Green's function is $G(x,t) = (x^2/t) - x$. We have the general solution to the normalized nonhomogeneous equation $y'' - (2/x)y' + (2/x^2)y = f(x)$ given by

$$y(x) = c_1x + c_2x^2 + \int_{x_0}^{x} \left(\frac{x^2}{t} - x\right) f(t)\,dt, \qquad x_0 \neq 0.$$

It's straightforward to show for instance that with $f(x) = x$ the integral term works out to $y_p(x) = \frac{1}{2}x^3 - x_0x^2 + \frac{1}{2}x_0^2x$. The last two terms can be combined with the homogeneous solution to give $y(x) = c_1x + c_2x^2 + \frac{1}{2}x^3$.

---

example 7

The equation $(x - 1)y'' - xy' + y = 1$ has a particular solution $y_p(x) = 1$, and the associated homogeneous equation has independent solutions $y_1(x) = x$ and $y_2(x) = e^x$ if $x \neq 1$. Initial conditions $y(x_0) = y_0$, $y'(x_0) = z_0$ are satisfied by solving for c_1 and c_2 in

$$y_0 = c_1y_1(x_0) + c_2y_2(x_0) + 1,$$
$$z_0 = c_1y_1'(x_0) + c_2y_2'(x_0).$$

If $x_0 = 0$, it turns out that $c_1 = z_0 - y_0 + 1$ and $c_2 = y_0 - 1$. However, Theorem 2.6 fails to apply at $x_0 = 1$, because the coefficient $(x - 1)$ of y'' is zero there. You are asked to show in Exercise 32 that the initial-value problem at $x_0 = 1$ has a solution only if $z_0 = y_0 - 1$ and that for this case the solution is not unique.

EXERCISES

For each of the differential equations in Exercises 1 through 4, try to find, maybe guess, a solution y_1 of the associated homogeneous equation. Then determine $u(x)$ so that $y(x) = y_1(x)u(x)$ is the general solution of the given equation.

1. $y'' - 4y' + 4y = e^x$.

2. $y'' + (1/x)y' = x$, $x > 0$.

3. $x^2y'' - 3xy' + 3y = x^4$, $x > 0$. [*Hint:* Try $y_1 = x^n$, for some n].

4. $xy'' - (2x + 1)y' + (x + 1)y = 3x^2e^x$, $x > 0$. [*Hint:* Try $y_1(x) = e^{rx}$.]

Find a particular solution y_p to Exercises 5 through 9 by solving Equation 2.1 for $u_1(x)$, $u_2(x)$ to get $y_p(x) = y_1(x)u_1(x) + y_2(x)u_2(x)$. Unless y_1 and y_2 are given, find them first. Then write down the general solution.

5. $y'' + y' - 2y = e^{2x}$.

6. $y'' + y = \tan x$, $-\pi/2 < x < \pi/2$.

7. $y'' + y = \sec x$, $-\pi/2 < x < \pi/2$.

8. $y'' - y = xe^x$.

9. $x^2y'' - 2xy' + 2y = 1$; $y_1(x) = x$, $y_2(x) = x^2$, $x > 0$.

Use Equation 2.2 to find a formula for a solution $y_p(x)$ to each of the equations in Exercises 10 through 15, with homogeneous solution $y_1(x)$ given in some cases. Don't forget to make sure that the equation is normalized.

10. $y'' - 2y' + y = e^x$.

11. $y'' + 3y' + 2y = 1 + e^x$.

12. $y'' + 3y' + 2y = (1 + e^x)^{-1}$.

13. $2y'' + 8y = e^x$.

14. $xy'' - y' = x^2$; $y_1(x) = 1$.

15. $x^2y'' - 2xy' + 2y = x^4$; $y_1(x) = x$.

Find the Green's function $G(x,t)$ for an initial-value problem for each of the equations in Exercises 16 through 19. Then express the solution y_p satisfying $y_p(x_0) = y_p'(x_0) = 0$ in terms of $g(x)$. [*Hints:* 18 is first order in y'; 19 has homogeneous solutions x^n for some n.]

16. $y'' + 3y' + 2y = g(x)$.

17. $2y'' + 4y = g(x)$.

18. $y'' + (1/x)y' = g(x)$, $x > 0$.

19. $x^2y'' - 2xy' + 2y = g(x)$.

Using the same equations as in Exercises 16 through 19 and the solutions of the initial-value problems found there, solve the corresponding initial-value problems in Exercises 20 through 23 with the given choice for $g(x)$. Don't forget to normalize $g(x)$.

20. $g(x) = \begin{cases} 0, & x < 1, \\ 1, & 1 \le x; \end{cases}$ $y(0) = 1$, $y'(0) = 2$.

21. $g(x) = \begin{cases} 1, & x < 1, \\ 0, & 1 \le x; \end{cases}$ $y(0) = -1$, $y'(0) = 1$.

22. $g(x) = 1/x$; $y(1) = 0$, $y'(1) = 2$.

23. $g(x) = \begin{cases} 0, & 0 < x < 2, \\ 3, & 2 \le x \le 4, \\ 1, & 4 < x; \end{cases}$ $y(3) = 0$, $y'(3) = 0$.

24. Show that the derivative of the Green's function Formula 2.3 for $y_p(x)$ can be written

$$y_p'(x) = \int_{x_0}^{x} \frac{y_1(t)y_2'(x) - y_2(t)y_1'(x)}{w(t)} f(t)\, dt.$$

Show then that $y_p'(x_0) = 0$. [*Hint:* Separate the integral in Equation 2.3 into two integrals, and then apply the product rule for differentiation.]

25. The **Leibnitz rule** for differentiating an integral states that

$$\frac{d}{dx} \int_{a(x)}^{b(x)} F(x,t)\, dt = \int_{a(x)}^{b(x)} \frac{\partial F}{\partial x}(x,t)\, dt + b'(x)F(x,b(x)) - a'(x)F(x,a(x)),$$

if $\partial F/\partial x$ is continuous. Use this result to establish the formula for $y_p'(x)$ in Exercise 24.

In Exercises 26 through 28, the constant coefficient equation $y'' + ay' + by = f(x)$ has homogeneous solutions $y_1(x) = e^{r_1x}$, $y_2(x) = e^{r_2x}$. These solutions are independent if $r_1 \ne r_2$; otherwise we consider $y_1(x) = e^{r_1x}$, $y_2(x) = xe^{r_1x}$. Find the Green's function for the equations in Exercises 26, 27, and 28 by using the approach of the present section.

26. $r_1 \neq r_2$.

27. $r_1 = r_2$.

28. $r_1 = \alpha + i\beta, r_2 = \alpha - i\beta, \beta \neq 0$.

Do the calculation of the Green's function integral in Example 5 of the text for the choices in Exercises 29 through 31.

29. $f(x) = x$.

30. $f(x) = x^2$.

31. $f(x) = 1/x$.

32. Consider the differential equation $(x-1)y'' - xy' + y = 1$ in Example 7 of the text.

(a) Show that the initial-value problem $y(1) = y_0$, $y'(1) = z_0$ has a solution only if $z_0 = y_0 - 1$, and that for this case there are infinitely many solutions. [*Hint:* x and e^x are homogeneous solutions.]

(b) Explain why the results of part (a) don't contradict Theorem 2.6.

3 BOUNDARY PROBLEMS

3A Two-Point Boundaries

Initial-value problems for second-order linear differential equations are those in which we specify the values $y(a)$ and $y'(a)$ of a solution $y = y(x)$ at a single point $x = a$. We have seen that, under the hypotheses of Theorem 1.1 in the introduction to this chapter, such problems always have a unique solution. If instead we try to specify just the solution values $y(a)$ and $y(b)$ at two distinct points $x = a$ and $x = b$, then asking about existence of a solution leads to a wider variety of outcomes: (i) there may be a unique solution, (ii) there may be infinitely many solutions, or (iii) there may be no solution at all. It is customary to call the search for a solution with values given at two points a **boundary-value problem,** in this case the two points being the boundary of an interval.

Motivation. Boundary problems have a physical interpretation for the oscillator equation

$$\ddot{y} = -k(t)\dot{y} - h(t)y$$

with specified boundary values $y(t_0) = \alpha$ and $y(t_1) = \beta$. The solutions $y(t)$ of such an equation may represent the displacements of a damped oscillating mechanism as a function of time t. Our question then becomes the following: over the time interval from t_0 to t_1 can the oscillator go from displacement $y = \alpha$ to displacement $y = \beta$, and if so, is there a unique way to do this? The next example illustrates the three possibilities.

example 1 The differential equation $y'' + y = 0$ has the general solution

$$y(x) = c_1 \cos x + c_2 \sin x.$$

To satisfy the boundary conditions $y(0) = 1$ and $y(\pi/2) = 2$, we must satisfy two conditions on c_1 and c_2, namely, $y(0) = c_1 = 1$ and $y(\pi/2) = c_2 = 2$. The only

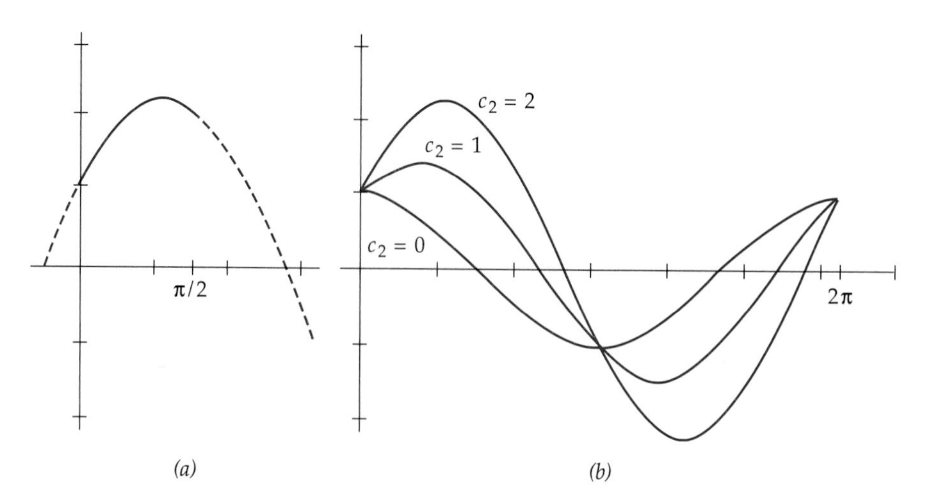

Figure 9.2 (*a*) Unique solution. (*b*) Multiple solutions.

solution is $y(x) = \cos x + 2\sin x$ since c_1 and c_2 are uniquely determined. The graph of the solution appears in Figure 9.2(a).

Keeping the same differential equation, but moving a boundary point from $\pi/2$ to 2π, we change the boundary conditions to $y(0) = 1$ and $y(2\pi) = 1$. Then the constants c_1 and c_2 in the general solution are subject to the requirement

$$y(0) = c_1 = 1, \qquad y(2\pi) = c_1 = 1,$$

while c_2 can evidently be chosen arbitrarily. The resulting 1-parameter family of solutions is infinite and is made up of all functions of the form

$$y(x) = \cos x + c_2 \sin x.$$

The graphs of three of these solutions appear in Figure 9.2(b). In Chapter 10 we'll see that multiple solutions to a boundary-value problem can be very desirable when we want to satisfy other conditions along with boundary conditions.

Still keeping the same differential equation, we can change the boundary conditions to

$$y(0) = 1, \qquad y(\pi) = 1.$$

These conditions require of c_1 in the general solution that

$$y(0) = c_1 = 1, \qquad y(\pi) = -c_1 = 1.$$

Since we cannot have both $c_1 = 1$ and $c_1 = -1$, it follows that there is no solution satisfying this third set of conditions.

The case of nonexistence and the case of multiple existence turn out to be independent of the precise solution values prescribed at the boundary points. It follows that both cases are distinguished from the case of existence of a unique solution for a given equation by conditions on the boundary points alone. The precise statement is as follows:

3.1 THEOREM

Assume the linear homogeneous differential equation

$$y'' + p(x)y' + q(x)y = 0$$

has continuous coefficients $p(x)$ and $q(x)$ on the interval $a \le x \le b$. For the boundary problem with $y(a) = \alpha$ and $y(b) = \beta$ to have a unique solution, either of the following equivalent conditions is sufficient:

(i) The only solution satisfying $y(a) = y(b) = 0$ is the identically zero solution.

(ii) For some pair of independent solutions, this 2-by-2 determinant condition holds:

$$y_1(a)y_2(b) - y_2(a)y_1(b) = \begin{vmatrix} y_1(a) & y_2(a) \\ y_1(b) & y_2(b) \end{vmatrix} \neq 0.$$

Proof. Let the general solution of the differential equation be

$$y(x) = c_1 y_1(x) + c_2 y_2(x).$$

The boundary conditions $y(a) = \alpha$ and $y(b) = \beta$ amount to

$$c_1 y_1(a) + c_2 y_2(a) = \alpha,$$
$$c_1 y_1(b) + c_2 y_2(b) = \beta.$$

It's easy to show (see Exercise 30) that these equations have a unique solution for c_1 and c_2 if and only if

$$y_1(a)y_2(b) - y_2(a)y_1(b) \neq 0.$$

But the condition that the determinant be nonzero is precisely the condition that the equations

$$c_1 y_1(a) + c_2 y_2(a) = 0,$$
$$c_1 y_1(b) + c_2 y_2(b) = 0$$

have only the trivial solution $c_1 = c_2 = 0$. ■

Theorem 3.1 shows that, given the differential equation, the existence of a unique boundary-value solution depends on the location of the boundary points. The term **conjugate points** is used to denote a pair of boundary points for which there *fails* to exist a unique solution. Thus a pair of points is conjugate relative to a homogeneous differential equation if there is a nontrivial solution that is zero at the two points. From the statement of Theorem 3.1, we see that, equivalently, a pair of points a and b is conjugate if and only if, for two linearly independent solutions y_1 and y_2, we have the relation

$$\begin{vmatrix} y_1(a) & y_2(a) \\ y_1(b) & y_2(b) \end{vmatrix} = 0.$$

example 2

Referring to Example 1, the pairs of points $\{0,2\pi\}$ and $\{0,\pi\}$ are conjugate for $y'' + y = 0$. More generally, $\{a,b\}$ is a conjugate pair of points if and only if

$$\begin{vmatrix} \cos a & \sin a \\ \cos b & \sin b \end{vmatrix} = 0,$$

that is, if $\sin b \cos a - \cos b \sin a = \sin(b-a) = 0$. We conclude for this example that $\{a,b\}$ is a conjugate pair of boundary points if and only if $b - a = n\pi$ for some integer n. In particular, it's easy to check that boundary-value problem

$$y'' + y = 0, \qquad y(a) = \alpha, \qquad y(a + n\pi) = \beta$$

has multiple solutions precisely when n is an integer and a unique solution otherwise.

example 3

For the equation $y'' - y = 0$, the general solution is expressible as

$$y(x) = c_1 e^x + c_2 e^{-x}.$$

The arbitrary pair $\{a,b\}$ of boundary points is conjugate precisely when the equations

$$y(a) = c_1 e^a + c_2 e^{-a} = 0,$$
$$y(b) = c_1 e^b + c_2 e^{-b} = 0$$

are satisfied by some c_1 and c_2 not both zero. But for real a and b,

$$\begin{vmatrix} e^a & e^{-a} \\ e^b & e^{-b} \end{vmatrix} = e^{(a-b)} - e^{-(a-b)} = 0$$

only when $a = b$. So there is always a unique solution to the boundary-value problem

$$y'' - y = 0, \qquad y(a) = \alpha, \qquad y(b) = \beta \qquad \text{if } a \neq b.$$

3B Nonhomogeneous Equations

We consider boundary-value problems for the nonhomogeneous equation with continuous coefficients $p(x)$, $q(x)$, and $f(x)$,

$$y'' + p(x)y' + q(x)y = f(x).$$

The key is first to find a solution satisfying the special **homogeneous boundary conditions**

$$y(a) = 0, \qquad y(b) = 0.$$

It turns out that this can always be done provided a and b are not conjugate points for the differential equation. Recall from Section 2 the variation of parameters formula,

$$y(x) = -y_1(x) \int \frac{y_2(x) f(x)}{w(x)}\, dx + y_2(x) \int \frac{y_1(x) f(x)}{w(x)}\, dx,$$

where y_1 and y_2 are independent solutions, and w is their Wronskian. By choosing the constants of integration properly in the two integrals, we can get the most general solution of the nonhomogeneous differential equation. In particular, we can specify these constants by using integrals with limits as follows:

3.2

$$\begin{aligned} y_p(x) &= -y_1(x)\int_b^x \frac{y_2(t)f(t)}{w(t)}\,dt + y_2(x)\int_a^x \frac{y_1(t)f(t)}{w(t)}\,dt \\ &= \int_a^x \frac{y_2(x)y_1(t)}{w(t)} f(t)\,dt + \int_x^b \frac{y_1(x)y_2(t)}{w(t)} f(t)\,dt, \end{aligned}$$

where we went from the first line to the second by interchanging limits in the first integral. Now suppose that the linearly independent solutions y_1 and y_2 have been chosen so that $y_1(a)=0$ and $y_2(b)=0$. It then follows by setting $x=a$ and $x=b$ that we get, respectively,

$$y_p(a) = 0 \quad \text{and} \quad y_p(b) = 0;$$

thus we have found a particular solution satisfying the specified homogeneous boundary condition.

The assumption that $x = a$ and $x = b$ are not conjugate points ensures that the desired independent solutions y_1 and y_2 can always be found, as stated in the following theorem.

3.3 THEOREM

If the equation $y'' + p(x)y' + q(x)y = 0$ has continuous coefficients, and $x = a$, $x = b$ is a pair of nonconjugate boundary points for the equation, then there are linearly independent solutions y_1 and y_2 satisfying $y_1(a) = 0$ and $y_2(b) = 0$.

Proof. All we have to do for y_1 is pick a nontrivial initial-value solution, with $y_1(a) = 0$, $y_1'(a) \neq 0$. Then automatically $y_1(b) \neq 0$, because otherwise having a and b not conjugate would imply that y_1 is necessarily identically zero. Similarly, choose y_2 so that $y_2(b)=0$, $y_2'(b)\neq 0$, with the result that $y_2(a)\neq 0$. Then y_1 and y_2 are linearly independent because, if one were a multiple of the other, each would have to be zero at both boundary points $x = a$ and $x = b$. But $y_1(a)y_2(b) \neq y_1(b)y_2(a)$, so this is impossible. ■

We denote by G the function of two variables defined for $a \le x \le b$ and $a \le t \le b$ by

3.4

$$G(x,t) = \begin{cases} \dfrac{y_1(t)y_2(x)}{w(t)}, & a \le t \le x, \\ \dfrac{y_2(t)y_1(x)}{w(t)}, & x \le t \le b, \end{cases}$$

where $y_1(a)=y_2(b)=0$, and $w(t)=y_1(t)y_2'(t) - y_2(t)y_1'(t)$ is the Wronskian of $y_1(t)$ and $y_2(t)$. The function $G(x,t)$ is called a **Green's function** for the boundary-value problem. [Note carefully the similarity and difference as compared with the Green's function for the initial-value problem.] Using Equation 3.2 we can write

3.5

$$y_p(x) = \int_a^b G(x,t)f(t)\,dt$$

for the solution of

$$y'' + p(x)y' + q(x)y = f(x), \qquad y(a) = y(b) = 0.$$

It is important to understand that while the independent homogeneous solutions y_1 and y_2 must satisfy the conditions $y_1(a) = 0$ and $y_2(b) = 0$, respectively, there is still some arbitrariness in the choice that we make of them. We illustrate this point with the next example. [Though division by $w(t)$ makes $G(x,t)$ turn out to be the same for all choices.]

example 4

Suppose we want to solve the equation $y'' = f(x)$, subject to $y(0) = 0$, $y(1) = 0$. (This can in principle be done by repeated integration.) The most general solution of the associated homogeneous equation is

$$y(x) = c_1 + c_2 x.$$

To satisfy $y_1(0) = 0$ we need to have $c_1 = 0$. Hence we can choose for simplicity $c_2 = 1$. At $x = 1$ we must have $c_1 + c_2 = 0$, so we can choose this time $c_1 = 1$, $c_2 = -1$. Thus we have selected

$$y_1(x) = x, \qquad y_2(x) = 1 - x$$

for our independent homogeneous solutions. The Wronskian $w[y_1, y_2](x)$ is

$$\begin{vmatrix} x & 1-x \\ 1 & -1 \end{vmatrix} = -1,$$

which happens to be a constant. Thus the Green's function is given by

$$G(x,t) = \begin{cases} \dfrac{t(1-x)}{-1}, & 0 \le t \le x, \\ \dfrac{(1-t)x}{-1}, & x \le t \le 1, \end{cases}$$

Figure 9.3 is a sketch of the graph of $G(x,t)$, showing different curves on the

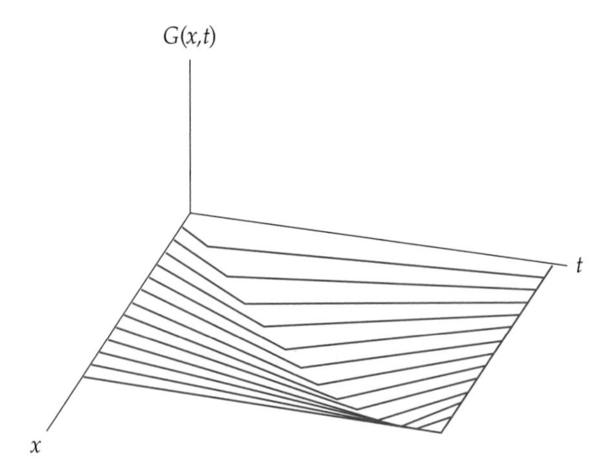

Figure 9.3 Green's function of y'' for $0 \le x \le 1,\ 0 \le t \le 1$.

graph for different fixed values of x. The formula for the particular solution of $y'' = 1$ is obtained by taking $f(t) = 1$ in the general Green's formula 3.5. Thus

$$y_p(x) = -(1-x)\int_0^x t\,dt - x\int_x^1 (1-t)\,dt$$
$$= -(1-x)\frac{x^2}{2} - x\left(\frac{1}{2} - x + \frac{x^2}{2}\right) = \frac{x^2}{2} - \frac{x}{2}.$$

Clearly, $y_p(0) = y_p(1) = 0$ and $y_p''(x) = 1$, so we have found the correct solution, though this problem was simple enough that we could have solved it without the Green's function.

The Green's function graph sketched in Figure 9.3 is typical in that it has a sharp crease running from one corner to the other. You're asked to show in Exercise 33 that for all such Green's functions the partial derivative $G_t(x,t)$ jumps up by 1 as t increases beyond x.

3C Nonhomogeneous Boundary Conditions

To solve the nonhomogeneous equation

$$y'' + p(x)y' + q(x)y = f(x)$$

with general **nonhomogeneous boundary conditions**

$$y(a) = \alpha, \qquad y(b) = \beta, \qquad \alpha,\ \beta \text{ not both zero.}$$

we continue to assume that a and b are not conjugate points of the associated homogeneous equation. Then first solve the equation with homogeneous boundary conditions

$$y(a) = 0, \qquad y(b) = 0,$$

for example by the method outlined in Section 3B, calling this solution y_p. Now the general solution of the nonhomogeneous equation is

$$y(x) = c_1y_1(x) + c_2y_2(x) + y_p(x),$$

where y_1 and y_2 are independent homogeneous solutions and c_1 and c_2 are constants. Since $y_p(a) = y_p(b) = 0$, all we have to do to satisfy $y(a) = \alpha,\ y(b) = \beta$ is choose c_1 and c_2 so that

$$c_1y_1(a) + c_2y_2(a) = \alpha$$
$$c_1y_1(b) + c_2y_2(b) = \beta.$$

This can always be done if a and b are not conjugate, because in that case the determinant of the system of linear equations for c_1 and c_2 is not zero. See Exercise 30, part (a).

example 5

The differential equation

$$y'' + y = 1$$

has linearly independent homogeneous solutions $\cos x$ and $\sin x$. To satisfy boundary conditions of the form

$$y(0) = \alpha, \qquad y\left(\frac{\pi}{2}\right) = \beta,$$

we can check that the points $x = 0$ and $x = \pi/2$ are not conjugate by observing that

$$\begin{vmatrix} \cos 0 & \cos \pi/2 \\ \sin 0 & \sin \pi/2 \end{vmatrix} = 1 \neq 0.$$

Rather than constructing a Green's function, we can observe by inspection that $y(x) = 1$ is a particular solution to the homogeneous equation, so

$$y(x) = c_1 \cos x + c_2 \sin x + 1$$

is the most general solution of the differential equation. To satisfy $y(0) = \alpha$, $y(\pi/2) = \beta$, we solve the equations

$$c_1 \cos 0 + c_2 \sin 0 + 1 = \alpha,$$
$$c_1 \cos \frac{\pi}{2} + c_2 \sin \frac{\pi}{2} + 1 = \beta$$

for c_1 and c_2 to get $c_1 = \alpha - 1$, $c_2 = \beta - 1$. Thus

$$y(x) = (\alpha - 1) \cos x + (\beta - 1) \sin x + 1$$

is the unique solution satisfying the boundary conditions.

However, the boundary points $x = 0$ and $x = \pi$ are conjugate for $y'' + y = 0$, because $\sin x$ is a nontrivial solution that is zero at both points. Thus an attempt to solve

$$c_1 \cos 0 + c_2 \sin 0 + 1 = \alpha,$$
$$c_1 \cos \pi + c_2 \sin \pi + 1 = \beta$$

leads to $c_1 = \alpha - 1$ *and* $-c_1 = \beta - 1$. Thus there is no solution to the boundary-value problem with boundary conditions

$$y(0) = \alpha, \qquad y(\pi) = \beta$$

unless $\alpha - 1 = -\beta + 1$ or $\alpha + \beta = 2$. In that case, there are infinitely many solutions, all of the form

$$y(x) = (\alpha - 1) \cos x + c_2 \sin x + 1,$$

where c_2 is arbitrary. These solutions all satisfy the rather special conditions

$$y(0) = \alpha, \qquad y(\pi) = 2 - \alpha.$$

EXERCISES

For each of the differential equations in Exercises 1 through 4, determine whether (i) there is a unique solution, (ii) there are infinitely many solutions, or (iii) there is no solution satisfying the given boundary conditions.

1. $y'' + y = 0,\ y(0) = 1,\ y(\pi/4) = 0.$

2. $y'' + y = 0,\ y(0) = 1,\ y(2\pi) = 0.$

3. $y'' + y = 0,\ y(0) = 1,\ y(2\pi) = 1.$

4. $y'' + 2y = 0,\ y(1) = 1,\ y(2) = 0.$

Find all pairs of conjugate points for the homogeneous equations in Exercises 5 through 8.

5. $y'' + 2y = 0.$

6. $y'' - 2y = 0.$

7. $y'' + y' + y = 0.$

8. $y'' + xy' + y = 0$. [*Hint:* Show $e^{-x^2/2}$ and $e^{-x^2/2}\int_0^x e^{t^2/2}\,dt$ are independent solutions.]

In Exercises 9 through 12, a pair of corresponding nonconjugate boundary points is given here for each of the differential equations in Exercises 5 through 8. In each case, find a pair of linearly independent solutions y_1 and y_2 that satisfy $y_1(a) = 0$ and $y_2(b) = 0$. Then use these solutions to construct a Green's function for the equation on the interval $[a, b]$.

9. $a = 0,\ b = \pi/(2\sqrt{2}).$

10. $a = 1,\ b = 2.$

11. $a = 0,\ b = \pi/\sqrt{3}.$

12. $a = 0,\ b = 1.$

Use the Green's functions found in Exercises 9 through 12 to represent the solution $y(x)$ of the corresponding equation $L[y] = f$ in Exercises 5 through 8, where f is defined as follows in Exercises 13 through 16, and where $y(a) = y(b) = 0$. Compute the solution by any method you choose.

13. $f(x) = x,\ 0 \le x \le \pi/(2\sqrt{2}).$

14. $f(x) = \begin{cases} -1, & 0 \le x < \frac{3}{2}, \\ 1, & \frac{3}{2} \le x \le 2. \end{cases}$

15. $f(x) = \cos x,\ 0 \le x \le \pi/\sqrt{3}.$

16. $f(x) = e^{-x},\ 0 \le x \le 1.$

Modify the solutions found in Exercises 13 through 16 so instead of satisfying the homogeneous conditions $y(a) = y(b) = 0$, they satisfy the following corresponding nonhomogeneous conditions.

17. $y(0) = 1,\ y(\pi/(2\sqrt{2})) = 0.$

18. $y(1) = 2,\ y(2) = -1.$

19. $y(0) = -1,\ y(\pi/\sqrt{3}) = 0.$

20. $y(0) = 1,\ y(1) = 1.$

21. Show that the points $a = 0$ and $b = \pi\sqrt{2}$ are conjugate for the differential equation $y'' + 2y = 0$, and find conditions on α and β under which there are solutions of this equation that satisfy $y(0) = \alpha$, $y(\pi\sqrt{2}) = \beta$. Find all such solutions. Does one of these solutions also satisfy $y'(0) = 1$?

22. For a homogeneous linear boundary problem, the only way to get a nontrivial (i.e., not identically zero) solution is to have more than one solution. In particular, consider the problem of finding a number λ and a nontrivial solution $y(x)$ to the equation $L[y] = \lambda y$, where L is a second-order linear operator. (Such a number λ is called an **eigenvalue,** and a corresponding nontrivial solution is called an **eigenfunction.**) Show that if $a < b$, the problem

$$y'' = \lambda y, \quad y(a) = y(b) = 0,$$

has nontrivial solutions if and only if $\lambda = -k^2\pi^2/(b-a)^2$, where $k = 0, 1, \ldots$. What are the corresponding solutions $y_k(x)$?

23. Consider the oscillator equation

$$my'' + ky' + hy = 0,$$

where m, k, and h are constant and $m > 0$.

(a) Show that if $k^2 \geq 4mh$, then there are no conjugate pairs a, b. (The case $k^2 = 4mh$ requires special attention.)

(b) Explain the physical significance of the conclusion in part (a).

(c) Find all conjugate pairs a, b in case $k^2 < 4mh$.

24. Buckling rod A uniform horizontal rod is subjected to a horizontal compressive force of magnitude F. The rod buckles an amount $y = y(x)$ at distance x from one end, where $0 \leq x \leq l$, in such a way that $y'' + (F/k)y = 0$; here k is a positive constant depending on the elasticity, the cross-sectional shape, and the density of the rod.

(a) Assuming boundary conditions $y(0) = y(l) = 0$ at the ends, show that vertical displacement takes place only when the applied force has one of the critical values $F_n = kn^2\pi^2/l^2$, where n is a positive integer.

(b) Since the differential equation is homogeneous, a constant multiple of a solution is also a solution, and the boundary-value problem doesn't have a unique solution. Sketch the general shapes of the critical displacement functions $y_n(x)$ corresponding to the three smallest critical forces.

25. (a) Show that the boundary conditions

$$A[y] = a_1 y(0) + a_2 y'(0) = 0,$$
$$B[y] = b_1 y(\pi) + b_2 y'(\pi) = 0,$$

when applied to the differential equation $y'' + y = 0$, can be satisfied by a solution $y(x) \neq 0$ precisely when $a_1 b_2 = a_2 b_1$.

(b) Show that $y'' + y = f(x)$ has a unique solution for each continuous f on $0 \leq x \leq \pi$ and satisfying $A[y] = \alpha$, $B[y] = \beta$, precisely when $a_1 b_2 \neq a_2 b_1$.

Beam Deflection. The downward deflection $y = y(x)$ of a uniform horizontal beam of length L satisfies the fourth-order differential equation

$$y^{(4)} = R, \qquad 0 \leq x \leq L,$$

where R is an appropriate constant and x represents distance from one end of the beam. If the beam has a simple support at an end $x = x_0$, then $y(x_0) = y''(x_0) = 0$. If the beam is held rigidly horizontal at x_0, say by embedding it in a concrete wall, then $y(x_0) = y'(x_0) = 0$. If the beam is unsupported at $x = x_0$, then $y''(x_0) = y'''(x_0) = 0$. Solve the differential equation for arbitrary $R > 0$ under the sets of conditions in Exercises 26 through 29. Give a physical explanation for what goes wrong under conditions in Exercise 29.

26. $y(0) = y''(0) = 0,\ y(L) = y''(L) = 0.$

27. $y(0) = y'(0) = 0,\ y(L) = y''(L) = 0.$

28. $y(0) = y'(0) = 0,\ y''(L) = y'''(L) = 0.$

29. $y(0) = y''(0) = 0,\ y''(L) = y'''(L) = 0.$

30. Consider the algebraic equations

$$Az + Bw = K,$$
$$Cz + Dw = L.$$

(a) Show that these equations have a unique solution for z, w if $AD - BC \neq 0$.

(b) Show that if $AD - BC = 0$, then these equations have either infinitely many solutions or else no solutions at all. [*Hint:* if $A, B, C,$ and D are all zero, the conclusion is obvious. If, for example, $B \neq 0$, then $C = AD/B$.]

31. In certain 1-dimensional diffusion processes the concentration $y(x)$ of the diffusing substance depends only on the distance x from one end of a medium of nominal width 1. In that case $y(x)$ can be shown to satisfy the equation

$$y'' = ky \exp\left[\gamma(1 - y)/\left(1 + \delta(1 - y)\right)\right],$$

with boundary conditions $y(0) = 0$, $y(1) = 1$. Here γ and δ are nonnegative constants, while k is positive.

(a) Solve the equation under the assumption $\gamma = 0$ and arbitrary $k > 0$.

(b) Sketch the graph of the solution found in part (a) for $k = 1$.

For general constants γ and δ see Exercise 7 of Section 4 on numerical methods.

32. Under circumstances different from those assumed in the previous exercise, a diffusion concentration may satisfy an equation of the form

$$y'' = py' - qy^2,$$

where p and q are positive constants, and there are boundary conditions of the form $y(0) = 0$, $y(1/p) = 1$.

(a) Solve the equation with $q = 0$ and $p = 1$.

(b) Sketch the graph of the solution found in part (a).

For more general choices of constants p and q, use numerical methods as in Exercise 8 of Section 4 on numerical methods.

33. (a) Verify that the Green's function $G(x,t)$ in Equation 3.4 is a continuous function of (x,t) for $a \leq t \leq b, a \leq x \leq b$.

(b) Verify that the partial derivative $G_t(x,t)$ is continuous for $a \leq t \leq b$, $a \leq x \leq b$ except when $t = x$, and that $G_t(x,t)$ jumps up by 1 as t increases past $t = x$.

(c) Rewrite Equation 3.4 for the Green's function to display it for fixed t over the intervals $a \leq t \leq x$ and $t \leq x \leq b$.

(d) Use the form of the definition from part (c) to show that the partial derivative $G_x(x,t)$ jumps up by 1 as x increases past $x = t$.

4 SHOOTING

Shooting is a method for applying numerical solving of initial-value problems to solution of boundary-value problems. Assuming that there is a solution to

$$y'' + p(x)y' + q(x)y = f(x), \qquad y(a) = \alpha, \qquad y(b) = \beta,$$

we can try to approximate the solution as follows. If we know how to solve initial-value problems numerically, for example, as described in the Chapter 4 Computing Supplement, we replace the condition $y(b) = \beta$ at b by an initial condition $y'(a) = z$ at a. If we are incedibly lucky in our choice of z it might then happen that the solution of the problem with initial value $y(a) = \alpha$, $y'(a) = z_0$ turns out to satisfy $y(b) = \beta$ within a tolerance for error that we find acceptable. In that case, we print or plot the result of our initial approximate computation. Normally the discrepancy between β and our computed $y(b)$ will be too big, but we can sometimes use the difference between them to guide us in adjusting z so as to reduce $|y(b) - \beta|$ to an acceptable size. The method we'll describe applies just as well to the general second-order problem

$$y'' = f(x, y, y'), \qquad y(a) = \alpha, \qquad y(b) = \beta.$$

The term "shooting" suggests the idea that from $y(a)$ we aim at $y(b)$ by adjusting $y'(a)$ and make a succession of shots, one after each adjustment of the initial value $z = y'(a)$. The values of each approximate solution should be recorded

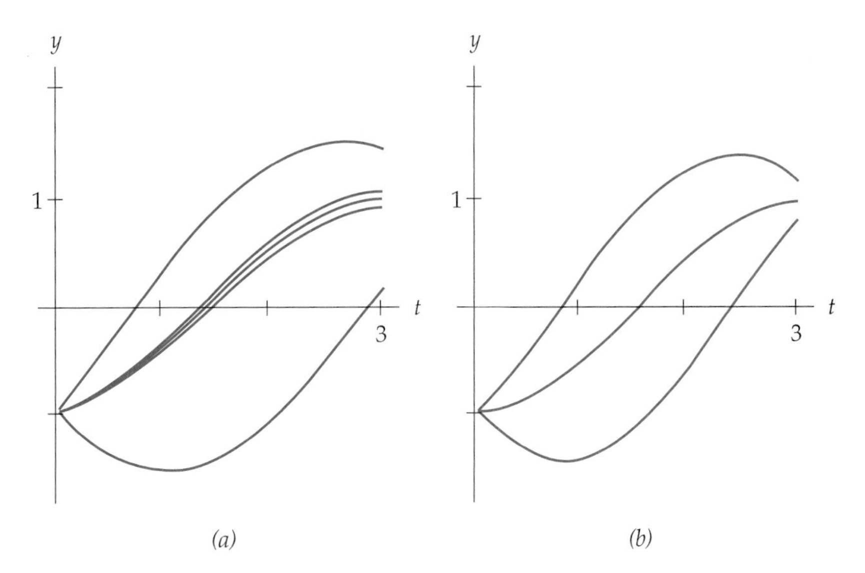

Figure 9.4 (*a*) $\ddot{y} = -\sin y,\ y(0) = -1,\ y(3) = 1.$
(*b*) $\ddot{y} = -y,\ y(0) = -1,\ y(3) = 1.$

numerically or graphically in case one of these solutions turns out to be acceptable. Figure 9.4(a) shows the graphs of six shots for a solution of the nonlinear pendulum equation, the last two graphs being indistinguishable at the scale used there.

To start the method, we use a computer to make two trial initial-value shots, using different initial slopes $y'(a) = z_1$ and $y'(a) = z_2$. We record the corresponding approximate values y_1 and y_2 that we get at $x = b$. Since $y(b) = \beta$ is what we want, we try to improve our accuracy by choosing the next slope $y'(a) = z_3$ to satisfy the linear interpolation equation

4.1 $$z_3 = z_2 + \frac{z_1 - z_2}{y_1(b) - y_2(b)}[\beta - y_2(b)].$$

This is just the equation in the yz plane of the line through (y_1, z_1) and (y_2, z_2) evaluated at $y = \beta$. If the shot aimed with the choice $z = z_3$ has an unacceptable value at $x = b$, then try again, replacing z_1 by z_2 and z_2 by z_3 in Equation 4.1 to get z_4. It is a good idea to try to choose z_1 and z_2 so that the corresponding y values $y_1(b)$ and $y_2(b)$ straddle the desired value $y = \beta$ and are not too close together. [The unlucky event that $y_1(b) = y_2(b)$ leads to big trouble in Equation 4.1. Why?] Aim each shot after that using a z value from Equation 4.1 along with a previous one that will lead to a straddle.

The central part of a computing routine to implement the method is a subroutine to approximate the solution of an initial-value problem with $y(a) = \alpha$, $y'(a) = z$. (See the Chapter 4 Computing Supplement.) The subroutine is called repeatedly until $|y(b) - \beta|$ is small enough. Except for the initial straddle, the entire process is automated in the Java Applet BOUNDARY2 at <http://math.dartmouth.edu/~rewn/>.

example 1

The boundary-value problem $\ddot{y} = -\sin y,\ y(0) = -1,\ y(3) = 1$, for an undamped pendulum equation requires us to pick two initial velocities; we try $y'(0) = z_1 = 1$, which swings the pendulum toward the position $y = 1$ that we want to attain at time $t = 3$, and $y'(0) = z_2 = -1$, which initially swings the pendulum away from $y = 1$. Using 200 equal steps from $t = 0$ to $t = 3$ gives the two extreme solution curves shown in Figure 9.4(a), where $y_1(3) \approx 1.46487$ and $y_2(3) \approx 0.173326$. Since these are quite far from the desired value 1, we apply the interpolation formula 4.1 to get a new slope

$$z_3 = -1 + \frac{1-(-1)}{1.46487 - 0.173326}(1 - 0.173326) \approx 0.280133.$$

Taking another shot with this slope gives an approximate solution y_3 with $y_3(3) \approx 1.04513$. To get more improvement, we note that y_2 and y_3 straddle the target value 1 at $x = 3$, so we interpolate the initial slopes $y_2(0)$ and $y_3(0)$ to get a slope $z_4 = 0.213875$ for the next shot. Continuing in this way gives $z_5 \approx 0.152445$, $z_6 \approx 0.150604$, with $y_6(3) = 1.00001$. We may choose to accept this solution without further improvement; its graph is the lowest of the intermediate ones in Figure 9.4(a). If we apply the same process to the linearized boundary-value problem $\ddot{y} = -y,\ y(0) = -1,\ y(3) = 1$, we discover that the third shot already gives $y_3(3) = 1.00000$ with $y_3'(0) = z_3 = 0.070893$, as shown in Figure 9.4(b). Of course, the solution of the linearized problem differs significantly from the nonlinear one.

Applying the shooting method to a *linear* equation, as at the end of the previous example, whether homogeneous or not, will in principle lead to the desired solution at the third shot. (See Exercise 5.) Furthermore, it's not necessary in a linear example for the terminal values of the first two shots to straddle the desired terminal value. (To prevent the method from going astray, the straddle requirement is typically quite important for a *nonlinear* equation.) Bear in mind though that if you're frustrated in a search for a solution to a boundary-value problem, you may be looking for something that doesn't exist. Nonexistence can occur even for linear equations, as you can see from Exercise 9.

example 2

The equation

$$\ddot{y} - 0.1t\dot{y} + y = 0$$

is linear with a single nonconstant coefficient. We try $z_1 = 1$ and $z_2 = 1.5$ for the first two shots. With this choice, and with 3000 equal steps in the improved Euler method, the result for the third shot turns out to give the value $y(3) = -1.00000$, along with $y'(0) = -0.070915$. The graphs of the three solutions are shown in Figure 9.5(a). To find numeric approximations to this solution, solve the initial-value problem with $y(0) = 1,\ y'(0) = -0.070915$.

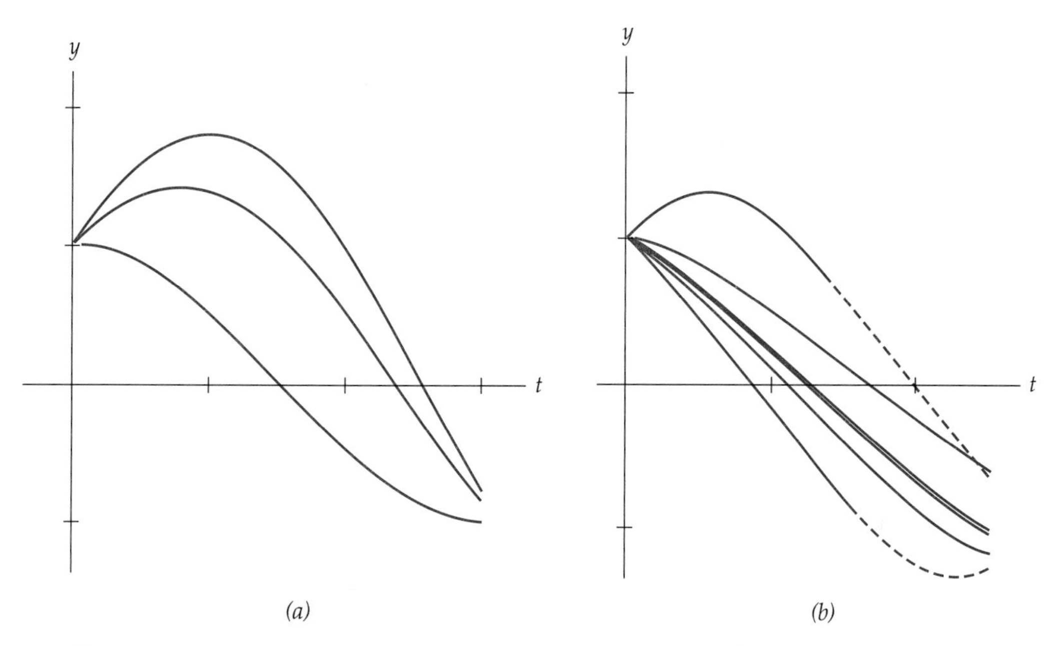

Figure 9.5 (a) $\ddot{y} - 0.1t\dot{y} + y = 0,\ y(0) = 1,\ y(3) = -1$. (b) $\ddot{y} = -y^3,\ y(0) = 1,\ y(2.5) = -1$.

example 3

The nonlinear boundary problem $\ddot{y} = -y^3,\ y(0) = 1,\ y(2.5) = -1$ requires us to straddle the value -1 with the terminal values of the first two shots with initial value $y(0) = 1$. It turns out that using $y'(0) = 1$ and $y'(0) = -1$ will give us the required straddle, and that on the sixth shot, having taken care to maintain a straddle each time, we get $y(2.5) = -1.00000$ with $y'(0) = -0.517353$. See Figure 9.5(b) for a graphical representation of the results. The exact solution can in principle be sought using the method of Chapter 3, Section 4B, but the required integration can't be carried out in terms of elementary functions.

The website <http://math.dartmouth.edu/~rewn/> contains the Applet BOUNDARY2 that finds approximate graphic solutions to two-point boundary problems using shooting; the required interpolations are calculated automatically. For numeric output, use the final endpoint value for y' given by the Applet in an initial-value numeric solver.

EXERCISES

1. (a) The differential equation $y'' - y = 1$ has the solution

$$y = (e+1)^{-1}(e^x + e^{-x+1}) - 1$$

satisfying $y(0) = y(1) = 0$. This exercise asks you to solve the same problem numerically. Find solutions y_1 and y_2 satisfying

$$y_1(0) = 0, \qquad y_1'(0) = z_1 = 1,$$
$$y_2(0) = 0, \qquad y_2'(0) = z_2 = -1.$$

Then find a third solution y_3 such that $y_3(0) = 0$, $y_3'(0) = z_3$, where z_3 is determined from Equation 4.1. Compare the result with the preceding solution.

(b) Derive the solution given at the beginning of part (a) by using a Green's function.

2. (a) The differential equation $y'' + y = 1$ has the solution

$$y = -\cos x + \sin x + 1$$

satisfying $y(0) = y(2\pi) = 0$. Find solutions y_1 and y_2 satisfying

$$y_1(0) = 0, \qquad y_1'(0) = 1,$$
$$y_2(0) = 0, \qquad y_2'(0) = -1.$$

Then explain why you cannot use Equation 4.1 to find a third solution.

(b) Find all solutions of the given boundary-value problem.

3. (a) Find graphical or numerical data for the solution of the boundary-value problem

$$y'' + \frac{2x}{1+x^2}y' = F(x), \qquad y(0) = -1, \qquad y(10) = 1.5,$$

for the case $F(x) = \frac{1}{5}$.

(b) Show that the general solution of the differential equation in part (a) has the form

$$y = c_1 + c_2 \arctan x + \int_0^x (1+t^2)(\tan^{-1} x - \tan^{-1} t)F(t)\,dt.$$

(c) Explain how to determine c_1 and c_2 in part (b) in order to find a solution to the boundary-value problem in part (a).

4. Consider the nonlinear pendulum problem

$$\ddot{y} = -\sin y - 0.1\dot{y}, \qquad y(0) = -1, \qquad y(3) = 1.$$

(a) Find a numerical solution for this problem, estimate the value of $y'(0)$ for your solution, and compare this value with the one given for the corresponding problem in Example 1 of the text.

(b) Give a physical interpretation of your comparison results in part (a).

5. If we apply the shooting method to a solvable linear, perhaps nonhomogeneous, boundary problem

$$y'' + p(x)y' + q(x)y = f(x), \qquad y(a) = \alpha, \qquad y(b) = \beta,$$

we will in principle arrive at a solution after just three shots. This exercise shows why, showing that the interpolation in Equation 4.1 of the text is just right for linear equations.

(a) Suppose that the first two shots give us solutions $y_1(x)$ and $y_2(x)$ that satisfy the initial conditions

$$y_1(a) = \alpha, \qquad y_2'(a) = z_1,$$
$$y_2(a) = \alpha, \qquad y_2'(a) = z_2,$$

where $z_1 \neq z_2$. Form the 1-parameter family of functions

$$y_3(x) = \lambda y_1(x) + (1 - \lambda) y_2(x),$$

where λ is a constant to be determined. [Note that for $0 \leq \lambda \leq 1$, $y_3(x)$ ranges from $y_2(x)$ if $\lambda = 0$ to $y_1(x)$ if $\lambda = 1$.] Show that, regardless of the value of λ, $y_3(x)$ satisfies the given differential equation and also satisfies the single boundary condition $y_3(a) = \alpha$.

(b) Assuming $y_1(b) \neq y_2(b)$, solve the equation $\beta = \lambda y_1(b) + (1 - \lambda) y_2(b)$ for λ to get

$$\lambda = \frac{\beta - y_2(b)}{y_1(b) - y_2(b)}.$$

If $y_1(b) - y_2(b)$ is uncomfortably small, adjust z_1 or z_2 to separate $y_1(b)$ and $y_2(b)$.

(c) Differentiate the equation for $y_3(x)$ in part (a), and show that using the value of λ found in part (b) yields Equation 4.1 with $z_3 = y_3'(a)$.

(d) Show that the choice of λ in part (b) yields $y_3(b) = \beta$.

6. Using the boundary conditions $y(0) = 0$, $y(10) = 0$, find a graphical solution of the **nonhomogeneous Airy equation**

$$y'' + xy = 1.$$

7. (a) Apply the shooting method to the boundary problem described in Exercise 31 of Section 3 under the assumptions $k = 1$, $\gamma = 0.3$, $\delta = 5$.

(b) Compare your solution numerically with the exact solution for the case $k = 1$, $\gamma = 0$, $\delta = 5$.

8. (a) Apply the shooting method to the boundary problem described in Exercise 32 of Section 3 under the assumptions $p = 1, q = 2$.

(b) Compare your solution numerically with the exact solution for the case $p = 1, q = 0$.

9. The purpose of this exercise is to see what can happen if you try to apply numerical solution to a boundary-value problem that has no solution. The last part of Example 1 of the previous section shows that the problem $y'' = -y,\ y(0) = 1,\ y(\pi) = 1$ fails to have a solution. However, note that any attempt to solve the problem numerically forces us to approximate π by a rational number p, perhaps in decimal or binary form.

(a) Show that if $p \neq \pi$ is a close approximation to π, the boundary-value problem to satisfy $y'' = -y,\ y(0) = 1,\ y(p) = 1$ has the unique solution

$$y(x) = \cos x + \frac{(1 - \cos p)}{\sin p} \sin x.$$

(b) Show that for the solution proposed in part (a), the slope at $x = p$, namely, $y'(p)$, tends to $+\infty$ if p approaches π from below and tends to $-\infty$ if p approaches π from above.

(c) Explain what the result will be of a naive attempt to find a nonexistent solution to the boundary-value problem by shooting. You may want to conduct some computer experiments.

10. Find a graphical solution to the boundary problem $\ddot{y} = -3y^3$, $y(0) = 1$, $y(2.5) = -1$, noting that some care is needed to straddle the value -1 with the terminal values of the first two shots.

11. Find a graphical solution to the boundary problem $\ddot{y} = -3y^3$, $y(0) = 1$, $y(6) = -1$, noting that some care is needed to straddle the value -1 with the terminal values of the first two shots.

12. The shape $y = y(x)$ of a chain suspended between two fixed supports at $y(a) = \alpha$ and $y(b) = \beta$ satisfies the **catenary equation** $y'' = k\sqrt{1 + (y')^2}$. The positive constant k depends on the length of the chain. Plot the solution assuming $k = 1$ and $y(0) = 1,\ y(6) = 2$.

5 ANALYTIC SOLUTIONS

A differential equation of the form $y' = f(x, y)$, if it has a solution $y = y(x)$ near $x = x_0$, determines the derivative $y'(x_0)$ by $y'(x_0) = f(x_0, y(x_0))$. If f is sufficiently differentiable, even higher derivatives are similarly determined at x_0. Our aim is to use this idea to find polynomial approximations to solutions of differential equations and, if possible, extend the approximations to power series representations

$$y(x) = \sum_{k=0}^{\infty} c_k (x - x_0)^k$$

for solutions convergent in some open interval symmetric about x_0. A function that can be represented by a power series is called an **analytic function,** and analytic functions have the property that they can be differentiated and integrated term by term as often as you like within the open interval of convergence. In addition, an analytic function $y(x)$ as displayed above is represented by its **Taylor series** at x_0, which simply means that the coefficients c_k can in principle be computed by the **Taylor coefficient formula**

$$c_k = \frac{1}{k!} y^{(k)}(x_0), \qquad k = 0, 1, 2, \ldots.$$

We'll see next that if we know a differential equation has an analytic solution we can sometimes calculate its Taylor coefficients at some point x_0 directly from the differential equation.

example 1 The differential equation $y' = y^2$ has $y(x) = (1 - x)^{-1}$ for a solution satisfying $y(0) = 1$. The power series expansion is the familiar geometric series formula

$$y = \frac{1}{1 - x} = 1 + x + x^2 + x^3 + \cdots, \qquad -1 < x < 1$$

This expansion is also the Taylor series of $y(x)$, so by the Taylor coefficient formulas we have

$$y = y(0) + \frac{y'(0)}{1!}x + \frac{y''(0)}{2!}x^2 + \frac{y'''(0)}{3!}x^3 + \cdots.$$

Comparison of coefficients of x^k in the two expansions shows that $y^{(k)}(0)/k! = 1$, so $y^{(k)}(0) = k!$, for $k = 0, 1, 2, \ldots$. Suppose, however, that we had no formula for the coefficients to begin with. (Most examples are of this sort.) Starting with the given differential equation, we can calculate successive derivatives using the chain rule, and then simplify by substitution from the earlier equations:

$$\begin{aligned}
y' &= y^2 = 1!y^2, \\
y'' &= 2yy' = 2y^3 = 2!y^3, \\
y''' &= 6y^2y' = 6y^4 = 3!y^4, \\
y^{(4)} &= 24y^3y' = 24y^5 = 4!y^5.
\end{aligned}$$

The general pattern is evidently $y^{(k)} = k!y^{k+1}$. The factorials cancel in a formal Taylor expansion for a solution $y = y(x)$ about x_0 with $y(x_0) = y_0$, so the Taylor expansion becomes a geometric series:

$$\begin{aligned}
y(x) &= y_0 + y_0^2(x - x_0) + y_0^3(x - x_0)^2 + \cdots \\
&= y_0\left[1 + y_0(x - x_0) + y_0^2(x - x_0)^2 + \cdots\right] \\
&= y_0 \sum_{k=0}^{\infty} \left[y_0(x - x_0)\right]^k = \frac{y_0}{1 - y_0(x - x_0)}.
\end{aligned}$$

This expansion contains the one we started with as the special case when $x_0 = 0$ and $y_0 = 1$. However, that expansion was valid for $-1 < x < 1$, while the more general one is valid for $|y_0(x - x_0)| < 1$, or $|x - x_0| < 1/|y_0|$.

Higher order equations, for example, $y'' = f(x, y, y')$, can be treated by the method illustrated above.

---

example 2 Suppose we want successive derivatives $y^{(k)}(0)$ to a solution of

$$y'' = yy',$$

given that $y(0) = 1$ and $y'(0) = -1$. First compute from the given equation some formulas for higher derivatives. Then simplify by substituting the given values $y(0) = 1$, $y'(0) = -1$. We find

$$\begin{aligned}
y'' &= yy'; \qquad y''(0) = -1, \\
y''' &= yy'' + (y')^2 = y^2y' + (y')^2; \qquad y'''(0) = 0, \\
y^{(4)} &= 2y(y')^2 + y^2y'' + 2y'y'' = 4y(y')^2 + y^3y'; \qquad y^{(4)}(0) = 3.
\end{aligned}$$

The first five terms of the Taylor expansion of $y(x)$ about $x = 0$ then add up to

$$y(0) + \frac{y'(0)}{1!}x + \frac{y''(0)}{2!}x^2 + \frac{y'''(0)}{3!}x^3 + \frac{y^{(4)}(0)}{4!}x^4 = 1 - x - \frac{1}{2}x^2 + \frac{1}{8}x^4.$$

In this example, no simple pattern emerges right away, but clearly we could compute as many terms as we had time and space for.

example 3

We rewrite the linear differential equation

$$y'' - xy' + y = 0 \quad \text{as} \quad y'' = xy' - y.$$

Then calculate, using substitution after differentiating,

$$\begin{aligned} y''' &= xy'' = x(xy' - y) \\ &= x^2y' - xy, \\ y^{(4)} &= x^2y'' + xy' - y = x^2(xy' - y) + xy' - y \\ &= (x^3 + x)y' - (x^2 + 1)y. \end{aligned}$$

If we let $y(0) = c_0$ and $y'(0) = c_1$, then from the previous calculations

$$y''(0) = -c_0, \qquad y'''(0) = 0, \qquad y^{(4)}(0) = -c_0.$$

The first few terms of the Taylor expansion of $y(x)$ about $x = 0$ has the form

$$\begin{aligned} y(x) &= c_0 + c_1x - \frac{c_0}{2!}x^2 + 0 - \frac{c_0}{4!}x^4 - \cdots \\ &= c_0\left(1 - \frac{1}{2}x^2 - \frac{1}{24}x^4 - \cdots\right) + c_1x. \end{aligned}$$

We verify directly that $y(x) = x$ is a solution of the differential equation. Thus we've identified two independent solutions,

$$y_1(x) = x \quad \text{and} \quad y_2(x) = 1 - \frac{1}{2!}x^2 - \frac{1}{4!}x^4 - \cdots.$$

If we try to find $y_2(x)$ by using $y_1(x)$ in Theorem 1.2 of Section 1, we discover that the solution is expressible only in power series form. You are asked to do this in Exercise 15.

EXERCISES

Find the first three nonzero terms in the Taylor expansion about $x = 0$ of $y = y(x)$ if y satisfies each of the equations in Exercises 1 through 4. Then use the resulting approximation near $x = 0$ to sketch the graph of the solution in a small neighborhood of $x = 0$.

1. $y' = y^2 + y, \; y(0) = 1.$

2. $y' = y^2 + x, \; y(0) = -1.$

3. $y' = xy, \; y(0) = 2.$

4. $y'' = xy, \; y(0) = 1, \; y'(0) = 0.$

5. **(a)** Find the first five nonzero terms in the Taylor expansion of $y = y(x)$ about $x = 0$ if $y'' = yy'$ and $y(0) = y'(0) = 1$.

(b) Derive the solution $y = \tan(x/2 + \pi/4)$ for the initial-value problem in part (a).

6. Find the first four nonzero terms in the Taylor expansion of $y = y(x)$ about $x = 1$ if $y''' = y$ and $y(1) = 2,\ y'(1) = 0,\ y''(1) = 1$.

7. Suppose $y'' = x^2 y$ while $y(0) = c_0$ and $y'(0) = c_1$. Show that $y = y(x)$ has the form

$$y = c_0\left(1 + \tfrac{1}{12}x^4 + \cdots\right) + c_1\left(x + \tfrac{1}{20}x^5 + \cdots\right).$$

8. Show that if $y''' = y^2 y'$ and $y(0) = y'(0) = y''(0) = 1$ then

$$y = 1 + x + \tfrac{1}{2}x^2 + \tfrac{1}{6}x^3 + \tfrac{1}{8}x^4 + \tfrac{3}{40}x^5 + \cdots.$$

9. In Example 2 of the text the computation stops with the fourth-degree term. Find the fifth-degree term.

10. Solve initial-value problem $y' = e^x y$, $y(0) = 1$ in terms of elementary functions. Then find the first three nonzero term in the Taylor expansion of that solution about $x = 0$.

11. The initial-value problem $y' = y$, $y(0) = 1$ has a well-known solution. Derive the power series form of that solution by the method of this section.

12. The initial-value problem $y'' = y$, $y(0) = y'(0) = 1$ has a well-known solution. Derive the power series form of that solution by the method of this section.

13. **(a)** Show that the equation $y' = |x|$ has solutions

$$y(x) = \begin{cases} x^2/2 + C, & x \geq 0, \\ -x^2/2 + C, & x < 0. \end{cases}$$

(b) Sketch the graph of the solution in part (a) for $-2 < x < 2$ assuming $C = 0$. Then show that none of the solutions in part (a) is analytic in an interval containing $x_0 = 0$.

14. Verify directly the assertion made at the end of Example 1 of the text that the expansion

$$\frac{y_0}{1 - y_0(x - x_0)} = y_0[1 + y_0(x - x_0) + y_0^2(x - x_0)^2 + \cdots]$$

is valid for $|x - x_0| < 1/|y_0|$.

15. The linear differential equation $y'' - xy' + y = 0$ has the solution $y_1(x) = x$ identified in text Example 3. Use Theorem 1.2 of Section 1 to find the general power series expansion for an independent solution $y_2(x)$. Compare the result with the solution $y_2(x)$ identified in Example 3. [*Hint:* Express the exponential factor in Theorem 1.2 as a power series.]

6 UNDETERMINED COEFFICIENTS

The solutions of many of the differential equations we have studied can be represented in terms of their Taylor expansions. For example, polynomials, the elementary transcendental functions $\cos x$, $\sin x$, e^x, and linear combinations of these have Taylor expansions that are valid for all x: In addition there is a large and important class of differential equations that has solutions that are *analytic*, that is, representable by power series. Even if an analytic solution is not a combination of elementary functions, its power series expansion may serve to define an important new function. Furthermore, the partial sums of a series expansion often give useful approximations to the true solution.

Recall that a Taylor expansion of a function $f(x)$ is a power series

$$\sum_{k=0}^{\infty} c_k(x - x_0)^k = c_0 + c_1(x - x_0) + c_2(x - x_0)^2 + \cdots,$$

where $c_k = f^{(k)}(x_0)/k!$. Such a series, unless it converges only for $x = x_0$, always converges absolutely in an interval $x_0 - r < x < x_0 + r$ that is symmetric about x_0; the number r is the **radius of convergence** of the series. For functions f that are analytic, the series actually converges to $f(x)$ for all x in that interval. Furthermore, Taylor expansions about the same point x_0 can be conveniently added and multiplied, and a Taylor expansion can be differentiated or integrated term by term to produce the Taylor expansion of a derivative or integral of the expanded function within the interval of convergence.

example 1 The Taylor expansions

$$e^x = \sum_{k=0}^{\infty} \frac{x^k}{k!}, \qquad -\infty < x < \infty,$$

$$\cos x = \sum_{k=0}^{\infty} \frac{(-1)^k x^{2k}}{(2k)!}, \qquad -\infty < x < \infty,$$

$$\sin x = \sum_{k=0}^{\infty} \frac{(-1)^k x^{2k+1}}{(2k+1)!}, \qquad -\infty < x < \infty,$$

$$\frac{1}{1-x} = \sum_{k=0}^{\infty} x^k, \qquad -1 < x < 1,$$

$$\ln(1 - x) = -\sum_{k=1}^{\infty} \frac{x^k}{k}, \qquad -1 < x < 1,$$

can all be computed directly from the general Taylor formula and can be shown to converge to the value of the function on the left for each x in the indicated interval.

example 2

To understand the series techniques that we'll need later, it's a good idea to follow carefully the steps in computing the derivative of $\cos 2x$ using term-by-term differentiation of its power series:

$$\begin{aligned}
\frac{d}{dx}\cos 2x &= \frac{d}{dx}\sum_{k=0}^{\infty}\frac{(-1)^k(2x)^{2k}}{(2k)!} \\
&= \sum_{k=0}^{\infty}\frac{d}{dx}\frac{(-1)^k(2x)^{2k}}{(2k)!} \\
&= \sum_{k=0}^{\infty}\frac{(-1)^k(2x)^{2k-1}(4k)}{(2k)!} \qquad \text{[Term is 0 when } k = 0.] \\
&= 2\sum_{k=1}^{\infty}\frac{(-1)^k(2x)^{2k-1}}{(2k-1)!} \qquad [2k \text{ canceled.}] \\
&= -2\sum_{k=0}^{\infty}\frac{(-1)^k(2x)^{2k+1}}{(2k+1)!} = -2\sin 2x.
\end{aligned}$$

In the last step, we simply replaced k by $k+1$ everywhere, including the limit equations $k = 0$ and $k = \infty$, to make the expansion look more like the corresponding expansion in the preceding examples.

The shift in the index k at the end of the previous example is the analogue of replacing x by $x+1$ in the integral

$$\int_{x=a}^{x=b} f(x-1)\,dx \quad \text{to get} \quad \int_{x=a-1}^{x=b-1} f(x)\,dx.$$

Such changes must be made carefully, with attention not only to the summand but also to the summation limits.

Many of the most important examples of series solutions of differential equations are expressible as power series about the point $x_0 = 0$, and most of our examples will be of that kind. Our main job in this section is efficient computation of coefficients in the expansion of solutions of linear differential equations. We begin with a familiar example for which we already know all solutions. The method involves treating the coefficients in a power series as "undetermined coefficients" to be determined recursively by requiring the coefficients of the same power of x on either side an equation to be zero.

example 3

To solve $y'' + y = 0$, we try to determine all the coefficients in a solution of the general form

$$y(x) = \sum_{k=0}^{\infty} c_k x^k.$$

This form of the expansion is particularly appropriate if we want to solve an initial-value problem with $y(0)$ and $y'(0)$ specified, because then $c_0 = y(0)$ and $c_1 = y'(0)$, assuming there is a Taylor expansion for the solution about $x = 0$. Proceeding under that assumption for the moment, we compute

$$y'(x) = \sum_{k=1}^{\infty} kc_k x^{k-1},$$

$$y''(x) = \sum_{k=2}^{\infty} (k-1)kc_k x^{k-2} \quad [k = 1 \text{ term is } 0.]$$

$$= \sum_{k=0}^{\infty} (k+1)(k+2)c_{k+2} x^k.$$

To simplify adding terms in x^k of the series for $y''(x)$ to terms in x^k for $y(x)$, we shifted the index of summation up by 2 in the series for $y''(x)$; to compensate, we also shift the summation limits down by 2. For $y(x)$ to represent a solution, we want

$$y''(x) + y(x) = \sum_{k=0}^{\infty} c_k x^k + \sum_{k=0}^{\infty} (k+1)(k+2)c_{k+2} x^k$$

$$= \sum_{k=0}^{\infty} \left[c_k + (k+1)(k+2)c_{k+2}\right] x^k = 0.$$

A fundamental fact about Taylor expansions is that for the series to be identically zero, all the coefficients must be zero. The reason is that the derivative formulas $f^{(k)}(x_0)/k!$ for the coefficients must all be zero. Thus from $c_k + (k+1)(k+2)c_{k+2} = 0$ we get a recurrence relation for the c_ks:

$$c_{k+2} = -\frac{c_k}{(k+1)(k+2)}, \qquad k = 0, 1, 2, \ldots.$$

Recalling that we can specify a particular solution by determining the numbers $c_0 = y(0)$, $c_1 = y'(0)$, it's natural to compute recursively in separate columns:

$$c_2 = -\frac{c_0}{1\cdot 2} = -\frac{c_0}{2!}, \qquad c_3 = -\frac{c_1}{2\cdot 3} = -\frac{c_1}{3!},$$

$$c_4 = -\frac{c_2}{3\cdot 4} = \frac{c_0}{4!}, \qquad c_5 = -\frac{c_3}{4\cdot 5} = \frac{c_1}{5!},$$

$$c_6 = -\frac{c_4}{5\cdot 6} = -\frac{c_0}{6!}, \qquad c_7 = -\frac{c_5}{6\cdot 7} = -\frac{c_1}{7!},$$

$$\vdots \qquad \qquad \vdots$$

$$c_{2k} = -\frac{c_{2k-2}}{(2k-1)(2k)} = \frac{(-1)^k c_0}{(2k)!}, \qquad c_{2k+1} = -\frac{c_{2k-1}}{(2k)(2k+1)} = \frac{(-1)^k c_1}{(2k+1)!}.$$

If we take $y(0) = c_0$ and $y'(0) = c_1 = 0$, then only the first column contains nonzero entries, and we get from the second formula in Example 1 the solution

$$y_0(x) = c_0 - \frac{c_0}{2!}x^2 + \frac{c_0}{4!}x^4 - \cdots + (-1)^k \frac{c_0}{(2k)!}x^{2k} + \cdots$$

$$= c_0 \cos x.$$

On the other hand, the choice $y(0) = c_0 = 0$ and $y'(0) = c_1$ makes the entries in the first column all zero, so we get an independent solution from the second column:

$$\begin{aligned} y_1(x) &= c_1 - \frac{c_1}{3!}x^3 + \frac{c_1}{5!}x^5 - \cdots + (-1)\frac{c_1}{(2k+1)}x^{2k+1} + \cdots \\ &= c_1 \sin x. \end{aligned}$$

Thus the general solution is $y(x) = c_0 \cos x + c_1 \sin x$ as we expected.

example 4

Solutions of **Airy's equation** $y'' + xy = 0$ are not obtainable in terms of the functions of elementary calculus, so we try

$$y(x) = \sum_{k=0}^{\infty} c_k x^k,$$

$$y''(x) = \sum_{k=2}^{\infty} (k-1)kc_k x^{k-2}.$$

Substitution into the differential equation gives

$$\begin{aligned} y''(x) + xy(x) &= \sum_{k=2}^{\infty}(k-1)kc_k x^{k-2} + \sum_{k=0}^{\infty} c_k x^{k+1} \\ &= \sum_{k=0}^{\infty}(k+1)(k+2)c_{k+2}x^k + \sum_{k=1}^{\infty} c_{k-1}x^k = 0, \end{aligned}$$

where this time we shifted the index in the first sum up by 2 and in the second sum down by 1 in order to make the powers of x agree. Note that the constant $2c_2$ is a term in the first sum that doesn't correspond to a constant term in the second sum. Separating out the constant enables us to combine the remaining terms in a sum over index k:

$$2c_2 + \sum_{k=1}^{\infty}[(k+1)(k+2)c_{k+2} + c_{k-1}]x^k = 0.$$

Setting all coefficients equal to zero gives the equations

$$c_2 = 0, \quad \text{and} \quad c_{k+2} = -\frac{c_{k-1}}{(k+1)(k+2)}, \quad k = 1, 2, 3, \ldots.$$

We find as a result that the terms are determined in sequences with indices differing by 3, and that

$$0 = c_2 = c_5 = c_8 = \cdots = c_{3k+2} = \cdots.$$

However, if c_0 or c_1 is not zero, we compute as follows:

$$
\begin{aligned}
c_3 &= -\frac{c_0}{2\cdot 3}, \\
c_6 &= -\frac{c_3}{5\cdot 6} &&= \frac{c_0}{2\cdot 3\cdot 5\cdot 6}, \\
c_9 &= -\frac{c_6}{8\cdot 9} &&= -\frac{c_0}{2\cdot 3\cdot 5\cdot 6\cdot 8\cdot 9}, \\
&\vdots &&\vdots \\
c_{3k} &= -\frac{c_{3k-3}}{(3k-1)3k} &&= \frac{(-1)^k c_0}{2\cdot 3\cdot 5\cdot 6\cdots(3k-1)3k}, \\
c_4 &= -\frac{c_1}{3\cdot 4}, \\
c_7 &= -\frac{c_4}{6\cdot 7} &&= \frac{c_1}{3\cdot 4\cdot 6\cdot 7}, \\
c_{10} &= -\frac{c_7}{9\cdot 10} &&= -\frac{c_1}{3\cdot 4\cdot 6\cdot 7\cdot 9\cdot 10}, \\
&\vdots &&\vdots \\
c_{3k+1} &= -\frac{c_{3k-2}}{3k(3k+1)} &&= \frac{(-1)^k c_1}{3\cdot 4\cdot 6\cdot 7\cdots 3k(3k+1)}.
\end{aligned}
$$

The solution determined by $y(0) = c_0 = 1,\ y'(0) = c_1 = 0$ is

$$y_0(x) = 1 + \sum_{k=1}^{\infty} \frac{(-1)^k x^{3k}}{2\cdot 3\cdot 5\cdot 6\cdots(3k-1)3k},$$

and the solution determined by $y(0) = c_0 = 0,\ y'(0) = c_1 = 1$ is

$$y_1(x) = x + \sum_{k=1}^{\infty} \frac{(-1)^k x^{3k+1}}{3\cdot 4\cdot 6\cdot 7\cdots 3k(3k+1)}.$$

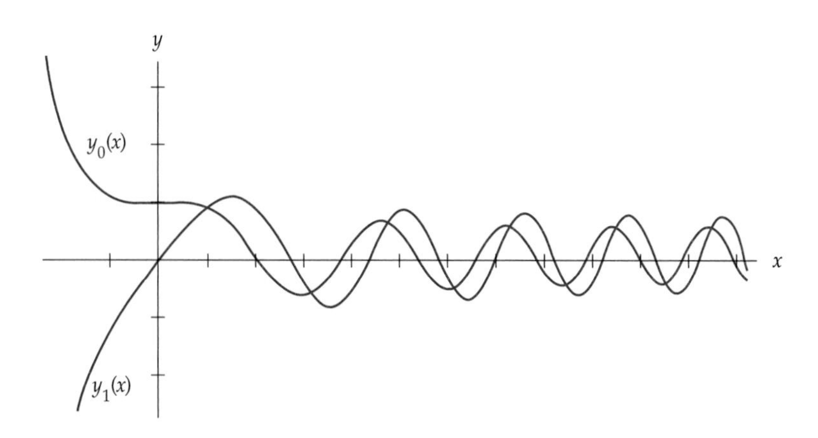

Figure 9.6 Solutions $y_0(x)$ and $y_1(x)$ to Airy's equation.

The respective power series for y_0 and y_1 both converge for all x because the denominators of the kth terms each contain increasing integer factors, $2k$ in number. Hence each series has terms dominated by those of a series that converges for all x, for example

$$\left|\frac{(-1)^k x^{3k}}{2\cdot 3\cdot 5\cdot 6\cdots(3k-1)3k}\right| \le \frac{|x|^{3k}}{k!},$$

since $(2\cdot 3)\cdot(5\cdot 6)\cdots\big((3k-1)\cdot(3k)\big) > 3\cdot 6\cdots(3k) > k!$.

Since the two solutions y_0 and y_1 are linearly independent, by Theorem 1.3 of Section 1 the general solution of $y'' + xy = 0$ has the form

$$y(x) = c_0 y_0(x) + c_1 y_1(x).$$

The sinusoidal solutions of the constant-coefficient equation $y'' + \alpha y = 0$ have infinitely many zero values if $\alpha > 0$ and at most one such value when $\alpha < 0$. Analogously (see Section 1B) a solution of the Airy equation is zero infinitely often when $x > 0$ but at most once when $x < 0$.

example 5

The **Legendre equation** of positive integer index n

$$(1-x^2)y'' - 2xy' + n(n+1)y = 0.$$

This equation appears also in Chapter 11, where a formula is given for polynomial solutions. To find solutions in the form of power series, we let

$$y(x) = \sum_{k=0}^{\infty} c_k x^k, \qquad y'(x) = \sum_{k=1}^{\infty} k c_k x^{k-1}, \qquad y''(x) = \sum_{k=2}^{\infty} (k-1)k c_k x^{k-2}.$$

Substitution of these sums into the differential equation and shifting the index k up by 2 in the first of the resulting sums shows that the c_k are determined by

$$\sum_{k=0}^{\infty}(k+1)(k+2)c_{k+2}x^k - \sum_{k=2}^{\infty}(k-1)k c_k x^k - 2\sum_{k=1}^{\infty} k c_k x^k + n(n+1)\sum_{k=0}^{\infty} c_k x^k = 0.$$

The indices $k = 0$ and $k = 1$ don't occur in all the sums so we get two special terms and a sum over k:

$$[2c_2 + n(n+1)c_0] + [2\cdot 3c_3 - 2c_1 + n(n+1)c_1]x$$
$$+ \sum_{k=2}^{\infty}[(k+1)(k+2)c_{k+2} - ((k-1)k + 2k - n(n+1))c_k]x^k = 0.$$

Setting the constant and the coefficient of each power of x equal to zero gives

$$c_2 = -\frac{n(n+1)}{2}c_0, \qquad c_3 = -\frac{(n-1)(n+2)}{2\cdot 3}c_1,$$

$$c_{k+2} = \frac{(k+1+n)(k-n)}{(k+1)(k+2)}c_k, \qquad k \ge 2.$$

Since the general recurrence relation contains a shift by 2, it is natural to split the coefficients into those of even and those of odd index. For example

$$c_4 = \frac{(3+n)(2-n)}{3\cdot 4}c_2 = \frac{(n-2)n(n+3)(n+1)}{4!}c_0,$$

$$c_6 = \frac{(5+n)(4-n)}{5\cdot 6}c_4 = -\frac{(n-4)(n-2)n(n+5)(n+3)(n+1)}{6!}c_0,$$

$$\vdots$$

and

$$c_5 = \frac{(4+n)(3-n)}{4\cdot 5}c_3 = \frac{(n-3)(n-1)(n+4)(n+2)}{5!}c_1,$$

$$c_7 = \frac{(6+n)(5-n)}{6\cdot 7}c_5 = -\frac{(n-5)(n-3)(n-1)(n+6)(n+4)(n+2)}{7!}c_1$$

$$\vdots$$

If $n = 2l$ is a positive even integer, then all even coefficients are zero beyond $2l$. Thus the series expansion with even powers reduces to an even polynomial. Similarly, if $n = 2l + 1$ is a positive odd integer the series expansion with odd powers reduces to an odd polynomial. For example, when $n = 4$ the independent solutions we derived are

$$P_4(x) = 1 - \frac{4\cdot 5}{2!}x^2 + \frac{2\cdot 4\cdot 7\cdot 5}{4!}x^4, \qquad \text{if } c_0 = 1,\ c_1 = 0,$$

$$Q_4(x) = x - \frac{3\cdot 6}{3!}x^3 + \frac{1\cdot 3\cdot 8\cdot 6}{5!}x^5 - \frac{-1\cdot 1\cdot 3\cdot 10\cdot 8\cdot 6}{7!}x^7 + \cdots, \qquad \text{if } c_0 = 0,\ c_1 = 1.$$

The ratio test for convergence (see Exercise 7) shows that for all values of n the infinite series solutions converges for $-1 < x < 1$. The two solutions, one with odd terms and one with even terms form the basis for the collection of all solutions of the homogeneous Legendre equation of index n on the interval $-1 < x < 1$, and the solutions that are not polynomials tend to infinity as x tends to 1 or -1. In general, the Legendre equation of nonnegative integer index n evidently has independent solutions such that one is a polynomial and the other is not. The importance of the polynomial solutions P_n is explained in Chapter 11.

EXERCISES

1. (a) Use the undetermined coefficient method to derive the general solution to the differential equation $y'' - y = 0$ in terms of a power series expansion about $x_0 = 0$.

(b) Show that the general series solution derived in part (a) represents $y = c_0 \cosh x + c_1 \sinh x$, where c_0 and c_1 are arbitrary constants.

2. (a) Show that, if $y = f(x)$ is a solution of $y'' + xy = 0$ for all x, then $y = f(-x)$ is a solution of

$$y'' - xy = 0.$$

(b) Use the result of part (a) together with the result of Example 4 of the text to find a power series expansion for the general solution of $y'' - xy = 0$.

3. (a) Apply the method of power series to solve the first-order differential equation

$$y' + 2xy = 0.$$

(b) Solve the differential equation in part (a) by finding an exponential multiplier and then integrating.

(c) Do the results of parts (a) and (b) agree for all x?

4. (a) Apply the power series method to find the general solution of the differential equation

$$y'' - xy' = 0$$

in the form $y(x) = c_0 y_0(x) + c_1 y_1(x)$.

(b) Solve the differential equation in part (a) by solving the equivalent system

$$y' = u,$$
$$u' = xu.$$

(c) Do the results of parts (a) and (b) agree for all x?

5. (a) Apply the power series method to find the general solution of the differential equation

$$y'' + xy' + y = 0$$

in the form $y(x) = c_0 y_0(x) + c_1 y_1(x)$.

(b) Show that a special case of the general solution found in part (a) is the solution $y(x) = e^{-(x^2/2)}$.

(c) Given that $y_0(x) = e^{-(x^2/2)}$, use the order reduction method of Section 1 to find an independent solution $y_1(x)$ in terms of the function $F(x) = \int_0^x e^{u^2/2}\,du$.

***(d)** Reconcile the result of part (c) with the first three terms of the solution of part (a). [*Hint:* Multiply power series.]

6. (a) Apply the power series method to find a single particular solution of the differential equation

$$y'' + xy' + y = x.$$

(b) Combine the result of part (a) with that of the previous exercise to write the general solution of $y'' + xy' + y = x$.

(c) Starting with the solution $y_0(x) = e^{-(x^2/2)}$ to the associated homogeneous equation, use variation of parameters to find the general solution to the equation in part (a).

7. Apply the ratio test for series convergence to the series solution found for the Legendre equation in Example 5 of the text. Show that the series converges for $-1 < x < 1$. [*Hint:* Split into even and odd parts, and show that each converges separately.]

8. The *Bessel equation* of integer index n is

$$x^2y'' + xy' + (x^2 - n^2)y = 0,$$

and it's discussed in more detail in Section 7B.

(a) Show that when $n = 0$ a solution of the form $\sum_{k=0}^{\infty} c_k x^k$ satisfies $(k+2)^2 c_{k+2} = -c_k$.

(b) Show that, if we choose $c_0 = 1,\ c_1 = 0$ in part (a), we get the solution denoted $J_0(x)$, where

$$J_0(x) = \sum_{k=0}^{\infty}(-1)^k 2^{-2k}(k!)^{-2}x^{2k},$$

called a Bessel function of index 0.

(c) Show that $J_1(x) = -J_0'(x)$ defines a solution of the Bessel equation of index 1. [*Hint:* Apply d/dx to the equation for J_0; then express J_0 in terms of J_0' and J_0''.]

9. A **Bessel function** of integer index n is defined by

$$J_n(x) = \left(\frac{x}{2}\right)^n \sum_{k=0}^{\infty}(-1)^k \frac{x^{2k}}{2^{2k}k!(n+k)!}.$$

(a) Show that J_n satisfies the Bessel equation of index n given in Exercise 8.

(b) Show that if y_n is a solution of the Bessel equation of index n and $u_n(x) = \sqrt{x}\,y_n(x)$, then u_n satisfies

$$u'' + \left(1 - \frac{4n^2 - 1}{4x^2}\right)u = 0.$$

10. Find a power series expansion about $x_0 = 1$ for the solution to initial-value problem $y' = 3(x-1)^2 y$, $y(1) = 1$ as follows.

(a) Use the undetermined coefficient method of this section.

(b) Find the solution to the problem in terms of elementary functions by working with the solution to part (a).

(c) Find some form of the solution to the related problem $y' = 3(x-1)^2 y$, $y(0) = 1$, by any method you choose.

11. (a) Show that the only power series expansion about $x_0 = 0$ that provides a solution for $x^2y'' - y = 0$ has all its coefficients equal to zero and so is the identically zero solution.

(b) Show nevertheless that this differential equation has solutions defined for $x > 0$ by determining those exponents r for which $y_r(x) = x^r$ is a solution.

12. (a) Show that the only power series expansion about $x_0 = 0$ that provides a solution for $x^2y'' + xy' + \left[x^2 - \left(\frac{1}{2}\right)^2\right]y = 0$ has all its coefficients equal to zero and so is the identically zero solution.

(b) Show nevertheless that this differential equation, the Bessel equation of index $\frac{1}{2}$, has solutions defined for $x > 0$ defined by series of the form

$$y = \sum_{k=0}^{\infty} c_k x^{k-1/2}.$$

(c) Show that the series solutions found in part (b) can be combined to give solutions of the form $y = x^{-1/2}(c_0 \cos x + c_1 \sin x)$.

13. Consider the solutions of Airy's equation $y'' = -xy$ in Example 4 of the text.

(a) Show that for $x > 0$ the graph of a solution $y(x)$ must be concave down when $y > 0$ and concave up when $y < 0$.

(b) Show that for $x < 0$ the graph of a solution $y(x)$ must be concave up when $y > 0$ and concave down when $y < 0$.

(c) Explain how the observations of parts (a) and (b) are consistent with the general nature of the graphs in Figure 9.6.

7 SINGULAR POINTS

The power series method of the previous section is usually applied just to second-order homogeneous equations of the form $a_0(x)y'' + a_1(x)y' + a_2(x)y = 0$, in which we assume that the coefficient functions $a_0(x)$, $a_1(x)$, and $a_2(x)$ are analytic at each point of their common domain of definition; in other words, each of the coefficient functions has a power series representation in some neighborhood of each point of that domain. Our special focus in this section will be on isolated points $x = x_0$ such that $a_0(x_0) = 0$; in that case the differential equation fails to define y'' in terms of x, y, and y', with the result that solutions may fail to be defined at x_0. Even if solutions are defined at x_0, satisfying the conventional initial conditions $y(x_0) = y_0$, $y'(x_0) = z_0$ is not possible in general. Nevertheless, solutions that exist in a neighborhood of such a point are sometimes very useful, and so deserve special study. We begin by putting the differential equation in **normal form** with leading coefficient 1:

$$y'' + a(x)y' + b(x)y = 0, \quad \text{where} \quad a(x) = \frac{a_1(x)}{a_0(x)} \quad \text{and} \quad b(x) = \frac{a_2(x)}{a_0(x)}.$$

If there is sufficient cancellation in the ratios $a(x)$ and $b(x)$ so that these two coefficients are analytic at x_0, then x_0 is called an **ordinary point** for the differential equation, and the method of the previous section can in principle be applied; otherwise x_0 is called a **singular point**, and a modification of our earlier technique may be necessary. If the coefficient $a_0(x)$ of y'' has not been divided out to produce the normalized form, a singular point may arise from a point $x = x_0$, where $a_0(x) = 0$. A standard bit of terminology is to refer to a point x_0 at which an analytic function has value zero as a *zero* of the function.

example 1

The normal form for $x^2y'' - xy' + x^2y = 0$ is $y'' - x^{-1}y' + y = 0$. In this example, $a(x) = -x^{-1}$ fails to be analytic just at $x_0 = 0$, though $b(x) = 1$ is analytic everywhere. Hence $x_0 = 0$ is the equation's only singular point.

Even though solutions $y(x)$ exist in a neighborhood of a singular point x_0 of a differential equation, the behavior of $y(x)$ as x approaches x_0 may be quite unruly; examples are given later. However, most of the examples that come up in practice have solutions expressible at their singularities by relatively mild modifications of ordinary power series. These equations have the property that their singular points are not too "bad" in the following precise sense. A singular point x_0 of the equation $y'' + a(x)y' + b(x)y = 0$ is called a **regular singular point** if the functions $(x - x_0)a(x)$ and $(x - x_0)^2b(x)$ can both be defined at x_0 in such a way as to be analytic there.

example 2

The equation $y'' - x^{-1}y' + y = 0$ has its only singular point at $x_0 = 0$. This point is a regular singular point, since both $xa(x) = -xx^{-1} = -1$ and $x^2b(x) = x^2$ are analytic for all x, in particular at the singular point $x_0 = 0$.

example 3

The **Legendre equation** of index 1 is $(1 - x^2)y'' - 2xy' + 2y = 0$. The normal form is $y'' - [2x/(1-x^2)]y' + [2/(1-x^2)]y = 0$, and the normal-form coefficients $a(x) = -2x/(1 - x^2)$ and $b(x) = 2/(1 - x^2)$ both have singular points at $x = \pm 1$. At $x_0 = 1$ we have

$$(x - 1)a(x) = \frac{-2(x - 1)x}{1 - x^2} = \frac{2x}{1 + x}.$$

Also

$$(x - 1)^2b(x) = \frac{2(x - 1)^2x}{1 - x^2} = \frac{-2(x - 1)}{1 + x}.$$

Thus $(x-1)a(x)$ and $(x-1)^2b(x)$ are both analytic at $x_0 = 1$, with respective values 1 and 0 there. Notice that to pass the regularity test at $x_0 = 1$, the product functions in question are not required to be analytic at the other singular point at $x_1 = -1$. Indeed, similar computations show that at $x_1 = -1$ the functions $(x + 1)a(x)$ and $(x+1)^2b(x)$ are analytic also, so both singular points are regular for the Legendre equation.

We'll consider two differential equations in detail, the Euler equation and the Bessel equation. Between them, their solutions exhibit most of the important features associated with equations having regular singular points.

7A Euler's Equation

The Euler differential equation

$$x^2y'' + axy' + by = 0, \qquad a, b \text{ constant}$$

arises naturally in the solution of the Laplace equation in Chapter 10, and is the simplest example of an equation with a regular singular point. The form of the equation suggests that there may be a solution of the form $y = x^\mu$ for $x > 0$. This is indeed the case, and we proceed to find the correct values for μ by first finding $y' = \mu x^{\mu-1}$ and $y'' = \mu(\mu - 1)x^{\mu-2}$. (The validity of these differentiation formulas for complex values of μ and $x > 0$ follows from the definition $x^{\alpha+i\beta} = x^\alpha x^{i\beta} = x^\alpha e^{i\beta \ln x}$.) Substitution into the differential equation gives

$$\mu(\mu - 1)x^\mu + \mu a x^\mu + bx^\mu = 0, \quad \text{or} \quad [\mu(\mu - 1) + \mu a + b]x^\mu = 0.$$

We assumed $x > 0$, so $x^\mu \neq 0$ and we can divide by it. This leaves us with an *indicial equation* that determines the **index** μ:

$$\mu(\mu - 1) + \mu a + b = 0 \quad \text{or} \quad \mu^2 + (a - 1)\mu + b = 0.$$

The roots of the indicial equation determine the solutions of the differential equation, as the roots of the characteristic equation do for a constant-coefficient equation. We'll assume that the coefficients a and b are real numbers. The indices are given by the quadratic formula: $\mu = \frac{1}{2}\left[(1 - a) \pm \sqrt{(a - 1)^2 - 4b}\,\right]$.

Distinct Real Roots: $\mu_1 \neq \mu_2$. This is the simplest case in every respect. Since the differential equation is linear, it has solutions $y = c_1x^{\mu_1} + c_2x^{\mu_2}$ for $x > 0$.

Equal Roots: $\mu_1 = \mu_2 = (1 - a)/2$. There is a solution $y_1(x) = x^{(1-a)/2}$. To find an independent solution, we resort to order reduction, using Theorem 1.2 of Section 1A. The normalized form of the differential equation is $y'' + (a/x)y' + (b/x^2)y = 0$, so an independent solution is

$$\begin{aligned} y_2(x) &= x^{(1-a)/2} \int \frac{e^{-\int (a/x)\,dx}}{x^{1-a}}\,dx \\ &= x^{(1-a)/2} \int \frac{e^{-a \ln x}}{x^{1-a}}\,dx = x^{(1-a)/2} \ln x. \end{aligned}$$

Since the equation is linear, we have solutions $y(x) = x^{(1-a)/2}(c_1 + c_2 \ln x)$ for $x > 0$.

Complex Roots: $4b > (a - 1)^2$. Denoting the distinct complex roots by $\mu_1 = \alpha + i\beta$, $\mu_2 = \alpha - i\beta$, $\beta > 0$, we have

$$x^{\alpha \pm i\beta} = x^\alpha x^{\pm i\beta} = x^\alpha e^{\pm i\beta \ln x} = x^\alpha\left[\cos(\beta \ln x) \pm \sin(\beta \ln x)\right].$$

Thus we have independent real solutions $y_1(x) = x^\alpha \cos(\beta \ln x)$ and $y_2(x) = x^\alpha \sin(\beta \ln x)$. As x increases from 1 to ∞, $\beta \ln x$ increases from 0 to ∞. Hence the sine and cosine of $\beta \ln x$ oscillate infinitely often, though with decreasing frequency because the rate of increase of $\ln x$ tends to 0. As x decreases from 1 to 0, $\beta \ln x$ decreases from 0 to $-\infty$, so the sine and cosine of $\beta \ln x$ oscillate infinitely often, with increasing frequency as x approaches 0.

example 4

The Euler equation $x^2y'' - 2xy' + 2y = 0$ has indicial equation $\mu^2 - 3\mu + 2 = (\mu - 1)(\mu - 2) = 0$, with roots $\mu_1 = 1,\ \mu_2 = 2$. The corresponding independent solutions are $y_1(x) = x$ and $y_2(x) = x^2$. The constants in the linear combination $y = c_1x + c_2x^2$ can be chosen so as to satisfy arbitrary initial conditions $y(x_0) = y_0,\ y'(x_0) = z_0$, *except* at the singular point $x_0 = 0$, where zero is the only possible value for the solution. Solving for the constants gives $c_1 = (2y_0 - x_0z_0)/x_0$, $c_2 = (x_0z_0 - y_0)/x_0^2$.

example 5

The family of equations $x^2y'' - \nu xy' + \nu y = 0$ with real parameter ν has $\mu^2 - (\nu + 1)\mu + \nu = (\mu - 1)(\mu - \nu) = 0$ for indicial equation with roots $\mu_1 = 1,\ \mu_2 = \nu$. If $\nu \neq 1$, we get x and x^ν for independent solutions. If $\nu = 1$, the double root leads to the solutions x and $x \ln x$. See Figure 9.7 for some graphs.

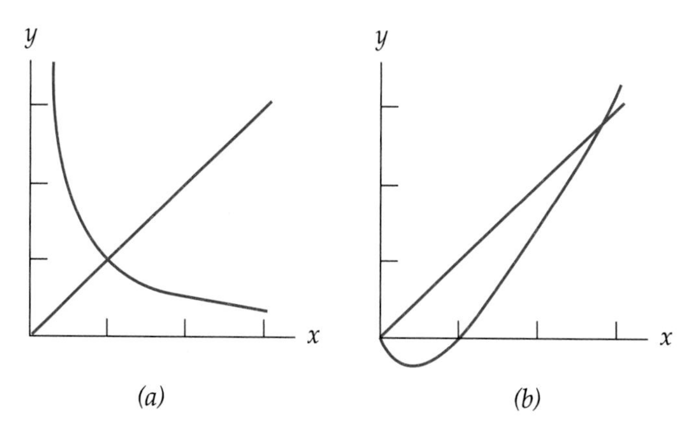

Figure 9.7 (a) $y_1 = x,\ y_2 = x^{-1};\ \mu_1 = 1,\ \mu_2 = -1.$
(b) $y_1 = x,\ \ y_2 = x \ln x;\ \mu_1 = \mu_2 = 1.$

example 6

The family of equations $x^2y' + m^2y = 0$ with real parameter $m^2 > \frac{1}{4}$ has $\mu^2 - \mu + m^2 = 0$ for indicial equation, with roots $\mu = \frac{1}{2} \pm \frac{i}{2}\sqrt{4m^2 - 1}$. With $m^2 = 401/4$, we get $x^{1/2}\sin(10 \ln x)$ and $x^{1/2}\cos(10 \ln x)$ for independent solutions. Now change the differential equation by including the first-order term $2xy'$, to get $x^2y'' + 2xy' + m^2y = 0$. The indicial equation becomes $\mu^2 + \mu + m^2 = 0$, with roots $\mu = -\frac{1}{2} \pm \frac{i}{2}\sqrt{4m^2 - 1}$. This just changes the factor $x^{1/2}$ in the solutions of the previous equation to $x^{-1/2}$. Figure 9.8 shows some typical graphs.

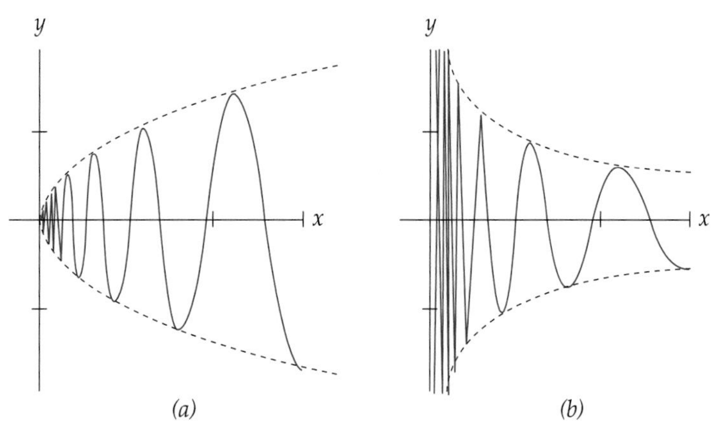

Figure 9.8 (*a*) $y = x^{1/2}\sin(10\ln x)$; $x^2y'' + \frac{101}{4}y = 0$.
(*b*) $y = x^{-1/2}\sin(10\ln x)$; $x^2y'' + 2xy' + \frac{101}{4}y = 0$.

EXERCISES

Find all singular points of each of the differential equations in Exercises 1 through 8, and state which of these points is regular and which is not.

1. $xy'' + xy' + y = 0$.

2. $x^2y'' + x^2y' + y = 0$.

3. $(1 - x)y'' + xy' + y = 0$.

4. $(1 - x^2)xy'' + xy' + y = 0$.

5. $y'' + xy' + y = 0$.

6. $x^3y'' + xy' + y = 0$.

7. $e^xy'' + xy' + y = 0$.

8. $x^2y'' + (\sin x)y' + y = 0$

Find the roots of the indicial equation for each of the equations in Exercises 9 through 16. Then find a pair of independent real-valued solutions defined for $x > 0$.

9. $x^2y'' + xy' + y = 0$.

10. $x^2y'' - xy' - y = 0$.

11. $x^2y'' + y = 0$.

12. $xy'' + y' = 0$.

13. $x^2y'' + 5xy' - 5y = 0$.

14. $x^2y'' - 3xy' + y = 0$.

15. $x^2y'' - 4xy' + y = 0$.

16. $x^2y'' - xy' + 7y = 0$.

17. For what positive values of x does the function $\sin(\ln x)$ in the examples equal zero?

18. For what positive values of x does the function $x\sin(\ln x)$ have relative maximum or minimum values?

19. The most general equation with a regular singular point at $x_0 = 0$ can be written $x^2y'' + xa(x)y' + b(x)y = 0$. Assume $x > 0$ and let $t = \ln x$ so that $x = e^t$.

(a) Use the chain rule for differentiation to show that

$$x\frac{dy}{dx} = \frac{dy}{dt} \quad \text{and} \quad x^2\frac{d^2y}{dx^2} = \frac{d^2y}{dt^2} - \frac{dy}{dt}.$$

(b) Use the result of part (a) to show that the differential equation is equivalent, for $x > 0$, to $\ddot{y} + \left[a(e^t) - 1\right]\dot{y} + b(e^t)y = 0$.

(c) Specialize the result of part (b) to show that the constant-coefficient Euler equation is equivalent to $\ddot{y} + (a - 1)\dot{y} + by = 0$. Show that the characteristic roots of this last equation are the same as the roots of the Euler equation's indicial equation.

20. Use the definition $x^{\alpha+i\beta} = x^\alpha x^{i\beta} = x^\alpha e^{i\beta \ln x}$ to verify that the derivative of $x^{\alpha+i\beta}$ is given by $(a + i\beta)x^{\alpha+i\beta-1}$ for $x > 0$.

7B Bessel's Equation

Bessel functions share certain properties with the sine function that allow them to play a part in the representation of oscillatory phenomena, as shown in Chapter 11, Section 3, as well as in many other areas of pure and applied mathematics. From our present point of view the definition of Bessel function starts most naturally with the *Bessel equation*

$$x^2y'' + xy' + (x^2 - \nu^2)y = 0.$$

This differential equation depends on the parameter ν, which for our purposes we will assume to be a real number. Integer indices ν, particularly $\nu = 0$ and $\nu = 1$ are the ones that occur most often in applications, but other choices for ν also occur in practice. In particular, choosing $\nu = n + \frac{1}{2}$ for integer n leads to elementary solutions.

A routine check shows that the Bessel equation never has a one-term solution of the form x^μ that work for the Euler equation of Section 7A. Nevertheless there is still an indicial equation to be satisfied by constants μ if we try to find a **Frobenius series** solution of index μ, which is just a power series multiplied through by x^μ:

7.1
$$y(x) = \sum_{k=0}^{\infty} c_k x^{k+\mu}.$$

Since μ need not be an integer, we'll assume for simplicity that $x > 0$ throughout the following discussion to avoid fractional powers of negative numbers. If $\mu = n$ is a nonnegative integer, the Frobenius series is an ordinary power series with first term c_0x^n.

We apply term-by-term differentiation to the series in Equation 7.1 and then substitute into the Bessel equation in the form $x^2y'' + xy' - \nu^2y = -x^2y$ to get

$$\sum_{k=0}^{\infty} \left[(k+\mu)(k+\mu-1) + (k+\mu) - \nu^2\right] c_k x^{k+\mu} = -\sum_{k=0}^{\infty} c_k x^{k+\mu+2}.$$

Now simplify the left side and shift k down by 2 on the right, replacing k by $k-2$ everywhere:

$$\sum_{k=0}^{\infty} \left[(k+\mu)^2 - \nu^2\right] c_k x^{k+\mu} = -\sum_{k=2}^{\infty} c_{k-2} x^{k+\mu}.$$

Next, equate coefficients of like powers $x^{k+\mu}$. When $k=0$ we get $(\mu^2-\nu^2)c_0=0$, and when $k=1$ we get $\left[(1+\mu)^2-\nu^2\right]c_1=0$. The constant c_0 is set up to be the first nonzero coefficient in the series, so, dividing it out, we're left with the equation $\mu^2-\nu^2=0$. This is the *indicial equation*, and it requires $\mu=\pm\nu$, where ν is the index of the Bessel equation. We'll assume $\mu=\nu\geq 0$ at first. The equation containing c_1 then reduces to $(1+2\nu)c_1=0$, so $c_1=0$ since $\nu\geq 0$. Equating coefficients of index $k\geq 2$ enables us to compute coefficients recursively using the **recurrence relation** we get by equating coefficients of the same power of x on the two sides of the previous displayed equation:

$$\left[(k+\nu)^2-\nu^2\right]c_k = -c_{k-2} \quad \text{or} \quad c_k = -\frac{c_{k-2}}{k(k+2\nu)}.$$

Since $c_1=0$, it follows that $c_k=0$ if k is odd. Since only even indices survive, we write $k=2j$ for the remaining ones. The recurrence relation says then that

7.2

$$c_{2j} = \frac{-c_{2j-2}}{2j(2j+2\nu)}.$$

As we mentioned earlier, the cases $\nu=0$ and $\nu=1$ are the most important, so we'll first consider nonnegative integer values $\nu=n\geq 0$. With $\nu=n$ in Equation 7.2, we get

$$c_2 = \frac{-c_0}{2^2(n+1)}, \quad c_4 = \frac{c_0}{2^4 2!(n+1)(n+2)}, \quad c_6 = \frac{-c_0}{2^6 3!(n+1)(n+2)(n+3)},$$

and in general

$$c_{2j} = \frac{(-1)^j c_0}{2^{2j} j!(n+1)(n+2)\cdots(n+j)}.$$

A conveniently simplifying choice for c_0 turns out to be $c_0 = 1/(2^n n!)$, so

7.3

$$c_{2j} = \frac{(-1)^j}{2^{2j+n} j!(n+j)!}.$$

We define the *Bessel function* J_n for $n=0,1,2,\ldots$ by

$$J_n(x) = \sum_{j=0}^{\infty} \frac{(-1)^j x^{2j+n}}{2^{2j+n} j!(n+j)!} = \sum_{j=0}^{\infty} \frac{(-1)^j (x/2)^{2j+n}}{j!(n+j)!},$$

a series that converges for all x, for example by comparison with the series for e^{x^2}.

example 7

Index $\nu = 0$.

A computer plot of

$$J_0(x) = \sum_{j=0}^{\infty} \frac{(-1)^j (x/2)^{2j}}{(j!)^2}$$

for $x \geq 0$ is shown in Figure 9.9. Since the expansion contains only even powers of x, J_0 is an even function, and the graph for $x < 0$ is obtained by reflecting in the vertical axis. It's easy to check that $J_0(0) = 1$ and $J_0'(0) = 0$.

It's proved in Section 1B that $J_0(x) = 0$ for an infinite sequence of positive numbers $\{x_n\}$, the *zeros* of the function. Furthermore, the successive differences $x_{n+1} - x_n$ between zeros tend to π. For example, the difference between the locations of the last two zeros shown in Figure 9.9 is about 3.14. This is just one among many analogies between solutions of the Bessel equation and solutions of damped harmonic oscillator equations. The first four zeros of J_0 are at approximately 2.40, 5.52, 8.65, and 11.79.

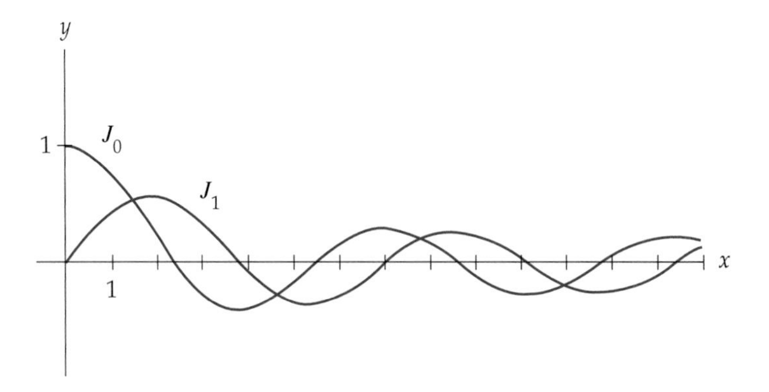

Figure 9.9 Graphs of J_0 and J_1 for $0 \leq x \leq 14$.

example 8

Index $\nu = 1$.

The Bessel function J_1 is

$$J_1(x) = \sum_{j=0}^{\infty} \frac{(-1)^j (x/2)^{2j+1}}{j!(j+1)!}.$$

As with J_0, the distance between successive zeros tends to π. It is left as an exercise using term-by-term differentiation of the series for J_0 to show that $J_0'(x) = -J_1(x)$ for all x. This last relation is analogous to $d/dx \cos x = -\sin x$, and its simplicity confirms the correctness of the choices 1 and $\frac{1}{2}$ for c_0 when $\nu = 0$ and 1 respectively. The graphical relationship between J_0 and J_1 is indicated in Figure 9.9, where it appears that the locations of the local maxima and minima of J_0 coincide with the zeros of J_1, the first four positive zeros of J_1 being at approximately 3.85, 7.02, 10.17, and 13.32.

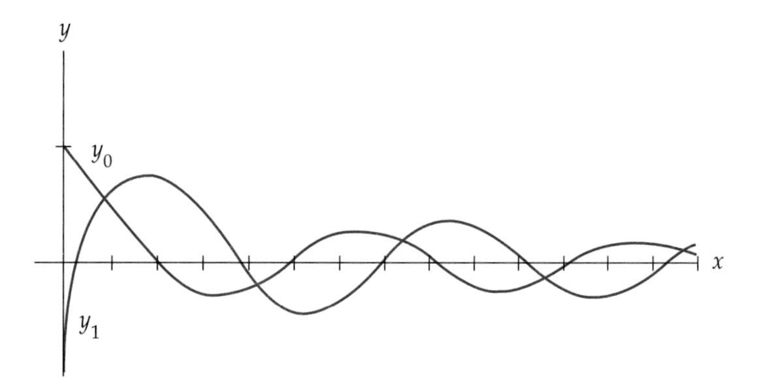

Figure 9.10 Bessel functions of the second kind:
$y_0(1) = 1,\ y_0'(1) = -1;\ y_1(1) = y_1'(1) = 1.$

Identifying series expansions for solutions of the Bessel equation of integer index n that are independent of J_n is relatively complicated. A **Bessel function of the second kind** is a solution of the index-n Bessel equation that is independent of $J_n(x)$. The functions J_n said to be of the first kind. Up to the addition of a constant multiple of J_n, these independent solutions are all given by Theorem 1.2 of Section 1 in the form

7.4
$$y_n(x) = C_n J_n(x) \int \frac{1}{x J_n^2(x)}\, dx, \qquad C_n = \text{const.} \neq 0.$$

Fortunately all we usually need know for our later applications is that all these independent solutions y_n are unbounded near zero, as indicated in Figure 9.10, and as Theorem 7.5 below shows. It follows from Theorem 2.6 of Section 2 that every solution y_n of the index-n Bessel equation has the form

$$y_n(x) = c_1 J_n(x) + c_2 Z_n(x),$$

where Z_n is a Bessel function of the second kind.

7.5 THEOREM

A solution y_n given by Equation 7.4 tends to $+\infty$ or to $-\infty$ as x tends to 0. In particular, for each n, there are numbers $b_n > 0$ and $\delta_n > 0$ such that for $0 < |x| \leq \delta_n$

$$|y_0(x)| \geq b_0 \ln|1/x| \quad \text{and} \quad |y_n(x)| \geq b_n/x^n,\ n = 1, 2, 3, \ldots.$$

Proof. For simplicity, take $C_n = 1/2^n n!$ in Equation 7.4, and assume first that $x > 0$. Note that $y_n(x)$ differs from

$$z_n(x) = x^n E_n(x) \int_x^{\delta_n} \frac{dt}{t^{2n+1} E_n^2(t)},$$

by only a constant multiple of the continuous function J_n. Here $E_n(x) = x^{-n} J_n(x)/2^n n!$ is a continuous even function such that $E_n(0) = 1$. Choose δ_n between 0 and

1 so that $\frac{1}{2} \le E_n(x) \le 2$ if $|x| \le \delta_n$. Then for $0 < x \le \delta_n$,

$$z_n(x) \ge \tfrac{1}{2}x^n \int_x^{\delta_n} \tfrac{1}{4}t^{-2n-1}\,dt \ge \begin{cases} -\frac{1}{8}\ln x + \frac{1}{8}\ln\delta_n, & n = 0, \\ \frac{1}{2n}x^{-n} - \frac{1}{2n}\delta_n^{-n}, & n \ge 1. \end{cases}$$

It follows that z_n plus a constant satisfies the claims made for y_n, so $y_n = z_n + C_n J_n$ also satisfies the claim. A similar argument with an integral over $-\delta_n \le x \le x < 0$ proves the result for $x < 0$. ■

Boundary-value problems for a Bessel equation are typically posed over an interval $0 \le x \le b$, in which the requirement at $x = 0$ is often only that the solution should remain bounded as x approaches 0 from the right. According to Theorem 7.5 this requirement rules out functions of the second kind. At $x = b$ we impose a more conventional condition $y(b) = \beta$.

example 9

The index-0 Bessel equation can be written $xy'' + y' + xy = 0$. A condition requiring that $y(x)$ remain bounded near $x = 0$ dictates that we should avoid using an unbounded solution and concentrate on solutions $c_1 J_0(x)$. A condition $y(b) = \beta$ requires $c_1 J_0(b) = \beta$, which can be satisfied by choosing $c_1 = \beta / J_0(b)$ if $J_0(b) \ne 0$. If $J_0(b) = 0$, then $c_1 J_0(b) = \beta$ has no solution unless $\beta = 0$, in which case c_1 is arbitrary. This is just one reason why it's important to know the location of the zeros of Bessel functions.

The Gamma Function. Finding an independent pair of solutions for $x > 0$ to the Bessel equation is somewhat simpler if ν is not an integer. The only additional complexity arises from the need to choose the initial constant c_0 in Equation 7.1 so that relations among the resulting Bessel functions are fairly simple. Just as in the case $\nu = n \ge 0$, the recurrence relation 7.2 leads to even-index coefficients c_k of the form

$$c_{2j} = (-1)^j \frac{c_0}{2^{2j} j!(\nu+1)(\nu+2)\cdots(\nu+j)}.$$

If $\nu \ge 0$ is an integer, the appropriate choice for c_0 turned out to be $c_0 = 1/2^\nu \nu!$, resulting in Equation 7.3. To make this work for noninteger ν, and even $\nu < 0$, it's customary to generalize the factorial function by using the **gamma function,** defined by an improper integral:

$$\Gamma(\nu) = \int_0^\infty t^{\nu-1}e^{-t}\,dt, \quad \nu > 0.$$

Some properties of $\Gamma(\nu)$ that are pertinent for us here are summarized in Theorem 7.6.

7.6 THEOREM

For n a nonnegative integer, $\Gamma(n+1) = n!$, and if $\nu > 0$ is a real number, $\Gamma(\nu + 1) = \nu\Gamma(\nu)$. This last relation can be used to extend the definition of $\Gamma(\nu)$ to noninteger values $\nu < 0$ so that the relation becomes an identity for the extended function. Defining $1/\Gamma(v) = 0$ at $\nu = 0, -1, -2, -3\ldots$ generalizes $1/n!$ to $1/\nu! = 1/\Gamma(v+1)$ for all real numbers ν. In particular, $(-\frac{1}{2})! = \sqrt{\pi}$ and $\frac{1}{2}! = \sqrt{\pi}/2$. (Relevant graphs are shown in Figure 9.11.)

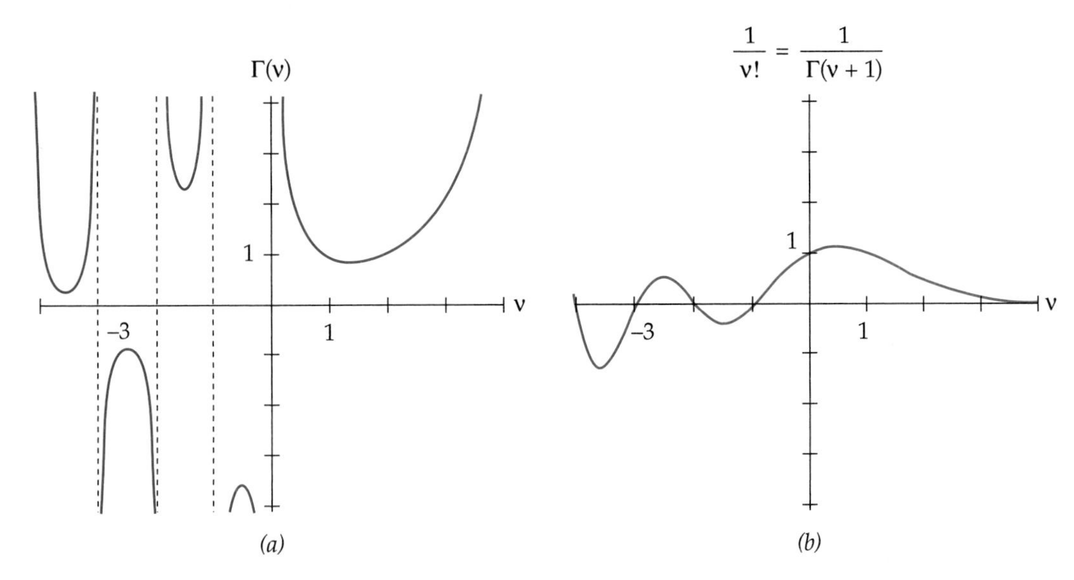

Figure 9.11 Graphs of (*a*) $\Gamma(\nu)$ and (*b*) $1/\Gamma(\nu+1) = 1/\nu!$ for $-4 < \nu < 4$.

Proof. For $\nu > 0$, integration by parts shows that

$$\begin{aligned}\Gamma(\nu+1) &= \int_0^\infty t^\nu e^{-t} dt \\ &= -t^\nu e^{-t}\Big|_0^\infty + \nu \int_0^\infty t^{\nu-1} e^{-t} dt = \nu\Gamma(\nu).\end{aligned}$$

In particular, if n is a positive integer,

$$\Gamma(n+1) = n\Gamma(n) = n(n-1)\Gamma(n-2) = \cdots = n(n-1)(n-2)\cdots 3\cdot 2\Gamma(1).$$

Direct computation gives $\Gamma(1) = 1$, so $\Gamma(n+1) = n!$. We define $\Gamma(\nu) = \Gamma(\nu+1)/v$ for $-1 < \nu < 0$ and more generally

$$\Gamma(\nu) = \frac{\Gamma(\nu+n)}{\nu(\nu+1)\cdots(\nu+n-1)} \quad \text{for} \quad -n < \nu < -n+1.$$

[Thus $\Gamma(\nu)$ is defined for $\nu < 0$, ν not equal to a negative integer, in terms of the original $\Gamma(\nu)$ for $0 < \nu < 1$.] Suppose $-n < \nu < -n+1$. Then $-n+1 < \nu+1 < -n+2$, and

$$\begin{aligned}\Gamma(\nu+1) &= \frac{\Gamma(\nu+1+n-1)}{(\nu+1)\cdots(\nu+1+n-2)} \\ &= \frac{\Gamma(\nu+n)}{(\nu+1)\cdots(\nu+n-1)} = \nu\Gamma(\nu).\end{aligned}$$

The quotient relation shows that $\Gamma(\nu)$ tends to $\pm\infty$ as ν approaches 0 or a negative integer from one side or the other, so it is natural to define $1/\nu! = 1/\Gamma(\nu+1)$ to be 0 at negative integers. Exercises 21 and 22 ask you to calculate $\Gamma(\nu)$ for $\nu = \pm\frac{1}{2}$. ■

Series Expansions for J_ν. At first sight the graphs shown in Figure 9.11 for $\nu < 0$ look like a somewhat arbitrary attempt to extend the definition of $1/n!$. The justification for this extension is that using these definitions allows us to extend the definition of $J_n(x)$ and other important functions in a convincing and memorable way to values of n that aren't nonnegative integers. Replacing the integer $n \geq 0$ by the real index ν in the definition of $J_n(x)$ gives

$$J_\nu(x) = \sum_{j=0}^{\infty} \frac{(-1)^j (x/2)^{2j+\nu}}{j!(\nu+j)!} = \sum_{j=0}^{\infty} \frac{(-1)^j (x/2)^{2j+\nu}}{j!\Gamma(\nu+j+1)}.$$

This series converges for all $x > 0$, and represents an analytic function if ν is an integer. When ν isn't an integer, x^ν isn't analytic at $x = 0$ but can be factored out so we see that $J_\nu(x)$ is x^ν times a function that is analytic for all x and so can be differentiated as often as we like. It's routine, and tedious, to verify that $J_\nu(x)$ satisfies the Bessel equation for $x > 0$, but this fact follows formally from knowing the result for an arbitrary integer $\nu = n$. We can then just replace n by ν in that term-by-term verification and use the fundamental relation $\nu! = \nu(\nu-1)!$ as we would for arbitrary positive integer index. The same remarks apply if we make a formal replacement of n by $-\nu < 0$, so that both $J_\nu(x)$ and $J_{-\nu}(x)$ are solutions of the Bessel equation of index ν. (Recall that the roots of the indicial equation are $\pm\nu$.) Here is a summary of some important relations involving Bessel functions.

7.7 THEOREM

The Bessel equation $x^2y'' + xy' + (x^2 - \nu^2)y = 0$ has $J_\nu(x)$ and $J_{-\nu}(x)$ for an independent pair of solutions precisely when $\nu > 0$ is not an integer, $J_{-\nu}(x)$ being unbounded like $x^{-\nu}$ near $x = 0$. If $\nu = n$ is an integer, then $J_n(x) = (-1)^n J_{-n}(x)$, so J_n and J_{-n} are not independent. For each real ν the **Bessel recurrence identities** hold:

(A)
$$J_{\nu+1}(x) = \frac{2\nu}{x} J_\nu(x) - J_{\nu-1}(x),$$

(B)
$$J_{\nu+1} = \frac{\nu}{x} J_\nu(x) - J'_\nu(x).$$

Proof. Assume $\nu > 0$. Then $J_\nu(x) = x^\nu C_1(x)$, where $C_1(x)$ is continuous, so $J_\nu(x)$ is bounded near $x = 0$. If $-\nu$ is not a negative integer, then none of the coefficients in $J_{-\nu}(x) = x^{-\nu} C_2(x)$ is zero, so the function is unbounded as x approaches 0 from the right. Hence, J_ν and $J_{-\nu}$ must be independent. On the other hand, if $-\nu = -n$ is a negative integer, then $1/(-n+j)! = 0$ for $j = 1, 2, \ldots, n-1$, so

$$J_{-n}(x) = \sum_{j=0}^{\infty} \frac{(-1)^j (x/2)^{2j-n}}{j!(-n+j)!} = \sum_{j=n}^{\infty} \frac{(-1)^j (x/2)^{2j-n}}{j!(-n+j)!}$$

$$= \sum_{j=0}^{\infty} \frac{(-1)^{j+n} (x/2)^{2j+n}}{(j+n)!j!} = (-1)^n J_n(x).$$

The proofs of the recurrence identities are Exercises 24 and 25 on series manipulation. ■

The functions $J_{n+1/2}(x)$, where n is an integer are exceptional in that they are elementary, expressible in terms of the functions of elementary calculus.

example 10 From the definition of $J_\nu(x)$ we have

$$J_{-1/2}(x) = \sum_{j=0}^{\infty} \frac{(-1)^j (x/2)^{2j-1/2}}{j!(j-1/2)!} = \sqrt{2}\,x^{-1/2} \sum_{j=0}^{\infty} \frac{(-1)^j x^{2j}}{2^{2j} j!(j-1/2)!}.$$

To simplify the terms in the sum, note that

$$(j-\tfrac{1}{2})! = (j-\tfrac{1}{2})(j-\tfrac{3}{2})\cdots[j-(j-\tfrac{1}{2})](-\tfrac{1}{2})! = 2^{-j}(2j-1)(2j-3)\cdots 3\cdot 1\sqrt{\pi}.$$

Hence $2^{2j} j!(j-\frac{1}{2})! = (2j)!\sqrt{\pi}$. It follows that the last sum displayed above is the Taylor expansion of $\cos x$ about 0, so $J_{-1/2}(x) = \sqrt{2/(\pi x)}\,\cos x$. A similar computation with $\nu = \frac{1}{2}$ would show that $J_{1/2}(x) = \sqrt{2/(\pi x)}\,\sin x$, but the same result can be had easily using the expression for $J_{\nu+1}$ in Theorem 7.7, Equation (A). (See also Exercise 12.)

example 11 Using the results of the previous example and Theorem 7.7, Equation (A), we find for $x > 0$

$$\begin{aligned} J_{3/2}(x) &= \frac{1}{x} J_{1/2}(x) - J_{-1/2}(x) \\ &= \frac{1}{x}\left(\sqrt{\frac{2}{\pi x}}\,\sin x\right) - \sqrt{\frac{2}{\pi x}}\,\cos x = \sqrt{\frac{2}{\pi x}}\left(\sin\frac{x}{x} - \cos x\right). \end{aligned}$$

You can verify that $J_{1/2}(x)$ is bounded near $x = 0$, as we know from Theorem 7.7.

EXERCISES

1. Differentiate the series for $J_0(x)$ to show that $J_0'(x) = -J_1(x)$ for all x.
2. Differentiate the series for $J_1(x)$ to show that $J_1'(x) = J_0(x) - J_1(x)/x$ for all $x > 0$.
3. Show that the Bessel equation of order ν does not have a solution of the form $y = x^\mu$ for $x > 0$.

A standard theorem on convergent alternating series $\sum_{k=0}^{\infty}(-1)^k a_k,\ a_k \geq 0$, states that the error e_n made in replacing the sum by a partial sum $s_n = \sum_{k=0}^{n}(-1)^k a_k$ satisfies $|e_n| \leq a_{n+1}$. Apply this theorem along with the conclusions of Exercises 1 and 2 to derive the estimates 4 through 7.

4. $J_0(1) \approx 0.7651977$.
5. $J_0'(1) \approx -0.4400506$.
6. $J_1(1) \approx 0.3251471$.
7. $J_1'(1) \approx 0.4400506$.

Use the appropriate numerical results from the Exercises 4 through 7 in a numerical differential equations solver to plot the graphs in Exercises 8 and 9. Recall that the initial-value problems at 0 are not uniquely determined, and that numerical approximation for the initial-value problem starting at $x = 1$ requires a negative step ($h < 0$) on the interval $0 \leq x \leq 1$.

8. Plot the graph of $J_0(x)$ for $0 \le x \le 10$.

9. Plot the graph of $J_1(x)$ for $0 \le x \le 10$.

Use the numerical estimates from Exercises 4 through 7 in a numerical DE solver to estimate the location of Exercises 10 and 11 to three decimal places.

10. The first five positive zeros of $J_0(x)$.

11. The first five positive zeros of $J_1(x)$.

12. The relationship between the Bessel equation and the harmonic oscillator equation can be illuminated as follows.

(a) Let $u(x) = x^{1/2}y(x)$ for $x > 0$, and show that $u(x)$ satisfies $u'' + \left[1 + (\frac{1}{4} - \nu^2)/x^2\right]u = 0$ if and only if $y(x)$ is a solution of $x^2y'' + xy + (x^2 - \nu^2)y = 0$ for $x > 0$. Note that for large x the differential equation for u is approximately $u'' + u = 0$.

(b) Use part (a) to show that the Bessel equation of index $\nu = \frac{1}{2}$ has for $x > 0$ the general solution

$$y(x) = c_1\frac{\cos x}{\sqrt{x}} + c_2\frac{\sin x}{\sqrt{x}}.$$

(It can be shown that for large values of x, solutions of Bessel equations have approximately this form, which explains the spacing of the zero values for large x.)

13. Weakening spring The harmonic oscillator equation $y'' + h_0 y = 0$, with constant $h_0 > 0$, represents the undamped oscillation of a spring. If the stiffness of the spring undergoes slow exponential decay over time, then h_0 should be replaced by $h(t) = h_0 e^{-at}$, for some small constant $a > 0$. (For a numerical analysis of the more general linear oscillator, see the Exercises for Chapter 4, Sections 7A and 7B, under the heading **time-dependent linear oscillator.**)

(a) Use the chain rule to show that the change of variable $u = ce^{-dt}$ transforms the differential equation into

$$d^2u^2\frac{d^2y}{du^2} + d^2u\frac{dy}{du} + h_0\left(\frac{u}{c}\right)^{a/d} y = 0.$$

(b) Show that the choices $c = 2/(a\sqrt{h_0})$ and $d = a/2$ give the zero-index Bessel equation $u^2y'' + uy' + u^2y = 0$.

(c) Show that bounded solutions of the original oscillator equation have the form

$$y(t) = C J_0\left[(2\sqrt{h_0}/a)e^{-at/2}\right].$$

(d) Show that the number of oscillations of the solution in part (c) is limited by the number of sign changes of $J_0(x)$ in the interval $0 < x \le 2\sqrt{h_0}/a$.

The index-ν **Bessel functions of the first kind** are the nontrivial constant multiples of $J_\nu(x)$. (For fixed ν these multiples obviously all have the same zeros, Theorem 1.4 of Section 1B shows that an independent index-ν Bessel function of the second kind can't have a zero in common with J_ν.) Use the modified Euler method to generate approximate solutions of the appropriate Bessel equation to the right and left of the given point to make computer graphics plots of the functions of the second kind in Exercises 14 through 17.

14. The index-0 function Z_0 such that $Z_0\left(\frac{1}{2}\right) = 0$ and $Z_0'\left(\frac{1}{2}\right) = 1$.

15. The index-1/2 function $Z_{1/2}$ such that $Z_{1/2}(\pi/2) = 0$ and $Z_{1/2}'(\pi/2) = -\sqrt{2/\pi}$. This is an elementary function; can you identify it?

16. The index-1 function Z_1 such that $Z_1\left(\frac{1}{2}\right) = 0$ and $Z_1'\left(\frac{1}{2}\right) = 1$.

17. The index $-\frac{1}{2}$ function $Z_{-1/2}$ such that $Z_{-1/2}(\pi/2) = 0$ and $Z_{-1/2}'(\pi/2) = -(2/\pi)^{3/2}$. This is an elementary function; can you identify it?

18. Find a solution of the boundary problem with $y(2) = 3$ and $y(x)$ bounded near 0, for the equation $x^2y'' + xy' + (x^2 - 1)y = 0$.

19. Consider the equation $x^2y'' + xy' + k^2x^2y = 0$, where k is constant.

(a) Show that the equation has solutions $y(x) = c_1 J_0(kx)$.

(b) Let $k = 2$, and solve the boundary problem with conditions $y(3) = 4$ and $y(x)$ bounded near 0.

20. Consider the equation $x^2y'' + xy' + (k^2x^2 - 1)y = 0$, where k is constant.

(a) Show that the equation has solutions $y(x) = c_1 J_1(kx)$.

(b) Let $k = 5$, and solve the boundary problem with conditions $y(2) = 1$ and $y(x)$ bounded near 0.

Gamma Function

21. Show that $(-\frac{1}{2})! = \Gamma(\frac{1}{2}) = \sqrt{\pi}$ by first showing that $\int_0^\infty t^{-1/2}e^{-t}\,dt = 2\int_0^\infty e^{-u^2}du$. Then compute this last integral by observing that it is the square root of the double integral $\int_0^\infty \int_0^\infty e^{-u^2-v^2}du\,dv$; this can be evaluated by changing to polar coordinates in the first quadrant of the uv plane.

22. Use $\Gamma(\nu + 1) = \nu\Gamma(\nu)$ together with the result of the previous exercise to show that $\frac{1}{2}! = \Gamma(\frac{3}{2}) = \sqrt{\pi}/2$.

23. The reciprocal factorial $1/\nu!$ can also be defined for all ν, consistent with the conventional interpretation of $k!$ for integer k, by

$$\frac{1}{\nu!} = \lim_{k\to\infty} \frac{(\nu+1)(\nu+2)\cdots(\nu+k)}{k!k^\nu}.$$

It can be proved that this limit exists and is finite for all ν. Under just that assumption, prove the following.

(a) $(\nu + 1) \cdot 1/(\nu + 1)! = 1/\nu!$.

(b) $1/\nu! = 0$ if ν is a negative integer.

(c) $1/\nu!$ has the conventional meaning when ν is a positive integer n. Note that

$$\frac{1}{n!} = \frac{(n+1)\cdots(n+k)}{n!(n+1)\cdots(n+k)} = \frac{(n+1)\cdots(n+k)k^n}{k!(k+1)\cdots(k+n)k^n}.$$

(It can be proved that the function $1/\nu!$ defined here is the same as the one defined in the text using $\Gamma(\nu)$.)

Bessel Functions of General Index ν. In Exercises 24 and 25, use the relevant infinite series to prove Equations (A) and (B) of Theorem 7.7.

24. $J_{\nu+1}(x) = (2\nu/x)J_\nu(x) - J_{\nu-1}(x),\ x > 0.$

25. $J_{\nu+1}(x) = (\nu/x)J_\nu(x) - J_\nu'(x),\ x > 0.$

26. Prove that $J_{\nu-1}(x) = (\nu/x)J_\nu(x) + J_\nu'(x),\ x > 0.$

27. Show that $J_{1/2}(x) = \sqrt{2/\pi x}\ \sin x$ in each of the following two ways.

(a) Using the appropriate identity in Theorem 7.7 together with $J_{-1/2}(x) = \sqrt{2/\pi x}\ \cos x$.

(b) Directly from the Frobenius series for $J_{1/2}(x)$.

***28.** The form of a recurrence identity stated for the Bessel functions J_ν in Theorem 7.7 can be used to generate, for fixed ν, a sequence of solutions to the Bessel equations of index $\nu,\ \nu+1,\ \nu+2, \ldots,$ in the following steps. Let Z_ν be an arbitrary solution of the index-ν equation, so that

$$Z_\nu'' + \frac{1}{x}Z_\nu' + \left(1 - \frac{\nu^2}{x^2}\right)Z_\nu = 0.$$

Define $Z_{\nu+1}$ by

$$Z_{\nu+1}(x) = \frac{\nu}{x}Z_\nu(x) - Z_\nu'(x).$$

(a) Show that $xZ_{\nu+1}' = (\nu+1)Z_\nu' + \left[x - (\nu^2+\nu)/x\right]Z_\nu$.

(b) Show that $(xZ_{\nu+1}')' = \left[x - (\nu+1)^2/x\right]Z_\nu' - \nu\left[1 - (\nu+1)^2/x^2\right]Z_\nu$.

(c) Show that $Z_{\nu+1}$ satisfies the Bessel equation of index $\nu+1$.

29. For noninteger ν define the **Weber function** Y_ν by

$$Y_\nu(x) = \frac{J_\nu(x)\cos\pi\nu - J_{-\nu}(x)}{\sin\pi\nu}.$$

For integer n it can then be shown that $Y_n(x) = \lim_{\nu\to n} Y_\nu(x)$ defines a solution independent of $J_n(x)$ for the index-n Bessel equation. You are asked to use the recurrence identities stated for $J_n(x)$ in Theorem 7.7 to show that $Y_n(x)$ satisfies the same relations.

(a) Show that $Y_{n+1}(x) = (2n/x)Y_n(x) - Y_{n-1}(x)$.

(b) Show that $Y_{n+1}(x) = (n/x)Y_n(x) - Y_n'(x)$.

30. (a) Apply l'Hôpital's rule to the limit definition of $Y_n(x)$ in the previous exercise to show that the Weber function Y_n of integer index satisfies

$$Y_n(x) = \frac{1}{\pi}\lim_{\nu\to n}\frac{\partial}{\partial\nu}\left[J_\nu(x) - (-1)^n J_{-\nu}(x)\right].$$

(b) Assuming that limits and order of partial differentiation can be interchanged as needed, use the result of part (a) to verify formally that $Y_n(x)$ satisfies the Bessel equation of index n. (The stated assumption isn't universally valid, but can be justified in this case.)

(c) Show that Y_n and J_n are independent by showing that $Y_n(x)$ is unbounded for x near zero. [*Hint:* Consider the series expansions for J_ν and $J_{-\nu}$.]

Prove successively Equations 31 through 34; use $J_{1/2}(x) = \sqrt{2/\pi x}\sin x$, $J_{-1/2}(x) = \sqrt{2/\pi x}\cos x$ and Theorem 7.7, Equation (A). [J_ν is an elementary function precisely when $\nu = n + \frac{1}{2}$ with integer n.]

31. $J_{3/2}(x) = \sqrt{2/\pi x}(\sin x/x - \cos x)$.

32. $J_{-3/2}(x) = \sqrt{2/\pi x}(-\cos x/x - \sin x)$.

33. $J_{5/2}(x) = \sqrt{2/\pi x}(3\sin x/x^2 - 3\cos x/x - \sin x)$.

34. $J_{-5/2}(x) = \sqrt{2/\pi x}(3\cos x/x^2 + 3\sin x/x - \cos x)$.

35. (a) Use Theorem 7.7, Equation (B), to prove that $(d/dx)x^{-\nu}J_\nu(x) = -x^{-\nu}J_{\nu+1}$ for $x > 0$.

(b) Use Rolle's theorem and part (a) to prove that between every two positive zeros of $J_\nu(x)$ there is a zero of $J_{\nu+1}(x)$.

CHAPTER 10

Partial Differential Equations

1 INTRODUCTION

All the phenomena discussed in the book up to this point have been modeled by functions of one real variable, though these functions have sometimes been vector valued. But many processes are governed by simultaneous changes in more than one independent variable, for example, time t and position x. If we can establish an equation relating rates of change with respect to these variables, then that equation will be expressible in terms of partial derivatives with respect to the independent variables and we'll be looking at what is called a **partial differential equation.** This is in contrast to the "ordinary" differential equations with 1-dimensional independent variables of the earlier chapters. The purpose of the first section is to highlight some prominent mathematical similarities and differences between ordinary and partial differential equations, and to motivate the introduction of solutions constructed from products of exponential and trigonometric functions.

1A Equations and Solutions

Differential equations that involve the partial derivatives of some real function $u(x, y, \ldots)$ of several variables arise just as often in pure and applied mathematics as do ordinary differential equations for functions of one variable. For example, the height $u(x, t)$ of a moving water wave is often treated as a function of position x in some direction and time t. The differential equation that $u(x, t)$ satisfies under various physical conditions may contain partial derivatives

$$\frac{\partial u}{\partial x}, \quad \frac{\partial u}{\partial t}, \quad \frac{\partial^2 u}{\partial x^2}, \quad \frac{\partial^2 u}{\partial t \partial x}, \quad \frac{\partial^2 u}{\partial t^2}, \ldots,$$

with respect to both x and t. (To make the formulas easier to write, we will sometimes use the alternative notation

$$u_x,\ u_t,\ u_{xx},\ u_{xt},\ u_{tt},\ \ldots$$

for these partial derivatives.) Our examples will be restricted mainly to linear equations of at most second order in two independent variables, say x and y. Such an equation has the form

1.1 $$au_{xx} + bu_{xy} + cu_{yy} + du_x + eu_y + fu = g,$$

where the coefficients a, b, c, d, e, f, and the function g are assumed to be continuous functions, perhaps constant, of x and y in some common region R of the xy plane. In addition, we'll look only for solutions $u(x, y)$, all of whose partial derivatives of order 1 and 2 are continuous in R. Under that assumption the equation $u_{xy} = u_{yx}$ will hold in the region R, a theorem from multivariable calculus called **Clairaut's theorem.**

example 1

We'll solve the equation $yu_{xy} = x$ in the region of the xy plane for which $y > 0$, called the "upper half-plane." Since $y > 0$ we can write the equation as $u_{xy} = x/y$ and can integrate both sides with respect to y, for each *fixed* x, getting

$$u_x(x, y) = x \ln y + c(x).$$

The integration "constant" $c(x)$ can be chosen to be an arbitrary differentiable function of x. Now integrate both sides with respect to x to get

$$u(x, y) = \tfrac{1}{2}x^2 \ln y + C(x) + D(y).$$

Here $C(x)$ is an integral of $c(x)$ and so is independent of y, and $D(y)$ is allowed to be an arbitrary twice-differentiable function of y. We have found the most general twice-differentiable solution of $yu_{xy} = x$ for $y > 0$.

For a second-order linear differential equation of ordinary type (i.e., with a single independent variable), we are used to having two arbitrary constants in the general solution formula. Notice, however, that in Example 1 the general solution contains two arbitrary twice-differentiable *functions*, allowing for a wide variety in the choice of a particular solution. This wide freedom of choice is an indication that we can expect to find particular solutions satisfying fairly complicated combinations of boundary and initial conditions. We will refer to boundary or initial conditions as **side conditions,** and they will be explored later in the chapter in the context of specific physical problems.

example 2

With $u = u(x, y)$, the differential equation

$$u_{xx} - y^2u = 0, \qquad y > 0,$$

is peculiar in that it contains a derivative only with respect to x, even though the solution will be of the form $u = u(x, y)$. (In a sense we're really dealing with

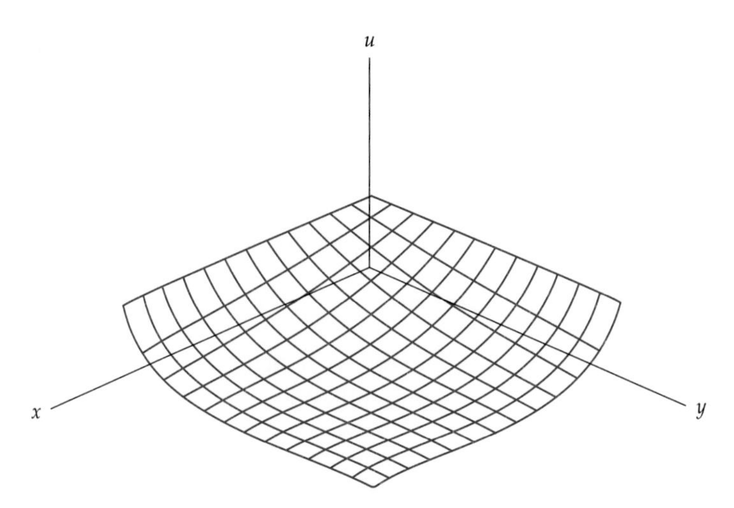

Figure 10.1 $u_{xx} - y^2 u = 0$, $u(0,y) = 1$, $u_x(0,y) = -y$;
solution $u(x,y) = e^{-xy}$.

an ordinary differential equation here, and solving the equation depends only on techniques for ordinary differential equations.) Thinking of y as held fixed, we find

$$u(x,y) = c_1(y)e^{xy} + c_2(y)e^{-xy}.$$

Some conditions that single out one solution are, for example,

$$u(0,y) = g(y), \qquad u_x(0,y) = h(y),$$

where g and h are preassigned functions of y. [Note that in this example the solution need not be even once differentiable as a function of y.] Imposing these conditions on $u(x,y)$ leads to the following linear equations for determining $c_1(y)$ and $c_2(y)$:

$$c_1(y) + c_2(y) = g(y), \qquad yc_1(y) - yc_2(y) = h(y).$$

The unique solution is $c_1(y) = \frac{1}{2}\left[g(y) + h(y)/y\right]$, $c_2(y) = \frac{1}{2}\left[g(y) - h(y)/y\right]$. Thus the particular solution satisfying the side conditions is

$$u(x,y) = \frac{1}{2}\left[g(y) + \frac{h(y)}{y}\right]e^{xy} + \frac{1}{2}\left[g(y) - \frac{h(y)}{y}\right]e^{-xy}.$$

In particular, if $g(y) = u(0,y) = 1$ and $h(y) = u_x(0,y) = -y$, we get $u(x,y) = e^{-xy}$. The graph of this solution appears in Figure 10.1 for $0 < x < 2$, $0 < y < 2$.

The partial differential operator

$$L = a\frac{\partial^2}{\partial x^2} + b\frac{\partial^2}{\partial y \partial x} + c\frac{\partial^2}{\partial y^2} + d\frac{\partial}{\partial x} + e\frac{\partial}{\partial y} + f$$

is linear in its action on twice-differentiable functions u and v:

$$L(\alpha u + \beta v) = \alpha L u + \beta L v, \qquad \alpha, \beta \text{ constant.}$$

The verification amounts to applying the operator to $\alpha u + \beta v$ and using the linearity of partial differentiation. Just as for ordinary differential operators, it follows that we can restate linearity as a **superposition principle** for homogeneous equations $Lu = 0$. This principle just says that if u and v are homogeneous solutions, then so is a linear combination $\alpha u + \beta v$. For partial differential operators, some of the most important solution techniques use the extension of the superposition principle to an arbitrary number of homogeneous solutions $u_1, \ldots, u_n$. Repeated application of the two-term formula gives

$$L(\alpha_1 u_1 + \cdots + \alpha_n u_n) = \alpha_1 L u_1 + \cdots + \alpha_n L u_n.$$

It's in this way that we'll start to construct solutions to general boundary problems from more elementary solutions.

1B Exponential Solutions

For a constant-coefficient homogeneous equation

1.2 $$a u_{xx} + b u_{xy} + c u_{yy} + d u_x + e u_y + f u = 0,$$

we can always find solutions of exponential form, just as in the case of a linear ordinary differential equation with constant coefficients. Guided by what worked for those linear differential equations, we assume a solution of the form $u(x, y) = e^{rx+sy}$, where r and s are to be determined so that $u(x, y)$ is a solution. We see that

$$u_x = ru, \qquad u_y = su, \qquad u_{xx} = r^2 u, \qquad u_{xy} = rsu, \qquad u_{yy} = s^2 u.$$

Thus a solution of the simple exponential form will have to have its parameters r and s chosen so that the equation

1.3 $$ar^2 + brs + cs^2 + dr + es + f = 0$$

is satisfied. This equation is the analogue of the *characteristic equation* introduced in Chapter 3 for an ordinary differential equation. The graph of this equation is an ellipse, a parabola or a hyperbola accordingly as $b^2 - 4ac$ is negative, zero, or positive, and the nature of the solutions to the corresponding differential equation depends significantly on which condition holds. Since the type of the differential equation depends just on the quadratic terms, $ar^2 + brs + cs^2$ is called the **characteristic polynomial** of the differential equation. Examples of each of the three types will be taken up in detail in Sections 4 and 5, where we'll see that the heat equation is parabolic, the wave equation is hyperbolic, and the Laplace equation is elliptic.

example 3

An example of a hyperbolic equation is

$$u_{xx} - u_{yy} = 0,$$

which has characteristic equation $r^2 - s^2 = 0$ so $s = \pm r$. Thus exponential solutions have the form

$$u(x, y) = e^{rx+ry}, \quad \text{and} \quad u(x, y) = e^{rx-ry}.$$

By forming linear combinations of these two families of basic exponential solutions, it's possible to obtain representations for solutions satisfying many different boundary conditions. It's easy to check directly that the equation $u_{xx} - u_{yy} = 0$ and the side conditions

$$u(x,0) = \cos\beta x, \quad \text{and} \quad u_y(x,0) = 0$$

are satisfied by $u(x,y) = \cos\beta x \cos\beta y$ for $-\infty < x < \infty$. On the other hand, using the exponential solutions, we have

$$\begin{aligned}\cos\beta x \cos\beta y &= \tfrac{1}{2}(e^{i\beta x} + e^{-i\beta x}) \cdot \tfrac{1}{2}(e^{i\beta y} + e^{-i\beta y}) \\ &= \tfrac{1}{4}e^{i\beta(x+y)} + \tfrac{1}{4}e^{i\beta(x-y)} + \tfrac{1}{4}e^{-i\beta(x+y)} + \tfrac{1}{4}e^{-i\beta(x-y)}.\end{aligned}$$

Hence $\cos\beta x \cos\beta y$ is a solution of the differential equation that we get by letting $r = i\beta$ and $r = -i\beta$ in the exponential solution formulas and forming a suitable linear combination. It turns out that all solutions of $u_{xx} - u_{yy} = 0$ can be constructed by generalizing the idea of linear combination. We'll use the techniques of *Fourier expansion* explained in Sections 2 and 3 to facilitate such constructions.

Most of the examples that we'll look at in detail have been studied for many years. The next example first appeared in print in 1973, and since then has been credited with having a profound effect on financial markets.

example 4

The **Black-Scholes differential equation** is

$$\frac{\partial V}{\partial t} + \frac{1}{2}\sigma^2 x^2 \frac{\partial^2 V}{\partial x^2} + rx\frac{\partial V}{\partial x} - rV = 0,$$

where $V(x,t)$ denotes the value V of the option to buy or sell a particular security at price x and time t. The parameter r is the current interest rate on a risk free investment such as a government bond, and the constant σ^2 is a measure of the volatility of the security's return to the investor. The option to buy or sell is called a *derivative* of the underlying security. Note that the differential equation is linear and has nonconstant coefficients depending on the variable x. With good estimates for σ^2, numerical solutions to this equation are widely used as a basis for establishing option prices. You're asked in Exercise 21 to investigate simple solutions of the form $V(x,t) = Ce^{-\alpha t}x^{\beta}$, where α, β, and C are positive constants.

EXERCISES

Verify by direct substitution that each of the differential equations in Exercises 1 through 5 is satisfied by the given function. Then verify that the function itself satisfies the given side condition.

1. $u_{xx} - u_y = 0$; $u(x,y) = e^{-y}\sin x$; $u(0,y) = 0$.

2. $u_{xx} + u_{yy} = 0$; $u(x,y) = e^{y}\sin x$; $u(0,y) = u(\pi,y) = 0$.

3. $u_{xx} - u_{yy} = 0$; $u(x,y) = \sin x \sin y$; $u(x,0) = u(x,\pi) = 0$.

4. $u_{xx} - 4u_y = 0$; $u(x,y) = y^{-1/2}e^{-x^2/y}$, $y > 0$; $\lim_{x\to\infty} u(x,y) = 0$.

5. $x^2u_{xx} + xu_x + u_{yy} = 0$; $u(x,y) = x\sin y$; $u(x,0) = u(x,\pi) = 0$.

6. (a) Solve the differential equation $u_{xy} = 0$ by integrating first with respect to y, then x.

(b) Solve the differential equation $u_{xy} = 1$ by integrating first with respect to y, then x.

(c) Verify that $u_p(x,y) = xy$ is a particular solution of $u_{xy} = 1$. Then use the linearity of the partial differential operator $L = \partial^2/\partial y\partial x$ to find the general twice-differentiable solution of $u_{xy} = 1$ by using the solution to part (a). [*Hint:* Use Theorem 3.1 of Chapter 3, Section 3.]

7. Find the general twice-differentiable solution $u(x,y)$ of the differential equation

$$u_{xx} + 2u_x + u = y.$$

8. To find the general solution in $u(x,y)$ of $u_{xx} - u_{yy} = 0$ for all x and y, define new variables z and w by

$$z = x + y,$$
$$w = x - y.$$

Define $\bar{u}$ as a function of two variables by

$$\bar{u}(z,w) = u\left(\tfrac{1}{2}(z+w), \tfrac{1}{2}(z-w)\right).$$

(a) Show that if (z,w) corresponds to (x,y) as above, then $\bar{u}(z,w) = u(x,y)$.

(b) Assuming that $u_{xx} - u_{yy} = 0$, use the chain rule to show that $\bar{u}_{zw} = 0$. [*Hint:* Compute $u_{xx} - u_{yy}$ in terms of $\bar{u}_{zz}$, $\bar{u}_{zw}$, and $\bar{u}_{ww}$.]

(c) Show that the general solution of $\bar{u}_{zw} = 0$ is $\bar{u}(z,w) = g(z) + h(w)$, where g and h are arbitrary twice-differentiable functions.

(d) Conclude that all solutions of the original equation have the form

$$u(x,y) = g(x+y) + h(x-y).$$

Classification. We consider here the geometric properties of the set of points in the rs plane that satisfy the characteristic equation

$$ar^2 + brs + cs^2 + dr + es + f = 0$$

of the associated operator

$$L = a\frac{\partial^2}{\partial x^2} + b\frac{\partial^2}{\partial y\partial x} + c\frac{\partial^2}{\partial y^2} + d\frac{\partial}{\partial x} + e\frac{\partial}{\partial y} + f;$$

these properties provide an important classification of the operators L. If a, b, and c are not all zero, the characteristic equation is that of a conic section, perhaps consisting only of lines or a point, of elliptic, parabolic, or hyperbolic type. The corresponding operator L is then said to be, respectively, **elliptic, parabolic,** or **hyperbolic.** For example, $\partial^2/\partial x^2 + \partial^2/\partial y^2$ is elliptic, because $r^2 + s^2 = 0$ is a degenerate ellipse; $\partial^2/\partial x^2 + \partial/\partial y$ is parabolic, because $r^2 + s = 0$ is a parabola; $\partial^2/\partial x^2 - \partial^2/\partial y^2$ is hyperbolic, because $r^2 - s^2 = 0$ is a degenerate hyperbola. Classify each of the operators in Exercises 9 through 12 as elliptic, parabolic, or hyperbolic.

9. $\dfrac{\partial^2}{\partial x^2} + \dfrac{\partial^2}{\partial y^2} - \dfrac{\partial}{\partial x}$.

10. $\dfrac{\partial^2}{\partial x \partial y}$.

11. $\dfrac{\partial^2}{\partial x^2} - 2\dfrac{\partial^2}{\partial y^2} + 4\dfrac{\partial}{\partial y}$.

12. $\dfrac{\partial^2}{\partial y^2} - \dfrac{\partial}{\partial x} - 3$.

13. The partial differential equation $u_{xx} + u_{yy} = u$ has solutions of the form $u(x,y) = e^{rx+sy}$ for certain constants r and s. Find the relationship that must hold between r and s for such a solution to exist. Then write a formula for such a solution under the assumption that $r = 1$.

14. The partial differential equation $u_{xx} + 2u_x + u = u_{yy}$ has solutions of the form $u(x,y) = e^{rx+sy}$ for certain constants r and s. Find the relationship that must hold between r and s for such a solution to exist.

15. The partial differential equation $u_{xx} - u_{yy} + u_y = 0$ has solutions of the form $u(x,y) = e^{rx+sy}$ for certain constants r and s. Find the relationship (i.e., characteristic equation) that must be satisfied by both r and s for such a solution to exist. Then write a formula for all such solutions in terms of x, y, and s.

16. Find all solutions of the **telegraph equation:** $u_{tt} + bu_t + cu = au_{xx}$, a, b, and c nonnegative constants, that are independent of t, that is, find the *steady-state* solutions. [*Hint:* $u_t = u_{tt} = 0$.]

17. **(a)** Find the characteristic equation of the telegraph equation $u_{tt} + bu_t + cu = au_{xx}$, a, b, and c nonnegative constants.

(b) Show that product solutions $u(x,t) = e^{rx+st}$ of the telegraph equation cannot have r and s both purely imaginary unless $b = 0$.

Partial differential equations have lots of solutions that are not of exponential form; some of these are developed later in the chapter. The following exercises give some indication of the possible variety of solutions.

18. **(a)** Show that the **wave equation** $u_{tt} = a^2 u_{xx}$ has solution $u(x,t) = f(x - at)$, where f is an arbitrary twice-differentiable function of one real variable and a is constant. For example, if $f(z) = 1/(1 + z^2)$, then $f(x - at) = 1/(1 + (x - at)^2)$.

(b) Use the linearity of the wave equation to show that it has solutions of the form $u(x,t) = f(x - at) + g(x - at)$.

19. **(a)** Show that the **Laplace equation** $u_{xx} + u_{yy} = 0$ has solutions $u(x,y) = f(x \pm iy)$, where $f(z)$ is an arbitrary **analytic function** given by a power series in z. [For example, if $f(z) = z^2$, then $u(x, y) = (x \pm iy)^2 = x^2 - y^2 \pm 2ixy$ are complex solutions.]

(b) Use the linearity of the Laplace equation to show that it has solutions of the form $u(x, y) = f(x + iy) + f(x - iy)$. What solution $u(x, y)$ results from choosing $f(z) = z^2$?

20. The third-order nonlinear **Korteweg-deVries** wave equation is

$$\frac{\partial u}{\partial t} + \beta u \frac{\partial u}{\partial x} + \frac{\partial^3 u}{\partial x^3} = 0, \qquad \beta \text{ constant.}$$

(a) Show that if $u = u(x, t)$ is a nonzero solution and c is a constant different from 0 or 1, then $cu(x, t)$ is not a solution of the differential equation with nonlinear term $\beta u u_x$, but satisfies the equation with β replaced by β/c.

(b) Show that the differential equation with constant $\beta = 12$ has solution $u(x, t) = \operatorname{sech}^2(x - 4t)$, where $\operatorname{sech} w = 1/\cosh w$ is the hyperbolic secant.

(c) Sketch the graphs of the function $u = u(x, t)$ in part (b) for $t = 0, 1$, and 2, and show that the solitary hump moves to the right above the x axis with velocity 4 as t increases.

21. The **Black-Scholes equation,** with positive constant parameters r and σ^2, and described in Example 4 of the text, is $\partial V/\partial t + \frac{1}{2}\sigma^2 x^2 (\partial^2 V/\partial x^2) + rx(\partial V/\partial x) - rV = 0$.

(a) Show that the differential equation has solutions $V(x, t) = Ce^{-\alpha t} x^{\beta}$, assuming $x > 0$, and α, β, and C are positive constants. Find the equation that relates these constants.

(b) Normally a solution $V(x, t)$ is concave up as a function of x if t is held fixed. Use the result of part (a) to show that the solutions identified there decrease over time if x is held fixed.

(c) Consider a simplification of the differential equation for which $r = 0$. Show directly from the simplified differential equation that a solution $V(x, t)$ is concave up as a function of x alone if and only if $V(x, t)$ is decreasing as a function of t alone.

2 FOURIER SERIES

2A Introduction

Finite linear combinations of the exponential solutions discussed in the previous section are not enough to satisfy the wide variety of side conditions that come up in practice. However, if we allow infinite series whose partial sums contain linear combinations of exponentials of the form $e^{i\beta x}$, or what amounts to the same thing, of $\cos \beta x$ and $\sin \beta x$, it turns out that we can produce solutions that satisfy many interesting side conditions. Disregarding any question of convergence for the moment, we define a **trigonometric series** to be a series of the form

2.1
$$\frac{a_0}{2} + \sum_{k=1}^{\infty}(a_k \cos kx + b_k \sin kx).$$

In the special circumstance that the coefficients a_k and b_k arise from an integrable function $f(x)$ by means of the **Euler formulas**

2.2
$$a_k = \frac{1}{\pi}\int_{-\pi}^{\pi} f(x)\cos kx\,dx, \qquad b_k = \frac{1}{\pi}\int_{-\pi}^{\pi} f(x)\sin kx\,dx,$$

then the trigonometric series is called the **Fourier series** of f. The coefficients a_k and b_k as given by Equations 2.2 are called the **Fourier coefficients** of f. The most fundamental question about a Fourier series is the extent to which the series represents the function f. The importance of such series for us stems from the inclusion in Sections 4 and 5 of individual terms of similar series in solutions of differential equations.

Our first examples illustrate the beautiful way in which the partial sums of a Fourier series attempt to mimic the function f that generates them. A partial sum

2.3
$$S_N(x) = \frac{a_0}{2} + \sum_{k=1}^{N}(a_k \cos kx + b_k \sin kx)$$

is called a **trigonometric polynomial,** and if the coefficients a_k and b_k are computed using Equations 2.2, then this trigonometric polynomial is called the Nth **Fourier approximation** to the function $f(x)$. Note that each term in S_N is a **periodic function** of period 2π, since $\cos kx$ and $\sin kx$ have period 2π. It follows that S_N is also periodic with period 2π, that is

$$S_N(x + 2\pi) = S_N(x), \qquad \text{for all } x.$$

Note also that the Fourier coefficients a_k and b_k are determined by integral formulas that use the values of $f(x)$ only for $-\pi \le x \le \pi$. The restriction to this interval will be removed in Section 3.

2B Computing Coefficients

The integrals in the Euler formulas deserve special consideration. If we let $f(x)$ equal $\cos lx$ or $\sin lx$ in Equations 2.2, it so happens that for nonnegative integer k and l we have

2.4
$$\frac{1}{\pi}\int_{-\pi}^{\pi} \cos kx\,\sin lx\,dx = 0,$$
$$\frac{1}{\pi}\int_{-\pi}^{\pi} \cos kx\,\cos lx\,dx = \begin{cases} 0, & k \ne l, \\ 1, & k = l \ne 0, \end{cases}$$
$$\frac{1}{\pi}\int_{-\pi}^{\pi} \sin kx\,\sin lx\,dx = \begin{cases} 0, & k \ne l \text{ or } k, l = 0, \\ 1, & k = l \ne 0. \end{cases}$$

The three Equations 2.4 are called **orthogonality relations** for reasons explained in Chapter 11, and they follow easily from trigonometric identities. (See Exercise 28.) As a sample application of the orthogonality relations, suppose that a trigonometric series 2.1 satisfies some condition that enables us to integrate it term by term on the interval $-\pi \leq x \leq \pi$. For example, the series might just be a finite sum, with all a_k and b_k equal to zero from some point on. We have the following theorem that connects Formulas 2.1 and 2.2.

2.5 THEOREM

If the trigonometric series 2.1 converges to a function $f(x)$ and can be integrated term by term over the interval $[-\pi, \pi]$, then the coefficients a_k and b_k must be given by the Euler Formulas 2.2.

Proof. We define $f(x)$ by

$$f(x) = \frac{a_0}{2} + \sum_{k=1}^{\infty} (a_k \cos kx + b_k \sin kx).$$

Then for a fixed integer $l > 0$, we multiply the equation by $(1/\pi) \cos lx$ and integrate from $-\pi$ to π to get

$$\begin{aligned}
&\frac{1}{\pi} \int_{-\pi}^{\pi} f(x) \cos lx \, dx \\
&= \frac{1}{\pi} \int_{-\pi}^{\pi} \left[\frac{a_0}{2} + \sum_{k=1}^{\infty} (a_k \cos kx + b_k \sin kx) \right] \cos lx \, dx \\
&= \frac{a_0}{2\pi} \int_{-\pi}^{\pi} \cos lx \, dx + \sum_{k=1}^{\infty} \left(\frac{a_k}{\pi} \int_{-\pi}^{\pi} \cos kx \cos lx \, dx + \frac{b_k}{\pi} \int_{-\pi}^{\pi} \sin kx \cos lx \, dx \right).
\end{aligned}$$

The first two relations in 2.4 show that all but one of the terms in parentheses is zero; the only term that survives is the term containing a_l, and we find, for $l \neq 0$,

$$\frac{1}{\pi} \int_{-\pi}^{\pi} f(x) \cos lx \, dx = \frac{a_l}{\pi} \int_{-\pi}^{\pi} \cos lx \cos lx \, dx = a_l.$$

When $l = 0$ the only nonzero term that survives in the sum is the first one. Since the integral of 1 over the interval is 2π, we get $2a_0$, which accounts for the term $\frac{1}{2}a_0$ in the expansion. A similar computation shows that $(1/\pi) \int_{-\pi}^{\pi} f(x) \sin lx \, dx = b_l$. ■

Theorem 2.5 shows that under fairly broad conditions on the coefficients of a trigonometric series, the coefficients must be given by the Euler Formulas 2.2. In particular, the Fourier series of a trigonometric polynomial is precisely equal to the polynomial. [For example, $f(x) = \sin x + \cos 2x$ is its own Fourier series.] It follows that Formulas 2.2 for determining the a_k and b_k appear to be quite appropriate if we want to represent integrable functions $f(x)$ by trigonometric series using Formula 2.1.

Odd and Even Functions. The first of the three Equations 2.4 follows, without any detailed computation, from the useful observation that the integral of an odd

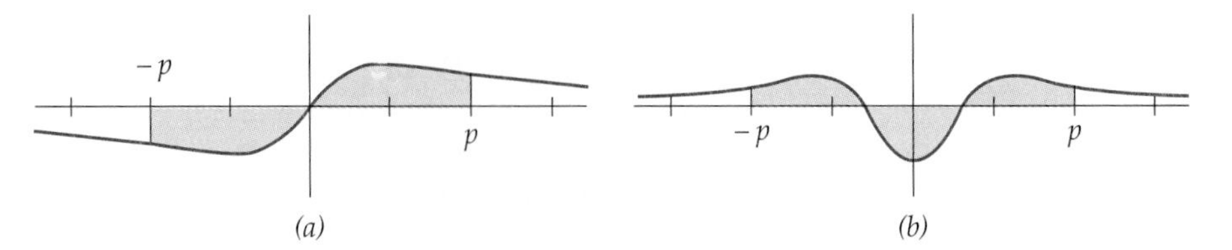

Figure 10.2 (*a*) Odd function. (*b*) Even function.

function, $g(x) = \cos kx \sin lx$ in this case, over an interval symmetric about the origin is necessarily zero. Recall that by definition an **odd function** $g(x)$ satisfies $g(-x) = -g(x)$ for $-p \le x \le p$ and that an **even function** satisfies $g(-x) = g(x)$ for $-p \le x \le p$. Figures 10.2(a) and 10.2(b) illustrate the two types. The pictures show geometrically, and it's easy to show analytically, that

2.6

$$\text{(a)} \quad \int_{-p}^{p} g(x)\,dx = 0 \quad \text{if } g \text{ is odd,}$$

$$\text{(b)} \quad \int_{-p}^{p} g(x)\,dx = 2\int_{0}^{p} g(x)\,dx \quad \text{if } g \text{ is even.}$$

---

example 1

Let $f(x) = |x|$ for $-\pi \le x \le \pi$. Then

$$a_k = \frac{1}{\pi}\int_{-\pi}^{\pi} |x| \cos kx\,dx, \qquad b_k = \frac{1}{\pi}\int_{-\pi}^{\pi} |x| \sin kx\,dx.$$

Clearly, $|x| \sin kx$ has integral zero over $[-\pi, \pi]$, because it is an odd function. Hence $b_k = 0$ for $k = 1, 2, \ldots$. On the other hand, the graph of $|x| \cos kx$ is symmetric about the y axis, so we can settle for twice its integral over $[0, \pi]$. For $k \ne 0$ we integrate by parts, getting

$$\begin{aligned}
a_k &= \frac{2}{\pi}\int_0^{\pi} x \cos kx\,dx \\
&= \frac{2}{\pi}\left[\frac{x \sin kx}{k}\right]_0^{\pi} - \frac{2}{\pi k}\int_0^{\pi} \sin kx\,dx \\
&= \left[\frac{2}{\pi k^2}\cos kx\right]_0^{\pi} = \frac{2}{\pi k^2}(\cos k\pi - 1) \\
&= \frac{2}{\pi k^2}[(-1)^k - 1] = \begin{cases} 0, & k = 2, 4, 6, \ldots, \\ -\dfrac{4}{\pi k^2}, & k = 1, 3, 5, \ldots. \end{cases}
\end{aligned}$$

When $k = 0$, we have $a_0 = 2/\pi \int_0^{\pi} x\,dx = \pi$. To summarize,

$$b_k = 0, \qquad k = 1, 2, 3, \ldots, \qquad a_0 = \pi, \qquad a_k = \begin{cases} 0, & k = 2, 4, 6, \ldots, \\ -\dfrac{4}{\pi k^2}, & k = 1, 3, 5, \ldots. \end{cases}$$

Hence, the Nth Fourier approximation is given for $N = 1, 3, 5, \ldots$ by the

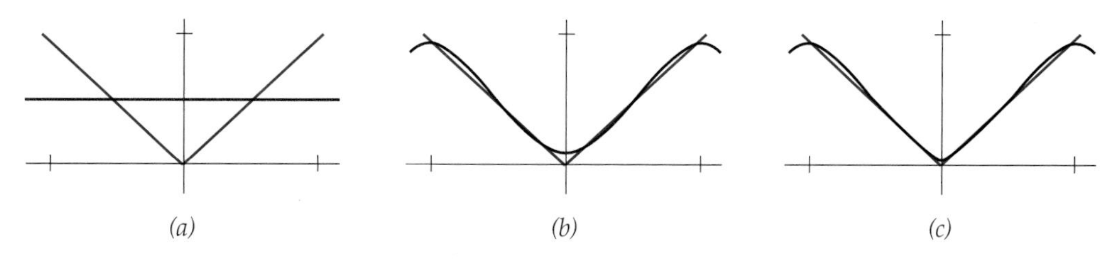

Figure 10.3 Fourier approximations to $f(x) = |x|$ on $[-\pi, \pi]$. (a) $s_0(x) = \pi/2$. (b) $s_1(x) = s_0(x) - (4/\pi)\cos x$. (c) $s_3(x) = s_1(x) - (4/9\pi)\cos 3x$.

trigonometric polynomial

$$s_N(x) = \frac{\pi}{2} - \frac{4}{\pi}\cos x - \frac{4}{\pi}\frac{\cos 3x}{3^2} - \cdots - \frac{4}{\pi}\frac{\cos Nx}{N^2}.$$

If N is even, we have $s_N(x) = s_{N-1}(x)$. Figure 10.3 shows how the graphs of s_0, s_1, and s_3 approximate that of $|x|$ on $[-\pi, \pi]$; it appears that adding more terms improves the approximation.

example 2

Let the step function $g(x) = \begin{cases} 1, & 0 \le x \le \pi, \\ -1, & -\pi \le x < 0. \end{cases}$ To compute a_k and b_k we break the interval of integration $[-\pi, \pi]$ at 0:

$$a_k = \frac{-1}{\pi}\int_{-\pi}^{0} \cos kx\, dx + \frac{1}{\pi}\int_{0}^{\pi} \cos kx\, dx.$$

Since cosine is an even function, $\cos(-kx) = \cos kx$, and the two integrals are equal, so we get $a_k = 0$. Similarly,

$$b_k = \frac{-1}{\pi}\int_{-\pi}^{0} \sin kx\, dx + \frac{1}{\pi}\int_{0}^{\pi} \sin kx\, dx.$$

Since sine is an odd function, $\sin(-kx) = -\sin kx$, the integrals themselves are negatives of each other, and we get, since we can assume $k \neq 0$,

$$\begin{aligned} b_k &= \frac{2}{\pi}\int_0^{\pi} \sin kx\, dx = \frac{2}{\pi}\left[\frac{-\cos kx}{k}\right]_0^{\pi} \\ &= \frac{2}{k\pi}\left[-(-1)^k + 1\right] = \begin{cases} 0, & k \text{ even}, \\ \dfrac{4}{k\pi}, & k \text{ odd}. \end{cases} \end{aligned}$$

In summary,

$$a_k = 0, \qquad k = 0, 1, 2, \ldots, \qquad b_k = \begin{cases} 0, & k = 2, 4, 6, \ldots, \\ \dfrac{4}{k\pi}, & k = 1, 3, 5, \ldots. \end{cases}$$

Hence, for odd N, the Nth Fourier approximation to g is given by

$$s_N(x) = \frac{4}{\pi}\sin x + \frac{4}{\pi}\frac{\sin 3x}{3} + \cdots + \frac{4}{\pi}\frac{\sin Nx}{N}.$$

The graphs of s_1, s_3, and s_5 are shown in Figure 10.4, together with that of $g(x)$.

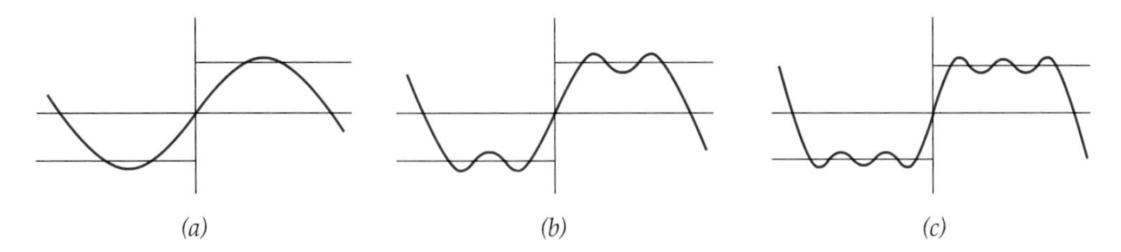

Figure 10.4 Fourier approximations to step function $g(x)$ on $[-\pi, \pi]$. (a) $s_1(x) = (4/\pi) \sin x$. (b) $s_3(x) = s_1(x) + (4/3\pi) \sin 3x$. (c) $s_5(x) = s_3(x) + (4/5\pi) \sin 5x$.

Sketching the graphs of a few partial sums along with the graph of the function being approximated is often a good way to tell whether you've made a mistake in computing the coefficients. The partial sums should appear to be trying to approximate the function, as in Figures 10.3 and 10.4.

2C Convergence

An important question is whether the Fourier approximations $s_N(x)$ converge as $N \to \infty$ to $f(x)$ for some values of x, where f is the function on $[-\pi, \pi]$ from which the Fourier coefficients are computed. Since the partial sums of a Fourier series are themselves periodic functions, a function to which they converge must also be periodic, and is called a **periodic extension** of f. This periodic extension may differ from the precise definition of $f(x)$ at some points in the interval $-\pi \leq x \leq \pi$. Indeed, changing a value $f(x_0)$ at a point x_0 has no effect on the integral formulas for the Fourier coefficients, so it's customary to alter the values of f at some points accordingly in defining a periodic extension of a function from an interval to the entire real number line. Figure 10.5 shows a sketch of a periodic extension from the interval $-\pi \leq x \leq \pi$.

The Fourier series of f is by definition the infinite series

$$\frac{a_0}{2} + \sum_{k=1}^{\infty} (a_k \cos kx + b_k \sin kx),$$

where a_k and b_k are given by the Euler Formulas 2.2. Theorem 2.7 states some conditions on f under which the Fourier series can be used to represent f. Thus we'll assume that the graph of f is not only bounded on $[-\pi, \pi]$ but **piecewise**

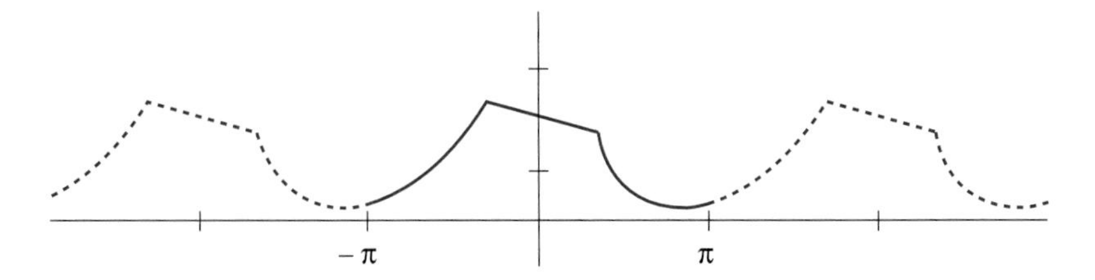

Figure 10.5 Periodic extension from $-\pi < x \leq \pi$.

monotone: This means that the interval $[-\pi, \pi]$ can be broken into finitely many subintervals, with endpoints $-\pi = x_1 < x_2 < \cdots < x_n = \pi$, such that f is either nondecreasing or nonincreasing on each open subinterval $x_k < x < x_{k+1}$. It can be proved that the Fourier series of a piecewise monotone function f will then converge to a 2π-periodic extension of f, also denoted by f. At a possible discontinuity at $x = x_0$ the series will converge to the *average* value

$$\tfrac{1}{2}[f(x_0-) + f(x_0+)].$$

Here $f(x_0-)$ stands for the left-hand limit of f at x_0, and $f(x_0+)$ is the right-hand limit. The graph of a representative piecewise monotone function is shown in Figure 10.5, with the average value indicated by a dot at each jump and the periodic extension by dots. A somewhat stronger version of the next theorem is called **Dirichlet's theorem,** in which the conclusion is the same but the hypotheses are weaker.

2.7 THEOREM

Let f be bounded and piecewise monotone on $[-\pi, \pi]$. Then the 2π-periodic Fourier series of f converges at every point x to a periodic extension of f. In particular, if f is continuous at x, then the series converges to $f(x)$. If f is discontinuous at x_0, the series converges to the average value $\frac{1}{2}[f(x_0-) + f(x_0+)]$.

Note. The precise value of $f(x)$ at a single point x has no effect on the Euler coefficient integrals, in particular the value assigned at a jump discontinuity has no effect on the Fourier series of $f(x)$, so we can think of $f(x)$ being defined for $-\pi \le x \le \pi$ or $-\pi < x < \pi$ as we please.

Examples 1 and 2 gave an indication of how partial sums of a Fourier series converge. In each of those examples, the function satisfies the conditions of boundedness and piecewise monotonicity; hence the series converges to the function claimed in Theorem 2.7.

example 3

The function g defined in Example 2 is rather arbitrarily defined to have the value 1 at $x = 0$. In spite of this arbitrariness, Theorem 2.7 enables us to conclude that the Fourier series of g converges as follows:

$$\sum_{k=0}^{\infty} \frac{4}{\pi} \frac{\sin(2k+1)x}{2k+1} = \begin{cases} 1, & 0 < x < \pi, \\ 0, & x = 0, \\ -1, & -\pi < x < 0. \end{cases}$$

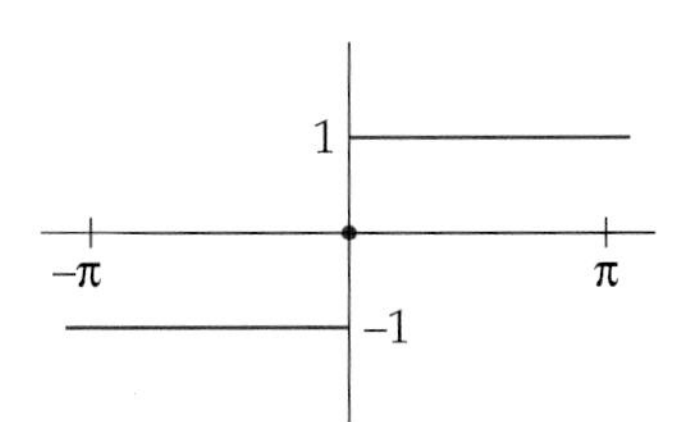

The graph of this discontinuous function is shown in the figure. To be very specific, we can set $x = \pi/2$ and arrive at the alternating series expansion

$$\sum_{k=0}^{\infty} \frac{(-1)^k}{2k+1} = \frac{\pi}{4}.$$

Theorem 2.7 shows to some extent the reason for choosing the coefficients in a trigonometric polynomial according to Formula 2.2. The reason is that, under favorable circumstances, the resulting sequence of trigonometric polynomials will converge to the function f. Figure 10.5 shows a function f extended periodically, with period 2π, from the interval $[-\pi, \pi]$ to other values of x. Since the partial sums of the Fourier series are also periodic with period 2π, whatever convergence takes place on $[-\pi, \pi]$ extends periodically to all values of x. At this point Theorem 2.7 has implications only for functions of period 2π, but we'll see in the next section that the statement can easily be modified to be valid for functions with any positive period $2p$.

The coefficient values a_k and b_k that we get from the Euler formulas are quite independent of the finite values assigned to $f(x)$ at isolated points; this is because the definite integrals in the Euler formulas don't distinguish between two functions that differ at finitely many points. For example, the two functions

$$f(x) = \begin{cases} 0, & -\pi \le x \le 0 \\ 1, & 0 < x \le \pi \end{cases} \quad \text{and} \quad g(x) = \begin{cases} 0, & -\pi \le x < 0 \\ 1, & 0 \le x \le \pi \end{cases}$$

differ only at $x = 0$, where $f(0) = 0$ and $g(0) = 1$. Intuitively speaking, the areas under the two graphs should be the same, namely, π. (That this is actually the case is a significant property of the integral.) Since the Fourier series of a piecewise monotone function converges to the average of the right and left limits at each point x, it makes sense simply to redefine such a periodic function to have the average value $\frac{1}{2}\left[f(x_0+) + f(x_0-)\right]$ at each jump discontinuity x_0. We refer to the end result as the **normalized function.**

example 4

If the function $f(x) = x + \pi$ is extended periodically from $-\pi < x < \pi$ to other values of x, its graph consists of parallel line segments of slope 1; it remains undefined at odd integer multiples of π since it's initially undefined at $\pm\pi$. To produce a normalized version of the function, defined for all x, all we have to do is define the function to have the average value π at odd integer multiples of π. The Fourier series can be computed for the function as originally defined, and the series will converge to the normalized function for real x. The periodically extended function is shown in Figure 10.6 together with the fourth partial sum of the Fourier expansion.

The Fourier coefficients can be computed as follows.

$$\begin{aligned} a_0 &= \frac{1}{\pi}\int_{-\pi}^{\pi} (x + \pi)\, dx \\ &= \frac{1}{\pi}\int_{-\pi}^{\pi} x\, dx + \frac{1}{\pi}\int_{-\pi}^{\pi} \pi\, dx = 0 + 2\pi = 2\pi. \end{aligned}$$

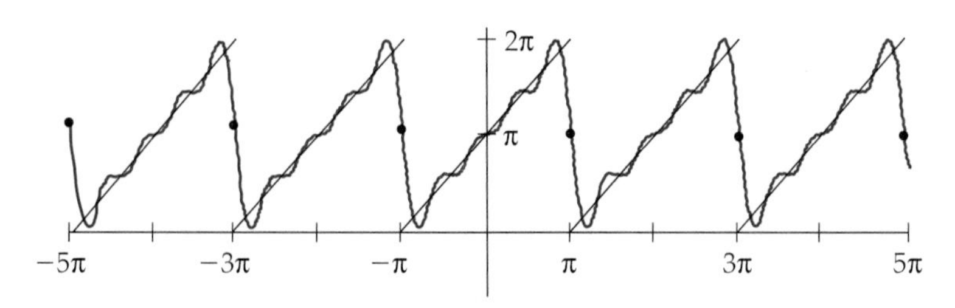

Figure 10.6 The fourth Fourier partial sum is superimposed. $f(x) = x + \pi$ extended periodically from $(-\pi, \pi)$ and normalized at jumps.

[Note that the integral of $f(x) = x$ over an interval symmetric about 0 is always 0.] When $k > 0$,

$$\begin{aligned} a_k &= \frac{1}{\pi}\int_{-\pi}^{\pi}(x+\pi)\cos kx\,dx \\ &= \frac{1}{\pi}\int_{-\pi}^{\pi}x\cos kx\,dx + \frac{1}{\pi}\int_{-\pi}^{\pi}\pi\cos kx\,dx = 0+0=0. \end{aligned}$$

The last integral above is zero because the indefinite integral is 0 at $\pm\pi$. The previous one is most easily seen to be zero by observing that the integrand, $x\cos kx$ is an odd function, so that the integral over $[-\pi, 0]$ is the negative of the integral over $[0, \pi]$. Now for the b_ks,

$$\begin{aligned} b_k &= \frac{1}{\pi}\int_{-\pi}^{\pi}(x+\pi)\sin kx\,dx \\ &= \frac{1}{\pi}\int_{-\pi}^{\pi}x\sin kx\,dx + \frac{1}{\pi}\int_{-\pi}^{\pi}\pi\sin kx\,dx \\ &= \frac{2(-1)^{k+1}}{k} + 0 = \frac{2(-1)^{k+1}}{k}. \end{aligned}$$

The last integral above is zero because the integrand is an odd function. The previous one can be computed using integration by parts, with

$$\begin{aligned} \frac{1}{\pi}\int_{-\pi}^{\pi}x\sin kx\,dx &= -\frac{1}{\pi k}x\cos kx\Big|_{-\pi}^{\pi} + \frac{1}{\pi k}\int_{-\pi}^{\pi}\cos kx\,dx \\ &= -\frac{1}{k}\cos k\pi - \frac{1}{k}\cos(-k\pi) + 0 = \frac{2(-1)^{k+1}}{k}. \end{aligned}$$

The full expansion, including the constant $a_0/2$, is then

$$f(x) = \pi + \sum_{k=1}^{\infty}\frac{2(-1)^{k+1}}{k}\sin kx = \pi + 2\sin x - \sin 2x + \frac{2}{3}\sin 3x - \frac{1}{2}\sin 4x + - \cdots.$$

The Java Applet FOURIER at website <http://math.dartmouth.edu/~rewn/> or via <http://www.mhhe.com/williamson> approximates Fourier coefficients using Simpson's rule and then plots graphs of partial sums.

EXERCISES

The following observations about values of sine and cosine are useful for computing Fourier coefficients; k is always an integer here: (1) $\sin k\pi = 0$, (2) $\cos k\pi = (-1)^k$, (3) $\sin(k+\frac{1}{2})\pi = (-1)^k$, and (4) $\cos(k+\frac{1}{2})\pi = 0$. Compute the Fourier coefficients of each of the functions in Exercises 1 through 10, and write the corresponding Fourier series in the form of Equation 2.1. Sketch the graph of each function extended to have period 2π on the interval $-2\pi \le x \le 2\pi$. Finally, sketch the graphs of the first three partial sums $S_0(x)$, $S_1(x)$, and $S_2(x)$ of the Fourier series.

1. $f(x) = x,\ -\pi < x < \pi.$

2. $f(x) = \begin{cases} -\pi - x, & -\pi < x < 0, \\ \pi - x, & 0 \le x \le \pi. \end{cases}$

3. $f(x) = x^2,\ -\pi < x < \pi.$

4. $f(x) = |x| + 1,\ -\pi \le x \le \pi.$

5. $f(x) = \begin{cases} 0, & -\pi < x \le 0. \\ 1, & 0 < x \le \pi. \end{cases}$

6. $f(x) = x + 1,\ -\pi < x < \pi.$

7. $f(x) = \begin{cases} -\pi, & -\pi < x < 0, \\ \pi, & 0 \le x \le \pi. \end{cases}$

8. $f(x) = 2x + 1,\ -\pi < x < \pi.$

9. $f(x) = -|x|,\ -\pi \le x \le \pi.$

10. $f(x) = \begin{cases} -1, & -\pi < x \le 0. \\ 2, & 0 < x \le \pi. \end{cases}$

11.–20. According to Theorem 2.7, the Fourier series of each function $f(x)$ in Exercise 1 converges to some function $F(x)$ whose graph may or may not differ at some points from the graph of $f(x)$. For Exercises 11 through 20 sketch the $F(x)$ corresponding in order to the $f(x)$ in Exercises 1 through 10 on the interval $-2\pi \le x \le 2\pi$.

21. Show that if $f(x)$ and $g(x)$ have the Fourier coefficients a_k, b_k and a'_k, b'_k, respectively, then $\alpha f(x) + \beta g(x)$, where α and β are constant, has Fourier coefficients $\alpha a_k + \beta a'_k,\ \alpha b_k + \beta b'_k$.

The Nth partial sum of a trigonometric series is called a *trigonometric polynomial* of degree N, and a trigonometric polynomial is necessarily the Fourier series of the function it represents. For example, the identity $\cos^2 x = \frac{1}{2} + \frac{1}{2}\cos 2x$ is the Fourier expansion of $\cos^2 x$. Find the Fourier series of each of the following functions by using appropriate identities, for example, the ones in Exercise 28.

22. $\sin^2 x.$

23. $\cos^3 x.$

24. $\sin 2x \cos x.$

25. $\sin^3 x.$

26. $\cos^4 x - \sin^4 x.$

27. Using elementary properties of integrals, prove Equations 2.6 (a) and 2.6 (b) of the text.

28. Establish the three orthogonality relations in Equations 2.4 of the text, as follows:

(a) Use the trigonometric identities

$$\cos\alpha\cos\beta = \tfrac{1}{2}\cos(\alpha+\beta) + \tfrac{1}{2}\cos(\alpha-\beta),$$
$$\sin\alpha\cos\beta = \tfrac{1}{2}\sin(\alpha+\beta) + \tfrac{1}{2}\sin(\alpha-\beta),$$
$$\sin\alpha\sin\beta = \tfrac{1}{2}\cos(\alpha-\beta) - \tfrac{1}{2}\cos(\alpha+\beta),$$

to compute the relevant integrals.

***(b)** Use instead the identities $\cos nx = \frac{1}{2}(e^{inx} + e^{-inx})$, $\sin nx = (1/2i)(e^{inx} - e^{-inx})$ together with the identity $e^{i(\alpha+\beta)} = e^{i\alpha}e^{i\beta}$ for the exponential function.

29. Carry out the details of the proof that $(1/\pi)\int_{-\pi}^{\pi} f(x)\sin lx\, dx = b_l$, parallel to the computation in Theorem 2.5.

30. Suppose $f(x) = x^2 - x$ for $-\pi \le x \le \pi$. What does the Fourier series of f converge to at $x = \pi$ and at $x = -\pi$?

31. Let $f(x) = \sqrt{|x|}$ for $-\pi \le x \le \pi$. Does $f(x)$ satisfy the hypotheses of Theorem 2.7? Explain your answer.

32. Is $f(x) = x \sin(1/x)$ piecewise monotone for $0 < x < \pi$? Explain your answer.

3 ADAPTED FOURIER EXPANSIONS

3A General Intervals

The direct application of Fourier methods to practical problems usually requires some modification of the standard formulation presented in the previous section. In the present section we describe some of these modifications and calculate some examples.

While the interval $[-\pi, \pi]$ is a natural one for Fourier expansions because it is a basic interval of definition for $\cos x$ and $\sin x$, it usually happens that a function encountered in an application needs to be approximated on some other interval. If the function f to be approximated is defined not on the interval $[-\pi, \pi]$ but on $[-p, p]$, a suitable change in the computation of the approximation can be made as follows. With f defined on $[-p, p]$, we define a function f_p restricted to $[-\pi, \pi]$ by

$$f_p(x) = f\left(\frac{px}{\pi}\right), \qquad -\pi \le x \le \pi.$$

Then we can compute the Fourier coefficients of f_p by Formula 2.2. The resulting Fourier sums $s_N(x)$ will converge to f_p on $[-\pi, \pi]$ under the assumptions of Theorem 2.7 of the previous section. To approximate f itself on $[-p, p]$, it's then natural to think of approximating $f(x)$, which is equal to, $f_p(\pi x/p)$ by the partial sums

$$s_N\left(\frac{\pi x}{p}\right) = \frac{a_0}{2} + \sum_{k=1}^{N}\left(a_k \cos\frac{k\pi x}{p} + b_k \sin\frac{k\pi x}{p}\right), \qquad -p \le x \le p.$$

The coefficients a_k and b_k can be computed directly in terms of f by making a change of variable as follows. We change variable in the integral over $[-\pi, \pi]$, replacing x by $\pi x/p$:

$$\begin{aligned} a_k &= \frac{1}{\pi}\int_{-\pi}^{\pi} f_p(x)\cos kx\,dx = \frac{1}{\pi}\int_{-\pi}^{\pi} f\left(\frac{px}{\pi}\right)\cos kx\,dx \\ &= \frac{1}{p}\int_{-p}^{p} f(x)\cos\left(\frac{k\pi x}{p}\right)dx. \end{aligned}$$

A similar computation holds for b_k, so the formulas in terms of f itself are

3.1 $$a_k = \frac{1}{p}\int_{-p}^{p} f(x)\cos\frac{k\pi x}{p}\,dx, \qquad b_k = \frac{1}{p}\int_{-p}^{p} f(x)\sin\frac{k\pi x}{p}\,dx$$

for the coefficients in the Fourier expansion

$$\frac{a_0}{2} + \sum_{k=1}^{\infty} \left(a_k \cos \frac{k\pi x}{p} + b_k \sin \frac{k\pi x}{p} \right)$$

for the $2p$-periodic extension of the function f, originally defined on $[-p, p]$. [Note that the trigonometric parts of these formulas have period $2p$, as they should.] We conclude from the foregoing results that π *can be replaced by* p *in the convergence Theorem 2.7* of Section 2C. Recall that the most important hypothesis about $f(x)$ in that theorem is piecewise monotonicity, and that will hold for all our examples.

example 1 We define a step function $h(x)$ by

$$h(x) = \begin{cases} 1, & 0 \le x \le p, \\ -1, & -p \le x < 0. \end{cases}$$

The function $h(x)$ differs from an odd function only at the single point $x = 0$, so the cosine coefficients a_k are all zero. The sine coefficients are given by

$$\begin{aligned} b_k &= \frac{1}{p} \int_{-p}^{0} (-1) \sin \frac{k\pi x}{p} \, dx + \frac{1}{p} \int_0^p (1) \sin \frac{k\pi x}{p} \, dx = \frac{2}{p} \int_0^p \sin \frac{k\pi x}{p} \, dx \\ &= \frac{2}{\pi} \int_0^{\pi} \sin kx \, dx = \begin{cases} 0, & k = 2, 4, 6, \ldots, \\ \dfrac{4}{\pi k}, & k = 1, 3, 5, \ldots. \end{cases} \end{aligned}$$

The Fourier series of $h(x)$ converges to zero at integer multiples of p, and at other points

$$h(x) = \frac{4}{\pi} \sin \frac{\pi x}{p} + \frac{4}{3\pi} \sin \frac{3\pi x}{p} + \frac{4}{5\pi} \sin \frac{5\pi x}{p} + \cdots.$$

For a function f defined on an arbitrary interval $a \le x \le b$, it is helpful to think of a periodic extension f_E of f having period $b - a$ and defined for all real numbers x. Such an extension is illustrated in Figure 10.7. We set $2p = b - a$ so that $p = (b - a)/2$ and $-p = -(b - a)/2$. We then compute the Fourier coefficients of f_E over the interval $[-p, p]$ according to Formula 3.1. Furthermore, because the integrands in Formula 3.1 have period $2p$, we can use the geometrically obvious fact that the integration can be performed over any interval of length $2p = b - a$ in particular, over $[a, b]$. (See Exercise 18.) Thus Formula 3.1 can be rewritten:

3.2 $$a_k = \frac{2}{b-a} \int_a^b f(x) \cos \frac{2k\pi x}{b-a} \, dx, \qquad b_k = \frac{2}{b-a} \int_a^b f(x) \sin \frac{2k\pi x}{b-a} \, dx.$$

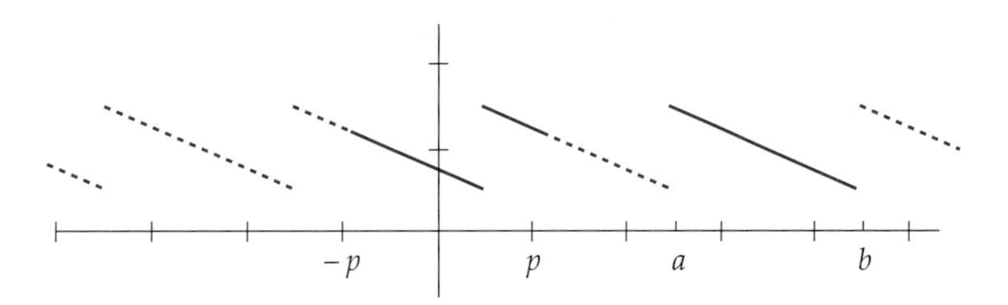

Figure 10.7 $p = (b-a)/2$.

The associated Fourier approximations are

$$s_N(x) = \frac{a_0}{2} + \sum_{k=1}^{N} \left(a_k \cos \frac{2k\pi x}{b-a} + b_k \sin \frac{2k\pi x}{b-a} \right).$$

Equations 3.2 are sometimes useful computationally because the definition of $f(x)$ on $[a,b]$ may make for easier computing than integrating over a symmetric interval $[-p,p]$.

---

example 2

Let $f(x) = x$, for $0 < x < 1$. We find, integrating by parts for $k \neq 0$,

$$\begin{aligned} a_k &= 2\int_0^1 x \cos 2k\pi x \, dx \\ &= 2\left[x \frac{\sin 2k\pi x}{2k\pi} \right]_0^1 - \frac{2}{2k\pi} \int_0^1 \sin 2k\pi x \, dx = 0, \\ b_k &= 2\int_0^1 x \sin 2k\pi x \, dx \\ &= 2\left[-x \frac{\cos 2k\pi x}{2k\pi} \right]_0^1 + \frac{2}{2k\pi} \int_0^1 \cos 2k\pi x \, dx \\ &= -\frac{\cos 2k\pi}{k\pi} = -\frac{1}{k\pi}. \end{aligned}$$

Since $a_0 = 2\int_0^1 x \, dx = 1$, then $a_0/2 = 1/2$, and the Fourier series is

$$\frac{1}{2} - \frac{1}{\pi}\left(\sin 2\pi x + \frac{\sin 4\pi x}{2} + \frac{\sin 6\pi x}{3} + \cdots \right).$$

3B Sine and Cosine Expansions

It sometimes happens that an expansion in terms only of cosines or only of sines is more convenient to use than a general Fourier expansion. We begin with the observation that the general cosine in a Fourier expansion is an even function [i.e., $\cos(-k\pi x/p) = \cos(k\pi x/p)$], and that the sine terms are odd [i.e., $\sin(-k\pi x/p) = -\sin k\pi x/p)$]. It follows that if f is an even periodic function, the product

$$f(x)\sin\frac{k\pi x}{p}$$

is odd, since *both* factors change sign when x changes sign. Therefore, for the Fourier sine coefficient b_k, we have by Equation 3.1,

$$b_k = \frac{1}{p}\int_{-p}^{p} f(x)\sin\frac{k\pi x}{p}\,dx = 0.$$

It follows that *an even function has only cosine terms in its Fourier expansion, plus a possible constant.* Similarly if f is an odd periodic function, the product

$$f(x)\cos\frac{k\pi x}{p}$$

is also odd; so for the Fourier cosine coefficient we have

$$a_k = \frac{1}{p}\int_{-p}^{p} f(x)\cos\frac{k\pi x}{p}\,dx = 0.$$

Thus *an odd function has only sine terms in its Fourier expansion.*

The facts in the preceding paragraph are the key to solving the following problem: given a function $f(x)$ defined just on the interval $0 \le x \le p$, find a trigonometric series expansion for f consisting only of cosine terms or, alternatively, only of sine terms. The trick is to extend the definition of f from the interval $0 \le x \le p$ to all real x in such a way that the extension is periodic of period $2p$ and either is even or is odd. We then compute the Fourier series of the extension. If f_e is an even periodic extension of f, then f_e will have only cosine terms in its Fourier series but is designed to represent f just on $0 \le x \le p$. Similarly, if f_o is an odd periodic extension of f, then f_o has only sine terms in an expansion designed to represent f for $0 \le x \le p$. For a **sine expansion** of $f(x)$ on $0 \le x \le p$ we have

3.3 $$\sum_{k=1}^{\infty} b_k \sin\frac{k\pi x}{p}, \qquad b_k = \frac{2}{p}\int_0^p f(x)\sin\frac{k\pi x}{p}\,dx.$$

For a **cosine expansion** of $f(x)$ on $0 \le x \le p$, we have

3.4 $$\frac{1}{2}a_0 + \sum_{k=1}^{\infty} a_k \cos\frac{k\pi x}{p}, \qquad a_k = \frac{2}{p}\int_0^p f(x)\cos\frac{k\pi x}{p}\,dx.$$

We illustrate the procedure with two examples.

example 3

We'll compute the cosine expansion for the function defined by $f(x) = 1 - x$ for $0 \le x \le 2$. We consider the even periodic extension shown in Figure 10.8. To find the extension we define f_e by $f_e(x) = f(-x)$ for $-2 \le x < 0$, and then extend periodically, with period 4, to the whole x axis. We use Formula 3.4 to compute the Fourier-cosine expansion of f_e. (Since f_e is even, we know that $b_k = 0$

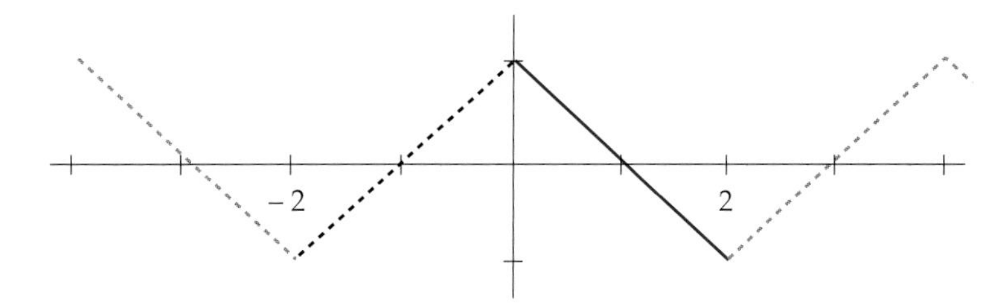

Figure 10.8 Even extension of $f(x) = 1 - x$ from $0 \le x \le 2$.

for all k.) The coefficient formula in 3.4 enables us to write

$$a_k = \frac{1}{2}\int_{-2}^{2} f_e(x)\cos\frac{k\pi x}{2}\,dx$$
$$= \int_0^2 f_e(x)\cos\frac{k\pi x}{2}\,dx.$$

Since on $0 \le x \le 2$, the function f_e is the same as the given function $f(x) = 1 - x$, we have, for $k > 0$,

$$\begin{aligned} a_k &= \int_0^2 (1-x)\cos\frac{k\pi x}{2}\,dx \\ &= \left[\frac{2}{k\pi}(1-x)\sin\frac{k\pi x}{2}\right]_0^2 + \frac{2}{k\pi}\int_0^2 \sin\frac{k\pi x}{2}\,dx \\ &= \frac{4}{\pi^2 k^2}[1-\cos k\pi] \\ &= \begin{cases} 0, & k \text{ even}, \\ \dfrac{8}{\pi^2 k^2}, & k \text{ odd}. \end{cases} \end{aligned}$$

Finally, $a_0 = \int_0^2 (1-x)\,dx = 0$. Thus the cosine expansion of f on $0 \le x \le 2$ has for its general nonzero term

$$\frac{8}{\pi^2 k^2}\cos\left(\frac{k\pi x}{2}\right), \qquad k \text{ odd}.$$

Written out, the expansion of the given function looks like

$$f(x) = \frac{8}{\pi^2}\left(\frac{\cos \pi x/2}{1} + \frac{\cos 3\pi x/2}{9} + \frac{\cos 5\pi x/2}{25} + \cdots\right), \qquad 0 \le x \le 2.$$

example 4

Starting with the same function as in Example 3, $f(x) = 1 - x$ for $0 \le x \le 2$, we compute a sine expansion by considering the odd periodic extension shown in Figure 10.9. We first define $f_o(x) = -f(-x)$ for $-2 \le x < 0$, and then extend periodically with period 4. (Since f_o is odd, we know that $a_k = 0$ for all k.) Also,

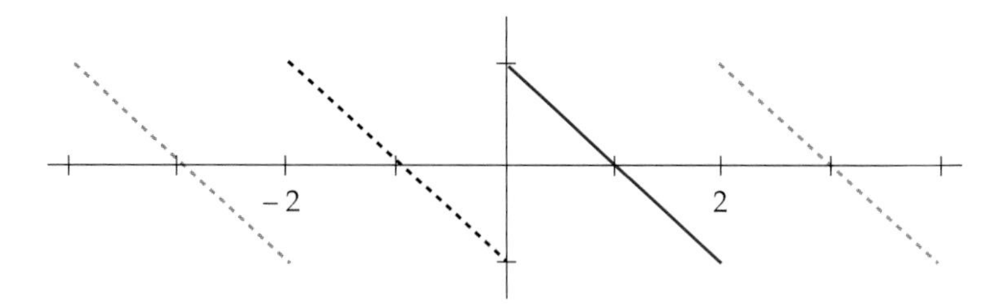

Figure 10.9 Odd extension of $f(x) = 1 - x$ from $0 \le x \le 2$.

by Equations 3.1 or 3.3,

$$b_k = \frac{1}{2}\int_{-2}^{2} f_o(x)\sin\frac{k\pi x}{2}\,dx$$
$$= \int_0^2 f_o(x)\sin\frac{k\pi x}{2}\,dx.$$

But $f_o(x) = 1 - x$ for $0 \le x \le 2$, so

$$b_k = \int_0^2 (1-x)\sin\frac{k\pi x}{2}\,dx$$
$$= \left[-\frac{2}{k\pi}(1-x)\cos\frac{k\pi x}{2}\right]_0^2 - \frac{2}{k\pi}\int_0^2 \cos\frac{k\pi x}{2}\,dx$$
$$= \frac{2}{\pi k}(-1)^k + \frac{2}{k\pi} - \frac{2}{k\pi}\left[\frac{2}{k\pi}\sin\frac{k\pi x}{2}\right]_0^2$$
$$= \begin{cases} 0, & k \text{ odd}, \\ \dfrac{4}{\pi k}, & k \text{ even}. \end{cases}$$

Thus the general nonzero term in the sine expansion is

$$\frac{4}{k\pi}\sin\left(\frac{k\pi x}{2}\right), \qquad \text{for even } k > 0.$$

A careful interpretation of this formula shows that the sine expansion of $f(x)$ is then

$$f(x) = \frac{2}{\pi}\left(\sin\pi x + \frac{\sin 2\pi x}{2} + \frac{\sin 3\pi x}{3} + \cdots\right), \qquad 0 < x < 2,$$

with convergence to zero at $x = 0$ and $x = 2$, since every term is zero there.

Finally, it's worth noting that Examples 1 and 2 of the previous section can be regarded as cosine and sine expansions respectively of the given functions restricted to $0 \le x \le \pi$.

The Java Applet FOURIER at website <http://math.dartmouth.edu/~rewn/> approximates adapted Fourier coefficients using Simpson's rule and then plots graphs of partial sums. An alternative is to use computer algebra software such as *Maple* to compute Fourier coefficients for elementary functions.

EXERCISES

1. Find the Fourier series for the function

$$f(x) = -x, \qquad -2 < x < 2.$$

To what values will the series converge at $x = 2$ and $x = -2$?

2. Find the Fourier series for the function

$$f(x) = 1 + x, \qquad 1 < x < 2.$$

To what values will the series converge at $x = 1$ and $x = 2$?

Find the sine expansion of each of the functions in Exercises 3 through 7. In each case, sketch the graph of the odd periodic extension of f after normalizing f to have the average value $\frac{1}{2}[f(x+) + f(x-)]$ at each jump discontinuity. Sketch also the graph of the sum of the first two nonzero terms of the sine expansion; this should exhibit some reasonable relation to your sketch of f.

3. $f(x) = 1, 0 < x < \pi.$

4. $f(x) = x^2, 0 < x < \pi.$

5. $f(x) = \cos x, 0 < x < \pi/2.$

6. $f(x) = \begin{cases} 1, & 0 < x < 1, \\ 0, & 1 \leq x < 2. \end{cases}$

7. $f(x) = \begin{cases} 0, & 0 < x < 2, \\ x - 2, & 2 \leq x < 3. \end{cases}$

Find the cosine expansion of each of the functions in Exercises 8 through 12. In each case, sketch the graph of the even periodic extension of f after normalizing f to have the average value at each jump discontinuity. Sketch also the graph of the sum of the first two nonzero terms of the cosine expansion; this should exhibit some reasonable relation to your sketch of f.

8. $f(x) = 1, 0 < x < \pi.$

9. $f(x) = x^2, 0 < x < \pi.$

10. $f(x) = \sin x, 0 < x < \pi/2.$

11. $f(x) = \begin{cases} 0, & 0 < x < 1, \\ 1, & 1 \leq x < 2. \end{cases}$

12. $f(x) = \begin{cases} 1, & 0 \leq x < 1, \\ 0, & 1 \leq x < 3. \end{cases}$

13. Find (a) the Fourier cosine expansion and (b) the Fourier sine expansion of the function

$$f(x) = x, \qquad 0 < x < \pi.$$

(c) Compare the results of (a) and (b) with the complete Fourier expansion of

$$g(x) = x, \qquad -\pi < x < \pi.$$

14. Prove the following:

(a) A product of even functions is even.

(b) A product of odd functions is even.

(c) The product of an even function and an odd function is odd.

(d) A linear combination of even functions is even, and of odd functions is odd.

15. Let f be a real-valued function defined on a symmetric interval $-a < x < a$.

(a) Show that f can be written as the sum of two functions, an even part P_e and an odd part P_o. [*Hint:* Let $P_e(x) = \frac{1}{2}[f(x) + f(-x)]$.]

(b) What are even and odd parts for a trigonometric polynomial?

(c) What are even and odd parts for $f(x) = e^x$?

16. Show that the even and odd functions P_e and P_o of the previous exercise are uniquely determined by f.

17. Let f be an odd function on $[-\pi, \pi]$ [i.e., $f(-x) = -f(x)$], and let g be an even function [i.e., $g(-x) = g(x)$]. Let a_k, b_k and a'_k, b'_k be the Fourier coefficients of f and g, respectively. Show that

$$a_k = 0, \qquad b_k = \frac{2}{\pi}\int_0^{\pi} f(x)\sin kx\,dx,$$

$$a'_k = \frac{2}{\pi}\int_0^{\pi} g(x)\cos kx\,dx, \qquad b'_k = 0.$$

18. Show that if $f(x + 2p) = f(x)$ for all real x, then the integral of f over every interval of length $2p$ is the same; specifically, show that

$$\int_t^{t+2p} f(x)\,dx = \int_0^{2p} f(x)\,dx.$$

(a) Do this by using $\int_t^{t+2p} f(x)\,dx = \int_t^0 f(x)\,dx + \int_0^{2p} f(x)\,dx + \int_{2p}^{t+2p} f(x)\,dx$.

(b) Do this for continuous f by defining $F(t) = \int_t^{t+2p} f(x)\,dx$ and showing that $F'(t) = 0$.

(c) Let $f(x)$ be a periodic function of period $2p$, and define $F(x) = \int_0^x f(u)\,du$. Show that $F(x)$ has period $2p$ if and only if $F(2p) = 0$.

19. **(a)** Show that if f is periodic and differentiable, then f' is periodic.

(b) Show by example that if f' is periodic, then f need not be periodic.

20. Show that if f is even and differentiable, then f' is odd. [*Hint:* Consider the limit of $[f(-x+h) - f(-x)]/h$ as $h \to 0$.]

21. Show that if f is odd and differentiable, then f' is even. See the hint for the previous exercise.

22. Show that the set of functions $\left\{\sqrt{2/p}\,\sin(k\pi x/p)\right\}_{k=1}^{\infty}$ is an **orthonormal set** on the interval $0 \le x \le p$. That is, show that

$$\frac{2}{p}\int_0^p \sin\frac{k\pi x}{p}\sin\frac{l\pi x}{p}\,dx = \begin{cases} 0, & k \neq l, \\ 1, & k = l \neq 0. \end{cases}$$

(a) Do this directly, using the identity $\sin\alpha\sin\beta = \frac{1}{2}\cos(\alpha-\beta) -\frac{1}{2}\cos(\alpha+\beta)$.

(b) Show how the result follows from Equations 2.4 of Section 2.

23. Show that the set of functions $\left\{\sqrt{2/p}\,\cos(k\pi x/p)\right\}_{k=1}^{\infty}$ is an *orthonormal set* on the interval $0 \le x \le p$, that is, show that

$$\frac{2}{p}\int_0^p \cos\frac{k\pi x}{p}\cos\frac{l\pi x}{p}\,dx = \begin{cases} 0, & k \neq l, \\ 1, & k = l \neq 0. \end{cases}$$

(a) Do this directly, using the identity $\cos\alpha\cos\beta = \frac{1}{2}\cos(\alpha-\beta) +\frac{1}{2}\cos(\alpha+\beta)$.

(b) Show how the result follows from Equations 2.4 of Section 2.

4 1-DIMENSIONAL HEAT AND WAVE EQUATIONS

In this section, we show how a Fourier series can be used to solve some problems in heat conduction and wave motion. Each of the pertinent partial differential equations is an example of an **evolution equation,** that is, an equation that determines how its solutions change as time increases. Each of these equations will have solutions described by a dynamical system whose states are themselves described by functions of a space variable.

4A 1-Dimensional Heat Equation

Suppose we are given a thin homogeneous wire of length p. Let $u(x,t)$ be the temperature, at time t, at a point x units from one end. Thus $0 \le x \le p$, and we'll let $t \ge 0$. We will assume that the only heat transfer is along the direction of the wire and that the temperature at the two ends is held fixed. For this reason we can, without loss of generality, represent the wire as a straight line segment along an x axis and picture the temperature at time t and position x as the graph of a function $u = u(x,t)$; an example is shown in Figure 10.10 with single temperature states traced for fixed time t as x varies. [Figure IS.9 in the Introductory Survey shows a sketch of the same temperature function $u(x,t)$ with emphasis on the variation of temperature at fixed points x while time t increases.]

The basic physical principle of heat conduction is that heat flow is proportional to, and in the direction opposite to, the temperature gradient ∇u. Recall that ∇u is the direction in which the temperature is increasing most rapidly, so it is reasonable that heat should flow in the opposite direction, from hotter to colder. Since the medium is 1-dimensional and is represented by a segment of the x axis, the gradient is represented by $\partial u/\partial x$. The rate of change of total heat in a segment $[x_1, x_2]$ with respect to time t is proportional to the flow rate in at the right end, $u_x(x_2, t)$, plus the flow rate in at the left end, $-u_x(x_1, t)$. Note,

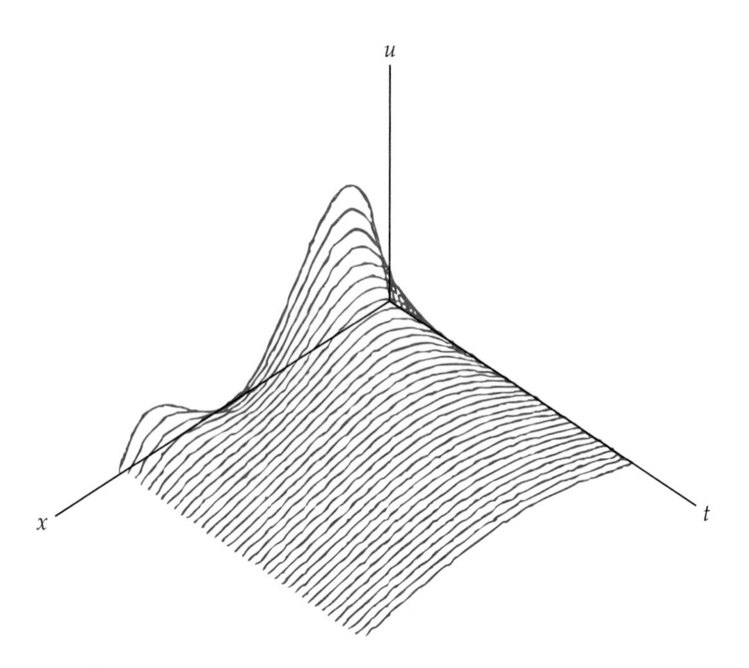

Figure 10.10 Temperature states $u(x,t)$ for fixed times t.

for example, that $u_x(x,t) > 0$ implies that heat flows to the left at x, and that $u_x(x,t) < 0$ implies that heat flows to the right at x. The rate of heat flow is

$$k\left[-\frac{\partial u}{\partial x}(x_1, t) + \frac{\partial u}{\partial x}(x_2, t)\right], \tag{1}$$

where the positive constant of proportionality k is the **conductivity** of the heat-conducting medium. This constant is arrived at empirically for each material. By the fundamental theorem of calculus, expression (1) can be written as

$$k\int_{x_1}^{x_2} \frac{\partial^2 u}{\partial x^2}(x,t)\,dx. \tag{2}$$

It is also true that the rate of change of heat in the segment is

$$\frac{d}{dt}\int_{x_1}^{x_2} \rho c u(x,t)\,dx = \rho c\int_{x_1}^{x_2} \frac{\partial u}{\partial t}(x,t)\,dx. \tag{3}$$

The factors ρ and c are, respectively, the linear **density** and **heat capacity** per unit of material. We can equate the expressions (2) and (3) for rate of change of total heat to give

$$a^2\int_{x_1}^{x_2} \frac{\partial^2 u}{\partial x^2}(x,t)\,dx = \int_{x_1}^{x_2} \frac{\partial u}{\partial t}(x,t)\,dx,$$

where $a^2 = k/(c\rho)$. Allowing x_2 to vary, we can differentiate both sides of this last equation with respect to the upper limit, getting

4.1

$$a^2\frac{\partial^2 u}{\partial x^2}(x,t) = \frac{\partial u}{\partial t}(x,t).$$

Equation 4.1 is the 1-dimensional **heat equation** or **diffusion equation;** it has the **qualitative interpretation** that if the graph of $u(x,t)$ is concave up as a function of x on some interval, so that $\partial^2 u/\partial x^2 > 0$ there, then $\partial u/\partial t > 0$ there also, so that temperature is increasing. Similarly, decreasing temperature goes with downward concavity of the graph of u as a function of x. Figure 10.10 supports these observations.

Equation 4.1 is linear in the usual sense that, if u_1 and u_2 are solutions, then so are linear combinations $c_1 u_1 + c_2 u_2$. To single out particular solutions, impose **boundary conditions** of the form

$$u(0,t) = 0 \quad \text{and} \quad u(p,t) = 0, \qquad t \geq 0. \tag{4}$$

These conditions hold temperature at zero for all time $t \geq 0$ at both ends of an x interval $[0, p]$. We also suppose we know the initial temperature at $t = 0$ throughout the x interval in terms of a function $h(x)$:

$$u(x,0) = h(x), \qquad 0 \leq x \leq p.$$

Separating Variables. In the standard method of solution we start by trying to find product solutions of the form

$$u(x,t) = X(x)T(t), \qquad X(0) = X(p) = 0,$$

the boundary conditions (4) hold for $t \geq 0$. If such product solutions exist, we'll try to find them by substitution into $a^2 u_{xx} = u_t$. We get

$$a^2 X''(x)T(t) = X(x)T'(t),$$

for $0 \leq x \leq p, 0 < t$. Assuming $X(x)T(t) \neq 0$, we divide by it to separate the variables:

$$a^2 \frac{X''(x)}{X(x)} = \frac{T'(t)}{T(t)}. \tag{5}$$

The left-hand side of this equation is independent of t so the right-hand side must be constant as a function of t as well as x. Similarly, the right-hand side is independent of x, so the left-hand side must be constant as a function of x as well as t. Thus both sides have the same constant value, which we'll denote by $-\lambda^2$; this notation for the constant is not essential, but is motivated by the convenient form of the resulting ordinary differential equations. (Note that $-\lambda^2$ can be positive if $\lambda = i\mu$.) Setting both sides of Equation (5) equal to $-\lambda^2$ gives the two equations:

$$a^2 X'' + \lambda^2 X = 0, \qquad T' + \lambda^2 T = 0. \tag{6}$$

The first of Equations (6), for X, is a harmonic oscillator equation, and has solutions

$$X(x) = c_1 \cos\left(\frac{\lambda x}{a}\right) + c_2 \sin\left(\frac{\lambda x}{a}\right).$$

But $u(0,t) = 0$ requires $X(0) = 0$, so $c_1 = 0$. Similarly, $u(p,t) = 0$ requires $X(p) = 0$ so $c_2 \sin(\lambda p/a) = 0$. Unless we trivialize the solution by making $c_2 = 0$ also, this last condition must be achieved by choosing λ so that $\lambda p/a = k\pi$, where

k is an integer. That is, we must take $\lambda = ka\pi/p$, with the result that $X(x)$ has the form

$$X(x) = c_k \sin\left(\frac{k\pi x}{p}\right), \qquad k = 1, 2, \ldots.$$

The second of Equations (6), for T, has solutions $T(t) = d_k e^{-\lambda^2 t}$, which, because now $\lambda^2 = (k^2a^2\pi^2)/p^2$, becomes

$$T(t) = d_k e^{-k^2(a^2\pi^2/p^2)t}.$$

The product solutions $u(x,t)$ are thus given, except for a constant factor $b_k = c_k d_k$, by

$$u_k(x,t) = e^{-k^2(a^2\pi^2/p^2)t} \sin\left(\frac{k\pi}{p}\right)x, \qquad k = 1, 2, \ldots.$$

Satisfying Initial Conditions. Equation 4.1 is linear, so linear combinations $\sum_{k=1}^{N} b_k u_k(x,t)$ are solutions also. Letting N tend to infinity, the formal result is

4.2
$$u(x,t) = \sum_{k=1}^{\infty} b_k e^{-k^2(a^2\pi^2/p^2)t} \sin\left(\frac{k\pi x}{p}\right).$$

Equation 4.2 represents dynamical systems in which, for fixed $t = t_0 \geq 0$, the states are functions $U(x) = u(x, t_0)$ to be determined by an initial state $h(x) = u(x,0)$.

To satisfy the initial condition $u(x,0) = h(x)$, we set $t = 0$ in Equation 4.2 and require

$$\sum_{k=1}^{\infty} b_k \sin\left(\frac{k\pi x}{p}\right) = h(x),$$

for suitable choices of the b_k. If $h(x)$ can be expressed in this form, we can expect that a solution to the problem is given by Equation 4.2. The boundary conditions, $u(0,t) = u(p,t) = 0$, require that the temperature remain zero at the ends, and the initial condition $u(x,0) = h(x)$ specifies the initial temperature at each point x of the wire between 0 and p. Thus we are naturally led to the problem of finding a Fourier sine series representation for $h(x)$, as in Equation 3.3 of Section 3. The coefficients in the expansion are

4.3
$$b_k = \frac{2}{p}\int_0^p h(x) \sin\frac{k\pi x}{p}\,dx, \quad k = 1, 2, 3, \ldots.$$

The matter of conditions under which the infinite series 4.2 actually represents a solution of Equation 4.1 is taken up briefly in Exercise 20.

example 1

Let's be more specific about solving the heat equation. For simplicity let $p = \pi$. Recall that to solve $a^2u_{xx} = u_t$ with boundary condition $u(0,t) = u(\pi,t) = 0$, and initial condition $u(x,0) = h(x)$, we want in general to be able to represent h

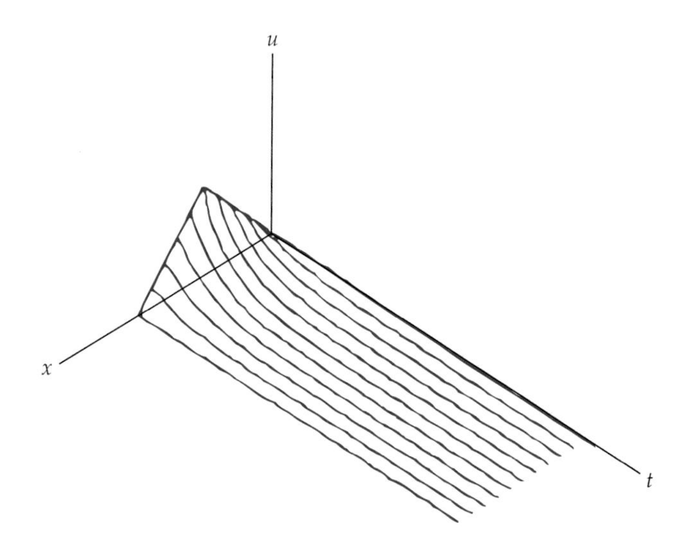

Figure 10.11 Temperature varying over time t for fixed x.

by an infinite series of the form

$$h(x) = \sum_{k=1}^{\infty} b_k \sin kx.$$

Suppose for example that h is given in $[0, \pi]$ by

$$h(x) = \begin{cases} x, & 0 \leq x \leq \pi/2, \\ \pi - x, & \pi/2 < x \leq \pi. \end{cases}$$

The graph of h is shown in Figure 10.11 in perspective above the x axis.

To find the Fourier sine expansion of h on $[0, \pi]$, we extend h to the interval $[-\pi, \pi]$ in such a way that the cosine terms in the expansion of h will all be zero, leaving only the sine terms to be computed. We do this by extending the graph of h asymmetrically about the vertical axis so as to represent an odd function. Then $a_k = 0$, because $h(x) \cos kx$ is an odd function on $[-\pi, \pi]$. Now use Equation 4.3:

$$\begin{aligned} b_k &= \frac{1}{\pi} \int_{-\pi}^{\pi} h(x) \sin kx \, dx \\ &= \frac{2}{\pi} \int_{0}^{\pi} h(x) \sin kx \, dx \\ &= \frac{2}{\pi} \int_{0}^{\pi/2} x \sin kx \, dx + \frac{2}{\pi} \int_{\pi/2}^{\pi} (\pi - x) \sin kx \, dx \\ &= \frac{4}{\pi k^2} \sin\left(\frac{k\pi}{2}\right). \end{aligned}$$

Hence

$$b_k = \begin{cases} 0, & k \text{ even}, \\ \dfrac{4}{\pi k^2}, & k = 1, 5, 9, \ldots, \\ \dfrac{-4}{\pi k^2}, & k = 3, 7, 11, \ldots. \end{cases}$$

Theorem 2.5 then implies that

$$h(x) = \frac{4}{\pi}\left(\frac{\sin x}{1^2} - \frac{\sin 3x}{3^2} + \frac{\sin 5x}{5^2} - \frac{\sin 7x}{7^2} + - \cdots\right)$$

for each x in $[0,\pi]$. Finally, from Equation 4.2 we expect the solution to the equation $a^2u_{xx} = u_t$ to be given by

$$u(x,t) = \frac{4}{\pi}\left(e^{-a^2t}\sin x - \frac{1}{3^2}e^{-3^2a^2t}\sin 3x + \frac{1}{5^2}e^{-5^2a^2t}\sin 5x - \frac{1}{7^2}e^{-7^2a^2t}\sin 7x + - \cdots\right).$$

Verification that $u(0,t) = u(\pi,t) = 0$ follows immediately from setting $x=0$ and $x=\pi$. By setting $t = 0$, we get the representation of h by its Fourier series, which is guaranteed by Theorem 2.5. That $u(x,t)$ satisfies the equation $a^2u_{xx} = u_t$ depends on term-by-term differentiation of the series for u. The graph of the solution is sketched in Figure 10.11.

4B Equilibrium Solutions

A solution $u(x,t) = v(x)$ of the heat equation that is independent of t is called an **equilibrium solution** because it doesn't vary with time. The heat equation for equilibrium solutions becomes simply $v'' = 0$, and all solutions are necessarily of the form $u(x,t) = v(x) = \alpha + \beta x$, where α and β are constant. Solutions of this type are useful for solving the time-dependent problem when we have **nonhomogeneous boundary conditions** of the form

$$u(0,t) = u_0, \qquad u(p,t) = u_1, \qquad t > 0,$$

where at least one of u_0 and u_1 is a nonzero constant. The idea is to choose α and β in the equilibrium solution $v(x) = \alpha + \beta x$ so that

$$v(0) = u_0, \qquad v(p) = u_1.$$

Then

$$u(x,t) = w(x,t) + v(x)$$

will satisfy $u(0,t) = u_0$, $u(p,t) = u_1$ if $w(x,t)$ is a solution of the heat equation satisfying **homogeneous conditions**

$$w(0,t) = w(p,t) = 0, \qquad t > 0.$$

Note that the function $w + v$ is indeed a solution of the heat equation, because both w and v are solutions and because the heat operator $a^2D_{xx} - D_t$ acts linearly.

To solve the problems of the form

$$a^2u_{xx} = u_t, \qquad u(0,t) = u_0, \qquad u(p,t) = u_1, \qquad u(x,0) = h(x),$$

we first find an equilibrium solution $v(x) = \alpha + \beta x$. We need $v(0) = u_0 = \alpha$ and $v(p) = u_1 = \alpha + \beta p$. Thus $\alpha = u_0$, and $\beta = (u_1 - u_0)/p$. Then solve the heat equation $a^2w_{xx} = w_t$ with boundary conditions $w(0,t) = w(p,t) = 0$ and initial condition $w(x,0) = h(x) - v(x)$. To understand this you can check immediately that the solution to the original problem has the form $u(x,t) = v(x) + w(x,t)$, or

4.4
$$u(x,t) = v(x) + \sum_{k=1}^{\infty} b_k e^{-k^2(\pi^2 a^2/p^2)t} \sin\frac{k\pi x}{p}, \quad \text{with}$$
$$b_k = \frac{2}{p}\int_0^p \left[h(x) - v(x)\right] \sin\frac{k\pi x}{p}\, dx.$$

example 2

Consider the problem

$$a^2 u_{xx} = u_t, \quad u(0,t) = 10, \quad u(5,t) = 30, \quad u(x,0) = h(x) = 10 - 2x, \quad 0 < x < 5.$$

We find the equilibrium solution $v(x) = 10 + 4x$. The desired solutions

$$u(x,t) = (10 + 4x) + w(x,t)$$

come from finding solutions $w(x,t)$ that satisfy homogeneous boundary conditions

$$w(0,t) = w(5,t) = 0, \qquad t \geq 0,$$

and an initial condition

$$u(x,0) = w(x,0) + v(x) = h(x);$$

this last condition is just

$$w(x,0) = h(x) - v(x) = -6x.$$

Thus our solution $u(x,t)$ has the form of Equation 4.4, where the b_k are Fourier sine coefficients

$$b_k = \frac{2}{5}\int_0^5 -6x \sin\frac{k\pi x}{5}\, dx = \frac{60(-1)^k}{k\pi}.$$

The solution to the problem is thus

$$u(x,t) = (10 + 4x) - \frac{60}{\pi}\sum_{k=1}^{\infty} \frac{(-1)^{k+1}}{k} e^{-k^2 a^2 \pi^2 t/25} \sin\frac{k\pi x}{5}.$$

EXERCISES

Solve the heat equation $a^2 u_{xx} = u_t$, given boundary and initial conditions in Exercises 1 through 6.

1. $u(0,t) = u(p,t) = 0$; $u(x,0) = \sin(\pi x/p)$, $0 < x < p$.
2. $u(0,t) = u(1,t) = 0$; $u(x,0) = x$, $0 < x < 1$.
3. $u(0,t) = u(1,t) = 0$; $u(x,0) = 1 - x$, $0 < x < 1$.
4. $u(0,t) = u(\pi,t) = 0$; $u(x,0) = x(\pi - x)$, $0 < x < \pi$.
5. $u(0,t) = u(\pi,t) = 0$; $u(x,0) = \sin x + \frac{1}{2}\sin 2x$, $0 < x < \pi$.
6. $u(0,t) = u(2,t) = 0$; $u(x,0) = \begin{cases} 1, & 0 < x < 1, \\ 0, & 1 < x < 2. \end{cases}$

Find equilibrium solutions $u(x, t) = v(x)$ of the heat equation $a^2 u_{xx} = u_t$ that satisfy each of the conditions in Exercises 7 through 10.

7. $u(0,t) = -1$, $u(2,t) = 1$.

8. $u(0,t) = 0$, $u(100,t) = 100$.

9. $u_x(0,t) = 1$, $u(1,t) = 2$.

10. $u_x(0,t) = -1$, $u(1,t) = 3$.

Solve the heat equation $a^2 u_{xx} = u_t$ given and initial conditions in Exercises 11 through 14.

11. $u(0,t) = 1$, $u(p,t) = 3$; $u(x,0) = \sin(\pi x/p)$, $0 < x < p$.

12. $u(0,t) = -1$, $u(2,t) = 1$; $u(x,0) = x$, $0 < x < 2$.

13. $u(0,t) = 0$, $u(1,t) = 1$; $u(x,0) = 1 - x$, $0 < x < 1$.

14. $u(0,t) = -1$, $u(1,t) = 2$; $u(x,0) = 3x - 1$, $0 < x < 1$.

15. Verify that the partial differential operator $L = a^2 D_{xx} - D_t$, defined by $L(u) = a^2 u_{xx} - u_t$ is linear in its action on twice differentiable functions.

Find all solutions of the form $u(x,t) = X(x)T(t)$ for each of the equations in Exercises 16 through 19:

16. $u_{xx} + u_x = u_t$.

17. $u_{xx} - u_x = u_t$.

18. $xu_x = 2u_t$.

19. $u_{xx} = u_{tt}$.

20. **(a)** Verify that the series expansion for $u(x,t)$ given by Equation 4.2 of the text is formally a solution of $a^2 u_{xx} = u_t$. Use term-by-term differentiation of the series.

*__(b)__ Show that the formal verification in part (a) is valid for $t > 0$, by showing that the differentiated series all converge uniformly for $t > \delta > 0$.

Note. If f is smooth enough it can be shown that $\lim_{t \to 0+} u(x,t) = f(x)$.

Find the equilibrium solution to each of the problems in Exercises 21 through 24.

21. $u_{xx} = u_t + 2$; $u(0,t) = 1$, $u(1,t) = 2$.

22. $u_{xx} = u_t + u$; $u(0,t) = 0$, $u(2,t) = 0$.

23. $u_{xx} = u_t + x$; $u(0,t) = 1$, $u(1,t) = 2$.

24. $u_{xx} = u_t - 2x + 1 - x$; $u(0,t) = 1$, $u(1,t) = 0$.

Insulated Endpoints. The 1-dimensional heat flow problem

$$u_{xx} = u_t, \qquad u_x(0,t) = u_x(p,t) = 0, \qquad u(x,0) = f(x)$$

for $0 \le x \le p$, can be interpreted as requiring the conducting medium to have insulated ends at $x = 0$ and $x = p$. The reason is that the temperature gradient u_x is always 0 at the endpoints, so there is no heat flow past those points. The solution involves Fourier cosine series. The next four exercises are about problems of this kind.

25. **(a)** Show that product solutions $u(x,t) = T(t)X(x)$ of the insulated endpoint problem have the form

$$u_k(x,t) = a_k e^{-k^2\pi^2 t/p^2} \cos \frac{k\pi}{p} x.$$

(b) Use the Fourier cosine expansion for $f(x)$ on $0 \le x \le p$ to solve the boundary value problem for a general initial temperature $f(x)$.

(c) What is the equilibrium temperature function $u(x,\infty)$ for $0 \le x \le p$?

26. Solve the heat equation $a^2 u_{xx} = u_t$ with insulated end conditions $u_x(0,t) = u_x(1,t) = 0$ and initial condition $u(x,0) = x$ for $0 < x < 1$.

27. Solve the heat equation $a^2 u_{xx} = u_t$ with insulated end conditions $u_x(0,t) = u_x(1,t) = 0$ and initial condition $u(x,0) = 1$ for $0 < x < 1$.

28. Solve the heat equation $a^2 u_{xx} = u_t$ with insulated end conditions $u_x(0,t) = u_x(\pi,t) = 0$ and initial condition $u(x,0) = \cos x$ for $0 < x < \pi$.

Suppose that $u(x,t)$ satisfies $u_{xx} = u_t$ as well as $u(0,t) = \alpha$, $u(p,t) = \beta$ for $t \ge 0$. Without doing any detailed computations, make a qualitative sketch of the graph of u relative to axes oriented as in Figures 10.10 and 10.11. Do this with each of the choices in Exercises 29 through 35 for $p = b - a$, for α and β, and for initial function $u(x,0)$.

29. $u(x,0) = x - x^2$, $0 \le x \le 1$, $\alpha = \beta = 0$.

30. $u(x,0) = x^2 - x^3$, $0 \le x \le 1$, $\alpha = \beta = 0$.

31. $u(x,0) = -x^2$, $0 \le x \le 1$, $\alpha = 0$, $\beta = -1$.

32. $u(x,0) = |x - 1|$, $0 \le x \le 2$, $\alpha = 0$, $\beta = 0$.

33. $u(x,0) = -x$, $0 \le x \le 1$, $\alpha = 0$, $\beta = -1$.

34. $u(x,0) = 1 - x$, $0 \le x \le 2$, $\alpha = 1$, $\beta = -1$.

35. The partial differential equation $tu_{xx} = u_t$ has product solutions of the form $u(x,t) = X(x)T(t)$.

(a) Find two ordinary differential equations satisfied by $X(x)$ and $T(t)$, respectively. [Note that the equation for X should not contain t and the equation for T should not contain x.]

(b) Solve the ordinary differential equations found in part (a), and use the results to specify the general form of the solutions $X(x)T(t)$.

36. The partial differential equation $txu_x = u_t$ has product solutions of the form $u(x,t) = X(x)T(t)$ for $x > 0$ and $t > 0$. Find the general form of all such solutions.

37. Suppose that $u(x,t)$ satisfies $a^2 u_{xx} = u_t$ and that $u(x, t_0)$ is concave up as a function of x for $x_0 < x < x_1$. Show that for each x with $x_0 < x < x_1$ there is a time $t(x)$ such that $u(x,t)$ increases as t increases from t_0 to $t(x)$. What if $u(x,t_0)$ is concave down?

38. The assumption that the separation constant C for the problem $a^2u_{xx} = u_{tt}$, $u(0,t) = u(p,t) = 0$ has the special form $-\lambda^2$ is convenient but not essential. Derive the product solutions $u_k(x,t) = e^{-k^2(a^2\pi^2/p^2)t}\sin(k\pi/p)x$ using C instead of $-\lambda^2$.

4C Wave Equation

Consider for physical motivation a stretched elastic string of length p and uniform density ρ placed along the x axis in 3-dimensional space. Suppose that the ends of the string are held fixed at $(x,y,z) = (0,0,0)$ and $(x,y,z) = (p,0,0)$ and that the absolute magnitude of the tension force τ is constant over the entire length of the string. If the string is somehow made to vibrate, our problem is to determine the position vector $\mathbf{x}(s,t)$ at time t of a point on the string a distance s from the end fixed at $x = 0$. Figure 10.12 shows a possible configuration. We imagine the string subdivided into short pieces of length Δs and then derive two different expressions for the total force acting on a typical segment of the subdivision. If $\mathbf{t}(s)$ is the unit tangent vector to the string at $\mathbf{x}(s)$, then the opposing forces at $\mathbf{x}(s_0)$ and $\mathbf{x}(s_0 + \Delta s)$ are

$$\tau\mathbf{t}(s_0 + \Delta s) \quad \text{and} \quad -\tau\mathbf{t}(s_0).$$

Hence the total force is

$$\tau[\mathbf{t}(s_0 + \Delta s) - \mathbf{t}(s_0)].$$

The mass of the short section is $\rho\,\Delta s$, density times length. Also, by Newton's law, the force equals mass, $\rho\,\Delta s$, times acceleration $\mathbf{a}(s_0)$. It follows on dividing by Δs that

$$\rho\mathbf{a} = \tau\left[\frac{\mathbf{t}(s_0 + \Delta s) - \mathbf{t}(s_0)}{\Delta s}\right].$$

The acceleration and tension vectors are $\mathbf{a} = (\partial^2\mathbf{x}/\partial t^2)(s,t)$ and $\tau\mathbf{t}(s) = (\partial\mathbf{x}/\partial s)(s,t)$. Making these substitutions in the previous equation for $\rho\mathbf{a}$ and letting $\Delta s \to 0$, we get

$$\rho\frac{\partial^2\mathbf{x}}{\partial t^2} = \tau\lim_{\Delta s\to 0}\frac{1}{\Delta s}\left[\frac{\partial\mathbf{x}(s_0 + \Delta s)}{\partial s} - \frac{\partial\mathbf{x}(s_0)}{\partial s}\right] = \tau\frac{\partial^2\mathbf{x}}{\partial s^2}.$$

This vector differential equation is equivalent to a system of three scalar

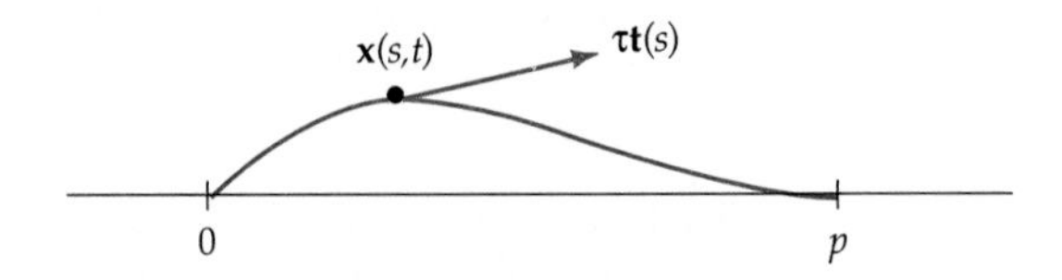

Figure 10.12 Stretched string and tension vector.

equations in which $\mathbf{x} = (x,y,z)$ and $a^2 = \tau/\rho$:

$$\frac{\partial^2 x}{\partial t^2} = a^2 \frac{\partial^2 x}{\partial s^2}, \qquad \frac{\partial^2 y}{\partial t^2} = a^2 \frac{\partial^2 y}{\partial s^2}, \qquad \frac{\partial^2 z}{\partial t^2} = a^2 \frac{\partial^2 z}{\partial s^2}.$$

Technically, the problem of finding solution functions $x(s,t)$, $y(s,t)$, and $z(s,t)$ is the same for all three equations. In practice, however, the equation for $x(s,t)$ is usually set aside when the longitudinal motion (along the x axis) is slight, and we make this assumption. Between the other two equations there is little difference in physical significance unless some other special assumption is made. From now on we'll use the neutral symbols $u(s,t)$ to stand for $y(s,t)$ or $z(s,t)$ as the case may be. We might in particular suppose that the string has been plucked in such a way that its spatial displacement takes place entirely in the u direction. Finally, we assume that the displacements are small enough that we can replace $\partial^2 y/\partial s^2$ by $\partial^2 y/\partial x^2$. This last assumption is one that requires experimental justification in actual practice. [Indeed Exercise 12 shows that the equation for $u(s,t)$ is nonlinear.] Thus, with some loss of generality, we consider, instead of the vector equation, a single linear scalar equation for one dimensional displacement $u(x,t)$:

4.5
$$\frac{\partial^2 u}{\partial t^2} = a^2 \frac{\partial^2 u}{\partial x^2}.$$

Equation 4.5 is the 1-dimensional *wave equation*; it has the **qualitative interpretation** that if $u(x,t)$ is concave down as a function of x ($u_{xx} < 0$), then the acceleration $u_{tt}(x,t)$ is negative, or alternatively that then the velocity $u_t(x,t)$ is decreasing. Similarly, upward concavity as a function of x implies increasing velocity. A moment's reflection shows that this interpretation is physically reasonable, since a curved string under tension will tend to straighten itself.

Equation 4.5 doesn't completely specify the vibration of a string unless we impose some initial conditions:

$$u(x,0) = f(x), \tag{1}$$

$$\frac{\partial u}{\partial t}(x,0) = g(x). \tag{2}$$

The first condition specifies the initial ($t = 0$) displacement in the u direction as a function of x, rather than s, and the second equation specifies the initial velocity in the u direction, also as a function of x. For string ends held fixed we use boundary conditions

$$u(0,t) = 0 \quad \text{and} \quad u(p,t) = 0. \tag{3}$$

The graph of a fairly complicated solution is sketched in Figure 10.13, where the individual curves on the graph represent string position states at equally spaced times. In this picture, the string has not only been released from an initial shape (i.e., "plucked"), but has simultaneously been given a downward initial velocity (i.e., "hammered") near its center. Plucking and hammering are the respective ways of starting a sound on a harpsichord and on a piano.

Recall that for Equation 4.5 to be physically realistic the amplitudes of vibrations of the string should be fairly small. In Figure 10.13, the graph of the solution is shown with $a = 1$, $p = \pi$, and the height of the initial plucked position chosen to be quite large to bring out some qualitative features of the picture.

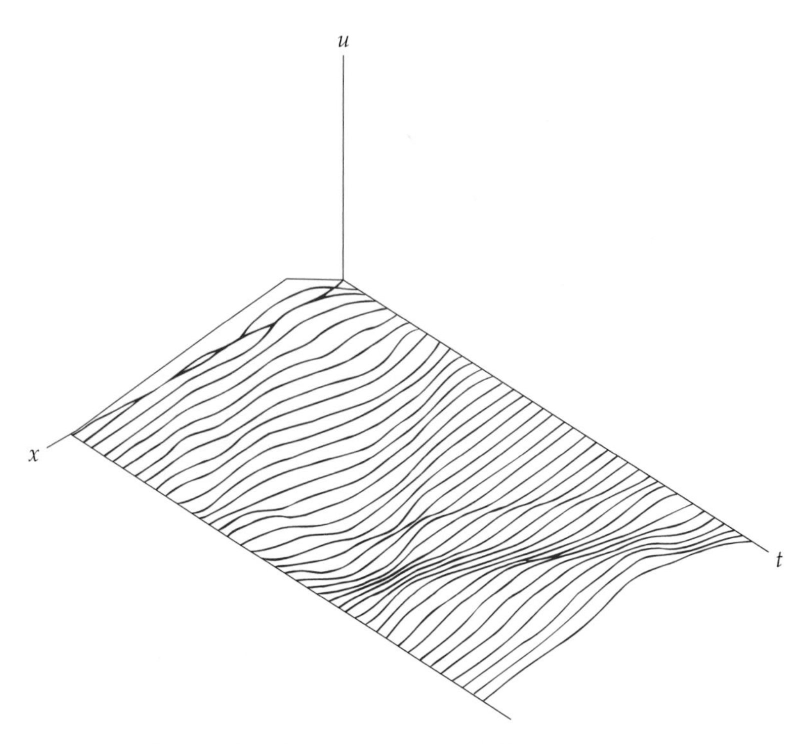

Figure 10.13 String shapes at equally spaced times.

As in the case of the heat equation, we use separation of variables and rely on the linearity of Equation 4.5 and the boundary conditions (3) for constructing solutions that satisfy the conditions (1) and (2) also.

Separating Variables. We try $u(x,t) = X(x)T(t)$, upon which Equation 4.5 becomes

$$X(x)T''(t) = a^2 X''(x)T(t) \quad \text{or} \quad \frac{T''(t)}{T(t)} = a^2\frac{X''(x)}{X(x)}.$$

Since X''/X and T''/T are independent of t and x, respectively, both sides of the last equation are constant. If λ is such a constant value, then

$$\frac{X''(x)}{X(x)} = \lambda \quad \text{and} \quad \frac{T''(t)}{T(t)} = a^2\lambda.$$

The first equation, $X''(x) = \lambda X(x)$, has solutions

$$X(x) = c_1 e^{\sqrt{\lambda}x} + c_2 e^{-\sqrt{\lambda}x}.$$

The boundary conditions (3) require for all $t \geq 0$ that

$$u(0,t) = X(0)T(t) = 0, \qquad u(p,t) = X(p)T(t) = 0,$$

so we set $X(0) = X(p) = 0$. Imposing these requirements on our expression for $X(x)$ gives

$$c_1 + c_2 = 0 \quad \text{and} \quad c_1 e^{\sqrt{\lambda}p} + c_2 e^{-\sqrt{\lambda}p} = 0.$$

Now solve for c_1 and c_2. We see that for solutions to exist we must have first that $c_1 = -c_2$, and hence that $e^{-\sqrt{\lambda}\,p} - e^{\sqrt{\lambda}\,p} = 0$. Thus we need $e^{2\sqrt{\lambda}\,p} = 1$. Allowing for complex exponents, we observe first that

$$e^{\alpha+i\beta} = e^{\alpha}(\cos\beta + i\sin\beta) = 1,$$

precisely when $\alpha = 0$ and $\beta = 2k\pi$ for some integer k. To make $e^{2\sqrt{\lambda}\,p} = 1$, we need $2\sqrt{\lambda}\,p = 2\pi k i$ for some integer k. We see that λ can't be positive, more specifically that

$$\sqrt{\lambda} = \frac{k\pi i}{p} \quad \text{and} \quad \lambda = \frac{-k^2\pi^2}{p^2}.$$

Solutions $X(x)$ are then of the form

$$\begin{aligned} X(x) &= c_1 e^{(\pi k i/p)x} - c_1 e^{(-\pi k i/p)x} \\ &= 2c_1 i \sin\frac{\pi k}{p}x. \end{aligned}$$

We can let $b_k = 2c_1 i$ and write $X_k(x) = b_k \sin(\pi k/p)x$.

The differential equation for T, namely, $T''(t) = a^2\lambda T(t)$, now takes the form

$$T''(t) + \frac{\pi^2k^2a^2}{p^2}T(t) = 0,$$

since we have determined that $\lambda = -\pi^2k^2/p^2$. Solutions of this equation are of the form

$$T_k(t) = c_2\cos\frac{\pi k a}{p}t + c_3\sin\frac{\pi k a}{p}t.$$

Writing $A_k = b_k c_2$ and $B_k = b_k c_3$, product solutions $u_k(x,t) = X_k(x)T_k(t)$ take the form

$$u_k(x,t) = \left[A_k\cos\frac{\pi k a}{p}t + B_k\sin\frac{\pi k a}{p}t\right]\sin\frac{\pi k}{p}x.$$

Satisfying Initial Conditions. We form finite or infinite sums of the type

4.6
$$u(x,t) = \sum_{k=1}^{\infty}\left[A_k\cos\frac{\pi k a}{p}t + B_k\sin\frac{\pi k a}{p}t\right]\sin\frac{\pi k}{p}x.$$

The initial conditions $u(x,0) = f(x)$ and $u_t(x,0) = g(x)$ become formally

4.6(a)
$$u(x,0) = \sum_{k=1}^{\infty} A_k\sin\frac{\pi k}{p}x = f(x),$$

and upon differentiating Equation 4.6 with respect to t,

4.6(b)
$$u_t(x,0) = \sum_{k=1}^{\infty}\frac{\pi k a}{p}B_k\sin\frac{\pi k}{p}x = g(x).$$

The coefficients A_k and $(\pi k a/p)B_k$ are then determined so that they are the Fourier sine coefficients of f and g, respectively. Thus Equation 4.6 represents a dynamical system in which, for fixed t, a state consists of a pair of functions $\big(U(x), V(x)\big) = \big(u(x,t), u_t(x,t)\big)$, where $U(x)$ describes the string's shape and $V(x)$ is its velocity at x. In particular, $\big(f(x), g(x)\big) = \big(u(x,0), u_t(x,0)\big)$ is an initial state for some trajectory in the state space.

example 3

For a simple example, we consider a string that is released with zero initial velocity, so that $u_t(x,0) = 0$. We assume the initial displacement is given by

$$u(x,0) = b \sin \frac{\pi x}{p},$$

and $u(0,t) = u(p,t) = 0$. It happens that f and g are so simple in this case that we can find their Fourier sine coefficients by inspection. We choose $B_k = 0$ in Equation 4.8 to make $u_t(x,t) = 0$. In Equation 4.6, set $A_1 = b$ and $A_k = 0$ for $k \neq 1$. The solution to Equation 4.5 then takes the form

$$u(x,t) = b \cos \frac{\pi a}{p} t \sin \frac{\pi}{p} x.$$

Again recall that for Equation 4.5 to be physically realistic the vibrations of the string should be fairly small; in other words, the coefficient b should be small.

In general, the computation of the coefficients A_k and B_k in Equation 4.6 is effected by applying the Fourier sine formulas to Equations 4.6(a) and 4.6(b):

4.7 $$A_k = \frac{2}{p} \int_0^p f(x) \sin \frac{k\pi x}{p}\, dx,$$

4.8 $$B_k = \frac{2}{k\pi a} \int_0^p g(x) \sin \frac{k\pi x}{p}\, dx.$$

example 4

The solution of $a^2 u_{xx} = u_{tt}$ that satisfies

$$u(0,t) = u(1,t) = 0, \qquad u(x,0) = x(1-x), \qquad u_t(x,0) = \sin \pi x$$

is found as follows. We compute from Equations 4.7 and 4.8, with $p = 1$,

$$A_k = 2 \int_0^1 x(1-x) \sin k\pi x\, dx = \frac{4}{k^3\pi^3}[1 - (-1)^k].$$

$$B_k = \frac{2}{k\pi a} \int_0^1 \sin \pi x \sin k\pi x\, dx = \begin{cases} \dfrac{1}{\pi a}, & k = 1 \\ 0, & k = 2, 3, \ldots. \end{cases}$$

The term corresponding to $k = 1$ in the general expansion is just

$$\left(\frac{8}{\pi^3}\cos \pi a t + \frac{1}{\pi a}\sin \pi a t\right)\sin \pi x.$$

All the other nonzero terms have odd index, so the solution formula is

$$u(x,t) = \left(\frac{8}{\pi^3}\cos \pi a t + \frac{1}{\pi a}\sin \pi a t\right)\sin \pi x$$
$$+ \sum_{j=1}^{\infty} \frac{8}{(2j+1)^3\pi^3}\cos(2j+1)\pi a t \sin(2j+1)\pi x.$$

Analysis of Solutions. The method of Fourier series and separation of variables is important for the wave equation not just because it provides solution formulas. (In fact, the d'Alembert method discussed in Exercises 6, 7, and 8 does this also, and does it in a way that is in some ways has more direct physical appeal.) What the Fourier method enables us to see is an analysis of vibratory motion into simpler vibrations. It is this analysis and its flexibility when applied to other kinds of wave propagation that makes the Fourier method so important.

example 5

Suppose $u(x,t)$ is the displacement at time t and position x of a taut string with ends fixed: $u(0,t) = u(p,t) = 0$. To simplify the discussion, suppose the initial velocity is zero: $u_t(x,0) = 0$. Then the Fourier solution has the form

$$u(x,t) = \sum_{k=1}^{\infty} A_k \cos \frac{k\pi a}{p} t \sin \frac{k\pi}{p} x,$$

where the amplitudes A_k of the individual terms are determined as Fourier sine coefficients in Equation 4.7 of the initial displacement $u(x,0) = f(x)$. A typical term

$$u_k(x,t) = A_k \cos \frac{k\pi a}{p} t \, \sin \frac{k\pi}{p} x$$

is periodic in the time variable t with period T determined by $k\pi a T/p = 2\pi$, so solving for T gives $T = 2p/ka$. The frequency, or number of vibrations per time unit, is thus $\nu = ka/2p$. The **fundamental frequency** is the frequency that corresponds to the first nonzero A_k. If $A_1 \neq 0$, the fundamental vibration has frequency $a/2p$ and all others have frequencies that are integer multiples of the fundamental frequency. If we think of each term $u_k(x,t)$ as generating an audible tone, it is usual to use the term **fundamental tone** and call the other of higher frequency tones **overtones** or **harmonics.**

The relative sizes of the amplitudes A_k give a complex vibration its recognizable quality. Some displacements of a fundamental tone and its overtones appear in Figure 10.14 with uniform amplitude 1. A point that remains stationary in one of these oscillations is called a **nodal point.** For example, $u_2(x,t)$ has a nodal point at $x = p/2$. Otherwise, for fixed x each term exhibits vertical motion as time t varies and is called a **standing wave** because it doesn't move longitudinally.

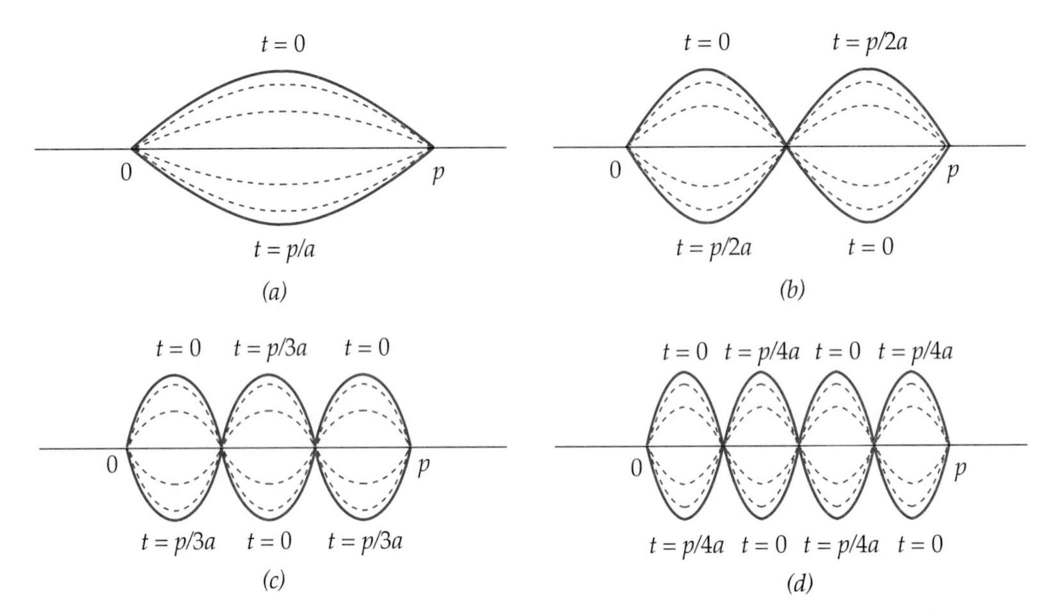

Figure 10.14 (*a*) $u_1(x,t) = \cos(\pi a t/p)\sin(\pi x/p)$. (*b*) $u_2(x,t) = \cos(2\pi a t/p)\sin(2\pi x/p)$. (*c*) $u_3(x,t) = \cos(3\pi a t/p)\sin(3\pi x/p)$. (*d*) $u_4(x,t) = \cos(4\pi a t/p)\sin(4\pi x/p)$.

Exercise 7 uses the d'Alembert solution to motivate the use of the term "wave" in connection with such vibrations.

EXERCISES

Solve the wave equation $a^2 u_{xx} = u_{tt}$ given the following boundary and initial conditions in Exercises 1 through 4.

1. $u(0,t) = u(\pi,t) = 0$, $u(x,0) = \sin x$, $u_t(x,0) = 0$.
2. $u(0,t) = u(\pi,t) = 0$, $u(x,0) = \begin{cases} x, & 0 < x \leq \pi/2, \\ \pi - x, & \pi/2 < x < \pi, \end{cases}$ $u_t(x,0) = 0$.
3. $u(0,t) = u(\pi,t) = 0$, $u(x,0) = 0$, $u_t(x,0) = \begin{cases} 0, & 0 < x \leq \pi/2, \\ 1, & \pi/2 < x < \pi. \end{cases}$
4. $u(0,t) = u(1,t) = 0$, $u(x,0) = x(1-x)$, $u_t(x,0) = \sin \pi x$.
5. The **nonhomogeneous wave equation**

$$a^2 u_{xx} = u_{tt} + g$$

incorporates g, the acceleration of gravity, into the vibrating string problem.

(a) Find the equilibrium solutions $u(x,t) = v(x)$ of the nonhomogeneous equation.

(b) Among the solutions found in part (a), select the solution that satisfies the boundary conditions $u(0,t) = u(p,t) = 0$.

(c) Explain how the Fourier solution method for $a^2 u_{xx} = u_{tt}$ can be modified to cover the solution of the nonhomogeneous problem.

6. The d'Alembert solution to the wave equation Let $U(x)$ and $V(x)$ be twice differentiable functions defined for all real x.

(a) Show that $u(x,t) = U(x + at) + V(x - at)$ defines a solution of $a^2 u_{xx} = u_{tt}$ that is valid for all real x and all real t.

(b) Assume that $f''(x)$ exists. Show that

$$u(x,t) = \tfrac{1}{2}[f(x + at) + f(x - at)]$$

is a solution to the wave equation having the form described in part (a) and also satisfies the initial conditions $u(x,0) = f(x)$, $u_t(x,0) = 0$.

(c) Assume that $g'(x)$ exists. Show that

$$u(x,t) = \frac{1}{2a} \int_{x-at}^{x+at} g(s)\, ds$$

is a solution to the wave equation having the form described in part (a) and also satisfies the initial $u(x,0) = 0$, $u_t(x,0) = g(x)$.

(d) Combine the results of parts (b) and (c) to find a solution formula for the wave equation subject to the initial conditions $u(x,0) = f(x)$, $u_t(x,0) = g(x)$.

7. (a) Show that the d'Alembert solution $U(x + at) + V(x - at)$ described in the previous exercise represents the sum of two wave motions, the first moving left with speed $a > 0$, the other moving right with the same speed.

(b) Show for the general term of Equation 4.6 that

$$\left[A \cos\left(\frac{k\pi a}{p}\right) t + B \sin\left(\frac{k\pi a}{p}\right) t\right] \sin\left(\frac{\pi k}{p}\right) x$$
$$= \sqrt{A^2 + B^2} \cos\left(\frac{k\pi a t}{p} - \theta\right) \sin \frac{k\pi x}{p},$$

where θ depends on A and B. This term is called a *standing wave*.

(c) Show that the term in part (b) can be written in the d'Alembert form

$$\frac{1}{2}\sqrt{A^2 + B^2} \left[\sin\left(\frac{k\pi}{p}(x + at) - \theta\right) + \sin\left(\frac{k\pi}{p}(x - at) + \theta\right)\right],$$

where θ depends on A and B. Use an identity from Exercise 28 of Section 2 to prove this.

8. This exercise imposes boundary conditions and initial *velocity* zero on d'Alembert's solution described in Exercise 6.

(a) Let $f(x)$ be twice differentiable for $0 \le x \le p$. Extend $f(x)$ to the interval $-p < x < 0$ by defining

$$f(x) = -f(-x), \qquad \text{if } -p < x < 0.$$

Then extend $f(x)$ to $-\infty < x < \infty$ to have period $2p$. Show that $f(x)$ so extended is not only periodic, but odd, that is, $f(-x) = -f(x)$ for $-\infty < x < \infty$.

(b) Sketch the odd periodic extension of $f(x) = x(1-x)$, as described in part (a), from $0 \le x \le 1$ to $-\infty < x < \infty$.

(c) Show that if $f(x)$ is odd and has period $2p$ for $-\infty < x < \infty$ then the function

$$u(x,t) = \tfrac{1}{2}[f(x+at) + f(x-at)]$$

is a d'Alembert solution to $a^2 u_{xx} = u_{tt}$ that satisfies the boundary conditions $u(0,t) = u(p,t) = 0$ and initial conditions $u(x,0) = f(x)$ and $u_t(x,0) = 0$.

9. This exercise imposes boundary conditions and initial *displacement* zero on d'Alembert's solution described in Exercise 6.

(a) Let $G(x)$ be twice differentiable for $0 \le x \le p$. with $G'(0) = G'(p) = 0$. Extend $G(x)$ to the interval $-p < x < 0$ by defining

$$G(x) = G(-x), \qquad \text{if } -p < x < 0.$$

Then extend $G(x)$ to $-\infty < x < \infty$ to have period $2p$. Show that $G(x)$ so extended is not only periodic, but even, that is, $G(-x) = G(x)$ for $-\infty < x < \infty$.

(b) Sketch the even periodic extension of $G(x) = x^2(1-x)^2$, as described in part (a), from $0 \le x \le 1$ to $-\infty < x < \infty$.

(c) Show that if $G(x)$ is even and has period $2p$ for $-\infty < x < \infty$ then the function

$$u(x,t) = \left(\tfrac{1}{2}a\right)[G(x+at) - G(x-at)]$$

is a d'Alembert solution to $a^2 u_{xx} = u_{tt}$ that satisfies the boundary conditions $u(0,t) = u(p,t) = 0$ and initial conditions $u(x,0) = 0$ and $u_t(x,0) = G'(x)$. Why did we assume $G'(0) = G'(p) = 0$?

10. The 1-dimensional wave equation with linear frictional drag is

$$a^2 u_{xx} = u_{tt} + k u_t,$$

where k is a positive constant. Find all product solutions $u(x,t) = X(x)T(t)$.

11. The partial differential equation $t u_{xx} = u_{tt}$ has product solutions of the form $u(x,t) = X(x)T(t)$. Find two ordinary differential equations, each dependent on a parameter λ, satisfied by $X(x)$ and $T(t)$, respectively. (The equation for X should be independent of t and the equation for T should be independent of x.)

***12.** Our derivation of the wave equation produced an equation of the form $a^2 u_{ss} = u_{tt}$. This is a linear equation if we keep arc length s as an independent space variable. However, introducing the more convenient coordinate variable x invariably leads to a **nonlinear wave equation** unless we accept a linearization. Here are three different nonlinear alternatives, each depending on a different assumption. The derivations depend on the chain rule for differentiation and the knowledge that arc length s and coordinate distance x are related by $ds/dx = \sqrt{1+u_x^2}$, so that $dx/ds = (1+u_x^2)^{-1/2}$. Observe that in each case the acceleration $|u_{tt}|$ decreases as the inclination $|u_x|$ increases.

(a) Replace $\partial^2 u/\partial s^2$ by $\partial(\partial u/\partial x)/\partial s$, and show that the wave equation becomes

$$a^2\left(1+u_x^2\right)^{-1/2} u_{xx} = u_{tt}.$$

(b) Replace $\partial^2 u/\partial s^2$ by $\partial(\partial u/\partial s)/\partial x$, and show that the wave equation becomes

$$a^2\left(1+u_x^2\right)^{-3/2} u_{xx} = u_{tt}.$$

Note that the left side is $a^2\kappa(x,t)u_{xx}$, where $\kappa(x,t)$ is the curvature of the string.

(c) Compute $\partial^2 u/\partial s^2$ with no special assumptions, and show that the wave equation is

$$a^2\left(1+u_x^2\right)^{-2} u_{xx} = u_{tt}.$$

5 THE LAPLACE EQUATION

The 2-dimensional **Laplace equation**

5.1
$$u_{xx} + u_{yy} = 0$$

arises naturally as the time-independent **equilibrium** versions of the 2-dimensional heat and wave equations

$$a^2(u_{xx} + u_{yy}) = u_t, \qquad a^2(u_{xx} + u_{yy}) = u_{tt}.$$

The derivation of these equations is similar to the ones given in Section 4. If R is a region in the xy plane whose boundary consists of finitely many smooth curves, we let $u(x,y,t)$ be the temperature at (x,y), taken at time t. Let the temperature at the boundary be maintained so that it does not change as time goes on, although the temperature may be different at different points of the boundary. Now let enough time elapse so that there is no detectable change in the function u as a function of t; in other words, wait long enough to be able to assume that $u_t(x,y,t) = 0$ throughout the interior of the region from that time on. In effect, we've assumed that $u(x,y,t) = u(x,y)$ is independent of t, in which case the heat equation reduces to the Laplace Equation 5.1, since $u_t = 0$. It is often convenient to abbreviate the Laplace operator by

$$\Delta = \frac{\partial^2}{\partial x^2} + \frac{\partial^2}{\partial y^2}.$$

The operator Δ acts linearly on twice-differentiable functions $u(x,y)$, and a solution of the partial differential equation $\Delta u = 0$ is called a **harmonic function.**

Solving the Laplace equation in two independent variables x and y is somewhat different from solving the heat or wave equation in the variables x and t. One important difference is that the boundary and initial conditions on $u(x,t)$

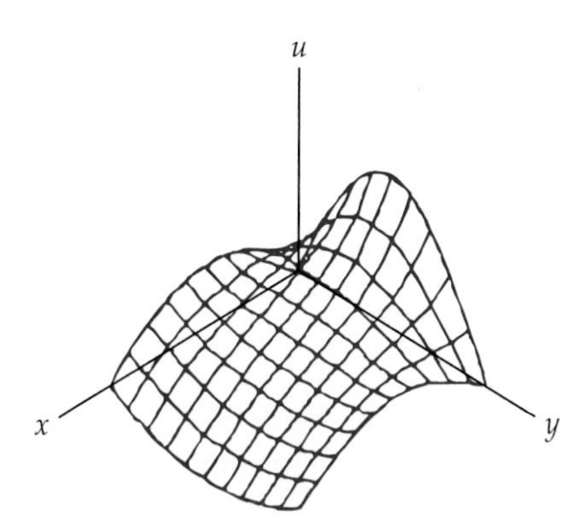

Figure 10.15 Solution to $\Delta u(x,y) = 0$.

that are appropriate for the heat and wave equations are replaced by a boundary condition on a solution $u(x,y)$ that is defined on some 2-dimensional region described by plane variables x and y. We'll consider rectangular regions, and in Exercise 19 circular ones, though there are techniques that can be applied to much more complicated regions.

A solution of the Laplace equation over a rectangle $0 \le x \le p, 0 \le y \le q$ is shown in Figure 10.15. On each of the four sides there are specified boundary values for $u(x,y)$:

$$u(x,0) = f_0(x), \qquad u(x,q) = f_q(x),$$
$$u(0,y) = g_0(y), \qquad u(p,y) = g_p(y).$$

To simplify matters for the moment, suppose that $f_0(x), g_0(y)$, and $g_p(y)$ are identically zero; we'll see that the general case will follow from this special case in a simple way.

Separating Variables. We look for product solutions in the form

$$u(x,y) = X(x)Y(y).$$

Substitution into $u_{xx} + u_{yy} = 0$ gives

$$X''Y + XY'' = 0$$

for $0 < x < p, 0 < y < q$. Assuming $X \neq 0,\ Y \neq 0$, we divide by XY, separate the variables, and write

$$\frac{Y''}{Y} = -\frac{X''}{X} = \lambda,$$

where λ is constant with respect to x and y because Y''/Y does not depend on x and X''/X does not depend on y. This gives us the two ordinary differential equations

$$X'' + \lambda X = 0, \qquad Y'' - \lambda Y = 0.$$

If $X_\lambda(x)$ and $Y_\lambda(y)$ are the solutions of these equations for the same λ, then $u_\lambda(x,y) = X_\lambda(x)Y_\lambda(y)$ will be a solution of $u_{xx} + u_{yy} = 0$.

If boundary conditions on $u(x,y)$ are $u(x,0) = 0$, $u(0,y) = 0$, and $u(p,y) = 0$, leaving the top edge at $y = q$ for later, then the corresponding implications for

X and Y are

$$u(x,0) = 0 \quad \text{requires} \quad X(x)Y(0) = 0, \qquad 0 < x < p,$$
$$u(0,y) = 0 \quad \text{requires} \quad X(0)Y(y) = 0, \qquad 0 < y < q,$$
$$u(p,y) = 0 \quad \text{requires} \quad X(p)Y(y) = 0, \qquad 0 < y < q.$$

We achieve these three conditions by making

$$X(0) = X(p) = 0, \qquad Y(0) = 0.$$

The resulting two-point boundary problem for X is then

$$X'' + \lambda X = 0, \qquad X(0) = X(p) = 0.$$

The basic solutions of the differential equation are $\sin\sqrt{\lambda}x$. We choose

$$X_k(x) = C_k \sin\frac{k\pi}{p}x, \qquad k = 1, 2, 3, \ldots.$$

with $\lambda = k^2\pi^2/p^2$ being the only λ values for which $X(0) = X(p) = 0$. Similarly, the equation for Y is now

$$Y'' - \frac{k^2\pi^2}{p^2}Y = 0, \qquad Y(0) = 0.$$

This differential equation has solutions that can be written either as

$$Y_k(y) = \alpha_k e^{k\pi y/p} + \beta_k e^{-k\pi y/p} \quad \text{or as} \quad Y_k(y) = A_k \cosh\frac{k\pi y}{p} + B_k \sinh\frac{k\pi y}{p}.$$

[Recall that $\cosh x = (e^x + e^{-x})/2$ and $\sinh x = (e^x - e^{-x})/2$.] Either form for $Y_k(y)$ will do, but writing the constants is a little simpler in terms of hyperbolic functions, so we'll use them. We satisfy $Y(0) = 0$ by making $A_k = 0$. Writing $b_k = C_k B_k$, we have

$$u_k(x,y) = X_k(x)Y_k(y) = b_k \sinh\frac{k\pi y}{p} \sin\frac{k\pi x}{p}.$$

Satisfying Boundary Conditions. The series representation

5.2
$$u(x,y) = \sum_{k=1}^{\infty} b_k \sinh\frac{k\pi y}{p} \sin\frac{k\pi x}{p}$$

will satisfy $u(x,q) = f_q(x)$, as well as $u(x,0) = 0$, provided that we can choose the b_k so

$$\sum_{k=1}^{\infty} b_k \sinh\frac{k\pi q}{p} \sin\frac{k\pi x}{p} = f_q(x).$$

So we take $b_k \sinh k\pi q/p$ to be the kth Fourier sine coefficient of $f(x) = f_q(x)$:

5.3
$$b_k = \left(\sinh\frac{k\pi q}{p}\right)^{-1} \frac{2}{p}\int_0^p f(x)\sin\frac{k\pi x}{p}\,dx.$$

If f_q is smooth enough that its Fourier series converges to it, Equation 5.2 will then satisfy the boundary condition at $y = q$. The second-order partial derivatives of $u(x,y)$ as defined by Equation 5.2 can be computed using term-by-term differentiation if the coefficients b_k tend to zero fast enough. Since each term satisfies the Laplace equation, the linearity of the equation would then guarantee that $u(x,y)$ would satisfy it also.

example 1

We want to solve the Laplace equation in the rectangle $0 < x < 2, 0 < y < 3$ (i.e., $p = 2, q = 3$) with boundary conditions shown in Figure 10.16(a):

$$u(x,0) = 0, \qquad u(x,3) = 1, \qquad 0 < x < 2,$$
$$u(0,y) = 0, \qquad u(2,y) = 0, \qquad 0 < y < 3.$$

Note that no condition is imposed at the corners of the rectangle. Equation 5.3 yields

$$\begin{aligned} b_k &= \left(\sinh \frac{3k\pi}{2}\right)^{-1} \int_0^2 1 \sin \frac{k\pi x}{2}\, dx \\ &= \frac{2[1-(-1)^k]}{k\pi \sinh(3k\pi/2)}, \qquad k = 1, 2, 3, \ldots. \end{aligned}$$

We remark that all even-order coefficients b_{2j} are zero, because $[1-(-1)^{2j}] = 0$. The solution Formula 5.2 gives us the particular solution

$$u_3(x,y) = \sum_{j=0}^{\infty} \frac{4}{(2j+1)\pi \sinh\left[3(2j+1)\pi/2\right]} \sinh \frac{(2j+1)\pi y}{2} \sin \frac{(2j+1)\pi x}{2}.$$

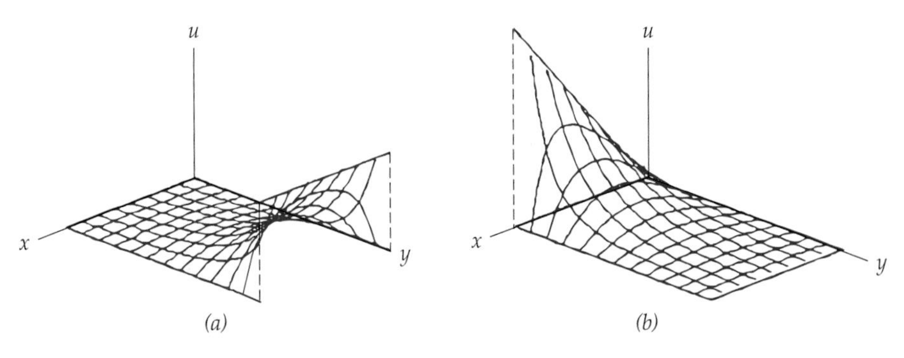

Figure 10.16 Solutions of $\Delta u = 0$ with boundary conditions mostly zero. (*a*) $u(x,3) = 1, 0 < x < 2$. (*b*) $u(x,0) = x, 0 < x < 2$.

Replacing $\sinh k\pi y/p$ by $\sinh k\pi(q-y)/p$ in Equation 5.2 gives a different solution $v(x,y)$ of the Laplace equation, a solution that satisfies $v(x,q) = 0$. The corresponding coefficients b_k are then determined by Equation 5.3 with $f_0(x) = v(x,0)$ plugged in for $f(x)$. This simple type change enables us to solve the Laplace equation with zero boundary values everywhere around the rectangle except along an edge where $y = 0$ or where $y = q$.

example 2

Using the technique of the previous example, we can solve the Laplace equation with boundary conditions

$$u(x,0) = x, \quad u(x,3) = 0, \quad 0 < x < 2,$$
$$u(0,y) = 0, \quad u(2,y) = 0, \quad 0 < y < 3.$$

The coefficients b_k in Equation 5.2 are determined by the requirement $u(x,0) = f_0(x) = x$ along the bottom edge of the rectangle. From Equation 5.3 we compute

$$b_k = \left(\sinh \frac{3k\pi}{2}\right)^{-1} \int_0^2 x \sin \frac{k\pi x}{2}\, dx$$
$$= \left(\sinh \frac{3k\pi}{2}\right)^{-1} \frac{4(-1)^{k+1}}{k\pi}.$$

The particular solution, graphed in Figure 10.16(b), is

$$u_0(x,y) = \sum_{k=1}^{\infty} \frac{4(-1)^{k+1}}{k\pi \sinh(3k\pi/2)} \sinh \frac{k\pi(3-y)}{2} \sin \frac{k\pi x}{2}.$$

The factor $\sinh k\pi(3-y)/2$ replaces $\sinh k\pi y/2$ as remarked earlier, so $u_0(x,3) = 0$ will hold. But each term in the series is a solution of the Laplace equation, and it only requires justification of term-by-term differentiation of the series to show that u_0 itself satisfies the Laplace equation.

example 3

To solve the Laplace equation with boundary conditions

$$u(x,0) = 5x, \quad u(x,3) = 4, \quad 0 < x < 2,$$
$$u(0,y) = 0, \quad u(2,y) = 0, \quad 0 < y < 3,$$

we just have to form a linear combination of the solutions u_3 and u_0 that we found in the two previous examples. The reason is, first, that $5u_0(x,y) + 4u_3(x,y)$ satisfies the Laplace equation because the Laplace operator Δ is linear:

$$\Delta(5u_0 + u_3) = 5\Delta u_0 + 4\Delta u_3 = 0 + 0.$$

Second,

$$5u_0(x,3) + 4u_3(x,3) = 5x + 0 = 5x,$$

and

$$5u_0(x,0) + 4u_3(x,0) = 0 + 4 = 4.$$

The reason this works is that each of u_3 and u_0 has a nonzero boundary value only where the other has a zero boundary value.

The previous examples have had nonzero boundary values imposed only on the top and bottom edges of rectangular regions in which a solution is sought. By reversing the roles of x and y in these examples, and in Equations 5.2 and 5.3, we can solve the Laplace equation with nonzero boundary values on the right and left edges of a rectangle, as in Exercises 3 and 4. See Exercise 19 for solutions of the Laplace equation in the interior of a circular disk.

EXERCISES

Each of the functions in Exercises 1 through 4 is a solution of the Laplace equation $u_{xx} + u_{yy} = 0$. Verify this, and verify that $u(x,y)$ satisfies the indicated boundary conditions. Then sketch the graph of $z = u(x,y)$ for (x,y) inside the indicated boundary.

1. $u(x,y) = (e^y - e^{-y}) \sin x$; $u(0,y) = u(\pi,y) = 0$ for $0 < y < \ln 3$, $u(x,0) = 0$, $u(x, \ln 3) = \left(\frac{8}{3}\right) \sin x$ for $0 < x < \pi$.

2. $u(x,y) = x^2 - y^2$; $u(0,y) = -y^2$, $u(1,y) = 1 - y^2$ for $0 < y < 1$, $u(x,0) = x^2$, $u(x,1) = x^2 - 1$ for $0 < x < 1$.

3. $u(x,y) = e^x \cos y$; $u(0,y) = \cos y$, $u(1,y) = e \cos y$ for $0 < y < \pi/2$, $u(x,0) = e^x$, $u(x, \pi/2) = 0$ for $0 < x < 1$.

4. $u(x,y) = (\sinh 2)^{-1} \sinh(2 - y) \sin x$; $u(0,y) = u(\pi,y) = 0$ for $0 < y < 2$, $u(x,0) = \sin x$, $u(x,2) = 0$ for $0 < x < \pi$.

Recall that $\cosh y = (e^y + e^{-y})/2$, $\sinh y = (e^y - e^{-y})/2$. Prove the following:

5. $d \cosh y/dy = \sinh y$.

6. $d \sinh y/dy = \cosh y$.

7. $\cosh^2 y - \sinh^2 y = 1$.

8. $e^y = \cosh y + \sinh y$.

9. **(a)** Solve the Laplace equation $u_{xx} + u_{yy} = 0$ subject to the boundary conditions $u(0,y) = u(\pi,y) = 0$, $u(x,0) = 0$, $u(x,\pi) = 0$, $0 < x < \pi$, $0 < y < \pi$.

(b) Sketch the graph of the solution.

Find a series solution for the Laplace equation subject to each of the sets of boundary conditions in Exercises 10 through 14.

10. $u(x,0) = u(x,1) = 1$, $0 < x < 1$, $u(0,y) = 0$, $u(1,y) = 0$, $0 < y < 1$.

11. $u(x,0) = 0$, $u(x,1) = x(1 - x)$, $0 < x < 1$, $u(0,y) = u(1,y) = 0$, $0 < y < 1$.

12. $u(x,0) = 1$, $u(x,1) = 2$, $0 < x < 1$, $u(0,y) = 0$, $u(1,y) = 0$, $0 < y < 1$.

13. $u(x,0) = u(x,1) = 0$, $0 < x < 1$, $u(0,y) = y$, $u(1,y) = 0$, $0 < y < 1$.

14. $u(x,0) = 1$, $u(x,2) = -1$, $0 < x < 1$, $u(0,y) = y$, $u(1,y) = y$, $0 < y < 2$.

15. **(a)** Assume that $f(x + iy)$ and $g(x - iy)$ are twice differentiable functions of x and y in some region R of the xy plane common to both f and g. Verify that

$$u(x,y) = f(x + iy) + g(x - iy)$$

defines a solution u of the Laplace equation in R.

(b) Let $f(x+iy) = g(x+iy) = e^{x+iy}$. What is the function $u(x,y)$ defined in part (a)?

16. Describe all solutions of the Laplace equation $u_{xx} + u_{yy} = 0$ that are independent of the variable y and are defined for (x,y) satisfying $a < x < b$ for some $a < b$.

17. **(a)** The derivation of Equation 5.3 for the coefficients b_k in Equation 5.2 was based on the assumption that $u(x,q)$ was specified for $0 < x < p$. If instead we specify the values of $u_y(x,q)$ for $0 < x < p$ by $u_y(x,q) = f(x)$, show that Equation 5.3 is replaced by

$$b_k = \left(\cosh \frac{k\pi q}{p}\right)^{-1} \frac{2}{k\pi} \int_0^p f(x) \sin \frac{k\pi x}{p}\, dx.$$

(b) Solve the Laplace equation under the boundary conditions

$$u(x,0) = 0, \qquad u_y(x,q) = 1, \qquad \text{for } 0 < x < p,$$
$$u(0,y) = u(p,y) = 0, \qquad \text{for } 0 < y < q.$$

18. **Graphs of harmonic functions.** Note that the Laplace equation $\Delta u = 0$ in two dimensions can be written $u_{xx} = -u_{yy}$. Also, recall that a "saddle point" for a function of two variables is one where the graph of the function has opposite concavity in two different directions. See Figures 10.15 and 10.16.

(a) Show that if $\Delta u(x,y) = 0$ in a region and $u_{xx}(x_0, y_0) \neq 0$, then (x_0, y_0) corresponds to a saddle point of the graph of u. [Thus a typical point on the graph of a nonlinear harmonic function $u(x,y)$ is a saddle point.]

(b) Show that every point of the graph of the harmonic function $u(x,y) = x^2 - y^2$ is a saddle point, paying special attention to $(x_0, y_0) = (0,0)$.

19. The **Laplace equation in polar coordinates** (r,θ), where $x = r\cos\theta$, $y = r\sin\theta$, is $r^2\bar{u}_{rr} + r\bar{u}_r + \bar{u}_{\theta\theta} = 0$. We define $\bar{u}(r,\theta) = u(r\cos\theta, r\sin\theta)$, where $u(x,y)$ satisfies the (x,y) Laplace equation $u_{xx} + u_{yy} = 0$.

(a) Compute $\bar{u}_r$, $\bar{u}_{rr}$, and $\bar{u}_{\theta\theta}$ by applying the chain rule to $u(r\cos\theta, r\sin\theta)$. Then verify that $\bar{u}$ satisfies the polar form of the Laplace equation.

(b) Verify that the polar form of the Laplace equation has solutions

$$\bar{u}(r,\theta) = r^n(a_n \cos n\theta + b_n \sin n\theta), \qquad n = 0, 1, 2, \ldots.$$

(c) Let $F(\theta) = u(R\cos\theta + a, R\sin\theta + b)$ be the boundary values of a harmonic function on the circular region $(x-a)^2 + (y-b)^2 \leq R^2$. Define

$$\bar{u}(r,\theta) = \frac{1}{2}a_0 + \sum_{n=1}^{\infty} r^n(a_n \cos n\theta + b_n \sin n\theta),$$

$$a_n = \frac{1}{R^n\pi}\int_{-\pi}^{\pi} F(\theta)\cos n\theta\, d\theta, \qquad b_n = \frac{1}{R^n\pi}\int_{-\pi}^{\pi} F(\theta)\sin n\theta\, d\theta.$$

Show that the series for $\bar{u}(r,\theta)$ and its partial derivatives converge uniformly if $0 \le r \le R_0$ for any $R_0 < R$, and hence that $\bar{u}(r,\theta)$ is a solution of the Laplace equation in polar coordinates for $0 \le r < R$ and $-\pi \le \theta < \pi$.

(d) Set $r = 0$ in the expansion of part (c) to show that a value $u(a,b)$ of a harmonic function u equals the average value of u over small circles centered at (a,b).

20. The Laplace equation $\Delta u = 0$ is the steady-state specialization of the wave equation $a^2\Delta u = u_{tt}$. In Exercises 6, 7, and 8 of Section 4C we discussed the *d'Alembert solution* of the wave equation assuming the space variable is 1-dimensional, in which case the wave equation is $a^2 u_{xx} = u_{tt}$. Here we consider the d'Alembert solution in two space dimensions, where the wave equation is $a^2(u_{xx} + u_{yy}) = u_{tt}$.

(a) Show that if $\alpha^2 + \beta^2 = 1$ and $f(z)$ is twice differentiable, then $u = f(\alpha x + \beta y - at)$ and $u = f(\alpha x + \beta y + at)$ are solutions of the 2-dimensional wave equation.

(b) Show that $u = f(\alpha x + \beta y - at)$ represents a wave with shape determined by $f(z)$ and moving relative to the xy plane in the direction of the unit vector (α,β) with speed $a > 0$. More specifically, show that the value $u = f(\alpha x + \beta y)$ along the line $\alpha x + \beta y = 0$ in $\mathcal{R}^2$ at time $t = 0$ gets translated after time $t > 0$ to the same value along the line $\alpha x + \beta y = at$.

(c) Show analogously to part (b) that $u = f(\alpha x + \beta y + at)$ represents a wave with shape determined by $f(z)$ and moving in the direction of the unit vector $(-\alpha,-\beta)$ with speed $a > 0$.

CHAPTER REVIEW AND SUPPLEMENT

What characteristic equation relating the constants r and s must be satisfied in order for $u(x,y) = e^{rx+sy}$ to be a solution of each of the partial differential equations in Exercises 1 through 4?

1. $u_{xx} - u_{yy} + 2u_x = 0.$

2. $u_{xx} + u_{xy} + u = 0.$

3. $u_{xx} + u_{yy} = u.$

4. $uu_{xxx} - u_y^2 = 0.$

5. Let $f(x) = \begin{cases} 1, & 0 \le x \le \pi/2, \\ 0, & \pi/2 < x \le \pi. \end{cases}$

(a) Sketch the graph of the even periodic extension of $f(x)$ with period 2π.

(b) Sketch the graph of the odd periodic extension of $f(x)$ with period 2π.

(c) Sketch the graph of the periodic extension of $f(x)$ with period π.

6. A function f is defined on the interval $0 \le x < 4$ by

$$f(x) = \begin{cases} x^2, & 0 \le x \le 1, \\ 2 - x, & 1 < x < 2, \\ 1, & 2 \le x < 3, \\ x - 3, & 3 \le x < 4. \end{cases}$$

(a) The function f has a unique periodic extension of period 4 to $-\infty < x < \infty$. Sketch the graph of this extension for $0 \leq x \leq 8$, indicating clearly the value of the function at points of discontinuity.

(b) To what respective values does the Fourier series of f converge at $x = 0$, $x = 1$, $x = 2$, $x = 3$, and $x = 4$?

(c) Sketch the graph of the odd function on $-4 \leq x \leq 4$ that agrees with f for $0 \leq x \leq 4$.

(d) Sketch the graph of the even function on $-4 \leq x \leq 4$ that agrees with f for $0 \leq x \leq 4$.

7. (a) Show that the Fourier sine expansion of $f(x) = x$ for $0 \leq x \leq \pi$ is

$$2\sum_{k=1}^{\infty}(-1)^{k+1}\frac{1}{k}\sin kx.$$

(b) Using the result of part (a), show that $\sum_{n=1}^{\infty}(-1)^{n-1}/(2n-1) = \pi/4$.

8. (a) Find the Fourier expansion of $f(x) = x + x^2$ for $-\pi \leq x \leq \pi$.

(b) Using the result of part (a), show that $\sum_{n=1}^{\infty} 1/n^2 = \pi^2/6$.

9. The partial differential equation $u_{xx} + 5u_x + u_t = 0$ has product solutions of the form $u(x,t) = X(x)T(t)$. Find two ordinary differential equations, each dependent on a parameter λ, satisfied by $X(x)$ and $T(t)$, respectively. (The equation for X should be independent of t and the equation for T should be independent of x.)

10. The heat equation $u_{xx} = u_t$ with $u(0,t) = u(1,t) = 0$ has solution

$$u(x,t) = \sum_{k=1}^{\infty} A_k e^{-k^2\pi^2 t}\sin k\pi x.$$

Find coefficients A_k so that (a) $u(x,0) = 2\sin 3\pi x$ and (b) $u(x,0) = x$ for $0 < x < 1$.

11. Mark each of the following statements T for true or F for false, as the case may be. It's possible to arrive logically at correct answers without doing any elaborate computation.

______ The cosine coefficients a_k of the full Fourier expansion of $f(x) = x/(1+x^2)$ for $-\pi < x < \pi$ are all zero.

______ The cosine coefficients a_k of the full Fourier expansion of $f(x) = \cos(1+x^2)$ for $-1 < x < 1$ are all zero.

______ The Fourier sine series of $f(x) = 2x^2,\ 0 \leq x \leq 1$ converges to 2 at $x = 1$.

______ The steady-state solution of $4u_{xx} + 4u_x + u = u_t$ that satisfies $u(0,t) = 1$ and $u(1,t) = 2$ is $u(x,t) = 1 + x$.

12. Match each item in the following list of functions with its Fourier series by writing an identifying letter in the appropriate blank. Do as little calculation as possible in reaching a decision.

(a) $f(x) = \begin{cases} -1, & -\pi < x < 0 \\ 1, & 0 < x < \pi \end{cases}$ ______ $\frac{\pi}{2} - \frac{4}{\pi}\sum_{k=1}^{\infty} \frac{\cos(2k-1)x}{(2k-1)^2}$

(b) $f(x) = x, -\pi < x < \pi$ ______ $\frac{2}{\pi} - \frac{4}{\pi}\sum_{k=1}^{\infty} \frac{\cos 2kx}{4k^2 - 1}$

(c) $f(x) = x^2, -\pi < x < \pi$ ______ $\frac{4}{\pi}\sum_{k=1}^{\infty} \frac{\sin(2k-1)x}{2k-1}$

(d) $f(x) = |\sin x|, -\pi < x < \pi$ ______ $2\sum_{k=1}^{\infty} \frac{(-1)^{k-1}}{k} \sin kx$

(e) $f(x) = |x|, -\pi < x < \pi$ ______ $\frac{\pi^2}{3} + 4\sum_{k=1}^{\infty} (-1)^k \frac{\cos kx}{k^2}$

Find product solutions $U(x,t) = X(x)T(t)$ for each of the equations in Exercises 13 through 16. Also, if possible, find all nonconstant steady state solutions for each equation, that is, solutions that are independent of t.

13. $U_{xx} = k^2 U_{tt}$, k const.

14. $U_{xx} = U_t + tU$.

15. $U_{xx} = U_t$.

16. $U_{xx} + U_{tt} = 0$.

17. Solve the wave equation $a^2 y_{xx} = y_{tt}$ given boundary conditions $y(0,t) = y(2,t) = 0$ and initial conditions $y(x,0) = 0$, $y_t(x,0) = 1 - |x - 1|$ for $0 \leq x \leq 2$.

18. Solve the heat equation $a^2 u_{xx} = u_t$ given boundary conditions $u(0,t) = 0$ and $u(p,t) = 1$, and initial condition $u(x,0) = x/p$ for $0 \leq x \leq p$.

19. Let L be a linear operator.

(a) Explain what "linear operator" means in general.

(b) If **u** and **v** are solutions of $L(\mathbf{x}) = 0$, prove that $\mathbf{x} = c\mathbf{u} + d\mathbf{v}$ is also a solution of $L(\mathbf{x}) = 0$ for constant c, d.

(c) Explain exactly what it means for λ and **u** to be an eigenvalue-eigenvector pair for L.

(d) Give three examples of linear operators and three of nonlinear operators.

20. The **Black-Scholes equation** for financial option pricing is described at the end of Section 1:

$$\frac{\partial V}{\partial t} + \frac{1}{2}\sigma^2 x^2 \frac{\partial^2 V}{\partial x^2} + rx\frac{\partial V}{\partial x} - rV = 0.$$

(a) Use separation of varibles to derive the following separated equations for the factors in product solutions $V(x,t) = T(t)X(x)$:

$$T' + \lambda T = 0, \qquad \tfrac{1}{2}\sigma^2 x^2 X'' + rxX' - (\lambda + r)X = 0.$$

(b) Find the form of the product solutions assuming $r = 0.05$ and $\sigma^2 = 0.10$.

CHAPTER 11 Sturm-Liouville Expansions

1 ORTHOGONAL FUNCTIONS

Many of the useful properties of Fourier expansions are shared by a large class of similar expansions in which the functions sin kx and cos kx are replaced by some sequence of functions $\{\phi_k\}_{k=1,2,3,\ldots}$, all of which are mutually orthogonal. Recall that orthogonality of the trigonometric system amounts to having an integral of the product of two distinct functions equal to zero. The same is true for other systems of orthogonal functions, but since the precise integral that is to be computed varies from system to system, it is convenient to introduce a general notation for treating all systems at once. To the extent that the different systems have common features, the uniform notation enables us to display some of their similarities more clearly. Furthermore, the notation fosters some geometric insight into the computations. To be specific, we define the **Fourier inner product** $\langle f,g\rangle$ of real-valued functions $f(x)$ and $g(x)$ for $-\pi \le x \le \pi$, by

1.1 $$\langle f,g\rangle = \int_{-\pi}^{\pi} f(x)g(x)\,dx.$$

It follows by direct computation (see Chapter 10, Section 2) that, if we set $\phi_1(x) = 1/\sqrt{2\pi}$, $\phi_{2n}(x) = (\cos nx)/\sqrt{\pi}$, and $\phi_{2n+1}(x) = (\sin nx)/\sqrt{\pi}$, then

$$\langle \phi_k,\phi_l\rangle = \begin{cases} 0, & k \ne l, \\ 1, & k = l. \end{cases}$$

Thus the sequence $\{\phi_k\}_{k=1,2,\ldots}$ is not only **orthogonal** in the sense that $\langle \phi_k,\phi_l\rangle = 0$ when $k \ne l$, but is **orthonormal.** The "normal" part of the terminology comes from having $\langle \phi_k,\phi_k\rangle = 1$. This normalization comes, in the case of the trigonometric system, from dividing by $\sqrt{\pi}$.

It turns out that many consequences of defining the Fourier inner product by Equation 1.1 follow from four defining properties of a general real-valued **inner product:**

1.2 POSITIVITY: $\langle f,f\rangle > 0$ except that $\langle 0,0\rangle = 0$.

SYMMETRY: $\langle f,g\rangle = \langle g,f\rangle$.

ADDITIVITY: $\langle f+g,h\rangle = \langle f,h\rangle + \langle g,h\rangle$, $\langle f,g+h\rangle = \langle f,g\rangle + \langle f,h\rangle$.

HOMOGENEITY: $\langle cf,g\rangle = c\langle f,g\rangle$, $\langle f,cg\rangle = c\langle f,g\rangle$.

These properties for the particular inner product defined in Equation 1.1 follow directly from properties of the integral. For example, symmetry is just the statement that

$$\int_{-\pi}^{\pi} f(x)g(x)\,dx = \int_{-\pi}^{\pi} g(x)f(x)\,dx.$$

One reason for singling out these four properties here is that they are shared by many other similar products, including the ordinary dot product of coordinate vectors. In addition, we'll see in the next section that inner products of the form

1.3 $$\langle f,g\rangle = \int_a^b f(x)g(x)r(x)\,dx,$$

with **weight function** $r(x) > 0$ arise in a natural way.

With any definition of $\langle f,g\rangle$ that satisfies the four properties 1.2, we can define a **norm** or **length** of a real-valued function f such that f^2 is integrable by

$$\|f\| = \langle f,f\rangle^{1/2}.$$

For our definition of Fourier inner product, with $r(x) = 1$, this amounts to

$$\|f\| = \left[\int_{-\pi}^{\pi} f^2(x)\,dx\right]^{1/2}.$$

More generally, we may consider functions $f(x)$ and $g(x)$ that are continuous for $a \le x \le b$. We denote the set of such functions by $C[a,b]$, and we define

$$\langle f,g\rangle = \int_a^b f(x)g(x)\,dx, \qquad \|f\| = \left[\int_a^b f^2(x)\,dx\right]^{1/2}$$

for f and g in $C[a,b]$.

example 1

Let $f(x)$ and $g(x)$ be bounded integrable functions. We could compute $\langle f+g, f-g\rangle$ directly from the specific definition of $\langle f,g\rangle$ in Equation 1.1. Alternatively, we can use the symmetry and additivity properties:

$$\begin{aligned}\langle f+g,f-g\rangle &= \langle f+g,f\rangle + \langle f+g,-g\rangle \\ &= \langle f,f\rangle + \langle g,f\rangle + \langle f,-g\rangle + \langle g,-g\rangle \\ &= \langle f,f\rangle + \langle f,g\rangle - \langle f,g\rangle - \langle g,g\rangle \\ &= \langle f,f\rangle - \langle g,g\rangle.\end{aligned}$$

This much is true regardless of how f and g are chosen. If it should happen that $f^2(x)=g^2(x)$, for example $f(x)=x$, $g(x)=|x|$, then $\langle f,f\rangle=\langle g,g\rangle$; so we conclude that $\langle f+g,\ f-g\rangle=0$. In other words, $f+g$ and $f-g$ are orthogonal. The conclusion is of course true also whenever $\langle f,f\rangle=\langle g,g\rangle$.

To see the importance of orthonormal sequences in general, we consider the following problem: Let $\langle f,g\rangle$ be an inner product, and let $\{\phi_k\}_{k=1,2,\ldots}$ be a sequence of elements, orthonormal with respect to the inner product. Using the norm defined by $\|f\|=\langle f,f\rangle^{1/2}$, we try to determine coefficients c_k, $k=1,2,\ldots,N$, such that the norm

$$\left\| g-\sum_{k=1}^{N} c_k\phi_k \right\|$$

is minimized for given g and N by the approximating sum $\sum_{k=1}^{N} c_k\phi_k$.

1.4 THEOREM

Let $\{\phi_k\}_{k=1,2,\ldots}$ be an orthonormal sequence in a space with an inner product in which numerical multiplication and addition are defined. Then, given an element g of the space, the distance between g and the approximating sum:

$$d_N=\left\| g-\sum_{k=1}^{N} c_k\phi_k \right\|$$

is minimized for $N=1,2,\ldots$ by taking $c_k=\langle g,\phi_k\rangle$.

Proof. The fact that the sequence ϕ_k is orthonormal makes the solution very simple; we get

$$\begin{aligned} 0\le\left\| g-\sum_{k=1}^{N} c_k\phi_k \right\|^2 &= \left\langle g-\sum_{k=1}^{N} c_k\phi_k,\ g-\sum_{k=1}^{N} c_k\phi_k \right\rangle \\ &= \|g\|^2-2\sum_{k=1}^{N} c_k\langle g,\phi_k\rangle+\sum_{k=1}^{N} c_k^2 \\ &= \|g\|^2-\sum_{k=1}^{N}\langle g,\phi_k\rangle^2+\sum_{k=1}^{N}\left[\langle g,\phi_k\rangle^2-2c_k\langle g,\phi_k\rangle+c_k^2\right] \\ &= \|g\|^2-\sum_{k=1}^{N}\langle g,\phi_k\rangle^2+\sum_{k=1}^{N}[\langle g,\phi_k\rangle-c_k]^2. \end{aligned} \tag{1}$$

We have added and subtracted $\sum_{k=1}^{N}\langle g,\phi_k\rangle^2$ at the next-to-last step. But the first two terms in the last expression are independent of the choice of the c_k and the last sum is then minimized by taking $c_k=\langle g,\phi_k\rangle$. ■

The numbers $c_k=\langle g,\phi_k\rangle$ in Theorem 1.4 are called the **Fourier coefficients** of g with respect to the orthonormal sequence $\{\phi_k\}$. An important remark about the conclusion of Theorem 1.4 is that the c_k are uniquely determined, independently of N. In other words, if we wanted to improve the closeness of the approximation to g by increasing N, then Theorem 1.4 says that the c_k already computed are to be left unchanged, and it is only necessary to compute additional coefficients $c_k = \langle g, \phi_k \rangle$, $k = N+1, \ldots$.

As a by-product of the proof of Theorem 1.4, we have the following.

1.5 BESSEL'S INEQUALITY

$$\|g\|^2 \geq \sum_{k=1}^{\infty} \langle g, \phi_k \rangle^2.$$

Proof. The inequality $0 \leq \|g\|^2 - \sum_{k=1}^{N} \langle g, \phi_k \rangle^2 + \sum_{k=1}^{N} [\langle g, \phi_k \rangle - c_k]^2$ was established in the last step of Equation (1). On taking $c_k = \langle g, \phi_k \rangle$, the inequality becomes

$$0 \leq \|g\|^2 - \sum_{k=1}^{N} \langle g, \phi_k \rangle^2.$$

Bessel's Inequality follows by letting N tend to infinity. ■

If the distance d_N in Theorem 1.4 always tends to 0 as $N \to \infty$, the orthonormal system $\{\phi_k\}$ is said to be **complete,** and the inequality in Bessel's Inequality 1.5 becomes the **Parseval equation:**

$$\|g\|^2 = \sum_{k=1}^{\infty} \langle g, \phi_k \rangle^2.$$

(The proof, assuming completeness, is left as Exercise 11.) Taken in the proper context, all the orthonormal sequences considered in this book turn out to be complete, and we'll assume that about each of them. Unfortunately the proofs of these facts lie considerably deeper than we're prepared to go here.

example 2

The approximation to g by a sum $\sum_{k=1}^{N} c_k \phi_k$ has been measured by a norm in Theorem 1.4. To see what this means for approximation by trigonometric polynomials, we use the inner product given by Equation 1.1 on the space $C[-\pi, \pi]$ of continuous functions on $[-\pi, \pi]$. Given the orthonormal sequence $\phi_1(x) = 1/\sqrt{2\pi}$, $\phi_{2n}(x) = (\cos nx)/\sqrt{\pi}$, $\phi_{2n+1}(x) = (\sin nx)/\sqrt{\pi}$, we try to minimize, for given g in $C[-\pi, \pi]$, the norm

$$\left\| g - \sum_{k=1}^{2N+1} c_k \phi_k \right\|.$$

We have seen that this is done by taking $c_k = \langle g, \phi_k \rangle$. By the definition of the inner product, we can take out constant factors to get

$$\langle g, \phi_k \rangle = \begin{cases} \frac{1}{\sqrt{2\pi}} \int_{-\pi}^{\pi} g(t)\, dt, & k = 1, \\ \frac{1}{\sqrt{\pi}} \int_{-\pi}^{\pi} g(t) \cos nt\, dt, & k = 2n, \\ \frac{1}{\sqrt{\pi}} \int_{-\pi}^{\pi} g(t) \sin nt\, dt, & k = 2n+1. \end{cases}$$

Hence the terms $c_k\phi_k$ become

$$c_1\phi_1 = \langle g,\phi_1\rangle\phi_1(x) = \frac{1}{2\pi}\int_{-\pi}^{\pi} g(t)\,dt,$$

$$c_{2n}\phi_{2n} = \langle g,\phi_{2n}\rangle\phi_{2n}(x) = \left[\frac{1}{\pi}\int_{-\pi}^{\pi} g(t)\cos nt\,dt\right]\cos nx,$$

$$c_{2n+1}\phi_{2n+1} = \langle g,\phi_{2n+1}\rangle\phi_{2n+1}(x) = \left[\frac{1}{\pi}\int_{-\pi}^{\pi} g(t)\sin nt\,dt\right]\sin nx.$$

Then

$$\sum_{k=1}^{2N+1} c_k\phi_k(x) = \frac{a_0}{2} + \sum_{k=1}^{N}(a_k\cos kx + b_k\sin\,kx),$$

where a_k and b_k are the trigonometric Fourier coefficients as defined in Section 2 of Chapter 10. The square of the norm to be minimized takes the form

$$\int_{-\pi}^{\pi}\left[g(x) - \frac{a_0}{2} - \sum_{k=1}^{N}(a_k\cos\,kx + b_k\sin\,kx)\right]^2 dx,$$

and Theorem 1.4 says that the minimum will be attained for a given fixed N by taking a_k and b_k to be the Fourier coefficients of g. Since the trigonometric system is complete, this minimum tends to zero as N tends to infinity, and the Parseval equation for the trigonometric system becomes, after division by π,

$$\frac{1}{\pi}\int_{-\pi}^{\pi} g(x)^2\,dx = \frac{1}{2}a_0^2 + \sum_{k=1}^{\infty}\left(a_k^2 + b_k^2\right).$$

This equation has an interesting interpretation in terms of sound intensity, as shown in Exercise 12.

The minimization of an integral of the form

$$\int_a^b \left[g(x) - \sum_{k=1}^{N} c_k\phi_k(x)\right]^2 dx$$

is called a best **mean-square approximation** to g. In this sense, we can say that the Fourier approximation provides the best mean-square approximation by a trigonometric polynomial.

EXERCISES

1. (a) Verify that

$$\langle f,g\rangle = \int_a^b f(x)g(x)\,dx$$

defines an inner product on the set $C[a,b]$ of continuous functions defined on $[a,b]$; that is, verify Equations 1.2 of the text.

(b) What condition is required of a continuous function $r(x)$ defined on $[a,b]$ in order that

$$\langle f,g\rangle = \int_a^b f(x)g(x)r(x)\,dx$$

satisfy $\langle f,f\rangle \geq 0$?

2. Define $\langle \mathbf{x},\mathbf{y}\rangle = x_1y_1 + x_2y_2 + x_3y_3$ for vectors $\mathbf{x} = (x_1, x_2, x_3)$, $\mathbf{y} = (y_1, y_2, y_3)$.

(a) Verify that $\langle \mathbf{x},\mathbf{y}\rangle$ is equal to the dot product $\mathbf{x}\cdot\mathbf{y}$ on $\mathcal{R}^3$.

(b) Verify that $\langle \mathbf{x},\mathbf{y}\rangle$ as defined above satisfies the four properties of a general inner product listed under 1.2.

3. Let $\{\phi_k\}$ be an orthonormal sequence of functions in $C[a,b]$, and let g be in $C[a,b]$. Show that if the real-valued function

$$\Delta(c_1,\ldots,c_N) = \int_a^b \left[g(x) - \sum_{k=1}^{N} c_k\phi_k(x)\right]^2 dx$$

has a local minimum as a function of $(c_1,\ldots,c_N)$, then

$$c_k = \int_a^b g(x)\phi_k(x)\,dx, \qquad k = 1, 2, \ldots, N.$$

(Theorem 1.4 asserts the existence of a global minimum.)
[*Hint:* Differentiate the formula for Δ under the integral sign. This is justified under the assumptions made here.]

Let $\{\phi_k\}$ be an orthonormal sequence of functions in $C[a,b]$ and let g be in $C[a,b]$. Use Bessel's inequality for Exercises 4, 5, and 6.

4. Show that $\sum_{k=1}^{\infty}\left[\int_a^b g(x)\phi_k(x)\,dx\right]^2$ converges.

5. Show that $\lim_{k\to\infty}\int_a^b g(x)\phi_k(x)\,dx = 0$.

6. Find an example of a function whose trigonometric Fourier series has coefficients a_k and b_k such that $\sum_{k=1}^{\infty}(a_k^2 + b_k^2)$ converges, but such that either $\sum_{k=1}^{\infty} a_k$ or $\sum_{k=1}^{\infty} b_k$ does not converge.

7. Define $\langle \mathbf{x},\mathbf{y}\rangle = r_1x_1y_1 + r_2x_2y_2 + r_3x_3y_3$, where $\mathbf{x} = (x_1,x_2,x_3)$, $\mathbf{y} = (y_1,y_2,y_3)$, for positive r_k. Verify that this definition of $\langle \mathbf{x},\mathbf{y}\rangle$ satisfies the relations 1.2 with x replacing f and y replacing g. Is this still true if $r_1 = 0$?

Use the properties 1.2 of a general inner product, together with its relation to length, to prove the equations in Exercises 8, 9, and 10. Also, assuming f and g are vectors in $\mathcal{R}^2$, give a geometric interpretation for each equation.

8. $\|f - g\|^2 = \|f\|^2 + \|g\|^2 - 2\langle f,g\rangle$.

9. $\|f + g\|^2 + \|f - g\|^2 = 2\|f\|^2 + 2\|g\|^2$.

10. $\|f - g\|^2 = \|f - h\|^2 + \|h - g\|^2$ if and only if $\langle f - h,\ h - g\rangle = 0$.

11. Use the displayed equations in the proof of Theorem 1.4 to show that the *Parseval equation* $\|g\|^2 = \sum_{k=1}^{\infty}\langle g,\phi_k\rangle^2$ holds for an orthonormal system $\{\phi_k\}$ if and only if the system is *complete*, that is, if and only if $\left\|g - \sum_{k=1}^{N}\langle g, \phi_k\rangle\phi_k\right\| \to 0$ as $N \to \infty$.

12. Periodic sound sources are perceived as pressure variations $g(t)$ representable by Fourier series: $g(t) = \frac{1}{2}a_0 + \sum_{k=1}^{\infty} \left[a_k \cos(k\pi t/p) + b_k \sin(k\pi t/p)\right]$. The values $g(t)$ are the deviations, plus or minus, from ambient atmospheric pressure at times t. The **intensity** of a pressure function of period $2p$ is defined to be the average value $(1/2p)\int_0^{2p} g^2(t)\,dt$.

(a) Show that $g(t) = a\cos(k\pi t/p) + b\sin(k\pi t/p)$ has intensity $\frac{1}{2}(a^2 + b^2)$.

(b) Assuming the validity of Parseval's equation for the general pressure function $g(t)$, show that the intensity of g equals the sum of the intensities of the individual terms in the Fourier expansion of g.

13. The inner product on continuous functions in $C[0,1]$ given by

$$\langle f, g \rangle = \int_0^1 f(x)g(x)x\,dx$$

arises naturally in problems involving Bessel functions. Verify the conditions listed in 1.2 of the text, paying particularly careful attention to the positivity condition.

Which of the definitions for the weight function $r(x)$ makes

$$\langle f, g \rangle = \int_{-1}^1 f(x)g(x)r(x)\,dx$$

a genuine inner product on $C[-1,1]$, the continuous real functions on $-1 \leq x \leq 1$? Look carefully at the conditions given by Equations 1.2, and for those choices of $r(x)$ in Exercises 14 through 19 that don't define inner products, state why.

14. $r(x) = x^2$.

15. $r(x) = x$.

16. $r(x) = (1 - x^2)^{-1/2}$.

17. $r(x) = x,\ x \geq 0,\ r(x) = 0,\ x < 0$.

18. $r(x) = |x|$.

19. $r(x) = x + 1$.

Exercises 20 through 24 relate to the nth **Chebychev polynomial** defined by

$$T_n(x) = \cos(n \arccos x), \qquad -1 \leq x \leq 1.$$

20. Compute the polynomial $T_n(x)$ for $n = 0, 1, 2, 3$.

21. Show that $|T_n(x)| \leq 1$ for $-1 \leq x \leq 1$.

22. Verify that the relation

$$\langle f, g \rangle = \int_{-1}^1 f(x)g(x)(1 - x^2)^{-1/2}\,dx$$

defines an inner product on $C[-1,1]$, that is, verify Properties 1.2.

23. Prove the orthogonality relations

$$\int_{-1}^1 T_k(x)T_l(x)(1 - x^2)^{-1/2}\,dx = \begin{cases} 0, & k \neq l, \\ \pi/2, & k = l \neq 0. \end{cases}$$

[*Hint:* Make the change of variable $x = \cos\theta$.]

24. Show that $T_n(x)$ is always a polynomial by setting $\arccos x = \theta$ and expanding the right side of $\cos n\theta = \text{Re}(e^{in\theta}) = \text{Re}(\cos\theta + i\sin\theta)^n$ by the binomial theorem.

Double Fourier Series

25. For continuous functions $f(x,y)$ and $g(x,y)$ defined on the rectangle $0 \le x \le p,\ 0 \le y \le q$ in $\mathcal{R}^2$, define

$$\langle f,g\rangle = \int_0^p \int_0^q f(x,y)g(x,y)\,dx\,dy.$$

(a) Show that $\langle f,g\rangle$ is an inner product on the set of continuous functions on the rectangle.

(b) Show that the functions

$$\phi_{jk}(x,y) = \frac{2}{\sqrt{pq}}\sin\frac{j\pi x}{p}\sin\frac{k\pi y}{q}, \qquad j,k = 1,2,3,\ldots,$$

form an orthonormal set relative to the above inner product:

$$\langle \phi_{jk}, \phi_{lm}\rangle = \begin{cases} 0, & (j,k) \ne (l,m) \\ 1, & (j,k) = (l,m). \end{cases}$$

26. Heat equation in a rectangular plate. The heat equation in rectangular x and y coordinates is $a^2(u_{xx} + u_{yy}) = u_t$.

(a) Assuming product solutions of the form $u(x,y,t) = X(x)Y(y)T(t)$, derive the solutions

$$u_{mn}(x,y,t) = e^{-\pi^2 a^2[(m/p)^2+(n/q)^2]t}\sin\frac{m\pi x}{p}\sin\frac{n\pi y}{q}$$

with boundary values identically zero on the edges of the rectangle with corners at $(0,0)$, $(p,0)$, (p,q), and $(0,q)$.

(b) Use the orthonormal system of the previous exercise to derive the formal expansion

$$u(x,y,t) = \sum_{m,n=1}^{\infty} b_{mn}e^{-\pi^2 a^2[(m/p)^2+(n/q)^2]t}\sin\frac{m\pi x}{p}\sin\frac{n\pi y}{q},$$

$$b_{mn} = \frac{2}{\sqrt{pq}}\int_0^p\int_0^q f(x,y)\sin\frac{m\pi x}{p}\sin\frac{n\pi y}{q}\,dx\,dy$$

for the zero-boundary problem with initial temperature function $f(x,y)$ on the rectangle.

(c) Suppose $w(x,y)$ solves the Laplace equation $u_{xx} + u_{yy} = 0$ on the rectangle with given boundary values on the edges of the rectangle. Verify that if $f(x,y)$ is replaced by $f(x,y) - w(x,y)$ in part (b) to produce solution $U(x,y,t)$, then $u(x,y,t) = w(x,y) + U(x,y,t)$ solves the heat equation subject to the same boundary conditions and with initial temperature function $f(x,y)$.

2 EIGENFUNCTIONS AND SYMMETRY

It is easy to check that the functions $\cos kx$ and $\sin kx$ are solutions of the harmonic oscillator equation $y'' = -k^2 y$. Standard terminology associated with these relations is that each of $\cos kx$ and $\sin kx$ is an **eigenfunction** of the differential operator d^2/dx^2, corresponding to the **eigenvalue** $-k^2$. More generally, let L be a linear differential operator. We say that a function $y = y_\lambda(x)$, not identically zero, is an **eigenfunction** of L corresponding to the **eigenvalue** λ if $Ly_\lambda = \lambda y_\lambda$ for some number λ. If you've found an eigenfunction y corresponding to an eigenvalue λ, then cy is also an eigenfunction for each constant $c \neq 0$; the reason is that, because L is linear, $L(cy) = cLy = c(\lambda y) = \lambda(cy)$.

To understand the connection between eigenfunctions and orthogonal sets of functions, we need one more definition. Suppose that $\mathcal{F}$ is a set of functions with an inner product $\langle f,g \rangle$. (We assume that sums and numerical multiples of functions in $\mathcal{F}$ are also in $\mathcal{F}$.) Let L be a linear operator from $\mathcal{F}$ to $\mathcal{F}$. Then L is **symmetric** with respect to the inner product if $\langle Lf,g \rangle = \langle f,Lg \rangle$ for all functions f and g for which Lf and Lg are defined as elements of $\mathcal{F}$.

In the following examples, we denote the set of continuous functions on $[a,b]$ by $C[a,b]$. Eventually we'll introduce a generalization of the Fourier inner product 1.1 to the **weighted inner product** on $C[a,b]$ with *positive weight function* $r(x)$:

2.1
$$\langle f,g \rangle_r = \int_a^b f(x)g(x)r(x)\,dx.$$

First, we'll continue to concentrate on the special case $r(x) = 1$.

2A Fourier Inner Product

We begin with the simplest example of a symmetric operator on $C[0,\pi]$. In all examples, it's important to understand that symmetry of an operator depends not just on a formula for a differential operator, but also on a precise specification of the functions on which the differential operator acts. This specification is usually made by imposing certain boundary conditions on the functions, and suitable conditions for a particular operator are called **symmetric boundary conditions.**

example 1

Let $\langle f,g \rangle$ be the Fourier inner product defined by Equation 1.1. If we let $Lf = d^2 f/dx^2$, then it is clear that Lf is in $C[0,\pi]$ only for those f in $C[0,\pi]$ that happen to have continuous second derivatives. Thus, for L to be symmetric, we must have $\langle Lf,g \rangle = \langle f,Lg \rangle$ for all twice continuously differentiable f and g on $[0,\pi]$. Equivalently, we must have, because of the definition of the inner product,

$$\int_0^\pi f''(x)g(x)\,dx = \int_0^\pi f(x)g''(x)\,dx.$$

This equation follows from integration by parts twice:

$$\int_0^\pi f''(x)g(x)\,dx = \Big[f'(x)g(x)\Big]_0^\pi - \int_0^\pi f'(x)g'(x)\,dx$$
$$= \Big[f'(x)g(x) - f(x)g'(x)\Big]_0^\pi + \int_0^\pi f(x)g''(x)\,dx.$$

Hence to make L symmetric we restrict its domain to some subset of $C[0,\pi]$ for which the bracketed terms will always add up to zero. This can be done in several ways. For example, we may restrict L to the subset consisting of those functions f in $C[0,\pi]$ for which $f(0) = f(\pi) = 0$, or to the subset for which $f'(0) = f'(\pi) = 0$. With either of these restrictions, L becomes symmetric. Notice that while a restriction of the required type is a boundary condition, in that it specifies the values of f at the endpoints 0 and π of its domain of definition, the symmetry follows from making rather special choices for these "symmetric" conditions.

The connection between orthogonal functions and symmetric operators is as follows.

2.2 THEOREM

Let L be a symmetric linear operator defined on a set of functions with an inner product. If y_1 and y_2 are eigenfunctions of L corresponding to distinct eigenvalues λ_1 and λ_2, then y_1 and y_2 are orthogonal.

Proof. We assume that

$$Ly_1 = \lambda_1 y_1, \qquad Ly_2 = \lambda_2 y_2,$$

and prove that $\langle y_1, y_2\rangle = 0$. We have

$$\langle Ly_1, y_2\rangle = \langle \lambda_1 y_1, y_2\rangle = \lambda_1\langle y_1, y_2\rangle$$

and

$$\langle y_1, Ly_2\rangle = \langle y_1, \lambda_2 y_2\rangle = \lambda_2\langle y_1, y_2\rangle.$$

Because L is symmetric, $\langle Ly_1, y_2\rangle = \langle y_1, Ly_2\rangle$, so $\lambda_1\langle y_1, y_2\rangle = \lambda_2\langle y_1, y_2\rangle$; equivalently, $(\lambda_1 - \lambda_2)\langle y_1, y_2\rangle = 0$. Since λ_1 and λ_2 are not equal, we must have $\langle y_1, y_2\rangle = 0$. ■

A differential operator of **Sturm-Liouville form** is one that can be written

2.3 $$Ly = (py')' + qy,$$

where $p(x)$ is a continuously differentiable function and $q(x)$ is assumed only continuous. [The operator d^2/dx^2 is a special case if we set $p(x) = 1,\ q(x) = 0$.] We want first to see what boundary conditions should be imposed on the domain of L in order to make L symmetric with respect to the Fourier inner product

$$\langle f, g\rangle = \int_a^b f(x)g(x)\,dx.$$

The following identity reduces the problem to its essence.

2.4 LAGRANGE IDENTITY

If L is given by $Ly = (py')' + qy$ on an interval $[a,b]$, then

$$\begin{aligned}\langle y_1, Ly_2\rangle - \langle Ly_1, y_2\rangle &= \left[p(x)\big(y_1(x)y_2'(x) - y_1(x)'y_2(x)\big)\right]_a^b \\ &= p(a)w(a) - p(b)w(b),\end{aligned}$$

where $w(x)$ is the Wronskian of y_1 and y_2. This identity shows that any conditions on the coefficient $p(x)$, or on the set $\mathcal{F}$ containing y_1 and y_2, that makes $p(a)w(a) = p(b)w(b)$ are symmetric conditions that will make L symmetric on $\mathcal{F}$.

Proof. Starting with the definition of L we rearrange $y_1(Ly_2) - (Ly_1)y_2$ as follows:

$$\begin{aligned}y_1(Ly_2) - (Ly_1)y_2 &= y_1[(py_2')' + qy_2] - y_2[(py_1')' + qy_1] \\ &= y_1(py_2')' - y_2(py_1')' \\ &= y_1[py_2'' + p'y_2'] - y_2[py_1'' + p'y_1'] \\ &= p'[y_1y_2' - y_1'y_2] + p[y_1y_2'' - y_1''y_2] \\ &= [p(y_1y_2' - y_1'y_2)]'.\end{aligned}$$

Integrating both sides from a to b gives the Lagrange identity. ■

example 2

The operator defined by $Ly(x) = y''(x)$ is in Sturm-Liouville form, where we take $p(x) = 1$ and $q(x) = 0$. Initially we'll consider L to be operating on twice continuously differentiable functions defined on $[-\pi, \pi]$. To ensure that L is symmetric we'll further restrict the domain of L by requiring that the right side of the Lagrange Identity 2.4 be zero. This we do by restricting attention to those functions $y(x)$ for which $y(0) = y(\pi) = 0$, or else to those for which $y'(0) = y'(\pi) = 0$. Solving the eigenvalue problem $y'' = \lambda y$, we find

$$y(x) = c_1 e^{\sqrt{\lambda}x} + c_2 e^{-\sqrt{\lambda}x}.$$

To satisfy the boundary conditions $y(0) = y(\pi) = 0$, we see that the coefficients c_1 and c_2 must satisfy

$$c_1 + c_2 = 0, \qquad c_1 e^{\sqrt{\lambda}\pi} + c_2 e^{-\sqrt{\lambda}\pi} = 0.$$

Solving for c_1 and c_2 shows that they are zero unless $e^{2\sqrt{\lambda}\pi} = 1$, in which case the only restriction is that $c_2 = -c_1$. The trivial choice $c_1 = c_2 = 0$ is conveniently ruled out as a part of the definition of eigenfunction. The equation $e^{2\sqrt{\lambda}\pi} = 1$ is satisfied by just those values of λ for which $2\sqrt{\lambda}\pi = 2k\pi i$, where k is an integer, that is $\sqrt{\lambda} = ki$. Thus the eigenvalues are $\lambda = -k^2$. Since $c_2 = -c_1$, the corresponding eigenfunctions are

$$y_k(x) = c_1 e^{kix} - c_1 e^{-kix} = 2ic_1 \sin kx.$$

Taking $c_1 = -\frac{1}{2}i$, we get a sequence of eigenfunctions $y_k(x) = \sin kx$, $k = 1, 2, 3, \ldots$, with corresponding eigenvalues $\lambda_k = -k^2$. Orthogonality now follows from Theorem 2.2:

$$\langle y_k, y_l \rangle = \int_0^\pi \sin kx \sin lx \, dx = 0, \qquad k \neq l.$$

example 3

A computation similar to the one in the previous example identifies the eigenfunctions of d^2/dx^2 that satisfy the symmetric boundary conditions $y'(0) = y'(\pi) = 0$; they are $z_k(x) = \cos kx$, $\lambda_k = -k^2$, $k = 0, 1, 2, \ldots$. Working through the details is a useful exercise.

example 4

The operator defined by $Ly(x) = (1 - x^2)y''(x) - 2xy'(x)$, is in Sturm-Liouville form if we set $p(x) = (1 - x^2)$ and $q(x) = 0$. We'll consider L to be operating on twice continuously differentiable functions defined on $[-1,1]$. As usual we make L symmetric by ensuring that the right side of the Lagrange Formula 2.4 is always zero for $a = -1$, $b = 1$. But since $p(x) = (1 - x^2)$, we find that $p(-1) = p(1) = 0$. Hence, L is symmetric on $C[-1,1]$ without further restriction on the domain, so the domain of the symmetric operator L consists of all twice continuously differentiable functions in $C[-1,1]$.

The symmetric operator $L(f)$ defined in Example 4 is usually associated with the differential equation

$$(1 - x^2)y'' - 2xy' + n(n + 1)y = 0.$$

This is called the **Legendre equation** of index n, and it is satisfied by the nth **Legendre polynomial** defined by

2.5
$$P_n(x) = \frac{1}{2^n n!} \frac{d^n}{dx^n}(x^2 - 1)^n, \qquad n = 0, 1, 2, \ldots.$$

That P_n satisfies the Legendre equation can be verified by repeated differentiation. (See Exercise 15.) The significance of having $P_n(x)$ satisfies the Legendre equation comes from writing the equation in the form $Ly = -n(n + 1)y$, where L is the symmetric operator $Ly = (1 - x^2)y'' - 2xy'$ on $C[-1,1]$. Then P_n can be looked at as an eigenfunction of L corresponding to the eigenvalue $-n(n + 1)$. Hence, by Theorem 2.2, the Legendre polynomials are orthogonal, that is,

$$\int_{-1}^{1} P_n(x) P_m(x) \, dx = 0, \qquad n \neq m.$$

Furthermore, a fairly complicated calculation (see Exercise 17) shows that

$$\int_{-1}^{1} P_n^2(x) \, dx = \frac{1}{n + \frac{1}{2}}, \qquad n = 0, 1, 2, \ldots$$

Therefore, the normalized sequence $\{(\sqrt{n + 1/2})P_n(x)\}$ $n = 0, 1, 2, \ldots$ is an orthonormal sequence in $C[-1,1]$. The graphs in Figure 11.1 of the first four

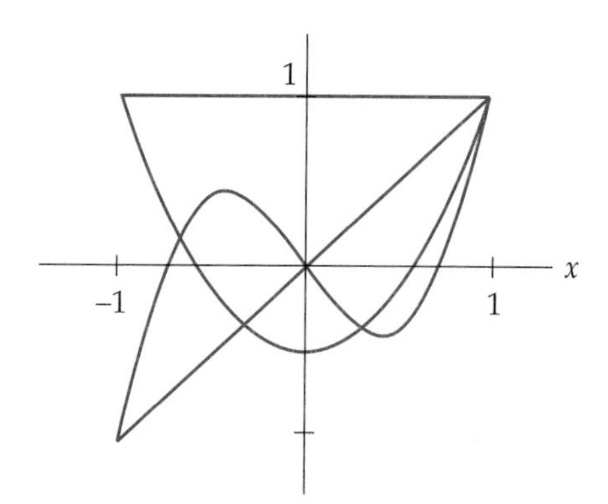

Figure 11.1 $P_n(x)$ for $-1 \leq x \leq 1$, $n = 0, 1, 2, 3$.

Legendre polynomials on $[-1,1]$ make it plausible that these functions are orthogonal to each other in the sense that the integral of the product of each pair is zero over $-1 \leq x \leq 1$. [Notice that in general the graphs of these functions don't intersect orthogonally.]

The Nth **Fourier-Legendre approximation** to a function g in $C[-1,1]$ is the finite sum

$$\sum_{k=0}^{N} c_k P_k(x),$$

where

2.6
$$c_k = \frac{2k+1}{2} \int_{-1}^{1} g(x) P_k(x)\, dx.$$

Theorem 1.4 then implies that the best mean-square approximation to g by a linear combination of Legendre polynomials is given by the Fourier-Legendre approximation. In other words,

$$\int_{-1}^{1} \left[g(x) - \sum_{k=0}^{N} c_k P_k(x) \right]^2 dx$$

is minimized by computing c_k by Equation 2.6. The precise details under which convergence takes place for particular values of x is generally of less interest than for trigonometric series, however, it's worth mentioning that Fourier-Legendre series converge just as trigonometric Fourier series do if $g(x)$ is piecewise monotone on $[-1,1]$.

2B Weighted Inner Products

By introducing the more general weighted inner product 1.2, we can extend the scope of the ideas in Section 2A considerably. In particular, we can include the following ordinary differential equations that arise from separating variables in partial differential equations. Each equation is followed by an equivalent equation in Sturm-Liouville form.

Bessel Equation of Order n

$$x^2y'' + xy' + (\lambda^2x^2 - n^2)y = 0; \qquad (xy')' + \frac{\lambda^2 x - n^2}{x}y = 0.$$

Chebychev Equation

$$(1 - x^2)y'' - xy' + \lambda^2 y = 0; \qquad \left(\sqrt{1-x^2}\, y'\right)' + \frac{\lambda^2}{\sqrt{1-x^2}\, y} = 0.$$

Hermite Equation

$$y'' - xy' + \lambda y = 0; \qquad (e^{-x^2/2}y')' + \lambda e^{-x^2/2}y = 0.$$

Laguerre Equation

$$xy'' + (1 - x)y' + \lambda y = 0; \qquad (xe^{-x}y')' + \lambda e^{-x}y = 0.$$

As is, the first equation in each pair is not in Sturm-Liouville form, but a simple modification enables us to put it in that form so we can apply the theory of Section 2A. Note that none of the equations above is written in the form $Ly = \lambda y$ that we used in Section 2A. However, in applying Theorem 2.2 the generic variable λ may be replaced by $-\lambda$ or even $\pm\lambda^2$ at our convenience. (Recalling our custom of writing $D^2y + \lambda^2 y = 0$, we continue to follow a useful tradition in writing these equations as we have.) The general idea is first to normalize the general eigenvalue-eigenfunction equation

$$a_0(x)y'' + a_1(x)y' + a_2(x)y + \lambda y = 0 \quad \text{to} \quad y'' + a(x)y' + b(x)y + \lambda c(x)y = 0$$

by dividing by $a_0(x)$. At this point we can apply the exponential integrating factor device of Chapter 1, Section 4, to the first two terms:

$$\text{if} \quad p(x) = e^{\int a(x)\,dx}, \quad \text{then} \quad p(x)y'' + p(x)a(x)y' = \frac{d}{dx}\left[p(x)y'\right].$$

After multiplication by $p(x)$, the normalized differential equation can then be written

$$\frac{d}{dx}\left[p(x)y'\right] + q(x)y + \lambda r(x)y = 0, \quad \text{where} \quad q(x) = p(x)b(x)$$
$$\text{and} \quad r(x) = p(x)c(x).$$

This last differential equation is in Sturm-Liouville form, and we can apply the Lagrange Identity 2.4 to it in proving the next theorem, which extends the related discussion of the Fourier inner product in Section 2A. First we look at an example.

example 5

Following the routine outlined above to put the nth order Bessel equation

$$x^2y'' + xy' + (\lambda^2x^2 - n^2)y = 0$$

in Sturm-Liouville form, we first normalize to get $y'' + (1/x)y' + (\lambda^2 - n^2/x^2)y = 0$.

Setting $p(x) = e^{\int (1/x)\,dx} = x$, we multiply this equation by $p(x) = x$ to get

$$xy'' + y' + \frac{\lambda^2 x - n^2}{x} y = 0, \quad \text{or} \quad (xy')' + \frac{\lambda^2 x - n^2}{x} y = 0.$$

Since not only $r(x) = x$ but also $p(x) = x$, we have $p(0) = 0$; thus to achieve symmetry an arbitrary condition such as boundedness is used at $x = 0$ along with a condition at another point $x = b$ that makes the Wronskian $w[y_1, y_2](b)$ equal zero.

2.7 THEOREM

The Sturm-Liouville equation

$$\left[p(x)y'\right]' + q(x)y + \lambda r(x)y = 0$$

is assumed to have $p(x)$ continuously differentiable and $q(x)$ and $r(x)$ continuous on an interval (a,b), which may be infinite. In addition, $r(x)$ is to be positive and integrable on (a,b). Suppose $y_1(x)$ and $y_2(x)$ are eigenfunctions that both satisfy the same symmetric boundary conditions. If the corresponding eigenvalues λ_1 and λ_2 are distinct, then y_1 and y_2 are orthogonal with respect to the $r(x)$-weighted inner product, for which $r(x) = 1$:

$$\langle y_1, y_2 \rangle_r = \int_a^b y_1(x) y_2(x) r(x)\,dx = 0.$$

Alternatively, the functions $w_1(x) = \sqrt{r(x)}\, y_1(x)$, $w_2(x) = \sqrt{r(x)}\, y_2(x)$ are orthogonal with respect to the Fourier inner product: $\langle w_1, w_2 \rangle = 0$.

Proof. Define the linear operator L_r acting on twice continuously differentiable functions in $C[a,b]$ by $L_r y = r^{-1}(x)\left[\left(p(x)y'\right)' + q(x)y\right]$. Note that the factors $r(x)$ and $r^{-1}(x)$ cancel in

$$\langle L_r y_1, y_2 \rangle_r = \int_a^b \left[L_r y_1(x)\right] y_2(x) r(x)\,dx = \langle L y_1, y_2 \rangle,$$

where $Ly = \left[p(x)y'\right]' + q(x)y$. Since L is a Sturm-Liouville operator, imposing symmetric boundary conditions and using the Lagrange Identity makes L a symmetric operator relative to the weighted inner product. Since L and L_r have the same eigenvalues-eigenvector pairs, our conclusion follows from Theorem 2.2 on orthogonality. The final statement follows just because $r \geq 0$ and $(\sqrt{r}\,)^2 = r$. ∎

example 6 The eigenfunction problem for the nth order Bessel equation requires us to solve the nth order Bessel equation $x^2 y'' + xy' + (\lambda^2 x^2 - n^2)y = 0$. To do this in terms of Bessel functions, we assume first that $\lambda \neq 0$ and change variable by $x = z/\lambda$. Thus $d/dx = (dz/dx)(d/dz) = \lambda\, d/dz$ and $d^2/dx^2 = \lambda^2 d^2/dz^2$. The differential equation transforms into the Bessel equation

$$z^2 \overline{y}'' + z\overline{y}' + (z^2 - n^2)\overline{y} = 0, \quad \text{where} \quad \overline{y}(z) = y\frac{z}{\lambda}.$$

By the results of Chapter 9, Section 7B, a solution that is finite at $z = 0$ is a constant multiple of $\overline{y}(z) = J_n(z)$, so $y(x) = J_n(\lambda x)$. In a typical application, boundary conditions consisting of finiteness at zero and $y(b) = 0$, $b > 0$ are

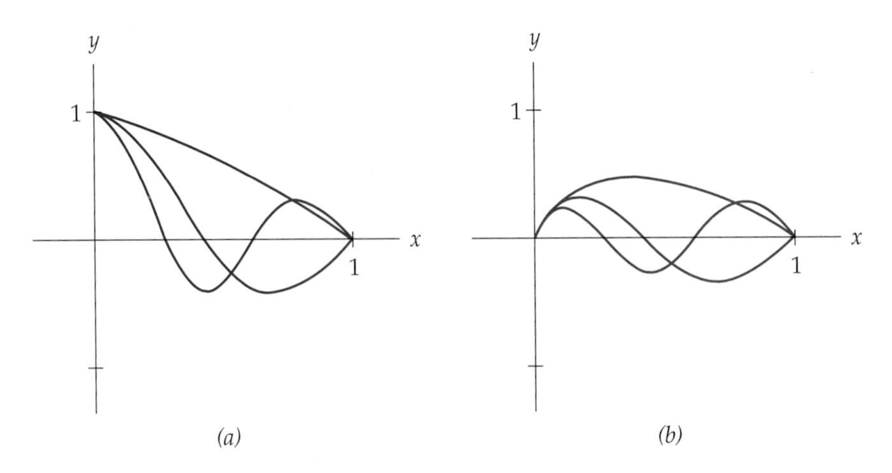

Figure 11.2 (*a*) $y = J_0(\lambda_k x)$. (*b*) $y = \sqrt{x}\, J_0(\lambda_k x)$.

required, so we want $y(b) = J_n(\lambda b) = 0$. In other words, the eigenvalues λ for the problem are those for which λb is a zero of $J_n(x)$. For example if $n = 0$ and $b = 1$, the first three eigenvalues in increasing order are $\lambda_1 \approx 2.4$, $\lambda_2 \approx 5.5$, and $\lambda_3 \approx 8.7$. The graphs of the first three corresponding eigenfunctions are shown in Figure 11.2(a) on the interval $0 \le x \le 1$. These are orthogonal with respect to the weight function $r(x) = x$. This orthogonality is hard to visualize, so the graphs in Figure 11.2(b) are used to show the eigenfunctions modified by the factor $\sqrt{r(x)} = \sqrt{x}$. The modified functions are orthogonal with respect to the uniformly weighted Fourier inner product. The graphs are the analogues for the Bessel equation of the harmonic-oscillator eigenfunctions $\sin \pi x$, $\sin 2\pi x$, and $\sin 3\pi x$.

EXERCISES

By solving $y'' - \lambda y = 0$ find all eigenvalues λ, and for each one a corresponding eigenfunction, of the differential operator d^2/dx^2 satisfying each of the sets of boundary conditions in Exercises 1 through 6.

1. $y(0) = y(\pi) = 0$.

2. $y'(0) = y'(\pi) = 0$.

3. $y(0) = y'(\pi) = 0$.

4. $y(0) = y'(\pi/2) = 0$.

5. $y(0) + y'(0) = y(1) = 0$.

6. $y'(0) = y(\pi) + y'(\pi) = 0$.

By solving $y'' + 2y' - \lambda y = 0$ find all eigenvalues λ, and for each one a corresponding eigenfunction, of the differential operator $d^2/dx^2 + 2d/dx$ satisfying each of the sets of boundary conditions in Exercises 7, 8, and 9.

7. $y(0) = y(\pi) = 0$.

8. $y(0) = y(1) = 0$.

9. $y'(0) = y'(\pi) = 0$.

10. Show that the Lagrange condition $\left[p(x)\left(f_1'(x) f_2(x) - f_1(x) f_2'(x)\right)\right]_a^b = 0$ of 2.4 can be written $p(a)w(a) = p(b)w(b)$, where $w(x)$ is the Wronskian determinant of $f_1(x)$ and $f_2(x)$.

11. **Periodic boundary conditions** A pair of boundary conditions of the form $y(a) = y(b)$, $y'(a) = y'(b)$, $a \neq b$, is called **periodic.**

(a) Show that a Sturm-Liouville operator $Ly = \left[p(x)y'\right]' + q(x)$ is symmetric if it is restricted by periodic boundary conditions and also satisfies $p(a) = p(b)$.

(b) Suppose a function $y = y(x)$ satisfies periodic boundary conditions at a and b and is continuously differentiable for $a \leq x \leq b$. Show that $y(x)$ can be extended to be continuously differentiable and periodic with period $b - a$ for all real x. (This result is the origin of the term "periodic" in the present context.)

Show that the differential operator d^2/dx^2 is symmetric with respect to the inner product

$$\langle y_1, y_2 \rangle = \int_{-1}^{1} y_1(x) y_2(x)\, dx,$$

and with each of the sets of boundary conditions in Exercises 12, 13, and 14.

12. $y(-1) = y(1) = 0$.

13. $y'(1) = y'(-1) = 0$.

14. $c_1 y(-1) + d_1 y'(-1) = c_2 y(1) + d_2 y'(1) = 0$, where c_k and d_k aren't both zero.

15. Verify that the Legendre polynomial P_n defined by Formula 2.5 satisfies the Legendre equation: $(1 - x^2)y'' - 2xy' + n(n+1)y = 0$. To do this let $u = (x^2 - 1)^n$. Then $(x^2 - 1)u' = 2nxu$. Differentiate both sides $(n + 1)$ times with respect to x. Use the Leibnitz rule for the derivative of a product:

$$(fg)^{(n)} = \sum_{k=0}^{n} \binom{n}{k} f^{(k)} g^{(n-k)}.$$

16. Compute the Legendre polynomials $P_0(x)$, $P_1(x)$, $P_2(x)$, and $P_3(x)$ from Equation 2.5. Then verify directly by integration that P_1 and P_3 are orthogonal with respect to the Fourier inner product over the interval $[-1,1]$.

17. **(a)** By using Formula 2.5 and repeated integration by parts, show that

$$\int_{-1}^{1} P_n^2(x)\, dx = \frac{(2n)!}{2^{2n}(n!)^2} \int_{-1}^{1} (1 - x^2)^n\, dx.$$

(b) Show that

$$\int_{-1}^{1} (1 - x^2)^n\, dx = 2 \int_{0}^{\pi/2} \sin^{2n+1}\theta\, d\theta.$$

(c) Show that

$$\int_{0}^{\pi/2} \sin^{2n+1}\theta\, d\theta = \frac{2 \cdot 4 \cdot 6 \cdots (2n)}{1 \cdot 3 \cdot 5 \cdots (2n+1)}.$$

(d) Show that

$$\int_{-1}^{1} P_n^2(x)\, dx = \frac{2}{2n+1}.$$

18. Prove that P_n, the nth Legendre polynomial, has n distinct roots in the interval $[-1,1]$. [*Hint:* Use Formula 2.5 and Rolle's theorem.]

19. Chebychev equation Show that $(1-x^2)y'' - xy' + \lambda^2 y = 0$ in Sturm-Liouville form is $(\sqrt{1-x^2}\, y')' + (\lambda^2/\sqrt{1-x^2}\,)y = 0$.

20. Hermite equation Show that $y'' - xy' + \lambda y = 0$ in Sturm-Liouville form is $(e^{-x^2/2}y')' + \lambda e^{-x^2/2}y = 0$.

21. Laguerre equation Show that $xy'' + (1-x)y' + \lambda y = 0$ in Sturm-Liouville form is $(xe^{-x}y')' + \lambda e^{-x}y = 0$.

22. The 3-dimensional Laplace equation in spherical coordinates (r,ϕ,θ) has the form

$$\frac{\partial}{dr}\left(r^2\frac{\partial u}{\partial r}\right) + \frac{1}{\sin\phi}\frac{\partial}{d\phi}\left(\sin\phi\frac{\partial u}{\partial\phi}\right) + \frac{1}{\sin^2\phi}\frac{\partial^2 u}{\partial\theta^2} = 0.$$

(a) Show that for solutions $u(r,\phi,\theta) = v(r,\phi)$ that are independent of θ, the equation has the form

$$r\frac{\partial^2}{dr^2}(rv) + \frac{1}{\sin\phi}\frac{\partial}{\partial\phi}\left(\sin\phi\frac{\partial v}{\partial\phi}\right) = 0.$$

(b) Show that the method of separation of variables applied to the equation of part (a) leads to the two ordinary differential equations

$$r^2R'' + 2rR' = \lambda R,$$

$$\frac{1}{\sin\phi}\frac{d}{d\phi}\left(\sin\phi\frac{d\Phi}{d\phi}\right) = -\lambda\Phi.$$

(c) Show that the equation of Euler type for $R(r)$ has solutions

$$r^{-(1/2)+\sqrt{\lambda+(1/4)}} \quad \text{and} \quad r^{-(1/2)-\sqrt{\lambda+(1/4)}}.$$

[*Hint:* Let $R = r^\mu$.]

(d) Show that the equation for Φ can be put in the form of the Legendre equation

$$(1-x^2)y'' - 2xy' + \lambda y = 0.$$

[*Hint:* Let $x = \cos\phi$, and use the chain rule.]

(e) By setting $\lambda = n(n+1)$, find a sequence of solutions to the partial differential equation of part (d).

23. Show that, in the case of the orthonormal sequence derived from the Legendre polynomials, the general expansion $\sum_{k=0}^{N}\langle g,\phi_k\rangle\phi_k$ reduces to $\sum_{k=0}^{N} c_k P_k(x)$, where c_k is given by Equation 2.6 of the text.

24. Let $a_0(x)$, $a_1(x)$, and $a_2(x)$ be continuous on an interval, with $a_0(x) \neq 0$. Show that the expression $a_0(x)y'' + a_1(x)y' + a_2(x)y$ can be written in the Sturm-Liouville form $(py')' + qy$ after multiplication by $m(x)/a_0(x)$, where $m(x) = \exp\left(\int (a_1(x)/a_0(x))\,dx\right)$. What are $p(x)$ and $q(x)$ in terms of $a_0(x)$, $a_1(x)$, and $a_2(x)$?

25. Write out a formal proof that if y is an eigenfunction of a linear operator L with eigenvalue λ, then so is cy for constant $c \neq 0$.

26. If λ is an eigenvalue of the linear operator L, the set of all linear combinations of corresponding eigenfunctions is called the **eigenspace** of λ for L.

(a) Let L be the linear differential operator d^2/dx^2 restricted to twice continuously differentiable functions $y = y(x)$ satisfying $y(0) = y(\pi)$. Show that the eigenvalues are all of the form $-n^2$, where n is a positive integer, and show that the eigenspace of L is 1-dimensional, consisting of all nonzero constant multiples $cy(x)$ of a single function.

(b) Let L be the linear differential operator d^2/dx^2 restricted to twice continuously differentiable functions $y = y(x)$ satisfying both $y(0) = y(2\pi) = 0$ and $y'(0) = y''(2\pi)$. Show that the eigenvalues are all of the form $-n^2$, where n is a positive integer, and show that the eigenspace of L is 2-dimensional, consisting of all nonzero linear combinations $c_1 y_1(x) + c_2 y_2(x)$ of two linearly independent functions.

3 VIBRATIONAL MODES

The purpose of this section is to apply the ideas in the previous two sections to interpreting solutions of boundary and initial-value problems for higher dimensional wave equations. Because its solutions are constructed using elementary trigonometric functions, we begin with a problem where separation of variables involves a **double Fourier sine series** of an integrable function $f(x, y)$:

3.1

$$\sum_{j,k=1}^{\infty} b_{jk} \sin \frac{j\pi x}{p} \sin \frac{k\pi y}{q}, \quad \text{where}$$

$$b_{jk} = \frac{4}{pq} \int_0^p \int_0^q f(x, y) \sin \frac{j\pi x}{p} \sin \frac{k\pi x}{q} \, dx \, dy.$$

We'll be specially concerned with the geometry of the individual terms in such expansions, in particular with regions in the xy plane in which their graphs lie above or below the plane and certain curves called *nodal curves* that separate these regions.

3A Rectangular Membrane

An ideal drumhead is a homogeneous 2-dimensional membrane that stays motionless along a boundary curve while the interior points move vertically in response to initial excitation. We consider small vertical displacements $u(x, y, t)$ of a membrane, depending on time t and position (x, y) in a plane rectangular region R: $0 \le x \le p,\ 0 \le y \le q$. For the boundary condition, the displacement u is held at value zero all around the edge of the rectangle, and natural initial conditions are $u(x, y, 0) = f(x, y)$ and $u_t(x, y, 0) = g(x, y)$, where $f(x, y)$ and $g(x, y)$ represent initial displacement and initial velocity respectively at (x, y). The appropriate partial differential equation for u turns out to be the **2-dimensional wave equation**

$$a^2(u_{xx} + u_{yy}) = u_{tt},$$

where a^2 is a positive constant depending on the uniform density and tension of the membrane. For functions $u(x,y,t) = u(x,t)$ that are independent of y, we have $u_{yy} = 0$, so the differential equation reduces to the 1-dimensional wave equation of Chapter 10, Section 4C. Recalling from Chapter 10, Section 5, the definition of the **2-dimensional Laplace operator** by $\Delta u = u_{xx} + u_{yy}$, the 2-dimensional wave equation can be written more concisely as $a^2 \Delta u = u_{tt}$.

We begin by finding product solutions $u(x,y,t) = X(x)Y(y)T(t)$ to the boundary-value problem, requiring $a^2 \Delta u = u_{tt}$ with $u(x,0,t)$, $u(p,y,t)$, $u(x,q,t)$, and $u(0,y,t)$ all to be zero on the boundary of the rectangle. These conditions are equivalent to requiring the factors in the product solution to satisfy $X(0) = X(p) = 0$ and $Y(0) = Y(q) = 0$. Substitution into the differential equation gives $a^2(X''YT + XY''T) = XYT''$ or $a^2(X''/X + Y''/Y) = T''/T = -\lambda^2$, where $-\lambda^2$ is just a convenient way to write the arbitrary constant that arises when we separate variables. Indeed, the resulting equation $T'' + \lambda^2 T = 0$ has the oscillatory solutions $T(t) = A\cos\lambda t + B\sin\lambda t$ that we associate with vibration. (The apparent prejudice here is purely formal; $-\lambda^2$ can assume positive values if λ is allowed to be imaginary.) The equation for X and Y is $X''/X + Y''/Y = -\lambda^2$. Separating variables again gives

$$\frac{X''}{X} = -\frac{Y''}{Y} - \lambda^2 = -l^2,$$

where $-l^2$ is constant because it can't vary with either x or y. The resulting ordinary differential equations are

$$X'' + l^2 X = 0 \quad \text{and} \quad Y'' + (\lambda^2 - l^2)Y = 0.$$

Mindful that the first of these equations has boundary conditions $X(0) = X(p) = 0$, we find solutions $X(x) = \sin j\pi x/p$, so $l^2 = j^2\pi^2/p^2$, where j and k are integers. Similarly the boundary problem for Y has solutions $Y(y) = \sin k\pi y/q$, so $\lambda^2 - l^2 = \lambda^2 - j^2\pi^2/p^2 = k^2\pi^2/q^2$. Note that as a result $\lambda^2 = \pi^2(j^2/p^2 + k^2/q^2)$. Thus our product solutions take the form

$$u_{jk}(x,y,t) = \left(A_{jk}\cos\lambda_{jk}t + B_{jk}\sin\lambda_{jk}t\right)\sin\frac{j\pi x}{p}\sin\frac{k\pi y}{q},$$

where $\lambda_{jk} = \pi\sqrt{j^2/p^2 + k^2/q^2}$.

We're now able to consider initial displacement $u(x,y,0)$ and velocity $u_t(x,y,0)$. To accommodate a significant amount of generality for these conditions, we use the linearity of the differential equation to try to put together solutions

3.2 $$u(x,y,t) = \sum_{j,k=1}^{\infty} \sin\frac{j\pi x}{p}\sin\frac{k\pi y}{q}\left(A_{jk}\cos\lambda_{jk}t + B_{jk}\sin\lambda_{jk}t\right).$$

Since we want $u(x,y,0) = \sum_{j,k=1}^{\infty} A_{jk}\sin(j\pi x/p)\sin(k\pi y/q)$, we determine the A_{jk} as double Fourier coefficients given in general form by Equation 3.1:

3.3 $$A_{jk} = \frac{4}{pq}\int_0^p\int_0^q u(x,y,0)\sin\frac{j\pi x}{p}\sin\frac{k\pi x}{q}\,dx\,dy.$$

Similarly, if $u_t(x,y,0)=\sum_{j,k=1}^{\infty}\lambda_{jk}B_{jk}\sin(j\pi x/p)\sin(k\pi y/q)$, we compute B_{jk} by

3.4 $$B_{jk}=\frac{4}{pq\lambda_{jk}}\int_0^p\int_0^q u_t(x,y,0)\sin\frac{j\pi x}{p}\sin\frac{k\pi y}{q}\,dx\,dy.$$

The convergence of the expansion in Equation 3.2 is quite complicated in general. We get some insight into this just by examining the multiplicity of individual terms. As in the 1-dimensional case, the t-dependent factors are just sinusoidal oscillations with respective circular frequencies $\lambda_{jk}=\pi\sqrt{j^2/p^2+k^2/q^2}$, and these frequencies are essentially pitches of certain vibrational "modes." Each factor $v_{jk}(x,y)=\sin(j\pi x/p)\sin(k\pi y/q)$ determines a basic **mode** of vibration. It is easy to verify directly, and it's implicit in the derivation given above, that each v_{jk} is an eigenfunction of the Laplace operator Δ corresponding to eigenvalue $-\lambda_{jk}^2=-\pi^2(j^2/p^2+k^2/q^2)$. Hence, in contrast with the case of one space dimension, there is frequently more than one independent eigenfunction for each eigenvalue, since a number can sometimes be represented as a sum of squares in more than one way. Beyond that, it's easy to verify that a linear combination of eigenfunctions with the same eigenvalue is also an eigenfunction, so the variety among eigenfunctions with a given eigenvalue can be considerable.

example 1

We continue to assume boundary values identically zero. If $p=q=\pi$, then $-\lambda^2=-j^2-k^2$, so the possible eigenvalues are the negatives of integers that can be represented as a sum of squares of two positive integers. For example, eigenvalue $-2=-1^2-1^2$ has the single eigenfunction $v_{11}(x,y)=\sin x\sin y$. But $-5=-1^2-2^2$ has the independent eigenfunctions $v_{12}(x,y)=\sin x\sin 2y$ and $v_{21}(x,y)=\sin 2x\sin y$. The eigenvalue -50 has three independent eigenfunctions: v_{17}, v_{71}, and v_{55}, so, for example, $v_{17}-v_{71}+2v_{55}$ is also an eigenfunction with eigenvalue -50. Each of these eigenfunctions is made to oscillate with the appropriate circular frequency $\lambda_{jk}=\sqrt{j^2+k^2}$ in the sense that the expansion in Equation 3.2 will contain, for example, the terms

$$(A_{17}\sin x\sin 7y+A_{71}\sin 7x\sin y+A_{55}\sin 5x\sin 5y)\cos\pi\sqrt{50}\,t,$$

where the constants A_{jk} depend on the initial displacement $u(x,y,0)$ as in Equation 3.3.

Figure 11.3 shows the graphs of three of the infinitely many different eigenfunctions with eigenvalue -50. A glance at Figure 11.3 suggests that there are

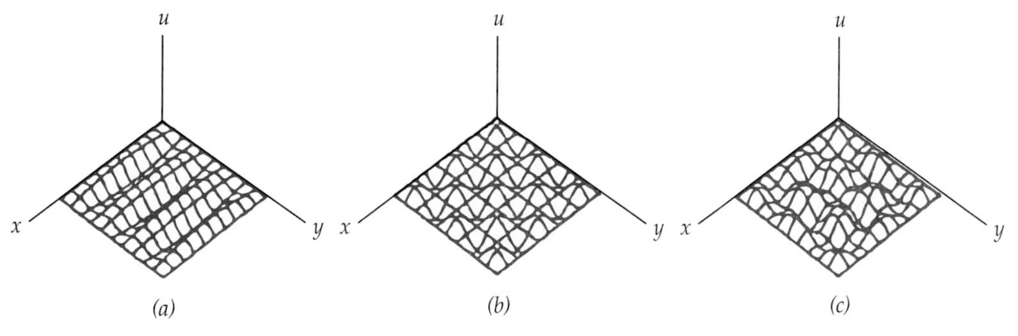

Figure 11.3 (a) $\frac{1}{3}v_{17}$. (b) $\frac{1}{3}v_{55}$. (c) $\frac{1}{6}v_{17}+\frac{1}{4}v_{71}+\frac{1}{4}v_{55}$.

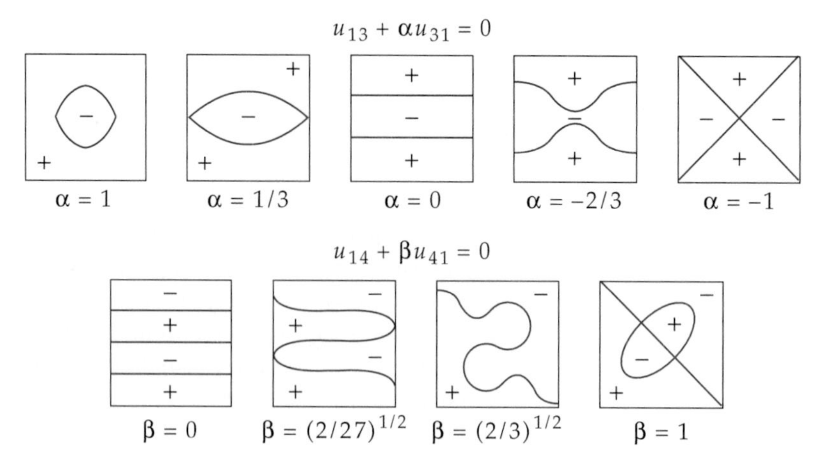

Figure 11.4 Nodal curves.

curves along which the displacement is always zero; such a curve is called a **nodal curve** and is analogous to a "nodal point" or **node** in a trigonometric oscillation, which also remains fixed during oscillation in one space dimension. Nodal curves are significant because they mark the boundaries of regions on which displacement actually takes place. Figure 11.4 shows a selection of nodal curves associated with the eigenvalues -10 and -17. Each of these drawings represents the level set at level zero of the corresponding eigenfunction, restricted in these examples to a π-by-π square. Each level set can be realized as the result of a gradual deformation of the adjacent ones. The signs show the regions in which the eigenfunction is positive or negative.

The orthogonality of the eigenfunctions $v_{jk}(x,y) = \sin(j\pi x/p)\sin(k\pi y/p)$ over the p-by-q square follows easily by elementary integration. It's left as an exercise to show that the Laplace operator Δ is symmetric with respect to the zero boundary values that we've quite naturally imposed. Orthogonality then follows from Theorem 2.2 without any additional computation.

3B Circular Membrane

We can think of a conventional drumhead as a homogeneous membrane bounded by a circular frame. The relevant wave equation is still $a^2\Delta u = u_{tt}$, but to adapt the method of Section 3A to a circular domain we need to use the chain rule to express the **Laplace operator in polar coordinates,** described in Exercise 19 of Chapter 10, Section 5.

$$\Delta u = u_{rr} + \frac{1}{r}u_r + \frac{1}{r^2}u_{\theta\theta}.$$

(The polar coordinate transformation is singular at the origin, so $r = 0$ is a singular point for the transformed operator.) The wave equation for a circular region becomes $a^2(u_{rr} + r^{-1}u_r + r^{-2}u_{\theta\theta}) = u_{tt}$. Looking for product solutions $u(r,\theta,t) = R(r)\Theta(\theta)T(t)$, we find

3.5
$$R''\Theta T + \frac{1}{r}R'\Theta T + \frac{1}{r^2}R\Theta''T = \frac{1}{a^2}R\Theta T''.$$

Division by $R\Theta T$ gives

$$\frac{R''}{R} + \frac{1}{r}\frac{R'}{R} + \frac{1}{r^2}\frac{\Theta''}{\Theta} = \frac{1}{a^2}\frac{T''}{T}.$$

Initial conditions specify $u(r,0)$ and $u_t(r,0)$ as initial displacement and velocity. One boundary condition for a membrane of radius b is $u(b,t) = 0$. The singularity at $r = 0$ gives rise to a new boundary point, where we impose just the condition that the solution remain bounded. We first look at solutions that are independent of θ; these can be generated by initial conditions that are independent of θ. (In practice, this might be done by striking a drum at its precise center.) A solution independent of θ is called a **radial function,** because position in the plane of equilibrium is uniquely determined by specifying r.

example 2

For solutions independent of θ, we have $\Theta'' = 0$ so radial solutions to the circular drum problem satisfy

$$\frac{R''}{R} + \frac{1}{r}\frac{R'}{R} = \frac{1}{a^2}\frac{T''}{T}.$$

Both sides of the equation must be constant, which we call $-\lambda^2$ for convenience. The two separated equations are

$$rR'' + R' + \lambda^2 rR = 0 \quad \text{and} \quad T'' + \lambda^2 a^2 T = 0.$$

The first of these equations is the Bessel equation of order 0, with solution $R(r) = J_0(\lambda r)$ bounded at $r = 0$. (We reject all solutions independent of this one, since they are guaranteed to be unbounded by Theorem 7.5 of Chapter 9.) The boundary condition at $r = b$ requires $R(b) = J_0(\lambda b) = 0$, so λ will be chosen from among the increasing sequence of positive values λ_j for which $J_0(\lambda_j b) = 0$. The key λs are then r_j/b, where the r_j are just the positive zeros of J_0, the first five of which are located approximately at $r_1 = 2.4$, $r_2 = 5.5$, $r_3 = 8.7$, $r_4 = 11.8$, and $r_5 = 14.9$. The corresponding solutions to the equation for T are the usual sinusoidal ones, so the θ-independent problem has solutions

$$u(r,t) = \sum_{j=0}^{\infty} J_0(\lambda_j r)(A_j \cos \lambda_j at + B_j \sin \lambda_j at).$$

The A_j and B_j are determined by the initial conditions as follows. In Section 2B, we saw that the functions $\sqrt{r}\,J_0(\lambda_j r)$ are orthogonal on the interval $0 \le r \le 1$, so that

$$\int_0^b r J_0(\lambda_j r) J_0(\lambda_k r)\, dr = 0, \qquad j \ne k.$$

Making a Fourier-Bessel expansion requires that the eigenfunctions be normalized by factors designed to make this integral equal 1 when $j = k$. This comes about if we multiply $J_0(\lambda_j r)$ by $N_j = \left[\int_0^b r J_0^2(\lambda_j r)\, dr\right]^{-1/2}$. Hence we can

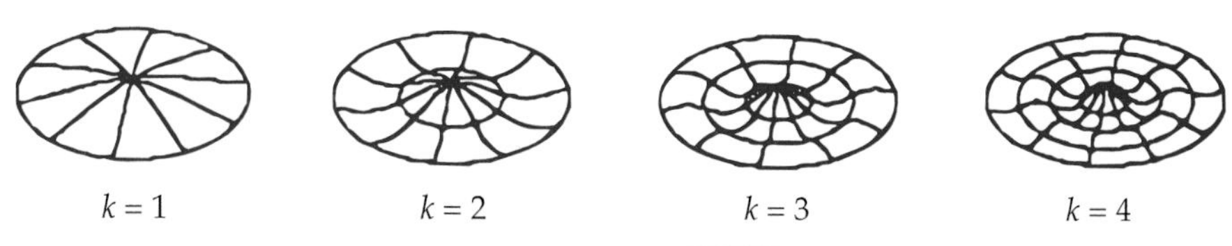

Figure 11.5 Graphs of $J_0(x_k r)$, $r = \sqrt{x^2 + y^2} \le 1$ with $J_0(x_k) = 0$.

compute the coefficients by

$$A_j = N_j^2 \int_0^b r J_0(\lambda_j r) u(r,0)\, dr, \qquad B_j = \frac{N_j^2}{a\lambda_j} \int_0^b r J_0(\lambda_j r) u_t(r,0)\, dr.$$

The coefficient formulas can be simplified somewhat for computing purposes, but such computations are complicated at best. We'll concentrate here on studying the eigenfunctions and their associated vibrational modes. Figure 11.5 shows in perspective the first four modes of the radially symmetric circular membrane of radius 1, starting with the "fundamental." The circles are the stationary nodal curves of an associated oscillation such as $J_0(r_k r) \sin a r_k t$, where r_k is the kth zero of J_0. The radii for fixed k are respectively $\rho_l = r_l / r_k$, $l = 1, 2, \ldots, k$. Note that the kth mode oscillates with circular frequency $a r_k$. Since none of these frequencies is a rational multiple of another, a linear combination of these oscillations will never be periodic, so a conventional drum will not sound "harmonious" by itself the way some musical instruments do. With some exceptions, the same remark applies to the rectangular drum.

---

example 3

Allowing dependence on the polar angle θ in the wave equation, we can first separate out the time-dependence in Equation 3.5 for product solutions such as

$$\frac{R''}{R} + \frac{1}{r}\frac{R'}{R} + \frac{1}{r^2}\frac{\Theta''}{\Theta} = \frac{1}{a^2}\frac{T''}{T} = -\lambda^2.$$

Of the two resulting equations

$$T'' + (\lambda a)^2 T = 0 \quad \text{and} \quad \frac{R''}{R} + \frac{1}{r}\frac{R'}{R} + \frac{1}{r^2}\frac{\Theta''}{\Theta} = -\lambda^2,$$

the second separates with an additional constant k^2 to give

$$r^2 \frac{R''}{R} + r\frac{R'}{R} + \lambda^2 r^2 = -\frac{\Theta''}{\Theta} = k^2.$$

Thus we have $\Theta'' + k^2\Theta = 0$ and $r^2 R'' + r R' + (\lambda^2 r^2 - k^2) R = 0$. To be well defined as a function of position in a plane, $\Theta(\theta)$ must have period 2π as a function of the angle θ. Hence k must be an integer in the general solution formula $\Theta_k(\theta) = \alpha_k \cos k\theta + \beta_k \sin k\theta$. The equation for $R(r)$ is a parametric Bessel equation with bounded solutions $R_k(r) = J_k(\lambda r)$. (Recall that by Theorem 7.5 of Chapter 9 the solutions independent of these are all unbounded near $r = 0$.) To make all solutions be zero on the boundary of a disk of radius b, we require $R_k(b) = J_k(\lambda b) = 0$; thus λb must be a zero r_{kj} of the Bessel function J_k. We define λ_{kj} by $\lambda_{kj} = r_{kj}/b$. Using these values for λ in the sinusoidal solutions $\cos \lambda a t$ and $\sin \lambda a t$ of $T'' + (\lambda a)^2 T = 0$, we find four types of product solution to the differential equation:

$$J_k(\lambda_{kj}r)\cos k\theta \cos \lambda_{kj}at, \qquad J_k(\lambda_{kj}r)\cos k\theta \sin \lambda_{kj}at,$$
$$J_k(\lambda_{kj}r)\sin k\theta \cos \lambda_{kj}at, \qquad J_k(\lambda_{kj}r)\sin k\theta \sin \lambda_{kj}at.$$

When $k = 0$, we get only the two terms $J_0(\lambda_{0j}r)\cos \lambda_{0j}at$ and $J_0(\lambda_{0j}r)\sin \lambda_{0j}at$ that appeared in the θ-independent solutions of the previous example.

Associated with each circular frequency $\lambda_{kj}a$ is a vibrational mode of the form

$$A_{kj}J_k(\lambda_{kj}r)\cos k\theta + B_{kj}J_k(\lambda_{kj}r)\sin k\theta = C_{kj}J_k(\lambda_{kj}r)\cos k(\theta - \phi_{kj}).$$

The A_{kj} and B_{kj}, or alternatively the C_{kj} and ϕ_{kj}, are determined as usual by initial displacement and velocity. It follows that the typical nodal curve tied to a given frequency has a fairly simple structure consisting of concentric circles, where the Bessel factor is zero, together with evenly spaced radial lines along which the trigonometric factor is zero. Figure 11.6 shows some nodal curves for some $u_{kj}(r,\theta) = J_k(\lambda_{kj}r)\cos k\theta$; comparison with Figure 11.4 shows that rectangular drums offer more variety in the way of regions that oscillate at a given circular frequency $\lambda_{kj}a$.

When an ideal drumhead of whatever shape is set in motion, the oscillatory modes are eigenfunctions $u_\lambda(x,y)$ of the Laplace operator: $\Delta u_\lambda = \lambda u_\lambda$. The edge of a drumhead is assumed to lie in a plane at level zero, so the sets of eigenfunctions that distinguish one drum shape from another have the common property that they all assume the value zero on the boundary. The corresponding eigenvalues determine the oscillation frequencies of the various modes and it is these frequencies that determine the quality of the sound produced by the drum. Can two differently shaped drums have exactly the same eigenvalues? (Roughly speaking, can you hear the shape of a drum by detecting these frequencies?) This question was much discussed but remained unanswered for about 25 years. It turns out that you can't hear the shape of a drum in this way; mathematicians Carolyn Gordon, David Webb, and Scott Wolpert collaborating in 1991 managed to find distinct nonconvex drum shapes that generate identical sequences of eigenvalues. For an elementary exposition see *American Scientist*, vol. 84, no. 1 (1996), pp. 46–55.

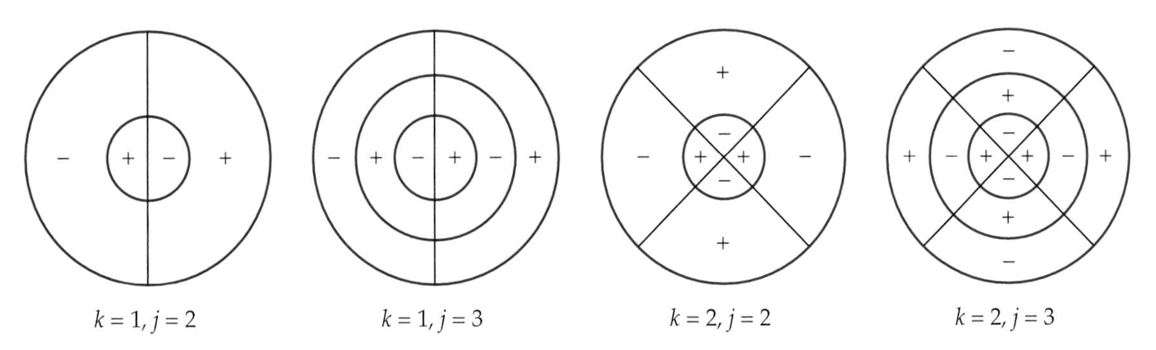

Figure 11.6 Nodal curves.

EXERCISES

1. Make the analogues of the sketches in Figure 11.4 for the eigenfunctions $u_{12}(x,y) = \sin x \sin 2y$ and $u_{21}(x,y) = \sin 2x \sin y$.

2. Make the analogues of the sketches in Figure 11.4 for the eigenfunctions $u_{13}(x,y) = \sin x \sin 3y$ and $u_{31}(x,y) = \sin 3x \sin y$.

3. Make the analogue of a sketch in Figure 11.4 for the eigenfunction $u_{12}(x,y) + u_{21}(x,y) = \sin x \sin 2y + \sin 2x \sin y$. [*Hint:* Aside from the boundary square, show that the level set satisfies $\cos x + \cos y = 0$.]

4. Make the analogues of the sketches in Figure 11.6 for the eigenfunctions $u_{11}(r,\theta) = J_1(r_{11}r)\cos\theta$ and $u_{21}(r,\theta) = J_2(r_{21}r)\cos 2\theta$. Here r_{k1} is the smallest positive zero of J_k.

5. Example 1 of the text shows that for a π-by-π membrane the mode of vibration linked to a given frequency can in general be decomposed into a number n of different sinusoidal products.

(a) For $n = 65$, there are four such products. What are they?

(b) How many such products are there when $n = 325$, and what are they?

6. The 2-dimensional heat equation in rectangular coordinates is $a^2(u_{xx} + u_{yy}) = u_t$. Find the product solutions that are identically zero on the boundary of the p-by-q rectangle with opposite corners at (0,0) and (p,q).

7. The 2-dimensional heat equation in polar coordinates is $a^2(u_{rr} + r^{-1}u_r + r^{-2}u_{\theta\theta}) = u_t$. Find the product solutions that are identically zero on the circle of radius b centered at (0,0).

8. Consider the following vibrational modes of a rectangular π-by-π membrane:

$$u_{jk}(x,y,t) = \sin(jx)\sin(ky)\sin\sqrt{j^2+k^2}\,t.$$

(a) Show that two modes u_{kj} and u_{lm} have a common time period of oscillation when the ratio $(j^2+k^2)/(l^2+m^2)$ is the square of a rational number.

(b) Give some examples of modes that have a common period. [*Hint:* Consider integer pairs $j = 2ab,\ k = a^2 - b^2$, where $a > b$.]

(c) Give some examples of two modes that don't have a common period, explaining why they don't.

9. **Green's theorem** for a plane region R with piecewise-smooth boundary curve ∂R asserts the validity of an equation relating an area integral for a gradient field $\nabla f = (f_x, f_y)$ to an arc-length integral of a directional derivative: $\int_R (f_{xx} + f_{yy})\,dx\,dy = \int_{\partial R}(\partial f/\partial \mathbf{n})\,ds$, where $\mathbf{n}$ is the outward-pointing unit normal vector on ∂R and f is continuously differentiable.

(a) Use Green's theorem to derive **Green's first identity** for the Laplace operator Δ:

$$\int_R f \Delta g \, dx\, dy + \int_R \nabla f \cdot \nabla g \, dx\, dy = \int_{\partial R} f \frac{\partial g}{\partial \mathbf{n}} \, ds.$$

(b) Use part (a) to prove **Green's second identity:**

$$\int_R (f \Delta g - g \Delta f) \, dx\, dy = \int_{\partial R} \left(f \frac{\partial g}{\partial \mathbf{n}} - g \frac{\partial f}{\partial \mathbf{n}} \right) ds.$$

10. Define an inner product $\langle f, g \rangle = \int_R fg \, dx\, dy$.

(a) Use Green's second identity of the previous exercise to show that the Laplace operator Δ is symmetric when acting on twice continuously differentiable functions on R with boundary values zero: $\langle \Delta f, g \rangle = \langle f, \Delta g \rangle$.

(b) Show that the symmetry result of part (a) holds if the condition that f and g be zero on the boundary is replaced by the condition that their normal derivatives be zero there.

11. The Bessel equation $x^2 y'' + xy' + (x^2 - n^2)y = 0$ with solutions $J_n(x)$ derived in Chapter 9, Section 7B, is somewhat less general than the parametric Bessel equation

$$x^2 y'' + xy' + (\lambda^2 x^2 - n^2)y = 0$$

that we need for solving the wave equation. Show that the more general parametric equation is satisfied by $y(x) = J_n(\lambda x)$.

12. Show that the Fourier-Bessel coefficients in Example 2 of the text can be computed by

$$A_k = \frac{1}{M_k} \int_0^1 x J_0(x_k x) u(bx, 0) \, dx, \qquad B_k = \frac{b}{a x_k M_k} \int_0^1 x J_0(x_k x) u_t(bx, 0) \, dx,$$

where x_k is the kth positive zero of $J_0(x)$ and $M_k = \int_0^1 x J_0^2(x_k x) \, dx$.

APPENDIX A

Complex Numbers

Complex numbers arise naturally when we try to extend our method for solving linear constant-coefficient differential equations of order two or more.

example 1

The differential equation $d^2y/dx^2 + y = 0$ has characteristic equation $r^2 + 1 = 0$, or $r^2 = -1$. Since there are no solutions r to this equation among the real numbers, the thing to do is work in the larger complex number field that contains solutions $r = \pm i$ having the property that $i^2 = -1$. (This is somewhat analogous to what we do when faced with the problem of finding a solution to the equation $r^2 = 2$ among the rational numbers; since it can be proved that there are no rational solutions, we enlarge the number system to include some irrational numbers such as $\pm\sqrt{2}$.) Following the routine that we use for real roots of a characteristic equation, we would expect solutions to the differential equation $d^2y/dx^2 + y = 0$ to look like

$$y(x) = c_1 e^{ix} + c_2 e^{-ix}.$$

Interpreting the formula at the end of the previous example in a useful way is the subject matter of Section 2 of Chapter 3. The rest of this appendix is taken up with the arithmetic of complex numbers, that is, the addition, multiplication, and division of numbers such as i, $-2i$, and $1 - \sqrt{3}i$. A **complex number** $z = x + iy$, with **real part** $\text{Re}(z) = x$ and **imaginary part** $\text{Im}(z) = y$, can be identified with the point, or vector, (x, y) in $\mathcal{R}^2$, as in Figure A.1. This identification not only gives us a useful pictorial representation for z, but makes $z = x + iy$ a clearly defined mathematical entity. In particular, the **purely real** complex numbers $x + i0$ are identified with the points $(x, 0)$ on the **real axis,** and the **purely**

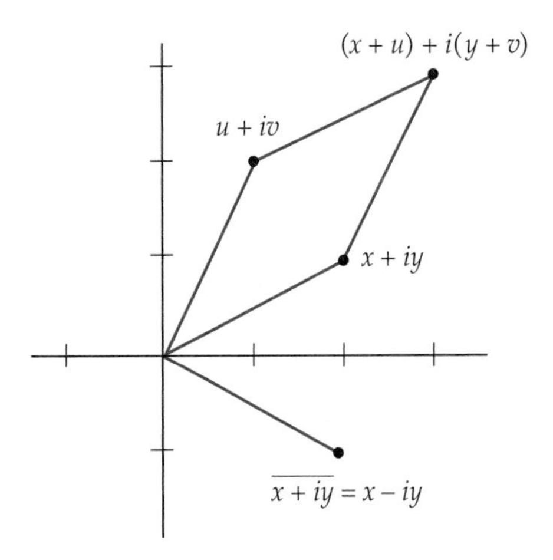

Figure A.1 Sum and conjugate.

imaginary complex numbers $0 + iy$ are identified with the points $(0, y)$ on the **imaginary axis.** Of course, to justify using the term *number,* we have to know how to operate arithmetically with complex numbers.

Complex Addition. We define addition analogously with vector addition in $\mathcal{R}^2$. Applying the usual rules of real number arithmetic to the real and imaginary parts of the numbers as well as to the symbol i we get

$$(x + iy) + (u + iv) = (x + u) + i(y + v).$$

Given our identification of $z = x + iy$ and $w = u + iv$ with the points (x, y) and (u, v), it follows that the sum $z + w$ is identified with the point $(x + u,\ y + v)$. Thus the location of the point for the sum can be arrived at by vector addition using the parallelogram law.

It's customary simply to write x for $x + i0$, iy for $0 + iy$, and 0 for $0 + i0$, as well as $-(x + iy)$ for $-x + i(-y)$.

Complex Multiplication. We define products using the usual rules of real number arithmetic except that i^2 is replaced by -1:

$$(x + iy)(u + iv) = (xu - yv) + i(xv + yu).$$

It follows that multiplication is commutative, that is, $(x + iy)(u + iv) = (u + iv)(x + iy)$. In particular $iy = yi$, and it's customary to write for example, $1 - 2i$ instead of $1 + i(-2)$.

example 2

Typical examples of complex addition are

$$4i + (1 - 2i) = 1 + 2i, \qquad (1 + 2i) + (1 - 2i) = 2, \quad \text{and} \quad (1 + 2i) - (1 - 2i) = 4i.$$

Typical examples of complex multiplication are

$$(1 + 2i)(1 - 2i) = 1 - 2i + 2i + 2i(-2i) = 1 - 4i^2 = 5,$$

and

$$4i(1-2i) = 4i + 4i(-2i) = 4i - 8i^2 = 8 + 4i.$$

The **conjugate** of a complex number $z = x + iy$ is the number $\bar{z} = x - iy$. For example, $\overline{2+3i} = 2 - 3i$, $\overline{2-i} = 2 + i$, and $\overline{-i} = i$. Geometrically, $\bar{z}$ is the reflection of z in the real number line, as indicated in Figure A.1.

Complex Division. To see how the **quotient** of two complex numbers should be defined, we can formally multiply the numerator and the *nonzero* denominator by the conjugate of the denominator:

$$\frac{u+iv}{x+iy} = \frac{(u+iv)(x-iy)}{(x+iy)(x-iy)} = \frac{(xu+yv)+i(xv-yu)}{x^2+y^2}.$$

Showing that this procedure always results in a uniquely defined quotient when $x + iy \neq 0$ is left as an exercise.

---

example 3 Computing a quotient is mostly multiplication:

$$\frac{2+3i}{2-i} = \frac{(2+3i)(2+i)}{(2-i)(2+i)} = \frac{(2\cdot 2 - 1\cdot 3) + i(2\cdot 3 + 1\cdot 2)}{2^2+1^2}$$
$$= \frac{1+8i}{5} = \frac{1}{5} + \frac{8}{5}i.$$

Note that the only division here in the more conventional sense is by a real number, in the last step.

The **absolute value** of $z = x + iy$ is defined to be $|z| = \sqrt{x^2+y^2}$. When we identify $x + iy$ with the point (x, y) in the plane, we see that $|x + iy|$ is just the distance from (x, y) to $(0,0)$, that is, to the complex number zero. Note that $z\bar{z} = |z|^2$.

---

example 4 For purely real complex numbers, the absolute value is the same as for real numbers, for example, $|-3| = 3$. Otherwise we have

$$|3i| = \sqrt{0^2+3^2} = 3,$$

$$|\sqrt{2} - i\sqrt{2}| = \sqrt{(\sqrt{2})^2 + (\sqrt{2})^2} = \sqrt{2+2} = 2, \quad \text{and}$$

$$\left|\frac{1}{2} + i\frac{\sqrt{3}}{2}\right| = \sqrt{\left(\frac{1}{2}\right)^2 + \left(\frac{\sqrt{3}}{2}\right)^2} = \sqrt{\frac{1}{4}+\frac{3}{4}} = 1.$$

Interpreting multiplication and division of complex numbers is done best by writing nonzero complex numbers $z = x + iy$ in **polar form,** with $x = |z| \cos\theta$

and $y = |z| \sin\theta$:

$$\begin{aligned} x + iy &= |z| \cos\theta + i|z| \sin\theta \\ &= |z|(\cos\theta + i \sin\theta). \end{aligned}$$

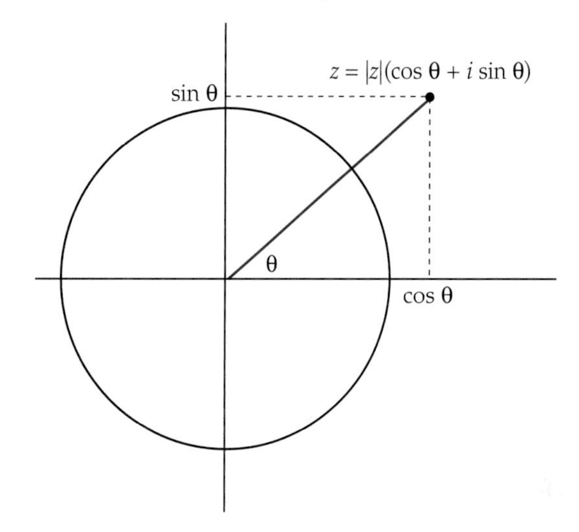

Figure A.2 Polar form.

Here $|z| = \sqrt{x^2 + y^2}$ is the radial distance of z from the origin and θ is an angle that the line joining z to 0 makes with the positive real axis. Because $\cos\theta$ and $\sin\theta$ are periodic with period 2π, the number z has infinitely many polar angles $\theta + 2\pi k$. Suitable angles can be found from the formula $\theta = \arctan(y/x)$. If $x = 0$, then $\theta = \pi/2$ when $y > 0$ and $\theta = -\pi/2$ when $y < 0$.

example 5

(a) The angle that the complex number $3i$ makes with the positive real axis is $\pi/2$, and $|3i| = 3$, so $3i = 3[\cos(\pi/2) + i\sin(\pi/2)]$.

(b) The angle that the number $\sqrt{2} - i\sqrt{2}$ makes with the positive real axis is $-\pi/4$, and $|\sqrt{2} - i\sqrt{2}| = 2$, so

$$\sqrt{2} - i\sqrt{2} = 2\left[\cos\left(-\frac{\pi}{4}\right) + i\sin\left(-\frac{\pi}{4}\right)\right] = 2\left(\cos\frac{\pi}{4} - i\sin\frac{\pi}{4}\right).$$

Here we've used the fact that $\cos\theta$ is an even function and $\sin\theta$ is an odd function.

(c) The angle that the number $1/2 + i\sqrt{3}/2$ makes with the positive real axis is $\pi/3$, and $|1/2 + i\sqrt{3}/2| = 1$, so

$$\frac{1}{2} + i\frac{\sqrt{3}}{2} = 1\left(\cos\frac{\pi}{3} + i\sin\frac{\pi}{3}\right).$$

As a reward for using the relatively complicated-looking polar form, we gain an advantage in interpreting products of complex numbers. If z and z' are complex numbers with polar angles θ and θ', respectively, their product can be written

$$\begin{aligned} zz' &= |z|(\cos\theta + i\sin\theta)\,|z'|(\cos\theta' + i\sin\theta') \\ &= |z||z'|\big[(\cos\theta\cos\theta' - \sin\theta\sin\theta') + i(\cos\theta\sin\theta' + \sin\theta\cos\theta')\big] \\ &= |z||z'|\big[\cos(\theta+\theta') + i\sin(\theta+\theta')\big]. \end{aligned}$$

The last step follows from the addition formulas for the cosine and sine functions. What we arrive at is a complex number zz' in polar form having absolute value $|z||z'|$ and polar angle $\theta + \theta'$. That is, if $\theta(w)$ stands for the polar angle of w, then

$$|zz'| = |z||z'| \quad \text{and} \quad \theta(zz') = \theta(z) + \theta(z').$$

These equations are illustrated in Figure A.3 and can be stated in words as follows:

(i) *The absolute value of a product of complex numbers is the product of the absolute values of the factors.*
(ii) *The polar angle of a product of complex numbers is the sum of the polar angles of the factors.*

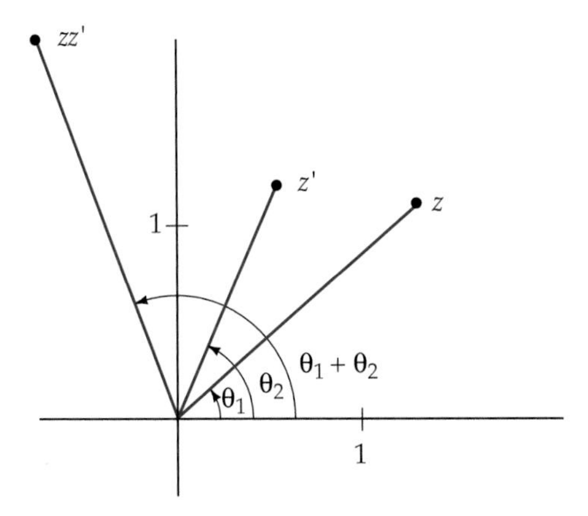

Figure A.3 Polar angle of zz' is $\theta_1 + \theta_2$.

These rules tell us geometrically how to locate a product in the plane when we have already plotted the individual factors.

example 6

Where is the product of $\sqrt{2} - i\sqrt{2}$ and $1/2 + i\sqrt{3}/2$ located in the complex plane? The absolute values of the two numbers are 2 and 1, respectively, and their polar angles are $-\pi/4$ ($-45°$) and $\pi/3$ ($60°$) according to the previous example. Hence the absolute value of the product is 2, and the polar angle is $-\pi/4 + \pi/3 = \pi/12$ or $15°$.

Polynomial roots have their existence guaranteed by the fundamental theorem of algebra, which states that an nth-degree polynomial equation with complex coefficients,

$$z^n + a_{n-1}z^{n-1} + \cdots + a_1 z + a_0 = 0, \qquad n \geq 1,$$

has a complex solution $z = r_1$. It then follows that there is a polynomial factorization

$$z^n + a_{n-1}z^{n-1} + \cdots + a_1 z + a_0 = (z - r_1)(z - r_2)\cdots(z - r_n),$$

corresponding to roots $r_1, \ldots, r_n$. The number of times a root r_k appears in the factorization is called the multiplicity of that root. Finding the roots, or equivalently the polynomial factorization, is immediate if $n = 1$ and quite simple in case $n = 2$ from the quadratic formula for roots r_1, r_2 of $z^2 + az + b$, namely, $r = \frac{1}{2}(-a \pm \sqrt{a^2 - 4b}\,)$. (See Exercise 23 for the derivation.) Analogous, but fairly complicated, formulas exist for polynomials of degrees 3 and 4, but it has been proved that no such general formulas exist for those of degree 5 or more. Certain particular equations can be solved quite easily, as the following examples show.

example 7 The 5th-degree polynomial $z^5 - z$ can be factored as follows:

$$z^5 - z = z(z^4 - 1) = z(z^2 - 1)(z^2 + 1) = z(z - 1)(z + 1)(z - i)(z + i).$$

Thus the solutions of the equation $z^5 - z = 0$ are evidently $0, 1, -1, i$, and $-i$.

example 8 If the 4th-degree equation $z^4 + z^2 - 6 = 0$ is regarded as quadratic in z^2, we find by the factorization $(z^2 - 2)(z^2 + 3)$, or by the quadratic formula, that either $z^2 = 2$ or $z^2 = -3$. Thus the four solutions to the equation are $\pm\sqrt{2}$ and $\pm\sqrt{3}i$.

EXERCISES

Locate the following complex numbers in a sketch showing real and imaginary axes.

1. $-2 + 2i$.

2. $(2 + i) + (-1 - 2i)$.

3. $i(1 + 2i)$.

4. $2[\cos(\pi/4) + i\sin(\pi/4)]$.

5. $(1 + i)/i$.

6. $(2 + i)/(-1 - 2i)$.

7. $i(1 + 2i)^2$.

8. $2[\cos(\pi/6) - i\sin(\pi/6)]$.

For each of the following complex numbers z, find $\bar{z}$ and $1/z$ in the form $x + iy$. Then find $|z|$ and write z in polar form $|z|(\cos\theta + i\sin\theta)$.

9. $1 + i$.

10. $-1 + 2i$.

11. $2i$.

12. $(2 + i)/i$.

13. $i(\cos t + i\sin t)$.

14. $(1 + 2i)(3 - i)$.

15. $(1 + i)(1 - i)$.

16. $(2 - i)(1 + i^{-1})$.

17. $(1 + i)/(1 - i)$.

18. Let z and w be complex numbers, $z \neq 0$. Show that the equation

$$\frac{w}{z} = \frac{1}{|z|^2}w\bar{z}$$

defines a *unique* quotient w/z such that $(w/z)z = w$. [*Hint:* Suppose that $qz = w$. Multiply by $\bar{z}$; then solve for q.]

19. Show that if $|z| = 1$, then $1/z = \bar{z}$.

20. For a given real number θ prove

(a) $(\cos\theta + i\sin\theta)^2 = \cos 2\theta + i\sin 2\theta$.

(b) $(\cos\theta + i\sin\theta)^n = \cos n\theta + i\sin n\theta$ for integers $n \geq 0$ (**deMoivre's formula**).

(c) the formula in part (b) for $n = -1$.

(d) the formula in part (b) for integers $n < 0$.

21. Prove the following:

(a) The equation $x^2 = -1$ has no solutions x in the real numbers $\mathcal{R}$.

(b) The equation $x^2 = 2$ has no solutions x in the rational numbers $\mathcal{Q}$, where $\mathcal{Q}$ is the set of all quotients m/n of integers. [*Hint:* Suppose $(m/n)^2 = 2$ and that all common factors in m and n have been canceled; then arrive at a contradiction.]

22. Verify for arbitrary complex numbers the commutative, associative, and distributive laws listed below using the analogous properties for real numbers.

(i) $z_1z_2 = z_2z_1$ and $z_1 + z_2 = z_2 + z_1$.

(ii) $z_1(z_2z_3) = (z_1z_2)z_3$ and $z_1 + (z_2 + z_3) = (z_1 + z_2) + z_3$.

(iii) $z_1(z_2 + z_3) = z_1z_2 + z_1z_3$.

23. **(a)** Show the equation $z^2 + az + b = 0$, where a and b are complex constants, can be rewritten in the form $(z + a/2)^2 + (4b - a^2)/4 = 0$. Hence show that all roots of the equation are given by the **quadratic formula** $z = \left(-a \pm \sqrt{a^2 - 4b}\,\right)/2$.

(b) Show that, for real a and b, the roots are complex conjugates if $a^2 < 4b$.

(c) If a and b are allowed to be complex, the expression $a^2 - 4b$ inside the square root symbol in the quadratic formula will itself in general be a complex number $u + iv$. To find the two distinct square roots of a nonzero complex number $u + iv$, let

$$x = \sqrt{\frac{\sqrt{u^2 + v^2} + u}{2}}, \qquad y = \sqrt{\frac{\sqrt{u^2 + v^2} - u}{2}}.$$

Show that the square roots of $u + iv$ are $\pm(x + iy)$ if $v \geq 0$ and $\pm(x - iy)$ if $v \leq 0$.

(d) Solve $z^2 + z + i = 0$.

24. There are n distinct solutions of $z^n = w$ if $w \neq 0$, that is, nth roots of a complex number $w \neq 0$. These roots can be found by first expressing ω in multiple polar form by allowing for the 2π periodicity of cosine and sine as follows:

$$w = |w|[\cos(\theta + 2k\pi) + i \sin(\theta + 2k\pi)], k = 0, 1, \ldots, n-1.$$

The desired roots are then

$$z_k = |w|^{1/n}\left[\cos\left(\frac{\theta}{n} + \frac{2k\pi}{n}\right) + i \sin\left(\frac{\theta}{n} + \frac{2k\pi}{n}\right)\right], k = 0, 1, \ldots, n-1.$$

(a) Verify that the numbers z_k satisfy $z_k^n = \omega, k = 0, 1, \ldots, n-1$.

(b) Find the cube roots of i, first in polar form, then in the form $x + iy$.

(c) Find the fourth roots of $-i$, first in polar form, then in the form $x + iy$.

APPENDIX B

Matrix Algebra

Matrices are useful in many areas of pure and applied mathematics, but their use in this book is for solving systems of linear differential equations and systems of linear algebraic equations. The systems

$$\frac{dx}{dt} = 2x + 5y \qquad\qquad 5x + y - z = 0$$
$$\text{and} \quad x + 2y + 3z = 0$$
$$\frac{dy}{dt} = 2x - y \qquad\qquad x \quad\; - z = 0$$

are examples of the two kinds of system we are concerned with. Once we know the general form of these two systems, it is clear that they are completely determined by the respective rectangular arrays of numbers

$$\begin{pmatrix} 2 & 5 \\ 2 & -1 \end{pmatrix} \quad \text{and} \quad \begin{pmatrix} 5 & 1 & -1 \\ 1 & 2 & 3 \\ 1 & 0 & -1 \end{pmatrix}.$$

Any rectangular array of numbers or functions is called a **matrix,** but we need matrix operations only with **column matrices** that we'll interpret as vectors, and **square matrices** as shown both above and below. Thus

$$\begin{pmatrix} 1 \\ 2 \\ 1 \end{pmatrix}, \quad \begin{pmatrix} 1 & 2 & 1 \\ 0 & 0 & 0 \\ 0 & 1 & 2 \end{pmatrix}, \quad \begin{pmatrix} 1 \\ e^t \\ e^{-t} \end{pmatrix}, \quad \begin{pmatrix} 1 & 1 & t \\ 1 & t & t^2 \\ 1 & t^2 & t^3 \end{pmatrix}$$

are examples of the type of matrix we'll be using. We'll sometimes want the entries in a matrix to be functions rather than constants as in the last two examples; but since the underlying algebra is the same in both instances, we'll restrict ourselves at first to examples with constant entries. (The first formula we'll encounter where functions as matrix entries make a notable difference is in Theorem 1.1, Formula 5, the product rule for matrix derivatives.) A horizontal line of entries in a matrix is called a **row** and a vertical line of entries is called a

column. The number of rows and columns in a matrix determine its **dimensions.** Thus the four previous examples have dimensions 3 by 1, 3 by 3, 3 by 1, and 3 by 3. Note that the number of rows is always listed before the number of columns. Two matrices are equal if they have the same dimensions and corresponding entries are equal. We'll use capital letters to denote matrices and corresponding small letters to denote a typical entry. Thus we may write

$$A = \begin{pmatrix} a_{11} & a_{12} \\ a_{21} & a_{22} \end{pmatrix} \quad \text{or} \quad A = (a_{ij}),\ i, j = 1, 2,$$

where a_{ij}, the ijth entry, is the entry in the ith row and jth column. If there is only one row or one column, we usually omit the unnecessary subscript and use a boldface letter to denote such a column. For example, the vector $\mathbf{x} = (x_1, x_2, x_3)$ in column format is

$$\mathbf{x} = \begin{pmatrix} x_1 \\ x_2 \\ x_3 \end{pmatrix}.$$

We'll denote by $\mathcal{R}^n$ the set of all vectors with n entries, whether written vertically or horizontally, separated by commas.

1 MATRIX OPERATIONS

Sums and Scalar Multiples. If A and B have the same dimensions, then the **sum** $A + B$ is defined to be the matrix having the same dimensions as A and B and with ijth entry equal to $a_{ij} + b_{ij}$, where $A = (a_{ij})$ and $B = (b_{ij})$.

example 1

Here are two matrix sums.

$$\begin{pmatrix} 1 & 1 \\ 0 & 2 \end{pmatrix} + \begin{pmatrix} -1 & 1 \\ 1 & 2 \end{pmatrix} = \begin{pmatrix} 0 & 2 \\ 1 & 4 \end{pmatrix},$$

$$\begin{pmatrix} 1 & 2 & 1 \\ -1 & 1 & 0 \\ -2 & 2 & -3 \end{pmatrix} + \begin{pmatrix} -1 & -2 & -1 \\ 1 & -1 & 0 \\ 2 & -2 & 3 \end{pmatrix} = \begin{pmatrix} 0 & 0 & 0 \\ 0 & 0 & 0 \\ 0 & 0 & 0 \end{pmatrix}.$$

Addition is not defined for matrices with different dimensions so we can't add

$$\begin{pmatrix} 1 & 1 \\ 0 & 2 \end{pmatrix} \quad \text{and} \quad \begin{pmatrix} 1 & 2 & 1 \\ -1 & 1 & 0 \\ -2 & 2 & -3 \end{pmatrix}.$$

For a matrix A and a real or complex number r, often called a **scalar,** the **scalar multiple** rA is defined to be the matrix with the same dimensions as A and with ijth the entry ra_{ij}.

example 2

Here are some examples of scalar multiplication, some in combination with sums:

$$-2\begin{pmatrix} 1 & 1 \\ 0 & 2 \end{pmatrix} = \begin{pmatrix} -2 & -2 \\ 0 & -4 \end{pmatrix},$$

$$\begin{pmatrix} 1 \\ 2 \\ 1 \end{pmatrix} + \begin{pmatrix} 2 \\ 0 \\ -1 \end{pmatrix} - 3\begin{pmatrix} 1 \\ 1 \\ 1 \end{pmatrix} = \begin{pmatrix} 0 \\ -1 \\ -3 \end{pmatrix}.$$

$$2\begin{pmatrix} 3 \\ -5 \end{pmatrix} - 3\begin{pmatrix} 3 \\ -5 \end{pmatrix} = \begin{pmatrix} 6 \\ -10 \end{pmatrix} + \begin{pmatrix} -9 \\ 15 \end{pmatrix} = \begin{pmatrix} -3 \\ 5 \end{pmatrix}.$$

A sum of scalar multiples of matrices is called a **linear combination** of matrices. Because of the way the matrix operations of addition and scalar multiplication are defined, a linear combination can of course always be written as a single matrix. However, it often happens that we want to expand a matrix into a linear combination. For example, if

$$\mathbf{e}_1 = \begin{pmatrix} 1 \\ 0 \\ 0 \end{pmatrix}, \quad \mathbf{e}_2 = \begin{pmatrix} 0 \\ 1 \\ 0 \end{pmatrix}, \quad \mathbf{e}_3 = \begin{pmatrix} 0 \\ 0 \\ 1 \end{pmatrix},$$

then any 3-dimensional column can be written as a linear combination of $\mathbf{e}_1$, $\mathbf{e}_2$, and $\mathbf{e}_3$:

$$\begin{pmatrix} c_1 \\ c_2 \\ c_3 \end{pmatrix} = c_1\begin{pmatrix} 1 \\ 0 \\ 0 \end{pmatrix} + c_2\begin{pmatrix} 0 \\ 1 \\ 0 \end{pmatrix} + c_3\begin{pmatrix} 0 \\ 0 \\ 1 \end{pmatrix}$$
$$= c_1\mathbf{e}_1 + c_2\mathbf{e}_2 + c_3\mathbf{e}_3.$$

Once the dimension you are working in has been identified, you can use the **e**-notation without ambiguity. Let $\mathbf{e}_k$ be the n-dimensional column with 1 in the kth entry and 0 elsewhere. Then any n-dimensional column can be represented uniquely as a linear combination of $\mathbf{e}_1, \ldots, \mathbf{e}_n$. For this reason, the set $\{\mathbf{e}_1, \ldots, \mathbf{e}_n\}$ is called the **standard basis** for the n-dimensional vectors.

example 3

With $\mathbf{e}_1$ and $\mathbf{e}_2$ of dimension 2, we have

$$\begin{pmatrix} 3 \\ 7 \end{pmatrix} = 3\begin{pmatrix} 1 \\ 0 \end{pmatrix} + 7\begin{pmatrix} 0 \\ 1 \end{pmatrix}$$
$$= 3\mathbf{e}_1 + 7\mathbf{e}_2.$$

example 4

A vector function can have a basis representation. For example,

$$\begin{pmatrix} t \\ e^t \\ e^{-t} \end{pmatrix} = t\begin{pmatrix} 1 \\ 0 \\ 0 \end{pmatrix} + e^t\begin{pmatrix} 0 \\ 1 \\ 0 \end{pmatrix} + e^{-t}\begin{pmatrix} 0 \\ 0 \\ 1 \end{pmatrix}$$
$$= t\mathbf{e}_1 + e^t\mathbf{e}_2 + e^{-t}\mathbf{e}_3.$$

Using both addition and scalar multiplication, we can write a linear combination of matrices with the same dimensions. For example, using 2-by-2 matrices,

$$2\begin{pmatrix} 1 & -1 \\ 2 & 3 \end{pmatrix} - 3\begin{pmatrix} 0 & 1 \\ 2 & 1 \end{pmatrix} = \begin{pmatrix} 2 & -2 \\ 4 & 6 \end{pmatrix} + \begin{pmatrix} 0 & -3 \\ -6 & -3 \end{pmatrix}$$
$$= \begin{pmatrix} 2 & -5 \\ -2 & 3 \end{pmatrix}.$$

We write $-A$ for $(-1)A$ and $A - B$ for $A + (-1)B$. Also, for each pair of dimensions, there is a **zero matrix,** denoted by 0, with all entries equal to zero, such that $A + 0 = A$.

Notational Warning. When just a single 0 is used to denote the zero vector or the zero matrix, it must be made clear from the context what dimension is intended. For example, if $A = \begin{pmatrix} 1 & 2 \\ 4 & 4 \end{pmatrix}$, then the 0 in $A + 0$ must mean $\begin{pmatrix} 0 & 0 \\ 0 & 0 \end{pmatrix}$, because no other dimensions would allow addition to a 2-by-2 matrix.

As a practical matter, it's important that if vectors $\mathbf{v}_1, \ldots, \mathbf{v}_n$ are used to express a vector $\mathbf{x}$ as a linear combination

$$\mathbf{x} = c_1\mathbf{v}_1 + \cdots + c_n\mathbf{v}_n,$$

then the coefficients $c_1, \ldots, c_n$ should be uniquely determined by the vector $\mathbf{x}$. In particular, the zero vector $\mathbf{x} = 0$ should be uniquely expressible, in which case $c_1 = c_2 = \cdots = c_n = 0$. Uniqueness follows in general from this special case because

$$c_1\mathbf{v}_1 + \cdots + c_n\mathbf{v}_n = d_1\mathbf{v}_1 + \cdots + d_n\mathbf{v}_n$$

is equivalent to $(c_1 - d_1)\mathbf{v}_1 + \cdots + (c_n - d_n)\mathbf{v}_n = 0$; thus $c_k - d_k = 0$ for $k = l$, $2, \ldots, n$. Another way of expressing the uniqueness condition on $\mathbf{v}_1, \ldots, \mathbf{v}_n$ is to say that no one of them should be expressible as a linear combination of the remaining ones. If the vectors $\mathbf{v}_1, \ldots, \mathbf{v}_n$ satisfy any one of these equivalent conditions, then the vectors are said to be **linearly independent;** otherwise they are **linearly dependent.**

---

example **5**

The vectors $\mathbf{e}_1$, $\mathbf{e}_2$, and $\mathbf{e}_3$ in $\mathcal{R}^3$ are linearly independent, because

$$c_1\begin{pmatrix} 1 \\ 0 \\ 0 \end{pmatrix} + c_2\begin{pmatrix} 0 \\ 1 \\ 0 \end{pmatrix} + c_3\begin{pmatrix} 0 \\ 0 \\ 1 \end{pmatrix} = \begin{pmatrix} 0 \\ 0 \\ 0 \end{pmatrix}$$

implies $c_1 = c_2 = c_3 = 0$. On the other hand

$$2\begin{pmatrix} 1 \\ 1 \\ 1 \end{pmatrix} + 3\begin{pmatrix} 0 \\ 1 \\ 2 \end{pmatrix} - \begin{pmatrix} 2 \\ 5 \\ 8 \end{pmatrix} = \begin{pmatrix} 0 \\ 0 \\ 0 \end{pmatrix}$$

so the three vectors on the left are linearly dependent; the third one is in fact 2 times the first plus 3 times the second.

Matrix Products. First consider an n-by-n matrix A and n-dimensional column $\mathbf{x}$. The **product** $A\mathbf{x}$ is defined to be the n-dimensional column vector equal to the linear combination of the columns of A with coefficients from $\mathbf{x}$, taken in the same order. For example,

$$\begin{pmatrix} a_{11} & a_{12} \\ a_{21} & a_{22} \end{pmatrix} \begin{pmatrix} x_1 \\ x_2 \end{pmatrix} = x_1 \begin{pmatrix} a_{11} \\ a_{21} \end{pmatrix} + x_2 \begin{pmatrix} a_{12} \\ a_{22} \end{pmatrix}$$

$$= \begin{pmatrix} a_{11}x_1 + a_{12}x_2 \\ a_{21}x_1 + a_{22}x_2 \end{pmatrix}.$$

Thus the product $A\mathbf{x} = \mathbf{b}$ represents the system of linear equations

$$a_{11}x_1 + a_{12}x_2 = b_1,$$

$$a_{21}x_1 + a_{22}x_2 = b_2,$$

where b_1 and b_2 are the entries in the column vector $\mathbf{b}$. Note also that if $\mathbf{x} = \mathbf{e}_j$, with 1 in the jth entry and 0 elsewhere, then $A\mathbf{e}_j$, is just the jth column of A:

$$A\mathbf{e}_j = \begin{pmatrix} a_{1j} \\ \cdot \\ \cdot \\ \cdot \\ a_{mj} \end{pmatrix}, \quad \text{where } A = (a_{ij}).$$

Another way to describe the product $A\mathbf{x}$ is that it is the column vector whose successive entries are the dot products of the rows of A with the vector $\mathbf{x}$.

If A and B are n-by-n matrices, the matrix product AB is the n-by-n matrix whose columns are the respective products of A with the columns of B. Thus matrix multiplication is often called *row-by-column multiplication*. For example,

$$\begin{pmatrix} a_{11} & a_{12} \\ a_{21} & a_{22} \end{pmatrix} \begin{pmatrix} b_{11} & b_{12} \\ b_{21} & b_{22} \end{pmatrix} = \begin{pmatrix} a_{11}b_{11} + a_{12}b_{21} & a_{11}b_{12} + a_{12}b_{22} \\ a_{21}b_{11} + a_{22}b_{21} & a_{21}b_{12} + a_{22}b_{22} \end{pmatrix}.$$

example 6

If

$$A = \begin{pmatrix} c_{11} & c_{12} \\ c_{21} & c_{22} \end{pmatrix} \quad \text{and} \quad B = \begin{pmatrix} 1 & 2 \\ 4 & 5 \end{pmatrix},$$

then

$$AB = \begin{pmatrix} c_{11} & c_{12} \\ c_{21} & c_{22} \end{pmatrix} \begin{pmatrix} 1 & 2 \\ 4 & 5 \end{pmatrix} = \begin{pmatrix} c_{11} + 4c_{12} & 2c_{11} + 5c_{12} \\ c_{21} + 4c_{22} & 2c_{21} + 5c_{22} \end{pmatrix}.$$

Note that the only requirement on the dimensions of matrices in a product is that the number of columns in the first matrix be the same as the number of rows in the second. Hence we can also form the product

$$BA = \begin{pmatrix} 1 & 2 \\ 4 & 5 \end{pmatrix} \begin{pmatrix} c_{11} & c_{12} \\ c_{21} & c_{22} \end{pmatrix} = \begin{pmatrix} c_{11} + 2c_{21} & c_{12} + 2c_{22} \\ 4c_{11} + 5c_{21} & 4c_{12} + 5c_{22} \end{pmatrix}.$$

Note that AB is not in general equal to BA.

example 7

The system of linear algebraic equations

$$\begin{aligned} 3x + 7y &= 1, \\ 2x + 5y &= 2, \end{aligned} \quad \text{or} \quad \begin{pmatrix} 3 & 7 \\ 2 & 5 \end{pmatrix} \begin{pmatrix} x \\ y \end{pmatrix} = \begin{pmatrix} 1 \\ 2 \end{pmatrix}$$

can be written $A\mathbf{x} = \mathbf{c}$, where

$$A = \begin{pmatrix} 3 & 7 \\ 2 & 5 \end{pmatrix}, \qquad \mathbf{x} = \begin{pmatrix} x \\ y \end{pmatrix}, \quad \text{and} \quad \mathbf{c} = \begin{pmatrix} 1 \\ 2 \end{pmatrix}.$$

Similarly, the system of differential equations

$$\begin{aligned} \frac{dx}{dt} &= 3x + 7y - 1, \\ \frac{dy}{dt} &= 2x + 5y - 2, \end{aligned} \quad \text{or} \quad \begin{pmatrix} \dot{x} \\ \dot{y} \end{pmatrix} = \begin{pmatrix} 3 & 7 \\ 2 & 5 \end{pmatrix} \begin{pmatrix} x \\ y \end{pmatrix} - \begin{pmatrix} 1 \\ 2 \end{pmatrix},$$

can be written $d\mathbf{x}/dt = A\mathbf{x} - \mathbf{c}$, since the vector derivative is computed in terms of coordinates by

$$\frac{d\mathbf{x}}{dt} = \frac{d}{dt} \begin{pmatrix} x \\ y \end{pmatrix} = \begin{pmatrix} \dot{x} \\ \dot{y} \end{pmatrix}.$$

It is worth remarking that the algebraic system becomes a special case of the differential system if we make the restriction that the solutions be constant, in which case $d\mathbf{x}/dt = 0$. It is easy to see that the algebraic system has solution $x = -9$, $y = 4$. It follows that the differential system has those constant solutions also. Of course, it has many nonconstant solutions in addition. (The constant solutions are called **equilibrium solutions,** and they are important in the discussion of stability in Chapter 6, Section 8, and Chapter 7, Section 5.)

For general questions about matrix products, it is sometimes useful to write the entries in the product of $A = (a_{ij})$ *and* $B = (b_{ij})$ by using the summation notation. The ijth entry in AB is

$$\sum_{k=1}^{n} a_{ik} b_{kj}.$$

Note that the summation index k runs over the column index of A (that is, along a row) and over the row index of B (that is, along the column).

The next theorem lists the four basic properties of matrix products, together with the rule for differentiation of matrix products. We assume that the various matrices have the proper shapes so that the combinations are well defined. The **derivative of a matrix,** and in particular of a vector, is computed entry by entry. For example,

$$\frac{d}{dt} \begin{pmatrix} t & t^2 \\ t^3 & 1 \end{pmatrix} = \begin{pmatrix} 1 & 2t \\ 3t^2 & 0 \end{pmatrix} \quad \text{and} \quad \frac{d}{dt} \begin{pmatrix} t \\ e^{-t} \end{pmatrix} = \begin{pmatrix} 1 \\ -e^{-t} \end{pmatrix}.$$

Matrices with variable entries are used extensively in Chapter 7, in particular products $A(t)\mathbf{x}(t)$ of matrices $A(t)$ and vectors $\mathbf{x}(t)$. In the latter case, the product rule, Formula 5 in Theorem 1.1 is sometimes written using overdots for time derivatives:

$$\frac{d}{dt}A(t)\mathbf{x}(t) = \dot{A}(t)\mathbf{x}(t) + A(t)\dot{\mathbf{x}}(t).$$

1.1 THEOREM

1. $(A + B)C = AC + BC$—Right distributivity.
2. $C(A + B) = CA + CB$—Left distributivity.
3. $(rA)B = r(AB) = A(rB)$—Associativity for scalar products.
4. $A(BC) = (AB)C$—Associativity for matrix products.
5. $d(AB)/dt = A(dB/dt) + (dA/dt)B$—Product rule for derivatives.

Proof. We prove only the last two since they are more complicated; the others are left as exercises. For property 4 let A, B, and C be n-by-n matrices. Then the ijth entry of BC is $\sum_{k=1}^{n} b_{ik}c_{kj}$, so the ijth entry of $A(BC)$ is

$$\sum_{l=1}^{n} a_{il}\left(\sum_{k=1}^{n} b_{lk}c_{kj}\right) = \sum_{l=1}^{n}\sum_{k=1}^{n} a_{il}b_{lk}c_{kj}.$$

Similarly, the ijth entry of $(AB)C$ is

$$\sum_{k=1}^{n}\left(\sum_{l=1}^{n} a_{il}b_{lk}\right)c_{kj} = \sum_{k=1}^{n}\sum_{l=1}^{n} a_{il}b_{lk}c_{kj}.$$

The two double sums on the right consist of the same terms added in different orders, so the ijth entries in $A(BC)$ and $(AB)C$ are the same.

For the product rule, we differentiate one entry at a time on the left-hand side. The derivative of the ijth entry is

$$\frac{d}{dt}\sum_{k=1}^{n} a_{ik}(t)b_{kj}(t) = \sum_{k=1}^{n}\frac{da_{ik}}{dt}(t)b_{kj}(t) + \sum_{k=1}^{n} a_{ik}(t)\frac{db_{kj}}{dt}(t).$$

But this is just the ijth entry in the sum of products of matrices on the right, so the formula is proved. ■

Formulas 1 and 2 in Theorem 1 are called the **distributive laws** for matrix multiplication. Two of them are needed because *matrix multiplication is not in general commutative;* that is, there are square matrices A and B, for which AB and BA both make sense but such that $AB \neq BA$, as we saw in Example 6. Properties 1, 2, and 3 of Theorem 1.1 express the **linearity** of matrix multiplication; in particular, if $\mathbf{x}$ and $\mathbf{y}$ are n-dimensional column vectors and A is an n-by-n matrix, the two equations

$$A(\mathbf{x} + \mathbf{y}) = A\mathbf{x} + A\mathbf{y}, \qquad A(r\mathbf{x}) = rA\mathbf{x}$$

express the fact that A acts as a linear function from $\mathcal{R}^n$ to $\mathcal{R}^n$. Formula 4 is the **associative law** for matrix multiplication, and it shows that it makes sense to write the product ABC since the result is independent of the order in which the products are formed. Formula 5 is the **product rule** for matrix multiplication, which mimics the familiar rule for differentiation of a product of scalar-valued functions. The order of the factors on the right-hand side must in general be preserved, once again because of the failure of commutativity in general.

Inverting Matrices. A square matrix of the form

$$I = \begin{pmatrix} 1 & 0 \\ 0 & 1 \end{pmatrix} \quad \text{or} \quad I = \begin{pmatrix} 1 & 0 & 0 \\ 0 & 1 & 0 \\ 0 & 0 & 1 \end{pmatrix} \quad \text{or} \quad I = \begin{pmatrix} 1 & 0 & \dots & 0 & 0 \\ 0 & 1 & \dots & 0 & 0 \\ \vdots & \vdots & \ddots & \vdots & \vdots \\ 0 & 0 & \dots & 0 & 1 \end{pmatrix}$$

that has ones on its **main diagonal** and zeros elsewhere is called an **identity matrix.** The matrix I satisfies $IA = A$ and $AI = A$ for matrices A such that the products are defined. Thus I is an identity element for matrix multiplication somewhat as the number 1 is an identity for multiplication of numbers. There is an n-by-n identity matrix for every value of n, but as with the zero matrices, it must be made clear from the context what the dimension of an identity matrix is.

example 8

You can check the following.

$$\begin{pmatrix} 1 & 0 \\ 0 & 1 \end{pmatrix} \begin{pmatrix} 1 & 2 \\ 4 & 5 \end{pmatrix} = \begin{pmatrix} 1 & 2 \\ 4 & 5 \end{pmatrix}$$

$$\begin{pmatrix} 1 & 0 & 0 \\ 0 & 1 & 0 \\ 0 & 0 & 1 \end{pmatrix} \begin{pmatrix} 1 & 2 & 1 \\ 3 & 4 & 1 \\ 5 & 6 & 1 \end{pmatrix} = \begin{pmatrix} 1 & 2 & 1 \\ 3 & 4 & 1 \\ 5 & 6 & 1 \end{pmatrix}.$$

If $\mathbf{x}$ is an n-dimensional column vector, then $I\mathbf{x} = \mathbf{x}$ looks like this when $n = 3$:

$$\begin{pmatrix} 1 & 0 & 0 \\ 0 & 1 & 0 \\ 0 & 0 & 1 \end{pmatrix} \begin{pmatrix} x_1 \\ x_2 \\ x_3 \end{pmatrix} = \begin{pmatrix} x_1 \\ x_2 \\ x_3 \end{pmatrix}.$$

If A is a given square matrix and A^{-1} is a square matrix of the same size such that

$$AA^{-1} = A^{-1}A = I,$$

then A is said to be an **invertible** matrix and A^{-1} the **inverse** of A. Exercise 46, part (c), shows that a matrix A can have at most one inverse, so we're justified in writing *the* inverse of A.

example 9

If $A = \begin{pmatrix} 1 & 2 \\ 3 & 7 \end{pmatrix}$, then $A^{-1} = \begin{pmatrix} 7 & -2 \\ -3 & 1 \end{pmatrix}$, because

$$AA^{-1} = \begin{pmatrix} 1 & 2 \\ 3 & 7 \end{pmatrix} \begin{pmatrix} 7 & -2 \\ -3 & 1 \end{pmatrix} = \begin{pmatrix} 1 & 0 \\ 0 & 1 \end{pmatrix}$$

and

$$A^{-1}A = \begin{pmatrix} 7 & -2 \\ -3 & 1 \end{pmatrix} \begin{pmatrix} 1 & 2 \\ 3 & 7 \end{pmatrix} = \begin{pmatrix} 1 & 0 \\ 0 & 1 \end{pmatrix}.$$

A general 2-by-2 matrix $\begin{pmatrix} a & b \\ c & d \end{pmatrix}$ is invertible if and only if $ad - bc \neq 0$. The real-valued function $\det A = ad - bc$ of the 2-by-2 matrix A is called the **determinant** of the matrix. In that case

1.2
$$\begin{pmatrix} a & b \\ c & d \end{pmatrix}^{-1} = \frac{1}{ad - bc}\begin{pmatrix} d & -b \\ -c & a \end{pmatrix}.$$

In particular,

$$\begin{pmatrix} 1 & 3 \\ -1 & 2 \end{pmatrix}^{-1} = \frac{1}{5}\begin{pmatrix} 2 & -3 \\ 1 & 1 \end{pmatrix} = \begin{pmatrix} \frac{2}{5} & -\frac{3}{5} \\ \frac{1}{5} & \frac{1}{5} \end{pmatrix}.$$

Formula 1.2 is worth remembering, and we leave it as an exercise to show that the formula is correct.

If A is an n-by-n matrix, then the matrix equation $A\mathbf{x} = \mathbf{b}$ is equivalent to a system of n linear equations in n unknowns. If A happens to be an invertible matrix with inverse A^{-1}, then we can solve the system in matrix form by multiplying both sides on the left by A^{-1} to get

$$A^{-1}A\mathbf{x} = A^{-1}\mathbf{b}.$$

Since $A^{-1}A = I$, we have $A^{-1}A\mathbf{x} = I\mathbf{x} = \mathbf{x}$, so the equation becomes

$$\mathbf{x} = A^{-1}\mathbf{b}.$$

In other words, $\mathbf{x} = A^{-1}\mathbf{b}$ is a solution to the given equation, and in fact is the only solution, because $\mathbf{x}$ could have been an arbitrary vector satisfying $A\mathbf{x} = \mathbf{b}$.

---

example 10 The system

$$\begin{aligned} x + 2y &= 3 \\ 3x + 7y &= -4 \end{aligned} \quad \text{is equivalent to} \quad \begin{pmatrix} 1 & 2 \\ 3 & 7 \end{pmatrix}\begin{pmatrix} x \\ y \end{pmatrix} = \begin{pmatrix} 3 \\ -4 \end{pmatrix}.$$

By Formula 1.2,

$$\begin{pmatrix} 1 & 2 \\ 3 & 7 \end{pmatrix}^{-1} = \begin{pmatrix} 7 & -2 \\ -3 & 1 \end{pmatrix},$$

so we multiply the left side of the equation by the inverse to get

$$\begin{pmatrix} 7 & -2 \\ -3 & 1 \end{pmatrix}\begin{pmatrix} 1 & 2 \\ 3 & 7 \end{pmatrix}\begin{pmatrix} x \\ y \end{pmatrix} = \begin{pmatrix} 1 & 0 \\ 0 & 1 \end{pmatrix}\begin{pmatrix} x \\ y \end{pmatrix} = \begin{pmatrix} x \\ y \end{pmatrix}.$$

Hence multiplying also on the right gives

$$\begin{pmatrix} x \\ y \end{pmatrix} = \begin{pmatrix} 7 & -2 \\ -3 & 1 \end{pmatrix}\begin{pmatrix} 3 \\ -4 \end{pmatrix} = \begin{pmatrix} 29 \\ -13 \end{pmatrix}.$$

Thus $(x, y) = (29, -13)$ is the unique solution.

Computing the inverse of a large matrix using determinants often requires lots of arithmetic, so the generally preferred method for matrices much larger than 2 by 2 is to use the the row operations that are useful for solving systems of linear equations by elimination. These operations, all of which leave the solutions of $A\mathbf{x} = 0$ unchanged are

1. Interchanging two rows.
2. Multiplying a row by a nonzero scalar.
3. Adding a multiple of one row to another row.

Computing A^{-1}. If a square matrix A can be converted to the identity matrix I by a sequence of row operations, then A is invertible, and applying the same sequence of row operations to I will produce A^{-1}. If the process of converting A to the identity matrix I fails, it will be because there is a sequence of row operations that results in a row or column consisting of nothing but zero entries. If there is a row or column of zeros the matrix equation $A\mathbf{x} = 0$ will have infinitely many solutions, so A^{-1} necessarily fails to exist. Reduced to essentials, the row operation process applied to A may be used to prove

1.3 THEOREM

Let A be a square matrix. Either there is a sequence of operations that reduces A to the identity matrix or, if that fails, there is a sequence of row operations that reduces A to a matrix containing a row or column of zeros.

Remark. Section 2 of this appendix shows that determinants of square matrices are computable by row operations. It will follow from Theorem 1.3, together with the row method for computing determinants, that *a square matrix A is invertible if and only if the determinant of A is not zero.*

---

example 11

We'll find the inverse of a 3-by-3 matrix A by applying the same row operations to I that we apply to A to reduce A to I. The example shows just one of several possible sequences of row operations that would produce 1s on the main diagonal and 0s elsewhere.

$$A = \begin{pmatrix} 2 & 4 & 8 \\ 1 & 0 & 0 \\ 1 & -3 & -7 \end{pmatrix}, \qquad I = \begin{pmatrix} 1 & 0 & 0 \\ 0 & 1 & 0 \\ 0 & 0 & 1 \end{pmatrix}.$$

Add -2 times the second row to the first and -1 times the second to the third:

$$\begin{pmatrix} 0 & 4 & 8 \\ 1 & 0 & 0 \\ 0 & -3 & -7 \end{pmatrix}, \qquad \begin{pmatrix} 1 & -2 & 0 \\ 0 & 1 & 0 \\ 0 & -1 & 1 \end{pmatrix}.$$

Multiply the first row by $\frac{1}{4}$, then add 3 times the first to the third:

$$\begin{pmatrix} 0 & 1 & 2 \\ 1 & 0 & 0 \\ 0 & 0 & -1 \end{pmatrix}, \qquad \begin{pmatrix} \frac{1}{4} & -\frac{1}{2} & 0 \\ 0 & 1 & 0 \\ \frac{3}{4} & -\frac{5}{2} & -1 \end{pmatrix}.$$

Multiply the third row by -1, then add -2 times the third to the first:

$$\begin{pmatrix} 0 & 1 & 0 \\ 1 & 0 & 0 \\ 0 & 0 & 1 \end{pmatrix}, \qquad \begin{pmatrix} \frac{7}{4} & -\frac{11}{2} & 2 \\ 0 & 1 & 0 \\ -\frac{3}{4} & \frac{5}{2} & -1 \end{pmatrix}.$$

Interchange first and second rows:

$$\begin{pmatrix} 1 & 0 & 0 \\ 0 & 1 & 0 \\ 0 & 0 & 1 \end{pmatrix}, \qquad \begin{pmatrix} 0 & 1 & 0 \\ \frac{7}{4} & -\frac{11}{2} & 2 \\ -\frac{3}{4} & \frac{5}{2} & -1 \end{pmatrix}.$$

You can check that the last matrix on the right is A^{-1} by multiplying A by it in either order.

---

example 12 An attempt to invert A using row operations might turn out to look like this:

$$A = \begin{pmatrix} 1 & 0 & 2 \\ 2 & 3 & 1 \\ -1 & -3 & 1 \end{pmatrix}, \qquad I = \begin{pmatrix} 1 & 0 & 0 \\ -2 & 1 & 0 \\ 1 & 0 & 1 \end{pmatrix}.$$

Subtract 2 times the first row from the second, and add the first row to the third:

$$A = \begin{pmatrix} 1 & 0 & 2 \\ 0 & 3 & -3 \\ 0 & -3 & 3 \end{pmatrix}, \qquad I = \begin{pmatrix} 1 & 0 & 0 \\ -2 & 1 & 0 \\ 1 & 0 & 1 \end{pmatrix}.$$

Adding the second row to the third creates a row with all zeros, so the system $A\mathbf{x} = 0$ is equivalent to the system

$$x + 2z = 0,$$

$$3y - 3z = 0.$$

Assigning z an arbitrary value gives $x = -2z$ and $y = z$, so there are infinitely many solutions. Hence A can't be invertible.

EXERCISES

Given that

$$A = \begin{pmatrix} -1 & 2 \\ 0 & 1 \end{pmatrix}, \quad B = \begin{pmatrix} 1 & 4 \\ 1 & 1 \end{pmatrix}, \quad \mathbf{x} = \begin{pmatrix} 1 \\ 2 \end{pmatrix}, \quad \mathbf{b} = \begin{pmatrix} 1 \\ 2 \end{pmatrix},$$

$$C = \begin{pmatrix} 1 & 1 & 1 \\ 0 & 1 & 1 \\ 0 & 0 & 1 \end{pmatrix}, \quad D = \begin{pmatrix} 1 & 0 & 1 \\ 1 & 0 & 1 \\ 2 & 2 & 2 \end{pmatrix}, \quad \mathbf{y} = \begin{pmatrix} 1 \\ 2 \\ 3 \end{pmatrix}, \quad \mathbf{d} = \begin{pmatrix} 1 \\ 1 \\ 1 \end{pmatrix},$$

compute

1. AB.

2. $A\mathbf{x} + \mathbf{b}$.

3. $BA + B^2$.

4. CD.

5. $C\mathbf{y} + \mathbf{d}$.

6. $DC + C^2$.

For each of the systems in Exercises 7 through 12 find a matrix A, a column vector $\mathbf{x}$, and a column vector $\mathbf{b}$ such that the system can be written in the form $A\mathbf{x} = \mathbf{b}$.

7. $x - y = 1,$
$x + y = 2.$

8. $2x + 3y = 0,$
$x + 3y = 0.$

9. $x + y = 1,$
$y = 1.$

10. $x + y + z = 0,$
$x + y - z = 1,$
$x - y - z = 0.$

11. $x + y = 0,$
$y + z = 0,$
$x - z = 1.$

12. $u + v - w = 1,$
$u - v + w = 2.$

For each of the systems in Exercises 13 through 16, find a matrix A, a column vector $\mathbf{x}$, and a column vector $\mathbf{b}$ such that the system can be written in the form $d\mathbf{x}/dt = A\mathbf{x} + \mathbf{b}$. Note that in general the entries in A, $\mathbf{b}$, and $\mathbf{x}$, will depend on t.

13. $dx/dt = 2x + 3y,$
$dy/dt = 2x - 4y.$

14. $dx/dt = x + y + t,$
$dy/dt = x - y - t.$

15. $dx/dt = x + y - z,$
$dy/dt = x - y + z,$
$dz/dt = x + ty.$

16. $du/dt = tu - tv + w,$
$dv/dt = u + v,$
$dw/dt = u + v + t.$

Find matrices A and vectors $\mathbf{b}$ such that each of the systems in Exercises 17 through 21 can be written in the form $\dot{\mathbf{x}} = A\mathbf{x} + \mathbf{b}$, where $\mathbf{x}$ is an n-by-1 column vector of the appropriate dimension.

17. $\dot{x} - 2y = 1,$
$x + 3\dot{y} = 2.$

18. $\dot{x} = 1,$
$\dot{y} = 2.$

19. $\dot{x} - \dot{y} = 0,$
$\dot{x} + \dot{y} = 0.$

20. $\dot{x} - y + z = 1,$
$x - \dot{y} - z = 0,$
$x + y + \dot{z} = 0.$

21. $\dot{x} - 2y = 0,$
$\dot{y} + z = 0,$
$x + \dot{z} = 1.$

Express each of the vector functions in Exercises 22 through 27 as a linear combination of $\mathbf{e}_1$ and $\mathbf{e}_2$ in the case of dimension 2, or $\mathbf{e}_1$, $\mathbf{e}_2$, and $\mathbf{e}_3$ in the case of dimension 3.

22. $f(t) = \begin{pmatrix} e^t \\ e^t \\ t^2 \end{pmatrix}.$

23. $f(t) = \begin{pmatrix} t^2 \\ t^3 \\ t^4 \end{pmatrix}.$

24. $f(t) = \begin{pmatrix} t-1 \\ t-1 \\ t+2 \end{pmatrix}.$

25. $f(t) = \begin{pmatrix} 1 \\ t \end{pmatrix}.$

26. $f(t) = \begin{pmatrix} \cos t \\ \sin t \end{pmatrix}.$

27. $f(t) = \begin{pmatrix} 1 \\ 1 \end{pmatrix}.$

The derivative of a vector-valued function is computed by differentiating each entry. The same is true for matrices, for example,

$$\frac{d}{dt}\begin{pmatrix} 1 & t \\ t & 2 \end{pmatrix} = \begin{pmatrix} 0 & 1 \\ 1 & 0 \end{pmatrix}.$$

Compute the derivative $dA(t)/dt$ for each of the matrices $A(t)$ in Exercises 28 through 31.

28. $A(t) = \begin{pmatrix} e^t & te^t \\ 0 & e^t \end{pmatrix}.$

29. $A(t) = \begin{pmatrix} t^2 & 2t \\ 3 & 4 \end{pmatrix}\begin{pmatrix} t & 2t \\ t & 3t \end{pmatrix}.$

30. $A(t) = \begin{pmatrix} e^t & 0 & 0 \\ 0 & e^{2t} & 0 \\ 0 & 0 & e^{3t} \end{pmatrix}.$

31. $A(t) = \begin{pmatrix} t & t^2 & t^3 \\ 1 & t^3 & t^4 \\ 1 & 1 & t^5 \end{pmatrix}.$

32. Let $U = \begin{pmatrix} -1 & 2 \\ 2 & -4 \end{pmatrix}$, $V = \begin{pmatrix} 2 & 6 \\ 1 & 3 \end{pmatrix}$. Compute UV and VU. Are they the same? Is it possible for the product of two matrices to be zero without either factor being the zero matrix?

33. **(a)** Using the matrix C of Exercise 1 compute the vectors $C\mathbf{e}_1$, $C\mathbf{e}_2$, and $C\mathbf{e}_3$, where the $\mathbf{e}_k$ are the standard basis vectors in $\mathcal{R}^3$.

(b) Show that if A is an n-by-n matrix, then $A\mathbf{e}_j$ is the jth column of A, where the $\mathbf{e}_k$ are the standard basis vectors in $\mathcal{R}^3$.

Compute the products.

34. $\begin{pmatrix} 1 & 2 & 3 \\ 4 & 5 & 6 \\ 7 & 8 & 9 \end{pmatrix}\begin{pmatrix} 0 \\ 1 \\ 0 \end{pmatrix}$. **35.** $\begin{pmatrix} 0 & 1 & 1 \\ 1 & 0 & 1 \\ 1 & 1 & 0 \end{pmatrix}\begin{pmatrix} 2 \\ 1 \\ 3 \end{pmatrix}$. **36.** $\begin{pmatrix} 2 & 0 & 0 \\ 0 & 4 & 0 \\ 0 & 0 & 5 \end{pmatrix}\begin{pmatrix} -1 \\ 1 \\ -1 \end{pmatrix}$.

37. Prove Equations 1, 2, and 3 of Theorem 1.1.

If A is a square matrix, it can be multiplied by itself, and we can define $A^2 = AA$, $A^3 = AAA = A^2A$, $A^n = AA\ldots A$(n factors). These powers of A all have the same dimension as A. Find A^2 and A^3 if

38. $A = \begin{pmatrix} 2 & 1 \\ 0 & 1 \end{pmatrix}$.

39. $A = \begin{pmatrix} 1 & 0 & -1 \\ -1 & 0 & 1 \\ 2 & 1 & -1 \end{pmatrix}$.

40. $A = \begin{pmatrix} 2 & 0 \\ 1 & 1 \end{pmatrix}$.

41. $A = \begin{pmatrix} 2 & 0 \\ 0 & 3 \end{pmatrix}$.

42. The numerical equation $a^2 = 1$ has $a = 1$ and $a = -1$ as its only solutions.

(a) Show that if $A = I$ or $-I$, then $A^2 = I$, where I is an identity matrix of any dimension.

(b) Show that $\begin{pmatrix} a & b \\ c & -a \end{pmatrix}^2 = \begin{pmatrix} 1 & 0 \\ 0 & 1 \end{pmatrix}$ if $a^2 + bc = 1$; so the equation $A^2 = I$ has infinitely many different solutions in the set of 2-by-2 matrices.

(c) Show that every 2-by-2 matrix A for which $A^2 = I$ is either I, $-I$, or one of the matrices described in part (b).

43. Show that if A is an n-by-n matrix and I is the n-by-n identity matrix, then

(a) $(A - I)(A + I) = A^2 - I$.

(b) $(A + I)^2 = A^2 + 2A + I$.

(c) Show that, if B is also n-by-n, it's not true in general that $(A - B)(A + B) = A^2 - B^2$ or that $(A + B)^2 = A^2 + 2AB + B^2$.

44. Find inverses for those of the following 2-by-2 matrices that have inverses.

(a) $\begin{pmatrix} 1 & 1 \\ 1 & 2 \end{pmatrix}$.

(b) $\begin{pmatrix} 3 & 6 \\ 2 & 4 \end{pmatrix}$.

(c) $\begin{pmatrix} \frac{1}{2} & \frac{1}{4} \\ \frac{1}{4} & \frac{1}{5} \end{pmatrix}$.

(d) $\begin{pmatrix} -7 & -5 \\ 12 & 9 \end{pmatrix}$.

45. Solve the matrix equation $A\mathbf{x} = \mathbf{b}$ by multiplying both sides by A^{-1} with

(a) $A = \begin{pmatrix} 2 & -1 \\ 3 & 4 \end{pmatrix}; \mathbf{b} = \begin{pmatrix} 1 \\ 1 \end{pmatrix}$. **(b)** $A = \begin{pmatrix} 7 & 2 \\ 1 & 1 \end{pmatrix}; \mathbf{b} = \begin{pmatrix} -2 \\ 4 \end{pmatrix}$.

46. (a) Show that if A and B are invertible matrices of the same dimension, then AB is invertible and $(AB)^{-1} = B^{-1}A^{-1}$.

(b) Show that, if $A_1, \ldots, A_n$ are invertible matrices with the same dimension, then the matrix product $A_1A_2 \cdots A_n$ is invertible and $(A_1A_2 \cdots A_n)^{-1} = A_n^{-1} \cdots A_2^{-1}A_1^{-1}$.

(c) Show that if $AB = BA = I$ and $AC = CA = I$, then $B = C$.

2 DETERMINANTS

For a square matrix A, the determinant is a numerical-valued function written det A. However, for a displayed matrix it is also customary just to replace the parentheses by vertical bars. Thus the notations

$$\begin{vmatrix} 1 & 4 & 5 \\ 6 & 7 & -3 \\ -2 & 1 & 0 \end{vmatrix}, \qquad \begin{vmatrix} a & b \\ c & d \end{vmatrix}$$

have the same meaning as

$$\det \begin{pmatrix} 1 & 4 & 5 \\ 6 & 7 & -3 \\ -2 & 1 & 0 \end{pmatrix}, \qquad \det \begin{pmatrix} a & b \\ c & d \end{pmatrix}.$$

Our definition of determinant will be inductive; that is, we will define det A first for 1-by-1 matrices, and then for each n define the determinant of an n-by-n matrix in terms of determinants of certain $(n-1)$-by-$(n-1)$ matrices called minors. For any matrix A, the matrix obtained by deleting the ith row and jth column of A is called the ijth minor of A and is denoted by A_{ij}. Recall that we use the small letter a, to denote the ijth entry of a matrix A. Thus the ijth minor A_{ij} corresponds to the entry a_{ij} in a natural way, because the minor is obtained by deleting the row and column containing a_{ij}.

---

example 1 Let

$$A = \begin{pmatrix} -5 & -6 & 7 \\ 8 & -9 & 0 \\ -3 & 4 & 2 \end{pmatrix}, \qquad B = \begin{pmatrix} 1 & 2 \\ 3 & 4 \end{pmatrix}.$$

Then some examples of entries and corresponding minors are

$$a_{11} = -5, \qquad A_{11} = \begin{pmatrix} -9 & 0 \\ 4 & 2 \end{pmatrix},$$

$$a_{23} = 0, \qquad A_{23} = \begin{pmatrix} -5 & -6 \\ -3 & 4 \end{pmatrix},$$

$$b_{11} = 1, \qquad B_{11} = (4),$$
$$b_{12} = 2, \qquad B_{12} = (3).$$

We can now give the definition of **determinant.** For a 1-by-1 matrix $A = (a)$ we define

$$\det A = a.$$

For an n-by-n matrix, $A = (a_{ij})$, $i, j = 1, \ldots, n$, we define

2.1 $$\det A = a_{11} \det A_{11} - a_{12} \det A_{12} + \cdots - (-1)^n a_{1n} \det A_{1n}.$$

The definition is inductive in the sense that defining the determinant of an n-by-n matrix A requires us to know the determinants of the $(n-1)$-by-$(n-1)$ minors A_{ij}. But the simple definition for the 1-by-1 case enables us to go on to 2 by 2, then 3 by 3, and so on. In words, the formula says that det A is the sum, with alternating signs, of the elements of the first row of A, each multiplied by the determinant of its corresponding minor. For this reason, the numbers

$$\det A_{11}, -\det A_{12}, \ldots, (-1)^{n+1} \det A_{1n}$$

are called the **cofactors** of the corresponding elements of the first row of A. In general, the **cofactor** of the entry a_{ij} in A is defined to be $(-1)^{i+j} \det A_{ij}$. Thus, in Example 1 the entry $a_{23} = 0$ in the matrix A has cofactor

$$(-1)^{2+3} \det \begin{pmatrix} -5 & -6 \\ -3 & 4 \end{pmatrix} = 38.$$

The factor $(-1)^{i+j}$ associates plus and minus signs with $\det A_{ij}$ according to the pattern.

$$\begin{pmatrix} + & - & + & - & \cdots \\ - & + & - & + & \cdots \\ + & - & + & - & \cdots \\ - & + & - & + & \cdots \\ \vdots & \vdots & \vdots & \vdots & \end{pmatrix}.$$

Note that the **main diagonal,** from upper left to lower right, contains only plus signs.

---

example 2

(a) $$\det \begin{pmatrix} 1 & 2 \\ 3 & 4 \end{pmatrix} = (1)(4) - (2)(3) = 4 - 6 = -2.$$

(b) $$\det \begin{pmatrix} -5 & -6 & 7 \\ 8 & -9 & 0 \\ -3 & 4 & 2 \end{pmatrix} = -5 \det \begin{pmatrix} -9 & 0 \\ 4 & 2 \end{pmatrix} - (-6) \det \begin{pmatrix} 8 & 0 \\ -3 & 2 \end{pmatrix}$$
$$+7 \det \begin{pmatrix} 8 & -9 \\ -3 & 4 \end{pmatrix}$$
$$= (-5)(-18 - 0) + (6)(16 - 0) + 7(32 - 27)$$
$$= 90 + 96 + 35 = 221.$$

(c) $$\det \begin{pmatrix} a & b \\ c & d \end{pmatrix} = ad - bc.$$

The result of the last example is worth remembering as a general rule of calculation. *The determinant of a 2-by-2 matrix is the product of the entries on the main diagonal minus the product of the other two entries.* Thus 2-by-2 determinants can often be computed mentally, and 3-by-3 determinants in one or two lines.

It is an important fact, the proof of which we omit, that if in Equation 2.1 the elements and cofactors of the first row are replaced by the elements and cofactors of any other row, or of any column, then the expansion is still valid. Formally, the statement can be expressed as follows.

2.2 THEOREM

If A is a square matrix, then

$$\det A = \sum_{j=1}^{n} (-1)^{i+j} a_{ij} \det A_{ij}, \qquad \textbf{expansion by } i\textbf{th row},$$

and

$$\det A = \sum_{i=1}^{n} (-1)^{i+j} a_{ij} \det A_{ij}, \qquad \textbf{expansion by } j\textbf{th column}.$$

Notice that Equation 2.1, which we used to define determinant, appears as a special case of the first equation in the theorem when we let $i = 1$. The alternating pattern of cofactor signs applies to all expansions by row or column.

example 3

The determinant computation in Example 1 can be done also by using the elements and cofactors of the second row:

$$\det \begin{pmatrix} -5 & -6 & 7 \\ 8 & -9 & 0 \\ -3 & 4 & 2 \end{pmatrix}$$

$$= -(8) \det \begin{pmatrix} -6 & 7 \\ 4 & 2 \end{pmatrix} + (-9) \det \begin{pmatrix} -5 & 7 \\ -3 & 2 \end{pmatrix} - (0) \det \begin{pmatrix} -5 & -6 \\ -3 & 4 \end{pmatrix}$$

$$= -8(-12 - 28) - 9(-10 + 21)$$

$$= 221.$$

Computing the same determinant using the elements and cofactors of the third column gives

$$\det \begin{pmatrix} -5 & -6 & 7 \\ 8 & -9 & 0 \\ -3 & 4 & 2 \end{pmatrix}$$

$$= 7 \det \begin{pmatrix} 8 & -9 \\ -3 & 4 \end{pmatrix} - (0) \det \begin{pmatrix} -5 & -6 \\ -3 & 4 \end{pmatrix} + 2 \det \begin{pmatrix} -5 & -6 \\ 8 & -9 \end{pmatrix}$$

$$= 7(32 - 27) + 2(45 + 48)$$

$$= 221.$$

The following theorem shows the effect on $\det A$ of the row operations we use to solve linear systems. The proof is an easy consequence of the definition of $\det A$. (See Exercise 16).

2.3 THEOREM
Let A be a square matrix. Then

1. Interchanging two rows or two columns changes the sign of $\det A$.
2. Multiplying a row, or a column, of A by a number r multiplies $\det A$ by r.
3. Adding a multiple of one row to another leaves $\det A$ unchanged; likewise for columns.

Thus while putting a matrix in another form B we just need to keep track of the row multipliers r and the sign changes that occur when rows are interchanged. Then $k \det A = \det B$, where k is plus or minus the product of the row multipliers. According to Theorem 1.3 in Section 1, the row operations just listed can be used to reduce a square matrix A to the identity, or, if that fails, to a matrix containing a row of zeros. In the first case, the determinant is not zero and A is invertible. In the second case, the determinant is zero and A is not invertible. Thus we have arrived at the following Theorem 2.4, which is stated as a remark following Theorem 1.3 in Section 1, and which is also proved again as a part of Theorem 2.8.

2.4 THEOREM
A square matrix A is invertible if and only if $\det A \neq 0$.

---

example 4

Let

$$A = \begin{pmatrix} 1 & 3 & -2 \\ 2 & -4 & 1 \\ 3 & 5 & -2 \end{pmatrix}, \qquad C = \begin{pmatrix} 1 & 3 & 0 \\ 2 & -4 & 5 \\ 3 & 5 & 4 \end{pmatrix}.$$

The third column of C is equal to the third column of A plus 2 times the first column. We compute

$$\det C = (1)(-16 - 25) - (3)(8 - 15) + (0)(10 + 12) = -20.$$

It follows that $\det A = -20$ also, and that both A and C are invertible.

---

example 5

Let

$$A = \begin{pmatrix} 2 & 4 & -1 & 0 \\ 3 & 0 & 2 & 3 \\ -1 & 2 & 3 & 1 \\ 0 & 1 & -2 & -1 \end{pmatrix}.$$

By adding 2 times column 3 to column 1, and 4 times column 3 to column 2, we obtain

$$B = \begin{pmatrix} 0 & 0 & -1 & 0 \\ 7 & 8 & 2 & 3 \\ 5 & 14 & 3 & 1 \\ -4 & -7 & -2 & -1 \end{pmatrix},$$

and by operation 2 of Theorem 2.3, $\det A = \det B$. The expansion of $\det B$ by the first row has only one nonzero term, and we get

$$\det B = (-1)\det\begin{pmatrix} 7 & 8 & 3 \\ 5 & 14 & 1 \\ -4 & -7 & -1 \end{pmatrix}$$

$$= -\det\begin{pmatrix} 7 & 1 & 3 \\ 5 & 9 & 1 \\ -4 & -3 & -1 \end{pmatrix}$$

$$= -\det\begin{pmatrix} 0 & 1 & 0 \\ -58 & 9 & -26 \\ 17 & -3 & 8 \end{pmatrix}.$$

Then $\det B = -(-1)[(-58)(8) - (17)(-26)] = -22$.

The product of two square matrices is again a square matrix. It's a remarkable fact that the determinant of the product equals the product of the determinants of the individual matrices. The theorem is called the **product rule** for determinants; we'll state it without proof.

2.5 PRODUCT RULE

If A and B are square matrices of the same size, then

$$\det(AB) = (\det A)(\det B).$$

A consequence of the product rule is that, for an invertible matrix A,

$$\det(A^{-1}) = \frac{1}{\det A}.$$

There is a determinant formula for the inverse of an invertible matrix that generalizes the very simple Equation 2.2 that holds for 2-by-2 matrices. The formula is reasonably efficient for simple 3-by-3 or even 4-by-4 matrices. For large matrices, it is usually preferable to use row operations to compute a determinant. However, the inversion formula enables us to see some general facts about determinants that are awkward to derive in other ways. We first prove a theorem from which the facts about inverses follow easily.

2.6 THEOREM

For an n-by-n matrix A,

$$\sum_{i=1}^{n}(-1)^{i+j}a_{ik}\det A_{ij} = \begin{cases} \det A & \text{if } k = j, \\ 0 & \text{if } k \neq j, \end{cases}$$

and

$$\sum_{j=1}^{n}(-1)^{i+j}a_{kj}\det A_{ij} = \begin{cases} \det A & \text{if } k = i, \\ 0 & \text{if } k \neq i. \end{cases}$$

Proof. Consider the expression

$$\sum_{i=1}^{n}(-1)^{i+j}x_i \det A_{ij},$$

where $x_1, \ldots, x_n$ may be any n numbers. We see that this is a certain determinant; it is the expansion, using the jth column, of the matrix obtained from A by replacing the jth column by $x_1, \ldots, x_n$. Now consider what happens if we take $x_1 = a_{1k}, x_2 = a_{2k}, \ldots, x_n = a_{nk}$, in other words, if we enter the elements from the kth column in the jth column. If $k = j$, we just get $\det A$. If $k \neq j$, we have the determinant of a matrix with two columns (the jth and kth) identical, so the result is zero. (See Exercise 14.) This proves the first equation; the second follows similarly by reversing the roles of row and column. ■

At this point we need the idea of transposing a square matrix, that is, reflecting it across its main diagonal. The matrix A and its **transpose** A^t are related as shown:

$$A = \begin{pmatrix} a_{11} & a_{12} & \ldots & a_{1n} \\ a_{21} & a_{22} & \ldots & a_{2n} \\ \vdots & \vdots & \ddots & \vdots \\ a_{n1} & a_{n2} & \ldots & a_{nn} \end{pmatrix}, \qquad A^t = \begin{pmatrix} a_{11} & a_{21} & \ldots & a_{n1} \\ a_{12} & a_{22} & \ldots & a_{n2} \\ \vdots & \vdots & \ddots & \vdots \\ a_{1n} & a_{2n} & \ldots & a_{nn} \end{pmatrix}.$$

Another way to state the relationship is to say that transposing interchanges the kth rows and the kth column.

Now the number $(-1)^{i+j} \det A_{ij}$ is the ijth cofactor of A; we'll abbreviate the cofactor $\tilde{a}_{ij}$ and write $\tilde{A}$ for the matrix with entries $\tilde{a}_{ij}$. Using this notation, we can interpret the previous theorem as a statement about the matrix products $\tilde{A}^t A$ and $A\tilde{A}^t$. In fact, we have

2.7 $$\tilde{A}^t A = A \tilde{A}^t = (\det A) I,$$

because the jkth entry in the product $\tilde{A}^t A$ is the product of the jth row of $\tilde{A}^t$ (i.e., the jth column of $\tilde{A}$) and the kth column of A, in other words, the sum $\sum_{i=1}^{n} \tilde{a}_{ij} a_{ik}$. But by the previous theorem this sum is equal to $\det A$ if $k = j$ and 0 otherwise. Hence $\tilde{A}^t A$ is equal to $(\det A)I$, a scalar multiple of the identity matrix. A similar calculation using the second equation in Theorem 2.6 shows that $A\tilde{A}^t = (\det A)I$ also. Thus we have proved Equation 2.7.

We can now easily write down a formula for A^{-1} when $\det A \neq 0$. A somewhat elaborated version of this theorem, described in Theorem 2.9, is called **Cramer's rule** and is used for solving $M\mathbf{x} = \mathbf{c}$ when M is an invertible square matrix.

2.8 THEOREM

If $\det A \neq 0$, then A is invertible, and

$$A^{-1} = \frac{1}{\det A} \tilde{A}^t,$$

where $\tilde{A}^t$ is the transpose of the matrix of cofactors of A. Conversely, if A is invertible, then $\det A \neq 0$.

Proof. If $\det A \neq 0$, then Formula 2.7 can be written

$$\left(\frac{1}{\det A} \tilde{A}^t \right) A = A \left(\frac{1}{\det A} \tilde{A}^t \right) = I.$$

Hence A^{-1} exists and equals $(\det A)^{-1} \tilde{A}^t$. Conversely, suppose $\det A = 0$. If A were invertible, then there would be a matrix A^{-1} such that $AA^{-1} = I$. By the product rule for determinants,

$$(\det A)(\det A^{-1}) = \det I = 1,$$

so $\det A$ can't be zero. ■

---

example 6

We can use Theorem 2.8 to compute the inverse of the matrix

$$A = \begin{pmatrix} 2 & 3 & 4 \\ 5 & 6 & 7 \\ 8 & 9 & 0 \end{pmatrix}.$$

The matrix with entries $\det A_{ij}$ is easily computed to be

$$\begin{pmatrix} -63 & -56 & -3 \\ -36 & -32 & -6 \\ -3 & -6 & -3 \end{pmatrix}.$$

To get the matrix of cofactors, we insert the factor $(-1)^{i+j}$. This changes the sign of every second entry, giving

$$\tilde{A} = \begin{pmatrix} -63 & 56 & -3 \\ 36 & -32 & 6 \\ -3 & 6 & -3 \end{pmatrix}.$$

The transposed matrix is

$$\tilde{A}^t = \begin{pmatrix} -63 & 36 & -3 \\ 56 & -32 & 6 \\ -3 & 6 & -3 \end{pmatrix}.$$

The cofactors in $\tilde{A}$ can be used to expand $\det A$ using the last column. We expand $\det A$ by its last column to get $\det A = 4(-3) + 7(6) + 0(-3) = 30$. Finally,

$$A^{-1} = \frac{1}{\det A}\tilde{A}^t = \begin{pmatrix} -\frac{63}{30} & \frac{36}{30} & -\frac{3}{30} \\ \frac{56}{30} & -\frac{32}{30} & \frac{6}{30} \\ -\frac{3}{30} & \frac{6}{30} & -\frac{3}{30} \end{pmatrix}.$$

The formula for A^{-1} of Theorem 2.8 occurs most often in the combination $A^{-1}\mathbf{b}$ that provides the solution to the vector equation $A\mathbf{x} = \mathbf{b}$ when there is a unique solution. The result is known as **Cramer's rule:**

2.9 THEOREM

Let A be an n-by-n invertible matrix. Then the kth coordinate x_k of the unique vector that satisfies $A\mathbf{x} = \mathbf{b}$ is given by

$$x_k = \frac{\det A_k}{\det A}, \qquad k = 1, 2, \ldots, n,$$

where A_k is the n-by-n matrix obtained by replacing the kth column of A by the column vector $\mathbf{b}$.

Proof. By Theorem 2.9, $A^{-1} = (1/\det A)\tilde{A}^t$, so the solution to $A\mathbf{x} = \mathbf{b}$ is $\mathbf{x} = (1/\det A)\tilde{A}^t\mathbf{b}$. Writing $\tilde{A} = (\tilde{a}_{ij})$, the kth coordinate of $\tilde{A}^t\mathbf{b}$ is

$$\sum_{i=1}^{n} \tilde{a}_{ij} b_i = \sum_{i=1}^{n} (-1)^{i+k} b_i \det A_{ik} = \det(A \text{ with column } k \text{ replaced by } \mathbf{b}) = \det A_k.$$

Hence $x_k = \det A_k / \det A$. ■

example 7

The matrix A and vector $\mathbf{b}$ of the system

$$\begin{aligned} 2x + 3y &= 5 \\ 4x - 2y &= 3 \end{aligned} \quad \text{are} \quad \begin{pmatrix} 2 & 3 \\ 4 & -2 \end{pmatrix} \quad \text{and} \quad \mathbf{b} = \begin{pmatrix} 5 \\ 3 \end{pmatrix}.$$

The matrices A_1 and A_2 of Theorem 2.9 are $A_1 = \begin{pmatrix} 5 & 3 \\ 3 & -2 \end{pmatrix}$ and $A_2 = \begin{pmatrix} 2 & 5 \\ 4 & 3 \end{pmatrix}$. Hence

$$x = \frac{\det\begin{pmatrix} 5 & 3 \\ 3 & -2 \end{pmatrix}}{\det\begin{pmatrix} 2 & 3 \\ 4 & -2 \end{pmatrix}} = \frac{19}{16}, \qquad y = \frac{\det\begin{pmatrix} 2 & 5 \\ 4 & 3 \end{pmatrix}}{\det\begin{pmatrix} 2 & 3 \\ 4 & -2 \end{pmatrix}} = \frac{7}{8}.$$

The following remarkable theorem appears in a crucial way in the validation of the general method for computing the exponential matrix in Chapter 7, Section 1.

2.10 CAYLEY-HAMILTON THEOREM

Let

$$P(\lambda) = \det(A - \lambda I) = (-1)^n \lambda^n + d_{n-1}\lambda^{n-1} + \cdots + d_1\lambda + d_0$$

be the **characteristic polynomial** of the n-by-n matrix A. Then $P(A)$ is equal to the zero matrix, that is, $(-1)^n A^n + d_{n-1}A^{n-1} + \cdots + d_1 A + d_0 I = 0$.

Proof. Apply Theorem 2.8 to the matrix $A - \lambda I$. It follows that if $P(\lambda) \neq 0$ then

$$P(\lambda)I = (A - \lambda I)\widetilde{(A - \lambda I)}^t,$$

where the entries in the cofactor matrix are polynomials of degree at most $n - 1$ in λ. Note that $\widetilde{(A - \lambda I)}^t$ can be written as a polynomial in λ with matrix coefficients B_k, so that

$$P(\lambda)I = (A - \lambda I)(B_{n-1}\lambda^{n-1} + \cdots + B_1\lambda + B_0),$$

where the n-by-n matrices B_k have constant entries constructed from the entries of A. Now $P(A) - P(\lambda)I$ can be written

$$\begin{aligned} &\left[(-1)^n A^n + d_{n-1}A^{n-1} + \cdots + d_0 I\right] - \left[(-1)^n \lambda^n I + d_{n-1}\lambda^{n-1} I + \cdots + d_0 I\right] \\ &= (-1)^n (A^n - \lambda^n I) + d_{n-1}(A^{n-1} - \lambda^{n-1} I) + \cdots + d_1(A - \lambda I) = (A - \lambda I)Q(A). \end{aligned}$$

We can factor out $(A - \lambda I)$ using the usual difference of powers factorization, because the powers of A commute. Now add our expressions above for $P(\lambda)I$

to both sides to get

$$P(A) = (A - \lambda I)\left[Q(A) + (\widetilde{A - \lambda I})^t\right].$$

On the right we have a polynomial with matrix coefficients of degree at least 1 in λ, unless the factor in brackets is zero. Since the left side doesn't contain λ, the factor in brackets must be the zero matrix, so $P(A)$ is the zero matrix. ■

It's worth pointing out that if a square matrix A is invertible, then the Cayley-Hamilton theorem can be used to write A^{-1} as a polynomial in A. For if

$$a_0 I + a_1 A + \cdots + a_{n-1}A^{n-1} \pm A^n = 0$$

is the characteristic equation of A with λ replaced by A, we can multiply by A^{-1} to get

$$a_0 A^{-1} + a_1 I + \cdots + a_{n-1}A^{n-2} \pm A^{n-1} = 0.$$

Now solve for A^{-1}, noting that $a_0 = \det A \neq 0$.

example 8

The matrix $A = \begin{pmatrix} 2 & 3 & 1 \\ 0 & 2 & 2 \\ 0 & 0 & 2 \end{pmatrix}$ has characteristic equation

$$(2 - \lambda)^3 = 8 - 12\lambda + 6\lambda^2 - \lambda^3 = 0,$$

so

$$8A^{-1} - 12I + 6A - A^2 = 0, \quad \text{or} \quad 8A^{-1} = 12I - 6A + A^2 = 0.$$

Hence we need only divide by 8 after computing

$$8A^{-1} = 12\begin{pmatrix} 1 & 0 & 0 \\ 0 & 1 & 0 \\ 0 & 0 & 1 \end{pmatrix} - 6\begin{pmatrix} 2 & 3 & 1 \\ 0 & 2 & 2 \\ 0 & 0 & 2 \end{pmatrix} + \begin{pmatrix} 4 & 12 & 10 \\ 0 & 4 & 8 \\ 0 & 0 & 4 \end{pmatrix} = \begin{pmatrix} 4 & -6 & 4 \\ 0 & 4 & -4 \\ 0 & 0 & 4 \end{pmatrix}.$$

EXERCISES

1. Find AB, BA, and the determinants of A, B, AB, and BA when

(a) $A = \begin{pmatrix} 1 & -2 \\ 3 & 1 \end{pmatrix}$, $B = \begin{pmatrix} 0 & 1 \\ 2 & -3 \end{pmatrix}$.

(b) $A = \begin{pmatrix} 2 & 0 & 0 \\ 0 & 3 & 0 \\ 0 & 0 & 4 \end{pmatrix}$, $B = \begin{pmatrix} -1 & 0 & 1 \\ 2 & -1 & -3 \\ 0 & 3 & 5 \end{pmatrix}$.

2. For the matrices in Exercise 1, part (b), what are $\det(2A)$ and $\det(2B)$?

3. A **diagonal matrix** D is a square matrix such that only the main diagonal entries $d_i = d_{ii}$, are allowed to be nonzero. We sometimes write $D = \text{diag}(d_1, \ldots, d_n)$. Show that $\det D = d_1 d_2 \ldots d_n$. In particular, $\det I = 1$.

4. What is the relation between $\det A$ and $\det(-A)$?

5. Verify the product rule for the pairs of matrices (a) and (b) in Exercise 1.

6. Apply the product rule to show that, if A is invertible, then $\det A \neq 0$ and $(\det A^{-1}) = (\det A)^{-1}$.

7. Let A be an m-by-m matrix and B an n-by-n matrix. Consider the $(m+n)$-by-$(m+n)$ matrix $\begin{pmatrix} A & 0 \\ 0 & B \end{pmatrix}$, which has A in the upper left corner, B in the lower right corner, and zeros elsewhere. Show that its determinant is equal to $(\det A)(\det B)$. [*Hint:* Consider the cases $A = I$ and $B = I$. Then use the product rule.]

8. Use the method of Example 5 of the text to evaluate

(a) $\det \begin{pmatrix} -1 & 0 & 1 & 2 \\ 0 & 1 & 2 & -1 \\ 1 & 2 & -1 & 0 \\ 2 & -1 & 0 & 1 \end{pmatrix}$. **(b)** $\det \begin{pmatrix} 1 & 1 & 1 & 1 \\ 1 & 2 & 4 & 8 \\ 1 & 3 & 9 & 27 \\ 1 & 4 & 16 & 64 \end{pmatrix}$.

9. **(a)** Compute

$$\det \begin{pmatrix} 1 & 2 & 3 & 4 \\ 0 & -1 & 5 & 6 \\ 0 & 0 & 3 & -1 \\ 0 & 0 & 0 & 4 \end{pmatrix}.$$

(b) A matrix A, like the one in part (a), in which every element below the diagonal is 0, is said to be **upper triangular.** Show that if A is any triangular matrix then $\det A$ is equal to the product of the diagonal elements.

10. Let A be the 3-by-3 matrix

$$\begin{pmatrix} 1 & 4 & 1 \\ 2 & 5 & 6 \\ -1 & 3 & 7 \end{pmatrix}.$$

(a) What is A^t?

(b) Compute $\det A$ and $\det A^t$.

11. Show that for an n-by-n matrix A

$$\det(A\tilde{A}^t) = \det(\tilde{A}^t A) = (\det A)^n.$$

12. **(a)** Show that if A, B, and C are n-by-n matrices such that

$$AB = CA = I$$

then $B = C$. [*Hint:* Multiply $CA = I$ by B on the right.]

(b) Show that if A and B are n-by-n matrices such that $AB = I$, then A and B are both invertible, with $A^{-1} = B$ and $B^{-1} = A$. [*Hint:* Use the product rule for determinants, together with part (a).]

13. Use Theorem 2.8 to determine which of the following matrices have inverses and then use either Theorem 2.8 or the remark following the

Cayley-Hamilton theorem to find the inverses of the ones that are invertible.

(a) $\begin{pmatrix} 1 & 0 & 0 \\ 3 & 1 & 5 \\ -2 & 0 & 1 \end{pmatrix}$. **(b)** $\begin{pmatrix} 1 & 2 & 3 \\ -1 & 1 & 0 \\ 0 & 3 & 3 \end{pmatrix}$. **(c)** $\begin{pmatrix} 2 & 4 & 8 \\ 1 & 0 & 0 \\ 1 & -3 & -7 \end{pmatrix}$.

(d) $\begin{pmatrix} t & 0 & 0 \\ 0 & 2 & 0 \\ 0 & 0 & 1 \end{pmatrix}$, t real. **(e)** $\begin{pmatrix} 1 & 2 & 1 \\ 0 & 0 & 1 \\ 0 & 0 & 3 \end{pmatrix}$. **(f)** $\begin{pmatrix} 1 & -1 & 1 \\ 0 & -1 & 1 \\ 0 & 0 & 1 \end{pmatrix}$.

(g) $\begin{pmatrix} 1 & 2 & -1 & 3 \\ 0 & 2 & 0 & 1 \\ 0 & 0 & 1 & 1 \\ 0 & 0 & 0 & 4 \end{pmatrix}$. **(h)** $\begin{pmatrix} 1 & 0 & 1 & 0 \\ 0 & 2 & 0 & 0 \\ 0 & 0 & 3 & 0 \\ 0 & 0 & 0 & 4 \end{pmatrix}$. **(i)** $\begin{pmatrix} 1 & 0 & 0 \\ 0 & e^t & te^t \\ 0 & 0 & e^t \end{pmatrix}$, t real.

14. Show that, if a square matrix A has two rows proportional, then $\det A = 0$. Show that the same result holds for columns.

15. Let A and B be identical square matrices except in some one row or column. Let C also be the same as A and B except that in that one row or column its entries are the sums of the corresponding entries in A and B. Show that $\det C = \det A + \det B$.

16. Prove parts 1, 2, and 3 of Theorem 2.3. The first two follow immediately from expansion by minors of the distinguished row. The third is immediate for 2-by-2 matrices and can be proved by induction using expansion by minors of a row different from the two to be interchanged.

17. Cramer's rule Theorem 2.9 says that if A is a square matrix with nonzero determinant, then the solution of the system $A\mathbf{x} = \mathbf{c}$ has its kth coordinate given by $x_k = \det A_k / \det A$, where the matrix A_k is obtained by replacing the entries in the kth column of A by the corresponding entries in the vector $\mathbf{c}$.

(a) Use Cramer's rule to solve $A\mathbf{x} = \mathbf{c}$, where

$$A = \begin{pmatrix} 1 & 3 & -2 \\ 2 & -4 & 1 \\ 3 & 5 & -2 \end{pmatrix} \quad \text{and} \quad \mathbf{c} = \begin{pmatrix} 1 \\ 2 \\ -2 \end{pmatrix}.$$

(b) Use Cramer's rule to solve for the functions $u(x)$ and $v(x)$ if

$$u(x) \sin x + v(x) \cos x = 0,$$

$$u(x) \cos x - v(x) \sin x = x \sin x.$$

18. Find coefficients α and β, depending on a, b, c, and d, that express the inverse of the 2-by-2 matrix $A = \begin{pmatrix} a & b \\ c & d \end{pmatrix}$ in the form $A^{-1} = \alpha I + \beta A$. [*Hint:* By Theorem 2.10, A satisfies its characteristic equation.]

APPENDIX C

Existence of Solutions

The Picard method described at the beginning of Section 1 is based on a fundamental, and very useful, idea, the iterative solution of equations. The general idea behind the method has important applications in many branches of pure and applied mathematics. Proofs of the theorems stated below are quite technical, and in that way exceed the demands made by the rest of the book. Instead of proving the theorems, the accompanying discussion aims simply at explaining in concrete terms the process of constructing solutions.

1 THE PICARD METHOD

In this section we take up an iterative method for finding approximate solutions to the vector differential equation with initial condition:

$$\frac{d\mathbf{x}}{dt} = \mathbf{F}(t,\mathbf{x}), \qquad \mathbf{x}(t_0) = \mathbf{x}_0.$$

The method has the advantage that it proves the existence of solutions under certain hypotheses.

The first step in applying the Picard method to the preceding vector differential equation is to replace it by an equivalent vector **integral equation:**

$$\mathbf{x}(t) = \mathbf{x}_0 + \int_{t_0}^{t} \mathbf{F}(u, \mathbf{x}(u))\, du.$$

It is easy to check (see Exercise 8) that, if $\mathbf{x} = \mathbf{x}(t)$ is a solution of either equation, then it is a solution of the other, in particular because the initial condition $\mathbf{x}(t_0) = \mathbf{x}_0$ is satisfied by the integral equation. The vector integral is, of course, computed by integrating the individual coordinate functions of the integrand. Next we compute a sequence of approximations to the desired solution by substituting each approximation into the next integrand:

$$\mathbf{x}_1(t) = \mathbf{x}_0 + \int_{t_0}^{t} \mathbf{F}(u, \mathbf{x}_0)\, du,$$

$$\mathbf{x}_2(t) = \mathbf{x}_0 + \int_{t_0}^{t} \mathbf{F}(u, \mathbf{x}_1(u))\, du,$$

$$\vdots$$

$$\mathbf{x}_{k+1}(t) = \mathbf{x}_0 + \int_{t_0}^{t} \mathbf{F}(u, \mathbf{x}_k(u))\, du.$$

To get some feeling for the kind of approximation that the method produces, we take up an example for we can show that a solution formula exists by routine calculus methods.

example 1

The real-valued differential equation $dx/dt = x + 1$ is easily seen to have the solution $x(t) = -1 + e^t$, satisfying the initial condition $x(0) = 0$. Let's arrive at the solution using the Picard method. Since $t_0 = 0,\ x_0 = 0$, and $F(t,x) = 1 + x$, we find

$$x_1(t) = \int_0^t 1\, du = t,$$

$$x_2(t) = \int_0^t (1 + u)\, du = t + \frac{t^2}{2},$$

$$x_3(t) = \int_0^t \left(1 + u + \frac{u^2}{2}\right) du = t + \frac{t^2}{2} + \frac{t^3}{3 \cdot 2},$$

$$\vdots$$

$$x_k(t) = \int_0^t \left[1 + u + \frac{u^2}{2} + \cdots + \frac{u^{k-1}}{(k-1)!}\right] du = t + \frac{t^2}{2} + \cdots + \frac{t^k}{k!}.$$

If we let $k \to \infty$, we get

$$\lim_{k\to\infty} x_k(t) = \lim_{k\to\infty} \left(t + \frac{t^2}{2} + \cdots + \frac{t^k}{k!}\right)$$

$$= \sum_{k=1}^{\infty} \frac{t^k}{k!} = -1 + e^t.$$

Thus, the kth Picard approximation happens in this example to be just the first k terms in the power series expansion of the solution.

example 2

The nonlinear system with initial conditions

$$\dot{x} = xy, \qquad x(0) = 1,$$
$$\dot{y} = x + y, \qquad y(0) = 1,$$

has Picard approximations given by the formulas

$$x_{k+1}(t) = 1 + \int_0^t x_k(u)y_k(u)\,du,$$

$$y_{k+1}(t) = 1 + \int_0^t [x_k(u) + y_k(u)]\,du.$$

The first approximation, made using $(x_0, y_0) = (1,1)$, is

$$x_1(t) = 1 + \int_0^t 1\,du = 1 + t,$$

$$y_1(t) = 1 + \int_0^t (1+1)\,du = 1 + 2t.$$

The next step gives

$$x_2(t) = 1 + \int_0^t (1+u)(1+2u)\,du = 1 + t + \frac{3t^2}{2} + \frac{2t^3}{3},$$

$$y_2(t) = 1 + \int_0^t (1+u) + (1+2u)\,du = 1 + 2t + \frac{3t^2}{2}.$$

We sketch the *trajectories* $\mathbf{x}_k(t) = \big(x_k(t), y_k(t)\big)$ of the two approximations, which are shown in Figure C.1 along with a computer approximation to the solution itself.

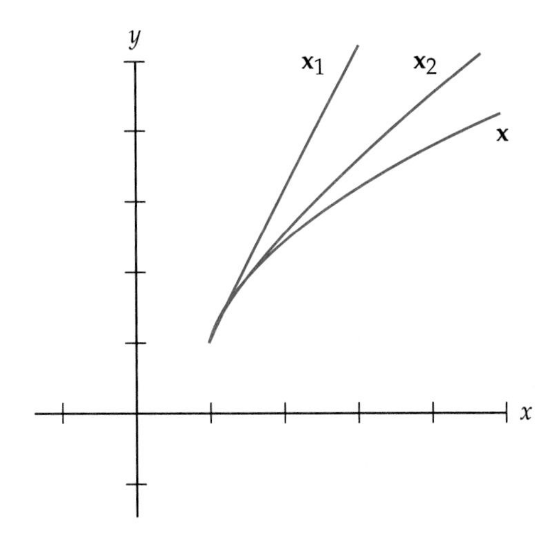

Figure C.1 Approximations $\mathbf{x}_1$, $\mathbf{x}_2$ to $\mathbf{x}$.

The Picard method applies to differential equations of order higher than one by first reducing such equations to first-order systems. Thus solving an initial-value problem for the system of Example 2 is equivalent to solving an initial-value problem for the second-order equation $\ddot{y} - \dot{y} - ty = 0$. The solutions to the nonlinear system are not available in terms of elementary functions. However, the Picard approximations to the associated initial-value problems converge to a solution, thus showing that solutions do exist. Proving this convergence is the standard method for proving existence in the first two theorems of Section 2.

2 EXISTENCE THEOREMS

Without some restriction on the nature of the function $\mathbf{F}(t,\mathbf{x})$, we cannot prove that the Picard iterates $\mathbf{x}_k(t)$ converge to a solution $\mathbf{x}(t)$ of $d\mathbf{x}/dt = \mathbf{F}(t,\mathbf{x})$. All such restrictions that have been used have the common feature that they prevent the values of $\mathbf{F}$ from changing too fast in response to changes in $\mathbf{x}$. The most widely used restriction is the assumption that the coordinate functions F_j have continuous partial derivatives with respect to the real coordinates x_k in $\mathbf{x}$. Under such a condition the Picard iterates converges to a unique solution of the differential equation with prescribed initial value. The precise statement follows.

2.1 THEOREM

Suppose that $\mathbf{F}(t,\mathbf{x})$ and its partial derivatives with respect to the variables x_k in $\mathbf{x}$ are continuous on an open rectangle in $\mathcal{R}^{n+1}$ containing $(t_0, \mathbf{x}_0)$. Then the initial-value problem

$$\dot{\mathbf{x}} = \mathbf{F}(t,\mathbf{x}), \qquad \mathbf{x}(t_0) = \mathbf{x}_0,$$

has a unique solution defined on some t interval containing t_0.

example 1

If F and x are real-valued, with $F(t,x) = 1 + x^2$, then $F_x(t,x) = 2x$. If we take $x(t) = \tan t$, we see that we get a solution for $-\pi/2 < t < \pi/2$, but in no larger interval containing $t = 0$.

example 2

If $\dot{\mathbf{x}} = \mathbf{F}(t,\mathbf{x})$ is linear, then $\mathbf{F}$ has the form

$$\mathbf{F}(t,\mathbf{x}) = A(t)\mathbf{x} + \mathbf{b}(t),$$

with coordinate functions

$$F_j(t,\ x_1, \ldots, x_n) = \sum_{k=1}^{n} a_{jk}(t)x_k + b_j(t).$$

The partial derivatives in question are $(\partial F_j/\partial x_k)(t,\mathbf{x}) = a_{jk}(t)$, so checking the hypothesis of Theorem 1.1 is just a matter of verifying that the entries in the matrix $A(t)$ are continuous on some common interval; if these entries are constant, then Chapter 7, Section 2, shows more, namely, that all solutions are expressible using elementary formulas.

For linear systems, the conclusion of Theorem 1.1 holds on the domain of continuity of the coefficients, rather than shrinking as may happen for a nonlinear equation. The result is fundamentally important if the coefficient matrix has variable coefficients.

2.2 THEOREM

If the matrix $A(t)$ and vector function $\mathbf{b}(t)$ have continuous entries on a common interval I, then for each t_0 in the interval the initial value problem

$$\dot{\mathbf{x}} = A(t)\mathbf{x} + \mathbf{b}(t), \qquad \mathbf{x}(t_0) = \mathbf{x}_0,$$

has a unique solution defined on the entire interval I.

example 3

The result of Theorem 1.1 appears very concretely in the case of dimension $n = 1$. The exponential integrating factor $M(t)$ of Chapter 1, Section 4, provides an explicit solution formula. The fundamental matrix $X(t)$ of Chapter 7, Section 3B, serves as a substitute in higher dimensions to $M(t)$, but is usually too hard to compute.

A **Euler polygon** for an initial-value problem is a concatenation of line segments joining successive Euler approximation points. See Figure 2.8 in Chapter 2, Section 6, for dimension 1. The definition applies to first-order normal-form systems of any dimension. An analysis replacing Picard approximations by Euler polygons shows that the assumption about $\mathbf{F}(t,\mathbf{x})$ can be weakened to continuity alone if all we want is existence of solutions on some restricted interval. Thus it might seem that the continuity requirement on partial derivatives of $\mathbf{F}$ is somewhat artificial. However, the stronger assumption is usually appropriate, because some such condition is necessary anyway to guarantee uniqueness.

It would be nice if continuity of $\mathbf{F}$ implied the convergence of Euler polygons to a solution graph of $\dot{\mathbf{x}} = \mathbf{F}(t, \mathbf{x})$, $\mathbf{x}(t_0) = \mathbf{x}_0$ as the time-difference between approximation points tends to zero. The truth is less pleasant, but in a way more interesting, as the following theorem shows.

2.3 PEANO'S THEOREM

Let $\mathbf{F}(t,\mathbf{x})$ be continuous for t in a neighborhood of t_0 and for $\mathbf{x}$ in some neighborhood of $\mathbf{x}_0$. Let $\mathbf{x}_k(t)$ be a sequence of Euler polygons for the initial-value problem

$$\dot{\mathbf{x}} = \mathbf{F}(t,\mathbf{x}), \qquad \mathbf{x}(t_0) = \mathbf{x}_0,$$

with maximum time difference tending to zero. Then a *subsequence* of $\mathbf{x}_k(t)$ converges to a solution on some neighborhood of t_0.

EXERCISES

For each initial-value problem in Exercises 1 through 4, write an equivalent integral equation.

1. $\dot{x} = tx + t^2$, $x(0) = 1$.

2. $\dot{x} = 1$, $x(0) = 2$.

3. $\dot{x} = x - t$, $x(0) = 0$.

4. $\dot{x} = e^{tx}$, $x(0) = 0$.

(5–8). For each equation in Exercises 1 through 4, find the Picard approximate solutions $x_1(t)$ and $x_2(t)$. Start with $x_0(t) = x(0)$ in each case.

9. (a) Find a vector integral equation equivalent to the system

$$\dot{x} = ty, \qquad x(0) = 1,$$
$$\dot{y} = tx, \qquad y(0) = 2.$$

(b) Find the Picard approximate solution $(x_1(t), y_1(t))$ by starting with $(x(0), y(0)) = (1,2)$.

(c) Find the Picard approximate solution $(x_2(t), y_2(t))$ from the approximation found in part (b).

(d) Sketch the trajectory of the approximate solution found in part (c)

10. (a) The second-order differential equation

$$\frac{d^2y}{dt^2} + ty = 0, \qquad y(0) = 0, \qquad \dot{y}(0) = 1$$

is equivalent to a first-order system of differential equations. What is that system?

(b) Find a pair of integral equations equivalent to the system found in part (a).

(c) Starting with the constant initial solution, find the next two Picard approximate solutions and pick out the relevant part as an approximate solution to the second-order differential equation in part (a).

11. Which of the following initial-value problems have solutions on the prescribed intervals? Justify your answers.

(a) $\dot{x} = x^2,\ x(0) = 1,\ 0 \le t \le 1.$

(b) $\dot{x} = t^2 + x^2,\ x(0) = 1,\ 0 \le t \le 1.$

(c) $\dot{x} = tx + y,\ x(0) = 1,\ \dot{y} = x + y,\ y(0) = 1,\ 0 \le t \le 1.$

12. Let A be an n-by-n constant matrix. Prove that the Picard iterates $\mathbf{x}_n(t)$ for the differential equation $d\mathbf{x}/dt = A\mathbf{x}$ are just the partial sums of the power series expansion of the matrix exponential e^{tA} applied to the initial vector $\mathbf{x}_0$.

13. Sketch the exact solution graph along with the graphs of the first three Picard approximations found in Example 1 of the text.

14. Verify that the initial-value problem and integral equation

$$\dot{\mathbf{x}} = \mathbf{F}(t,\mathbf{x}), \qquad \mathbf{x}(t_0) = \mathbf{x}_0, \quad \text{and} \quad \mathbf{x}(t) = \mathbf{x}_0 + \int_{t_0}^{t} \mathbf{F}(u, \mathbf{x}(u))\, du$$

have the same solutions if $\mathbf{F}$ is continuous.

(a) Show by using one version of the fundamental theorem of calculus that a solution of the initial-value problem is also a solution of the integral equation.

(b) Show by using the other version of the fundamental theorem of calculus that a solution of the integral equation is necessarily differentiable and solves the initial-value problem.

15. Just by looking at the direction field of $dx/dt = \sqrt{1 - x^3}$ it's hard to guess whether the graph actually touches the horizontal line $x = 1$ or is only asymptotic to it.

(a) Show that, for a given value of x, a solution $x(t)$ of $dx/dt = \sqrt{1 - x^3}$ always increases at least as fast as a solution to $dx/dt = \sqrt{1 - x}$. [*Hint:* Note that $1 - x^3 = (1 + x + x^2)(1 - x)$.]

(b) Show that $dx/dt = \sqrt{1 - x}$ has solutions $x(t) = 1 - (t - c)^2/4$.

(c) Show that all solution graphs of $dx/dt = \sqrt{1 - x^3}$ have a point on the line $x = 1$. [*Hint:* Compare, without computing, the integrals of $(1 - x)^{-1/2}$ and $(1 - x^3)^{-1/2}$.]

16. You can verify that the initial-value problem $dx/dt = \sqrt{x}$, $x(0) = 0$ has two different solutions, $x(t) = \frac{1}{4}t^2$ and $x(t) = 0$ for all t and that Picard iteration sequences are generated, for various choices of the initial estimate $x_0(t)$, by $x_{n+1}(t) = \int_0^t \sqrt{x_n(u)}\,du$. So far we've considered starting the iteration only with the constant $x_0(t) = x(0)$ but here we'll make another choice in parts (b) through (f).

(a) Show that starting with $x_0(t) = 0$ for all t makes $x_n(t) = 0$ for all t and $n = 1, 2, 3, \ldots$, so that the limit of this sequence is the identically zero solution.

(b) Show that starting with $x_0(t) = t$ for all $t \geq 0$ produces a sequence of the form $x_n(t) = c_n t^{r_n}$.

(c) Show that the exponents in part (b) are $r_n = (2^{n+1} - 1)/2^n$, and that $\lim_{n\to\infty} r_n = 2$.

(d) Show that the coefficients c_n in part (b) satisfy $c_n = \left[2^n/(2^{n+1} - 1)\right]\sqrt{c_{n-1}}$.

(e) Show that the coefficients c_n in part (b) satisfy $\frac{1}{4} \leq c_n \leq c_{n-1}$, $n = 2, 3, 4, \ldots$, and so conclude that $\lim_{n\to\infty} c_n$ exists and is at least $\frac{1}{4}$. [*Hint:* Use induction and part (d).]

(f) Show that the sequence c_n has limit $\frac{1}{4}$, and hence that the sequence $c_n t^{r_n}$ tends to the corresponding value of the nontrivial solution $\frac{1}{4}t^2$. [*Hint:* Let n tend to infinity in the equation of part (d).]

17. Carry out the analogues of parts (b) through (f) of the previous problem for the initial choice $x_0(t) = t^2$. All other assumptions are to remain the same. [This problem is really easier than the previous one, except that you're not given as much detailed guidance here.]

18. The 1-dimensional problem $\dot{x} = x|x|^{-3/4} + t\sin(\pi/t)$, $x(0) = 0$, has Euler polygons that fail to converge as the time steps tend to zero. (See Chapter 1, Exercise 12, in Coddington and Levinson, *Theory of Ordinary Differential Equations*, Krieger, 1984.)

(a) Show that Theorem 1.3 implies the existence of a solution in a neighborhood of $t = 0$ if we define $0|0|^{-3/4} = 0\sin(\pi/0) = 0$.

(b) Show directly that the hypotheses of Theorem 1.1 fail to hold using the part (a) definitions at $x = 0$ and $t = 0$.

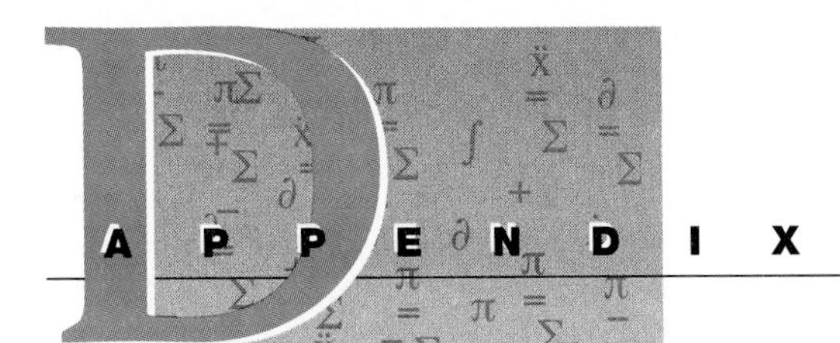

APPENDIX D

Indefinite Integrals

The table at the end of this appendix lists some frequently occurring integrals. As a supplement to the table it's often useful to use a symbolic calculator that provides some indefinite integrals. If you don't see how to compute an indefinite integral directly and don't find it in a table or from a calculator you can try one of the following techniques. Integration constants are omitted, since they're a side issue here.

Identity Substitutions. Rewriting the integrand using an algebraic, trigonometric, exponential, or logarithmic identity will sometimes convert an apparently intractable integrand into an amenable one.

example 1

The integral $\int (e^x + e^{3x})^2\,dx$ can be rewritten by squaring out the binomial to get

$$\int (e^x + e^{3x})^2\,dx = \int (e^{2x} + 2e^{4x} + e^{6x})\,dx$$
$$= \tfrac{1}{2}e^{2x} + \tfrac{1}{2}e^{4x} + \tfrac{1}{6}e^{6x}.$$

example 2

To integrate $\cos^2 x$ recall the trigonometric identity $\cos 2x = 2\cos^2 x - 1$, which is equivalent to $\cos^2 x = \frac{1}{2}(1 + \cos 2x)$. Thus Formula 30 in the table follows from

$$\int \cos^2 x\,dx = \frac{1}{2}\int (1 + \cos 2x)\,dx = \frac{1}{2}x + \frac{1}{4}\sin 2x.$$

Identities to facilitate the integration of rational functions $P(x)/Q(x)$, where $P(x)$ and $Q(x)$ are polynomials, are derived by **partial fraction decomposition,** described in detail in Chapter 8, Section 1 on Laplace transforms.

example 3

A partial fraction decomposition is used to compute Formula 6 in the table:

$$\int \frac{1}{(x-a)(x-b)}\,dx = \int \left[\frac{1}{(a-b)(x-a)} - \frac{1}{(a-b)(x-b)}\right] dx$$

$$= \frac{\ln|x-a|}{a-b} - \frac{\ln|x-b|}{a-b} = \frac{1}{a-b}\ln\left|\frac{x-a}{x-b}\right|.$$

Substitution for the Integration Variable. Awkwardness in an integrand can sometimes be circumvented by a substitution. If the given integration variable is x, a substitution $x = g(u)$, $dx = g'(u)\,du$ may simplify the integrand enough that the corresponding integral in u can be computed. Then replace u by its equivalent in terms of x using the inverse relation $u = h(x)$, where $g\big(h(x)\big) = x$.

example 4

An awkward occurrence of $\sqrt{x}$ can be circumvented by letting $x = u^2$, $dx = 2u\,du$. For example,

$$\int \frac{dx}{\sqrt{x}+1} = \int \frac{2u\,du}{u+1} = 2\int \left(1 - \frac{1}{u+1}\right) du.$$

The last step comes from division of u by $u+1$. Now integrate with respect to u and reintroduce x using the inverse relation $u = \sqrt{x}$ to get

$$\int \frac{dx}{\sqrt{x}+1} = 2[u - \ln(u+1)] = 2[\sqrt{x} - \ln(\sqrt{x}+1)].$$

example 5

To integrate $\sqrt{1-x^2}$, set $x = \sin u$, $dx = \cos u\,du$. Then

$$\int \sqrt{1-x^2}\,dx = \int \sqrt{1-\sin^2 u}\,\cos u\,du$$

$$= \int \cos^2 u\,du = \frac{1}{2}u + \frac{1}{4}\sin 2u,$$

by Example 2. Since $\sin 2u = 2\sin u\cos u$, we find $\int \sqrt{1-x^2}\,dx = \frac{1}{2}\arcsin x + \frac{1}{2}x\sqrt{1-x^2}$.

Substitution for a Part of the Integrand. Here we try to write a given integral in the form $\int f\big(g(x)\big)g'(x)\,dx$ and set $u = g(x)$, $du = g'(x)\,dx$ in the hope that we can compute the indefinite integral $F(u) = \int f(u)\,du$. If $F'(u) = f(u)$, the result is

$$\int f(g(x))g'(x)\,dx = \int f(u)\,du = F(g(x)).$$

example 6

In the integral $\int \cos^2 x \sin x\,dx$ we note the square of a function, namely, $g(x) = \cos x$ multiplied by a function $\sin x$, which is easily modified to be the derivative $g'(x) = -\sin x$. By including the constant factor -1 in the integrand and compensating with a "$-$" before the integral, we rewrite the integral as

$$\int \cos^2 x \sin x\,dx = -\int (\cos x)^2(-\sin x)\,dx.$$

It's now natural to think of substituting u for $g(x) = \cos x$ and du for $g'(x) = (-\sin x)\,dx$ to get

$$\begin{aligned}\int \cos^2 x \sin x\,dx &= -\int u^2\,du \\ &= -\tfrac{1}{3}u^3 = -\tfrac{1}{3}\cos^3 x.\end{aligned}$$

Integration by Parts. This technique is one of the most important, because of its frequent use in deriving other general formulas; it is embodied in the formula

$$\int f(x)g'(x)\,dx = f(x)g(x) - \int f'(x)g(x)\,dx,$$

which follows from the product rule for differentiation. To apply the method, you need to recognize the integrand of a given integral as a product of two functions; one of them, $f(x)$, you differentiate and the other one, $g'(x)$, you try to identify as a function you can integrate easily. If one choice for $f(x)$ and $g'(x)$ fails to work you may want to try another. Formulas 17, 19, and 23 in the table can be computed by a single application of integration by parts, and 21, 24, and 29 by repeated integration by parts.

example 7

In $\int x \sin x\,dx$ a good choice is $f(x) = x$ and $g'(x) = \sin x$, because $f'(x) = 1$ simplifies the remaining integration, and $g(x) = -\cos x$ is easy to integrate. Thus

$$\begin{aligned}\int x \sin x\,dx &= (x)(-\cos x) - \int (1)(-\cos x)\,dx \\ &= -x\cos x + \sin x.\end{aligned}$$

Integral table

1. $\displaystyle\int (ax+b)^n\,dx = \frac{1}{a(n+1)}(ax+b)^{n+1}, \quad n \neq -1$

2. $\displaystyle\int \frac{dx}{ax+b} = \frac{1}{a}\ln|ax+b|$

3. $\displaystyle\int \frac{x\,dx}{ax+b} = \frac{x}{a} - \frac{b}{a^2}\ln|ax+b|$

4. $\displaystyle\int \frac{x\,dx}{(ax+b)^2} = \frac{b}{a^2(ax+b)} + \frac{1}{a^2}\ln|ax+b|$

5. $\displaystyle\int \frac{x\,dx}{(ax+b)^3} = \frac{b}{2a^2(ax+b)^2} - \frac{1}{a^2(ax+b)}$

6. $\displaystyle\int \frac{1}{(x-a)(x-b)}\,dx = \frac{1}{a-b}\ln\left|\frac{x-a}{x-b}\right|$

7. $\displaystyle\int \frac{dx}{(ax+b)(cx+d)} = \frac{1}{ad-bc}\ln\left|\frac{ax+b}{cx+d}\right|, ad-bc \neq 0$

8. $\displaystyle\int \frac{dx}{(a^2-x^2)} = \frac{1}{2a}\ln\left|\frac{a+x}{a-x}\right|,\ a \neq 0$

9. $\displaystyle\int \frac{dx}{x^2(x^2-a^2)} = \frac{1}{a^2x} = \frac{1}{2a(a^2-b^2)}\ln\left|\frac{x-a}{x+a}\right|,\ a \neq 0$

10. $\displaystyle\int \frac{dx}{(x^2-a^2)(x^2-b^2)} = \frac{1}{2a(a^2-b^2)}\ln\left|\frac{x-a}{x+a}\right| + \frac{1}{2b(b^2-a^2)}\ln\left|\frac{x-b}{x+a}\right|,$
$a, b \neq 0$

11. $\displaystyle\int \frac{dx}{x^2(x^2-a^2)} = \frac{1}{a^2x} = \frac{1}{2a(a^2-b^2)}\ln\left|\frac{x-a}{x+a}\right|,\ a \neq 0$

12. $\displaystyle\int \frac{dx}{(x^2-a^2)(x^2-b^2)} = \frac{1}{2a(a^2-b^2)}\ln\left|\frac{x-a}{x+a}\right| + \frac{1}{2b(b^2-a^2)}\ln\left|\frac{x-b}{x+a}\right|,$
$a, b \neq 0$

13. $\displaystyle\int x\sqrt{ax+b}\,dx = \frac{2(3ax-2b)}{15a^2}(ax+b)^{3/2}$

14. $\displaystyle\int \frac{x\,dx}{\sqrt{ax+b}} = \frac{2(ax-2b)}{3a^2}\sqrt{ax+b}$

15. $\displaystyle\int \frac{dx}{(x+a)^2+b^2} = \frac{1}{b}\arctan\left(\frac{x+a}{b}\right)$

16. $\displaystyle\int \frac{x\,dx}{(x+a)^2+b^2} = \frac{1}{2}\ln\left[(x+a)^2+b^2\right] + \frac{a}{b}\arctan\left(\frac{x+a}{b}\right)$

17. $\displaystyle\int \frac{dx}{x(ax^2+b)} = \frac{1}{2b}\ln\left|\frac{x^2}{ax^2+b}\right|$

18. $\displaystyle\int \sqrt{a^2-x^2}\,dx = \frac{1}{2}x\sqrt{a^2-x^2} + \frac{1}{2}a^2\arcsin(x/a)$

19. $\displaystyle\int \frac{dx}{\sqrt{a^2-x^2}} = \arcsin\frac{x}{a}$

20. $\displaystyle\int e^{ax}\,dx = \frac{1}{a}e^{ax}$

21. $\displaystyle\int xe^{ax}\,dx = \frac{1}{a^2}(ax-1)e^{ax}$

22. $\displaystyle\int x^2 e^{ax}\,dx = \frac{1}{a^3}(a^2x^2 - 2ax + 2)e^{ax}$

23. $\displaystyle\int \ln ax\,dx = x\ln ax - x$

24. $\displaystyle\int x\ln ax\,dx = \frac{1}{2}x^2\ln ax - \frac{1}{4}x^2$

25. $\displaystyle\int x^2\ln ax\,dx = \frac{1}{3}x^3\ln ax - \frac{1}{9}x^3$

26. $\displaystyle\int \sin ax\,dx = -\frac{1}{a}\cos ax$

27. $\displaystyle\int x\sin ax\,dx = \frac{1}{a^2}\sin ax - \frac{1}{a}x\cos ax$

28. $\displaystyle\int x^2\sin ax\,dx = \frac{2}{a^2}x\sin ax + \frac{2}{a^3}\cos ax - \frac{1}{a}x^2\cos ax$

29. $\displaystyle\int \sin^2 ax\,dx = \frac{x}{2} - \frac{\sin 2ax}{4a}$

30. $\displaystyle\int \sin^3 ax\,dx = -\frac{1}{a}\cos ax + \frac{\cos^3 ax}{3a}$

31. $\displaystyle\int \cos ax\,dx = \frac{1}{a}\sin ax$

32. $\displaystyle\int x\cos ax\,dx = \frac{1}{a^2}\cos ax + \frac{1}{a}x\sin ax$

33. $\displaystyle\int x^2\cos ax\,dx = \frac{2}{a^2}x\cos ax - \frac{2}{a^3}\sin ax + \frac{1}{a}x^2\sin ax$

34. $\displaystyle\int \cos^2 ax\,dx = \frac{x}{2} + \frac{\sin 2ax}{4a}$

35. $\displaystyle\int \cos^3 ax\,dx = \frac{1}{a}\sin ax - \frac{\sin^3 ax}{3a}$

36. $\displaystyle\int \sin ax\sin bx\,dx = \frac{\sin(a-b)x}{2(a-b)} - \frac{\sin(a+b)x}{2(a+b)},\ |a| \neq |b|$

37. $\displaystyle\int \cos ax\cos bx\,dx = \frac{\sin(a-b)x}{2(a-b)} + \frac{\sin(a+b)x}{2(a+b)},\ |a| \neq |b|$

38. $\displaystyle\int \sin ax\cos bx\,dx = -\frac{\cos(a-b)x}{2(a-b)} - \frac{\cos(a+b)x}{2(a+b)},\ |a| \neq |b|$

39. $\displaystyle\int \sin b(c-x)\cos bx\,dx = -\frac{1}{4b}\sin b(2x-c) + \frac{\cos bc}{2}x$

40. $\displaystyle\int \sin b(c-x)\sin bx\,dx = \frac{1}{4b}\sin b(2x-c) - \frac{\cos bc}{2}x$

41. $\displaystyle\int \cos b(c-x)\cos bx\,dx = -\frac{1}{4b}\sin b(2x-c) + \frac{\cos bc}{2}x$

42. $\displaystyle\int e^{ax}\sin bx\,dx = \frac{e^{ax}}{a^2+b^2}(a\sin bx - b\cos bx)$

43. $$\int e^{ax}\cos bx\,dx = \frac{e^{ax}}{a^2+b^2}(a\cos bx + b\sin bx)$$

44. $$\int \tan ax\,dx = -\frac{1}{a}\ln\cos ax$$

45. $$\int \tan^2 ax\,dx = \frac{1}{a}\tan ax - x$$

46. $$\int \tan^3 ax\,dx = \frac{1}{2a}\tan^2 ax + \frac{1}{a}\ln\cos ax$$

47. $$\int \sec ax\,dx = \frac{1}{a}\ln|\tan ax + \sec ax| = \frac{1}{2a}\ln\left(\frac{1+\sin ax}{1-\sin ax}\right)$$

48. $$\int \sec^2 ax\,dx = \frac{1}{a}\tan ax$$

49. $$\int \sec^3 ax\,dx = \frac{1}{2a}\tan ax\sec ax + \frac{1}{a}\ln|\tan ax + \sec ax| = \frac{1}{2a}\left[\frac{\sin ax}{\cos^2 ax} + \ln\left(\frac{1+\sin ax}{1-\sin ax}\right)\right]$$

50. $$\int \tan ax\sec ax\,dx = \frac{1}{a}\sec ax$$

Answers to Selected Exercises

Answers to exercises are provided as a spot check on the correctness and accuracy of work, and not as definitive validations. It's good to maintain the habit of checking results both for general reasonableness and for precision; the latter can usually be done routinely after solving a differential equation by substitution into the equation, while the former depends on the development of sound reasoning, particularly when it comes to the interpretation of a formal solution. Many exercises in this book include as integral parts of their statements what amount to answers; for many others, brief answers are provided here as the outcome of more elaborate work. For numbered exercises with several similar parts, the general, though not invariable, practice is to give an answer for part (a) and then for alternate parts thereafter.

Chapter 1

Section 1

1. Yes **3.** Yes **5.** Yes **7.** Yes **9.** Yes **11.** Yes **13.** Yes **15.** Yes **17.** $y = 7e^{3x}$
19. $x = -2 + \ln t$ **21.** $y = -2\cos x$ **23.** $y = -\sqrt{2 - x^2}$ **25.** $y = -\sqrt{3x}$ **27.** $y' = 3y$
29. $x' = 1/t, \quad t > 0$ **31.** $y' = -y\tan x$ **33.** $y' = -x/y$ **35.** $y' = y/2x$ **37.** $r = -2$
39. $r = 1$ **41.** **(b)** $k = (\ln 3)/2$ **43.** $y' = \sqrt{x^2 + y^2}$ **45.** $y' = kx/y$

Section 2AB

19. **(b)** $y = \begin{cases} 0, & x < 0 \\ (1/2)x^2 + 1, & x \geq 0 \end{cases}$
23. **(c)** The solution graphs are concave up where $y > 0$ and concave down where $y < 0$
39. $y = \pm x$ **41.** All solutions are increasing
43. Solutions are increasing in the region bounded above by $y = |x|$ and below by $y = -|x|$, and decreasing in the complementary region
45. Solutions are increasing in the regions inside the parabola $y = x^2$ to the left of the y-axis, and outside $y = x^2$ to the right of the y-axis. They are decreasing in the complementary region.
53. $y' = y/x$ **55.** $y' = -y/x$ **57.** $y' = 3y/x$ **59.** $y' = \cot x$ **61.** $y' = 0$ **63.** $y' = -y/x$

Section 2C

1. $S = \boldsymbol{R}^2$; $R = S$
3. S is the union of the 1st and 3rd quadrants of $\boldsymbol{R}^2$; $R = \{(x, y) \mid x, y > 0\}$ or $R = \{(x, y) \mid x, y < 0\}$
5. $S = \{(x, y) \mid x > 0\}$ or $S = \{(x, y) \mid x < 0\}$; $R = S$ **7.** $S = \boldsymbol{R}^2$; $R = S$
9. $S = \boldsymbol{R}^2$; $R = S$ **13.** **(b)** $s(t) = 0$ is a second solution **17.** **(a)** $R = \{(x, y) \mid x > 0, \ y > 0\}$
19. **(c)** No. $\frac{\partial F}{\partial y}$ is not defined for $x^2 + y^2 = 1$
21. **(c)** $y = \begin{cases} x^2 - 1, & x \leq 0 \\ (1/2)x^2 - 1, & x > 0 \end{cases}$. **(d)** Yes. $p(x)$ and $\frac{\partial}{\partial y}P(x) = 0$ are continuous for all (x, y).

Section 3A

1. $y = \frac{1}{2}x^2 - \frac{1}{3}x^3 + c$; $y = \frac{1}{2}x^2 - \frac{1}{3}x^3 + 1$ **3.** $y = -\frac{1}{2}\ln(1 - x^2) + c$; $y = -\frac{1}{2}\ln(1 - x^2) + 1$
5. $y = -\sin x + c_1 x + c_2$; $y = -\sin x + 2x + 1$ **7.** $z = (t - 1)e^t + c$; $z = (t - 1)e^t + 2$
9. $x = e^t + c_1 t + c_2$; $x = e^t - t$ **17.** $t_{\max} \approx 155$ seconds; $h_{\max} \approx 388{,}199$ feet
19. **(a)** $t = 1$ second; height ≈ 4984 feet; **(b)** $v_0 \approx 401$ ft/sec **21.** $v_0 \approx 80.25$ ft/sec
25. **(a)** $t \approx 2.9$ sec; **(b)** stopping distance ≈ 129 feet;
(c) dec. rate ≈ 25 ft/sec^2; stopping distance ≈ 155 ft
27. **(a)** $v_0 \approx 5.67$ ft/sec; **(b)** $y_{\max} \approx 3.1$ feet

35. **(b)** Other solution $s(t) = 0$. $\frac{\partial}{\partial s} = g/\sqrt{2gs}$ is not defined at $s = 0$.
37. **(a)** $y(x) = -\frac{1}{24}Px^4 + c_1x^3 + c_2x^2 + c_3x + c_4$; **(b)** Maximum deflection is about $(.005)L^4P$
39. **(a)** $y(x) = -\frac{1}{24}\left(Px^4 + \frac{(48a-2PL^4)}{L^3}x^3 + \frac{(pL^4-72a)}{L^2}x^2\right)$
41. **(a)** Boundary conditions lead to inconsistent equations when $R, L > 0$;
(b) Boundary conditions describe a beam simply supported at the left end and hanging free at the right end, which is impossible.

Section 3B

1. $y = \pm\sqrt{\frac{2}{3}x^3 + c}$; $c = -13/3$ **3.** $y = (3e^{-x} + c)^{1/3}$; $c = -2$ **5.** $v = \arccos(-\sin u + c)$; $c = 1$
15. **(b)** The direction field is horizontal on the lines $y = \pm 1$
19. **(b)** $h(t) = \left(\frac{-ka\sqrt{2g}}{2A}t + \sqrt{h_0}\right)^2$; **(c)** $(A/(ka))\sqrt{2h_0/g}$ time units; 7.4 minutes; **(d)** $k \approx 0.74$
23. **(a)** $A = \pi(2h - h^2)$; **(b)** 33.6 seconds **25.** $r = (3Ct/(4\pi))^{1/3}$
27. **(a)** $y = 4/(x+c)^2$; **(d)** The solution $y = 4/x^2$.
29. **(a)** $y(x) \geq x$, graph increasing; **(b)** $y(x) \geq x^2/2$, graph increasing and concave up;
(c) $y(x) \geq e^x$ if $x \geq 0$, $y(x) \leq e^x$ if $x \leq 0$; **(d)** $x^3/6 + x \leq y(x) \leq x^3 + 3 + x$;
(e) $y(x) \geq e^x$ if $x \geq 0$, $y(x) \leq e^x$ if $x \leq 0$; **(f)** $y(x) \leq x$, graph concave down.
31. **(a)** $y = (Ke^{2x} - 1)/(Ke^{2x} + 1)$, $K \neq 0$;
(c) Allowing $K = 0$ gives $y(x) = -1$. No value of K gives $y(x) = 1$

Section 3C

1. $y = ce^x - 2x - 5$ **3.** $\arctan\left(\frac{1}{\sqrt{3}}\left(\frac{2y}{x} - 1\right)\right) = \frac{\sqrt{3}}{2}\ln|x| + c$ **5.** $\ln(x^2 + y^2) + 2\arctan(y/x) = c$
7. $(x - y + 1)^2 = 2x + c$ **9.** $\ln|x - 1 + \sqrt{2}(y+2)| - \ln|x - 1 - \sqrt{2}(y+2)| - \sqrt{2}\ln|(x-1)^2 - 2(y+2)^2| = c$
11. $2\arctan\left(\frac{x+1/2}{y+1/2}\right) = \ln\left((x+1/2)^2 + (y+1/2)^2\right) + c$ **13.** $7y^2 - 4x^2 - 6xy - 8y + 14x = c$
15. $y^2 - x^2 + 2xy + 2y + 1/2 = c, c \neq 0$ **17.** **(b)** $(x + 2y + 1)^2 - 6y = c$
27. **(b)** Derivatives for positive and negative x don't agree as $x \to 0$; **(c)** (ii) $C = 0$.

Section 4A

1. $M(x) = e^{2x}$ **3.** $M(x) = x^2$ **5.** $M(x) = e^{-\frac{1}{2}e^{-2x} - x}$ **7.** $s = 1 + Ce^{-t^2/2}$; $s = 1 - e^{-t^2/2}$
9. $y = Ce^{x^2/4}$; $y = 0$ **11.** $x = 1 + Ce^{-y}$; $x = 1$ **13.** $y = \frac{1}{2}e^x + Ce^{-x}$; $y = \frac{1}{2}(e^x - e^{-x})$
15. $y = \frac{1}{2} + xe^{-2x} + Ce^{-2x}$; $y = \frac{1}{2} + xe^{-2x} + \frac{1}{2}e^{-2x}$ **17.** $x = \frac{1}{4}te^{-2t} + \frac{1}{16}e^{-2t} + Ce^{2t}$; $x = \frac{1}{4}te^{-2t} + \frac{1}{16}e^{-2t} + \frac{15}{16}e^{2t}$
19. $x = -y - 1 + Ce^y$ **21.** $y = \begin{cases} x - 1 + Ce^{-x}, & 0 \leq x < 1 \\ x^2/3 - 1/(3x) + C/xe, & x \geq 1 \end{cases}$ **23.** $y = x - C_1e^{-x} + C_2$
25. $y = C_1\ln|x| + x^2/4 + C_2$ **27.** $y = C_1e^{-x} + C_2x + C_3$ **29.** $y = C_1x^{1/2} + C_2x + C_3$

Section 4B

1. $S = 10e^{-t/50}$ **5.** $u_{\min} = 47.5°$ at about 1 hr 23 min; $\lim_{t\to\infty} = 50°$ **7.** ≈ 36.8 lb. **9.** ≈ 14.35 lb
11. **(a)** 155 lb. **(b)** 200 lbs; about 29.4 min. **13.** ≈ 34.66 lb. **15.** ≈ 112.5 lb.
17. $u(t) = 20(e^{-t} - e^{-2t})$; $b = 5$ **19.** $u(t) = 10(1 - 2e^{-t} + e^{-2t})$; $b = 10$ **21.** $u(t) = 50(1 + e^{-t})$
23. $u(t) = 100e^{-t}$ **25.** $u(t) = 50(4 - e^{-t} - e^{-2t})$ **27.** **(a)** $k \approx 0.0924$; **(b)** $t = 22.5$ min
29. **(a)** $u(t_1) = \frac{sa + u_1 + s(u_0 - a)e^{-kt_1}}{s+1}$; **(b)** $v(t_1) = \frac{sa + a + \left(s(u_0-a) + (u_1-a)\right)e^{-kt_1}}{s+1}$; **(c)** The routine in part (b).
31. **(a)** $dx/dt = rx/(\ell + rt)$; **(b)** $x(t) = x_0(\ell + rt)/\ell$

Section 4C

1. $y' = y$ **3.** $y' = 2y$ **5.** $y' = -y$ **7.** $y' = -4x^2(y+1)$
9. **(a)** $y_1 = 6/(2 - 3x^2)$; **(b)** $y_2 = (3/2)(1 + e^{3x^2})$; **(c)** $y_1(1/2) = 4.8$; $y_2(1/2) \approx 4.6755$
13. **(a)** $dv/dt = \alpha g(1 - (k/gm)^{1/\alpha}v)$; **(b)** $v(t) = (gm/k)^{1/\alpha}$
17. **(b)** As $t \to \infty$, $u(t) \to 1$ for the non-linear solution and $u(t) \to 1/2$ for the linearized solution.
19. $y = (1 + 7e^{-3x})^{-1/3}$ **21.** $(y + y_0)^2v^2 = \frac{2}{3}\left((y + y_0)^3 - y_0^3\right)$

Section 5AB

1. Exact, separable; $F(x.y) = xy + x^2y^2 = C$; $xy + x^2y^2 = 2$
3. Exact, separable; $F(x, y) = x(y + e^y) = C$; $x(y + e^y) = 1$ **5.** Not exact, separable; $ye^y = Ce^{x^2/2}$; $y = 0$
7. Exact, separable; $F(x, y) = xy + \ln|xy| = C$; $xy + \ln|xy| = 1$
9. Exact, not separable; $F(x, y) = x^2y + xy^3 = C$; $x^2y + xy^3 = 2$ **11.** $k = 1$; $xy + y + y^3/3 = C$
13. Exact for all k; $x + \ln|x + ky| = C$ or $(x + ky)e^x = C$
17. $Q(x, y) = -x \sin y + g(y)$, where g is any integrable function of y
19. **(b)** $y - y/(x^2 + y^2) = C$; **(d)** $\left(1 + \frac{y^2-x^2}{(x^2+y^2)^2}\right) dx - \frac{2xy}{(x^2+y^2)^2} dy = 0$; $x + x/(x^2 + y^2) = C$

Section 5C

23. $x/y + x = C$ **25.** $x^3 + 2x^3y - x^2y^2 = C$ **27.** **(b)** $F(x, y) = H(x) - G(y) = C$ **29.** $x/y - y = C$
31. $xy^2 = C$ **33.** $y^2 \sin x = C$ **35.** $(x - 1)y^2 = C$ **37.** $M(x) = x$; $x^2y^2 + 2x^3 = C$
39. $M(x) = e^x$; $y^2e^x + 2ye^{2x} = C$

Chapter Review and Supplement

1. $xy - x^2/2 = C$ **3.** $x = (t + C)e^{t^2/2}$ **5.** $\ln(y^4 + 1) = 4e^x + C$ **7.** $y = \begin{cases} x^3(1/2 + C_1e^{-2x}), & x \geq 0; \\ x^3(1/2 + C_2e^{-2x}), & x < 0. \end{cases}$
9. $x = (1/5)(t^3 - 12/t^2)$, $t > 0$ **11.** $x = 1/(3t + C)$, $x = 0$ **13.** $x = \tan(t + C) - t$ **15.** $y = -xe^x + (1 + e)x$
17. $x = \ln(-1 + Ce^t)$ **19.** $y = x$ **21.** $zy^3 + 2zy - z^3 = C$ **23.** $x^3/3 + xy - \cos y = C$
25. $y = e^{-x^2/2}\left(\int e^{-x^2/2} dx + C\right)$ **27.** $y^4 = 1 + Ce^{-2x^2}$, $y \neq 0$ **29.** $2x^2 + y^2 + 3xy + 2x - y = C$
31. $x + e^{-x} \sin y = C$ **33.** $x^2 - y^2 + xy - 1 = Cx$ **35.** $ye^{xy} = C$ **37.** $xy^2 + y = C$, $y \neq 0$
39. $y = (1 - Ce^{-x/3})^3$ **41.** **(a)** P is the entire xy plane; **(b)** Q is the region above the line $y = x$; **(c)** Yes; **(d)** Yes; **(e)** $y = \ln(e^x + C)$
45. $S(t) = 200(1 - e^{-t/100})$; $S(t) = 200 - 150e^{-t/100}$
47. **(d)** Solution of linearized equation can be solved for u as a function of t.
49. (ii) The contrapositive statement to (i) is "If a solution $f(x)$ is 0 for at least one value of x then it is not the case that the $f(x)$ is not 0 for all x; i.e., $f(x) = 0$ for all x if it is 0 for at least one x.

Chapter 2

Section 1

1. $r = (1/10)\ln(3/2)$; $P_0 = 200/3$; $P(-10) = 400/9$ **3.** $(1/r)\ln 3$ **7.** $P' = \beta P(1 - P/(\beta/\delta))$
9. $a = r$; $b = r/K$ **11.** **(b)** $\ln P - (1+c)\ln(K-P) = rt + C$; **(c)** $\ln(P/P_0) - (1+c)\ln[(K-P)/(K-P_0)] = rt$
13. **(a)** $D = (1/r)\ln[2(K - P_0)/(K - 2P_0)]$

Section 2

1. **(a)** $\approx$99.5 ft/sec; **(b)** $\approx$7.6 miles **3.** **(a)** $k = 0.36$; **(b)** $k = 1296$; **(c)** Yes
7. **(a)** $v = 4v_0/(2 + 10^{-3}\sqrt{v_0}t)^2$; **(b)** $x = 2v_0t/(2 + 10^{-3}\sqrt{v_0}t)$
9. **(a)** $\approx$4.05 seconds; **(b)** $\approx$304.4 feet; **(c)** $\approx$4.7 sec; ≈ -121 ft/sec;
   (d) Air resistance prevents the velocity at impact with the ground from being as large as v_0;
   (e) Maximum height $= 402.5$ at $t = 5$ seconds, and final velocity equals the negative of initial velocity after an additional 5 seconds. Compared with initial assumptions, the ball remains moving upward longer and therefore goes higher without air resistance. Speed on the way down is also higher, so it takes only slightly more time to reach the ground while traveling approximately 4/3 as far.
11. **(b)** $v = -gt + v_0$, and $t_{max} = v_0/g$, the same as in part (a).
15. **(b)** $v = \sqrt{mg/k}(Ce^{2bt} - 1)(Ce^{2bt} + 1)$, where $b = \sqrt{kg/m}$; **(c)** $v_\infty = \sqrt{mg/k}$

Section 3

1. Equilibrium solution: $y(t) = 0$. All solutions approach 0 as $t \to \infty$.
3. Equilibrium solution is $y(t) = 0$. No nonconstant solution approaches 0.
5. Equilibrium solutions: $y_1(t) = 1$, $y_2(t) = -1$. Solutions with values < -1 approach -1 as $t \to -\infty$;

solutions with values >1 approach 1 as $t \to -\infty$; solutions with values in $(-1, 1)$ approach -1 as $t \to -\infty$, and approach 1 as $t \to \infty$.

7. Equilibrium solutions are $u_1(t) = 0$, $u_2(t) = 1$. Solutions with values <0 approach 0 as $t \to \infty$; solutions with values >1 approach 1 as $t \to -\infty$; solutions with values in $(0, 1)$ approach 0 as $t \to \infty$, and approach 1 as $t \to -\infty$.

9. Equilibrium solution: $y(t) = 0$. All solutions approach 0 as $t \to \pm\infty$.

11. **(a)** $v_1(t) = -1 + \sqrt{1 + mg/k}$, $v_2(t) = -1 - \sqrt{1 + mg/k}$;
(b) Only v_1, the terminal velocity, is physically meaningful.

13. $v_\infty = e^{mg/k} - 1$ **15.** **(b)** $v_\infty = mg/k$ if $m \le k/g$, and $v_\infty = \sqrt{2mg/k - 1}$ if $m > k/g$.

19. **(a)** $0 < h < kc^2/4$; **(b)** $y(t) = (c \pm \sqrt{c^2 - 4h/k})/2$;
(c) Solutions approach $(c + \sqrt{c^2 - 4h/k})/2$ as $t \to \infty$, and approach $(c - \sqrt{c^2 - 4h/k})/2$ as $t \to -\infty$;
(d) Solutions approach $(c - \sqrt{c^2 - 4h/k})/2$ as $t \to -\infty$;
(e) Solutions approach $(1 + \sqrt{1 - 4h})/2$ approach as $t \to \infty$;
(f) If $h = kc^2/4$ there is a unique equilibrium point, and if $h > kc^2/4$ there are no equilibrium points and the D.E. does not model the history of a real population.

21. **(b)** $0 < h < k\alpha^\alpha \beta^\beta \left(c/(\alpha + \beta)\right)^{\alpha+\beta}$;
(c) Solutions approach the larger value as $t \to \infty$, and approach the smaller value as $t \to -\infty$.

Section 4

1. **(a)** $v = \frac{\alpha v_e}{k} + c(m_1 - \alpha t)^{k/\alpha}$ **3.** **(b)** $m/m_0 = e^{-c/v_e}$; **(c)** $m/m_0 \approx 10^{-37,000}$ **5.** **(a)** $v = v_0 + v_e \ln(m_0/m)$

9. **(a)** $v = 800 \ln\left(\frac{20}{20-t}\right) - 32t$, $a = \frac{800}{20-t} - 32$ ft/sec^2;
(b) $h(10) \approx 854.8$ ft, $v = -32t + 235.5$, $h = -16t^2 + 235.5t + 854.8$, $h_{\text{max}} \approx 1714$ ft, speed at impact ≈ 331.2 ft/sec

13. $v = \left(\frac{(c+v_0)-(c-v_0)(m/m_0)^{2v_e/c}}{(c+v_0)+(c-v_0)(m/m_0)^{2v_e/c}}\right)$ **21.** Fuel consumed $= (m_0 - m_1)(1 - e^{(v_2 - v_1)/v_e})$

23. **(a)** $m\frac{dv}{dt} = -(v + s)\frac{dm}{dt}$; **(b)** $v = \frac{m_0(v_0+s)}{m_0+st} - s$ **25.** **(b)** $x = v_0\sqrt{m_0/a} \arctan \sqrt{\frac{a}{m_0}}t$, $s_0 = \frac{v_0 \pi}{2}\sqrt{m_0/a}$

27. **(a)** $v = \frac{16t(20-t)}{10-t}$; **(b)** $v = \frac{3200}{19}\left[\left(1 - \frac{t}{10}\right)^{-0.9} - \left(1 - \frac{t}{10}\right)\right]$;
(c) $v(7) \approx 485$ ft/sec (part(a)), $v(7) \approx 447$ ft/sec (part (b))

29. **(a)** $v = 32t$ **31.** **(a)** $(75 - t)\frac{dv}{dt} = 10 - 0.0005v^2$, $(0 \le t \le 25)$;
(b) $v(25) \approx 102.0$ ft/sec; **(c)** ≈ 67.5 ft/sec.

Section 5AB

1. $y = e^x$ **7.** $y = e^x$ **9.** $y = e^x + x - 1$ **15.** **(b)** $(r, q) \approx (0.5, 1.4)$

19. **(a)** 100 minutes; **(b)** $dS/dt = 3(1 - e^{-t}) - 4S/(100 - t)$;
(c) $S_{\max} \approx 48.4$ and occurs at about $t = 35.5$;
(d) Using the I.F. $(100 - t)^{-4}$ requires integrating $e^{-t}/(100 - t)^4$.

21. ≈ 18 seconds

Section 5C

1. $x \approx 23.50393$ **3.** $x \approx 0.56714$ **5.** $x = 1$ **7.** $x_1 \approx -0.72449$, $x_2 \approx 1.122074$ **9.** $x \approx 1.3098$

11. $x_1 = -1$, $x_2 \approx 0.405564$ **17.** The slope of the line is steeper than the tangent if $0 < h < 1$.

21. **(b)** Gen. sol.: $y = Cx/(x - 1)$; **(c)** $x_n \to \infty$ for $|x_0| > 1$, $x_n \to 0$ for $|x_0| < 1$, $x_n \to 1$ for $|x_0| = 1$.

Chapter 3

Section 1AB

1. $r^2 + r - 6 = 0$; $y(x) = c_1 e^{-3x} + c_2 e^{2x}$ **3.** $r^2 + 2r + 1 = 0$; $y(x) = c_1 e^{-x} + c_2 x e^{-x}$ **5.** $r^2 - r = 0$; $y(x) = c_1 + c_2 e^x$

7. $2r^2 - 3r + 1 = 0$; $y(x) = c_1 e^{x/2} + c_2 e^x$ **9.** $y(x) = (2/5)e^{-3x} + (8/5)e^{2x}$ **11.** $y(x) = (3x + 1)e^{-x}$

13. $y(x) = (e - e^x)/(e - 1)$ **15.** $y(x) = 0$ **17.** $y'' + 2y' + y = 0$ **19.** $y'' = 0$ **21.** $y'' + 2y' + y = 0$

23. $-e^{-2x}$ **25.** $27e^{3x}$ **27.** $-2 \sin x$ **29.** $D^2 + 2D + 1 = (D + 1)^2$ **31.** $2D^2 - 1 = (\sqrt{2}D - 1)(\sqrt{2}D + 1)$

33. D^2 **35.** $z = C_1$, $y(x) = c_1 + c_2 e^{3x}$ $(c_1 = -C_1/3)$ **37.** $z = -1 + C_1 e^x$, $y(x) = -1 + c_1 e^x + c_2 e^{-x}$ $(c_1 = C_1/2)$

39. **(a)** $y(x) = c_1 e^{-x} + c_2 x e^{-x}$; **(b)** $y(x) = e^{-x} + 3xe^{-x}$;
(c) $y(x) \to 0$ as $x \to \infty$, $y(x) \to -\infty$ as $x \to -\infty$; **(d)** $(2/3, 3e^{-2/3})$
41. **(b)** $y(x) = c_1 x + c_2/x$; **(d)** $y(x) = c_1 \ln|x| + c_2$
47. $y(x) = (1-2x)e^x$ $(y'(0) = -1)$; $y(x) = (1-x)e^x$ $(y'(0) = 0)$; $y(x) = e^x$ $(y'(0) = 1)$
51. **(b)** $g \approx 32.2$ ft/sec^2; **(c)** $y(t) = y_0 \cosh \sqrt{g/\ell}\, t + v_0 \sqrt{\ell/g} \sinh \sqrt{g/\ell}\, t, 0 \le y \le \ell$.
53. **(a)** $y(x) = \cosh x + \alpha \sinh x$

Section 2AB

1. $\pi/2$ **3.** $-\pi/4$ **5.** $c_1 = c_2 = 1/2$ **7.** $c_1 = -i\pi/2$, $c_2 = i\pi/2$ **9.** **(b)** Smallest positive period $= 2\pi/\beta$
13. $y = e^{-x}(c_1 \cos x + c_2 \sin x)$ **15.** $y = x^2/2 - x + c_1 + c_2 e^{-x}$ **17.** $r_{1,2} = \pm i\sqrt{2}$; $y = c_1 \cos\sqrt{2}x + c_2 \sin\sqrt{2}x$
19. $r_{1,2} = 1 \pm i$; $y = e^x(c_1 \cos x + c_2 \sin x)$ **21.** $r_1 = 1/2, r_2 = -1$; $y = c_1 e^{x/2} + c_2 e^{-x}$
23. $r_{1,2} = (-1 \pm i\sqrt{7})/4$; $y = e^{-x/4}\left(c_1 \cos \frac{\sqrt{7}}{4}x + c_2 \sin \frac{\sqrt{7}}{4}x\right)$ **25.** $y = (1/\sqrt{2}) \sin\sqrt{2}x$
27. $y = 0$ **29.** $y = \frac{4}{3}(e^{x/2} - e^{-x})$ **31.** $y = 0$ **35.** **(b)** $A = 2$; $\theta = \pi/6$
41. **(a)** The solution corresponds to the pendulum at rest but pointing vertically upward, a particularly difficult feat to accomplish in practice.
(b) The constant solutions are $\theta = n\pi$, n an integer. Even values of n correspond to the pendulum at rest pointing downward, odd values of n correspond to the pendulum at rest pointing upward.
45. $y'' + 4y = 0$ **47.** $y'' - 2y' + 5y = 0$ **49.** $4y'' + y = 0$ **51.** $y'' + 9y = 0$
53. They are not independent on any interval if $\beta = 0$

Section 2C

1. (i) $p = 2\pi/3$; (ii) $\nu = 3/2\pi$; (iii) $A = 1$ **3.** (i) $p = 2\pi/3$; (ii) $\nu = 3/2\pi$; (iii) $A = \sqrt{5}$
5. (i) $p = 2\pi/3$; (ii) $\nu = 3/2\pi$; (iii) $A = \sqrt{2}$ **7.** $y'' + y = 0$ **9.** $y'' + 4y = 0$ **11.** $y'' + 2y' + 2y = 0$
13. Not a solution **15.** $A = 2$; $\phi = \pi/3$; $y = 2\cos(x - \pi/3)$ **17.** $A = 3$; $\phi = \pi/2$; $y = 3\cos(2x - \pi/2)$
19. $A = 1$; $\phi = \pi$; $y = \cos(4x - \pi)$ **21.** $A = 6$; $\phi = 2\pi/3$; $y = 6\cos(2x - 2\pi/3)$
23. $y'' + 4y = 0$, $y(0) = 0$, $y'(0) = 6$ **25.** $y'' + y = 0$, $y(0) = 2$, $y'(0) = -2\sqrt{3}$
27. $y'' + 2y' + 5y = 0$, $y(0) = -1$, $y'(0) = 1$ **29.** $y'' - 2y' + 2y = 0$, $y(0) = 0$, $y'(0) = 1$

Section 2D

1. $y = c_1 e^x + e^{-x/2}\left(c_2 \cos \frac{\sqrt{3}}{2}x + c_3 \sin \frac{\sqrt{3}}{2}x\right)$ **3.** $y = c_1 + c_2 e^{\sqrt{2}x} + c_3 e^{-\sqrt{2}x}$
5. $y = c_1 e^{2x} + e^{-x}(c_2 \cos\sqrt{3}x + c_3 \sin\sqrt{3}x)$ **7.** $y = c_1 e^{2x} + c_2 e^{-2x} + c_3 e^x + c_4 e^{-x}$
9. $y = c_1 + c_2 x + c_3 e^{2x}$ **11.** $y'' + 16y = 0$ **13.** $y^{(6)} + 48y^{(4)} + 768y'' + 4096y = 0$ or $(D^2 + 16)^3 y = 0$
15. $y^{(4)} - 4y''' + 8y'' - 8y' + 4y = 0$ **17.** $y^{(6)} = 0$ **19.** $y'' + 2y' + 2y = 0$ **21.** $y''' + y'' = 0$
23. $y''' - y'' + y' - y = 0$ **25.** $y''' = 0$ **27.** $y^{(4)} - y = 0$ **29.** $y'' + y = 0$ **31.** $c_1 = 0, c_2 = c_3 = 1$
33. $c_1 = c_3 = 1, c_2 = -1$ **35.** $c_1 = c_2 = c_3 = 1$ **37.** $c_1 = c_3 = 1, c_2 = c_4 = -1$
39. $y = c_1 + c_2 x + c_3 x^2 + c_4 x^3 + c_5 e^x + c_6 e^{-x}$ **41.** $y = c_1 + c_2 e^x + c_3 e^{-x} + c_4 \cos x + c_5 \sin x$
43. $y = c_1 e^{3x} + c_2 e^{-2x} + c_3 e^{4x} + c_4 e^{-4x}$ **45.** $y^{(4)} + 5y'' + 4y = 0$ **47.** **(a)** $y = c \sin \pi x$
51. **(a)** $y_h = c_1 + c_2 x + c_3 x^2 + c_4 x^3$

Section 3AB

5. Linear **7.** Not linear **9.** $y = x + c_1 \cos x + c_2 \sin x$ **11.** $y = e^x + c_1 e^{x/\sqrt{2}} + c_2 e^{-x/\sqrt{2}}$
13. $y'' - 3y' + 2y = 0$ **15.** $y'' = 0$ **17.** $y^{(4)} + 18y'' + 81y = 0$ **19.** $y^{(4)} - 4y''' + 8y'' - 8y' + 4y = 0$
21. $y_p = A\cos x + B \sin x$ **23.** $y_p = Axe^x$ **25.** $y_p = Ax^2 e^x + Bx^3 e^x$
27. **(a)** $y_p = Ae^{2x}$; **(b)** $y = \frac{1}{3}e^{2x} + c_1 e^x + c_2 e^{-x}$; **(c)** $y = \frac{1}{3}e^{2x} - \frac{1}{3}e^{-x}$
29. **(a)** $y_p = Ae^x$; **(b)** $y = \frac{1}{4}e^x + c_1 e^{-x} + c_2 x e^{-x}$; **(c)** $y = \frac{1}{4}e^x - \frac{1}{4}e^{-x} + \frac{1}{2}xe^{-x}$
31. **(a)** $y_p = A + Bx + Cxe^x$; **(b)** $y = \frac{1}{2}xe^x - x + c_1 e^x + c_2 e^{-x}$; **(c)** $y = \frac{1}{2}xe^x - x + \frac{3}{4}e^x - \frac{3}{4}e^{-x}$
33. **(a)** $y_p = Ax\cos x + Bx \sin x$; **(b)** $y = \frac{1}{2}x \sin x + c_1 \cos x + c_2 \sin x$; **(c)** $y = \frac{1}{2}x \sin x + \sin x$
35. **(a)** $y_p = Ax^2 \cos x + Bx^2 \sin x + Cx \cos x + Dx \sin x$; **(b)** $y = \frac{1}{4}x^2 \sin x + \frac{1}{4}x \cos x + c_2 \cos x + c_2 \sin x$;
(c) $y = \frac{1}{4}x^2 \sin x + \frac{1}{4}x \cos x + \frac{3}{4} \sin x$

37. **(a)** $y_p = A + Bx$; **(b)** $y = -\frac{3}{4} + \frac{3}{4}x + c_1e^{-2x} + c_2xe^{-2x}$; **(c)** $y = -\frac{3}{4} + \frac{3}{4}x + \frac{3}{4}e^{-2x} + \frac{7}{4}xe^{-2x}$
39. **(a)** $y_p = Ae^x$; **(b)** $y = \frac{1}{5}e^x + e^{-x}(c_1 \cos x + c_2 \sin x)$; **(c)** $y = \frac{1}{5}e^x - \frac{1}{5}e^{-x}(\cos x - 3\sin x)$
41. $y = c_1e^x + e^{-x/2}\left(c_2 \cos \frac{\sqrt{3}}{2}x + c_3 \sin \frac{\sqrt{3}}{2}x\right) + \frac{1}{3}xe^x$ **43.** $y = c_1 + c_2e^{\sqrt{2}x} + c_3e^{-\sqrt{2}x} - 2x$
45. $y = c_1e^{2x} + e^{-x}(c_2 \cos \sqrt{3}x + c_3 \sin \sqrt{3}x) + \frac{1}{12}xe^{2x}$ **47.** $y = c_1e^{2x} + c_2e^{-2x} + c_3e^x + c_4e^{-x} - \frac{1}{6}\sinh x$
49. $y = c_1 + c_2x + c_3e^{2x} - \frac{3}{16}x^2 - \frac{1}{8}x^3 - \frac{1}{16}x^4 - \frac{1}{40}x^5$

Section 3C

1. $y_p = \frac{1}{2}xe^x$ **3.** $y_p = 1$ **5.** $y_p = -e^x \ln|x|$ **7.** $y_p = \cos x \ln|\cos x| + x \sin x$
9. $y_p = -\cos x \ln|\sec x + \tan x|$ **11.** **(a),(b)** $y_p = -\frac{1}{8}x \sin 2x + \frac{1}{8}$

Section 3D

1. $g(x-t) = \frac{1}{4}\left(e^{2(x-t)} - e^{-2(x-t)}\right)$; $y(x) = \frac{1}{4}\int_{x_0}^{x}\left(e^{2(x-t)} - e^{-2(x-t)}\right)f(t)\,dt$
3. $g(x-t) = e^{3(x-t)} - e^{2(x-t)}$; $y(x) = \int_{x_0}^{x}\left(e^{3(x-t)} - e^{2(x-t)}\right)f(t)\,dt$
5. $g(x-t) = (x-t)e^{x-t}$; $y(x) = \int_{x_0}^{x}(x-t)e^{x-t}f(t)\,dt$
7. $g(x-t) = \frac{4}{\sqrt{7}}e^{3(x-t)/4}\sin\left(\frac{\sqrt{7}}{4}(x-t)\right)$; $y(x) = \frac{2}{\sqrt{7}}\int_{x_0}^{x}e^{3(x-t)/4}\sin\left(\frac{\sqrt{7}}{4}(x-t)\right)f(t)\,dt$
9. $y = -x + \sinh x$ **11.** $y = -\frac{1}{3}xe^{-x} - \frac{1}{9}e^{-x} + \frac{1}{9}e^{2x}$ **13.** $y = -\frac{1}{3}e^x + \frac{1}{4}e^{2x+1} + \frac{1}{12}e^{-2x-3}$
15. $y = \frac{1}{24}(3\sin x - \sin 3x)$ **17.** $y = \begin{cases}(1/4)(-1 + \cosh 2x), & x \ge 0;\\ 0, & x < 0.\end{cases}$
19. $y = \begin{cases}(1/4)(-1 + \cosh 2x), & x \ge 0;\\ (1/4)(1 - \cosh 2x), & x < 0.\end{cases}$ **21.** $y = \begin{cases}(1/8)(-2x + 2 + \sinh(2x-2)), & x \ge 1;\\ 0, & x < 1.\end{cases}$
23. **(b)** $x(t) = \frac{3}{2}\sin t - \frac{1}{2}t\cos t$ **25.** $y = (e^{-x} + e^{-2x})\ln(1 + e^x) - (1 + \ln 2)\,e^{-x} + (1 - \ln 2)\,e^{-2x}$
27. **(a)** $y = -\frac{1}{2}e^x + c_1e^{2x} + c_2e^{-x}$; **(b)** $y = \frac{1}{5}e^{3x} + c_1e^{2x} + c_2e^{-2x}$ **35.** $y(x) = \int_0^x e^{r(x-t)}f(t)\,dt$

Section 4AB

1. $y = c_1t^2 + c_2$ **3.** $y = \ln|t + c_1| + c_2$ **5.** $y = t^4/16 + c_1 \ln|t| + c_2$ **7.** $y = c_1e^{c_2t}$
9. $y = c_1 \pm \sqrt{c_2 - 2t}$, and $y =$ constant. **11.** $y = \ln|c_1e^t + c_2e^{-t}|$, ($c_1, c_2$ not both 0) **13.** $y = t^3/3 + t$
15. $y = \sqrt{2}\tan(t/\sqrt{2})$, ($|t| < \pi\sqrt{2}/2$) **19.** **(a)** $y = (1/k)\cosh kx$
23. **(a)** $y = c_1\tan(c_1t + c_2)$; $y = -c_1\tanh(c_1t + c_2)$; $y = -c_1\coth(c_1t + c_2)$

Chapter Review and Supplement

1. $y = \frac{1}{2}x^2e^{-x} + \frac{3}{4}e^x + c_1e^{-x} + c_2xe^{-x}$ **3.** $y = -\frac{1}{2}\sin x + c_1e^x + c_2e^{-x}$
5. $y = 1/3 + e^{-x}(c_1\cos\sqrt{2}x + c_2\sin\sqrt{2}x)$ **7.** $y = -\frac{1}{6}x\cos 3x + \cos 3x + \frac{1}{18}\sin 3x$
9. $y = -\frac{1}{5}\cos 3x + c_1\cos 2x + c_2\sin 2x$ **11.** $y = -\cos x + c\sin x$ **13.** $y_p = Ax^2e^{2x} + Bxe^{2x}$
15. $y_p = Axe^{2x} + Bx^2e^{2x} + Cxe^{3x}$ **17.** $y_p = Axe^{2x} + B\cos x + C\sin x$ **19.** $y_p = Ax + Bx^2 + Cx^3 + Dxe^x$
21. $y_p = Ax^3 + Bx^4 + Cx^5 + Dx^6$ **29.** $A = 1/R$, $\alpha = \pi/2$
31. **(c)** $x = \ln\sqrt{3/2} + i\left(n + \frac{1}{2}\right)\pi$, $n = 0, \pm 1, \pm 2, \ldots$.
35. $z = 2y_1 - 3y_2$ **37.** $z = -y_1 - y_2$
43. **(d)** 322 ft/sec; **(e)** ≈2.65 seconds
45. **(b)** $\dot{y}(t) \to gt + v_0$ as $k \to 0$
47. $A = \sqrt{(\theta(0))^2 + (\ell/g)(\dot{\theta}(0))^2}$; period $= 2\pi\sqrt{\ell/g}$
49. **(a)** $\ddot{y} + (k/m)\dot{y} + (g/l)y = 0$;
(b) $k < 2m\sqrt{g/\ell}$; **(c)** The period decreases as m increases.
55. **(a)** $\theta(t) = 0.1\cos 4t$; **(b)** $\pi/8 \approx 0.4$ seconds; **(c)** 0.4 rad/sec.
57. Period $= \pi/4$ seconds; amplitude ≈ 0.12 radians; maximum angular velocity ≈ 0.96 rad/sec.

Chapter 4

Section 1

1. Critically damped **3.** Harmonic **5.** Underdamped **7.** $x(t) = c_1 e^{-t} + c_2 t e^{-t}$
9. $x(t) = c_1 \cos 3t + c_2 \sin 3t$ **11.** $x(t) = e^{-at/2}\left(c_1 \cos \frac{a\sqrt{3}}{2}t + c_2 \sin \frac{a\sqrt{3}}{2}t\right)$
13. $x(t) = \frac{1}{2}\sin 2t;\ \ddot{x} + 4x = 0$ **15.** $x(t) = -te^{-2t};\ \ddot{x} + 4\dot{x} + 4x = 0$
17. $x(t) = -\frac{5}{2}e^{-2t} + \frac{3}{2}e^{-4t};\ \ddot{x} + 6\dot{x} + 8x = 0$ **19.** $x_p(t) = \frac{1}{10}\cos t + \frac{3}{10}\sin t$; $\approx$5.01 seconds
21. **(a)** $h = 6$ lb/ft; **(b)** $h \approx 8.9$ kg/m; **(c)** 80 pounds;
(d) Hooke's law is not valid for this range of compression.
27. $k^2 < 8h$ **29.** $k = \sqrt{3}$ **31.** $\omega = 0, \pm 1$ **33.** $k = 1/2$
35. $A = \sqrt{13}$; frequency $= 1/(2\pi)$; $\phi = \arctan(3/2) \approx 0.9828$ radians
37. $A = \sqrt{5}$; frequency $= 1/2$; $\phi = \arctan(1/2) \approx 0.4635$ radians **39.** $\pi/3$ radians
45. **(a)** $h = 5/2$; $2/\pi \approx 0.64$ oscillations per second;
(b) The amplitude of the quake's motion is 0.25 feet and the amplitude of the forced oscillation of the weight is 0.08 feet.
49. **(c)** $\ddot{x} + hx = \sin\sqrt{h}t$
51. **(a)** $\ddot{y} + 125\pi y = 0,\ y(0) = -1/2,\ \dot{y}(0) = 0$; **(b)** Amplitude $= 1/2$ feet; period $= \frac{2}{5}\sqrt{\pi/5} \approx 0.32$ seconds
55. **(a)** $(\alpha - \beta)/\omega$; **(b)** $\alpha - \beta - \pi/2$
57. **(a)** $k = 6$; $h = 45$; **(b)** $m = 1/26$; $k = 1/13$; **(c)** $m = 1$; $h = 1/2$; **(d)** No such m exists.
59. **(a)** $v = \sqrt{2g\ell}$

Section 2

1. **(a)** $Q(t) = e^{-5t}(c_1 \cos 80t + c_2 \sin 80t)$; **(b)** $Q(t) = e^{-5t}\left(\cos 80t + (1/16)\sin 80t\right)$;
(c) $Q(t) = \frac{5}{6416}e^{-t} + e^{-5t}\left(\frac{6411}{6416}\cos 80t + \frac{1603}{25664}\sin 80t\right)$;
(d) $Q(t) = \frac{-8}{290725}\cos 20t + \frac{241}{290725}\sin 20t + e^{-5t}\left(\frac{290733}{290725}\cos 80t + \frac{289769}{4651600}\sin 80t\right)$
5. **(b)** $I(t) = \frac{E_0 C}{\sqrt{1+R^2C^2\omega^2}}\cos(\omega t - \alpha) + Ke^{-t/(RC)}$, $\alpha = \arctan(RC\omega)$;
(c) $I(t) = \frac{E_0 C}{\sqrt{1+R^2C^2\omega^2}}\left(\cos(\omega t - \alpha) - \frac{e^{-t/(RC)}}{\sqrt{1+R^2C^2\omega^2}}\right)$
11. **(a)** $Q_\infty = 1.5 \cdot 10^{-4}$ coulombs; **(b)** ≈ 0.001945 seconds; **(c)** $I_{\max} \approx 0.018899$ amperes
13. $Q(0.01) \approx 0.5650$ coulombs; $I(0.01) \approx -8.37534$ amperes

Section 3A

13. Two families of parallel lines with slopes 1 and -1, respectively. **15.** $y = 0, \pm 1$
19. Equilibrium points: $(-1, 0)$, $(0, 0)$, $(1, 0)$. Solutions move away from $(-1, 0)$ for $y_0 \in (-\infty, -1) \cup (-1, 0)$, toward $(0, 0)$ for $y_0 \in (-1, 0) \cup (0, 1)$, and away from $(1, 0)$ for $y_0 \in (0, 1) \cup (1, \infty)$.
21. Equilibrium points: $(-1, 0)$, $(0, 0)$, $(1, 0)$. Solutions move toward $(-1, 0)$ for $y_0 \in (-\infty, -1) \cup (-1, 0)$, away from $(0, 0)$ for $y_0 \in (-1, 0) \cup (0, 1)$, and toward $(1, 0)$ for $y_0 \in (0, 1) \cup (1, \infty)$.
23. Equilibrium points: $(k\pi, 0)$, $k = 0, \pm 1, \pm 2, \ldots$. Solutions move away from equilibrium points with k even and toward equilibrium points with k odd.
29. **(e)** $y = 2^{-1/3}(4 - 3t)^{2/3}$ **45.** **(a)** $C = \frac{1}{2}z_0^2 - (g/\ell)\cos y_0$; **(b)** $|z_0| > \sqrt{2(g/\ell)(1 + \cos y_0)}$
47. **(a)** $\dot{y}(10) \approx \pm 7.5$; **(b)** Yes. **(c)** Maximum velocity ≈ 4 and minimum velocity ≈ -4.
(d) Maximum position ≈ 11.5 and minimum position ≈ 6.5. **(e)** Yes. **(f)** $\approx 0, \pm 9, \pm 18$
(g) Clockwise.

Section 3B

1. $y = 0, 1$; Yes, periodic solutions exist near the point $(0, 0)$.

Section 4AB

1. **(a)** Zeros are $\approx$ 0.000, 2.68, 4.35, 5.75, 6.99, 8.14, 9.20, 10.22, 11.17.
Zeros are not altered for $a \neq 0$, but $a = 0$ gives the zero solution.

(b) There are no negative zeros.
(c) Maxima: $\approx 1.52, 5.07, 7.57, 9.71, 11.64$; Minima $\approx 3.54, 6.38, 8.67, 10.70$.
3. Non-negative zeros of the nonlinear equation are $\approx 3.37k$, $k = 0, 1, 2, \ldots$.
11. **(a)** $k = 8/9$; **(b)** $\approx$9.21 seconds; **(c)** $k \approx 0.6212$; **(d)** $\approx$8.44 seconds
15. **(a)** $\approx 0.1572, 0.1549, 0.1526$; **(b)** $\approx 0, 2.48, 4.96, 7.44, 9.92, 12.40, 14.88$; **(c)** $\approx 0, 3.1, 6.18, 9.25, 12.3$.

Section 4C

9. $f(t) = H(t+1) - 2H(t-1) + 2H(t-2), (-1 \le t < 3)$ **13.** **(a)** $c(t) = (\cos t)H(\pi - t) - H(t - \pi), (t \neq \pi)$
15. **(a)** $d(t) = (1-t)H(t) + tH(t-1)$
17. $\text{sqw}(t) = 1 + 2\sum_{k=2[t/2]+1}^{\infty}(-1)^k H(t-k)$, where $[\,\cdot\,]$ is the greatest integer function. **23.** **(b)** $A \approx 1$

Chapter 5

Section 1ABC

1. $x(t) = c_1e^t - 1, y(t) = c_2e^t$; $x(t) = 2e^t - 1, y(t) = 2e^t$
3. $x(t) = c_1e^t, y(t) = c_2e^{t/2}, z(t) = c_3e^{t/3}$; $x(t) = 0, y(t) = e^{t/2}, z(t) = -e^{t/3}$
5. **(i)** $\mathbf{F}(t, \mathbf{x}) = (x+1, y)$; **(ii)** $|\mathbf{F}(t, \mathbf{x})| = \sqrt{(x+1)^2 + y^2}$
7. **(i)** $\mathbf{F}(t, \mathbf{x}) = (x, y/2, z/3)$; **(ii)** $|\mathbf{F}(t, \mathbf{x})| = \sqrt{x^2 + y^2/4 + z^2/9}$
19. **(a)** $\dot{x} = -tx + t, \dot{y} = x$; **(b)** $x(0) = 1, y(0) = 0$; **(c)** $y(t) = t$ **21.** $\dot{x} = y, \dot{y} = xy$
23. $\dot{x} = y, \dot{y} = z, \dot{z} = 12xy$ **25.** $dx/dt = -x + 2y, dy/dt = x - y$ **27.** $dx/dt = \cosh t, dy/dt = \sinh t$
29. $dy/dx = x + y$; $y = ce^x - (x+1)$; constant solutions: $x = y =$ constant.
31. $dy/dx = x/y$; $y^2 - x^2 = c$; constant solution: $x = y = 0$ **33.** **(a)** $x = z_0t, y = w_0t - \frac{1}{2}gt^2$
35. $\theta = \arctan\big(h(1/a + 1/b)\big)$; $v_0 = \sec\theta\sqrt{abg/(2h)}$
37. **(a)** $\ddot{x} = 0, \ddot{y} = -g$; **(b)** $\theta \approx 54.4^\circ$, $v_0 \approx 3659$ ft/sec, $y_{\text{max}} \approx 26.2$ miles;
(c) $m\ddot{x} = -a(y)\dot{x}, m\ddot{y} = -32m - a(y)\dot{y}$
41. **(b)** $\ddot{\phi} = -2\dot{\theta}\dot{\phi}\cot\theta$; **(c)** $\ddot{\theta} = -(g/\ell)\sin\theta + \dot{\phi}^2\sin\theta\cos\theta$
45. **(a)** $x(t) = \frac{1}{a} - \frac{1}{a^2+\gamma^2}\left(\frac{\gamma^2}{a}e^{-at} + a\cos\gamma t + \gamma\sin\gamma t\right)$
$y(t) = \frac{1}{b}(1 - e^{-bt}) - \frac{a}{a^2+\gamma^2}\left[\frac{\gamma^2}{a(b-a)}(e^{-at} - e^{-bt}) + \frac{\gamma^2 - ab}{b^2+\gamma^2}e^{-bt} - \frac{(\gamma^2-ab)\cos\gamma t - \gamma(a+b)\sin\gamma t}{b^2+\gamma^2}\right]$
47. **(a)** $a^3\omega^2 = G(m_1 + m_2)$; **(b)** $x(0) = a, \dot{x}(0) = 0, y(0) = 0, \dot{y}(0) = \sqrt{G(m_1+m_2)/a}$.

Section 1D

1. $\mathbf{X}_0 = (-1/5, 1/5)$ **3.** $\mathbf{X}_0 = (0, a)$, a arbitrary. **5.** $\mathbf{X}_0 = (0, 0, 0), (-1/2, 1/2, -1/2)$
7. $\mathbf{X}_0 = (c/\sqrt{1+c^2}, -1/\sqrt{1+c^2}), (-c/\sqrt{1+c^2}, 1/\sqrt{1+c^2})$; They all lie on the unit circle.
9. $(0, 0), (d/c, a/b)$ **11.** $\mathbf{X}_0 = (0, 0, 0), (0, 2, 3), (3, 3, 0)$ **13.** No
15. Equilibrium solutions are all points on the xz plane, the line $y = -x$ in the xy plane, and the line parametrized by $(t, t, 2t)$
17. $\dot{x} = 1, \dot{y} = \sin(xy)$ **19.** $\dot{x} = wy, \dot{y} = wz, \dot{z} = wx, \dot{w} = 1$

Section 1E

7. **(a)** $H(x, y) = -\frac{1}{2}(x^2 + y^2) + C$;
11. **(b)** $J_t = e^{3t(x^2+y^2)}$;
(d) No. $\text{div}(F, G) = 2(x, y)$ can take on both positive and negative values.

Section 2ABC

1. Nonlinear **3.** Linear **5.** Linear **7.** $x = c_1e^{-2t} + c_2e^{2t}, \ y = -c_1e^{-2t} - \frac{1}{2}c_2e^{2t}$
9. $x = c_1e^{(1+\sqrt{2})t} + c_2e^{(1+\sqrt{2})t} + 4 - 2t, \ y = \frac{\sqrt{2}}{2}c_1e^{(1+\sqrt{2})t} - \frac{\sqrt{2}}{2}c_2e^{(1+\sqrt{2})t} - 3 + t$
11. $x = -c_1t^3 + (2c_1 - c_2)t, \ y = c_1t^2 + c_2$ **13.** $dx/dt = -x + 2y, \ dy/dt = x - y$
15. $dx/dt = \frac{1}{2}(\sin t + \cos t), \ dy/dt = \frac{1}{2}(\sin t - \cos t)$
17. $x = 5\sqrt{2}\sinh((-1+\sqrt{2})t), \ y = 5\cosh((-1+\sqrt{2})t)$
19. $x = \frac{1}{2}(-\cos t + \sin t + 1), \ y = \frac{1}{2}(-\cos t + \sin t + 3)$ **21.** $x = 1, \ y = 0$

23. $x = \frac{1}{2}e^t + \frac{1}{4}e^{-t/\sqrt{2}}[(-1+\sqrt{2})\cos(t/\sqrt{2}) - \sin(t/\sqrt{2})] + \frac{1}{4}e^{t/\sqrt{2}}[-(1+\sqrt{2})\cos(t/\sqrt{2}) + \sin(t/\sqrt{2})]$,
$y = -\frac{1}{2}e^t + \frac{1}{4}e^{-t/\sqrt{2}}[\cos(t/\sqrt{2}) + (-1+\sqrt{2})\sin(t/\sqrt{2})] + \frac{1}{4}e^{t/\sqrt{2}}[\cos(t/\sqrt{2}) + (1+\sqrt{2})\sin(t/\sqrt{2})]$
25. $\dot{x} = u,\ \dot{y} = v,\ \dot{u} = v,\ \dot{v} = -u,\ \ y(0) = u(0) = v(0) = 0,\ x(0) = 1$
27. $\dot{x} = u,\ \dot{y} = v,\ \dot{u} = y + e^t,\ \dot{v} = -x,\ \ x(0) = y(0) = u(0) = v(0) = 0$
29. Solutions: $x = 0,\ \ y = c_1$ **31.** Solutions: $x = c_1e^{t/2},\ \ y = c_1e^{t/2}$
33. **(a),(b)** $\dot{x} = u,\ \dot{y} = v,\ \dot{u} = -2x + y,\ \dot{v} = 3x + 2y$; **(c)** $x = c_1e^{at} + c_2e^{-at} + c_2\cos at + c_4\sin at$,
$y = (2+a^2)(c_1e^{at} + c_2e^{-at}) + (2-a^2)(c_3\cos at + c_4\sin at)$, where $a = 7^{1/4}$.

Section 3AB

1. Is an eigenvector, $\lambda = 7$ **3.** Is an eigenvector, $\lambda = 7$ **5.** Not an eigenvector **7.** $\lambda = -2, 2$
9. $\lambda = 1, 2$ **11.** $\mathbf{u}_1 = (-2, 1)$ for $\lambda = -2$; $\mathbf{u}_2 = (2, 1)$ for $\lambda = 2$
13. $\mathbf{u}_1 = (0, 1, -1)$ for $\lambda = 1$; $\mathbf{u}_2 = (0, 0, 1)$ for $\lambda = 2$ **15.** $\mathbf{u}_1 = (0, 1)$ for $\lambda_1 = 2$; $\mathbf{u}_2 = (1, 0)$ for $\lambda_2 = 3$
17. $\mathbf{u}_1 = (i, 1)$ for $\lambda_1 = 2 + i$; $\mathbf{u}_2 = (-i, 1)$ for $\lambda_2 = 2 - i$ **19.** $\mathbf{x}(t) = (c_1e^{3t},\ c_2e^{2t})$
21. $\mathbf{x}(t) = e^{2t}(c_1\cos t - c_2\sin t,\ c_2\cos t + c_1\sin t)$ **23.** $\mathbf{x}(t) = (-e^{3t},\ 0)$ **25.** $\mathbf{x}(t) = (0, 0)$
29. **(b)** $A = \begin{pmatrix} -3 & 4 \\ -6 & 7 \end{pmatrix}$; **(d)** $\dot{x} = -3x + 4y,\ \ \dot{y} = -6x + 7y$
31. $\lambda = \pm i$; Solutions are periodic oscillations.
33. $\lambda = \frac{1}{2}(3 \pm \sqrt{13})$; Solutions do not oscillate but some can tend to 0 and some can tend to ∞, depending on the initial conditions.
39. **(a)** $\lambda = \cos\theta \pm i\sin\theta,\ \mathbf{u} = (\pm i,\ 1)$
43. **(b)** $y = c_1e^{at},\ \ x = c_2e^{at} + c_1bte^{at}$; **(c)** Generalized eigenvector: $\mathbf{v} = (0, 1)$
45. **(a)** $\mathbf{u} = (2, 1)$; **(b)** $\mathbf{v} = (1, 1)$; $\mathbf{x}(t) = c_1te^t\begin{pmatrix} 2 \\ 1 \end{pmatrix} + c_2e^t\begin{pmatrix} 1 \\ 1 \end{pmatrix}$ **49.** **(a)** $x = c_1e^t,\ \ y = c_2e^t + c_3te^t$, $z = c_3e^t$

Section 4A

1. $\lambda = -1, 1$; saddle **3.** $\lambda = 2 \pm i$; unstable spiral **5.** $\lambda = \pm\sqrt{2}i$; stable center
7. $\lambda = 0, 1$; degenerate—not categorized **9.** $\lambda = -1, 1$; saddle
11. $\lambda = \frac{1}{2}(-1 \pm \sqrt{3}i)$; asymptotically stable spiral
13. $k = 0 \Rightarrow$ stable center; $0 < k < 2 \Rightarrow$ unstable spiral; $k = 2 \Rightarrow$ asymptotically stable star; $k > 2 \Rightarrow$ asymptotically stable node.
15. $k = 0 \Rightarrow$ stable center; $0 < k < 2\sqrt{2} \Rightarrow$ asymptotically stable spiral; $k = 2\sqrt{2} \Rightarrow$ asymptotically stable star; $k > 2\sqrt{2} \Rightarrow$ asymptotically stable node.
23. **(a)** B; **(b)** C; **(c)** A; **(d)** D
25. **(a)** $x = e^{-t/2}\left[c_1\sin\left(\frac{\sqrt{3}}{2}t\right) + c_2\cos\left(\frac{\sqrt{3}}{2}t\right)\right], y = \frac{1}{2}e^{-t/2}\left[-(c_1+\sqrt{3}c_2)\sin\left(\frac{\sqrt{3}}{2}t\right) + (\sqrt{3}c_1 - c_2)\cos\left(\frac{\sqrt{3}}{2}t\right)\right]$,
$z = c_3e^{-t} + \frac{1}{4}e^{-t/2}\left[(c_1+\sqrt{3}c_2)\sin\left(\frac{\sqrt{3}}{2}t\right) + (-\sqrt{3}c_1 + c_2)\cos\left(\frac{\sqrt{3}}{2}t\right)\right]$;
(b) $z = c_3e^t - \frac{1}{6}e^{-t/2}\left[(3c_1 - \sqrt{3}c_2)\sin\left(\frac{\sqrt{3}}{2}t\right) + (\sqrt{3}c_1 + 3c_2)\cos\left(\frac{\sqrt{3}}{2}t\right)\right]$

Section 4B

1. $(0, 0)$: $\dot{x} = y,\ \ \dot{y} = x$; saddle.
3. $(0, 0)$: $\dot{x} = -y,\ \ \dot{y} = x$; no conclusion. $(-1, 0)$: $\dot{x} = -y,\ \ \dot{y} = -x$; saddle
5. $(0, 0)$: $\dot{x} = x,\ \ \dot{y} = x$; no conclusion. $(1, 1)$: $\dot{x} = -2y + 2,\ \ \dot{y} = x - 3y + 2$; asymptotically stable node. $(-1, -1)$: $\dot{x} = -2y - 2,\ \ \dot{y} = x - 3y - 2$; asymptotically stable node.
7. $(0, 0)$: $\dot{x} = x - y,\ \ \dot{y} = x + y$; unstable spiral
9. $(m\pi, 1)$: $\dot{x} = -(y - 1),\ \ \dot{y} = (-1)^m(x - m\pi)x$; saddle ($m$ odd), no conclusion (m even)
15. $B > (A+1)^2 \Rightarrow$ unstable node, $B = (A+1)^2 \Rightarrow$ unstable star, $B < (A+1)^2 \Rightarrow$ unstable spiral.
17. Equilibrium points are $(0, 0)$ and all points on the unit circle.
19. Equilibrium points: $(0, 0)$ (unstable saddle), $(\pm 1, 0)$ (asymptotically stable star)
25. **(a)** $\dot{x} = y,\ \ \dot{y} = -x + \alpha y$ **27.** **(b)** $\dot{x} = \alpha x + y,\ \ \dot{y} = -x$

Section 4C

1. Strict Liapunov function for $a, b > 0$ **3.** Strict Liapunov function for $a, b > 0$
5. Liapunov function (not strict) for $a, b > 0$ **7.** Strict Liapunov function for $a, b, c > 0$
9. Not a Liapunov function for any choice of a, b, c **11.** Liapunov function (not strict) for $a, b, c > 0$
15. **(a)** All points $(x, y, 0)$ on the xy plane; **(b)** $z = z_0 e^{-t}$ **25.** **(a)** $\dot{x} = y,\ \dot{y} = -ky - hx - \alpha x^2 - \beta x^3$

Chapter Review and Supplement

1. $x = \tan(t + \pi/4),\ y = 2e^t\ (-3\pi/4 < t < \pi/4)$ **3.** $x = 2e^t - e^{-t},\ y = -1 + 2e^t + e^{-t}$
5. $x = e^{-5t},\ y = 2e^{-5t}$ **7.** $x = e^{2t},\ y = 2e^{2t} - t$ **9.** $x = 2\cos t + \cos 2t,\ y = 4\cos t - \cos 2t$
11. $x = -\sin t,\ y = \cos t,\ z = -e^t$
13. For each solution of one of the systems there is a solution of the other system with the same trajectory but traced out in the opposite direction.

Chapter 6

Section 1

1. **(a)** $\dot{y} = -\frac{y}{25} + \frac{z}{25},\ \dot{z} = \frac{y}{100} - \frac{z}{25}$; **(b)** $y = \frac{25}{2}e^{-t/50} + \frac{15}{2}e^{-3t/50},\ z = 25e^{-t/50} - 15e^{-3t/50}$
3. **(a)** $\dot{x} = -ax + ay,\ \dot{y} = ax - ay$
5. **(a)** 200 gallon tank empties in 200 minutes. **(b)** $\dot{x} = -\frac{x}{50} + \frac{2y}{200-t}\ (x(0) = 0),\ \dot{y} = \frac{x}{100} - \frac{2y}{200-t}\ (y(0) = 10)$
7. **(a)** 50 minutes; **(b)** $\dot{x} = -\frac{2x}{50-t} + \frac{y}{100+t}\ (x(0) = 0),\ \dot{y} = \frac{2x}{50-t} - \frac{y}{100+t}\ (y(0) = 10)$;
(c) $x(50) = 0,\ y(50) = 150$
9. **(a)** $t_1 = 100$ minutes; **(b)** $\dot{x} = -\frac{2x}{100-t} + 1,\ \dot{y} = \frac{2x}{100-t} - \frac{3y}{100-t}$; **(c)** $x = \frac{t}{100}(100-t),\ y = \left(\frac{t}{100}\right)^2(100-t)$
11. **(a)** $t_1 = 100$ minutes; **(b)** $\dot{x} = -\frac{2x}{100-t} + \frac{3y}{100-t}\ (x(0) = x_0),\ \dot{y} = \frac{2x}{100-t} - \frac{3y}{100-t}\ (y(0) = y_0)$;
(d) $x = 18 + 8(1 - t/100)^5,\ y = 12 - 8(1 - t/100)^5\ (0 \le t < 100)$

Section 2

1. **(a)** $P(t) = \frac{1}{2}\left(P_0 - \frac{\sqrt{6}}{3}Q_0\right)e^{(1+\sqrt{6})t} + \frac{1}{2}\left(P_0 + \frac{\sqrt{6}}{3}Q_0\right)e^{(1-\sqrt{6})t}$,
$Q(t) = -\frac{\sqrt{6}}{4}\left(P_0 - \frac{\sqrt{6}}{3}Q_0\right)e^{(1+\sqrt{6})t} + \frac{\sqrt{6}}{4}\left(P_0 + \frac{\sqrt{6}}{3}Q_0\right)e^{(1-\sqrt{6})t}$
(b) P prospers and Q dies out if $P_0 > \frac{\sqrt{6}}{3}Q_0$, while Q prospers and P dies out if $P_0 < \frac{\sqrt{6}}{3}Q_0$.
19. **(a)** $(K, P) = (0, 0),\ (d/c_2, a_2/b_2)$ if H is absent and $(H, P) = (0, 0),\ (d/c_1, a_1/b_1)$ if K is absent.
(b) Conditions: (i) $a_1b_2 = a_2b_1$, (ii) $H(0) = (d - c_2K(0))/c_1$, (iii) $0 < K(0) < d/c_2$. Equilibrium solutions with all positive coordinates: $(H, K, P) = ((d - c_2K(0))/c_1,\ K(0),\ a_1/b_1)$.
21. **(b)** $H + P = (c/b)\ln|H| + C$

Section 3AB

3. $\mu = \sqrt{\frac{5+\sqrt{5}}{2}},\ \sqrt{\frac{5-\sqrt{5}}{2}}$ **5.** They are the same. **9.** $x_e = \frac{1}{3}(\ell + 2\ell_1 - \ell_2 - \ell_3),\ y_e = \frac{1}{3}(2\ell + \ell_1 + \ell_2 - 2\ell_3)$
11. **(b)** $x_e = 2h/5,\ y_e = 3h/5$; **(c)** $\mu = \sqrt{h},\ \sqrt{5h}$
13. $\dot{z} = u/m_1,\ \dot{u} = -(h_1 + h_2)z + h_2w + h_1\ell_1 - h_2\ell_2,\ \dot{w} = v/m_2,\ \dot{v} = h_2z - (h_2 + h_3) + h_2\ell_2 - h_3\ell_3 + h_3b$
15. $m\ddot{\mathbf{x}} = h_1(|\mathbf{a} - \mathbf{x}| - \ell_1)\frac{(\mathbf{a}-\mathbf{x})}{|\mathbf{a}-\mathbf{x}|} + h_2(|\mathbf{b} - \mathbf{x}| - \ell_2)\frac{(\mathbf{b}-\mathbf{x})}{|\mathbf{b}-\mathbf{x}|} + h_3(|\mathbf{c} - \mathbf{x}| - \ell_3)\frac{(\mathbf{c}-\mathbf{x})}{|\mathbf{c}-\mathbf{x}|}$

Section 4

3. $\dot{x} = z,\ \dot{z} = -kx/(x^2 + y^2)^{3/2},\ \dot{y} = w,\ \dot{w} = -ky/(x^2 + y^2)^{3/2}$
5. $\ddot{\mathbf{x}}_1 = \frac{Gm_2}{r_{21}^3}\mathbf{r}_{21} + \frac{Gm_3}{r_{31}^3}\mathbf{r}_{31} + \frac{Gm_4}{r_{41}^3}\mathbf{r}_{41},\ \ddot{\mathbf{x}}_2 = \frac{Gm_1}{r_{12}^3}\mathbf{r}_{12} + \frac{Gm_3}{r_{32}^3}\mathbf{r}_{32} + \frac{Gm_4}{r_{42}^3}\mathbf{r}_{42}$,
$\ddot{\mathbf{x}}_3 = \frac{Gm_1}{r_{13}^3}\mathbf{r}_{13} + \frac{Gm_2}{r_{23}^3}\mathbf{r}_{23} + \frac{Gm_4}{r_{43}^3}\mathbf{r}_{43},\ \ddot{\mathbf{x}}_4 = \frac{Gm_1}{r_{14}^3}\mathbf{r}_{14} + \frac{Gm_2}{r_{24}^3}\mathbf{r}_{24} + \frac{Gm_3}{r_{34}^3}\mathbf{r}_{34}$
7. $\ddot{\mathbf{x}} = -\frac{G(m_1+m_2)}{|\mathbf{x}|^{p+1}}\mathbf{r}$

9. **(a)** $G \approx 6.67 \times 10^{-11}$; **(b)** $g \approx 9.805\ \text{m/sec}^2$

25. $\ddot{x}_1 = -\frac{Gm_2}{r^3}(x_1 - x_2)$ $\ddot{y}_1 = -\frac{Gm_2}{r^3}(y_1 - y_2)$ $\ddot{z}_1 = -\frac{Gm_2}{r^3}(z_1 - z_2)$
$\ddot{x}_2 = -\frac{Gm_1}{r^3}(x_2 - x_1)$ $\ddot{y}_2 = -\frac{Gm_1}{r^3}(y_2 - y_1)$ $\ddot{z}_2 = -\frac{Gm_1}{r^3}(z_2 - z_1)$

Section 5AB

1. **(b)** $x = \frac{1}{2}(3e^t - e^{-t}),\ y = \frac{1}{2}(3e^t + e^{-t})$ **3.** **(c)** $x = 1 + e^{-t}(1 - 4t),\ y = t - 2 + e^{-t}(3 + 4t)$

11. **(a)** $\frac{dx_1}{dt} = -\frac{7x_1}{50+t} + \frac{3x_2}{100+t},\ \frac{dx_2}{dt} = \frac{3x_1}{50+t} - \frac{4x_2}{100+t} + 2\ \ (0 \le t \le 50)$; **(b)** $x_1(8.60) \approx 7.45$ lbs.

19. $m\ddot{\mathbf{x}} = h_1(|\mathbf{a} - \mathbf{x}| - \ell_1)\frac{(\mathbf{a}-\mathbf{x})}{|\mathbf{a}-\mathbf{x}|} + h_2(|\mathbf{b} - \mathbf{x}| - \ell_2)\frac{(\mathbf{b}-\mathbf{x})}{|\mathbf{b}-\mathbf{x}|} + h_3(|\mathbf{c} - \mathbf{x}| - \ell_3)\frac{(\mathbf{c}-\mathbf{x})}{|\mathbf{c}-\mathbf{x}|}$

Chapter 7

Section 1

1. $e^{t\mathbf{A}} = \begin{pmatrix} e^{-t} & 0 \\ 0 & e^t \end{pmatrix}$ **3.** $e^{t\mathbf{A}} = \begin{pmatrix} \cos t + t\sin t & 0 \\ 0 & \cos t - t\sin t \end{pmatrix}$ **5.** **(d)** $\mathbf{x}(t) = e^t \begin{pmatrix} -1 + 2t \\ 2 \end{pmatrix}$

11. Yes, it is always the most general solution

13. **(a)** $e^{t\mathbf{A}} = \begin{pmatrix} 2e^{-t} - e^t & -e^{-t} + e^t \\ 2e^{-t} - 2e^t & -e^{-t} + 2e^t \end{pmatrix}$; **(b)** $e^{t\mathbf{A}} = \begin{pmatrix} e^{3t} & 0 \\ 0 & e^{2t} \end{pmatrix}$; **(c)** $e^{t\mathbf{A}} = \begin{pmatrix} e^t & e^{5t} - e^t \\ 0 & e^{5t} \end{pmatrix}$;

(d) $e^{t\mathbf{A}} = e^{2t}\begin{pmatrix} \cos t & -\sin t \\ \sin t & \cos t \end{pmatrix}$

Section 1D

1. **(a)** $\mathbf{x}(t) = c_1\begin{pmatrix} 1 \\ 0 \end{pmatrix} + c_2\begin{pmatrix} 0 \\ e^t \end{pmatrix}$; **(b)** $\mathbf{x}(t) = c_1\begin{pmatrix} e^t \\ 0 \end{pmatrix} + c_2\begin{pmatrix} 0 \\ e^{2t} \end{pmatrix}$;

(c) $\mathbf{x}(t) = c_1\begin{pmatrix} e^t \\ 0 \\ 0 \end{pmatrix} + c_2\begin{pmatrix} 0 \\ e^{2t} \\ 0 \end{pmatrix} + c_3\begin{pmatrix} 0 \\ 0 \\ e^{3t} \end{pmatrix}$; **(d)** $\mathbf{x}(t) = c_1\begin{pmatrix} e^t \\ 0 \\ 0 \end{pmatrix} + c_2\begin{pmatrix} te^t \\ e^t \\ 0 \end{pmatrix} + c_3\begin{pmatrix} 0 \\ 0 \\ e^{2t} \end{pmatrix}$

7. $\mathbf{A} = \begin{pmatrix} 2 & 3 \\ 2 & -3 \end{pmatrix}$

Section 1E

5. **(a)** $\mathbf{X}(t) = \begin{pmatrix} 1 & 0 \\ 1 & (1-t)^{-1} \end{pmatrix}$; **(b)** $\det \mathbf{X}(t) = (1-t)^{-1}$

Section 2

1. $e^{t\mathbf{A}} = \begin{pmatrix} -5e^{2t} + 6e^{3t} & 3e^{2t} - 3e^{3t} \\ -10e^{2t} + 10e^{3t} & 6e^{2t} - 5e^{3t} \end{pmatrix}$, $\mathbf{x}(t) = c_1\begin{pmatrix} -5e^{2t} + 6e^{3t} \\ -10e^{2t} + 10e^{3t} \end{pmatrix} + c_2\begin{pmatrix} 3e^{2t} - 3e^{3t} \\ 6e^{2t} - 5e^{3t} \end{pmatrix}$

3. $e^{t\mathbf{A}} = e^t\begin{pmatrix} 1 & t & e^t + t - 1 \\ 0 & 1 & 1 - e^t \\ 0 & 0 & e^t \end{pmatrix}$, $\mathbf{x}(t) = c_1e^t\begin{pmatrix} 1 \\ 0 \\ 0 \end{pmatrix} + c_2e^t\begin{pmatrix} t \\ 1 \\ 0 \end{pmatrix} + c_3e^t\begin{pmatrix} e^t + t - 1 \\ 1 - e^t \\ e^t \end{pmatrix}$

5. $e^{t\mathbf{A}} = \frac{1}{2}\begin{pmatrix} 1 + e^{2t} & -1 + e^{2t} & -1 + e^{2t} & 0 \\ -1 + 2e^t - e^{2t} & 1 + 2e^t - e^{2t} & 1 - e^{2t} & 0 \\ -2e^t + 2e^{2t} & -2e^t + 2e^{2t} & 2e^{2t} & 0 \\ -1 + 2e^t(1+t) - e^{2t} & 1 + 2te^t - e^{2t} & 1 - e^{2t} & 2e^t \end{pmatrix}$,

$\mathbf{x}(t) = c_1\begin{pmatrix} 1 + e^{2t} \\ -1 + 2e^t - e^{2t} \\ -2e^t + 2e^{2t} \\ -1 + 2e^t(1+t) - e^{2t} \end{pmatrix} + c_2\begin{pmatrix} -1 + e^{2t} \\ 1 + 2e^t - e^{2t} \\ -2e^t + 2e^{2t} \\ 1 + 2te^t - e^{2t} \end{pmatrix} + c_3\begin{pmatrix} -1 + e^{2t} \\ 1 - e^{2t} \\ 2e^{2t} \\ 1 - e^{2t} \end{pmatrix} + c_4\begin{pmatrix} 0 \\ 0 \\ 0 \\ 2e^t \end{pmatrix}$

7. $x = -19e^{2t} + 21e^{3t},\ y = -38e^{2t} + 35e^{3t}$ **9.** $x = te^t,\ y = e^t,\ z = 0$

11. $x = \frac{1}{2}(-1 + e^{2t}),\ y = \frac{1}{2}(1 + 2e^t - e^{2t}),\ z = -e^t + e^{2t},\ w = \frac{1}{2}(1 + 2e^t(1+t) - e^{2t})$

13. $\mathbf{A} = \begin{pmatrix} 2 & 0 & 1 \\ -1 & 3 & 1 \\ -1 & 0 & 4 \end{pmatrix}$; $x = e^{3t}(c_1(1-t) + c_3 t),\ y = e^{3t}(-c_1 t + c_2 + c_3 t),\ z = e^{3t}(-c_1 t + c_3(1+t))$

15. **(c)** $e^{t\mathbf{A}} = \begin{pmatrix} e^{\alpha t} & \frac{e^{\alpha t} - e^{\beta t}}{\alpha - \beta} \\ 0 & e^{\beta t} \end{pmatrix}$ $(\alpha \neq \beta)$, $e^{t\mathbf{A}} = \begin{pmatrix} e^{\alpha t} & te^{\alpha t} \\ 0 & e^{\alpha t} \end{pmatrix}$ $(\alpha = \beta)$

17. **(a)** $b_0(t) = 2e^t - e^{2t},\ b_1(t) = e^{2t} - e^t;\ e^{t\begin{pmatrix} 1 & \beta \\ 0 & 2 \end{pmatrix}} = \begin{pmatrix} e^t & \beta(e^{2t} - e^t) \\ 0 & e^{2t} \end{pmatrix},\ e^{t\begin{pmatrix} 2 & \beta \\ 0 & 1 \end{pmatrix}} = \begin{pmatrix} e^{2t} & \beta(e^{2t} - e^t) \\ 0 & e^t \end{pmatrix}$;

(b) $e^{t\mathbf{A}} = e^{\alpha t}\begin{pmatrix} 1 & \beta t \\ 0 & 1 \end{pmatrix}$

21. **(a)** $\dot{\mathbf{x}} = \begin{pmatrix} z \\ w \\ -2x + y \\ x - 2y \end{pmatrix}$; **(b)** $\mathbf{A} = \begin{pmatrix} 0 & 0 & 1 & 0 \\ 0 & 0 & 0 & 1 \\ -2 & 1 & 0 & 0 \\ 1 & -2 & 0 & 0 \end{pmatrix}$; **(c)** $\lambda_{1,2} = \pm i$ and $\lambda_{3,4} = \pm\sqrt{3}i$;

(d) $\mathbf{x}(t) = \begin{pmatrix} c_1 \cos t + c_2 \sin t + c_3 \cos\sqrt{3}t + \frac{1}{\sqrt{3}} c_4 \sin\sqrt{3}t \\ c_1 \cos t + c_2 \sin t - c_3 \cos\sqrt{3}t - \frac{1}{\sqrt{3}} c_4 \sin\sqrt{3}t \\ c_2 \cos t - c_1 \sin t + c_4 \cos\sqrt{3}t - \sqrt{3}c + 3\sin\sqrt{3} \\ c_2 \cos t - c_1 \sin t - c_4 \cos\sqrt{3}t + \sqrt{3}c_3 \sin\sqrt{3} \end{pmatrix}$

Section 3ABC

1. **(a)** $x = c_1(-e^t + 2e^{-t}) + c_2(e^t - e^{-t}) - 2 - e^{2t}/3,\ y = c_1(-2e^t + 2e^{-t}) + c_2(2e^t - e^{-t}) - 3 - 4e^{2t}/3$;
(b) $x = c_1 e^{3t} - e^t/2 - 1/3,\ y = c_2 e^{2t} - e^{-t}/3$; **(c)** $x = c_1 e^t + c_2(e^{5t} - e^t) - 1 - te^t - e^t/4,\ y = c_2 e^{5t} - e^t/4$;
(d) $x = e^{2t}(c_1 \cos t - c_2 \sin t - 2),\ y = e^{2t}(c_2 \cos t + c_1 \sin t + 1)$

9. **(b)** $d\mathbf{x}/dt = \begin{pmatrix} -1 & e^{-t} \\ -2e^t & 3 \end{pmatrix}\mathbf{x}$; Yes, the solutions are linearly independent.

Section 4AB

1. $\lambda = \frac{1}{2}(-1 \pm \sqrt{29})$; Unstable (saddle). **3.** $\lambda = 1, 1, -1$; Unstable (saddle).

5. For $\mathbf{x}_0 = \begin{pmatrix} -1 \\ 0 \end{pmatrix}$: $\mathbf{F}'(\mathbf{x}_0) = \begin{pmatrix} 1 & 0 \\ -2 & 0 \end{pmatrix}$; $\lambda = 1, 0$; Unstable.

For $\mathbf{x}_0 = \begin{pmatrix} 1 \\ \sqrt{2} \end{pmatrix}$: $\mathbf{F}'(\mathbf{x}_0) = \begin{pmatrix} 1 & -2\sqrt{2} \\ 2 & 0 \end{pmatrix}$; $\lambda \approx \frac{1}{2} \pm 2.325i$; Unstable.

For $\mathbf{x}_0 = \begin{pmatrix} -1 \\ \sqrt{2} \end{pmatrix}$: $\mathbf{F}'(\mathbf{x}_0) = \begin{pmatrix} 1 & 0 \\ 2 & 0 \end{pmatrix}$; $\lambda \approx 2.93, -1.93$; Unstable (saddle).

7. For $\mathbf{x}_0 = \begin{pmatrix} 0 \\ 0 \end{pmatrix}$: $\mathbf{F}'(\mathbf{x}_0) = \begin{pmatrix} 0 & 1 \\ 1 & 0 \end{pmatrix}$; $\lambda = \pm 1$; Unstable (saddle).

9. For $\mathbf{x}_0 = \begin{pmatrix} 0 \\ 0 \\ 0 \end{pmatrix}$: $\mathbf{F}'(\mathbf{x}_0) = \begin{pmatrix} -1 & 0 & 0 \\ 0 & -1 & 0 \\ 0 & 0 & -1 \end{pmatrix}$; $\lambda = -1, -1, -1$; Asymptotically stable.

11. $P(\lambda) = -((\lambda + 1)^3 + 1)$; $\lambda = -2, \frac{1}{2}(-1 \pm \sqrt{3}i)$; Asymptotically stable.

13. **(b)** $\dot{x} = z,\ \dot{y} = w,\ \dot{z} = -4x + y^2,\ \dot{w} = -y + 2xy$; $\mathbf{x}_0 = \begin{pmatrix} 0 \\ 0 \end{pmatrix}, \begin{pmatrix} 1/2 \\ \sqrt{2} \end{pmatrix}, \begin{pmatrix} 1/2 \\ -\sqrt{2} \end{pmatrix}$

17. **(a)** All types except type V are hyperbolic. Type II is of saddle type.
(b) Types I, IV, VII are sources; types III, VI, VIII are sinks.

Chapter Review and Supplement

1. **(a)** $e^{t\mathbf{A}} = e^t\begin{pmatrix} \cos t & -\sin t \\ \sin t & \cos t \end{pmatrix}$; **(b)** Unstable.

3. **(a)** $e^{tA} = \frac{1}{2}\begin{pmatrix} 3e^t - e^{-t} & -e^t + e^{-t} \\ 3e^t - 3e^{-t} & -e^t + 3e^{-t} \end{pmatrix}$; **(b)** Unstable (saddle).

5. **(a)** $e^{tA} = \frac{1}{2}\begin{pmatrix} 3e^t - e^{3t} & -e^t + e^{3t} \\ 3e^t - 3e^{3t} & -e^t + 3e^{3t} \end{pmatrix}$; **(b)** Unstable.

7. **(a)** $e^{tA} = e^t \begin{pmatrix} 1 & 0 & 0 \\ 0 & 1 & t \\ 0 & 0 & 1 \end{pmatrix}$; **(b)** Unstable. **9.** **(a)** $e^{tA} = e^t \begin{pmatrix} 1 & 0 & 0 \\ t + \frac{1}{2}t^2 & 1 & t \\ t & 0 & 1 \end{pmatrix}$; **(b)** Unstable.

11. $x = -\frac{1}{2}(1+t) + e^t(c_1 \cos t + c_2 \sin t),\ y = -\frac{1}{2}t + e^t(-c_2 \cos t + c_1 \sin t)$

Chapter 8

Section 1A

3. $\mathcal{L}[f](s) = \frac{4s}{s^2+16}$ **5.** $\mathcal{L}[f](s) = \frac{2}{s^3} + \frac{2}{s^2} - \frac{1}{s}$ **7.** $\mathcal{L}[f](s) = \frac{1}{s-3} + \frac{2}{(s-3)^2}$ **9.** $y(t) = \frac{1}{2}(e^t - e^{-t})$
11. $y(t) = \sin 2t$ **13.** $y(t) = \frac{1}{3}e^{2t}\sin 3t$ **15.** $y = -1 - t + 3e^t$ **17.** $y = \frac{1}{13}\left(2\sin 2t + 3\cos 2t - 3e^{-3t}\right)$
19. $y = -4 + \frac{2}{111}\left(228e^{t/2} - 6\cos 3t - \sin 3t\right)$ **21.** $x = 2e^{2t} - 2e^{-2t},\ y = -e^{2t} + 2e^{-2t}$
23. $x = 4 - 2t + (-2+\sqrt{2})e^{(1+\sqrt{2})t} - (2+\sqrt{2})e^{(1-\sqrt{2})t},\ y = -3 + t + (1+\sqrt{2})e^{(1+\sqrt{2})t} + (1-\sqrt{2})e^{(1-\sqrt{2})t}$
25. $x = 2t + \frac{4}{3}e^{-t} - \frac{10}{3}e^{t/2} + 2e^t, y = 1 - t + \frac{1}{2}t^2 - \frac{4}{3}e^{-t} - \frac{5}{3}e^{t/2} + 2e^t$

Section 1BC

1. $\mathcal{L}[f](s) = \frac{1}{s}(e^{-s} - e^{-2s})$ **3.** $\mathcal{L}[f](s) = \frac{1}{s}(1 + e^{-s} - 2e^{-2s})$ **5.** $\mathcal{L}[f](s) = \frac{1}{s^2} - 2\left(\frac{1}{s^2} + \frac{1}{s}\right)e^{-s} + \left(\frac{1}{s^2} + \frac{2}{s}\right)e^{-2s}$
7. $g(t) = H(t-1) - H(t-2),\ \mathcal{L}[g](s) = \frac{1}{s}(e^{-s} - e^{-2s})$
9. $g(t) = \sin t H(t) + (1 + \cos t - \sin t)H(t-\pi) - (1+\cos t)H(t-2\pi)$,
$\mathcal{L}[g](s) = \frac{1}{s^2+1}(1 + e^{-\pi s}) - \frac{s}{s^2+1}(e^{-\pi s} + e^{-2\pi s}) + \frac{1}{s}(e^{-\pi s} - e^{-2\pi s})$
11. $g(t) = (t-1)^2 H(t-1) + \left(e^{t-1} - (t-1)^2\right)H(t-2),\ \mathcal{L}[g](s) = \frac{2}{s^3}e^{-s} + \frac{1}{s-1}e^{-2s+1} - \left(\frac{2}{s^3} + \frac{2}{s^2} + \frac{1}{s}\right)e^{-2s}$
13. $y(t) = \begin{cases} 1 - e^{-t}, & 0 \le t < 1; \\ t - 1 + (e-1)e^{-t}, & 1 \le t. \end{cases}$
15. **(a)** $y(t) = \frac{1}{6}(t-1)^3 H(t-1) - \frac{1}{3}(t-2)^3 H(t-2) + \frac{1}{6}(t-3)^3 H(t-3)$
17. $y(t) = \frac{t}{\pi^2} - \frac{\sin \pi t}{\pi^3} - \left(\frac{t-1}{\pi^2} + \frac{\sin \pi t}{\pi^3}\right)H(t-1)$

Section 2

1. $(f * g)(t) = -1 + t + e^{-t}$ **3.** $(f * g)(t) = t^2/2$ **5.** $-1 + t + e^{-t}$ **7.** $\frac{1}{2}(t-2)^2 H(t-2)$
9. $\frac{1}{9} - \frac{1}{9}(1+3t)e^{-3t}$ **11.** $e^{-t}\sin t$ **13.** $1 + H(t-1)$ **15.** $y(t) = -1 + \frac{1}{10}(17e^t + 3e^{-t} - 2\sin 2t)$
17. $y(t) = 5 - t + t^2/2 - (3+t)e^{-t}$ **19.** $y(t) = -1 + 2e^{-t/2}\cosh \frac{\sqrt{5}}{2}t$

Section 3AB

3. -1 **5.** 0 **7.** $\pi^2 + 1$ **9.** $e^{-\pi s}$ **11.** $\frac{1}{s}e^{-\pi s}$ **13.** $e^{-\pi s}$ **19.** $y(t) = (t - t_0)H(t - t_0)$
21. $y(t) = (t-1)e^{-t+1}H(t-1)$ **23.** $y(t) = (1 - e^{-t+1})H(t-1)$ **25.** $y(t) = 2\sinh(t-2)H(t-2)$
27. $y(t) = \frac{2}{\sqrt{15}}e^{-t/4}\left(2\sin\frac{\sqrt{15}}{4}t + e^{1/4}\sin\frac{\sqrt{15}}{4}(t-1)H(t-1) - e^{1/2}\sin\frac{\sqrt{15}}{4}(t-2)H(t-2)\right)$
31. $c_1 = 2,\ c_2 = 1$
33. **(a)** $\tau_j = \frac{2\pi j}{\sqrt{4-k^2}}$; **(b)** $y(t) = \frac{2a}{\sqrt{4-k^2}}e^{-kt/2}\sin\frac{\sqrt{4-k^2}}{2}t\left[\sum_{j=1}^{\infty} e^{k\pi j/\sqrt{4-k^2}}H\left(t - 2\pi j/\sqrt{4-k^2}\right)\right]$

Chapter 9

Section 1A

1. Independent **3.** Independent **5.** Dependent ($|x|^5 = -x^5$ for $x < 0$) **9.** $y_2(x) = \frac{1}{2}x^2 e^x$
11. $y_2(x) = x^2$ **13.** $y(x) = c_1 \int e^{x^2/2}\,dx + c_2$ **15.** $y_2(x) = xe^{\alpha x}$
17. **(a)** $y_2(x) = 1 + \frac{x}{2}\ln\left|\frac{x-1}{x+1}\right|$; **(b)** $y(x) = 2x + 1 + \frac{x}{2}\ln\left(\frac{1-x}{1+x}\right)$ **19.** $y(x) = x^2\ln|x| - x^2$
23. $y(x) = c_1 x + c_2 x^{-1}$ **25.** $y(x) = c_1/x + c_2(\ln x)/x$
27. **(c)** $y_2(x) = \frac{1}{1-a-2\mu_0}x^{1-a-\mu_0}$ for $\mu_0 \ne \frac{1}{2}(1-a)$, $y_2(x) = x^{\mu_0}\ln x$ for $\mu_0 = \frac{1}{2}(1-a)$.

29. **(b)** $y'' - (2\cot x)y' + (2\cot^2 x + 1)y = 0$; **(b)** $(x-1)y'' - xy' + y = 0$
31. $W[f,g](x) = 0$ for all x. **33.** **(b)** $y_1(x) = x^{-1/2}\sin x,\ y_2(x) = x^{-1/2}\cos x$

Section 2ABC

1. $y = e^x + c_1 xe^{2x} + c_2 e^{2x}$ **3.** $y = x^4/3 + c_1x^3 + c_2 x$ **5.** $y_p = \frac{1}{4}e^{2x},\ y = c_1e^x + c_2e^{-2x} + \frac{1}{4}e^{2x}$
7. $y_p = \cos x\ln(\cos x) + x\sin x,\ y = c_1\cos x + c_2\sin x + \cos x\ln(\cos x) + x\sin x$
9. $y_p = 1/2,\ y = 1/2 + c_1x + c_2x^2$ **11.** $y_p = 1/2 + e^x/6$ **13.** $y_p = e^x/10$ **15.** $y_p = x^4/6$
17. $G(x,t) = \frac{1}{\sqrt{2}}\sin\left(\sqrt{2}(x-t)\right),\ y_p = \frac{1}{2\sqrt{2}}\int_{x_0}^{x}\sin\left(\sqrt{2}(x-t)\right)g(t)\,dt$
19. $G(x,t) = \frac{x}{t}(x-t),\ y_p = \int_{x_0}^{x}\frac{x}{t^3}(x-t)g(t)\,dt$

21. $y = \frac{1}{4}\begin{cases}-5\cos\sqrt{2}x + 2\sqrt{2}\sin\sqrt{2}x + 1, & x < 1;\\ -5\cos\sqrt{2}x + 2\sqrt{2}\sin\sqrt{2}x + \cos\sqrt{2}(x-1), & 1 \le x.\end{cases}$ **23.** $y = \begin{cases}x/2 - 5x^2/24, & 0 < x < 2;\\ 3/2 - x + x^2/6, & 2 \le x \le 4;\\ 1/2 - x/2 + 5x^2/48, & 4 < x.\end{cases}$

27. $G(x,t) = (x-t)e^{r_1(x-t)}$ **29.** $y_p = \frac{1}{2}x(x-x_0)^2$ **31.** $y_p = -x + x^2/x_0 - x\ln(x/x_0)$

Section 3ABC

1. Unique solution **3.** Infinitely many solutions **5.** $\{a, a + n\pi/\sqrt{2}\}$, a arbitrary, n an integer.
7. $\{a, a + 2n\pi/\sqrt{3}\}$, a arbitrary, n an integer.

9. $y_1(x) = \sin\sqrt{2}x,\ y_2(x) = \cos\sqrt{2}x;\ G(x,t) = \begin{cases}-\frac{1}{\sqrt{2}}\cos(\sqrt{2}x)\sin(\sqrt{2}t), & 0 \le t \le x;\\ -\frac{1}{\sqrt{2}}\sin(\sqrt{2}x)\cos(\sqrt{2}t), & x \le t \le \pi/2\sqrt{2}.\end{cases}$

11. $y_1(x) = e^{-x/2}\sin\frac{\sqrt{3}}{2}x,\ y_2(x) = e^{-x/2}\cos\frac{\sqrt{3}}{2}x;\ G(x,t) = \begin{cases}-\frac{2}{\sqrt{3}}e^{(t-x)/2}\cos\frac{\sqrt{3}}{2}x\sin\frac{\sqrt{3}}{2}t, & 0 \le t \le x;\\ -\frac{2}{\sqrt{3}}e^{(t-x)/2}\sin\frac{\sqrt{3}}{2}x\cos\frac{\sqrt{3}}{2}t, & x \le t \le \frac{\pi}{\sqrt{3}}.\end{cases}$

13. $y(x) = \frac{1}{2}x - \frac{\sqrt{2}}{8}\pi\sin\sqrt{2}x$ **15.** $y(x) = \sin x - e^{-(x-\pi/\sqrt{3})/2}\sin\frac{\pi}{\sqrt{3}}\sin\frac{\sqrt{3}}{2}x$
17. $y(x) = \cos\sqrt{2}x + \frac{1}{2}x - \frac{\sqrt{2}}{8}\pi\sin\sqrt{2}x$ **19.** $y(x) = \sin x - e^{-x/2}\left(\cos\frac{\sqrt{3}}{2}x + e^{\pi/(2\sqrt{3})}\sin\frac{\pi}{\sqrt{3}}\sin\frac{\sqrt{3}}{2}x\right)$
21. $\alpha = \beta;\ y(x) = \alpha\cos\sqrt{2}x + c\sin\sqrt{2}x$, c constant; $y'(0) = 1$ for $c = 1/\sqrt{2}$.
23. **(b)** There is no oscillation if $k^2 \ge 4mh$ and therefore the equilibrium point will be attained, at most, once.
(c) $\{a, a + 2mn\pi/\sqrt{4mh - k^2}\},\ n = 1, 2, 3, \ldots$.
27. $y(x) = \frac{R}{48}x^2(3L^2 - 5Lx + 2x^2)$
29. No solution. The beam has a simple support at one end and is unsupported at the other end. This can happen only if $R = 0$.

31. **(a)** $y(x) = \frac{\sinh\sqrt{k}x}{\sinh\sqrt{k}}$ **33.** **(c)** $G(x,t) = \begin{cases}\frac{y_1(t)}{w(t)}y_2(x), & a \le t \le x;\\ \frac{y_2(t)}{w(t)}y_1(x), & x \le t \le b.\end{cases}$

Section 4

1. **(a)** $y_1 = e^x - 1,\ y_2 = e^{-x} - 1,\ y_3 = (e+1)^{-1}(e^x + e^{x+1}) - 1$

Section 5

1. $y(x) = 1 + 2x + 3x^2 + \cdots$ **3.** $y(x) = 2 + x^2 + \frac{1}{4}x^4 + \cdots$ **5.** **(a)** $y(x) = 1 + x + \frac{1}{2}x^2 + \frac{1}{3}x^3 + \frac{5}{24}x^4 + \cdots$
9. $\frac{1}{20}x^5$ **11.** $y(x) = \sum_{k=0}^{\infty}\frac{!}{k!}x^k\ (= e^x)$ **15.** $y_2(x) = -1 + \sum_{k=1}^{\infty}\frac{1}{2^k(2k-1)k!}x^{2k}$

Section 6

1. **(a)** $y(x) = c_0\sum_{n=0}^{\infty}\frac{x^{2n}}{(2n)!} + c_1\sum_{n=0}^{\infty}\frac{x^{2n+1}}{(2n+1)!}$ **3.** **(a)** $y(x) = c_0\sum_{n=0}^{\infty}\frac{(-1)^n x^{2n}}{n!}$; **(b)** $y(x) = Ce^{-x^2}$; **(c)** Yes.
5. **(a)** $y(x) = c_0\sum_{n=0}^{\infty}\frac{(-1)^n}{2^n n!}x^{2n} + c_1\sum_{n=0}^{\infty}\frac{(-1)^n 2^n n!}{(2n+1)!}x^{2n+1}$; **(c)** $y_1(x) = F(x)/F'(x)$ **11.** **(b)** $r = (1 \pm \sqrt{5})/2$

Section 7A

1. $x = 0$ (regular) **3.** $x = 1$ (regular) **5.** No singular points. **7.** No singular points.
9. $\mu = \pm i;\ y_1 = \cos(\ln x),\ y_2 = \sin(\ln x)$ **11.** $\mu = \frac{1}{2} \pm i\frac{\sqrt{3}}{2};\ y_1 = x^{1/2}\cos\left(\frac{\sqrt{3}}{2}\ln x\right),\ y_2 = x^{1/2}\sin\left(\frac{\sqrt{3}}{2}\ln x\right)$
13. $\mu - 1, -5;\ y_1 = x,\ y_2 = x^{-5}$ **15.** $\mu = (5 \pm \sqrt{21})/2;\ y_1 = x^{(5+\sqrt{21})/2},\ y_2 = x^{(5-\sqrt{21})/2}$
17. $x = e^{m\pi},\ m = 0, \pm 1, \pm 2, \ldots$

Section 7B

11. 3.832, 7.016, 10.174, 13.324, 16.471 **15.** $Z_{1/2}(x) = \frac{\cos x}{\sqrt{x}}$ **17.** $Z_{-1/2}(x) = \frac{2\cos x}{\pi\sqrt{x}}$ **19. (b)** $y(x) = \frac{4J_0(2x)}{J_0(6)}$

Chapter 10

Section 1AB

7. $u = y + c_1(y)e^{-x} + c_2(y)xe^{-x}$ **9.** Elliptic **11.** Hyperbolic **13.** $r^2 + s^2 = 1$; $u = e^x$
15. $r^2 - s^2 + s = 0$; $u = c_1 e^{sy+\sqrt{s(s-1)}x} + c_2 e^{sy-\sqrt{s(s-1)}x}$ **17. (a)** $ar^2 - s^2 - bs - c = 0$
19. (b) $u = 2(x^2 - y^2)$ **21. (a)** $(\beta - 1)(\sigma^2\beta + 2r) = 2\alpha$ with $C > 0$ arbitrary.

Section 2ABC

1. $\sum_{k=1}^{\infty} \frac{2(-1)^{k+1}}{k} \sin kx$ **3.** $\frac{\pi^2}{3} + \sum_{k=1}^{\infty} \frac{4(-1)^k}{k^2} \cos kx$ **5.** $\frac{1}{2} + \sum_{k=1}^{\infty} \frac{2}{(2k-1)\pi} \sin(2k-1)x$ **7.** $\sum_{k=1}^{\infty} \frac{4}{2k-1} \sin(2k-1)x$
9. $-\frac{\pi}{2} + \sum_{k=1}^{\infty} \frac{4}{(2k-1)^2\pi} \cos(2k-1)x$ **23.** $\frac{3}{4}\cos x + \frac{1}{4}\cos 3x$ **25.** $\frac{3}{4}\sin x - \frac{1}{4}\sin 3x$
31. Yes, since f is decreasing on $(-\pi, 0)$ and increasing on $(0, \pi)$.

Section 3AB

1. $\sum_{k=1}^{\infty} \frac{4(-1)^k}{k\pi} \sin \frac{k\pi x}{2}$; Series converges to 0 at $x = \pm 2$. **3.** $\sum_{k=0}^{\infty} \frac{4}{(2k+1)\pi} \sin(2k+1)x$ **5.** $\sum_{k=1}^{\infty} \frac{8k}{(4k^2-1)\pi} \sin 2kx$
7. $\sum_{k=1}^{\infty} \left(\frac{2}{k\pi}(-1)^{k+1} - \frac{6}{k^2\pi^2} \sin \frac{2k\pi}{3}\right) \sin \frac{k\pi x}{3}$ **9.** $\frac{\pi^2}{3} + \sum_{k=1}^{\infty} \frac{4(-1)^k}{k^2} \cos kx$ **11.** $\frac{1}{2} + \sum_{k=0}^{\infty} \frac{2(-1)^{k+1}}{(2k+1)\pi} \cos \frac{(2k+1)\pi x}{2}$
13. (a) $\frac{\pi}{2} + \sum_{k=0}^{\infty} \frac{-4}{(2k+1)^2\pi} \cos(2k+1)x$; **(b)** $\sum_{k=1}^{\infty} \frac{2(-1)^{k+1}}{k} \sin kx$;
(c) Since $g(x) = x$ is odd, the complete Fourier expansion of g contains only sine terms and is therefore the sine expansion given in part (b).
15. (b) The odd part is the sum of the terms involving sine, and the even part is the sum of the remaining terms.
(c) $P_o = \sinh x$, $P_e = \cosh x$.
19. (b) $f(x) = x + \sin x$

Section 4AB

1. $u(x,t) = e^{-(\pi^2a^2/p^2)t} \sin \frac{\pi x}{p}$ **3.** $u(x,t) = \sum_{k=1}^{\infty} \frac{2}{k\pi} e^{-k^2\pi^2a^2t} \sin k\pi t$ **5.** $u(x,t) = e^{-a^2t}\sin x + \frac{1}{2}e^{-4a^2t}\sin 2x$
7. $v(x) = -1 + x$ **9.** $v(x) = 1 + x$ **11.** $u(x,t) = 1 + \frac{2x}{p} + \frac{2}{\pi}\sum_{k=1}^{\infty} \frac{2[3(-1)^k - 1]}{k} e^{-(k^2\pi^2a^2/p^2)t} \sin \frac{k\pi x}{p}$
13. $u(x,t) = x + \frac{2}{\pi}\sum_{k=1}^{\infty} \frac{1}{k} e^{-4k^2\pi^2a^2t} \sin 2k\pi x$
17. $u(x,t) = \left(c_1 e^{(1/2+\alpha)x} + c_2 e^{(1/2-\alpha)x}\right) e^{(\alpha^2-1/4)t}$ $(\alpha \neq 0)$,
$u(x,t) = (c_1 + c_2x)e^{x/2-t/4}$, $u(x,t) = \left(c_1\cos\beta x + c_2\sin\beta x\right)e^{x/2-(\beta^2+1/4)t}$ $(\beta \neq 0)$.
19. $u(x,t) = \left(c_1e^{\alpha x} + c_2e^{-\alpha x}\right)\left(c_3e^{\alpha t} + c_4e^{-\alpha t}\right)$ $(\alpha \neq 0)$, $u(x,t) = (c_1 + c_2x)(c_3 + c_4t)$,
$u(x,t) = \left(c_1\cos\beta x + c_2\sin\beta x\right)\left(c_3\cos\beta t + c_4\sin\beta t\right)$ $(\beta \neq 0)$
21. $u(x,t) = x^2 + 1$ **23.** $u(x,t) = x^3/6 + 5x/6 + 1$
25. (b) $u(x,t) = \frac{1}{p}\int_0^p f(u)\,du + \frac{2}{p}\sum_{k=1}^{\infty}\left[\left(\int_0^p f(u)\cos\frac{k\pi u}{p}du\right)e^{-k^2\pi^2a^2t/p^2}\cos\frac{k\pi x}{p}\right]$;
(c) $u(x,\infty) = \frac{1}{p}\int_0^p f(u)\,du$
27. $u(x,t) = 1$
35. (a) $X'' + \lambda^2X = 0$, $T' + \lambda^2tT = 0$; **(b)** $X(x)T(t) = e^{-\lambda^2t^2/2}(c_1\cos\lambda x + c_2\sin\lambda x)$ and $X(x)T(t) = c_1 + c_2x$.
37. If $u(x,t_0)$ is concave down on an x-interval I then, for each $x \in I$, there is a time $t(x)$ such that $u(x,t)$ decreases as t increases from t_0 to $t(x)$.

Section 4C

1. $u(x,t) = \cos at \sin x$
3. $u(x,t) = \frac{1}{\pi a}\left[2\sin at\sin x - \sin 2at\sin 2x + \frac{2}{9}\sin 3at\sin 3x + \frac{2}{25}\sin 5at\sin 5x - \frac{1}{9}\sin 6at\sin 6x + \cdots\right]$
5. (a) $v(x) = \frac{g}{2a^2}x^2 + c_1x + c_2$; **(b)** $v(x) = \frac{g}{2a^2}x(x-p)$;
(c) Solve the associated homogeneous wave equation $a^2u_{xx} = u_{tt}$ with homogeneous boundary conditions. Then, since the equation is linear, add the homogeneous and equilibrium solutions to obtain a solution of the given nonhomogeneous equation.
9. (c) $G'(0) = G'(p) = 0$ is assumed in order to be consistent with the assumption of fixed endpoints.

11. $X'' - \lambda X = 0,\ T'' - \lambda t T = 0$

Section 5

9. **(a)** $u(x, y) = 0$ **11.** $u(x, y) = \sum_{k=0}^{\infty} \frac{8 \sin(2k+1)\pi x \sinh(2k+1)\pi y}{(2k+1)^3\pi^3 \sinh(2k+1)\pi}$

13. $u(x, y) = \sum_{k=1}^{\infty} \frac{2(-1)^{k+1}}{k\pi \sinh k\pi} \sinh[k\pi(1-x)] \sin k\pi y$ **15.** **(b)** $u(x, y) = 2e^x \cos y$

17. **(b)** $u(x, y) = \sum_{k=0}^{\infty} \frac{4p}{\pi^2(2k+1)^2 \cosh\left((2k+1)\pi q/p\right)} \left[\sinh \frac{(2k+1)\pi y}{p} \sin \frac{(2k+1)\pi x}{p}\right]$

19. **(a)** $\bar{u}_r = \cos\theta\, u_x + \sin\theta\, u_y,\ \ \bar{u}_{rr} = \cos^2\theta\, u_{xx} + 2\cos\theta \sin\theta\, u_{xy} + \sin^2\theta\, u_{yy},$
$\bar{u}_{\theta\theta} = r^2\left(\cos^2\theta\, u_{yy} - 2\cos\theta\sin\theta\, u_{yx} + \sin^2\theta\, u_{xx}\right) - r\left(\cos\theta\, u_x + \sin\theta\, u_y\right)$

Chapter Review and Supplement

1. $r^2 - s^2 + 2r = 0$ **3.** $r^2 + s^2 = 1$ **9.** $X'' + 5X' - \lambda X = 0,\ T' + \lambda T = 0$ **11.** T, F, F, F

13. Product solution: $U(x, t) = (c_1 e^{k\sqrt{\lambda}x} + c_2 e^{-k\sqrt{\lambda}x})(c_3 e^{\sqrt{\lambda}t} + c_4 e^{-\sqrt{\lambda}t})$; Non-constant steady state solutions: $U(x, t) = c_1 + c_2 x,\ c_2 \neq 0$.

15. Product solution: $U(x, t) = (c_1 e^{\sqrt{\lambda}x} + c_2 e^{-\sqrt{\lambda}x})e^{\lambda t}$; Non-constant steady state solutions: $U(x, t) = c_1 + c_2 x$, $c_2 \neq 0$.

17. $y(x, t) = \frac{16}{\pi^3 a} \sum_{k=1}^{\infty} \frac{(-1)^{k-1}}{(2k-1)^3} \sin \frac{(2k-1)\pi at}{2} \sin \frac{(2k-1)\pi x}{2}$

19. **(a)** L acts on functions (producing new functions) in such a way that $L(\alpha f + \beta g) = \alpha L(f) + \beta L(g)$ for any functions f, g and any constants α, β. **(c)** $L(\mathbf{u}) = \lambda\mathbf{u}$

Chapter 11

Section 1

1. **(b)** $r(x)$ must be non-negative on $[a, b]$ with only isolated zeros.
7. No, since positivity fails if $r_1 = 0$.
9. The sum of the lengths of the diagonals of a parallelogram is equal to its perimeter.
15. $r(x)$ does not define an inner product (positivity fails).
17. $r(x)$ does not define an inner product (positivity fails). **19.** $r(x)$ defines an inner product.

Section 2AB

1. $\lambda = -n^2$; $y_\lambda(x) = \sin nx$, for $n = 1, 2, 3, \ldots$. **3.** $\lambda = -\left(n + \frac{1}{2}\right)^2$; $y_\lambda(x) = \sin\left(n + \frac{1}{2}\right)x$, for $n = 0, 1, 2, \ldots$.

5. $\lambda = -a^2$; $y_\lambda(x) = \sin a(x-1)$, where $a\ (> 0)$ satisfies $\sin 2a = 2a/(1+a^2)$.

7. $\lambda = -(n^2+1)$: $y_\lambda(x) = e^{-x} \sin nx$, for $n = 1, 2, 3, \ldots$.

9. $\lambda = -(n^2+1)$: $y_\lambda(x) = e^{-x} \cos nx$, for $n = 0, 1, 2, \ldots$.

Section 3AB

5. **(a)** $v_{18}(x, y) = \sin x \sin 8y,\ v_{81}(x, y) = \sin 8x \sin y, v_{47}(x, y) = \sin 4x \sin 7y,\ v_{74}(x, y) = \sin 7x \sin 4y;$
(b) $v_{1,18}(x, y) = \sin x \sin 18y,\ v_{18,1}(x, y) = \sin 18x \sin y,\ v_{6,17}(x, y) = \sin 6x \sin 17y,$
$v_{17,6}(x, y) = \sin 17x \sin 6y,\ v_{10,15}(x, y) = \sin 10x \sin 15y,\ v_{15,10}(x, y) = \sin 15x \sin 10y.$

Appendix A

9. $\bar{z} = 1 - i;\ 1/z = \frac{1}{2} - \frac{1}{2}i;\ |z| = \sqrt{2};\ z = \sqrt{2}\left(\cos(\pi/4) + i\sin(\pi/4)\right)$

11. $\bar{z} = -2i;\ 1/z = -\frac{1}{2}i;\ |z| = 2;\ z = 2\left(\cos(\pi/2) + i\sin(\pi/2)\right)$

13. $\bar{z} = -\sin t - i\cos t;\ 1/z = -\sin t - i\cos t;\ |z| = 1;\ z = 1\left(\cos(t + \pi/2) + i\sin(t + \pi/2)\right)$

15. $\bar{z} = 2;\ 1/z = \frac{1}{2};\ |z| = 2;\ z = 2(\cos 0 + i\sin 0)$

17. $\bar{z} = -i;\ 1/z = -i;\ |z| = 1;\ z = 1\left(\cos(\pi/2) + i\sin(\pi/2)\right)$ **23.** **(d)** $z_{1,2} = \frac{1}{2}\left(-1 \pm \sqrt{\frac{\sqrt{17}+1}{2}} - i\sqrt{\frac{\sqrt{17}-1}{2}}\right)$

Appendix B

Section 1: Matrix Operations

1. $\begin{pmatrix} 1 & -2 \\ 1 & 1 \end{pmatrix}$ **3.** $\begin{pmatrix} 4 & 14 \\ 1 & 8 \end{pmatrix}$ **5.** $\begin{pmatrix} 7 \\ 6 \\ 4 \end{pmatrix}$ **7.** $A = \begin{pmatrix} 1 & -1 \\ 1 & 1 \end{pmatrix}$, $\mathbf{x} = \begin{pmatrix} x \\ y \end{pmatrix}$, $\mathbf{b} = \begin{pmatrix} 1 \\ 2 \end{pmatrix}$

9. $A = \begin{pmatrix} 1 & 1 \\ 0 & 1 \end{pmatrix}$, $\mathbf{x} = \begin{pmatrix} x \\ y \end{pmatrix}$, $\mathbf{b} = \begin{pmatrix} 1 \\ 1 \end{pmatrix}$ **11.** $A = \begin{pmatrix} 1 & 1 & 0 \\ 0 & 1 & 1 \\ 1 & 0 & -1 \end{pmatrix}$, $\mathbf{x} = \begin{pmatrix} x \\ y \\ z \end{pmatrix}$, $\mathbf{b} = \begin{pmatrix} 0 \\ 0 \\ 1 \end{pmatrix}$

13. $A = \begin{pmatrix} 2 & 3 \\ 2 & -1 \end{pmatrix}$, $\mathbf{x} = \begin{pmatrix} x \\ y \end{pmatrix}$, $\mathbf{b} = \begin{pmatrix} 0 \\ 0 \end{pmatrix}$ **15.** $A = \begin{pmatrix} 1 & 1 & -1 \\ 1 & -1 & 1 \\ 1 & t & 0 \end{pmatrix}$, $\mathbf{x} = \begin{pmatrix} x \\ y \\ z \end{pmatrix}$, $\mathbf{b} = \begin{pmatrix} 0 \\ 0 \\ 0 \end{pmatrix}$

17. $A = \begin{pmatrix} 0 & 2 \\ -1/3 & 0 \end{pmatrix}$, $\mathbf{b} = \begin{pmatrix} 1 \\ 2/3 \end{pmatrix}$ **19.** $A = \begin{pmatrix} 0 & 0 \\ 0 & 0 \end{pmatrix}$, $\mathbf{b} = \begin{pmatrix} 0 \\ 0 \end{pmatrix}$ **21.** $A = \begin{pmatrix} 0 & 2 & 0 \\ 0 & 0 & -1 \\ -1 & 0 & 0 \end{pmatrix}$, $\mathbf{b} = \begin{pmatrix} 0 \\ 0 \\ 1 \end{pmatrix}$

23. $f(t) = t^2\mathbf{e}_1 + t^3\mathbf{e}_2 + t^4\mathbf{e}_3$ **25.** $f(t) = \mathbf{e}_1 + t\mathbf{e}_2$ **27.** $f(t) = \mathbf{e}_1 + \mathbf{e}_2$ **29.** $dA(t)/dt = \begin{pmatrix} 3t^2 + 4t & 6t^2 + 12t \\ 7 & 18 \end{pmatrix}$

31. $dA(t)/dt = \begin{pmatrix} 1 & 2t & 3t^2 \\ 0 & 3t^2 & 4t^3 \\ 0 & 0 & 5t^4 \end{pmatrix}$ **33. (a)** $C\mathbf{e}_1 = \begin{pmatrix} 1 \\ 0 \\ 0 \end{pmatrix}$, $C\mathbf{e}_2 = \begin{pmatrix} 1 \\ 1 \\ 0 \end{pmatrix}$, $C\mathbf{e}_3 = \begin{pmatrix} 1 \\ 1 \\ 1 \end{pmatrix}$ **35.** $\begin{pmatrix} 4 \\ 5 \\ 3 \end{pmatrix}$

39. $A^2 = \begin{pmatrix} -1 & -1 & 0 \\ 1 & 1 & 0 \\ -1 & -1 & 0 \end{pmatrix}$, $A^3 = \begin{pmatrix} 0 & 0 & 0 \\ 0 & 0 & 0 \\ 0 & 0 & 0 \end{pmatrix}$ **41.** $A^2 = \begin{pmatrix} 4 & 0 \\ 0 & 9 \end{pmatrix}$, $A^3 = \begin{pmatrix} 8 & 0 \\ 0 & 27 \end{pmatrix}$

45. (a) $\mathbf{x} = \begin{pmatrix} 5/11 \\ -1/11 \end{pmatrix}$; **(b)** $\mathbf{x} = \begin{pmatrix} -2 \\ 6 \end{pmatrix}$

Section 2

1. (a) $AB = \begin{pmatrix} -4 & 7 \\ 2 & 0 \end{pmatrix}$, $BA = \begin{pmatrix} 3 & 1 \\ -7 & -7 \end{pmatrix}$, $\det A = 7$, $\det B = -2$, $\det AB = \det BA = -14$;

(b) $AB = \begin{pmatrix} -2 & 0 & 2 \\ 6 & -3 & -9 \\ 0 & 12 & 20 \end{pmatrix}$, $BA = \begin{pmatrix} -2 & 0 & 4 \\ 4 & -3 & -12 \\ 0 & 9 & 20 \end{pmatrix}$, $\det A = 24$, $\det B = 2$, $\det AB = \det BA = 48$

9. (a) -12 **13. (a)** $\begin{pmatrix} 1 & 0 & 0 \\ -13 & 1 & -5 \\ 2 & 0 & 1 \end{pmatrix}$; **(b)** Not invertible; **(c)** $\begin{pmatrix} 0 & 1 & 0 \\ 7/4 & -22/4 & 2 \\ -3/4 & 10/4 & -1 \end{pmatrix}$;

(d) $\begin{pmatrix} 1/t & 0 & 0 \\ 0 & 1/2 & 0 \\ 0 & 0 & 1 \end{pmatrix}$, for $t \neq 0$; **(e)** Not invertible; **(f)** $\begin{pmatrix} 1 & -1 & 0 \\ 0 & -1 & 1 \\ 0 & 0 & 1 \end{pmatrix}$;

(g) $\begin{pmatrix} 1 & -1 & 1 & -3/4 \\ 0 & 1/2 & 0 & -1/8 \\ 0 & 0 & 1 & -1/4 \\ 0 & 0 & 0 & 1/4 \end{pmatrix}$; **(h)** $\begin{pmatrix} 1 & 0 & -1/3 & 0 \\ 0 & 1/2 & 0 & 0 \\ 0 & 0 & 1/3 & 0 \\ 0 & 0 & 0 & 1/4 \end{pmatrix}$; **(i)** $\begin{pmatrix} 1 & 0 & 0 \\ 0 & e^{-t} & -te^{-t} \\ 0 & 0 & e^{-t} \end{pmatrix}$, for all t.

17. (a) $\mathbf{x} = \begin{pmatrix} -1/4 \\ -5/4 \\ -5/2 \end{pmatrix}$; **(b)** $u(x) = x \sin x \cos x$, $v(x) = -x \sin^2 x$

Appendix C

Sections 1 & 2

1. $x(t) = 1 + \int_0^t (ux + u^2)\, du$ **3.** $x(t) = \int_0^t (x - u)\, du$ **5.** $x_1(t) = 1 + \frac{1}{2}t^2 + \frac{1}{3}t^3$, $x_2(t) = 1 + \frac{1}{2}t^2 + \frac{1}{3}t^3 + \frac{1}{8}t^4 + \frac{1}{15}t^5$

7. $x_1(t) = -\frac{1}{2}t^2$, $x_2(t) = -\frac{1}{6}t^3 - \frac{1}{2}t^2$

9. (a) $\mathbf{x}(t) = \begin{pmatrix} 1 \\ 2 \end{pmatrix} + \int_0^t \begin{pmatrix} uy \\ ux \end{pmatrix} du$; **(b)** $(x_1(t), y_1(t)) = (1 + t^2, 2 + t^2/2)$

(c) $(x_2(t), y_2(t)) = (1 + t^2 + t^4/8, 2 + t^2/2 + t^4/8)$

11. (a) No; **(b)** No; **(c)** Yes

INDEX